Porsche 911 Carrera

Service Manual

Carrera, Carrera S, Carrera 4, Carrera 4S
(Type 993)
1995, 1996, 1997, 1998

BentleyPublishers®
.com

BENTLEY PUBLISHERS® | Automotive Reference™

Bentley Publishers, a division of Robert Bentley, Inc.
1734 Massachusetts Avenue
Cambridge, MA 02138 USA
800-423-4595 / 617-547-4170

Information that makes the difference®

BentleyPublishers®.com

Technical contact information

We welcome your feedback. Please submit corrections and additions to our technical discussion forum at:

`http://www.BentleyPublishers.com`

Updates and corrections

We will evaluate submissions and post appropriate editorial changes online as updates or tech discussion. Appropriate updates and corrections will be added to the book in future printings. Check for updates and corrections for this book before beginning work on your vehicle. See the following web address for additional information:

`http://www.BentleyPublishers.com/updates/`

WARNING—Important Safety Notice

Do not use this manual unless you are familiar with automotive repair procedures and safe workshop practices. This manual illustrates the workshop procedures required for some maintenance and service work. It is not a substitute for full and up-to-date information from the vehicle manufacturer or for proper training as an automotive technician. Note that it is not possible for us to anticipate all of the ways or conditions under which vehicles may be serviced or to provide cautions as to all of the possible hazards that may result.

We have endeavored to ensure the accuracy of the information in this manual. Please note, however, that considering the vast quantity and the complexity of the service information involved, we cannot warrant the accuracy or completeness of the information contained in this manual.

FOR THESE REASONS, NEITHER THE PUBLISHER NOR THE AUTHOR MAKES ANY WARRANTIES, EXPRESS OR IMPLIED, THAT THE INFORMATION IN THIS BOOK IS FREE OF ERRORS OR OMISSIONS, AND WE EXPRESSLY DISCLAIM THE IMPLIED WARRANTIES OF MERCHANTABILITY AND OF FITNESS FOR A PARTICULAR PURPOSE, EVEN IF THE PUBLISHER OR AUTHOR HAVE BEEN ADVISED OF A PARTICULAR PURPOSE, AND EVEN IF A PARTICULAR PURPOSE IS INDICATED IN THE MANUAL. THE PUBLISHER AND AUTHOR ALSO DISCLAIM ALL LIABILITY FOR DIRECT, INDIRECT, INCIDENTAL OR CONSEQUENTIAL DAMAGES THAT RESULT FROM ANY USE OF THE EXAMPLES, INSTRUCTIONS OR OTHER INFORMATION IN THIS BOOK. IN NO EVENT SHALL OUR LIABILITY WHETHER IN TORT, CONTRACT OR OTHERWISE EXCEED THE COST OF THIS MANUAL.

Before attempting any work on your Porsche, read **00 Warnings and Cautions** and any **WARNING** or **CAUTION** that accompanies a procedure in this manual. Review the **WARNINGS** and **CAUTIONS** each time you prepare to work on your Porsche.

Your common sense and good judgment are crucial to safe and successful service work. Read procedures through before starting them. Think about whether the condition of your car, your level of mechanical skill, or your level of reading comprehension might result in or contribute in some way to an occurrence which might cause you injury, damage your car, or result in an unsafe repair. If you have doubts for these or other reasons about your ability to perform safe repair work on your car, have the work done at an authorized Porsche dealer or other qualified shop.

Part numbers listed in this manual are for identification purposes only, not for ordering. Always check with your authorized Porsche dealer to verify part numbers and availability before beginning service work that may require new parts.

Special tools required to perform certain service operations are identified in the manual and are recommended for use. Use of tools other than those recommended in this manual may be detrimental to the car's safe operation as well as the safety of the person servicing the car.

The vehicle manufacturer will continue to issue service information updates and parts retrofits after the editorial closing of this manual. Some of these updates and retrofits will apply to procedures and specifications in this manual. We regret that we cannot supply updates to purchasers of this manual.

This manual is prepared, published and distributed by Bentley Publishers, 1734 Massachusetts Avenue, Cambridge, Massachusetts 02138, USA. All information contained in this manual is based on the information available to the publisher at the time of editorial closing. Porsche has not reviewed and does not vouch for the accuracy or completeness of the technical specifications and work procedures described and given in this manual.

ISBN 978-0-8376-1719-0 **Editorial closing 07/13** Job code: P998-02

Library of Congress Cataloging-in-Publication Data
Porsche 911 Carrera service manual : Carrera, Carrera 4, Carrera 4S (Type 993) 1995, 1996, 1997, 1998.
pages cm
Summary: "Service and repair information for Porsche 911 vehicles, Type 993, applicable to model years 1995, 1996, 1997, and 1998"--Provided by publisher.
Includes index.
ISBN 978-0-8376-1719-0 (alk. paper)
1. Porsche 911 automobile--Maintenance and repair--Handbooks, manuals, etc. I. Title: Porsche 911 (Type 993) service manual : Carrera, Carrera 4, Carrera 4S 1995, 1996, 1997, 1998.
TL215.P75P632 2013
629.28'722--dc23
2013019297

 The paper used in this publication is acid free and meets the requirements of the National Standard for Information Sciences–Permanence of Paper for Printed Library Materials. ∞

Manufactured in the United States of America.

202605R001

Foreword

For the Porsche owner with basic mechanical skills and for independent auto service professionals, this book includes many of the specifications and procedures that were available in an authorized Porsche dealer service department as this manual went to press. The Porsche owner with no intention of working on his or her car will find that owning and referring to this manual makes it possible to be better informed and to discuss repairs with a professional automotive technician more knowledgeably.

If you are a Porsche owner intending to do maintenance and repair work, make sure you have screwdrivers, a set of metric wrenches and sockets, and metric Allen and Torx wrenches, since these basic hand tools are needed for much of the work described in this manual. Many procedures also require a torque wrench to ensure that fasteners are tightened properly and in accordance with specifications. In some cases, the text refers to special tools that are recommended or required to accomplish adjustments or repairs. These tools are often identified by their Porsche special tool number and illustrated.

Disclaimer

We have endeavored to ensure the accuracy of the information in this manual. When the vast array of data presented in the manual is taken into account, however, no claim to infallibility can be made. We therefore cannot be responsible for the result of any errors that may have crept into the text. Please also read **WARNING–Important Safety Notice** on the copyright page at the beginning of this book.

Prior to starting a repair procedure, read the procedure, **00 Warnings and Cautions** and the warnings and cautions that accompany the procedure. Reading a procedure before beginning work helps you determine in advance the need for specific skills, identify hazards, prepare for appropriate capture and handling of hazardous materials, and the need for particular tools and replacement parts such as gaskets.

Bentley Publishers encourages comments from the readers of this manual with regard to errors, and suggestions for improvement of our product. These communications have been and will be carefully considered in the preparation of this and other manuals. If you identify inconsistencies in the manual, you may have found an error. Please contact the publisher and we will endeavor to post applicable corrections on our website. Review updates and additions that we have posted before beginning work. Please see the following web address:

http://www.BentleyPublishers.com/updates/

Porsche offers extensive warranties, especially on components of the fuel delivery and emission control systems. Therefore, before deciding to repair a Porsche that may be covered wholly or in part by any warranties issued by Porsche Car North America, Inc., consult your authorized Porsche dealer. Regardless of its age, or whether it is under warranty, your Porsche is both an easy car to service and an easy car to get serviced. So if at any time a repair is needed that you feel is too difficult to do yourself, a trained Porsche technician is ready to do the job for you.

Bentley Publishers

00 Warnings and Cautions

Please read these warnings and cautions before proceeding with maintenance and repair work.

WARNINGS—
*See also **CAUTIONS** on next page.*

- Read the important safety notice on the copyright page at the beginning of the book.
- Some repairs may be beyond your capability. If you lack the skills, tools and equipment, or a suitable workplace for any procedure described in this manual, we suggest you leave such repairs to an authorized Porsche dealer service department or other qualified shop.
- Thoroughly read each procedure and the **WARNINGS** and **CAUTIONS** that accompany the procedure. Also review posted corrections at `www.BentleyPublishers.com/updates/` before beginning work.
- If any procedure, tightening torque, wear limit, specification or data presented in this manual does not appear to be appropriate for a specific application, contact the publisher or the vehicle manufacturer for clarification before using the information in question.
- Porsche is constantly improving its cars. Sometimes these changes, both in parts and specifications, are made applicable to earlier models. Therefore, before starting any major jobs or repairs to components on which passenger safety may depend, consult your authorized Porsche dealer about technical bulletins that may have been issued.
- Do not reuse any fasteners that are worn or deformed in normal use. Many fasteners are designed to be used only once and become unreliable and may fail when used a second time. This includes, but is not limited to, nuts, bolts, washers, self-locking nuts or bolts, circlips and cotter pins. Replace these fasteners with new parts.
- Do not work under a lifted car unless it is solidly supported on stands designed for the purpose. Do not support a car on cinder blocks, hollow tiles or other props that may crumble under continuous load. Do not work under a car that is supported solely by a jack. Do not work under the car while the engine is running.
- If you are going to work under a car on the ground, make sure that the ground is level. Block the wheels to keep the car from rolling. Disconnect the battery negative (–) terminal (ground strap) to prevent others from starting the car while you are under it.
- Do not run the engine unless the work area is well ventilated. Carbon monoxide kills.
- Remove finger rings, bracelets and other jewelry so that they cannot cause electrical shorts, get caught in running machinery or be crushed by heavy parts.
- Tie long hair behind your head. Do not wear a necktie, a scarf, loose clothing, or a necklace when you work near machine tools or running engines. If your hair, clothing, or jewelry were to get caught in the machinery, severe injury could result.
- Do not attempt to work on your car if you do not feel well. You increase the danger of injury to yourself and others if you are tired, upset or have taken medication or any other substance that may keep you from being fully alert.
- Illuminate your work area adequately but safely. Use a portable safety light for working inside or under the car. Make sure the bulb is enclosed by a wire cage. The hot filament of an accidentally broken bulb can ignite spilled fuel, vapors or oil.
- Catch draining fuel, oil or brake fluid in suitable containers. Do not use food or beverage containers that might mislead someone into drinking from them. Store flammable fluids away from fire hazards. Wipe up spills at once but do not store the oily rags, which can ignite and burn spontaneously.
- Observe good workshop practices. Wear goggles when you operate machine tools or work with battery acid. Wear gloves or other protective clothing whenever the job requires working with harmful substances.
- Greases, lubricants and other automotive chemicals contain toxic substances, many of which are absorbed directly through the skin. Read the manufacturer's instructions and warnings carefully. Use hand and eye protection. Avoid direct skin contact.
- Friction components (such as brake pads or shoes or clutch discs) contain asbestos fibers or other friction materials. Do not create dust by grinding, sanding, or by cleaning with compressed air. Avoid breathing dust. Breathing any friction material dust can lead to serious diseases and may result in death.
- Disconnect the battery negative (–) terminal (ground strap) whenever you work on the fuel system or the electrical system. Do not smoke or work near heaters or other fire hazards. Keep an approved fire extinguisher handy.
- Batteries give off explosive hydrogen gas during charging. Keep sparks, lighted matches and open flame away from the top of the battery. If hydrogen gas escaping from the cap vents is ignited, it will ignite gas trapped in the cells and cause the battery to explode.
- Battery acid (electrolyte) can cause severe burns. Flush contact area with water and seek medical attention.
- Do not quick-charge the battery (for boost starting) for longer than one minute. Wait at least one minute before boosting the battery a second time.
- Connect and disconnect a battery charger only with the battery charger switched OFF.
- Connect and disconnect battery cables, jumper cables or a battery charger only with the ignition switched OFF. Do not disconnect the battery while the engine is running.
- Do not allow the battery charging voltage to exceed 16.5 volts. If the battery begins producing gas or boiling violently, reduce the charging rate. Boosting a sulfated battery at a high rate can cause an explosion.

continued on next page

Please read these warnings and cautions before proceeding with maintenance and repair work.

WARNINGS (continued)

- The air-conditioning system is filled with a hazardous refrigerant. Make sure the system is serviced only by a trained technician using approved refrigerant recovery / recycling equipment, trained in related safety precautions, and familiar with regulations governing the discharging and disposal of automotive chemical refrigerants.
- Do not expose any part of the air-conditioning system to high temperatures such as open flame. Excessive heat increases system pressure and may cause the system to burst.
- Some aerosol tire inflators are highly flammable. Be extremely cautious when repairing a tire that may have been inflated using an aerosol tire inflator. Keep sparks, open flame or other sources of ignition away from the tire repair area. Inflate and deflate the tire at least four times before breaking the bead from the rim. Completely remove the tire from the rim before attempting any repair.
- Cars covered by this manual are equipped with a supplemental restraint system (SRS) that automatically deploys airbags in case of a frontal or side impact. These are explosive devices. Handled improperly or without adequate safeguards, they can be accidently activated and cause serious injury.
- The ignition system produces high voltages that can be fatal. Avoid contact with exposed terminals and use extreme care when working on a car with the engine running or the ignition switched ON.
- Place jack stands only at locations specified by the manufacturer. The vehicle lifting jack supplied with the vehicle is intended for tire changes only. Use a heavy duty floor jack to lift the vehicle before installing jack stands. See **03 Service and Maintenance**.
- Aerosol cleaners and solvents may contain hazardous or deadly vapors and are highly flammable. Use only in a well ventilated area. Do not use on hot surfaces (engines, brakes, etc.).
- Due to risk of personal injury, be sure the engine is cold before beginning any service or maintenance work.
- Do not quick-charge the battery (for boost starting) for longer than one minute. Wait at least one minute before boosting the battery a second time.

CAUTIONS—
*See also **WARNINGS** on previous page.*

- If you lack the skills, tools and equipment, or a suitable workshop for any procedure described in this manual, we suggest you leave such repairs to an authorized Porsche dealer or other qualified shop.
- Porsche is constantly improving its cars and sometimes these changes, both in parts and specifications, are made applicable to earlier models. Therefore, part numbers listed in this manual are for reference only. Always check with your authorized Porsche dealer parts department for the latest information.
- Before starting a job, make certain that you have the necessary tools and parts on hand. Read the instructions thoroughly and do not attempt shortcuts. Use tools appropriate to the work and use replacement parts meeting Porsche specifications. Makeshift tools, parts and procedures do not make good repairs
- Do not use pneumatic and electric tools to tighten fasteners, especially on light alloy parts. Use a torque wrench to tighten fasteners to the tightening torque specification listed.
- Be mindful of the environment. Before you drain the crankcase, find out the proper way to dispose of the oil. Do not pour oil onto the ground, down a drain, or into a stream, pond or lake. Dispose of in accordance with Federal, State and Local laws.
- Do not expose electronic control modules to temperatures from a paint-drying booth or a heat lamp in excess of 203°F (95°C). Do not expose control modules to temperatures in excess of 185°F (85°C) for more than two hours.
- Before doing any electrical welding on cars equipped with ABS, disconnect the battery negative (–) terminal (ground strap) and the ABS control module connector.
- Make sure ignition is OFF before disconnecting the battery.
- Label battery cables before disconnecting. On some models, battery cables are not color coded.
- Disconnecting the battery may erase fault code(s) stored in control module memory. If the Check Engine light (malfunction indicator light or MIL) is illuminated or any other system faults are detected (indicated by an illuminated warning light), be sure to retrieve and record fault codes using a Porsche scan tool or equivalent prior to disconnecting the battery
- If a normal or rapid charger is used to charge battery, disconnect the battery and remove from the vehicle in order to avoid damaging paint and upholstery.
- Do not quick-charge the battery (for boost starting) for longer than one minute. Wait at least one minute before boosting the battery a second time.
- Slow-charge a sealed or "maintenance free" battery at an amperage rate that is approximately 10% of the battery's ampere-hour (Ah) rating.

01 Vehicle Identification and VIN Decoder

Vehicle Identification Number (VIN), decoding

Some of the information in this manual applies only to cars of a particular model year or range of years. For example, 1995 refers to the 1995 model year but does not necessarily match the calendar year in which the car was manufactured or sold. To be sure of the model year of a particular car, check the Vehicle Identification Number (VIN) on the car.

The VIN is a unique sequence of 17 characters assigned by Porsche to identify each individual car. When decoded, the VIN tells the country and year of manufacture; make, model and serial number; assembly plant and some equipment specifications.

The Porsche VIN is on a plate mounted on top of the dashboard, on the driver's side where the number can be seen through the windshield. The 10th character is the model year code. The letters I, O, Q and U are not used for model year designation. S for 1995, T for 1996, V for 1997, and W for 1998. The table below explains some of the codes in the VIN for 1995 through 1998 Porsche 993 models covered by this manual.

Sample VIN: WP0AA2999TS32----

position: 1 2 3 4 5 6 7 8 9 10 11 12 13 14 - 17

VIN position	Description	Decoding information	
1	Manufacturing Country	W	Germany
2	Manufacturer	P	Dr. Ing. h.c. F. Porsche AG
3	Vehicle Type	0	Sports car
4	Body code	A C D	Coupe Cabriolet Targa
5	Engine code	A C	Normally aspirated Turbocharged
6	Restraint system	2	Driver and passenger airbags
7-8	Model type	99	First 2 digits of Porsche model (993)
9	Check digit		0 - 9 or X, calculated by NHTSA
10	Model year	S T V W	1995 1996 1997 1998
11	Assembly plant	S	Stuttgart, Germany
12	Model type	3	Third digit of Porsche model (993)
13	Engine/body type	2 4 7 8	Coupe Cabriolet Turbo Targa
14-17	Serial Number		Sequential production number for specific vehicle

02 Product Familiarization

TABLES

General

This product familiarization section contains introductory and technical information for the 1995 through 1998 Porsche 911, including a year-by-year model summary.

This model was internally designated by Porsche as the Type 993. The 993 represented the final configuration of the original air-cooled 911 series of 1963. Coinciding with the 30 year anniversary, the 911 Carerra was developed in 1993 and launched in the US in 1994 as a 1995 model.

During the 4 year production run, a small number of model variants and technical systems were introduced. The Coupe was available first, and soon there after the Cabriolet was introduced. Rear wheel drive (C2) and all wheel drive (C4) versions were available. The glass-roofed Targa was added to the lineup in 1996. Additional special models included the Turbo, the Turbo S, the Carrera S and the Carerra 4S.

The most dramatic changes to the 993 when compared to the previous 911 (type 964) were the sleek redesign of the body and heavily modified suspension systems.

The 993 exterior featured a smoother front and rear, with a noticeably more aerodynamic front end. This restyle was carried out by Tony Hatter, under the supervision of design chief Harm Lagaay.

Along with the revised bodywork, the Porsche 993 received a multi-link rear suspension that dramatically improved ride comfort and handling. The new suspension, along with chassis refinements, improved the 993's handling over the previous model.

The original 1995 3.6 liter air-cooled boxer engine developed an impressive 272 hp owing to improved engine management and exhaust design. Beginning with model year 1996, the upgraded 3.6 liter engine's power increased to 286 hp.

Engine

This section covers general information on engine mechanical components.

The Porsche 993 was fitted with an air-cooled horizontally-opposed ("boxer") engine. Two basic versions of the M64 normally-aspirated engine were used in the 1995 - 1998 cars covered by this manual. General engine specifications are listed in **Table a**.

See also:

- **03 Service and Maintenance** for drive belt service, oil change, air filter service, spark plug replacement, as well as other routine maintenance item procedures.
- **10a Engine Removal and Installation**.
- **10b Engine Disassembly and Assembly**.
- **13a Engine Crankshaft**.
- **13b Pistons and Cylinders**.
- **15a Cylinder Head**.
- **15b Camshafts**.
- **17 Lubrication System** for lubrication system description and crankcase seal service.

NOTE–

- *Although there is brief mention of the Turbo models in this section, Turbo models are not covered in this repair manual.*

02

Table a. Porsche 993 engine applications and specifications

Year model	Engine code	Displacement cc (cid)	Stroke x bore mm	Comp. ratio	Horsepower hp @ rpm	Torque lb-ft @ rpm	Engine management, features
Normally aspirated models							
1995 Carerra, Carrera 4	M64.07 M64.08 (Tiptronic)	3600 (219.7)	100 x 76.4	11.3 : 1	272 @ 6100	243 @ 5000	Bosch Motronic M 2.10.1 Hot film mass air-flow sensor with adaptive knock control
1996 - 1998 Carerra, Carrera 4, Carerra S, Carerra 4S	M64.23 M64.24 (Tiptronic)	3600 (219.7)	100 x 76.4	11.3 : 1	286 @ 6100	250 @ 5250	Bosch Motronic M 5.2 (OBD II) with Varioram induction
Turbo models							
1996 - 1998 Turbo	M64.60	3600 (219.7)	100 x 76.4	8.0 : 1	400 @5750	400 @4500	Bosch Motronic M 5.2 (OBD II) Twin Turbo
1997 Turbo S	M64.60	3600 (219.7)	100 x 76.4	8.0 : 1	424 @ 5750	400 @ 4500	Bosch Motronic M 5.2 (OBD II) modified Twin Turbo

M64 air-cooled boxer engine

M64 3.6 liter air-cooled engine features:

- Hydraulic valve lifters
- Dry sump lubrication
- Intake resonance charging
- DME with hot film mass air flow sensor
- Twin ignition with adaptive knock control
- Electronic engine idle air regulation
- Three-way catalytic converter with Lambda control

Technical updates for 1996-1998 engine (M64.23/24):

- Revised cylinder heads with larger valves
- Revised pistons
- Modified cam lobe shape
- Modified hydraulic valve lifter element
- Varioram intake system

The two-piece crankcase is made of aluminum-silicon alloy. The crankshaft is an eight-bearing design.

The cylinders are made of high-temperature resistant light alloy. The Nikasill bore surface coating functions to compensate for varying temperatures across the cylinder and to maintain a constant running clearance. The bore is machined with a slight taper.

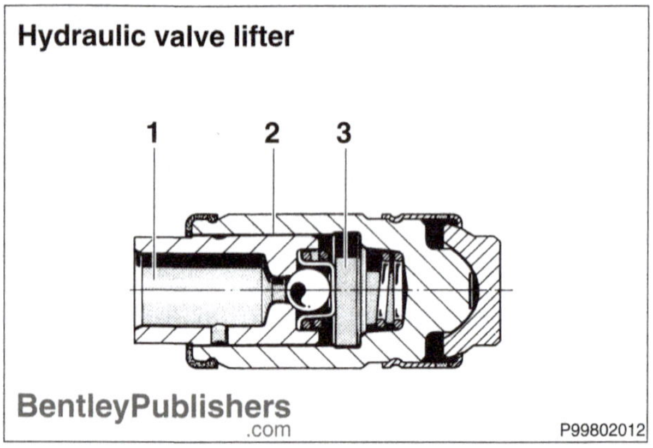

The M64 engine features hydraulic valve lifters. The lifters are inserted in the rocker arms. The reservoir chamber (**1**) is supplied with pressurized engine oil. When the valve clearance is readjusted, the oil flows past the ball into the pressure chamber (**3**). The oil escapes slowly via the drain gap (**2**) to ensure proper closing of the valve.

The camshafts are driven by the intermediate shaft via two duplex roller chains. The chains are guided by tensioning and guide rails. Oil-fed chain tensioners are used to tension the chains and to dampen chain oscillations.

Valvetrain

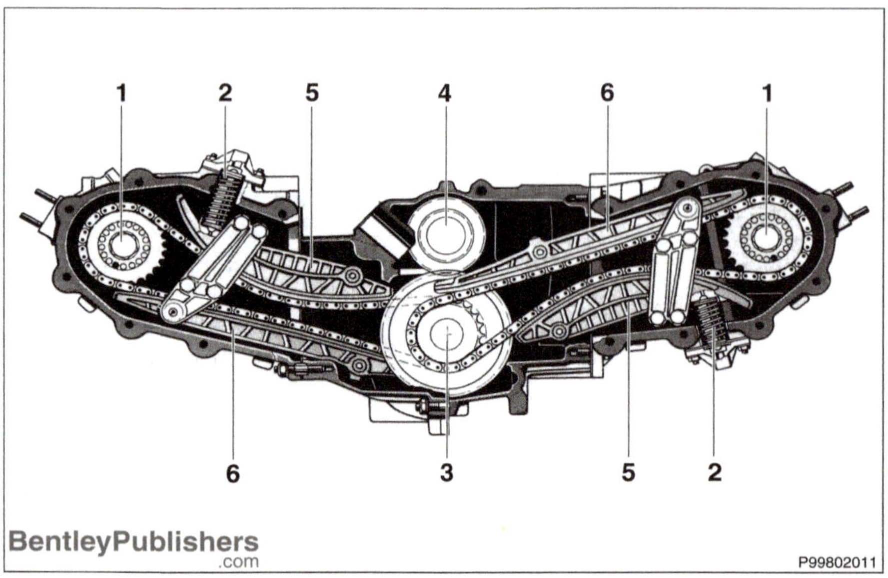

1. Camshaft
2. Chain tensioner
3. Intermediate shaft
4. Crankshaft
5. Chain tensioning rail
6. Chain guide rail

1996 - 1998 M64 full load curves

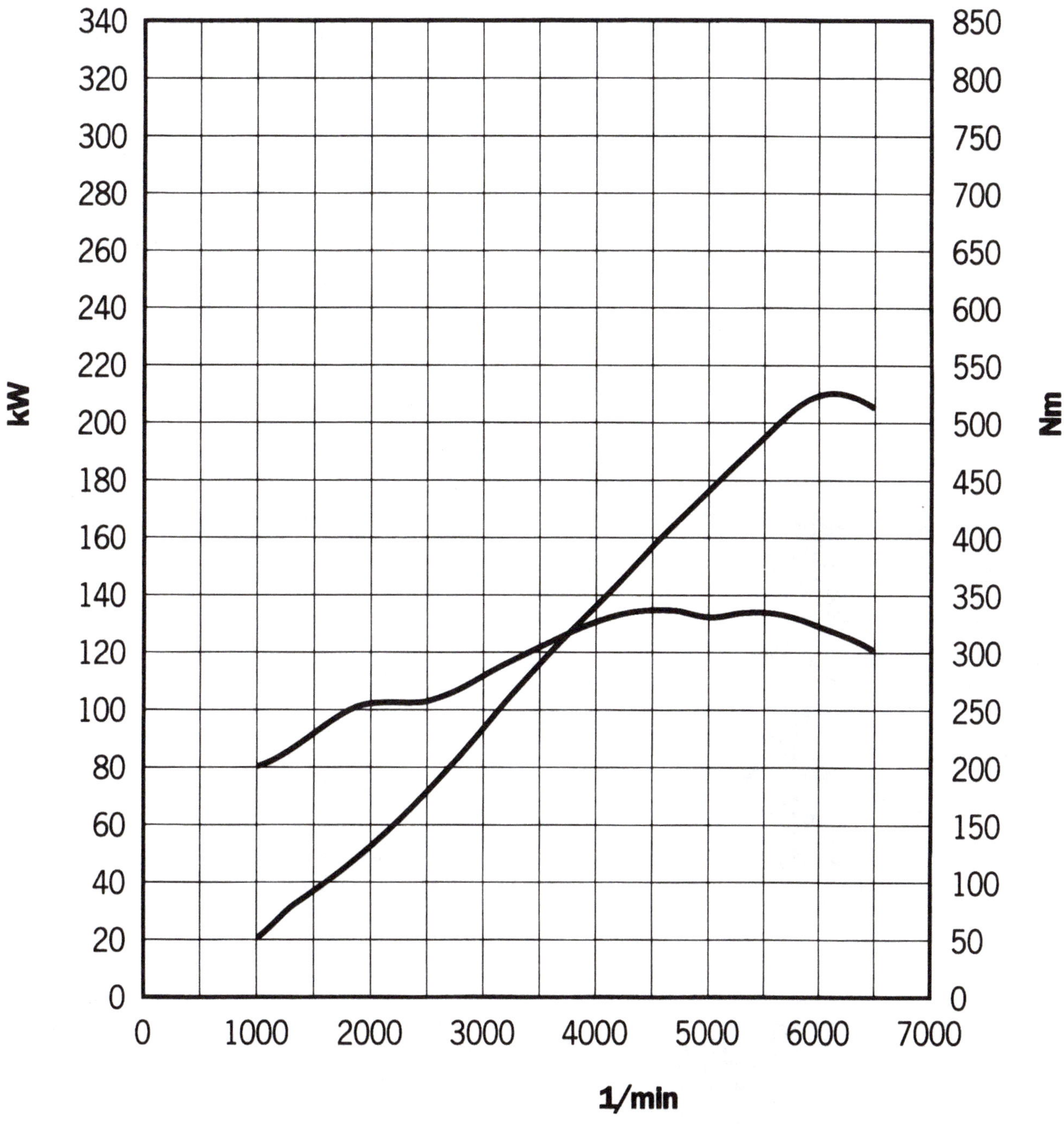

Displacement	3,600 cm^3	Max. torque	251 ft-lb (340 Nm) @ 5250 rpm
Stroke	76.4 mm		
Bore ∅	100 mm	Compression ratio	11.3:1
Engine power	285 hp (210 kW) @ 6100 rpm	Fuel grade	98 RON / 85 MON

P99802009

Engine management

There are two versions of engine management systems used on the 1995 through 1998 911 Carrera models, depending on engine application/model year. These systems, that control fuel injection and ignition functions, are referred to as DME (for digital motor electronics) or Motronic. See **Table b**.

Table b. Engine management systems	
Year, engine	**DME system**
1995 (M64.07, M64.08)	Bosch M 2.10.1
1996 - 1998 (M64.23, M64.24)	Bosch ME 5.2

The engine control module (ECM) uses signals from the following sensors as the primary inputs to electronically control fuel delivery and ignition timing:

- Oxygen sensors
- Mass air flow sensor
- Knock sensors
- Crankshaft sensor
- Camshaft sensors
- Air and coolant temperature sensors
- Throttle position sensor

Each DME system has the same basic components and operating principles. The most notable difference is that the 1996 and later cars (DME 5.2) have advanced diagnostic capabilities (OBD II), Varioram intake system, and stereo air/fuel control with 4 oxygen sensors (1995 models use one oxygen sensor).

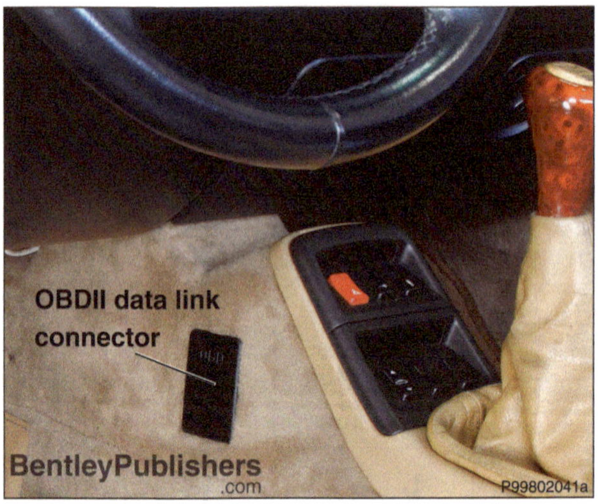

DME 5.2 engine management is compliant with second generation on-board diagnostics (OBD II). OBD II diagnostic software is capable of recognizing and electronically storing hundreds of diagnostic troubles codes (DTCs) in the engine control module (ECM). DTCs can be accessed using special test equipment (scan tool) via the DLC (data link connector). See **OBD On-Board Diagnostics** for OBD II fault codes.

Varioram intake system

To achieve high engine output and increased torque in the middle rpm range, the M64.23/24 engines are fitted with a timed tuning flap and variable-length intake pipes integrated into the intake manifold.

One section of the intake manifold is bolted to the cylinder head while the other section slides in or out to double the overall intake pipe length. One vacuum-controlled slide is used on each cylinder bank. The slides are actuated by solenoids that are connected to engine vacuum.

CLUTCH AND TRANSMISSION

The Carrera uses a longitudinal, rear-engine drivetrain. The transmission is bolted directly to the rear of the engine, behind the passenger compartment.

See also:

- **30 Clutch**
- **34 Manual Transmission**
- **37 Automatic Transmission**

Manual transmission

The clutch on manual transmission cars uses a single dry disc and dual-mass flywheel.

BentleyPublishers.com P99802026

The Porsche 993 was fitted with one of two different 6-speed manual transmissions for the US market, depending on the drive line configuration. See **Table c**.

Table c. Manual transmission applications (993 models)	
Carerra, Carrera S (RWD)	G50.20, 6-speed
Carrera 4, Carrera 4S (AWD)	G64.20, 6-speed

See **34 Manual Transmission** for additional information.

Tiptronic (automatic) transmission

The 993 automatic is equipped with a 4-speed Tiptronic type A50 transmission. See **Table d**.

BentleyPublishers.com P99802025

The Porsche Tiptronic is an electronically controlled automatic transmission that can be operated in automatic or manual mode. When in manual mode, automatic gear changes are only made at pre-defined engine speed limits to protect the engine against overrevving or underrevving. See **37 Automatic Transmission for additional information**.

Table d. Automatic transmission applications (993 models)	
1995 - 1996	A50.05
1997 - 1998	A50.07

Suspension, Brakes and Steering

The rear engine design, MacPherson strut front suspension and multi-link rear suspension ensure optimal driving safety, maneuverability and handling. Lightweight and high performance components, including aluminum control arms, crossmembers, and wheel bearing carriers are used throughout.

Type 993 cutaway

BentleyPublishers.com

P99802001

Front suspension

The Porsche 993 uses a MacPherson strut front suspension with coil springs. It is a compact, lightweight design that provides outstanding handling and ride comfort. The MacPherson strut design improves steering precision and suspension dynamics.

The front suspension is supported by an aluminum crossmember, two aluminum side members, and lower control arms mounted to individual wheel bearing carriers at each side. See **40 Front Suspension, Front Drive Axles** for additional information.

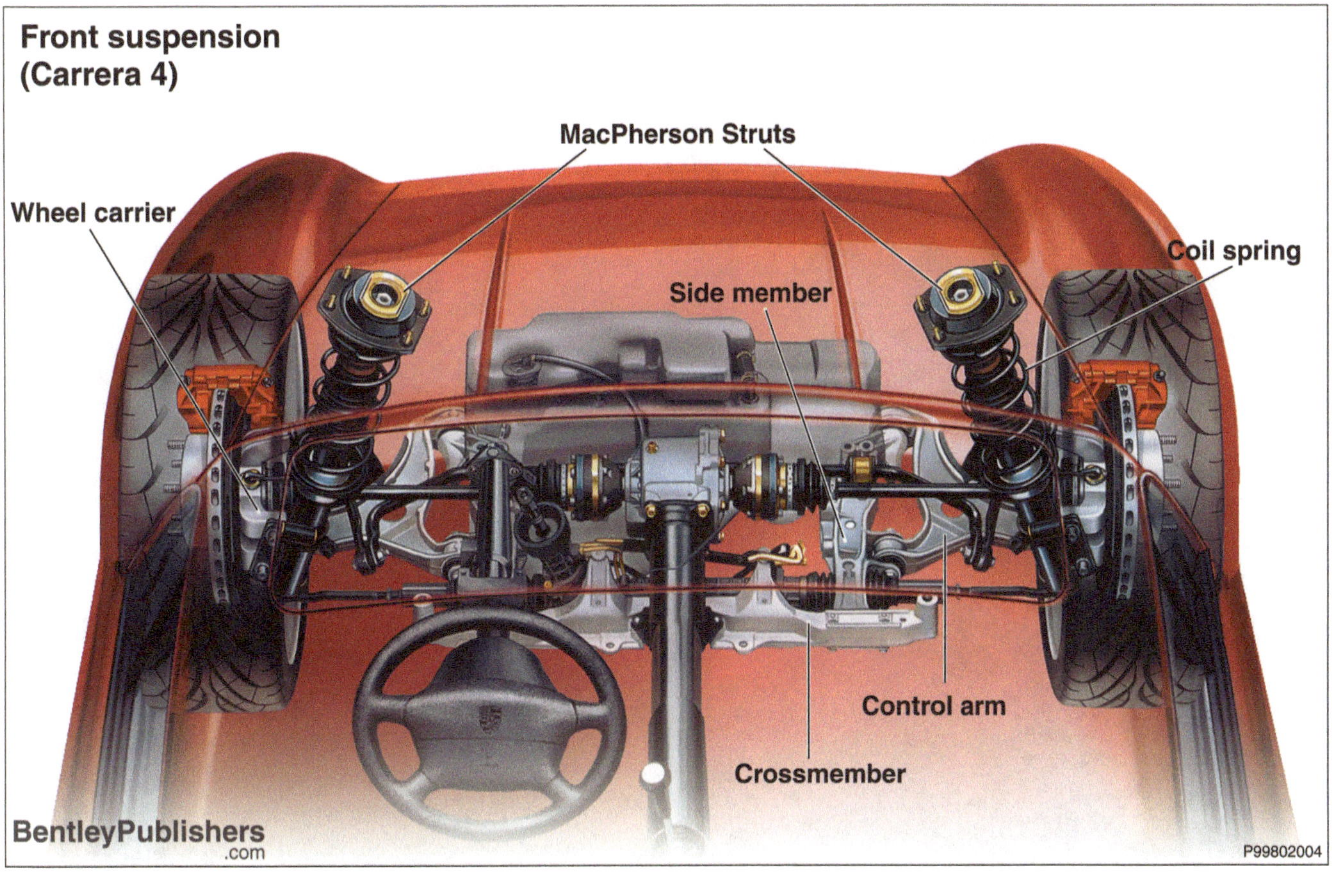

02

Rear suspension

The rear suspension uses a weight-optimized LSA (light, stabile, agile) design that enhances the dynamics of the vehicle. The multi-link LSA rear suspension is a race-proven, compact design that utilizes a rear subframe assembly consisting of two upper members and a crossmember.

Wheel carriers are mated to the subframe via two upper control arms, a lower toe control arm and a lower A-arm. The entire subframe assembly uses four rubber body mounts to reduce noise and vibration. See **42 Rear Suspension, Rear Drive Axles** for additional information.

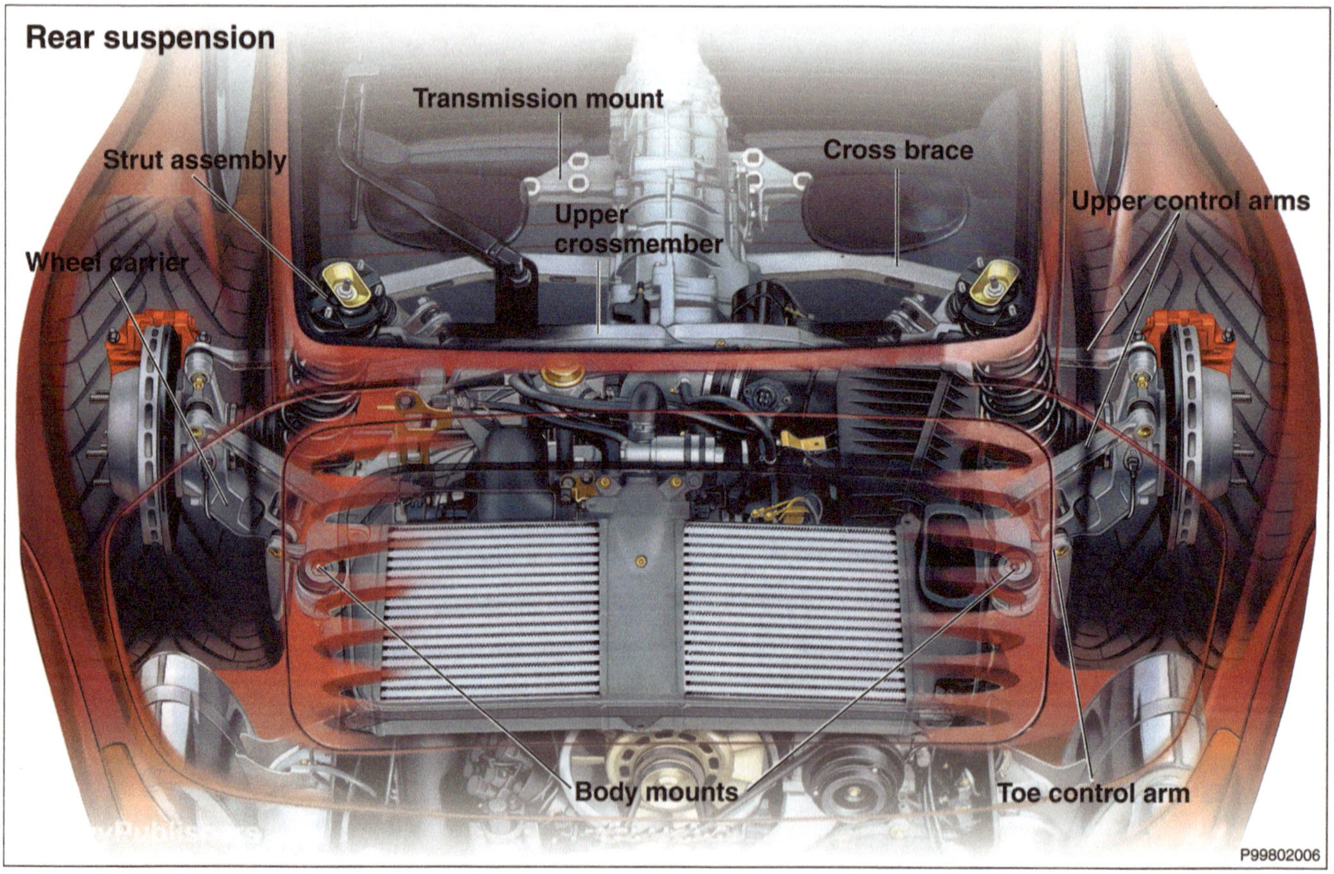

Steering

The power-assisted rack and pinion steering combined with lightweight aluminium components creates quick response and excellent road feel. The power steering rack mounts to the front subframe assembly. The 993 has a steering ratio of 16.48:1.

The engine-mounted power steering pump is driven off of the camshaft via a belt. The hydraulic steering fluid reservoir is located on the right side of engine bay.

NOTE–

- *The 993 power steering system used Dexron II D ATF on cars built before March 1996 and Pentosin (CHF 11S) power steering fluid on cars built March 1996 and later. See* **48 Steering** *for more information.*

Brakes

Porsche 993 models are equipped with an ABS system with an optional ABD (automatic brake differential) system.
See **45 Anti-lock Brakes (ABS)** for additional application information.

The ABS brake system is a dual circuit design with either a vacuum brake booster (3-channel ABS system) or a hydraulic booster (4-channel ABS system). See **Table e**.

Table e. Brake system applications	
Vacuum boost	Carerra, Carerra S
Hydraulic boost	Carrera 4, Carrera 4S, Turbo, Turbo S

The brake circuits are split for the front and rear axles and use a proportioning valve (two proportioning valves if equipped with ABD) to distribute pressure.

Ventilated and cross-drilled front and rear brake rotors help reduce braking temperatures at the wheels. Light alloy four-piston brake calipers are used at each corner.

Body

Coupe

The 993 Coupe bodyshell was largely based on that of its predecessor and consists of a steel body with a welded box frame and bolted-on front fenders. Optimum corrosion protection was achieved by combining cathodic electrophoretic dip priming with undercoating and a high quality topcoat.

To reduce weight while simultaneously retaining the required fatigue strength, energy consumption potential in a crash and full corrosion resistance, a number of structural body components were made from hot dip galvanized high tensile strength (HSS) steel panels.

When compared to its predecessor, the 993's torsional strength was increased by 20% at the same weight.

At the time of manufacturing, passenger safety requirements were exceeded with the 993, resulting in a rigid passenger compartment and an efficient crumple zone for the occupants.

Rollover safety was provided by strong body pillars, roof frame cross sections, and structural joints.

1995 Carrera Cabriolet
P99802028

Cabriolet

The 993 Carrera was designed from the start with a Cabriolet version in mind. As a result, the only differences between the bodies of the Cabriolet and the Coupe are a few reinforcements and the modifications required to mount the convertible top.

In detail, the following modifications were made:

- Reinforcements are incorporated in the A-pillars for improved rollover protection.
- The upper cowl panel is modified for mounting the convertible top lock.
- The central rear panel between the convertible top and the engine compartment lid is modified.
- The convertible top mounts and the upper safety belt mounts for the front seats are installed in the B-pillars.
- The inside rear wall is reinforced for mounting the convertible top frame.
- A longer gusset plate is installed in the sill to improve torsional stiffness.

See **61 Convertible Top** for additional information.

P99802007

Targa

The 993 Targa body design is largely identical to that of the 993 Cabriolet and features the same body shell reinforcements.

The 993 Targa offers full 2+2 seating. Compared with the Coupe, headroom is the same in the rear and is slightly more generous in the front.

The roof assembly is light and rigid. It is an attractive complement to the classic lines of the 911 and, when seen from the side, features a body styling that reflects the lines of the Coupe.

The roof assembly includes the following components:

- Roof rail with side window glass.
- Wind deflector.
- Glass roof.
- Rear window.
- Drive motor.

The wind deflector and the glass roof are made of durable twin-layer safety glass. The glass is tinted, offering UV protection and is opaque from the outside. The rear window is also made of tinted safety glass and is electrically heated. See **60 Sunroof, Targa Top** for additional information.

Model Year News

The 911 Carrera (993) was produced for 4 years for the US market. The information given here details the 993's annual model introductions, product features, and technical updates.

1995 model year

For the 1995 model, the Coupe and the Cabriolet were available, although the Cabriolet was introduced later in the model year.

The C4 Coupe and C4 Cabriolet with all-wheel-drive were also available.

The 3.6 liter 272 hp (M64.07/08) engine was unique to the 1995 model year.

1996 model year

The 1996 model introduced additional body styles and models, as well as many significant technical updates, including:

- Carrera 4S introduction.
- Targa introduction.
- Turbo introduction.
- A more powerful 285 hp normally-aspirated engine with Varioram and OBD II (On-Board Diagnostics, generation II).
- Revised Tiptronic tuning.
- Redesigned vehicle key with integrated remote for vehicles with immobilizer.
- New line of Becker radios with optional 6-disc CD changer.

◄ Optional Litronic (xenon) low-beam headlights.

◄ The 911 Carrera 4S is based on the body of the 911 Turbo. It is fitted with a naturally-aspirated engine, combined with the Varioram system and features the wider Turbo body work and an electrically retractable spoiler instead of a fixed Turbo spoiler. The suspension was slightly lower than that of the Carrera. Wheel spacers (31 mm / 1.2 in) were used in the rear for wheel fitment in the wider rear fenders.

◄ The 911 Targa is based on the 911 Cabriolet construction and features an extended glass roof.

Turbo

P99802038

Turbo

A Turbo version of the Porsche 993 was launched in 1995 as a 1996 model. The Turbo was the first standard production Porsche with twin turbochargers and the first 911 Turbo to be equipped with standard all-wheel drive (not available with rear-wheel drive). This flagship model featured a 3.6 liter turbocharged engine producing 400 hp. The Turbo was fitted with twin turbochargers, air-to-air intercoolers, DME 5.2 (OBD II) engine management, redesigned cylinder heads, as well as many other modified engine components.

The Turbo bodywork differed from the Carrera body in that it was by widened at rear wheel arches (approximately 60 mm / 2.4 in), The Turbo also featured a fixed rear spoiler that housed the intercoolers. Special 18-inch (460 mm) alloy wheels with weight-reducing hollow spokes were standard issue.

Turbo

P99802037

1997 model year

In 1997, Porsche introduced a limited run (182 units) of the 911 Turbo S. This exclusive model featured higher performance (additional 24 hp, or 424 hp), a Turbo S vented rear spoiler, engine cooling ducts in the rear quarters, yellow calipers, and carbon fiber interior appointments.

Carerra S

The 1997 Carrera S was similar to the Carerra 4S from the previous model year. Like the 4S, the S shared the Turbo model's basic body, but used the naturally aspirated Carrera engine.

1998 model year

The 1998 models were carried over from the 1997 models

Option Codes

Table f. Option codes	
Code	Description
M001	Carrera Cup
M002	911 Carrera RS road version
M003	911 Carrera RS Club Sport
M005	Carrera RSR
M007	Carrera RS for Japan (with rear seats, electric windows, airbags)
M018	sports steering wheel with elevated hub 30 mm
M020	speedometer with two scales mph-km/h
M024	version for Greece
M030	sport suspension
M032	touring suspension
M033	lowered suspension
M034	version for Italy
M058	impact absorbers front and rear
M061	version for Great Britain
M062	version for Sweden
M063	version Luxemburg
M064	version for Netherlands
M065	version for Denmark
M066	version for Norway
M067	version for Finland
M068	version for Thailand
M069	other country version
M071	EU country version
M092	special model Turbo S - 1997
M111	version for Austria
M113	version for Canada
M114	version for Taiwan
M119	version for Spain
M124	version for France
M126	control and indications in French
M127	control and indications in Swedish
M130	control and indications in English
M139	seat heating system, left seat
M150	without catalyst, operates with leaded gas
M159	engine sound package
M185	rear safety belts 2-point, automatic
M186	rear safety belts 2-point, manual
M193	version for Japan
M197	stronger battery
M215	version for Saudi Arabia
M219	differential
M220	locking differential 40%
M224	automatic limited slip differential
M225	version for Belgium
M249	Tiptronic transmission
M261	outside mirror - plain - passenger's side, electrically adjustable and heatable

Table f. Option codes	
Code	Description
M277	version for Switzerland
M288	headlamp washer
M320	Porsche/Becker CR-11 cassette radio
M324	Porsche/Becker CR-10 cassette radio
M325	version for South Africa/New Zealand
M326	Porsche/Becker CR-21 radio
M329	Blaupunkt Toronto SQR 48 or Porsche/Becker CR-210 cassette radio
M330	Blaupunkt Toronto SQR 46 or Porsche/Becker CR-31 cassette radio - 1997
M331	Porsche/Becker CR-1 cassette radio
M333	Blaupunkt Paris RCR 41 cassette radio
M334	Blaupunkt Bremen RCM 42 cassette radio
M335	rear automatic safety belt, 3-point,
M336	Blaupunkt Duesseldorf RCR 84 cassette radio
M338	rear wheel drive
M339	4 wheel drive
M340	seat heating system, right seat
M371	sport seat, left manually adjustable
M372	sport seat, right manually adjustable
M373	sport seat, left, electrically height adjustable
M374	sport seat, right, electrically height adjustable
M379	standard seat, left, electrical vertical adjustment
M380	standard seat, right, electrical vertical adjustment
M381	standard seat left, manually adjustable
M382	standard seat right, manually adjustable
M383	sport seat left, electrical vertically adjustable
M384	bucket seat, left
M385	bucket seat, right
M387	sport seat right, electrical vertically adjustable
M393	17" Targa wheel
M397	16" wheels
M398	cup wheel, 17"
M401	cup wheel, 16"
M405	RS wheel, 18"
M408	18" techno wheel
M413	18" turbo-look-wheel
M419	rear luggage dump
M423	box for cassettes and coins
M425	rear window wiper
M426	without rear window wiper
M432	steering wheel with Tiptronic control
M433	leather steering wheel Atiwe
M435	leather steering wheel MOMO
M437	comfort seat left, electrically adjustable
M438	comfort seat right, electrically adjustable
M439	electrical roof catch
M441	radio preparation
M445	silver wheel trim with full-color Porsche crest
M446	concave hub cape with full-color Porsche crest

02

Table f. Option codes	
Code	Description
M451	reduced radio preparation
M454	automatic speed control
M455	wheel locks
M459	strut bar
M467	outside mirror for driver's side, convex, electrically heatable and adjutable
M470	fixed spoiler
M471	fixed spoiler with additional wing
M473	power left seat (all)
M479	version for Australia
M480	6-speed manual transmission
M484	version for USA
M487	control for fog lights with side marker lights
M488	inscriptions in German language
M490	sound system
M491	turbo-look Carrera 4S
M492	headlights for left-hand traffic
M493	narrow bodywork
M495	Carrera S
M498	without model designation on rear end
M499	version for Germany
M513	back-support (lordose) right seat
M520	passenger compartment monitoring
M529	anti-drive-off feature 315 mhz
M530	anti-drive-off feature to 1996, anti-drive-off feature 433 mhz 1997
M534	theft security system
M544	75 liter fuel tank
M545	92 liter fuel tank
M551	wind deflector
M553	version for USA, Canada
M561	airbag driver's side
M562	airbag driver's side and front passenger's side
M564	without airbags
M566	fog headlamps, white
M567	windshield tinted, upper part darker colored
M573	air conditioner
M576	without rear fog lamp
M586	back-support (lordose) left seat
M599	roof cover molding
M601	Litronic headlamp
M602	araised stop lamp
M605	headlight vertical aim control
M611	preparation of telephone installation Nokia (D-network)
M614	preparation of telephone installation Motorola (D-network)
M615	facility for the installation of a Nokia mobile phone
M631	child's seat detection
M650	electrical sliding roof
M651	electric window opener
M656	without power-assisted steering

Table f. Option codes	
Code	Description
M657	power assisted steering
M659	on-board computer
M660	OBD 2
M661	stricter emission-control concept
M680	-mj 01 digital sound package/ mj 02- sound system Bose analog
M685	rear seats, divided
M686	Porsche/Becker CDR-21 CD radio
M687	Blaupunkt Muechen RD 104 CD radio
M688	Porsche/Becker CDR-210 CD radio
M692	Porsche CD changer
M693	Blaupunkt London RDM 42 CD radio
M694	Turbo
M936	seat covers rear, leather
M937	seat covers rear, leatherette
M938	seat covers rear, cloth
M939	seat covers rear, draped leather
M946	seat covers front, leather/leather/leatherette
M975	luggage boot trim with velour carpet
M981	leather equipment without seat covers
M982	seat covers front, draped leather/leather/leather
M983	seat covers front, leather
M990	seat covers front, cloth/cloth/leatherette
P02	stereo with CD changer (Coupe)
P04	stereo with CD Changer (Targa/Cabriolet)
P08	limited slip differential (RWD)
P14	heated seat pkg includes adjustable heating range
P15	power seat pkg includes dual power seats
P31	sport chassis (includes 17" 5-spoke wheels)
P32	sport chassis (includes 18" technology wheels)
P36	special chassis with M393 (17" Targa wheels)
P49	digital sound package
PCX	Porsche cellular phone handset kit
S11	cellular telephone with mounting bracket (includes installation kit)
SAC	AC rapid charger
SBE	battery eliminator-cellular phone
SLB	large battery-cellular phone
SUL	ultra light battery-cellular phone
SVD	voice dialer kit-cellular phone
X09	console with special leather extended center
X17	light rootwood dashboard
X18	dark rootwood dashboard
X19	leather lower dashboard
X26	steering wheel-special leather
X28	light rootwood steering wheel
X30	dark rootwood steering wheel
X31	light rootwood parking brake lever
X32	dark rootwood parking brake lever
X34	painted instrument dials includes leather trim on instrument rings

Table f. Option codes	
Code	Description
X40	leather sun visors with passenger reading light
X43	leather console switch frame
X52	console CD storage for 5 discs
X53	console CD storage for 8 discs
X54	oval chrome exhaust pipes
X56	lower carbon dashboard
X57	carbon door trim panels
X68	tonneau cover (Cabriolet)
X70	metal door sill with insignia
X71	aluminum instrument dials
X76	flared rocker panels
X77	carbon steering wheel
X86	light rootwood door trim panels
X87	dark rootwood door trim panels
X89	painted Porsche crest rim caps
X97	aluminium/leather shift knob
X98	aluminium/leather parking break handle
XC3	hardtop (Cabriolet)
XC8	light rootwood gear shift knob
XC9	dark rootwood gear shift knob
XD1	Aero II Turbos S front air dam
XD2	Aero II Turbo S rear deck lid
XD3	rain sensing windshield
XD4	Porsche crest rim caps
XD9	wheels painted to match vehicle's color
XE3	automatic dimming rear view mirror
XF5	leather covered instrument rings
XF6	leather gear box tunnel
XF7	leather covered tray located behind parking brake
XJ4	leather ignition lock rosette
XJ6	leather steering wheel casing
XK7	leather wrapped shift knob and cover (deviating colors or to sample)
XMA	roof, A and B pillar with leather of choice
XM5	leather control knobs
XM7	leather covered glove box knob
XM8	leather sun visor with mirror and reading light
XN1	leather power window switches
XN2	leather inside door handles
XN3	leather fresh air side dash vents
XN4	leather fresh air center vent
XN7	leather parking brake lever
XN8	leather sun visors
XN9	leather sun visors with mirror
XP3	leather Cabriolet boot
XP6	leather seat belt locks & housing
XP9	leather radio speaker trim rings
XR3	leather seat hinges

Table f. Option codes	
Code	Description
XR6	leather front backrests lock controls
XR7	rear window washer
XRC	brake calipers painted black
XRD	brake calipers painted red
XRF	brake calipers painted yellow
XSA	sport seat
XTA	carbon fiber inside door latch
XV1	leather covered ventilating cover
XV2	leather fresh air side vents
XV3	leather air conditioning switches
XV4	leather heat adjuster switches
XV5	leather seat adjuster switches
XV6	leather ring plates for seat adjuster
XV7	leather fuel tank pull knob
XV9	leather B-pillar seat belt covers
XW2	leather rear seat belt locks & housing
XW4	leather light switch
XW5	leather wiper/instrument light knobs
XW6	leather door lock pin rosettes
XW8	leather door entrance panel sills
XW9	leather entrance panel covers
XX1	embroidered floor mats (requires leather interior trim)
XX2	footwell lights
XX3	map pocket lights
XX6	leather shift lever knob pattern insert with deviating colors
XY5	leather Tiptronic shift plate
XZ7	leather OS mirror adjuster frame
Y01	spoiler-aero kit I with front duct
Y02	spoiler-aero kit II with rockers
Y03	shift lever/brake handle-carbon
Y04	carbon pkg I - lower dash, door trim, gauge panel
Y05	shift lever/brake handle-carbon and aluminum
Y06	shift lever/brake handle-aluminum
Y07	shift lever/brake handle-aluminum and light wood
Y08	shift lever/brake handle-aluminum and dark wood
Y09	light rootwood package Tiptronic, X17 dash, X25 shifter, X28 steering wheel, X31 brake handle (requires color)
Y10	dark rootwood package Tiptronic, X18 dash, X24 shifter, X30 steering wheel, X32 brake handle (requires color)
Y17	exclusive pkg II - leather seat belt locks and housings, shifter tunnel, parking brake, tray behind parking brake
Y23	aluminum Tiptronic selector includes aluminum brake lever
Y24	carbon Tiptronic selector includes carbon brake lever
Y25	carbon package II (Y04 plus: steering wheel, shifter, brake handle)
Y26	carbon Tiptronic package X56 lower dash, X57 door panel, X48 shifter (handbrake), X77 steering wheel (requires color)
Y29	aluminum/chrome pkg (instruments, tailpipe, door sill)
Y59	light wood pkg (steering wheel, shifter, brake lever, dash)
Y60	dark wood pkg (steering wheel, shifter, brake lever, dash)
Y61	carbon Tiptronic shifter includes carbon hand brake
Y62	light rootwood/aluminum Tiptronic shift selector & brake handle (requires color brake base)
Y63	dark rootwood/aluminum Tiptronic shift selector & brake handle (requires color brake base)

Table f. Option codes	
Code	Description
Y65	dash pkg (leather vent covers)
Y66	door pkg (leather sills, pwr switches, locks, speaker trim, etc)
Y67	leather seat pkg (switches, hinges, adjusters, etc)
Y75	front and rear spoilers with aero kit
Z02	leather console tray
Z04	shift lever knob & hand brake in leather single seam stitch (requires color)
Z05	leather headliner (custom color) includes A and B pillars and visors
Z06	steering wheel-deviating color
Z07	trunk carpeting (matches interior color, without CD changer)
Z08	trunk carpeting (Matches interior color, with CD changer)
Z09	custom leather carpet welting (color matched to sample interior)
Z10	carpet welting leatherette prior model year color (subject to availability)
Z11	leather carpet welting standard color or matching interior color
Z21	seat stitching-alternate standard colors
Z23	door handle stitching in deviating color
Z24	dashboard stitching in deviating color
Z25	center console stitching in deviating color
Z26	door storage top stitching, deviating (requires color) available on full leather interior only
Z27	headliner, leatherette (deviating current color; visor & pillars)
Z27	headliner, standard leather
Z27	leather headliner (Porsche colors) includes A and B pillars and visors
Z28	roofliner, A&B pillars, XN8 in current deviating color in leatherette (requires color) N/A cashmere
Z31	leather knee bar
Z35	leather knee bar (requires all-leather interior)
Z41	pearl white metallic paint (requires metallic paint to sample)
Z42	door trim panels-leatherette
Z43	door trim panels-leather (includes armrests, requires full leather interior)
Z44	leatherette door panels, armrest/storage compartment lids & handles-leatherette interior, deviating standard color (requires color)
Z45	front seat inlays (deviating current colors; requires full leather seats)
Z46	rear seat inlays (deviating current colors; requires full leather seats)
Z47	rear seat inlays in leatherette, deviating color (requires color)
Z49	front seat belts in deviating color
Z50	seatbelts, red or yellow front
Z51	front/rear seat inlays (deviating current color; requires leather)
Z53	head rests stamped with Porsche crest (requires leather upholstery)
Z65	leatherette dash, knee bar, window sills, back wall top section-deviating standard color not available with full leather interior or light colored dashboard (requires color)
Z66	leatherette belt line (Cabriolet) in deviating standard color
Z6x	leatherette trunk in deviating color
Z69	carpet-standard color (deviating from interior colors)
Z70	carpet colors from prior model years-subject to availability (requires color)
Z79	deviating standard color A & B Pillars
Z80	leather A-pillar and window frame
Z81	leather A/B pillar
Z86	dashboard in deviating standard color

03 Service and Maintenance

General

Carry out the maintenance work described in this repair group at the factory specified time or mileage interval. Following these intervals will help ensure safe and dependable operation.

NOTE–

- *Porsche, at times, changes their recommended maintenance procedures and requirements. The information contained here may not include updates or revisions made by Porsche since the publication of the documents supplied with the car. If there is any doubt about what procedures apply to a specific model or model year, or what intervals should be followed, remember that an authorized Porsche center has the latest maintenance information.*

Emergencies

The following is a list of emergency procedures and where they are covered in this manual:

- Jump starting. See **Jump starting** in this repair group.
- Tire change. See **Flat tire, changing** in this repair group.
- Towing. See **Towing** in this repair group.
- Sunroof manual closing. See **60 Sunroof, Targa top**.
- Targa roof manual closing. See **60 Sunroof, Targa Top**.
- Convertible top manual closing. See **61 Convertible Top**.
- Deactivating immobilizer. See **96 Interior Lights, Anti-theft**.

Indicator and warning lights

Many vehicle systems are self-monitored for faults while driving. Generally, a red warning light that illuminates while driving should be considered serious. If you cannot immediately determine the seriousness of the warning light, stop the vehicle in a safe place and switch engine OFF as soon as possible. Consult the owner's manual for additional information on the warning light and the recommended action.

P99890017

1996 and later models: if the malfunction indicator light (**arrow**) (MIL) (Check Engine warning light) comes on or flashes, it indicates an emissions-related fault in the engine management system. Faults such as a bad oxygen sensor or a faulty fuel injector can cause exhaust or evaporative emissions to exceed a specified limit. When these limits are exceeded, the MIL illuminates. The engine can be safely driven with the light on, but the faults should be interrogated with a scan tool as soon as possible.

If the MIL flashes, it indicates that an emissions-related fault has occurred along with a catalyst damaging misfire. Pull the vehicle a safe distance off the road, set the parking brake, and turn the emergency flashers on. Use appropriate warning device (flares, warning triangles, etc.) to alert other drivers. Arrange to have vehicle towed to a Porsche workshop. Failure to do so could result in catalytic converter damage. See **OBD On-Board Diagnostics** for more information on the MIL and the on-board diagnostic system.

Jump starting

Jump starting is not a recommended practice on vehicles equipped with sophisticated electronic systems. Voltage spikes associated with jump starting have the potential to damage sensitive electronics. The preferred method to charge a discharged battery is to remove the battery from the vehicle and slow-charge it using a 12V battery charger.

If jump starting is the only available option, the method described below can be used.

– Open trunk and pull back trunk floor carpet to expose battery.

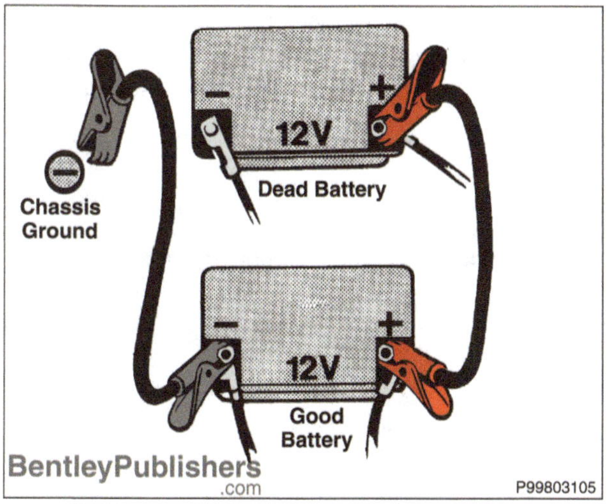

◄ Connect one end of positive jumper cable (**red**) to positive battery terminal (+). Connect other end of positive cable to positive terminal (+) of good battery.

– Connect one end of negative cable to negative terminal (–) on good battery. Connect other end of negative cable to grounding point on vehicle with discharged battery.

NOTE–

- *The grounding point should be connected to ground as far away from battery as possible. For example, a solid unpainted metal part or bolt head is a good grounding point.*

> ***WARNING—***
>
> - *When one end of jumper cables is attached to a battery, do not allow cable clamps to contact each other. This could result in electric shock.*

– On both vehicles, put transmission in neutral and engage parking brake.

– Start and run engine of vehicle with good battery. Raise idle speed.

– Start engine of vehicle with discharged battery. Do not attempt starting for more than 15 seconds. Allow a waiting period of at least one minute between attempts.

– To reduce voltage spikes, switch on one or two electrical loads (heated rear window and heating fan blower, for example) before disconnecting jumper cables.

– With engine running, carefully remove jumper cables.

Flat tire, changing

Prior to changing a flat tire, position the vehicle a safe distance off of the road, set the parking brake, and turn the emergency flashers on. Use appropriate warning devices (flares, warning triangles, etc.) to alert other drivers.

◄ Use the proper jacking points to raise the car safely and avoid damage. The jack supplied with the car can only be used at the front and rear side jacking lugs (**arrows**).

> ***WARNING—***
> - *Do not work under a raised car unless it is solidly supported on jack stands that are intended for the purpose.*
> - *Watch jack closely. Make sure it stays stable and does not shift or tilt. As the car is raised, the car may roll slightly and jack may shift.*

◄ Working in trunk, pull back trunk floor carpet. Remove spare tire, jack, and jack handle.

> ***WARNING—***
> - *Do not park vehicle where it could contact dry tall grass, brush or other flammable materials. The hot exhaust system could ignite such materials and cause fire.*

– Firmly set parking brake. Turn emergency flashers ON. Chock wheel diagonally opposite to tire being changed.

> ***WARNING—***
> - *Do not rely on transmission or parking brake to keep the vehicle from rolling. They are not a substitute for positively blocking opposite wheel.*

– Loosen lug bolts while vehicle is on ground, but leave them snug.

◄ Place jack in position, making sure raised part on jack fits into hole in jacking lug. Make sure jack is resting on flat, solid ground. Use a board or other support to provide a firm surface for jack, if necessary. Raise vehicle only far enough so that the wheel is fully off ground and then remove lug bolts and wheel.

– Install temporary collapsible emergency spare tire. Install lug bolts and tighten hand-tight using lug wrench.

– Inflate emergency spare tire using supplied air compressor in trunk. Inflate tire to 36 psi.

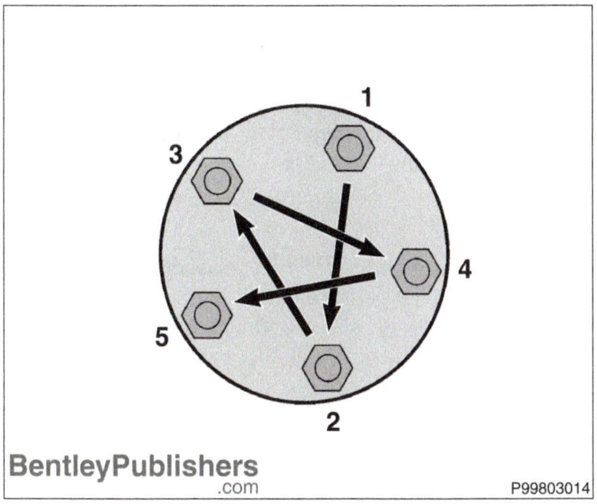

Lower vehicle. With all wheels on ground, fully tighten lug bolts in a crisscross pattern. Check inflation pressure of installed spare tire.

> ***WARNING—***
> - *Avoid high speed driving, hard cornering, and hard acceleration and braking when using the temporary spare.*

Tightening torque	
Wheel to wheel hub	130 Nm (96 ft-lb)

Towing

The recommended technique for towing vehicles covered by this manual (Carrera 2 and Carrera 4 models with automatic and manual transmission) is flat bed towing.

To access front threaded towing eye socket, carefully pry out round cover in front bumper (**arrow**).

- Working in trunk, remove threaded towing eye from tool pouch. Tool pouch is stored with spare tire and jack.
- Fully screw towing hook into tow eye.
- Use towing eye and hook for winching vehicle on flat bed.

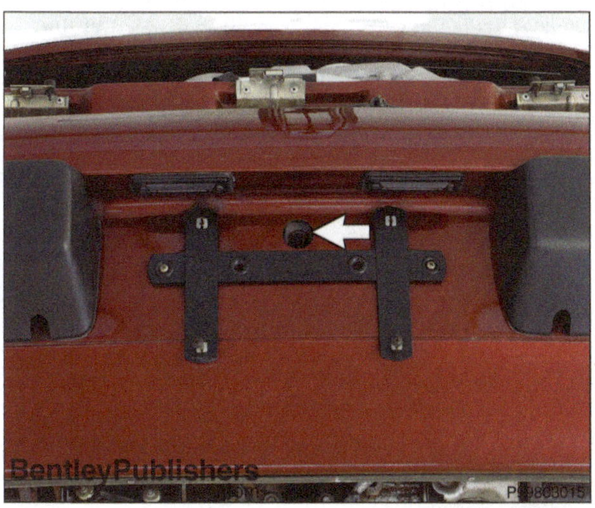

Rear tow eye (**arrow**) is located behind license plate.

Basic Service Information

Much of the necessary maintenance and minor repair described in the manual can be done with ordinary tools. Below is some important information on how to work safely, a discussion of what tools will be needed and how to use them.

Safety

Although an automobile presents many hazards, common sense and good equipment can help ensure safety. Many accidents happen because of carelessness. Pay attention and stick to safety rules in this manual. Read the cautions and warnings given in **00 Warnings and Cautions** prior to working on the vehicle.

Working under vehicle safely

WARNING—

- *A jack is a temporary lifting device. Do not use a jack alone to support vehicle while you are under it.*
- *Do not work under a lifted vehicle unless it is solidly supported on jack stands that are intended for that purpose.*
- *Do not use wood, concrete blocks or bricks to support a vehicle. Wood may split. Blocks and bricks, while strong, are not designed for that kind of load and may break or collapse.*
- *Use care when removing major (heavy) components from one end of the vehicle. The sudden change in weight and balance can cause vehicle to tip off the lift or jack stands.*
- *Do not support vehicle at engine oil pan, transmission, or on front or rear axle components. Serious damage may result.*

– Disconnect negative (–) cable from battery so that vehicle cannot be started. Let others know what you are doing.

CAUTION—

- *Prior to disconnecting the battery, read the battery disconnection cautions given in* **00 Warnings and Cautions**.

– Use proper jacking points (**arrows**) to prevent undercarriage damage. See **Raising vehicle safely using floor jack** in this repair group.

◀ When lifting car using an automotive lift, raise vehicle slowly, Once car is off ground, stop lift and check that vehicle is stable by gently rocking car at front or rear bumper.

– Use at least two jack stands to support vehicle. Use jack stands designed for the purpose of supporting vehicle.

– Place jack stands on firm, solid surface. If necessary, use a flat board or similar solid object to provide a firm footing.

– Lower vehicle slowly until its weight is fully supported by jack stands. Watch to make sure jack stands do not tip or lean as the vehicle settles on them.

– Observe jacking precautions again when raising vehicle to remove jack stands.

Raising vehicle safely using floor jack

The procedure given below describes how to raise the car using a floor jack and how to support it with jack stands.

WARNING—

- *A jack is a temporary lifting device. Do not use a jack alone to support the vehicle while you are under it.*
- *Do not work under a lifted vehicle unless it is solidly supported on jack stands that are intended for that purpose.*
- *Do not use wood, concrete blocks or bricks to support a vehicle. Wood may split. Blocks and bricks, while strong, are not designed for that kind of load and may break or collapse.*
- *Watch the jack closely. Make sure it stays stable and does not shift or tilt. As the car is raised, it may roll slightly and the jack may shift.*
- *Use at least two jack stands to support car. Use automotive jack stands designed for supporting a car.*
- *Place jack stands on a firm solid surface.*

CAUTION—

- *When raising the car using a floor jack or hydraulic lift, carefully position the jack pad to prevent damaging the car body. If possible, use a rubber pad between jack and body.*

– Disconnect negative (–) cable from battery so that no one can start car. Let others know what you are doing.

CAUTION—

- *Prior to disconnecting the battery, read the battery disconnection cautions given in* **00 Warnings and Cautions**.

P99803058

◄ To raise front of car, position floor jack on raised jacking lug (**arrow**) allowing room to place jack stand.

– Raise vehicle until jack stands can fit under front jacking lug. If possible, place a rubber pad between body and jack stand.

P99803056

◄ To raise rear of car, position floor jack on raised jacking lug (**arrow**) allowing room to place jack stand.

– Raise vehicle until jack stands can fit under rear jacking lug. If possible, place a rubber pad between body and jack stand.

– Lower car slowly until its weight is fully supported by jack stands. Watch to make sure jack stands do not tip or lean as car settles on them.

– Observe jacking precautions when raising car to remove jack stands.

Underbody splash shields, removing

The underbody splash shields are best removed with the vehicle raised in the air on a lift. Plastic nuts, rivets and screws retain the underbody panels to underside of the vehicle.

Leave room around the lift arms if removing the rear shields to access fasteners. Front and rear splash shields can be removed independently. If removing the center shield, remove either the front or rear shield first.

For more information see:

- **50 Fenders** for inner fender liner removal.

Splash shields

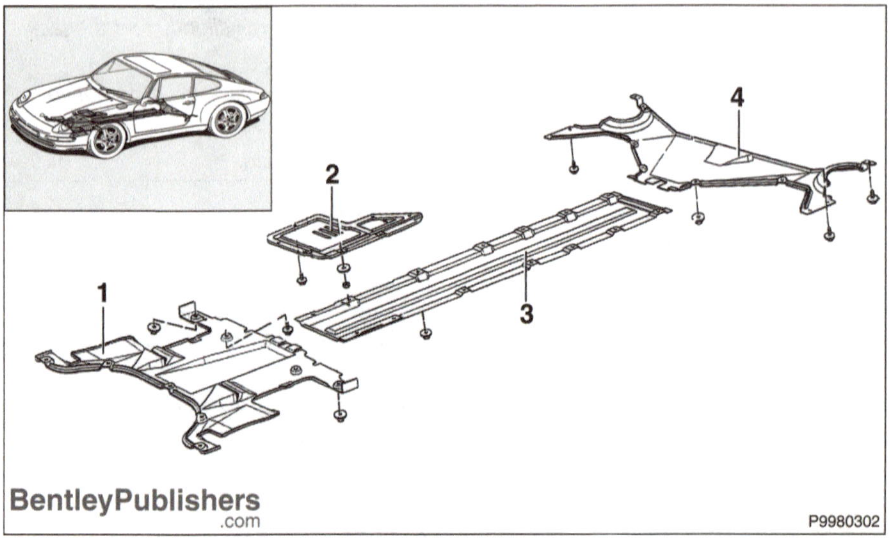

1. Front splash shield
2. Fuel pump cover
3. Center splash shield
4. Rear splash shield

Engine protection panel, removing

- Turn engine protection panel fasteners (**arrows**) counterclockwise then pull off panel.
- To install, turn fasteners clockwise to lock into place.

Transmission protection panel, removing

- Turn transmission protection panel fasteners (**arrows**) counterclockwise then remove panel.
- To install, turn fasteners clockwise to lock into place.

Non-reusable fasteners

Many fasteners used on the cars covered by this manual must be replaced with new ones once they are removed. These include but are not limited to: bolts, nuts (self-locking, nylock, etc.), cotter pins, studs, brake fittings, roll pins, clips and washers. Use genuine Porsche parts for this purpose.

Some bolts are designed to stretch during assembly and are permanently altered, rendering them unreliable once removed. These are know as torque-to-yield fasteners. Replace fasteners where instructed to do so. Failure to replace these fasteners could cause vehicle damage and personal injury. See an authorized Porsche center for applications and ordering information.

Tightening fasteners

It is good practice to tighten fasteners gradually and evenly to avoid misalignment or over-stressing any one portion of the component. For components sealed with gaskets, this method helps to ensure that the gasket will seal properly.

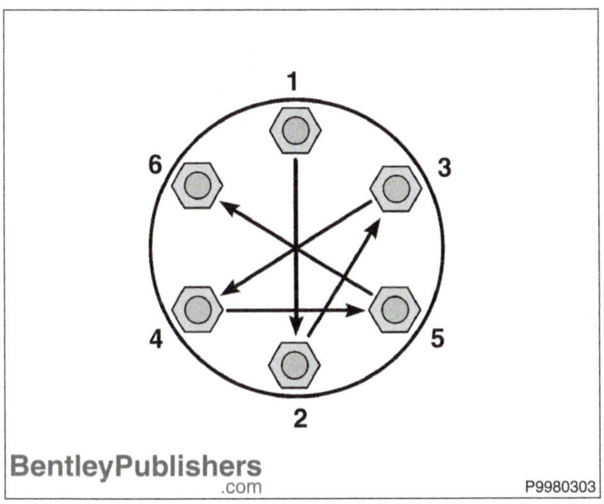

- Where there are several fasteners, tighten them in a sequence alternating between opposite sides of the component. Repeat the sequence until all the bolts are evenly tightened to the proper specification.

For some repairs a specific tightening sequence is necessary, or a particular order of assembly is required. Such special conditions are noted in the text, and the necessary sequence is described or illustrated. Where no specific torque is listed, use **Table a** as a general guide for tightening fasteners.

> ***WARNING—***
>
> - **Table a** *is a general reference only. The values listed are not intended to be used as a substitute for torques specifically called out in text.*

NOTE–

- *Metric bolt classes or grades are marked on the bolt head.*
- *Do not confuse wrench size with bolt diameter. As an example, a 13 mm wrench normally drives an M8 bolt.*

Table a. Bolt tightening torques–general (in Nm) (max. permissible)

Bolt diameter	Bolt class (according to DIN 267)					
	5.6	**5.8**	**6.8**	**8.8**	**10.9**	**12.9**
M5	2.5	3.5	4.5	6	8	10
M6	4.5	6	7.5	10	14	17
M8	11	15	18	24	34	40
M10	23	30	36	47	66	79
M12	39	52	62	82	115	140
M14	62	82	98	130	180	220
M16	94	126	150	200	280	340
M18	130	174	210	280	390	470

Buying parts

Some of the maintenance and repair tasks in this manual call for the installation of new parts, or the use of new gaskets and other materials when reinstalling parts. In most cases, make sure needed parts are on hand before beginning the job. Read the introductory text and the complete procedure to determine which parts are needed.

For bigger jobs, partial disassembly and inspection are required to determine a complete parts list. Read the procedure carefully and, if necessary, make other arrangements to get the necessary parts while your car is disassembled.

Genuine Porsche parts

Genuine Porsche replacement parts from an authorized Porsche center are designed and manufactured to the same high standards as the original parts. They are guaranteed to fit and work as intended by the engineers who designed the car.

Many independent repair shops make a point of using genuine Porsche parts. They know the value of doing the job right with the right parts. Parts from aftermarket sources can be as good, particularly if manufactured by one of Porsche's original equipment suppliers, but it is often difficult to know.

Porsche is constantly updating and improving their cars, often making improvements during a given model year. Porsche may recommend a newer, improved part as a replacement, and your authorized center's parts department will know about it. The Porsche parts organization is best equipped to deal with any Porsche parts needed.

Non-returnable parts

Some parts cannot be returned. The best example is electrical parts. Buy electrical parts carefully, and be as sure as possible that a replacement is needed, especially for expensive parts such as electronic control modules. It may be wise to let an authorized Porsche center or other qualified shop confirm your diagnosis before replacing a non-returnable part.

Information you need to know

Model: Porsche 911 normally aspirated models covered by this manual are known as the Carrera, the Carrera 4 and the Carrera 4S. The official platform designation is model type 993.

The type 993 was available on the U.S. market from 1995 through 1998. If necessary, confirm that the vehicle is a type 993 by decoding the vehicle identification number. See **01 Vehicle Identification and VIN Decoder**.

Model year: This is not necessarily the same as date of manufacture or date of sale. A 1995 model may have been manufactured in 1994. It is still a 1995 model. Model years covered by this manual are 1995 to 1998.

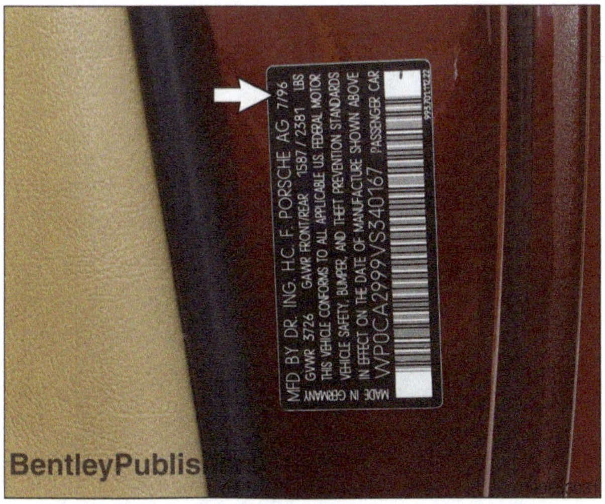

◀ **Date of manufacture**: This information may be necessary when ordering replacement parts or determining if any recalls are applicable to your car. The label located on the left lower B-pillar specifies the month and year (**arrow**) that the car was built.

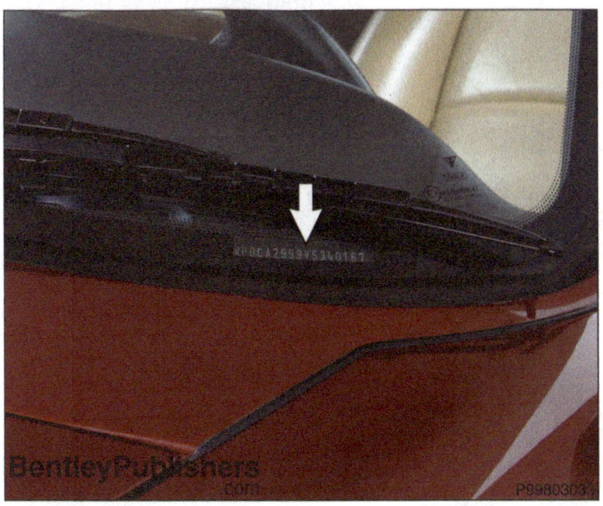

◀ **Vehicle identification number (VIN)**: This is a combination of seventeen letters and numbers that identify the particular car. The VIN appears on the state registration document and on the car itself. One location is in the lower left corner of the windshield (**arrow**). See **01 Vehicle Identification and VIN Decoder**.

Engine code: Cars covered in this manual are powered by a 6-cylinder engine. For information on engine codes and engine applications, see **02 Product Familiarization**.

Transmission code: The transmission type with its identifying code may be important when buying clutch parts, seals, gaskets and other transmission-related parts. For information on transmission codes and applications, see **02 Product Familiarization**.

Porsche service

Porsche centers are uniquely qualified to provide service for Porsche cars. Their authorized relationship with the Porsche service organization means that they are constantly receiving special tools and equipment, together with the latest and most accurate repair information and training.

Authorized Porsche centers are committed to supporting the Porsche product. On the other hand, there are select independent shops that specialize in Porsche service and are also capable of doing high quality repair work. Checking with other Porsche owners for recommendations on service facilities is a good way to learn of reputable Porsche shops in your area.

Tools

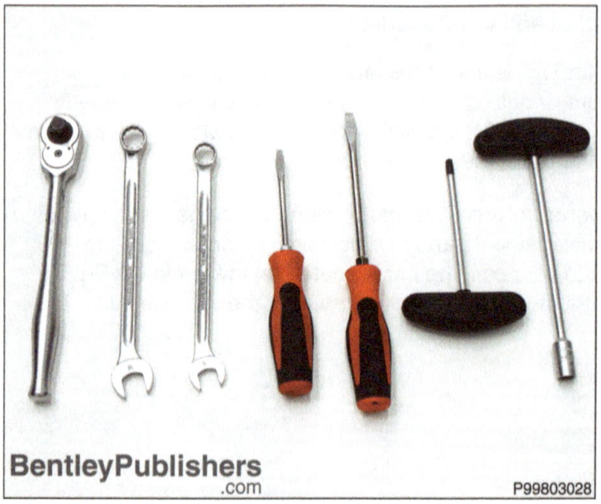

Most maintenance can be accomplished with a small selection of the right tools. Tools range in quality from inexpensive junk, which may break at first use, to very expensive and well-made tools for the professional. The best tools for most do-it-yourself Porsche owners lie somewhere in between.

Many reputable tool manufacturers offer good quality, moderately priced tools with a lifetime guarantee. These are your best buy. They will cost a little more, but they are good quality tools that will do what is expected of them. Sears Craftsman ® line is one such source of good quality tools.

Some of the repairs covered in this manual require the use of special tools, such as a custom puller or specialized electrical test equipment. These special tools are called out in the text and can be purchased through an authorized Porsche center or other Porsche specialty automotive tool manufacturers.

MAINTENANCE TABLES

Except where noted, the maintenance items listed apply to all models and model years covered by this manual. Porsche recommends a yearly service for vehicles that accumulate low yearly mileage between service intervals. Cars driven regularly should follow the service schedule of maintenance every 12,000 miles.

Yearly maintenance and inspections are listed in **Table b**. Minor maintenance (every 12,000 miles) items are listed in **Table c**. Additional maintenance (every 24,000 miles) items are listed in **Table d**. Additional mileage services are listed in **Table e.**

Table b. Yearly service for vehicles that accumulate low yearly mileage between required service intervals

Maintenance item	Tools Required	New parts required	Warm engine required	Porsche service recommended	Additional repair information
Change engine oil	•	•	•		
Oils and fluids: Visually inspect for leaks					
On-board diagnostics: Check fault memory using Porsche diagnostic scan tool for codes	•			•	OBD
Parking brake: Check free travel of parking brake lever					
Brake fluid: Check level					
Steering gear: Visually inspect bellows for damage					
Tie rods: Check play and dust boots					
Ball joints: Visually inspect dust boots for damage					
Drive shafts: Visually inspect boots for leaks and damage					
Exhaust system: Check for leaks and damage, check tightness					
Tires and spare wheel: Check condition and correct tire pressure	•				
Seals for doors, hood and roof: Remove rubber residue from body seal contact areas, apply suitable lubricant					
Vehicle lights: Check operation and headlight alignment	•				94
Windshield washer, headlight washer: Check fluid level, nozzle aim and washer fluid anti-freeze protection level for the winter season. Check wiper system operation.					92
Battery and electrical: Check electrolyte level and hydrometer test. Check for any present warning lights.	•				

Table c. Required maintenance and lubrication service every 12,000 miles

Maintenance item	Tools Required	New parts required	Warm engine required	Porsche service recommended	Additional repair information
Change engine oil (no filters required, filters changed every other oil change)	•	•	•		
Cabin air filter: Replace filter	•	•			
Engine air filter: Inspect and replace if necessary	•	•			
Crankcase ventilation: Check tightness of hose connections	•				
Fuel system: Visually inspect for damage, clearance and tightness of fuel line connections	•				
Intake air: Check hoses, lines and connections for tightness	•				
Brake hoses and lines: Visually inspect for clearance, damage and corrosion					
Clutch: Check play or final position of clutch pedal, check fluid level and condition					
Throttle linkage: Check for smooth movement and full throttle opening					
Resonance flap and Varioram (if applicable): Check operation using Porsche scan tool. Visually inspect system.	•				
Power steering: Check fluid levels and inspect system for leaks					
Manual transmission/axle drives: Check fluid level, inspect for leaks					
Automatic transmission: Check fluid level, inspect for leaks					
Drive shafts: Visually inspect for damage, check tightness					
Door hinges: Lubricate all door hinges and latches					
Door locks, hood locks and safety hook on front door: Check adjustment and function.					
Oils and fluids: Visually inspect for leaks					
On-board diagnostics: Check fault memory using Porsche diagnostic scan tool for codes	•			•	
Parking brake: Check free travel of parking brake lever					
Brake fluid: Check level					
Steering gear: Visually inspect of bellows for damage					
Tie rods: Check play and dust boots					
Ball joints: Visually inspect dust boots for damage, check tightness of wheel alignment adjustment bolts	•				
Drive shafts: Visually inspect boots for leaks and damage					
Exhaust system: Check for leaks and damage, check tightness					
Tires and spare wheel: Check condition and correct tire pressure	•				
Seals for doors, hood and roof: Remove rubber residue from body seal contact areas, apply suitable lubricant					
Vehicle lights: Check operation and headlight alignment	•				94
Windshield washer, headlight washer: Check fluid level, nozzle aim and washer fluid anti-freeze protection level for the winter season. Check wiper system operation.					92
Battery and electrical: Check electrolyte level and hydrometer test. Check for warning lights.	•				

Table d. Additional maintenance requirements every 24,000 miles, including all Table c items.					
Maintenance item	**Tools Required**	**New parts required**	**Warm engine required**	**Porsche service recommended**	**Additional repair information**
V-belt and Polyrib belt: Check condition and tension					
Engine oil filters: Replace both	•	•	•		
Drive belt for power steering pump: Inspect condition	•				

Table e. Additional maintenance requirements.					
Maintenance item	**Tools Required**	**New parts required**	**Warm engine required**	**Porsche service recommended**	**Additional repair information**
Spark plugs: Replace (every 30,000 miles, or no later than 4 years)	•	•			
Manual transmission and axle drives: Change fluids (every 60, 000 miles)	•	•	•		34/39
Air filter: Replace filter element (every 30,000 miles)	•	•			
Fuel filter: Replace (every 60,000 miles)	•	•			
Automatic transmission: Change fluid and filter (every 30,000 miles)	•	•	•		37
Automatic transmission: Change fluid in differential (every 60, 000 miles)	•	•	•		37
Brake system: Flush and change brake fluid (every 2 years)	•	•			47
Air bag supplemental restraint system: Scheduled maintenance after 4 years, then continue to check every two years				•	69

Diagnosis System

Fault memory, reading

A Porsche diagnostic scan tool can be used to access fault codes and OBD data.

Porsche 993 systems capable of on-board self-diagnosis:

- Engine management (DME M 2.10 and DME M 5.2)
- Airbag
- ABS/ABD
- Automatic (Tiptronic) transmission
- Alarm system
- Heater system

NOTE–

- *1996 - 1998 models are equipped with an advanced OBD II (DME M 5.2) emissions-related diagnostic system. 1995 models are equipped with OBD I (DME M 2.10) .*
- *For OBD I vehicles, emissions-related fault information is displayed with blink codes through the Check Engine light. See* **OBD On-Board Diagnostics.**
- *For OBD II vehicles, a generic OBD II compatible diagnostic scan tool can be used to retrieve basic emissions-related DTCs. A list of DTCs can be found in* **OBD On-Board Diagnostics**.

P99890017

◄ OBD II: The Check Engine light (**arrow**), also know as malfunction indicator light or MIL, alerts the driver of a problem and the need to have the system checked for fault codes.

P99803311

◄ The diagnostic scan tool is connected via the diagnostic port. Remove cover (**arrow**) to access port. Port is located in driver's footwell on left side of center console.

Fluids and Lubricants

Engine oil level, checking

The Porsche 993 air-cooled engines are equipped with a dry sump lubrication system. Oil is stored in a remote supply tank and a pressure oil pump circulates oil through the engine. Oil level checking is done with the engine running at operating temperature.

NOTE–

- *See* **17 Lubrication System** *for additional information.*
- A small amount of oil consumption is normal. The rate of consumption depends on quality and viscosity of the oil and operating conditions. Regularly check the oil level on the instrument panel. The oil level may also be checked using the dipstick in the engine compartment.

– Engine oil level should be checked under the following conditions
  - Vehicle level.
  - Engine fully warm (oil temperature in area of upper third marking on oil temperature gauge).
  - Engine idling.

◀ Remove dipstick (**arrow**) located at right upper corner of engine bay. Wipe dipstick with clean cloth and fully reinsert into tube.

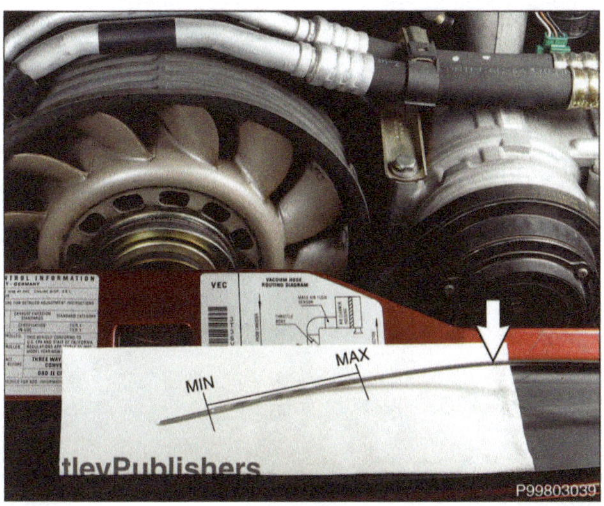

◀ Remove dipstick (**arrow**) again and read oil level. Make sure oil is between MIN and MAX positions. Twisted portion of dipstick signifies MIN and MAX positions. The difference between marks corresponds to approximately 1.5 lters (1.6 quarts).

> ***CAUTION—***
>
> - *Avoid overfilling. There is a risk of damaging the catalytic converter if oil level is too high.*
> - *Lubricant specifications are subject to change. Consult your authorized Porsche center for the latest information regarding approved lubrication applications and specifications.*

Engine oil and filter, changing

Requirements for changing engine oil are:

- Engine oil warm
- Vehicle raised and level

> ***WARNING—***
>
> - *Do not drain engine oil when engine is hot. Hot engine oil can scald.*
> - *Wear hand and face protection when draining warm engine oil.*

– Loosen right rear wheel bolts.

– Raise car and support safely. Remove right rear wheel.

> ***WARNING—***
>
> - *Make sure car is firmly supported on jack stands designed for the purpose. Place jack stands underneath structural chassis points, not under suspension parts.*

< Turn engine protection panel fasteners (**arrows**) counterclockwise to release.

– Remove panel.

< Turn transmission protection panel fasteners (**arrows**) counterclockwise to release.

– Remove panel.

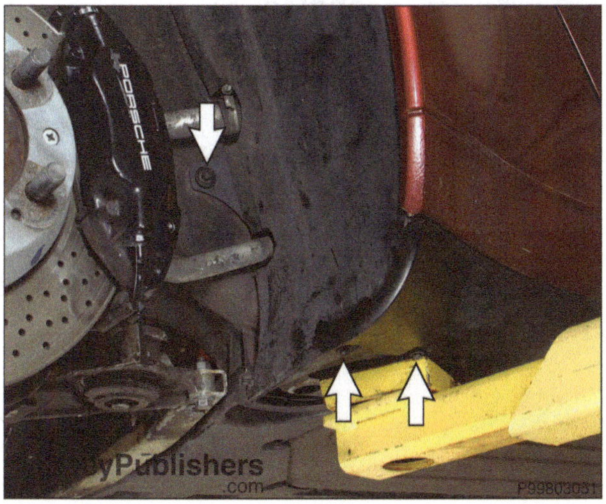

< Remove right rear inner wheel liner fasteners (**arrows**).

< Pull back fender liner (**arrow**).

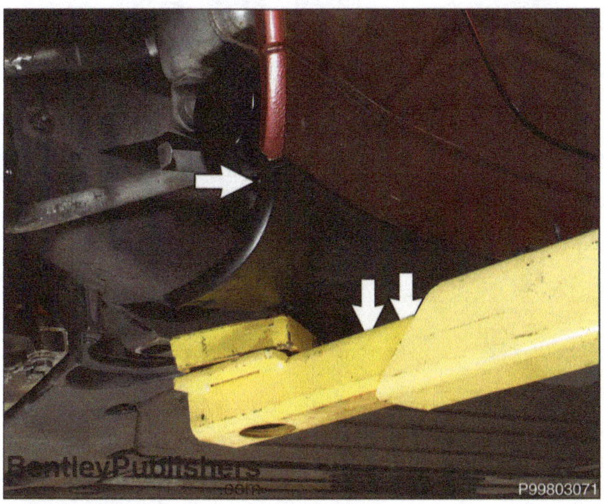

< Remove right side rocker panel guard fasteners (**arrows**).

– Pull back guard enough to expose oil thermostat housing.

< Remove driver side heater flap clamps (**arrows**).

– Remove driver side heater flap.

NOTE–

- *Drain plug can be removed with flap in place, but oil flow will coat heater flap and flow erratically off of heater flap.*

< Remove crankcase oil drain plug (**arrow**). Drain engine oil into suitable container.

NOTE–

- *Oil drains from engine at a rapid rate. The oil drain container should have a large opening to prevent over flow.*
- *Use flexible plastic shield to guide oil into container if necessary.*

– After oil has stopped draining, install new drain plug seal and reinstall drain plug.

Tightening torque	
Oil drain plug to crankcase	50 + 5 Nm (37 + 3.7 ft-lb)

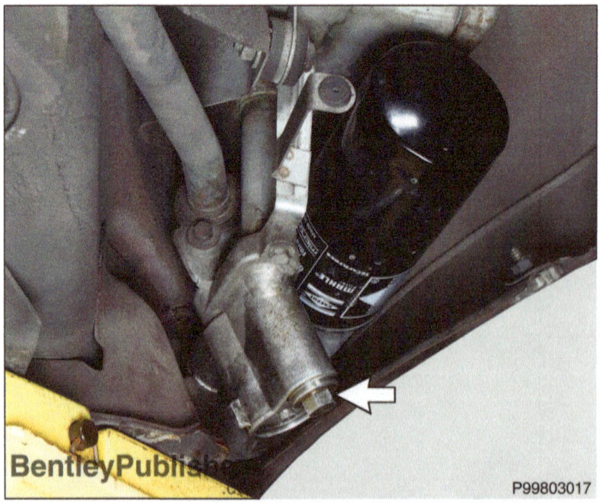

< Remove oil thermostat housing drain plug (**arrow**).

– Drain engine oil into suitable container.

– After oil has stopped draining, install new drain plug seal and reinstall drain plug.

Tightening torque	
Oil drain plug to thermostat housing	50 + 5 Nm (37 + 3.7 ft-lb)

Remove heater air pipe clamps (**A**) from right side of engine.

- Remove heater air pipe (**B**).

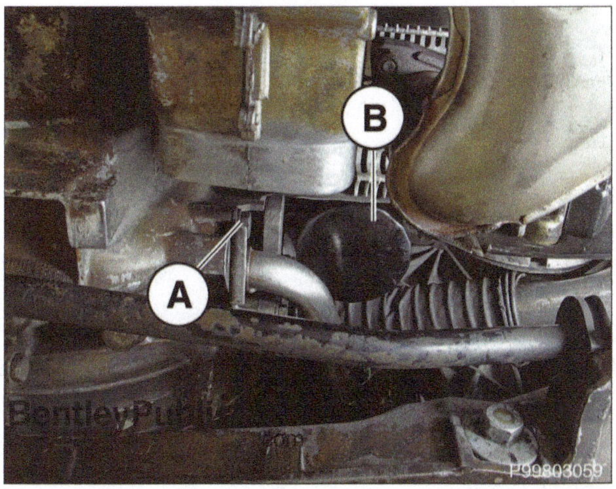

Remove auxiliary oil filter:

- Remove bolt (**A**) from crankcase oil pipe.
- Place drain pan under crankcase oil pipe and oil filter.
- Pull oil pipe from crankcase. Move pipe to one side.
- Using Porsche special tool 9204 or equivalent, remove auxiliary oil filter (**B**). Allow to drain fully.

- Lubricate seal on new oil filter with fresh motor oil.
- Install new oil filter.

NOTE–

- *Tighten oil filter by hand until seal is seated firmly. Do not use oil filter wrench to install new filter or seal damage could occur.*

- Using a new O-ring, install crankcase oil pipe and tighten bolt.

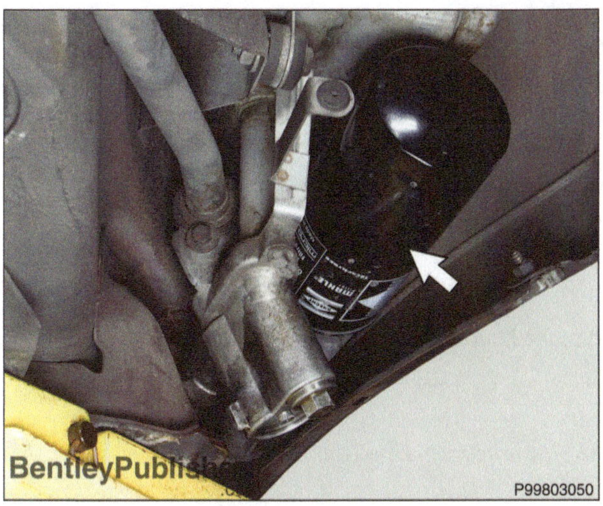

Remove main oil filter (**arrow**):

- Place drain pan under oil filter.
- Using Porsche special tool 9204 or equivalent, remove oil filter from thermostat housing.
- Allow to drain fully.

- Lubricate seal on new oil filter with fresh motor oil before installing.
- Install new oil filter.

NOTE–

- *Tighten oil filter by hand until seal is seated firmly. Do not use oil filter wrench to install new filter or seal damage could occur.*

- Reinstall heater pipe.
- Reinstall right inner wheel liner and right rocker panel guard.
- Reinstall all underbody protection panels.
- Reinstall right rear wheel.

Tightening torque	
Wheel to wheel hub	130 Nm (96 ft-lb)

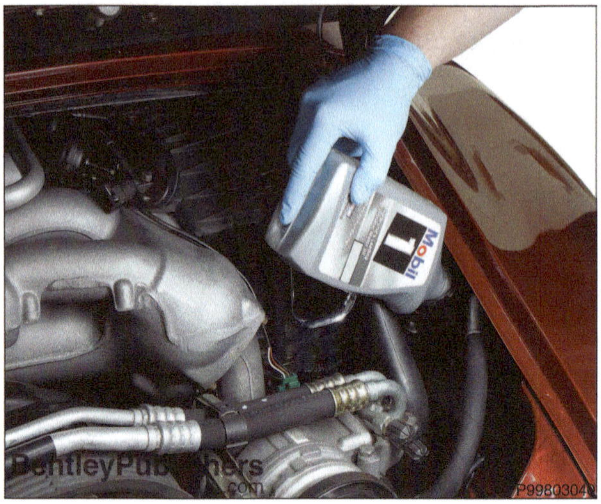

◄ Fill engine oil to capacity from table below.

Engine oil capacities	
Oil change with filters	9.5 liters (10 qt)
Oil change without filters	8 liters (8.5 qt)

NOTE–

- *Engine oil capacity specifications given are approximate. Always check that the oil level is correct after an oil change. See* **Engine oil level, checking** *given earlier in this repair group.*

Engine oil specifications	
Specification: Fully synthetic	Oils meeting Porsche A40 oil specification
Oil viscosity (recommended)	0W-40

NOTE–

- *Oil specifications are subject to change. Consult your authorized Porsche center for the latest information regarding approved engine oils.*

Brake fluid, checking

WARNING—

- *Brake fluid is poisonous. DO NOT ingest brake fluid. Wash thoroughly with soap and water if brake fluid comes in contact with skin.*

CAUTION—

- *DO NOT let brake fluid come in contact with paint. Wash immediately with soap and water.*
- *Brake fluid absorbs moisture from the air. Store in an airtight container.*
- *DO NOT allow brake fluid to exceed the maximum level in the fluid reservoir.*

– Inspect system for fluid leaks or other damage. See **Under Car Maintenance** in this repair group for brake pad inspection and brake system inspection.

– Working in trunk, pull back carpet on driver side.

◄ Check that brake fluid level in reservoir is between MIN and MAX marks.

NOTE–

- *The brake fluid reservoir is also the reservoir for the hydraulically operated clutch.*
- *A slight decrease in the brake fluid level due to brake pad wear is normal.*
- *Some early cars (MY 1995) had a black, non-transparent reservoir, removal of cap is necessary to check level.*

– Top off brake fluid as necessary so fluid level is at or below MAX mark.

Brake fluid specifications	
Porsche brake fluid	Super DOT 4

WARNING—
- *Do not mix DOT 5 (silicone) brake fluid with DOT 4 or Super DOT 4 brake fluid. Component corrosion results and may lead to brake system failure.*

Brake fluid, changing

Brake fluid absorbs moisture from the atmosphere. This moisture can cause brake system corrosion and adversely affect braking performance. Therefore, flush out old brake fluid and add new, fresh fluid at least every 3 years, or 60,000 miles. See **47 Brakes–Hydraulic** for brake fluid replacement and brake system bleeding procedures.

Power steering fluid, checking

Damage to the power-assisted steering is often the result of running a system while low on fluid. Groaning noises when the steering is turned lock-to-lock or foam formation in the fluid reservoir can indicate low fluid level or that air bubbles are present in the fluid. Before topping up reservoir, remedy any leaks on the suction side and replace faulty parts on the pressure side.

The reservoir is located in the upper right side of engine compartment.

NOTE–
- *Check power steering fluid level with engine running at idle.*

– Open engine hood.

< Unscrew power steering reservoir cap and remove.

– Wipe attached dipstick clean.

– Screw cap back on fully, then remove and check level.

NOTE–
- *Fluid level should be between upper and lower marks on dipstick.*

– Top up with Pentosin (CHF 11 S) power steering fluid as necessary. Do not overfill.

Power steering fluid specification	
Power steering fluid	Pentosin CHF 11 S

Manual transmission oil, checking

Requirements for checking manual transmission oil are:

- Transmission oil warm
- Vehicle raised and level
- 10 mm Allen socket

– Raise vehicle and support safely. Make sure vehicle is level.

> ***WARNING–***
>
> - *Make sure the car is firmly supported on jack stands designed for the purpose. Place jack stands underneath structural chassis points, not underneath suspension parts.*

◄ Using a 10 mm Allen socket, remove transmission oil fill plug (**arrow**).

– Add Hypoid gear oil SAE 75W90 until it runs out of fill hole.

– Install and tighten fill plug.

Tightening torque	
Drain and fill plug to manual transmission	30 Nm (22 ft-lb)

Manual transmission oil capacity - (Gear oil SAE 75W90)	
Carrera (RWD)	3.6 liters (3.8 US qt)
Carrera 4 (AWD)	3.8 liters (4.0 US qt)

See **34 Manual Transmission** for fluid changing procedure and more information.

Automatic transmission fluid (ATF), checking and topping off

Requirements for checking automatic transmission fluid (ATF) are:

- Fluid warmed to 80° C.
- Vehicle on level ground.
- Engine running at idle.
- Car in park with emergency brake engaged.

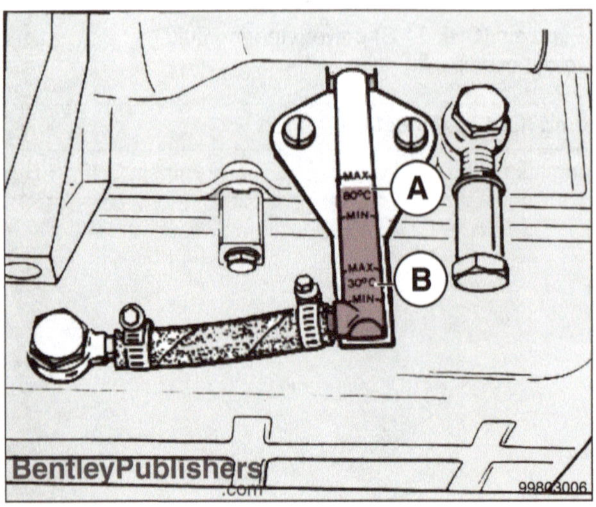

◄ Level of automatic transmission fluid can be checked via sight tube located on side of transmission. Level should be at or below MAX reading.

- Testing range (**A**) when fluid is warmed to 80° C.
- Testing range (**B**) when fluid temperature is 30° C.

NOTE–

- *Porsche system tester 9288 can determine precise ATF temperature.*

See **37 Automatic Transmission** for fluid change procedure and more information.

Differential oil

Porsche specifies a 60,000 mile change interval for the front (Carrera 4 only) and rear differential oil.

On cars with manual transmission, the rear differential oil is shared with the transmission gear oil. See **34 Manual Transmission** for information on changing the transmission oil.

On cars with automatic transmission, the rear differential oil is separate from the ATF. See **39 Differentials, Driveshaft** for procedures on changing the rear differential oil.

Windshield wipers and washer fluid, checking

Washer fluid reservoir filler (**arrow**) is located on left front fender behind fuel filler lid. Use washer fluid with correct temperature range for anticipated driving conditions.

Washer fluid capacity (approximate)	
Washer fluid with headlight cleaning system	7 liters (7.4 qt)

- Check spray patter of front windshield washer nozzles. Adjust as necessary. See **92 Wipers and Washers**.
- Inspect windshield wiper blades, front and rear (if applicable). Replace damaged or deteriorated parts. See **92 Wipers and Washers**.

NOTE–

- *In cases where cleaning of wiper blades and replacement of wiper blades fails to eliminate streaking conditions, thoroughly clean outside of windshield with a non-abrasive cleaner, water, and a soft cloth. Rub until windshield is completely clean of all foreign material. Rinse windshield thoroughly.*

ENGINE COMPARTMENT MAINTENANCE

Drive belts

The Porsche 993 is equipped with four drive belts. There is an engine cooling fan belt, alternator belt, and an air-conditioner compressor belt at the front of engine. The camshaft-driven power steering toothed belt is located on the rear passenger side of engine. See **48 Steering** for power steering belt replacement.

A/C compressor belt, removing and installing

- Remove A/C line bracket bolt (**arrow**).

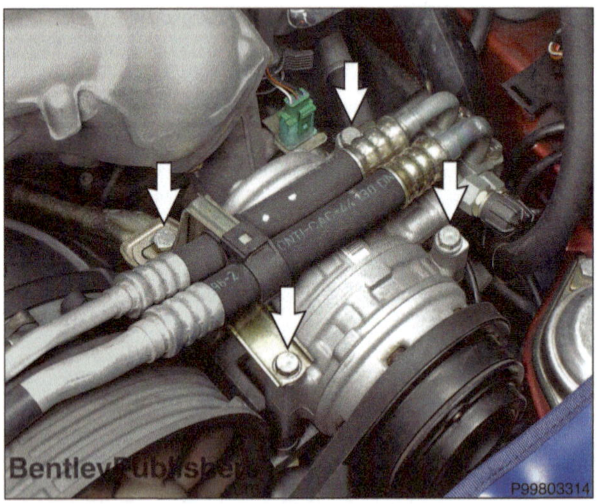

- Loosen compressor mounting fasteners (**arrows**) but do not remove.

Working at lower right side of compressor, loosen lock nut then tensioning bolt (**inset**) located just below compressor on passenger side.

- Slide compressor toward engine.
- Remove belt from compressor pulley then crank pulley.

If belt will not come off compressor pulley, remove compressor bolts.

- Tilt compressor toward rear of car until belt can be removed.

NOTE–

- *When removing compressor bolts, hold square nuts below to prevent from dropping.*

- Installation is reverse of removal. Note the following:
  - Replace belt if signs of cracking, fraying or layer separation are visible.

Turn tensioning bolt clockwise to increase belt tension.

NOTE–

- *Check belt tension by applying thumb pressure at belt midpoint. Belt should deflect approximately 5 mm (0.2 in).*

Tightening torque	
Compressor mounting bolts (M8)	28 Nm (20.5 ft-lb)

Alternator belt, removing and installing

– Remove A/C compressor belt. See **A/C compressor belt, removing and installing** in this repair group.

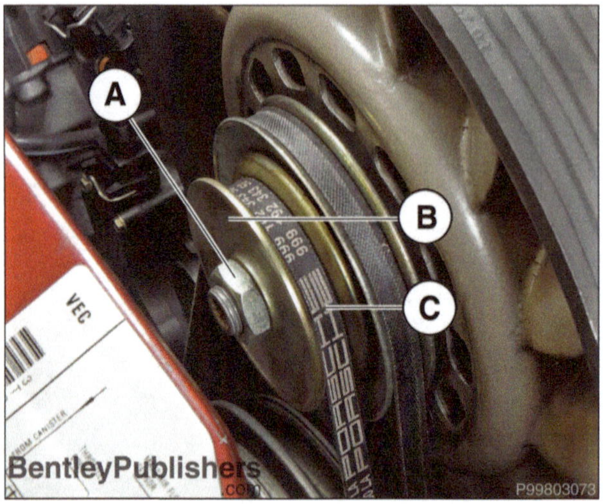

◄ A securing nut (**A**) holds alternator belt pulley (**B**) and alternator belt (**C**) onto shaft.

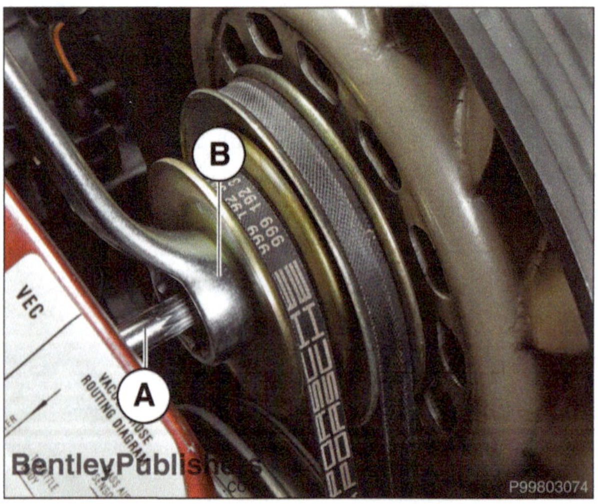

◄ Counterhold alternator shaft with 12 point Allen wrench (**A**).

– Using 24 mm wrench (**B**), loosen securing nut.

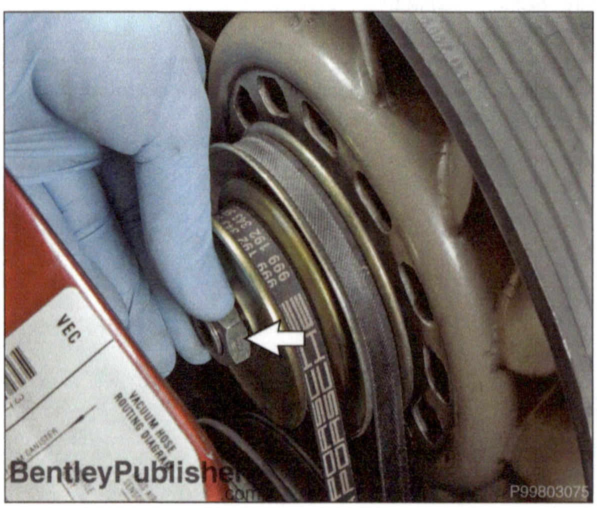

◄ Remove securing nut (**arrow**).

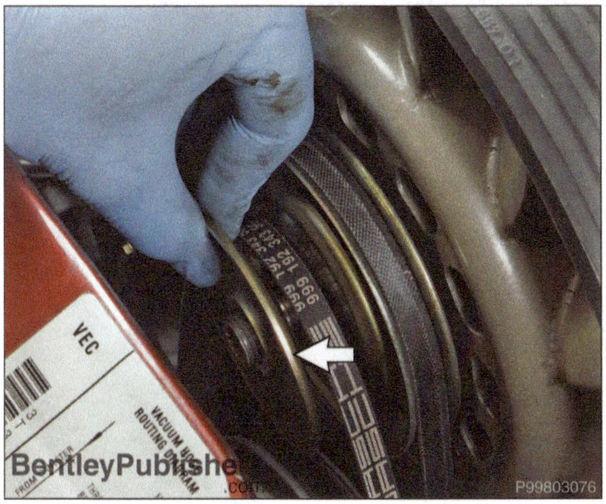

∢ Remove front alternator belt pulley half (**arrow**).

– Remove belt.

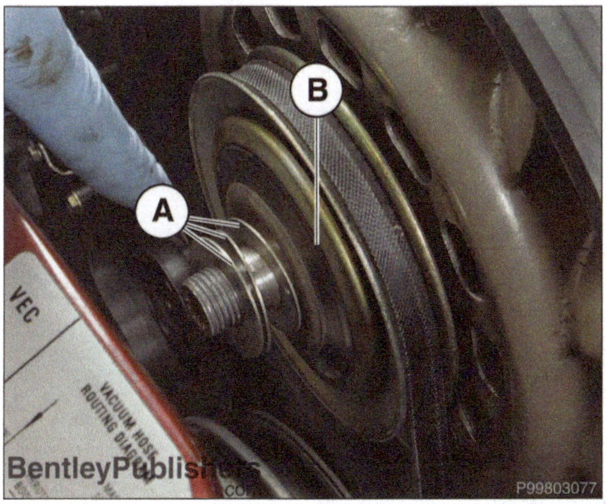

∢ Remove shims (**A**).

NOTE–

- *Shims may be present between pulley halves, behind pulley or in front of pulley. Note position upon removal.*

– Remove rear alternator belt pulley half (**B**) and inspect for wear.

– Installation is reverse of removal. Note the following:

- Install shims (if used) into same position.
- Replace belt if signs of cracking, fraying or layer separation is visible.
- Check belt tension by applying thumb pressure at belt midpoint. Belt should deflect approximately 5 mm (0.2 in).
- If tension adjustment necessary, reposition or add shims. Run engine for a few minutes (at least 1,500 rpm), then recheck belt tension.
- Observe torque specs.
- Counterhold alternator shaft with 12 point Allen socket when refitting securing nut.

Tightening torque	
Nut to alternator shaft	50 Nm (37 ft-lb)

Engine cooling fan belt, removing and installing

– Remove A/C compressor belt. See **A/C compressor belt, removing and installing** in this repair group.

– Remove alternator belt, shims and alternator pulley halves. See **Alternator belt, removing and installing** in this repair group.

◄ Using 5 mm Allen socket, remove fan pulley bolts.

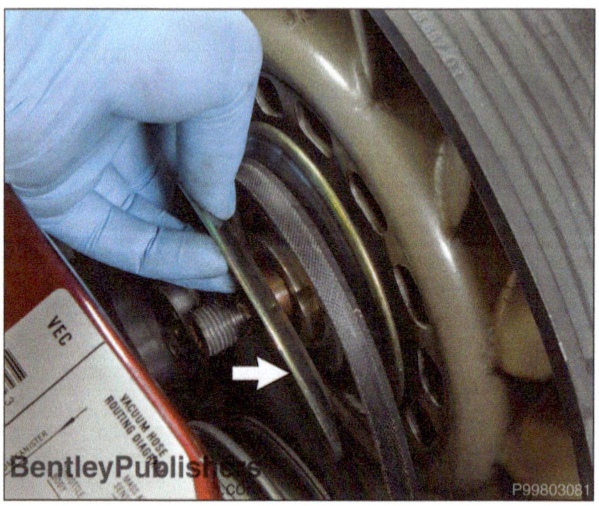

◄ Pull front fan belt pulley half (**arrow**).

– Remove belt.

NOTE–

- *Shims may be present between pulley halves, behind pulley or in front of pulley. Note position upon removal.*

– Installation is reverse of removal. Note the following:

- Remove rear pulley half and inspect for wear.
- Install shims (if used) into same position.
- Replace belt if signs of cracking, fraying or layer separation is visible.
- Check belt tension by applying thumb pressure at belt midpoint. Belt should deflect approximately 5 mm (0.2 in).
- If tension adjustment necessary, reposition shims. Run engine for a few minutes (at least 1,500 rpm), then recheck belt tension.
- Observe torque specs.
- Counterhold alternator shaft with 12 point Allen socket when tightening bolts.

Tightening torque	
Nut to alternator shaft	50 Nm (37 ft-lb)

Air filter (engine), replacing

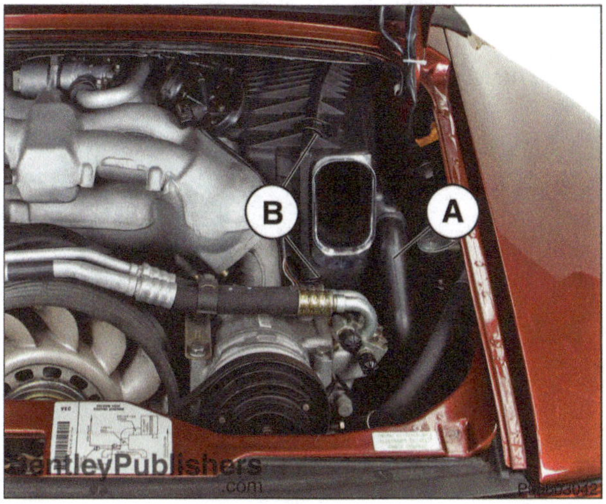

◄ Working in engine compartment, remove secondary air injection inlet tube (**A**).

– Release air filter housing cover retaining clips (**B**).

– Lift air filter housing cover out of engine bay.

◄ Remove air filter in direction of **arrow**.

– Clean air filter housing and housing cover.

– Install new air filter.

– Refit air filter housing cover and fasten clips.

– Reinstall inlet tube.

Fuel filter, replacing

- Fuel filter (**arrow**) is located in right side of engine bay behind power steering reservoir and engine oil dipstick.
- Remove air intake housing. See **24 Fuel Injection**.

- Counter hold fuel filter with wrench. Disconnect fuel lines (**arrows**) from front and back of filter.

NOTE–

- *Place rag or towel below fuel filter to catch fuel when removing fuel lines.*

WARNING—

- *Gasoline is dangerous to your health. Wear suitable hand and skin protection when working on the fuel system. Do not breathe fuel vapors. Always work in a well-ventilated area.*
- *Fuel and fuel vapor are present when the fuel system is opened. Do not smoke or create sparks. Be aware of pilot lights in gas operated equipment (heating systems, water heaters, etc.). Have an approved fire extinguisher handy.*
- *Gasoline fuel supply systems are designed to maintain system pressure after the engine is turned off. Fuel is expelled under pressure as fuel lines are disconnected. This can be a fire hazard, especially if engine is warm. Always wrap a clean shop towel around fuel line fitting before removing.*

- Loosen clamp (**arrow**).
- Remove fuel filter.
- Installation is reverse of removal. Note the following:

NOTE–

- *Fuel filter marked with direction of flow arrow. Be sure to install in facing correct direction.*

Under Car Maintenance

Many of the following procedures must be performed with the vehicle on a lift or safely positioned on jack stands.

> ***WARNING—***
> - *Never work under a raised car unless it is solidly supported on jack stands that are intended for that purpose.*
> - *When raising the car using a floor jack or hydraulic lift, carefully position the jack pad to prevent damaging the car body.*
> - *Watch the jack closely. Make sure it stays stable and does not shift or tilt. As the car is raised, it may roll slightly and the jack may shift.*
> - *A jack is a temporary lifting device. Do not use alone to support the car while you are under it.*
> - *Do not use wood, concrete blocks or bricks to support a car. Wood may split. Blocks and bricks, while strong, are not designed for that kind of load and may break or collapse.*

Spark plugs, replacing

– Raise car and support safely.

> ***WARNING–***
> - *Make sure the car is firmly supported on jack stands designed for the purpose. Place jack stands underneath structural chassis points, not underneath suspension parts.*

◄ Turn engine protection panel fasteners (**arrows**) counterclockwise to release.

– Remove engine protection panel.

Left spark plugs, lower

◄ Working under car near left rear muffler, remove heat exchanger tube clamp (**arrow**).

– Remove tube from heat exchanger.

< Remove spark plug wires (**arrows**).

< Using a spark plug socket on long extension, remove spark plugs.

– Installation is reverse of removal.

Tightening torque	
Spark plugs to cylinder head	30 Nm (22 ft-lb)

Left spark plugs, upper

– Working in engine compartment, remove rear blower fan assembly. See **80 Heating**.

NOTE–

- *Plugs can be removed with rear blower fan assembly in place. Removing rear blower fan assembly allows for easier access.*

< Remove spark plug wires (**arrows**).

- Using spark plug socket with 4 inch extension and U-joint adapter, remove spark plugs.
- Installation is reverse of removal.

Tightening torque	
Spark plugs to cylinder head	30 Nm (22 ft-lb)

Right side spark plugs

- Raise car and support safely.

> ***WARNING–***
> - *Make sure the car is firmly supported on jack stands designed for the purpose. Place jack stands underneath structural chassis points, not underneath suspension parts.*

- Remove right rear muffler. See **26 Exhaust System**.

- Loosen heat exchanger tube hose clamps (**arrows**). Remove heat exchanger tube.
- Remove spark plug wires from right lower bank.

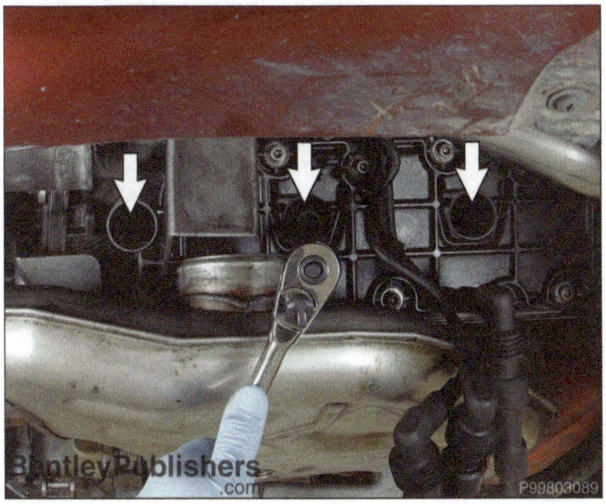

- Using a spark plug socket and a long extension, remove lower right side spark plugs (**arrows**).

◀ Push spark plug wire grommet (**arrows**) through right side engine shroud.

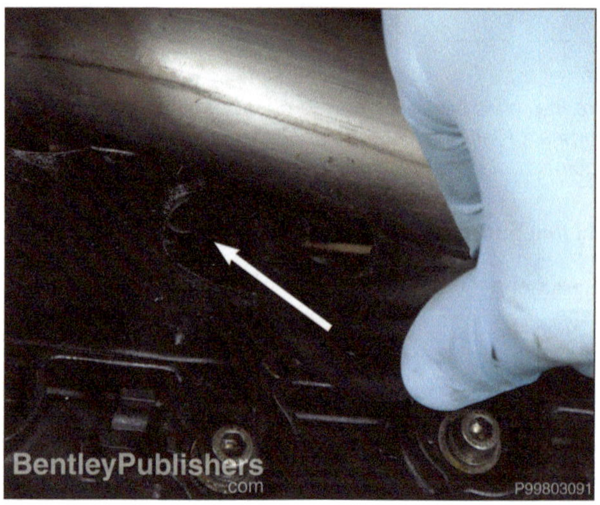

◀ Feed each plug wire through cutout in engine shroud.

◀ Loosen lower muffler bracket bolt (**arrow**).

◄ Remove upper muffler bracket fastener (**arrow**).

– Remove muffler bracket (**A**).

◄ Vehicles with manual transmission: Remove clutch vent tube fasteners (**arrows**).

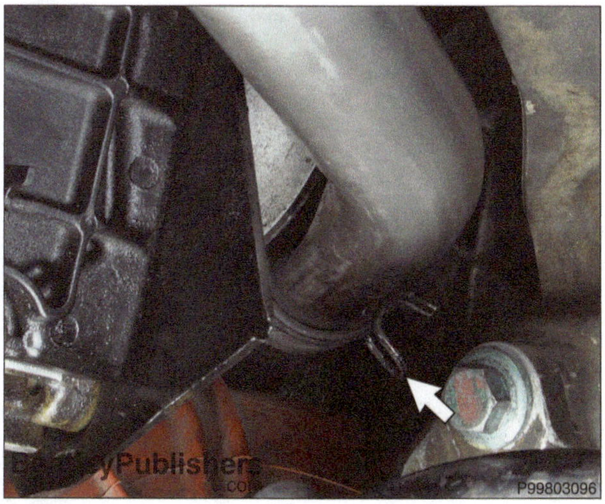

◄ Vehicles with manual transmission: Remove clamp from clutch vent tube (**arrow**).

Vehicles with manual transmission: Remove forward-most clutch vent tube fastener (**arrow**).

Remove remaining engine shroud fasteners (**arrows**).

NOTE–

- *An additional engine shroud fastener is located at rear of engine shroud near muffler bracket mount location.*

– Lower car and remove air intake housing assembly. See **24 Fuel Injection**.

– Vehicles with manual transmission: Remove hose clamp from clutch vent tube above shroud.

Raise car, lower engine shroud with clutch vent tube (on manual transmission vehicles) out of car.

◀ Working from below, remove spark plug wires (**arrows**)

- Remove spark plugs.
- Installation is reverse of removal.

Brake pads and rotors, checking

◀ Replace brake pads when brake pad warning indicator (**arrow**) lights up, or when there is a residual pad thickness of 2 mm (0.08 in) or less.

- To avoid replacing pad wear sensors, replace brake pads when pad thickness is about 2.5 mm (0.1 in). Replace sensors if wire core is worn. There are two sensors at each caliper.
- With vehicle on ground, loosen but do not remove wheel lug bolts.
- Raise car and support safely.
- Remove wheel lug bolts then remove wheel.

> ***WARNING–***
>
> • *Make sure car is firmly supported on jack stands. Place jack stands underneath structural chassis points, not underneath suspension parts.*

◀ Inspect brake pad thickness (**A**) and rotor thickness (**B**) for wear.

NOTE–

• *Brake pad and rotor replacement is covered in* **46 Brakes Mechanical**.

Brake wear specifications	
Brake pad wear limit (dimension **A**):	2.0 mm (0.078 in)
New brake rotor (dimension **B**): • Front Carrera, Carrera 4S • Rear Carrera Carrera 4S	 32.0 mm (1.259 in) 24.0 mm (0.944 in) 28.0 mm (1.102 in)
Brake rotor minimum thickness, after machining (dimension **B**): • Front Carrera, Carrera 4S • Rear Carrera Carrera 4S	 30.6 mm (1.204 in) 22.6 mm (0.899 in) 26.6 mm (1.047 in)
Brake rotor wear limit (dimension **B**): • Front Carrera, Carrera 4S • Rear Carrera Carerra 4S	 30.0 mm (1.181 in) 22.0 mm (0.866 in) 26.0 mm (1.024 in)

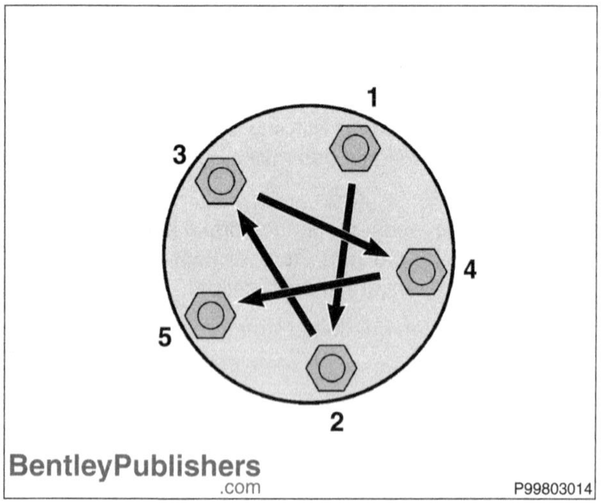

- Install wheel and tighten lug bolts by hand, evenly in a diagonal pattern.
- Repeat brake inspection procedure for each wheel.
- Lower vehicle to ground. Tighten lug bolts evenly to specification in a diagonal pattern.

Tightening torque	
Wheel to wheel hub	130 Nm (96 ft-lb)

Exhaust system, checking

> ***WARNING—***
> - *The exhaust system operates at high temperatures. Do not touch exhaust system while engine is running.*
> - *Allow exhaust system to cool at least one hour before touching.*

- Inspect exhaust system for leaks or damage. Inspect exhaust system mounts. Replace faulty, missing or deteriorated parts.

Underbody, checking

Inspect underbody sealant for damage. Also check underbody, wheel housings and splash shields. Repair or replace missing or damaged parts.

Fluid lines and hoses, checking

- Check brake, fuel, power steering, oil and other fluid lines for leaks, damage, corrosion and loose connections.

Constant velocity (CV) joint boots, checking

Clean inner and outer CV joint (drive axle) boots prior to inspection.

– Raise car and support safely.

> ***WARNING–***
>
> *• Make sure car is firmly supported on jack stands. Place jack stands underneath structural chassis points, not underneath suspension parts.*

< Inspect inner and outer CV joint boots (**arrows**) on both drive axles (including front axles on Carrera 4 models). Press down on rubber boot with fingers to reveal hidden cracks.

NOTE–

• The boot can be replaced separately as long as the joint is not damaged or worn. If the joint is worn, replace the complete drive axle assembly. See **42 Rear Suspension, Rear Drive Axles**, *or* **40 Front Suspension, Front Drive Axles**.

Suspension and steering components, checking

– Check ball joint dust boots for damage and correct seating. Check for play in ball joints.

– Check inner and outer tie rod end boots for damage and correct seating. Check for play by moving tie rods and wheels.

– If a rubber boot is found to have rips or holes, replace corresponding component. Dirt or moisture that has entered joint can destroy component. See **40 Front Suspension, Front Drive Axles**, **42 Rear Suspension, Rear Drive Axles** or **48 Steering** for component replacement.

– Check that suspension mountings (nuts, bolts and fasteners) at front and rear are properly torqued.

Front suspension, tightening torques

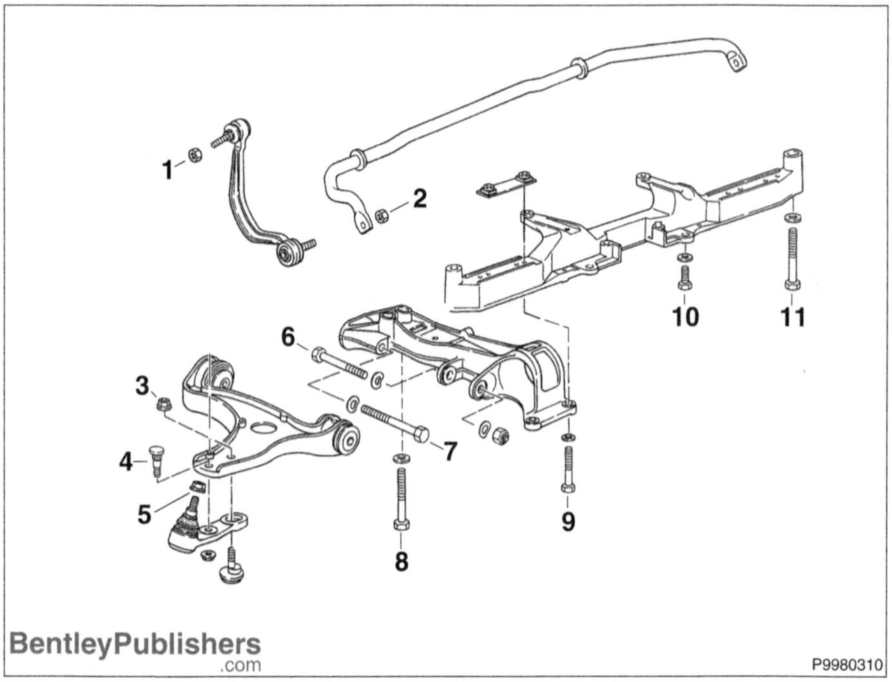

1. Stabilizer bar link to wheel carrier
   - 46 Nm (34 ft-lb)
2. Stabilizer bar link to stabilizer bar
   - 46 Nm (34 ft-lb)
3. Control arm to ball joint (inner)
   - M10, 80 Nm (59 ft-lb)
4. Control arm to ball joint (outer)
   - M12, 140 Nm (103 ft-lb)
5. Ball joint to wheel carrier
   - 75 Nm (55 ft-lb)
6. Control arm to side member (rear)
   - M12, 85 Nm (63 ft-lb)
   - Tighten with rear suspension in normal loaded position (car on ground, suspension settled)
7. Control arm to side member (front)
   - M12, 105 Nm (77 ft-lb)
   - Tighten with rear suspension in normal loaded position (car on ground, suspension settled)
8. Side member to body
   - M12, 90 Nm (66 ft-lb)
9. Side member to cross member
   - M10, 46 Nm (34 ft-lb)
10. Crossmember to body
    - M10, 45 Nm (33 ft-lb)
11. Crossmember to body
    - M12, 105 Nm (77 ft-lb)

For more information and procedures, see **40 Front Suspension, Front Drive Axles** and **44 Wheels, Tires and Alignment**.

Rear suspension, tightening torques

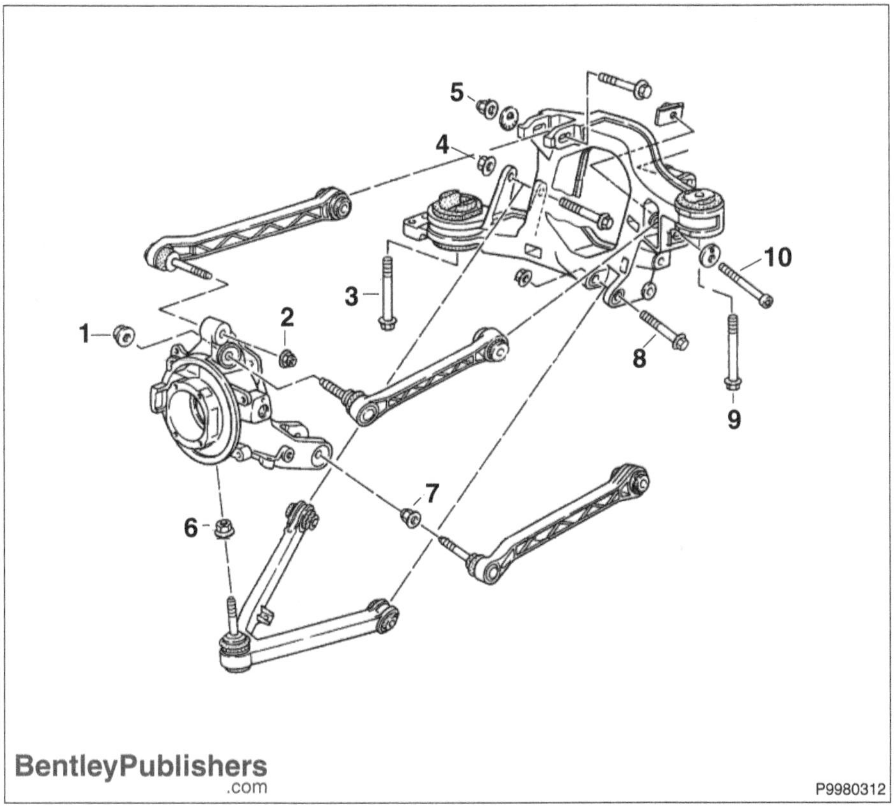

1. Camber control arm to wheel carrier
   - M12, 75 Nm (56 ft-lb)
2. Caster control arm to wheel carrier
   - M12, 75 Nm (56 ft-lb)
3. Subframe to body
   - M12, 120 Nm (89 ft-lb)
4. Lower control arm to subframe
   - M14, 200 Nm (147 ft-lb)
5. Camber control arm to subframe
   - M12, 85 Nm (63 ft-lb)
6. Lower control arm ball joint to wheel carrier
   - M12, 75 Nm (55 ft-lb)
7. Toe control arm to wheel carrier
   - M12, 85 Nm (63 ft-lb)
8. Lower control arm to subframe
   - M12, 85 Nm (63 ft-lb)
9. Subframe (rear) to body
   - M12, 120 Nm (89 ft-lb)
10. Caster control arm to subframe
    - M12, 85 Nm, 63 (ft-lb)

For more information on rear suspension see **42 Rear Suspension, Rear Drive Axles**.

Exterior and Interior Body Maintenance

Airbags, inspecting

Porsche 993 models are fitted with a driver airbag and a passenger airbag.

> ***WARNING—***
>
> - *Do not cover the padded airbag covers on the steering wheel and instrument panel nor affix any objects to them.*
> - *Do not apply chemical treatment to airbag unit covers. Clean with a dry or water moistened cloth only.*

– Inspect padded airbag unit covers on steering wheel and dashboard for signs of external damage or excessive wear. Check with an authorized Porsche dealer if faults are found.

– Switch ignition ON to perform a functional self-test of airbag warning light:

- Airbag warning light illuminates for a short period of time then goes off.
- If warning light does not illuminate, check bulb and power supply.

Cabin micro filters, replacing

The passenger compartment cabin (pollen) micro filters are located under the front cowl panel in the trunk. They each have their own housings and mount directly in front of cabin fan blower motors.

< Remove rubber weather strip (**A**).

– Remove cowl panel cover (**B**).

⮜ Disconnect connector (**arrow**) from right side blower fan housing and slide out of bracket.

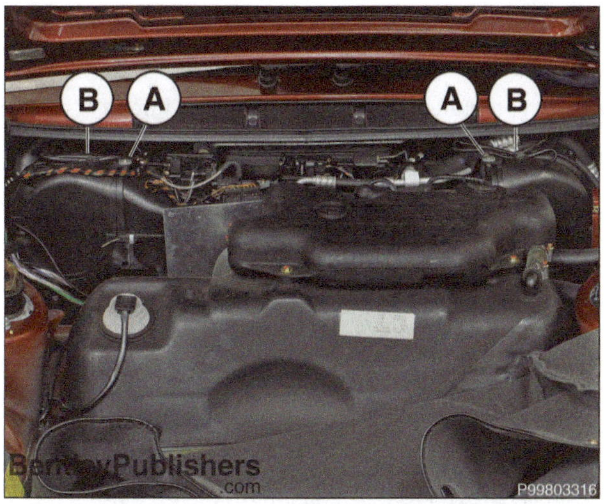

⮜ Remove cabin filter housing retaining nuts (**A**).

– Remove clip (**B**) and wiring from cabin filter housings.

– Remove cabin filter housings by lifting straight up.

NOTE–

• *Cross panel and central electric panel shown removed for clarity.*

⮜ Remove cabin micro filters (**arrows**) by lifting up out of slots in housing.

NOTE–

• *Cross panel and central electric panel shown removed for clarity.*

– Installation is reverse of removal. Note the following:

NOTE–

• *Ensure proper fitting of micro filter and housing before securing.*

Battery service

The battery is in the trunk, under floor carpeting. Under normal operating conditions, the battery is maintenance free.

Some batteries have a charge indicator (magic eye) on top that displays electrolyte level and charge condition.

- Green: Charge and electrolyte level are OK
- Black: There is insufficient or no charge.
- Yellow or colorless: Electrolyte level is critically low. Top off with distilled water immediately.

NOTE–

- *If the battery age is in excess of 5 years and the charge indicators are colorless, replace battery.*
- *Air bubbles that occur normally during battery charging or during vehicle operation may adversely affect charge indicator reading. To obtain an accurate reading, gently tap the charge indicator with a screwdriver handle to displace air bubbles that have formed.*
- *Some batteries are equipped with sealing plugs with plastic foil. These batteries are considered maintenance free.*

– At high outside temperatures, check electrolyte level through translucent battery housing. Make sure electrolyte level is just above battery plates and separators.

- Remove filler caps to see battery plates. Electrolyte level should align with internal electrolyte level indicator (lip). This equates to external marking on battery case.

– For a battery with removable electrolyte caps: If electrolyte level is low, replenish it by adding distilled water only. Use of non-distilled water may contain electrolyte impurities, which cause self-discharging.

CAUTION—

- *DO NOT overfill battery. Overfilled batteries can boil over.*
- *Too little electrolyte reduces the service life of a battery.*
- *Extract excess electrolyte using a hydrometer.*
- *Dispose of electrolyte (sulfuric acid) properly. Refer to local regulations pertaining to electrolyte disposal.*

– During battery service, check the following:

- Battery securely bolted down.
- Battery cable clamping bolts tight.
- Battery vent hose correctly routed.
- Battery cell caps equipped with O-ring seals.

NOTE–

- *For battery testing, service and replacement, see* **27 Battery, Starter, Alternator**.

Battery disconnection

If the battery is disconnected and reconnected, some systems and components may need to be reset.
See **27 Battery, Starter, Alternator** for more information.

Body exterior

Automotive finishes are subject to deterioration from industrial fumes, corrosive road salt, acid rain and other damaging air-borne elements. Regular body finish care contributes to maintaining and preserving the exterior of your Porsche.

The best protection against environmental influences is frequent washing and waxing. How often this is required depends on the environment where the vehicle is used.

– Protect paint with a coat of hard wax at least twice a year.

– Check paint for chips and scratches. Touch up paint defects soon after they occur to prevent corrosion.

– After winter, thoroughly wash underside of vehicle.

– Keep exterior plastic and vinyl clean. Occasionally apply a colorless vinyl or leather preservative. Do not wax plastic or vinyl.

Clutch, checking

The clutch actuation is through a hydraulic master cylinder and slave cylinder. It is therefore not possible to check clutch free travel at the pedal.

For a quick functionality check, pull pedal up toward driver seat. Make sure pedal does not move upward. If there is any play, check boost spring or pedal assembly.

Proper clutch operation is dependent on:

- Correct bleeding of clutch hydraulics. See **30 Clutch**.
- No leaks in hydraulic system.
- Pedal return to starting position.
- Proper installation and position of pedals.
- Checking the pedal end position (pedal return/ return force).

Door check straps and hinges, lubricating

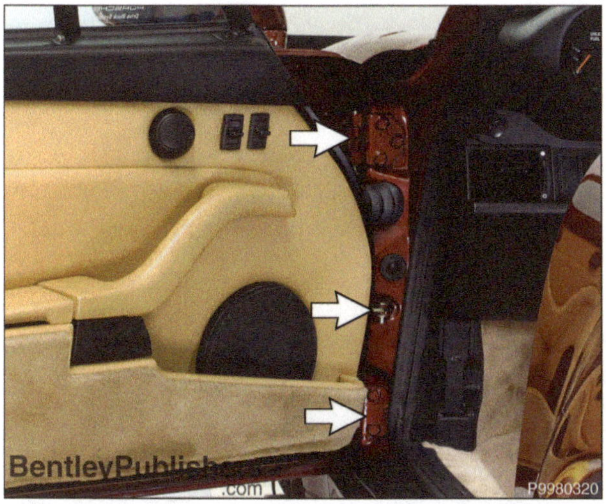

◀ Lubricate door check straps and door hinges periodically with a lubricant designed for door hinges. Clean hinges before applying lubricant.

– Check sheet metal where door check meets body. Look for signs of fatigue or sheet metal cracking. If any damage is found, consult a body repair shop for proper repair methods.

Locks, front lid safety hook, checking

– Check mounting bolt torques for door lock, front lid lock and rear lid lock, as well as retaining nuts for upper parts of locks of front and rear lids.

Tightening torque	
Door lock to door	10 Nm (7 ft-lb)
Trunk latch to body	10 Nm (7 ft-lb)
Hood latch to body	10 Nm (7 ft-lb)
Upper lock sections of hood and trunk	10 Nm (7 ft-lb)

Door lock, inspection

– Check that door lock engages when doors are closed and disengages again when door handle is operated. Check operation inside and outside.

Trunk lid and engine hood lock, inspection

– Check that lower lid locks engage by insertion of lock upper parts when lids (trunk and hood) are closed and disengage again when lid releases are pulled.

Trunk lid safety hook, inspection

◄ Check that trunk lid is held down by safety hook (**arrow**) after lid lock is opened. Check that safety hook engages in retaining plate at its lowest point.

– When lid is open, check that safety hook return spring loads hook toward base plate of lock upper part.

Exterior lights, checking

– Check operation of headlights (high and low beam), marker lights, taillights, turn signal lights, brake lights, back-up lights and emergency flasher lights. See **94 Exterior Lights** for bulb replacement.

NOTE–

- *Keep new bulbs clean and free from grease. Do not touch bulb surface with hands. Use a cloth or soft paper towel when replacing bulbs. Grease from fingerprints may lead to premature bulb failure.*

– Use soapy water to clean exterior lights and plastic headlight lenses. Never use chemical cleaning agents. To avoid scratches, do not rub with dry or slightly damp cloths or paper towels.

Check that headlight washer nozzles are present and do not need adjustment.

Headlights, adjusting

Headlight aim is adjustable. See **94 Exterior Lights** for more information.

Interior lights, checking

- Check operation of indicator and instrument cluster warning bulbs. Check operation of interior cabin illumination bulbs. See **96 Interior Lights, Anti-theft**.

Parking brake, checking and adjusting

If the parking brake lever can be pulled up by more the 4 clicks without engaging, adjust parking brake.

- Raise vehicle and support safely. Remove rear wheel.

> ***WARNING—***
>
> - *Make sure vehicle is stable and well supported at all times. Use a professional automotive lift or jack stands. Place jack stands underneath structural chassis points, not underneath suspension parts. A floor jack is not adequate support.*

- Position rotor bore at approximately 4 o'clock.

- ◄ Using a screwdriver, advance parking brake adjuster by prying star wheel toward front of vehicle through bore until rotor cannot be turned. Next, turn adjuster back 5 notches (3 notches until the wheel can be turned freely, then another 2).
- Repeat procedure on opposite wheel.
- Pull up parking brake lever by 2 clicks to test engagement.
- Release parking brake lever and check that wheels can be turned freely.

Tightening torque	
Wheel to wheel hub	130 Nm (96 ft-lb)

Seat belts, checking

- Check that it is possible to pull out seat belt smoothly and evenly from belt retractor.
- Check that tongue of seat belt engages audibly in buckle.
- Check that abrupt pull on belt strap locks belt retractor.
- Inspect seat belt for damage. If cuts, fraying or seam tears are found, replace seat belt.

Tire service

- Check tire pressure on a regular basis when tires are cold. Refer to tire pressure sticker on left front door jamb. Be sure to also check spare tire pressure.
- Check that all tires are the same type and tread pattern. Measure tread depth. If tread wear has exceeded minimum specification listed, replace tire.

Tire specifications	
Minimum tread depth	1.6 mm

> ***CAUTION—***
> - *Check wheel alignment after replacing tires to ensure maximum tire life.*

- Check tires for abnormal wear patters. Tread wear pattern of front tires in an indication of whether the toe and camber settings need to be checked:
  - Feathering on tread: Incorrect toe
  - Tread worn on one side: Incorrect camber

For more information on tires, see **44 Wheels, Tires and Alignment**.

Test drive

- Upon completion of maintenance work, road test vehicle checking operation of the following:
  - Key fob remote control
  - Front power seats
  - Brake pedal and parking brake free play
  - Brake operation and noise
  - ABS function
  - Engine
  - Steering play
  - Transmission operation
  - Cruise control (if equipped)
  - Heating and air-conditioning system
  - Instruments, radio and horn operation
  - Vehicle handing dynamics such as pulling, vibrations, noises
- Manual transmission:
  - Gearshift smooth operation
  - Clutch operation, smoothness and normal pedal pressure
- Automatic transmission:
  - Neutral safety switch: Make sure engine only starts in PARK or NEUTRAL.
  - Shift lock: Make sure shift lever does not move out of PARK or NEUTRAL unless foot brake is applied.
  - Kick-down: Depress accelerator pedal full to floor while driving. Depending on vehicle and engine speed, upshift is either delayed or transmission shifts down into next lower gear.

1 Engine–General

General

This section lists engine specifications for engines used in the 1995 through 1998 Porsche 911 Carerra.

The Porsche 993 is fitted with an air cooled horizontally-opposed ("boxer") engine.

There are subtle differences between the 1995 (OBD I) engines and the 1996 through 1998 (OBD II) engines. See **02 Product Familiarization** for more information.

Key engine features:

- Six-cylinder horizontally opposed light-alloy construction
- Air cooled, twin-valve, hydraulic valve lifters
- Dry sump lubrication
- DME Bosch engine management with twin ignition

General engine specifications are listed in **Table a**.

Engine Overview

Porsche 993 3.6 liter air-cooled "boxer" engine. 1997 engine shown.

1. Ignition coils
2. Twin ignition distributor
3. Engine cooling fan
4. Alternator
5. Idle speed control
6. Varioram actuator
7. Power steering pump
8. Belt tension monitor

Table a. Porsche 993 engine applications and specifications

Year model	Engine code	Displacement cc (cid)	Stroke x bore mm	Comp. ratio	Horsepower hp @ rpm	Torque lb-ft @ rpm	Engine management, features
Normally aspirated models							
1995 Carerra, Carrera 4	M64.07 M64.08 (Tiptronic)	3600 (219.7)	100 x 76.4	11.3 : 1	272 @ 6100	243 @ 5000	Bosch Motronic M 2.10.1, Hot film mass air-flow sensor with adaptive knock control
1996 - 1998 Carerra, Carrera 4, Carerra S, Carerra 4S	M64.23 M64.24 (Tiptronic)	3600 (219.7)	100 x 76.4	11.3 : 1	286 @ 6100	250 @ 5250	Bosch Motronic M 5.2 (OBD II) with Varioram induction
Turbo models							
1996 - 1998 Turbo	M64.60	3600 (219.7)	100 x 76.4	8.0 : 1	400 @5750	400 @4500	Bosch Motronic M 5.2 (OBD II) Twin Turbo
1997 Turbo S	M64.60	3600 (219.7)	100 x 76.4	8.0 : 1	424 @ 5750	400 @ 4500	Bosch Motronic M 5.2 (OBD II) modified Twin Turbo

10a Engine Removal and Installation

General

This repair group covers engine removal and installation.

Porsche recommends removing the engine and transmission as one unit, then separating the two. Engine removal is described in detail, followed by a reference to the procedure for splitting and rejoining the engine and transmission.

See also:

- **03 Service and Maintenance** for drive belt removal.
- **1 Engine-General** for engine codes and applications.
- **27 Battery, Starter, Alternator** for starter removal.
- **34 Manual Transmission** for shift cables and manual transmission removal.
- **37 Automatic Transmission** for shift cables and automatic transmission removal.
- **39 Differentials, Driveshaft** for driveshaft removal.
- **42 Rear Suspension** for drive axle removal.

Warnings and Cautions

WARNING–

- *The fuel system is designed to retain pressure even when the ignition is OFF. When working with the fuel system, loosen fuel lines slowly to allow residual fuel pressure to dissipate. Avoid spraying fuel.*
- *Before beginning any work on the fuel system, place a fire extinguisher in the vicinity of the work area.*
- *Fuel is highly flammable. When working around fuel, do not disconnect any wires that could cause electrical sparks. Do not smoke or work near heaters or other fire hazards.*
- *Loosen the fuel tank cap to release pressure in the tank before working on the tank or lines.*
- *When disconnecting a fuel hose, wrap a shop towel around the end of the hose to prevent fuel spray.*
- *Due to risk of personal injury, be sure the engine is cold before beginning the removal procedure.*
- *Do not use a work light with an incandescent bulb near any fuel. Fuel may spray on the hot bulb causing a fire.*
- *Make sure the work area is properly ventilated.*

10a

CAUTION–

- *If the MIL (malfunction indicator light) is illuminated, see* **OBD On-Board Diagnostics** *for DME fault code information.*
- *If other system faults are indicated by an illuminated ABS, SRS or other warning light, see the appropriate repair group in this manual or consult a Porsche repair shop for more information on fault codes.*
- *Cover painted surfaces before beginning engine removal.*
- *As an aid to installation, label components, wires and hoses before removing them.*
- *Do not reuse gaskets, O-rings or seals during reassembly.*

ENGINE REMOVAL AND INSTALLATION

The photos in this section are from a 1997 Carrera with a manual transmission, unless otherwise noted.

Engine and transmission, removing

– Disconnect negative (–) battery cable and cover battery terminal to keep cable from accidentally contacting terminal.

CAUTION–

- *Prior to disconnecting the battery, read the battery disconnection cautions given in* **00 Warnings and Cautions.**

– Raise vehicle and support safely.

WARNING–

- *Make sure the car is stable and well supported at all times. A floor jack is not adequate support.*

◄ Place fender cover over engine hood opening and engine hood latch.

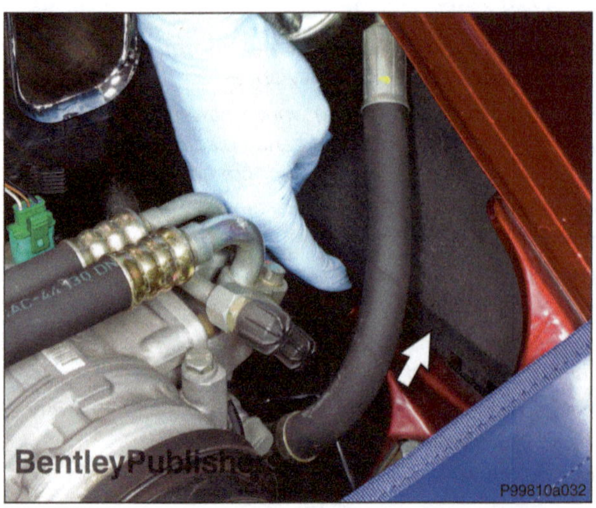

◄ Working in engine compartment on right side, remove plastic cover (**arrow**) by unclipping bottom and pulling up.

< Disconnect A/C compressor electrical connector (**arrow**).

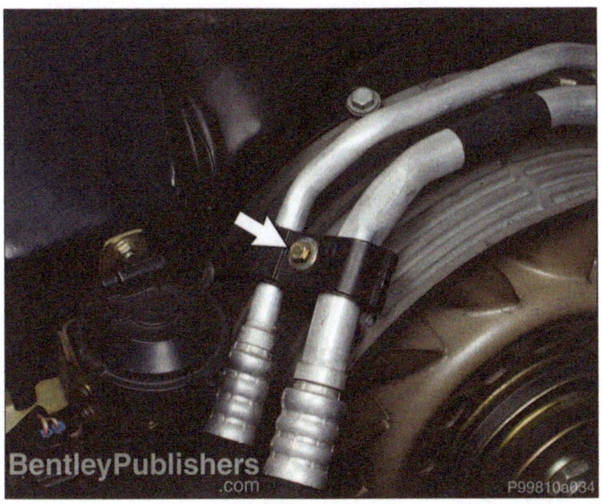

< Working at top of cooling fan shroud, remove A/C line mounting bracket fastener (**arrow**).

– Remove drive belts. See **03 Service and Maintenance.**

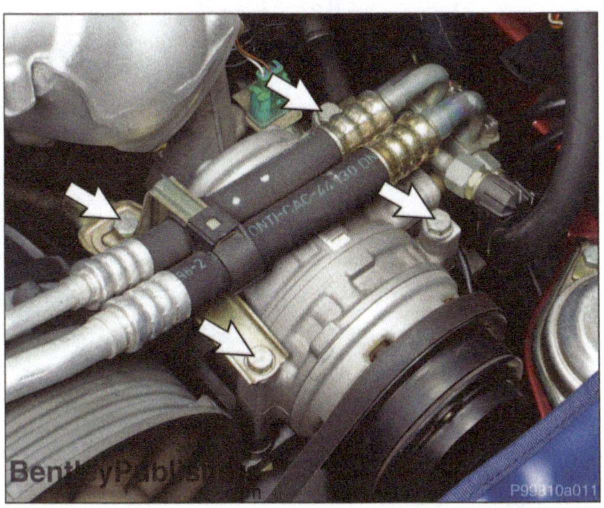

< Remove A/C compressor fasteners (**arrows**).

NOTE–

- *Be prepared to catch threaded lugs from bottom of bracket when bolts are removed.*

– Remove A/C compressor from mounting bracket and lay compressor on edge of engine compartment.

NOTE–

- *Use pad on top of bumper to protect from damage.*
- *Do not disconnect refrigerant lines.*

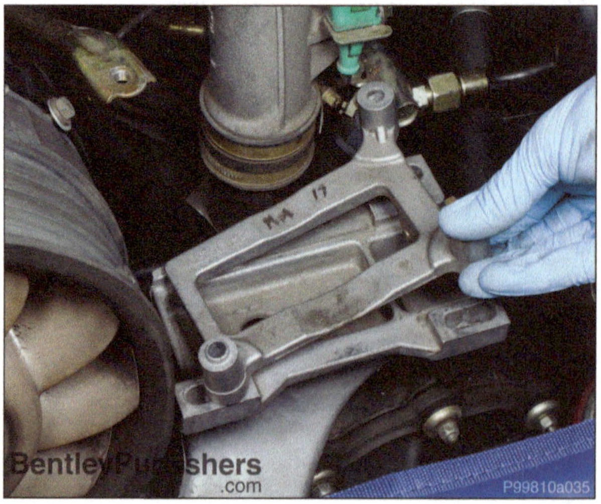

◄ Remove A/C compressor adjusting bracket from mounting bracket and store in a safe place.

– Remove intake air housing. See **24 Fuel Injection.**

– Remove heater blower motor. See **80 Heating**.

◄ Working in engine compartment on right side, disconnect oxygen sensor electrical connectors (**arrows**) by opening sliding lock to separate.

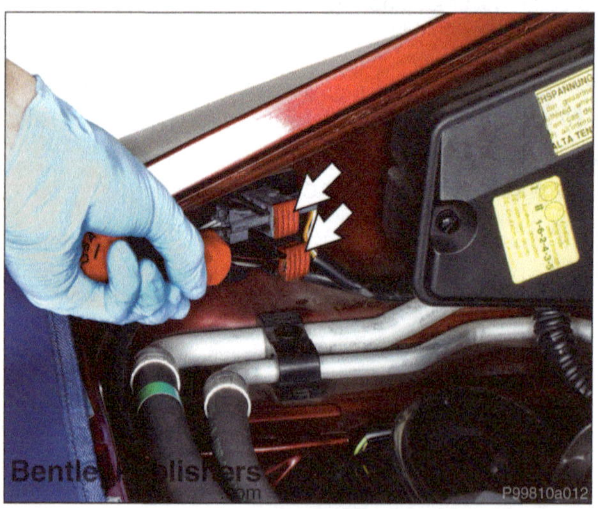

◄ Working in engine compartment on left side, disconnect oxygen sensor electrical connectors (**arrows**) by opening sliding lock to separate.

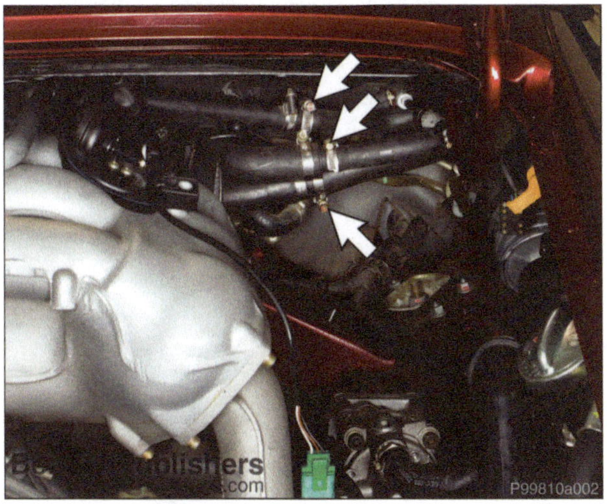

- Working at right side rear of engine, disconnect crankcase breather hoses (**arrows**).

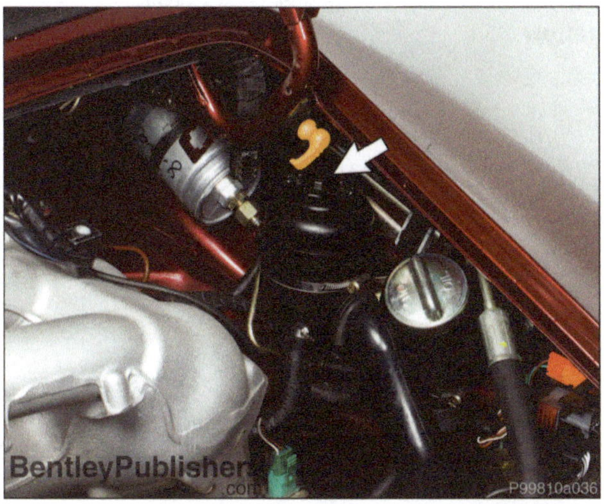

- Working at power steering fluid reservoir, use a clean syringe to siphon power steering fluid out of reservoir (**arrow**).

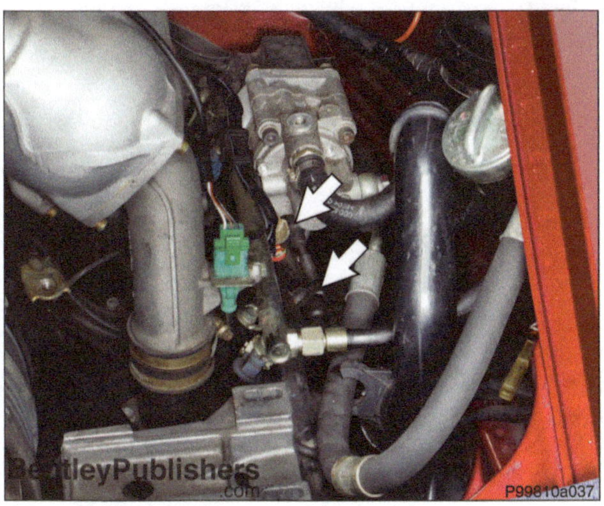

- Working at right side of engine below power steering pump, disconnect cylinder 4 and 5 spark plug cables (**arrows**).

Place a catch container under engine. Remove power steering lines (**arrows**) from power steering pump and allow fluid to drain.

Working at intake manifold, disconnect idle valve electrical connector (**arrow**).

Working at right front of intake manifold, disconnect electrical connector (**arrow**).

< Working at right rear of intake manifold, disconnect mass air flow and throttle housing electrical connectors (**arrows**).

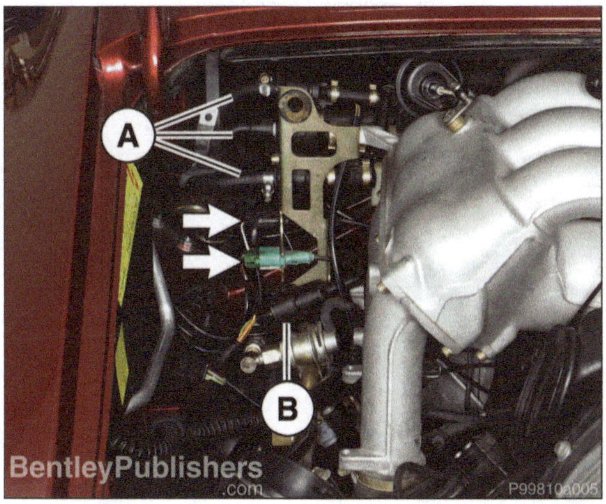

< Working at left side of engine compartment:

- Loosen hose clamps and disconnect hoses (**A**).
- Disconnect electrical connectors (**arrows**).
- Open electrical connector housing and disconnect (**B**).

CAUTION–

- *Be careful not to lose main engine harness plug seal when disconnecting.*

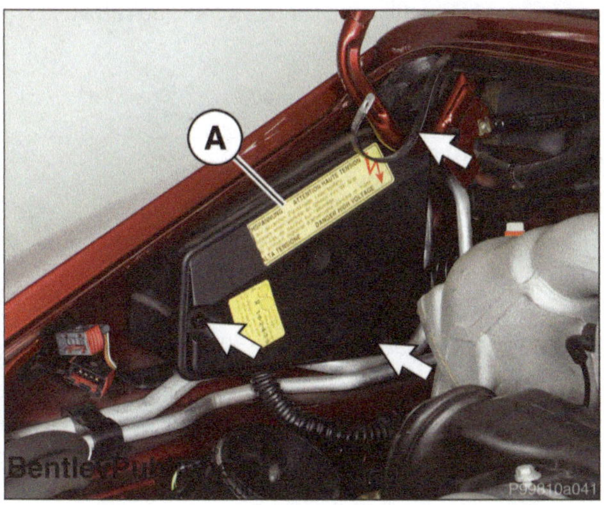

< Working in engine compartment on left side, loosen E-box cover fasteners (**arrows**) then remove E-box cover (**A**).

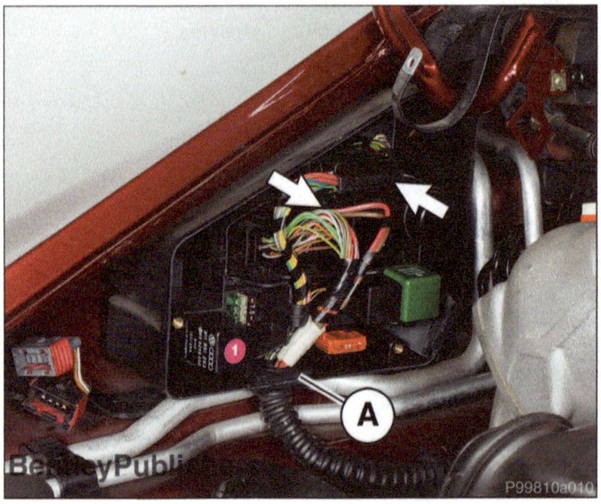

< Working inside E-box, pull wiring harness with grommet (**A**) out of E-box, then disconnect electrical connectors (**arrows**).

< Open sliding lock and separate electrical connector (**arrow**).

< Working at fuel pressure regulator, disconnect fuel line (**arrow**).

NOTE–

- *When separating lines, be prepared to catch dripping fuel.*

WARNING–

- *Fuel will be expelled under pressure.*
- *Before disconnecting the fuel hose, wrap a shop cloth around the hose to absorb any leaking fuel.*
- *Do not smoke or work near heaters or other fire hazards.*
- *Keep a fire extinguisher handy.*
- *Immediately plug open fuel hoses and ports.*

⮜ Working at right side of engine, disconnect fuel line (**arrow**).

NOTE–

- *When separating lines, be prepared to catch dripping fuel.*

> ***WARNING–***
>
> - *Fuel will be expelled under pressure.*
> - *Before disconnecting the fuel hose, wrap a shop cloth around the hose to absorb any leaking fuel.*
> - *Do not smoke or work near heaters or other fire hazards.*
> - *Keep a fire extinguisher handy.*
> - *Immediately plug open fuel hoses and ports.*

– Carrera 4: Working at rear of center console, remove storage compartment trim piece .

⮜ Carrera 4: Working at rear of center console, remove screws (**arrows**).

– Carrera 4: Lift rear storage compartment up and remove from center console.

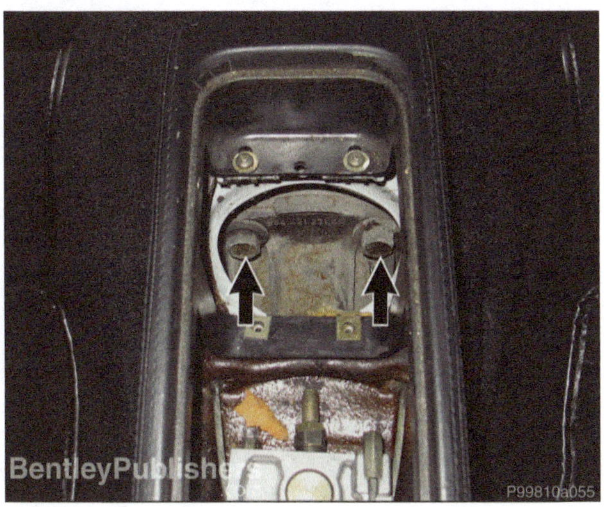

⮜ Carrera 4: Peel back sound insulation, then remove upper torque tube fasteners (**arrows**).

– Working under vehicle, remove engine protection plate fasteners then remove engine protection plate. See **03 Service and Maintenance**.

– Working under vehicle, remove underbody protection panels. See **03 Service and Maintenance**.

– Remove mufflers. See **26 Exhaust System**.

◄ Remove left and right rear CV joint (drive axle) mounting fasteners (**arrows**). See **42 Rear Suspension**.

– Detach CV joints from transmission drive flanges. Cover CV joints to protect from damage and secure aside using stiff wire.

◄ Manual transmission: Slide shift linkage selector rod rubber boot toward front of vehicle (**arrow**). Remove selector rod fastener (**A**), then slide selector rod off transmission shaft.

– Carrera 4: Working above shift rod, pull rubber cover off torque tube.

◄ Carrera 4: Working at torque tube, rotate front wheels until Allen fastener (**arrow**) is visible.

– Remove Allen head fastener.

– Carrera 4: Slide retaining collar toward front of vehicle until rear input shaft (**A**) is free.

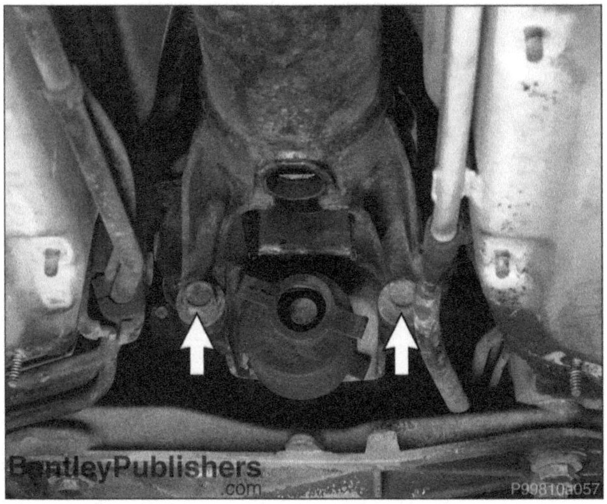

◄ Carrera 4: Remove lower torque tube fasteners (**arrows**).

– Carrera 4: Separate torque tube from transmission.

◄ Automatic transmission: Remove electrical connector (**arrow**) from mounting bracket, then disconnect electrical connector and remove from transmission housing.

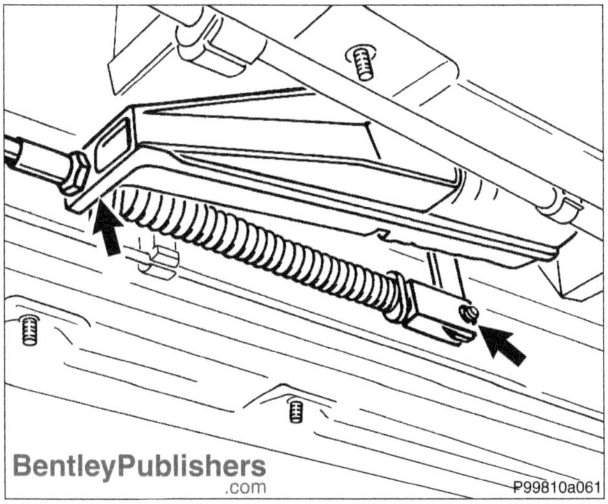

◄ Automatic transmission: Remove cable end from transmission range switch (**arrows**).

– Automatic transmission: Pull cable out of deflection lever and remove cable from guide tube (**arrows**).

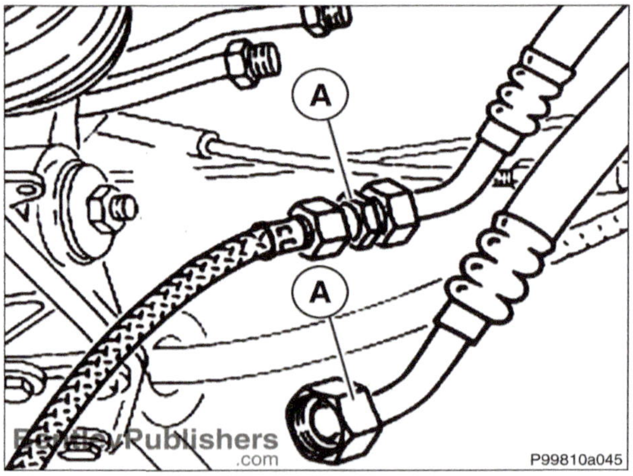

◀ Automatic transmission: Remove transmission oil cooler lines (**A**).

◀ Working under vehicle , loosen heat exchange flap hose clamps (**arrows**).

NOTE–

- *Left side shown, right is similar.*

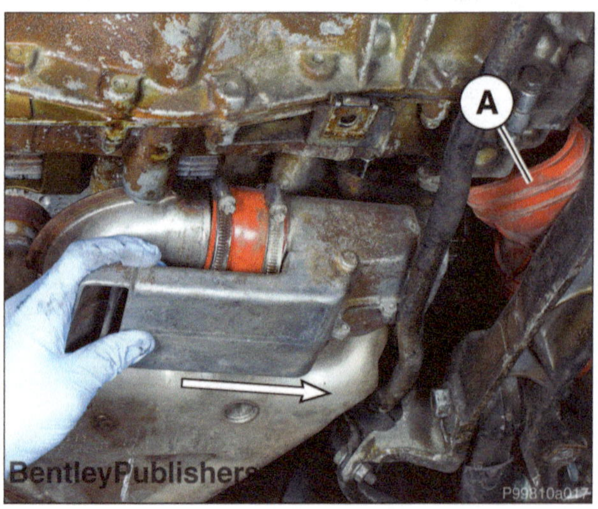

◀ Pull rear hose (**A**) off heat exchange flap, then slide heat exchanger out of front hose (**arrow**) and remove.

NOTE–

- *Left side shown, right is similar.*

Working at heat exchange pipes, loosen hose clamps (**arrows**).

NOTE–

- *Right side shown, left is similar.*

– Remove heat exchange pipe mounting fasteners (**A**) then remove pipe (**B**) from vehicle.

NOTE–

- *Right side shown, left is similar.*

– Remove rear sway bar. See **42 Rear Suspension, Rear Drive Axles**.

Working at left and right side toe control arms, mark eccentric washer position (**arrow**).

– Remove left and right side toe control arms. See **42 Rear Suspension, Rear Drive Axles**.

Remove crossmember fasteners (**arrows**), then remove crossmember (**A**).

10a

< Working at cross brace, remove fasteners (**arrows**), then remove cross brace (**A**) from vehicle.

< Remove engine oil drain plug (**arrow**) and drain oil into suitable container.

< Remove oil line mounting bracket fastener (**arrow**).

- Working at left side of engine, counterhold oil return line fitting hex (**A**) while loosening oil line (**B**).

- Working at starter motor, remove ground strap fastener (**arrow**) then remove ground strap from starter.
- Remove starter from engine. See **27 Battery, Starter, Alternator**.

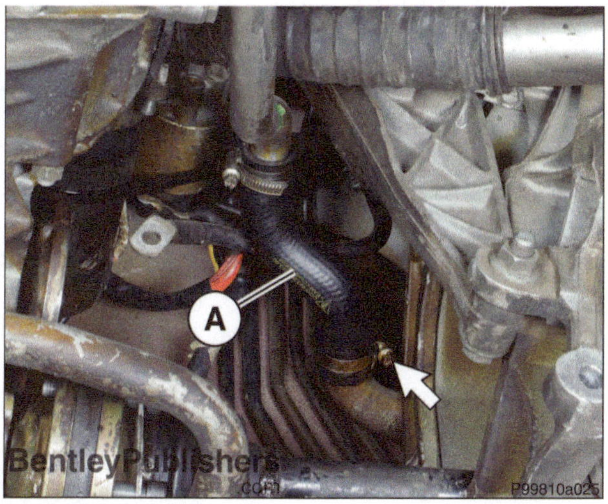

- Working at right side of engine, remove hose clamp (**arrow**) from oil hose and remove hose (**A**). Be prepared to catch leaking oil in a suitable container.

- Manual transmission: Disconnect reverse light switch electrical connector (**arrow**).
- Place engine jack under engine and secure in place.

◄ Remove transmission crossmember fasteners (**arrows**).

◄ Remove engine mount fasteners (**inset**).

– Lower engine enough to access fastener for center transmission mount.

– Remove center transmission mounting fastener.

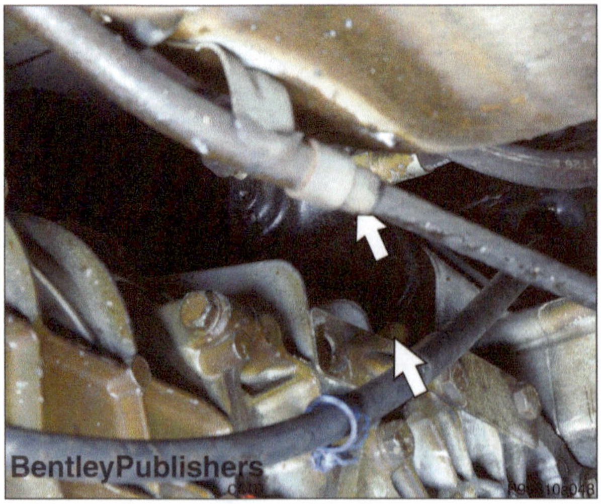

◄ Manual transmission: Lower engine enough to access to clutch slave cylinder mounting fasteners (**arrows**). Remove slave cylinder mounting fasteners.

– Manual transmission: Pull clutch slave cylinder out of transmission and tie aside without disengaging line.

> ***CAUTION—***
>
> *• Do not press on clutch pedal while the slave cylinder is unbolted from transmission.*

– Disconnect throttle cable from pedal return throttle housing.

– Lower engine-transmission assembly carefully out of vehicle. Check for wiring, fuel hoses or mechanical parts that might become snagged as assembly is lowered. Be sure not to damage transmission vent hose or cylinder #3 spark plug cable.

> ***CAUTION—***
>
> *• If vehicle is to be placed on its wheels and moved, reinstall and torque rear crossmember and toe control arms.*

Engine and transmission, installing

Engine transmission assembly reinstallation is done in reverse order of removal. See **Engine and transmission, removing** in this repair group.

With vehicle securely raised on suitable lift, raise engine and transmission assembly into place using a suitable jack.

– Replace all fasteners when reassembling.

– Install toe control arm and torque fastener with suspension in zero position. Zero position is when toe arm and crossmember are in horizontal line.

> ***CAUTION–***
> - *Check grade of toe control arm eccentric bolts. Original equipment used 8.8 grade bolts. An update from Porsche recommends 10.9 grade.*

– Install throttle cable, be sure rubber bushing is properly seated.

– Place engine compartment rubber seal between engine and body before moving engine and transmission into final position.

– Install suspension assembly hole lids, if applicable. Be sure lids are orientated properly.

Tightening torques	
Toe control arm to cross member (eccentric), M12, 10.9 grade*	100 Nm (74 ft-lb)
Toe control arm to wheel carrier, M12	85 Nm (63 ft-lb)
Wheel to wheel hub, M14	130 Nm (96 ft-lb)
Selector rod mounting bolt (automatic transmission) (threadlocker 270)	18 Nm (13 ft-lb)
Engine to engine mount (M12)	85 Nm (63 ft-lb)
Transmission mount to body (M12)	46 Nm (34 ft-lb)
Drive axle to transmission (M10)	81 Nm (60 ft-lb)
Oil drain plug to thermostat housing	50 + 5 Nm (37 + 3.7 ft-lb)
Oil drain plug to crankcase	50 + 5 Nm (37 + 3.7 ft-lb)
Stabilizer bar retaining brackets to crossmember (M8)	23 Nm (17 ft-lb)
Stabilizer bar link to strut (M10)	46 Nm (34 ft-lb)
Crossmember to side subframe (M12)	120 Nm (88 ft-lb)
Cross brace to side subframe (M10)	65 Nm (48 ft-lb)
Muffler exhaust clamp	25 Nm (18 ft-lb)
Catalytic converter clamp	25 Nm (18 ft-lb)

*If 8.8 grade bolt installed, replace with 10.9 grade bolt (part supersession). Bolt grade marked on bolt head.

10a

Engine and Transmission Separating and Rejoining

Engine and transmission, separating and rejoining (manual transmission)

◀ Working at left side of transmission, remove clutch vent tube fasteners (**arrows**) then remove vent tube (**A**) from transmission.

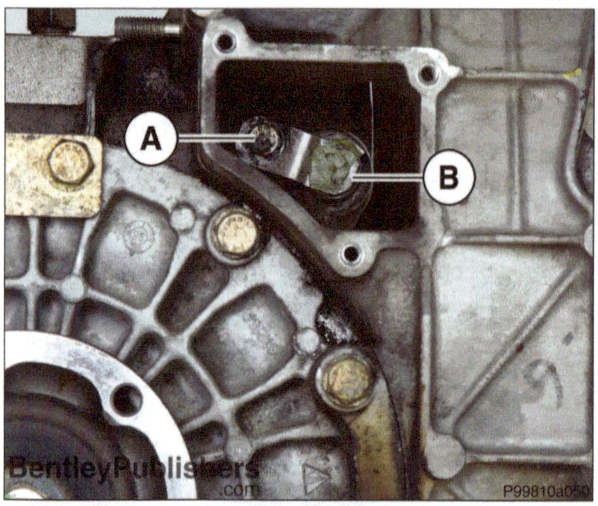

◀ Working at clutch release fork mounting bracket, remove fastener (**A**) then remove cover (**B**).

◀ Using an M6 x 40 mm bolt (**arrow**), extract clutch release fork shaft.

◄ Loosen and remove bell housing fasteners (**arrows**) and separate engine from transmission.

– Check the following while engine and transmission are separated:

- Clutch, clutch release bearing and flywheel replacement. See **30 Clutch**.
- Crankshaft seal replacement. See **13a Crankcase and Crankshaft**.

– Clean transmission input shaft splines and lubricate with a thin coat of light grease.

◄ Use a piece of tape (**arrow**) to secure clutch release fork in proper position for engine transmission rejoining.

– Ensure that bell housing centering dowel sleeves are seated correctly in crankcase flange.

– Slide transmission into place, aligning transmission input shaft carefully with clutch disc splines.

– Install two bell housing bolts finger tight.

– Install clutch release fork shaft with new seals into transmission. Check that shaft properly engages with clutch release fork.

– Remove tape from clutch release fork.

– Install clutch release fork mounting bracket and tighten fastener.

– Install bell housing bolts, torquing each to correct specification.

Tightening torques	
Dual-mass flywheel to crankshaft	85 Nm (63 ft-lb)
Pressure plate to dual-mass flywheel	23 Nm (17 ft-lb)
Transmission bell housing to engine	23 Nm (17 ft-lb)

Engine and transmission, separating and rejoining (automatic transmission)

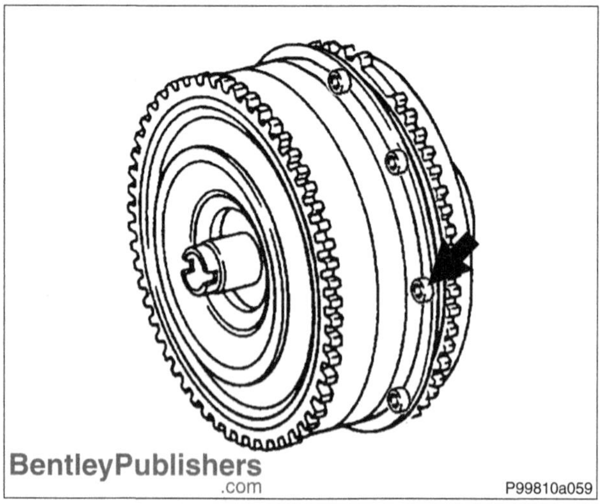

< Working through starter motor opening, remove torque converter fasteners:

- Rotate engine until one of 8 torque converter bolts (**arrow**) is visible. Remove bolt then rotate engine 45° and remove next bolt. Repeat for remaining torque converter fasteners.

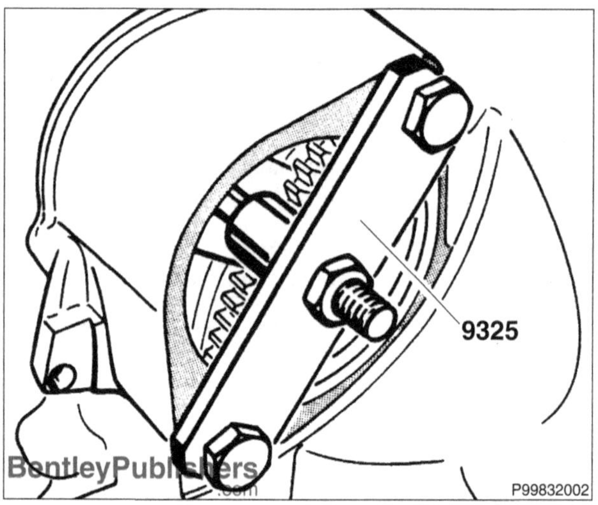

< Install Porsche special tool **9325** to secure torque converter in place.

– Loosen and remove transmission bell housing bolts.

– Separate transmission from engine.

– When rejoining make sure torque converter is secured in place by Porsche special tool **9325**.

– Carefully mate up transmission bell housing to engine flange.

– Install transmission bell housing bolts and torque to correct specification.

– Working through starter motor opening, reinstall torque converter fasteners. Install all fasteners finger tight before torquing.

– Tighten torque converter fasteners in stages.

Tightening torques	
Drive plate to crankshaft (use new bolts)	90 Nm (70 ft-lb)
Torque converter to drive plate	16 Nm (12 ft-lb)
Transmission bell housing to engine	23 Nm (17 ft-lb)

10b Engine Disassembly and Assembly

General

This repair group covers engine disassembly and reassembly. Engine removal procedures, including separating the transmission from the engine, are covered in **10a Engine Removal and Installation**.

The repair information given here requires that the engine first be removed from the car. The disassembly procedure should be carried out in a clean environment on an engine stand designed for the Porsche flat six air cooled engine.

Reconditioning procedures and specifications are not covered here. Once the engine is disassembled to the component level, refer to the appropriate repair group for applicable reconditioning information.

- **13a Crankcase and Crankshaft**
- **13b Pistons and Cylinders**
- **15a Cylinder Heads**
- **15b Camshafts**

Some photos in this repair group are from an earlier air cooled Porsche engine similar to the M64 engine. These photos are intended to help illustrate the specific procedure.

> ***CAUTION—***
> - *If the engine is disassembled because of bearing failure, a thorough flushing of the lubrication system must be carried out prior to reassembly.*

Engine, Disassembling

Some special Porsche tools are required to disassemble and assemble the engine. Read the procedure through prior to beginning the job.

The disassembly procedure is divided into subsections as follows:

- **Engine peripherals, removing**
- **Camshaft timing chains and camshafts, removing**
- **Camshaft cases and cylinder heads, removing**
- **Pistons and cylinders, removing**
- **Crankcase, disassembling**
- **Crankshaft, disassembling**

NOTE–

- *The pulley end of the engine is referred to as the front. The flywheel end of the engine is referred to as the rear.*

Engine peripherals, removing

The V-belts are driven by the crankshaft pulley and power the alternator and air conditioning compressor (if equipped). Alternator V-belt tension is maintained by pulley shims, and is electrically monitored. The air conditioning V-belt tension is maintained by an adjustable bracket.

– Remove engine speed sensor and sensor bracket from right side of crankcase. See **28 Ignition System**.

P99810b176a

< Manual transmission: Remove clutch pressure plate fasteners (**arrows**) and remove pressure plate and clutch disc from flywheel.

P99810b175a

< Manual Transmission: Hold flywheel stationary with flywheel locking device (Porsche special tool P 238) and loosen flywheel fasteners (**arrows**). Remove flywheel from engine.

– Automatic transmission: Hold drive plate stationary with locking device. Remove drive plate fasteners and remove drive plate from crankshaft flange.

– Remove metal shrouding from flywheel (or drive plate) end of engine. Mount engine on engine stand.

> **WARNING—**
> • *A 4-arm or 5-arm engine stand adapter (Porsche special tool* **P 201***) should be used to mount the engine. 3-arm or universal engine stands do not provide proper support and may damage engine block.*

– Remove metal shrouding from pulley end of engine.

– Remove V-belts. See **03 Service and Maintenance**.

◄ Remove engine mounting bracket fasteners (**arrows**) then remove bracket (**A**) from engine.

– Rotate engine on stand so exhaust system is on top. Lock engine stand in position.

– Remove exhaust manifolds. See **26 Exhaust System**.

◄ Loosen muffler bracket fastener (**arrow**) then remove bracket (**A**) from engine. Repeat on opposite side of engine.

– Rotate engine so that intake side is up. Lock engine stand in position.

- ◀ Counterhold crankshaft pulley using Porsche special tool **9290** while loosening pulley fastener.
- Remove crankshaft pulley from engine.

- ◀ Remove A/C compressor bracket fasteners (**arrows**), then remove A/C compressor bracket (**A**) from engine.
- Reinstall crankshaft pulley fastener and tighten nut. Counterhold crankshaft pulley using Porsche special tool **9290** while tightening.
- Remove intake manifold. See **24 Fuel Injection**.

- ◀ Disconnect electrical connectors (**arrows**) from oil pressure gauge sensor and remove sensor.
- Working at top of engine cooling fan, remove ignition wire supports.
- Remove cooling fan / alternator housing. See **27 Battery, Starter, Alternator**.
- Pull alternator out partially and unbolt alternator ground strap from top of engine crankcase.

NOTE–

- *Once drive belt is removed, reinstall alternator belt pulley and shims and tighten nut finger tight.*

- Disconnect spark plug connectors. Unclip and remove distributor caps with wires.
- Remove left and right side metal shrouding from engine.
- Remove intermediate shaft end cover from engine.

Camshaft timing chains and camshafts, removing

Set crankshaft to cylinder 1 compression TDC (**arrow**).

Remove oil feed tube fasteners (**arrows**), then lift tube (**A**) straight up to remove. Remove and discard O-rings. Repeat on opposite side of engine.

Remove intake and exhaust valve cover fasteners (**arrows**), then remove valve covers from engine. Repeat on opposite side of engine.

– Remove camshafts. See **15b Camshafts**.

– Remove left timing chain housing fasteners.

◄ Remove left side upper and lower timing chain guide rail fasteners (**arrows**), then remove timing chain housing with guide rails (**A**).

– Repeat procedures for right side chain guide rails and timing chain housing. Note lower right guide rail is different from all other chain guide rails.

Camshaft cases and cylinder heads, removing

◄ Remove 18 fasteners (**arrows**) from left camshaft case.

– Remove camshaft case and mark for left side.

– Rotate engine on stand. Working on left side bottom of crankcase, remove oil pressure line and all brackets holding it to crankcase.

– Remove cylinder head nuts from each cylinder head.

– Remove cylinder heads, marking installed locations.

– Repeat procedures for right side.

NOTE—

- *Check for broken cylinder head studs when removing cylinder heads. If any are found, check area around cylinder head and base of stud for signs of blow-by. If blow-by is present, mark that area on both cylinder and head for further inspection once parts are thoroughly cleaned.*
- *Various head studs have been used on the M64 engine. Always check with an authorized Porsche parts department for the latest in parts information.*

– Disassemble cylinder heads:

- Using Porsche special tool **P 200** or equivalent valve spring compressor to remove valves. See **15a Cylinder Heads** for reconditioning information

NOTE–

- *Keep parts for each valve assembly (keepers, springs, spring retainer etc.) together.*

Pistons and cylinders, removing

- Set crankshaft to cylinder 1 compression TDC (**arrow**). If necessary, temporarily reinstall crankshaft pulley.
- Working at cylinders, remove cylinder air shroud plates and knock sensor bridge.

- Mark cylinder locations on each cylinder barrel. Remove cylinders. If necessary, use a soft-faced hammer to tap cylinders off studs.

NOTE—

- *Label all parts during disassembly. Used parts must be reinstalled in their exact location and position.*

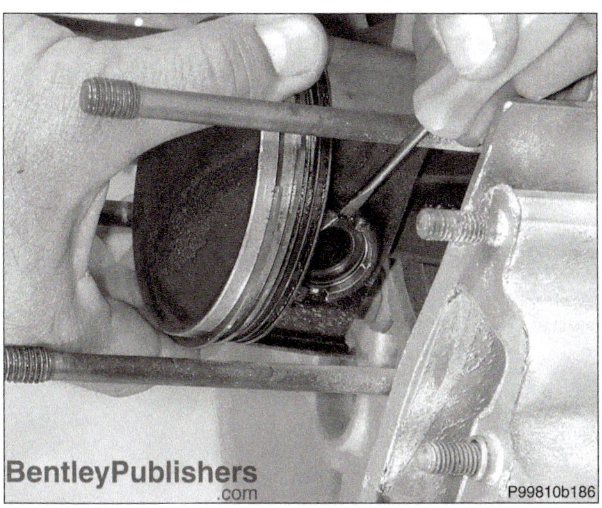

- Remove wrist pin circlips from piston 1.

10b

- Using a soft drift, tap wrist pin out of piston 1.
- Remove piston, marking its position.
- Rotate crankshaft 240°. Remove piston 2, marking its position.

> ***CAUTION—***
>
> - *When rotating crankshaft, manually guide pistons in and out of crankcase bores in order to avoid contacting piston skirts or rings. Use shop towels around the pistons for protection.*

- Rotate crankshaft appropriate amount and remove remaining pistons, marking positions.

NOTE—

- *The firing order is 1 - 6 - 2 - 4 - 3 - 5.*

Crankcase, disassembling

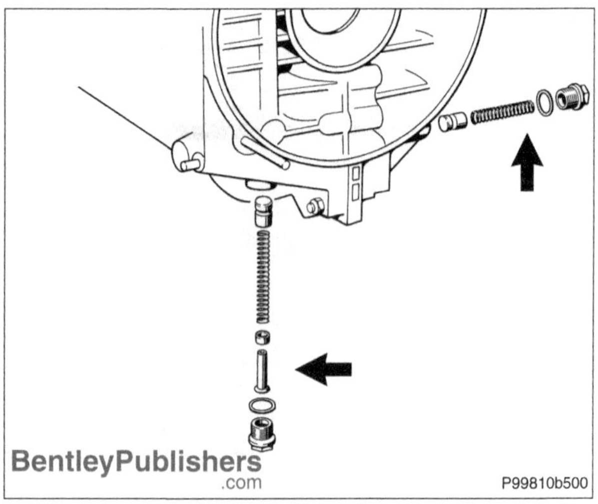

- Remove both oil pressure relief valve assemblies (**arrows**) in bottom of crankcase.
- Remove crankshaft pulley bolt using Porsche special tool **9290** or equivalent to counterhold pulley.
- Remove pulley.
- Remove oil filter housing.

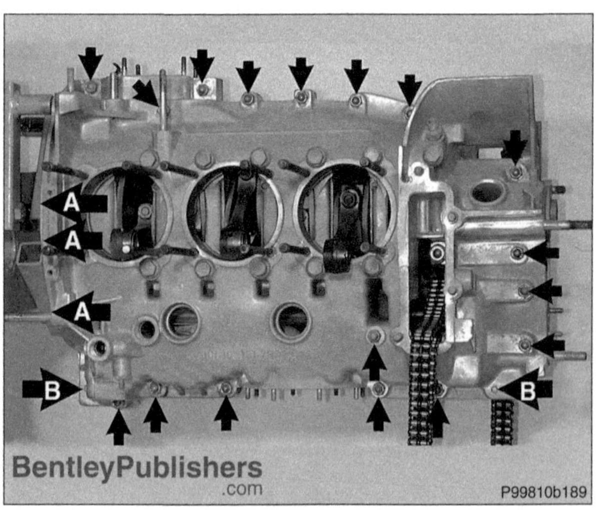

- Remove M8 nuts (**arrows**) joining case halves together.
  - 3 nuts (**A**) are in flywheel housing.
  - 2 nuts (**B**) face right (bottom of case).

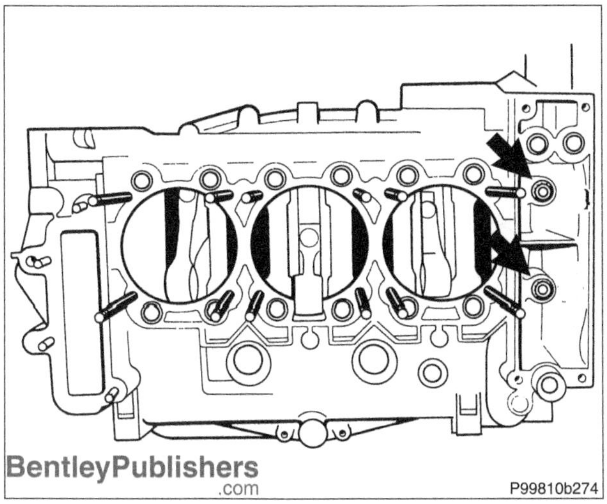

◀ Remove case nuts (**arrows**).

◀ Remove M10 case nut (**arrow**) and washer from left timing chain housing.

- Remove remaining M10 case nuts (12-point, quantity 11) on left side, counterholding bolt heads on right side of crankcase.
- Carefully separate crankcase halves.

> ***CAUTION—***
> - *Do not pry case apart at sealing surfaces. A damaged surface will result in an oil leak.*

- Lift left half of crankcase off and set aside.
- Lift crankshaft assembly out of crankcase and set aside.
- Remove oil pump mounting nuts.
- Remove and disassemble intermediate shaft by removing timing chains, oil pump and oil pump drive shaft.

Crankshaft, disassembling

- Mount crankshaft to crankshaft holding tool (Porsche special tool **P 209**) and clamp tool with crankshaft in shop vise.

NOTE—
- *An adequate substitute for special tool* **P 209** *can be made from an old flywheel. The plate mounts to the crankshaft flange so that it can be securely clamped to the vise.*

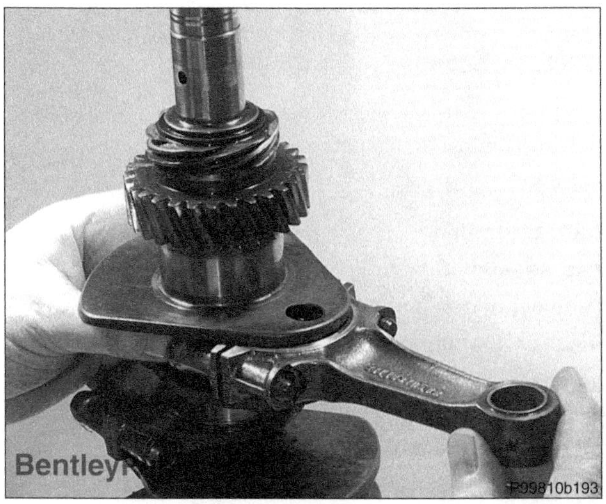

◀ Mark installed position of connecting rods and remove from crankshaft.

- Slide crankshaft nose bearing off crankshaft.
- Remove circlip holding gear assembly to crankshaft.

Remove gear assembly on front of crankshaft using suitable gear puller.

Engine, Assembling

All parts must be clean and ready for reassembly at this point. All machine work should be completed and all new replacement parts should be on hand. Use suitable engine assembling lubricant during assembly.

> ***CAUTION—***
> - *If the engine failed mechanically, it is important to check the intake system for foreign objects.*
> - *If the engine is disassembled because of bearing failure, a thorough flushing of the lubrication system must be carried out prior to reassembly.*

NOTE—
- *Various head studs have been used on the M64 engine. Always check with an authorized Porsche parts department for the latest in parts information.*
- *If either the crankcase, the crankshaft drive gear or the intermediate shaft drive gear are replaced, special installation instructions apply. See* **13a Engine Crankshaft**.

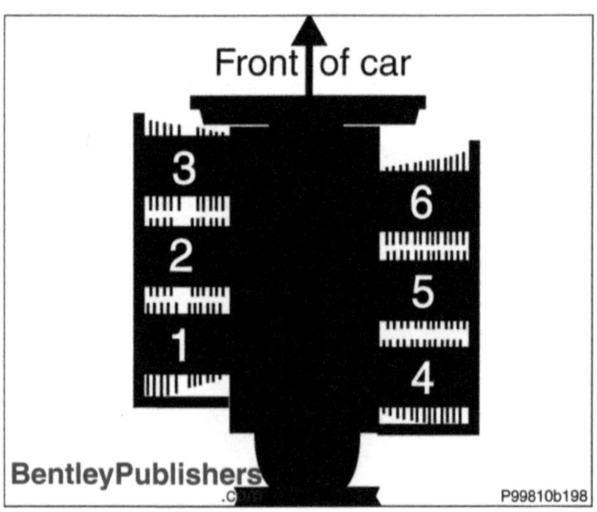

Cylinder numbering is shown in illustration. Cylinder position will be referred to during the assembly process.

The assembly procedure is divided into subsections, as follows:
- **Crankshaft, assembling**
- **Crankcase, assembling**
- **Pistons and cylinders, installing**
- **Cylinder heads, installing**
- **Camshaft cases, camshafts and camshaft timing chains, installing**
- **Camshaft sprocket assembly**
- **Timing chain parallelism**

- **Camshaft timing, adjusting**
- **Rocker shaft and arm, installing**
- **Engine peripherals, installing**

Crankshaft, assembling

◄ Crankshaft reconditioning information is given in **13a Engine Crankshaft**.

1. Crankshaft nose bearing
2. Circlip
3. Drive gear (for distributor)
4. Spacer
5. Drive gear (with mark 0 or 1)

NOTE—

- *Check crankshaft drive gear for wear prior to installation. If gear is worn or damaged, replacement gear must be matched to crankcase and to gear on intermediate shaft.*

– Mount crankshaft to Porsche special tool **P 209** or equivalent and clamp tool with crankshaft in shop vise.

◄ Install drive gear to crankshaft.

- Install woodruff key to crankshaft
- Heat gear to 150°C (302°F) in oil bath.
- With shoulder (**arrow**) on gear facing in, slide drive gear fully onto crankshaft.

WARNING—

- *Use protective gloves or shop rag when handling the hot gear.*

10b

– Install spacer to crankshaft.

– Install distributor drive gear to crankshaft.

- Heat gear to 100°C (212°F) in water bath.

WARNING—

- *Use protective gloves or shop rag when handling the hot gear.*

◄ Install gear retaining circlip to crankshaft.

- Circlip used to hold gears in place on crankshaft comes in different thicknesses.
- Select circlip to best obtain near zero clearance.

Porsche P/N	Thickness	Mark
901 102 148 00	2.4 mm	0
901 102 148 01	2.3 mm	1
901 102 148 02	2.2 mm	2
901 102 148 03	2.1 mm	3

– Install new connecting rod bolts in each connecting rod.

– Install connecting rod bearing shells in connecting rods.

◀ Install connecting rods on crankshaft in their original locations. Matching numbers on side of connecting rod and rod cap should all be facing same side of engine (i.e., all numbers face exhaust side of engine).

– Install new connecting rod nuts on all rod bolts.

Tightening torques	
Connecting rod cap to connecting rod • Stage 1 • Stage 2	 30 Nm (22 ft-lb) 90° ± 2°

Crankcase, assembling

– Mount right side crankcase half on engine stand.

– Install and tighten oil drain plug using new sealing ring.

Tightening torque	
Oil drain plug to crankcase	50 + 5 Nm (37 + 3.7 ft-lb))

◀ Install intermediate shaft bearing shells in crankcase halves.

NOTE—

- *Due to the high loads the timing chains put on the intermediate shaft, it is recommended that the intermediate shaft bearings be replaced during engine overhaul.*

◀ Place new oil pump sealing O-rings (**arrows**) in right crankcase half.

– Lubricate oil pump gears with engine assembling lubricant.

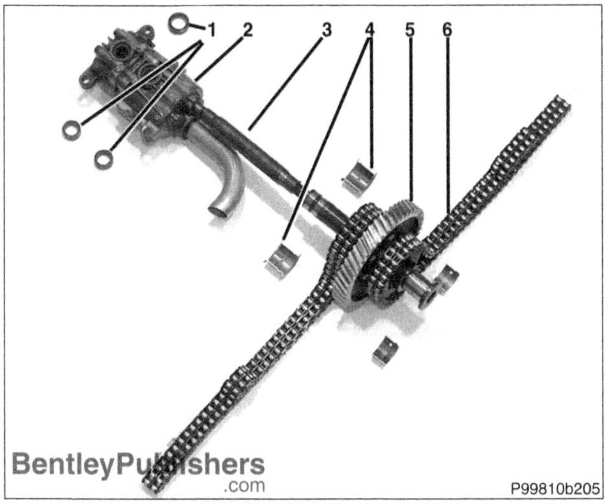

Place both timing chains on intermediate shaft sprockets. Install oil pump drive shaft with oil pump to intermediate shaft.

1. Oil pump sealing O-rings
2. Oil pump
3. Oil pump drive shaft
4. Intermediate shaft bearings
5. Drive gear (with 0 or 1 mark)
6. Timing chain

NOTE—

- *Check intermediate shaft gear for wear prior to installation. If the intermediate shaft gear is worn or damaged, the replacement gear must be matched to the crankcase and crankshaft drive gear.*
- *The intermediate shaft is hollow and should be cleaned out whenever the engine is being reconditioned and in cases of engine bearing failure.*

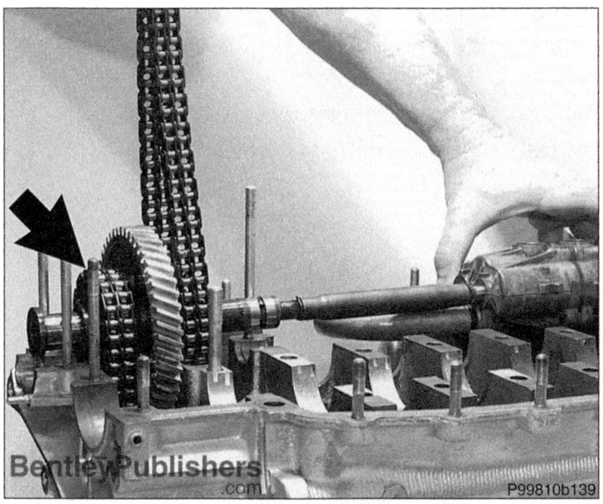

Lower complete intermediate shaft assembly into right crankcase half, guiding front timing chain (**arrow**) through opening in case.

– Install new oil pump with new self-locking nuts.

Tightening torque	
Oil pump to crankcase (replace with new)	25 Nm (18 ft-lb)

Lubricate crankshaft main bearing shells with engine oil and install in crankcase halves.

– Install crankshaft nose bearing locating pin in crankcase half.

◀ Without nose bearing outer sealing O-ring in place, lay crankshaft nose bearing in crankcase so that it locks into locating pin (**A**). Draw a line (**B**) on each side of bearing to use as a reference when installing crankshaft assembly.

– Remove nose bearing.

– Install nose bearing outer O-ring.

– Apply Curil T sealant to crankcase to nose bearing sealing surface.

– Lubricate nose bearing and install on crankshaft.

◀ Place crankshaft assembly into right crankcase half.

- Make sure right side connecting rods for cylinders 4, 5, and 6 are pointing down.
- Check that nose bearing (**A**) is correctly seated in case.
- Slowly turn crankshaft back and forth to check for binding or interference.

NOTE—

- *Porsche special tool* **P 221** *can be used to support the left side connecting rods during crankcase assembly.*
- *Porsche special tool* **P 222** *can be used to support the left timing chain during crankcase assembly.*

◀ Place two new sealing O-rings (**A**) in oil pump.

- Check to see that oil passage sealing O-ring previously placed in right crankcase half (**B**) remains undisturbed.
- Lubricate all O-rings lightly with engine oil.

NOTE—

- *Different size seals are used. Be sure to install the correct parts.*
- *Oil pump nut locking tabs not used on M64 engine.*

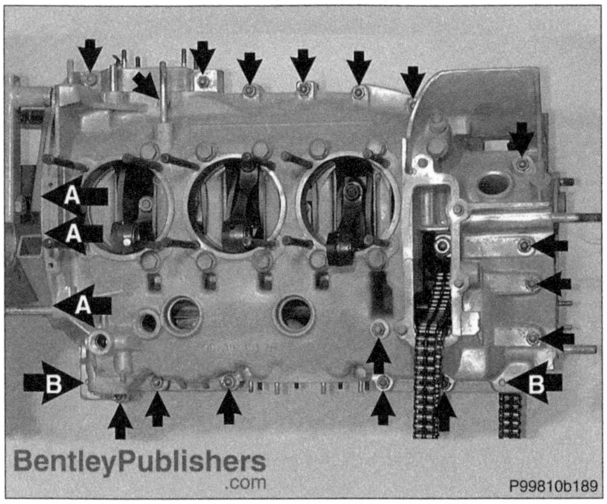

– Using Loctite® 574 sealant, apply even thin coat to the mating surface of the left crankcase half.

NOTE–

- *The preferred method of applying Loctite 574 is to use a smooth paint roller. Once sealant is applied, crankcase should be bolted together within ten minutes. Use a paint brush to apply sealant around the bolt holes.*

– Place left case half in position over right case half. Tap case halves together lightly with a rubber mallet.

< Install 22 M8 nuts around outer edge of crankcase. Tighten nuts until case halves make solid contact (not final torque).

- Three nuts are in flywheel housing (**A**).
- Two nuts face right side (**B**).

> ***CAUTION—***
>
> - *Turn crankshaft back and forth periodically to check for binding. Protect connecting rod small ends (use short length of rubber tube) prior to turning crankshaft.*

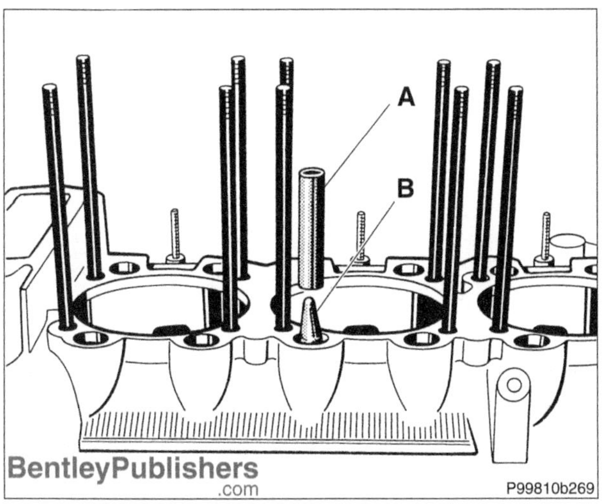

– Lubricate 11 case bolts and 24 O-ring seals lightly with oil.

NOTE–

- *Use of Porsche special tool* **P 9511** *(tapered adapter (***A***) and cylindrical sleeve (***B***)) is strongly recommended when installing sealing O-rings to case bolts. The two-part tool is designed to install the O-rings without damage.*

– Install insulating washers to all bolts.

10b

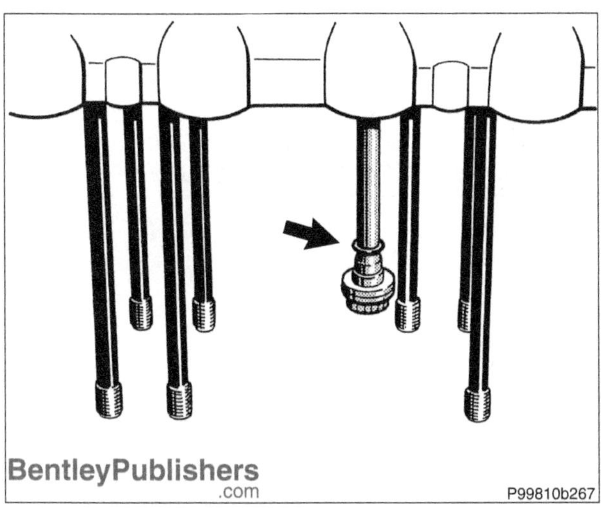

< Install sealing O-rings to shouldered-end of bolts:

- Install tapered adapter (**P 9551**) to threaded end of bolt. Install O-ring to tapered adapter and slide O-ring onto bolt. Slide O-ring up to shoulder (**arrow**). Repeat for all bolts.

– Apply Curil T sealant to crankcase bolt washer surface.

– Install case bolts into right half of crankcase and press firmly until seated.

– Working at left half of crankcase, install O-ring onto tapered adapter (**P 9511**). Install adapter over threaded end of bolt. Using cylindrical sleeve (big end), press O-ring off adapter until seated in crankcase. Install insulating washer. Using cylindrical sleeve (small end), press washer firmly until seated in crankcase. Repeat for all bolts.

NOTE–

- *If necessary, hold case bolt stationary from opposite side when installing sealing O-rings and insulating washers.*

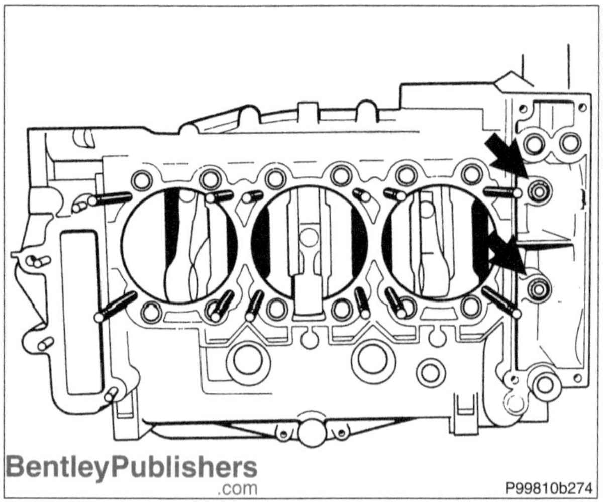

– Install 12-point case bolt nuts (threads lubricated with oil) finger tight.

◄ Using O-ring install tools (**P 9551**), install O-rings and insulating washers on case studs (**arrows**) in crankcase near bearing no. 1. Install case stud nuts (threads lubricated with oil) finger tight.

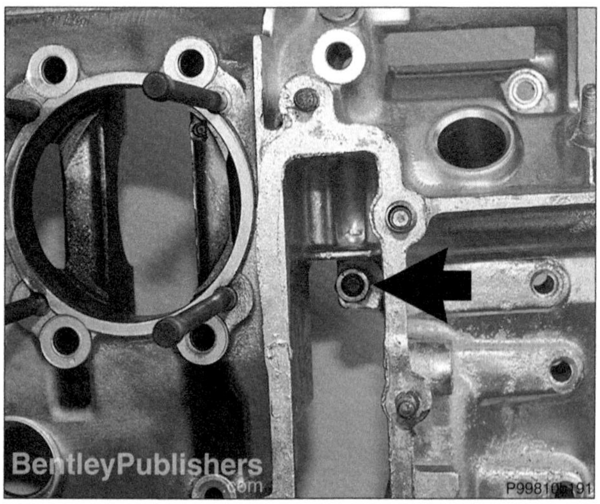

◄ Working in left timing chain cutout, install M10 nut (**arrow**) with wavy washer facing up in timing chain cutout.

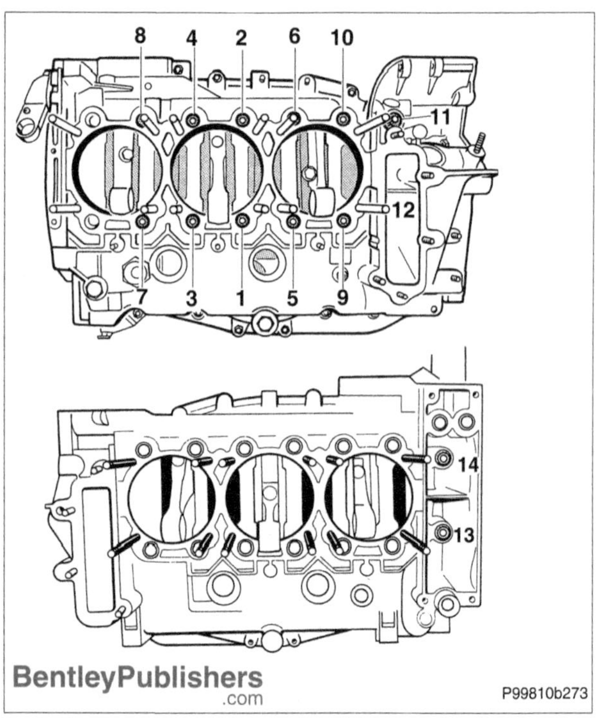

◄ While counterholding case bolts, torque nuts using M14 12-point socket in sequence shown.

Tightening torques	
Case bolts and nuts (12-point M10)	50 Nm (37 ft-lb)
Outer studs/nuts (M8)	25 Nm (18 ft-lb)

CAUTION—

- *To avoid damaging sealing O-rings, always counterhold bolts with help from an assistant when tightening nuts.*

– Install crankshaft pulley seal and rear main seal. See **13a Engine Crankshaft**.

– Install crankshaft pulley, counterholding with Porsche special tool **9290** or equivalent when torquing bolt.

Tightening torque	
Crankshaft pulley to crankshaft	235 Nm (173 ft-lb)

NOTE—

- *If crankshaft pulley locating hole or pin is damaged, valve timing may be affected. Replace a pulley with elongated / damaged hole.*

– Install oil pressure sending units.

Tightening torques	
Oil warning light sending unit to crankcase	20 Nm (15 ft-lb)
Oil pressure sensor to crankcase	35 Nm (26 ft-lb)

– Install breather cap with a new gasket.

Tightening torque	
Breather cap to crankcase	10 Nm (7 ft-lb)

– Install oil temperature sensor.

Tightening torque	
Oil temperature sensor to crankcase	25 Nm (18 ft-lb)

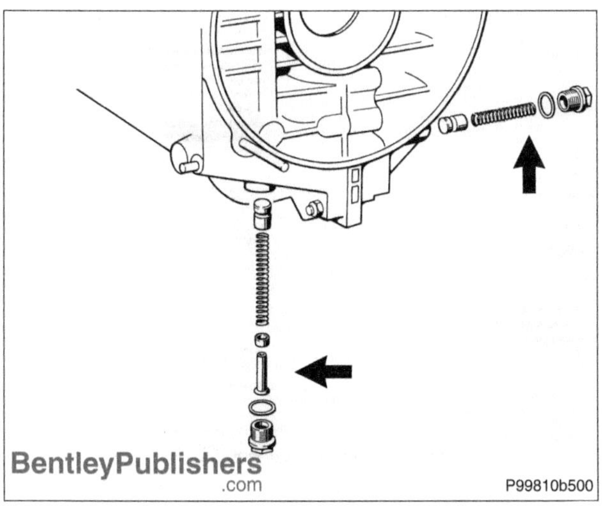

◄ Install oil pressure relief valve assemblies (**arrows**) in crankcase. Coat relief valves with assembly lube before installing.

Tightening torque	
Oil pressure relief valve plugs to crankcase	60 Nm (44 ft-lb)

10b

Pistons and cylinders, installing

Measurement and reconditioning of pistons and cylinders is covered in **13b Pistons and Cylinders**.

> ***CAUTION—***
> - *Be sure to keep each piston and cylinder sets clearly marked and together.*
> - *Be sure used pistons and cylinders are reinstalled in the exact same locations and positions.*

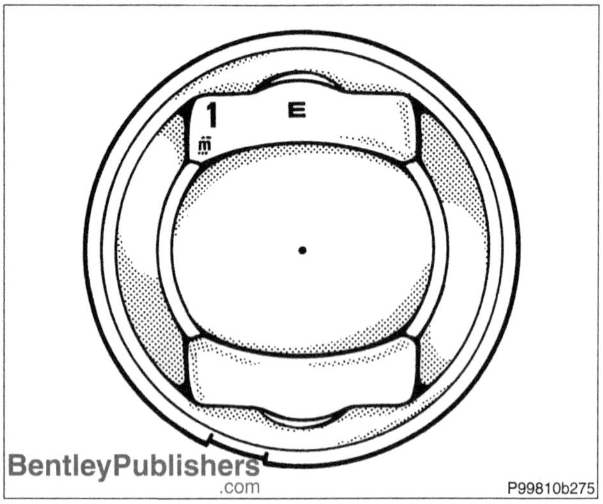

P99810b275

◀ Installed piston (orientation of stampings on upper piston crown):
- E stamp in center (faces intake side)
- Tolerances group (0, 1, 2, or 3) to left of E stamp
- Weight class is next tolerance group

NOTE–
- *Markings to right of E stamp for internal use by piston manufacturer (Mahle).*

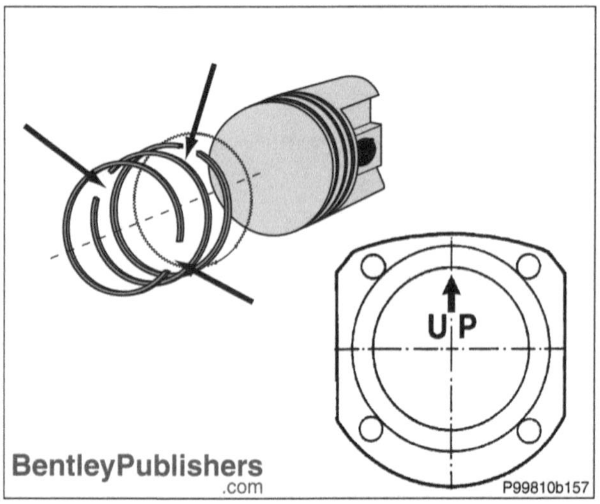

P99810b157

◀ Set piston ring gaps to clock positions 4, 10 and 2 (**arrows**).

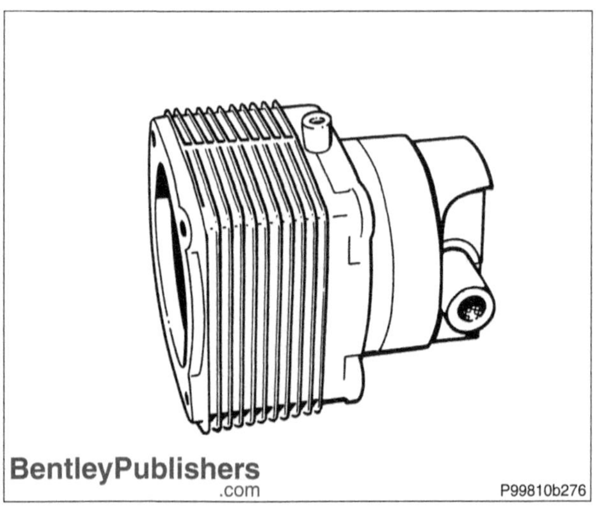

P99810b276

◀ Working on a wooden or plastic work surface, lubricate cylinder walls and preassemble cylinders to pistons as shown (wrist pin partially installed).

– Install one of two wrist pin retaining circlips to piston (opposite end of partially installed wrist pin). Be sure anti-rotation tang on circlip is properly engaged in notch on piston.

> ***CAUTION—***
> - *If installed, wrist pin circlips on new replacement pistons are for transportation purposes only. Replace with circlips provided separately with pistons.*
> - *Always check that both piston pin circlips are in place before sliding cylinder into position.*

- Reset crankshaft to cylinder 1 compression TDC.
- Install cylinder and piston assembly no 1.
  - Place new cylinder base O-ring on cylinder 1.
  - Position cylinder / piston through head studs and align connecting rod. Deeper cooling fins point down.
  - Push wrist pin into position and install remaining circlip. Check carefully that circlip is properly installed.
  - Fully seat cylinder and temporarily install Porsche special tool to hold cylinder in place.
- Slowly rotate crankshaft and install cylinders / pistons 2 and 3 with new base O-rings.

> ***CAUTION—***
> - *When turning crankshaft, manually guide connecting rods pistons to avoid contact.*

< Install new head gaskets (**arrow**) in groove at top of cylinders 1, 2 and 3.

< Temporarily hold cylinders in position using Porsche special tools **P 140** (**arrows**).

NOTE–
- *As a substitute for the special tools, slide two sockets over the cylinder head studs for each cylinder and install two cylinder head nuts finger tight.*

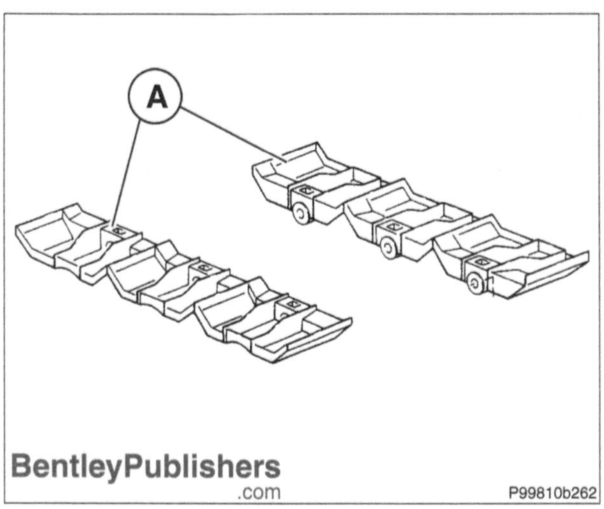

< Install left side cylinder lower air shrouds (**A**).
- Install right side pistons and cylinders (4, 5, 6) with base O-rings and head gaskets.

Cylinder heads, installing

Repairs to cylinder heads and valves are described in **15a Cylinder Head**.

- With engine on engine stand and pistons, cylinders, gaskets, and air guides installed, rotate engine on stand so that left side cylinders are up.

NOTE—

- *Remove temporary cylinder hold down nuts prior to cylinder head installation.*

- Lubricate cylinder head studs, nuts and washers lightly with Optimoly®HT copper paste (Porsche part number 901 104 382 02) or equivalent. Starting with cylinder head 3, install left side cylinder heads.

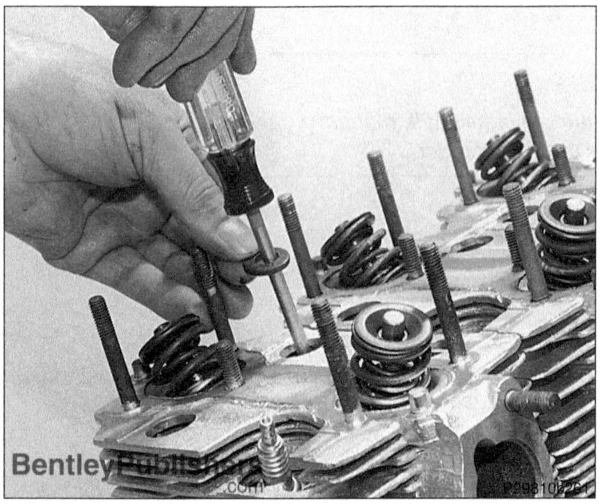

- < Using screwdriver as guide, slide head washers over head studs Install head nuts and hand tighten.
- Temporarily fit camshaft case in position on heads.
  - Alignment pins on cylinder heads must line up with alignment holes in camshaft case. Position cylinder heads as required until machined sealing surfaces meet.
  - Temporarily install a few camshaft case mounting nuts (hand tight) to hold case firmly in position.
- Install knock sensor bridge and tighten fasteners finger tight.

NOTE—

- *Sealing surfaces on heads and camshaft cases must be clean, flat, and without burrs.*

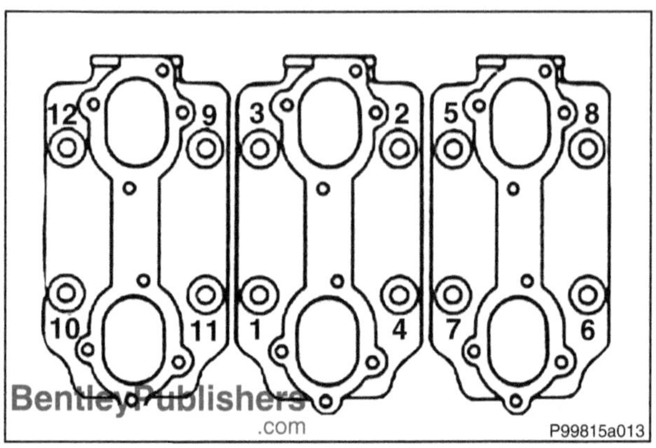

- < Use cross pattern to torque cylinder head nuts.

Tightening torque	
Cylinder head to crankcase • Stage 1 • Stage 2	 20 Nm (15 ft-lb) 90° ± 2°

- Remove camshaft case from cylinder heads 1 - 3. Rotate engine on stand so that right side cylinders 4 - 6 are up.
- Starting with cylinder 6, install right side cylinder heads.
- Slide head washers over head studs. Install head nuts and hand tighten.
- Temporarily fit camshaft case in position on heads.
  - Alignment pins on cylinder heads must line up with alignment holes in camshaft case. Position cylinder heads as required until machined sealing surfaces meet.
  - Temporarily install a few camshaft case mounting nuts (hand tight) to hold case firmly in position.
- Use cross pattern to torque cylinder head nuts.
- Remove camshaft case from cylinder heads 4 - 6.

- Rotate engine on stand. Install left side bottom oil pressure line.
- Tighten knock sensor bridge bolts.

Tightening torques	
Knock sensor bridge to cylinder heads	10 Nm (7 ft-lb)
Knock sensor to knock sensor bridge	20 ± 5 Nm (15 ± 3.5 ft-lb)

Camshaft cases, camshafts and camshaft timing chains, installing

- Use Loctite® 574 to coat machined sealing surfaces on camshaft cases.
- Place new sealing O-rings on oil return tubes.
- Holding oil return tubes, guide right camshaft case into place.

NOTE—

- *Two piece oil return tube replacement is covered in* **17 Lubrication System**.

◀ Install 18 retaining nuts (**arrows**) and tighten after making sure oil return tubes are correctly positioned.

Tightening torque	
Camshaft case to cylinder head	25 Nm (18 ft-lb)

- Install timing chain guide rails with timing chain housing to engine.

NOTE—

- *The short guide rail is used to maintain tension on the timing chain via the timing chain tensioner.*

- Tighten right timing chain housing. Repeat for left side.

Tightening torque	
Timing chain housing to crankcase	25 Nm (18 ft-lb)

◀ Tighten timing chain guide rail (**A**) retaining bolts (**arrow**).

Tightening torque	
Timing chain guide rail retaining bolt to crankcase	25 Nm (18 ft-lb)

- Slide right camshaft into camshaft case.
- Install camshaft end cover with new O-ring and gasket. Install end cover retaining bolts and tighten.
- Install right camshaft sprocket to timing chain. Slide sprocket over sprocket flange.
- Install sprocket washer and mounting bolt finger tight.
- Install right timing chain tensioner and idler sprocket assembly using new O-ring. Tighten tensioner mounting nuts.
- Install left camshaft, camshaft case, timing chain guide rail, camshaft sprocket assembly and timing chain as described in previous steps.

Camshaft sprocket assembly

Camshaft sprocket assembly

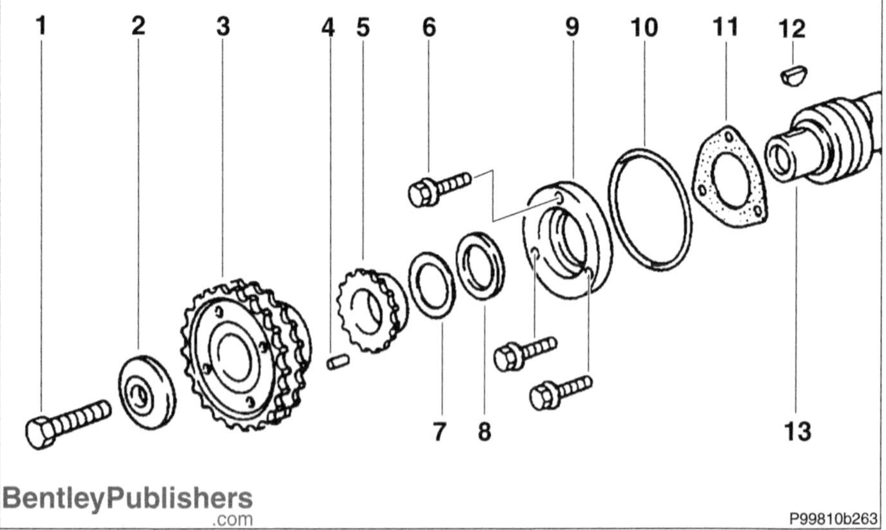

P99810b263

1. Sprocket mounting bolt
2. Lock washer
3. Camshaft sprocket
4. Sprocket lockpin
5. Sprocket flange
6. Camshaft retaining bolt
7. Sprocket spacer shims
8. Thrust spacer
9. Camshaft end cover
10. Sealing O-ring
11. Gasket
12. Camshaft
13. Woodruff key

– Check parallelism of intermediate shaft and camshaft sprockets by measuring distance from front of intermediate shaft to front edge of gear wheels. Use quality straight edge when making measurement.

NOTE—

- *Normally, if reinstalling camshafts and sprockets, 3 sprocket spacer shims will be necessary under the left sprocket and 4 shims under the right sprocket.*
- *If new camshafts or sprockets are being installed, camshaft sprocket spacing must be adjusted by adding or subtracting spacer shims so that each sprocket is parallel with the corresponding intermediate shaft sprocket.*

Timing chain parallelism

Timing chain sprocket parallelism

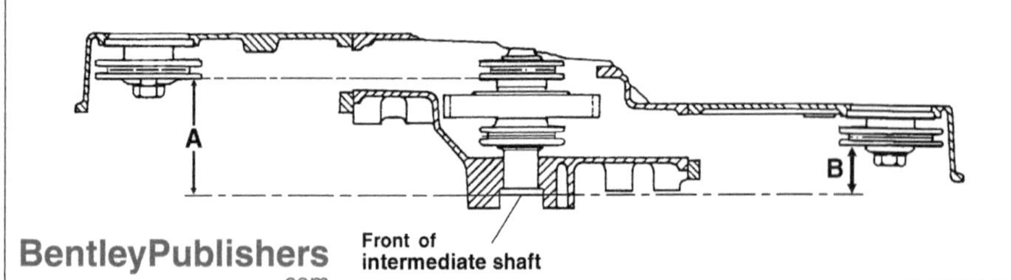

P99810b165

- **A** Factory distance from end of intermediate shaft to face of cylinder 1 - 3 (left) timing chain sprocket = 98.07± 0.25 mm (3.86 ± 0.01 in.)
- **B** Factory distance from end of intermediate shaft to face of cylinder 4 - 6 (right) timing chain sprocket = 43.27 ± 0.25 mm (1.70 ± 0.01 in.)

Camshaft timing, adjusting

For camshaft timing adjusting, see **15b Camshafts**.

Rocker shaft and arm, installing

◀ Set crankshaft to cylinder 1 compression TDC (**arrow**).

◀ When installed, camshaft punch mark (**arrow**) should face up.

NOTE—

- *Camshaft punch marks or woodruff key hole will point up when at cylinder 1 TDC.*

– Assemble rocker arms with shaft.

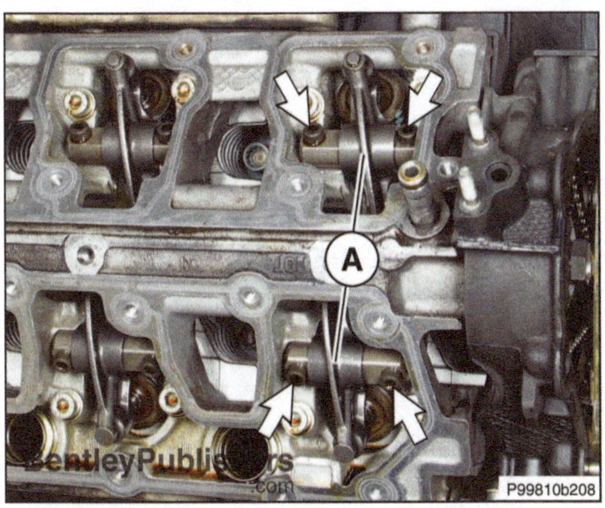

◀ Install cylinder 1 intake and exhaust rocker shafts with rocker arms (**A**).

– Install rocker arm fasteners (**arrows**) and torque.

– Install remaining rockers arms.

– Check hydraulic lifter height, see **15b Camshafts**.

Engine peripherals, installing

< With crankshaft at TDC, install distributor so that mark on distributor rotor aligns with mark on distributor housing (**arrows**).

- The marks line up approximately, not precisely.
- There is only one way to install the distributor housing.

Tightening torque	
Distributor base clamp (M8)	18 Nm (14 ft-lb)

— Install new spark plugs.

Tightening torque	
Spark plug to cylinder head	30 Nm (22 ft-lb)

— Install engine mount carrier and cross member.

— Install A/C compressor mounting brackets.

Tightening torques	
Engine mount carrier to crankcase	40 Nm (30 ft-lb)
Engine mount crossmember to carrier	40 Nm (30 ft-lb)

— Install oil filter housing.

> ***CAUTION—***
> *• Always use new O-ring seals.*

Tightening torque	
Oil filter housing to crankcase	23 Nm (17 ft-lb)

— Install all sheet metal shrouding around engine.

— Set complete fuel system, intake manifold, alternator and cooling shroud in place.

- Reattach alternator ground to top of crankcase under shroud.
- Install intake runner washers and nuts.

Tightening torque	
Intake manifold to cylinder heads	25 Nm (18 ft-lb)

— Connect all spark plug wires and install all wire holders.

— Rotate engine on stand and install heat exchangers, catalytic converter and muffler. See **26 Exhaust System**.

Tightening torque	
Heat exchangers to cylinder heads	25 Nm (18 ft-lb)

— Remove engine from engine stand.

— Install flywheel, clutch assembly and transmission as described in **10a Engine Removal and Installation**.

Tightening torques	
Clutch to flywheel	25 Nm (18 ft-lb)
Flywheel to crankshaft (use new bolts)	85 Nm (63 ft-lb)
Transmission to engine	45 Nm (33 ft-lb)
Drive plate to flywheel (Tiptronic) (use new bolts)	90 Nm (66 ft-lb)

– Install engine. See **10a Engine Removal and Installation**.

Tightening torques	
Crankshaft pulley to crankshaft	235 Nm (173 ft-lb)

13a Engine Crankshaft

GENERAL

This repair group covers engine pulley removal and installation. Also covered is crankshaft seal replacement and crankshaft specifications.

Engine drive belt service is covered in **03 Service and Maintenance.**

CRANKSHAFT PULLEY

Pulley system

The V-belts are driven by the crankshaft pulley and power the alternator and air conditioning compressor (if equipped). Alternator and fan V-belt tension is maintained by pulley shims.

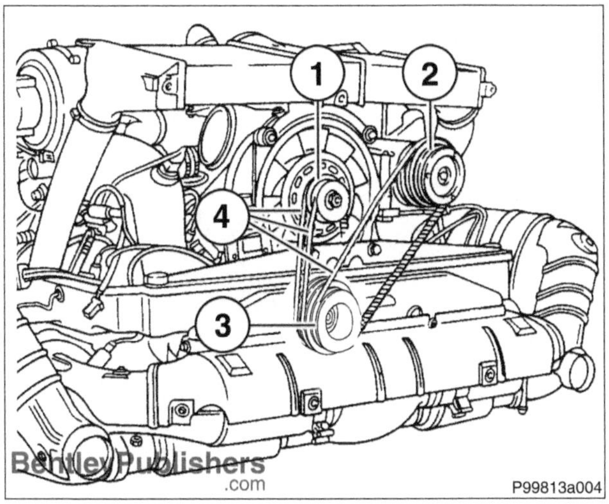

Belt routing

1. Alternator pulley
2. Air conditioning pulley
3. Crankshaft pulley
4. V-belts

Crankshaft pulley, removing and installing

The crankshaft pulley mounting bolt is tightened to a high torque. Be sure to have the Porsche crankshaft counterholding tool (special tool **9548**) available before starting the job.

– Open engine hood and remove intake air housing. See **24 Fuel Injection**.

– Mark direction of drive belts using tire crayon before removing.

– Remove drive belts. See **03 Service and Maintenance**.

– Remove ignition coils from front engine support bracket and lay aside. See **28 Ignition System**.

– Place engine jack under engine and secure in place.

Working from below, remove engine mount fasteners (**inset**).

Remove engine mounting bracket fasteners (**arrows**) then remove bracket (**A**) from engine.

NOTE–

- *Engine shown removed to better show fastener location.*

- Counterhold crankshaft pulley using Porsche special tool **9548** while loosening pulley fastener.
- Remove crankshaft pulley from engine.

- Before installing crankshaft pulley, clean crankshaft (**arrow**) and pulley mating surfaces. Both should be free of oil and debris.
- Apply a thin coat of oil to crankshaft pulley fastener threads and pulley fastener mating surface.
- Install pulley fastener and torque. Counterhold crankshaft pulley using Porsche special tool 9548 while torquing.

Tightening torques	
Crankshaft pulley to crankshaft	235 Nm (173 ft-lb)

- Inspect crankshaft pulley runout using a dial indicator. Maximum runout of 0.5 mm (0.01 in) is permissible. Replace a bent pulley.
- Reinstall drive belts, reassemble engine, start engine and allow to idle for a few minutes. Then recheck pulley and drive belt tension. See **03 Service and Maintenance**.

CRANKSHAFT SEALS

Crankshaft pulley seal, removing and installing

- Remove crankshaft pulley. See **Crankshaft pulley, removing and installing** in this repair group.

- Using a flat blade screwdriver, penetrate rubber sealing lip, then insert screwdriver behind seal face and lever out.

> ***CAUTION—***
> - *Be careful not to damage sealing surface of crankshaft when prying seal out.*

- Before installing new seal, lightly coat seal with oil.

◄ Install seal using Porsche special tool **P 216**.

– Install crankshaft pulley.

◄ Install pulley fastener and torque. Counterhold crankshaft pulley using Porsche special tool 9548 while torquing.

Tightening torques	
Crankshaft pulley to crankshaft	235 Nm (173 ft-lb)

– Inspect crankshaft pulley runout. See **Crankshaft pulley, removing and installing** in this repair group.

– Recheck crankshaft pulley fastener torque after straightening.

– Reinstall drive belt, reassemble engine, start engine and allow to idle for a few minutes. Then recheck pulley and drive belt.

– Idle engine and inspect for oil leaks.

Rear main seal, removing and installing

Rear main seal problems are frequently the cause of engine oil leaks. Periodically inspect the seam between the engine crankcase and transmission bellhousing for signs of fresh oil drips.

– Remove transmission. See **34 Manual Transmission** or **10 Engine Removal and Installation**.

– **Manual transmission**: Remove clutch pressure plate and flywheel. To avoid warping flywheel, loosen bolts in diagonal pattern and in several stages. See **30 Clutch**.

– **Automatic transmission**: Remove drive plate. To avoid warping drive plate, loosen bolts in diagonal pattern and in several stages. See **37 Automatic Transmission**.

– Drill a 3 mm hole in the face of rear main seal.

> ***CAUTION—***
>
> *• Do not exceed a depth of 5 mm (¼ inch) when drilling hole in rear main seal.*

– Thread sheet metal screw with washer approximately 5 mm (¼ in) into drilled hole. Use Porsche puller (special tool **9237**) to remove seal.

– Inspect crankshaft sealing surface. Clean crankcase bore seating surface.

– Lightly coat outer surface of seal with oil.

< Install seal using Porsche special tool **9517** and **9517 / 1**.

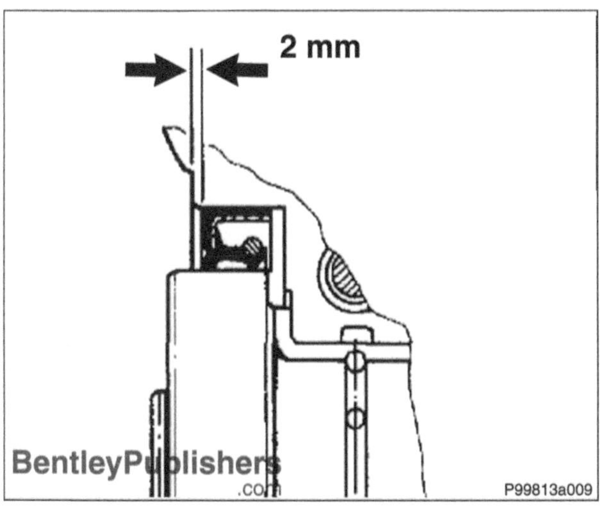

< Install seal until it is recessed 2 mm beyond crankcase flange.

– Reinstall drive plate or clutch.

Tightening torques	
Dual-mass flywheel to crankshaft (use new bolts)	85 Nm (63 ft-lb)
Pressure plate to dual-mass flywheel	23 Nm (17 ft-lb)
Drive plate to crankshaft (use new bolts)	90 Nm (66 ft-lb)
Torque converter to drive plate	16 Nm (12 ft-lb)
Transmission bell housing to engine	23 Nm (17 ft-lb)

– Reinstall transmission. See **10a Engine Removal and Installation**.

Crankshaft Assembly

Crankshaft assembly

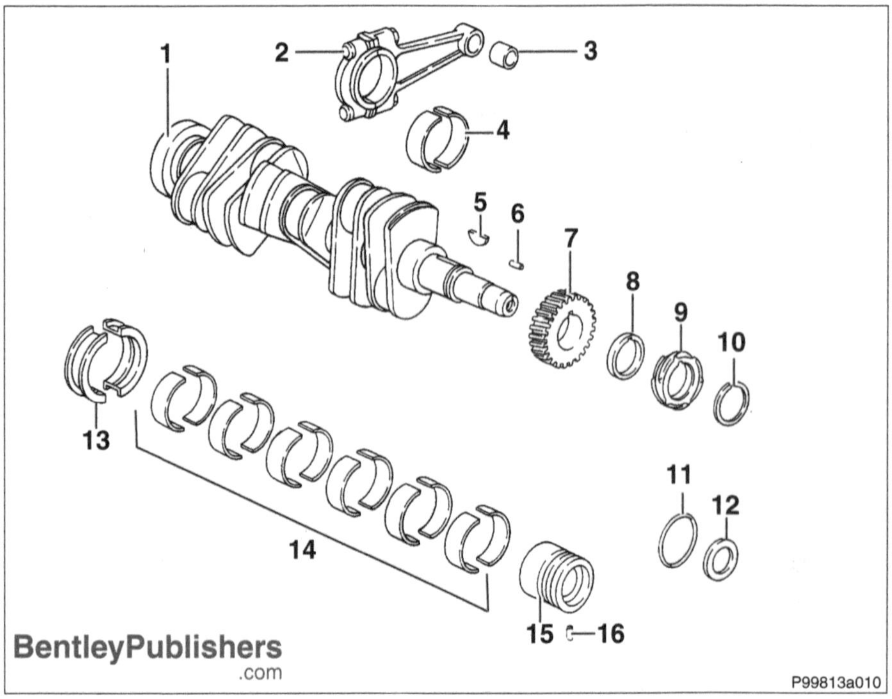

1. Crankshaft
2. Connecting rod
   - Connecting rod nut
   - Connecting rod bolt
3. Connecting rod bushing
4. Connecting rod bearing
   - Apply thin coat of oil
5. Woodruff key
6. Tensioning sleeve (dowel)
7. Timing gear
8. Intermediate ring
9. Drive wheel
10. Snap ring
11. O-ring
12. Crankshaft seal (pulley end)
13. Main thrust bearing
    - Apply thin coat of oil
14. Main bearings
    - Apply thin coat of oil
15. Nose bearing
16. Tensioning sleeve (dowel)

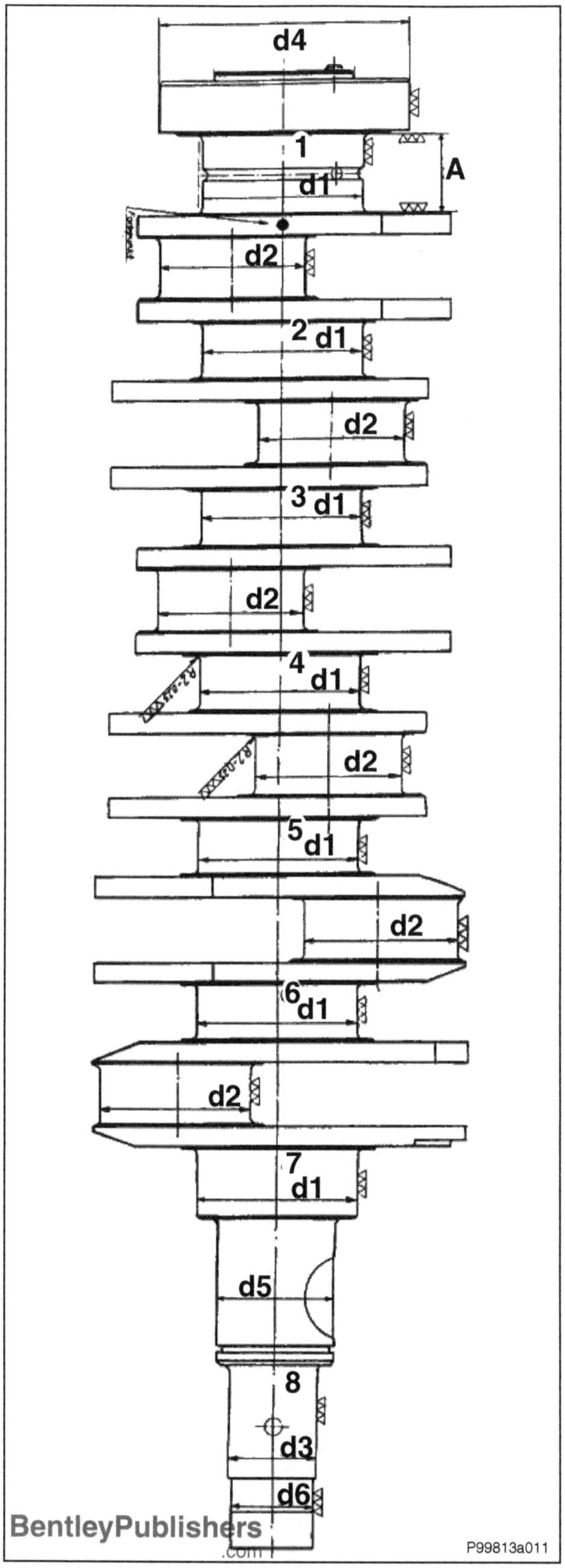

Crankshaft specifications

- Polish oil seal running surfaces, RT = 3.
- Machine oil seal running surfaces to 29.670 mm (crankshaft seal) and 89.580 mm (rear main seal) only if score marks are too deep.
- Oil bores are rounded to R 0.5 mm after machining.
- Remove sharp edges with R = 0.2 mm to 0.5 mm.
- Maximum permissible radial runout relative to support in -.-.- maximum 0.04 mm.
- Surface treatment using gas carbonitrided PN 2063.
- Do not straighten main bearings 3 and 5 after nitriding treatment.
- Straightening other main bearings by levering at the radii if needed.
- Polish running surfaces of main and big end bearing journals after nitriding treatment.

Repair size color coding:

- 1st repair size - blue dot
- 2nd repair size - green dot

Bearing size	Main bearing d1	Big end bearing d2	Main bearing d3
Standard	59.971 - 59.990	54.971 - 54.990	30.980 - 30.993
0.25	59.721 - 59.740	54.721 - 54.740	30.730 - 30.743
0.50	59.471 - 59.490	54.471 - 54.490	30.480 - 30.493

Bearing size	Flange 0 d4	TIming gear fit d5	Support d6
Standard	89.780 - 90.000	42.002 - 42.013	29.960 - 29.993
Wear diameter	89.580	n/a	29.670

Bearing size	Guide bearing width A	Crankcase diameter
Standard	28.000 - 28.060	65.000 - 65.019
Oversize	n/a	65.250 - 65.269

13a

13b Pistons and Cylinders

GENERAL

This repair group provides reconditioning information for the pistons and cylinders. Engine disassembly is described in **10b Engine Disassembly and Assembly**.

> ***CAUTION—***
> - *The information contained here is intended to be used as a reconditioning guide for the experienced automotive mechanic.*

Pistons and Cylinders

Pistons must be matched by weight. When replacing pistons and cylinders, always check with an authorized dealer for the latest in replacement parts information.

Mark all parts during engine disassembly. All used parts must be reinstalled in their exact installation position. Do not interchange between cylinders.

Piston and cylinder components

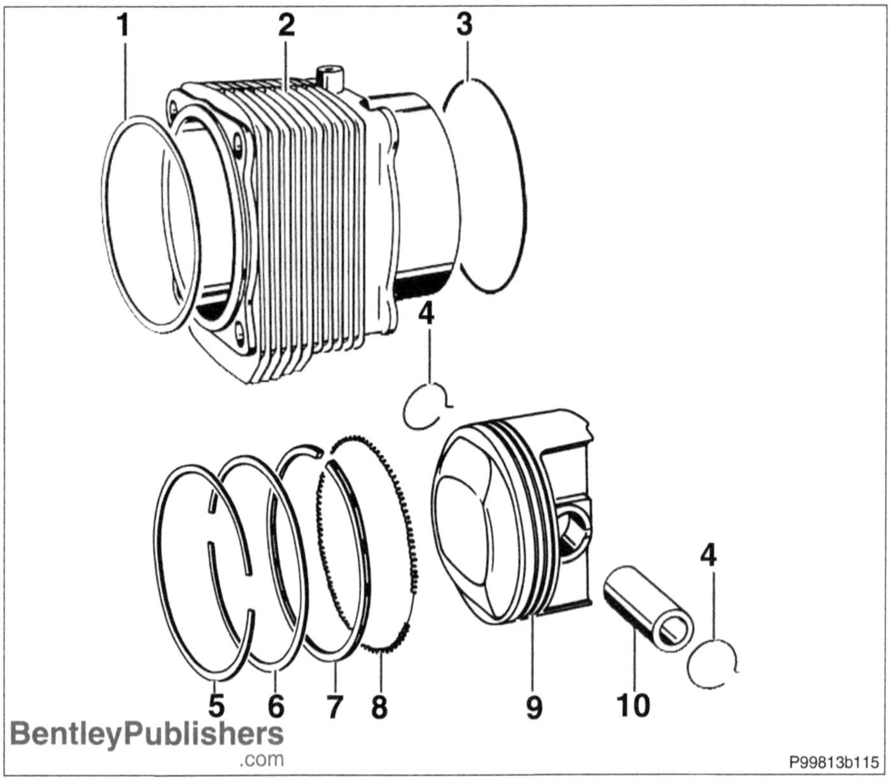

1. Cylinder head seal
   - replace with new, install dry
2. Cylinder
3. Base O-ring (102 x 2)
4. Wrist pin circlip
   - replace with new
5. Tapered face ring, groove 1
6. Stepped tapered faced ring, groove 2
7. Double beveled oil control ring, groove 3
8. Tubular spring, groove 3
9. Piston
10. Wrist pin

Cylinders

P99813b118

Cylinders have a bore diameter of 100 mm and are made from a high-temperature-resistant light alloy. The bore surface coating (Nikasil) is applied galvanically. The bore is machined with a slight taper to compensate for differing temperatures across the height of the cylinder and to maintain a constant running clearance.

◄ Cylinder manufacturer identification and dimensions are stamped on the cylinder base opposite from the knock bridge mounting lug.

1. Cylinder height tolerance group (5 or 6)
2. Manufacturer identification
3. Cylinder diameter tolerance group

Cylinders are divided into two height tolerance groups and four diameter tolerance groups.

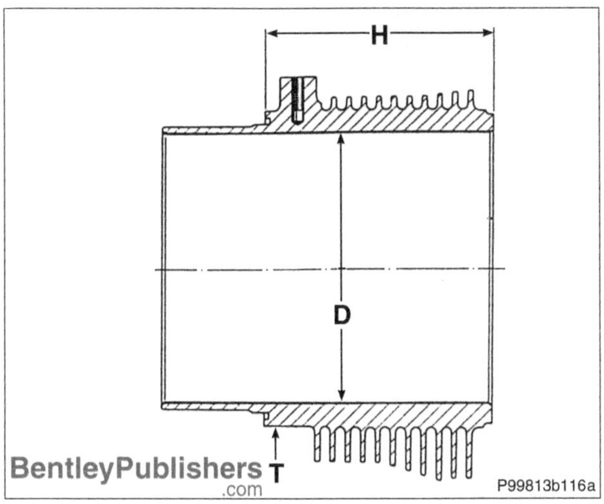

P99813b116a

Height tolerance group (stamp)	**Diameter tolerance group (stamp)**	**Height size (H) (-0.20)**	**Cylinder diameter (D) (+0.007)**
5	0	82.750 mm	100.000 mm
	1	82.750 mm	100.007 mm
	2	82.750 mm	100.014 mm
	3	82.750 mm	100.021 mm
6	0	82.770 mm	100.000 mm
	1	82.770 mm	100.007 mm
	2	82.770 mm	100.014 mm
	3	82.770 mm	100.021 mm

Cylinder roundness, checking

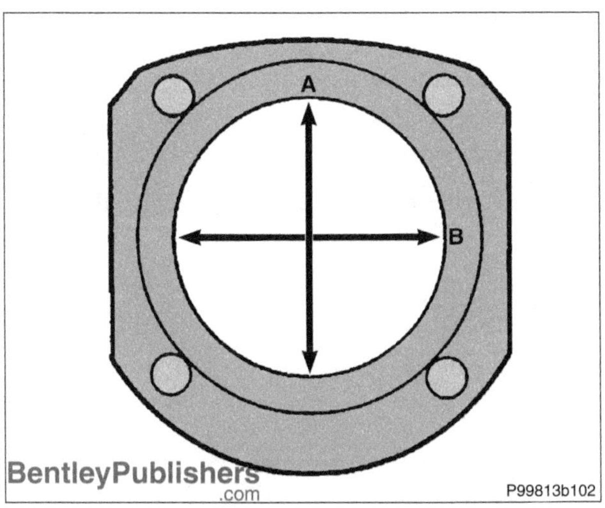

P99813b102

◄ Check cylinder roundness by measuring bore in two directions perpendicular to stud holes (**A** and **B**).

Cylinder roundness wear limit	
Difference between **A** and **B**	0.04 mm (0.0016 in)

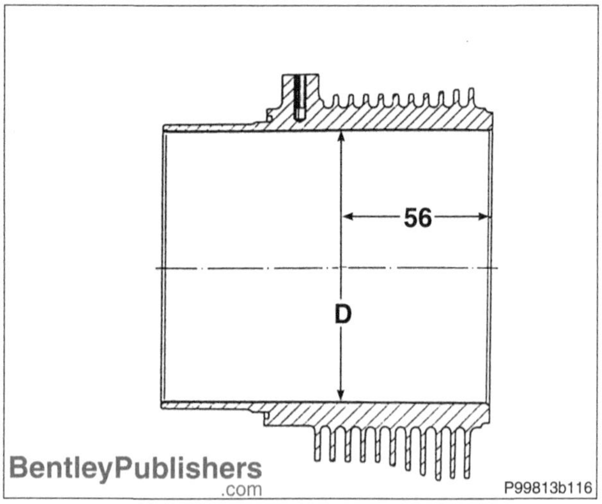

⊲ Check bore diameter 56 mm (2.2 in) below upper edge.

Replace cylinders if measured diameter exceeds 0.08 mm (0.0031 in) from installed size.

Replace pistons and cylinders if running clearance exceeds 0.12 mm (0.005 in).

Piston wear, checking

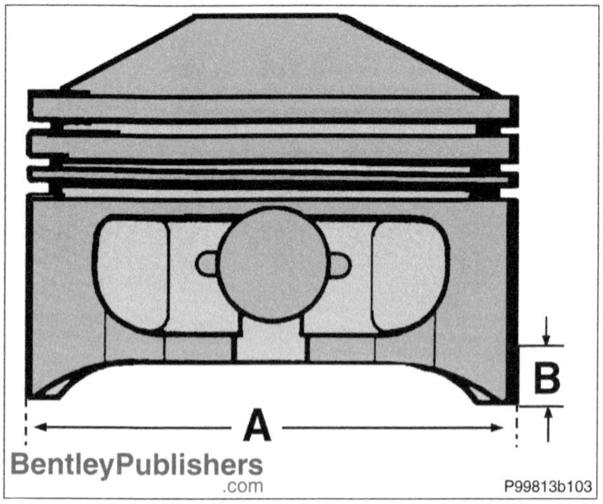

⊲ Check piston for wear by measuring across piston skirt (**A**).

- Measure Mahle® pistons 18 mm (0.71 in) from bottom (**B**).
- Measure KS® pistons 10 mm (0.39 in) from bottom (**B**).

Piston weight

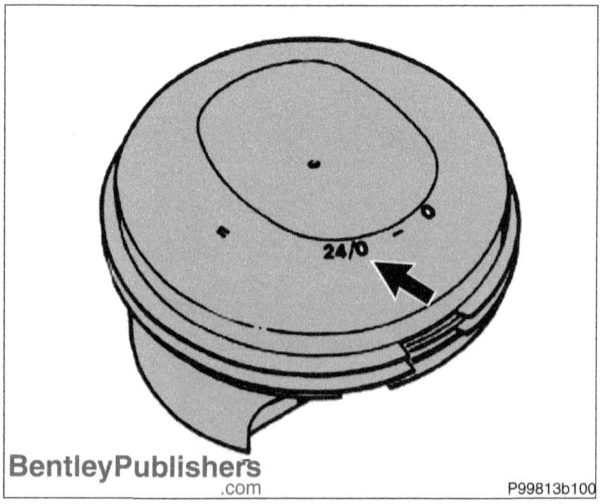

⊲ To check piston weight, clean top of piston using wire brush to expose weight marking (**arrow**). If weight marking is worn, weigh pistons using an accurate scale.

- New pistons: match to within 4 grams (0.14 oz)
- Replacement piston: match to within 8 grams (0.28 oz)

Piston is weighed with wrist pin, wrist pin clips and rings installed.

Piston rings

Check and adjust piston ring end gaps if reusing old pistons and cylinders.

Measure ring end clearance by pushing ring into bottom of cylinder to approximate height of cylinder base O-ring (**arrow**).

Piston ring end gap clearance

Ring (groove)	Gap at installation	Wear limit
Compression (1)	0.2 – 0.4 mm (0.008 – 0.016 in)	0.8 mm (0.031 in)
Compression (2)	0.2 – 0.4 mm (0.008 – 0.016 in)	1.0 mm (0.039 in)
Oil control (3)	0.3 – 0.6 mm (0.012 – 0.024 in)	2.0 mm 0.078 in)

Piston ring side clearance

Ring (groove)	Gap at installation	Wear limit
Compression (1)	0.070 – 0.102 mm (0.0028 – 0.0040 in)	0.2 mm (0.0079 in)
Compression (2)	0.040 – 0.072 mm (0.0016 – 0.0028 in)	0.2 mm (0.0079 in)
Oil control (3)	0.020 – 0.052 mm (0.0008 – 0.0020 in)	0.1 mm (0.0039 in)

13b

Piston and cylinder, servicing notes

Piston and cylinder, removing

- Remove camshafts and cylinder heads. See **15a Cylinder Heads** and **15b Camshafts**.
- Begin at front of engine, slide cylinder up and pull off piston to remove from engine.

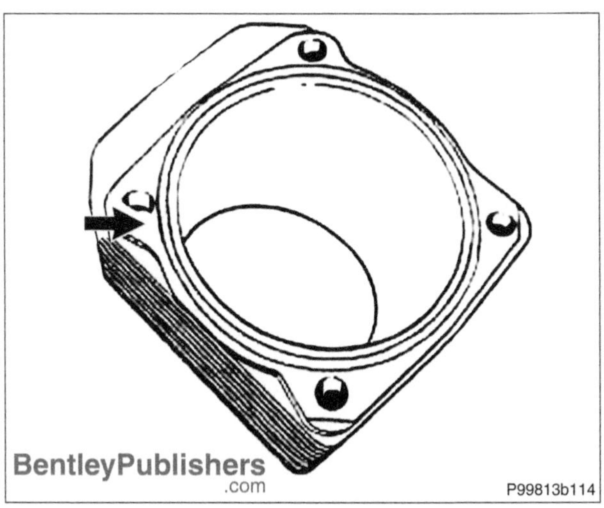

Place cylinder on plastic or wood work surface, then using a small flat blade screwdriver, remove cylinder seal by prying in reinforced area (**arrow**). Clean sealing surface.

Piston and cylinder, servicing notes

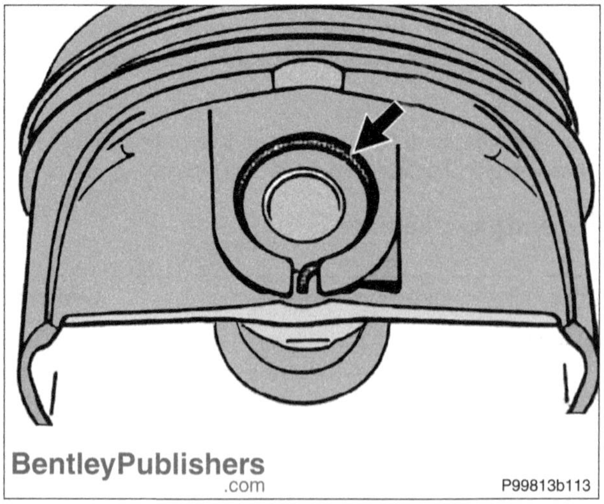

< Using a small pick, remove wrist pin circlip (**arrow**) and discard.

– Push or gently drive wrist pin out of piston and rod, then remove piston from crankcase.

– Repeat for remaining cylinders.

NOTE—

- *Note wrist pin cylinder location and orientation. Wrist pins are not interchangeable between cylinders.*

Piston and cylinder, installing

< Place new snap ring manually into tapered assembly sleeve, Porsche special tool 9500 / 2.

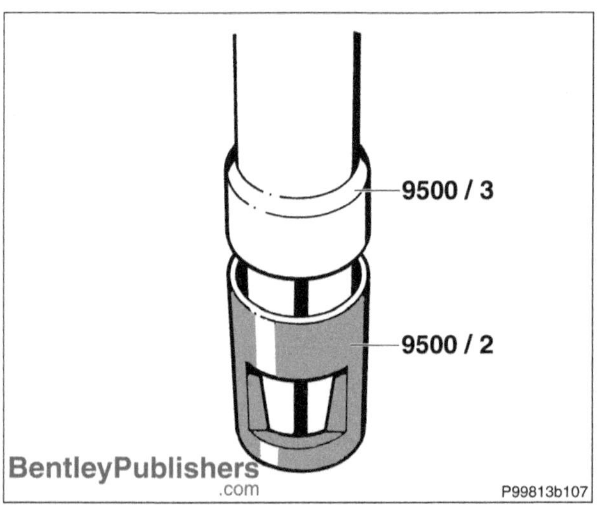

< Using Porsche special tool 9500 / 3, push snap ring down into groove of assembly sleeve 9500 / 2 and listen for an audible click to confirm engagement.

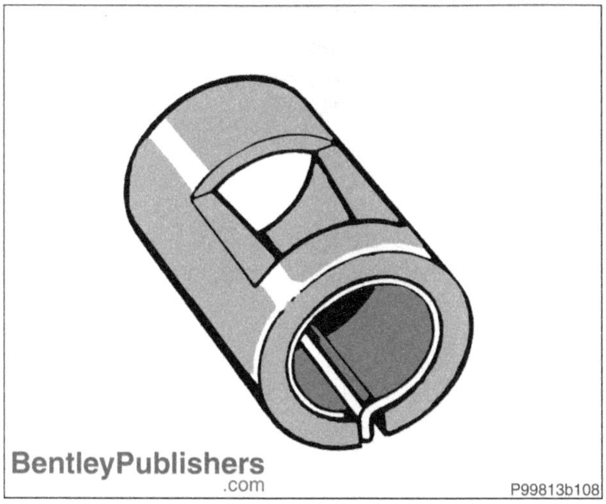

◀ Remove assembly tool 9500 / 3 and visually check snap ring for correct seating in groove of tapered assembly sleeve 9500 / 2.

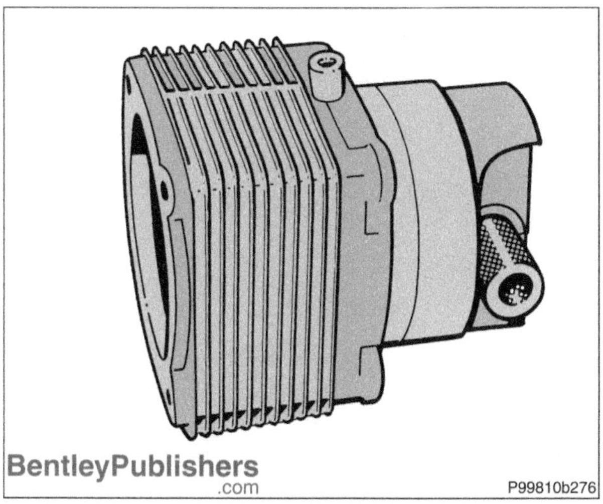

◀ Working on a wood or plastic surface, preassemble pistons and cylinders.

– Install new cylinder base O-ring onto cylinder base. Push wrist pin into position through connecting rod.

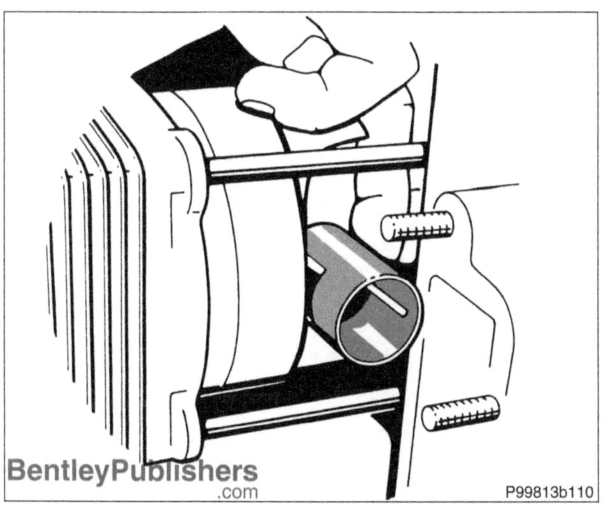

◀ Place preassembled cylinder on cylinder head studs. Place tapered assembly sleeve 9500 / 2 on piston wrist pin bore.

◀ Insert plunger (**arrow**) into tapered assembly sleeve.

– Smaller diameter of plunger provides guidance for wrist pin. Use plunger to press snap ring into wrist pin snap ring groove. Confirm snap ring engagement with an audible click.

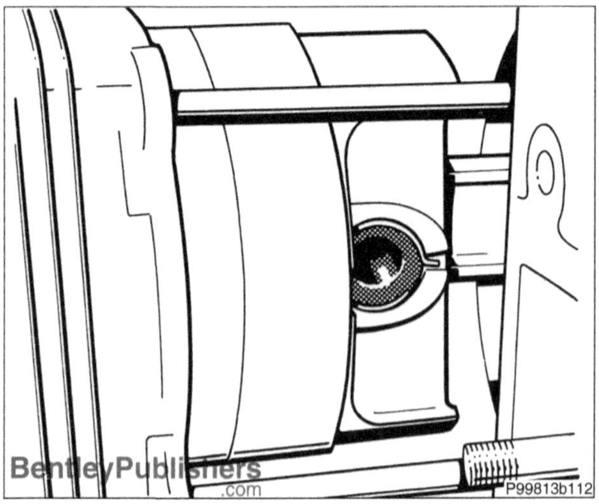

◀ Confirm wrist pin circlip is properly seated (**arrow**).

– Slide cylinder down onto crankcase.

– Install cylinder heads and camshafts. See **15a Cylinder Heads** and **15b Camshafts**.

15a Cylinder Head

General

This repair group provides reconditioning information for the cylinder heads, including valvetrain reconditioning specifications.

Cylinder head repair and reconditioning require that the engine be removed from the car and disassembled on an appropriate engine stand. Engine disassembly is described in **10b Engine Disassembly and Assembly**.

CAUTION—

- *The information contained here is intended to be used as a reconditioning guide for the experienced automotive mechanic.*

CYLINDER HEADS

All six cylinder heads are identical.

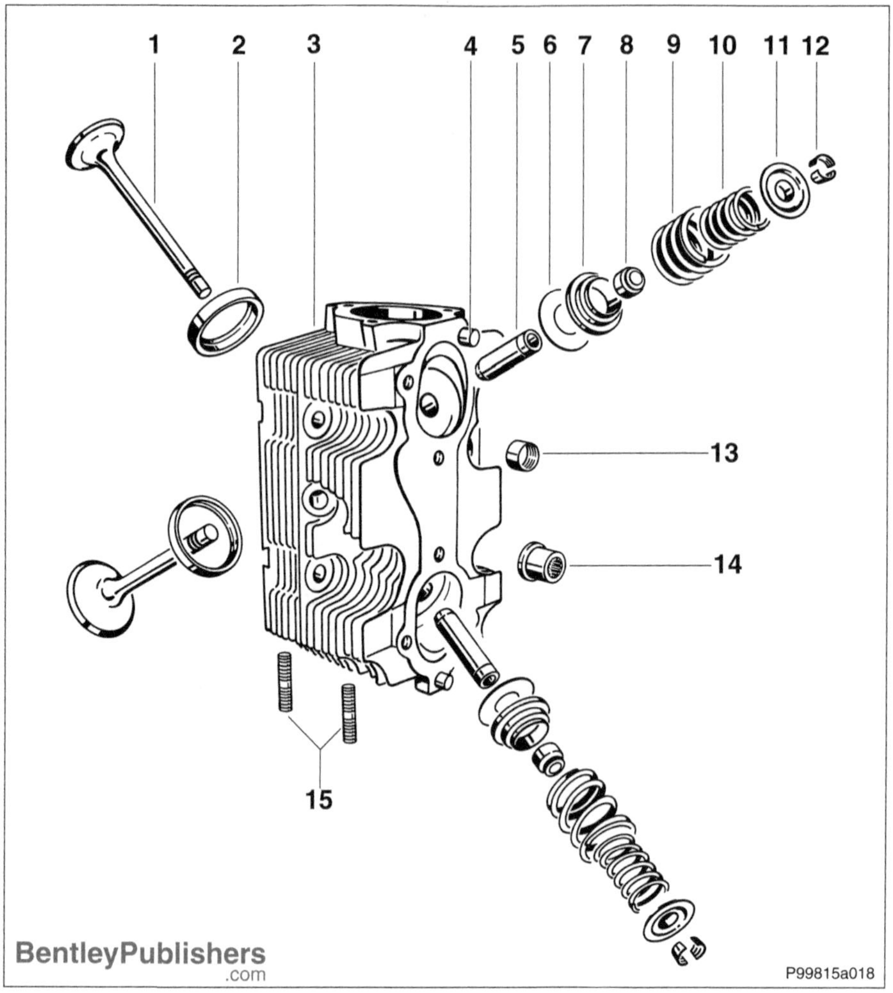

1. Valve
2. Valve seat
3. Cylinder head
4. Roll pin
5. Valve guide
6. Valve spring seat shim
7. Valve spring seat
8. Valve stem seal
9. Outer valve spring
10. Inner valve spring
11. Valve spring retainer
12. Valve keeper
13. Spark plug threaded insert
14. Cylinder head nut
15. Exhaust studs

Cylinder head service

Always check hydraulic lifter lift whenever hydraulic lifters are replaced or valve or valves seats are machined. See **15b Camshafts**.

< Remove and install valve springs using Porsche special tool P 200 or equivalent.

NOTE--

- *Keep parts for each valve assembly (keepers, springs, spring retainer etc.) together.*

– On cylinder head no. 3: remove temperature sensor using Porsche special tool 9222 / 1 or equivalent.

– Wash cylinder head and clean thoroughly by bead blasting.

NOTE--

- *Check sealing surfaces on cylinder head prior to cleaning. If any head studs were found to be broken during disassembly, check carefully for damage to the castings caused by blow-by.*

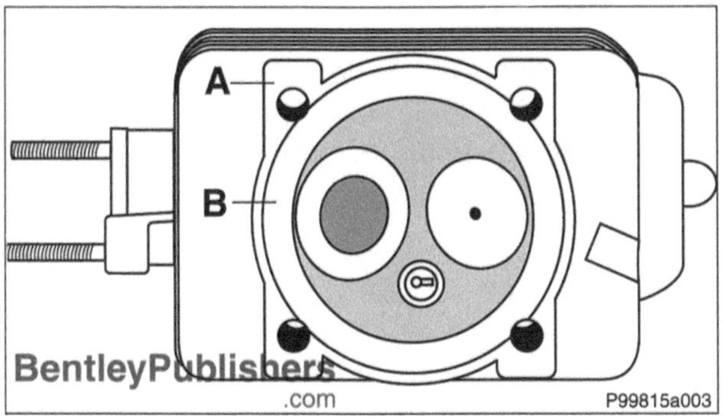

Inspect cylinder head sealing surface. Cylinder heads can be machined, provided the following:

- Mill surfaces **A** and **B** equal amounts.
- Mill all three heads on each side an equal amount.
- Bevel edges of machined surfaces slightly.
- Mark cylinder head with "–25", indicating that head was machined.

Cylinder head machining limit	
Mill surfaces **A** and **B**	0.25 ± 0.02 mm (0.01 ± 0.001 in)

CAUTION—
- *Cylinder head surface may only be machined once.*

Valve Seats and Guides

Valve guide, checking

Use Porsche special tool P 206 (valve guide go / no go gauge) or bore gauge / micrometer to check guide for wear. Check against specification.

– Thoroughly clean valve guides.

– Temporarily install a new valve to measure tilt play.

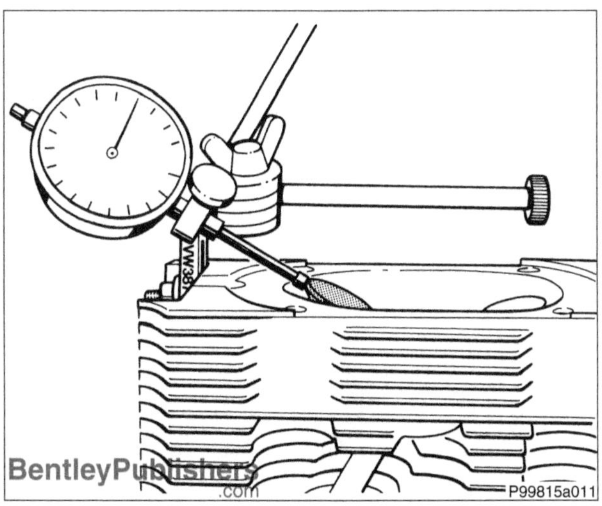

Secure VW 387 dial indicator with holder to cylinder head. Align dial indicator parallel to valve head.

– Tilting play is measured at a valve lift of 10 mm (0.39 in) (clearance from valve head to valve seat).

NOTE--
- *Ovality in the entry area of the valve guides may be disregarded.*

Valve guide wear limit (valve tilt)	
Intake valve	0.80 mm (0.0315 in)
Exhaust valve	0.80 mm (0.0315 in)

Valve guide wear limit (stem to guide clearance)	
Intake valve guide inside diameter to intake valve stem clearance	0.20 mm (0.008 in)
Exhaust valve guide inside diameter to exhaust valve stem clearance	0.20 mm (0.008 in

Valve guide, replacing

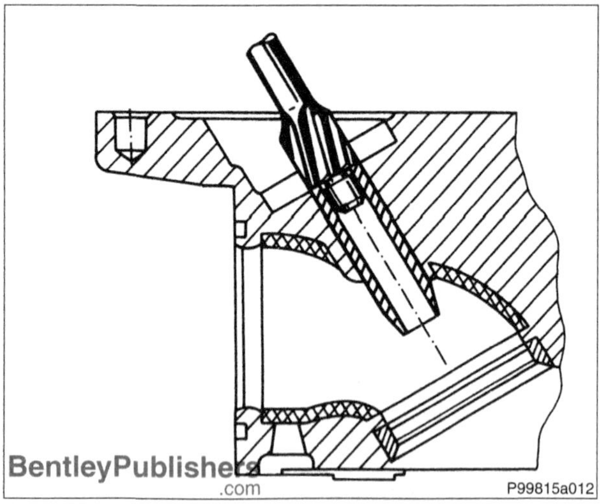

- Working from camshaft side, machine away top of valve guide flush with top of cylinder head.
- Dislodge each valve guide with sharp blow from a hammer.
- Press guide out through combustion chamber using an appropriate drift.

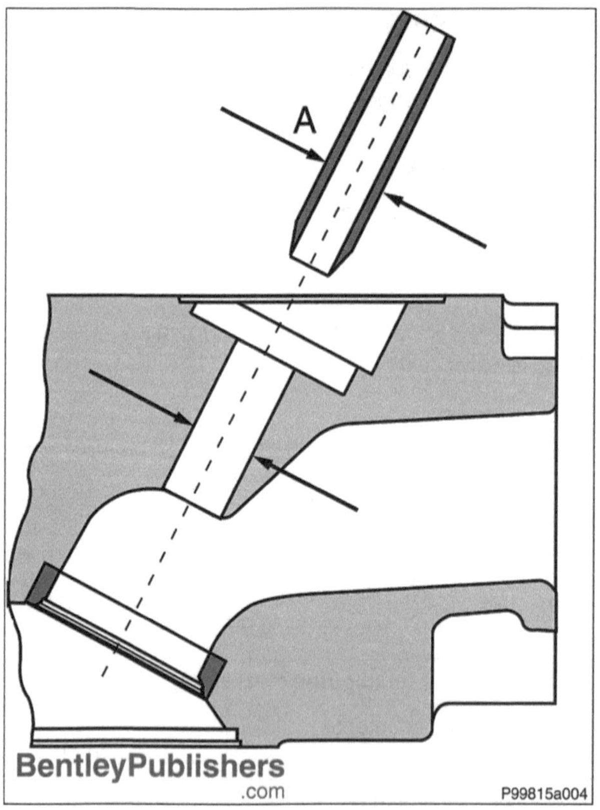

- Measure guide bores in cylinder head. Machine replacement valve guides so that valve guide diameter is 0.06 - 0.08 mm (0.0023 - 0.0031 in.) larger than valve guide hole.

NOTE--

- *When valve guides are removed, bores in the cylinder head will be widened. Replacement valve guides from Porsche are oversized and must be machined (turned on a lathe).*

Valve guide outside diameters	
Valve guide	**Guide outside diameter (A)**
Standard size	13.060 mm (0.514 in)
Oversize 1	13.260 mm (0.522 in)

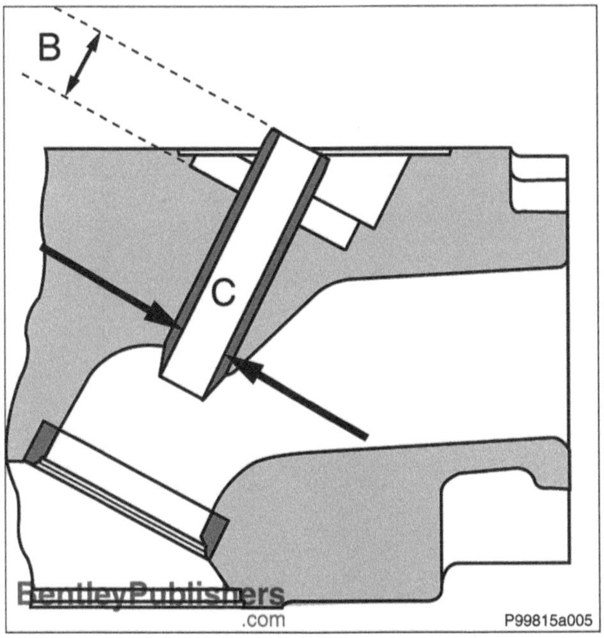

Apply lubricant (tallow-based) to valve guide and press new guides in to specified installed height using stepped drift working from the camshaft side.

Valve guide Installed height	
Height above cylinder head (B)	16.5 mm (0.649 in)
Reamed diameter (C)	8.00 - 8.015 mm (0.3149 - 0.3155 in)

– Bore inner valve guide diameter using precision boring equipment.

NOTE–

- *Intake and exhaust valve guides differ. The exhaust guides have a stepped end.*

Valve seat, replacing

– Use rough cut stone to carefully grind away valve seat until seat becomes loose in cylinder head. Remove seat.

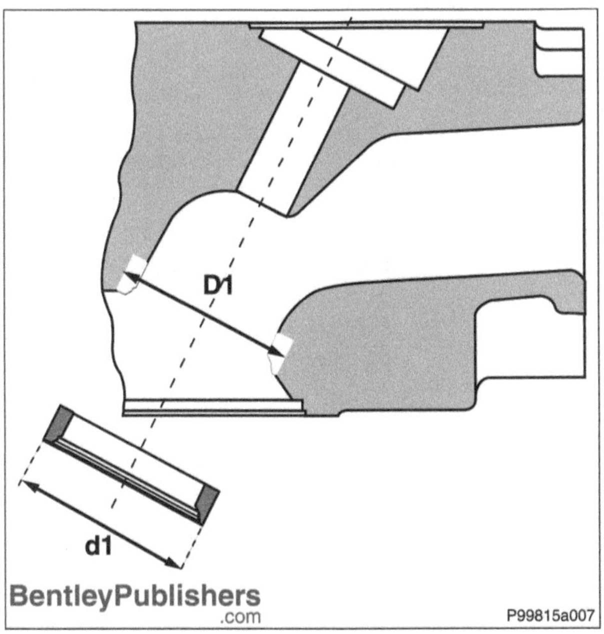

Measure valve seat outer diameter d1. Measure cylinder head bore diameter D1.

– Machine replacement valve seats so that seat diameter is larger than valve seat hole according to fit specifications. See **Valve seat dimensions** in this repair group.

– Heat cylinder head slowly to 200° C (392° F).

– Install valve seat using appropriate drift.

– Allow cylinder head to slowly cool to room temperature.

– Reheat cylinder head to 200° C (392° F) and maintain temperature for 2 hours.

– Allow cylinder head to slowly cool to room temperature.

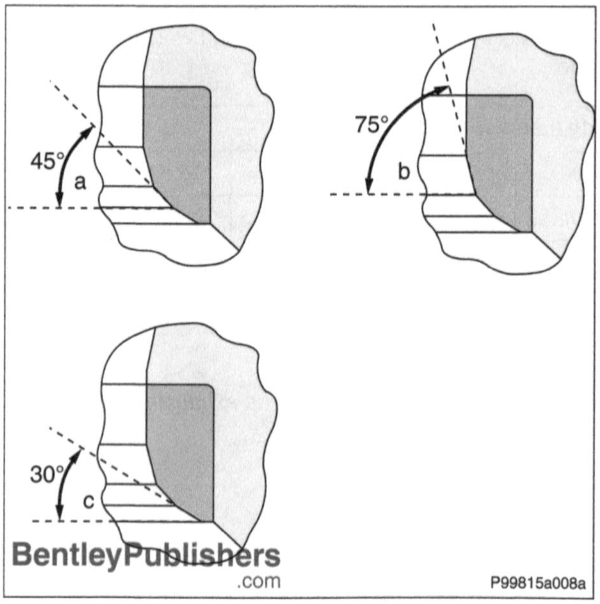

Use illustration and table below as guide in cutting valve seat.

Valve seat cutting guide (intake, exhaust valve)	
Valve seat angle (**a**)	45°
Valve seat lower edge (**b**)	75°
Valve seat upper edge (**c**)	30°

Valves

The intake valves are sodium filled. The exhaust valve stems are tapered.

Valves

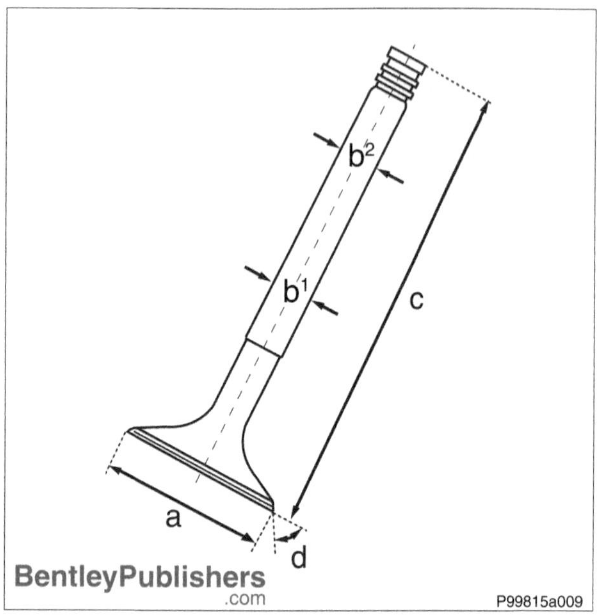

Use illustration and table below as a guide for valve specifications and measurements.

Valve Specifications

Dimension	Intake	Exhaust
Valve head diameter (**a**) M64.07/08 engines M64.23/24 engines	 49.0 mm ± 0.10 mm (1.929 ± 0.004 in) 50.0 mm ± 0.10 mm (1.968 ± 0.004 in)	 42.5 ± 0.10 mm (1.673 ± 0.004 in) 43.5 ± 0.10 mm (1.712 ± 0.004 in)
Valve stem diameter ($\mathbf{b^1}$)	7.97 – 0.012 mm (0.3137 – 0.00047 in)	7.95 – 0.012 mm (0.3129 – 0.0004 in)
Valve stem diameter ($\mathbf{b^2}$)	n-a	7.97 – 0.012 mm (0.3137 – 0.00047 in)
Total valve length (**c**)	110.1 ± 0.1 mm (4.33 ± 0.004 in)	109 ± 0.1 mm (4.291 ± 0.004 in)
Valve seat angle (d)	45°	

Valve spring height, setting

Keep each set of valves, springs, keepers and spring retainers with the cylinder head that they came out of. Check both inner and outer springs for damage.

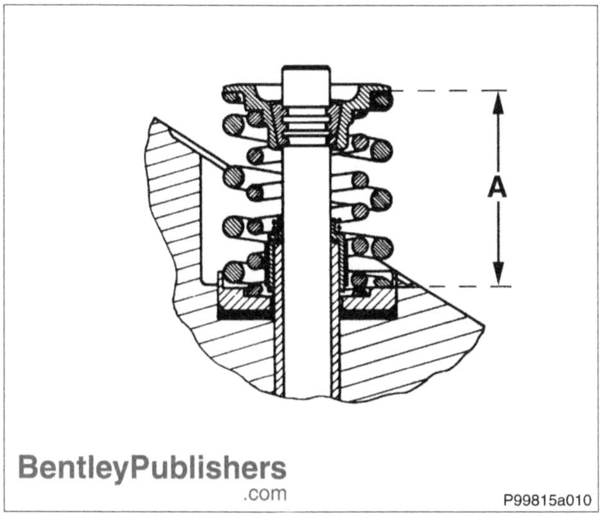

◄ Use Porsche special tool P 10 c or equivalent to measure distance **A**.

- Adjust valve spring height by adding or subtracting shims under valve spring.
- Correct valve spring height to specifications listed.

Valve spring height specification	
Intake (**A**)	36.7 + 0.3 mm (1.445 + 0.012 in)
Exhaust (**A**)	35.7 + 0.3 mm (1.406 + 0.012 in)

Valve stem seals, replacing

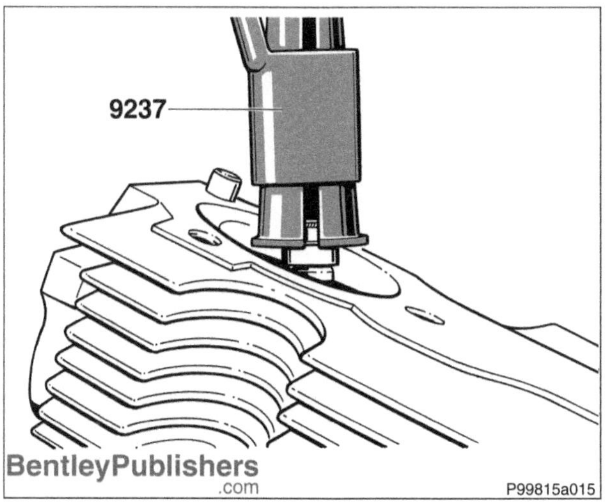

◄ With valve spring removed, hook Porsche special tool 9237 into recess in valve seal and pull valve seal off valve guide.

– Install washers and valve spring seat.

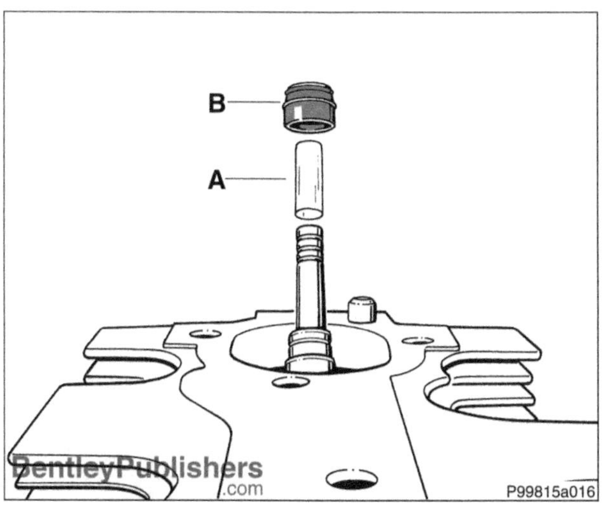

◄ Push plastic installation sleeve (8 mm diameter) onto valve stem (A).

– Lubricate valve stem seal (**B**) lip and seat surface with clean engine oil.

– Install valve seal by hand.

15a

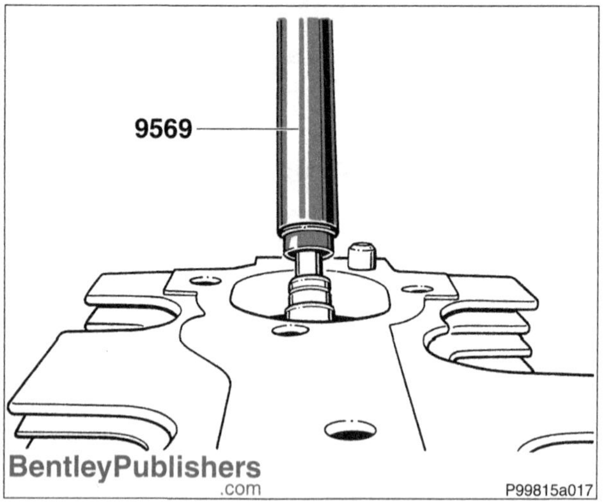

< Using Porsche special tool **9569** carefully tap seal onto valve guide.

– Remove plastic sleeve from valve stem.

15b Camshafts

GENERAL

This section covers information and repairs to the camshaft cases, camshafts, rocker shafts and rocker arms. Engine disassembly, including camshaft case removal and timing chain replacement, is covered in **10b Engine Disassembly and Assembly**.

CAMSHAFTS

> ***CAUTION—***
> • *The information contained here is intended to be used as a reconditioning guide for the experienced automotive mechanic.*

Camshafts, removing

The camshaft removal procedure shown here is carried out the engine removed. It is possible to remove camshafts from the camshaft cases with the engine installed, although working space is limited.

When removing camshafts with engine installed, remove rear bumper and rear bumper heat shield, then lower engine slightly when sliding camshaft out of camshaft housing. See **63 Bumpers**.

Read complete procedure before beginning job. Multiple Porsche special tools are required.

– Raise car and support safely.

> ***WARNING—***
> • *Make sure car is stable and well supported at all times. Use a professional automotive lift or jack stands designed for the purpose. A floor jack is not adequate support.*

– Disconnect negative (–) battery cable and cover battery terminal to keep cable from accidentally contacting terminal.

– Remove engine and transmission from vehicl. See **10a Engine Removal and Installation.**

Camshafts, removing

Specials tools are needed when servicing the camshafts. Below are the basic set of tools for camshaft and timing chain removal and installation.

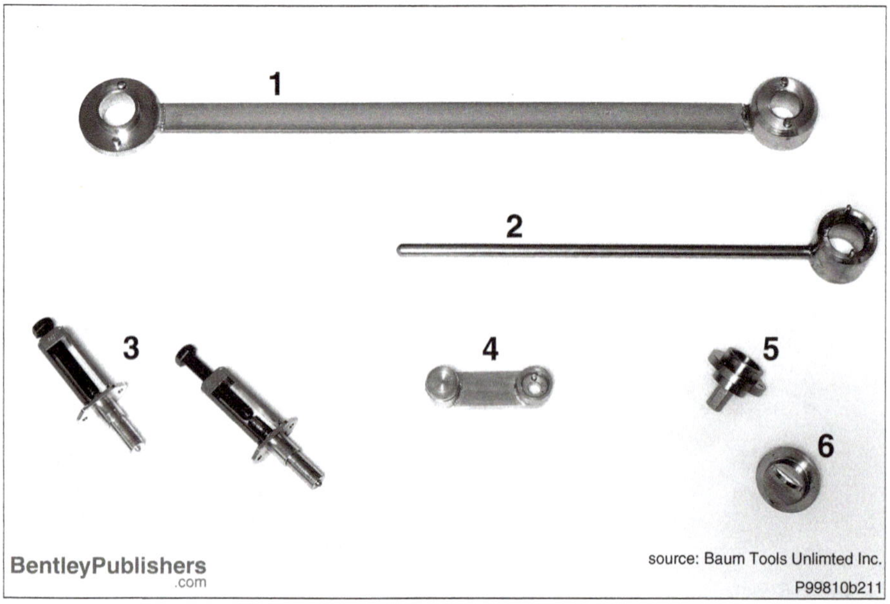

1. Porsche special tool **9548**
   - For holding crankshaft while loosening and tightening center pulley bolt
2. Porsche special tool **9582**
   - For holding camshaft while loosing and tightening center sprocket bolt
3. Porsche special tool **9401**
   - Auxiliary chain tensioners for setting camshaft timing
4. Porsche special tool **9553**
   - For locking crankshaft at TDC position
5. Porsche special tool **9551**
   - For locking and turning camshaft at TDC (left side)
6. Porsche special tool **9552**
   - For locking and turning camshaft at TDC (right side)

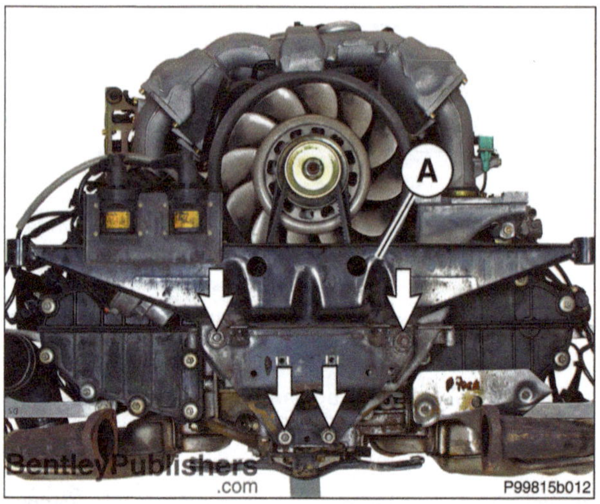

< Remove engine mounting bracket fasteners (**arrows**) then remove bracket (**A**) from engine.

< Set crankshaft to cylinder 1 compression TDC (**arrow**).

◀ Counterhold crankshaft pulley using Porsche special tool **9548** while loosening pulley fastener.

– Remove crankshaft pulley from engine.

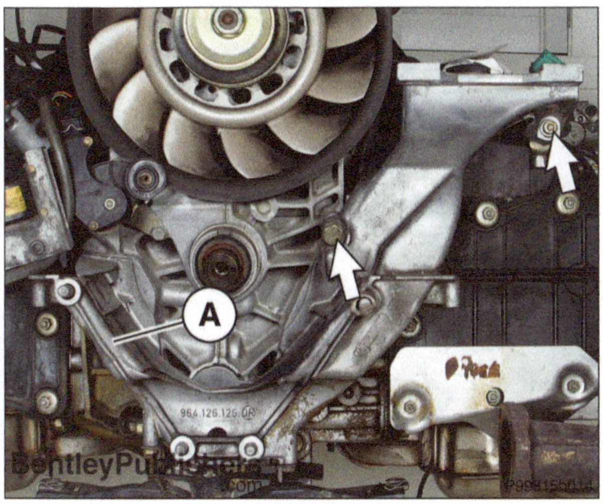

◀ Remove A/C compressor bracket fasteners (**arrows**), then remove A/C compressor bracket (**A**) from engine.

– Lock crankshaft in TDC position using Porsche special tool **9553**, if available.

◀ Loosen muffler bracket fastener (**arrow**) then remove bracket (**A**) from engine. Right side is shown, left is similar.

15b

Remove oil feed pipe fasteners (**arrows**), then lift pipe (**A**) with O-ring straight up to remove.

NOTE–

- *It is recommended that the plastic oil feed pipe and all O-rings be replaced once removed to prevent oil leaks.*

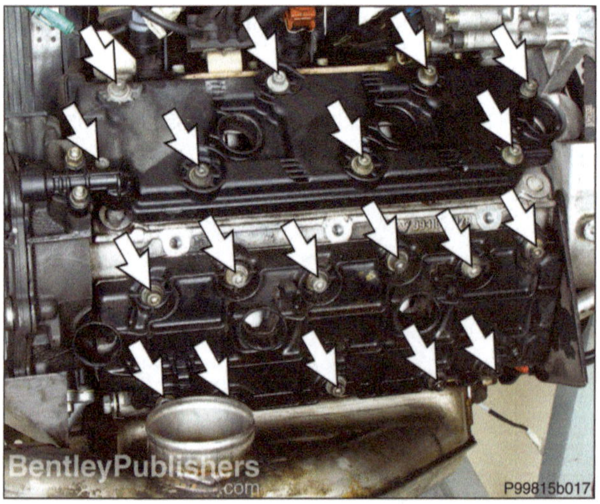

Remove intake and exhaust valve cover fasteners (**arrows**), then remove valve covers. Right side shown, left is similar.

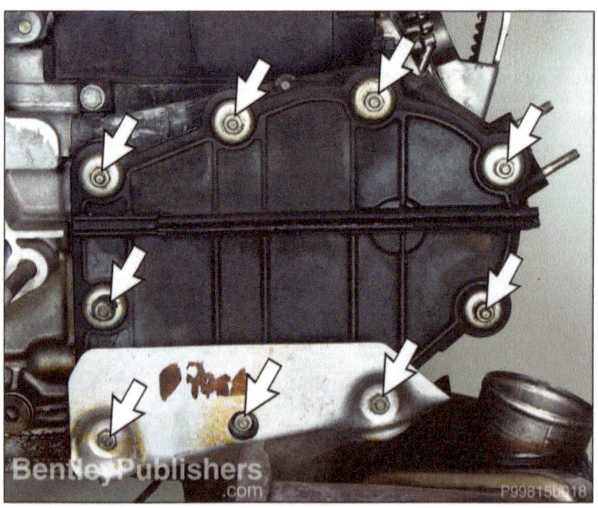

Remove timing cover fasteners (**arrows**), then remove timing cover. Right side shown, left is similar.

- Starting with cylinder 1, remove rocker arm fasteners (**arrows**) then remove rocker arms (**A**) from engine.
- Remove all remaining rocker arms.

NOTE—

- *Note location of each rocker arm for reinstallation.*

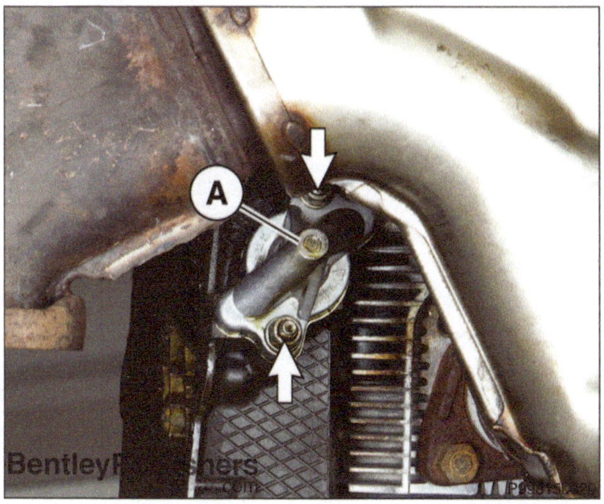

- Working at bottom of engine, remove timing chain tensioner fasteners (**arrows**) then remove tensioner cover (**A**) from engine. Right side shown, left is similar.

- Remove timing chain tensioner (**arrow**) from engine. Right side shown, left is similar.

Using Porsche special tool **9582,** counterhold camshaft sprocket while loosening camshaft sprocket fastener. Right side shown, left is similar.

Remove timing chain rail guide fasteners (**arrows**) then remove guide from engine.

- Remove timing chain sprocket fastener (**B**).

Remove camshaft sprocket from camshaft and lay sprocket with timing chain aside.

< Remove thrust plate fasteners (**arrows**).

< Remove thrust plate from engine.

< Slide camshaft out of camshaft housing.

Camshafts, installing

– Remove exhaust manifolds. See **26 Exhaust**.

– Remove power steering pump, pulley and bracket. See **48 Steering**.

– Coat camshaft bearing contact surfaces and lobes with assembly lube.

◄ Slide camshaft into camshaft housing.

NOTE—

- *Camshaft should slide into housing smoothly with little resistance. If camshaft binds at any point, do not proceed. Inspect associated parts for improper installation or damage.*

◄ When installed, both camshaft punch marks should face up with engine in normal installed position (**arrow**).

NOTE—

- *Camshaft punch marks or woodruff key hole will point up when at cylinder 1 TDC.*

◄ Camshaft locking tool **9551** (left camshaft) or **9552** (right camshaft), can be used to rotate camshafts from flywheel side of engine during installation.

Install camshaft locking tools into flywheel side of engine. Bolt tools into position. Right side shown, left is similar.

Install thrust plate then torque fasteners (**arrows**).

Tightening torque	
Thrust plate to camshaft housing	15 Nm (11 ft-lb)

NOTE—

- *Always replace thrust place gasket and O-ring.*

Install timing chain sprocket (**arrow**) onto camshafts.

Install timing chain rail guide (**A**) and torque fasteners (**arrows**).

Tightening Torque	
Timing chain guide rail to crankcase	15 Nm (11 ft-lb)

– Set camshaft timing. See **Camshaft timing, adjusting** in this repair group.

Camshaft timing, adjusting

Porsche special camshaft timing tools are necessary to time engine.

Camshaft locking tools **9551** (left camshaft) and **9552** (right camshaft) are used to rotate camshafts from flywheel side of engine during timing procedure.

– Remove exhaust manifolds. See **26 Exhaust**.

– Remove power steering pump, pulley and bracket. See **48 Steering**.

Install Porsche special tool **9401** (auxiliary timing chain tensioner).

When adjusting or checking camshaft timing, preload auxiliary camshaft tensioner until machined groove (**arrow**) on thrust piece is barely visible.

– Place Porsche special camshaft locking tool **9551** on left camshaft and install two bolts to secure to camshaft housing (**arrows**).

NOTE—

- *Due to tolerances of bolt hole pattern on camshaft housing, you may have to use one bolt from special tool kit and one M6 bolt to secure Porsche camshaft locking tools in place.*

Place Porsche special camshaft locking tool **9552** on right camshaft and install two bolts to secure to camshaft housing (**arrow**).

NOTE—

- *Due to tolerances of bolt hole pattern on camshaft housing, you may have to use one bolt from special tool kit and one M6 bolt to secure Porsche camshaft locking tools in place.*

With both camshafts locked in position and crankshaft at cylinder 1 TDC, torque camshaft pulley bolt. Use Porsche special tool **9582** to counterhold camshaft sprocket while torquing.

Tightening torque	
Camshaft sprocket to camshaft • Stage 1 • Stage 2	 20 Nm (15 ft-lb) Additional 90°

– Remove auxiliary timing chain tensioner tools and install hydraulic timing chain tensioner.

– Remainder of installation is reverse of removal.

15b

Hydraulic Lifter Service

Rocker arm, removing

- Remove valve covers. See **Valve covers, removing and installing** in this repair group.

- Starting with cylinder 1, remove rocker arm fasteners (**arrows**) then remove rocker arms (**A**) from engine.
- Remove all remaining rocker arms, as necessary.

NOTE—

- *Note location of each rocker arm for reinstallation.*

Hydraulic lifter, replacing

The hydraulic lifters can be removed from rocker arms and replaced. There are three sizes available, standard, size 1 and size 2. When replacing, measure valve lift and adjust lifter size as needed. See **Measuring hydraulic lifter**, in this repair group.

- Rocker arms can be identified by hydraulic lifter bore:
  - Intake rocker arm lifter bore is drilled and has cover (**A**).
  - Exhaust rocker arm lifter bore is solid (**B**).

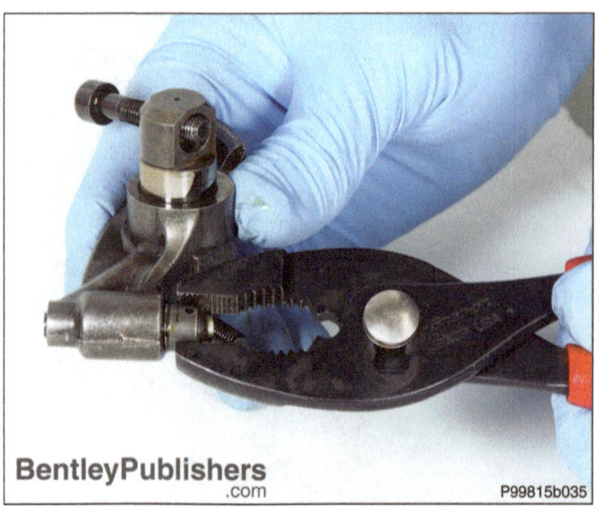

- Using a pair of pliers, pull old hydraulic lifter out of rocker arm.
- Thoroughly clean hydraulic lifter bore.
- Lubricate hydraulic lifter O-ring using clean engine oil.

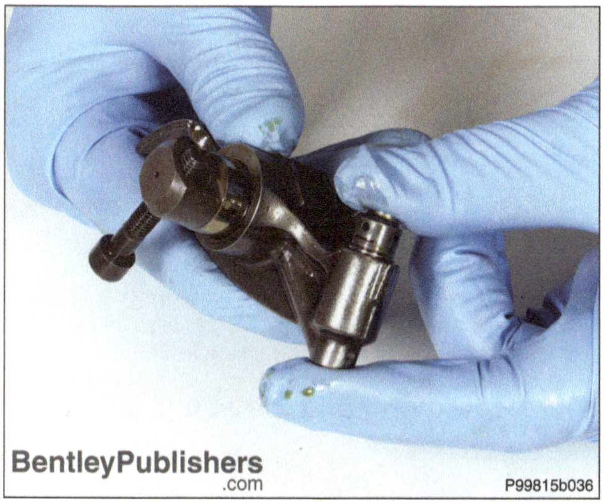

Install new hydraulic lifter into rocker arm by pressing firmly until fully seated.

Tightening torque	
Rocker arm to camshaft housing	13 Nm (9.5 ft-lb)

Measuring hydraulic lifter height

Measure hydraulic lifter height whenever hydraulic lifters are replaced or valve or valves seats are machined.

The procedure given here assumes that the engine is removed from the car.

The following parts are needed to measure hydraulic lifter lift:

- one spare intake rocker arm with hydraulic lifter.
- one spare exhaust rocker arm with hydraulic lifter.

NOTE—

- *Hydraulic lifters used when measuring lift should be free from engine oil and residue.*
- *Do not use rocker arms or hydraulic lifters from engine to measure hydraulic valve lift, use spare parts.*

– Remove all intake and exhaust rockers from engine.

For each cylinder, set piston to TDC, both valves are closed. If camshaft timing chain is not installed, turn camshaft until rocker arms are within the circle of the cam. Use Porsche special tools **9551** (left camshaft) and **9552** (right camshaft) to rotate camshafts.

> ***CAUTION—***
>
> - *If chain is installed, be sure to turn then engine over at the crankshaft pulley bolt. Do not turn the engine over using the camshaft special tools.*

Install dial indicator on top of hydraulic lifter bore on rocker arm. Then push rocker arm to measure lift.

– Lift must be at least 0.2 mm on the intake rocker arm and 0.6 mm on the exhaust rocker arm.

– If valve lift measured is lower than specified amount, install a shorter replacement hydraulic lifter.

Replacement valve lifters are available in two sizes:

- Size 1: 0.5 mm shorter
- Size 2. 1.0 mm shorter

> ***CAUTION—***
> • *Do not exceed a valve lift of 1.85 mm (0.073 in) on intake rocker arm and 2.25 mm (0.089 in) on exhaust rocker arm.*

Valve Covers

The M64 engine has two gasketed valve covers on either side of the engine: an upper cover and lower cover. The lower covers are relatively easy to access and remove once the car is raised in the air. The upper covers are not as easy and space is limited.

Valve covers, removing and installing

– Raise car and support safely.

> ***WARNING–***
> • *Make sure car is firmly supported on jack stands designed for the purpose. Place jack stands underneath structural chassis points, not underneath suspension parts.*

◄ Working beneath engine, turn engine protection panel fasteners (**arrows**) counterclockwise and remove engine protection panel.

Left lower valve cover

◄ Working near left rear muffler, remove heat exchanger tube clamp (**arrow**).

– Remove tube from heat exchanger.

∢ Unplug spark plug wire ends from plugs. (**arrows**).

∢ Remove valve cover bolts (**arrows**).

- Remove valve cover.
- Using a new gasket (**inset**), install valve cover. Tighten bolts from the middle out. The remainder of installation is the reverse of removal.

Tightening torque	
Valve cover to camshaft housing	10 Nm (7 ft-lb)

Left upper valve cover, removing and installing

- Working in engine bay, remove rear blower fan assembly. See **80 Heating**.

∢ Unplug spark plug wire ends from plugs (**arrows**).

- ◀ Remove valve cover bolts (**arrows**).
- Remove valve cover.
- Using three new gaskets, install valve cover. Tighten bolts from the middle out. The remainder of installation is the reverse of removal.

Tightening torque	
Valve cover to camshaft housing	10 Nm (7 ft-lb)

Right side valve covers

- Raise car and support safely.

> ***WARNING–***
> - *Make sure car is firmly supported on jack stands. Place jack stands underneath structural chassis points, not underneath suspension parts.*

- Remove right rear muffler. See **26 Exhaust System**.

- ◀ Loosen heat exchanger tube hose clamps (**arrows**) and remove heat exchanger tube.
- Unplug spark plug wire ends from plugs on right lower bank.

◀ Push spark plug wire grommet (**arrows**) through right side engine shroud.

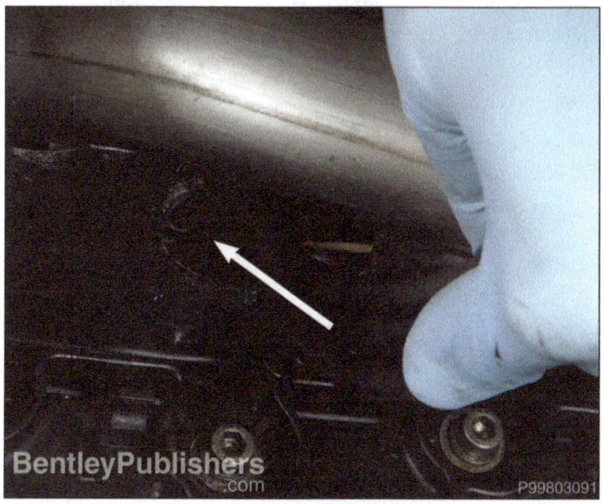

◀ Feed each plug wire through cutout in engine shroud.

◀ Remove upper muffler bracket fastener (**arrow**).

– Remove muffler bracket.

Remove lower muffler bracket bolt (**arrow**).

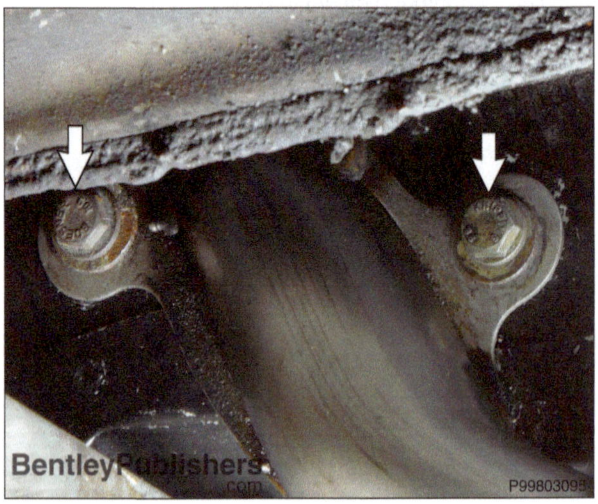

Vehicles with manual transmission: Remove clutch vent tube fasteners (**arrows**).

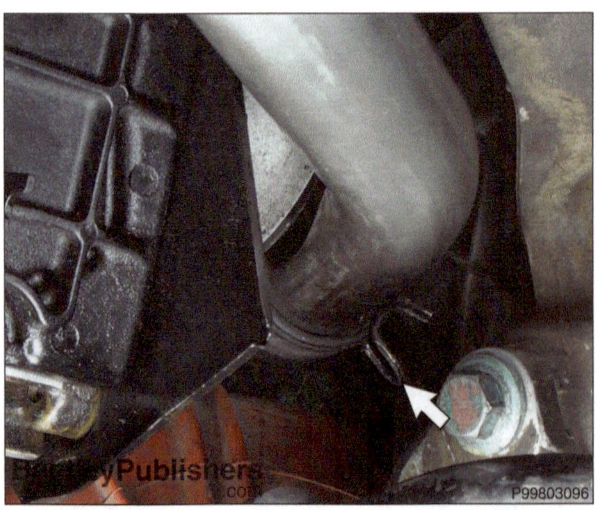

Vehicles with manual transmission: Remove clamp from clutch vent tube (**arrow**).

⊲ Vehicles with manual transmission: Remove forward-most clutch vent tube fastener (**arrow**).

⊲ Remove remaining engine shroud fasteners (**arrows**).

NOTE–

- *An additional engine shroud fastener is located at rear of engine shroud near muffler bracket mount location.*

– Lower car and remove air filter housing assembly. See **24 Fuel Injection**.

– Vehicles with manual transmission: Remove hose clamp from clutch vent tube above shroud.

⊲ Raise car, lower engine shroud with clutch vent tube (on manual transmission vehicles) out of car.

◀ From below, unplug upper bank spark plug wire ends (**arrows**).

◀ Remove lower valve cover bolts (**arrows**).

– Remove lower valve cover.

◀ Remove accessible upper valve cover bolts (**arrows**) from below.

– Lower car and remove oil feed pipe fasteners (**arrows**), then lift pipe (**A**) with O-ring straight up to remove.

NOTE–

- *It is recommended that the plastic oil feed pipe and all O-rings be replaced once removed to prevent oil leaks.*

– Remove remaining upper valve cover bolts from above.

– Remove upper valve cover.

◄ Installation is reverse of removal. Using new gaskets, install valve covers. Tighten bolts from the middle out. Install exhaust system as described in **26 Exhaust System**.

NOTE–

- *Right side upper and lower valve covers shown with engine removed for clarity.*

Tightening torque	
Valve cover to camshaft housing	10 Nm (7 ft-lb)

15b

17 Lubrication System

General

This section covers the engine lubrication system, including troubleshooting and component replacement. If the engine is disassembled because of bearing failure, a thorough flushing of the lubrication system must be carried out prior to reassembly.

CAUTION—
- *The information contained here is intended to be used as a reconditioning guide for the experienced automotive mechanic.*

Lubrication system

The dry sump lubrication system contains the tandem oil pump, oil supply tank, oil cooler, pressure safety system (including oil pressure gauge), filter, and connecting oil lines.

The tandem oil pump is in the base of the crankcase. As the name implies, the tandem oil pump contains two pumps. One pump draws oil from the oil tank and pumps it to the main oil circuit and main bearings. The second pump draws circulated oil from the crankcase and delivers it back to the oil tank via the oil filter. Oil pump removal requires that the crankcase halves be separated. See **10b Engine Disassembly and Assembly**.

The main bearings, rod bearings and camshaft bearings are supplied with pressurized oil. In addition, the undersides of the pistons are cooled using oil spray jets. Oilways in the camshaft cases provide splash lubrication for the rocker shafts and valve stems.

A thermostat regulates the flow of oil to the engine oil cooler. The oil cooler is mounted behind the front right fender.

If the oil temperature is below 83°C (181°F), a thermostatically-operated flap is held closed to prevent oil from circulating through the cooler. When the flap is open, oil is circulated through the cooler before reaching the bearings.

Porsche 993 engines are equipped with an oil pressure warning system and an oil level warning system to help prevent engine damage. Other safety features include a filter bypass to guard against bursting the filter due to over pressure and three different oil pressure relief valves.

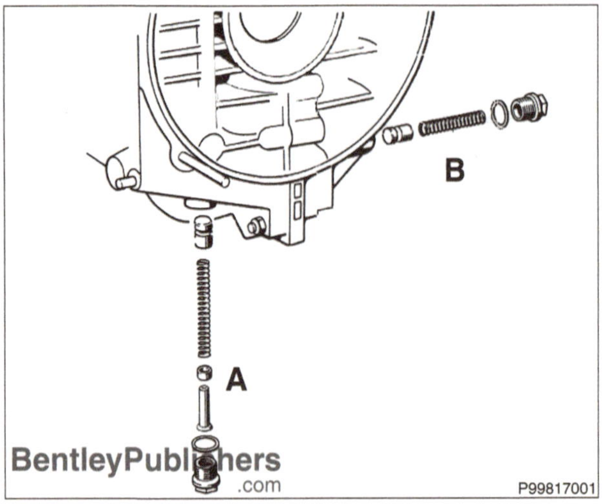

The crankcase incorporates two different oil pressure relief valves. The main oil pressure relief valve (**A**) is mounted at the bottom of the crankcase. The safety relief valve (**B**) is mounted in the left side of the crankcase.

The oil pressure relief valve opens and allows oil to pass directly into the crankcase when oil pressure exceeds 5.3 bar. (76.9 psi). The oil pressure relief valve spring is 89 mm (3.51 in) long.

In the event that the oil pressure relief valve does not open when oil pressure is excessive, the safety relief valve will open to prevent damage to the oil cooler, oil filter and oil lines. The oil pressure safety valve spring is 70 mm (2.76 in) long. The safety valve is regulated to open at approximately 9.0 bar (130.5 psi).

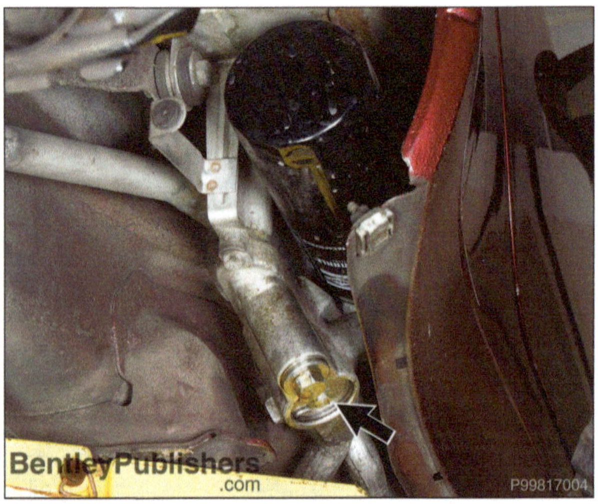

The oil cooler thermostat / pressure relief valve (**arrow**) controls oil flow to the oil cooler at the front of the car. The thermostat / pressure relief valve is located under the car ahead of the right rear wheel.

The thermostat / pressure relief valve contains two spring-loaded valves. One valve is thermostatically controlled and opens to allow pressurized oil to enter the front cooler when the temperature of the returning oil exceeds approximately 83° C (181° F). The other valve is pressure controlled and will open to bypass the cooler when the oil pressure is excessive.

System monitoring

The engine oil lubricating system has sensors monitoring engine oil temperature, oil pressure and oil level.

Engine oil temperature

Engine oil temperature is monitored via a temperature sensor mounted on the right rear of engine crankcase. Temperature sensor sends input directly to the oil temperature gauge on dashboard.

A second engine oil temperature sensor (**arrow**) is mounted to the oil cooler. Oil cooler is located behind the right front wheel housing liner.

This temperature sensor monitors oil temperature at the oil cooler and sends input to climate control unit. The climate control unit activates oil cooler blower fan at either high or slow speed based on oil temperature at oil cooler.

Oil pressure

Engine oil pressure is monitored via an oil pressure sensor/switch mounted on the right rear of engine crankcase. This sensor/switch sends info to the oil pressure gauge on instrument panel. An additional function of the oil pressure sensor/switch is to turn off the oil pressure warning light when oil pressure is present at the sensor/switch after engine startup.

Oil level

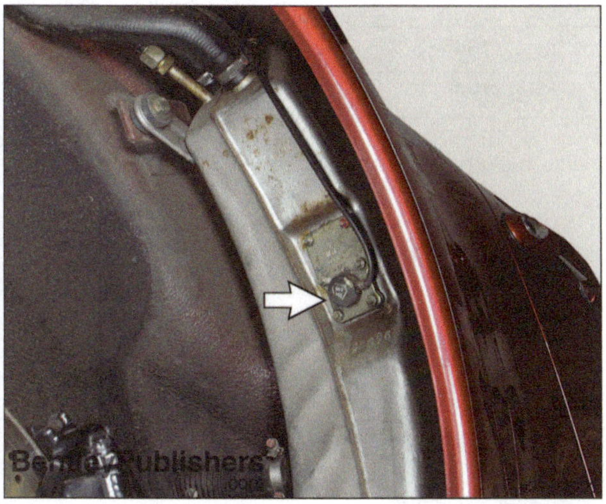

The oil level sensor is mounted in the oil tank, behind right rear wheel housing liner. The oil level sensor is a float-style sensor that rides atop oil in the tank. As oil level drops or rises, the float drops and rises sending appropriate signal to oil level gauge.

Troubleshooting

Oil pressure, checking

> ***WARNING—***
> - *If red oil pressure warning light comes on or flashes while driving, always assume that the oil pressure is low. Turn engine off as soon as possible and have vehicle towed.*

To test oil pressure, the intake manifold has to be removed to install pressure gauge, then reinstalled to run engine and check oil pressure.

– Remove intake manifold. See **24 Fuel Injection**.

– Remove secondary air injection pump. See **24 Fuel Injection**.

< Disconnect electrical connectors (**arrows**) from oil pressure sensor and unscrew sensor.

– Thoroughly clean around the oil pressure sensor before removing.

– Install pressure gauge in place of oil pressure sensor.

NOTE—
- *The pressure gauge will need to be adapted to a 22 mm male thread.*

– Install intake manifold. See **24 Fuel Injection**.

NOTE—
- *Route oil pressure gauge hose on top of intake manifold and away from moving engine parts.*

– With gauge installed, start engine and allow to reach operating temperature.

NOTE–
- *As a general guide, oil pressure should rise approximately 10 psi for every 1,000 rpm.*

Oil pressure specifications	
idle @90° C (194° F)	2 to 2.5 bar (29 to 36 psi)
2,500 rpm @90° C (194° F)	> 5 bar (> 73 psi)
5,000 rpm @90° C (194° F)	6.5 bar (94 psi)

– Remove pressure gauge and reinstall sensor.

NOTE—
- *Lubricate sensor threads prior to installation.*

Tightening torque	
Oil pressure sensor to crankcase	35 Nm (26 ft-lb)

If testing shows low oil pressure, one or more of the following conditions may be indicated:

- Worn or faulty oil pump assembly
- Worn or faulty engine bearings
- Severe engine wear

Lubrication warning system

The lubrication warning system consists of the following:

- Oil pressure warning light
- Oil pressure gauge
- Oil level gauge
- Oil temperature gauge

Oil pressure warning light, testing

When the ignition is switched ON, the oil pressure warning light illuminates. When the engine is started and the oil pressure rises slightly, the oil pressure warning sensor/switch contacts open and the warning light goes out. Make sure the oil level is correct before starting tests.

If the oil warning light remains on while the engine is running, check engine oil pressure. See **Oil pressure, checking** in this repair group. Do not drive the car until the problem is corrected. Severe engine damage may occur.

– Turn ignition switch ON.
  - Oil pressure warning light located in instrument panel should illuminate.

– If warning light does not illuminate when ignition is switched ON, disconnect green/white wire electrical connector (**arrow**) from oil pressure warning sensor/switch and ground terminal to a clean metal surface.
  - If warning light illuminates, check sensor/switch as described below.
  - If warning light does not illuminate, wiring to instrument cluster or to light itself is faulty.

NOTE–

- *The oil pressure sensor/switch also functions as the oil pressure sensor (green/red wire terminal) for the oil pressure gauge.*

– If warning light illuminates with key on but does not go out with engine started, disconnect green/white wire connector from oil pressure sensor/switch.
  - Warning light on instrument panel should go out. Check oil pressure. See **Oil Pressure, checking**.
  - If warning light does not go out, wiring to sensor /switch is most likely grounded somewhere between switch terminal and warning light. See **EWD Electrical Wiring Diagrams**.

– To test sensor/switch, connect ohmmeter between sensor terminal (green/white wire) and ground.
  - With engine switched OFF, there should be continuity.
  - With engine running, oil pressure should open switch and there should be no continuity.

– Replace sensor/switch if needed. Lubricate switch threads using clean engine oil during installation.

Tightening torque	
Oil pressure sensor/switch to crankcase	35 Nm (26 ft-lb)

17

Oil level sensor, testing

The oil level sensor is similar in operation to the fuel level sensor in the fuel tank. Check that the oil level is correct before testing the sensor.

The oil level sensor is mounted in the oil tank behind the right rear wheel housing liner ahead of wheel. It is accessed from below.

– Raise car and support safely.

> ***WARNING—***
>
> - *Make sure the car is stable and well supported at all times. Use a professional automotive lift or jack stands designed for the purpose. A floor jack is not adequate support.*

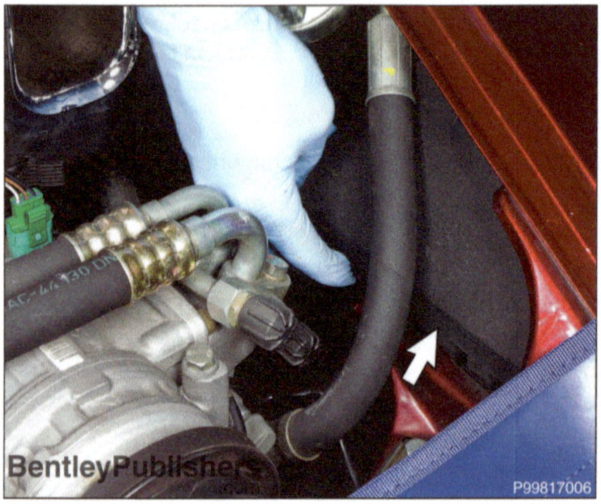

◄ Working in engine compartment on right side, remove plastic cover (**arrow**) by unclipping bottom and pulling up.

◄ Disconnect oil level sensor electrical connector (**arrow**).

– Check electrical resistance between sensor terminals in connector. If sensor is electrically open, it is faulty and should be replaced.

– If no faults are found, ground disconnected wire. Switch ignition ON and check that gauge moves to full position.

If gauge reacts correctly, sensor is faulty and should be replaced. If gauge does not respond, check electrical wiring to gauge. See **EWD Electrical Wiring Diagrams**.

NOTE–

- *Oil level sensor can be removed and installed with the oil tank in installed. It is not necessary to drain the engine oil to replace the oil level sensor.*

Lubrication system, flushing

After an engine bearing failure, the oil passages in the crankcase and crankshaft, cylinder heads and camshaft cases must be thoroughly flushed. See **13a Crankcase and Crankshaft**. In addition, the entire lubrication system must be flushed clean.

It is impossible to remove all metal particles from the engine oil cooler. It must be replaced after an engine bearing failure.

– Remove oil filter and oil tank See **Oil tank, removing** in this repair group. Rinse oil tank with engine cleaner.

– Remove oil pressure relief valve for oil cooler. See **Oil pressure relief valve, replacing** in this repair group.

◄ Disassemble and clean oil pressure relief valve.

NOTE–

- *In stubborn conditions, it will help to apply heat to the pressure fitting nuts to avoid damaging the threads on the relief valve.*

– Remove oil cooler and replace with new. See **Oil cooler, removing and installing** in this repair group.

– Flush oil lines that run front to rear with engine cleaner and carefully blow clean and dry with compressed air.

– Flush oil lines from oil tank to engine and from engine to oil cooler pressure regulator.

Component Replacement

NOTE–

- *Engine oil pump replacement is covered in* **10b Engine Disassembly and Assembly**. *Access to the oil pump requires that the engine / transmission be removed and separated. Engine can then be disassembled.*

Oil pressure and temperature sensor, replacing

To replace the oil pressure or oil temperature sensors, the intake manifold has to be removed.

– Remove intake manifold. See **24 Fuel Injection**.

– Remove secondary air pump. See **24 Fuel Injection**.

– Thoroughly clean around oil pressure sensor.

◄ Disconnect electrical connectors (**arrows**) from sensors.

– Unscrew oil pressure sensor (**A**) from engine.

– Unscrew oil temperature sensor (**B**) from engine.

– Installation is reverse of removal, use new sealing washers.

Tightening torque	
Oil pressure sensor to crankcase	35 Nm (26 ft-lb)
Oil temperature sensor to crankcase	20 Nm (15 ft-lb)

Engine oil tank, removing and installing

The engine oil supply tank is located in the right rear fender well and can be accessed from below.

- Raise car and support safely.

> ***WARNING—***
>
> - *Make sure car is stable and well supported at all times. Use a professional automotive lift or jack stands. A floor jack is not adequate support.*

- Remove right rear wheel. Remove right rear wheel housing liner.
- Remove rocker panel trim. See **66 Exterior Equipment.**

Remove oil tank drain plug (**arrow**) and drain oil into suitable container.

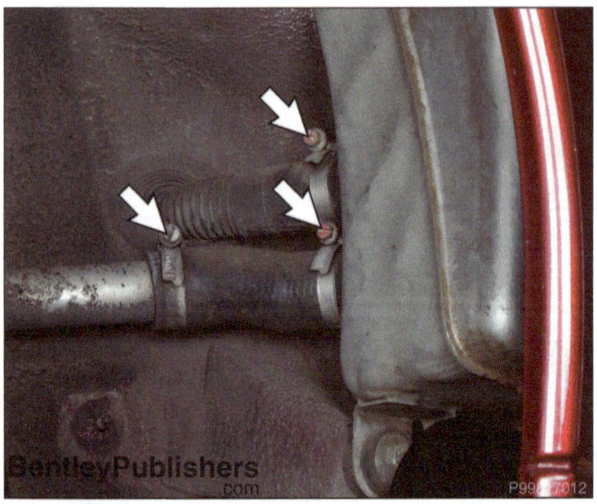

Loosen oil tank hose clamps (**arrows**).

- Disconnect both oil hoses from tank. Be prepared to catch residual oil in a suitable container.

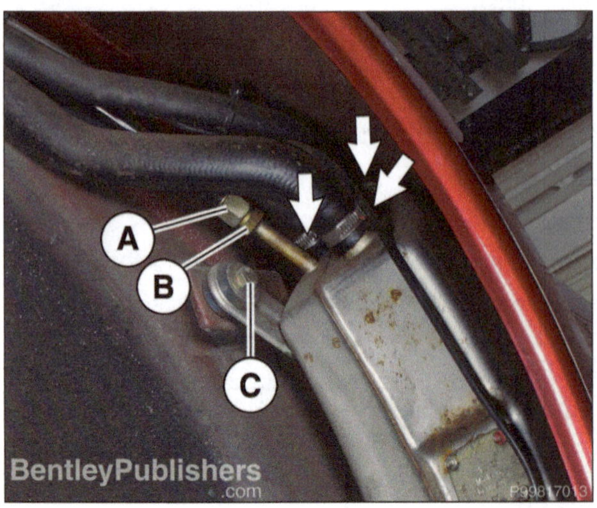

Loosen hose clamps (**arrows**) then remove hoses from oil tank.

- Loosen oil line fitting (**A**) while counterholding hex (**B**).
- Remove oil tank mounting fastener (**C**).

◄ Remove oil line retaining bracket fastener (**arrow**), then remove retaining bracket from oil lines.

◄ Remove oil line mounting bracket fasteners (**arrows**).

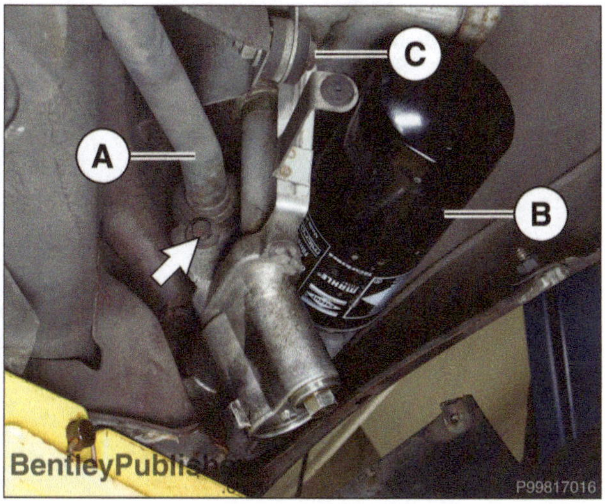

◄ Remove oil line fastener (**arrow**), then pull oil line (**A**) out of oil tank.

– Remove oil filter (**B**).

– Remove oil tank mounting fastener (**C**).

– Remove oil tank by pulling it back and down.

– Installation is reverse of removal.

– Top up engine and fill tank with oil.

  • Run engine and check for leaks.

  • Check oil level. See **03 Service and Maintenance**.

Tightening torque	
Engine oil drain plug to oil tank	50 + 5 Nm (37 + 5 ft-lb)

Oil cooler thermostat / pressure relief valve, removing and installing

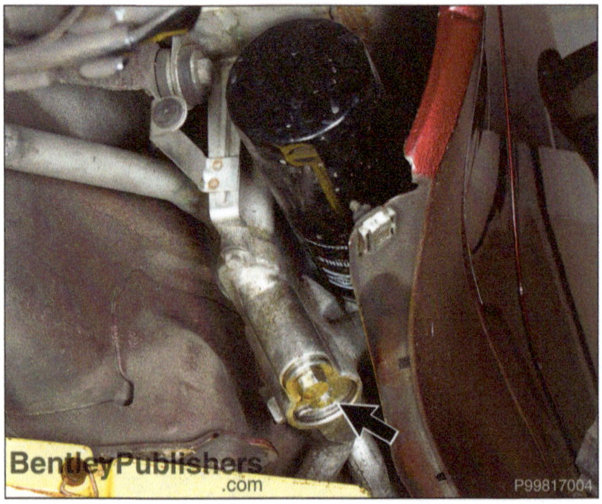

◀ The oil cooler thermostat / pressure relief valve (**arrow**) is located under the car, ahead of the right rear wheel.

– Raise car and support safely.

> ***WARNING—***
> - *Make sure car is stable and well supported at all times. Use a professional automotive lift or jack stands. A floor jack is not adequate support.*

– Remove right rear wheel.

– Remove rocker panel trim. See **66 Exterior Equipment.**

◀ Remove oil tank drain plug (**arrow**) and drain engine oil tank into a suitable container.

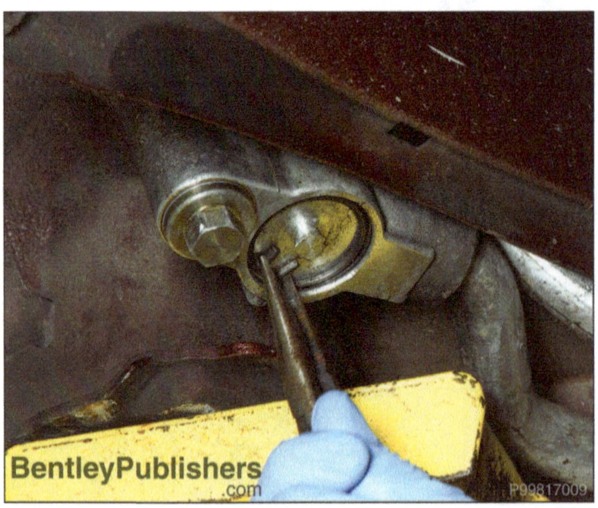

◀ Remove oil cooler thermostat / pressure relief valve snap ring.

◄ Using a pair of pliers, pull cap out of oil manifold.

◄ Using a pair of pliers, pull oil cooler thermostat / pressure relief valve out of oil manifold.

◄ Disassemble and clean valve if reusing.

– Installation is reverse of removal. Replace O-ring each time valve is removed.

Tightening torque	
Engine oil drain plug to oil tank	50 + 5 Nm (37 + 5 ft-lb)

Oil cooler, removing and installing

The oil cooler is mounted in front of the right front wheel housing, just behind the front bumper. Oil flow to the cooler is controlled via the oil cooler thermostat/pressure relief valve.

NOTE–

- *Shown with bumper cover removed for clarity.*

– Raise car and support safely.

> ***WARNING—***
>
> - *Make sure car is stable and well supported at all times. Use a professional automotive lift or jack stands. A floor jack is not adequate support.*

– Remove right front wheel.

– Remove right front lower spoiler. See **63 Bumpers.**

Remove lower support bracket rubber trim (**A**) and lower support bracket bolts (**arrows**).

– Remove lower support bracket.

– If necessary, remove cooling fan from cooler. See **Oil cooler fan, replacing** in this repair group.

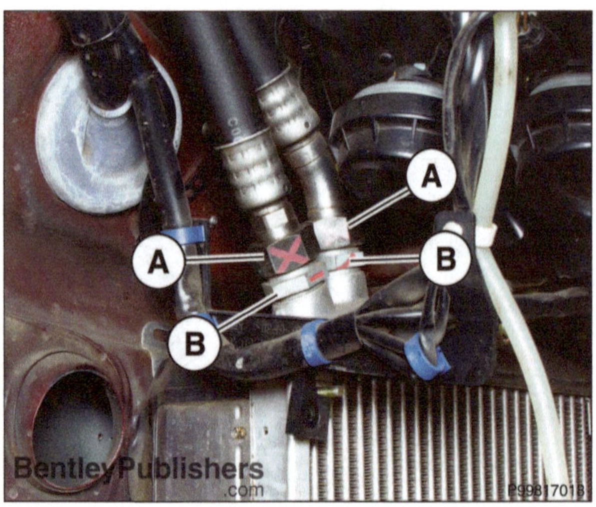

Loosen oil cooler lines (**A**) while counterholding oil cooler line fittings (**B**).

– Unscrew and remove oil lines from oil cooler. Be prepared to catch any dripping oil with a rag.

– Plug disconnected oil lines to keep debris out of system.

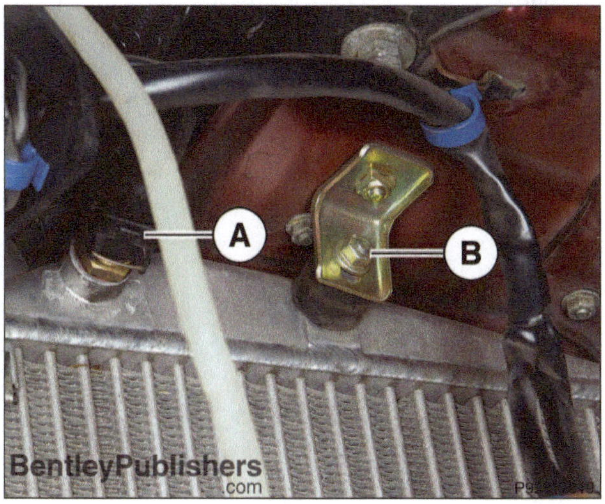

◀ Disconnect oil cooler temperature sensor electrical connector (**A**).

– Remove oil cooler mounting nut (**B**).

– Lower oil cooler from vehicle.

– Installation is reverse of removal.

- Clean threads on oil cooler fittings before reinstalling oil lines.
- Coat oil cooler line fittings with Optimoly® TA.
- Always check the oil cooler and its connections for leaks after allowing the engine to reach operating temperature. The oil cooler thermostat / pressure relief valve opens for oil to flow to the cooler.

Tightening Torques	
Oil cooler temperature sensor to oil cooler	30 Nm (22 ft-lb)
Oil cooler lines to oil cooler	80-100 Nm (59-74 ft-lb)

Oil cooling fan, replacing

◀ The oil cooler fan is thermostatically controlled. A temperature sensor (**arrow**) on the oil cooler reports oil cooler temperature to the climate control unit. When the oil temperature reaches 110° C (230° F), the climate control unit via the oil cooler fan relay and ballast resistor turn the fan on (low speed, stage 1). The fan operation can be quick-checked by applying battery voltage to the fan harness connector.

NOTE–

- *Shown with bumper cover removed for clarity.*

– Raise car and support safely.

> ***WARNING—***
>
> - *Make sure car is stable and well supported at all times. Use a professional automotive lift or jack stands. A floor jack is not adequate support.*

– Remove right front wheel.

– Remove right front lower spoiler. See **63 Bumpers**.

◀ Disconnect oil cooler cooling fan electrical connector (**arrow**).

17

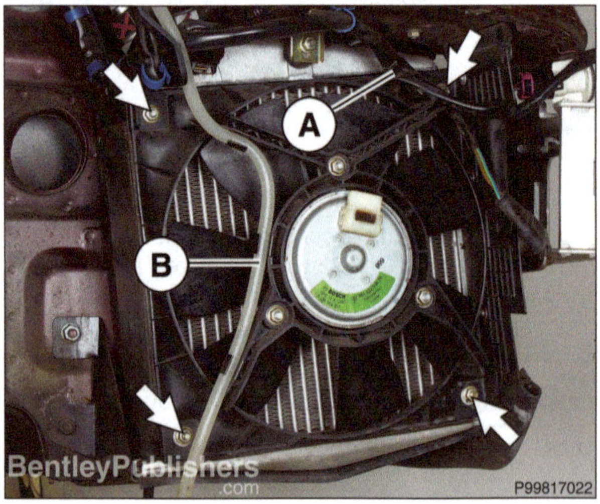

- Detach wiring harness (**A**) and drain hose (**B**) from cooling fan.
- Remove cooling fan mounting fasteners (**arrows**), then remove fan from oil cooler.
- Installation is reverse of removal.

Oil cooler fan ballast resistor, replacing

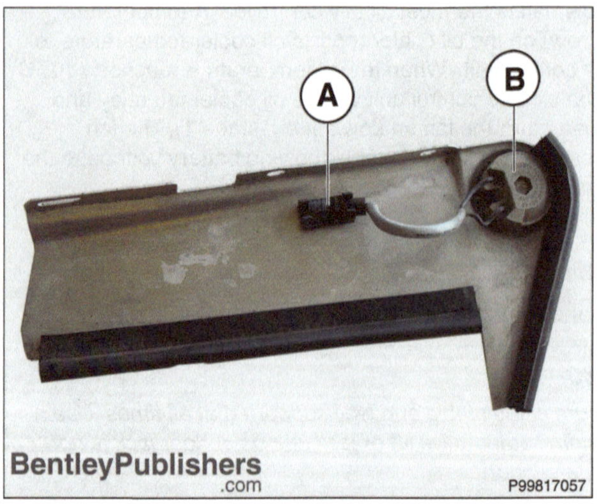

- The oil cooler blower fan resistor (**B**) controls low speed fan operation. The oil cooler fan resistor is located on an aluminum panel above oil cooler and oil cooler fan. Ballast resistor connector (**A**) sits atop the panel.
- Raise car and support safely.

> ***WARNING—***
>
> - *Make sure car is stable and well supported at all times. Use a professional automotive lift or jack stands. A floor jack is not adequate support.*

- Remove right front wheel.
- Remove right front lower spoiler. See **63 Bumpers**.

- Remove lower support bracket rubber trim (**A**) and lower support bracket bolts (**arrows**).
- Remove lower support bracket.

- Remove oil cooler mounting nut (**arrow**).
- Lower oil cooler enough to access ballast resistor.

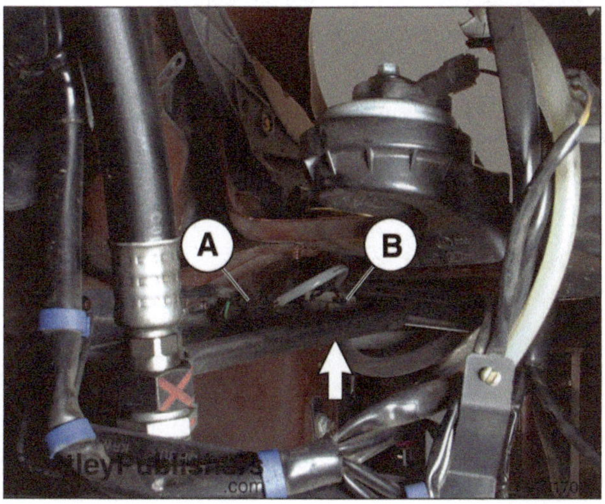

- Disconnect ballast resistor connector (**A**). Ballast resistor (**B**) is held to panel with a nut (**arrow**) under panel. Remove nut and remove resistor from panel.
- Installation is reverse of removal.

Oil return tube, replacing

The procedure below describes replacing the one-piece oil return tube with an expandable two-piece tube. This procedure can be done with the engine installed.

Oil return tubes can be individually replaced, although replacing in pairs is recommended.

- Remove exhaust manifold. See **26 Exhaust System**.
- Disconnect heater hoses from front and rear of heat exchanger.

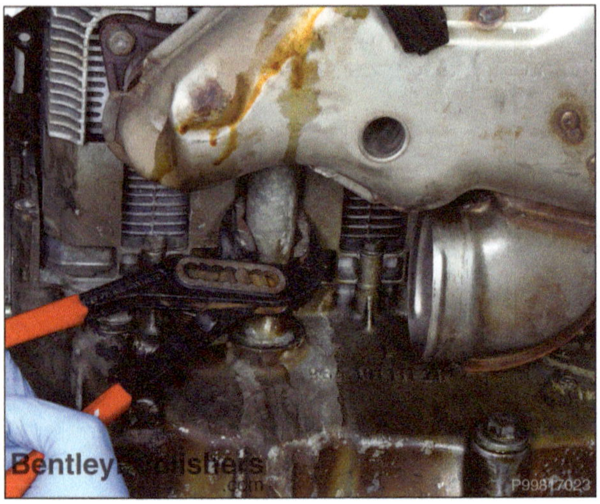

< Crush center of oil return tube with large pliers and bend.

- Have a drain pan ready to catch any engine oil that may spill.
- Remove destroyed return tube and clean sealing surfaces on crankcase and camshaft case.
- Mount O-ring seals to oil return tube sections and assemble.
- Push tube locking ring out of locking groove and slide approximately 40 mm (1.6 in) away from locking groove.
- Apply a light film of tire mounting paste to seals and collapse sectioned oil return tube.
- Align telescoping end of tube with crankcase and camshaft housing holes. Expand tube until it fits tightly into both holes.
- Slide locking ring into locking groove.
- Reinstall exhaust manifold.
- Run engine and check for oil leaks.

20 Fuel Supply

TABLES

20

General

This repair group covers repairs to the fuel tank, fuel pump, fuel level sender and connecting lines. For additional information, see:

- **24 Fuel Injection** for information on the Motronic (DME) fuel injection and emission control systems.
- **03 Service and Maintenance** for fuel filter replacement.

Warnings and Cautions

Please observe the following cautions and warnings when performing any service or repair on the fuel system.

> ***WARNING—***
> - *When working on an open fuel system, wear suitable hand protection, as prolonged contact with fuel can cause illnesses and skin disorders.*
> - *Gasoline is highly flammable and its vapors are explosive. Do not smoke or work on a car near heaters or other fire hazards when diagnosing and repairing fuel system problems. Have a fire extinguisher available in case of an emergency.*

CAUTION—

- *Disconnecting power from the engine control module (ECM) can erase its adaptive memory. It may be necessary to drive the car for approximately 10 minutes after reconnecting the battery to re-set ECM adaptive memory and restore normal performance.*
- *Always unscrew the fuel tank cap to release pressure in the tank before working on the tank or lines.*
- *Before performing any electrical tests with the ignition turned on, disable the ignition system. See* **28 Ignition System**.
- *To prevent damage to the ignition system or the electronic fuel system components, including the ECM, always connect and disconnect wires and test equipment with the ignition off.*
- *Fuse and relay locations are subject to change. Use care when troubleshooting the electrical system at the fuse/relay panels. To resolve problems in identifying a relay, consult an authorized Porsche dealer for the latest information*
- *Absolute cleanliness is essential when working with the fuel system. Even a minute particle of dirt can cause trouble if it reaches an injector. Thoroughly clean the fuel line unions before disconnecting any of the lines.*
- *Use only clean tools. Keep removed components clean and sealed or covered with a clean, lint-free shop towel, especially if completion of the repair is delayed.*
- *When replacing parts, install only new, clean components.*
- *Avoid nearby use of compressed air, and do not move the car while the fuel system is open.*
- *Do not use compressed air above 40 psi to blow out lines or components. Internal damage may result to components.*
- *Always replace seals and O-rings.*

Fuel Tank and Lines

The fuel tank is located in the trunk. The fuel level sending unit is mounted through the top of the tank. Connecting lines for the evaporative emission control system and fuel supply and return lines are also attached to the tank.

Fuel Tank Capacity	
Carrera 2 and Carrera 4	19 gallon (72 liters)

Fuel tank, draining

– Remove front trunk carpet. Remove spare tire.

– Disconnect negative (–) battery cable and cover battery terminal to keep cable from accidentally contacting terminal.

> ***CAUTION–***
> - *Prior to disconnecting the battery, read the battery disconnection cautions given in* **01 Warnings and Cautions.**

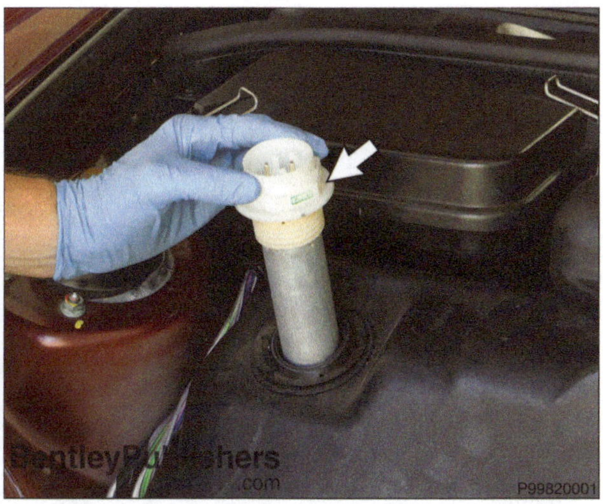

◄ Remove fuel level sending unit (**arrow**). See **Fuel level sending unit, replacing** in this repair group.

◄ Insert fuel transfer pump hose (**arrow**) into fuel tank opening and transfer fuel to gasoline designated container.

NOTE—
- *Be sure the container is large enough to hold the volume of fuel that is to be pumped out of the tank.*
- *The gasoline storage container (gas can) must comply with all state and local fuel container laws and regulations.*

20

Fuel tank, removing and installing

– Drain fuel tank. See **Fuel tank, draining** in this repair group.

– Disconnect battery negative (–) cable.

> ***CAUTION–***
> • *Prior to disconnecting the battery, read the battery disconnection cautions given in* **01 Warnings and Cautions.**

– Raise car and support safely.

> ***CAUTION–***
> • *Make sure the car is stable and well supported at all times. Use a professional automotive lift or jack stands designed for the purpose. A floor jack is not adequate support.*

◀ Working underneath right front of car, remove fuel pump cover fasteners (**arrows**), then remove fuel pump cover (**A**).

◀ Using hose clamping pliers, pinch off fuel feed line (**arrow**) from tank.

– Lower car to ground.

– Remove trunk carpet.

– Remove spare tire.

◄ Working inside trunk, loosen hose clamps (**A**), then remove vapor expansion tank fasteners (**arrows**).

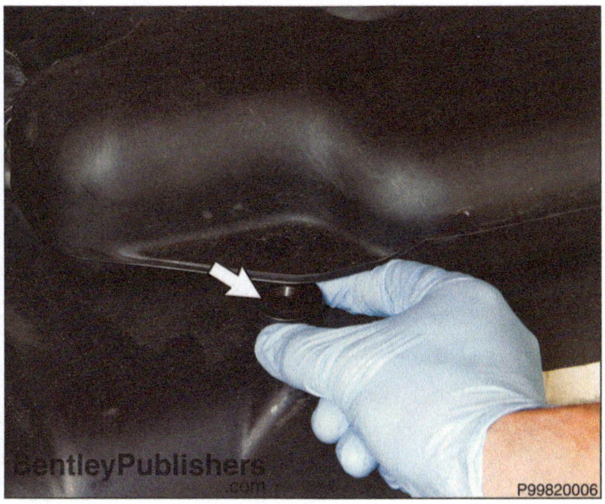

◄ Be careful not to lose vapor expansion tank fastener spacers (**arrow**) when removing tank.

◄ Pull tank out of hoses (**arrows**).

– Pull check valve (**A**) straight out of tank with grommet toward bulkhead.

20

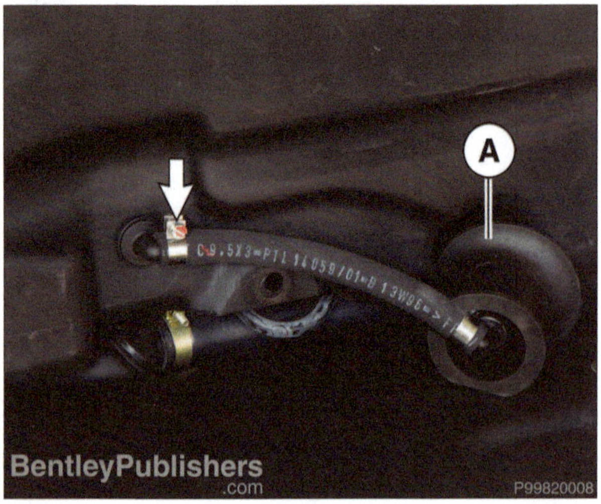

◄ Working at bottom of expansion tank, remove check valve (**A**) then hose (**arrow**).

– Remove expansion tank from vehicle.

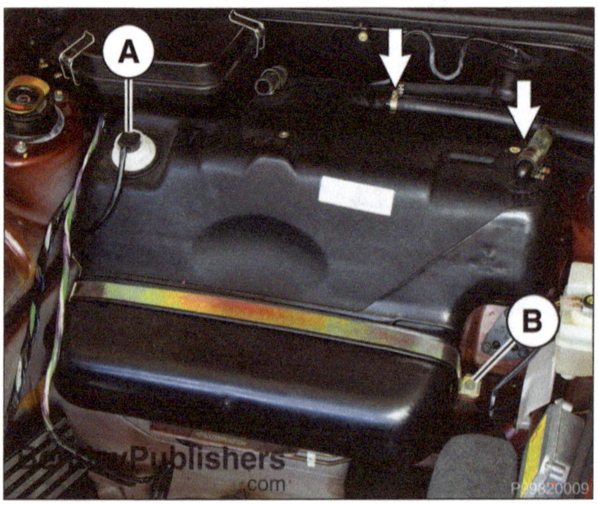

◄ Loosen filler neck hose clamps (**arrows**) then remove hoses from tank.

– Disconnect fuel level sender electrical connector (**A**).

– Remove fuel tank strap fastener (**B**), then remove fuel tank strap.

– Lift fuel tank up enough to disconnect remaining hoses.

– Remove fuel tank by lifting it up and out through trunk.

– Installation is reverse of removal.

- Always use new tank mounting gasket and hose clamps.
- Inspect all rubber parts including hoses and replace any that are worn, chafed, or cracked.

Fuel level sending unit, replacing

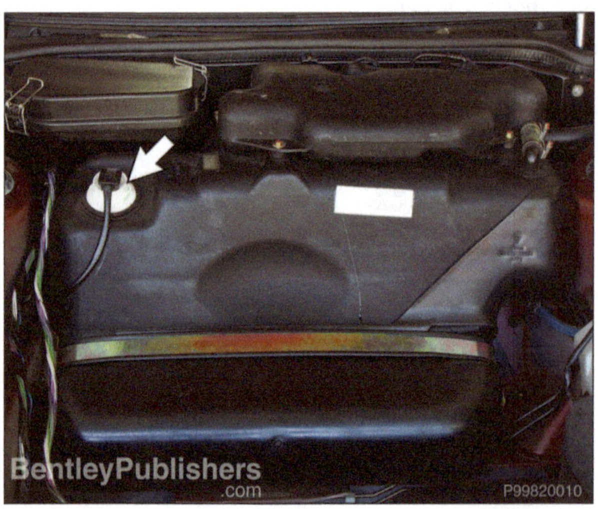

◄ Fuel level sending unit (**arrow**) is mounted in top of fuel tank. The unit has a float connected to a variable resistor for fuel level, and a switch that closes to turn on the low fuel light. When replacing the sending unit always replace hose clamps, gaskets and O-rings.

– Remove trunk floor carpet.

– Disconnect negative (–) battery cable and cover battery terminal with insulating material to keep cable from accidentally contacting terminal.

> ***CAUTION–***
>
> - *Prior to disconnecting the battery, read the battery disconnection cautions given in* **00 Warnings and Cautions.**

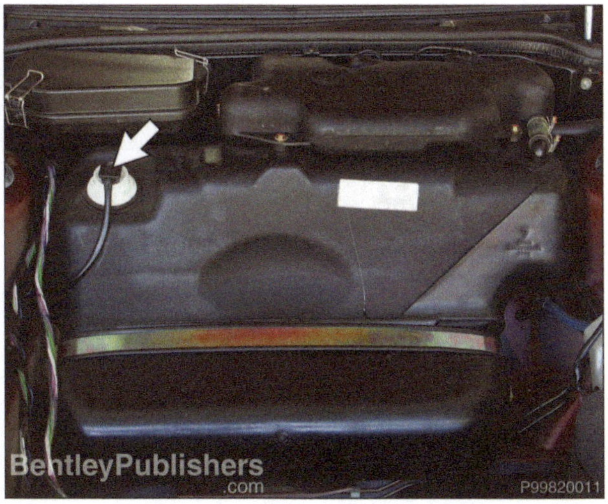

◀ Disconnect sending unit electrical connector (**arrow**).

◀ Using large pliers, unscrew sending unit from tank.

◀ Slowly withdraw sending unit from fuel tank, allowing it to drain as it is being removed.

– Installation is reverse of removal, noting the following:

- New replacement fuel level sending units may be fitted with a guard for protection during shipping. This guard must be removed prior to installing the unit.
- Install new sealing O-ring ring (**arrow**).
- Reconnect electrical connector.
- Check for leaks.

Tightening torque	
Fuel level sender to fuel tank	30 Nm (22 ft-lb)

Fuel Delivery Tests

Checking fuel delivery is a fundamental part of troubleshooting and diagnosing the engine management system. Fuel pressure directly influences fuel delivery. An accurate fuel pressure gauge will be needed to make the tests.

Fuel Pressure Checks

The fuel pump delivers fuel at high pressure to the fuel injection system. A pressure regulator maintains system pressure. The quantity of fuel supplied exceeds demand, so excess fuel returns to the fuel tank via a return line.

There are three significant fuel pressure and delivery checks:

- System pressure created by the fuel pump and maintained by the pressure regulator.
- Fuel delivery volume created by the fuel pump and affected by restrictions, such as clogged fuel filter.
- Residual pressure maintained in the closed system after the engine and fuel pump are shut off.

Relieving fuel pressure and connecting fuel pressure gauge

WARNING—

- *The fuel system is designed to retain pressure even when the ignition is OFF. When working with the fuel system, loosen the fuel lines slowly to allow residual fuel pressure to dissipate. Avoid spraying fuel. Use shop towels to capture leaking fuel.*
- *Before beginning work on the fuel system, place a fire extinguisher in the vicinity of the work area.*

– Remove blower motor. See **80 Heating**.

– Tightly wrap a shop towel around fuel rail test port and loosen fitting. Dispose of shop towel properly.

< Connect fuel pressure gauge to fuel rail test port (**arrow**) and tighten to 12 Nm (9 ft-lb).

CAUTION—

- *When removing test port cap, note small sealing ball under cap. Be careful not to drop sealing ball.*

NOTE–

- *The fuel pressure gauge should have a range of 0 to 5 bar (0 to 75 psi).*

System pressure, testing

Fuel system pressure is created by the fuel pump and regulated by the fuel pressure regulator. System pressure is not adjustable.

< Relieve fuel pressure and connect pressure gauge to fuel rail (**arrow**). See **Relieving fuel pressure and connecting fuel pressure gauge** in this repair group.

> ***CAUTION—***
>
> - *The fuel pump is capable of developing a higher pressure than that regulated by the pressure regulator. If the fuel pump check-valve malfunctions (stuck closed), make sure the fuel pressure does not rise above 6.0 bar (87 psi). Damage to the fuel lines or fuel system components could result.*

– Operate fuel pump without running engine using a fuel pump relay jumper. See **Operating fuel pump for tests** in this repair group.

– Check that fuel pressure corresponds to specifications listed in **Table a**.

– If system pressure is low:

- Repeat test while gradually pinching off return hose. Pressure should rise rapidly. If not, fuel pump is most likely faulty. If pressure does rise, fuel pressure regulator is faulty. See **24 Fuel Injection** for replacement information.

– If system pressure is too high:

- Check return line from pressure regulator to tank.
- Check for kinks in hose.
- Blow compressed air through line to check for blockages.

> ***WARNING—***
>
> - *Do not use compressed air above 40 psi to blow out lines or components. Internal damage to components may result.*

– Remove fuel pump relay jumper and reattach relay to harness.

– Start engine and recheck fuel pressure.

– Remove vacuum hose from fuel pressure regulator.

– Recheck fuel pressure. Compare to specifications in **Table a**.

NOTE–

- *If any faults are found, either fuel pressure regulator or vacuum hose is faulty and should be replaced.*

- When finished testing, disconnect pressure gauge and reinstall fuel pressure test port cap with sealing ball.

Tightening torque	
Fuel pressure test port cap	12 Nm (9 ft-lb)

Table a. Fuel pump pressure specifications	
Condition	**Fuel pressure**
Engine OFF	3.8 bar (55 psi)
Engine running	3.3 bar (48 psi)
Engine running, vacuum hose removed from pressure regulator	4.0 - 4.2 bar (58 - 61 psi)

Fuel delivery volume, testing

< Disconnect return line from fuel rail (**arrow**) on rear left side of engine.

> ***WARNING—***
> - *To prevent fuel from spraying on a hot engine, system pressure should be relieved before disconnecting fuel lines. One method is to tightly wrap a shop towel around a fuel line fitting and loosen or disconnect the fitting.*

- Connect a length of hose with threaded female fitting (14 mm x 1.5) to fuel rail.
- Place open end of hose in a suitable container for catching fuel (2 quart capacity).
- Run fuel pump for exactly 30 seconds. See **Operating fuel pump for tests** in this repair group.
- Measure fuel collected. Compare to specification in **Table b**.
- When finished testing, reconnect fuel line.
- Tighten all hose clamps.

Table b. Fuel pump delivery volume	
Delivery condition	**Delivery rate**
30 seconds @ 12V	0.850 liter (0.90 qt)

Fuel Pump Electrical Checks

Fuel pump, troubleshooting

During cranking and engine running, the DME main/fuel pump relay is energized and the pump runs. If an electrical system fault interrupts power to the fuel pump, the engine will not run.

– To quickly check if fuel pump is coming on, actuate starter while an assistant listens for a running pump.

– If pump is not running, check for faulty fuse or faulty DME main / fuel pump relay.

– Check fuel pump fuses. See **97 Electrical Component Locations**.

– Test DME main / fuel pump relay and circuit. See **DME main/fuel pump relay, testing** in this repair group.

– Check if voltage is reaching pump. See **EWD Electrical Wiring Diagrams**.

Operating fuel pump for tests

To operate fuel pump for testing purposes without having to run engine, the DME main / fuel pump relay can be bypassed to power pump directly.

◄ The DME main / fuel pump relay (**arrow**) is located in trunk in central electric panel.

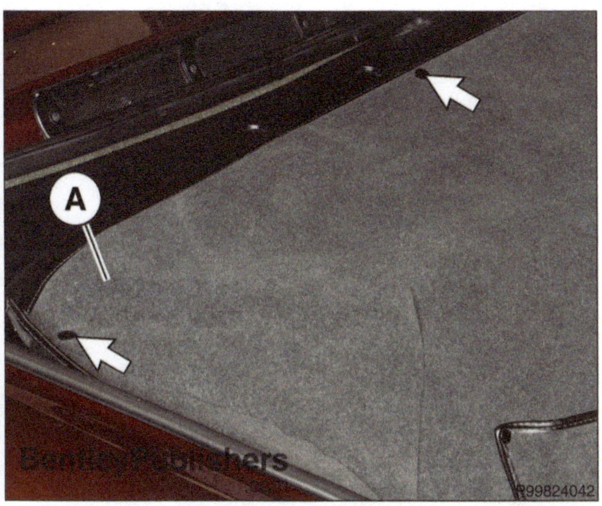

◄ Working in right side of trunk, disconnect snaps (**arrows**) and peel back carpet (**A**).

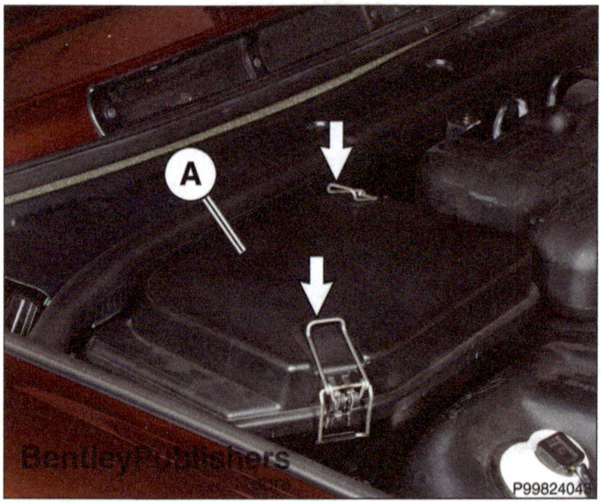

◀ Open locks (**arrows**), then remove central electric panel cover (**A**).

◀ Pull DME main / fuel pump relay (**arrow**) out of central electric panel.

NOTE–

- *Disconnecting power from the engine control module (ECM) erases its adaptive memory. It may be necessary to drive the car for approximately 10 minutes after reconnecting the battery to reset ECM adaptive memory and restore normal performance.*

◀ Connect socket for relay terminal **30** to socket for relay terminal **87b** with fused jumper wire to run pump. Ignition key does not need to be turned ON.

– After completing tests, remove jumper wire.

NOTE–

- *The jumper wire should be 1.5 mm (14 gauge) thick and include an in-line fuse holder with a 15 amp fuse. To avoid socket damage from repeated connecting and disconnecting, also include a toggle switch.*

DME main / fuel pump relay, testing

◀ The DME main / fuel pump relay (**arrow**) is located in trunk in central electronics panel.

The DME main/fuel pump relay contains two relays within a single housing. One relay is the DME main relay and the other is the fuel pump relay.

The Motronic DME main relay is energized when the key is turned on. The closed main relay supplies positive battery voltage (via terminal **87**) to the fuel injectors, the engine control module (ECM), the idle speed control valve, and the fuel pump relay. The fuel pump relay is then energized (closed) by the ECM. The switched ground for this relay (terminal **85b**) is supplied only if certain parameters are met, such as the presence of an ignition pulse and a starter input signal. This is a safety feature to prevent the pump from operating if the engine is not running.

On cars with a factory alarm, the alarm system interrupts power to terminal **86** of the main/fuel pump relay when the system is armed. When diagnosing a no-start condition, be sure to check for key-on power at terminal **86**.

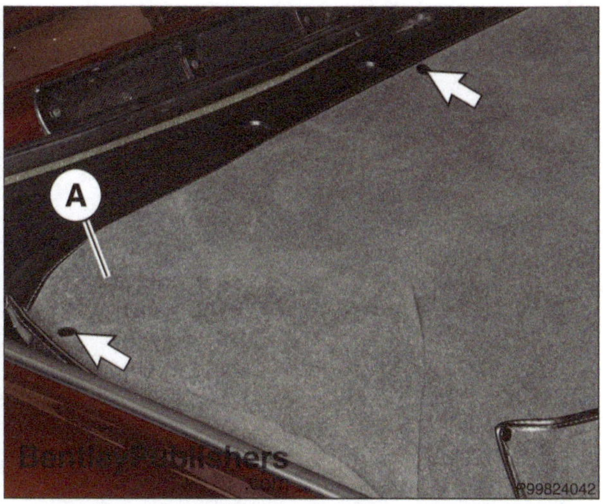

◀ Working in right side of trunk, disconnect snaps (**arrows**) and peel back carpet (**A**).

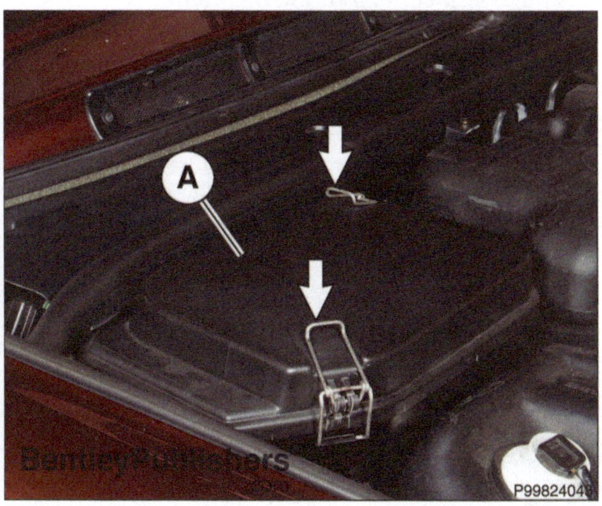

◀ Open locks (**arrows**), then remove central electric panel cover (**A**).

Pull DME main / fuel pump relay (**arrow**) out of central electric panel.

With key OFF, check for battery voltage at terminal **30**.

- If battery voltage is not present, check for breaks in large red wire in relay harness. This wire is a direct (non-fused) connection to battery. See **97 Electrical Wiring Diagrams**.

With ignition switch in ON position check for battery voltage at terminal **86**.

- If battery voltage is not present, check for break in black wire in relay harness. See **97 Electrical Wiring Diagrams**.

NOTE–

- *The vehicle alarm system interrupts power to terminal* **86**. *When diagnosing a no-start condition be sure to test key-ON power to terminal* **86**.

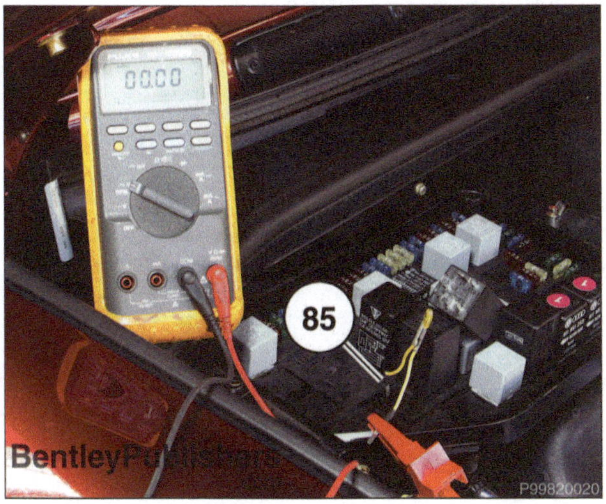

Check for ground at terminal **85**.

- If ground is present continue testing.
- If ground is not present, check connections at ground point on intake runner 1.

– Turn ignition OFF and install relay.

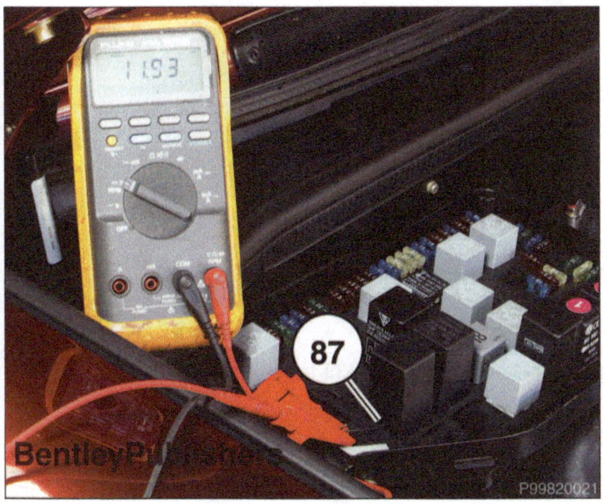

With ignition switch in ON position and relay connected, check for battery voltage at terminal **87**.

- If battery voltage is present, first relay has energized and is functioning correctly.
- If battery voltage is not present and all earlier tests are OK, replace DME main / fuel pump relay.
- If DME main relay has energized but fuel pump still will not run, test fuel pump relay ground and output. Relay must be plugged into harness for test.

– With engine cranking, test for ground at terminal **85b**.

– With engine cranking check for battery voltage at terminal **87b**.

- If battery voltage is not present at terminal **87b** but is present at terminal **87**, replace DME main/fuel pump relay.
- If battery voltage is present but fuel pump does not run, check fuel pump fuse in luggage compartment fuse box.
  See **97 Fuses, Relays, Component Locations**.

Fuel pump voltage, checking

If fuel pump fuse and DME main/fuel pump relay are OK, but fuel pump will not run, check for voltage at pump.

– Raise car and support safely.

> ***CAUTION—***
>
> *• Make sure the car is stable and well supported at all times. Use a professional automotive lift or jack stands designed for the purpose. A floor jack is not adequate support.*

< Working underneath right front of car, remove fuel pump cover fasteners (**arrows**).

– Remove fuel pump cover (**A**).

< Remove fuel pump mounting bracket fasteners (**arrows**).

– Pull fuel pump down out of body and allow to hang.

◄ Working at fuel pump, pull back protective rubber boots (**arrows**).

◄ Bypass fuel pump relay. See **Operating fuel pump for tests** in this repair group.

– Check for voltage between fuel pump connectors.

- If battery voltage is present and connectors are not loose or corroded, fuel pump is faulty.
- If there is no voltage, look for a faulty wire or connector.
- Also, check fuel pump ground wire (brown wire) for continuity to ground.

20

Fuel pump current draw, checking

– Raise car and support safely.

> ***CAUTION—***
>
> - *Make sure the car is stable and well supported at all times. Use a professional automotive lift or jack stands designed for the purpose. A floor jack is not adequate support.*

◄ Working underneath right front of car, remove fuel pump cover fasteners (**arrows**).

– Remove fuel pump cover (**A**).

Fuel pump current draw, checking

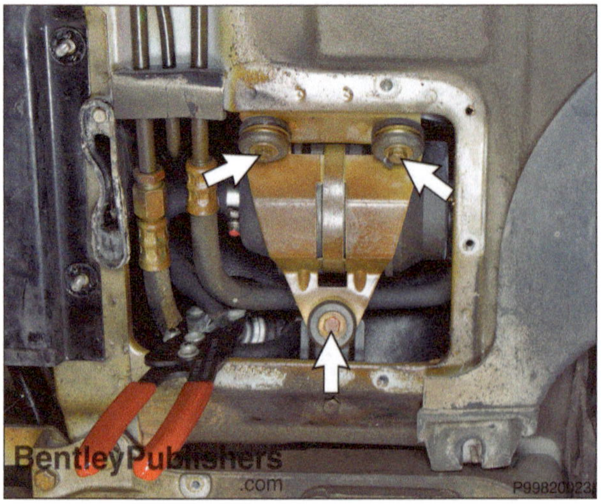

< Remove fuel pump mounting bracket fasteners (**arrows**).

- Pull fuel pump down and allow to hang.

< Pull back two protective rubber boots from pump connectors. Disconnect 2.5 mm green wire from pump.

< Connect ammeter between disconnected green wire and fuel pump terminal as shown. The other wire remains connected to pump.

> ***CAUTION—***
>
> *• Do not allow the test leads to short to ground.*

- Bypass fuel pump, see **Operating fuel pump for tests** in this repair group.
- Fuel pump current should be in the range of 4 to 6 amps.

NOTE–

• A higher than normal power consumption usually indicates a worn fuel pump, which may cause intermittent fuel starvation due to pump overheating and seizure. The only remedy is to replace the pump.

Fuel Pump Replacement

The fuel pump is mounted underneath right front of vehicle, behind a metal cover (**arrow**).

Fuel pump, replacing

– Disconnect negative (–) battery cable and cover battery terminal to keep cable from accidentally contacting terminal.

> ***CAUTION–***
>
> • *Prior to disconnecting the battery, read the battery disconnection cautions given in* **00 Warnings and Cautions.**

– Raise car and support safely.

> ***CAUTION—***
>
> • *Make sure the car is stable and well supported at all times. Use a professional automotive lift or jack stands. A floor jack is not adequate support.*

Working underneath right front of car, remove fuel pump cover fasteners (**arrows**).

– Remove fuel pump cover (**A**).

◀ Using hose clamping pliers, pinch off fuel feed line (**arrow**) from tank.

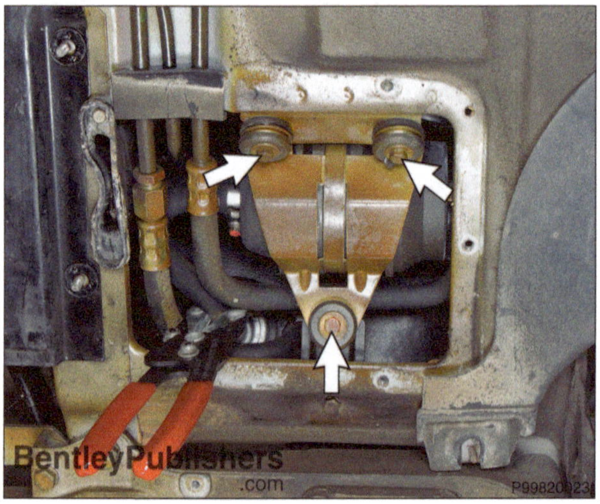

◀ Remove fuel pump mounting bracket fasteners (**arrows**).

– Pull fuel pump away from floorboard and allow to hang.

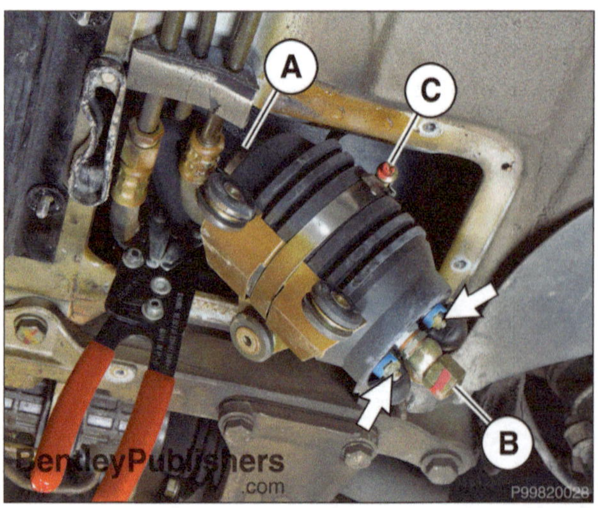

◀ Disconnect electrical connectors (**arrows**) at fuel pump.

– Loosen clamp and remove fuel supply line (**A**) from fuel pump.

– Remove pressure side fuel line cap nut (**B**) while counter holding pump fitting with wrench.

– Remove fuel line from fuel pump.

– Loosen pump mounting clamp (**C**) and remove pump.

– Installation is reverse of removal. Note the following:

- Always replace sealing washers.
- Start engine and check for leaks after tightening fittings, before reinstalling fuel pump cover.

Tightening torque	
Electrical connector nuts	1.5 Nm (13 in-lb)
Hose clamps	2 Nm (18 in-lb)
Fuel pump clamp bolt	7 Nm (62 in-lb)

Fuel pump check valve, replacing

– Disconnect negative (–) battery cable and cover battery terminal to keep cable from accidentally contacting terminal.

> ***CAUTION***
>
> • *Prior to disconnecting the battery, read the battery disconnection cautions given in* **00 Warnings and Cautions.**

– Raise car and support safely.

> ***CAUTION—***
>
> • *Make sure the car is stable and well supported at all times. Use a professional automotive lift or jack stands. A floor jack is not adequate support.*

◄ Working underneath right front of car, remove fuel pump cover fasteners (**arrows**).

– Remove fuel pump cover (**A**).

◄ Using hose clamping pliers, pinch off fuel feed line (**arrow**) from tank.

◄ Remove fuel pump mounting bracket fasteners (**arrows**).

– Pull fuel pump down away from floorboard and allow to hang.

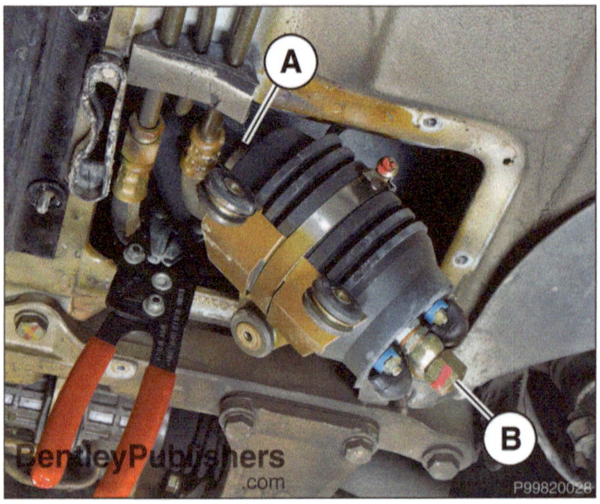

◄ Remove pressure side fuel line cap nut (**A**) while counter holding pump fitting (**B**) with wrench.

– Remove fuel line from fuel pump.

◄ Remove check valve (**A**) while counter holding pump fitting (**B**) with wrench, then remove check valve from fuel pump.

– Installation is reverse of removal. Note the following:

- Always replace sealing washers.
- Start engine and check for leaks after tightening fittings, before reinstalling cover.

Tightening torque	
Electrical connector nuts	1.5 Nm (13 in-lb)
Hose clamps	2 Nm (18 in-lb)
Fuel pump clamp bolt	7 Nm (62 in-lb)

24 Fuel Injection

TABLES

General

This repair group covers the fuel injection components of the DME (Motronic) system.

- Ignition System information is covered in **28 Ignition System**.
- For fuel pump and related fuel supply component testing and repair see **20 Fuel Storage and Supply**.
- Wiring diagrams for the DME engine management system can be found in **EWD Electrical Wiring Diagrams**.
- Fuel filter replacement is covered in **03 Service and Maintenance**.

Digital Motor Electronics (DME) system

The 911 Carrera models covered in this manual are equipped with an electronic engine management system, controlling fuel delivery and ignition through a single engine control module (ECM). This system is known as the Motronic system or Digital Motor Electronics (DME).

The fuel injection is completely electronic in operation. The DME engine control module (ECM) controls fuel delivery and ignition timing using signals from the following sources:

- Air flow sensor
- Intake air temperature (IAT) sensor
- Cylinder head temperature sensor
- Reference position (TDC) sensor
- Crankshaft speed (RPM) sensor
- Throttle position switches
- Oxygen sensor

There are two variants of the DME systems used on the cars covered by this manual. **Table a**.

Table a. DME system application	
• 1995 Carrera, Carrera 4	DME M 2.10 (55-pin ECU)
• 1996 - 1998 Carrera, Carrera S, Carrera 4, Carrera 4S	DME M 5.2 (88-pin ECU)

Bosch M 2.10 is installed on all normally-aspirated 1995 Porsche 993 models for U.S. market. The main features of the M 2.10 system are:

- 55-pin connector, OBD I compliant.
- Cylinder sequential fuel injection
  (separate fuel mixture control for each cylinder)
- Warming up and acceleration enrichment
- Adaptive Lambda control (single oxygen sensor)
- Adaptive selective cylinder control for knock regulation
- Tiptronic: Torque reduction via ignition retarding at gearshifts
- Engine idle speed control with adaptive air correlation
- Carbon canister purge
- Hot film mass air flow sensor
- Separate air temperature sensor
- Throttle position potentiometer
- Secondary air injection pump control
- Resonance flap control

Bosch M 5.2 is installed on all normally aspirated 1996-1998 Porsche 993 models for the U.S. market. The main features of the M 5.2 system are:

- 88-pin connector, OBD II compliant
- Cylinder sequential fuel injection (separate fuel mixture control for each cylinder).
- Electronically mapped twin ignition system (two spark plugs per cylinder)
- Adaptive heated oxygen sensor control (post- and pre-sensors for each catalytic converter).
- Varioram air induction system control.
- Hot film mass air flow sensor.
- Adaptive knock control.
- Hall-effect sensor installed in distributor to detect firing Top Dead Center (TDC) of cylinder 1.
- Idle air control using a twin-coil rotary adjuster.
- Electronically controlled and monitored secondary air injection system.
- Carbon canister with electronic venting.
- Throttle potentiometer.
- Engine temperature sensing using NTC resistor.

Emission controls

The emission control system used in the 911 Carrera includes the oxygen sensor system, catalytic converters, secondary air injection system and the evaporative emission system.

Oxygen sensor: This system provides the fuel injection system with feedback information about combustion efficiency.

Catalytic converters: This system reduces tailpipe emissions.

Secondary air injection system: This system reduces tail pipe and engine emissions during cold starting.

Evaporative emission controls: This system provides venting for the fuel tank while at the same time trapping the fuel vapors that would otherwise be vented into the atmosphere. The evaporative emission system collects fuel vapors from the fuel tank in a charcoal canister. During certain engine operating conditions, the canister is purged, drawing the fuel vapors into the engine to be burned.

DME system components (Bosch M 2.10)

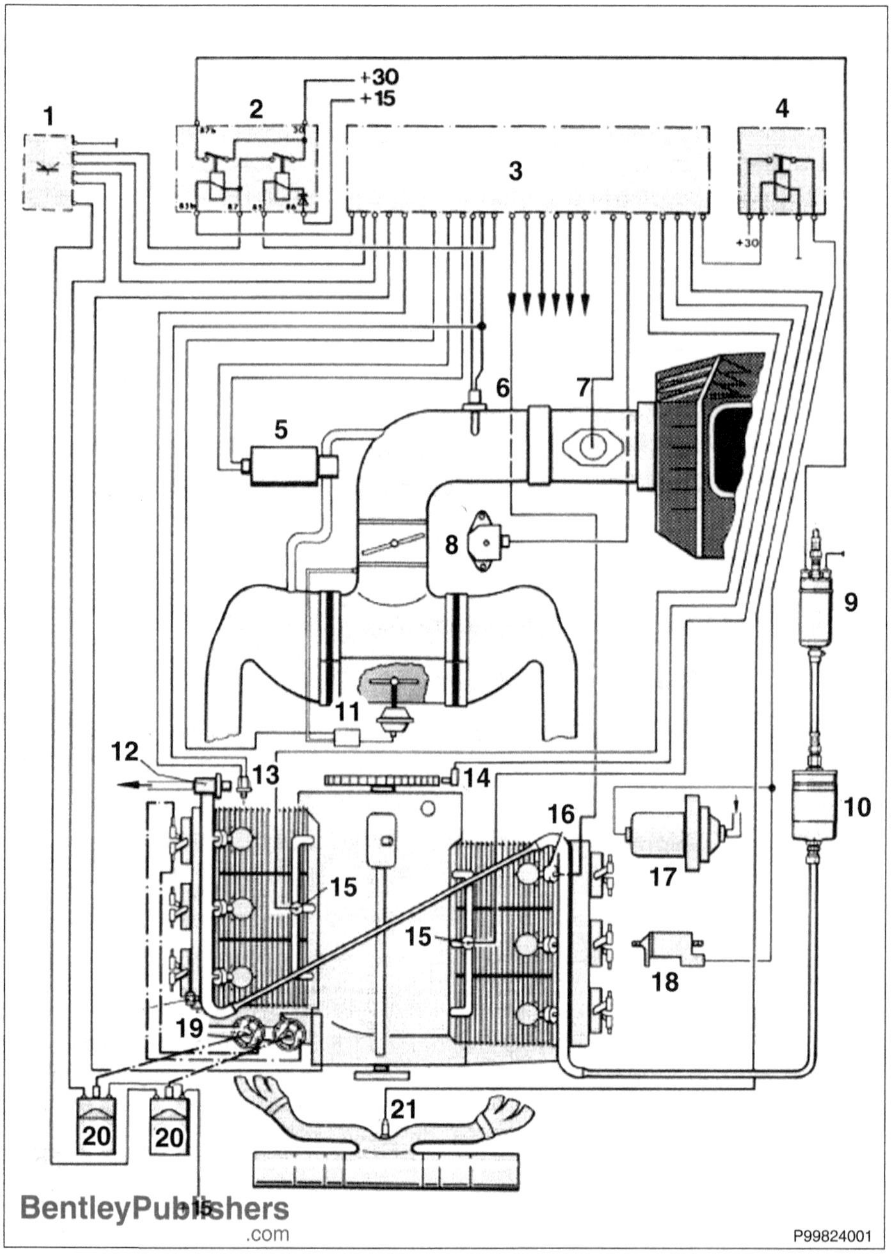

1. Ignition switching unit (final stage)
2. DME relay
3. DME control unit
4. Secondary air injection pump relay
5. Idle air control (IAC) valve
6. Intake air temperature sensor
7. Mass air flow sensor
8. Throttle position sensor
9. Fuel pump
10. Fuel filter
11. Resonance flap on - off valve
12. Fuel pressure regulator
13. Engine temperature sensor
14. Engine speed sensor
15. Knock sensor
16. Fuel injectors
17. Secondary air injection pump
18. Secondary air injection vacuum solenoid
19. Distributor
20. Ignition coils
21. Oxygen sensor

DME system components (Bosch M 5.2)

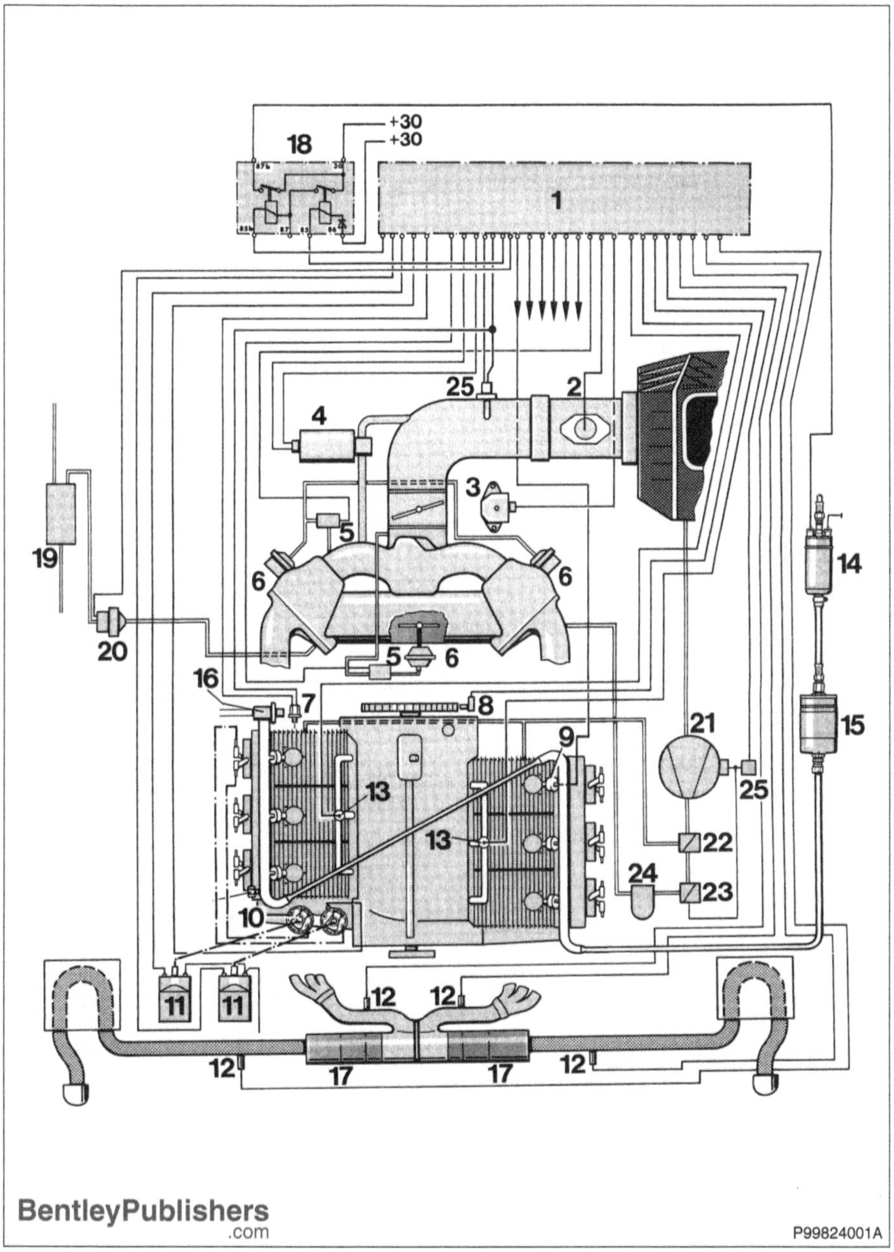

1. DME control unit
2. Mass air flow sensor
3. Throttle position sensor
4. Idle air control (IAC) valve
5. Solenoid valve (Varioram)
6. Diaphragm valves (Varioram)
7. Engine temperature sensor
8. Engine speed sensor
9. Injection valve
10. Distributor (twin with Hall-effect sensor)
11. Ignition coils
12. Oxygen sensors
13. Knock sensor
14. Fuel pump
15. Fuel filter
16. Fuel pressure regulator
17. Catalytic converter
18. DME relay
19. Carbon canister
20. Tank venting valve
21. Secondary air injection pump
22. Pneumatic valve
23. Solenoid valve
24. Vacuum tank
25. Intake air temperature sensor

Basic engine settings

The DME system compensates automatically for changes in the engine due to age, minor wear or small vacuum leaks. As a result, idle speed and mixture do not need adjustment as part of routine maintenance.

Secondary Air Injection (SAI)

The secondary air injection (SAI) system uses an electric pump to inject ambient air into the exhaust stream after a cold engine start. This strategy effectively reduces the warm up time of the catalytic converter, resulting in a reduction of HC and CO emissions. The DME engine control module (ECM) controls and monitors the secondary air injection system.

NOTE–

- *993 cars sold in the US were fitted with secondary air injection (SAI).*
- *On 1996 and later cars (DME M 5.2), the SAI system is monitored by OBD II software. On DME M 5.2 cars, insufficient SAI air flow faults can turn on the Check Engine light.*
- *1995 cars are OBD I compliant which does not monitor the SAI system.*

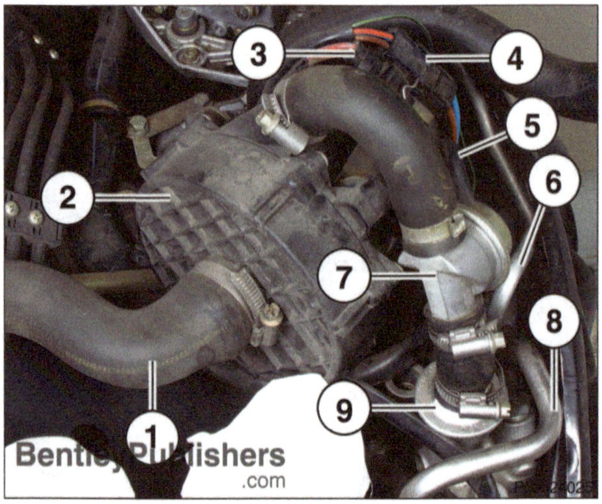

◀ The SAI pump is mounted to the right rear of engine behind intake manifold.

1. SAI pump air intake hose
2. SAI pump
3. SAI pump vacuum solenoid
4. SAI pump electrical connector
5. SAI pump control valve vacuum hose
6. Air pipe to left camshaft housing
7. SAI control valve
8. Air pipe to right camshaft housing
9. SAI check valve

When the SAI relay is energized by the DME control unit, power is supplied to the SAI pump and the vacuum switching solenoid. When the switching solenoid is energized, engine vacuum opens the control valve to allow SAI pump air to be injected through the SAI check valve and into the left and right camshaft housings.

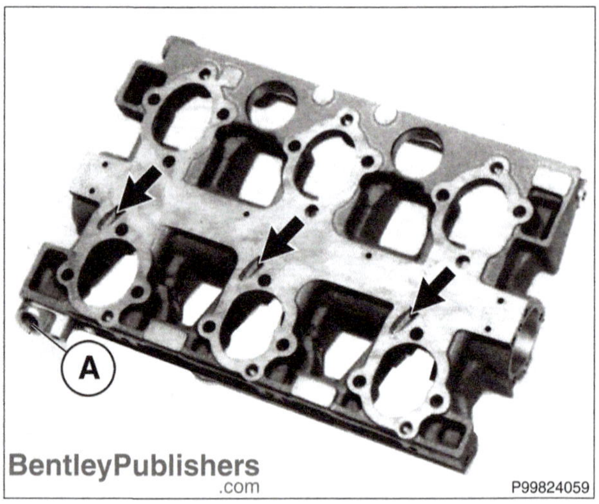

P99824059

◀ From the port in the camshaft housings (**A**), air flows through the bores in the camshaft housings (**arrows**) into the exhaust ports in the cylinder heads.

P99824060

◀ Air flow from camshaft housing is directed into transverse passage (**A**) in cylinder head in exhaust port.

Common SAI issues

Owing to the harsh operating environment, it is not uncommon for the SAI air passages and check valve to become clogged with combustion by-products. This clogging results in insufficient flow or no flow of secondary air into exhaust system.

DME M 5.2 OBD II SAI fault codes:

- P0410 - Secondary air injection, insufficient flow (cyl. 1-3)
- P1411 - Secondary air injection, insufficient flow (cyl. 4-6)

OBD II SAI monitor: When the SAI pump is commanded by the DME control unit to run (only during cold start conditions), the oxygen sensor senses the excess oxygen condition. If the SAI ports are blocked, additional air will not be detected in exhaust, and an insufficient flow fault will be set in the DME control unit.

If the blockage is simply at the one-way check valve, the valve can be replaced first to see if air flow is restored. But if the check valve is severely blocked, the passages in the camshaft housing and cylinder heads may also blocked. Properly cleaning blocked passages requires engine disassembly to access the ports in the camshaft housings and cylinder heads.

Secondary air injection components

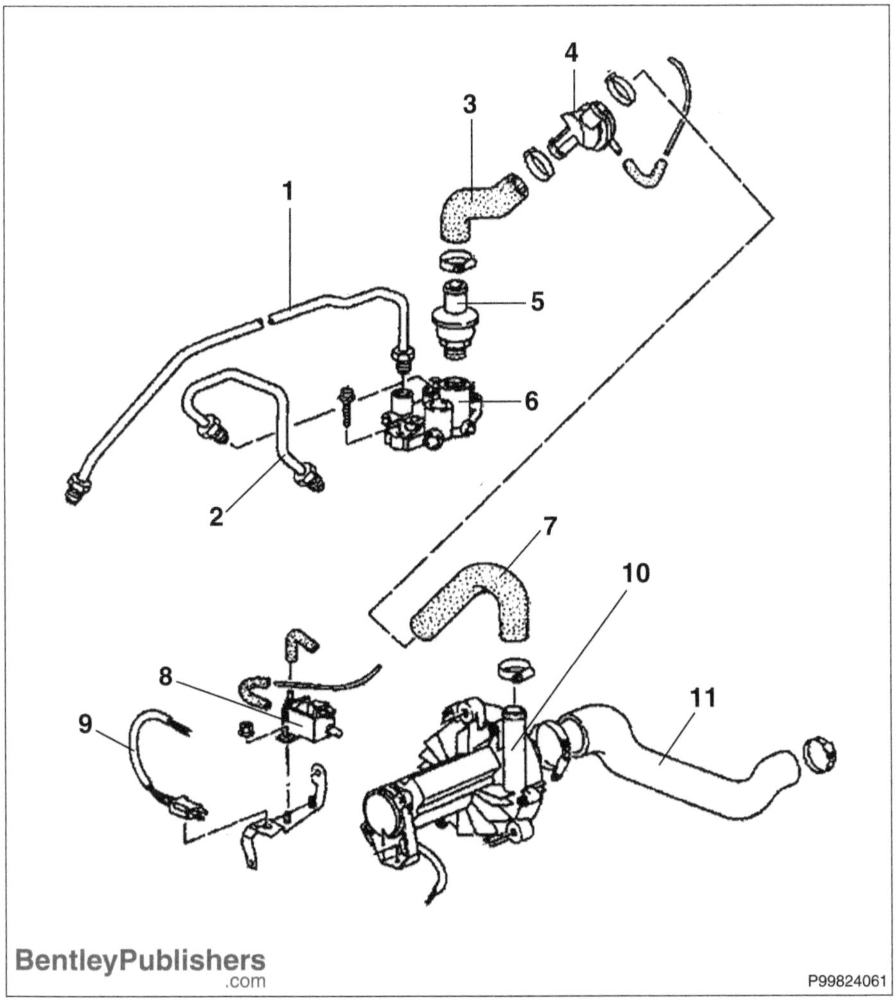

1. Air pipe to left side camshaft housing
2. Air pipe to right side camshaft housing
3. Hose
4. SAI control valve (shut-off valve)
5. SAI check valve
6. Air distributor housing
7. Hose
8. SAI vacuum solenoid
9. Connector
10. SAI pump
11. SAI air intake hose

FUEL INJECTION COMPONENT TESTING

Warnings and Cautions

WARNING—

- *When working on an open fuel system, wear suitable hand protection, as prolonged contact with fuel can cause illnesses and skin disorders.*
- *The fuel system is designed to retain pressure even when the ignition is off. When working with the fuel system, loosen the fuel lines very slowly to allow residual fuel pressure to dissipate gradually. Avoid spraying fuel.*
- *Fuel is highly flammable. When working around fuel, do not disconnect any wires that could cause electrical sparks. Do not smoke or work near heaters or other fire hazards. Keep an approved fire extinguisher handy.*
- *The ignition system produces high voltages that can be fatal. Avoid contact with exposed terminals. Use extreme caution when working on a car with the ignition switched on or the engine running.*
- *Do not touch or disconnect any high voltage cables from the coil, distributor, or spark plugs while the engine is running or being cranked by the starter.*

CAUTION—

- *Use a high impedance digital multimeter for all voltage and resistance tests.*
- *An LED test light should be used in place of an incandescent-type test lamp.*
- *Avoid damaging harness connectors or relay panel sockets. Jumper wires should be fabricated with flat blade end connectors that are the same size as the connector or relay terminal being tested.*
- *Always switch the test meter to the appropriate function and range before making test connections. Connect and disconnect the test equipment only with the ignition switched off.*
- *Wire colors may vary from those given in text and schematics. Always check component terminals to verify wire colors.*

Fuel injectors, testing

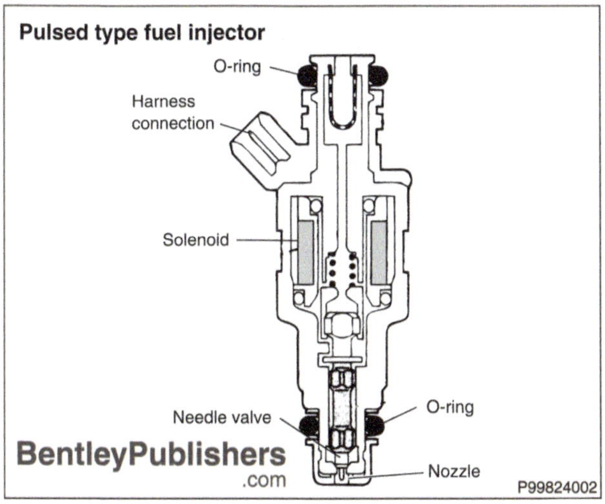

The fuel injectors are turned on and off by the DME engine control module (ECM). The quantity of fuel delivered to the engine is determined by the amount of time the injectors remain open.

Injector tests include electrical checks for resistance, power and ground signal from the ECM.

Injector flow tests require specialized equipment that opens each injector for a precise amount of time to measure fuel delivery. Most Porsche work shops can perform an injector balance test to locate clogged or faulty injectors.

To quickly check if an injector is electrically functioning, start engine and place screwdriver or finger on injector body. A slight vibration or buzzing sound indicates that injector is functioning electrically.

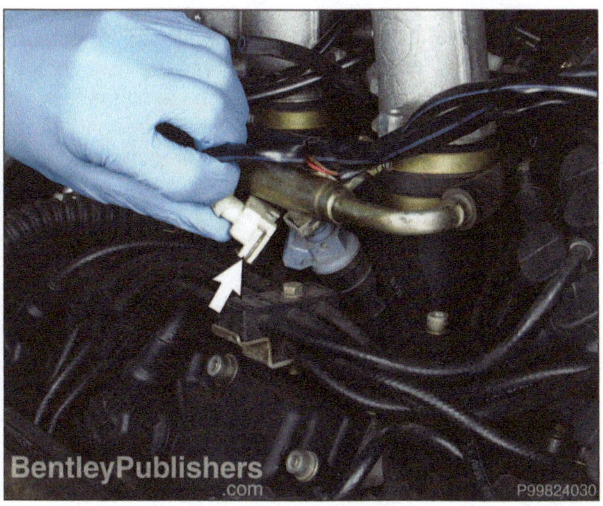

With engine running, remove injector electrical connectors, one at a time.

- Engine RPM should drop each time one injector is disconnected.
- If disconnecting one injector does not affect engine RPM, continue testing that injector and harness connector.

– If any faults are found, turn engine OFF and disconnect electrical connector from suspected injector.

– Using an ohmmeter, check injector resistance. Compare to specification listed in **Table b**.

Table b. Fuel injector resistance specification	
Electrical connector disconnected • Resistance at 20°C (68°F)	16 Ω

– If injector resistance is within specification, check that ECM signal is present at injector connector.
  - Connect LED injector tester or back probe injector harness connector using digital voltmeter (select ms-pulse width range if available).
  - Operate starter or run engine and check for pulsed voltage signal.
  - Repeat for each injector.

– If no signal is present, check for power to injectors.
  - There should be battery voltage (+) at red/white wire of each injector connector with ignition key on.
  - If voltage is not present, check main/fuel pump relay output at terminal 87.

– If there is positive (+) battery voltage at the injectors but no pulsed response at the voltmeter, check the wire(s) from the ECM to the injectors. If no wiring faults can be found, the pulsed ground signal from the control module may be missing. Check the outputs from the ECM.

– If no wiring faults can be found after the above checks, the ECM may be faulty. ECM harness electrical tests and fuel injector replacement are covered later in this repair group.

DME engine control module (ECM)

When troubleshooting engine management problems, check the wiring and confirm that all input signals are correct before assuming the ECM is faulty.

ECM pin assignments are given in **DME engine control module (ECM) M 2.10, pin assignments** and **DME engine control module (ECM) M 5.2, pin assignments** in this repair group. This information can be helpful when diagnosing faults to or from the ECM. If all inputs and wiring are OK but operational problems still exist, the ECM itself may be faulty.

Test results with incorrect values do not necessarily mean that a component is faulty. Check for loose, broken or corroded connections and wiring before replacing components. If the results are still incorrect, test the component itself.

When making checks at the ECM, a pinout test harness (breakout box) should be used to allow tests to be made with the connector attached to the ECM. As an alternative, the electrical connector housing can be separated so that electrical tests can be made from the back of the connector with the connector still plugged in.

DME engine control module (ECM), accessing for tests

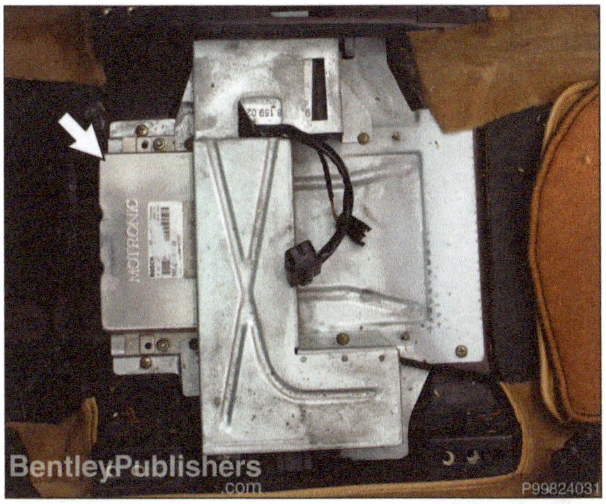

The DME engine control module (**arrow**) is located under driver seat.

- Disconnect negative (–) battery cable. Wait approximately 1 minute.

NOTE—

- *Prior to disconnecting the battery, read the battery disconnection cautions, see* **00 Warnings and Cautions**.

- Remove driver seat. See **72 Seats**.

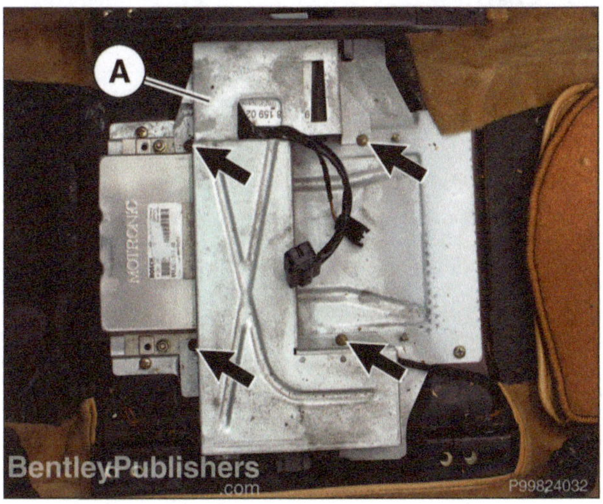

Remove cover plate (**A**) shear bolts (**arrows**).

- Remove cover plate from vehicle.
- Disconnect ECM electrical connector and connect pinout test harness to ECM for electrical tests of DME components.

24

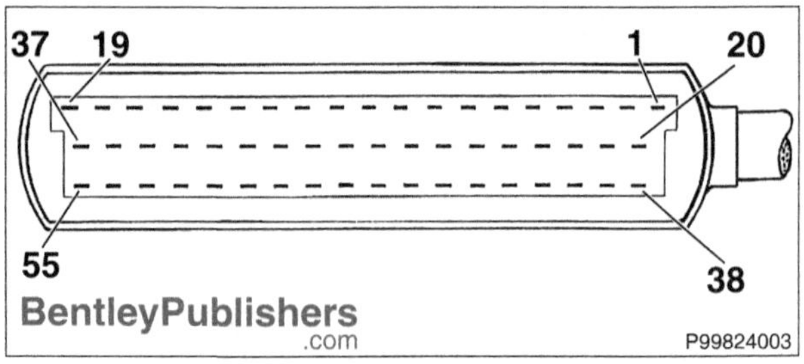

DME engine control module (ECM) (Bosch M 2.10) pin assignments

DME M 2.10 control module pin assignments (55-pin)	
1	Resonance flap
2	Not used
3	Ground DME relay, (Terminal **85b**)
4	Intake manifold flap open
5	Cylinder 3 fuel injector
6	RPM
7	Mass air flow signal
8	Hall sensor signal
9	Vehicle speed
10	Oxygen sensor ground
11	Knock sensor 1
12	Positive supply: throttle potentiometer, hall sender
13	Diagnosis L wire
14	Ground supply: throttle potentiometer, NTC sensors
15	Not used
16	Cylinder 6 fuel injector
17	Cylinder 1 fuel injector
18	Battery positive (+)
19	Ground: electronics shield
20	Oxygen sensor, heating
21	Check engine light
22	Intake manifold flap close
23	Cylinder 5 fuel injector
24	Ground: final stages
25	Final stage, ignition circuit 1
26	Ground, mass air flow sensor
27	Tiptronic: terminal 15
28	Oxygen sensor
29	Knock sensor
30	Ground, knock sensors
31	Final stage, ignition circuit 2
32	Secondary air pump
33	Fuel tank vent

DME M 2.10 control module pin assignments (55-pin)	
34	Cylinder 4 fuel injector
35	Cylinder 2 fuel injector
36	Ground, DME relay, (Terminal **85**)
37	DME relay, (Terminal **85**)
38	Not used
39	Not used
40	A/C compressor clutch
41	Heater control unit
42	Tiptronic: selector lever switch
43	Not used
44	Tiptronic: coding
45	Engine NTC II
46	Not used
47	Intake air temperature sensor
48	Engine speed sensor –
49	Engine speed sensor +
50	Engine knock: yes or no
51	Transmission input
52	Tiptronic: throttle position signal to transmission
53	Throttle position sensor
54	Characteristic map switch
55	Diagnosis K wire

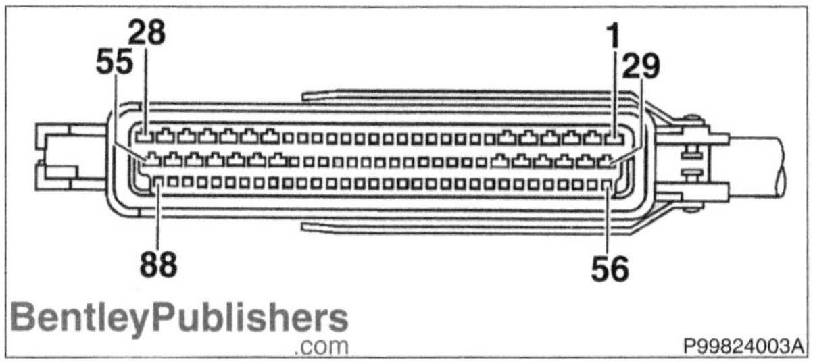

DME M 5.2 control module pin assignments (88-pin)	
1	Oxygen sensor heating, downstream
2	Intake manifold flap open
3	Cylinder 1 fuel injector
4	Cylinder 2 fuel injector
5	Cylinder 3 fuel injector
6	Ground
7	Not used
8	Check engine light
9	Not used
10	Vacuum coasting signal
11	A/C compressor clutch
12	Not used
13	Not used
14	Not used
15	Not used
16	Intake air temperature sensor
17	Mass air flow signal
18	Oxygen sensor, upstream (cylinder 4,5,6)
19	Oxygen sensor, upstream (cylinder 1,2,3)
20	Engine speed sensor +
21	Hall sensor
22	Final stage, ignition circuit 2
23	Not used
24	Not used
25	Not used
26	Battery positive (+)
27	Terminal 85b ground
28	Ground
29	Intake manifold flap close
30	Oxygen sensor heating, upstream
31	Cylinder 6 fuel injector
34	Cylinder 4 fuel injector
33	Cylinder 5 fuel injector

DME engine control module (ECM) (Bosch M 5.2) pin assignments

DME M 5.2 control module pin assignments (88-pin)	
34	Ground, intake temperature sensor
35	Varioram switch over
36	Not used
37	Secondary air injection pump
38	Compressor on information
39	Fuel level information
40	Knock sensor (cylinder 4,5,6)
41	Not used
42	Transmission input
43	Not used
44	Throttle position sensor
45	Ground, mass air flow sensor
46	Ground, oxygen sensor
47	Not used
48	Not used
49	Final stage, ignition circuit 1
50	Not used
51	Not used
52	Not used
53	Positive supply: throttle potentiometer, hall sensor
54	DME relay, (Terminal **87**)
55	Ground, ignition
56	Tiptronic: terminal 15
57	Timing valve, secondary air injection pump
58	Not used
59	Resonance flap
60	Not used
61	Fuel tank vent
62	A/C compressor control
63	Ground, DME relay, (Terminal **85**)
64	Not used
65	Not used
66	Not used
67	Engine knock: yes or no
68	Tiptronic: selector lever switch
69	Not used
70	Knock sensor (cylinder 1,2,3)

DME M 5.2 control module pin assignments (88-pin)	
71	Ground, knock sensor
72	Not used
73	Not used
74	Engine temperature
75	Not used
76	Oxygen sensor, downstream (cylinder 1,2,3)
77	Oxygen sensor, downstream (cylinder 4,5,6)
78	Engine speed sensor –
79	Vehicle speed
80	RPM
81	Not used
82	Tiptronic: throttle position signal to transmission
83	Not used
84	Not used
85	Not used
86	Not used
87	Diagnosis L line
88	Diagnosis K line

DME main / fuel pump relay

P99828021

The DME main / fuel pump relay (**arrow**) is located in central electronics panel in trunk. If the DME main / fuel pump relay is not functioning correctly, the engine will not run. Testing of the DME main / fuel pump circuit is covered in **20 Fuel Storage and Supply**.

Cylinder head temperature sensor, testing

The cylinder head temperature sensor (**arrow**) is installed in cylinder head 3. The cylinder head temperature sensor signal is the main temperature signal to the ECM.

– Remove blower motor. See **80 Heating**.

Disconnect white electrical connector on left side of engine (**arrow**).

NOTE—

- *Cylinder head temperature sensor electrical connector is white, located behind green knock sensor connector shown in photo.*

Connect ohmmeter across connector terminals (**inset**).

- Resistance should correspond to specifications listed in **Table c**.
- If any faults are found, replace sensor. Replacement procedure is covered later in this group.
- An open circuit in the temperature sensor will cause the DME system to run rich. A short circuit in the temperature sensor causes leaner mixtures.

Table c. Cylinder head temperature sensor resistance

Sensor temperature	Resistance
0° C (32° F)	4.4 - 6.8 Ω
15 - 30° C (59 - 86° F)	1.4k - 3.6k Ω
40° C (104° F)	1k - 1.3k Ω
80° C (176° F)	250 - 390 Ω
100° C (212° F)	160 - 210 Ω

24

Mass air flow sensor, testing

The amount of air entering the engine is electrically measured by the mass air flow (MAF) sensor (**arrow**). When the engine is running, intake air is drawn past the sensor hot film and registered as a input parameter for the DME.

– Disconnect MAF sensor electrical connector.

– Remove DME main / fuel pump relay and use a jumper wire to bypass circuit. See **Operating fuel pump for tests** in **20 Fuel Supply**.

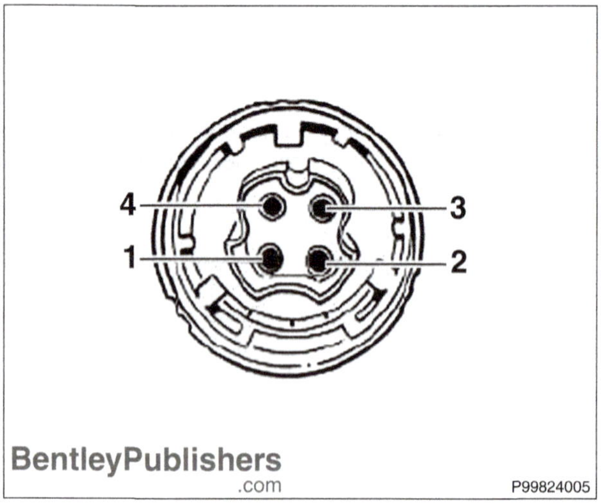

Connect voltmeter to terminals 1 and 3.

- Voltage reading of 10.8 Volts.

– Reconnect electrical connector

– Connect Voltmeter to DME terminals 7 and 26 (ground).

- Voltage reading of 160 mV

– Blow air through hot film in mass air flow sensor.

- A voltage increase should occur.

Intake air temperature (IAT) sensor, testing

To reduce engine knock at high intake temperatures, the 993 engine is fitted with a intake air temperature sensor (IAT).

Model year 1995, the intake air temperature sensor (**arrow**) is mounted between the intake air duct between the mass air flow sensor and the throttle housing.

From model year 1996, the intake air temperature sensor (**arrow**) is mounted on the intake air housing between the mass air flow sensor and the throttle housing.

This signal is used to continuously modify the ignition timing angle (timing retard direction) at an intake temperature starting at 25° C (77° F).

NOTE–

- *The value of the temperature sensor may be read in ° C using a Porsche scan tool in the Actual Values menu item.*

– Disconnect electrical connector from intake air temperature sensor.

– Connect ohmmeter across terminals 1 and 2 of intake air temperature sensor connector. Compare readings to specifications in **Table d**.

Table d. Intake air temperature sensor resistance	
Sensor temperature	**Resistance**
0° C (32° F)	4.4 - 6.8 Ω
15 - 30° C (59 - 86° F)	1.4k - 3.6k Ω
40° C (104° F)	1k - 1.3k Ω

Throttle position sensor, testing

The throttle position sensor is used to determine the position of the throttle shaft. The 0° position (engine idle position) of the potentiometer is determined in an adaptive process by the DME control unit.

If the battery has been disconnected or the DME has been disconnected, restart the engine with the throttle housing fully closed. The DME uses the smallest opening angle of the throttle as the engine idle position. If the throttle moves to 'open', the engine idle settings are switched off by the DME as soon as an opening angle of 1° is reached.

– Connect test lead VW 1501 between throttle potentiometer and disconnected electrical connector.

– Connect voltmeter between terminals 1 and 2.

– Ignition ON: 5 V (throttle position sensor reference voltage)

- No voltage: Check circuit using wiring diagram.

– Connect voltmeter to lead No. 2 and No. 3.

- Voltage reading of 0.5 V.
- Operate throttle: Voltage should increase to 4.3 V.

24

Engine speed sensor, testing

The engine speed sensor is an inductive sender for engine speed and reference mark detection. For this purpose, a toothed ring gear is machined onto the flywheel. This ring has a total of 60 teeth. Two of these teeth are replaced by cutouts to generate the reference mark signal. The reference signal is set at 84° BTDC.

The air gap between the sender and the flywheel ring gear is adjustable and must be 1.0 - 0.2 mm.

Test engine speed sensor using an oscilloscope. Connect and oscilloscope according to manufacturer's instructions.

- Disconnect DME electrical connector and install break out box.
- Connect oscilloscope test lead with terminal **49** and terminal **48** of the breakout box.

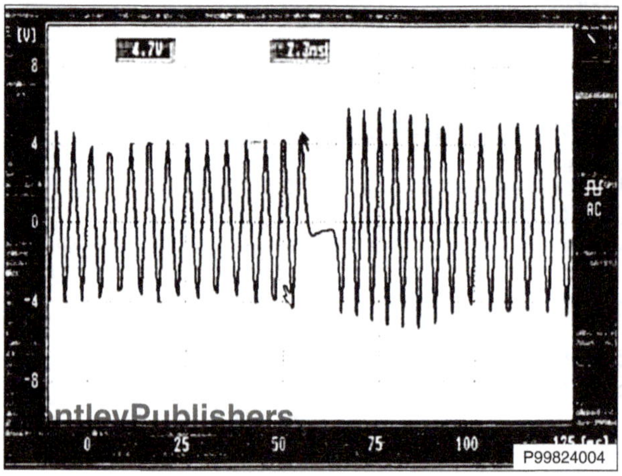

Start engine.

- Sine wave fluctuations of 3 V should be displayed.
- A signal less than 3 V indicates improper air gap. Adjust sensor air gap to 1.0 - 0.2 mm.
- An intermittently higher amplitude indicates the reference mark signal.

Oxygen sensor system

The oxygen sensor system provides the ECM with feedback information about combustion efficiency. The signal is used to continuously control the air-fuel mixture.

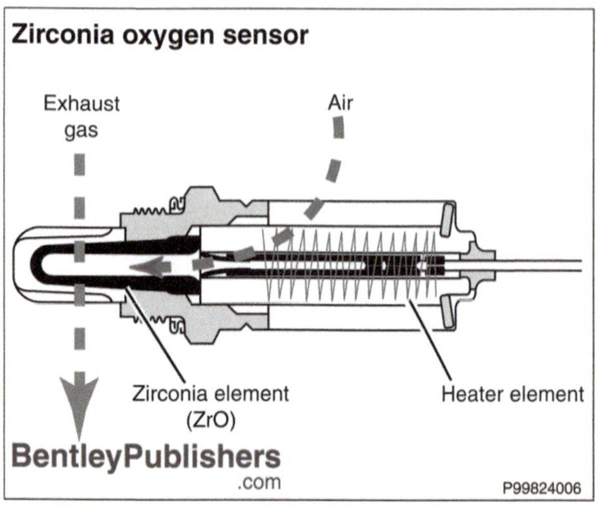

One surface of the exhaust-mounted oxygen sensor is exposed to the exhaust gas, while the other surface is exposed to the atmosphere. The difference in oxygen content between the two surfaces causes a chemical reaction that generates a low voltage electrical signal (100 - 900 mV).

The low voltage signal of the oxygen sensor is not reliably produced until the sensor reaches a temperature of approximately 300° C (572° F). For this reason, a separate heater circuit is used to bring the oxygen sensor up to operating temperature rapidly after a cold start.

– **1995 models**: A single heated oxygen sensor is mounted in the center of the catalytic converter assembly.

◄ **1996 - 1998 models**: four heated oxygen sensors are mounted before (**A**) and after (**B**) the left and right catalytic converters.

– The heater circuit gets power from terminal 87b of the DME main / fuel pump relay and is not fused.

CAUTION—

- *Lubricating sprays containing silicone can contaminate the oxygen sensor.*
- *Handle the oxygen sensor with care. Hitting or dropping will damage it.*

Oxygen sensor signal, testing

Basic checking of oxygen sensor efficiency using a pinout test harness (Porsche special tool **9543** or equivalent) is described below.

NOTE–

- *For the most accurate test results, the engine must be fully warmed, the exhaust system must be free of leaks, and all electrical consumers must be off.*

WARNING—

- *Exhaust system parts can be hot enough to cause serious burns. When working near hot pipes or mufflers, use heavy gloves and other appropriate protection.*

– Disconnect ECM harness connector and connect breakout box.

– Connect a digital voltmeter (millivolt scale) to terminals. Start engine and let idle. After a maximum of two minutes, oxygen sensor output voltage should begin fluctuating.

Oxygen sensor test terminals	Test conditions and test results
1995 **28** (sensor signal) and **10** (ground)	Engine warm 0.1 – 0.9 VDC fluctuating
1996 - 1998 **Bank 1, Sensor 1** **19** (sensor signal) and **46** (ground) **Bank 1, Sensor 2** **18** (sensor signal) and **46** (ground) **Bank 2, Sensor 1** **76** (sensor signal) and **46** (ground) **Bank 2, sensor 2** **77** (sensor signal) and **46** (ground)	Engine warm 0.1 – 0.9 VDC fluctuating

– With engine running, loosen oil filler cap to admit unmeasured intake air and simulate a lean running condition.
  - Oxygen sensor output voltage should drop.

  If voltage signal does not fluctuate or if there is no change when oil filler cap is removed, check for a faulty sensor wire in harness or connector in engine compartment. If no wiring faults can be found, sensor is probably faulty and should be replaced. See **26 Exhaust System**.

Oxygen sensor heater circuit, testing

Check oxygen sensor heater circuit at harness side of oxygen sensor connector.

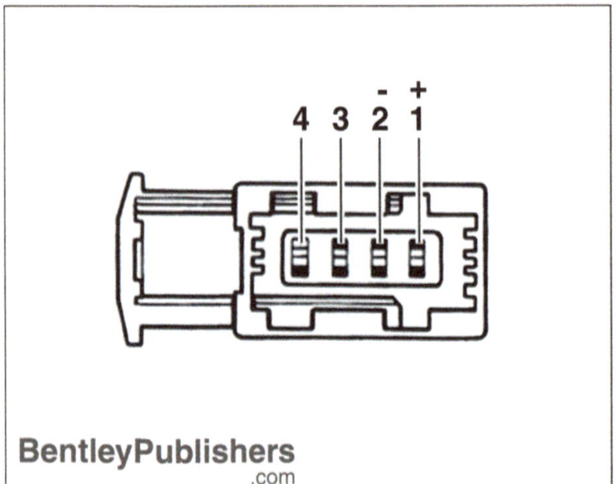

P99824007

Attach voltmeter (set on volts) to terminals 2 and 3.
  - With engine running, battery voltage (12 V) should be present to sensor heater.

– If not, check output at terminal 87b of DME main / fuel pump relay.

– Check oxygen sensor harness ground wire for continuity to ground.

NOTE–

- *Wire colors leading to the sensor plug are different from wire colors leading to the sensor itself. Use wiring diagram to identify wire colors. See* **EWD Electrical Wiring Diagrams**.

Component Replacement

DME main / fuel pump relay, replacing

P99828021

DME main / fuel pump relay (**arrow**) is located in trunk in central electric panel.

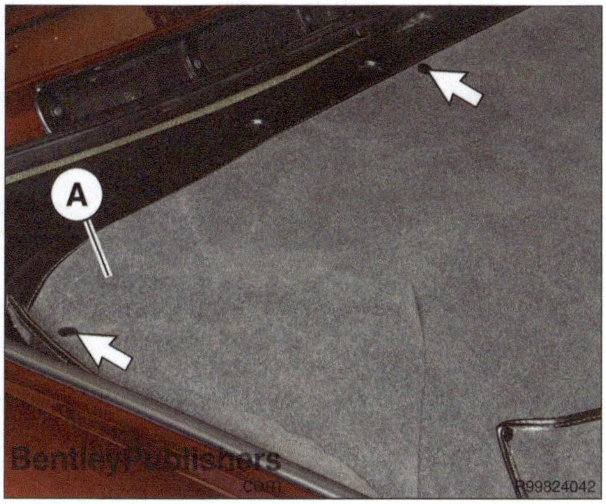

< Working in right side of trunk, disconnect snaps (**arrows**) and peel carpet back (**A**) enough to access central electric panel.

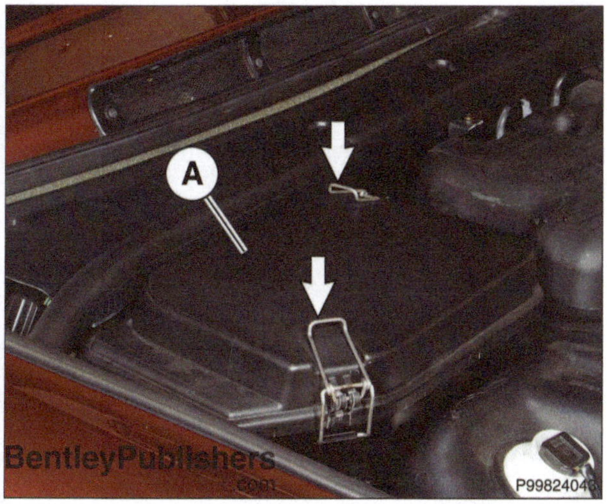

< Open locks (**arrows**), then remove central electric panel cover (**A**).

< Pull DME main / fuel pump relay (**arrow**) out of central electric panel.

– Installation is reverse of removal.

DME engine control module (ECM), replacing

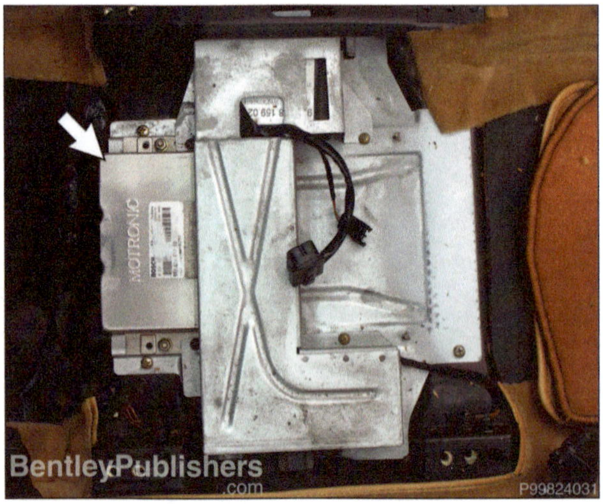

The DME engine control module (**arrow**) is located under the driver seat.

- Disconnect negative (–) battery cable. Wait approximately 1 minute. Prior to disconnecting the battery, read battery disconnection cautions. See **00 Warning and Cautions**.
- Remove driver seat. See **72 Seats**.

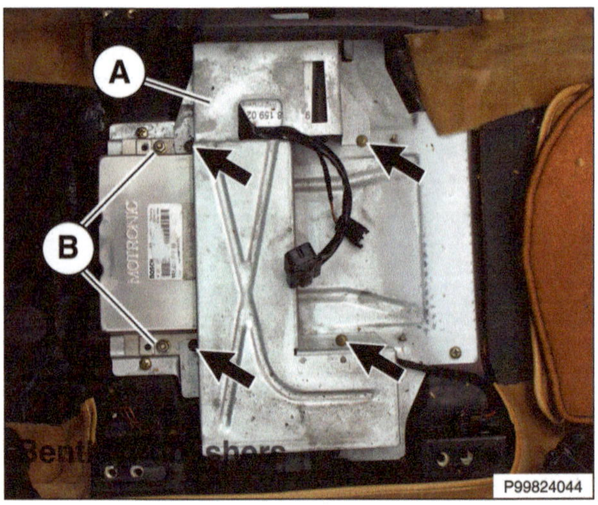

Remove cover plate (**A**) shear bolts (**arrows**), then remove cover plate from vehicle.

- Remove DME fasteners (**B**).
- Disconnect ECM electrical connector and remove DME from vehicle.
- Installation is reverse of removal.

Intake air housing, removing

Procedure shows intake air housing from 1995 model year 993. Other models are similar.

Working at intake air duct, disconnect intake air temperature and mass air flow sensor electrical connectors (**A**).

- Working at side of intake air housing, remove breather hose (**B**).
- Remove intake air housing fastener (**C**), then remove intake air housing from engine.
- Installation is reverse of removal.

Mass air flow sensor, replacing

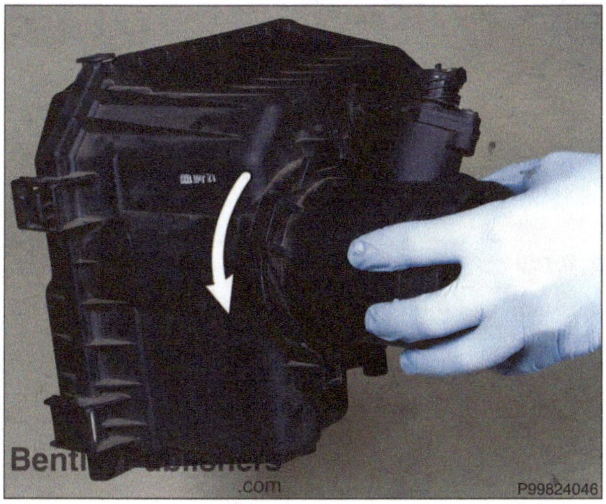

– Remove intake air housing. See **Intake air housing, removing** in this repair group.

– 1995: Loosen hose clamp at MAF sensor duct, then remove sensor from engine.

< From 1996: Rotate mass air flow sensor counterclockwsie (**arrow**), then pull straight out of intake air housing to remove.

– Installation is reverse of removal.

Cylinder head temperature sensor, replacing

– Raise car and support safely.

> ***CAUTION—***
>
> *• Make sure the car is stable and well supported at all times. Use a professional automotive lift or jack stands designed for the purpose. A floor jack is not adequate support.*

– Remove rear blower motor assembly. See **80 Heating**.

< Disconnect white electrical connector on left side of engine (**arrow**).

NOTE—

• Cylinder head temperature sensor electrical connector is white,located behind green knock sensor connector show in photo.

< Working underneath engine remove rubber grommets (**arrows**) at left rear (flywheel side) engine shroud.

> ***WARNING—***
>
> *• Allow engine to cool fully before working near the exhaust system or cylinder heads.*

24

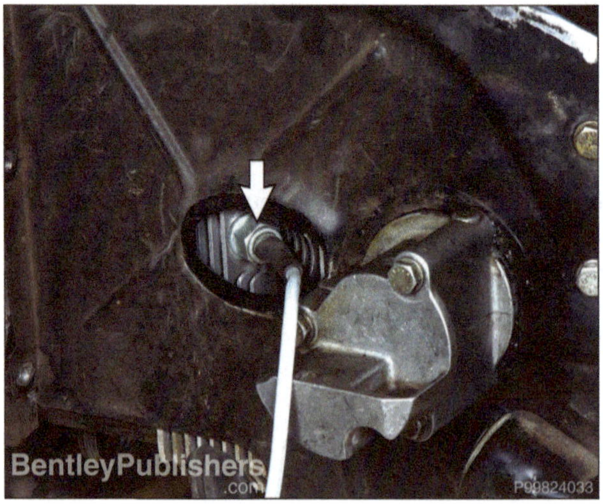

- Using Porsche special tool **9222** or equivalent (a deep socket with a slit cut in the side), remove temperature sensor (**arrow**) from cylinder head 3.
- Pull wire with temperature sensor attached through rubber grommet openings.
- Installation is reverse of removal.
  - Before installing, apply a light coat of Molykote® HTP to the threads.

Tightening torque	
Cylinder head temperature sensor to cylinder head	15 Nm (11 ft-lb)

Idle speed control valve, replacing

- Working at top of intake manifold, disconnect idle speed control valve electrical connector (**arrow**).

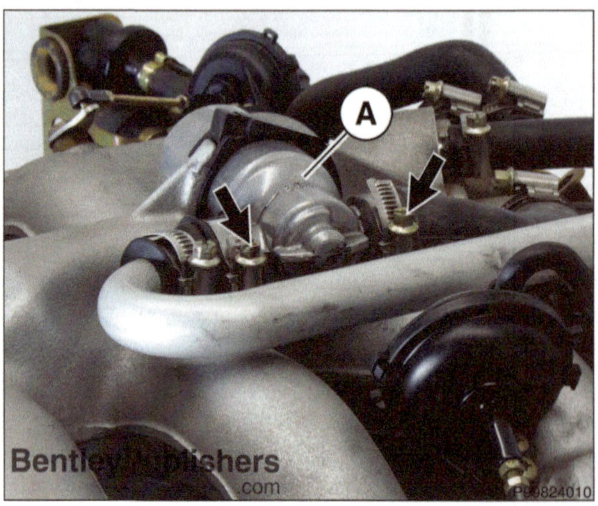

- Loosen hose clamps (**arrows**) then remove hoses from valve.
- Slide idle air control valve (**A**) out of mounting insulator and remove from engine.
- Installation is reverse of removal.

Fuel injectors, replacing

> ***WARNING—***
> - *Gasoline is highly flammable and its vapors are explosive.*
> - *Do not smoke or work on a car near heaters or other fire hazards when diagnosing and repairing fuel system problems.*
> - *Have a fire extinguisher available in case of an emergency.*

Left side injectors, replacing

– Remove rear blower motor assembly. See **80 Heating**.

◄ Working at rear of fuel rail, disconnect fuel line (**A**) and remove vacuum hose (**B**) from fuel pressure regulator.

◄ Working at fuel rail, disconnect fuel injector electrical connectors (**arrows**).

– Remove fuel rail mounting fasteners (**A**).

– Carefully pull up on fuel rail and remove with injectors.

24

- Remove locking clip (**A**) holding injector to fuel rail.
- Pull injector from rail (**arrow**).
- Installation is reverse of removal. Note the following:
  - Replace injector sealing O-rings.
  - Observe proper tightening torques.

Tightening torque	
Fuel rail to intake manifold	7.5 Nm (5.5 ft-lb)

Right side injectors, replacing

- Remove intake air housing. See **Intake air housing, removing** in this repair group.

- Place fender cover over engine hood opening and engine hood latch.

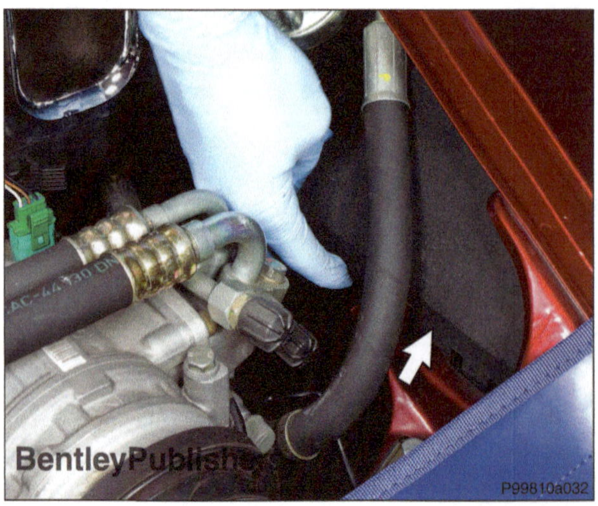

- Working in engine compartment on right side, remove plastic cover (**arrow**) by unclipping bottom and pulling up.

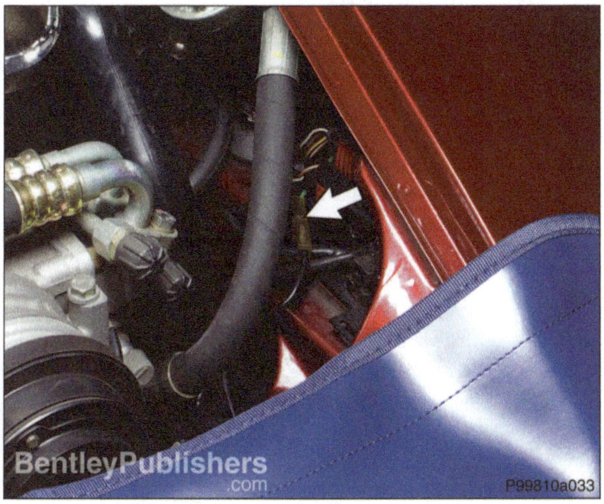

Disconnect A/C compressor electrical connector (**arrow**).

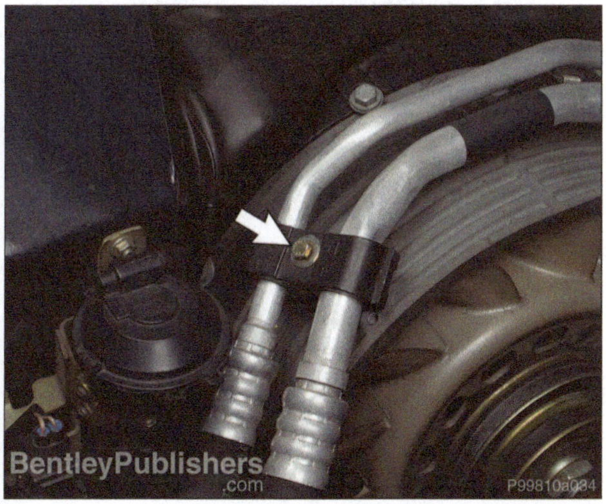

Working at top of cooling fan shroud, remove A/C line mounting bracket fastener (**arrow**).

– Remove drive belts. See **03 Service and Maintenance**.

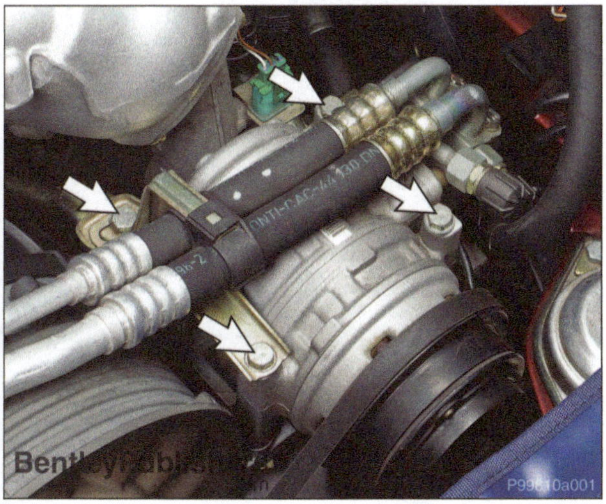

Remove A/C compressor fasteners (**arrows**).

NOTE–

- *Be prepared to catch square threaded lugs from bottom of bracket when bolts are removed.*

– Remove A/C compressor from mounting bracket and lay compressor on edge of engine compartment.

NOTE–

- *Use pad on top of bumper to protect from damage.*
- *Do not disconnect refrigerant lines.*

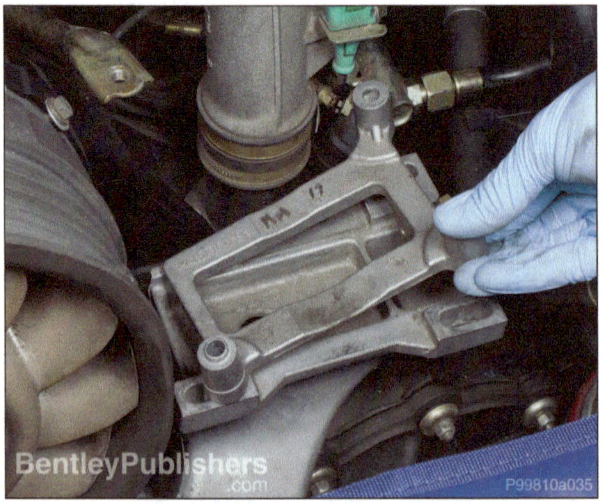

< Remove A/C compressor adjusting bracket from mounting bracket and store in a safe place.

< Working at right side of engine, disconnect fuel line (**arrow**).

NOTE–

- *When separating lines, be prepared to catch dripping fuel.*

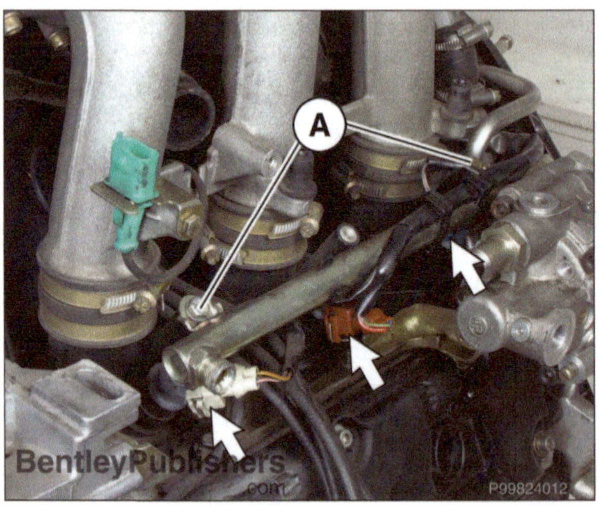

< Working at fuel rail, disconnect fuel injector electrical connectors (**arrows**).

– Remove fuel rail mounting fasteners (**A**).

– Carefully pull up on fuel rail and remove with injectors.

- Remove locking clip (**A**) holding injector to fuel rail.
- Pull injector from rail (**arrow**).
- Installation is reverse of removal. Remember to replace injector sealing O-rings.

Tightening torque	
Fuel rail to intake manifold	7.5 Nm (5.5 ft-lb)

Fuel pressure regulator, replacing

The fuel pressure regulator is located at the rear left of the engine, at the end of the left fuel rail.

> ***WARNING—***
>
> - *Gasoline is highly flammable and its vapors are explosive.*
> - *Do not smoke or work on a car near heaters or other fire hazards when diagnosing and repairing fuel system problems.*
> - *Have a fire extinguisher available in case of an emergency.*

- Remove rear blower motor assembly. See **80 Heating.**

- Working at rear of fuel rail, disconnect fuel line (**A**) and remove vacuum hose (**B**) from fuel pressure regulator.

< Carefully pry up and remove fuel pressure regulator retainer (**arrow**).

< Remove fuel pressure regulator (**arrow**) from fuel rail.

– Remove and discard sealing O-rings.

– Installation is reverse of removal.

Intake manifold, removing and installing

Procedure below shows the intake manifold from a 1997 model. Other models are similar.

– Disconnect negative (–) battery cable and cover battery terminal with insulating material to keep cable from accidentally contacting terminal.

> ***CAUTION—***
>
> *• Prior to disconnecting the battery, read the battery disconnection cautions given in* **00 Warnings and Cautions.**

– Raise vehicle and support safely.

> ***WARNING—***
>
> *• Make sure the car is stable and well supported at all times. A floor jack is not adequate support.*

< Place fender cover over engine hood opening and engine hood latch.

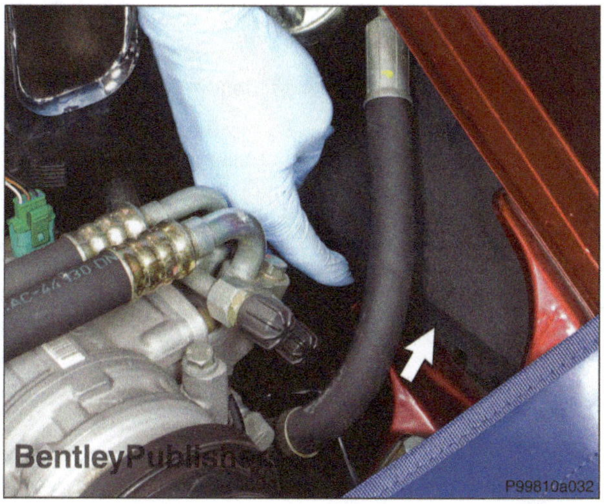

< Working in engine compartment on right side, remove plastic cover (**arrow**) by unclipping bottom and pulling up.

< Disconnect A/C compressor electrical connector (**arrow**).

< Working at top of cooling fan shroud, remove A/C line mounting bracket fastener (**arrow**).

– Remove drive belts. See **03 Service and Maintenance**.

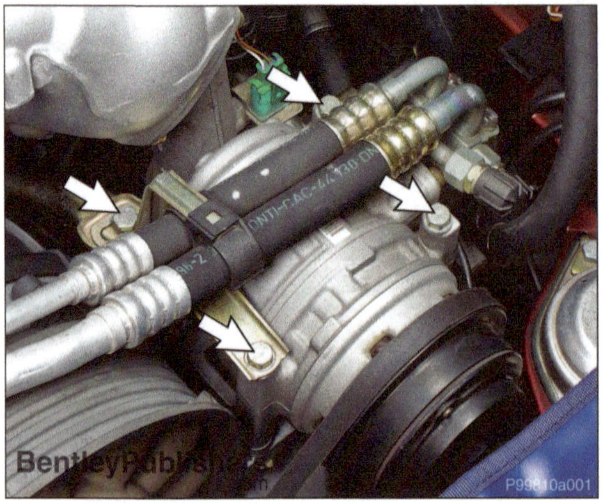

< Remove A/C compressor fasteners (**arrows**).

NOTE–

- *Be prepared to catch threaded lugs from bottom of bracket when bolts are removed.*

– Remove A/C compressor from mounting bracket and lay compressor on edge of engine compartment.

NOTE–

- *Use pad on top of bumper to protect from damage. Do not disconnect refrigerant lines.*

< Remove A/C compressor adjusting bracket from mounting bracket and store in a safe place.

– Remove intake air housing. See **24 Fuel Injection**.

– Remove rear blower motor assembly. See **80 Heating**.

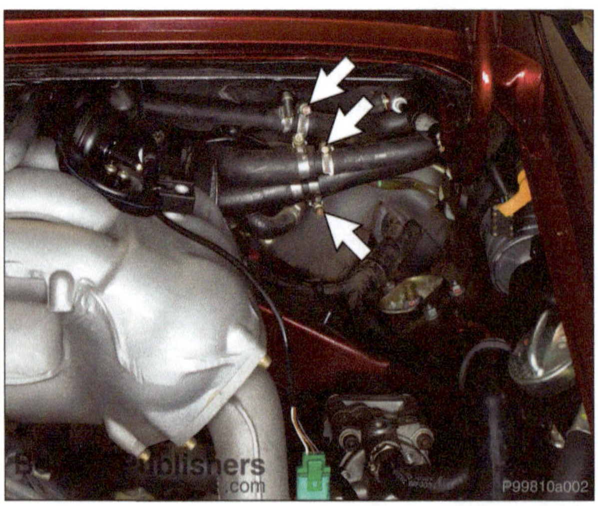

< Working at right side rear of engine, disconnect crankcase breather hoses (**arrows**).

⮜ Working at intake manifold, disconnect idle valve electrical connector (**arrow**).

⮜ Working at right front of intake manifold, disconnect electrical connector (**arrow**).

⮜ Working at right rear of intake manifold, disconnect electrical connectors (**arrows**).

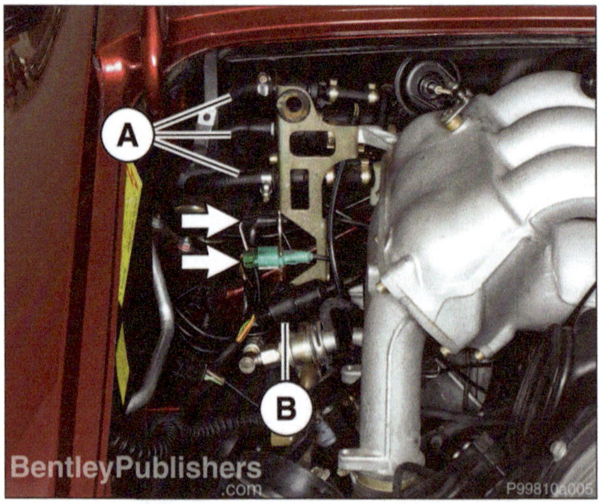

Working at left side of engine compartment:

- Loosen hose clamps and disconnect hoses (**A**).
- Disconnect electrical connectors (**arrows**).
- Open electrical connector housing and disconnect (**B**).

NOTE–

- *Be careful not to lose main engine harness plug seal when disconnecting.*

Working at left rear of intake manifold, disconnect vacuum hose (**arrow**).

Working at right rear of intake manifold, disconnect vacuum hose (**arrow**).

◀ Working at left side of intake manifold between runners, disconnect vacuum solenoid electrical connectors (**arrows**).

◀ Remove vacuum solenoids (**arrows**) from intake manifold.

◀ Working at left rear of intake manifold, disconnect fuel pressure regulator vacuum hose (**arrow**).

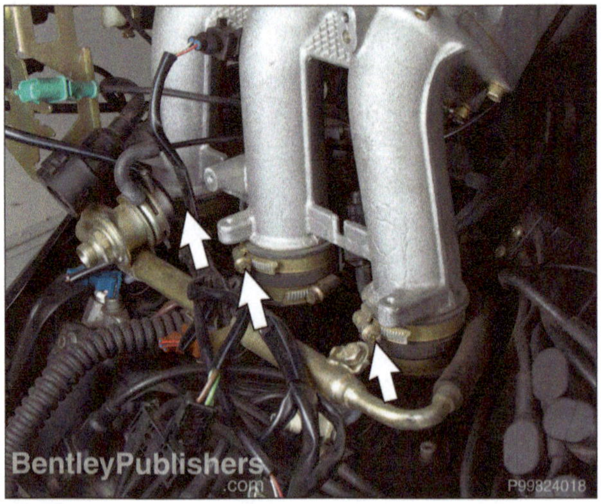

⋖ Working at left side of intake manifold runners, loosen hose clamps (**arrows**).

⋖ Working at right side of intake manifold runners, loosen hose clamps (**arrows**).

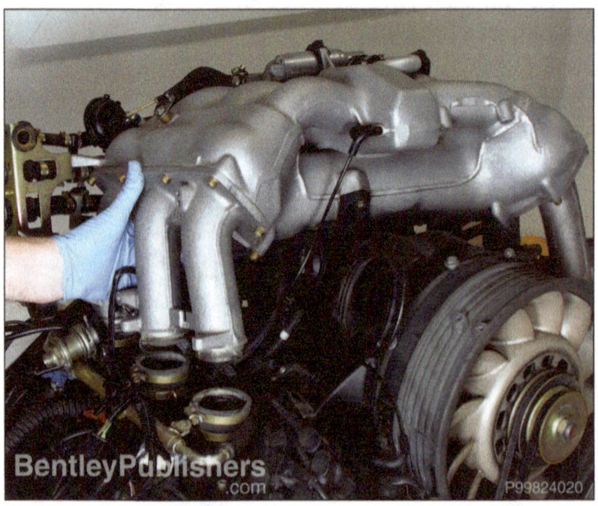

⋖ Lift intake manifold up and out of intake manifold runners to remove. Lift slowly, checking for wires of hoses that may snag.

– Plug intake manifold openings to prevent debris from entering engine.

To replace intake manifold runner gaskets use the following steps.

– Remove fuel rail. See **Fuel injectors, replacing** in this repair group.

⮜ Loosen intake manifold connecting hose clamp (**arrow**), then remove hose (**A**) from engine.

⮜ Remove intake manifold runner fasteners (**arrows**).

– Remove intake manifold runner from engine.

– Remove gasket and discard.

– Installation is reverse of removal.

  • Replace all gaskets and sealing O-rings.

Tightening torques	
Intake manifold runner to engine	10 Nm (7 ft-lb)
Fuel rail to intake manifold	7.5 Nm (5.5 ft-lb)

Throttle housing, replacing

Replacing the throttle housing can be done with the engine installed, however, it can be easier to perform with the intake manifold removed. Photos show throttle housing replacement with intake manifold removed for clarity.

– Remove mass air flow sensor. See **Mass air flow sensor, replacing** in this repair group.

– Disconnect throttle position sensor electrical connector.

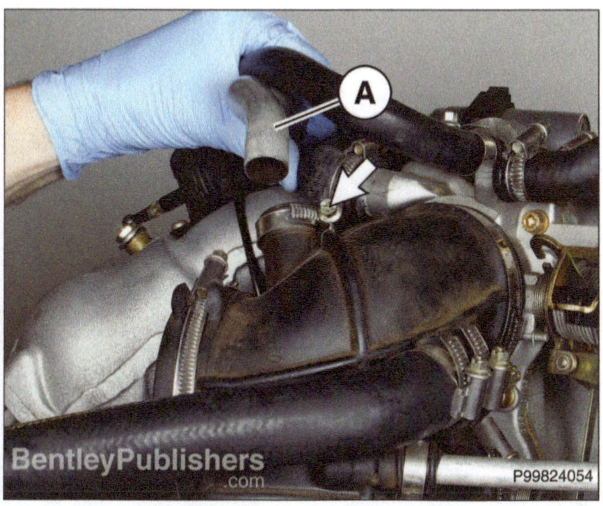

⮜ Loosen hose clamp (**arrow**), then remove intake air pipe (**A**) from throttle housing duct.

Throttle housing, replacing

◄ Loosen hose clamps (**arrows**).

– Remove intake air duct from throttle housing.

◄ Disconnect throttle and cruise control cables (**arrows**).

– Feed throttle cable through bracket (**A**) to remove.

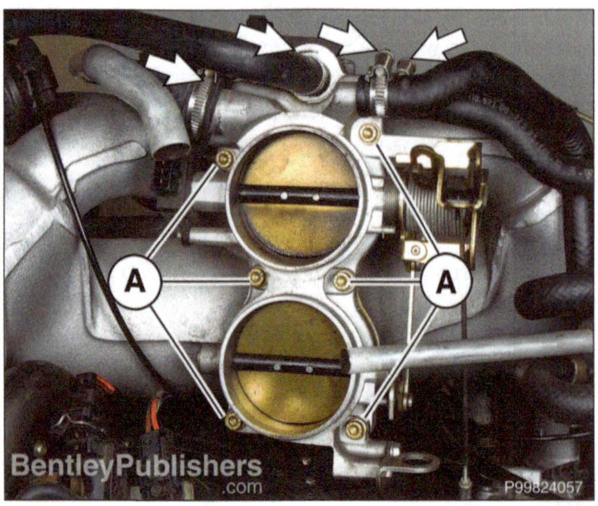

◄ Loosen hose clamps (**arrows**).

– Remove hoses from throttle housing.

– Remove throttle housing fasteners (**A**).

– Remove throttle housing from intake manifold.

– Installation is reverse of removal. Note the following:

- Replace throttle housing gasket.

Varioram actuator, replacing

Right side actuator shown in procedure, left side is similar.

– Remove intake air housing. See **Intake air housing, replacing** in this repair group.

< Disconnect actuator vacuum line (**A**).

– Pull actuator socket (**B**) off ball joint.

– Remove actuator fasteners (**arrows**).

– Remove actuator from intake manifold.

– Installation is reverse of removal.

< To replace center Varioram actuator, remove intake manifold. See **Intake manifold, removing and installing** in this repair group. Actuator is located at bottom rear of intake manifold (**arrow**).

Secondary air injection pump, replacing

– Remove intake manifold. See **Intake manifold, removing and installing** in this repair group.

< Working at right rear of engine, disconnect secondary air injection component electrical connectors (**arrows**).

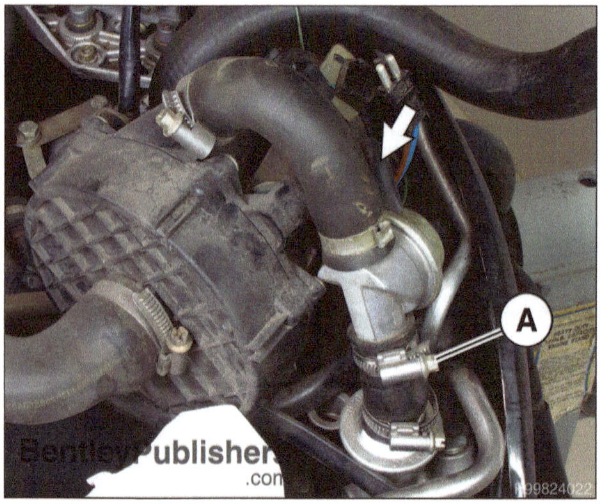

< Working at secondary air injection pump, loosen hose clamp (**A**) and remove vacuum hose (**arrow**).

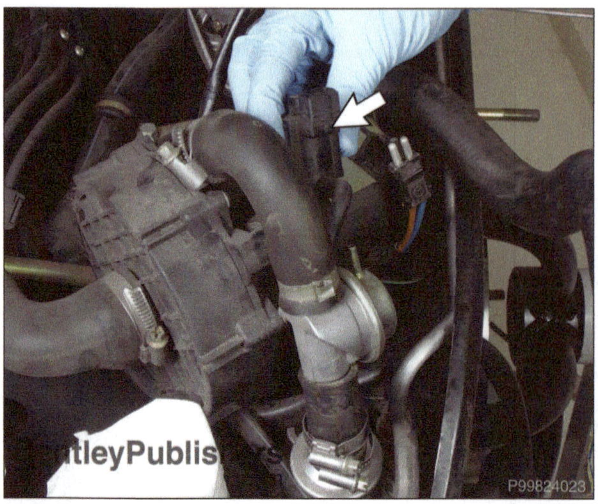

< Remove secondary air injection vacuum control solenoid from pump bracket (**arrow**).

- Remove secondary air injection pump mounting bracket fasteners (**arrows**).
- Remove secondary air injection pump from engine.
  - Hoses and secondary air injection control valve can be serviced once pump is removed.
- Installation is reverse of removal.

Secondary air injection check valve, replacing

- Remove secondary air pump. See **Secondary air injection pump, replacing** in this repair group.

- Loosen hose clamp (**A**) then remove hose (**B**) from check valve.

- Loosen secondary air injection pipe nuts (**arrows**).
- Remove secondary air injection valve manifold fasteners (**A**).
- Remove secondary air injection valve with manifold from engine.
- Installation is reverse of removal.

26 Exhaust System

General

The exhaust system is designed to be maintenance free, but regular inspection is warranted due to harsh operating conditions.

This section covers maintenance, troubleshooting, and repair of the exhaust system. For information on testing the oxygen sensors, see **24 Fuel Injection**.

> ***WARNING—***
> - *To avoid injury, wear eye protection and heavy gloves when working on rusty parts of an old exhaust system.*
> - *Do not use flammable workshop chemicals near a hot catalytic converter. They may ignite.*

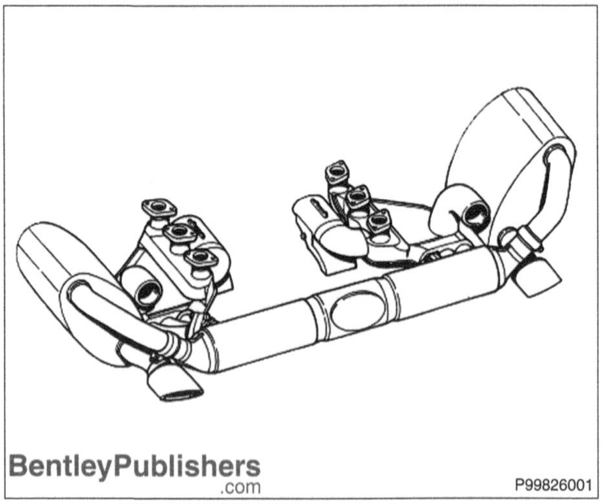

Exhaust system

The basic exhaust system components are the exhaust headers that are encased within the heat exchangers, a cross-over pipe, a catalytic converter assembly and a muffler.

Separate exhaust headers/heat exchangers are mounted to each bank of three cylinders. Exhaust gases are channelled from each exhaust port to a single collector for each cylinder bank. The exhaust headers from each bank of cylinders are housed in a heat exchanger that supplies heated air for passenger compartment heating and windshield defrosting.

A cross-over (connector) pipe brings the right side exhaust gases across to the left side. Both the cross-over pipe and left exhaust header connect to the catalytic converter assembly. The muffler connects directly to the rear of the catalytic converter assembly to complete the exhaust system.

Exhaust manifold components

1. Exhaust manifold
2. Flange gasket
3. Joining adapter
4. Clamp
5. Catalytic converter

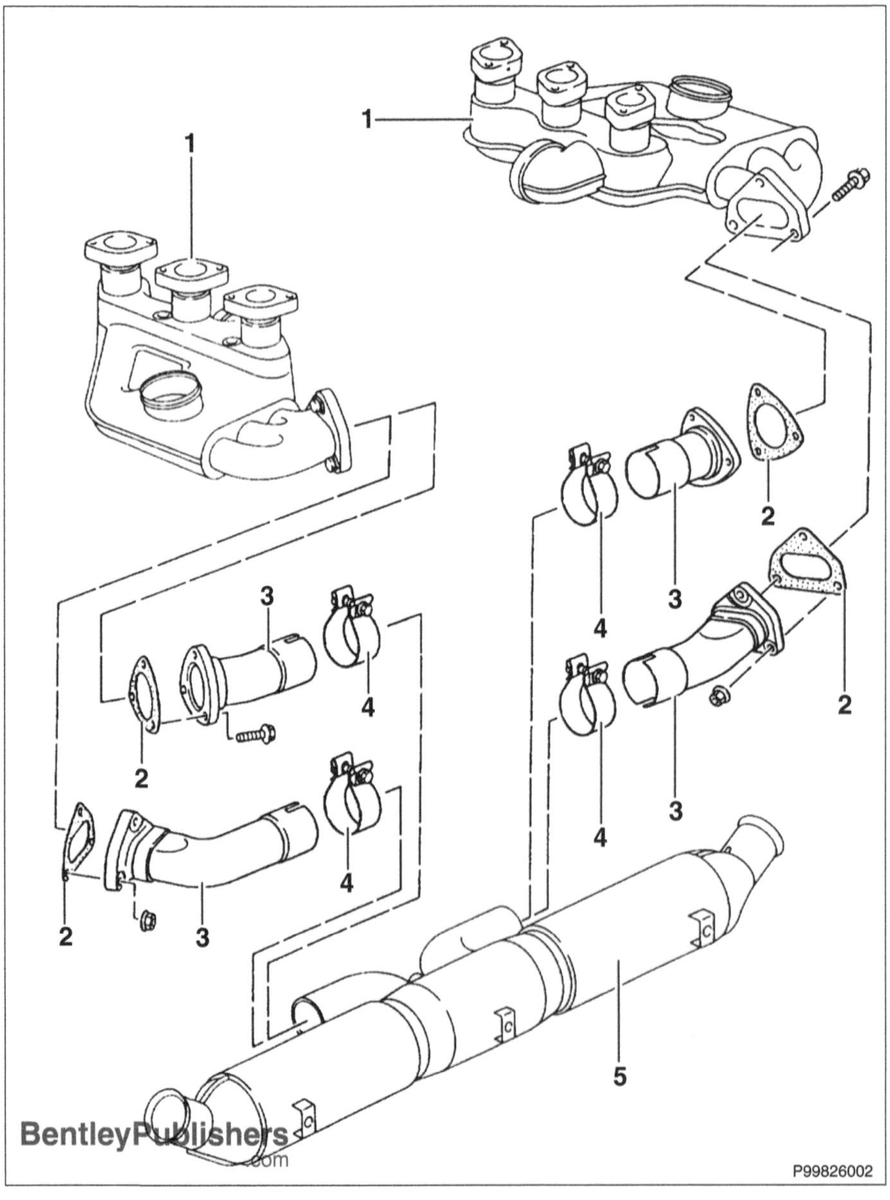

Exhaust system components

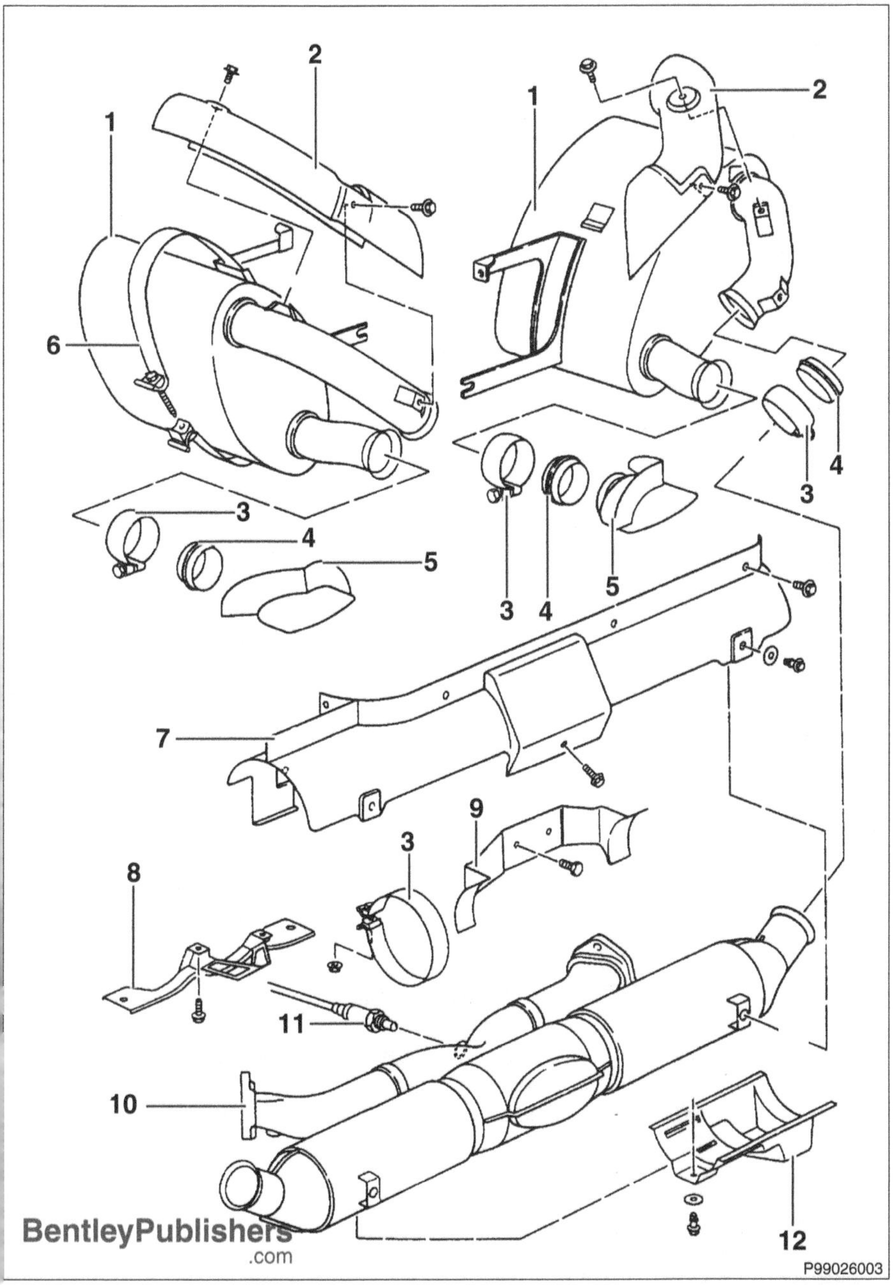

1. Muffler
2. Heat shield
3. Clamp
4. Sealing donut
5. Tail pipe
6. Muffler strap
7. Heat shield
8. Support bracket
9. Support bracket
10. Catalytic converter assembly
11. Oxygen sensor
    - 1995: single sensor (as shown)
    - 1996 - 1998: 2 sensors per catalytic converter, total of 4
12. Heat shield

Heater control components

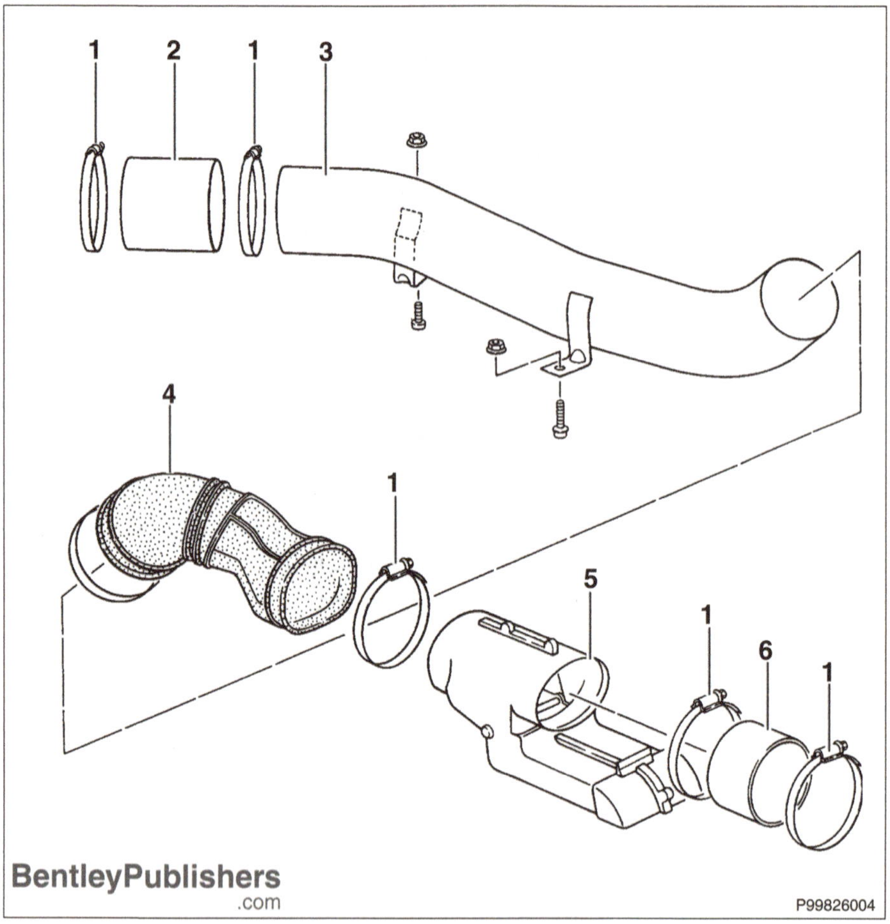

1. Clamp
2. Hose
   - 80 mm
3. Heating tube
4. Connecting hose
5. Heat control box
6. Hose
   - 70 mm

Component Replacement

Use of non-original exhaust system components can change system back pressure and may affect fuel mixture. It may be necessary to check and adjust the fuel injection system after installing a non-original muffler.

NOTE—

- *Use new gaskets and mounting hardware whenever removing and installing exhaust system parts.*
- *When mounting exhaust system components, tighten the mounting nuts in two stages.*

Muffler, replacing

– Raise car and support safely.

> ***WARNING—***
>
> - *Make sure the car is stable and well supported at all times. Use a professional automotive lift or jack stands designed for the purpose. A floor jack is not adequate support.*

◄ Remove exhaust tip fastener (**A**).

– Spread clamp (**B**) and remove tip from exhaust.

◀ Working at catalytic converter, loosen exhaust clamp (**A**).

– Spread clamp (**B**) to relieve tension.

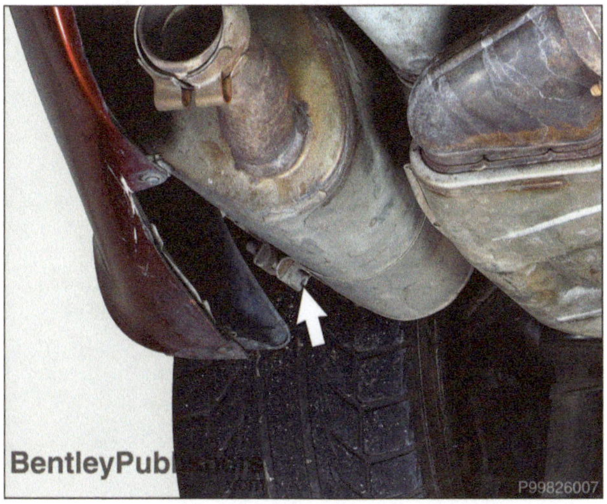

◀ Working at muffler, remove muffler clamp fastener (**arrow**).

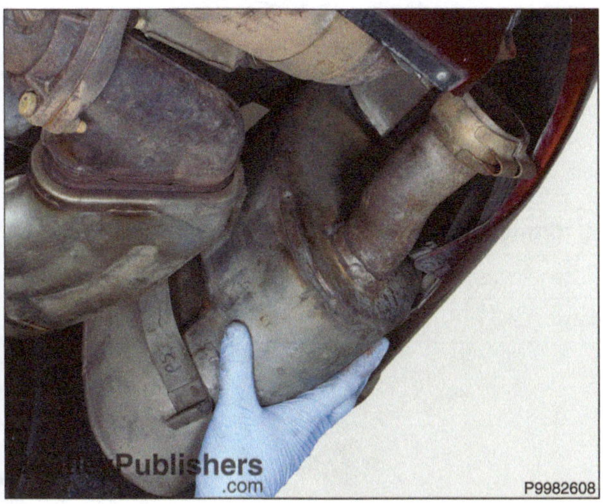

◀ Lift muffler up to disengage from upper mounting tab until muffler drops down.

– Remove muffler from vehicle.

– Installation is reverse of removal. Note the following:

- Replace all hardware and gaskets.
- Observe torque specifications when tightening hardware.

Tightening torques	
Exhaust strap bolt	20 Nm (15 ft-lb)
Exhaust clamp bolt	25 Nm (18 ft-lb)

Catalytic converter assembly, replacing

– Raise car and support safely.

> ***WARNING—***
> *• Make sure the car is stable and well supported at all times. Use a professional automotive lift or jack stands designed for the purpose. A floor jack is not adequate support.*

– Remove muffler. See **Muffler, replacing** in this repair group.

– Remove oxygen sensor(s). See **Oxygen sensors, replacing** in this repair group.

NOTE–

• If catalytic converter assembly is not being replaced with new part, oxygen sensor(s) can remain installed in exhaust system when removing.

< Remove exhaust flange fasteners (**arrows**) from both sides.

– Remove catalytic converter heat shield fasteners.

< Working at catalytic converter, loosen muffler clamp (**arrow**).

– Open clamps to aid removal of catalytic converter.

– Lower catalytic converter from engine.

– Installation is reverse of removal. Note the following:

• Replace all hardware and gaskets.

Tightening torques	
Muffler to catalytic converter	25 Nm (18 ft-lb)
Exhaust clamp bolt	25 Nm (18 ft-lb)

Exhaust manifold, removing

– Raise car and support safely.

> ***WARNING—***
>
> • *Make sure the car is stable and well supported at all times. Use a professional automotive lift or jack stands designed for the purpose. A floor jack is not adequate support.*

– Remove catalytic converter. See **Catalytic converter, replacing** in this repair group.

– Remove heat exchange flap. See **Heat exchange components, replacing** in this repair group.

◄ Working at exhaust manifold, remove fasteners (**arrows**).

– Remove exhaust manifold from engine.

– Installation is reverse of removal. Note the following:

• Replace all hardware and gaskets.

Tightening torque	
Exhaust manifold to cylinder head	25 Nm (18 ft-lb)

Oxygen sensors, replacing

On 1995 modes (DME M 2.10), a single oxygen sensor is installed in the center of the catalytic converter assembly.

On 1996 - 1998 models (DME M 5.2), there are a total of four oxygen sensors, two per catalytic converter, as part of OBD II regulations. The procedure given here describes the four sensor setup.

◄ **1996 - 1998**: Oxygen sensors are located at the front (**A**) and rear (**B**) of the catalytic converters.

> ***CAUTION—***
>
> • *Handle oxygen sensors with care. Oxygen sensors are easily damaged.*
>
> • *To avoid possible engine damage, do not swap oxygen sensor connectors. Mark before disconnecting.*

Oxygen sensors electrical connectors, right side

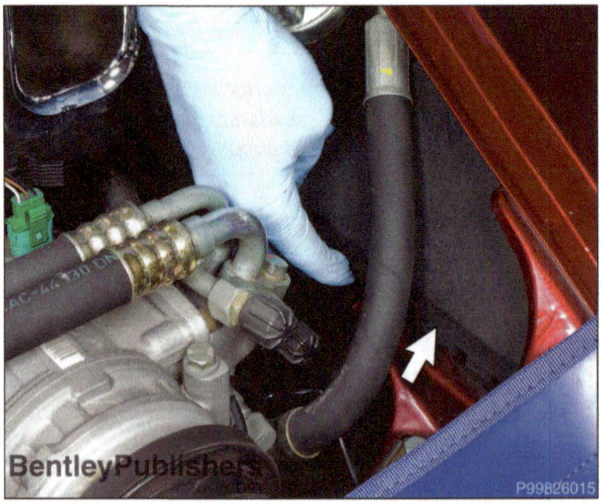

< Working in engine compartment on right side, remove plastic cover (**arrow**) by unclipping bottom and pulling up.

< Working in engine compartment on right side disconnect oxygen sensor electrical connectors by opening sliding locks (**arrows**) to separate.

Oxygen sensors electrical connectors, left side

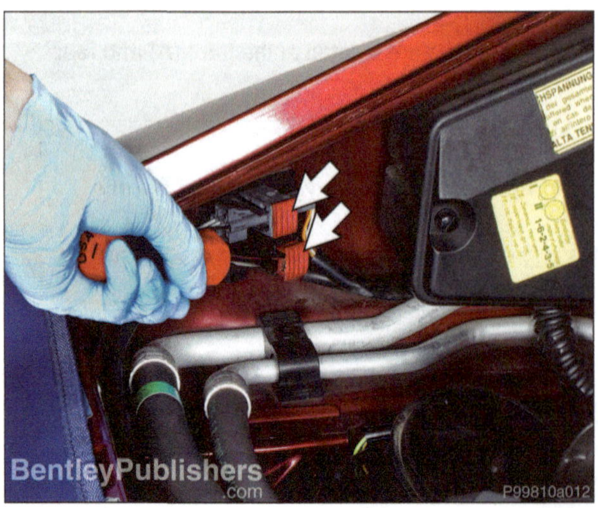

< Working in engine compartment on left side disconnect oxygen sensor electrical connectors by opening sliding locks (**arrows**) to separate.

- Push wire with rubber grommet through engine compartment bulkhead.
- Raise car and support safely.

> ***WARNING—***
>
> • *Make sure the car is stable and well supported at all times. Use a professional automotive lift or jack stands. A floor jack is not adequate support.*

Pre-catalyst oxygen sensors

- Using an oxygen sensor socket, remove oxygen sensor (**arrow**) from exhaust.
- Installation is reverse of removal. Note the following:
  - Apply a small amount of anti-sieze compound to sensor threads.
  - Make sure electrical connectors are routed as before and connected properly.

> ***CAUTION—***
> - *Do not overuse anti-sieze compound. Do not contaminate tip of sensor with anti-sieze compound or lubricants.*

Tightening torque	
Oxygen sensor to exhaust	47 Nm (35 ft-lb)

Post-catalyst oxygen sensors

- Using an oxygen sensor socket, remove oxygen sensor (**arrow**) from exhaust.
- Installation is reverse of removal. Note the following:
  - Apply a small amount of anti-sieze compound to sensor threads.
  - Make sure electrical connectors are routed as before and connected properly.

> ***CAUTION—***
> - *Do not overuse anti-sieze compound. Do not contaminate tip of sensor with anti-sieze compound or lubricants.*

Tightening torque	
Oxygen sensor to exhaust	47 Nm (35 ft-lb)

Heat exchange components, replacing

- Raise car and support safely.

> ***WARNING—***
> - *Make sure the car is stable and well supported at all times. Use a professional automotive lift or jack stands. A floor jack is not adequate support.*

Heat control box

Inspect rubber flap in heat control box for damage.

- Working under vehicle, loosen heat control box hose clamps (**arrows**).

26

< Pull rear hose (**A**) off heat control box, then slide heat control box out of front hose and remove.

– Installation is reverse of removal. Note the following:
  - Replace worn clamps and silicone hose couplers.

Heat exchange pipe

– Remove heat control box.

< Working at heat exchange pipes, loosen hose clamps (**arrows**).

– Remove heat exchange pipe mounting fasteners (**A**).

– Remove pipe (**B**) from vehicle.

– Installation is reverse of removal. Note the following:
  - Replace worn clamps and silicone hose couplers.

27 Battery, Starter, Alternator

TABLES

General

This section covers starter and alternator replacement and troubleshooting as well as battery load testing and charging. For additional electrical troubleshooting information that may apply to these components, see

- **03 Service and Maintenance** for battery replacement.
- **97 Fuses, Relays, Component Locations**.
- **EWD Electrical Wiring Diagrams**.

The battery, starter, and alternator are the three major components of the electrical system. The battery supplies electrical power when the engine is not running. The starter is an electric motor that rotates the engine via a flywheel when the key is in the start position. The alternator converts some of the engine's mechanical energy into electric energy whenever the engine is running.

CAUTION—

- *The starter and alternator are not protected by fuses. To prevent accidental shorts that might blow a fuse or damage wires and electrical components, the negative (–) battery cable should always be disconnected before working on the electrical system.*

Voltage and polarity

The cars covered by this manual have a 12-volt, direct current (DC), negative ground electrical system. The voltage regulator maintains the voltage in the system at approximately the 12-volt rating of the battery. All circuits are grounded by direct or indirect connection to the negative (–) terminal of the battery.

Battery

The battery is located in the trunk on left side below trunk carpet.

The battery stores electricity chemically. Lead plates inside of the battery interact with a mixture of sulfuric acid and water, called electrolyte, to create a flow of current.

Starter

The starter and its attached solenoid are mounted on the right side of the transmission just above the right drive axle.

Charging system

The charging system consists of a belt-driven alternator and voltage regulator. The voltage regulator is mounted on the back side of the alternator and also serves as the alternator brush holder.

Maintenance

The following electrical system maintenance is described in **03 Service and Maintenance**.

- Checking battery level.
- Checking alternator belt tension.
- Changing alternator V-belt.

TROUBLESHOOTING

The cause of a dead battery or an inoperative starter are often related. Before troubleshooting any battery-related problems, check that all ground connections are tight and free from corrosion.

Table a lists symptoms of trouble with the battery, starter and charging system, their probable causes, and suggested corrective actions. Most corrective actions are described later in this group.

ble a. Battery, starter and charging system troubleshooting

ymptom	Probable Cause	Corrective Action
ngine turns slowly or not at all or olenoid clicks when starter is operated	a. Battery cables loose, dirty, or corroded b. Battery discharged c. Body ground strap loose, dirty, or corroded d. Poor connection or no voltage at starter motor terminal 30 e. Starter motor or solenoid faulty	a. Clean or replace cables. b. Charge battery, test and replace if necessary. c. Inspect ground strap. Clean, tighten and replace if necessary. d. Check connection and test for voltage at starter. e. Test starter.
attery will not stay charged nore than a few days)	a. Short circuit draining battery b. Alternator V-belt loose or damaged c. Battery faulty d. Battery cables loose, dirty, or corroded e. Alternator or voltage regulator faulty	a. Test for static current draw. b. Inspect alternator V-belt. c. Load test battery and replace if necessary. d. Clean or replace cables. e. Test alternator output and regulator.
attery losing water	a. Battery being overcharged	a. Test voltage regulator.
ghts dim or light intensity varying with ngine speed	a. Alternator V-belt loose or damaged b. Alternator or voltage regulator faulty c. Body ground strap loose, dirty, or corroded	a. Inspect alternator V-belt. b. Test alternator and voltage regulator. Replace as required. c. Inspect ground strap. Clean, tighten, and replace if necessary.

Grounding points

The battery to chassis ground cable connection (**arrow**) is located in left of trunk behind battery.

Engine / transmission to body ground strap connection (**arrow**) is located on right side of transmission near starter motor.

Battery

Electrolyte contains sulfuric acid and can cause severe burns.

> ***WARNING—***
>
> - *Wear goggles, rubber gloves, and a rubber apron when working with battery electrolyte.*
> - *If electrolyte is spilled on your skin or clothing, flush the area at once with large quantities of water.*
> - *If electrolyte gets into your eyes, bathe them with large quantities of clean water for several minutes and call a physician.*
> - *Batteries generate explosive gasses. Keep sparks and open flame away. Do not smoke.*

Replace a battery if the casing is cracked or leaking. If electrolyte is spilled, clean the area with a solution of baking soda and water.

Low-maintenance batteries have been supplied as original equipment in the cars covered by this manual. These batteries have vented removable filler caps. The battery requires distilled water added periodically to maintain electrolyte level.
The six-cell, 12-volt lead-acid battery capacity is rated by ampere/ hours (Ah) and cold cranking amps (CCA). The Ah rating is determined by the average amount of current the battery can deliver over time without dropping below a specified voltage. The CCA is determined by the battery's ability to deliver starting current at 0°F (–18°C).

Temperature affects the efficiency of the battery. The current-producing capacity of a battery at 5°F (–15°C) is only half its capacity at 68°F (20°C), and partly-discharged batteries can freeze due to the higher proportion of water in the electrolyte. A frozen battery produces no current, but can usually be restored when thawed.

For more information on battery maintenance, see **03 Service and Maintenance**.

Battery testing

Battery testing determines the state of battery charge. The most common methods are open-circuit and load voltage testing. Batteries that have filler caps can also be tested by checking the specific gravity of the electrolyte. The specific gravity test checks the amount of acid in the electrolyte as an indication of battery charge. Inexpensive gravity testers are available at most auto supply stores.

Open-circuit voltage test

An open-circuit voltage test checks battery voltage by connecting an accurate digital voltmeter to the battery posts. Before making any open circuit voltage test on the battery, first load the battery with 15 amps for one minute, for example by turning on the headlights without the engine running. Open-circuit voltage levels and their corresponding percentages of charge are shown in **Table b**.

> ***WARNING—***
>
> - *Never operate the engine with the battery disconnected.*
> - *Prior to disconnecting the battery, read the battery disconnection cautions in this repair group.*

Table b. Open circuit voltage and battery charge	
Open-circuit voltage	**State of charge**
12.6 volts or more	Fully charged
12.4 volts	75% charged
12.2 volts	50% charged
12.0 volts	25% charged
11.7 volts	Fully discharged

The battery is in satisfactory condition if the open-circuit voltage is at least 12.4 volts. If the open-circuit voltage is at this level or above, but the battery still lacks power for starting, perform a load voltage test to determine the battery's service condition. If the open-circuit voltage is below 12.4 volts, recharge the battery. If the battery cannot be recharged to at least 75%, it should be replaced.

Battery load voltage test

A battery load tester is required for a load voltage test. The test is made by applying a high resistive load to the battery terminals and then measuring battery voltage. The battery should be fully charged for the most accurate results. The battery cables must be disconnected before making the test. If the voltage is below that listed in **Table c**, the battery should be replaced.

> ***WARNING—***
>
> - *Always wear protective goggles and clothing when performing a load test.*

Table c. Battery load test–minimum voltage (apply 200 amp load for 15 seconds)	
Ambient temperature	**Voltage**
80°F (27°C)	9.6 V
60°F (16°C	9.5 V
40°F (4°C)	9.3 V
20°F (–7°C)	8.9 V
0°F (-18°C)	8.5 V

Static current draw test

If the battery discharges over time, there may be a constant drain or current draw on the battery. A small static drain on the battery is normal, but a large drain will cause the battery to quickly discharge. Make a static current draw test the first step when experiencing battery discharge.

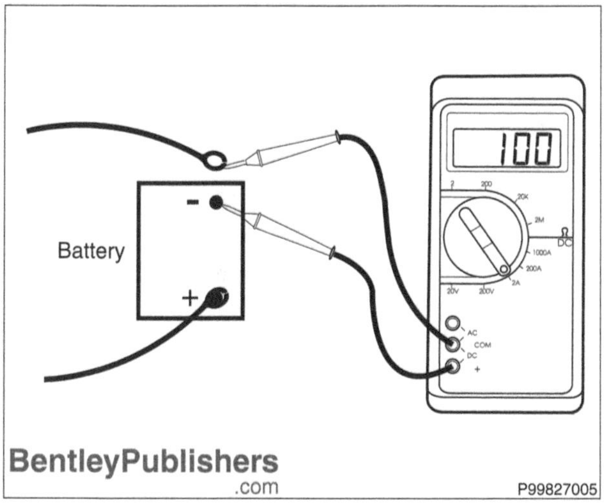

- Make sure ignition and all electrical accessories are switched off.
- Disconnect battery negative (–) cable.

< Connect a digital ammeter between negative battery post and negative battery cable to measure current. Wait at least one minute to get an accurate reading.

A range of about 0 to 100 milliampere is normal, depending on the number of accessories that need constant power. A current of 400 milliampere (0.4 amp) or more may indicate a problem.

- To determine the circuit or component causing the problem, remove one fuse at a time until the current drops to a normal range.

Battery charging

Discharged batteries can be recharged using a battery charger, but a battery can never be charged to a voltage in excess of that which it is capable of producing electro-chemically. Prolonged charging causes gassing that will evaporate the electrolyte to a level that can damage the battery.

Always read and follow the instructions provided by the battery charger's manufactures. Do not use a charger if the instructions are not available.

STARTING SYSTEM

When the ignition key is in the start position, battery power energizes a solenoid switch on the starter motor. The solenoid then engages the starter drive pinion with the ring gear on the engine flywheel/clutch assembly. To minimize wear and stress on the drive pinion and ring gear, the solenoid does not switch current to the starter until the drive pinion is fully engaged. Also the drive pinion has an overrunning clutch to prevent the starter from being driven by the engine.

Starting system troubleshooting

The battery and its cables should be in good condition when troubleshooting the starter. If in doubt about the battery, test as described earlier. Starter efficiency is affected by engine oil viscosity. This is especially true in cold weather.

Table d lists symptoms of trouble with the starting system, their probable causes, and suggested corrective actions, most described later in this group.

See **EWD Electrical Wiring Diagrams** for a wiring schematic of the starter system that can be used as a guide for troubleshooting.

Table d. Starting system troubleshooting		
Symptom	**Probable Cause**	**Corrective Action**
Starter does not operate when ignition switch is turned to start	a. Ignition switch or wire leading from ignition switch to solenoid faulty (less than 7.5 volts to solenoid switch) b. Solenoid switch faulty (less than 7.5 volts to starter motor) c. Starter motor faulty	a. Test for voltage at terminal 50 of solenoid switch with ignition switch at start. If not at least 7.5 volts, test for voltage at terminal 50 of ignition switch with switch at start. Replace ignition switch or eliminate open circuit between ignition switch and solenoid switch. b. Test for voltage at field-winding connector with ignition at start. If not at least 7.5 volts, replace solenoid. c. Test for voltage at field-winding connector with ignition at start. If at least 7.5 volts, repair or replace starter motor.
Starter turns slowly or fails to turn engine	a. Dirty, loose, or corroded starter connections b. Dirty, loose, or corroded ground strap between engine / transmission an body c. Starter worn or faulty	a. Remove, clean, and tighten connections. b. Remove and clean or replace ground strap. c. Repair or replace starter.
Starter makes unusual noises, turns erratically, or fails to turn	a. Drive pinion defective b. Flywheel damaged	a. Repair or replace starter motor. b. Replace flywheel.
Starter operates, but does not turn engine	a. Starter drive pinion or shaft faulty b. Solenoid switch mechanism faulty	a. Repair or replace starter. b. Replace starter solenoid switch.

Starter, removing and installing

– Disconnect negative (–) battery cable and cover battery terminal with insulating material to keep cable from accidentally contacting terminal.

– Raise vehicle and support safely.

> ***CAUTION—***
>
> • *Make sure car is stable and well supported at all times. Use a professional automotive lift or jack stands. A floor jack is not adequate support.*

– Remove engine and transmission protective panels. See **03 Service and Maintenance**.

◄ Working at heat exchange pipes, loosen hose clamps (**arrows**).

– Remove heat exchange pipe mounting fasteners (**A**).

– Remove pipe (**B**) from vehicle.

– Automatic transmission: Place jack under transmission.

– Automatic transmission: Loosen transmission mount fasteners then lower transmission 10 mm (0.40 in)

◄ Remove right rear CV joint (drive axle) mounting fasteners (**arrows**). See **42 Rear Suspension**.

– Detach CV joint from transmission drive flange.

NOTE—

- *Cover CV joints to protect from damage and secure aside using stiff wire.*

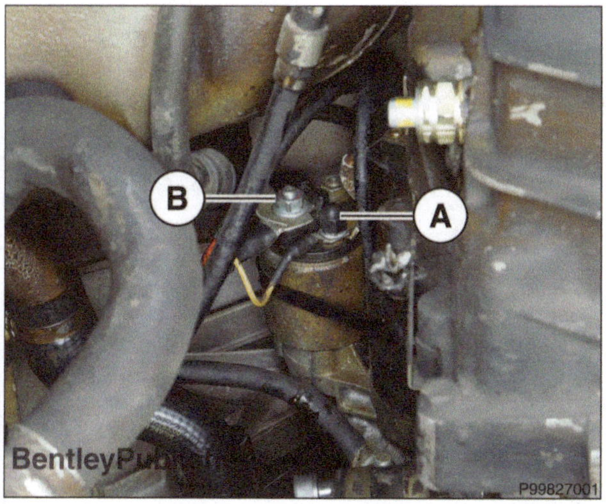

◄ Working at starter, remove plastic nut cover (**A**) then remove nut beneath it.

– Remove terminal **50** wire from starter.

– Remove nut (**B**).

– Remove terminal **30** wire from stater.

◄ Working at starter motor, remove ground strap fastener (**arrow**).

– Remove ground strap from starter.

– Next working at top of starter motor, remove starter fastener.

– Remove starter from engine.

– Installation is reverse of removal.

Tightening torque	
Starter to transmission	40 Nm (30 ft-lb)
Drive axle to transmission flange	81 Nm (60 ft-lb)
Transmission mount to body	46 Nm (34 ft-lb)

27

Charging System

The charging system provides the current necessary to keep the battery charged and to operate the car's electrical accessories. The system includes an alternator driven from the engine crankshaft by a V-belt to generate the charging current. A voltage regulator controls the charging current.

To prevent damage to the alternator or regulator when making tests or repairs, make all connections with negative (–) to negative, and positive (+) to positive unless directed otherwise. Even momentary contact with a conductor of the wrong polarity can damage the alternator's diodes. Make certain that the battery negative (–) cable is securely fastened, and that the cable to terminal **30** on the starter is connected to the battery's positive (+) terminal.

Rated output of the alternator when measured at the battery with the engine running can be 10 to 15 amps less than specified. The difference is accounted for by electrical consumers and is a normal condition.

WARNING—

- *Disconnect the battery negative (–) cable before disconnecting any wires from the alternator. Battery voltage is wired directly to the alternator without fuse protection. Reconnect the battery cable after all wires have been safely disconnected and insulated.*

CAUTION—

- *Never operate the engine with the battery disconnected.*
- *Never operate the alternator with its output terminal (B+ or 30) disconnected and the other terminals connected.*
- *Never short, bridge, or ground any terminals of the charging system except as specifically described in* **Charging system troubleshooting** *in this repair group.*

The voltage regulator (**arrow**) is mounted to the back of the alternator housing. Voltage regulators and alternators are available as replacement parts from an authorized Porsche dealer.

Charging system troubleshooting

Charging system trouble is indicated by an illuminated alternator warning light (battery warning light) on the instrument panel, or by an under or overcharged battery.

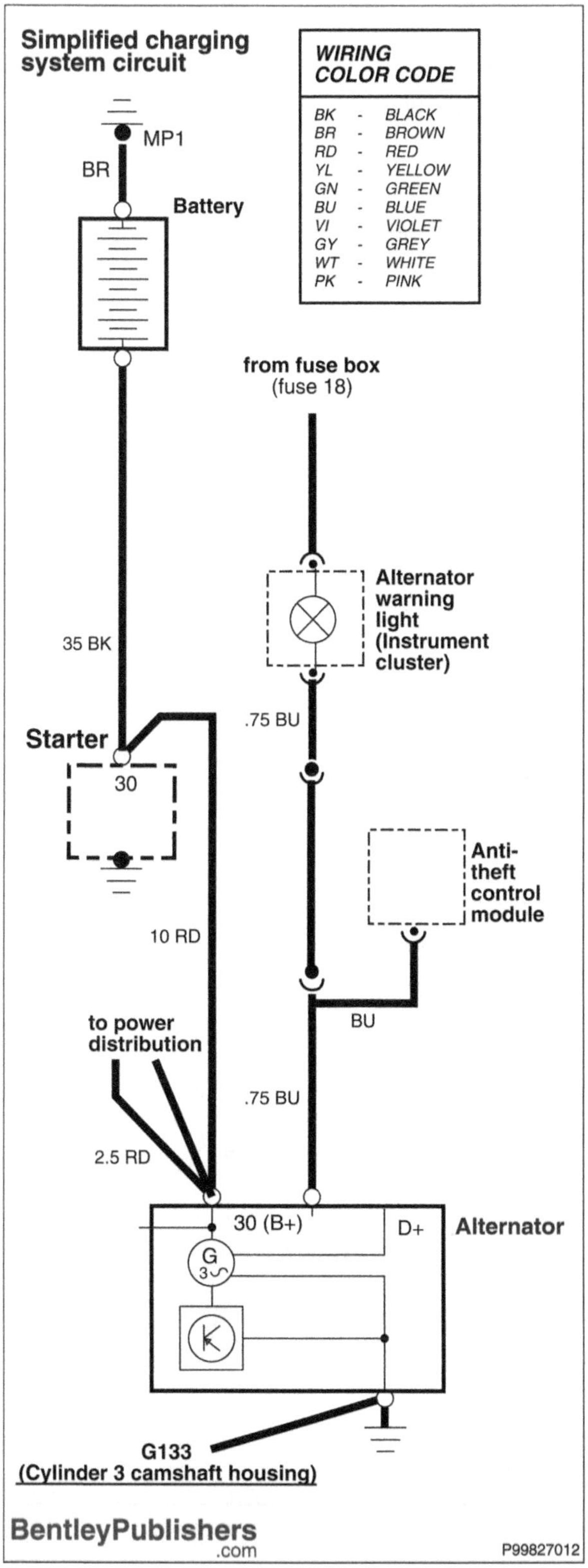

The alternator generates electrical current by electrical induction. That is, a magnetic field is placed in motion to induce a current in a stationary coil. When the engine is running and the alternator is spinning, part of the current it produces energizes its electromagnetic field. When starting, some other current must be provided to initially energize the field and begin the current generating process. This current is provided by the battery through the alternator warning light in the instrument panel.

As soon as the alternator output equals battery voltage, the warning light goes out. Normally, the warning light should be off when the ignition is off and the engine is stopped. The light should only come on when the ignition is turned on (current to the alternator rotor) and go out again when the engine is started and the battery is being charged.

An undercharged battery is usually associated with starting trouble. Again, make sure that the battery is in good condition and capable of accepting a full charge before looking at the charging system.

Causes of an undercharged battery are:

- The simultaneous use of many electrical accessories for long periods of time.
- Leaving accessories or lights in operation with the engine stopped.
- Frequent long periods of starter usage.
- Frequent short-trip driving.
- Improper alternator V-belt condition or tension.
- Broken or frayed charging system wiring.
- Corroded connectors at the D+ and B+ terminals of the alternator.
- Worn, corroded, or loose battery cable connections at the chassis.

Charging system troubleshooting

Table e lists symptoms of trouble indicated by the alternator warning light, their probable causes, and suggested corrective actions, most described later in this group.

Table e. Warning light troubleshooting

Symptom	Probable Cause	Corrective Action
Ignition OFF, engine OFF, warning light ON	Test: Disconnect blue wire (D+) from alternator. a. Light goes out: Faulty alternator diodes. b. Light does not go out: short to ground in wiring harness or wiring connector.	a. Repair or replace alternator. b. Repair or replace faulty wiring.
Ignition ON, engine OFF, warning light OFF	a. Battery fully discharged. b. Bulb burned out. Test: Disconnect blue wire (D+) from alternator. With battery connected and ignition on, touch blue wire to ground. c. Light does not come on: Faulty bulb socket, open circuit between socket and terminal 15 of ignition switch, or open circuit between blue wire (D+) on alternator and instrument cluster. d. Light comes on: Loose connection between regulator and alternator or loose connection between brushes and regulator. e. Light comes on, no faults with regulator: Internal alternator faults or faulty regulator.	a. Charge battery. b. Remove and test bulb. Replace faulty bulb. c. Replace instrument cluster bulb holder. Repair wiring to alternator. d. Inspect brushes. Correct loose connections. e. Repair or replace alternator or voltage regulator.
Engine running at any speed, warning light stays ON	a. Loose or broken alternator V-belt. Test: Disconnect red wire (B+) from alternator. Do not short wire. Using a voltmeter, test between red wire and ground with battery terminal connected. b. No voltage to alternator: Open circuit between red wire on alternator and starter or between starter and positive (+) pole. c. Exciter diodes burned out. d. Faulty regulator or faulty alternator windings. e. High voltage drop between red alternator positive (+) wire and starter due to broken, loose, or corroded wires.	a. Replace or adjust V-belt. b. Repair wire or connections between alternator and starter. c. Repair or replace alternator. d. Test charging system and replace faulty components as needed. e. Repair wires or connectors.

Charging system, testing

The tests described here will help determine the cause of charging system trouble. The battery should be fully charged and the alternator V-belt correctly tensioned.

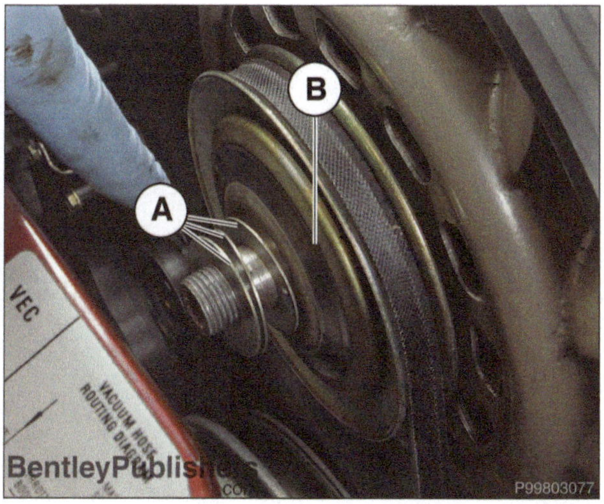

V-belt adjusting shims (**A**) may be present in front of, behind or between pulley halves (**B**). See **03 Service and Maintenance**.

The charging system warning light should be operating correctly. All electrical connections should be clean and tight. Replace wires that are hard and cracked.

If tests show that the alternator and regulator are operating correctly, but the battery still continually runs down, there may be a short in the electrical system causing a continuous current drain. Remember that the clock and some other components require constant current. See **Static current draw test** in this repair group.

A general test of charging system output can be made with an accurate digital voltmeter. Such a test is given below. The most accurate testing is done using an alternator and regulator tester that applies a high-current load to the alternator. Conclusive tests using this equipment can be made inexpensively by an authorized Porsche dealer or other qualified shop.

Alternator and regulator, checking

Attach digital voltmeter to negative (–) and positive (+) battery terminals.

- Make sure tester is connected to clean areas of terminals.
- Set voltmeter to DCV scale.

– Start engine and run it at about 1500 rpm with all electrical accessories turned off. Measure voltage at battery.

– A reading much higher than 14.2 volts most likely indicates a faulty voltage regulator.

– A reading below 13.5 volts means that the battery is not adequately charged. This may be due to:

- Failed V-belt
- Improper V-belt tension
- Faulty voltage regulator
- Faulty alternator

– To determine which component is faulty:

- Visually check V-belt. See **03 Service and Maintenance**.
- Keep engine running at about 1500 rpm and turn on all electrical accessories.
- Check voltage across battery terminals.
- A reading that is the same as the first most likely indicates a faulty voltage regulator.
- A reading that is lower most likely indicates a faulty alternator.

Noisy alternator

Alternator mechanical noises can be the result of misalignment between the V-belt and the pulley, a loose or broken pulley, worn bearings, or a bent rotor shaft. Check for bad bearings by removing the V-belt as described below and rotating the alternator pulley by hand. If the shaft grinds, the alternator should be replaced.

Alternator, replacing

– Disconnect negative (–) battery cable and cover battery terminal with insulating material to keep cable from accidentally contacting terminal.

– Remove drive belts. See **03 Service and Maintenance**.

Place fender cover over engine hood opening and engine hood latch.

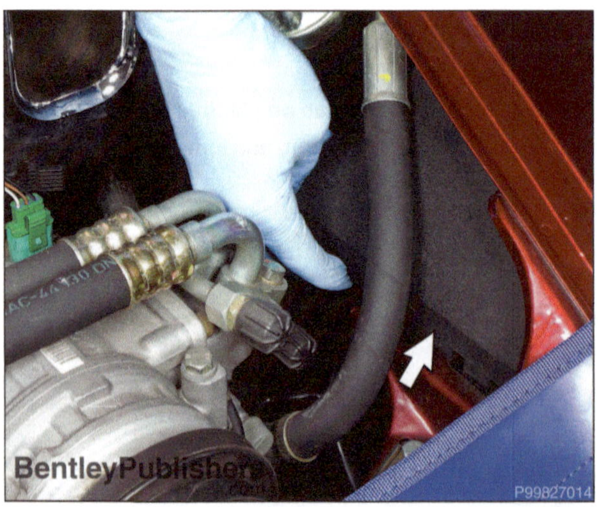

Working in engine compartment on right side, remove plastic cover (**arrow**) by unclipping bottom and pulling up.

Disconnect A/C compressor electrical connector (**arrow**).

Working at top of cooling fan shroud, remove A/C line mounting bracket fastener (**arrow**).

Remove A/C compressor fasteners (**arrows**). Be prepared to catch threaded lugs from bottom of bracket when bolts are removed.

– Remove A/C compressor from mounting bracket and lay compressor on edge of engine compartment. Use pad on top of bumper to protect from damage.

NOTE–

- *Do not disconnect refrigerant lines.*

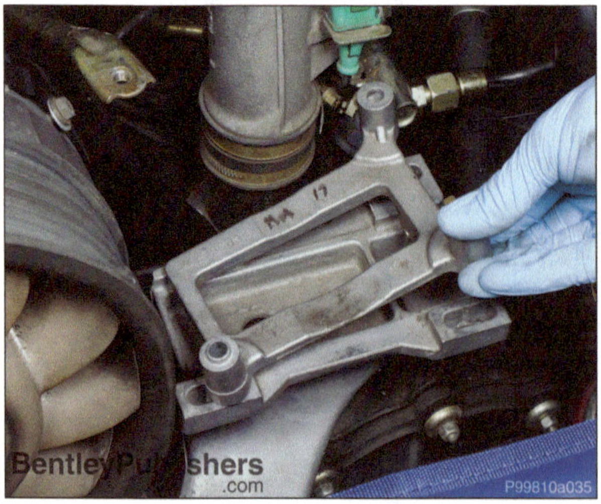

◀ Remove A/C compressor adjusting bracket from mounting bracket and store in a safe place.

◀ Remove fasteners (**arrows**) from engine cooling shroud.

– Remove duct (**A**) from engine cooling shroud by lifting straight up.

◀ Remove Allen bolt (**A**) from cooling housing strap.

– Pull alternator and fan out (toward rear of car) partially.

◀ Remove nuts (**arrows**) holding alternator air guide and slide guide back over wiring harness.

◀ If air guide is stuck to alternator, lever out air guide bushing (**arrow**) using a flat blade screwdriver.

◀ Detach electrical connections (**arrows**) in back of alternator.

- Remove alternator / cooling fan assembly.
- Working at bench, remove alternator to fan housing mounting nuts.
- Push alternator out of fan housing.
- Pull fan off alternator shaft using a puller.
- Installation is reverse of removal.

Tightening torque	
Fan housing mounting strap allen bolt	8 Nm (6 ft-lb)
Fan/alternator pulley to alternator shaft	14 Nm (10 ft-lbs)

Voltage regulator, replacing

The brushes contact the alternator's slip rings to supply the current that magnetizes the alternator rotor. The voltage regulator maintains a nominal voltage in the electrical system by feeding excess output from the alternator back to ground. In addition, it regulates the amount of current supplied to the alternator rotor, turning it on and off as needed. If either component is faulty, the result will be an over or undercharged battery. The brushes wear under normal use, and will eventually need to be replaced.

– Remove alternator. See **Alternator, replacing** in this repair group.

P99827023

< With alternator on a bench, remove two screws (**arrows**) holding regulator to alternator housing.

– Slide regulator away from shaft and lift out of alternator housing.

– Installation is reverse of removal.

28 Ignition System

General

The ignition system installed in the cars covered by this manual incorporates many components that maximize engine performance and reliability in the face of increasing restrictions on engine exhaust emissions. The ignition components are part of the engine management system known as digital motor electronics (DME) or Motronic. For more information on the DME system, see **24 Fuel Injection**.

See **03 Service and Maintenance** for spark plug replacement.

DME ignition system

In the DME system, all ignition and fuel injection control functions are controlled by a single engine control module (ECM). Ignition timing and fuel control are based on inputs the control module receives for engine load, engine speed, ignition quality, engine temperature, intake air temperature, and altitude sensor (barometric pressure). The only function that the distributor serves is to distribute the high voltage to the individual spark plugs.

The DME ignition system provides each spark plug with a precisely timed high-voltage charge to ignite the air fuel mixture in the combustion chamber. The system also makes adjustments to the ignition timing in response to changes in engine speed and load.

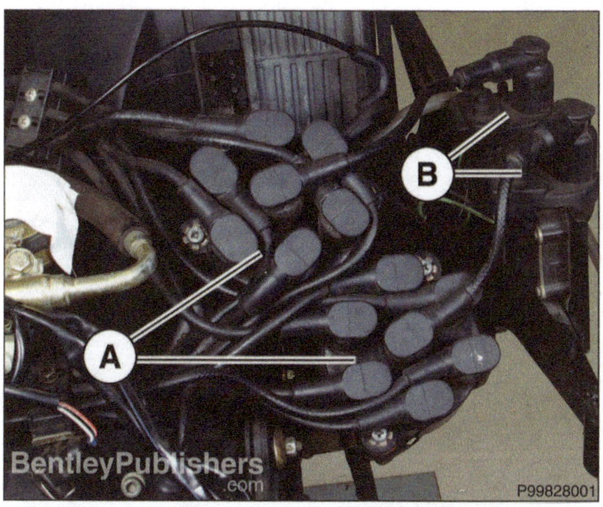

The 993 normally-aspirated engine is fitted with a twin ignition system. There are two distributors (**A**), two ignition coils (**B**) and two spark plugs per combustion chamber to ignite the air-fuel mixture. This shortens the spark travel in the cylinder and reduces combustion time while producing a faster pressure increase and a more stable combustion.

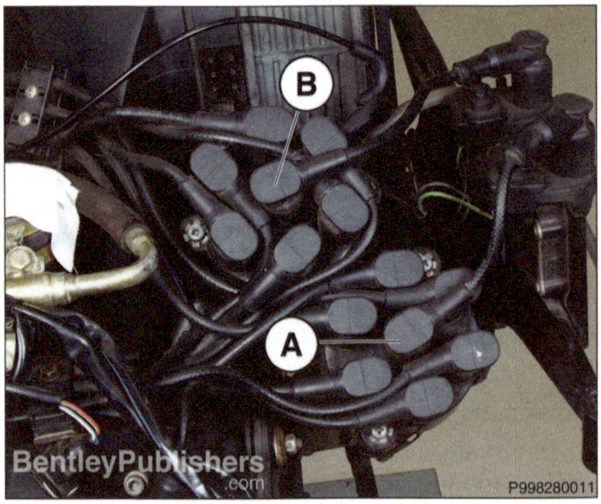

The high-voltage charge is created by the ignition coils. In the primary circuit, battery voltage is applied to the ignition coils to charge them. When the primary circuit is opened, the coils discharge their high voltage. The secondary ignition circuit, ignition rotors, distributor caps, spark plug wires and spark plugs distribute the high voltage to the cylinders to ignite the air fuel mixture. Ignition timing refers to the position of the piston in the cylinder when the coil discharges.

◀ The lower distributor (**A**) fires the upper three spark plugs on both sides of engine.
The upper distributor (**B**) fires the lower three spark plugs on both sides of the engine.

NOTE–

- *For firing order information, see* **Firing order** *in this repair group.*

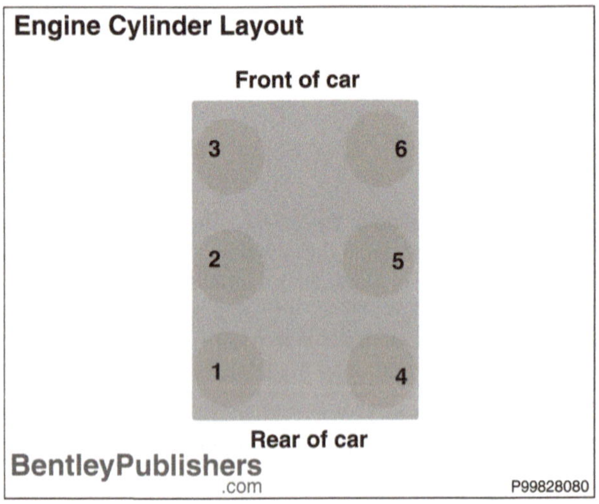

◀ Cylinders 1 through 3 are located on the left (driver) side of the car. Cylinder 1 being the rear most cylinder.

Cylinders 4 through 6 are located on the right (passenger) side of car. Cylinder 4 being the rear most cylinder.

Knock sensors

◀ In order to perform adaptive knock regulation, the electronic ignition map is subdivided up into 8 engine speed and 4 load ranges. If engine knock occurs, the ignition angle is retarded by 3° for the respective cylinder, and is then retuned to the optimized ignition angle in small steps dependent on time.

The engine is equipped with two knock bridges (**A**) with a knock sensor (**B**) screwed into each knock bridge. The knock bridges connect cylinders 1 to 3 and 4 to 6. The knock detection circuit is matched to the bridges which is why no other components are attached to the bridges.

If knocking is detected, ignition timing for the respective cylinder is retarded by 3°. If knocking continues, each knock detected retards ignition timing in 3° stages up to a maximum of 9°.

Maintenance

Porsche specifies the maintenance steps below to be carried out at particular time or mileage intervals for proper maintenance of the ignition system.

- Replace spark plugs.
- Inspect distributor cap, ignition rotor and spark plug wires.
- Information on other ignition system maintenance and on the prescribed maintenance intervals can be found in **03 Service and Maintenance**.

Firing order

◄ To determine start of firing order, inspect lip of ignition distributor body for cylinder 1 scribe mark (**arrows**).

The ignition rotors turn in a clockwise direction.

◄ Install spark plug wires so the spark plugs fire in the correct order.

Ignition Firing Order	
Porsche 911 (993) Carrera	1-6-2-4-3-5

Cylinder 1 is at left rear when viewing engine from rear of engine compartment.

Ignition timing

Ignition timing is electronically controlled and is not adjustable on DME systems. The initial (baseline) ignition point is determined by crankshaft position during starting. This is signalled by a crankshaft reference sensor. Once the engine is running, the ignition point is continually changed based on the various inputs to the control module. Engine speed is signalled by a crankshaft speed sensor.

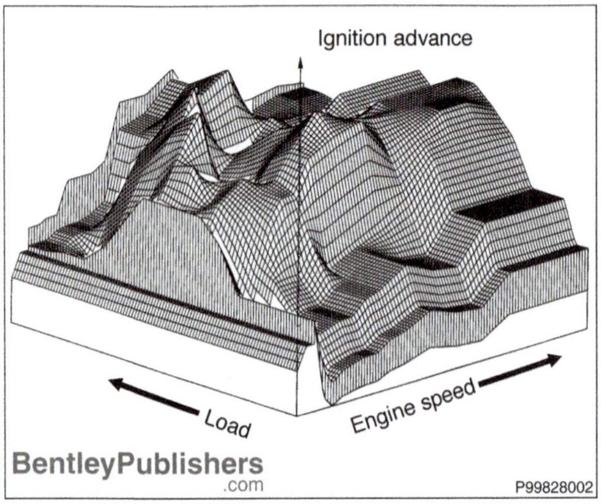

A DME ignition characteristic map illustrating all the possible ignition points is shown. A map similar to the one shown is digitally stored in the ECM.

If It is necessary to check ignition timing for testing purposes, use a strobe type timing light. Power to the light should come from 110V wall plug, but it is possible to get 12V power from the small fuse box in the left side of the engine compartment. Attach signal probe to spark plug wire 1 (left rear engine compartment).

Disabling ignition system

The ignition system should be disabled when performing certain tests such as compression test or starter current draw test.

Disconnect electrical connector to DME main relay (**arrow**) located in trunk to disable ignition system. See **97 Fuses, Relays, Component Locations**.

> ***WARNING—***
> - *Do not touch or disconnect any of the high tension cables from the coil, distributor cap or spark plugs while the engine is running or being cranked by the starter motor. Lethal voltage is present.*

CAUTION—
- *Connect or disconnect ignition system wires, multiple connectors, and ignition test equipment leads only while the ignition is off.*
- *A high impedance digital multimeter should be used for all voltage and resistance tests. An LED test light should be used in place of an incandescent-type test lamp.*
- *Switch multimeter functions or measurement ranges only with the test probes disconnected.*
- *Do not disconnect terminal 4 (center terminal) from the coil or remove the distributor cap to disable the ignition.*
- *Do not disconnect battery while engine is running.*
- *Do not connect a test instrument with a 12-volt supply to terminal 15 (+) of the ignition coil. The voltage back-flow can damage the ECM. In general, make test connections only as specified by Porsche, this manual or the instrument manufacturer.*
- *Before operating the starter without starting the engine (e.g., doing a compression test), always disable the ignition, as described above.*
- *Do not quick charge the battery (for boost starting) for longer than one minute, and do not exceed 16.5 volts at the battery with the boosting cables attached. Wait at least one minute before boosting the battery a second time.*
- *Do not wash engine while it is running, or any time the ignition is switched on.*
- *Disconnect the battery when doing any electric welding on the vehicle or when charging the battery.*
- *Do not start the engine if the car has been heated above 176°F (80°C). Wait for the car to cool to normal temperature.*
- *Do not connect terminal 1 of the coil to ground as a means of preventing the engine from starting.*

TROUBLESHOOTING

The DME ignition system contains sensitive electronic components. To protect the system, and for general safety, the cautions in this group should be observed during ignition system troubleshooting, maintenance or repair work.

Poor driveability may have a variety of causes. The fault may lie with the ignition system, the fuel system, parts of the emission control system, or a combination of the three. Because of the interrelated functions of these systems and their effects on each other, it is often difficult to know where to begin looking for problems. For this reason, effective troubleshooting should always consider these systems in unison, as one major system.

This troubleshooting section applies to starting and running problems caused specifically by faults in the ignition system, including the coil, the distributor cap, and rotor, and the spark plug wires. A complete failure of the ignition system to produce spark at the spark plugs is self-evident. For other problems such as rough idle, misfiring, or poor starting, however, the cause is not so clear. For troubleshooting engine management and the way the car runs, see **24 Fuel Injection**.

Basic troubleshooting principles

The function of the ignition system is to provide a properly timed high-voltage spark to the combustion chamber.

On DME ignition systems, the crankshaft position sensor (in the transmission bellhousing) determines crankshaft position and gives the electronic system its baseline ignition timing. Ignition timing is then controlled by the ECM based on engine load, engine speed, engine temperature, and throttle position.

An engine that starts and runs indicates that the ignition system is fundamentally working—delivering voltage to the spark plugs. A hard starting or poor-running engine, however may indicate a problem with how well the spark is delivered. A faulty coil, cracked or deteriorated spark plug wires, worn or cracked distributor cap or rotor, and worn or fouled spark plugs are all causes of reduced spark intensity and inefficient combustion.

Inefficient combustion can cause serious problems in the catalytic converter. The poorly burned mixture can overload the converter with raw fuel, leading to converter damage. An overheated catalytic converter can be a fire hazard.

An engine that has good cranking speed but will not begin to start may indicate a complete failure of the system to produce spark. Inspect the ignition system visually. Make sure the spark plug wires have not been interchanged. See **Firing order** in this repair group.

If no faults are located, make a basic check to see if spark is being produced. See **Ignition system quick-check** in this repair group. This is the most important troubleshooting step. If a strong spark is observed, then the failure to start is due to another cause, such as no fuel being delivered to the engine.

Ignition system quick-check

If the engine does not start, the first step in troubleshooting the ignition system is to determine whether or not the system is making a spark at the spark plug. If no spark is present, then more detailed testing of the ignition system is necessary.

– To check for spark, turn ignition OFF and remove ignition wire from a spark plug.

– Connect to known good spare spark plug.

< Attach jumper wire with alligator ends to spark plug threads and a good engine compartment ground.

> ***WARNING—***
>
> - *Do not hold the spark plug or its connector, even if using insulated pliers. The ignition system is a high-energy system operating in a dangerous voltage range which could prove to be fatal if exposed terminals or live parts are contacted.*

> ***CAUTION—***
>
> - *Any test set-up other than the one described above may lead to damage or inconclusive results.*

– While a helper actuates starter, look and listen for spark in spark plug gap.

- Bright blue spark indicates healthy ignition system.
- Yellow-orange spark is weaker and indicates that, while spark is present and system is functioning, it is not operating at peak efficiency.

– Check condition of ignition system components, see **Ignition system visual inspection** in this repair group.

> ***CAUTION—***
>
> - *If ignition system failure is not the problem, the engine may start during this test. Be prepared to turn OFF the ignition immediately. Running the engine with a spark plug wire disconnected may damage the catalytic converter.*
> - *Before checking the ignition system when there is no spark or a weak spark, make sure that the battery is fully charged.*

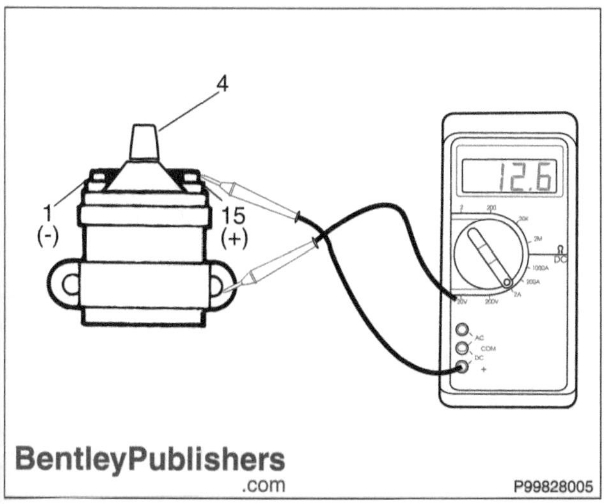

< If there is no spark, test for primary voltage at ignition coil. Connect digital voltmeter between terminal **15** (+) of ignition coil and ground (a clean, bare metal part).

– When ignition is turned ON, there should be battery voltage at terminal **15**.

– If battery voltage is not present, there is either a fault in wire between terminal **15** and ignition switch, in ignition switch itself, or in wiring from battery to ignition switch.

– Check ignition coil fuse in luggage compartment fuse panel.

- See **97 Fuses, Relays, Component Locations** for engine management schematics.
- If no faults have been detected up to this point but there is still no spark or a weak spark, see **Ignition system troubleshooting** in this repair group.
- If coil is receiving voltage, or if a strong spark is observed but engine still will not start, see **20 Fuel Supply** and **24 Fuel Injection** for fuel system testing.

Ignition system troubleshooting

Symptom	Probable cause	Corrective action
No spark or weak spark observed during spark test	Wet or damp distributor cap and/or spark plug wires	Remove cap and wires. Dry and reinstall.
	Faulty wires or connectors (primary circuit)	Inspect and repair wiring as needed.
	Weak or faulty coil	Test and replace as needed.
	Defective spark plug wires	Test and replace as needed.
	Worn or fouled spark plugs.	Replace spark plugs. See **03 Service and Maintenance.**
	Faulty engine speed sensor	Test and replace as needed.
	Faulty ECM	Test and replace as needed.

Ignition system visual inspection

The spark plug wires, ignition coil, distributor cap and ignition rotor are subject to wear and electrical breakdown which will impair their ability to deliver a crisply timed and powerful spark. Many of these conditions are most easily detected by a thorough visual inspection. Dirt and moisture on any of these components are also potential causes of poor spark at the spark plugs.

- To inspect cap and rotor, remove cap from distributor. See **Distributor cap and ignition rotor, removing and installing** in this repair group.
  - Inspect contacts inside cap and at tip of the rotor for corrosion, wear, or pitting.
  - Parts with corroded contacts can be cleaned and reused, but if there is wear, pitting, or heavy corrosion, replacement is recommended.
  - Center black carbon brush inside the cap should spring back when compressed.
  - Cracks or carbon tracks in distributor cap may cause shorts to ground.
  - Cracks may be fine and difficult to see. Check carefully, especially around the contacts. Carbon tracks are faint lines, usually running between two contacts or to ground, left over from high-voltage arcing.
  - A distributor cap that shows any sign of cracks or carbon tracking should be replaced.

P99828013

< Inspect black dust shields (**arrows**) mounted under rotor.

- Excessive oil residue on shield, rotor or cap can lead to engine misfire and poor running under load.
- If any signs of oil residue are found on shield, carefully inspect distributor shaft seal.

P99828014

< The coils (**arrows**) should be closely examined for cracks, burns, carbon tracks, or any leaking fluid.

- Coil tower, terminal 4, should be clean and dry.
- If necessary, remove coil for cleaning and close examination.
- Check that wiring at top of coil is routed as shown. Loosen nuts and reposition wires if necessary.
- Inspect all primary wires and connections for any corrosion or damage.

– Clean or repair any faults found. Corroded or loose connections may interfere with ignition function.

– A failed engine control module (ECM) can sometimes damage the ignition coil. Be sure to check the ignition coil anytime a faulty ECM is replaced and a failure to start condition still exists.

Component Replacement

The cap, rotor and spark plug wires all carry high voltage to the spark plugs. Proper engine performance depends on getting the best possible spark at the spark plug. Worn and corroded contacts or poor insulation which allows the spark to short to ground are the primary reasons for replacement of these components.

Distributor cap and ignition rotor, removing and installing

Porsche recommends the cap and rotor be replaced as a set during a major maintenance.

– If replacing cap, temporarily leave spark plug wires in old cap.

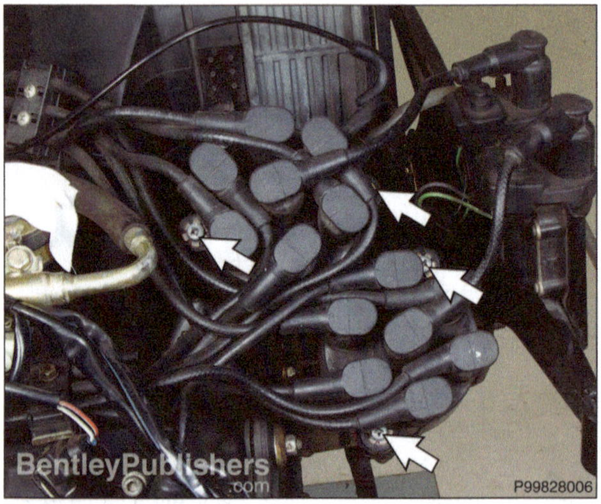

◀ To replace cap and rotor, loosen distributor cap fasteners (**arrows**).

– Lift cap off distributor.

– Install new cap on distributor and tighten fasteners.

NOTE—

- *The cap can only be mounted in one position. Make sure the cap is properly seated before locking the mounting pins.*

– Change over one spark plug wire at a time from old cap to new cap.

> ***CAUTION—***
>
> - *To avoid damaging the distributor cap, do not wiggle the connectors when removing the spark plug wires. If necessary, twist to loosen. Then, pull straight out from the cap.*

◀ To remove rotor, first remove distributor cap. Next, pull rotor (**arrow**) straight up to remove it from distributor shaft.

NOTE—

- *The rotor can only be mounted in one position. Align tab in the rotor shaft with the notch in the distributor shaft and push the rotor into place.*

– Installation is reverse of removal.

Spark plug wires, replacing

When removing spark plug wires, label their positions so that they can be reinstalled in the proper places. If the wires get mixed up, see **Firing order** in this repair group. See spark plug removal procedure in **03 Service and Maintenance** to remove wires.

Ignition coils, replacing

Ignition coils are replaced as a pair.

◄ Working at left side of engine compartment, remove ignition wires (**arrows**) from coils.

NOTE—

- *Label wire location for reinstalling.*

◄ Remove ignition coil protective covers (**arrows**).

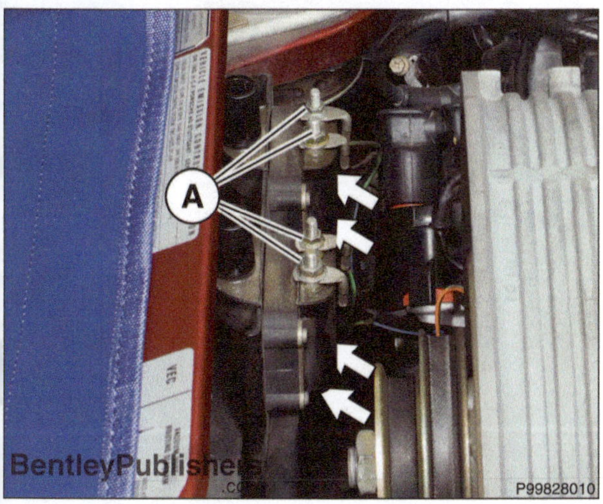

◄ Remove ignition coil electrical connector fasteners (**A**).

– Remove ignition coil fasteners (**arrows**).

– Remove ignition coils from engine brace.

– Installation is reverse of removal.

Engine speed sensor, replacing

The engine speed sensor is located at the top left side of transmission bell housing.

– Remove rear blower motor. See **80 Heating**.

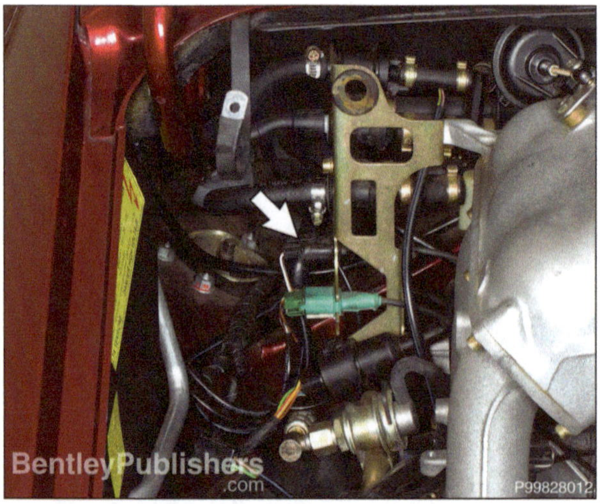

∢ Working at left side of engine compartment, disconnect engine speed sensor electrical connector (**arrow**).

– Remove electrical connector from mounting bracket.

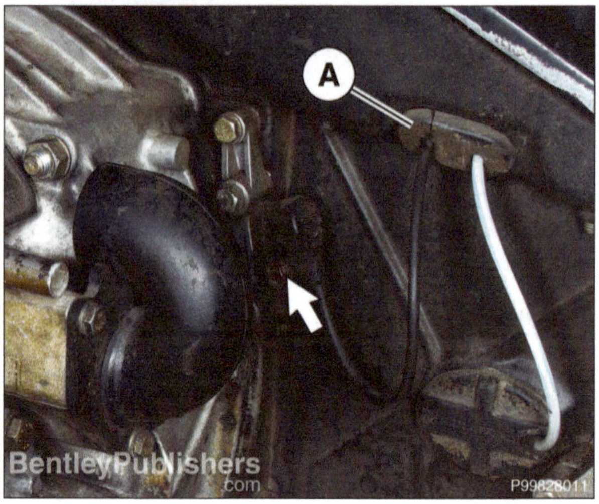

∢ Working at left side of transmission bell housing, remove engine speed sensor fastener (**arrow**).

– Remove rubber grommet (**A**) and feed wiring harness through hole in engine cover.

– Installation is reverse of removal.

Tightening torque	
Sensor to bracket	8 Nm (6 ft-lb)

Distributor, replacing

The dual distributor is serviced as one unit. The dual distributor utilizes a rubber connecting belt to synchronize the distributor rotors. This belt is available from aftermarket parts resources and can be replaced by disassembling distributor.

- Set crankshaft to cylinder 1 compression TDC (**arrow**).
- Remove distributor cap See **Distributor cap and ignition rotor, removing and installing** in this repair group.

- With crankshaft at TDC, distributor rotor aligns with mark on distributor housing (**arrows**).
  - The marks line up approximately, not precisely.

- Remove distributor mounting fastener (**arrow**).
- Remove distributor from engine.

◄ When installing, with crankshaft at TDC, install distributor so that mark on distributor rotor aligns with mark on distributor housing (**arrows**).

- The marks line up approximately, not precisely.
- There is only one way to install the distributor housing.
- Remainder of installation is reverse of removal.

Tightening torque	
Distributor base clamp (M8)	18 Nm (13 ft-lb)

Knock sensors, replacing

Knock sensor installation position and tightening torque are vital. Be sure to note orientation before removing.

– Remove alternator and alternator fan shroud from engine. See **27 Battery, Starter, Alternator**.

– Right side knock sensor: Remove intake air housing. See **24 Fuel Injection**.

◄ Right side knock sensor: Working behind A/C compressor, disconnect knock sensor electrical connector (**arrow**).

– Left side knock sensor: Remove rear blower motor. See **80 Heating**.

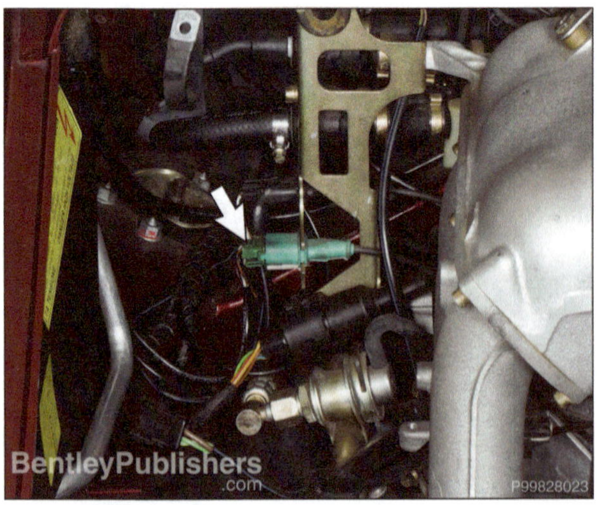

◄ Left side knock sensor: Working at rear of engine compartment, disconnect knock sensor electrical connector (**arrow**).

- Working at top of cylinder heads, remove knock sensor fastener (**arrow**).
- Remove knock sensor from engine while feeding wiring harness through engine shroud.
- Installation is reverse of removal.

Tightening torques	
Knock sensor bridge to cylinder head	10 Nm (7 ft-lb)
Knock sensor to knock sensor bridge	20 Nm (14 ft-lb)

Ignition switching unit, replacing (1995 models only)

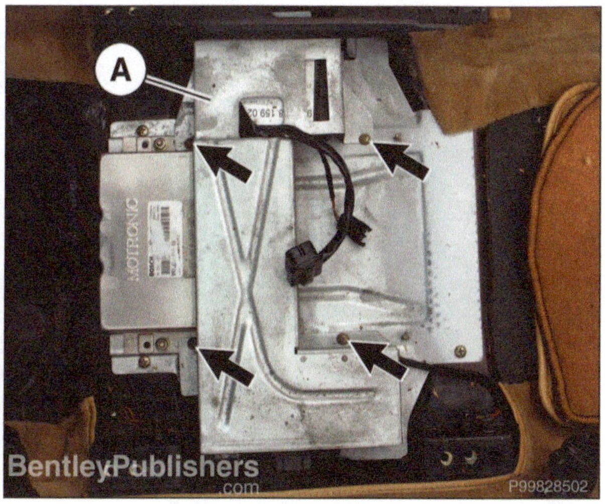

The ignition switching unit, also called the ignition final stage unit, is located under the driver seat.

- Disconnect negative (–) battery cable. Wait approximately 1 minute.

NOTE–

- *Prior to disconnecting the battery, read the battery disconnection cautions, see* **00 Warnings and Cautions**.

- Remove driver's seat. See **72 Seats**.
- Remove cover plate (**A**) shear bolts (**arrows**).
- Remove cover plate from vehicle.

28

- Pull back carpet to access ignition switching unit (**arrow**) and connector.
- Disconnect ignition switching unit connector.
- Remove two fasteners then remove unit.
- Installation is reverse of removal.

30 Clutch

General

This repair group covers replacement of clutch mechanical and hydraulic components.

Special tools are required for some procedures described in this repair group. Read procedures through before beginning a job.

Mechanical clutch service requires engine and transmission removal and separation. See **10a Engine Removal and Installation**.

Clutch system

The clutch mechanical system is made up of three major components: double-mass flywheel, clutch disc and pressure plate.

The clutch is hydraulically operated by the master and slave cylinders. Clutch disc wear is automatically taken up by the pushrod travel of the slave cylinder, making periodic adjustment unnecessary.

Double-mass flywheel

The double-mass flywheel reduces harmonic disturbances and torsional vibrations transmitted to the driveline. The damper mechanism that is traditionally part of the clutch disc is integrated into the double-mass flywheel. This works to reduce the engine resonance point to a speed lower than engine idle, minimizing driveline vibrations.

The flywheel is split into two sections:

- Primary section: Bolted to crankshaft.
- Secondary section: Makes contact with clutch disc and secures pressure plate.

The primary section of the flywheel contains springs to isolate engine vibrations and a torque limiter to prevent engine torque spikes from exceeding engine and transmission component strength.

Troubleshooting

A soft or spongy feel to the clutch pedal, excessive pedal free-play or grinding noises from the gears while shifting can all indicate problems with the clutch hydraulics. If any of those symptoms exist, perform clutch fluid flush. If fault persists, replace failing hydraulic component.

Failure of the clutch to fully disengage (leading to difficult gear shifting) may be caused by one or more of the following:

- Leaking or faulty clutch slave or master cylinder.
- Inadequate travel or misalignment of the slave cylinder pushrod.
- Faulty or bent master cylinder pushrod.
- Binding clutch disc.
- Seized pilot bearing.

Inspect bottom of transmission bellhousing for oil. If oil is present, remove flywheel to inspect crankshaft rear main and intermediate shaft seals. See **13a Crankcase and Crankshaft** for more information.

CLUTCH MECHANICAL

Clutch assembly

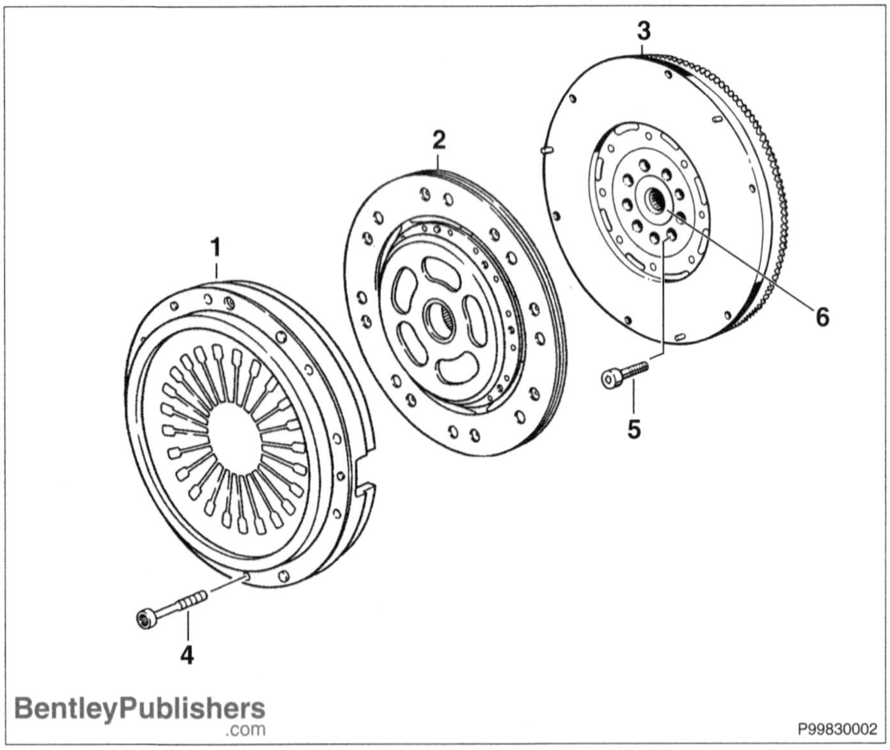

1. Pressure plate
2. Clutch disc
3. Double-mass flywheel
   - Make sure alignment pins are firmly seated.
   - Flywheel is not serviceable. Replace if damaged.
4. Screw, M8
   - To avoid damaging pressure plate, loosen and tighten bolts in stages in crisscross pattern.
   - Tighten to 23 Nm (17 ft-lb)
5. Screw, M10
   - Replace with new.
   - Tighten to 85 Nm (63 ft-lb)
6. Pilot bearing

Clutch operation

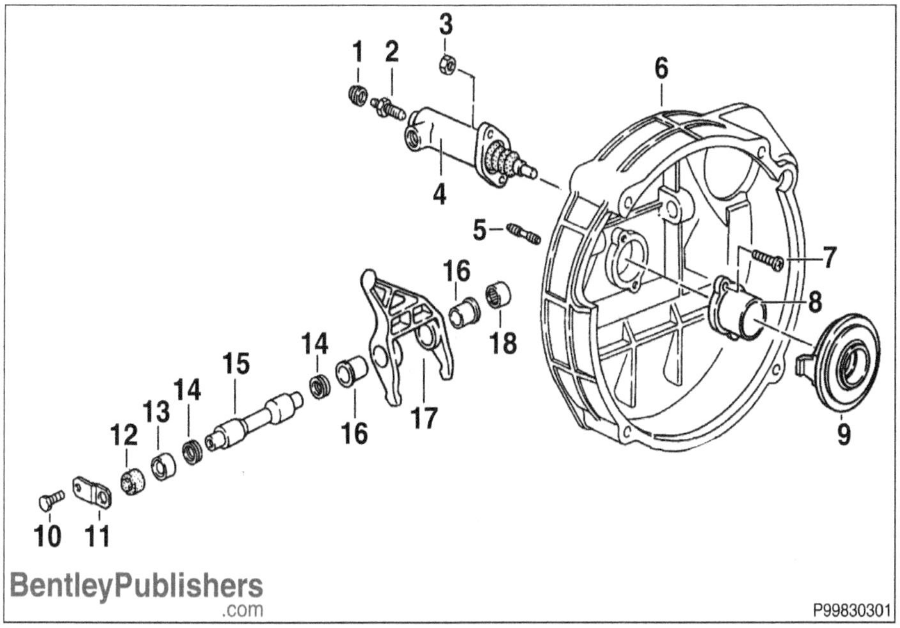

1. Dust boot
2. Bleeder valve
3. Nut
4. Slave cylinder
5. Stud
6. Transmission bellhousing
7. Screw, M6
8. Guide tube
9. Release bearing
10. Bolt, M6
11. Retaining bracket
12. Cover
13. Bushing
14. Gasket
15. Clutch control shaft
16. Bushing
17. Release fork
18. Articulated bushing

Clutch, replacing

The engine and transmission must be removed as a unit then separated in order to service the clutch. It is recommended to replace pressure plate and release bearing when installing a new clutch disc.

Carefully inspect flywheel and transmission pilot shaft bearing when replacing the clutch. Replacing the clutch disc without replacing other worn components may accelerate clutch disc wear.

– Raise car and support safely.

> ***WARNING–***
>
> *• Make sure the vehicle is stable and well supported at all times. Use a professional automotive lift or jack stands designed for the purpose. A floor jack is not adequate support.*

– Remove engine with transmission. See **10a Engine Removal and Installation**.

– Separate engine from transmission. See **10a Engine Removal and Installation**.

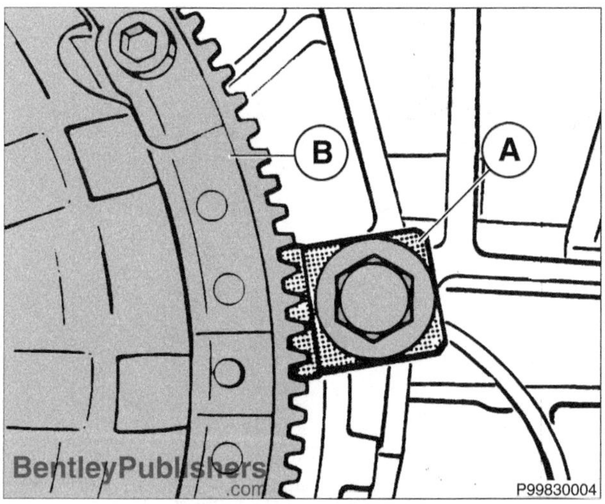

⮜ Using Porsche special tool **9538/1** (**A**) or equivalent, lock flywheel (**B**) in place.

◄ Loosen clutch pressure plate mounting bolts (**arrows**).

> ***CAUTION—***
>
> - *To avoid warping the pressure plate, loosen bolts evenly until clutch pressure is relieved.*

– Remove pressure plate and clutch disc.

> ***WARNING—***
>
> - *Clutch disc friction material produces dangerous dust. Do not breathe in dust. Use water to wet down components and collect dripping mixture in a shop towel.*

◄ Remove flywheel mounting bolts (**arrows**) and remove flywheel from crankshaft for inspection, pilot bearing replacement, or rear main seal replacement.

Oil residue, especially at base of bellhousing, usually indicates a faulty crankshaft rear main or intermediate shaft seal. For seal replacement, see **13a Crankcase and Crankshaft**.

◄ Inspect flywheel for damage or scoring.

- If flywheel is scored or damaged, replace.
- If flywheel shows signs of mild glazing, clean surface using a drill with abrasive pad.

Due to construction details, the double-mass flywheel should not be machined or commercially resurfaced.

Clutch pilot bearing is set in (**arrow**) from rear of flywheel. Measure pilot bearing depth before driving bearing out from flywheel. Install new pilot bearing to appropriate depth.

Depth of pilot bearing should match circular flange (**arrow**) of crankshaft allowing flywheel to sit flush on crankshaft.

Install flywheel on crankshaft.

- Align flywheel on crankshaft locating pin (**arrow**).

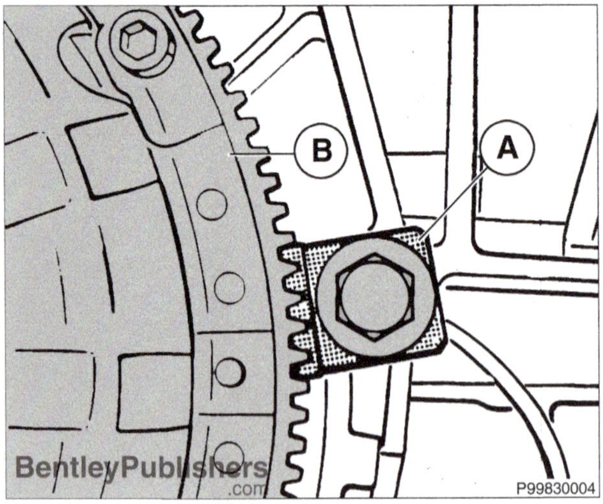

Use flywheel locking tool (**A**) or equivalent to counterhold flywheel (**B**).

Install new flywheel bolts. Torque bolts to specification in crisscross pattern in several stages.

Tightening torque	
Flywheel to crankshaft (use new bolts)	85 Nm (63 ft-lb)

Center clutch disc on flywheel using clutch disc centering tool (**arrow**).

Install pressure plate bolts (**arrows**) in crisscross pattern and in several stages.

Tightening torque	
Pressure plate to flywheel	23 Nm (17 ft-lb)

– Remove clutch disc centering tool.

Clutch Hydraulic System

The hydraulic clutch system shares a common fluid reservoir (**arrow**) with the brake system. The reservoir is located in the trunk on the left side. Make sure the reservoir is topped up with fluid.

The clutch hydraulic system is not adjustable. The clutch slave cylinder automatically compensates for normal wear of the clutch disc.

Air in the lines and moisture in the fluid can affect clutch operation.

Warnings and Cautions

WARNING—

- *Brake fluid is poisonous, highly corrosive and dangerous to the environment. Dispose of old brake fluid properly.*
- *Wear safety glasses and rubber gloves when working with brake fluid. Do not siphon brake fluid with your mouth.*

CAUTION—

- *Brake fluid is corrosive and dissolves paint. Immediately clean away any brake fluid spilled on painted surfaces and wash with water.*
- *Use brake fluid from a fresh, unopened container. Brake fluid absorbs moisture from the air. This can lead to corrosion problems in the clutch and brake hydraulic system, and also lowers the brake fluid boiling point.*

Clutch pedal assembly

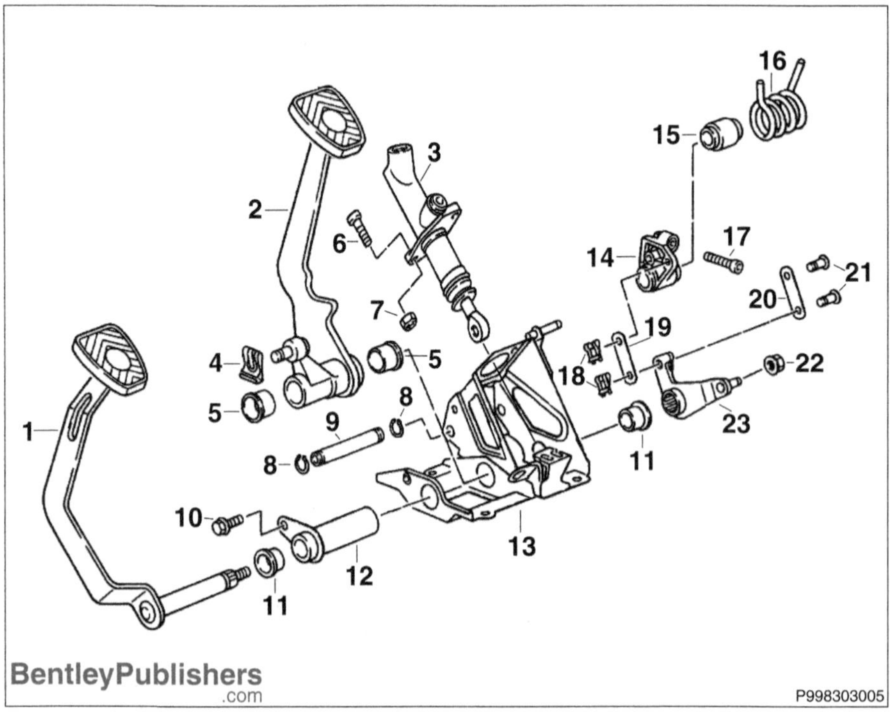

1. Clutch pedal
2. Brake pedal
3. Clutch master cylinder
4. Safety clip
5. Bushing
6. Bolt
7. Lock nut
8. Circlip
9. Shaft
10. Bolt, M6
11. Bushing
12. Bearing tube
13. Pedal bracket/assembly
14. Lever
15. Bearing sleeve
16. Clutch power spring
17. Screw, M5
18. Lock clip
19. Brace
20. Brace
21. Bolt
22. Nut, M10
23. Lever

Clutch hydraulic system, bleeding

– Working at brake fluid reservoir in trunk, fill reservoir up to MAX line with new Super DOT 4 brake fluid.

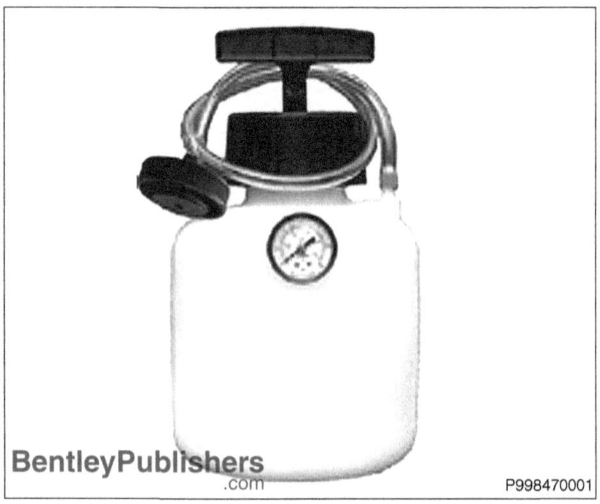

◄ Fill brake system pressure bleeder with new Super DOT 4 brake fluid and connect bleeder to brake fluid reservoir. Pressurize bleeder to approximately 1.5 bar (22 psi).

> ***CAUTION—***
> - *Do not exceed 2 bar (29 psi) pressure at the fluid reservoir when bleeding or flushing hydraulic system.*

– Press and hold clutch pedal to floor. Use pedal depressing tool or equivalent if necessary.

– Raise vehicle and support safely.

> ***WARNING–***
> - *Make sure the vehicle is stable and well supported at all times. Use a professional automotive lift or jack stands designed for the purpose. A floor jack is not adequate support.*

– Remove left rear wheel.

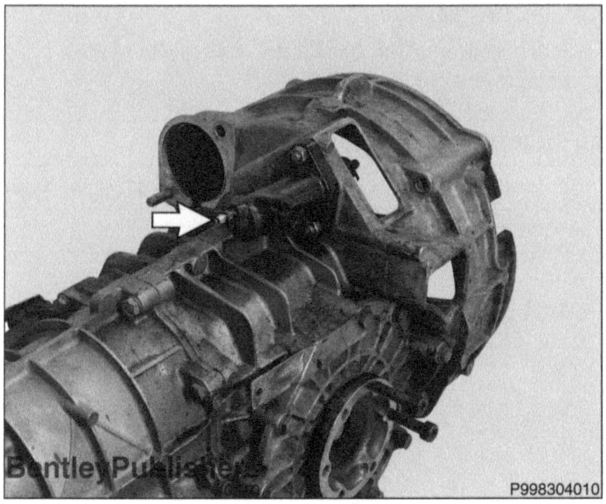

P998304010

- Working under vehicle, remove rear underbody splash shield. See **03 Service and Maintenance**.
- ◄ Working at upper left side of transmission, attach clear tube to bleeder valve (**arrow**) on clutch slave cylinder and route hose into collecting bottle.

NOTE–

- *Transmission shown removed for clarity.*

- Open bleeder valve until clear, bubble-free brake fluid emerges. Close bleeder valve.
- Slowly pull clutch pedal from floor to normal rest position.
- Depressurize and disconnect bleeding device. Add brake fluid to reservoir if level is below max.
- Check clutch operation. If necessary, repeat bleeding procedure.

Clutch slave cylinder, removing and installing

If the clutch hydraulic system is opened at any point, be sure to bleed system thoroughly. See **Clutch hydraulic system, bleeding** in this repair group.

- Raise vehicle and support safely.

> ***WARNING–***
>
> - *Make sure the vehicle is stable and well supported at all times. Use a professional automotive lift or jack stands designed for the purpose. A floor jack is not adequate support.*

- Remove left rear wheel.
- Working under vehicle, remove rear underbody splash shield. See **03 Service and Maintenance**.
- Remove left heater pipe.
- Using fluid line clamps, pinch off line to clutch slave cylinder to prevent fluid loss.

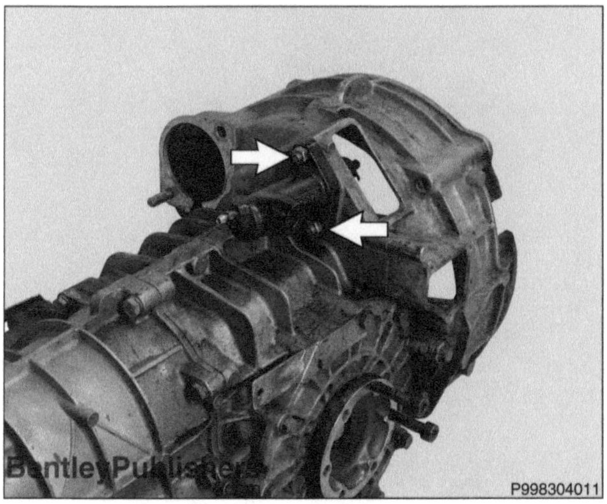

P998304011

- ◄ Working at upper left side of transmission, loosen slave cylinder mounting nuts (**arrows**).

NOTE–

- *Transmission shown removed for clarity.*

- Place rag under slave cylinder and remove hydraulic line from slave cylinder.
- Remove slave cylinder mounting bolts and slide slave cylinder out towards front of car off of transmission studs.

> ***CAUTION—***
>
> - *Do not press clutch pedal once clutch slave cylinder is removed from transmission. Hydraulic pressure could force actuating rod from slave cylinder assembly.*

- Installation is reverse of removal. Note the following:
  - If clutch actuating rod needs to be pushed back into slave cylinder, open bleeder valve first.
  - Inspect slave cylinder for fluid leaks.

- Inspect dust boot for tearing.
- Bleed clutch hydraulic system, see **Clutch hydraulic system, bleeding** in this repair group.

Tightening torques	
Slave cylinder to transmission	23 Nm (16 ft-lb)
Hydraulic line to slave cylinder	28 Nm (20 ft-lb)

Clutch master cylinder

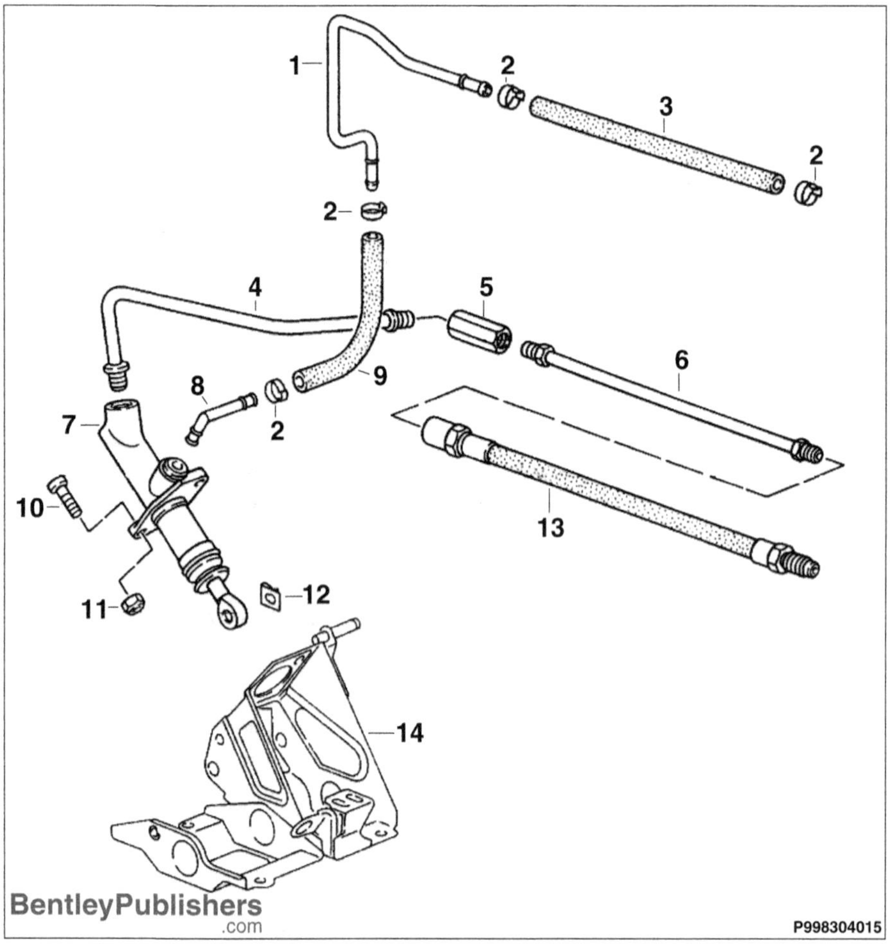

1. Clutch line
2. Clamp
3. Hose
4. Clutch line
5. Connecting piece
6. Clutch line
7. Master cylinder
8. Elbow
9. Hose
10. Bolt
11. Lock nut
12. Lock washer
13. Clutch hose
14. Pedal bracket assembly

Clutch master cylinder, replacing

When the clutch hydraulic system is opened, be sure to bleed system thoroughly. See **Clutch hydraulic system, bleeding** in this repair group.

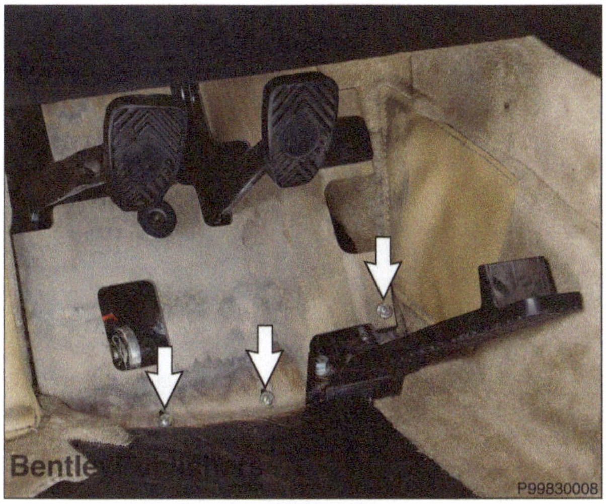

< Remove floor board:

- Working in pedal area, remove floor mat and peel off carpet from pedal floor board.
- Remove three retaining screws at base of floor board (**arrows**).
- Pull accelerator pedal from top and off of linkage ball mount.
- Remove floor board.

– Empty brake fluid reservoir using a syringe or other suction device.

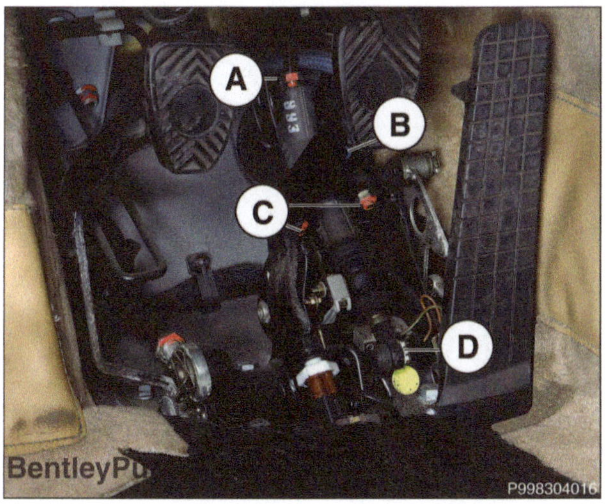

< To remove clutch master cylinder:

- Clamp off supply line to master cylinder and remove hydraulic lines (**A**) and (**B**) from cylinder.

> ***CAUTION—***
>
> - *Be prepared to catch dripping brake fluid.*
> - *Brake fluid is corrosive and dissolves paint. Immediately clean away any brake fluid spilled on painted surfaces and wash with water.*
> - *Place shop towel underneath open fluid port.*

- Loosen master cylinder mounting bolts (**C**).
- Remove clip (**D**) and pin from pushrod. Free pushrod from pedal assembly.
- Remove master cylinder mounting bolts (**C**) and pull master cylinder from pedal assembly.

– Installation is reverse of removal, noting the following:

- After loosely installing floor board, hold gas pedal lever with locking pliers and firmly push pedal back onto ball socket to refit.
- Once installation is complete, fill fluid reservoir with fresh brake fluid and bleed hydraulic system. See **Clutch hydraulic system, bleeding** in this repair group.
-

30

32 Torque Converter

GENERAL

This repair group covers removal and installation of the torque converter as well as torque converter seal replacement.

See also:

- **10a Engine Removal and Installation** for engine and transmission removal as a unit and separating transmission from engine.
- **37 Automatic Transmission** for more information on the Tiptronic transmission.

TORQUE CONVERTER

Torque converter, removing and installing

Removing

– Remove engine and transmission from vehicle as a unit. See **10a Engine Removal and Installation** for detailed instructions.

– Separate transmission from engine. See **10a Engine Removal and Installation**.

- Working through starter motor opening, unbolt torque converter perimeter bolts (qty. 8).
- Secure torque converter in place by installing Porsche special tool **9325**.
- Loosen and remove transmission bell housing bolts.
- Separate transmission from engine.

– Support transmission and ensure it is level.

– Remove torque converter.

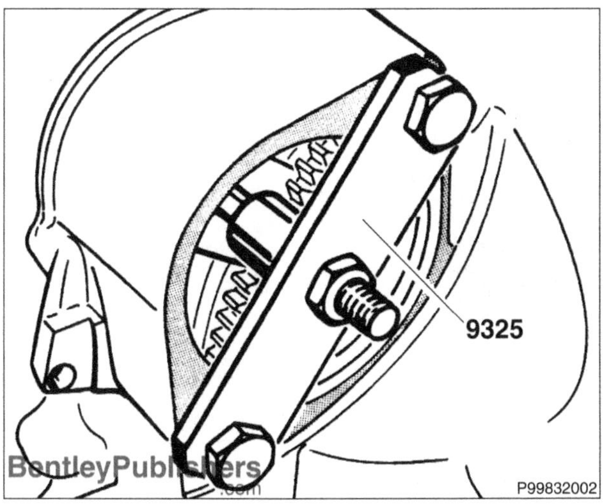

> **CAUTION—**
> - *Use care not to damage torque converter bearing assembly and torque converter seal when removing torque converter. Try to pull it off carefully in a straight line as to not stress the bearing or seal.*

Installing

> ***CAUTION—***
>
> *• If torque converter is not correctly installed, both ATF pump and torque converter may be damaged when reinstalling transmission to engine.*

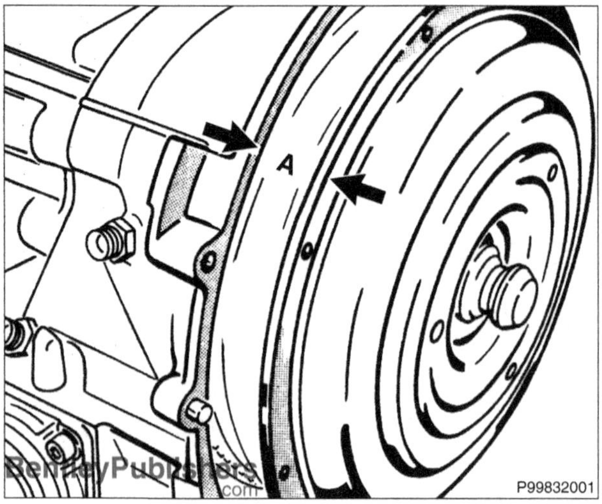

- With transmission supported and level, push torque converter onto transmission input shaft until dimension **A** is reached:
  - Rotate torque converter back and forth as necessary until gear toothing engages and it reaches installation position.
  - Dimension **A** = approximately 25 mm (0.98 in)
- Secure torque converter in place by installing Porsche special tool **9325,** then install transmission. See **10a Engine Removal and Installation** for detailed instructions and tightening torques.

Torque converter seal, removing and installing

Removing

- Remove transmission. See **10a Engine Removal and Installation.**

NOTE–

• Porsche recommends removing the engine and transmission as a unit. For engine and transmission removal and for separating transmission from engine, see **10a Engine Removal and Installation.**

- Remove torque converter. See **Torque converter, removing and installing**, in this repair group.

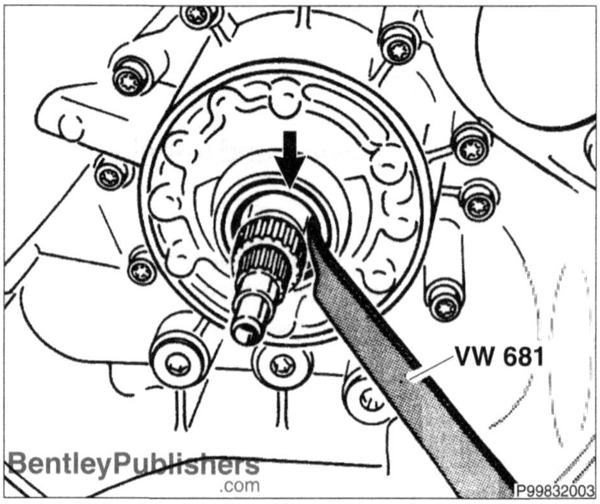

- Using special tool VW 681 or equivalent, carefully lever out torque converter sealing ring (**arrow**).

Installation

- Moisten new sealing ring with fresh ATF.

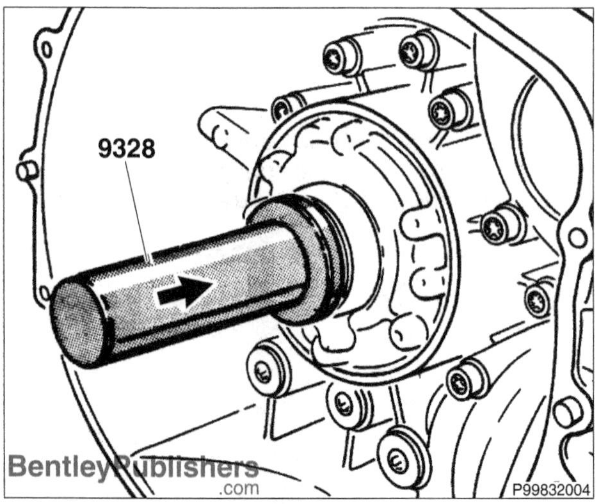

- Using special tool 9238 or equivalent, press in sealing ring to stop.
- Remainder of installation is reverse of removal.

34 Manual Transmission

GENERAL

This repair group covers external transmission service only. Internal repairs are not covered in this manual.

Transmission removal is covered in **10a Engine Removal and Installation**. Porsche recommends removing engine and transmission as one unit.

Separating transmission from engine is also covered in **10a Engine Removal and Installation.**

See also:

- **03 Service and Maintenance** for transmission fluid service.
- **10a Engine Removal and Installation**.
- **39 Differentials, Driveshaft** for driveshaft removal (Carrera 4).

Manual transmission overview

Two different manual transmissions were available for the 993 in the US market. Carrera models were fitted with the G50.20 transmission, while the Carrera 4/4S (AWD) models were fitted with the G64.20, transmission.

Table a. Manual transmission applications (993 models)	
Carrera (RWD)	G50.20, 6-speed
Carrera 4 (AWD)	G64.20, 6-speed

Table b. G50.20 manual transmission gear ratios	
Gear	**Gear ratio**
1st gear	3.818:1
2nd gear	2.048:1
3rd gear	1.407:1
4th gear	1.118:1
5th gear	0.921:1
6th gear	0.775:1
Reverse	2.857:1
Final drive ratio	3.444:1

Table c. G64.20 manual transmission gear ratios	
Gear	**Gear ratio**
1st gear	3.818:1
2nd gear	2.048:1
3rd gear	1.407:1
4th gear	1.118:1
5th gear	0.928:1
6th gear	0.775:1
Reverse	2.857:1
Final drive ratio	3.444:1

Manual Transmission Shift Assembly

Manual transmission shift knob, removing

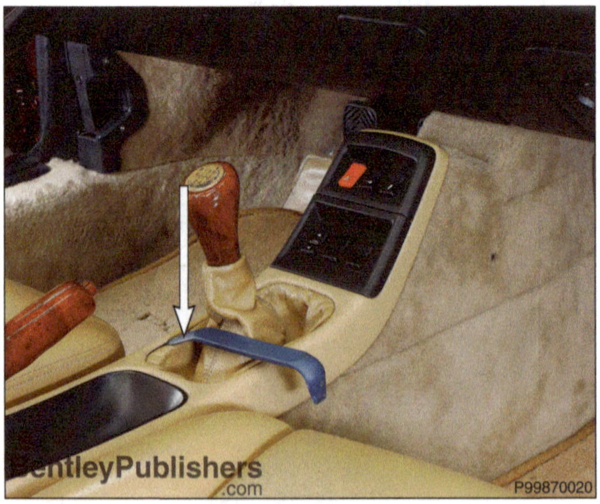

- Using plastic prying tool, press down on left side of shifter boot to disengage from center console.
- Pull leather boot up to shift knob base and remove Allen bolt holding knob to shift lever.
- Pull off shift knob with boot.
- Installation is reverse of removal.

Shift linkage components

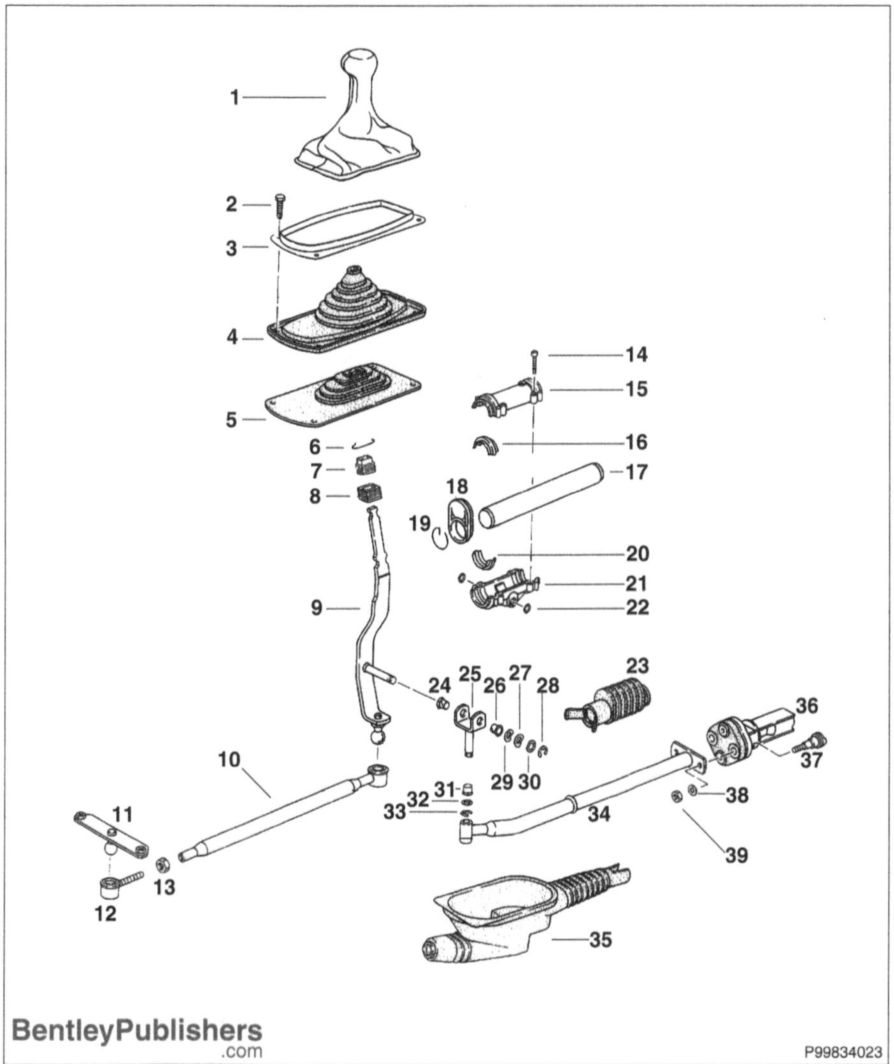

1. Shift knob with boot
2. Bolt, M6
3. Frame
4. Bellows
5. Plate
6. Clamp
7. Damper
8. Damper
9. Gearshift lever
10. Guide tube
11. Bearing plate (C2 only)
12. Ball joint
13. Lock nut
14. Screw
15. Slide
16. Insert
17. Pipe
18. Bearing sleeve
19. Circlip
20. Insert
21. Slide
22. O-ring
23. Rubber boot
24. Bearing
25. Fork
26. Bearing
27. Support washer
28. Lock washer
29. Support washer
30. Spring washer
31. Bearing
32. Spring washer
33. Lock washer
34. Gear shift rod
35. Bellows
36. Bearing body
37. Bolt
38. Washer
39. Nut, M8

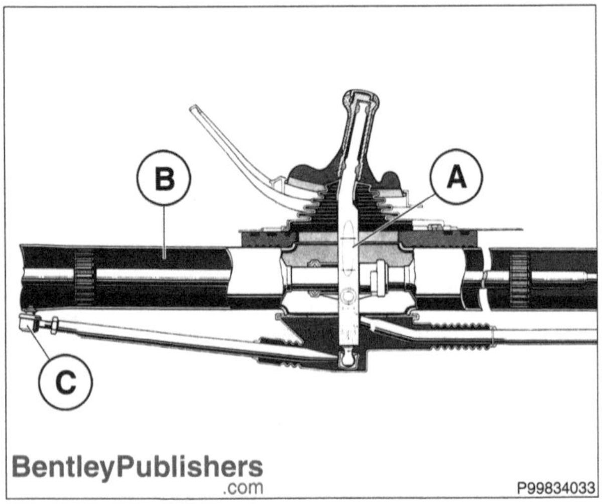

Shift lever, adjusting

To adjust shift lever, move to neutral position.

- Lower portion of gearshift lever (**A**) should be perpendicular to driveshaft tube (**B**) or transmission tunnel on rear wheel drive models. (Carrera 4 model shown)
- From inside car, shift through gears and check shift lever distance to surrounding trim. Should be evenly spaced all around.
- To adjust, loosen guide tube lock nut and remove ball socket (**C**) from driveshaft tube on Carrera 4 models or retaining plate on Carrera models. Turn ball socket to adjust length. One turn equals 3 to 5 mm in shifter angle. Adjust until shift lever is perpendicular as shown.

Transmission Mounts

Transmission mount (Carrera)

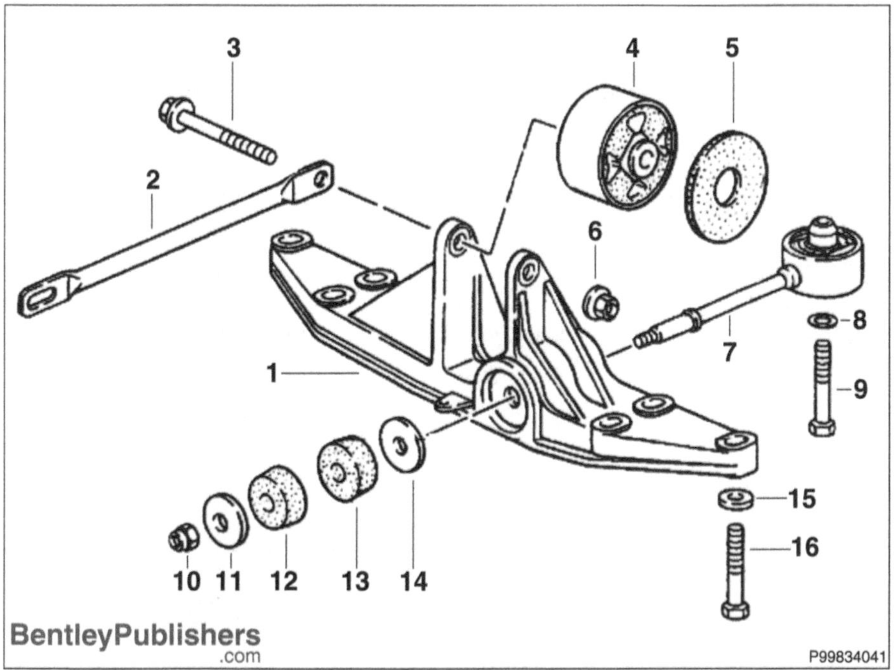

1. Transmission mount
2. Strut
3. Bolt, M12
4. Bushing
5. Stop
6. Lock nut
7. Transmission carrier
8. Washer
9. Bolt, M12
10. Lock nut
11. Washer
12. Rubber buffer
13. Rubber buffer
14. Washer
15. Washer
16. Bolt, M10

Transmission mount (Carrera 4)

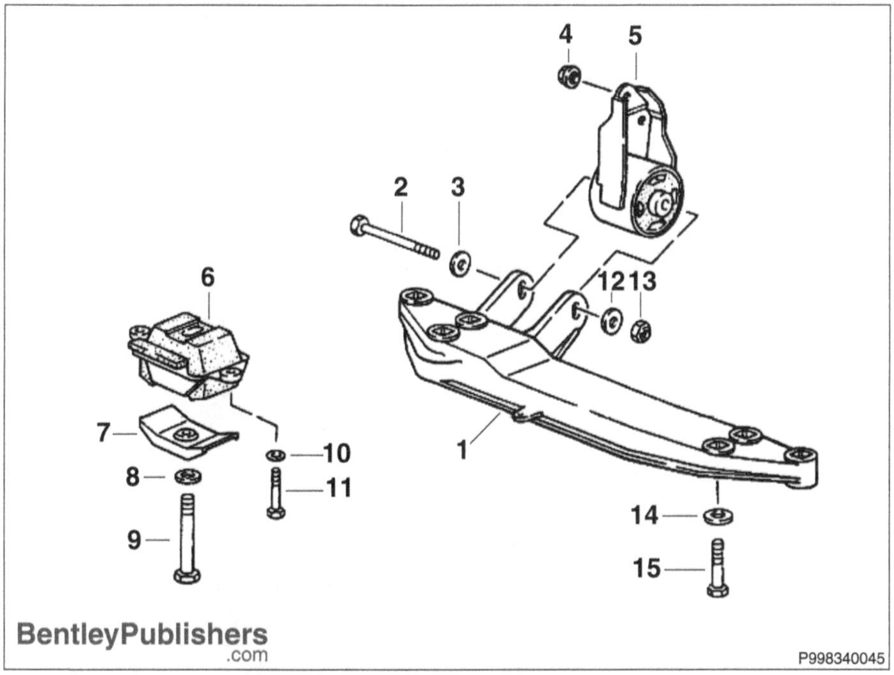

1. Transmission mount
2. Bolt, M10
3. Washer
4. Nut
5. Bushing
6. Bearing bracket
7. Stop
8. Washer
9. Bolt, M10
10. Washer
11. Bolt, M8
12. Washer
13. Lock nut
14. Washer
15. Bolt, M10

34

Transmission Service and Repairs

Manual transmission oil, checking

When checking manual transmission oil, drive vehicle for at least a few minutes to bring transmission oil up to operating temperature.

– Raise vehicle and support safely. Make sure vehicle is level.

> ***WARNING–***
> • *Make sure car is firmly supported on automotive jack stands. Place jack stands underneath structural chassis points, not underneath suspension parts.*

◄ Remove transmission oil fill plug (**arrow**). Use 10 mm Allen socket.

– Add Hypoid gear oil SAE 75W90 until it runs out of fill hole.

– Install and tighten fill plug.

Tightening torque	
Fill plug to manual transmission	30 Nm (22 ft-lb)

Manual transmission oil, changing

– Raise vehicle and support safely. Make sure vehicle is level.

> ***WARNING–***
> • *Make sure the car is firmly supported on jack stands designed for the purpose. Place jack stands underneath structural chassis points, not underneath suspension parts.*

– Working under vehicle, remove transmission protection panel. See **03 Service and Maintenance**.

– Place drain pan under transmission.

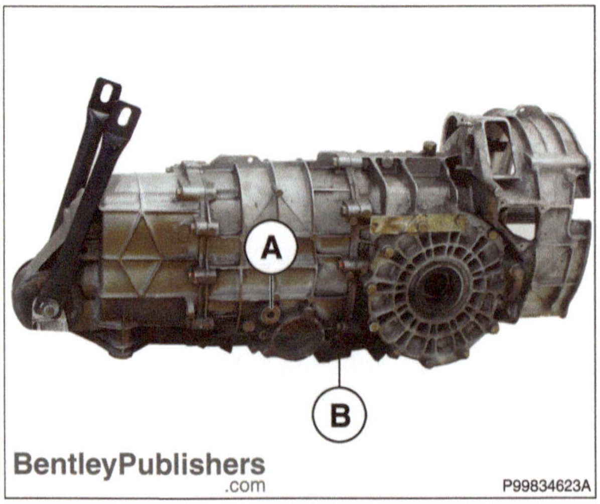

◄ Using 10 mm Allen socket, remove transmission oil fill plug (**A**) then remove transmission drain plug (**B**) and drain oil into pan.

– Clean drain and fill plugs. When oil is fully drained, install oil drain plug.

– Using fluid pump, add transmission fluid through fill hole until fluid flows out.

– Install and tighten fill plug.

Tightening torque	
Drain and fill plug to manual transmission	30 Nm (22 ft-lb)

Manual transmission oil capacity - (Gear oil SAE 75W90)	
Carrera (RWD)	3.6 liters (3.8 US qt)
Carrera 4 (AWD)	3.8 liters (4.0 US qt)

Transmission input shaft seal, replacing

If transmission fluid is pooled inside the bell housing or dripping underneath, the input shaft seal may be faulty.

- Remove engine and transmission assembly. See **10a Engine Removal and Installation**.
- Separate transmission from engine. See **10a Engine Removal and Installation**.

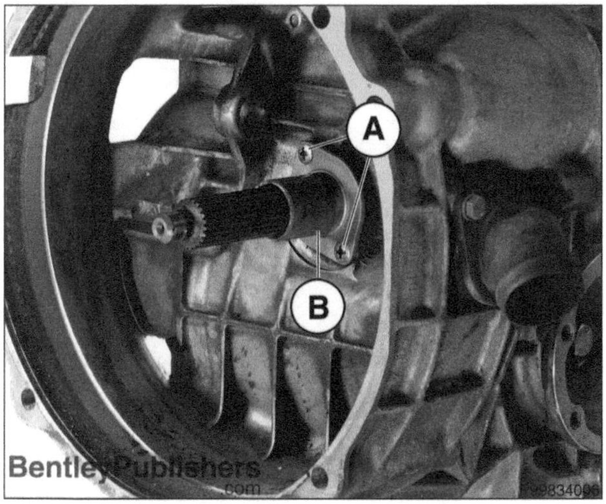

◄ Remove release bearing guide tube screws (**A**) and pull out guide tube (**B**).

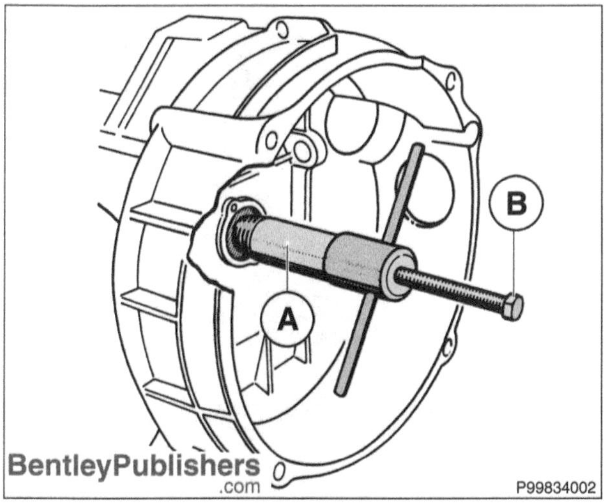

◄ Note installed depth of seal. Then screw Porsche special tool **9251** (**A**) or equivalent into oil seal. Tighten bolt (**B**) on tool to extract seal.

NOTE–

- *Use hooked pick tool to pull tension spring from input shaft if tension spring does not come out with input shaft seal.*

Installing

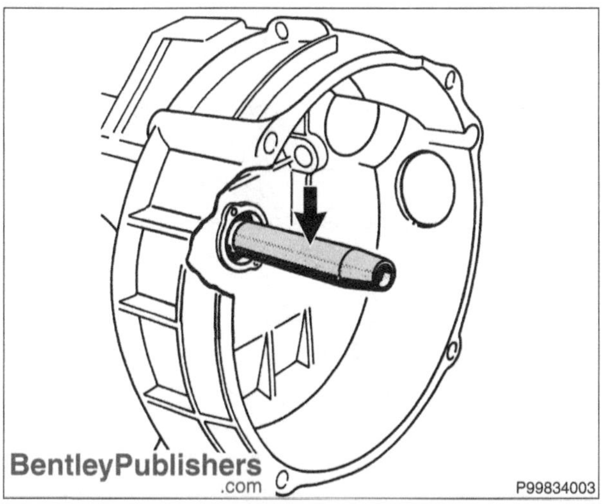

◄ Install Porsche assembling sleeve/special tool **9255** (**arrow**) onto input shaft splines.

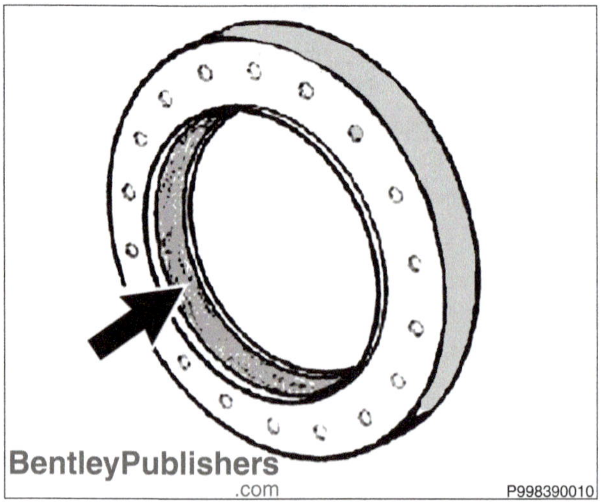

Fill gap (**arrow**) in output shaft seal lips with Optimal HT 2+2 grease.

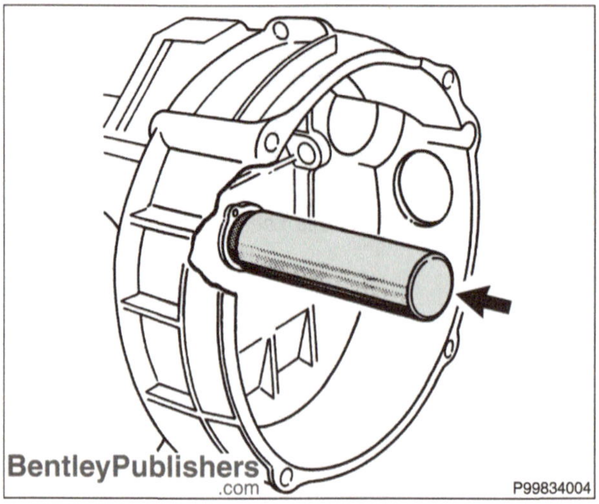

Slide input shaft seal over assembling sleeve and drive into place using Porsche special tool **9256** (**arrow**).

- Refit guide tube and install screws.

Tightening torque	
Guide tube to transmission	10 Nm (7 ft-lb)

Back-up light switch, replacing

- Working under vehicle, remove rear underbody splash shield. See **03 Maintenance**.

Remove connector (**arrow**) on passenger side of transmission and inspect terminals.

NOTE–

- *Corroded or damaged terminals are a common cause of back-up light malfunction.*

- Unscrew switch to remove.
- Installation is reverse of removal. Use a new sealing rig.

37 Automatic Transmission

General

This repair group covers basic maintenance of the automatic transmission. Internal repairs are not covered.

See also:

- **10a Engine Removal and Installation** for engine and transmission removal as a unit and separating transmission from engine.
- **32 Torque Converter**.
- **39 Differentials, Driveshaft**.

Tiptronic basics

P99837002

The 993 automatic is equipped with a Tiptronic type A50 transmission. The Porsche Tiptronic transmission is electronically controlled and can be operated in automatic or manual mode.

In automatic mode, gear shifts are automatically selected electronically based on operating conditions. The electronic control unit includes special features such as prevention of unintended upshifts or shifting in a curve.

The transmission control unit is capable of self-diagnosis and fault code storage. Tiptronic fault code retrieval requires the use of a diagnostic scan tool.

The transmission control unit is supplied with constant voltage to retain adaptive (learned) settings. If the control unit or vehicle battery is disconnected, adaptive settings are erased. An extended drive will be required to reestablish adaptive settings. On 1995 models, fault code memory is also erased when power is removed from the transmission control unit.

When operated in manual mode, the gears are engaged manually. Upshift by pushing the lever forward and downshift by pulling the lever back. Tapping the lever twice in rapid succession allows the driver to downshift by two gears at once.

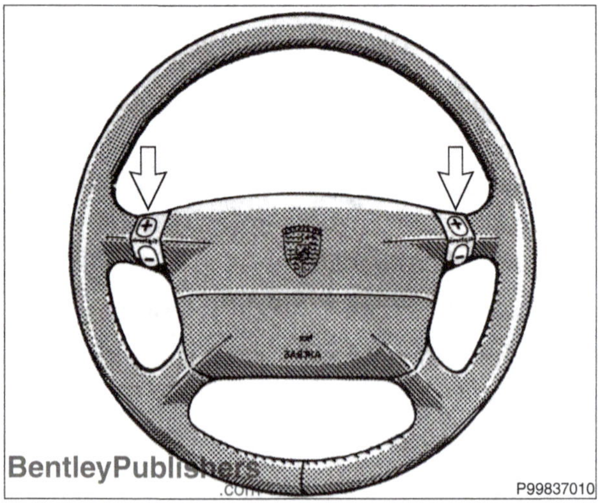

If so equipped, it is also possible to upshift and downshift via the buttons (**arrows**) on the steering wheel when in the manual mode.

If there is a Tiptronic fault, a warning light will illuminate (**arrow**), the transmission will shift to fourth gear (regardless of selector lever position), and any relevant faults will be stored in the transmission control unit. Tiptronic fault code information can be accessed using a Porsche diagnostic scan tool.

Gear ratio information is in **Table b**.

Table a. Gear ratios	
Gear	**Gear ratios**
Spur gear	1.100
1st gear	2.479
2nd gear	1.479
3rd gear	1.000
4th gear	0.728
Reverse	2.086

Automatic transmission electrical components

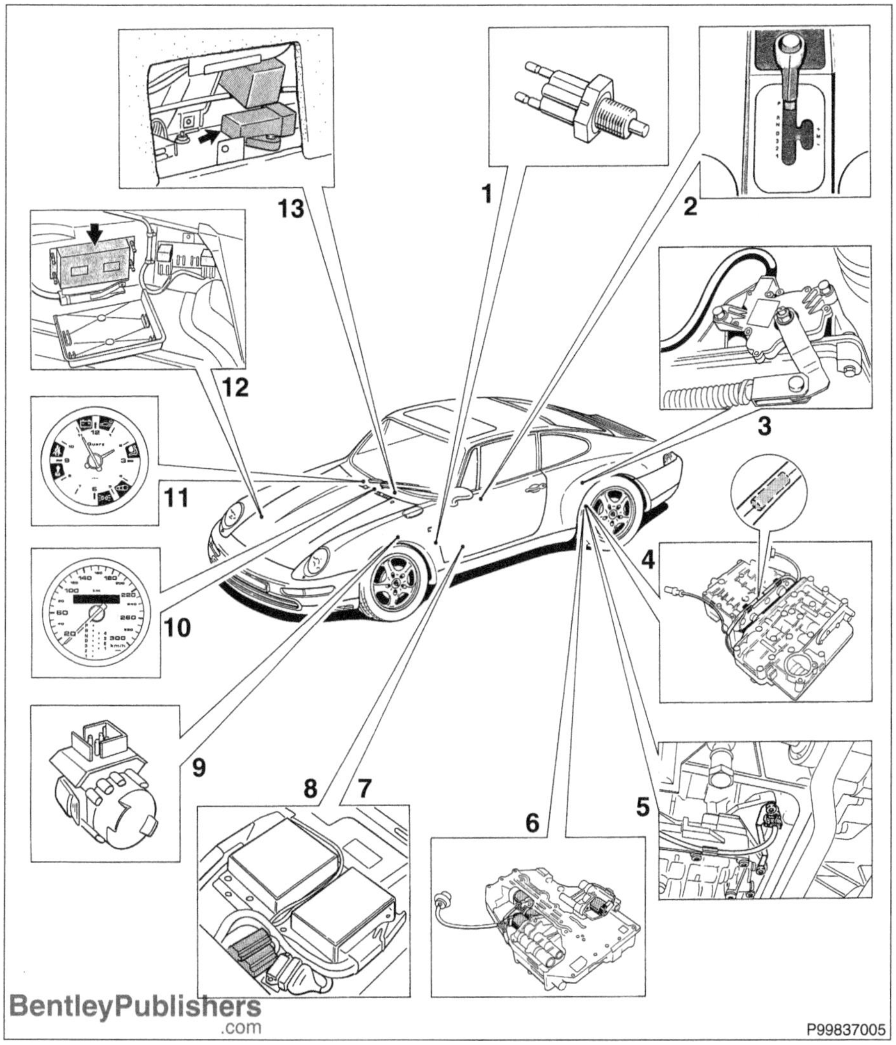

1. Stop light switch
   - in pedal cluster
2. Selector lever
3. Position switch
   - on transmission
4. ATF temperature sending unit
   - in transmission wiring harness
5. Inductive rpm pickup
   - in transmission valve body
6. Solenoid valves
   - in transmission valve body
7. Transmission control unit
   - under driver's seat
8. DME control unit
   - under driver's seat
9. Kickdown switch
   - at accelerator pedal
10. Selector lever position and gear indicator
    - in speedometer
11. Emergency mode warning light
    - in clock
12. ABS control unit
    - right side of trunk
13. Transverse acceleration sensor (1995 only)
    - underneath center console

AUTOMATIC TRANSMISSION SERVICE

Automatic transmission fluid, checking level

Correct fluid level is critical for proper functioning of the Tiptronic transmission. Prior to checking fluid level:

- Remove underbody paneling below transmission.
- Be sure vehicle is level.
- Be sure engine is running at idle speed.
- Apply hand brake.
- Move transmission selector lever to "P".
- Run engine until transmission fluid temperature reaches 50° C (122° F).

NOTE–

- *Exact fluid temperature can be determined with Porsche scan tool.*

– Raise vehicle and support safely.

> ***WARNING—***
>
> - *Make sure vehicle is stable and well supported at all times. Use a professional automotive lift or jack stands. A floor jack is not adequate support.*

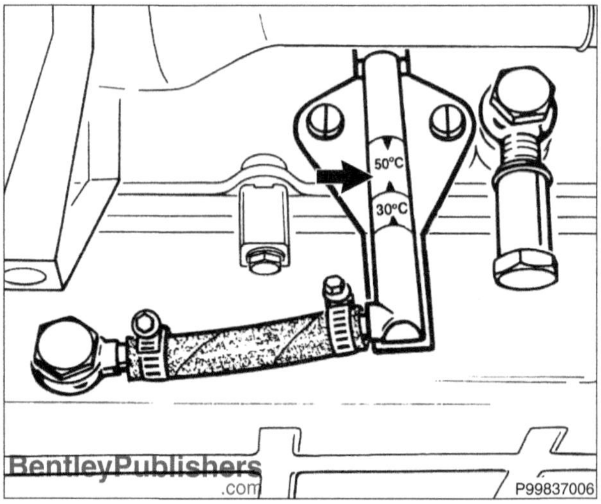

◄ Working at right side of transmission, check fluid level at level indicator (sight tube). Fluid level must be within 50° C range (**arrow**). Add or drain fluid as necessary.

Automatic transmission fluid and filter, changing

NOTE–

- *The original fill for the A50 Porsche Tiptronic was Dexron IID. Check with an authorized Porsche store or a Porsche repair specialist for the latest in fluid compatibility for the A50 transmission.*

– Raise vehicle and support safely.

> ***WARNING—***
>
> - *Make sure vehicle is stable and well supported at all times. Use a professional automotive lift or jack stands. A floor jack is not adequate support.*

– Place drain pan under transmission.

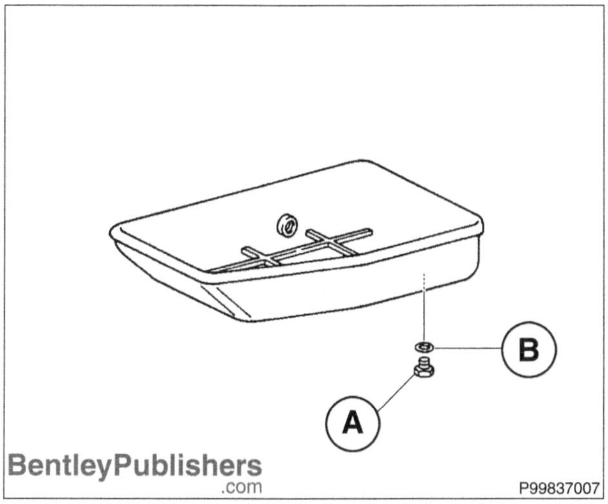

< Remove transmission drain plug (**A**) and drain fluid. Replace drain plug (**A**) and sealing washer (**B**).

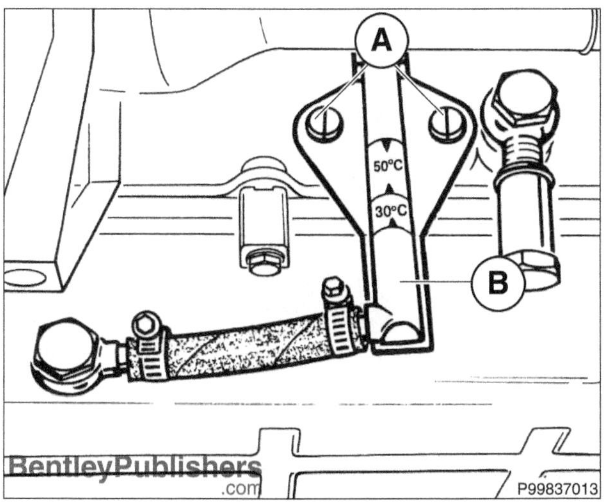

< Remove screws (**A**) and disconnect transmission fluid level indicator (**B**) from transmission housing.

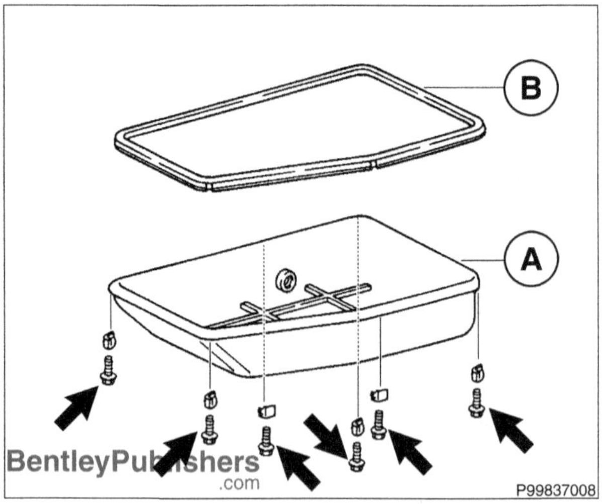

< Remove bolts and retaining brackets (**arrows**). Remove transmission oil pan (**A**) and gasket (**B**).

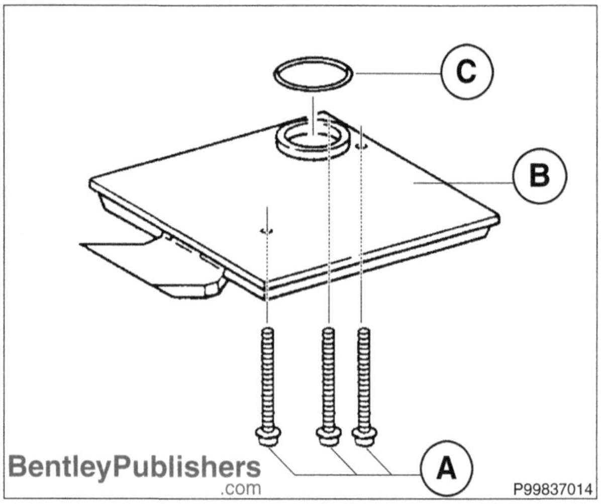

< Remove Torx bolts (**A**) and remove filter (**B**).

- Install new filter and O-ring (**C**).
- Install transmission oil pan with new gasket.

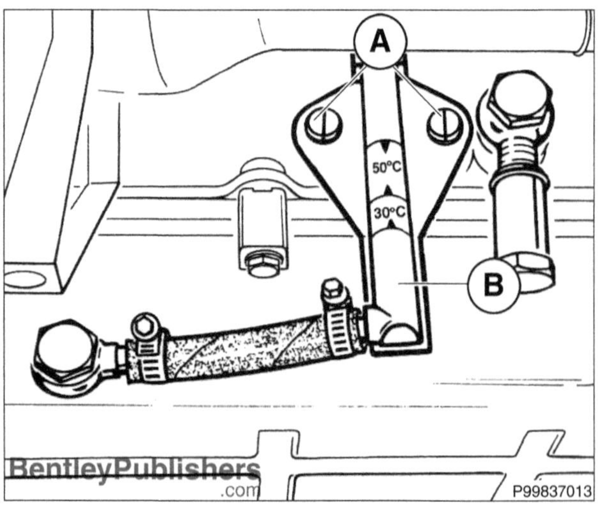

< Install screws (**A**) securing transmission fluid level indicator (**B**) to transmission housing.

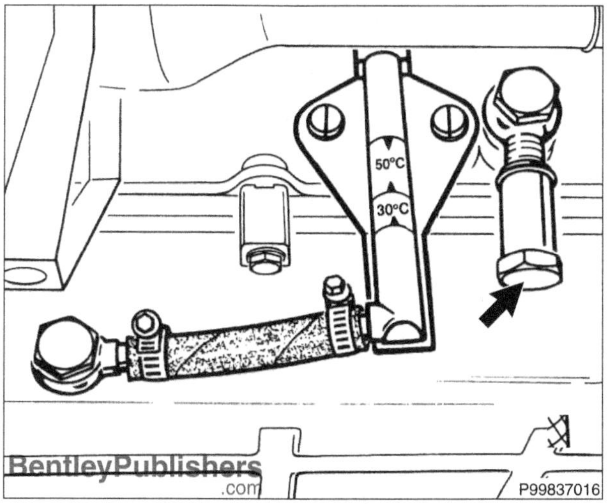

◀ Remove transmission quick fill connector plug (**arrow**).

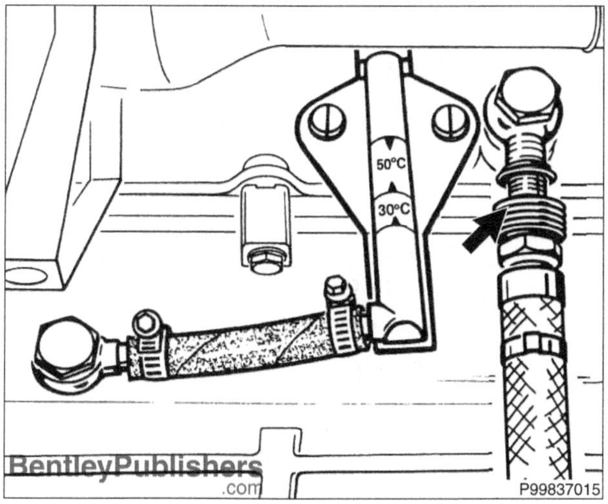

◀ With engine off, connect hose to quick fill connection (**arrow**). Using a fluid pump (ATF filling tool), pump ATF into transmission until fluid level is within 30° C mark on sight tube.

- Disconnect filling tool and reinstall quick fill connector plug.
- Start car and drive until transmission is warm.
- Check fluid level when transmission is at 50° C (122° F) .

NOTE–

- *Exact fluid temperature can be determined with Porsche scan tool.*

- Add or drain off fluid as necessary.

Tightening torques	
Drain plug to transmission oil pan	40 Nm (30 ft-lb)
Transmission fluid filter to transmission	8 Nm (6 ft-lb)
Transmission fluid quick fill connector plug	30 Nm (22 ft-lb)
Transmission oil pan to transmission	8 Nm (6 ft-lb)

Automatic transmission fluid capacity	
Fluid change quantity	3.5 liters (4.2 US qt)
Transmission fill capacity (including torque converter)	9.5 liters (10 US qt)

Automatic Transmission Removal and Installation

Automatic transmission, removing and installing

Porsche recommends removing the engine and transmission as a unit. For engine and transmission removal and for separating engine and transmission, see **10a Engine Removal and Installation**.

NOTE–

- *A special tool is recommended to hold the torque converter in place prior to separating the transmission from the engine. See also* **32 Torque Converter**.

Automatic Transmission Oil Cooler

The ATF cooler is mounted with the engine oil cooler, in the front right wheel housing.

Automatic transmission oil cooler assembly

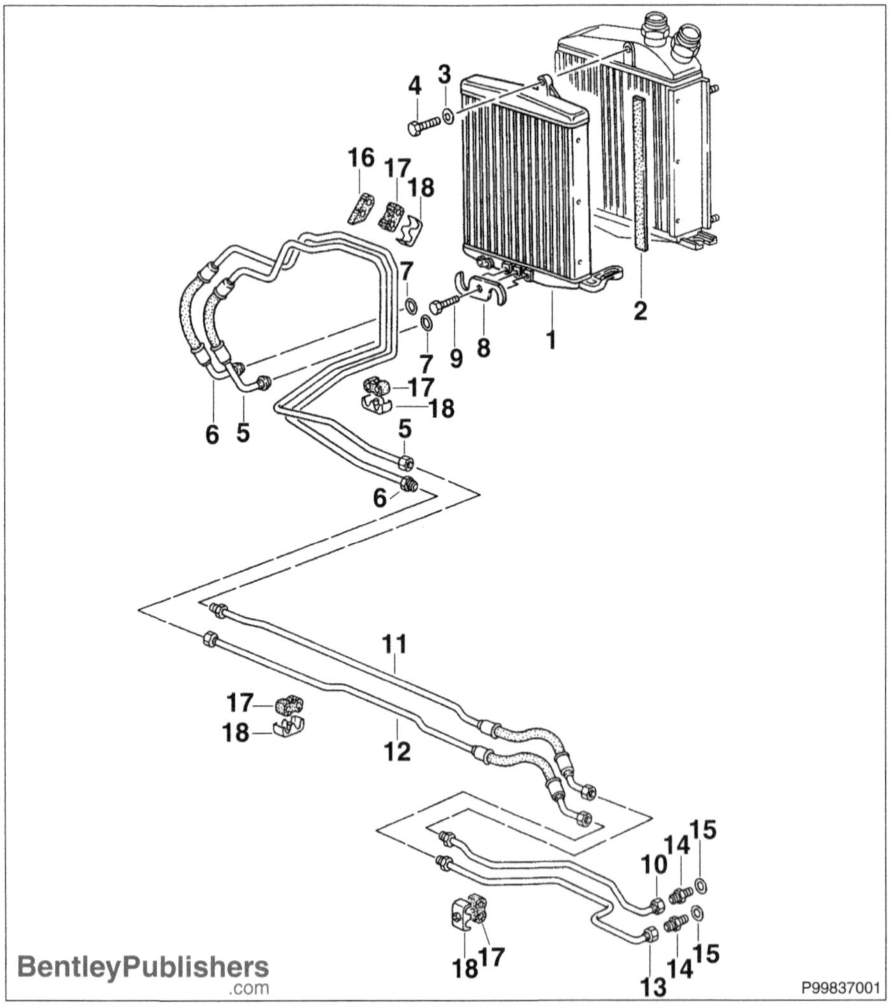

1. Automatic transmission fluid cooler
2. Sealing strap
3. Washer
4. Bolt, M6 x 22 mm
5. Oil pipe
6. Oil pipe
7. O-ring
8. Latch
9. Bolt, M6 x 20 mm
10. Oil pipe
11. Oil pipe
12. Oil pipe
13. Oil pipe
14. Connection piece
15. Sealing ring
16. Gasket
17. Rubber mounting
18. Clamp

Automatic Transmission Shift Assembly

Shift lever cable, removing and installing

- Move gear selector to position P.
- Raise vehicle and support safely.

> ***WARNING—***
>
> - *Make sure the vehicle is stable and well supported at all times. Use a professional automotive lift or jack stands designed for the purpose. A floor jack is not adequate support.*

- Remove underbody panels beneath transmission.

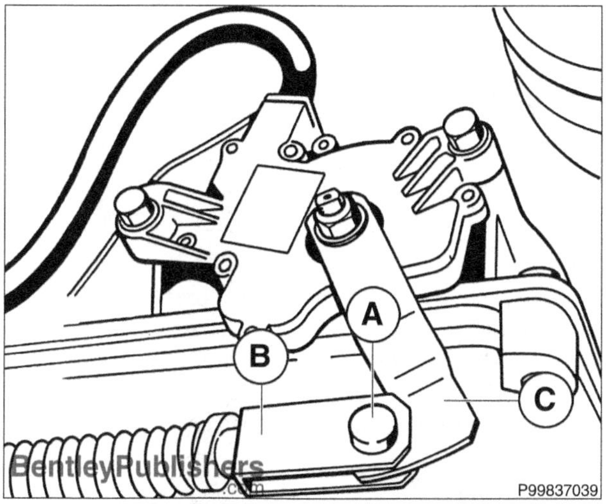

◄ Remove circlip and pin (**A**). Disconnect cable (**B**) from actuating lever (**C**).

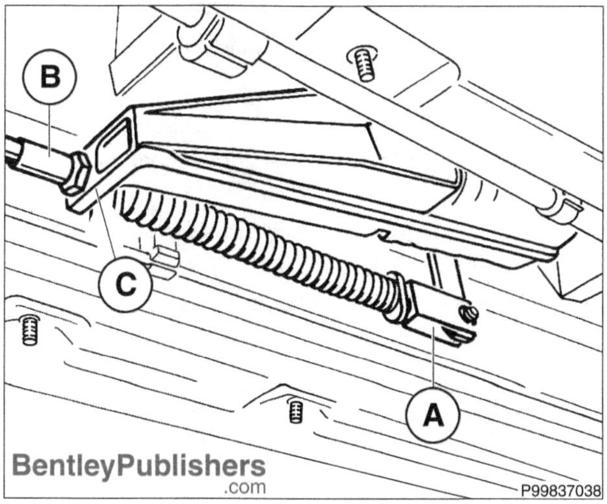

◄ Remove clevis (**A**) from cable. Disconnect guide tube (**B**) from bracket (**C**).

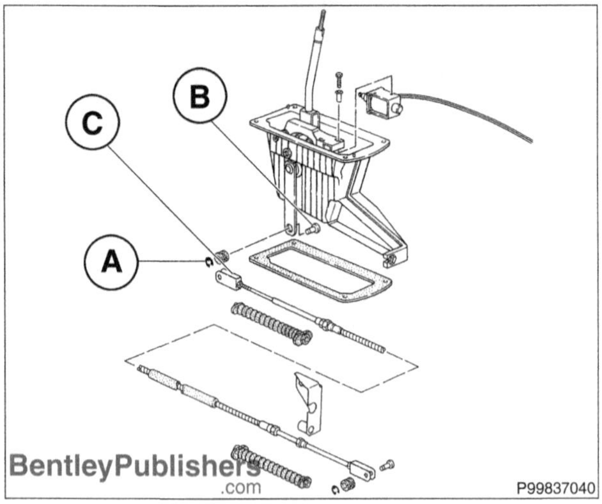

- Working underneath car, remove circlip (**A**) and pin (**B**).
- Remove clevis (**C**) from cable.
- Installation is reverse of removal. After installing, adjust cable. See **Shift lever cable, adjusting**, in this repair group.

Shift lever cable, adjusting

- Move gear selector to position P.
- Raise vehicle and support safely.

> ***WARNING—***
>
> - *Make sure the vehicle is stable and well supported at all times. Use a professional automotive lift or jack stands designed for the purpose. A floor jack is not adequate support.*

- Remove underbody panels beneath transmission.
- Remove left rear heater pipe.

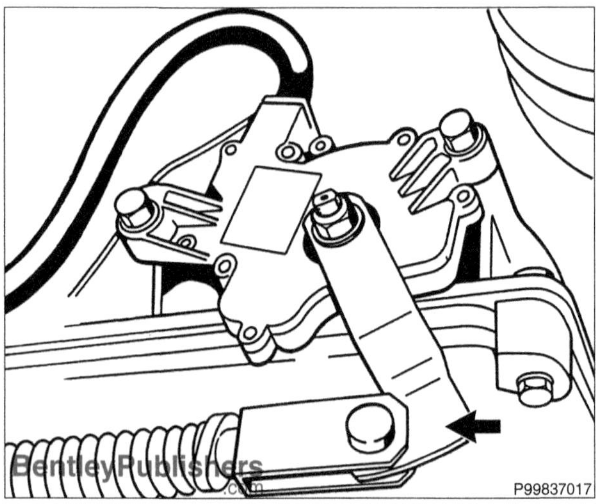

- Set multi-function switch to position P:
  - Press actuating lever (**arrow**) back up to stop.
- Set cable length at clevis so that bolt can be installed free from stress.
- Check cable adjustment:
  - Shift through all gears and confirm that correct gear position indicator light in speedometer lights up.
  - When in position "D", it must be possible to switch to manual shift gate in one smooth motion.
- Mount snap ring for bolt on actuator lever.

Automatic transmission gear position switch, removing and installing

A Porsche special alignment tool is required to adjust the switch during installation. Read the procedure through before beginning the procedure.

- Move gear selector to position N.
- Raise vehicle and support safely.

> ***WARNING—***
>
> - *Make sure vehicle is stable and well supported at all times. Use a professional automotive lift or jack stands. A floor jack is not adequate support.*

- Remove underbody panels beneath transmission.
- Remove left rear heater pipe.

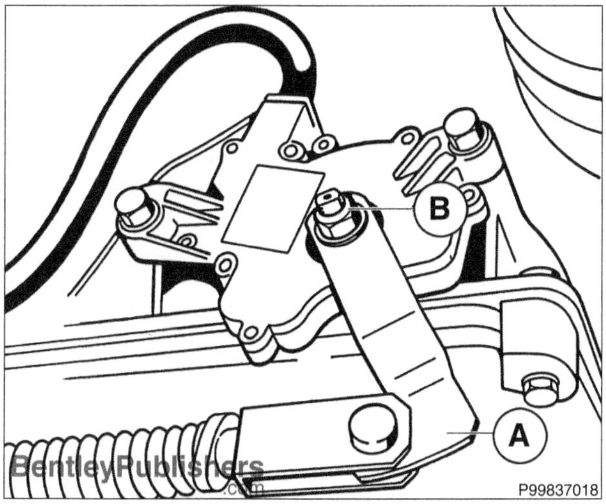

◄ Disconnect selector lever cable from actuating lever (**A**).

- Remove nut (**B**) and remove actuating lever (**A**).

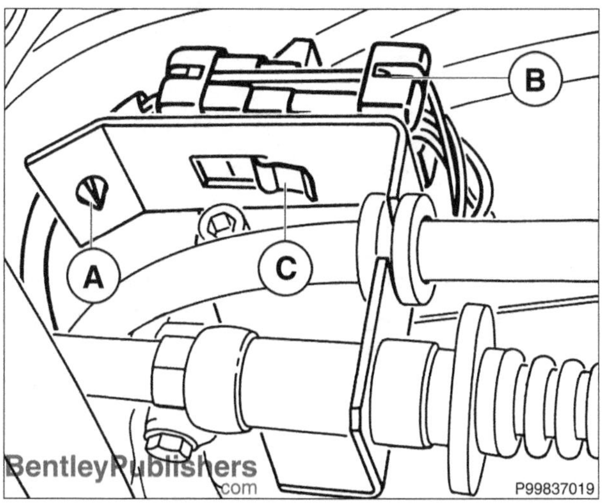

◄ Press wire harness retainer **(A)** out of bracket.

- Slide electrical connector (**B**) out bracket at **(C)** and disconnect electrical connector.

Automatic transmission gear position switch, removing and installing

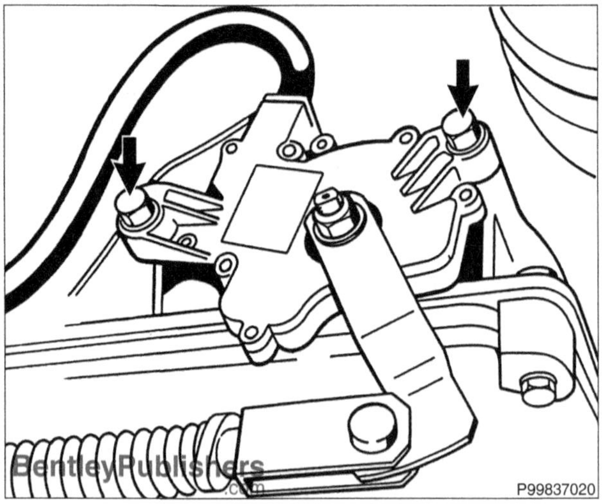

Remove bolts (**arrows**) and remove gear position switch.

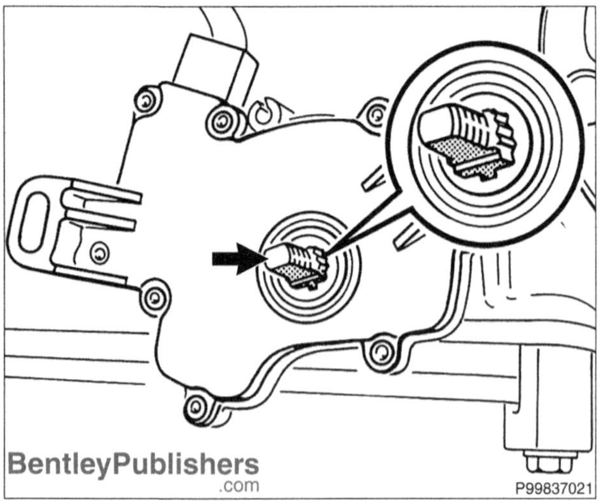

When installing gear position switch, turn shaft (**arrow**) counterclockwise to stop, then back two clicks to set selector shaft to position "N":

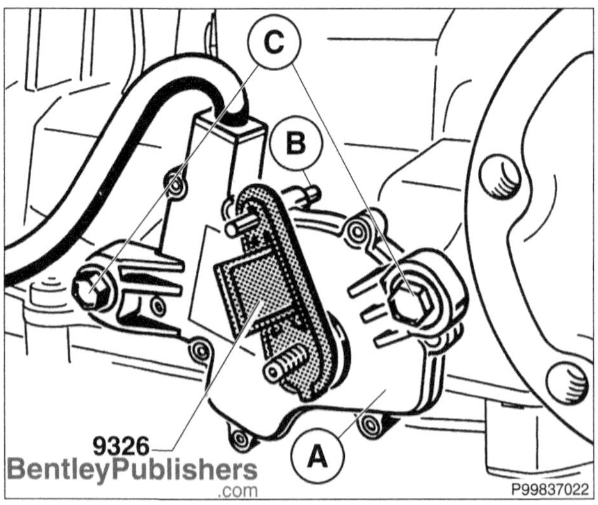

Adjust gear position switch.

- Push Porsche tool 9326 onto selector shaft and rotate switch (**A**) until locating pin (**B**) can be pushed into locating hole of switch.
- Tighten bolts (**C**).

– Remainder of installation is reverse of removal.

Tightening torques	
Actuating lever to selector shaft	15 Nm (11 ft-lb)
Gear position switch to transmission	10 Nm (7 ft-lb)

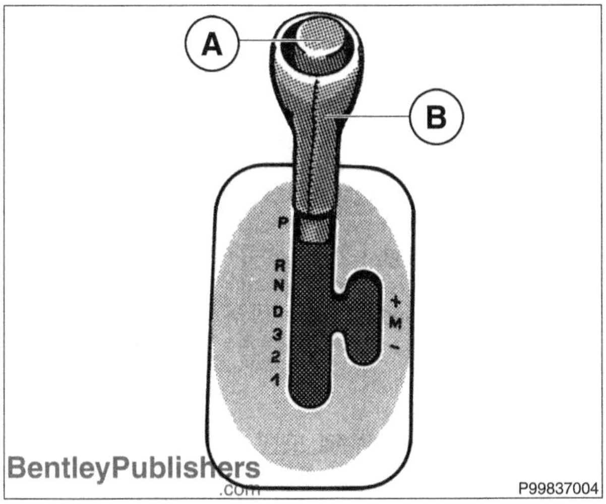

Automatic transmission shift knob, removing and installing

◀ Unscrew and remove button (**A**) in center of shift knob.

– Pull shift knob (**B**) up to remove.

– Installation is reverse of removal.

Coding transmission control unit (vehicles from 1996)

Starting with model year 1996, the transmission control unit is coded with the country of operation. Coding the control unit requires a Porsche scan tool. Replacement control units are pre-coded for USA vehicles.

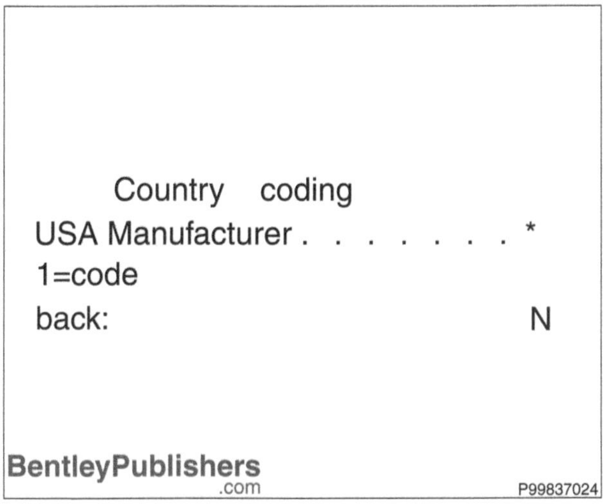

◀ Connect Porsche scan tool and select Country coding menu.

NOTE–

- *Country code and where control unit was coded are displayed on first line of country coding screen.*

– Press 1 for coding.

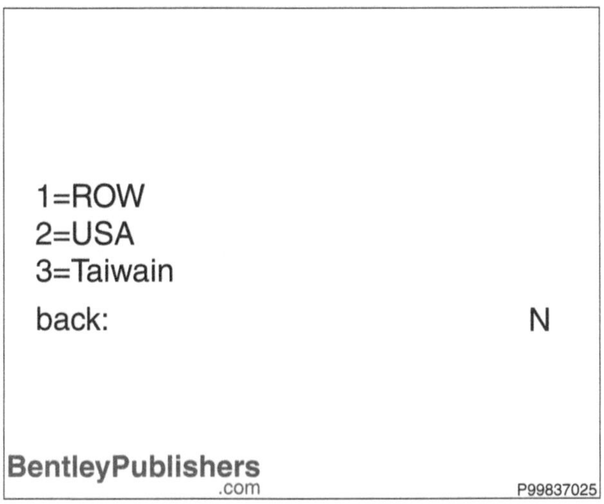

◀ At Country selection menu, press the button corresponding to the correct country.

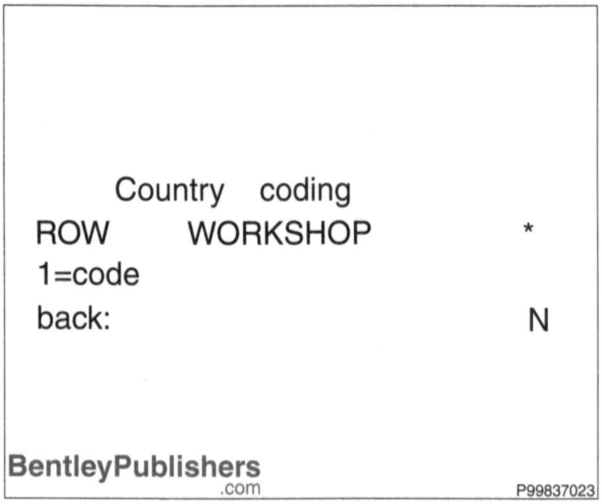

Screen will then display new coding information. Exit menu and disconnect scan tool.

39 Differential, Driveshaft

General

This repair group covers front and rear differential fluid service and seal replacement. Also covered is Carrera 4 front differential and center driveshaft replacement and repair information.

Differential internal repair is not covered in this manual.

NOTE–

- *On cars with manual transmission, the rear differential is integral with the manual transmission (transaxle). See* **34 Manual Transmission** *for manual transmission fluid service.*

Warnings and Cautions

WARNING—

- *Do not reuse self-locking nuts or bolts. They are designed to be used only once and may fail if reused. Always replace self-locking fasteners any time they are loosened or removed.*
- *Do not attempt to weld or straighten any suspension components. Replace damaged parts.*

CAUTION—

- *Due to the aluminum construction of suspension components, observe the following precautions:*
  When replacing damaged suspension components, always check the condition of control arms.

Front Differential and Driveshaft C4 Models

Front differential components

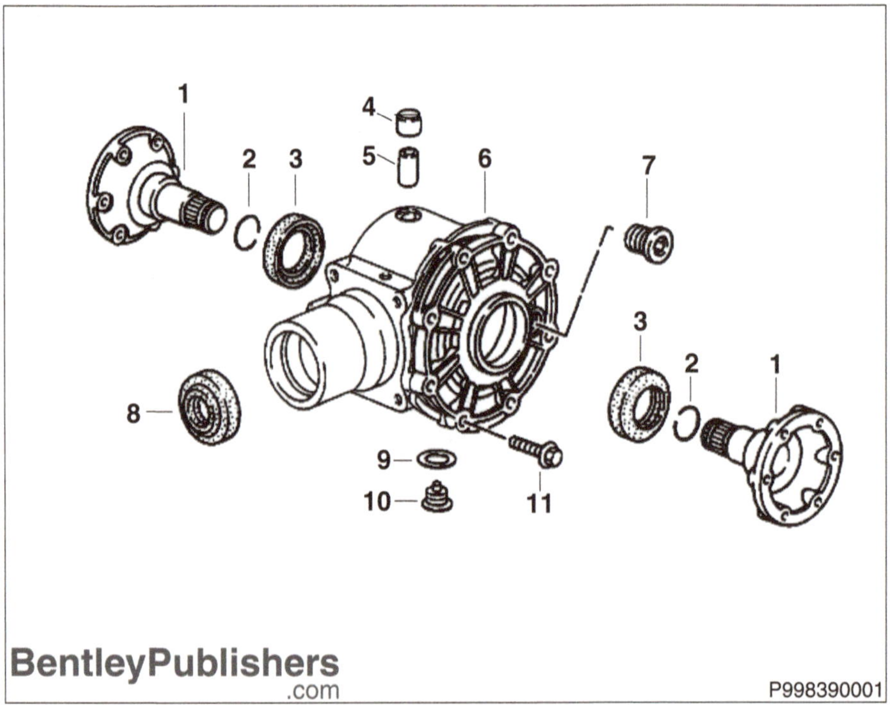

1. Axle flange
2. Circlip
3. Drive axle flange seal
4. Cap
5. Guide sleeve
6. Front differential assembly
7. Fill plug
   - Tighten to 22 Nm (16 ft-lb)
8. Shaft sealing ring
9. Sealing ring
10. Drain plug
    - Tighten to 30 Nm (22 ft-lb
11. Bolt, M8

Front differential fluid level, checking

– Raise car and support safely.

> ***WARNING—***
> • *Make sure the car is stable and well supported at all times. Use a professional automotive lift or jack stands designed for the purpose. A floor jack is not adequate support.*

– Working beneath car, remove plastic underbody shields. See **03 Service and Maintenance**.

◄ Remove front differential fill plug (**arrow**).

– Differential oil level should slightly drip from fill hole if level is correct. Top off oil level if necessary by filling until it drips from fill plug.

Front differential oil	
Transaxle hypoid gear oil SAE 75W90	0.6 L (0.63 qt)

– Refit fill plug and tighten.

Tightening torque	
Front differential fill plug	22 Nm (16 ft-lb)

Front differential fluid, replacing

– Raise car and support safely.

> ***WARNING—***
> *• Make sure the car is stable and well supported at all times. Use a professional automotive lift or jack stands designed for the purpose. A floor jack is not adequate support.*

– Working beneath car, remove plastic underbody shields. See **03 Service and Maintenance**.

◄ Remove front differential fill plug (**A**). Remove front differential drain plug (**B**) and drain fluid into suitable container.

– Once drained, replace sealing ring and reinstall drain plug.

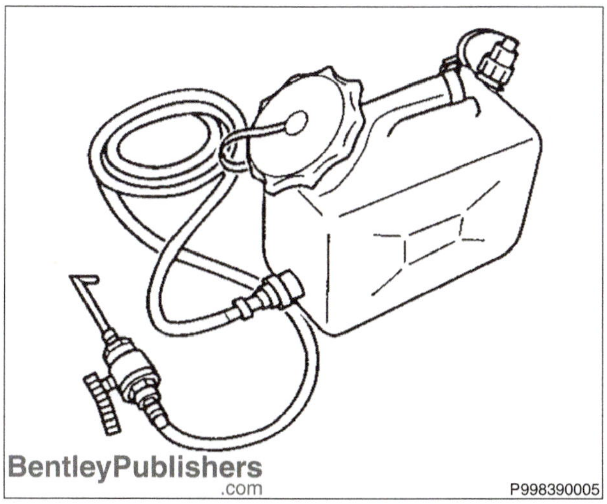

◄ Use fluid pump to fill differential oil until fluid runs out of fill hole.

Front differential oil	
Transaxle hypoid gear oil SAE 75W90	0.6 L (0.63 qt)

NOTE–

• Fill front differential slowly to avoid spilling.

– Refit fill plug and tighten.

Tightening torques	
Front differential drain plug	30 Nm (22 ft-lb)
Front differential fill plug	22 Nm (16 ft-lb)

39

Front drive axle flange seal, replacing

– Raise car and support safely.

> ***WARNING—***
>
> • *Make sure car is stable and well supported at all times. Use a professional automotive lift or jack stands. A floor jack is not adequate support.*

– Working beneath car, remove plastic underbody shields. See **03 Service and Maintenance**.

– Remove front drive axle. See **40 Front Suspension**.

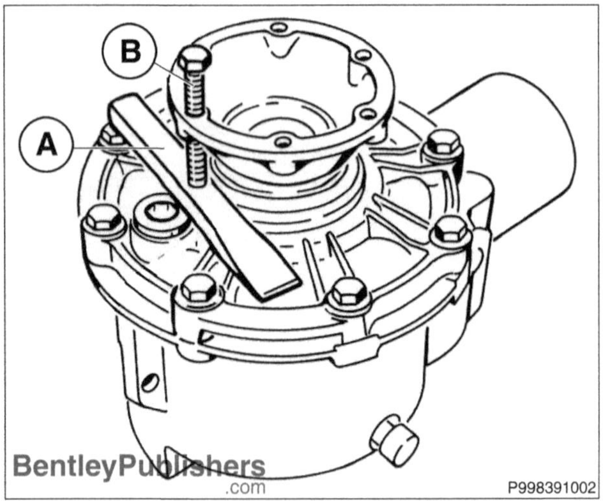

◄ Place a chisel or spacer (**A**) under drive axle flange and press out flange by tightening bolt (**B**). If drive axle flange tilts, place additional bolt into opposite hole and tighten.

– Using a seal puller, remove seal from cover.

Installing

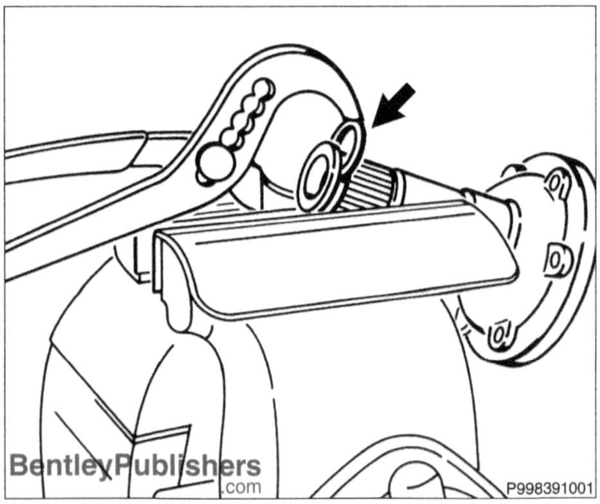

◄ Clamp flange in vise using protective jaws, then replace flange retaining snap ring (**arrow**).

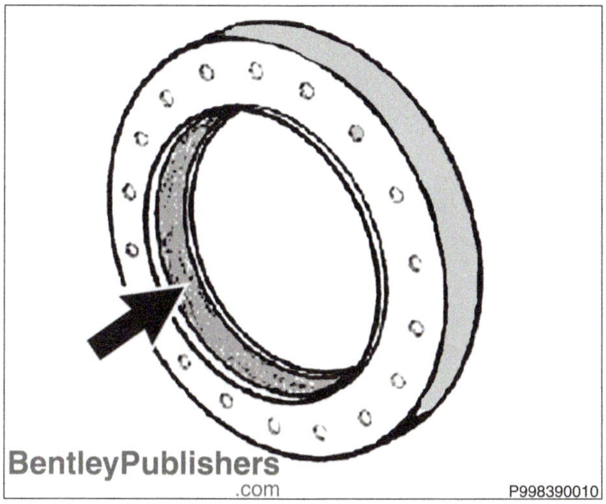

◀ Fill space (**arrow**) between dust and flange seal lips with Liqui-Moly grease.

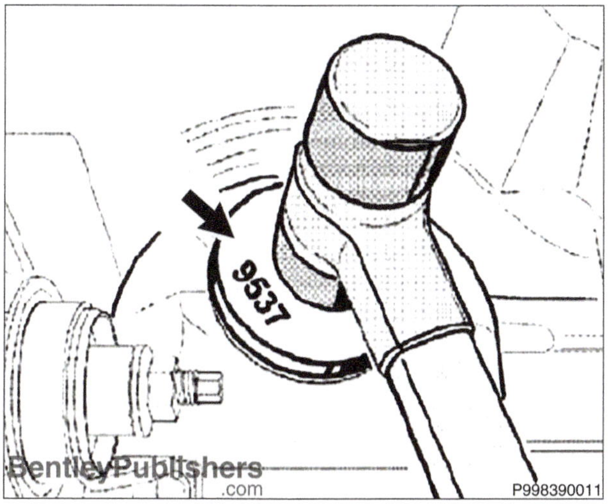

◀ Using Porsche seal installer **9537** (**arrow**) or equivalent, install drive axle flange seal ring fully until stops.

– Remainder of installation is reverse of removal.

Tightening torque	
Drive axle to output flange (M8)	42 Nm (31 ft-lb)
Drive axle to wheel hub (use new nut)	460 Nm (340 ft-lb)

Front differential, removing and installing

– Raise car and support safely.

> ***WARNING—***
> - *Make sure the car is stable and well supported at all times. Use a professional automotive lift or jack stands designed for the purpose. A floor jack is not adequate support.*

– Working beneath car, remove plastic underbody shields. See **03 Service and Maintenance**.

◀ Disconnect power steering lines from clip (**arrow**) on steering rack assembly.

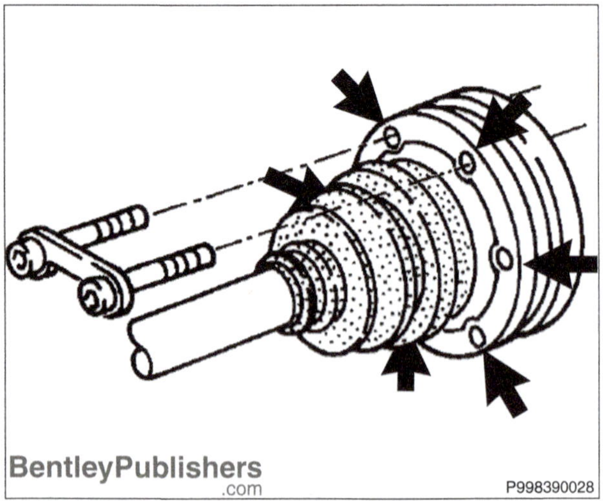

– Remove front drive axle bolts (**arrows**). Remove both front drive axles. See **40 Front Suspension, Front Drive Axles**.

◄ Remove torque tube/differential mount bolts (**arrows**).

- Lift and support differential slightly and remove mount.

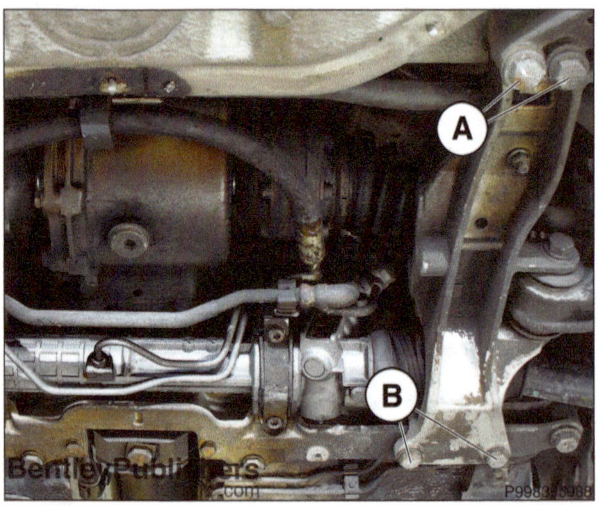

◄ Remove front side member mounting bolts (**A**) on both sides. Loosen rear side member mounting bolts (**B**) by approximately 5 mm, but do not remove.

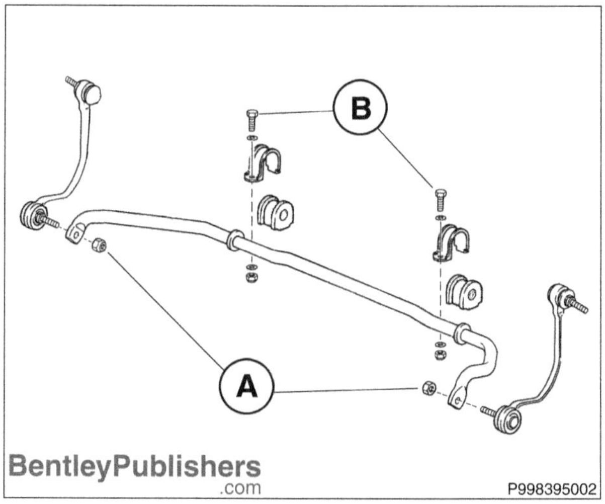

- Disconnect stabilizer bar links (**A**) from bar. Remove bolts (**B**) holding bar to side members.
- Push bar forward but to not remove.

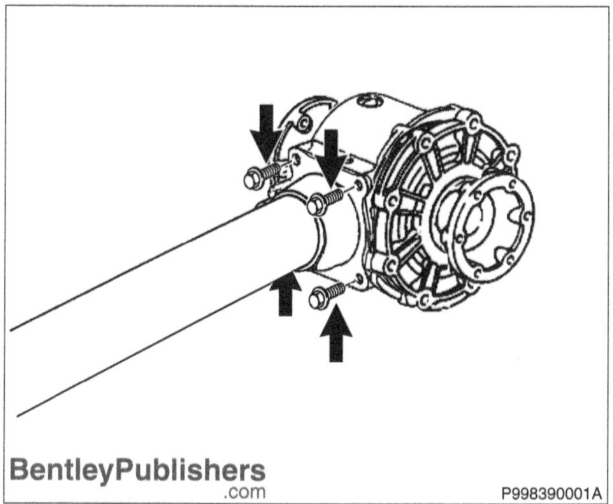

- Remove front differential to torque tube retaining bolts (**arrows**) and support torque tube with transmission jack or equivalent.
- Push front differential out of driveshaft splines. Lower and remove front differential from vehicle.
- Installation is reverse of removal. Note the following:
  - When installing, tighten and torque front fasteners, then rear.
  - Check fluid level once installed. See **Front differential fluid level, checking** in this repair group.

Tightening torque	
Front differential mount to torque tube (M12)	85 Nm (63 ft-lb)
Front differential mount to front axle cross member (M8)	23 Nm (17 ft-lb)
Front differential to torque tube (M10)	46 Nm (34 ft-lb)
Drive axle to output flange (M8)	42 Nm (31 ft-lb)
Stabilizer bar to side member (M8)	23 Nm (17 ft-lb)
Stabilizer bar to stabilizer link (M10)	46 Nm (34 ft-lb)
Side member to body and cross member (M12) front facing (M10) rear facing	 90 Nm (66 ft-lb) 46 Nm (34 ft-lb)
Cross member to body (M12) outer (M10) inner	 105 Nm (77 ft-lb) 48 Nm (35 ft-lb)
Drive axle to wheel hub (use new nut)	460 Nm (340 ft-lb)

Center driveshaft, removing and installing

Removing

- Remove center console. See **70 Interior Trim**.
- Raise car and support safely.

> ***WARNING–***
> - *Make sure vehicle is stable and well supported at all times. Use a professional automotive lift or jack stands. A floor jack is not adequate support.*

- Working beneath car, remove plastic underbody shields. See **03 Service and Maintenance**.

◄ Loosen lock bolt (**A**) on shift guide rod. Disconnect ball joint (**B**) from socket.

◄ Push back shift linkage bellows at transmission end. Remove locking bolt (**A**) holding shift linkage to transmission. Slide shift linkage off of shaft toward front of car.

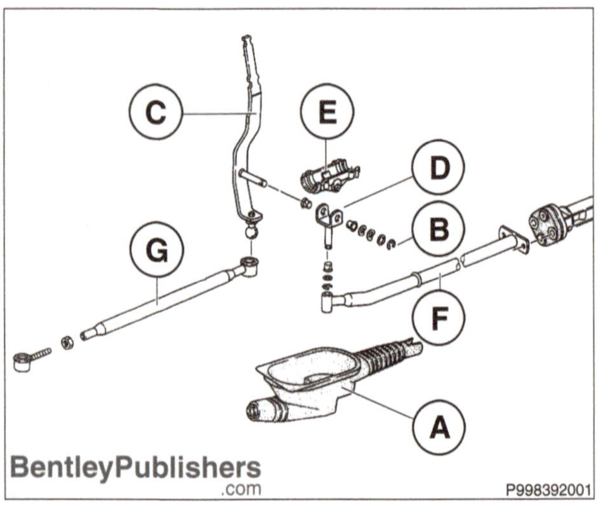

◄ Pull back bellows (**A**) over shift lever base under center of car. Remove C-clip (**B**) from shift lever retaining rod. Push shift lever (**C**) out of fork (**D**) and slide (**E**). Lower shift lever (**C**) with shift linkage (**F**) and shift guide rod (**G**) from car.

NOTE–
- *When removing C-clip (**B**), note position of washers between clip and shift lever retaining rod.*

- Remove front differential. See **Front differential, removing and installing** in this repair group.

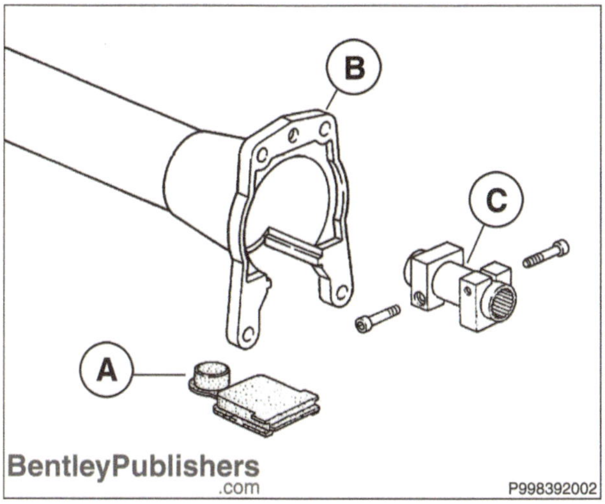

< Remove cover (**A**) at transmission end of torque tube (**B**). Once removed, clamping sleeve (**C**) and bolts will be visible.

39

< Remove clamping sleeve bolt (**A**). Rotate driveshaft until bolt (**B**) is visible through circular opening. Remove bolt (**B**). Slide clamping sleeve toward front of car to disengage transmission shaft from driveshaft.

< Support driveshaft torque tube with transmission jack or equivalent. Remove lower bolts (**arrows**) securing torque tube to transmission.

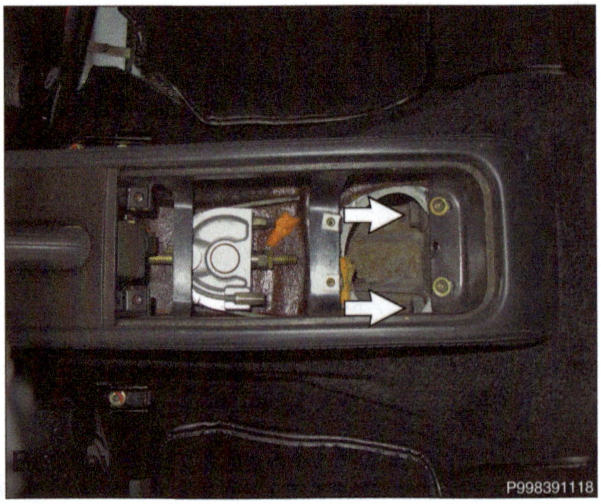

< Working inside car behind emergency brake lever, remove two upper bolts (**arrows**) securing torque tube to transmission.

– With helper, carefully push torque tube forward, then remove at angle from car.

– Installation is reverse of removal. Note the following:
  - Ensure torque tube is properly positioned on centering pins on transmission.
  - Check gearshift adjustment.

Tightening torque	
Shift linkage rod to transmission (M8)	23 Nm (17 ft-lb)
Torque tube to transmission (M12)	85 Nm (63 ft-lb)
Clamping sleeve to driveshaft (M8)	40 Nm (29 ft-lb)
Front differential mount to torque tube (M12)	85 Nm (63 ft-lb)
Front differential mount to front-axle crossmember (M8)	23 Nm (17 ft-lb)
Front differential to torque tube (M10)	46 Nm (34 ft-lb)
Drive axle to output flange (M8)	42 Nm (31 ft-lb)
Stabilizer bar to side member (M8)	23 Nm (17 ft-lb)
Stabilizer bar to stabilizer link (M10)	46 Nm (34 ft-lb)
Side member to body and cross member (M12) front facing (M10) rear facing	 90 Nm (66 ft-lb) 46 Nm (34 ft-lb)
Cross member to body (M12) outer (M10) inner	 105 Nm (77 ft-lb) 48 Nm (35 ft-lb)

Viscous coupler, function test

The AWD viscous coupler is the end housing of the manual transmission. When the rear wheels slip, the viscous coupling functions to transfers torque via the center driveshaft to the front differential.

Perform test on vehicle at room temperature.

– Raise car and support safely.

> ***WARNING—***
> - *Make sure car is stable and well supported at all times. Use a professional automotive lift or jack stands. A floor jack is not adequate support.*

– Set parking brake and shift transmission to neutral.

– With help from an assistant, hold right front wheel still while attempting to rotate left side wheel.

Left side wheel should resist rotating. If left wheel rotates with minimal force, inspect condition of viscous coupler.

Rear Differential

Rear differential components (automatic transmission)

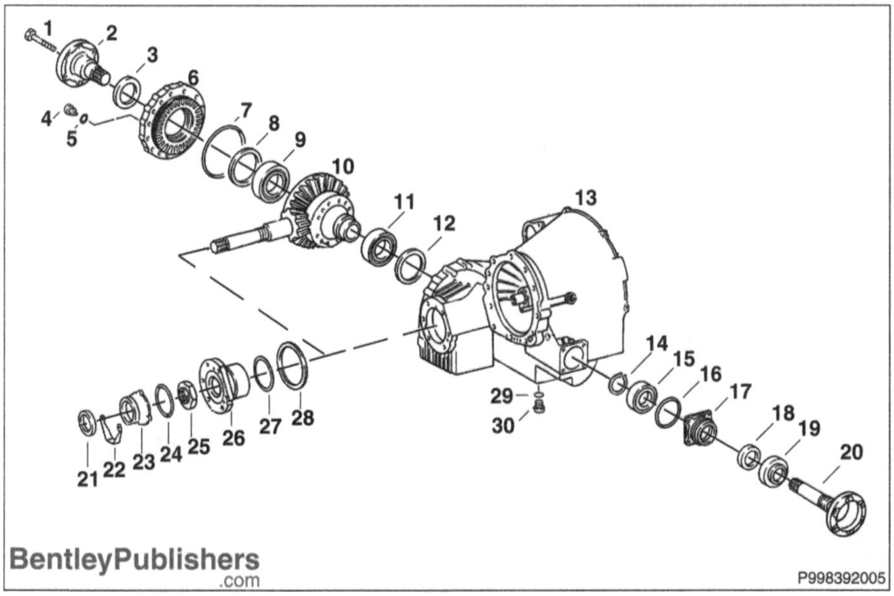

1. Bolt, M8
2. Drive axle flange
3. Rear drive axle flange seal
4. Fill plug
   - Tighten to 50 Nm (37 ft-lb)
5. Fill plug seal
6. Cover
7. Seal
8. Adjusting shim
9. Bearing
10. Differential
11. Bearing
12. Adjusting shim
13. Rear axle/differential housing
14. Circlip
15. Bearing
16. Seal
17. Housing
18. Rear drive axle flange seal
19. Protective ring
20. Drive axle flange
21. Shaft sealing ring
22. Support
23. Housing
24. Seal
25. Lock nut
26. Bearing sleeve
27. Seal
28. Adjusting shim
29. Drain plug seal
30. Drain plug
    - Tighten to 50 Nm (37 ft-lb)

Rear differential fluid level, checking (automatic transmission)

This procedure is for vehicle with Tiptronic automatic transmission. The rear differential on manual cars is integral the manual transmission. See **34 Manual Transmission**.

– Raise car and support safely.

> ***WARNING—***
> *• Make sure car is stable and well supported at all times. Use a professional automotive lift or jack stands. A floor jack is not adequate support.*

– Remove right rear wheel.

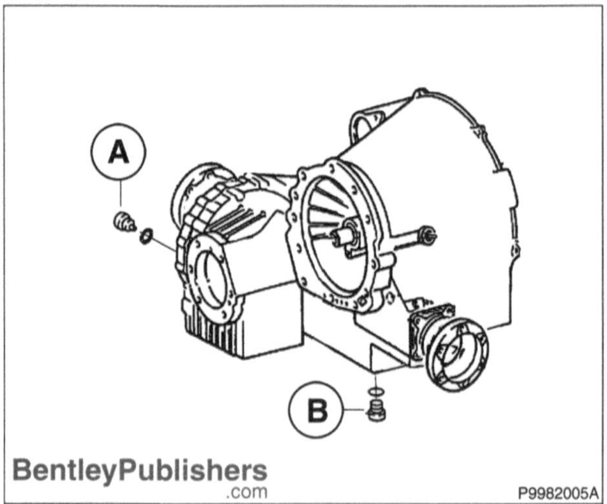

◄ Remove rear differential fill plug (**A**) located just in front of right rear drive axle flange on differential cover.

– Differential oil level should drip slightly from fill hole if level is correct. Top off oil if necessary. To change fluid see **Rear differential fluid replacing (automatic transmission)** in this repair group.

Rear differential oil (automatic transmission)	
Transaxle hypoid gear oil SAE 75W90	0.9 L (1 qt)

– Refit fill plug with new seal and tighten.

Tightening torque	
Rear differential fill plug	50 Nm (22 ft-lb)

Rear differential fluid, replacing (automatic transmission)

This procedure is for vehicles with Tiptronic automatic transmissions.

– Raise car and support safely.

> ***WARNING—***
> *• Make sure the car is stable and well supported at all times. Use a professional automotive lift or jack stands designed for the purpose. A floor jack is not adequate support.*

– Remove right rear wheel.

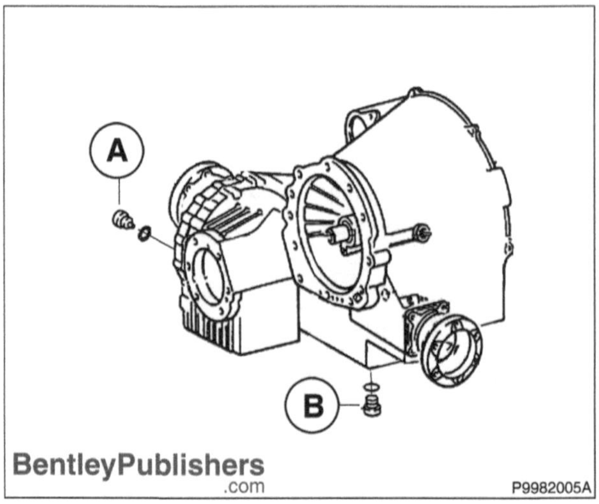

◄ Remove rear differential fill plug (**A**). Remove rear differential drain plug (**B**) and drain fluid into suitable container.

– Once drained, replace sealing ring and reinstall drain plug.

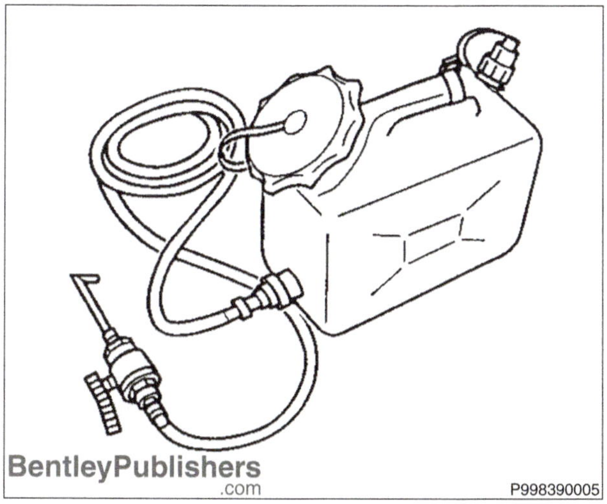

◀ Use fluid pump to fill rear differential oil until fluid runs out of fill hole.

Rear differential oil	
Transaxle hypoid gear oil SAE 75W90	0.9 L (1 qt)

NOTE–

- *Fill rear differential slowly to avoid spilling.*

– Refit fill plug with new seal and tighten.

Tightening torques	
Rear differential fill plug	50 Nm (37 ft-lb)
Rear differential drain plug	50 Nm (37 ft-lb)

Rear drive axle flange seal, replacing (automatic transmission, passenger side)

– Raise car and support safely.

> ***WARNING–***
>
> - *Make sure vehicle is stable and well supported at all times. Use a professional automotive lift or jack stands. A floor jack is not adequate support.*

– Remove right rear wheel.

– Working beneath car, remove plastic underbody shields for engine and transmission. See **03 Service and Maintenance**.

– Remove right rear heater pipe hose clamps and lift out pipe.

◀ Remove drive axle bolts (**arrows**) from drive axle flange. Separate axle from drive axle flange.

◀ Disconnect sway bar link (**A**) from sway bar. Remove toe control arm (**B**). See **42 Rear Suspension, Rear Drive Axles**. Remove bolt (**C**) holding both lower control arms to subframe and pry arms from subframe. Pull strut base (**D**) from wheel carrier.

NOTE–

- *Leave axle nut in place and leave axle attached to wheel carrier.*

Rear drive axle flange seal, replacing(automatic transmission, passenger side)

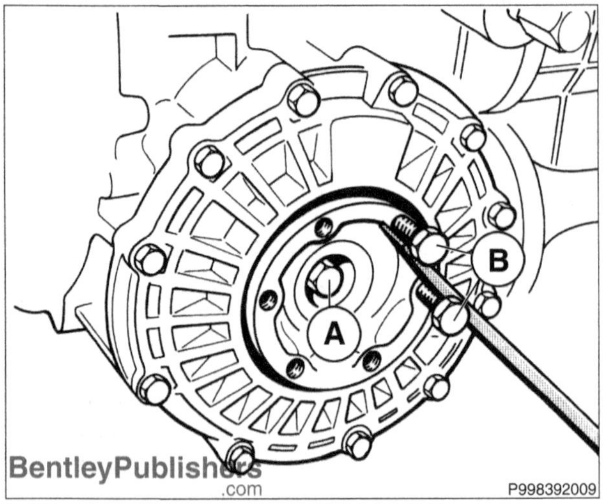

◀ Remove center drive axle flange bolt (**A**). Counterhold flange by installing two drive axle to drive axle flange bolts (**B**) and bracing with pry bar when loosening center nut.

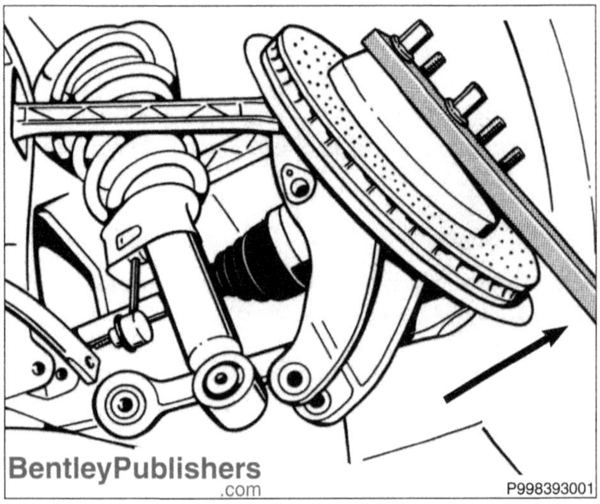

◀ With helper, pull back suspension and remove drive axle flange from differential housing.

– Lever out rear drive axle flange seal, noting position, using special tool VW 681 or equivalent.

– Installation is reverse of removal. Note the following:

- Press in rear axle drive flange seal flush with differential housing.
- Observe tightening torques.

Tightening torque	
Center flange bolt to differential	46 Nm (34 ft-lb)
Drive axle to drive axle flange	81 Nm (60 ft-lb)
Stabilizer link to stabilizer bar, M10	46 Nm (34 ft-lb)
Toe control arm to wheel carrier, M12 (use new nut)	85 Nm (63 ft-lb)
Toe control arm to subfame (eccentric), M12* grade 10.9, (use new nut)	100 Nm (74 ft-lb)
Lower control arm to subframe, use new nuts M14 M12	 200 Nm (147 ft-lb) 85 Nm (63 ft-lb)
Wheel to wheel hub, M14	130 Nm (96 ft-lb)

*If 8.8 grade bolt installed, replace with 10.9 grade bolt (part supersession). Bolt grade marked on bolt head.

Rear drive axle flange seal, replacing (automatic transmission, driver side)

- Raise car and support safely.

> ***WARNING–***
>
> - *Make sure vehicle is stable and well supported at all times. Use a professional automotive lift or jack stands. A floor jack is not adequate support.*

- Remove left rear wheel.
- Working beneath car, remove plastic under shields for engine and transmission. See **03 Service and Maintenance**.

◀ Remove drive axle bolts (**arrows**) from drive axle flange. Separate axle from drive axle flange.

- Remove drive axle. See **42 Rear Suspension, Rear Drive Axles** for more information.

39

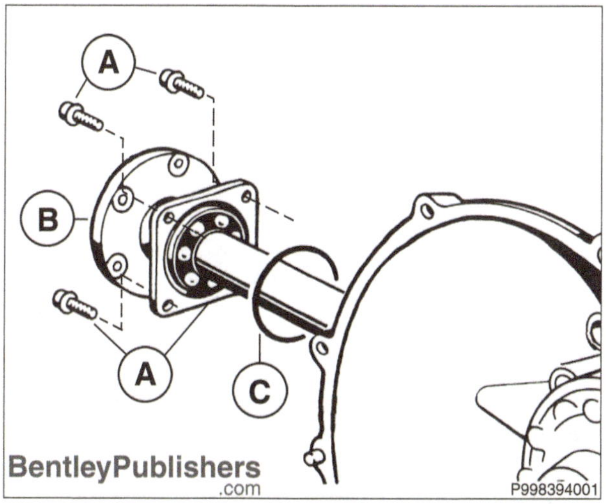

◀ Remove fasteners (**A**) holding flange to casing. Pull drive axle flange (**B**) from differential using Porsche special tool 9330 or equivalent. Ensure O-ring (**C**) is removed with drive axle flange.

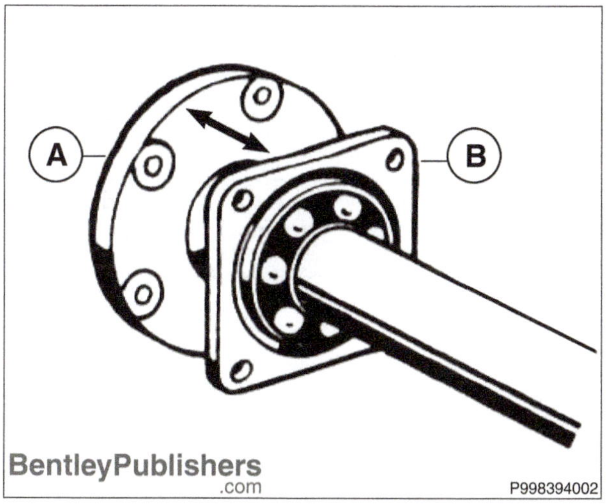

◀ Press drive axle flange (**A**) from housing (**B**) and remove drive axle flange seal.

- Installation is reverse of removal. Note the following:
  - Observe tightening torques.
  - Replace O-ring/seal from housing to differential.

Tightening torques	
Drive axle to drive axle flange	81 Nm (60 ft-lb)
Flange screws to differential casing	23 Nm (17 ft-lb)
Wheel to wheel hub, M14	130 Nm (96 ft-lb)

Rear drive axle flange seal, replacing (manual transmission)

This procedure is the same for both rear drive axle flange seals on cars with manual transmissions.

– Raise car and support safely.

> ***WARNING–***
>
> • *Make sure vehicle is stable and well supported at all times. Use a professional automotive lift or jack stands. A floor jack is not adequate support.*

– Remove rear wheel.

– Working beneath car, remove plastic under shields for engine and transmission. See **03 Service and Maintenance**.

– Remove drive axle. See **42 Rear Suspension, Rear Drive Axles**.

◄ Install two drive axle bolts (**B**) into adjacent holes in drive axle flange.

NOTE–

• *Transmission shown removed for clarity. This procedure can be completed with transmission installed in car.*

◄ Counterhold drive axle flange by bracing drive axle bolts with pry bar. Remove center flange bolt and pull out drive axle flange.

– Lever out rear drive axle flange seal, noting position, using special tool VW 681 or equivalent.

– Installation is reverse of removal. Note the following:

• Press in rear axle drive flange seal.

• Observe tightening torques.

Tightening torque	
Center flange bolt to transmission	44 Nm (32 ft-lb)
Drive axle to drive axle flange	81 Nm (60 ft-lb)

40 Front Suspension, Front Drive Axles

General

This repair group covers front suspension service. Special press tools and procedures are required to disassemble and service many of the suspension components, including the front struts and wheel bearing assemblies. Read the procedure through before beginning a repair. For additional information, see:

- **44 Wheels, Tires and Alignment.**
- **48 Steering** for tie rod end replacement.

Warnings and Cautions

WARNING—

- *Replace self-locking fasteners any time they are loosened or removed. They are designed to be used only once and may fail if reused.*
- *Do not attempt to weld or straighten any suspension components. Replace damaged parts.*

CAUTION—

- *Due to the aluminum construction of suspension components, observe the following precautions: When replacing damaged suspension components, always inspect the condition of the control arms. Do not expose aluminum components to temperatures exceeding 80°C (176°F) unless noted.*

Front suspension components

The 993 front suspension is comprised of an aluminum crossmember, two aluminum side members, and lower control arms mounted to individual wheel bearing carriers at each side. McPherson struts mount from the wheel carrier to the body. The tubular sway bar increases suspension roll stiffness.

The strut assemblies use twin-tube, gas pressure shock absorbers. The shock absorber body is threaded which allows the lower spring perch to be adjusted to increase or decrease front ride height.

Front suspension system

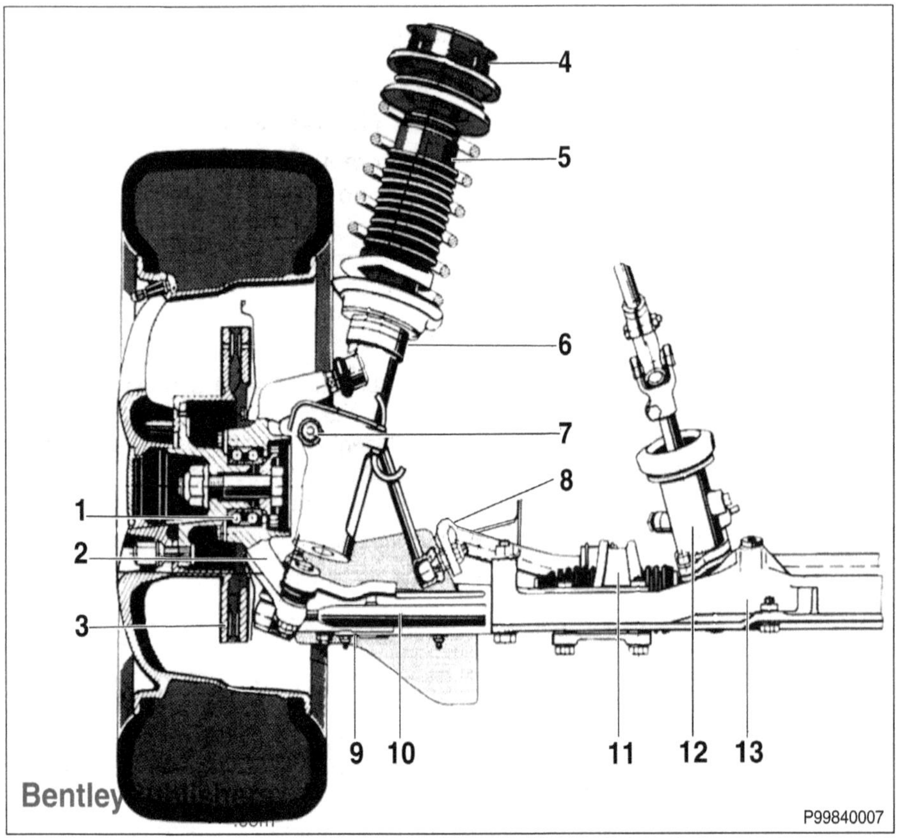

1. Wheel bearing
2. Wheel carrier
3. Brake disc
4. Strut mount
5. Strut
6. Spring retainer/height adjuster
7. Camber adjusting bolt
8. Stabilizer bar
9. Ball joint
10. Tie rod
11. Steering gear mount
12. Power steering rack
13. Crossmember

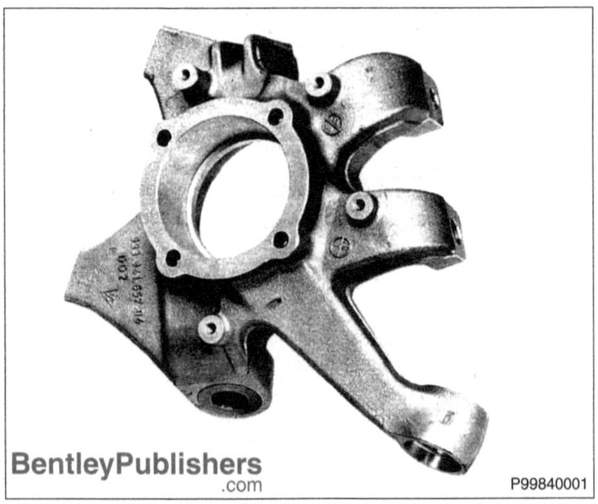

◀ The die-cast aluminum wheel carrier supports the wheel bearing, brake caliper, strut assembly, tie rod end, control arm ball joint, stabilizer bar link and ABS sensor.

Front suspension overview

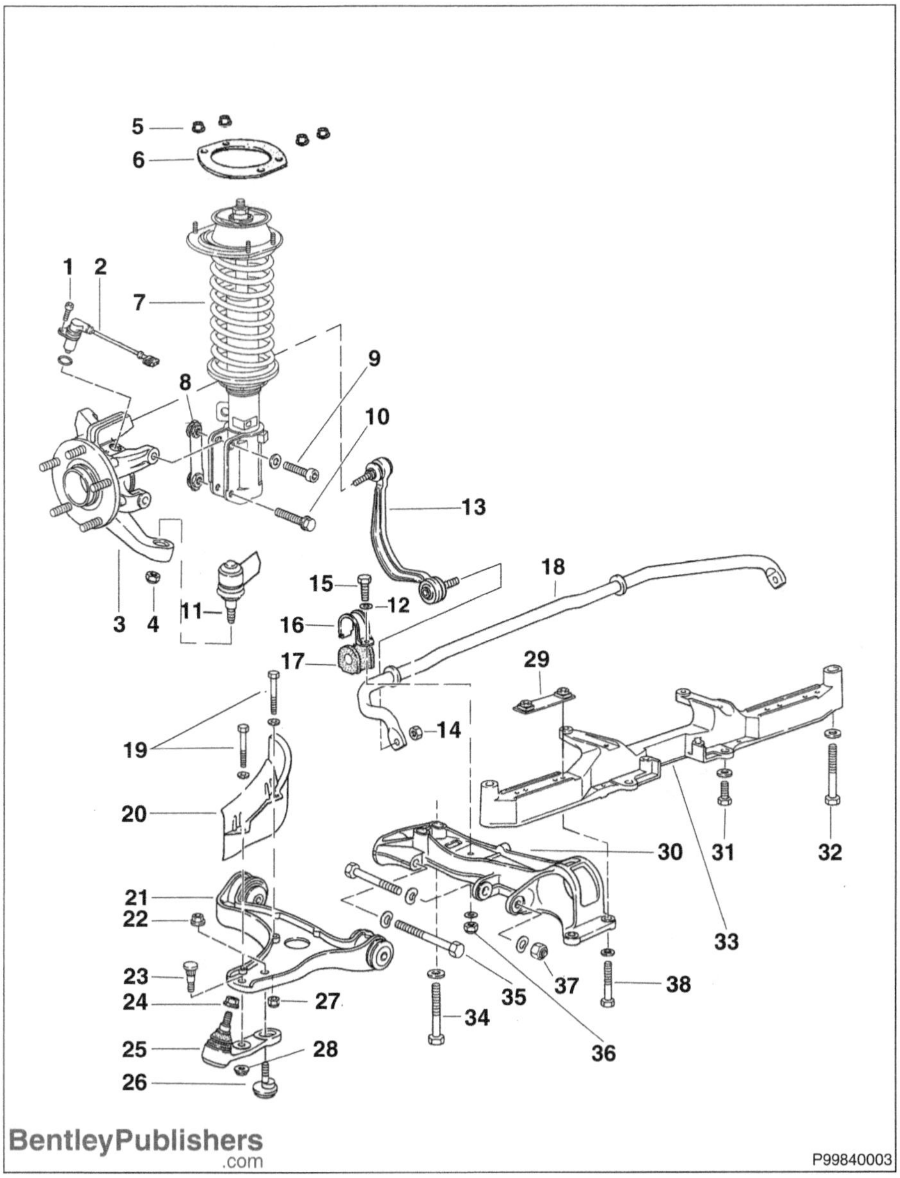

1. Bolt (Allen head)
   - 10 Nm (7 ft-lb)
2. Wheel speed sensor
3. Wheel carrier
   - If loosened or removed, front wheel alignment must be checked
4. Lock nut*
   - 65 Nm (48 ft-lb)
5. Lock nut *
   - 33 Nm (24 ft-lb)
6. Gasket*
7. Strut
8. Cage (collar nuts)*
9. Bolt (Allen-head)*
   - M12, 120 Nm (88 ft-lb)
   - Camber adjusting bolt
10. Bolt*
    - M14, 200 Nm (148 ft-lb)
11. Tie rod
12. Washer
13. Stabilizer link
14. Lock nut*
    - 46 Nm (34 ft-lb)
15. Bolt
    - M8, 23 Nm (17 ft-lb)
16. Stabilizer bushing retainer
17. Stabilizer bushing
    - Apply tire lube when assembling
18. Stabilizer bar
19. Screw
20. Air guide
21. Control arm
    - If loosened or removed, front wheel alignment must be checked
22. Lock nut*
    - M10, 80 Nm (59 ft-lb)
23. Bolt (knurled)
    - Press into control arm

*** Replace with new**

Front suspension overview (cont.)

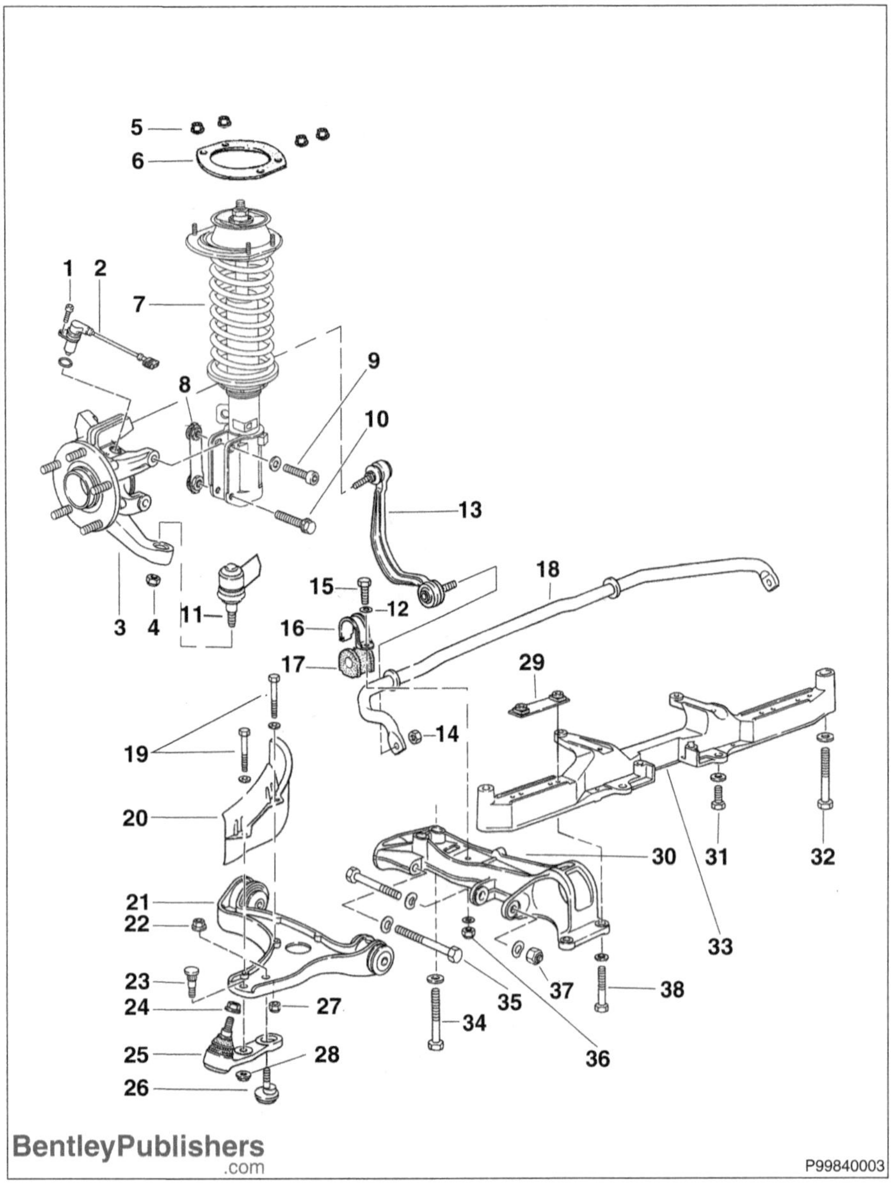

24. Lock nut*
- 75 Nm (55 ft-lb)

25. Ball joint
- Press off with ball joint removal tool
- If loosened or removed, front wheel alignment must be checked

26. Bolt (caster eccentric)

27. Lock nut (for air guide)*
- 10 Nm (7 ft-lb)

28. Lock nut*
- M12, 140 Nm (103 ft-lb)

29. Threaded plate

30. Side member

31. Bolt, crossmember to body
- M10, 45 Nm (33 ft-lb)

32. Bolt, cross member to body
- M12, 105 Nm (77 ft-lb)

33. Crossmember

34. Bolt
- M12, 90 Nm (66 ft-lb)

35. Bolt
- M12, 105 Nm (77 ft-lb)
- Tighten with rear suspension in normal loaded position (car on ground, suspension settled)

36. Lock nut*
- 23 Nm (17 ft-lb)

37. Lock nut (collar)*
- M12, 85 Nm (63 ft-lb)
- Tighten with rear suspension in normal loaded position (car on ground, suspension settled)

38. Bolt
- M10, 46 Nm (34 ft-lb)

*** Replace with new**

Front Suspension

Front suspension components on rear wheel drive and all-wheel drive models are similar. Differences will be noted within individual procedures.

Control arm, removing and installing

– Raise vehicle and support safely. Remove wheel.

> ***WARNING—***
>
> • *Make sure vehicle is stable and well supported at all times. Use a professional automotive lift or jack stands. Place jack stands underneath structural chassis points, not underneath suspension parts. A floor jack is not adequate support.*

– Remove front plastic underbody splash shield. See **03 Service and Maintenance.**

> ***CAUTION—***
>
> • *Wear eye protection and use care when removing underbody splash shield as dirt and debris collects above.*

◄ Remove nut (**arrow**) from control arm ball joint. If necessary, counterhold ball joint stud with Torx T40 driver while loosening nut.

◄ Using ball joint press tool (**inset**), press ball joint from wheel carrier.

> ***CAUTION—***
>
> • *To prevent damage to ball joint rubber boots, coat press tool and rubber boots with tire mounting paste prior to removal.*

- Remove nut (**A**) and bolt (**B**) where control arm attaches to side member. Carefully pry out remaining bolt and remove control arm.

- Installation is reverse of removal. Note the following:
  - Replace fasteners with new.
  - Ball joint taper must be free of grease prior to installation.
  - When tightening control arm ball joint to wheel carrier nut, counterhold with Torx 40 driver.
  - Tighten bushing-end of control arm bolts with rear suspension in normal loaded position (car on ground, suspension settled).

Tightening torque	
Control arm ball joint to wheel carrier (replace with new)	75 Nm (55 ft-lb)
Control arm to side member Rear (**A**) (replace with new) Front (**B**)	 85 Nm (63 ft-lb) 105 Nm (77 ft-lb)

Control arm ball joint, removing and installing

- Raise vehicle and support safely. Remove wheel.

> ***WARNING—***
> - *Make sure vehicle is stable and well supported at all times. Use a professional automotive lift or jack stands. Place jack stands underneath structural chassis points, not underneath suspension parts. A floor jack is not adequate support.*

> ***CAUTION—***
> - *Removal of the control arm ball joint affects wheel alignment. Wheel alignment must be checked any time the control arm ball joint is removed or loosened.*

- Remove front plastic underbody splash shield. See **03 Service and Maintenance.**

- Remove nut (**arrow**) from control arm ball joint. If necessary, counterhold ball joint stud with Torx T40 driver while loosening nut.

- Using ball joint press tool (**inset**), remove ball joint from wheel carrier.

> ***CAUTION—***
> - *To prevent damage to ball joint rubber boots, coat press tool and rubber boots with tire mounting paste prior to removal.*

- Mark position of eccentric bolt to control arm (**A**). Counterhold nut atop control arm and remove eccentric caster bolt. Remove outer fastener (**B**). Remove ball joint.

– Installation is reverse of removal. Note the following:
  - Replace fasteners with new.
  - Ball joint taper must be free of grease prior to installation.
  - When tightening control arm ball joint to wheel carrier nut, counterhold with Torx T40 screwdriver.
  - Check, and if necessary, adjust wheel alignment. See **44 Wheels, Tires and Alignment**.

Tightening torque	
Ball joint to wheel carrier (replace with new nut)	75 Nm (55 ft-lb)
Ball joint to control arm M12, outer, knurled (replace with new nut) M10, inner, eccentric (replace with new nut)	 140 Nm (103 ft-lb) 80 Nm (59 ft-lb)

Stabilizer bar assembly

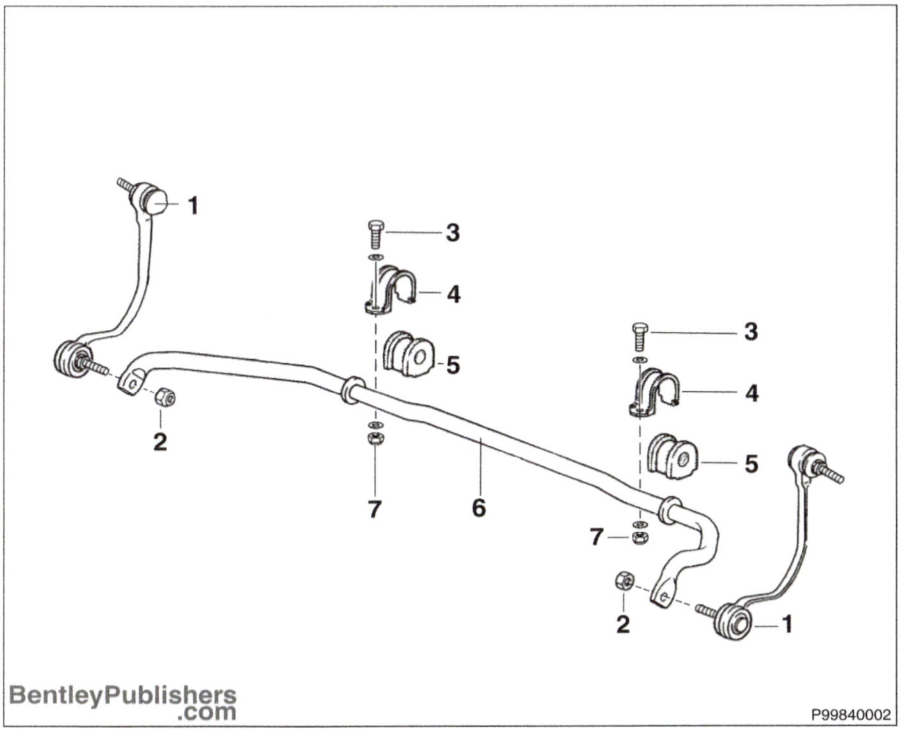

1. Stabilizer bar link
   - Upper joint threads into wheel carrier Use 16 mm open end wrench to remove
2. Lock nut
   - M10, 46 Nm (34 ft-lb)
   - Counterhold joint hex with wrench to loosen
   - Replace with new
3. Bolt
   - M8, 23 Nm (17 ft-lb)
4. Retaining bracket
5. Stabilizer bushing
   - Apply tire lube when installing
6. Stabilizer bar
7. Nut
   - M8, 23 Nm (17 ft-lb)
   - Replace with new

40

Front Struts

Front strut assembly

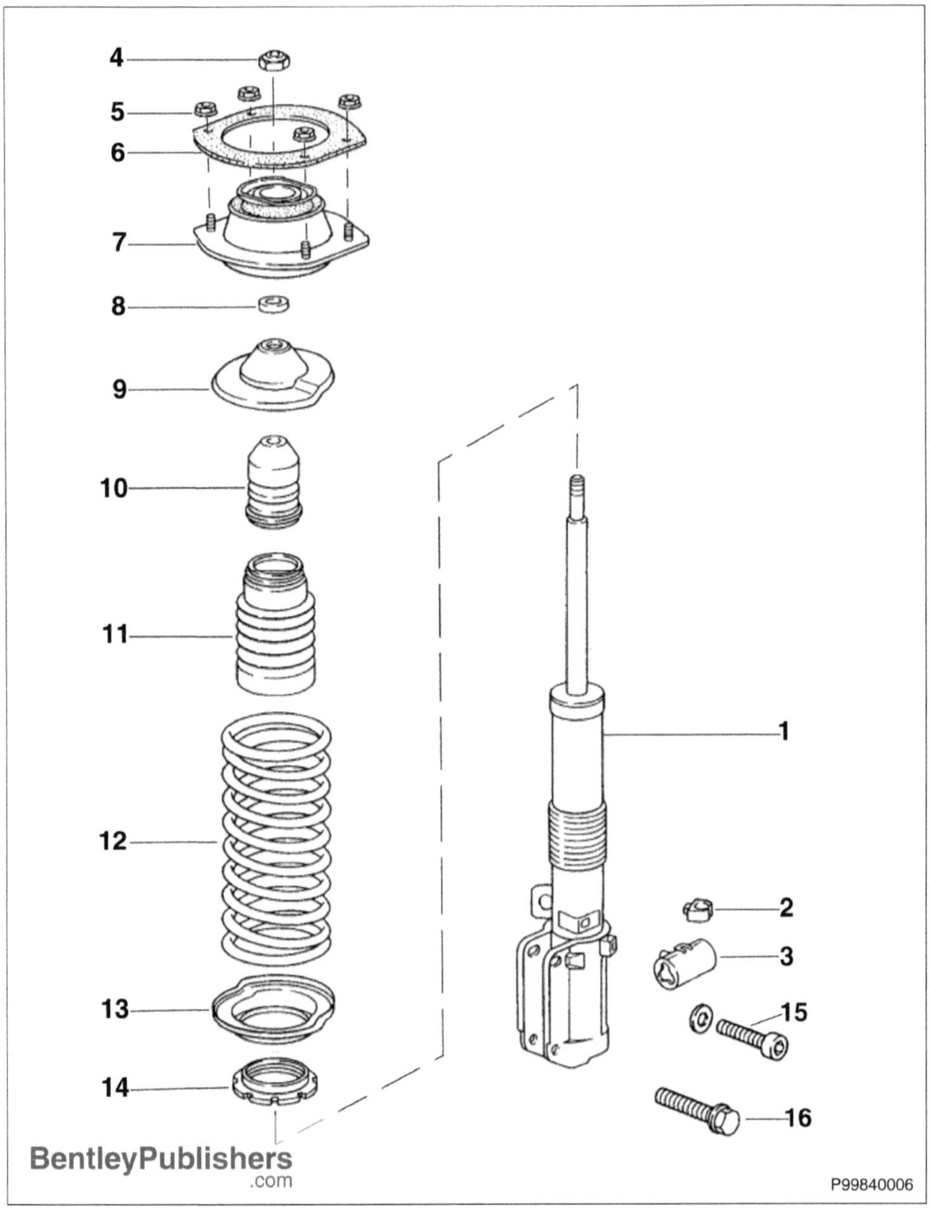

1. Shock absorber
2. Clip
3. Push-on connector
4. Nut
   - M14, 80 Nm (59 ft-lb)
   - Relieve spring tension before removing
   - Counterhold piston using 7 mm Allen wrench
   - Replace with new
5. Nut
   - M8, 33 Nm (24 ft-lb)
   - Replace with new
6. Gasket
7. Strut mount
8. Spacer sleeve
9. Upper spring perch
10. Bump stop
11. Bellows
12. Coil spring
13. Lower spring perch
14. Adjusting nut
    - See **44 Wheels, Tires and Alignment** for ride height adjusting specifications
15. Bolt (Allen-head)
    - M12, 120 Nm (88 ft-lb)
    - Camber adjusting bolt
    - Replace with new
16. Bolt
    - M14, 200 Nm (148 ft-lb)
    - Replace with new

Front strut, removing and installing

NOTE–

- *If replacing front strut, measure and record front ride height prior to strut removal. See* **44 Wheels, Tires and Alignment**.

– Raise vehicle and support safely. Remove front wheel.

> ***WARNING—***
>
> - *Make sure vehicle is stable and well supported at all times. Use a professional automotive lift or jack stands. Place jack stands underneath structural chassis points, not underneath suspension parts. A floor jack is not adequate support.*

◄ Open connector cover (**A**) and disconnect electrical connectors (**B**) for wheel speed sensor. Disconnect harness from strut bracket (**C**) and move aside.

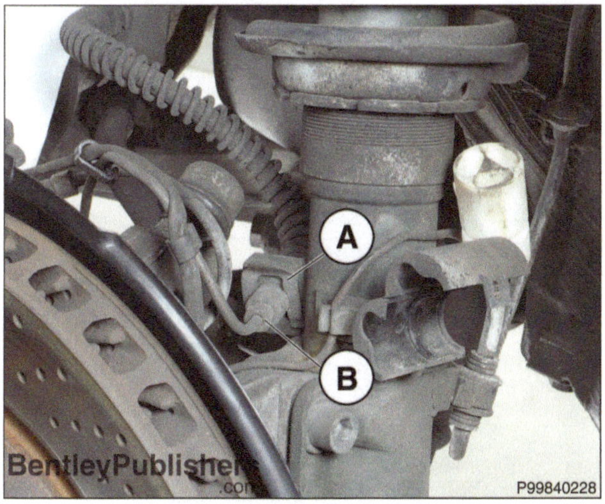

◄ Remove retaining clip (**A**) from brake line. Separate brake line (**B**) at strut bracket and pull out of bracket.

> ***CAUTION—***
>
> - *Thoroughly clean surfaces contaminated with brake fluid.*
> - *Bleed brakes when refitting. See* **47 Brakes-Hydraulic**.

– Mark position of spring strut base to wheel housing (front wheel camber position).

NOTE–

- *Front wheel camber is set via the upper bolt on the strut base. Camber is altered by changing the position of the strut in relation to the wheel carrier. The strut upper bolt hole is elongated.*

◄ Remove two retaining bolts (**arrows**) securing strut to wheel carrier.

While supporting strut from below, remove four nuts (**arrows**) at upper strut mount. Lower strut assembly out of vehicle.

> ***CAUTION—***
> - *Avoid overextending suspension during strut removal to prevent damage to tie rod ball joint.*

- Support suspension with stiff wire after removing strut.
- Installation is reverse of removal. Note the following:
  - Use new self-locking fasteners.
  - Paint mark on strut should face forward (strut mount is not symmetrical, strut piston is offset towards rear).
  - Bleed brakes. See **47 Brakes-Hydraulic**.
  - If replacing front strut, match position of adjusting nut on original strut to new strut to maintain proper ride height.
  - Inspect condition of coil spring and spring perches.
  - Follow tightening torques when refitting bolts and fasteners.
  - Check front ride height and front wheel alignment. See **44 Wheels, Tires and Alignment**.

> ***CAUTION—***
> - *If work was performed or parts replaced that affect vehicle ride height, a complete wheel alignment is necessary.*

Tightening torque	
Strut to wheel carrier M12 Allen-head (replace with new) M14 hex-head (replace with new)	 120 Nm (88 ft-lb) 200 Nm (148 ft-lb)
Lock nut, strut mount to body (replace with new)	33 Nm (24 ft-lb)

Front Wheel Bearings And Carrier

Front wheel carrier assembly

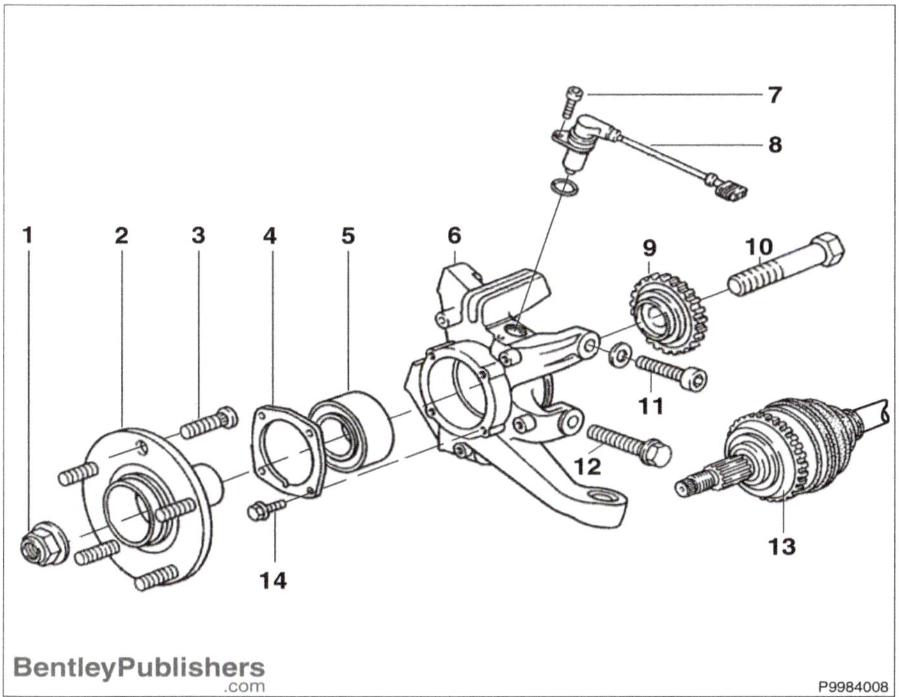

1. Lock nut
   - M22, 460 Nm (340 ft-lb)
   - Replace with new
2. Front wheel hub
3. Wheel bolt (stud)
   - Wheel nut : M14, 130 Nm (96 ft-lb)
4. Cap
5. Wheel bearing
6. Wheel carrier
7. Screw (Allen-head)
   - 10 Nm (7 ft-lb)
8. Wheel sensor
9. ABS pulse wheel (Carrera 2 models)
10. Tension bolt (Carrera 2 models)
11. Bolt (Allen-head)
    - M12, 120 Nm (88 ft-lb)
    - Replace with new
12. Bolt (hex-head)
    - M14, 200 Nm (148 ft-lb)
    - Replace with new
13. Front drive axle (Carrera 4 models)
14. Bolt (hex-head)
    - M8, 23 Nm (17 ft-lb)

40

Front wheel carrier, removing and installing

< Carrera 4 models: Using a breaker bar with a 32 mm socket, loosen drive axle nut. Discard nut.

- *If planning on servicing the wheel bearing on Carrera 2 models, loosen the tension bolt nut (use 32 mm socket). The tension bolt will need to be removed as part of replacing the wheel bearing.*

– Raise vehicle and support safely. Remove wheel.

> ***WARNING—***
>
> - *Make sure vehicle is stable and well supported at all times. Use a professional automotive lift or jack stands. Place jack stands underneath structural chassis points, not underneath suspension parts. A floor jack is not adequate support.*

– Remove front plastic underbody splash shield. See **03 Service and Maintenance.**

Front wheel carrier, removing and installing

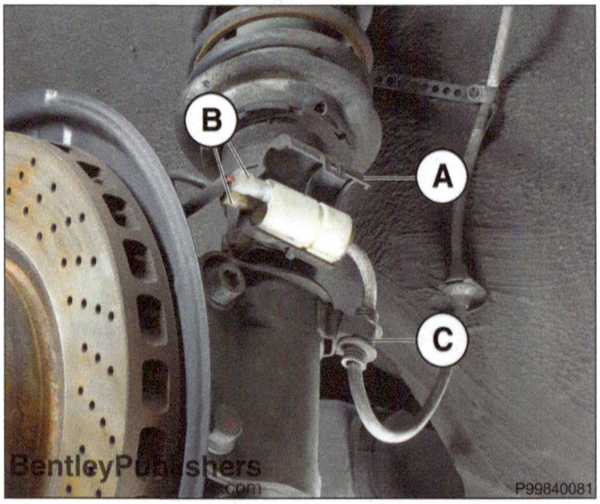

- Open connector cover (**A**).
- Disconnect electrical connectors (**B**) for wheel speed sensor (ABS sensor).
- Disconnect harness from strut bracket (**C**) and move aside.
- Remove ABS wheel speed sensor bolt then remove sensor from wheel carrier.

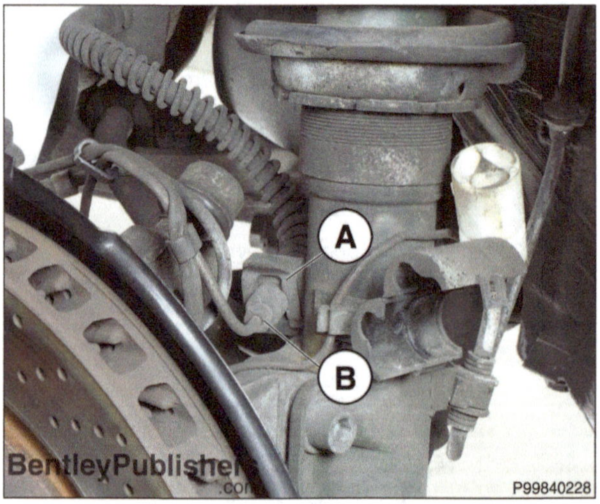

- Remove retaining clip (**A**) from brake line.
- Separate brake line (**B**) at strut bracket then pull out bracket.

> ***CAUTION—***
> - *Thoroughly clean surfaces contaminated with brake fluid.*
> - *Bleed brakes when refitting. See* **47 Brakes-Hydraulic**.

- Remove brake caliper bolts (**arrows**) and lift off caliper.
- Hang caliper out of way with stiff wire.
- Remove brake rotor. See **46 Brakes–Mechanical**.
- Remove two bolts securing brake cooling duct to control arm and remove duct.
- Unscrew stabilizer link from wheel carrier.

- Remove nut securing control arm ball joint. Using ball joint press tool (**inset**), remove ball joint from wheel carrier.

> ***CAUTION—***
>
> - *To prevent damage to ball joint rubber boots, coat press tool and rubber boots with tire mounting paste prior to removal.*

- Remove nut securing tie rod end ball joint (**arrow**).
- Using ball joint press tool, remove tie rod end from wheel carrier.

> ***CAUTION—***
>
> - *To prevent damage to ball joint rubber boots, coat press tool and rubber boots with tire mounting paste prior to removal.*

- Mark position of spring strut base to wheel housing (front wheel camber position).

NOTE–

- *Front wheel camber is set via the upper bolt on the strut base. Camber is altered by changing the position of the strut in relation to the wheel carrier. The strut upper bolt hole is elongated.*

- Remove two retaining bolts (**arrows**) securing strut to wheel carrier.
- All wheel drive (Carrera 4) vehicles: Remove nut from drive axle. Press drive axle from wheel carrier.

> ***CAUTION—***
>
> - *Do not place load on wheel hub with axle removed. Damage to wheel bearing will result.*

– Installation is reverse of removal. Note the following:
  - Use new self-locking fasteners.
  - Observe tightening torques.
  - Check front wheel alignment. See **44 Wheels, Tires and Alignment.**

Tightening torques	
Control arm ball joint to wheel carrier (use new nut)	75 Nm (55 ft-lb)
Spindle nut to tension bolt (C2 vehicles) (use new nut)	460 Nm (340 ft-lb)
Brake caliper to wheel carrier (M12, replace with new)	85 Nm (63 ft-lb)
Drive axle to wheel hub (C4 vehicles) (use new nut)	460 Nm (340 ft-lb)
Stabilizer link to stabilizer bar (M10) (use new nut)	46 Nm (34 ft-lb)
Strut to wheel carrier M12 Allen-head (replace with new) M14 hex-head (replace with new)	 120 Nm (88 ft-lb) 200 Nm (148 ft-lb)
Tie rod end to wheel carrier (use new nut)	65 Nm (48 ft-lb)
Wheel to wheel hub	130 Nm (96 ft-lb)
Wheel speed sensor to wheel carrier	10 Nm (7 ft-lb)

Front wheel bearing, removing

– Remove front wheel carrier. See **Front wheel carrier, removing and installing**.

NOTE–

– On Carrera 2 models, remove tension bolt nut and remove tension bolt together with ABS pulse wheel.

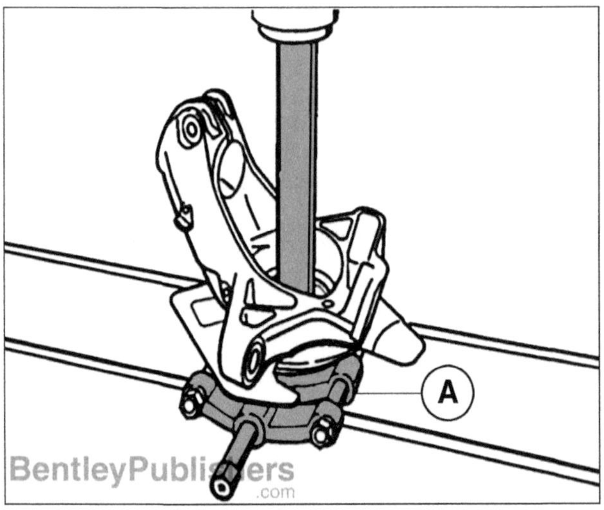

◀ Install bearing separator (**A**) between wheel carrier and wheel hub. With wheel carrier supported, press hub out of wheel bearing.

NOTE–

- *Wheel carrier shown for purposes of illustrating press tool use. Wheel carrier used on the Porsche 993 model is similar.*

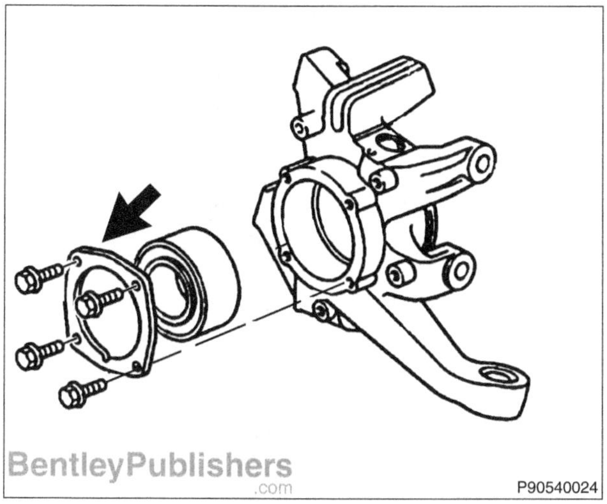

◄ Remove brake dust shield. Remove fasteners and remove bearing retainer plate (**arrow**).

– Heat wheel carrier to 100°C (212°F). Do not exceed 120°C (248°F).

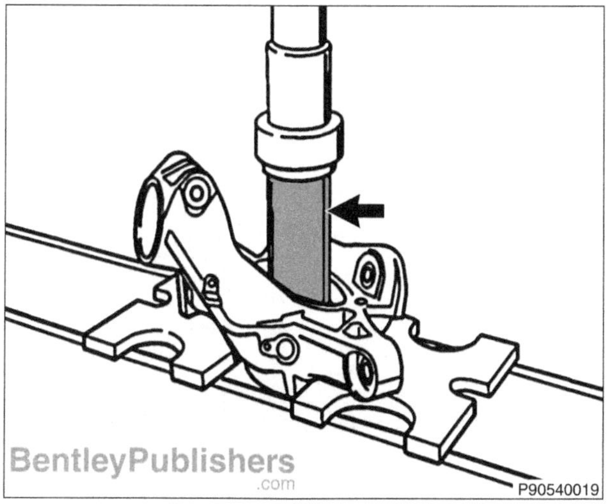

◄ Using press tool (**arrow**), press wheel bearing out of wheel carrier.

NOTE–

- *Wheel carrier shown for purposes of illustrating press tool use. Wheel carrier used on the Porsche 993 model is similar.*

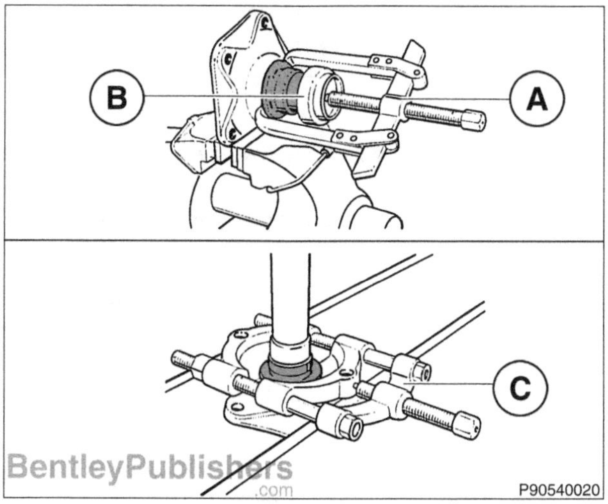

◄ Using two-jaw puller (**A**) in combination with protective cup (**B**), pull inner bearing race several millimeters off wheel hub. Install bearing separator (**C**) and fully separate inner bearing race from wheel hub.

Front wheel bearing, installing

Before installing wheel bearing, clean bore in wheel bearing housing and remove any surface imperfections with a polishing cloth.

– Heat wheel carrier to 100°C (212°F). Do not exceed 120°C (248°F). Once heated, insert new wheel bearing into front of wheel carrier. If necessary, lightly press bearing into carrier until fully seated.

– Install brake dust shield and bearing retainer plate.

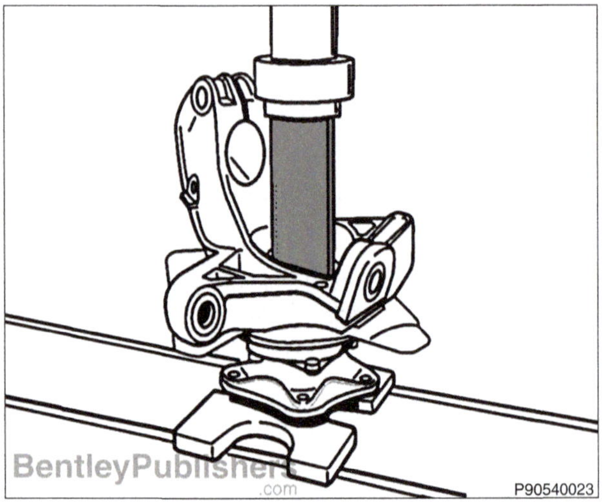

◄ Place wheel carrier on wheel hub. While pressing on bearing inner race, press until hub is fully seated in bearing.

– Remainder of installation is reverse of removal. Tighten spindle nut (Carrera 2) or drive axle nut (Carrera 4) to final torque with wheel installed and car on ground.

Tightening torques	
Control arm ball joint to wheel carrier (use new nut))	75 Nm (55 ft-lb)
Spindle nut to tension bolt (Carrera 2) (use new nut)	460 Nm (340 ft-lb)
Brake caliper to wheel carrier (M12, replace with new)	85 Nm (63 ft-lb)
Drive axle nut to drive axle (Carrera 4) (use new nut)	460 Nm (340 ft-lb)
Stabilizer bar link to stabilizer bar (M10) (use new nut)	46 Nm (34 ft-lb)
Tie rod end to wheel carrier (use new nut)	65 Nm (48 ft-lb)
Wheel bearing retainer plate to wheel carrier (M8)	23 Nm (17 ft-lb)
Wheel to wheel hub	130 Nm (96 ft-lb)
Wheel speed sensor to wheel carrier	10 Nm (7 ft-lb)

FRONT DRIVE AXLES (CARRERA 4)

Front drive axle (Carrera 4), removing and installing

– With vehicle on ground, remove wheel center cap from wheel.

◄ Using a breaker bar with a 32 mm socket, loosen drive axle nut.

– Raise vehicle and support safely. Remove wheel.

> ***WARNING–***
>
> • *Make sure the vehicle is stable and well supported at all times. Use a professional automotive lift or jack stands designed for the purpose. A floor jack is not adequate support.*

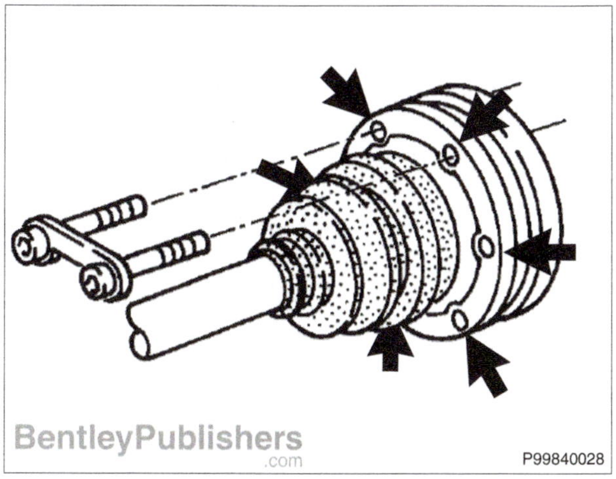

- Working beneath car, remove front plastic underbody splash shield. See **03 Service and Maintenance**.

> ***CAUTION—***
> - *Wear eye protection and use care when removing underbody splash shield as dirt and debris collects above.*

- ◄ Remove CV joint (drive axle) mounting fasteners (**arrows**).
- Detach CV joint from drive flange.
- Temporarily install two 8 mm bolts with nuts to hold grease cap in place.
- Remove brake cooling guide from control arm.

- ◄ Remove nut (**arrow**) from control arm ball joint. If necessary, counterhold ball joint stud with Torx T40 driver while loosening nut.

- ◄ Using ball joint press tool, remove ball joint from wheel carrier (**inset**).

> ***CAUTION—***
> - *To prevent damage to ball joint rubber boots, coat press tool and rubber boots with tire mounting paste prior to removal.*

- Detach stabilizer bar link from stabilizer bar. Counterhold link using open-ended wrench to remove fastener.

40

Front drive axle (Carrera 4), removing and installing

◄ Remove nut from tie rod end. Using ball joint press tool, remove tie rod end from wheel carrier.

> ***CAUTION—***
> - *To prevent damage to ball joint rubber boots, coat press tool and rubber boots with tire mounting paste prior to removal.*

◄ Use press tool (**arrow**) to press drive axle from wheel carrier.

– Pull wheel carrier out and carefully remove drive axle from vehicle.

– Installation is reverse of removal, note the following:

- Inspect all parts before assembly. If drive axle CV boots are damaged, replace boots and repack CV joint grease before installing drive axle.
- Lubricate drive axle splines lightly with Optimoly HT grease.
- Carefully pull wheel carrier out to insert drive axle into wheel hub.
- Tighten drive axle nut with car on ground. Use new nut and new self-locking fasteners.

Tightening torques	
Control arm ball joint to wheel carrier (use new nut)	75 Nm (55 ft-lb)
Drive axle to output flange (M8)	42 Nm (31 ft-lb)
Drive axle to wheel hub (use new nut)	460 Nm (340 ft-lb)
Stabilizer link to stabilizer bar (M10) (use new nut)	46 Nm (34 ft-lb)
Tie rod end to wheel carrier (use new nut)	65 Nm (48 ft-lb)
Wheel to wheel hub	130 Nm (96 ft-lb)

42 Rear Suspension, Rear Drive Axles

General

This repair group covers rear suspension service. Special press tools and procedures are required to disassemble and service the wheel bearings. Read through the procedure before beginning any repair.

Warnings and Cautions

WARNING—

- *Do not reuse self-locking nuts or bolts. They are designed to be used only once and may fail if reused. Always replace self-locking fasteners any time they are loosened or removed.*
- *Do not attempt to weld or straighten any suspension components. Replace damaged parts.*

CAUTION—

- *Due to the aluminum construction of suspension components, observe the following precautions: When replacing damaged suspension components, always check the condition of the control arms. Do not expose aluminum components to temperatures exceeding 80°C (176° F) unless noted.*

Rear suspension components

The 993 rear suspension assembly is a light weight multi-link assembly. There are two upper control arms, a lower toe control arm and a lower A-arm connecting the wheel carrier to the subframe. The entire subframe assembly uses four rubber body mounts to reduce noise and vibration.

The rear strut assemblies are comprised of twin-tube, gas pressure shock absorbers. The shock absorber body provides the lower perch for a cylindrical spring. The top of the spring is retained by the strut mount.

Rear suspension system

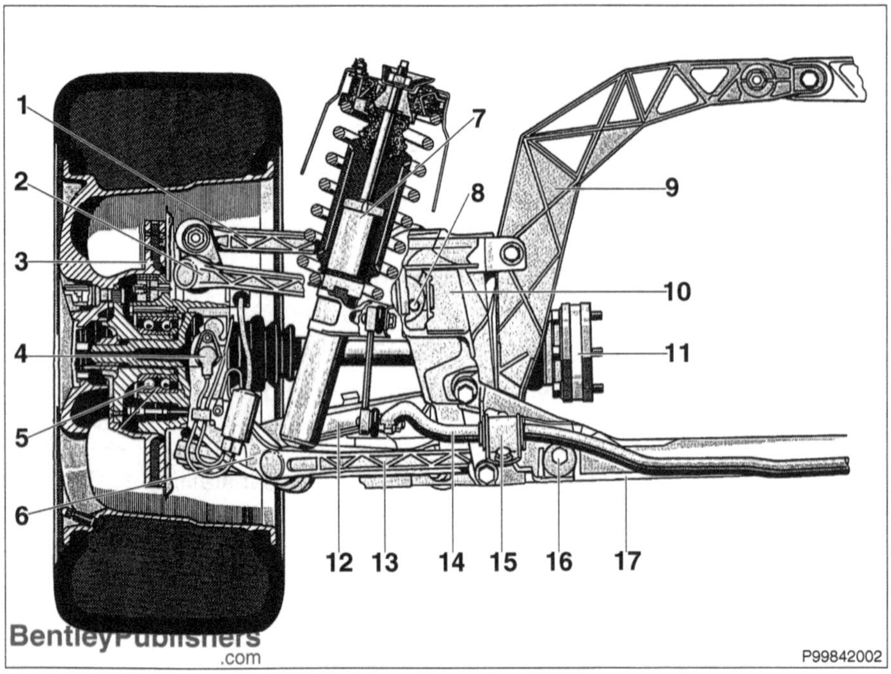

1. Camber control arm
2. Caster control arm
3. Brake rotor
4. Wheel speed sensor
5. Wheel bearing
6. Wheel carrier
7. Strut
8. Caster eccentric
9. Upper cross member
10. Rear axle side subframe
11. Drive axle
12. Lower control arm (A-arm)
13. Toe control arm
14. Stabilizer bar
15. Stabilizer bar retainer
16. Toe eccentric
17. Cross member

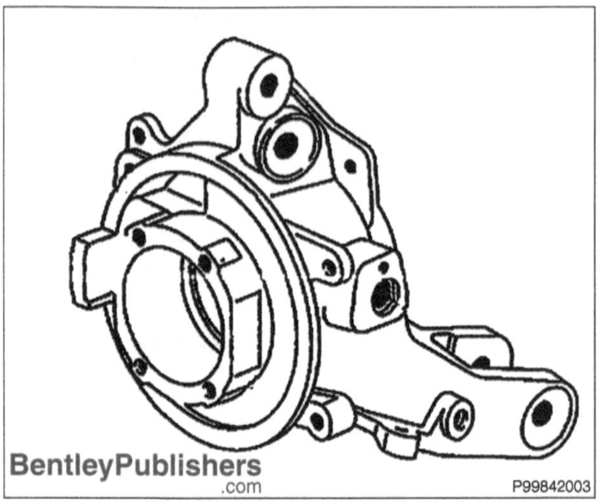

◄ The die-cast aluminum wheel carrier provides mounting for the wheel bearing, brake caliper, strut assembly, parking brake, control arms and wheel speed sensor.

Rear suspension overview

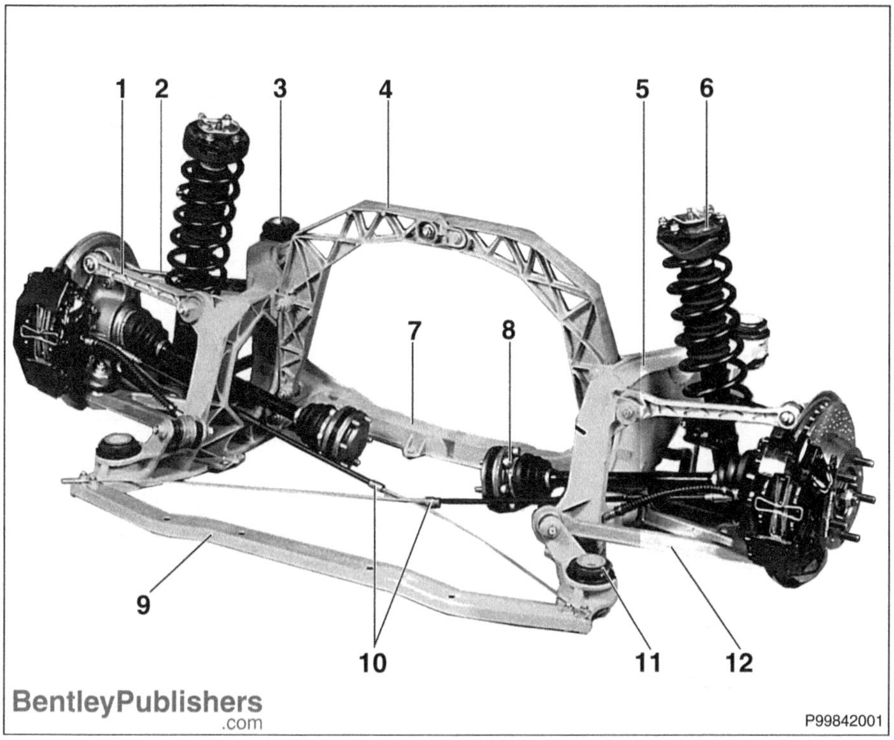

1. Camber control arm
2. Caster control arm
3. Suspension mount
   - Rubber
4. Upper crossmember
   - Two sections
5. Rear axle side subframe
6. Strut and coil spring assembly
7. Crossmember
8. Rear drive axle
9. Front cross brace
10. Emergency brake cables
11. Suspension mount
    - Rubber
12. Lower control arm

REAR STABILIZER BAR AND STRUTS

Rear stabilizer bar, removing and installing

– Raise vehicle and support safely.

> ***WARNING—***
>
> *• Make sure vehicle is stable and well supported at all times. Use a professional automotive lift or jack stands. Place jack stands underneath structural chassis points, not underneath suspension parts. A floor jack is not adequate support.*

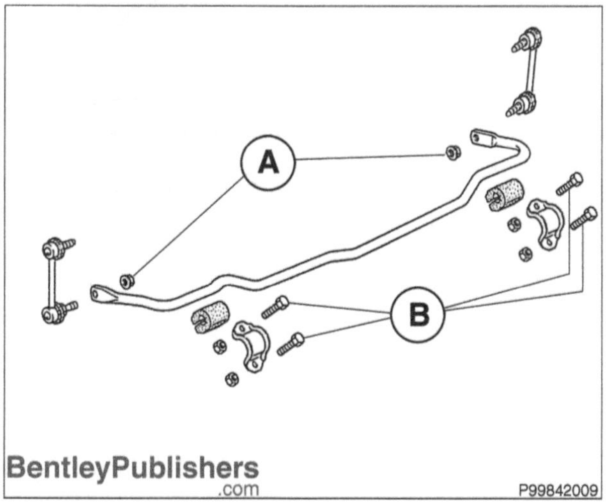

◀ Counter hold and remove stabilizer bar link to stabilizer bar nuts (**A**). Remove stabilizer bar to crossmember bolts (**B**).

– Remove stabilizer bar.

– Installation is reverse of removal.

Tightening torques	
Stabilizer bar link to stabilizer bar, M10 (use new nut)	46 Nm (34 ft-lb)
Stabilizer bar link to strut, M10 (use new nut)	46 Nm (34 ft-lb)
Stabilizer bar retaining bracket to cross member, M8 (use new nuts)	23 Nm (17 ft-lb)

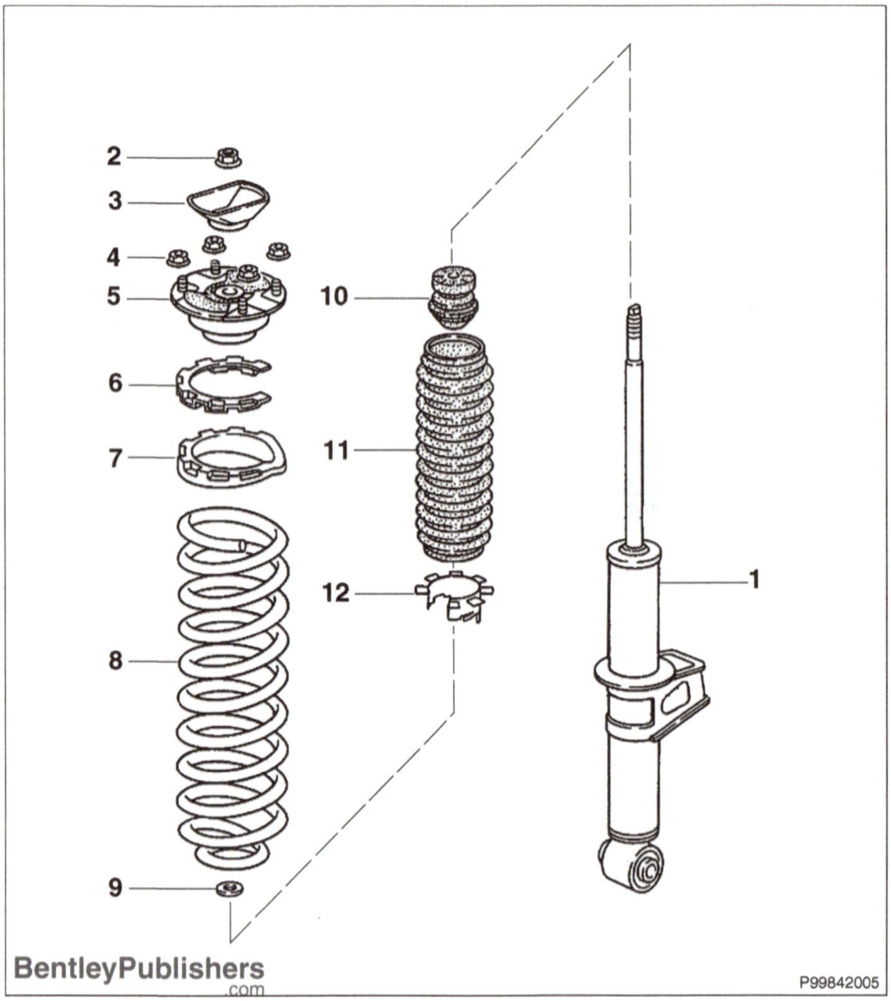

Rear strut assembly

1. Shock absorber
2. Nut, M12
   - Compress coil spring with spring compressor and counter hold shock absorber piston rod with 7 mm wrench before removing.
   - 58 Nm (43 ft-lb)
   - Replace with new
3. Stop plate
4. Nut, M8
   - 33 Nm (24 ft-lb)
   - Replace with new
5. Strut mount
   - Mounts are identical. Markings R for right and L for left for positioning only.
6. Intermediate section
   - Thickness depending on coil spring installed: 3.0 mm, 5.5 mm and 8 mm.
7. Spring support
8. Coil spring
   - Compress coil spring with spring compressor before disassembling strut.
9. Washer
10. Bump stop
11. Bellows
12. Mount (support clip)

Rear strut, removing and installing

– Measure and record rear ride height. See **44 Wheels. Tires, and Alignment**.

– Raise vehicle and support safely. Remove rear wheel.

> ***WARNING—***
> • *Make sure vehicle is stable and well supported at all times. Use a professional automotive lift or jack stands. Place jack stands underneath structural chassis points, not underneath suspension parts. A floor jack is not adequate support.*

– Remove engine and transmission protection panels. See **03 Service and Maintenance**.

– Remove stabilizer bar link from stabilizer bar. Keep link attached to strut.

◄ Mark position of toe control arm eccentric bolt (**arrow**).

– Remove toe control arm eccentric retaining bolt while counterholding retaining nut.

< Loosen toe control arm bolt (**arrow**) at wheel carrier. Do not remove.

- Counter hold fastening bolt with Torx 40.

> ***CAUTION—***
> - *Avoid damage to drive shaft by placing protective tube over if available. Avoid over extending drive shaft.*

– For left side strut removal:

- Remove heater blower from engine compartment. See **80 Heating**.

– For right side strut removal:

- Remove air filter housing from engine compartment. See **03 Service and Maintenance**.

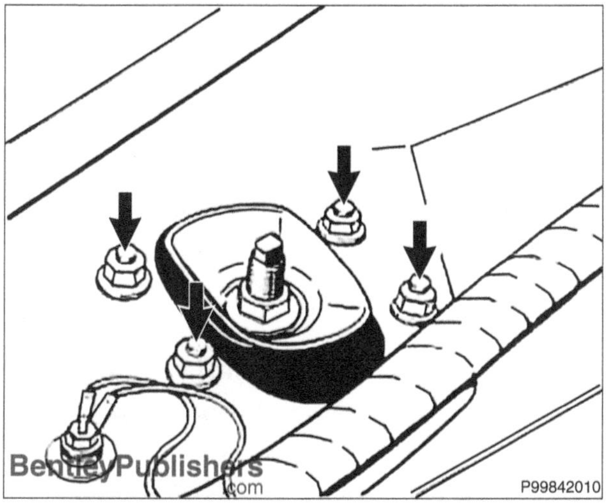

< Remove four strut mount nuts (**arrows**), noting installed position of stop plate.

– Remove toe control arm bolt at wheel carrier and remove toe control arm.

– Remove strut assembly.

– Installation is reverse of removal. Note the following:

- If coil spring removal is necessary, compress coil spring with spring compressor and counter hold shock absorber piston rod with 7 mm wrench before removing. When installing, replace upper retaining nut with new and tighten to 58 Nm (43 ft-lb). See **Rear strut assembly** in this repair group for more information.
- Replace fasteners with new.
- Do not grease fasteners coated in Dacromet finish (aluminum color).
- Refit toe control arm eccentric bolt to position marked at removal.
- Fully tighten toe control arm eccentric bolt with rear suspension in normal loaded position (car on ground, suspension settled).
- Check ride height as recorded earlier. Perform full suspension alignment.

Tightening torques	
Toe control arm to cross member (eccentric), M12, 10.9 grade*, use new nut	100 Nm (74 ft-lb)
Toe control arm to wheel carrier, M12, use new nut	85 Nm (63 ft-lb)
Strut mount to body, M8, use new nuts	33 Nm (24 ft-lb)
Stabilizer link to stabilizer bar, M10, use new nut	46 Nm (34 ft-lb)
Wheel to wheel hub, M14	130 Nm (96 ft-lb)

*If 8.8 grade bolt installed, replace with 10.9 grade bolt (part supersession). Bolt grade marked on bolt head.

42

Rear Control Arms

Multi-link axle assembly (LSA system)

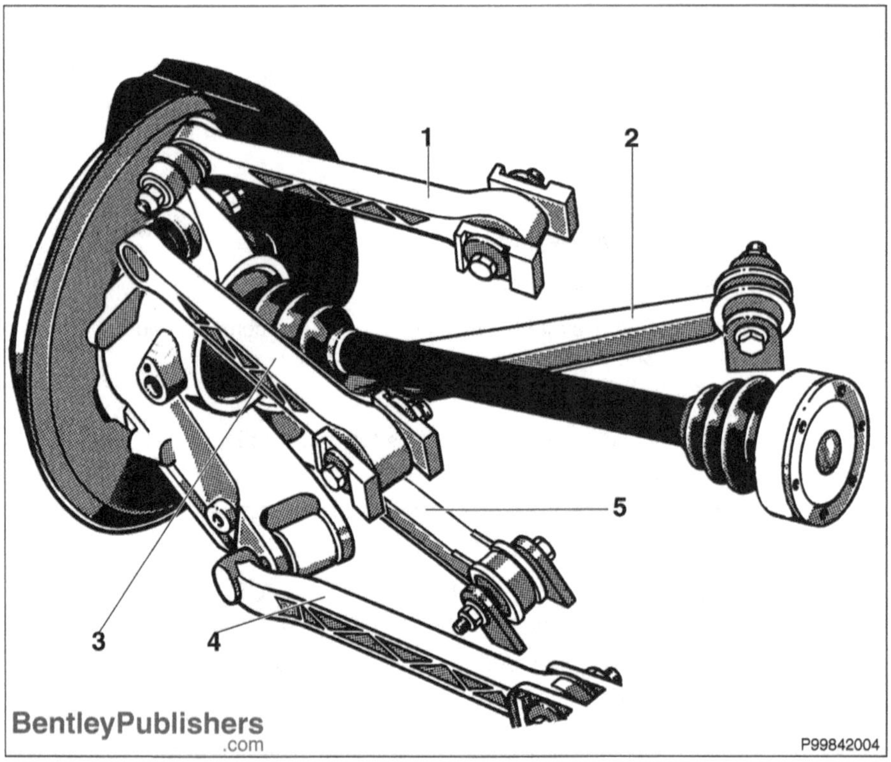

1. Camber control arm
2. Lower control A-arm (front facing)
3. Caster control arm
4. Toe control arm
5. Lower control A-arm (rear facing)

Upper control arms, removing and installing

Procedures for removing upper control arms (caster control arm and camber control arm) are identical. Fully tighten control arm eccentric bolts with rear suspension in normal loaded position (car on ground, suspension settled).

– Raise vehicle and support safely.

> ***WARNING—***
>
> - *Make sure vehicle is stable and well supported at all times. Use a professional automotive lift or jack stands. Place jack stands underneath structural chassis points, not underneath suspension parts. A floor jack is not adequate support.*

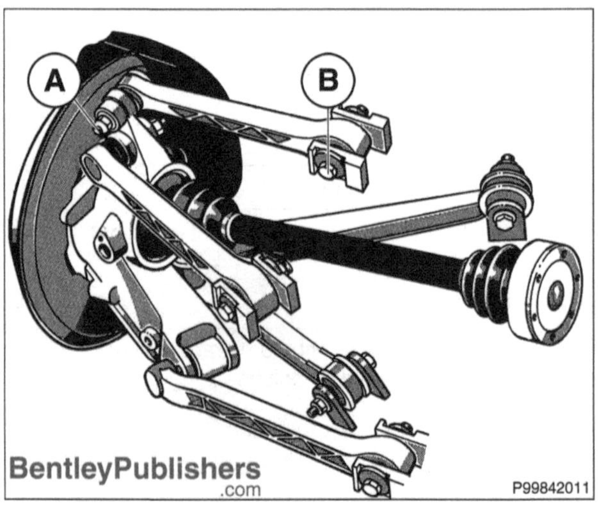

< Counter hold nut and loosen bolt (**A**).

– Mark position of eccentric bolt (**B**).

– Remove bolts (**A**) and (**B**).

– Remove control arm from side subframe.

– Press ball joint end from wheel carrier and remove control arm.

– Installation is reverse of removal. Note the following:
  - Refit eccentric bolt to marked position.
  - Use new self-locking nuts. Tighten control arm eccentric bolt with rear suspension in normal loaded position (car on ground, suspension settled).
  - Check suspension alignment.

Tightening torques	
Camber control arm to wheel carrier, M12, use new nut	75 Nm (55 ft-lb)
Camber control arm to subframe (eccentric), M12, use new nut	85 Nm (63 ft-lb)
Caster control arm to wheel carrier, M12, use new nut	75 Nm (55 ft-lb)
Caster control arm to subframe (eccentric), use new nut	85 Nm (63 ft-lb)
Wheel to wheel hub, M14	130 Nm (96 ft-lb)

Toe control arm, removing and installing

– Raise vehicle and support safely. Remove wheel.

> ***WARNING—***
>
> - *Make sure vehicle is stable and well supported at all times. Use a professional automotive lift or jack stands. Place jack stands underneath structural chassis points, not underneath suspension parts. A floor jack is not adequate support.*

◀ Mark position of eccentric bolt (**arrow**) then remove.

– Swing toe control arm from crossmember.

◀ Remove toe control arm nut (**arrow**) at wheel carrier.

- Counter hold toe control arm ball joint with Torx 40.

– Remove toe control arm from wheel carrier.

– Installation is reverse of removal. Note the following:

- Replace fasteners with new.
- Observe correct tightening torques.
- Refit eccentric bolt to marked position.
- Use new self-locking nut. Tighten control arm eccentric bolt with rear suspension in normal loaded position (car on ground, suspension settled).
- Check suspension alignment.

Tightening torques	
Toe control arm to cross member (eccentric), M12, 10.9 grade* (use new nut)	100 Nm (74 ft-lb)
Toe control arm to wheel carrier, M12 (use new nut)	85 Nm (63 ft-lb)
Wheel to wheel hub, M14	130 Nm (96 ft-lb)

*If 8.8 grade bolt installed, replace with 10.9 grade bolt (part supersession). Bolt grade marked on bolt head.

42

Lower control arm (A-arm), removing and installing

– Raise vehicle and support safely. Remove wheel.

> ***WARNING—***
>
> - *Make sure vehicle is stable and well supported at all times. Use a professional automotive lift or jack stands. Place jack stands underneath structural chassis points, not underneath suspension parts. A floor jack is not adequate support.*

◄ Remove nut (**arrow**) from lower control arm ball joint. If necessary, counter hold ball joint stud with Torx T40 driver while loosening nut.

– Using a press tool, remove ball joint from wheel carrier.

> ***CAUTION—***
>
> - *To prevent damage to ball joint rubber boots, coat tool and rubber boots with tire mounting paste prior to removal.*

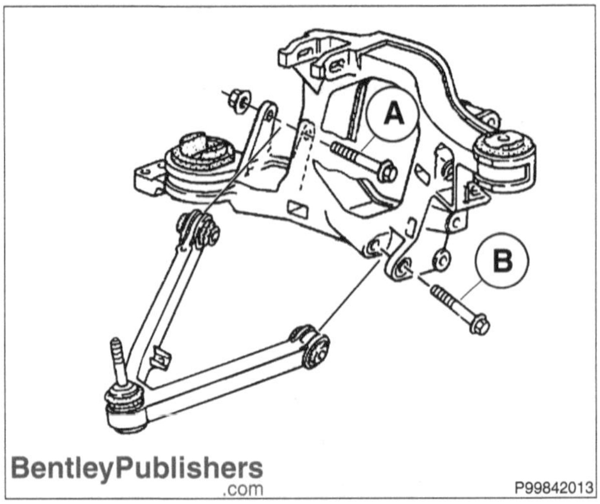

◄ Remove fastener (**A**) securing front arm of lower control arm to subframe. Counter hold nut while removing.

– Remove fastener (**B**) securing rear arm of lower control arm to subframe. Counter hold nut while removing.

– Carefully pry lower control arm from rear subframe..

– Installation is reverse of removal. Note the following:

- Use new self-locking nuts. Tighten control arm to side subframe bolts with rear suspension in normal loaded position (car on ground, suspension settled).
- Check suspension alignment.

Tightening torque	
Lower control arm to subframe, use new nuts M14 M12	 200 Nm (147 ft-lb) 85 Nm (63 ft-lb)
Lower control arm ball joint to wheel carrier, M12, use new nut	75 Nm (55 ft-lb)
Wheel to wheel hub, M14	130 Nm (96 ft-lb)

Rear Wheel Bearings

Wheel carrier assembly

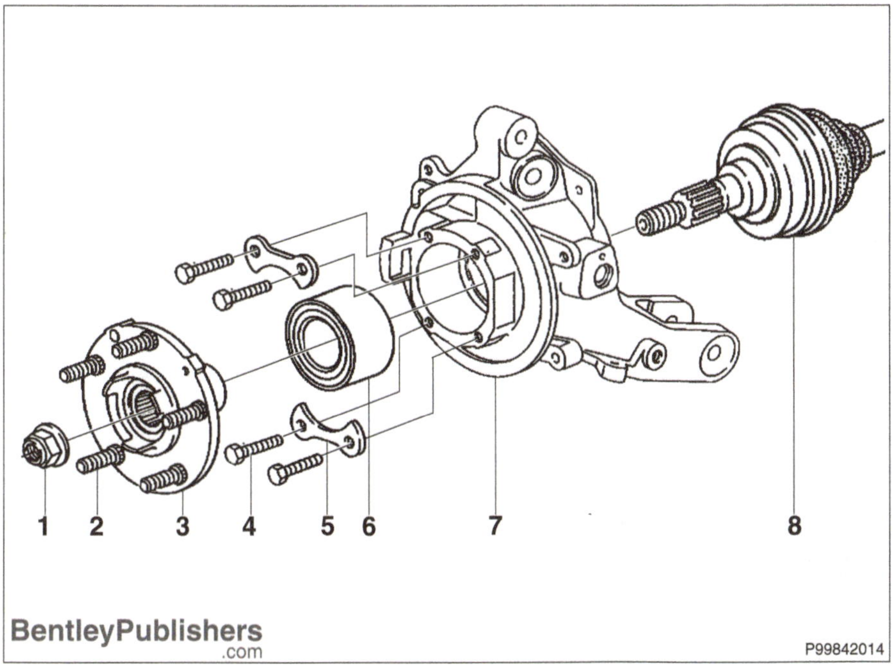

1. Axle nut
   - M22, replace with new
   - 460 Nm (340 ft-lb)
2. Wheel stud
3. Wheel hub
4. Fastener
   - M8
   - 23 Nm (17 ft-lb)
5. Bearing retainer
6. Wheel bearing
7. Wheel carrier
8. Rear drive axle

Wheel carrier, removing and installing

◀ With vehicle on ground, use a breaker bar with a 32 mm socket to loosen drive axle hex nut at wheel. If necessary, apply brakes.

– Raise vehicle and support safely. Remove wheel and axle nut.

> ***WARNING—***
>
> - *Make sure vehicle is stable and well supported at all times. Use a professional automotive lift or jack stands. Place jack stands underneath structural chassis points, not underneath suspension parts. A floor jack is not adequate support.*

– Remove engine and transmission protection panels. See **03 Service and Maintenance**.

- Working at wheel carrier, remove wheel speed sensor mounting screw (**A**).
- Remove wheel speed sensor and brake pad sensor connectors (**B**) from wheel carrier.
- Remove wheel speed sensor.
- Remove bracket fastener (**C**) for connector.
- Remove wiring from retainers (**D**).
- Unbolt brake caliper from wheel carrier and hang out of way using stiff wire. See **46 Brakes–Mechanical**.

- Counter hold stabilizer link (**B**) and remove stabilizer link nut (**A**).
- Remove brake rotor and parking brake assembly. See **46 Brakes–Mechanical**.
- Remove rear drive axle. See **Rear drive axle, removing and installing** in this repair group.

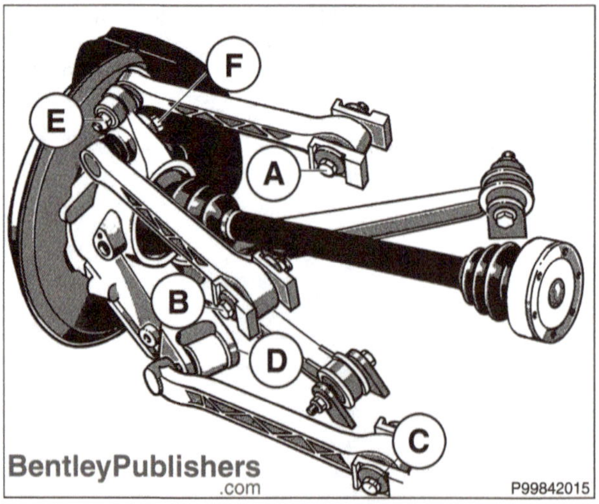

- Mark eccentric bolts (**A**), (**B**), and (**C**), then loosen.
- Remove toe control arm nut (**D**) at wheel carrier.
  - Counter hold toe control arm ball joint bolt with Torx 40.
  - Remove ball joint from wheel carrier.
- Carefully pry strut base from wheel carrier assembly.
- Remove camber control arm nut (**E**) at wheel carrier.
  - Remove ball joint from wheel carrier.
- Remove caster control arm nut (**F**) at wheel carrier.
  - Remove ball joint from wheel carrier.
- Remove lower control arm (A-arm) bolt at wheel carrier.
  - Remove ball joint from wheel carrier.
- Remove wheel carrier assembly.

Installing

– Installation is reverse of removal. Note the following:

- Replace self-locking fasteners with new.
- If replacing wheel carrier, replace with identical, or replace in pairs.
- Lubricate drive axle splines lightly with Optimoly HT grease.
- Tighten control arm eccentric fasteners and lower control arm bushing fasteners (using new self-locking nuts) with rear suspension in normal loaded position (car on ground, suspension settled).
- Use markings made earlier when installing eccentric bolts.
- Tighten drive axle nut only with car on ground. Replace drive axle hex nut.
- Check suspension alignment.

Tightening torque	
Brake caliper to wheel carrier, M12 (replace with new)	85 Nm (63 ft-lb)
Camber control arm to wheel carrier, M12 (use new nut)	75 Nm (55 ft-lb)
Camber control arm to subframe (eccentric), M12 (use new nut)	85 Nm (63 ft-lb)
Caster control arm to wheel carrier, M12 (use new nut)	75 Nm (55 ft-lb)
Caster control arm to subframe (eccentric) (use new nut)	85 Nm (63 ft-lb)
Drive axle to transmission, M10	81 Nm (60 ft-lb)
Drive axle to wheel hub, M22 (use new nut)	460 Nm (340 ft-lb)
Lower control arm to wheel carrier, M12 (use new nut)	75 Nm (55 ft-lb)
Lower control arm to subframe (use new nut) M14 M12	 200 Nm (147 ft-lb) 85 Nm (63 ft-lb
Stabilizer link to stabilizer bar, M10	46 Nm (34 ft-lb)
Toe control arm to cross member (eccentric), M12, 10.9 grade*, use new nut	100 Nm (74 ft-lb)
Toe control arm to wheel carrier, M12, use new nut	85 Nm (63 ft-lb)
Wheel to wheel hub, M14	130 Nm (96 ft-lb)

*If 8.8 grade bolt installed, replace with 10.9 grade bolt (part supersession). Bolt grade marked on bolt head.

Rear wheel bearing, replacing

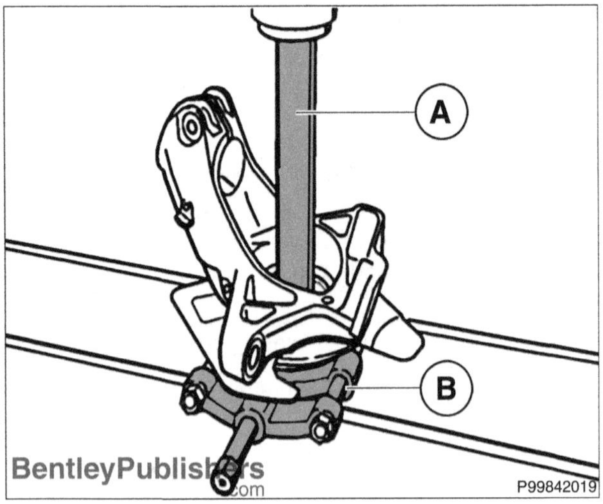

- Remove wheel carrier. See **Wheel carrier, removing and installing** in this repair group.
- Remove brake dust shield.

Support wheel carrier assembly with bearing separator (**B**) and press hub out of wheel carrier with pressing tool (**A**).

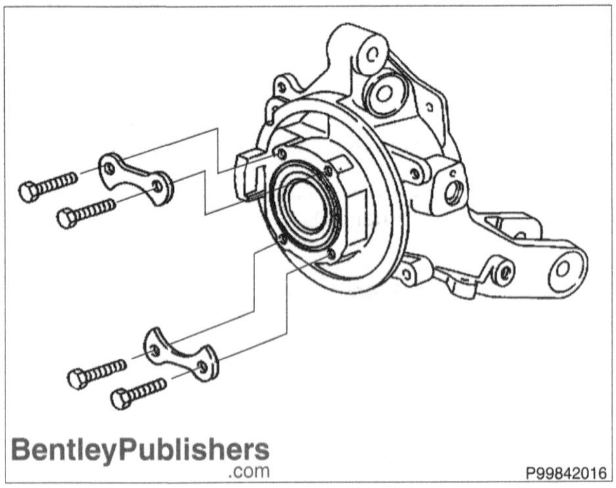

Remove fasteners and remove bearing retainer plates (**arrows**).

- Heat wheel carrier to 100°C (212°F). Do not exceed 120°C (248°F).

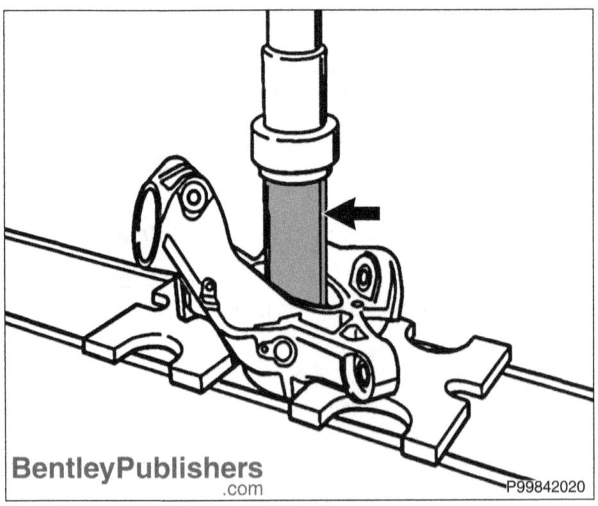

Using press tool (**arrow**), press wheel bearing out of wheel carrier.

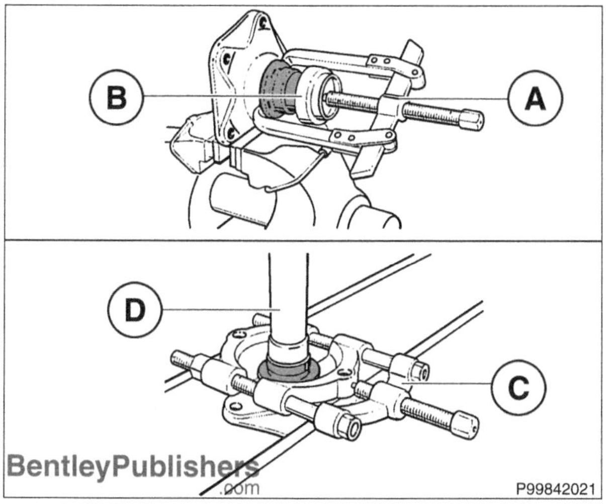

◀ Using two-jaw puller (**A**) in combination with protective cup (**B**), pull inner bearing race several millimeters off wheel hub and install bearing separator.

– Using bearing separator (**C**), press inner bearing race off wheel hub.

Installing

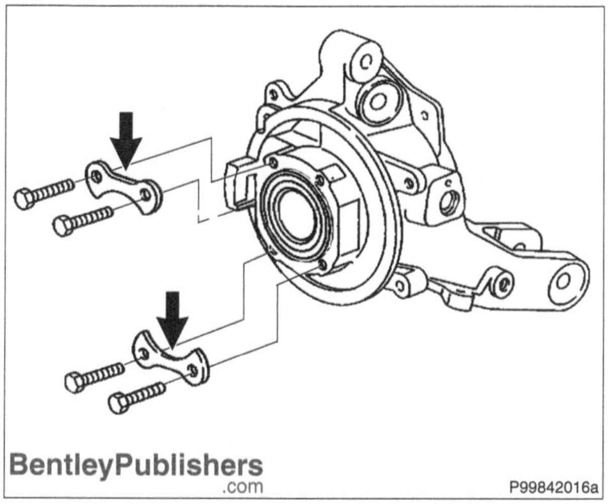

Before installing wheel bearing, clean bore in wheel bearing housing and remove any surface imperfections with a polishing cloth.

– Heat wheel carrier to 100°C (212°F). Do not exceed 120°C (248°F)

– Insert new wheel bearing into wheel carrier. If necessary, press wheel bearing into carrier until fully seated.

◀ Install bearing retainer plates (**arrows**).

Tightening torque	
Wheel bearing retainer plates to wheel carrier, M8	33 Nm (24 ft-lb)

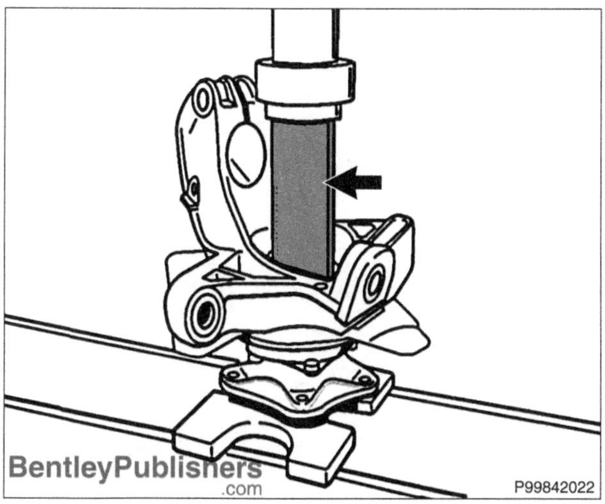

◀ Place wheel carrier on wheel hub and press on bearing inner race until hub is fully seated.

– Remainder of installation is reverse of removal. Use new self-locking nuts and new caliper mounting bolts.

- Install wheel carrier. See **Wheel carrier, removing and installing** in this repair group.
- Install parking brake, brake rotor and brake caliper. See **46 Brakes Mechanical**.

Rear Drive Axles

Rear drive axle, removing and installing

◄ With vehicle on ground, use a breaker bar with a 32 mm socket to loosen drive axle hex nut at wheel. If necessary, apply brakes.

– Raise car and support safely. Remove wheel and axle nut.

> ***WARNING—***
>
> *• Make sure vehicle is stable and well supported at all times. Use a professional automotive lift or jack stands. Place jack stands underneath structural chassis points, not underneath suspension parts. A floor jack is not adequate support.*

– Remove engine and transmission protection panel. See **03 Service and Maintenance**.

◄ Remove heater pipe (**A**) by loosening hose clamps (**arrows**) at each end. Unbolt from retaining fasteners and remove.

– Unbolt brake caliper from wheel carrier and hang out of way using stiff wire. See **46 Brakes–Mechanical**.

◄ Counter hold stabilizer link (**B**) and remove nut (**A**).

– Disconnect link from stabilizer bar.

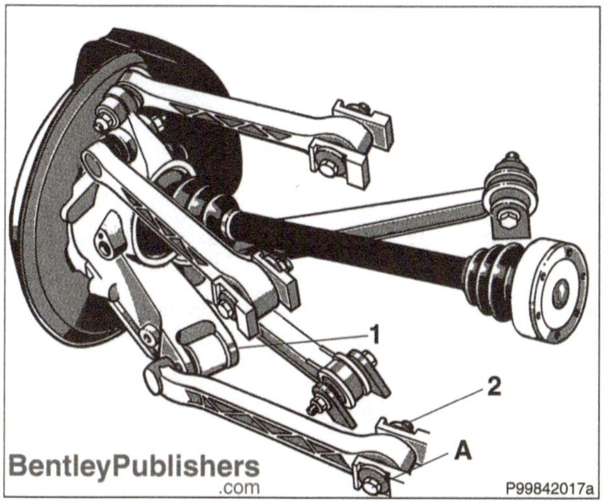

P99842017a

< Remove toe control arm

- Mark toe control arm eccentric bolt (**A**), then remove nut (**2**) and bolt.
- Remove toe control arm nut (**1**) at wheel carrier.
- Counter hold toe control arm ball joint bolt with Torx 40.
- Remove ball joint from wheel carrier.
- Remove toe control arm.

P998423202

< Remove nut (**arrow**) from lower control arm ball joint. If necessary, counter hold ball joint stud with Torx T40 driver while loosening nut.

- Using a ball joint press tool, remove ball joint from wheel carrier.

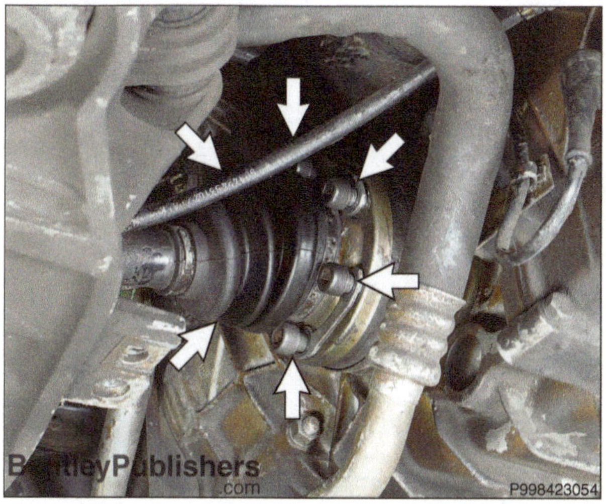
P998423054

< Remove CV joint (drive axle) mounting bolts (**arrows**).

– Separate CV joint from transmission flange.

Rear drive axle, removing and installing

< Using drive axle press tool (**arrow**), press drive axle out of wheel carrier.

– Lift wheel hub assembly to allow room for removal of axle assembly.

– Installation is reverse of removal, noting the following:

- Use new self-locking nuts.
- Inspect all parts before reassembly. If drive axle CV boots are damaged, replace boots and repack CV joint grease.
- If necessary, use a helper when fitting drive axle to wheel bearing carrier. Lubricate drive axle splines lightly with Optimoly HT grease.
- Tighten drive axle nut only with car on ground. Replace drive axle hex nut.
- Install eccentric bolts to marked positions. Tighten control arm bushing end with rear suspension in normal loaded position (car on ground, suspension settled).

Tightening torque	
Brake caliper to wheel carrier, M12 (replace with new)	85 Nm (63 ft-lb)
Drive axle to transmission, M10	81 Nm (60 ft-lb)
Drive axle to wheel hub, M22 (use new nut)	460 Nm (340 ft-lb)
Stabilizer link to stabilizer bar, M10	46 Nm (34 ft-lb)
Toe control arm to wheel carrier, M12 (use new nut)	85 Nm (63 ft-lb)
Toe control arm to subfame (eccentric), M12* grade 10.9, (use new nut)	100 Nm (74 ft-lb)
Lower control arm to wheel carrier, M12 (use new nut)	75 Nm (55 ft-lb)
Wheel to wheel hub, M14	130 Nm (96 ft-lb)
Wheel speed sensor to wheel hub, M6	10 Nm (7 ft-lb)
Crossmember to side subframe (M12)	120 Nm (88 ft-lbs)
Cross brace to side subframe (M10)	65 Nm (48 ft-lbs)

*If 8.8 grade bolt installed, replace with 10.9 grade bolt (part supersession). Bolt grade marked on bolt head.

44 Wheels, Tires and Alignment

WHEELS AND TIRES

Wheel and tire combinations

Original-equipment Porsche wheel and tire size combinations for 993 models, as delivered from the factory:

Wheel and tire sizes
16 Inch wheels Front 7 x 16 (ET 55) 205/55 ZR 16 Rear 9 x 16 (ET 70) 245/45 ZR 16
17 Inch wheels Front 7 x 17 (ET 55) 205/50 ZR 17 Rear 9 x 17 (ET 55) 255/40 ZR 17
18 Inch wheels Front 8 x 18 (ET 52) 225/40 ZR 18 Rear 10 x 18 (ET 65) 265/35 ZR 18 Front 8 x 18 (ET 52) 225/40 ZR 18 Rear 10 x 18 (ET 40) 285/30 ZR 18

WARNING—

- *Always replace tires in pairs (front and rear axles).*
- *Tires with high speed ratings (ZR, for example) should be replaced after 6 years, regardless of tread condition.*
- *When balancing wheels, use adhesive weights mounted to the inside of the wheel rim.*
- *Maximum radial run out for a tire and rim is 1.0 mm.*
- *Apply a light coating of Optimoly TA or equivalent aluminum paste anti-seize compound to lug bolt threads before installing.*
- *Lug bolts with damaged threads or heads should be replaced. Only Porsche original equipment lug nuts should be used.*
- *Start lug bolts by hand and tighten evenly, observing specified pattern. Never use an impact wrench to tighten lug nuts.*
- *16 inch wheels are standard. 17 and 18 inch wheels are available as an option.*

44

Wheel Alignment

Front alignment adjustments

Alignment is measured with full tank of fuel, spare tire and emergency tools loaded, without driver.

Tightening torque	
Wheel to wheel hub	130 Nm (96 ft-lb)

Front camber

Front wheel camber is adjusted via the upper bolt on the strut base. Camber is altered by changing the position of the strut in relation to the wheel carrier. The strut upper bolt hole is elongated at the strut base.

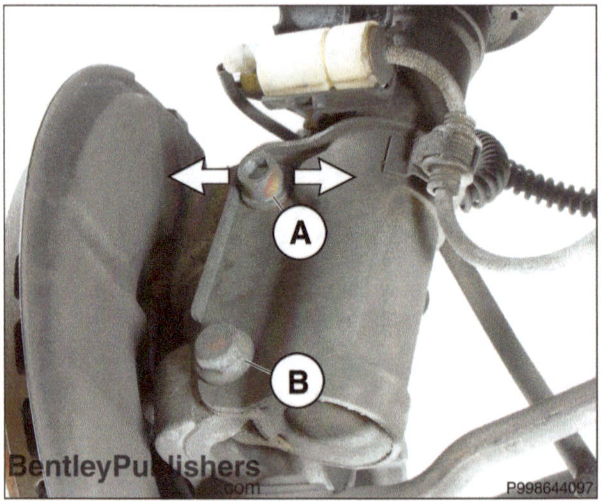

- Loosen upper strut to wheel bearing carrier mounting bolt (**A**).
- Loosen lower strut to wheel bearing carrier mounting bolt (**B**).
- Pivot wheel carrier (**arrows**) at upper bolt to adjust camber to specifications.
- Tighten bolts and recheck adjustment.

Tightening torque	
Strut to wheel carrier, upper bolt (camber adjustment), M12	120 Nm (88 ft-lb)
Strut to wheel carrier, lower bolt (hex-head), M14	200 Nm (148 ft-lb)

Front caster

Front wheel caster is adjusted via the eccentric bolt at the control arm ball joint.

- Bounce vehicle 2-3 times and allow car to reach normal ride height.

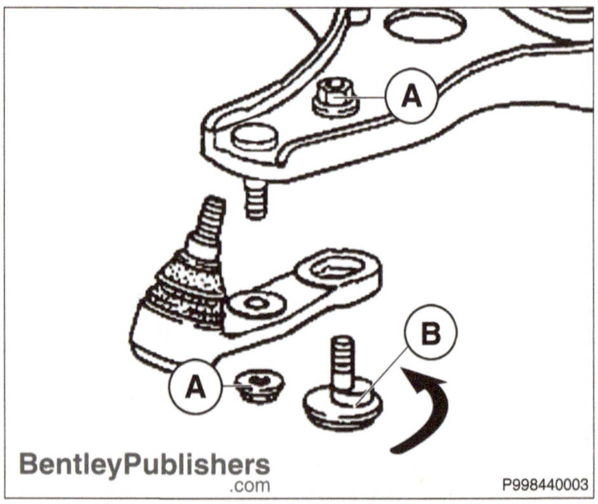

- Loosen ball joint lock nuts (**A**).
- Turn caster eccentric (**B**) to desired position then secure lock nuts (**A**).

Tightening torque	
Ball joint to control arm lock nut, M12	140 Nm (103 ft-lb)
Ball joint to control arm (caster eccentric), M10	80 Nm (59 ft-lb)

Front toe

- Bounce vehicle 2 - 3 times and allow car to reach normal ride height.
- Brace steering wheel in center position (wheels straight ahead).

◀ Loosen tie rod adjuster lock nut (**A**).

- Turn tie rod (**B**) to desired position.
- Tighten lock nut (**A**).

Front ride height

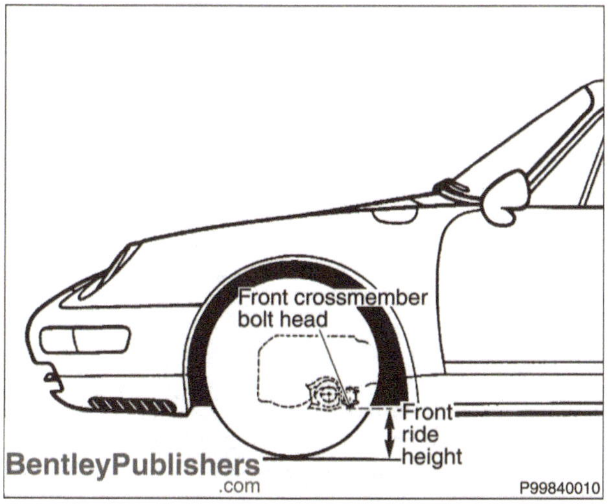

Ride height is measured with full tank of fuel, spare tire and emergency tools loaded, without driver.

◀ Ride height is measured from the ground to outer (hex-head) bolt connecting cross member to body.

> ***WARNING—***
>
> - *Changing ride height at one wheel will result in a change in wheel loads at the other wheels. If the ride height has been altered, wheel load scales should be used to be sure wheel load differences are within tolerance. The left-to-right tolerance on the front and rear axles must be below 20 kg (44 lb).*

◀ Turning lower adjusting nut (**arrow**) at front strut spring perch raises and lowers ride height for front of vehicle.

> ***WARNING—***
>
> - *Do not exceed +/- 10mm (0.39 in) from factory setting.*

- Raise vehicle to reduce compressed load on coil spring.
- Remove front wheel.
- Turn adjusting nut in small increments using hook wrench.
- Refit wheel and allow suspension to settle.

Ride height (front)	
USA (all models)	174 mm (6.85 in)
RoW	154 mm (6.06 in)
RoW (sport suspension and turbo)	144 mm (5.67 in)
Maximum difference left to right	5 mm (0.20 in)
Maximum difference front to rear	10 mm (0.39 in)

44

Rear alignment adjustments

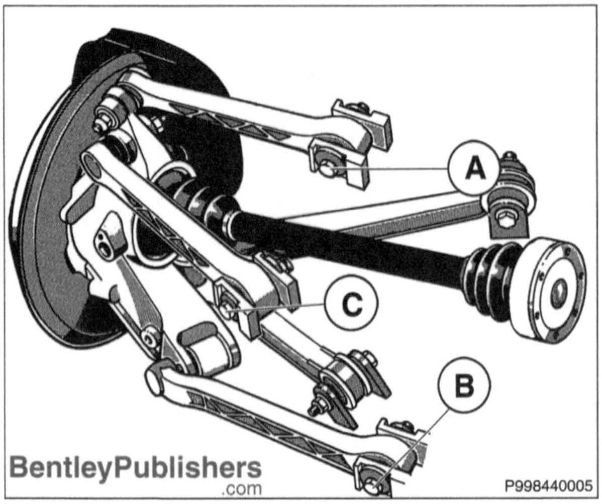

Alignment is measured with full tank of fuel, spare tire and emergency tools loaded, without driver.

◄ Rear suspension alignment is adjustable via eccentrics.

– To adjust camber, loosen lock nut and turn eccentric (**A**) to desired position. Secure lock nut.

– To adjust toe, loosen lock nut and turn eccentric (**B**) to desired position. Secure lock nut.

– Kinematic toe adjustment may be made using eccentric (**C**).

- To adjust kinematic toe, loosen lock nut and turn eccentric (**C**) to desired position. Secure lock nut.

Rear ride height

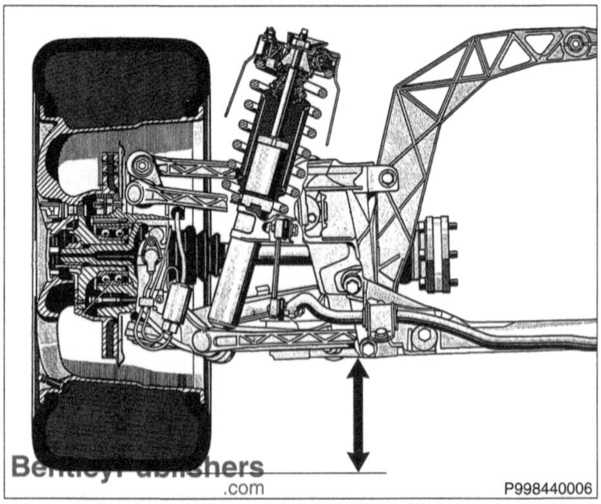

Rear ride height is controlled by the suspension springs. Adjustment to rear ride height is not normally required unless the rear springs are worn or other parts are damaged. Ride height is measured with full tank of fuel, spare tire and emergency tools loaded, without driver.

◄ Rear ride height is measured from the ground to mating face on rear subframe between lower control arm (A-arm) mounts.

> ***WARNING—***
>
> - *Changing ride height at one wheel will result in a change in wheel loads at the other wheels. If the ride height has been altered, wheel load scales should be used to be sure wheel load differences are within tolerance. The left-to-right tolerance on the front and rear axles must be below 20 kg (44 lb).*

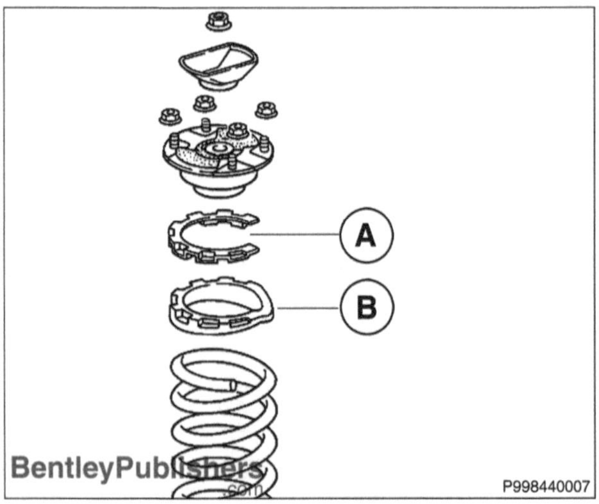

◄ If rear springs were replaced, a thicker or thinner intermediate section (**A**) may be installed to achieve proper ride height, depending on the spring classification installed.

– Strut must be disassembled to change intermediate section (**A**).

- 3.0 mm, 5.5 mm, and 8.0 mm intermediate sections (**A**) are available through Porsche parts.
- Only one intermediate section can be used per strut.
- Press intermediate section (**A**) into spring support (**B**) and reassemble strut.

Ride height (rear)	
USA (all models)	157 mm (6.18 in)
RoW	147 mm (5.79 in)
RoW (sport suspension and turbo)	127 mm (5.00 in)
Maximum difference left to right	5 mm (0.20 in)
Maximum difference front to rear	10 mm (0.39 in)

Alignment specifications

Table a. lists the alignment specifications for the Porsche 993.

Table a. Alignment specifications			
	USA (including sport)	**Rest of World (ROW)**	**Rest of World (ROW) Sport**
Front axle			
Toe unpressed (total)	+5' ± 5'	+5' ± 5'	+5' ± 5'
Toe difference angle (at 20° lock)	- 40' ± 30'	- 1° ± 30'	- 1° 45' ± 30'
Camber with wheels forward Max difference left to right	- 20° ± 10' 10'	- 20° ± 10' 10'	- 20° ± 10' 10'
Caster Max difference left to right	5° 20' + 15' / - 30' 15'	5° 20' + 15' / - 30' 15'	5° 20' + 15' / - 30' 15'
Rear Axle			
Toe per wheel Max difference left to right	+10' ± 5' 10'	+10' ± 5' 10'	+10' ± 5' 10'
Camber Max difference left to right	- 1° 10' ± 15' 20'	- 1° 10' ± 15' 20'	- 1° 10' ± 15' 20'

45 Antilock Brakes (ABS)

GENERAL

This repair group covers identification and replacement of components of the antilock brake system (ABS).

For more information on brakes, see the following repair groups:

- **46 Brakes–Mechanical**
- **47 Brakes–Hydraulic**

WARNING—

- *Brake fluid is poisonous, corrosive and dangerous to the environment. Wear safety glasses and rubber gloves when working with brake fluid. Do not siphon brake fluid with your mouth. Dispose of brake fluid properly.*
- *Do not reuse self-locking nuts, bolts or fasteners. They are designed to be used only once and may fail if reused. Replace with new self-locking fasteners.*
- *Vehicles fitted with ABD require a Porsche scan tool to properly bleed the brake hydraulic system. See* **47 Brakes–Hydraulic** *for more information.*
- *Immediately clean brake fluid spilled on painted surfaces and wash with water, as brake fluid removes paint.*
- *Use new brake fluid from an unopened container. Brake fluid absorbs moisture from the air. This can lead to corrosion problems in the braking system, and also lowers the fluid boiling point.*
- *Plug open lines and brake fluid ports to prevent contamination.*
- *If carrying out electric welding work, be sure to disconnect electrical harness connector from electronic control module.*
- *Do not expose electronic control modules to high sustained heat, such as in a paint drying booth. Maximum heat exposure: 95° C (203°F) for short periods of time, 85°C (185°F) for long periods (approximately 2 hours).*

ABS, ABS / ABD Systems

Electronically controlled ABS maintains vehicle stability and control during emergency braking by preventing wheel lock-up. ABS provides optimum deceleration and stability during adverse conditions by automatically adjusting brake system hydraulic pressure at each wheel to prevent wheel lock-up.

ABS and ABS / ABD systems are designed to be maintenance free. There are no adjustments that can be made to the system. The system is self-tested by the electronic control module each time the car is started. Once the test is complete, the warning light(s) turns OFF. If light remains ON or comes ON at any time during driving, a system fault has occurred and the system is electronically disabled. The conventional braking system remains fully functional.

The system's main components are:

- Wheel speed sensors
- Electronic control module
- Hydraulic unit

Wheel speed sensors continuously send wheel speed signals to the electronic control module. The control module compares these signals to determine, in fractions of a second, whether any of the wheels are about to lock. If any wheel is nearing a lock-up condition, the module signals the hydraulic unit to maintain or reduce pressure at the appropriate wheel(s). Pressure is modulated by electrically operated solenoid valves in the hydraulic unit.

System identification

Three ABS systems were available in the 993:

- **ABS 5, 3-channel, vacuum booster** - standard system on Carrera 2.
- **ABS 5 / ABD, 4-channel, vacuum booster (with ABD option)** Added Automatic Brake Differential (ABD) to the standard system.
- **ABS 5 / ABD, 4-channel, hydraulic booster** is standard on Carrera 4, 4S, and Turbo.

ABS 3-channel system (Carrera 2)

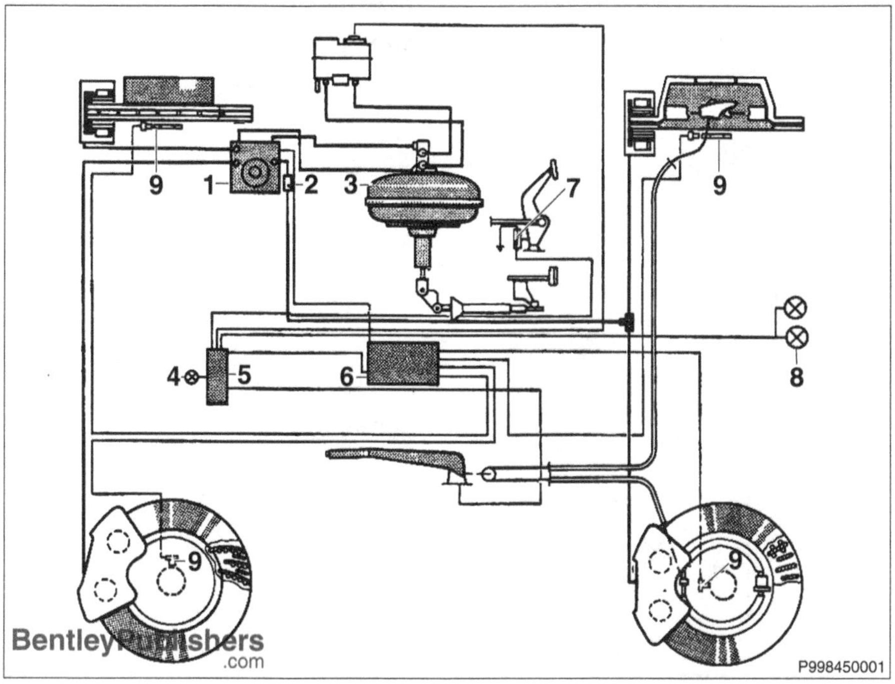

1. ABS hydraulic unit
2. Brake proportioning valve
3. Brake master cylinder with vacuum booster
4. ABS warning lamp
5. Central indicator controller
6. ABS control unit
7. Brake light switch
8. Brake lights
9. ABS wheel speed sensors

ABS 5 is a 3-channel, antilock brake system using a single proportioning valve for the rear brakes. This system has a master cylinder attached to a brake vacuum booster, mounted under the car parallel to the steering rack.

ABS 5 / ABD 4-channel system (Carrera 2 with ABD option)

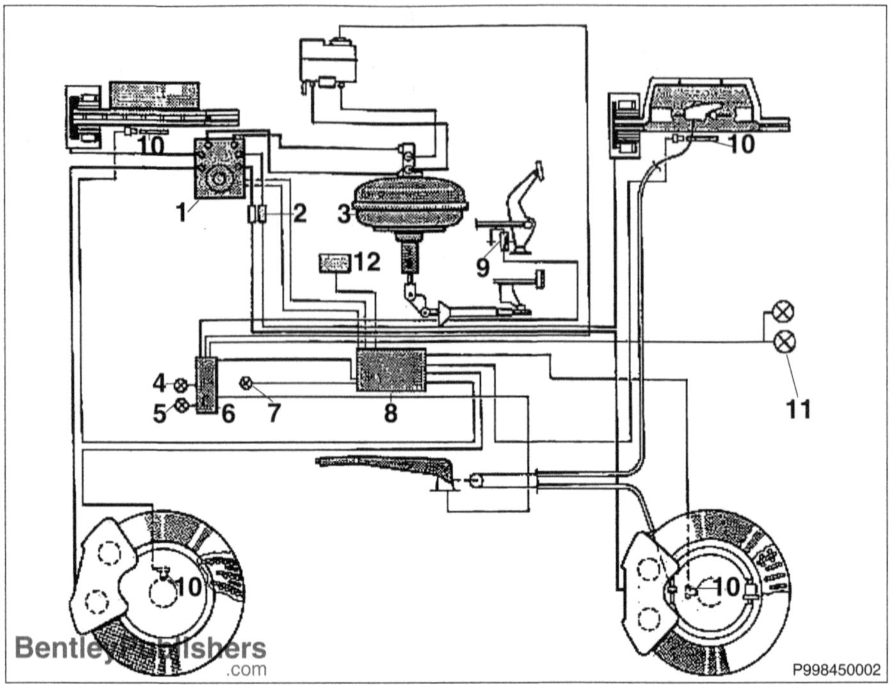

1. ABS/ABD hydraulic unit
2. Brake proportioning valve
3. Brake master cylinder with booster
4. ABS warning lamp
5. ABD warning lamp
6. Central indicator controller
7. ABD operation lamp
8. ABS/ABD control unit
9. Brake light switch
10. ABS wheel speed sensors
11. Brake lights
12. DME control module

ABS 5 / ABD 4-channel system was an option on two-wheel drive cars. Automatic Brake Differential (ABD) functions similarly to a locking or limited slip differential to help the vehicle retain traction.

Using the ABS wheel speed sensors to detect wheel spin during acceleration, the ABS control module will modulate the hydraulic pressure at the brake caliper of the spinning wheel to assist the vehicle in gaining traction. This 4-channel system uses two brake proportioning valves to individually control brake pressure at rear wheels.

ABS 5 / ABD 4-channel system (Carrera 4, 4S and Turbo)

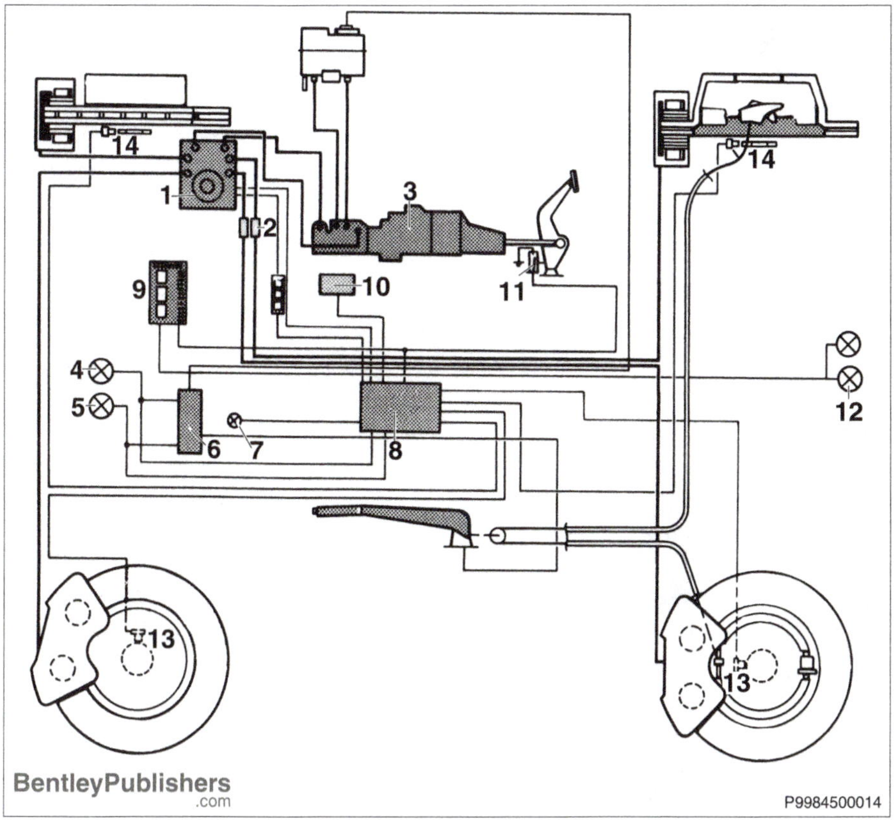

1. ABS hydraulic control unit
2. Brake proportioning valves
3. Brake master cylinder with hydraulic booster
4. ABS warning lamp
5. ABD warning lamp
6. Central indicator controller
7. ABD operation lamp
8. ABS electronic control unit
9. Relay / fuse block
10. DME control unit
11. Brake light switch
12. Brake lights
13. ABS wheel speed sensor

ABS 5 / ABD 4-channel with hydraulic booster is used on Carrera 4, 4S and Turbo cars. This uses a hydraulic brake booster instead of a vacuum booster. The hydraulic booster is mounted to the front of the master cylinder at the bulkhead on left side of trunk.

This system utilizes a brake boost hydraulic pump (located above battery) and a hydraulic pressure accumulator (on right hand side of trunk near ABS relays).

–NOTE

- *See* **47 Brakes–Hydraulic** *for additional information regarding the hydraulic brake boost system.*

A larger fluid reservoir (approximately 1.5 quarts) is used on the Carrera 4 cars owing to the requirements of the hydraulic brake boost system.

Component descriptions

Hydraulic unit

P998456182a

The ABS hydraulic unit (**arrow**) uses high speed electromagnetic valves and a pump to "pulse" braking force in ABS functions.

The hydraulic unit is located in the left front side of trunk beside battery.

Control unit (electronic)

P998451667

The 55-pin electronic control unit (**arrow**) receives signals from the brake light switch and the four wheel speed sensors. Wheel speed signals are processed independently of each other to calculate wheel slip.

The electronic control unit is located in the right hand side of the trunk. A sealed cover protects the control unit from dirt. Slide the protective cover up to remove.

Wheel speed sensors

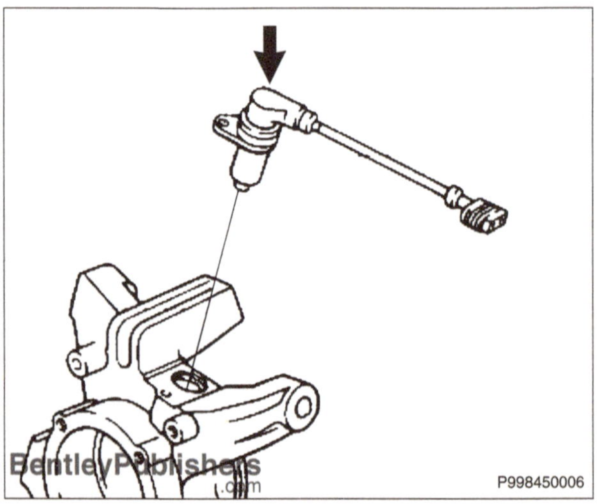
P998450006

Each wheel carrier has a wheel speed sensor attached. The sensors send A/C voltage signals to the ABS control unit when the wheels are rotating.

Power supply

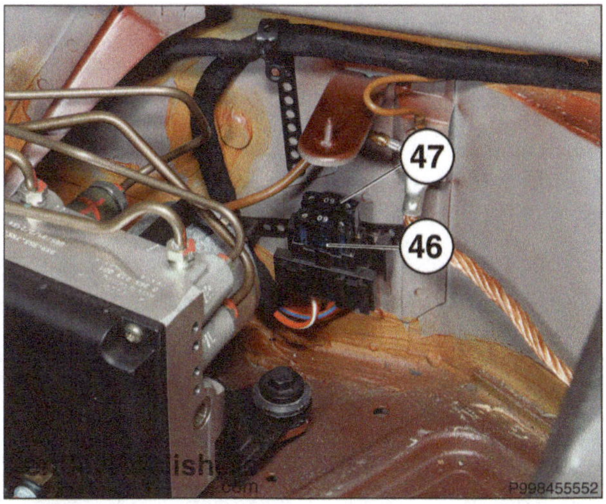

60 amp maxi fuse **46** supplies power to the ABS solenoids and the ABS pump. Fuse 46 is located just behind the ABS hydraulic unit.

60 amp maxi fuse **47** supplies power to the DME system and the hydraulic brake booster epump (C4 / C4S cars with hydraulic brake boost).

> ***CAUTION—***
>
> - *Fuse positions can vary. Confirm fuse identification by inspecting wire colors at the fuse block. Fuse 47 should have red wires. Fuse 46 should have red/white and red/blue wires.*

NOTE–

- *See* **47 Brakes–Hydraulic** *for additional information regarding the hydraulic brake boost system on Carrera 4 / 4S models.*

A 10 amp fuse (#16) in the central electric panel supplies power to the ABS electronic control unit.

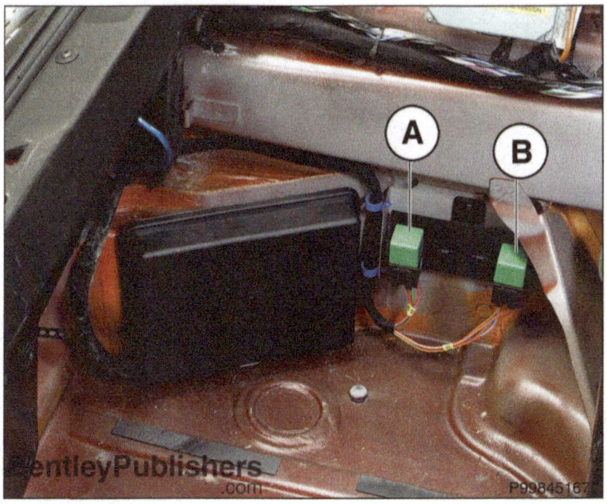

Relay R65 (**A**) is a 50 amp power relay located beside ABS electronic control unit in trunk. This relay supplies power to the ABS hydraulic unit pump.

Relay R68 (**B**) is a 50 amp power relay. This relay supplies power to the ABS hydraulic unit solenoids.

On Carrera 4 cars with hydraulic brake boost, relay R66 is located in the center position of the relay block. This relay supplies power to the brake booster electrohydraulic pump.

> ***CAUTION—***
>
> - *Relay positions can vary. Confirm relay identification by inspecting wire colors at the relay block. See EWD for ABS or ABS / ABD relay wire colors (EWD-62, for example)*

ABS / ABS /ABD system inspection and troubleshooting

A visual inspection of the ABS system may help to locate system faults. Carefully inspect the entire ABS wiring harness, particularly the pulse sensor harnesses and connectors near each wheel. Look for chafing or damage due to incorrectly routed wires.

Carefully remove the wheel speed sensors. See **Wheel speed sensor, removing and installing** in this repair group. Clean the sensor tips. Inspect the toothed pulse wheel at the wheel hubs. Check for missing, clogged or corroded teeth, or other damage that could alter the signal between the sensor tip and the pulse wheel.

– Remove and inspect ABS / ABD relay contacts for corrosion or poor contact surfaces. Clean contacts on relay and in relay block.

– Check maxi fuses (60 amp), located behind ABS hydraulic unit.

– Check fuse no. 16 (10 amp) located in central electric panel in engine compartment.

If no visual faults can be found and the ABS light remains ON, system diagnosis should be carried out using a Porsche scan tool.

Component Replacement

A Porsche scan tool should be used to clear system fault codes after replacing faulty components.

Wheel speed sensor, removing and installing

Front and rear wheel speed sensors are not interchangeable. The tip of the speed sensor is at a different angle and will not read pulse wheel properly if reversed.

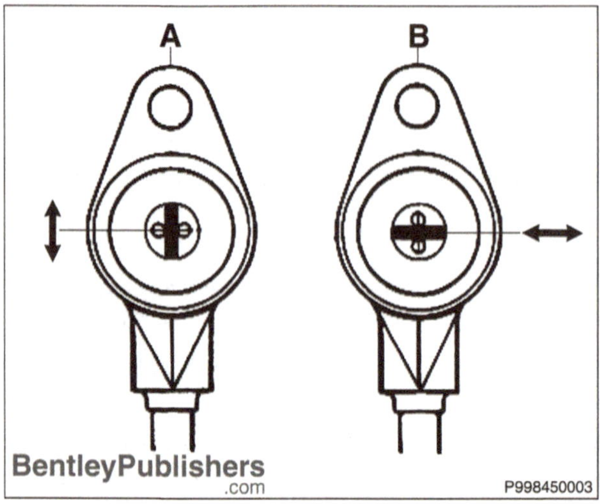

- Front wheel speed sensor (**A**). Rear wheel speed sensor (**B**).
- Raise and support vehicle. Remove wheel.

> ***WARNING—***
> - *Make sure vehicle is stable and well supported at all times. Use a professional automotive lift or jack stands. Place jack stands underneath structural chassis points, not underneath suspension parts. A floor jack is not adequate support.*

- Separate sensor electrical connector (**B**). Remove sensor fastener (**A**) from wheel carrier and pull sensor from wheel carrier.

NOTE–
- *Rear wheel speed sensor shown, front is similar. Connector bracket is bolted to strut housing.*

- Installation is reverse of removal. Note the following:
  - Clean debris from magnetic tip of sensor before installing.
  - Coat sensor shaft and sensor bore with Molykote Longterm 2 or alternate lightweight grease.
  - If sensor was replaced due to ABS faults or accident damage, check system function using Porsche scan tool.

NOTE–
- *Do not allow grease on sensor tip.*

Tightening torque	
Wheel speed sensor to wheel carrier	10 Nm (7 ft-lb)

Hydraulic unit, removing and installing

> ***WARNING—***
>
> - *This procedure requires use of a Porsche scan tool. Read the procedure through before beginning the job to determine the scope of the repair.*
> - *Vehicles fitted with ABD require a Porsche scan tool to cycle the ABS valves when bleeding the brake hydraulic system. See* **47 Brakes–Hydraulic** *for more information.*

– Remove trunk carpet and set aside.

– Remove spare tire and tool kit from front trunk.

– Remove hydraulic unit cover held in place by two plastic nuts.

– Using fluid extractor, empty brake fluid reservoir.

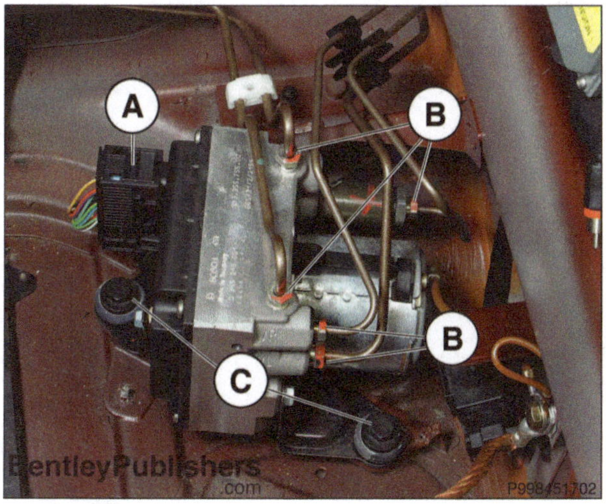

◄ To remove hydraulic unit:

- Mark location of hydraulic lines in relation to hydraulic unit.
- Disconnect hydraulic lines (**B**) and seal with suitable plugs.
- Disconnect electrical connector (**A**) from control module.
- Remove mounting fasteners (**C**) and remove hydraulic unit.

> ***CAUTION—***
>
> - *Brake fluid is corrosive. Protect painted surfaces from spilled brake fluid. Immediately clean up any spilled fluid and wash thoroughly with water.*

NOTE–

- *ABD-equipped models have additional lines and an additional proportioning valve on hydraulic unit.*

– Installation is reverse of removal, note the following:

- Bleed brakes. See **47 Brakes–Hydraulic**.
- Check brake lines and connections for leaks.
- Read and clear fault codes using Porsche scan tool.
- Start engine and ensure ABS warning lamps are not illuminated.
- Road test brakes, performing at least one ABS controlled operation.
- Read ABS fault memory. No faults should be stored.

45

Electronic control unit, removing and installing

> ***CAUTION—***
> - *This procedure requires use of a Porsche scan tool. Read the procedure through before beginning the job to determine the scope of the repair.*

– Remove trunk carpet and set aside.

– Remove spare tire and tool kit from trunk.

< Remove control unit cover by sliding up off of ball pins.

< Remove four ball pins (**arrows**) retaining control unit.

< Disconnect electrical connector (**arrow**) and remove electronic control unit.

– Installation is reverse of removal, note the following:
- Rear and clear fault codes using Porsche scan tool.
- Start engine and ensure ABS warning lamps are not illuminated.
- Road test brakes, performing at least one ABS controlled operation.
- Read ABS faulty memory. No faults should be stored.

46 Brakes–Mechanical

General

This repair group covers replacement of brake pads, brake rotors, and adjusting the parking brake assembly.

Also see the following repair groups for more information on brakes:

- **45 Antilock Brakes (ABS)**
- **47 Brakes–Hydraulic**

Warnings and Cautions

WARNING—

- *Brake friction materials such as brake linings or pads may contain asbestos fibers which can lead to illnesses. Do not create dust by grinding, sanding or cleaning the pads with compressed air. Avoid breathing any asbestos fibers or dust.*
- *Brake fluid is poisonous, highly corrosive and dangerous to the environment. Wear safety glasses and rubber gloves when working with brake fluid. Immediately clean away any fluid spilled on painted surfaces and wash with water. Dispose of brake fluid properly.*
- *Do not reuse self-locking nuts or fasteners. They are designed to be used only once and may fail if reused. Always replace them with new self-locking nuts.*

CAUTION—

- *Always machine or replace brake rotors in pairs.*
- *Always replace brake pads in sets (one set = one axle).*
- *To prevent brake fluid overflow when caliper pistons are pushed back, siphon some fluid out of brake master cylinder reservoir.*

Brake Pads and Rotors

Inspect brake pad thickness (**A**) and rotor thickness (**B**) for wear.

Brake wear specifications	
Approximate new brake pad thickness (dimension **A**): • Front • Rear	 11.0 mm (0.433 in) 11.0 mm (0.433 in)
Brake pad wear limit (dimension **A**):	2.0 mm (0.078 in)
New brake rotor (dimension **B**): • Front • Rear	 32.0 mm (1.259 in) 24.0 mm (0.944 in)
Brake rotor minimum thickness, after machining (dimension **B**): • Front • Rear	 30.6 mm (1.204 in) 22.6 mm (0.899 in)
Brake rotor wear limit (dimension **B**): • Front • Rear	 30.0 mm (1.181 in) 22.0 mm (0.866 in)

Brake pads, replacing

The following procedure applies to both front and rear brake pads. The brake calipers do not need to be removed in order to replace brake pads.

– Raise vehicle and support safely. Remove wheel.

> ***WARNING—***
>
> • *Make sure vehicle is stable and well supported at all times. Use a professional automotive lift or jack stands. Place jack stands underneath structural chassis points, not underneath suspension parts. A floor jack is not adequate support.*

Remove brake pad spring clip retainer by squeezing upper portion and pulling retainer off (**arrow**).

◀ Remove both brake wear indicators (**arrows**) by pulling away from brake pad.

◀ Release brake pad spring clip (**arrow**) by squeezing in center and lifting up and out of bottom bracket.

– Remove clip from caliper by squeezing upper portion to free from caliper detents.

◀ Depress caliper pistons with pliers or piston press tool for easier removal of brake pads.

CAUTION—

- *Remove brake fluid reservoir cap before resetting pistons.*
- *To prevent brake fluid overflow when caliper pistons are pushed back, siphon some fluid out of brake master cylinder reservoir if necessary. Clean up any spilled fluid and wash area with water.*

- Remove brake pads from caliper. If necessary, separate circular vibration dampers from back of pad with a small flat blade screwdriver before removing.
- Press pistons fully into calipers using a piston press tool or equivalent.
- Installation is reverse of removal, note the following:
  - Replace damping plates for caliper pistons if necessary.
  - Clean guide surface for brake pads before installing new pads. Be sure not to damage dust caps on caliper pistons.
  - Replace brake pad wear sensors, if necessary.
  - Firmly press brake pedal several times with vehicle stationary to ensure pads adjust to operating position before driving.
  - Check brake fluid level and secure reservoir cap.

NOTE —

- *New brake pads require approximately 150 miles to "break in" and attain their best friction and wear coefficient. During this period, do not brake hard except in an emergency.*

Brake rotors, removing and installing

The Porsche 993 uses a four wheel disc brake system with ventilated brake rotors at each wheel. New caliper bolts are specified by Porsche when remounting the caliper.

- Raise vehicle and support safely. Remove wheel.

WARNING—

- *Make sure vehicle is stable and well supported at all times. Use a professional automotive lift or jack stands. Place jack stands underneath structural chassis points, not underneath suspension parts. A floor jack is not adequate support.*

Brake rotor diameter	
Front	304 mm (11.968 in)
Rear	299 mm (11.710 in)

- Remove brake pads. See **Brake pads, replacing** in this repair group.

- Remove two brake caliper mounting bolts (**arrows**) and remove caliper.

NOTE–

- *Front caliper shown, rear caliper bolts are on inside edge of caliper.*

- Hang caliper from suspension using stiff wire.

Working at brake rotor, remove screws (**arrows**) holding rotor to hub. If necessary, tap screw heads using a drift to loosen.

Prior to removing rear rotors, parking brake must be released. Working at rear wheel, using a screwdriver through bore, retract parking brake adjuster by prying on star wheel toward rear of vehicle.

– Remove rotor from hub. If necessary, tap rotor using a soft face hammer to release rotor from hub.

– Installation is reverse of removal. Note the following:

- Clean rotor mounting surface of wheel hub. Apply a thin coating of Optimoly TA or equivalent aluminum anti-seize compound.
- Replace brake caliper mounting screws with new.
- Firmly press brake pedal several times with vehicle stationary to ensure pads adjust to operating position before driving.
- Check brake fluid level and secure reservoir cap.
- When refitting rear rotors, adjust parking brake. See **Parking brake shoes, adjusting.**

Tightening torques	
Brake rotor to hub, M6 (front or rear)	5 Nm (3.7 ft-lb)
Front brake caliper to wheel carrier, M12 (replace with new)	85 Nm (63 ft-lb)
Rear brake caliper to wheel carrier, M12 (replace with new)	85 Nm (63 ft-lb)
Wheel to wheel hub	130 Nm (96 ft-lb)

Parking Brake

Parking brake components

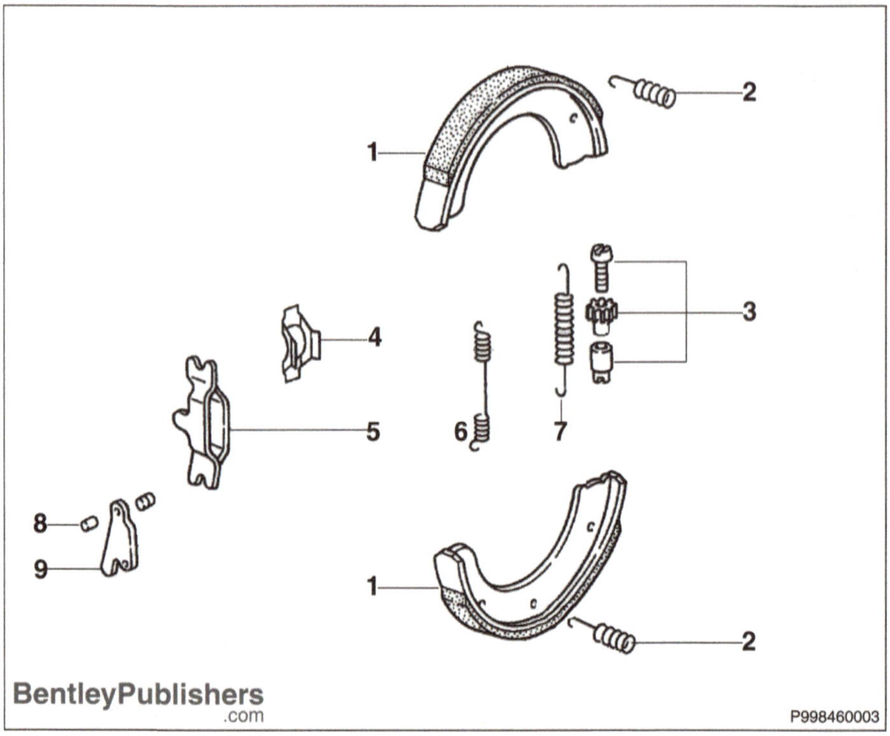

1. Parking brake shoes
2. Compression spring
3. Star wheel adjuster
4. Support plate
5. Clip
6. Lower return spring
7. Return spring
8. Screw
9. Control lever

Parking brake shoes, adjusting

If the parking brake lever can be pulled up by more than 4 clicks without engaging, adjust parking brake shoes.

– Raise vehicle and support safely. Remove rear wheel.

> ***WARNING—***
> • *Make sure vehicle is stable and well supported at all times. Use a professional automotive lift or jack stands. Place jack stands underneath structural chassis points, not underneath suspension parts. A floor jack is not adequate support.*

Using a screwdriver, advance parking brake adjuster by prying star wheel towards front of vehicle through bore until rotor cannot be turned. Next, turn adjuster back 5 notches (3 notches until the wheel can be turned freely, then another 2).

– Repeat procedure on opposite wheel.

– Pull up parking brake lever by 2 clicks to test engagement.

– Release parking brake lever and check the wheels can be turned freely.

Tightening torque	
Wheel to wheel hub	130 Nm (96 ft-lb)

Parking brake shoes, replacing

– Raise vehicle and support safely. Remove rear wheel.

> ***WARNING—***
>
> • *Make sure vehicle is stable and well supported at all times. Use a professional automotive lift or jack stands. Place jack stands underneath structural chassis points, not underneath suspension parts. A floor jack is not adequate support.*

– Remove rear brake rotors. See **Brake rotors, removing and installing** in this repair group.

< Remove parking brake shoes:

- Compress compression spring (**A**), then working from rear of wheel carrier, unhook compression spring end from wheel carrier and remove from front. Repeat for other compression spring.
- Unhook return spring (**B**) from brake shoe and remove and star wheel adjuster (**C**).
- Roll brakes shoes out and off. If necessary, unhook front spring from shoes before rolling shoes out. Note direction and orientation of shoes during removal as an aide to reassembly.

Brake shoe compression spring installed position

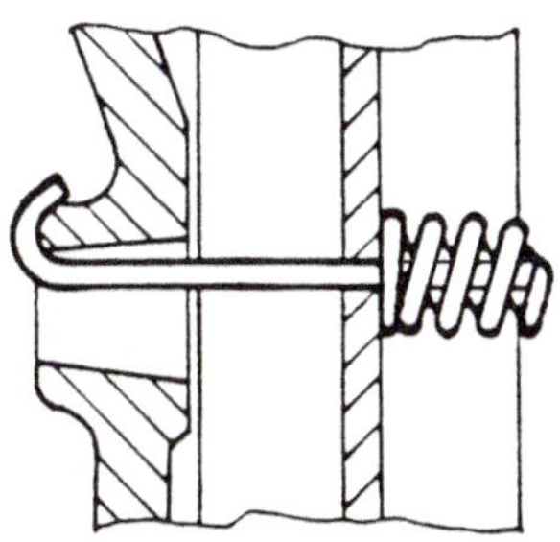

P99846004

– Installation is reverse of removal. Note the following:

- Grease adjuster, actuating lever pin, and brake shoe sliding surfaces.
- Check that parking brake shoes, adjuster, return springs, compression springs and expanding lever are correctly positioned. Adjust as necessary.
- Clean brake disc and hub mounting surface. Apply anti-seize compound to centering surface of wheel hub.
- Adjust parking brake shoes. See **Parking brake shoes, adjusting**.
- Fit brake caliper, replacing caliper mounting hardware.

Tightening torques	
Rear brake rotor to hub, M6	5 Nm (3.7 ft-lb)
Rear brake caliper to wheel carrier, M12 (replace with new)	85 Nm (63 ft-lb)
Wheel to wheel hub	130 Nm (96 ft-lb)

46

47 Brakes–Hydraulic

General

This repair group covers brake bleeding, as well as replacement procedures for hydraulic brake system components.

Also see the following repair groups:

- **03 Service and Maintenance** for brake fluid level check.
- **45 Antilock Brakes (ABS)** for antilock brake system.
- **46 Brakes–Mechanical** for brake pads, rotors and parking brake assembly.

System description

Porsche 993 models are equipped with power assisted disc brakes. Depending on model, either a vacuum or hydraulic brake booster is used to provide power assist to the dual-circuit master cylinder.

Brake booster system application	
Carrera 2	Vacuum brake booster
Carrera 4, Carrera 4S	Hydraulic brake booster

WARNING—
- *The hydraulic brake booster system operates under high pressure (up to 2600 psi). Depressurize the booster circuit before servicing brake system or bleeding brakes. Wear eye protection and protective gloves when depressurizing the booster circuit.*

CAUTION—
- *Brake fluid is highly corrosive and dangerous to the environment. Dispose of it properly.*
- *When adding or replacing brake fluid always add new brake fluid from an unopened container.*
- *It is important to bleed the entire brake system whenever any part of the brake hydraulic system has been opened.*

Brake fluid reservoir (**arrow**) for Carrera models with vacuum booster system.

Larger brake fluid reservoir (**arrow**) for Carrera 4 and 4S models with hydraulic booster system.

Troubleshooting

Brake performance is mainly affected by three things:

- Level and condition of brake fluid
- Brake system's ability to create and maintain hydraulic pressure
- Condition of friction components

Air in the brake fluid will make the brake pedal feel spongy during braking or will increase the brake pedal force required to stop. Fluid contaminated by moisture or dirt can corrode components. Inspect the brake fluid. If fluid is dirty or murky, or over two years old, fluid should be replaced.

Visually check hydraulic system starting with the master cylinder. To check the function of master cylinder, hold brake pedal down hard with car stopped and engine running. The pedal should feel solid and stay solid. If pedal slowly falls to floor, either fluid is escaping from other points or a the master cylinder may be faulty. Check all brake fluid lines and couplings for leaks, kinks, chafing or corrosion.

On models with vacuum boost, check brake vacuum booster by pumping brake pedal approximately 10 times with engine off. Hold pedal down and start engine. Pedal should fall slightly. If not, check for any visible faults before suspecting a faulty brake booster, and check the one-way check valve at the hose connection on the vacuum booster.

On models with hydraulic boost, the pedal should not feel overly hard or sink to the floor. Turn ignition key ON, but do not start engine. Check that the red brake pressure warning light goes out once system pressure is built up. The hydraulic boost pump may audibly run until pressure is built if the pedal is depressed with the key in the on position.

If ABS or ABS / ABD warning lights are illuminated, interrogate the system for faults using a Porsche scan tool.

–NOTE

- *See* **45 Antilock Brakes (ABS)** *for additional fuse and relay electrical troubleshooting information.*

Brake Hydraulic System

Replacing brake fluid

Porsche recommends replacing the brake fluid every two years. This is due to the fact that brake fluid readily absorbs moisture. Moisture in the brake fluid can adversely affect braking performance and may also damage the system, leading to costly repairs.

Replace the brake fluid using one of the bleeding procedures described later to expel the old fluid. Remove the filter/strainer from the brake fluid reservoir and clean it in new, unused brake fluid. Using new, unused brake fluid, pump at least 250 ml (0.53 pints) of brake fluid through each caliper to flush the system and expel the old fluid. Refill the reservoir and bleed the brakes as described below. Use only Porsche recommended Super DOT 4 brake fluid.

CAUTION—

- *Do not rely on flushing alone to clean a system contaminated with dirt or corrosion. The flushing procedure may actually force dirt from the lines into the calipers. To do the job thoroughly, disassemble the system and individually clean the parts. Use only brake fluid to flush the lines. Do not use alcohol since it encourages the accumulation of water in the system.*

Brake fluid	
Porsche brake fluid	Super DOT 4

WARNING—

- *Brake fluid is poisonous, highly corrosive and dangerous to the environment. Wear safety glasses and rubber gloves when working with brake fluid. Immediately clean away fluid spilled on painted surfaces and wash with water. Dispose of brake fluid properly.*

Bleeding brakes (with vacuum brake booster)

Read the procedure through before bleeding the brake system.

On Carrera models cars fitted with the optional ABD feature, a factory scan tool will be required to bleed the system if the ABS/ABD hydraulic unit was replaced or if the system requires a significant volume of fluid after repairs are carried out. See **Additional brake bleeding procedure (after replacing ABS/ABD hydraulic unit)**.

NOTE—

- *For information on ABS and ABS/ABD system identification, see* **45 Antilock Brakes (ABS)**.

P998470001

◄ Brake bleeding on vehicles with vacuum brake booster can be performed using a brake pressure bleeder, or through conventional methods.

– Top off brake fluid reservoir with fresh brake fluid.

– Connect brake pressure bleeder to fluid reservoir. Fill bleeder fluid tank to sufficient level and pressurize bleeder.

NOTE—

- *On models built before October 1995, connect hose to reservoir overflow fitting and block off hose end to prevent brake fluid from exiting reservoir overflow.*

Brake bleeding pressure (with vacuum brake boost)	
Approximate pressure (ABS)	1.5 bar (21.76 psi)
Approximate pressure (ABS / ABD)	1.5-2.0 bar (21.76 psi - 29 psi)

CAUTION—

- *Do not exceed 2.0 bar (29 psi) at the fluid reservoir when bleeding or flushing hydraulic system.*

– Connect transparent hose and collection bottle to bleeder valve at caliper. Open each bleeder valve until clear, bubble-free brake fluid appears. Bleed both valves at each brake caliper.

Tightening torques	
Brake bleeder valve to brake caliper, M10	8 -12 Nm (6-9 ft-lb)
Connecting brake pipe to brake caliper, M10	12 Nm (9 ft-lb)
Wheel to wheel hub	130 Nm (96 ft-lb)

– Bleed brakes using following sequence:

1. Right rear
2. Left rear
3. Right front
4. Left front

– If brake master cylinder or ABS hydraulic unit has been replaced, or brake hydraulic system required a significant volume of fluid, use following steps to remove air bubbles from primary master cylinder circuit:
  • Open right rear bleeder valve. Depress brake pedal several times.

> ***CAUTION—***
> *• Unless the master cylinder is new, it is recommended that the full stroke of the brake pedal not be used to avoid damaging master cylinder seals. Instead, use shorter pedal strokes and double the pumping cycles.*

  • Hold pedal depressed for 2-3 seconds, then release slowly. Close bleeder valve.

Tightening torque	
Brake bleed valve to caliper, M10	8 -12 Nm (6-9 ft-lb)

  • Repeat this procedure in sequence; left rear, right front, left front.
  • Depressurize bleeding device and disconnect.
  • Top up brake fluid as needed.

NOTE—
• If vehicle is equipped with ABS/ABD and hydraulic unit was replaced, continue to **Additional brake bleeding procedure for ABD circuit** *in this repair group.*

Additional brake bleeding procedure (after replacing ABS/ABD hydraulic unit)

To bleed trapped air from the ABS/ABD hydraulic unit, a Porsche scan tool is needed to systematically activate the return pump and solenoid valves in the hydraulic unit.

– Perform brake bleed as described earlier under **Bleeding brakes (with vacuum brake booster)** in this section.
– Leave pressure bleeding device connected and pressurized.
– Connect Porsche scan tool. Turn ignition ON. Select bleed menu.
– Open right rear caliper bleed valve with hose and collection bottle attached.
– Press start button on scan tool.

NOTE —
• Pressing start button initiates operation of return pump and hydraulic solenoid valves.

– Allow fluid to flow from bleeder until clear and bubble-free. Throughout bleeding process, depress brake pedal at least 10 times.

> ***CAUTION—***
> *• Unless the master cylinder is new, it is recommended that the full stroke of the brake pedal not be used to avoid damaging master cylinder seals. Instead, use shorter pedal strokes and double the pumping cycles.*

– Close right rear bleed valve and immediately press stop button on scan tool.

Tightening torque	
Brake bleed valve to caliper, M10	8-12 Nm (6-9 ft-lb)

– Switch ignition OFF and disconnect tester. Disconnect bleeding device and top off brake fluid as necessary.

Hydraulic brake booster circuit, depressurizing

On cars fitted with a hydraulic brake booster system, the high-pressure brake booster circuit must be depressurized before servicing the brake system.

- *For information on brake system identification, see* **45 Antilock Brakes (ABS)**.

> ***WARNING—***
> - *The hydraulic brake boost system operates under high pressure (up to 2600 psi). Wear eye protection and protective gloves when depressurizing the booster circuit.*

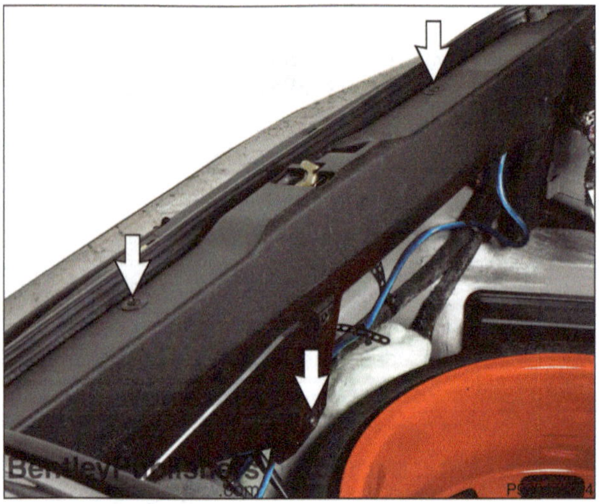

– Remove trunk carpet.

< Remove fasteners (**arrows**) holding trunk latch trim in place. Lift trim up and out to expose brake hydraulic booster pump.

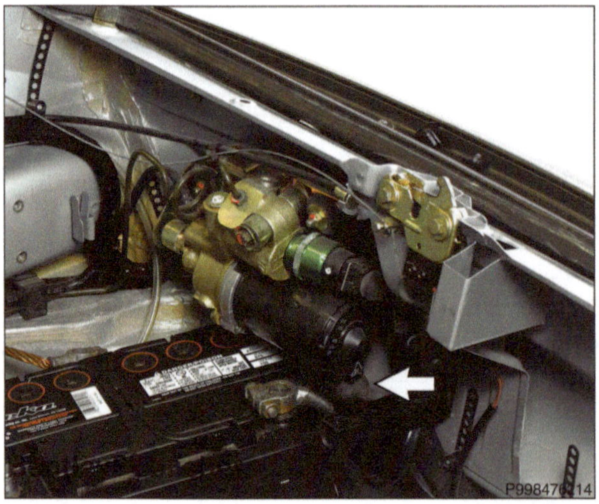

< Be sure ignition key is removed from ignition cylinder. Remove electrical connector (**arrow**) from brake booster pump.

– Depressurize booster circuit by pressing brake pedal 25 times until pedal feels hard.

– To keep hydraulic boost system depressurized, keep pump connector disconnected and ignition key OFF.

Bleeding brakes (with hydraulic brake booster)

Read the procedure through before bleeding the brakes. Note that the hydraulic brake booster circuit is a separate circuit from the hydraulic master cylinder and brake caliper circuit. The bleeding procedure given here includes bleeding both circuits.

◄ When bleeding the master cylinder and brake caliper circuit, a brake pressure bleeder should be used at the brake fluid reservoir.

- In addition, a factory scan tool will be required to bleed the system if the ABS/ABD hydraulic unit was replaced or if the system requires a significan volume of fluid after repairs are carried out. See **Additional brake bleeding procedure (after replacing ABS/ABD hydraulic unit)**.

> ***WARNING—***
>
> - *The hydraulic brake boost system operates under high pressure (up to 2600 psi). Wear eye protection and protective gloves when servicing the booster circuit.*

Bleeding hydraulic booster circuit

– Remove trunk carpet.

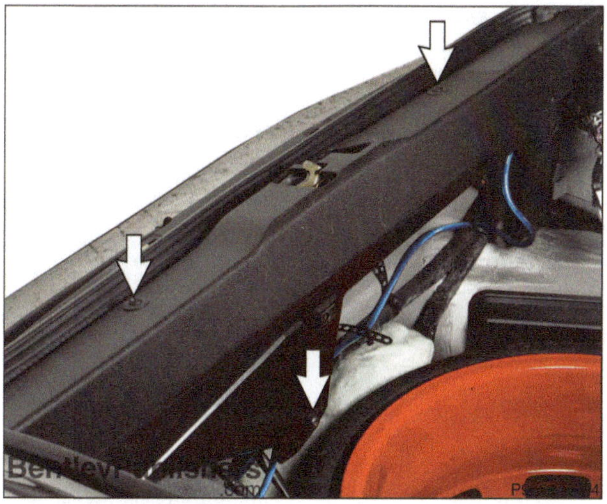

◄ Remove fasteners (**arrows**) holding trunk latch trim in place. Lift trim up and out to expose hydraulic booster pump.

– Charge pressure accumulator by turning ignition on and actuating brake pedal until hydraulic booster pump runs.

◄ Once hydraulic booster pumps shuts off, remove electrical connector from booster pump (**arrow**). Turn ignition key OFF and remove from ignition cylinder.

Bleeding brakes (with hydraulic brake booster)

◀ Working at pressure accumulator in right rear of trunk, connect a transparent hose and collection bottle to bleeder valve (arrow). Open bleeder valve slowly and allow fluid to escape into container. Close bleeder valve. Leave hose connected.

> ***WARNING—***
>
> *• The hydraulic brake boost system operates under high pressure (up to 2600 psi). Wear eye protection and protective gloves when servicing the booster circuit.*

– Top off brake fluid reservoir with fresh brake fluid.

> ***WARNING—***
>
> *• Brake fluid is poisonous, highly corrosive and dangerous to the environment. Wear safety glasses and rubber gloves when working with brake fluid. Immediately clean away fluid spilled on painted surfaces and wash with water. Dispose of brake fluid properly.*

– Connect brake pressure bleeder to fluid reservoir. Fill bleeder fluid tank to sufficient level and pressurize bleeder.

NOTE—

• On models built before October 1995, connect hose to reservoir overflow fitting and block off hose end to prevent brake fluid from exiting reservoir overflow.

Brake bleeding pressure (with hydraulic brake boost)	
Approximate pressure (ABS / ABD)	1.5 bar (21.75 psi)

> ***CAUTION—***
>
> *• Do not exceed 2.0 bar (29 psi) at the fluid reservoir when bleeding or flushing hydraulic system.*

– Open bleeder valve at accumulator.

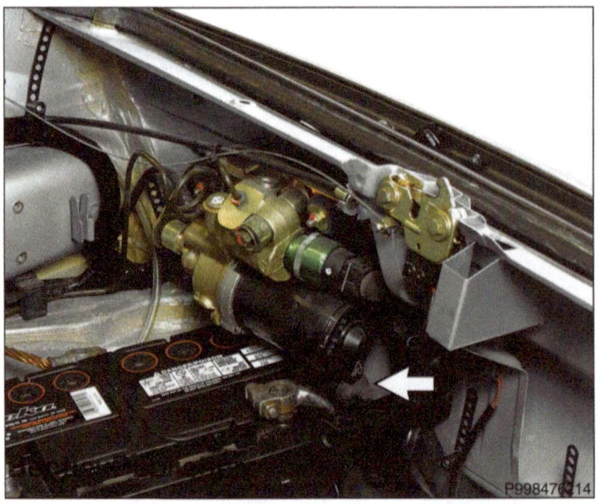

◀ Turn ignition on and reconnect electrical connector at booster pump (arrow). Allow fluid to pump into container. When fluid is clear and free of air bubbles, close bleeder valve and disconnect connector at pump. Turn ignition off.

Bleeding master cylinder and caliper circuit

– Connect a transparent hose and collection bottle to bleeder valve at caliper. With pressure bleeder connected, open each bleeder valve. Close bleeder valve when bubble-free brake fluid appears. Bleed both valves at each brake caliper.

– Bleed brakes using following sequence:

1. Right rear
2. Left rear
3. Right front
4. Left front

Tightening torque	
Brake bleed valve to caliper, M10	8 - 12 Nm (6-9 ft-lb)

– If brake master cylinder was replaced or brake hydraulic system required a significant volume of fluid, use following steps to remove air bubbles from primary master cylinder circuit:

- With pressure bleeder connected, open right rear bleeder valve. Depress brake pedal several times. Hold pedal depressed for 2-3 seconds, then release slowly. Close bleeder valve.

> ***CAUTION—***
>
> - *Unless the master cylinder is new, it is recommended that the full stroke of the brake pedal not be used to avoid damaging master cylinder seals. Instead, use shorter pedal strokes and double the pumping cycles.*

– Repeat above procedure in following sequence: left rear, right front, left front.

– Depressurize bleeding device and disconnect.

– Top up brake fluid as needed.

– Reconnect electrical connector to booster pump. Turn ignition ON to charge pressure accumulator. When pump turns off, turn ignition OFF and check brake fluid level. Top off as necessary.

Additional brake bleeding procedure (after replacing the ABS/ABD hydraulic unit)

To bleed trapped air from the ABS/ABD hydraulic unit, a Porsche scan tool is needed to activate the return pump and solenoid valves in the hydraulic unit.

– Perform brake bleed as described earlier under **Bleeding brakes (with vacuum brake booster)** in this section.

◄ Depressurize brake booster circuit. See **Hydraulic brake booster circuit, depressurizing**. Leave harness connector disconnected from hydraulic booster pump (**arrow**).

– Leave pressure bleeding device connected and pressurized.

– Connect Porsche scan tool. Turn ignition ON. Select bleed menu.

– Open right rear caliper bleeder valve with hose and collection bottle attached.

– Press start button on scan tool.

NOTE —

- *Pressing start button initiates operation of return pump and hydraulic solenoid valves.*

– Allow fluid to flow from bleeder until clear and bubble-free Throughout bleeding process, depress brake pedal at least 10 times.

> ***CAUTION—***
>
> - *Unless the master cylinder is new, it is recommended that the full stroke of the brake pedal not be used to avoid damaging master cylinder seals. Instead, use shorter pedal strokes and double the pumping cycles.*

– Close right rear bleeder valve and immediately press stop button on scan tool.

– Switch ignition OFF and disconnect tester. Disconnect bleeding device and top off brake fluid as necessary.

– Reconnect electrical connector to hydraulic booster pump. Switch ignition ON to charge pressure accumulator. When pump turns off, turn ignition OFF and check brake fluid level. Top off as necessary.

Brake master cylinder (with vacuum booster), removing and installing

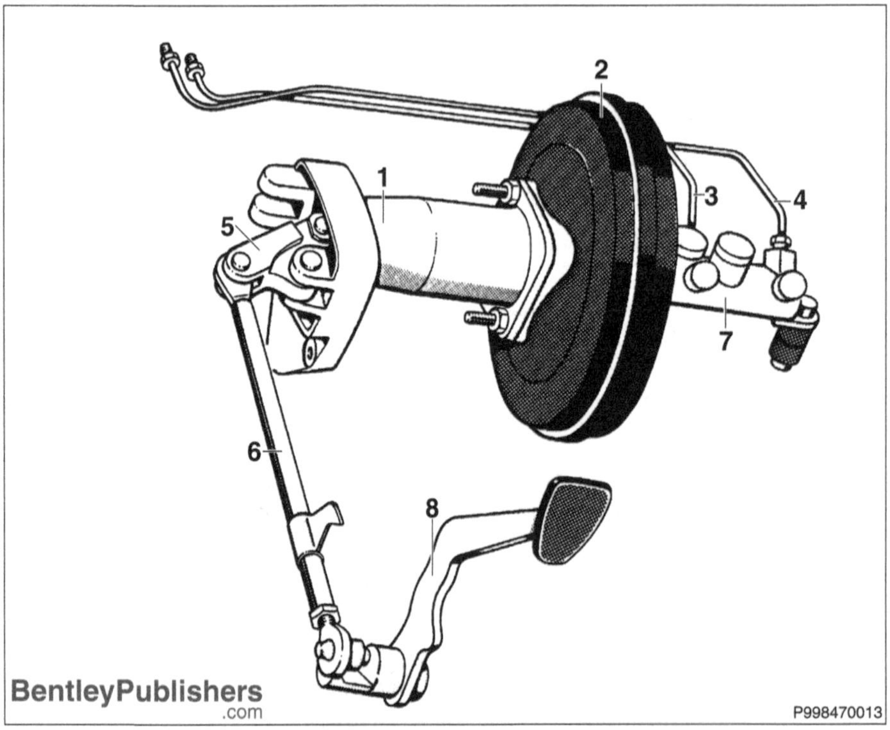

1. Bearing bracket
2. Brake booster
3. Brake line
4. Brake line
5. Reverse lever
6. Push rod
7. Master cylinder
8. Pedal assembly

– Using fluid extractor, drain brake fluid reservoir.

– Raise vehicle and support safely.

> ***WARNING—***
> - *Make sure vehicle is stable and well supported at all times. Use a professional automotive lift or jack stands. Place jack stands underneath structural chassis points, not underneath suspension parts. A floor jack is not adequate support.*

– Remove plastic splash shield from front of car. See **03 Service and Maintenance.**

◄ Disconnect brake lines at master cylinder (**arrows**) and plug to prevent leaking.

> ***WARNING—***
> - *Brake fluid is poisonous, highly corrosive and dangerous to the environment. Wear safety glasses and rubber gloves when working with brake fluid. Immediately clean away fluid spilled on painted surfaces and wash with water. Dispose of brake fluid properly.*

◄ Remove fastener (**arrow**) from master cylinder mounting bracket.

◄ Remove nuts (**arrows**) holding master cylinder to brake booster and pull master cylinder from brake booster.

– Installation is reverse or removal, note the following:

- Top off brake fluid reservoir.
- Bleed brakes as usual. See **Bleeding brakes** in this repair group.
- Check brake function.

Tightening torques	
Master cylinder to vacuum brake booster, M8	23 Nm (17 ft-lb)
Master cylinder to mounting bracket, M8	23 Nm (17 ft-lb)
Brakes lines to master cylinder (copper pipes) • M10 • M12	 12 - 14 Nm (9 - 10 ft-lb) 20 Nm (15 ft-lb)

Brake master cylinder (with hydraulic booster), removing and installing

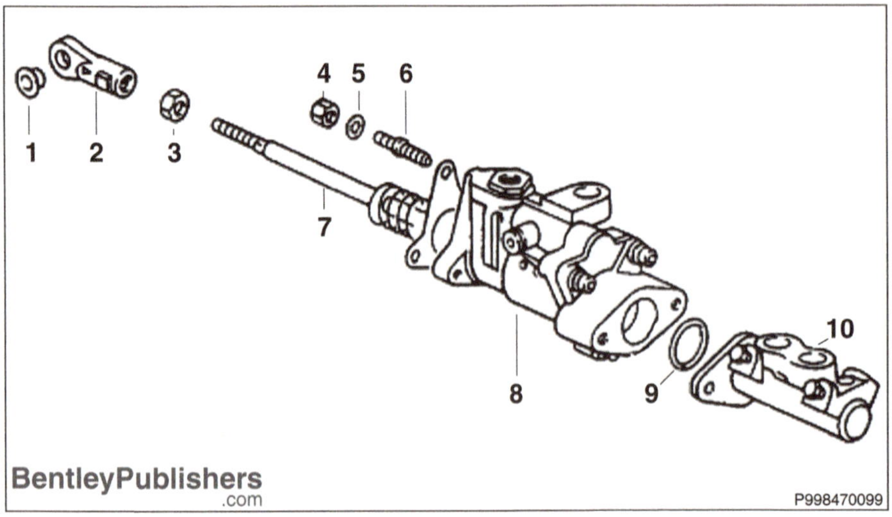

1. Bushing
2. Swivel head
3. Nut
4. Lock nut
   - Tighten to 23 Nm (17 ft-lb)
5. Washer
6. Stud
7. Rod
8. Brake booster
9. O-ring
10. Brake master cylinder

– Using fluid extractor, drain brake fluid reservoir.

– Remove trunk carpet.

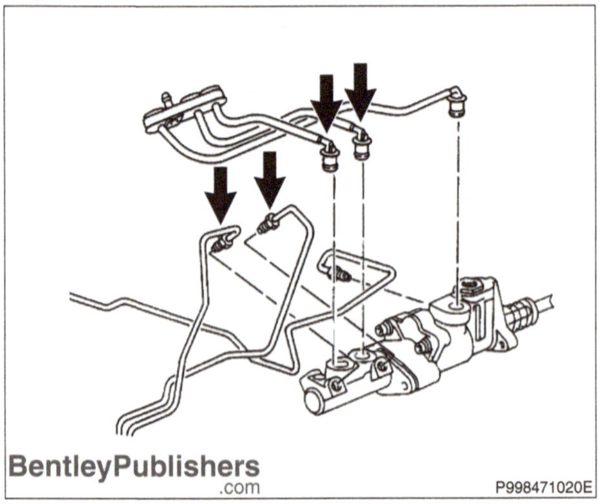

◄ Disconnect brake lines and hose fittings (**arrows**) at master cylinder.

> ***WARNING—***
> - *Brake fluid is poisonous, highly corrosive and dangerous to the environment. Wear safety glasses and rubber gloves when working with brake fluid. Immediately clean away fluid spilled on painted surfaces and wash with water. Dispose of brake fluid properly.*

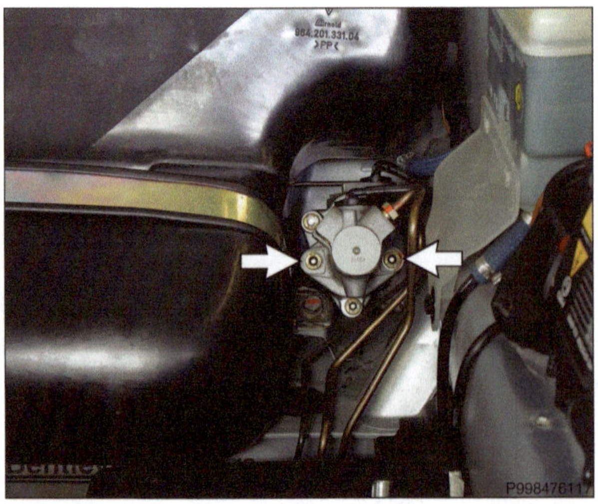

◄ Remove bolts (**arrows**) securing brake master cylinder to hydraulic brake booster assembly.

– Remove master cylinder from brake booster.

– Installation is reverse or removal, noting the following:
- Replace sealing O-ring with new.
- Top off brake fluid reservoir.
- Bleed master cylinder and caliper circuit. See **Bleeding brakes** in this repair group.
- Check brake function.

Tightening torques	
Master cylinder to hydraulic brake booster	23 Nm (17 ft-lb)
Brakes lines to master cylinder (copper pipes) • M10 • M12	 12 - 14 Nm (9 - 10 ft-lb) 20 Nm (15 ft-lb)

Vacuum brake booster, removing and installing

– Raise vehicle and support safely.

> ***WARNING—***
>
> • *Make sure vehicle is stable and well supported at all times. Use a professional automotive lift or jack stands. Place jack stands underneath structural chassis points, not underneath suspension parts. A floor jack is not adequate support.*

– Remove plastic splash shield from front of car. See **03 Service and Maintenance**.

< Remove vacuum line (**arrow**) from booster.

– Remove brake master cylinder. See **Brake master cylinder, (with vacuum booster), removing and installing** in this repair group.

< Loosen bearing bracket boot hose clamp (**arrow**). Slide boot off bearing bracket and out of the way.

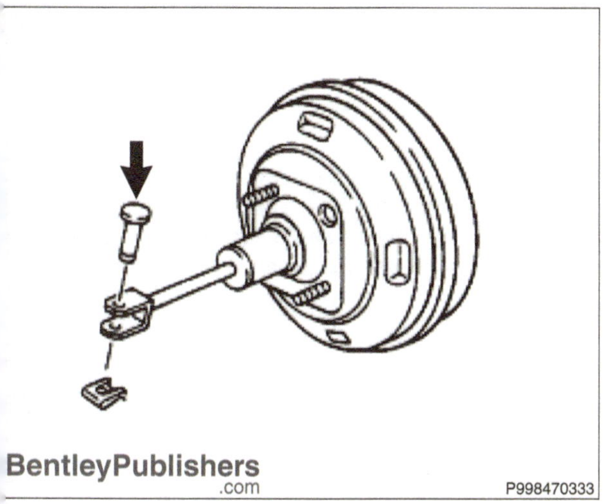

< Working inside boot, release clip and remove pin (**arrow**) holding push rod to reverse lever.

47

- Remove two nuts (**arrows**) securing vacuum booster to bearing bracket.
- Separate vacuum booster from bearing bracket, turn sideways and pull from car.
- Installation is reverse or removal. Note the following:
  - Top off brake fluid reservoir.
  - Bleed brakes as usual. See **Bleeding brakes** in this repair group.
  - Check brake function.

Tightening torques	
Vacuum brake booster bearing bracket to side member, M8	23 Nm (17 ft-lb)
Vacuum brake booster to bearing bracket, M8	23 Nm (17 ft-lb)

Hydraulic brake booster, removing and installing

- Depressurize hydraulic booster circuit. See **Hydraulic brake pressure circuit, depressurizing**.

> ***WARNING—***
> - *The hydraulic brake boost system operates under high pressure (up to 2600 psi). Wear eye protection and protective gloves when servicing the booster circuit.*

- Be sure electrical connector (**arrow**) at brake booster pump remains disconnected.
- Remove brake master cylinder from hydraulic brake booster. See **Brake master cylinder (with hydraulic booster), removing and installing** in this repair group.

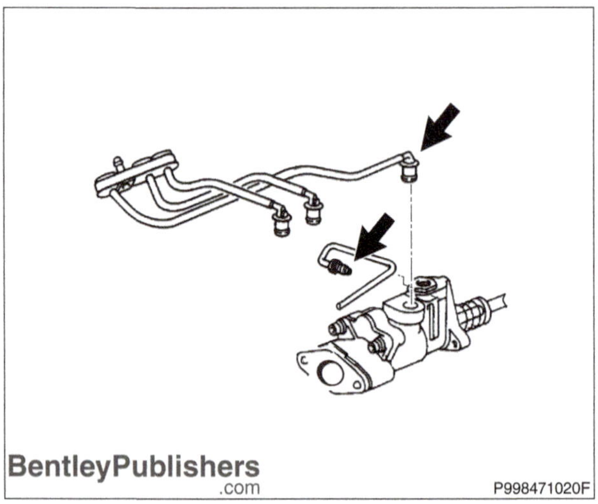

- Remove supply line and brake line (**arrows**) from brake booster. Plug open connections to prevent contamination.

> ***WARNING—***
> - *Brake fluid is poisonous, highly corrosive and dangerous to the environment. Wear safety glasses and rubber gloves when working with brake fluid. Immediately clean away fluid spilled on painted surfaces and wash with water. Dispose of brake fluid properly.*

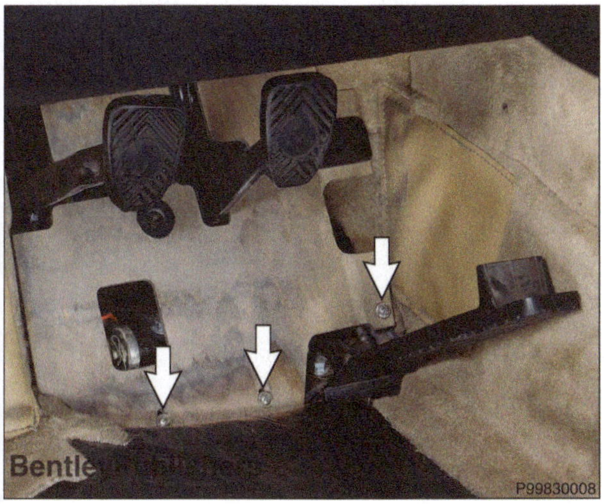

< Remove floor board:

- Working in pedal area, remove floor mat and peel carpet from pedal floor board.
- Remove retaining screws at base of floor board (**arrows**).
- Pull accelerator pedal from top and off of linkage ball mount.
- Remove floor board.

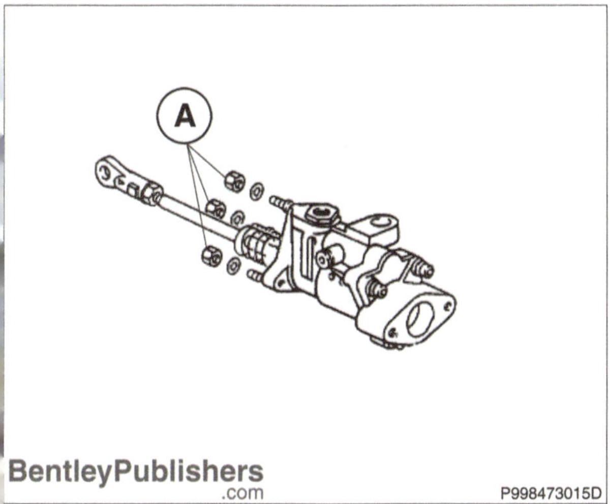

< Working above brake pedal, remove retaining clip and clevis pin from pushrod. Remove nuts (**A**) securing brake booster to bulkhead. Remove booster from trunk side.

– Installation is reverse or removal. Note the following:

- Support accelerator pedal push rod with locking pliers when reconnecting rod to pedal.
- Top off brake fluid reservoir.
- Bleed brakes. See **Bleeding brakes** in this repair group.
- Check brake function.
- Observe torque tightening specifications.

Tightening torques	
Brakes line to hydraulic brake booster, M10	14 - 16 Nm (10 - 12 ft-lb)
Hydraulic brake booster to bulkhead, M8	23 Nm (17 ft-lb)
Master cylinder to hydraulic brake booster, M8	23 Nm (17 ft-lb)

Hydraulic brake booster pump, removing and installing

– Depressurize hydraulic booster circuit. See **Hydraulic brake pressure circuit, depressurizing**.

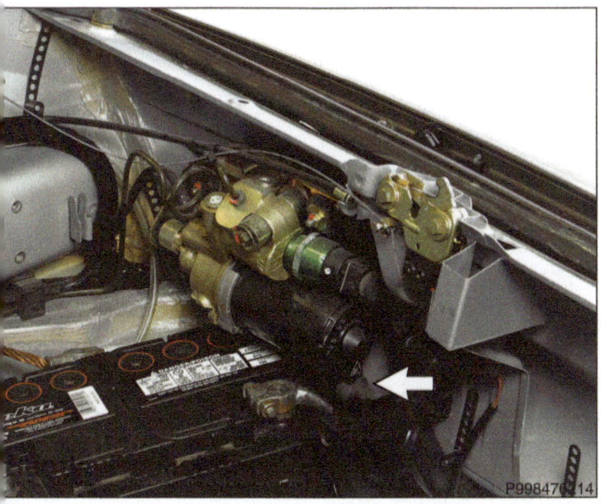

< Be sure electrical connector (**arrow**) at brake booster pump remains disconnected.

> ***WARNING—***
> - *The hydraulic brake boost system operates under high pressure (up to 2600 psi). Wear eye protection and protective gloves when servicing the booster circuit.*

Hydraulic brake booster pump, removing and installing

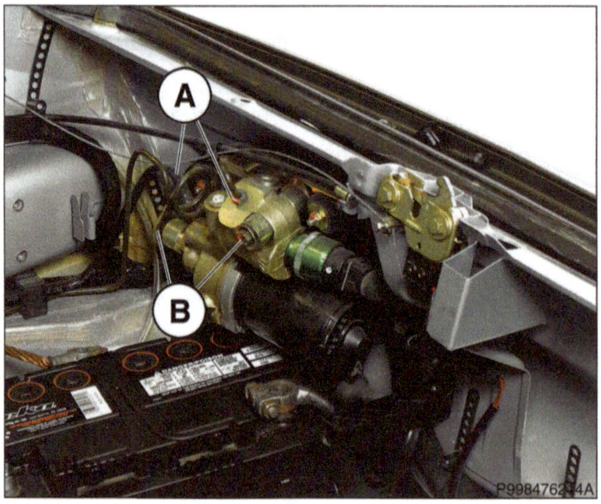

◄ Remove brake lines (**A**) from brake booster pump. Remove brake booster pump bracket retaining bolts (**B**) and remove booster pump.

> ***WARNING—***
>
> - *Brake fluid is poisonous, highly corrosive and dangerous to the environment. Wear safety glasses and rubber gloves when working with brake fluid. Immediately clean away fluid spilled on painted surfaces and wash with water. Dispose of brake fluid properly.*

– Installation is reverse of removal, noting the following:
- Top off brake fluid reservoir.
- Bleed hydraulic booster circuit. See **Bleeding brakes** in this repair group.
- Check brake function.
- Observe torque tightening specifications.

Tightening torques	
Brakes line to pump, M10	14 - 16 Nm (10 - 12 ft-lb)
Pressure warning switch to booster pump, M25	26 Nm (19 ft-lb)
Booster pump mounting, M6	10 -13 Nm (7 - 10 ft- lb)

Hydraulic brake pressure accumulator, removing and installing

– Depressurize hydraulic booster circuit. See **Hydraulic brake pressure circuit, depressurizing**.

◄ Be sure electrical connector (**arrow**) at brake booster pump remains disconnected.

> ***WARNING—***
>
> - *The hydraulic brake boost system operates under high pressure (up to 2600 psi). Wear eye protection and protective gloves when servicing the booster circuit.*

< Remove cap and attach bleeding device to bleeder nipple on accumulator (**arrow**). Open nipple to allow residual fluid to evacuate into container. Close bleed nipple.

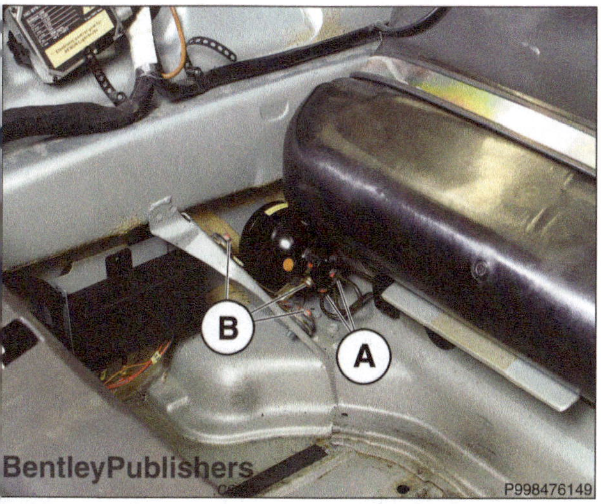

< Remove two brake lines (**A**) from accumulator. Remove two retaining nuts (**B**) and remove accumulator.

> ***WARNING—***
> - *Brake fluid is poisonous, highly corrosive and dangerous to the environment. Wear safety glasses and rubber gloves when working with brake fluid. Immediately clean away fluid spilled on painted surfaces and wash with water. Dispose of brake fluid properly.*

– Installation is reverse of removal. Note the following:
- Top off brake fluid reservoir.
- Bleed hydraulic booster circuit. See **Bleeding brakes** in this repair group.
- Check brake function.
- Observe torque tightening specifications.

Tightening torques	
Brakes line to pressure accumulator, M10	14 - 16 Nm (10 - 12 ft-lb)
Screw on fitting to pressure accumulator	14 - 16 Nm (10 - 12 ft-lb)

Brake caliper, removing and installing

– Raise vehicle and support safely. Remove wheel.

> ***WARNING—***
> - *Make sure vehicle is stable and well supported at all times. Use a professional automotive lift or jack stands. Place jack stands underneath structural chassis points, not underneath suspension parts. A floor jack is not adequate support.*

– Remove brake pads. See **46 Brake–Mechanical**

◄ Remove brake line (**arrow**) and plug line to prevent leakage.

> ***WARNING—***
> - *Brake fluid is poisonous, highly corrosive and dangerous to the environment. Wear safety glasses and rubber gloves when working with brake fluid. Immediately clean away fluid spilled on painted surfaces and wash with water. Dispose of brake fluid properly.*

NOTE–
- *On rear calipers, brake line location is slightly lower (front caliper pictured).*

◄ Remove two brake caliper mounting bolts (**arrows**) and remove caliper.

NOTE–
- *On rear calipers, mounting bolts are on inner side of caliper (front brakes pictured).*

– Installation is reverse of removal. Note the following:
- Replace brake caliper mounting screws with new.
- Bleed braking system when completed if caliper is removed. See **Bleeding brakes** in this repair group.
- Firmly press brake pedal several times with vehicle stationary to ensure pads adjust to operating position before driving.
- Check brake fluid level and secure reservoir cap.

Tightening torques	
Brake bleed valve to caliper, M10	8 -12 Nm (6-9 ft-lb)
Brake caliper to wheel carrier, M12 (replace with new)	85 Nm (63 ft-lb)
Brake line to brake caliper (copper pipes) • M10 • M12	 12 - 14 Nm (9 - 10 ft-lb) 20 Nm (15 ft-lb)
Connecting pipe to brake caliper, M10	12 Nm (9 ft-lb)
Wheel to wheel hub	130 Nm (96 ft-lb)

48 Steering

General

This repair group covers servicing of the steering system, including steering rack, power steering pump and pump belt, and tie rod replacement. Removal and installation of steering wheel is also included here.

See also the following repair groups:

- **40 Front Suspension**
- **44 Wheels, Tires and Alignment**

The Porsche 993 is equipped with power-assisted rack and pinion steering. The steering rack is bolted to the front cross member. There are no provisions for lubrication of the steering rack or linkage.

Wear and play in the steering system will cause sloppy, loose steering. Before checking for any steering problems, check tires, tire inflation pressures and front suspension components. Inspect ball joint and steering rack boots for tears or damage and replace if necessary. Tie rod ends are subject to wear. In general, the steering system is serviced by replacement of worn parts.

Warnings

WARNING—

- *The 993 is equipped with an airbag mounted to the steering wheel. The airbag is an explosive device and should be treated with extreme caution. See* **69 Seat belts, Airbags** *for additional airbag information.*
- *Do not reuse self-locking nuts. They are designed to be used only once and may fail if reused. Replace with new locking nuts.*
- *Do not install bolts and nuts coated with undercoating wax, as correct tightening torque cannot be assured. Clean threads with solvent before installation, or install new parts.*
- *Do not attempt to weld or straighten any steering components. Replace damaged parts.*

Steering Wheel

Steering wheel, removing and installing

– Disconnect negative (–) battery cable. Insulate battery terminal to keep cable from accidentally contacting terminal.

> ***CAUTION—***
> - *Prior to disconnecting the battery, read the battery disconnection cautions in* **00 Warnings and Cautions**.

> ***WARNING—***
> - *After disconnecting the battery, wait 1 minute before beginning work on airbag components.*

< Working from rear of steering wheel, use T30 wrench to remove Torx bolts (**arrows**) holding drivers airbag to steering wheel.

< Carefully pry off connector (**arrow**) from airbag. Remove airbag and store safely.

> ***WARNING—***
> - *Improper handling of the airbag can cause serious injury. Store the airbag with the horn pad facing up. If stored facing down, accidental deployment can propel it violently into the air, causing injury.*

NOTE–
- *Turning the ignition ON with the airbag removed will set an airbag system fault code. Resetting airbag fault memory requires a Porsche scan tool.*

< Turn wheels to straight ahead position and mark position of steering wheel splines (**A**) in relation to shaft splines. Disconnect horn wires (**B**).

- Remove center nut securing steering wheel to steering shaft.
- Carefully pull off steering wheel while feeding wiring through wheel.

NOTE–

- *To prevent damage caused by unintended rotation, airbag contact spring locks when steering wheel is removed. Contact spring unlocks automatically when steering wheel is installed.*

- Installation is reverse of removal, note the following:
  - Fit steering wheel using previously made spline alignment marks and make sure upper spokes of wheel are horizontal.
  - Install and tighten steering wheel mounting nut and washer.
  - Reattach airbag and horn harness connectors. Remount airbag.

CAUTION—

- *Use care not to pinch horn and airbag harness when mounting airbag.*

  - Using a Porsche scan tool, check and reset fault memory.
  - Test operation of horn. Start car and check that airbag warning lamp goes out.

Tightening torque	
Airbag to steering wheel (use new screws)	10 Nm (7 ft-lb)
Steering wheel to steering shaft	45 Nm (33 ft-lb)

Steering column stalk switch, removing and installing

The steering wheel stalk switches include wiper switch, directional light switch, and cruise control switch.

- Remove steering wheel. See **Steering wheel, removing and installing** in this repair group.

- Remove steering column cover fasteners (**arrows**) and remove three cover pieces.

Remove fasteners (**arrows**) holding contact unit in place and slide contact unit from steering shaft.

Remove knee bar (**arrow**) to access wiring for contact unit. See **70 Interior Trim**.

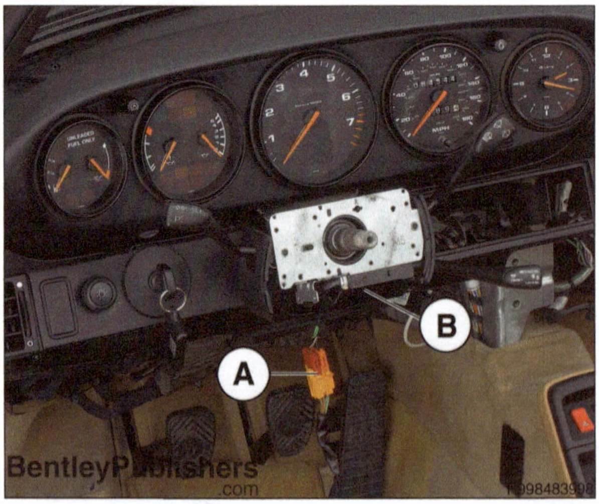

Pull out wiring. Disconnect connectors (**A**) and remove contact unit. Loosen fastener (**B**) holding stalk switch mount in place.

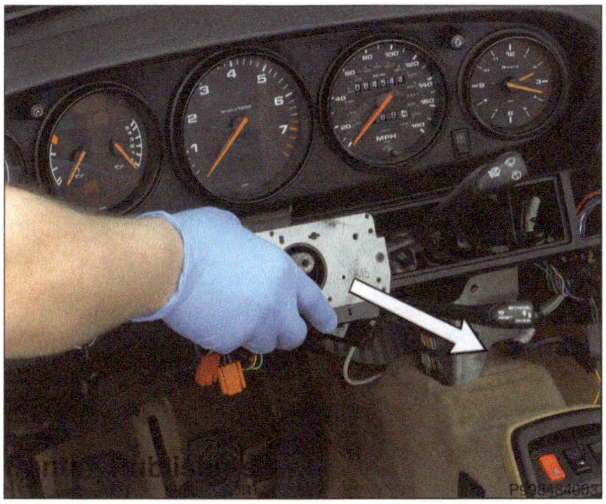

< Pull stalk switch mount with switches from steering column and disconnect electrical connector.

– Installation is reverse of removal, noting the following:
  - Ensure steering wheel aligns to marks made on spline.
  - Take care not to pinch airbag or contact unit wiring.

Tightening torque	
Steering wheel to steering shaft	45 Nm (33 ft-lb)
Airbag to steering wheel	10 Nm (7 ft-lb)

Tie Rods and Steering Rack

Tie rod, removing and installing

48

– Loosen front wheel lug bolts. Raise car and support safely. Remove front wheel.

> ***WARNING—***
> - *Make sure the car is stable and well supported at all times. Use a professional automotive lift or jack stands. A floor jack is not adequate support.*

– Working at wheel carrier, remove nut from tie rod end ball joint. If necessary, counter hold ball joint stud with a T-40 Torx if stud spins while loosening nut.

< Using ball joint press tool (**inset**), press ball joint out of wheel carrier.

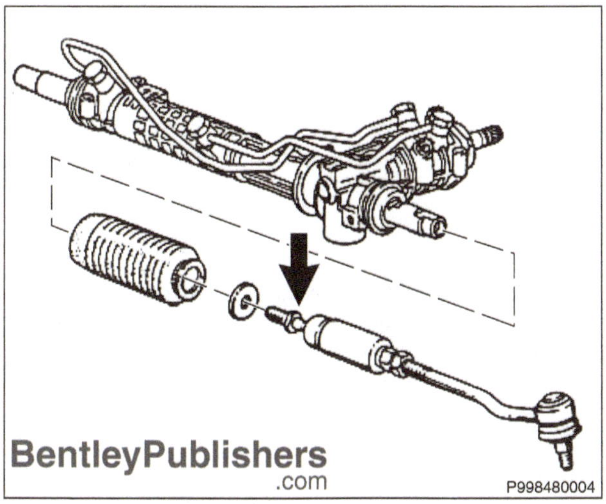

< Remove bellows clamp and push back bellows. Remove tie rod from steering rack using open-ended wrench on hex fitting (**arrow**). Counter hold steering rack shaft when loosening joint from rack.

– Installation is reverse of removal. Note the following:
  - Crimp new boot clamp using clamp crimping pliers.
  - Use new self-locking nut at wheel carrier end.

Tightening torques	
Tie rod inner joint to steering rack	70 Nm (52 ft-lb)
Tie rod end to wheel carrier (use new)	65 Nm (48 ft-lb)
Tie rod to tie rod inner joint (jam nut)	45 Nm (33 ft-lb)
Wheel to wheel hub	130 Nm (96 ft-lb)

NOTE–
- *When installing a new tie rod assembly, adjust length of new rod to match length of old tie rod until a wheel alignment can be performed. Carry out wheel alignment to ensure proper tracking and to minimize tire wear.*

Steering rack, removing and installing

– Set wheels in straight ahead position then loosen front wheel lug bolts. Raise car, and support safely. Remove front wheels.

> ***WARNING—***
> *• Make sure the car is stable and well supported at all times. Use a professional automotive lift or jack stands. A floor jack is not adequate support.*

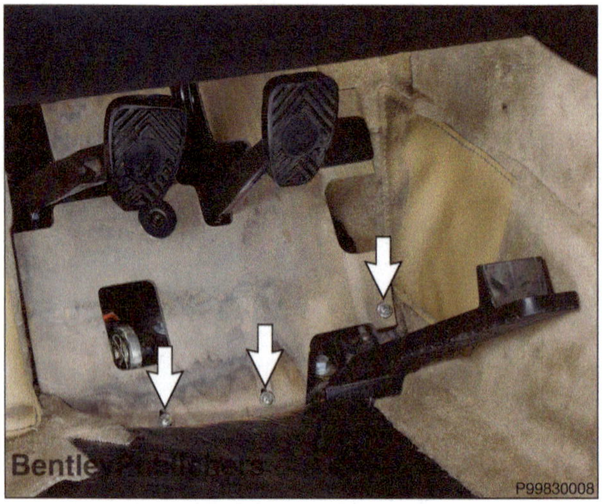

◄ Working inside car, remove pedal floor board:

- Remove floor mat and peel carpet from pedal floor board.
- Remove retaining screws at base of floor board (**arrows**).
- Pull accelerator pedal from top and off of linkage ball mount.
- Remove floor board.

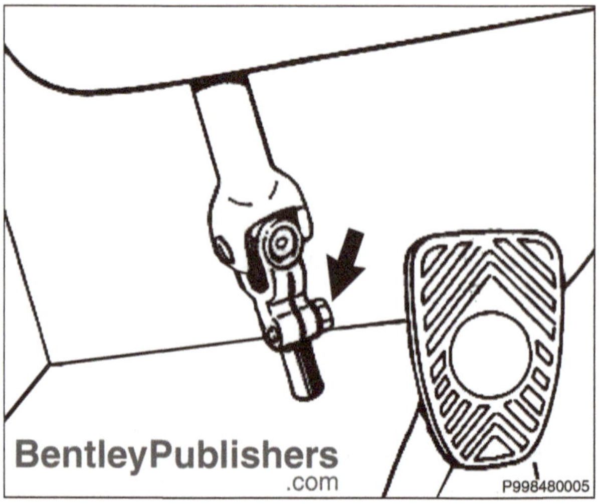

◄ With steering wheel pointed straight ahead, disconnect steering rack universal joint at column behind pedal assembly. Remove clamping bolt (**arrow**) and slide joint upward.

NOTE –
• Be sure steering wheels is in straight ahead position before removing U-joint clamping bolt.

– Remove left and right tie rods from steering rack and wheel carrier. See **Tie rods, removing and installing** in this repair group.

– Working in right side of engine compartment, siphon fluid from power steering reservoir.

◄ Disconnect power steering lines from the steering rack (**arrows**). Disconnect pipes from bracket on right side of rack. Clean up spilled fluid.

> ***CAUTION—***
> *• When removing and installing steering rack, be sure not to damage or score sliding parts. Use protective caps to prevent particles from entering rack when lines are removed.*

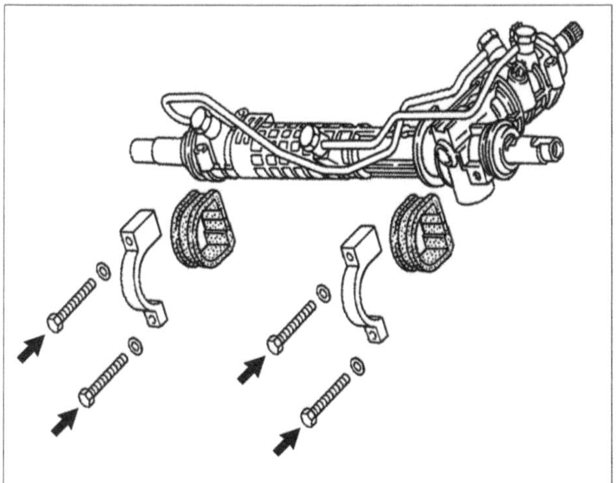
P998480009

< Remove mounting bolts (**arrows**) securing power steering rack to subframe.

– Lower steering rack from car. Take care not to damage assembly.

– Installation is reverse of removal. Note the following:

- Replace rack mounting and universal joint fasteners with new, original equipment. Clean away left over locking compound in threaded fastener bores.
- With rack fully extended, pull back boots and grease with VW steering gear grease AOF 063 000 04. Also grease outside of steering gear pinion shaft at teeth of universal joint.
- With steering rack and steering wheel in center position, push shaft onto universal joint. Secure new fastening bolt.
- After fitting hydraulic lines to steering rack, fill reservoir with appropriate power steering fluid and bleed system.
- When refitting pedal floor board, support accelerator pedal push rod with locking pliers when reconnecting rod to pedal.
- Observe torque tightening specifications.

Tightening torques	
Hydraulic lines to steering rack, M12	20 Nm (15 ft-lb)
Steering rack to cross member, M8 (replace with new)	45 Nm (33 ft-lb
Tie rod inner joint to steering rack	70 Nm (52 ft-lb)
Tie rod end to wheel carrier (replace with new)	65 Nm (48 ft-lb)
Tie rod to tie rod inner joint (jam nut)	45 Nm (33 ft-lb)
Universal joint to steering rack, M8 (replace with new)	23 Nm (17 ft-lb)
Wheel to wheel hub	130 Nm (96 ft-lb)

Power Steering

> ***CAUTION—***
> *• Do not overfill power steering reservoir.*
> *• Replace visibly swollen hoses.*

Power steering system

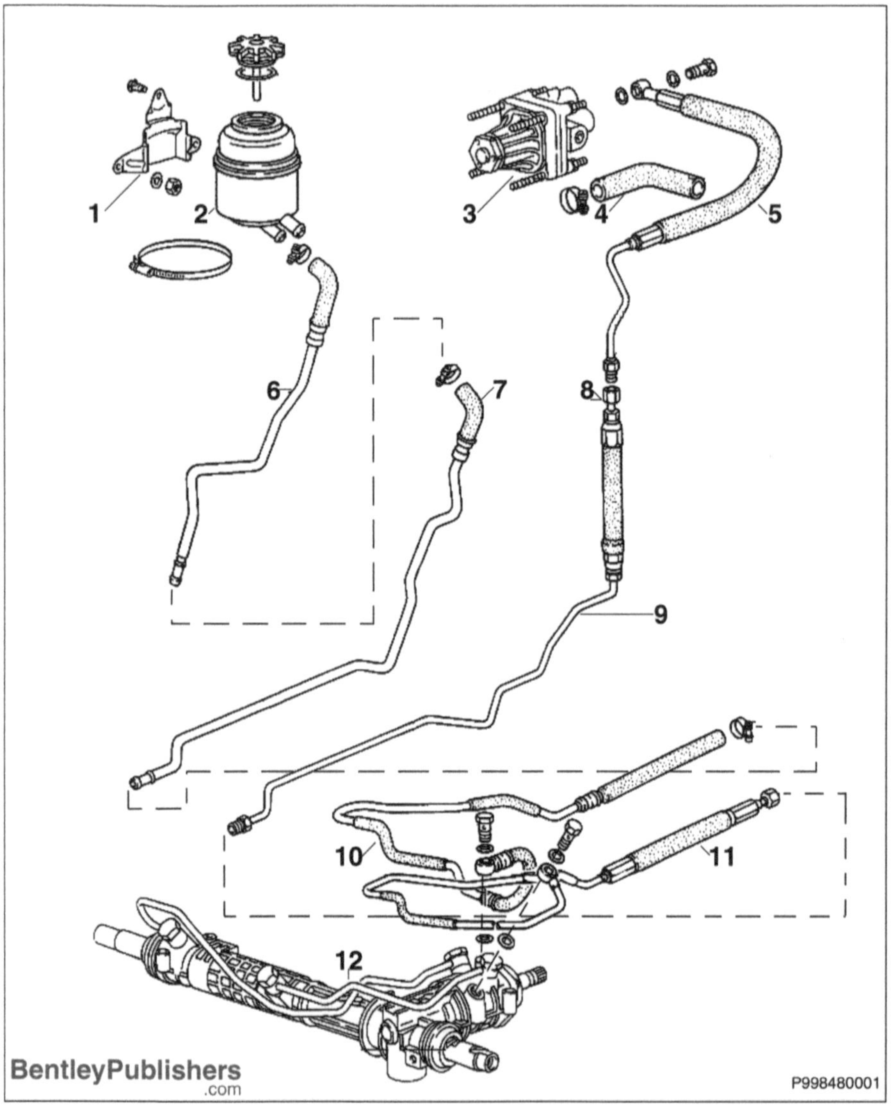

1. Support bracket
2. Fluid reservoir
3. Power steering pump
4. Hose
5. High pressure line
6. Return line
7. Return line
8. High pressure line fastener
9. High pressure line
10. Lower return line
11. Lower pressure line
12. Power steering rack/gear

Power steering system, checking fluid level

Damage to the power-assisted steering is often the result of running the system while low on oil. Insufficient oil level will damage the servo pump as a result of the high pressure in the hydraulic circuit.

Groaning noises when the steering is to the end (locked) position or foam formation in the reservoir indicates a low oil level and/ or that air has been sucked into system. Before topping up the reservoir, remedy any leaks on the suction side and replace any faulty parts on the pressure side.

The reservoir is located on the right side of the engine compartment. Check power steering fluid level with engine idling and car on level ground. Do not operate steering while checking fluid level.

There are two different power steering reservoirs that came on the 993. Early 993's had a transparent reservoir. It has a min and max mark on side. Fill level between the min and max marks.

< Later 993's have a black, non-transparent reservoir. Remove cap with attached dipstick. Wipe dipstick clean and reinsert to check fluid level.

NOTE–

- *Power steering system in the 993 uses Dexron II D ATF on cars built before March 1996 and Pentosin (CHF 11S) power steering fluid on cars built March 1996 and later.*
- *Pentosin and ATF can be mixed. Pentosin can be added to Dexron to top up if needed.*
- *If a vehicle is still using Dexron II, Porsche recommends flushing system and using Pentosin (CHF 11S) instead.*

Power steering system, bleeding

After installation of new steering components or after a substantial loss of fluid, bleed power steering system.

– Briefly start engine a few times (when it fires, quickly switch it off). This procedure causes fluid level in reservoir to fall rapidly. Continuously have helper check and maintain reservoir level using this procedure. Be sure reservoir is not run dry.

CAUTION—

- *Do not overfill power steering reservoir.*

– If power steering level does not fall when engine is turned over, start engine and let run at idle.

– Turn steering wheel steadily and quick from lock to lock several times to allow air to escape. Do not turn steering wheel with more force than necessary (to prevent unnecessary build-up of pressure).

– Have helper observe fluid level during procedure. If level continues to fall, add fluid until level remains constant and air bubbles stop rising.

Power steering pump, removing and installing

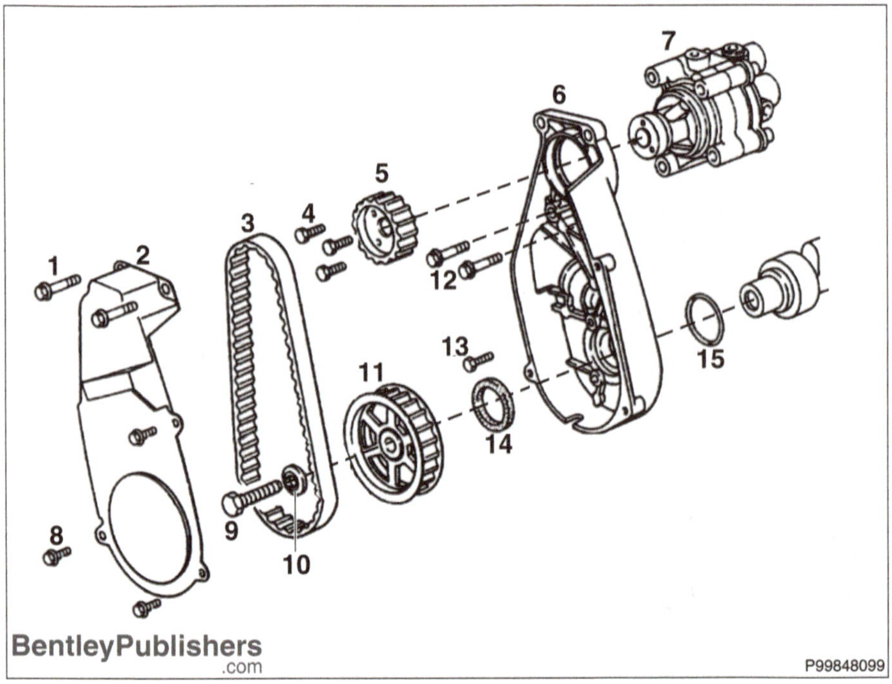

1. Bolt, M8
2. Cover
3. Toothed belt
4. Bolt, M6
5. Gear wheel (upper)
6. Power steering bracket
7. Power steering pump
8. Bolt, M6
9. Bolt, M12
10. Washer
11. Gear wheel (lower)
12. Bolt, M8
13. Bolt, M6
14. Shaft sealing ring
15. O-ring

NOTE–

- *Power steering pump removal can be completed with engine in vehicle, though access is severely limited making it a difficult task. Images shown here are with engine removed from vehicle.*

– Remove engine air filter box and intake housing assembly from passenger side of engine. See **24 Fuel injection**.

– Raise and support vehicle. Remove right rear wheel and inner fender liner.

> ***WARNING—***
>
> - *Make sure vehicle is stable and well supported at all times. Use a professional automotive lift or jack stands. Place jack stands underneath structural chassis points, not underneath suspension parts. A floor jack is not adequate support.*

◄ Remove power steering pump cover fasteners (**arrows**) then remove cover.

NOTE–

- *Two upper cover bolts are also retaining bolts for power steering pump.*

⋖ Remove bolts (**A**) from upper belt sprocket.

⋖ Pull off sprocket with belt in direction of **arrow**.

⋖ Remove remaining lower bolts (**arrows**).

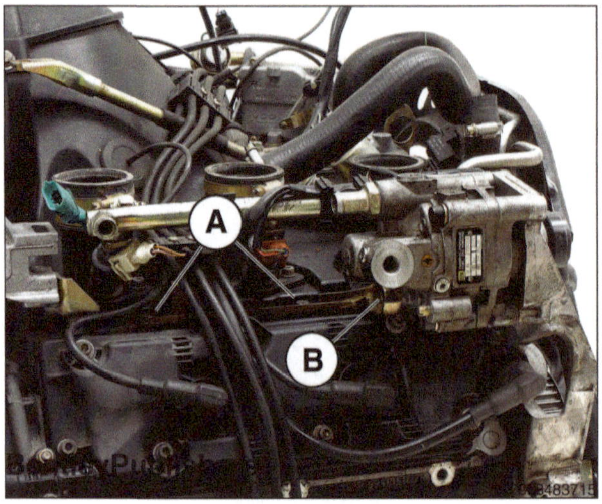

- Remove bolts (**A**) holding power steering pump bracket to engine and bolt (**B**) holding pump to bracket.
- Push bracket aside.

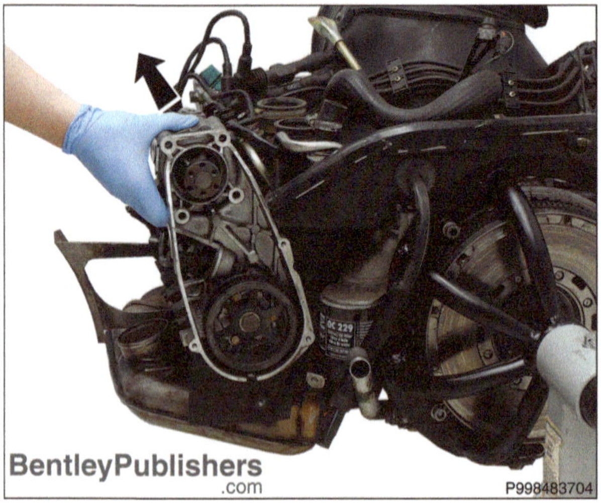

- Pull power steering pump out towards rear of car in direction of **arrow**.

- Installation is reverse of removal. Note following:
  - Ensure lettering on upper gear wheel is facing away from power steering pump (towards front of car).
  - There is no tension adjustment on power steering belt.
  - Ensure belt is properly seated on sprockets.

NOTE–

- *Engine shown with pump removed for clarity.*

Power steering belt, replacing

NOTE–

- *Power steering belt replacement can be completed with engine in vehicle. Pictures show engine out of car.*

– Remove engine air filter box and intake housing assembly from passenger side of engine. See **24 Fuel injection**.

– Support vehicle and remove right rear wheel and inner fender liner.

> *WARNING—*
>
> - *Make sure vehicle is stable and well supported at all times. Use a professional automotive lift or jack stands. Place jack stands underneath structural chassis points, not underneath suspension parts. A floor jack is not adequate support.*

◄ Remove power steering pump cover fasteners (**arrows**), then remove cover.

◄ Remove bolts (**A**) from upper sprocket.

◄ Pull off sprocket together with belt in direction of **arrow**.

– Installation is reverse of removal. Note the following:

- Ensure lettering on upper gear wheel is facing away from power steering pump (towards front of car).
- There is no tension adjustment on power steering belt.
- Ensure belt is properly seated on sprockets

Ignition Switch and Cylinder

Ignition switch and cylinder components

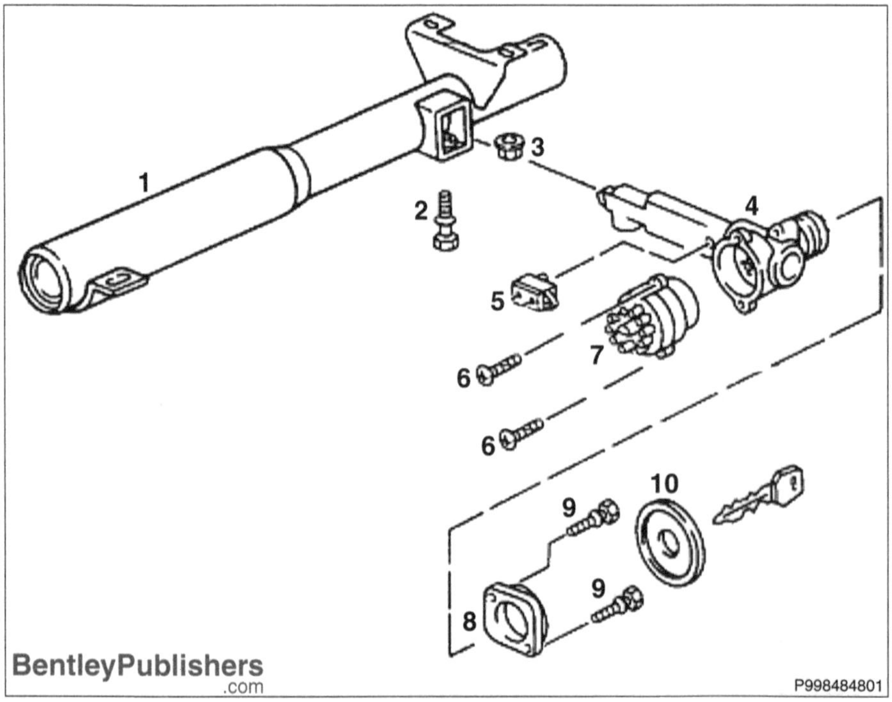

1. Steering outer tube
2. Bolt, M8
3. Threaded bushing
4. Steering column lock with lock cylinder
5. Buzzer contact
6. Screw
7. Ignition switch
8. Reinforcement plate
9. Shear bolt, M6
10. Cover

Ignition cylinder, removing and installing

– Remove knee bar. See **70 Interior trim**.

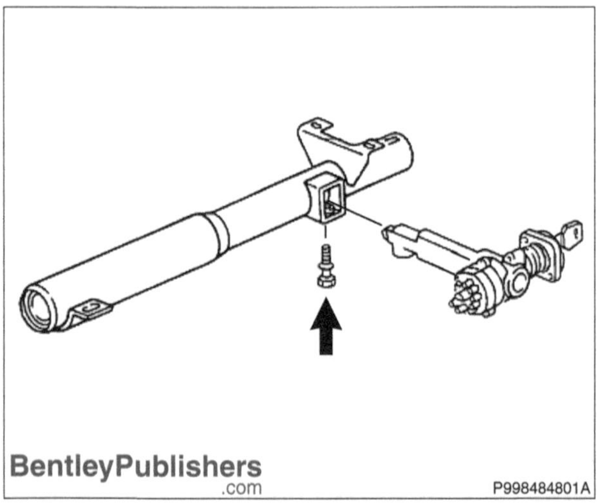

◄ Loosen fastener (**arrow**) holding steering column lock to steering outer tube.

– Working at rear of steering column lock, disconnect harness connector from ignition switch.

⋖ Remove plastic protective disc by inserting pick tool in hole on edge and threading disc in a counter clockwise direction.

⋖ Drill out shear bolts (**arrows**) mounting lock to dashboard. Use a screw extractor to remove shear bolts from steering column lock.

– Push steering column lock through dash and remove from behind.

– Installation is reverse of removal. Note the following:
  - Use new shear bolts and tighten until bolt heads shear off.

Ignition switch, removing and installing

– Remove knee bar. See **70 Interior Trim**.

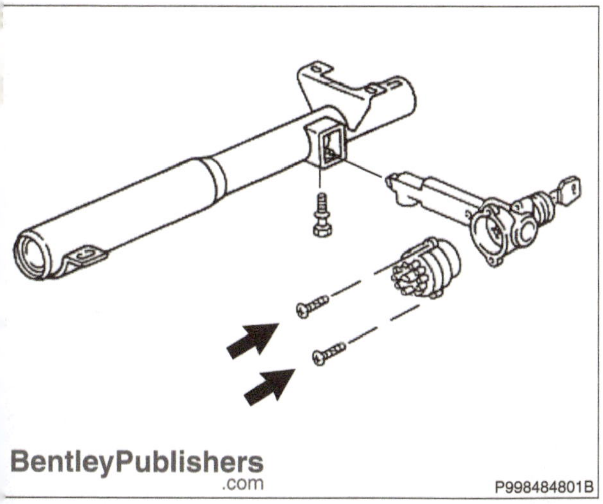

⋖ Remove connector from back of ignition switch and remove two fasteners (**arrows**) holding ignition switch to cylinder housing.

– Remove ignition switch.

– Installation is reverse of removal.

50 Fenders

General

This repair group covers service procedures for the front fenders. For more information on the 993 body, also see:

- **55 Trunk Lid, Engine Hood**
- **57 Doors and Locks**
- **63 Bumpers**
- **66 Exterior Equipment**

Front Fenders

Front fender, removing and installing

Front fenders are bolted and seam-sealed to the body.

– Raise vehicle and support safely. Remove front wheel.

> ***WARNING—***
>
> *• Make sure vehicle is stable and well supported at all times. Use a professional automotive lift or jack stands. Place jack stands underneath structural chassis points, not underneath suspension parts. A floor jack is not adequate support.*

– Remove headlight assembly. See **94 Exterior Lights**.

– Remove front bumper. See **63 Bumpers**.

– Remove front wheel housing liner. See **Front wheel housing liner, removing and installing** in this repair group.

– Remove fuel tank filler flap. See **Fuel tank filler flap, removing and installing** in this repair group.

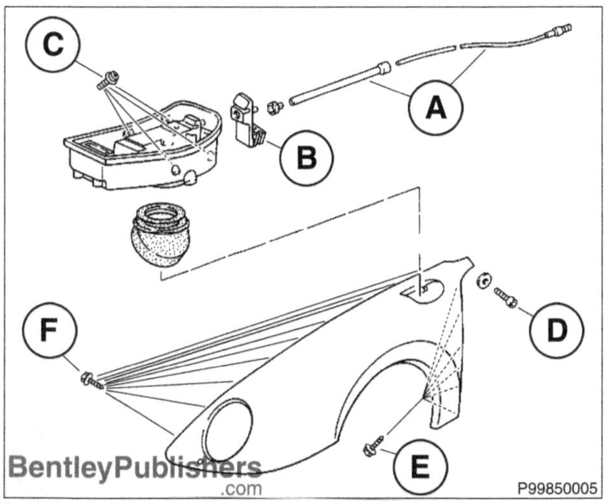

◄ Disconnect fuel tank filler flap release cable (**A**) from latch (**B**) and feed cable out through grommet on fender.

– Remove screws (**C**) securing tray to fender.

– Open door and remove bolt and washer (**D**).

– Working in wheel housing, remove bolts and remove plastic nut (**E**) securing fender to rocker cover.

– Open trunk. Peel back trunk gasket and remove bolts (**F**).

– Carefully cut away seam sealer at fender to body joint.

– Disconnect cable clips as necessary and remove fender.

– Installation is reverse of removal.

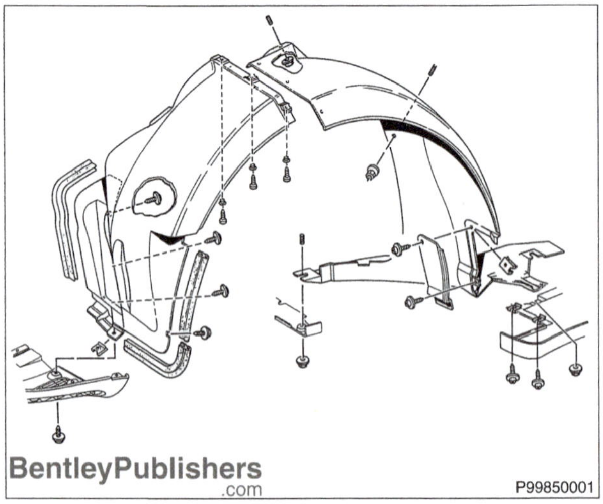

Front wheel housing liner, removing and installing

– Raise vehicle and support safely. Remove front wheel.

> ***WARNING—***
>
> - *Make sure vehicle is stable and well supported at all times. Use a professional automotive lift or jack stands. Place jack stands underneath structural chassis points, not underneath suspension parts. A floor jack is not adequate support.*

◀ Remove front wheel housing liner fasteners.

- Remove plastic nuts.
- Remove screws.

– Remove rear section of wheel housing liner from vehicle.

– Remove front section of wheel housing liner from vehicle.

– Installation is reverse of removal. Replace broken or damaged fasteners.

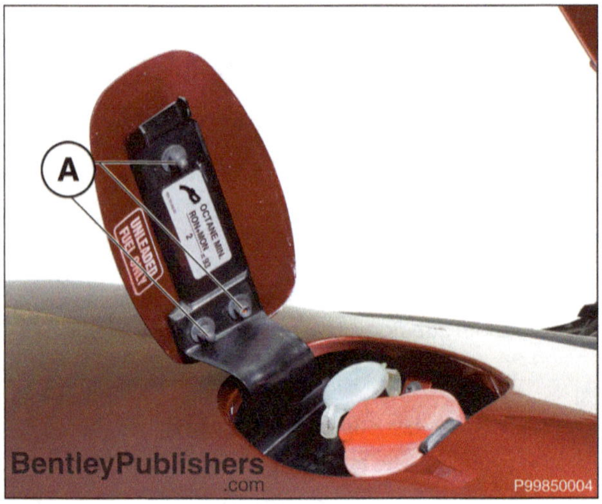

Fuel tank filler flap, removing and installing

◀ Open fuel filler flap.

– Remove fasteners (**A**) and remove fuel filler flap.

– Installation is reverse of removal.

Front fender components

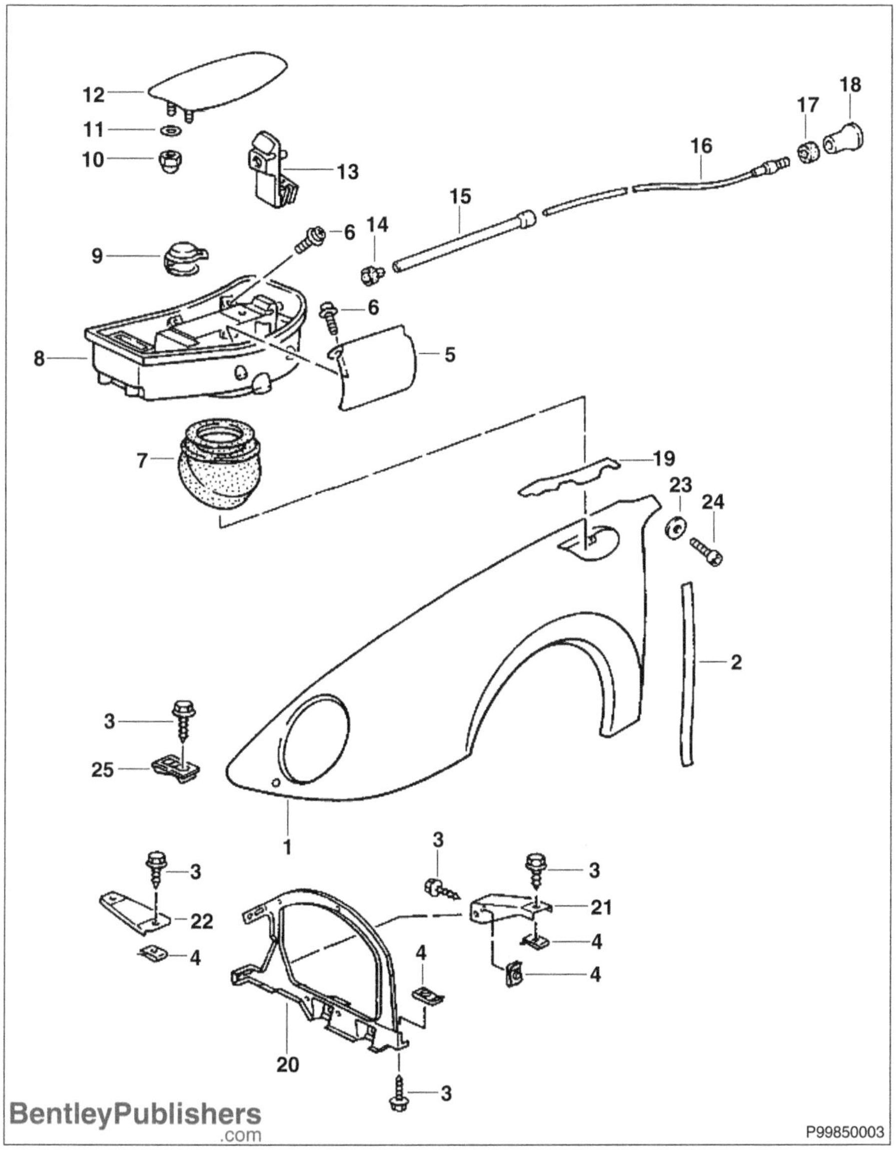

1. Fender
2. Sealing strip
3. Fastener
4. Speed nut
5. Protective flap
6. Fastener
7. Sleeve
8. Tray
9. Washer fluid reservoir filler neck cap
10. Fastener, M5
11. Washer
12. Fuel filler flap
13. Latch
14. Grommet
15. Guide tube
16. Fuel filler flap release cable
17. Grommet
18. Fuel filler flap release knob
19. Seal
20. Support plate
21. Support
22. Support
23. Washer
24. Fastener, M6
25. Speed nut

55 Trunk Lid, Engine Hood

General

This repair group covers service procedures for trunk lid, engine hood, lid release mechanisms, and rear spoiler.

Trunk

Trunk lid components

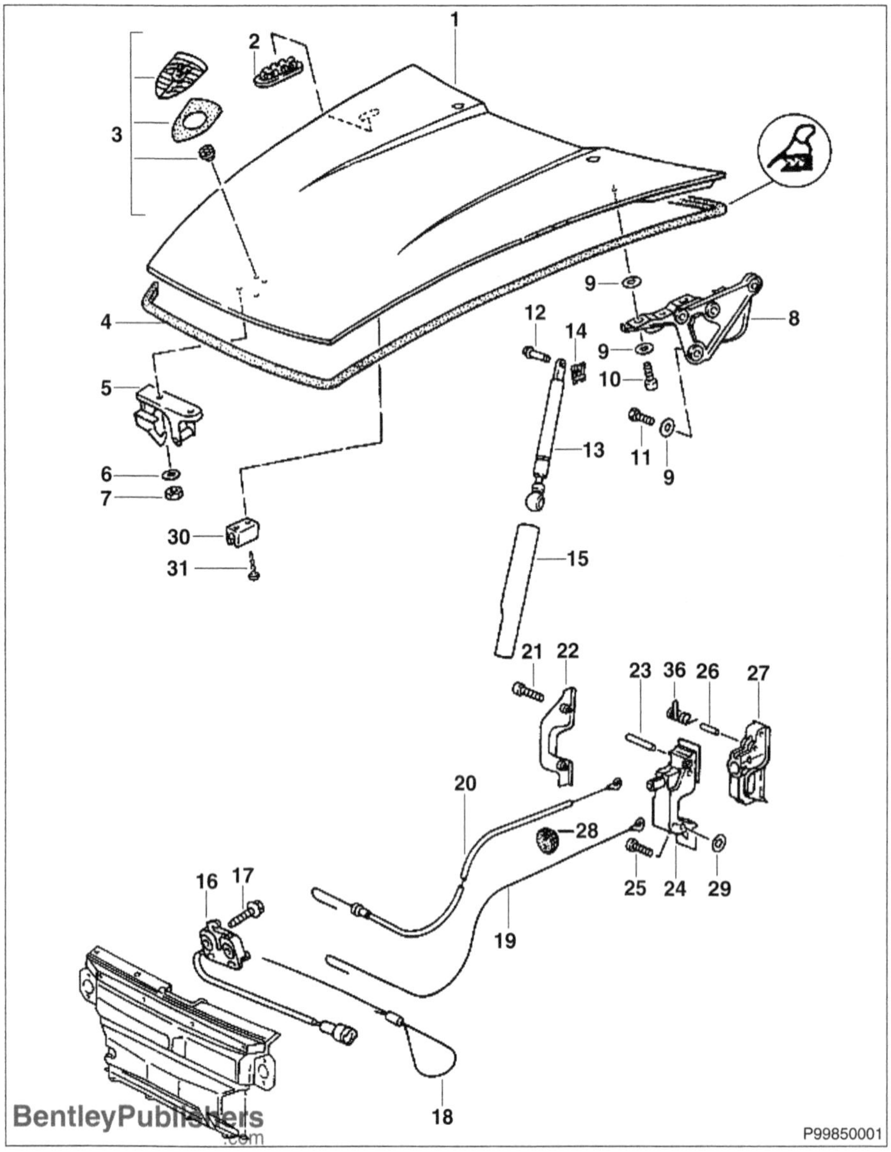

1. Trunk lid
2. Cap
3. Trunk lid emblem
4. Gasket
5. Lock (upper)
6. Washer
7. Nut, M6
8. Hinge
9. Washer
10. Fastener, M8 x 22 mm
11. Fastener, M8 x 16 mm
12. Fastener
13. Gas strut
14. Safety clip
15. Protective tube
16. Lock (lower)
17. Fastener, M6 x 30 mm
18. Trunk lid emergency release cable
19. Bowden cable
20. Fastener, M6 x 16 mm
21. Cover
22. Pin
23. Handle housing
24. Fastener, M6 x 15 mm
25. Pin
26. Handle
27. Rubber sleeve
28. Washer
29. Retainer
30. Fastener
31. Lock nut

Trunk lid strut, removing and installing

– Raise and support trunk lid.

◄ Pry out and slide off retainer clip (**arrow**). Push out retainer pin to release top of gas strut.

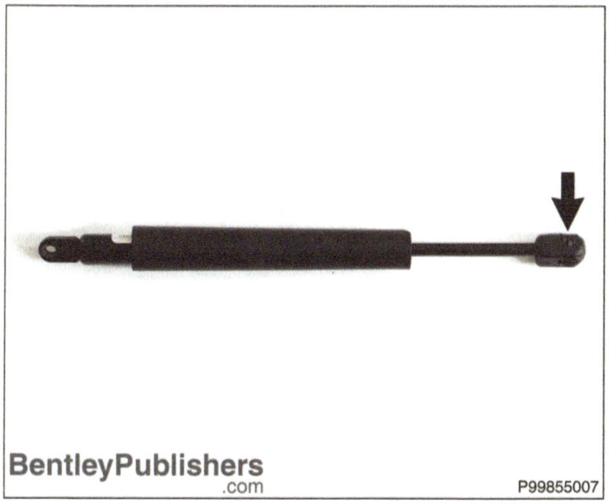

◄ Release bottom of gas strut from body by prying off spring clip (**arrow**) and removing strut from ball joint.

– Installation is reverse of removal. Press strut socket onto ball joint until socket is fully seated and spring clip locks in place.

Trunk lid, removing and installing

– Raise and support trunk lid.

◄ Using plastic prying tool, lever out cover (**arrow**).

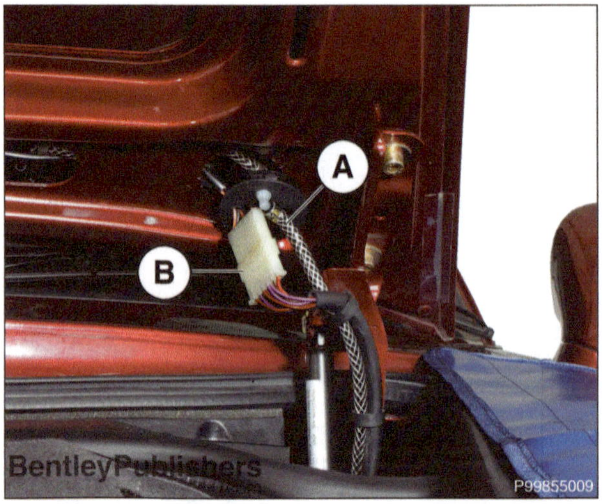

◄ Disconnect windshield washer fluid hose (**A**).

– Feed electrical connector (**B**) through hole in cover and disconnect.

◄ Mark hinge position to trunk lid. With an assistant supporting trunk lid, remove hinge fasteners (**arrows**) from left and right sides then remove trunk lid.

– Installation is reverse of removal. Tighten hinges in previously installed position to maintain trunk lid alignment. Check lid alignment and lid body gaps.

Trunk lid, aligning

If trunk lid was removed, make sure hinges are aligned as close to original mounting points as possible. Use touch up paint on any exposed surfaces.

In addition to hinge fasteners, front trunk lid alignment adjustments can be made at the lid striker and rubber bump stops.

- Adjust lid striker by loosening fasteners (**arrows**) and moving striker.

- Screw rubber bump stops (**arrow**) in or out, if necessary.
- Before shutting trunk lid, check striker-to-lock alignment. Adjust striker as needed.

- Close trunk lid and check alignment. Align lid so that gap dimensions (**arrows**) are even.

55

Engine Hood

> ***CAUTION—***
> - *The engine hood is heavy. Before removing engine hood supports, have an assistant support the lid.*
> - *Realignment of body panels may expose unpainted metal. Paint all exposed metal surfaces once work is complete.*

Engine hood components

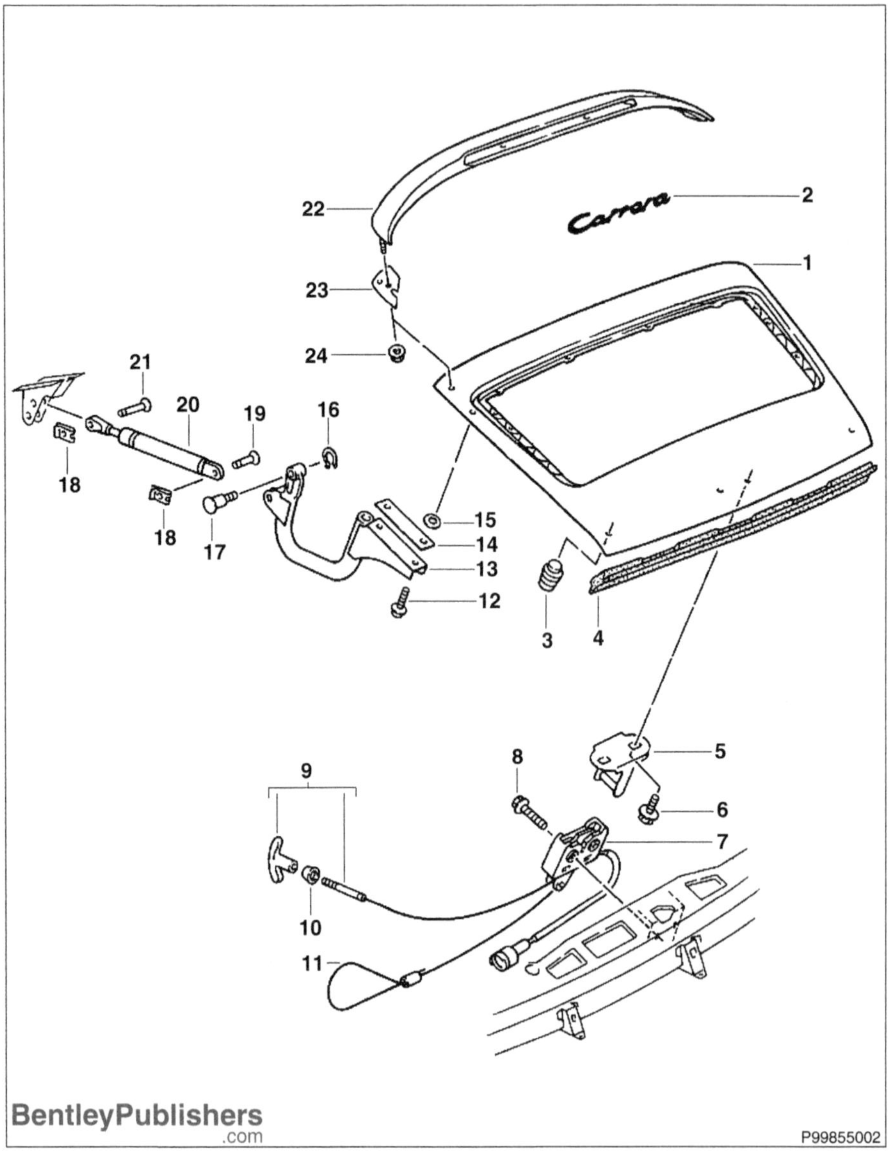

1. Engine hood
2. Emblem
3. Rubber buffer
4. Rubber seal
5. Lock (upper)
6. Fastener, M6 x 18 mm
7. Lock (lower)
8. Fastener, M6 x 30 mm
9. Engine hood release cable
10. Spacer
11. Emergency engine hood release cable
12. Fastener, M6 x 16 mm
13. Engine hood hinge
14. Insulator pad
15. Washer
16. Circlip
17. Hinge pin
18. Safety clip
19. Pin
20. Gas strut
21. Pin
22. Bracket
23. Protective pad
24. Lock nut

Engine hood strut, removing and installing

– Raise and support engine hood.

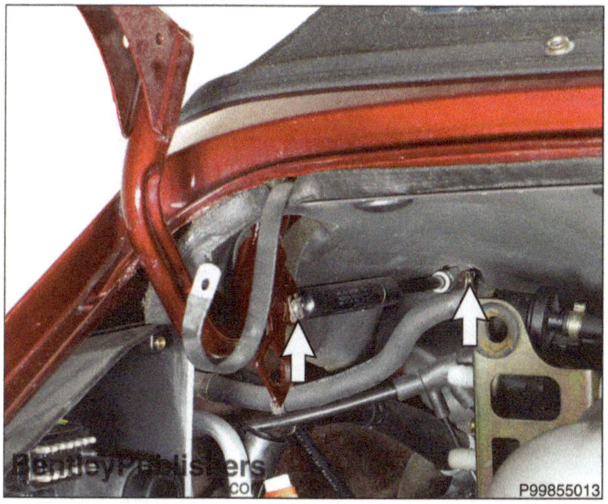

◄ Release gas strut from hinge by prying retainer clips (**arrows**) and pushing out retainer pins. Remove gas strut.

– Installation is reverse of removal.

Engine hood, removing and installing

– Raise and support engine hood.

– Mark hinge position.

– If applicable, disconnect spoiler and engine cooling fan electrical connectors.

◄ With an assistant supporting engine hood, remove hinge attachment fasteners (**arrows**) from left and right sides and remove engine hood.

– Installation is reverse of removal:

- Tighten hinges in previously installed position to maintain engine hood alignment.

Engine hood, aligning

If engine hood was removed, be sure hinges are aligned as close to original mounting points as possible. Use touch up paint on any exposed surfaces.

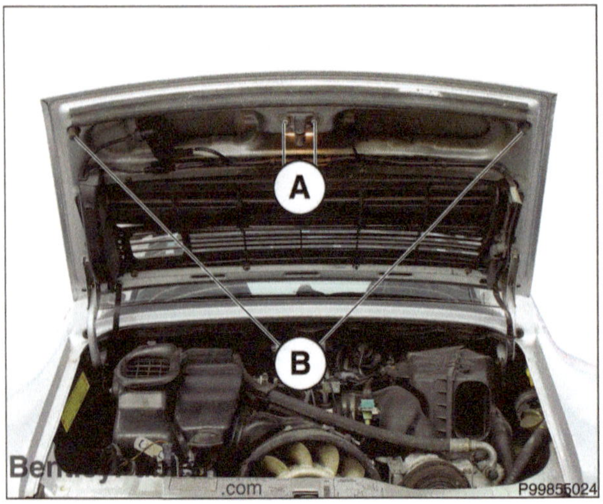

In addition to hinge bolts, engine hood alignment can be made at lid striker and rubber bump stops.

- Adjust hood striker by loosening bolts (**A**) and moving striker.
- Screw rubber bump stops (**B**) in or out.

– Before shutting engine hood, check striker-to-lock alignment. Adjust striker as needed.

Close engine hood and check alignment. Align lid so that gap dimensions (**arrows**) are even.

Trunk Lid and Engine Hood Release

Trunk lid release, servicing

– Open trunk.

– Remove trunk carpeting.

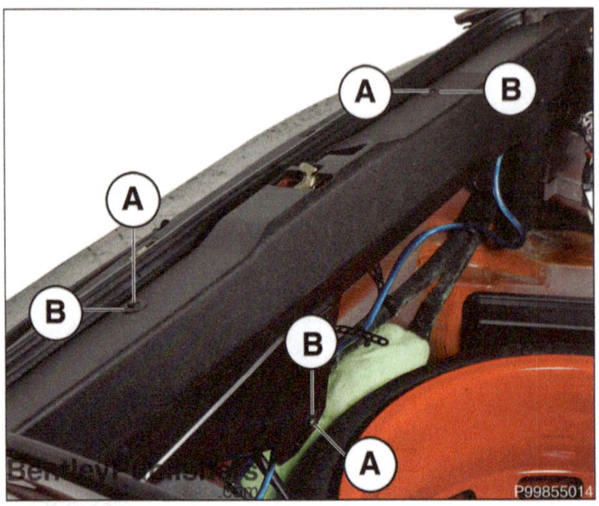

Remove screws (**A**). Using a plastic prying tool, lever out expander nuts (**B**).

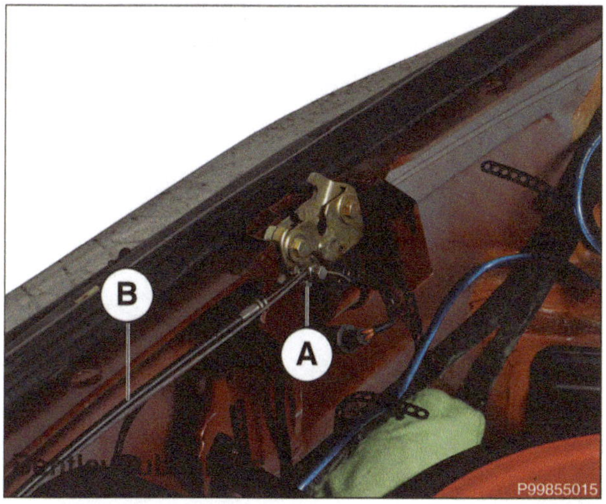

P99855015

◀ Loosen lock bolt (**A**) and disconnect cable (**B**) from trunk lid release.

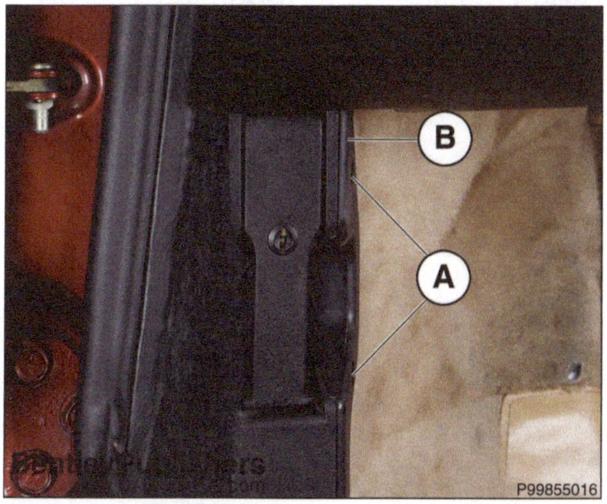

P99855016

◀ Working inside driver's side footwell, remove fasteners (**A**) and remove cover plate (**B**).

P99855017

◀ Insert Allen wrench through cover plate bolt holes and remove trunk lid release handle fasteners.

– Remove trunk lid release handle.

– Pull footwell carpeting toward rear of vehicle to remove.

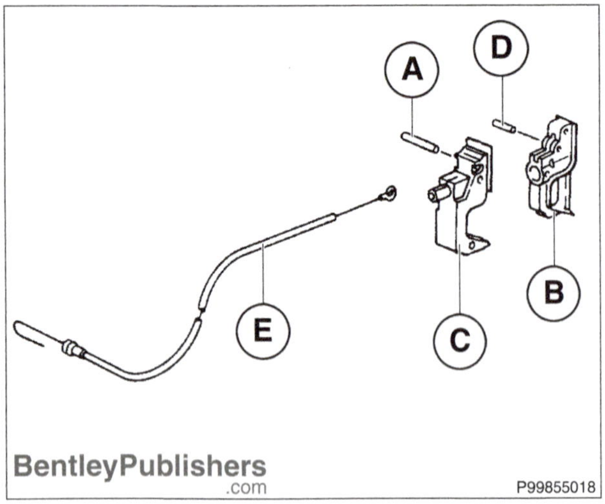

- Drive out pin (**A**) to disconnect handle (**B**) from handle housing (**C**).
- Drive out pin (**D**) to disconnect cable from handle.
- Installation is reverse of removal.

Engine hood release cable, replacing

- Open engine compartment.

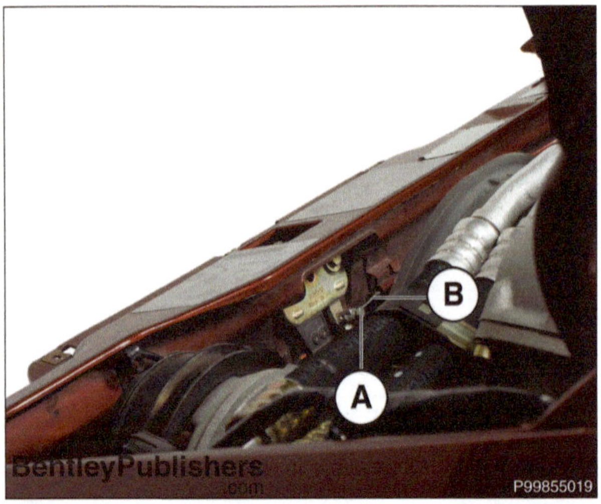

- Loosen lock bolt (**A**) and disconnect engine hood release cable (**B**).

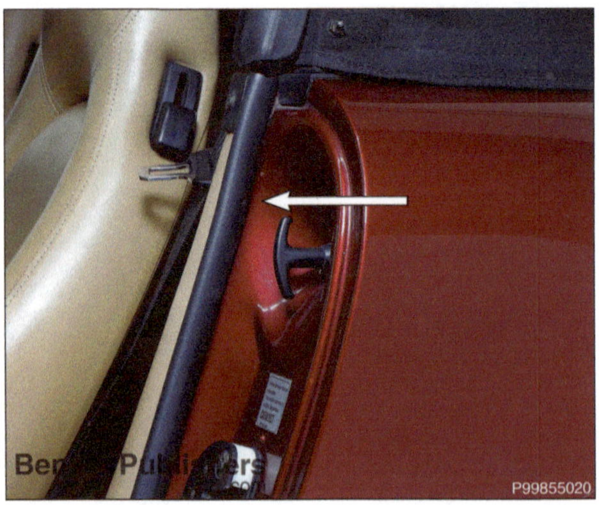

- Pull engine hood release handle in direction of arrow to remove cable.
- Installation is reverse of removal.

Spoiler

The motorized rear spoiler is comprised of the painted frame and the air inlet grille. The frame and the air inlet grille are bonded together. If the automatic function fails, the spoiler can be extended or retracted manually using the spoiler switch in the center console. Raise and lower the spoiler via the rocker switch only when the vehicle is at a full stop.

Rear spoiler motor drive

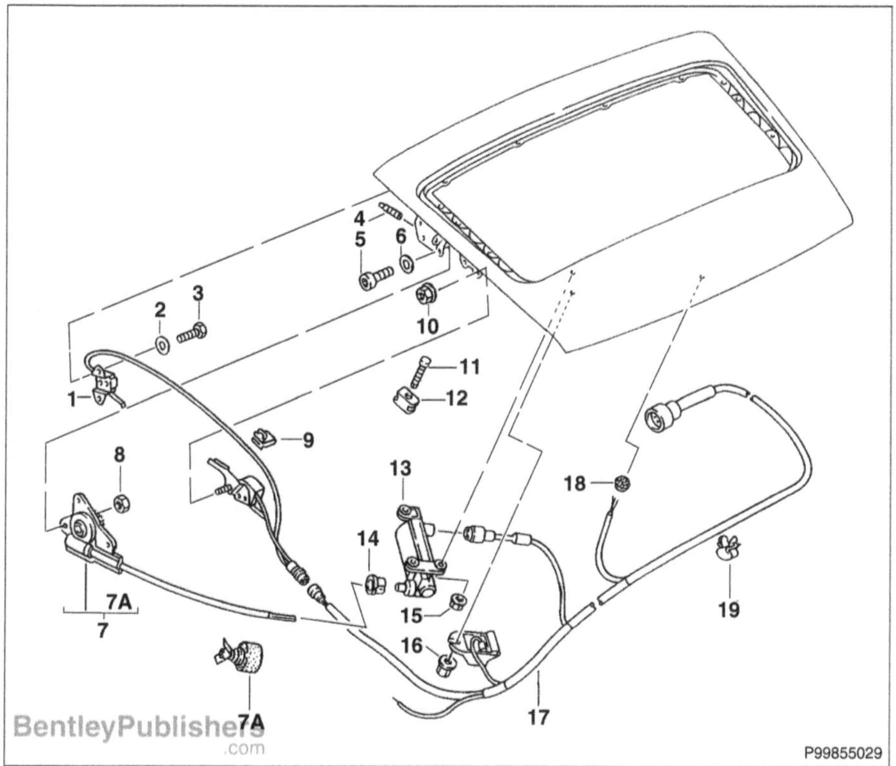

1. Microswitch harness (2 microswitches)
2. Washer
3. Bolt (M5 x10)
4. Threaded pin
5. Screw (self-locking, mircorencapsulated)
6. Washer
7. Transmission
   7A. Cap (for drive cable at transmission)
8. Nut (M6)
9. Cable clip
10. Nut (M5)
11. Threaded pin
12. Nut
13. Drive motor
14. Bushing
15. Nut (M6)
16. Nut, plastic
17. Wiring harness
18. Rubber sleeve
19. Cable clip

55

Rear spoiler mechanical components

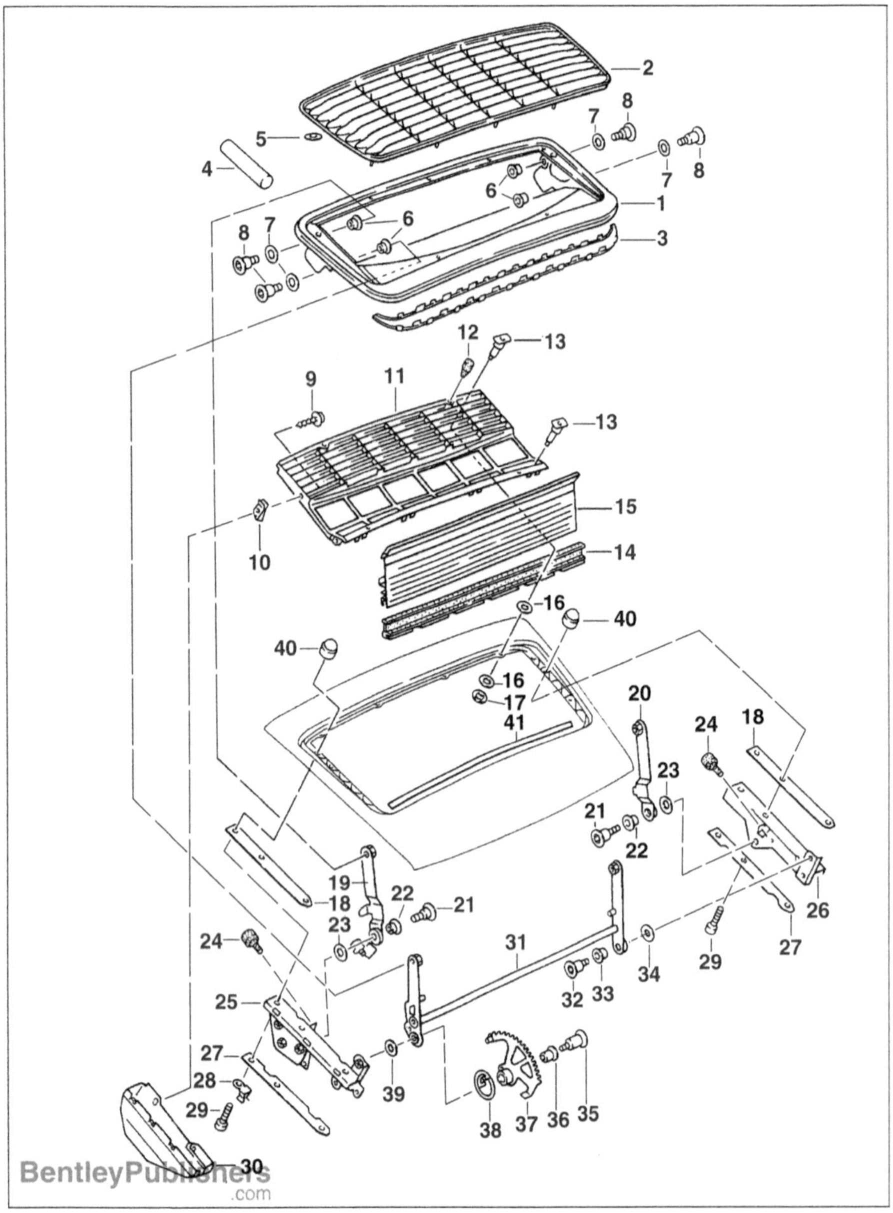

1. Spoiler
2. Air inlet grill
3. Cover
4. Adhesive kit
5. Clamping washer
6. Bearing sleeve
7. Washer
8. Screw (self-locking, mircorencapsulated)
9. Screw (self tapping)
10. Speed nut
11. Ventilation grill
12. Stop
13. T-head screw
14. Seal
15. Rear wall
16. Washer (9 x 15 x 2)
17. Clamping washer
18. Deck pad
19. Lever
20. Lever
21. Screw (self-locking, mircorencapsulated)
22. Bearing sleeve (17 x 7 x 10.1)
23. Washer (16 x 3.6 x 10.5)
24. Buffer
25. Mounting plate
26. Mounting plate
27. Deck pad
28. Cable holder
29. Screw (pan head, M6 x 16)
30. Cover
31. Linkage
32. Screw (self- locking, mircorencapsulated)
33. Bearing sleeve (17 x 7 x 10.1)
34. Washer (16 x 1.6 x 10.5)
35. Screw (self-locking, mircorencapsulated)
36. Bearing sleeve (16x18.5x10.1)
37. Toothed drive
38. Torsion spring
39. Washer (24x12.5x1.6)
40. Protective cap
41. Weather strip

57 Doors and Locks

General

This repair group covers removal and installation of the doors, door trim panels and door lock components. For related information see:

- **64 Door Windows** for power door windows and door glass removal.
- **96 Interior Lights, Anti-theft** for central locking and anti-theft system theory and operation.

Doors

Door, removing and installing

– Disconnect negative (–) battery cable.

> ***CAUTION—***
>
> - *Prior to disconnecting the battery, read the battery disconnection cautions given in* **00 Warnings and Cautions**.

– Remove door panel. See **Door trim panel, removing and installing** in this repair group.

– Label and disconnect wiring from all switches. Arrange wiring so it can be withdrawn once door is unbolted.

< Pry locking clip (**arrow**) off door check pin. Then drive pin out from below.

- Support door and remove upper and lower hinge bolts on body (**arrows**). Carefully withdraw wiring harnesses when removing door.

– Installation is reverse of removal, noting the following:

  - Align door so that panel gaps are equal on either side. See **Door, adjusting**, in this repair group.
  - Adjust door striker so that trailing edge of door is slightly higher than front edge.
  - Realignment of body panels may expose unpainted metal. Repair any paint damage and paint any exposed metal.

Door, adjusting

- Loosen door hinge bolts (**arrows**) to adjust door.

NOTE–

- *Realignment of body panels may expose unpainted metal. Paint all exposed surfaces once work is complete.*

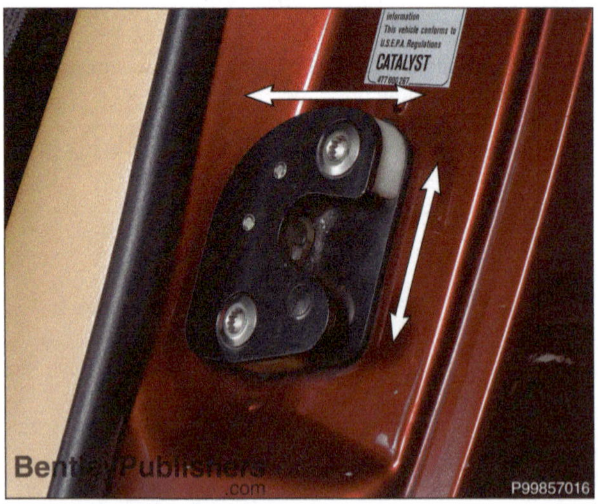

- Loosen striker bolts to adjust striker.

Door Trim Panel

Door trim panel, removing and installing

– Disconnect negative (–) cable from battery.

> ***CAUTION—***
> - *Prior to disconnecting the battery, read the battery disconnection cautions given in* **00 Warnings and Cautions.**

NOTE–
- *Trim panel plastic clips are easily damaged. Have extra clips on hand before beginning procedure.*

◄ Remove screws around bottom and side of door panel (**arrows**).

– Remove fastener inside door pocket and remove door pocket.

– Remove door speaker.

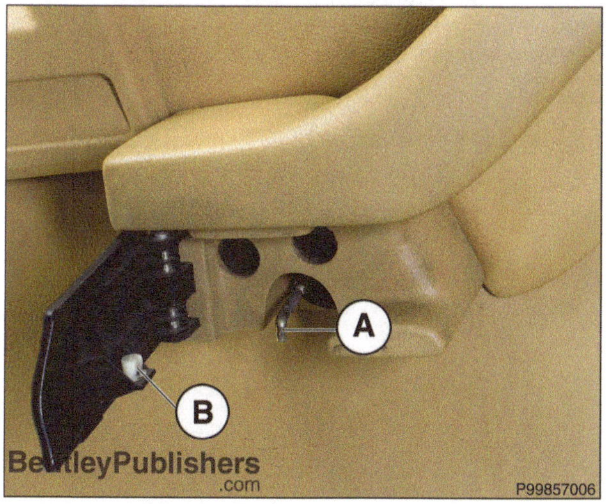

◄ Unclip and disconnect actuating rod (**A**) from door handle (**B**).

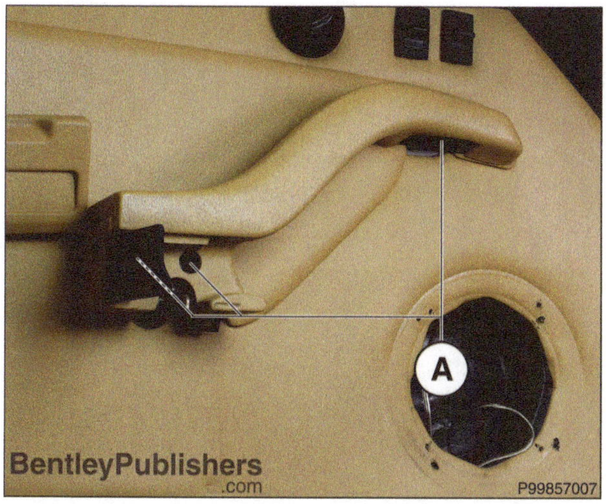

◄ Remove fasteners (**A**).

57

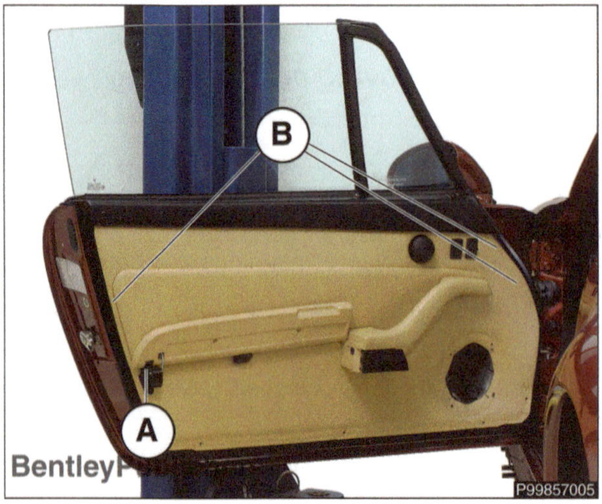

- ⮜ Remove fastener (**A**).
- Using plastic prying tool, lever out door trim panel clips (**B**).
- Disconnect power window switch and any other wiring remove door trim panel.
- Installation is reverse of removal. Remember to:
  - Make sure all electrical connectors are securely fastened.
  - Replace missing or broken trim clips.

Door Locks

Door striker, replacing

- Open door and mark position of striker on body.

- ⮜ Remove fasteners (**arrows**) securing striker to body.
- Remove striker and striker shims (if applicable).
- Installation is reverse of removal.

Door handle and lock cylinder, removing and installing

- Close window.
- Remove door panel. See **Door trim panel, removing and installing** in this repair group.
- Remove door vapor barrier. See **64 Door Windows**.

- ⮜ Working inside door, remove fasteners (**arrows**).
- Carefully remove door handle from door.
- Installation is reverse of removal. Use new gaskets between door and body.

Central locking switch, removing and installing

- Using a plastic prying tool, lever off lower switch panel, then upper switch panel (**arrow**) from center console.
- Disconnect electrical connectors.

- Unclip tabs (**arrows**) and remove central locking switch from switch panel.
- Installation is reverse of removal.

Door latch, removing and replacing

- Remove door panel. See **Door trim panel, removing and installing** in this repair group.
- Remove door vapor barrier and window glass. See **64 Door Windows**.

- Remove fasteners (**arrows**) and remove rear section of door window frame.
- Remove door handle. See **Door handle and lock cylinder, removing and installing** in this repair group.

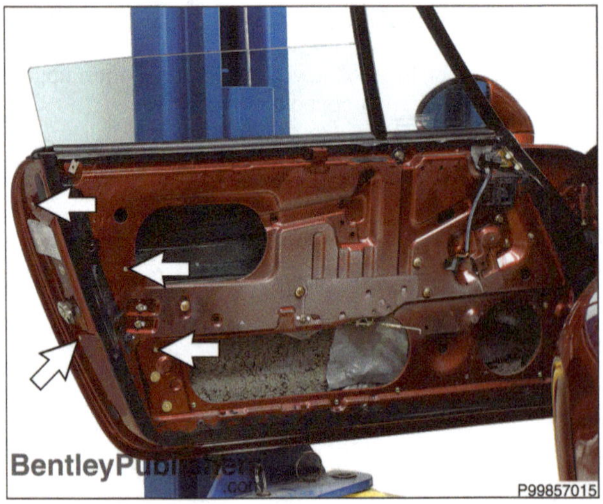

◀ Remove screws (**arrows**) and remove plastic covers inside door.

- Working inside door, disconnect release cable and power door lock control rod from latch assembly.

◀ Working at door exterior, remove latch bolts (**arrows**) and remove latch from inside door.

- Installation is reverse of removal. If necessary, adjust striker position and tighten bolts. See **Door, adjusting**, in this repair group.

Door window frame, removing and installing

- Remove door panel. See **Door trim panel, removing and installing** in this repair group.
- Remove door vapor barrier. See **64 Door Windows**.

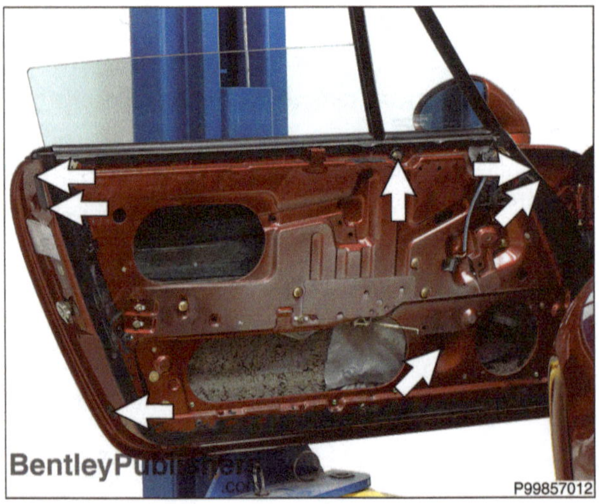

◀ Remove window frame mounting fasteners (**arrows**).

- Working inside door, remove bracket from lower end of window channel.
- Slide window frame up and out of door.
- Installation is reverse of removal. Note the following:
  - Align marks on window frame with door.
  - Check that window seats squarely in opening and frame closes evenly on the door gasket.
  - Adjust window frame if necessary. The frame tilt (in-out) can be adjusted at rear and middle channel bracket. Frame height (up-down) can be adjusted at two upper bolts at front of door.

60 Sunroof, Targa Top

General

This repair group covers service procedures for the sunroof and Targa top.

Porsche 993 models were available with an optional sliding metal sunroof, or a sliding glass roof (Targa top). Both the sunroof and Targa top are cable driven and electrically operated.

Sunroof

Sunroof components

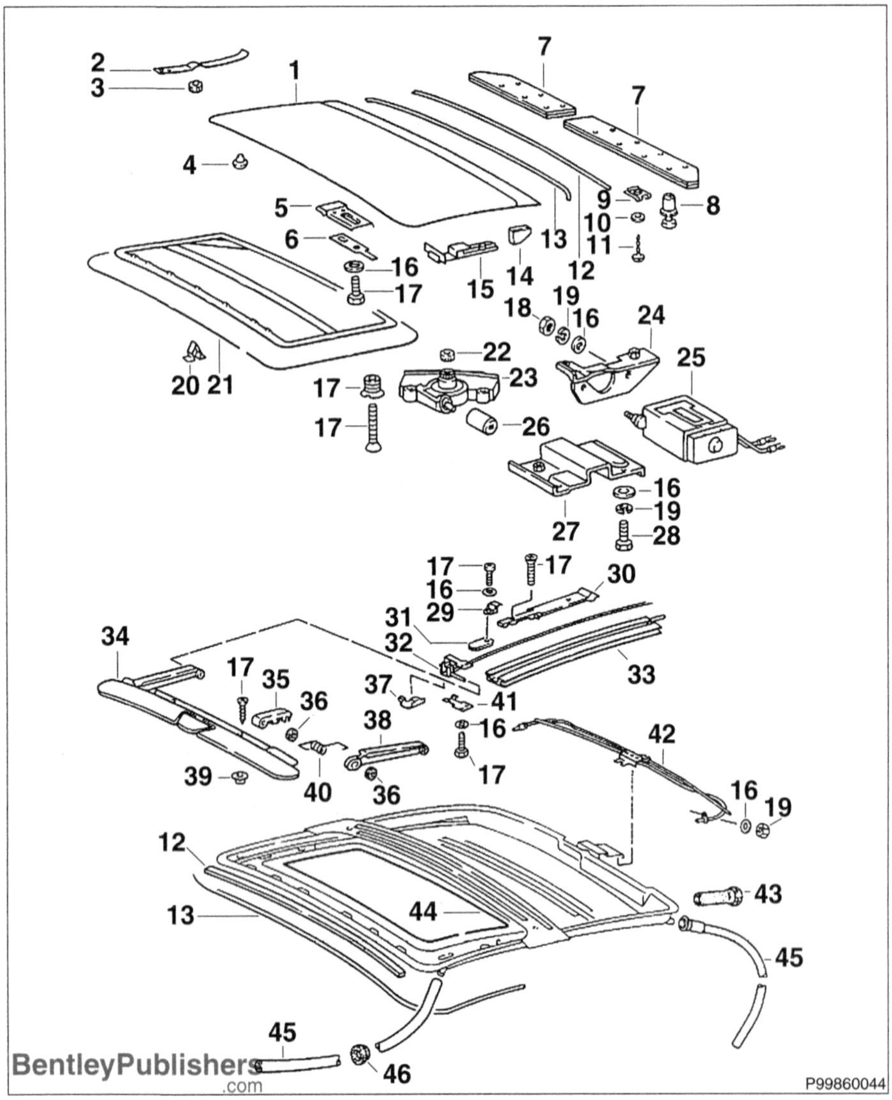

1. Sunroof panel
2. Tension spring
3. Knurled nut
4. Rubber stop
5. Guide
6. Tab washer
7. Cover
8. Expansion rivet
9. Speed nut
10. Cone washer
11. Tapping screw
12. Sealing rubber
13. Fabric strip
14. Guide shoe
15. Support
16. Washer
17. Screw
18. Hexagon nut
19. Lock ring
20. Clip
21. Sunroof interior panel
22. Drive pinion
23. Transmission
24. Motor support
25. Motor
26. Clutch
27. Support bracket
28. Hex head bolt
29. Terminal piece
30. Intermediate piece
31. Threaded plate
32. Drive cable (x2)
33. Guide rail
34. Wind deflector
35. Bearing clip
36. Spacer ring
37. Slide
38. Hinge

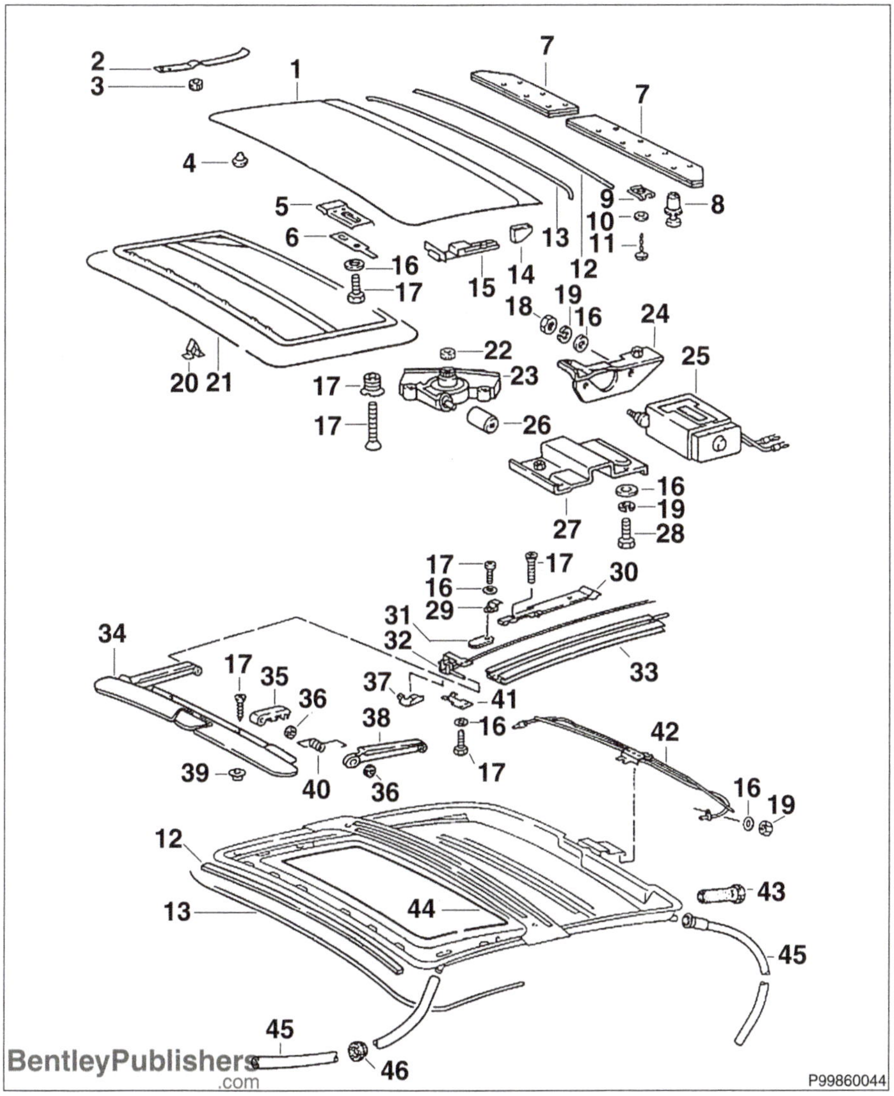

39. Protective cap

40. Spring

41. Reinforcement

42. Drive motor mounting track

43. Grommet

44. Beading

45. Drain tube

46. Rubber sleeve

Sunroof controls

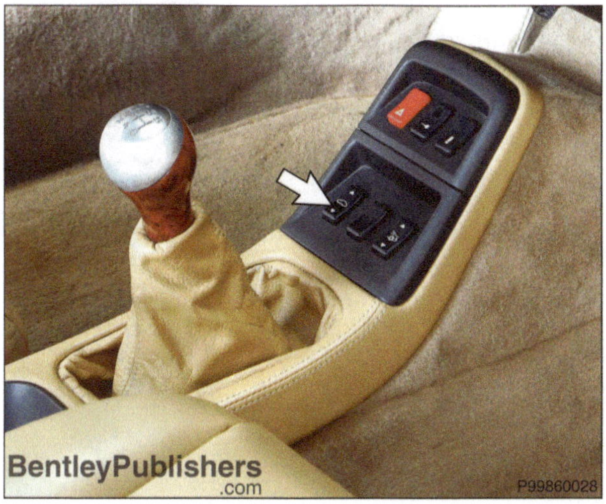

The power sunroof is operated by a rocker switch located on the center console (**arrow**). The sunroof can be operated with the ignition in the ON or accessory position.

> ***WARNING—***
>
> - *To avoid injury, keep fingers, hands, hair, etc. clear of power sunroof when opening or closing.*

> ***CAUTION—***
>
> - *Do not operate sunroof at vehicle speeds exceeding 60 mph as wind force may cause damage to sunroof headliner or mechanism.*

Sunroof emergency closing

In case of electrical fault, the sunroof can be closed manually.

– Open zipper in headliner above rear window.

◀ Remove plastic expanding rivet fasteners (**arrows**) from insulating panel and remove panel.

NOTE–

- *Sunroof drive access panel fasteners are a two-piece expanding rivet (**inset**). Use a Torx driver to remove center threaded pin, then pull rivet from panel.*

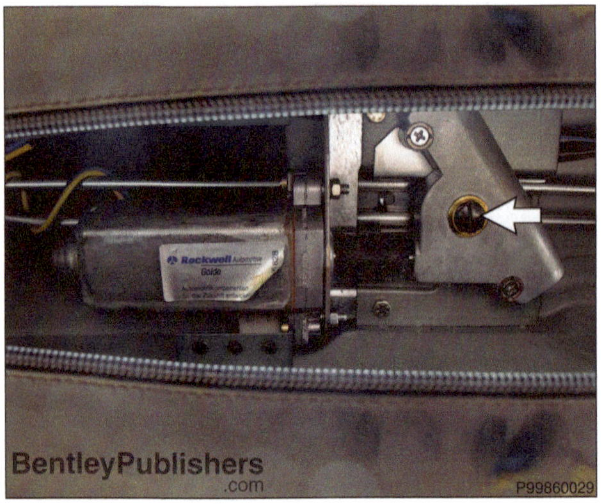

◀ Using a flat blade screwdriver, remove sunroof drive unit tensioning screw (**arrow**).

◀ Insert flat blade screwdriver or hand crank from vehicle tool kit into sunroof drive unit (**arrow**) and turn until sunroof is closed.

– Reinstall and fully tighten sunroof drive unit tensioning screw. Reinstall access panel. Close headliner.

Sunroof panel, adjusting height

The sunroof panel should be adjusted:

- If top of closed sunroof becomes misaligned with roof of car.
- If it does not close squarely.
- If there are wind noises at speed.
- If there are water leaks.
- If sunroof has been removed.

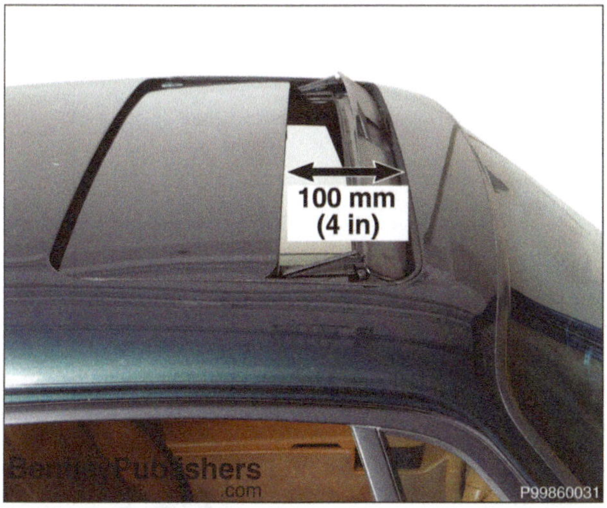

Open sunroof approximately 100 mm (4 in).

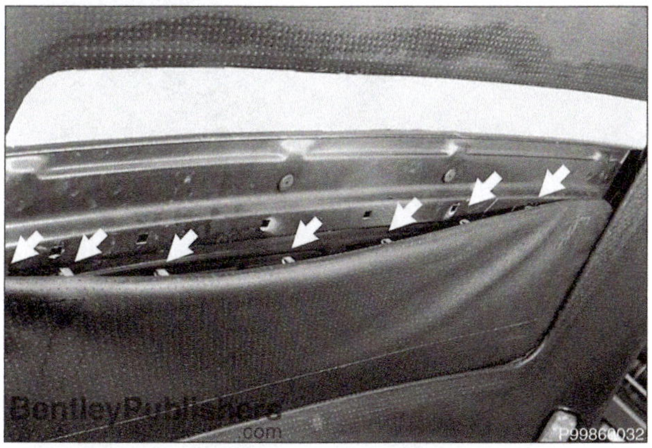

Using plastic prying tool, carefully lever down front of trim panel (**arrows**) until free from sunroof panel.

- Slide headliner panel back into roof cavity.
- Close sunroof.

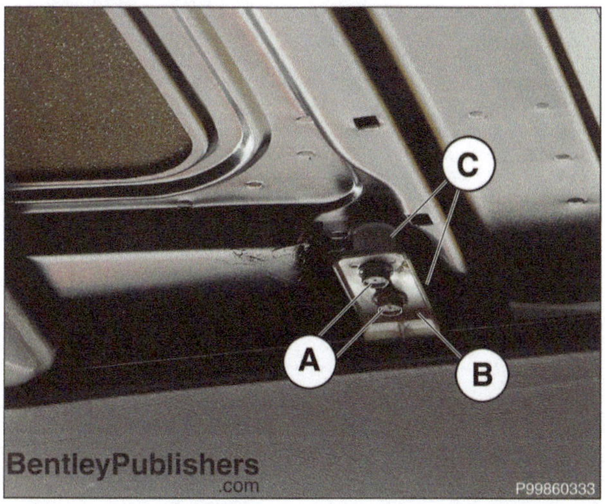

Remove bolts (**A**) and remove front guides (**B**).

- Turn knurled nuts (**C**) to adjust front height.

NOTE–

• Left side shown. Right side is similar.

- Reinstall guides and bolts. Check roof adjustment.

NOTE–

• Adjust front height so that panel is even with roof or slightly lower.

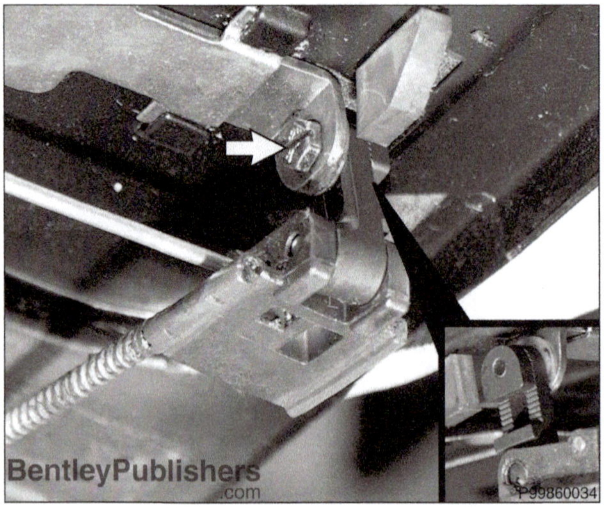

To adjust rear height of sunroof panel:

- Loosen adjustment screws (**arrow**).
- Adjust ratchets (**inset**) so that sunroof panel is even with roof or slightly higher.
- Tighten screws.

NOTE–

• Left side shown. Right side is similar.

– Open sunroof approximately 100 mm (4 in).

– Slide headliner trim panel forward and snap clips into place.

– Check operation of sunroof.

Sunroof panel, removing and installing

CAUTION—

• Use masking tape around sunroof opening to protect painted surfaces when working on sunroof.

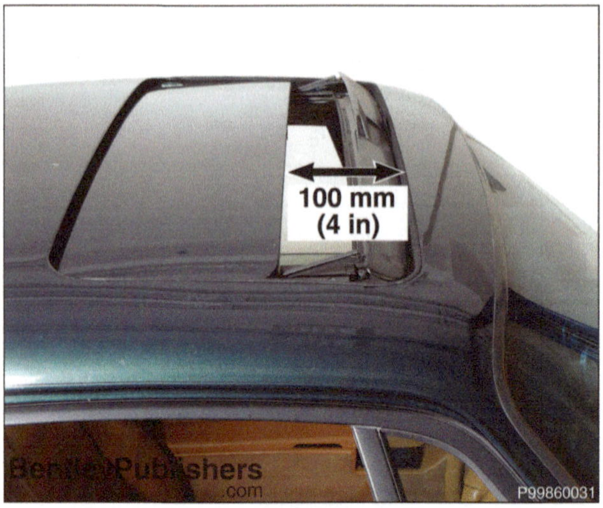

Open sunroof lid approximately 100 mm (4 in).

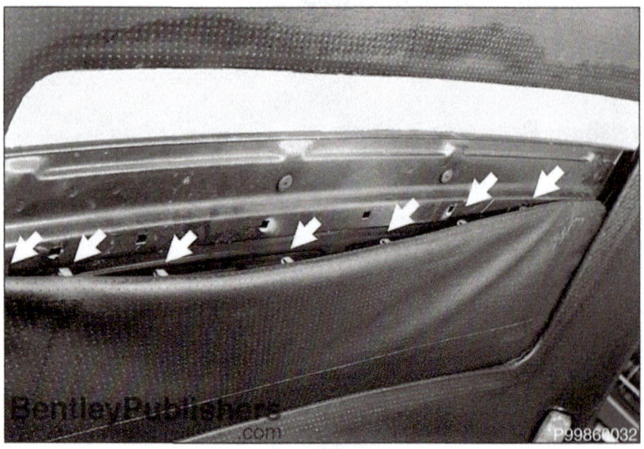

Using plastic prying tool, carefully lever down on front of trim panel to free clips (**arrows**).

– Slide headliner panel back into roof cavity.

– Close sunroof.

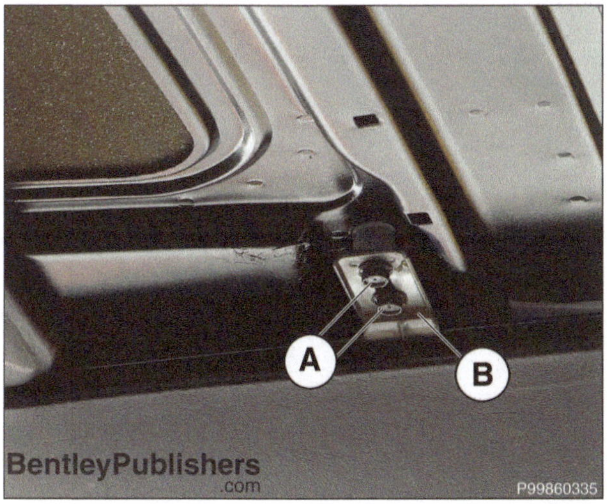

< Remove bolts (**A**), then remove front guides (**B**).

< Working at rear of panel, lift spring off pin and pivot to side (**arrow**).

– Remove bolts (**A**) from rear guide brackets.

NOTE–

• Left side shown. Right is similar.

– Carefully push front of sunroof panel up and slide forward and out of body.

– Installation is reverse of removal. Check sunroof adjustment. See **Sunroof panel, adjusting height** in this repair group.

Wind deflector, removing and installing

CAUTION—

• Use masking tape around sunroof opening to protect painted surfaces when working on sunroof.

– Open sunroof.

< Remove intermediate plate screws (**arrows**), then remove plates.

NOTE–

• Left side shown. Right side is similar.

< Remove trim cover mounting screws from (**arrows**).

< Using a plastic prying tool, carefully lever up rear of wind deflector clip (**arrow**) then and slide lip forward to disconnect from trim cover.

– Carefully lift out wind deflector with clips.

– Installation is reverse of removal.

Sunroof cables, replacing

– Remove sunroof panel. See **Sunroof panel, removing and installing** in this repair group.

– Remove sunroof motor and drive unit. See **Sunroof motor and drive unit, removing and installing** in this repair group.

– Remove wind deflector. See **Wind deflector, removing and installing** in this repair group.

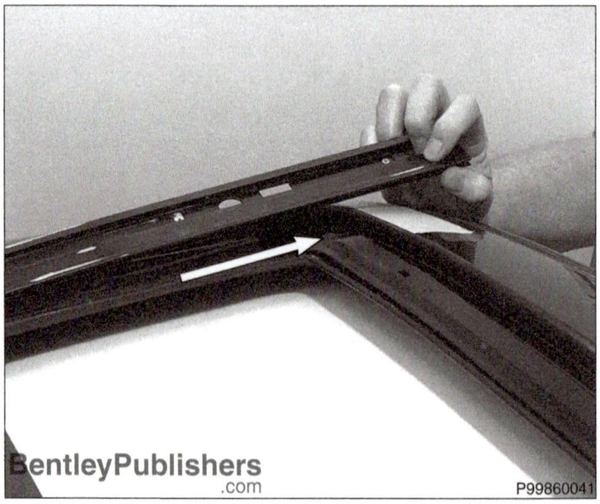

< Pull sunroof guide rails forward and out to remove.

NOTE–

• Use tape (as shown) to protect painted surfaces prior to removing rail.

– Pull sunroof cables from guides.

– Clean guide rails and wipe dry.

NOTE–

• Do not use a lubricant on sunroof cables.

– Installation is reverse of removal, noting the following:

- Inspect all parts for wear and replace as necessary.
- Adjust sunroof panel height. See **Sunroof panel, adjusting height** in this repair group.

Sunroof motor and drive unit, removing and installing

– Open zipper in headliner above rear window.

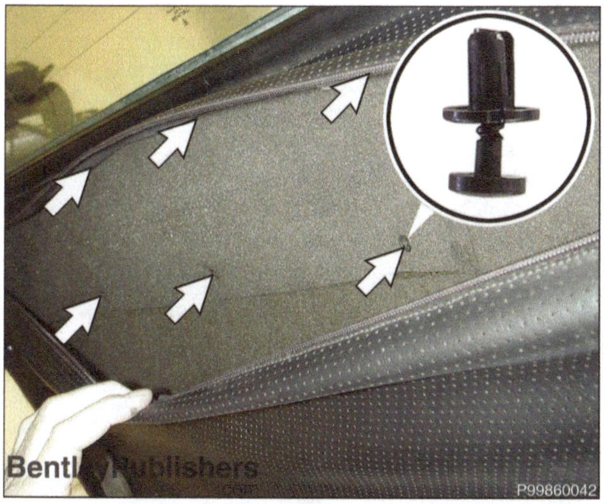

< Remove sunroof drive assembly access panel fasteners (**arrows**) and remove panel.

NOTE–

- *Sunroof drive access panel fasteners are a two-piece expanding rivet (**inset**). Use a Torx driver to remove center threaded pin, then pull rivet from panel.*

– Disconnect sunroof drive motor electrical connector.

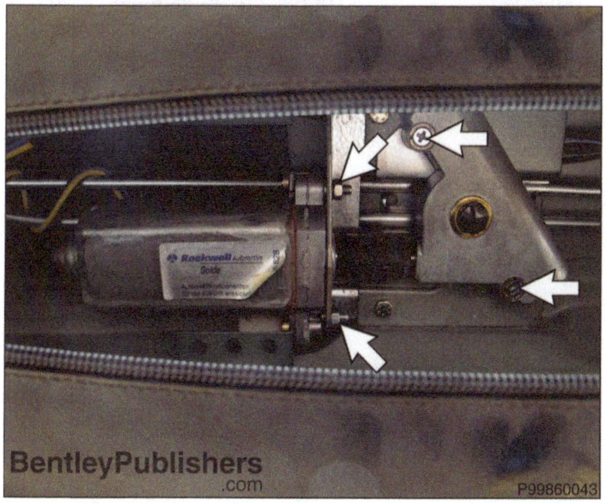

< Remove sunroof drive assembly mounting fasteners (**arrows**). Lower drive assembly with motor.

– Installation is reverse of removal.

Sunroof guide rails, replacing

> ***CAUTION—***
>
> - *Place strip of masking tape across front of sunroof opening to protect paint when removing sunroof parts.*

Remove sunroof panel. See **Sunroof panel, removing and installing** in this repair group.

– Remove sunroof motor and drive unit. See **Sunroof motor and drive unit, removing and installing** in this repair group.

– Remove wind deflector. See **Wind deflector, removing and installing** in this repair group.

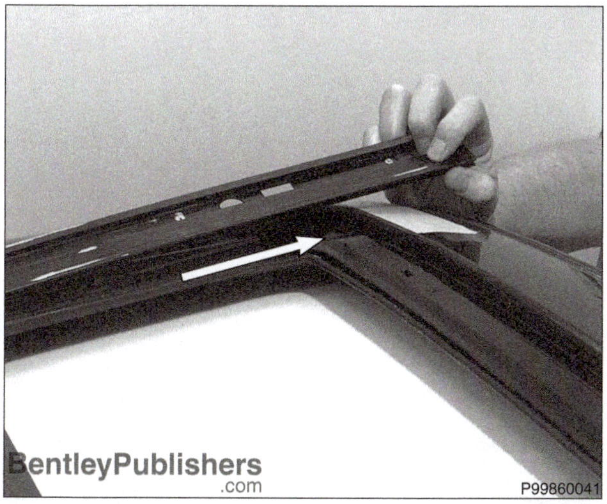

< Pull sunroof guide rails straight forward and out to remove.

– Installation is reverse of removal, noting the following:

- Inspect parts and replace as necessary.

Targa Top

Targa top overview

Porsche 993 Targa models were equipped with an electrically-operated sliding glass roof, wind deflector and sun blind. Separate rocker switches on the center console control the Targa top, wind deflector, and sun blind.

The rocker switches can be operated with the ignition in the ON or accessory position when the doors are closed. When the doors are open, the rocker switches can be operated with the ignition in any position. The sun blind can only be operated when the Targa top is closed.

The Targa is styled to reflect the lines of the Coupe. Though structurally it is based on the Cabriolet, featuring the Cabriolet's structural reinforcements.

The Targa roof was pre-assembled and installed as a single unit when the car was built. The wind deflector and glass roof are made of durable, twin-layer tinted safety glass and offer UV protection.

The fabric sun blind is mounted below the wind deflector in the front crossbow (**arrow**). An electric drive motor operates the rolling blind.

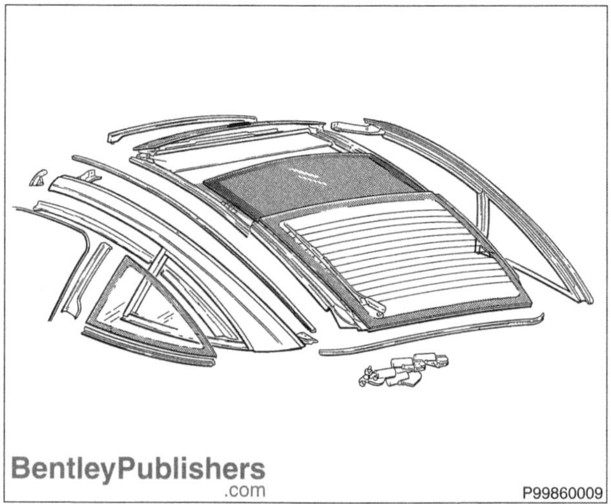

The electrically heated rear window is also made of tinted safety glass.

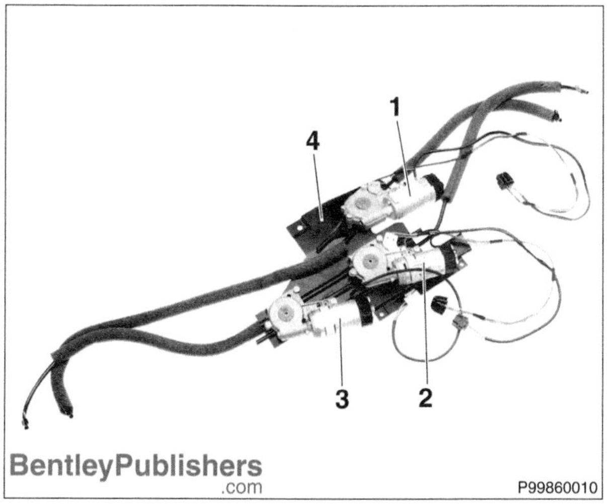

The Targa top, wind deflector, and sun blind are cable-driven by a gang of three electric motors.

1. Sun blind motor
2. Targa top motor / wind deflector
3. Targa top motor /wind deflector
4. Mounting plate

The wind deflector and Targa top drive motors (2 and 3) are operated in parallel. On motor 2, a pair of microswitches (**arrows**) operated by a drive gear control the switch off points (end travel) of the wind deflector motor and Targa top.

When working on the drive mechanism, never remove both motors from the support plate at the same time. One of the two motors should remain installed to keep the mechanism synchronized.

If only one drive motor is fitted to the support plate, it must not be operated electrically. Do not operate the top unless both motors are installed.

60

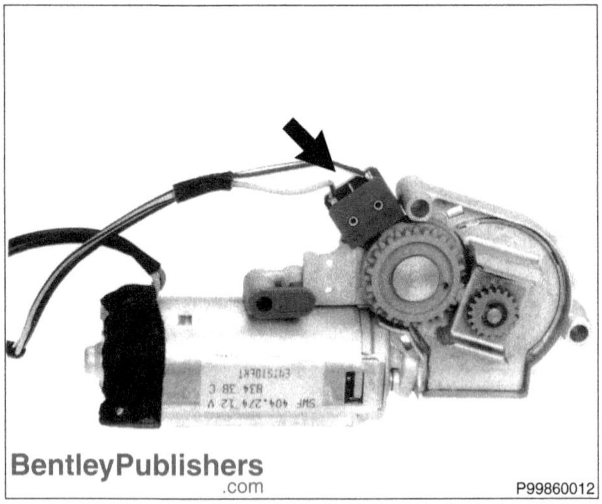

The sun blind motor simultaneously actuates both the left and right drive cables. A microswitch (**arrow**) operated by a drive gear switches the sun blind motor off at the end positions.

Targa top components

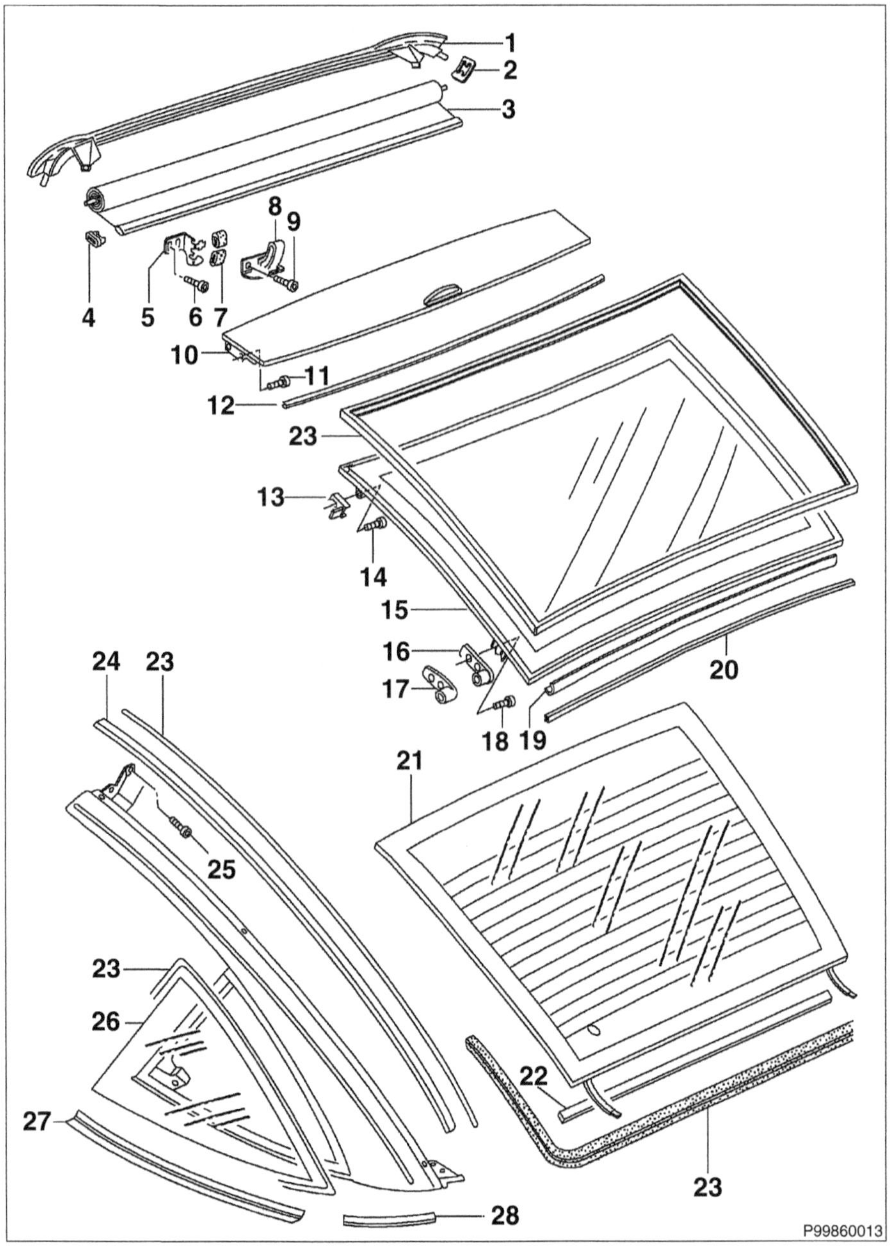

1. Gasket
2. Clamping washer
3. Shade
4. Bearing piece
5. Guide
6. Fastener, M5 x 12 mm
7. Sliding piece
8. Guide
9. Fastener, M5 x 12 mm
10. Wind deflector
11. Fastener, M5 x 12 mm
12. Seal
13. Axle bearing
14. Fastener, M6 x 12 mm
15. Targa roof
16. Axle plate
17. Axle plate
18. Torx fastener
19. Sealing strip
20. Gasket
21. Rear window
22. Retaining strip
23. Gasket
24. Cover strip
25. Torx fastener
26. Rear window
27. Molding
28. Retaining strip

Targa top drive components

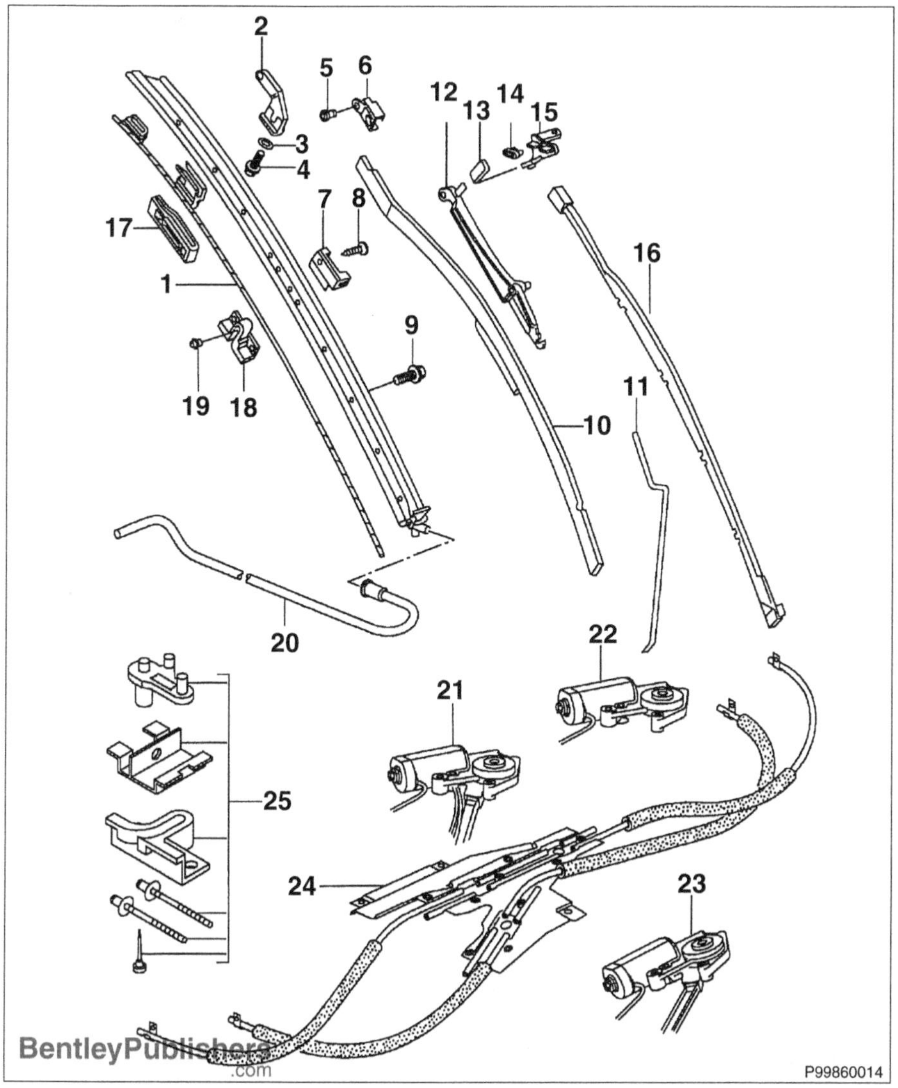

1. Sun blind drive shaft
2. Bearing bracket
3. O-ring
4. Torx fastener
5. Friction ring
6. Bearing bracket
7. Support
8. Fit bolt
9. Combination screw
10. Rail
11. Spring
12. Knock-out lever
13. Slide
14. Slide
15. Supporting bow
16. Guide
17. Guide
18. Guide
19. Buffer
20. Drain tube
21. Sun blind drive motor
22. Targa top drive motor
23. Targa top drive motor
24. Drive motor mounting plate
25. Knock-out lever

Targa top controls

Opening Targa top

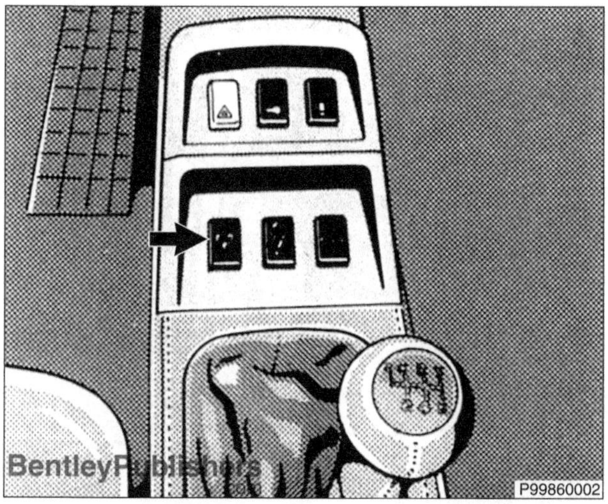

- ◄ Press and hold rocker switch (**arrow**) in center console until wind deflector is fully raised. Release rocker switch.
- Press and hold rocker switch one more time. Targa top opens and will remain in desired position as soon as rocker switch is released. Sun shade will roll in before top opens.

Closing Targa top

- Press rocker switch until Targa top has closed completely.
- Press and hold rocker switch one more time until the wind deflector is fully lowered.

> ***WARNING—***
>
> - *To avoid injury, keep fingers, hands, hair, etc. clear of Targa top when opening or closing.*

Targa top emergency closing

In case of drive motor failure, the Targa top, wind deflector and sun blind can be closed manually. Fold down rear seat backrests.

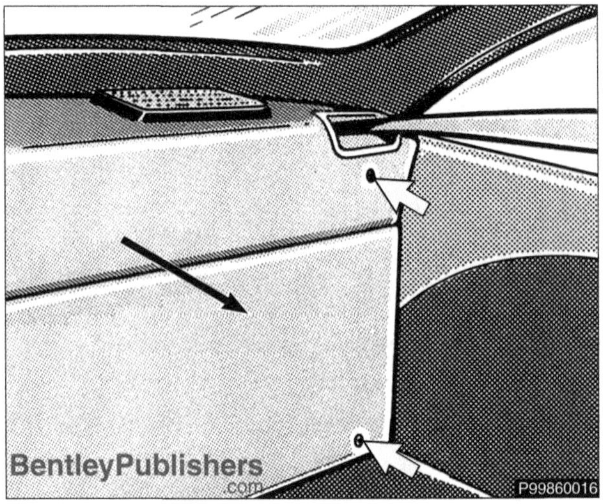

- Remove screws (**white arrows**) from rear parcel shelf on left and right sides.
- Carefully pull rear parcel shelf forward (**black arrow**) until drive motors are accessible.

- Using a flat blade screwdriver, pry off red plastic drivers (**arrows**) from drive motors.

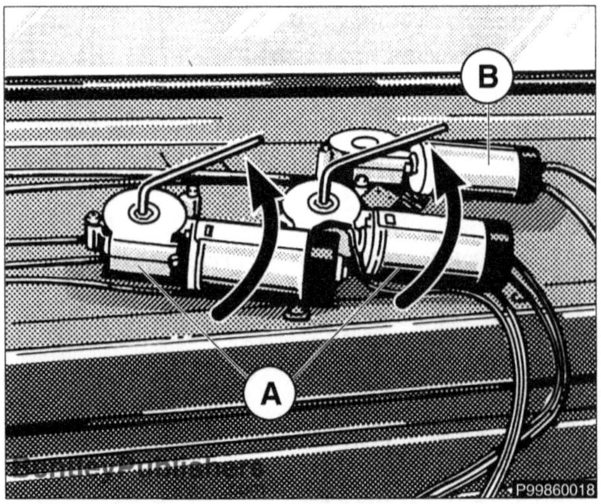

- Use Allen wrench(es) to manually close top or sun blind.
  - **Sun blind motor** (**B**), rotate motor counterclockwise until sun blind is rolled in.
  - **Top drive motors** (**A**): rotate both motors counterclockwise simultaneously until roof and wind deflector are fully closed.
- Reinstall plastic drivers. Clip Allen wrenches back in place and refit rear parcel shelf.

Targa top drive motors, basic adjustment

Basic adjustment of the drive cables is required if both motors or the drive cables have been removed.

NOTE–

- *Drive motors remain mounted to bracket for procedure.*

– Fold down rear seat backrests.

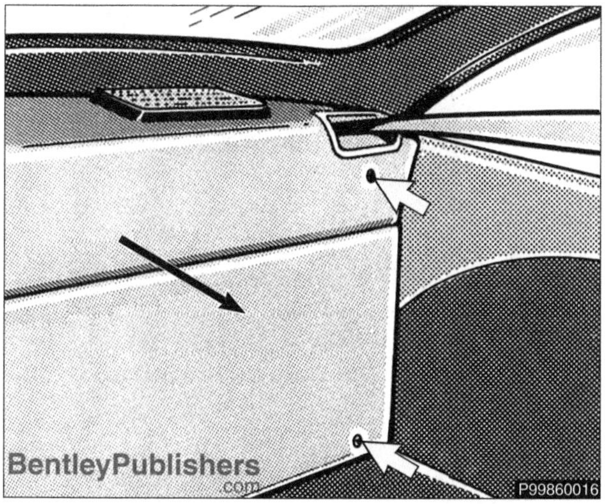

< Remove screws (**white arrows**) from rear parcel shelf on left and right sides.

– Carefully pull rear parcel shelf forward (**black arrow**) until drive motors are accessible.

< Disconnect both Targa top drive motor electrical connectors from control unit (**arrow**).

60

< Using a flat blade screwdriver, pry off red plastic drivers (**arrows**) from both drive motors.

Targa top drive motors, basic adjustment

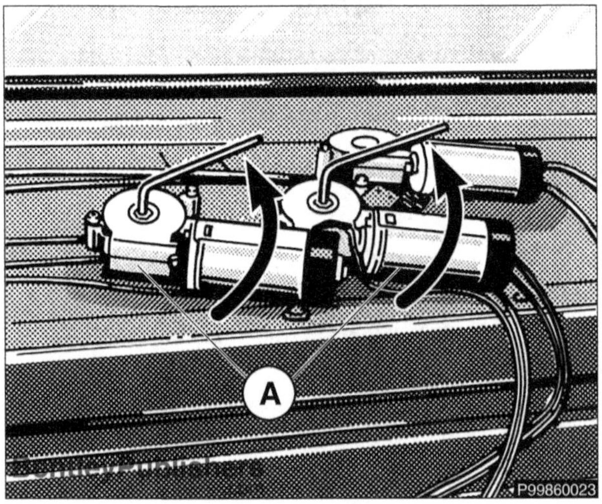

< Using Allen wrench, rotate drive motors (**A**) counterclockwise simultaneously until roof and wind deflector are fully closed.

- Front gate slide (connected to cable) must meet mounting supports on left and right sides.

< Reconnect drive motor (**A**) electrical connector to control unit (**C**).

NOTE–

- *Drive motor (**B**) connector remains disconnected from control unit and plastic drivers remain removed from motors.*

– Operate glass roof/wind deflector rocker switch in the close position until motor (**A**) turns off. Operate switch again until motor turns off (wind deflector closed).

NOTE–

- *When motor turns off, it has reached the end positions (microswitches signal end positions).*

– Reconnect drive motor (**B**) to control unit and refit plastic drivers.

– Operate roof and wind deflector to check basic adjustment.

– Remainder of installation is reverse of removal.

Sun blind motor, basic adjustment

- Fold down rear seat backrests.

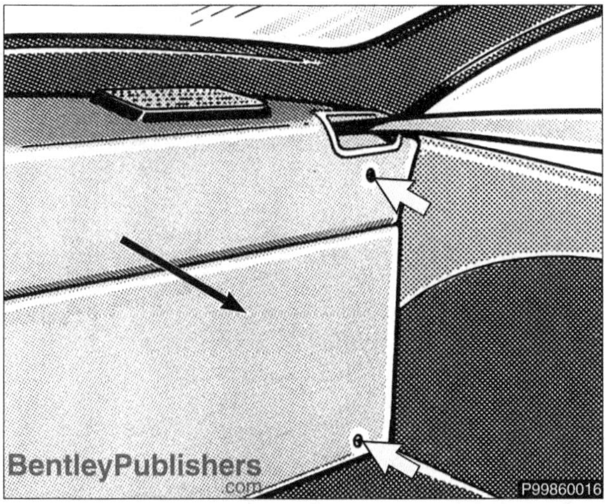

- Remove screws (**white arrows**) from rear parcel shelf on left and right sides.
- Carefully pull rear parcel shelf forward (**black arrow**) until drive motors are accessible.

- Remove fasteners and disconnect sun blind drive motor (**arrow**) from bracket.
- Pull left and right sun blind drive cables forward up to guide rail stop screw so that the sun blind is fully rolled in.
- Operate sun blind rocker switch in sun blind rolled in direction until motor turns off.

NOTE–

- *When motor turns off, it has reached the end position (microswitch signals end position).*

- Reinstall sun blind drive motor to bracket.
- Operate sun blind using rocker switch to check basic adjustment.
- Remainder of installation is reverse of removal.

60

61 Convertible Top

General

This repair group covers convertible top troubleshooting, emergency closing, component locations and repair procedures. Replacement of the soft top itself is not included in this manual. See also **EWD Electrical Wiring Diagrams** for circuit details.

The convertible top drive motor assembly is linked to the convertible top by flexible drive cables connected to planetary gearboxes (transmissions). The planetary gearboxes are connected to the convertible top via link arms to the roof frame assembly.

The convertible top can only be opened with vehicle stationary, parking brake lever set and ignition key in accessory or ON position.

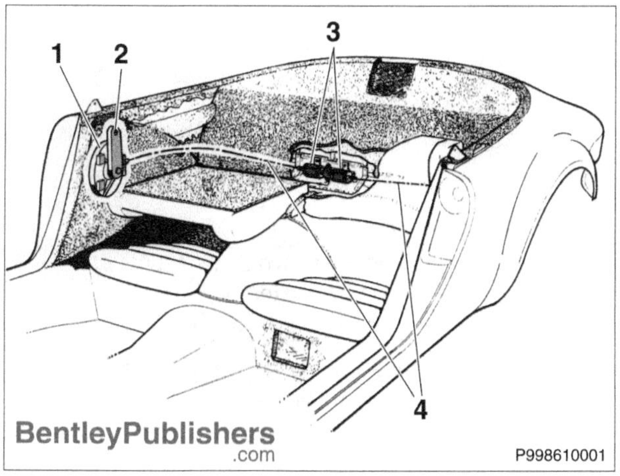

The drive motor(s) are mounted to a bracket behind the rear seats. The gearboxes are mounted on the sides of the vehicle.

1. Planetary gearbox
2. Link arm
3. Convertible top drive motor(s)
4. Drive cables

NOTE–

- *1995 models used two drive motors. 1996 and later models used a single drive motor.*

Two small motors are used for unlocking and locking the convertible top to the windshield frame. The motors are located in the front of the top (header bow).

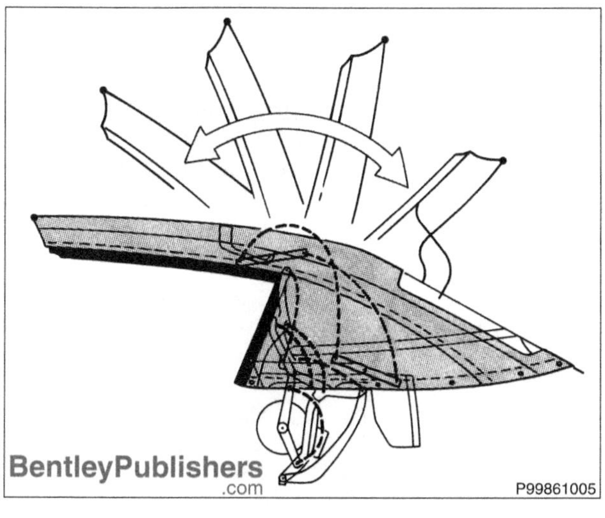

As the automatic convertible top is opened, the front roof frame swings inward, the header bow is lowered behind the rear seat backrests, and the roof frame comes to rest on the header bow.

61

Warnings and Cautions

> ***WARNING—***
> - *Do not operate convertible top while performing repairs.*
> - *Remove key from ignition before proceeding with repairs.*

> ***CAUTION—***
> - *Open or close convertible top with vehicle level and stationary and parking brake engaged.*
> - *Do not operate top with vehicle jacked up.*
> - *Do not operate top unless there is sufficient clearance above vehicle.*
> - *To prevent battery drain, it is advised to operate top with engine running.*
> - *Do not attempt to drive vehicle unless top is completely open or closed.*
> - *Do not modify convertible top control system to bypass built-in safety features.*
> - *Do not open top with temperatures below freezing.*

Convertible soft top

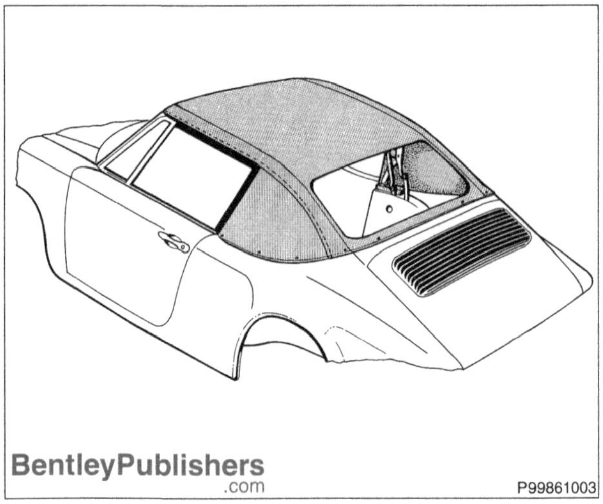

The convertible soft top is comprised of a 3-layer structure;

- Top layer: synthetic fiber provides resistance to weathering.
- Middle layer: rubber for waterproofing.
- Bottom layer: cotton fabric.

To improve the lifespan of the soft top:

- Park vehicle in shade when possible.
- Remove bird droppings immediately.
- Open convertible top only when dry to avoid mold spots and scratches. Leave convertible top closed to dry.
- Brush top with a soft brush to remove dust prior to washing.
- Wash top with clean water only. If top is very dirty, use a mild soap solution such as a detergent for cleaning wool .
- Do not wash vehicle in an automatic car wash.
- Leave convertible top closed to dry.
- Do not use sharp objects to remove snow and ice.

Convertible top frame

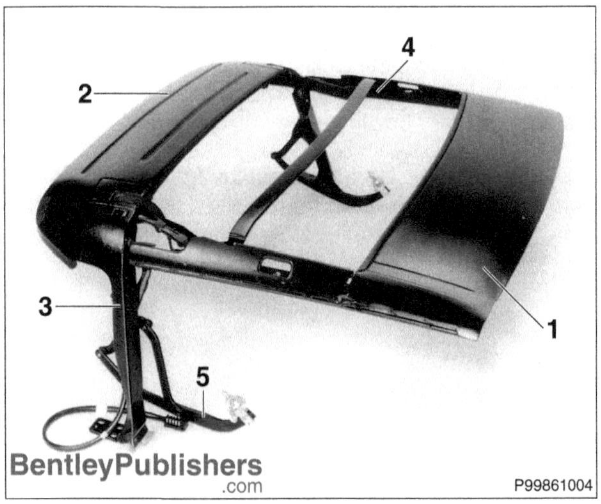

The convertible top frame is comprised of a steel roof frame, a stamped aluminum header bow, connecting members, and linkage. This structure provides a large area of solid roof surface.

1. Front roof frame
2. Header bow
3. Main bow
4. Drag bow
5. Links

Convertible top, troubleshooting

The automatic power top is electrically operated via the convertible top electronic control unit based on inputs from the rocker switch, microswitches, motor limit switches, and the parking brake lever switch.

◀ The convertible top control module (**arrow**) is located behind dashboard on driver's side (identified by four multi-pin connectors).

NOTE—

- *The convertible top control unit is mounted directly above the central informer control unit. Access to the control unit is greatly improved by removing the knee bar. See* ***70 Interior Trim****.*

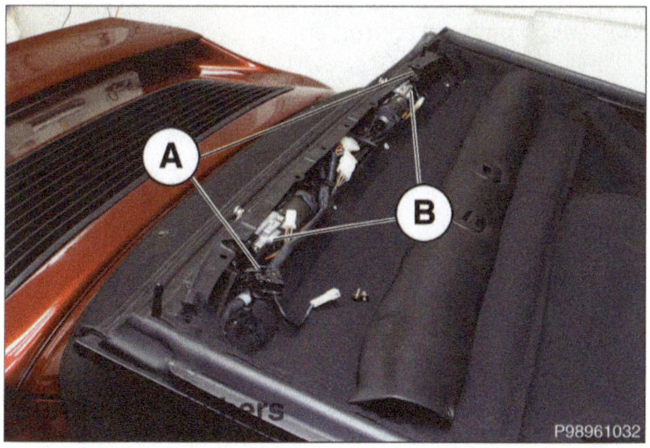

◀ Convertible top lock microswitches and limit switches

- Located in top head bow, beneath the trim cover.
- Top lock microswitches (**A**). These switches confirm top has engaged windshield frame, enabling top lock motors to run. They also function to turn convertible top drive motor off.
- Top lock motor limit switches (**B**). These switches signal control unit that top lock motors are in closed/locked position.

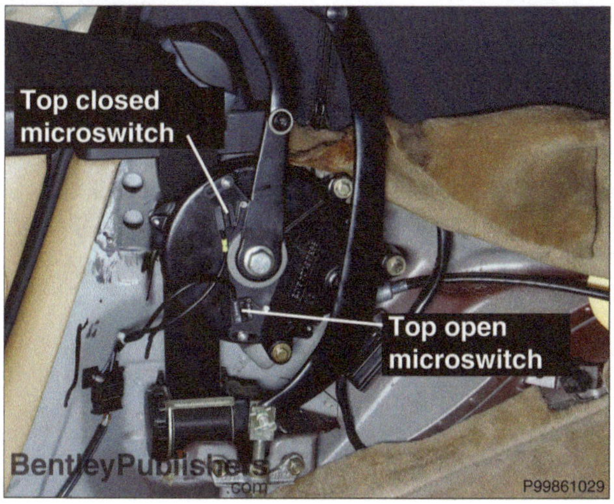

◀ Convertible top open/closed microswitches:

- Located on passenger side convertible top transmission.
- These switches interrupt power to the drive motor(s) when the switch is engaged. Top microswitch is contacted when top is fully closed. Bottom microswitch is contacted when top is fully open. Switch position is adjustable at the mounting plate.

When troubleshooting convertible top faults, first check to see that convertible top fuse (F13, 30A) is good and intact. Inspect the wiring and connections at the convertible top control module, top drive motor(s), top lock motors, and all related switches. See **97 Fuses, Relays, Component Locations** for fuse information.

Check limit switches and microswitches by verifying inputs to the control unit while actuating switches by hand or spinning motors by hand. See **EWD Electrical Wiring Diagrams** for detailed electrical schematics (switch logic and wiring identification).

If the switches are operating correctly, check to see if power is reaching the motor(s) when the rocker switch is actuated with the parking brake set and the ignition ON. Check that power (blue/white and yellow wires) and ground (orange wire) are present at the switch with the parking brake set and ignition ON. If no faults can be found the control module may be faulty.

Convertible top, emergency closing

In the event of an electrical fault, the convertible top can be closed manually.

< Remove cover caps from access points on left and right rear side panels.

– Insert wheel nut wrench from vehicle tool kit into access point and loosen side drive arm bolt (**arrow**) approximately four turns. Loosen bolt on opposite side.

NOTE–

- *The transmission drive arm bolt head is 18 mm. The wheel nut wrench is 19 mm. For a better fit, use an 18 mm socket if available.*

– Pull convertible top closed by hand until it makes contact with windshield frame.

< Press and hold convertible top rocker switch until convertible top locks to windshield frame and convertible top indicator light (**arrow**) goes out.

< If convertible top lock motors are not working, insert flat blade screwdriver into lock motor access points in windshield frame (**arrows**).

– Alternating between both locks, turn each lock motor clockwise approximately ten turns until convertible top indicator light goes out and top is firmly locked into windshield frame.

Convertible top drive components

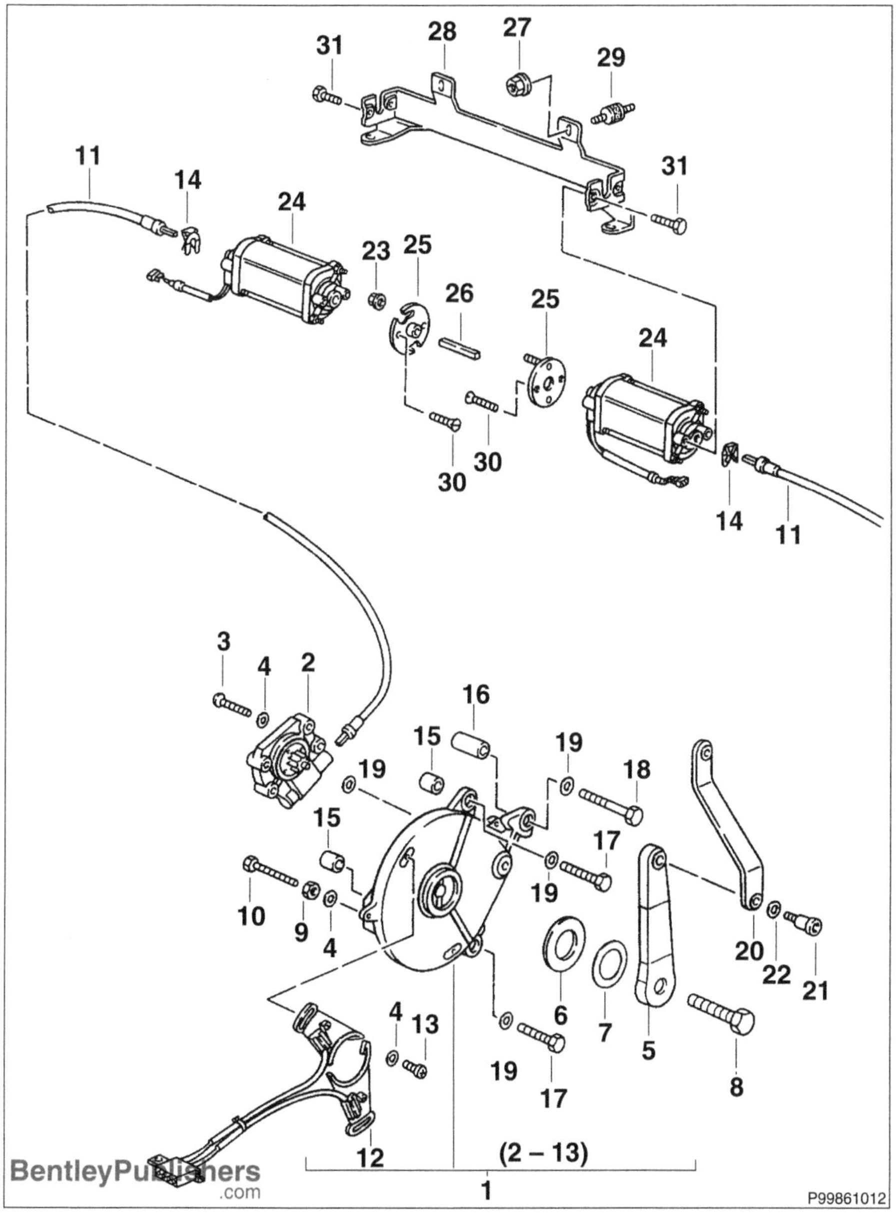

1. Convertible top transmission
2. Planetary gearbox
3. Fastener, M4 x 18 mm
4. Washer
5. Drive arm
6. Washer
7. Shim
8. Hex bolt, M12 x 30 mm
9. Hex nut
10. Hex bolt, M4 x 22 mm
11. Flexible cables
12. Convertible top microswitch
13. Fastener, M4 x 8 mm
14. Locking clip
15. Spacer
16. Spacer
17. Hex bolt, M6 x 35 mm
18. Hex bolt, M6 x 50 mm
19. Washer
20. Link arm
21. Shouldered bolt, M8 x 20 mm
22. Washer
23. Hex nut
24. Convertible top drive motor(s)*
25. Cover washer
26. Connecting piece
27. Hex nut
28. Support bracket
29. Bonded rubber mounting
30. Self-locking fastener, M5 x 16 mm
    Replace with new
31. Self-locking fastener, M5 x 16 mm
    Replace with new

** Dual motor drive on 1995 models. Single motor drive on 1996 and later models.*

61

Convertible top lock components

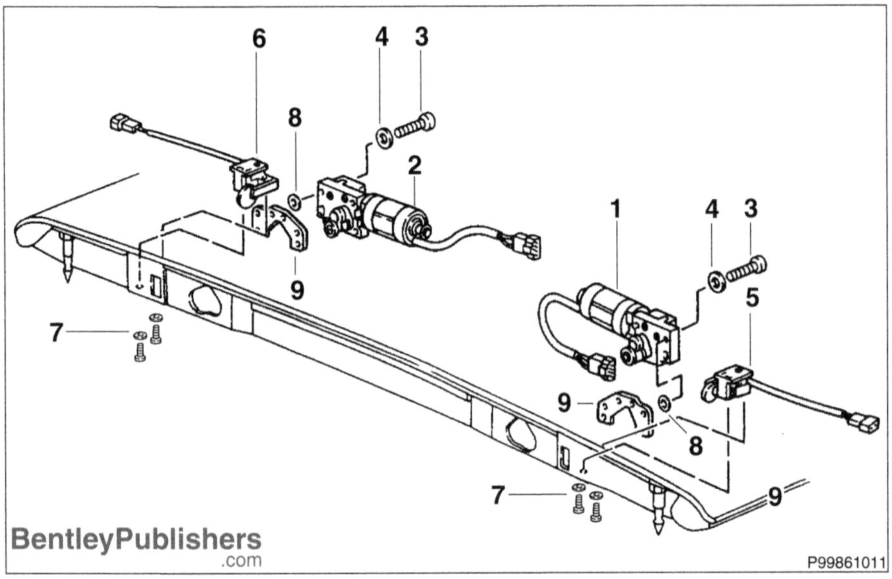

1. Convertible top lock motor assembly (left side)
2. Convertible top lock motor assembly (right side)
3. Fastener, M6 x 30 mm
4. Washer
5. Convertible top lock microswitch (left side)
6. Convertible top lock microswitch (right side)
7. Screw, M4 x 10 mm
8. Shim
9. Threaded plate

Convertible top frame components

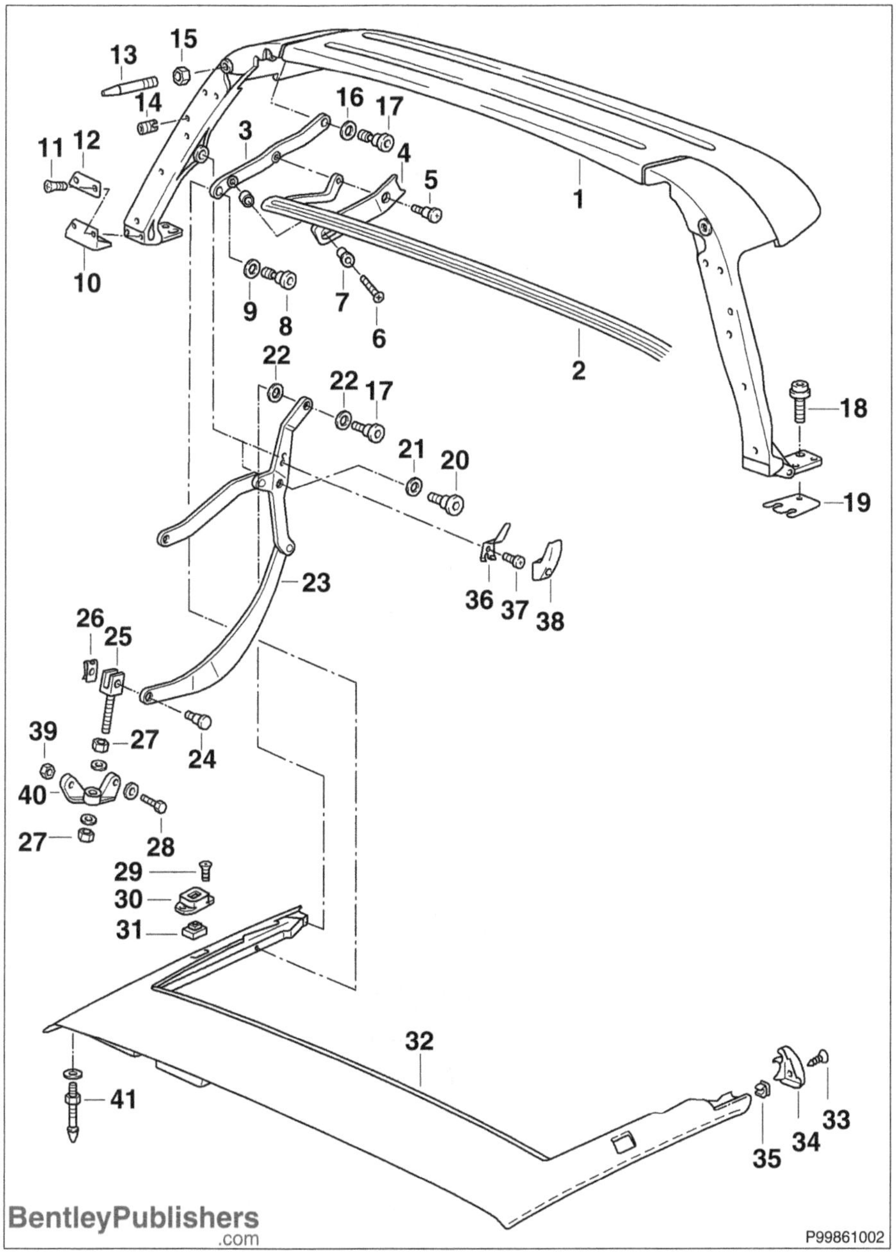

1. Main folding top bow
2. Intermediate folding top bow
3. Link arm
4. Protective cover
5. Shouldered bolt, M5 x 12 mm
6. Screw, M4 x 22 mm
7. Eccentric washer
8. Shouldered bolt
9. Washer
10. Gasket
11. Screw, M4 x 10 mm
12. Plate
13. Guide pin
14. Threaded insert
15. Hex nut, M8
16. Washer
17. Shouldered bolt
18. Fastener, M8 x 20 mm
19. Plate
20. Shouldered bolt
21. Washer
22. Washer
23. Link arm assembly
24. Clevis pin
25. Threaded yoke
26. Locking clip
27. Hex nut, M10
28. Hex bolt, M5 x 15 mm
29. Screw, M4 x 10 mm
30. Housing
31. Buffer
32. Roof frame
33. Screw
34. End piece
35. Nut
36. Coat hook
37. Bolt
38. Cover
39. Lock nut, M8
40. Bearing bracket
41. Locating pin

Convertible top cover components

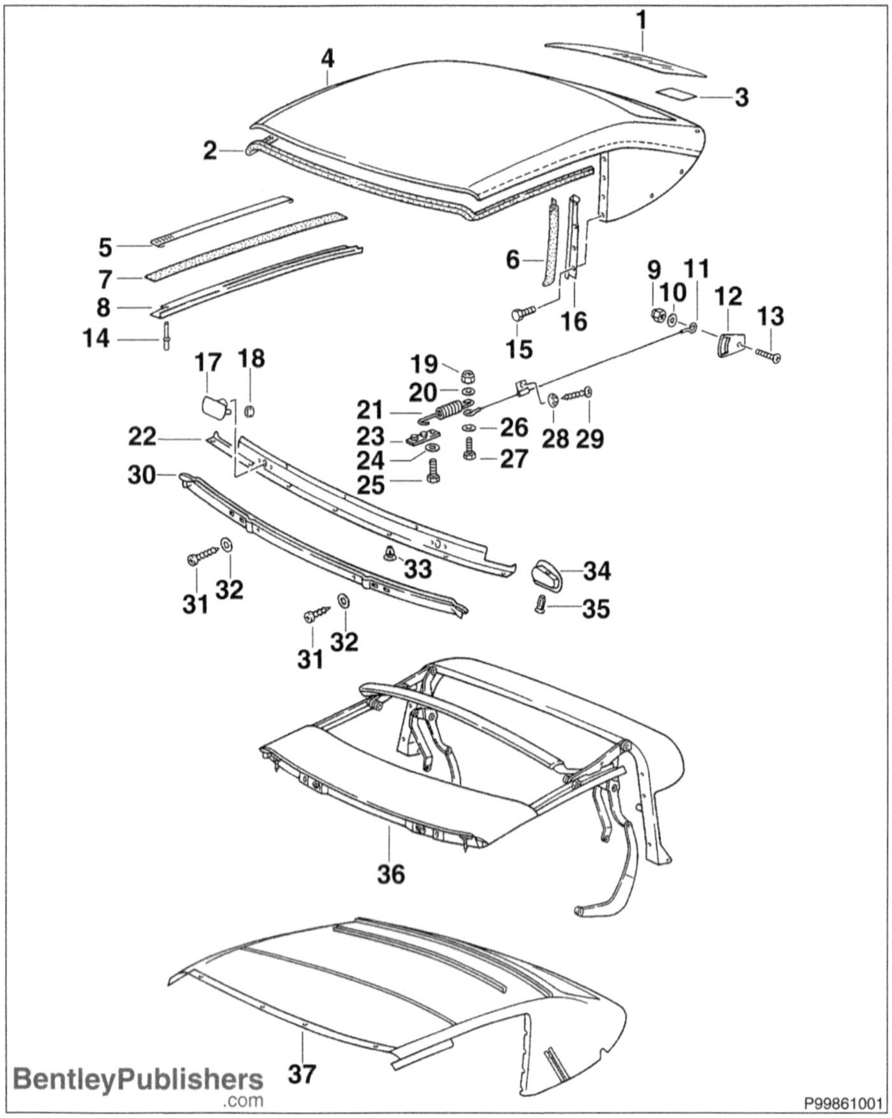

1. Rear window
2. Roof frame seal
3. Insulating plate
4. Convertible top cover
5. Retaining strap
6. Gasket
7. Sealing strap
8. Rain channel
9. Lock nut, M4
10. Washer
11. Tension cable
12. Trim cover
13. Fastener, M4 x 12 mm
14. Blind rivet
15. Screw
16. Rail
17. Cap
18. Clamping washer
19. Lock nut
20. Washer
21. Tension spring
22. Padded cover
23. Support
24. Washer
25. Hex bolt, M6 x 12 mm
26. Washer
27. Hex bolt, M4 x 12 mm
28. Washer
29. Self-tapping screw
30. Cover
31. Self-tapping screw
32. Washer
33. Clip
34. Handle
35. Expansion rivet
36. Convertible top frame
37. Roof liner

Convertible Top Drive Service

Convertible top motor, removing and installing

The convertible top electric drive motor is mounted behind rear seats.

NOTE–

- *1995 model year Cabriolet models were equipped with two convertible top drive motors. 1996 and later models were equipped with a single convertible top motor. Removal and installation procedures are similar. Single motor vehicle shown.*

– Remove rear seat backrests and seat belt frame cover. See **72 Seats**.

< Loosen rear carpet cover fastener (**arrow**).

– Lift out rear carpet cover.

< Lever off clips securing drive cables to motor (**arrow**).

Convertible top motor, removing and installing

Disconnect electrical connector (**arrow**).

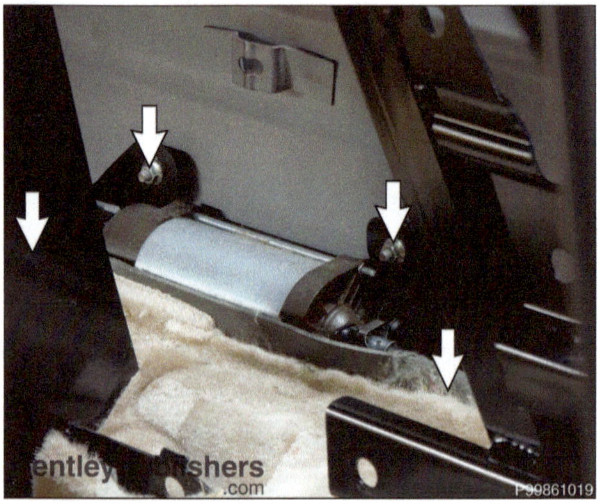

Remove nuts (**arrows**) and remove convertible top motor together with bracket.

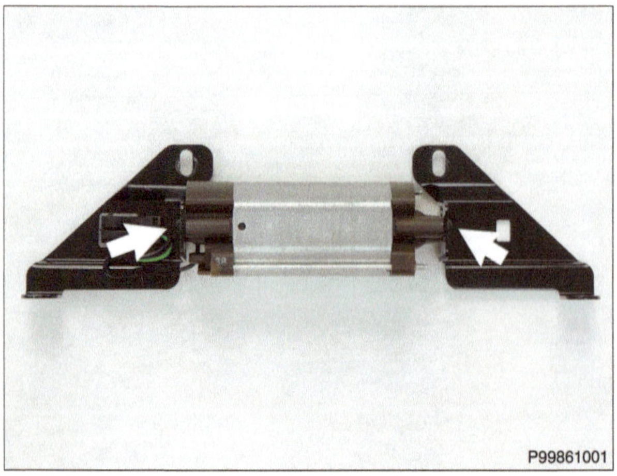

Remove fasteners (**arrows**) and separate convertible top drive motor from bracket.

- Installation is reverse of removal.

Convertible top transmission, removing and installing

- Remove rear seat backrests and seat belt frame cover. See **72 Seats**.

< Loosen rear carpet cover fastener (**arrow**).

- Lift out rear carpet cover.

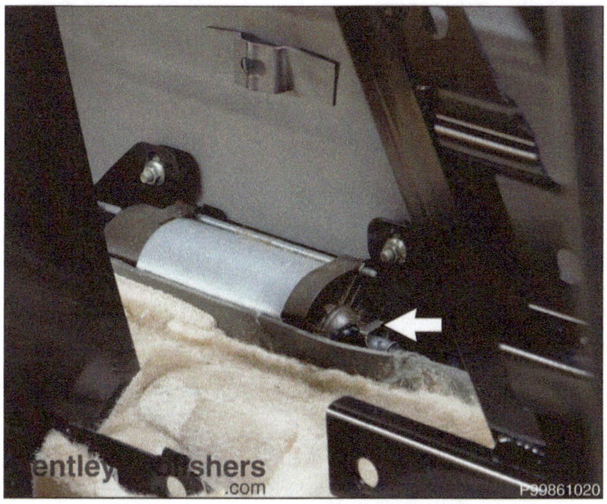

< Lever off clips securing drive cables to motor (**arrow**).

NOTE–

- *1995 model year Cabriolet models were equipped with two convertible top drive motors. 1996 and later models were equipped with a single convertible top motor. Single motor vehicle shown.*

- Remove rear interior side panel. See **70 Interior Trim**.

61

< Peel back carpeting (**arrow**) from side panel.

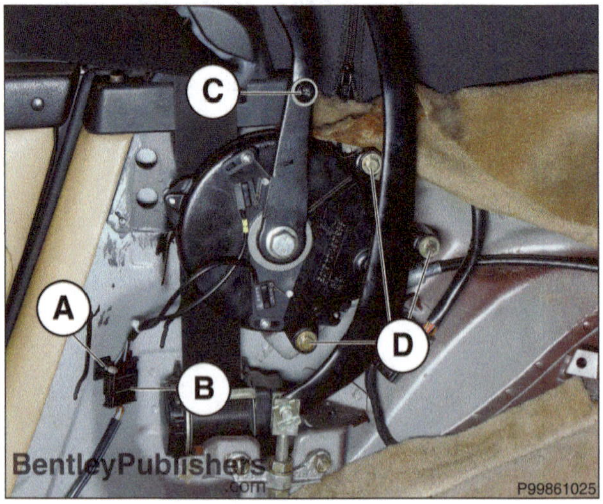

- Remove nut (**A**) and disconnect electrical connector (**B**).
- Remove Allen bolt (**C**) from drive arm.
- Remove transmission bolts (**D**) with spacer sleeves.
- Remove convertible top transmission with drive cable.
- Installation is reverse of removal.

Convertible top drive cable, replacing

- Remove convertible top transmission. See **Convertible top transmission, removing and installing** in this repair group.

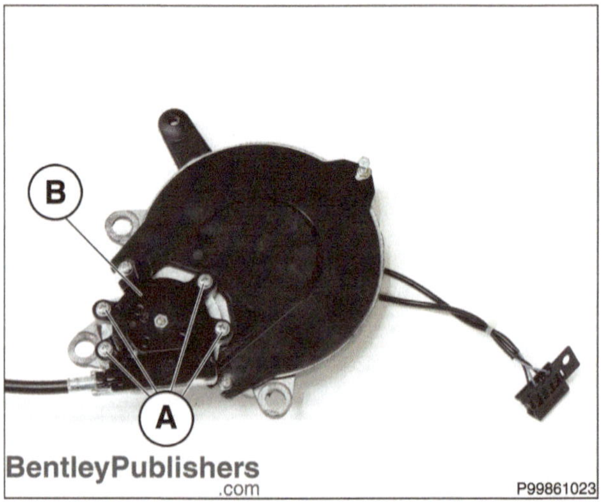

- Working from back side of the transmission, remove screws (**A**) and then remove cover plate (**B**).

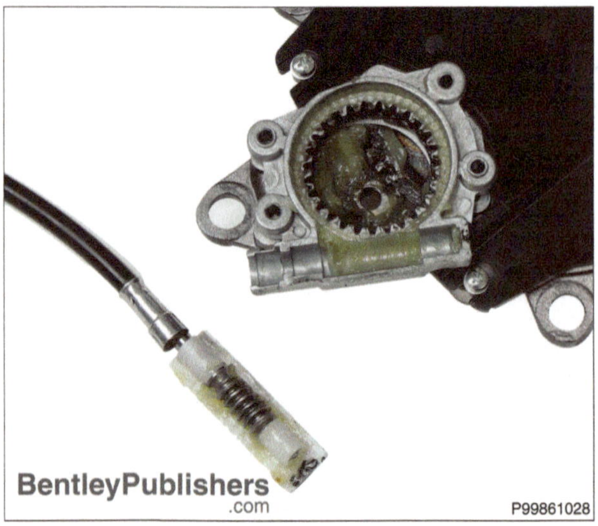

- Remove drive cable from gearbox with plastic piece.
- Installation is reverse of removal.

63 Bumpers

General

This repair group covers front and rear bumper removal and installation. For more information see:

- **66 Exterior Equipment**
- **94 Exterior Lights** for turn signal and sidemarker information

Bumper construction

The front bumper and rear bumper are made of extruded aluminum with a thickness of 4 mm (0.16 in). The bumpers are attached to the body via telescoping impact absorbers, one on each side.

A threaded bore on the right side (**arrow**) allows a towing lug to be screwed into the front bumper.

A threaded lug bore for the rear bumper is located in the center of the bumper, behind the license plate.

A multi-part heat shield is attached between the rear bumper and bumper cover to protect cover from exhaust heat.

63

Front Bumper

Front bumper components (Carrera and Carrera 4)

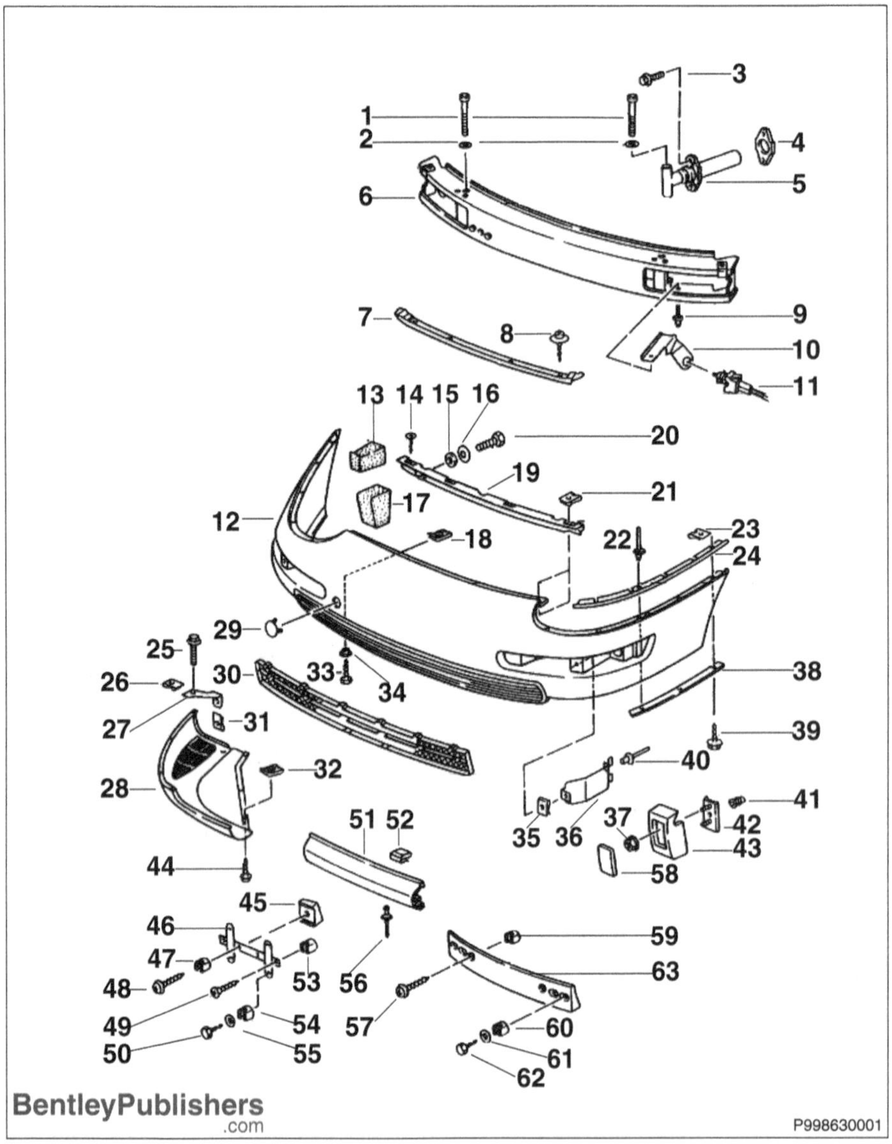

1. Fastener, M12
2. Washer
3. Fastener, M8
4. Pad
5. Impact absorber
6. Bumper
7. Sealing strip
8. Self-tapping screw
9. Rivet
10. Support
11. Outside temperature sensor
12. Bumper cover
13. Sealing wedge
14. Self-tapping screw
15. Spacer
16. Washer
17. Sealing wedge
18. Speed nut
19. Retaining strip
20. Bolt, M6
21. Speed nut
22. Rivet
23. Speed nut
24. Piping
25. Screw
26. Speed nut
27. Support
28. Lower spoiler
29. Cap
30. Ventilation grill
31. Speed nut

Front bumper components (Carrera and Carrera 4)

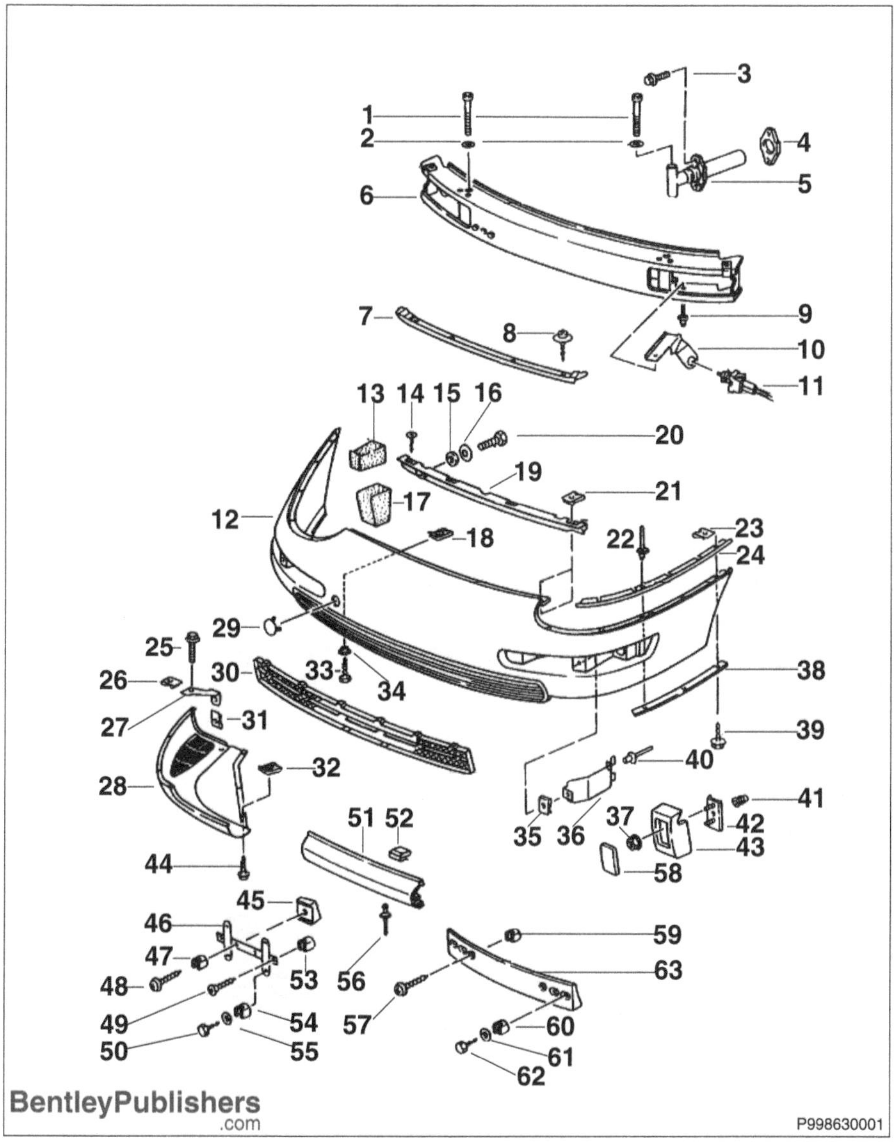

32. Speed nut
33. Self-tapping screw
34. Spacer
35. Speed nut
36. Retaining bow
37. Nut, M5
38. Retaining strip
39. Screw
40. Rivet
41. Nut, M5
42. Mounting
43. Bumper guard
44. Screw
45. Spacer
46. Support
47. Expanding nut
48. Self-tapping screw
49. Self-tapping screw
50. Self-tapping screw
51. Retaining strip
52. Cable clamp
53. Expanding nut
54. Special nut
55. Washer
56. Rivet
57. Self-tapping screw
58. Cap
59. Expanding nut
60. Expanding nut
61. Washer
62. Self-tapping screw

63

Front bumper components (Carrera S and Carrera 4S)

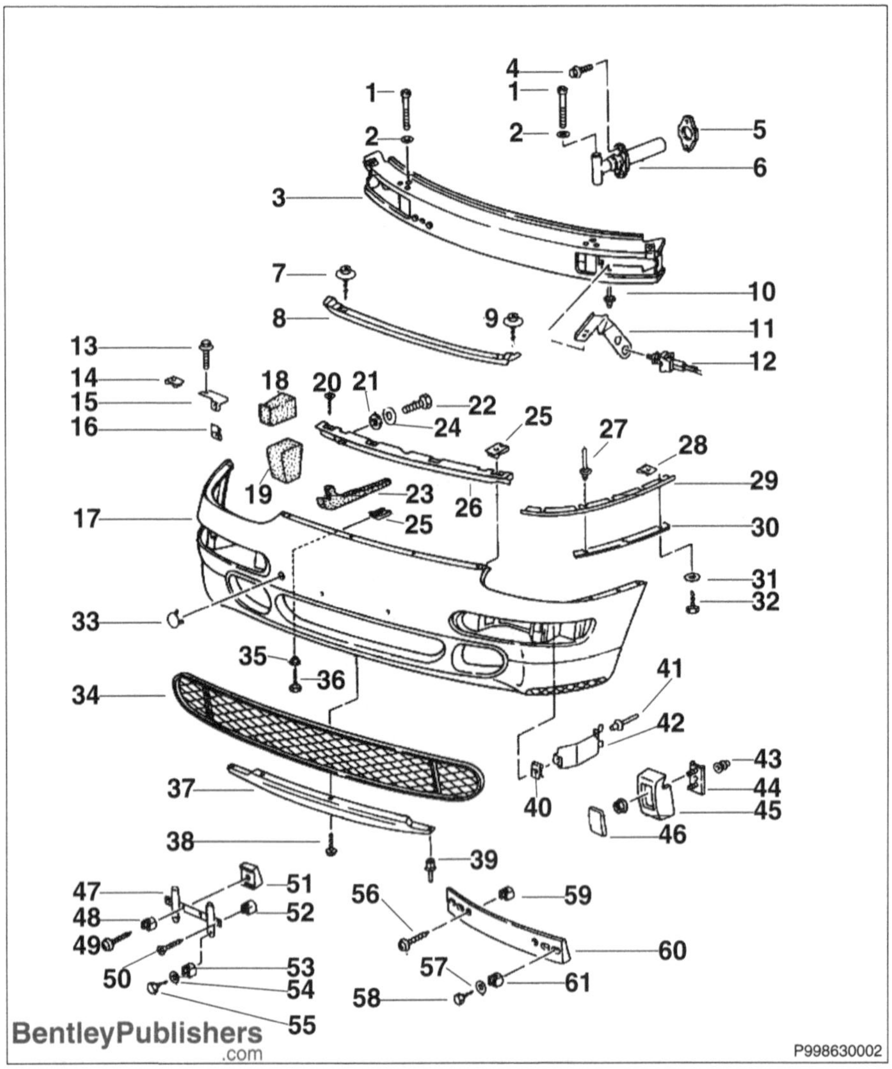

1. Fastener, M12
2. Washer
3. Bumper
4. Fastener, M8
5. Pad
6. Impact absorber
7. Self-tapping screw
8. Sealing strip
9. Self-tapping screw
10. Rivet
11. Support
12. Outside temperature sensor
13. Screw
14. Speed nut
15. Support
16. Speed nut
17. Bumper cover
18. Sealing wedge
19. Sealing wedge
20. Self-tapping screw
21. Spacer
22. Bolt, M6
23. Sealing wedge, lower
24. Washer
25. Speed nut
26. Retaining strip
27. Rivet
28. Speed nut
29. Piping
30. Retaining strip
31. Washer

Front bumper components (Carrera S and Carrera 4S)

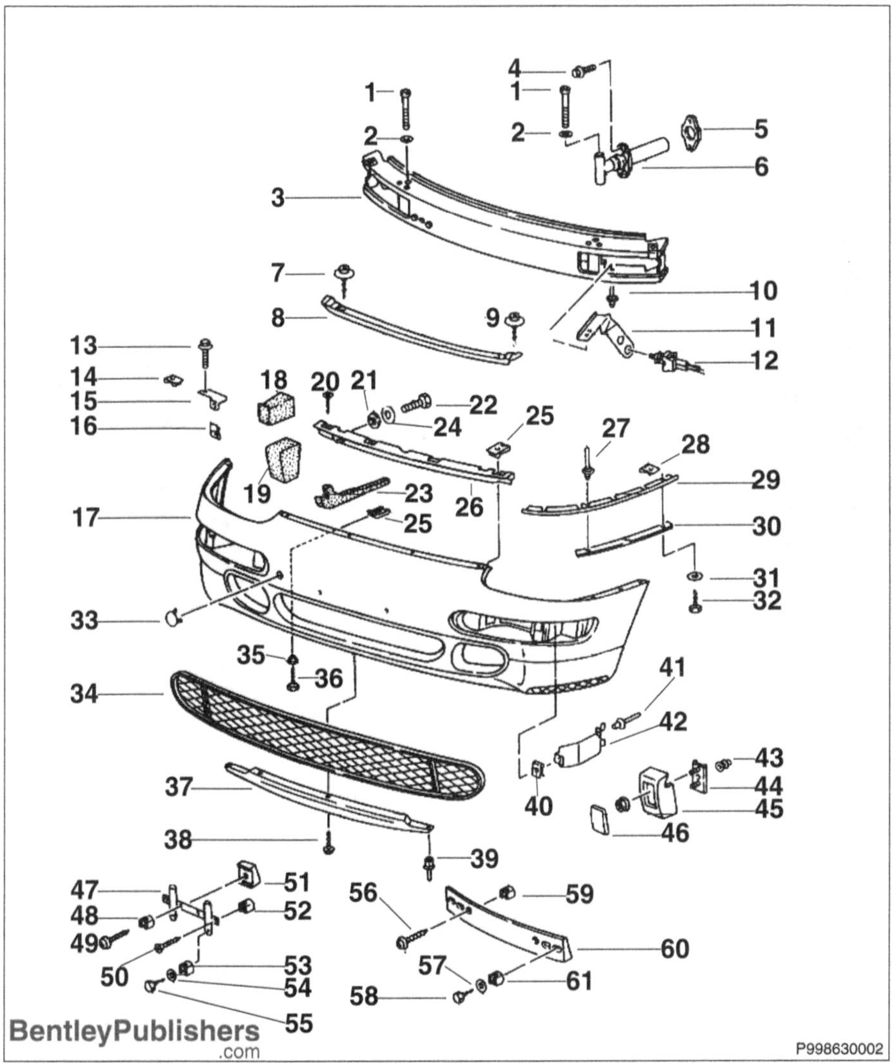

32. Self-tapping screw
33. Cap
34. Ventilation grill
35. Spacer
36. Self-tapping screw
37. Spoiler
38. Self-tapping screw
39. Expanding rivet
40. Speed nut
41. Rivet
42. Retaining bow
43. Nut, M5
44. Mounting
45. Bumper guard
46. Cap
47. Support
48. Expanding nut
49. Self-tapping screw
50. Self-tapping screw
51. Spacer
52. Expanding nut
53. Spacer nut
54. Washer
55. Screw
56. Self-tapping screw
57. Washer
58. Screw
59. Expanding nut
60. Support
61. Expanding nut

63

Front bumper cover, removing and installing (Carrera and Carrera 4)

– Raise vehicle and support safely. Remove front wheels.

> ***WARNING—***
>
> *• Make sure car is stable and well supported at all times. A floor jack is not adequate support.*

– Remove front wheel housing liners. See **50 Fenders**.

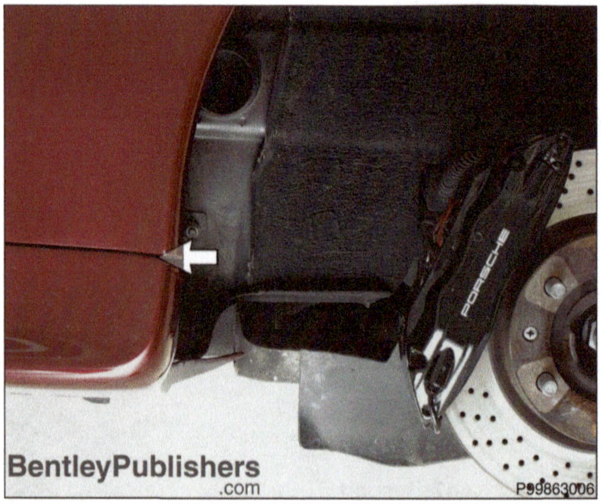

◄ Working inside front wheel well, remove side lower spoiler fastener (**arrow**). Repeat on opposite wheel well.

◄ Working from below, remove lower spoiler fasteners (**arrows**).

NOTE–

• Car shown fitted with factory Aerokit. Standard trim is similar.

– Disconnect drip hoses from mount on lower spoiler trim.

– Starting at wheel well, peel front lower spoiler off bumper cover.

NOTE–

• Plastic locking tabs may need to be pried out from behind to release.

◄ Open trunk and remove bumper strip screws (**arrows**).

– Remove bumper strip.

◀ Remove bumper cover fastener (**arrow**) from left and right sides.

◀ Working inside fender wells, remove retaining strip fasteners (**arrows**) with strip from left and right sides.

- Remove turn signal and fog lights. See **94 Exterior Lights**.
- With help from assistant, carefully pull front bumper cover off at bottom first, then down slightly to clear inner corner of fenders. Once clear, pull bumper cover straight off.
- Installation is reverse of removal.

Front bumper, removing and installing

– Raise vehicle and support safely. Remove front wheels.

> ***WARNING—***
>
> *• Make sure car is stable and well supported at all times. A floor jack is not adequate support.*

– Remove front bumper cover. See **Front bumper cover, removing and installing** in this repair group.

◄ Remove bumper mounting bolts (**arrow**) from left and right sides.

– Remove headlight washer hose from clips if applicable.

– Lift bumper from impact absorbers.

◄ If necessary, remove impact absorber mounting bolts (**arrows**).

– Remove absorber.

– Installation is reverse of removal.

Rear Bumper

Rear bumper components

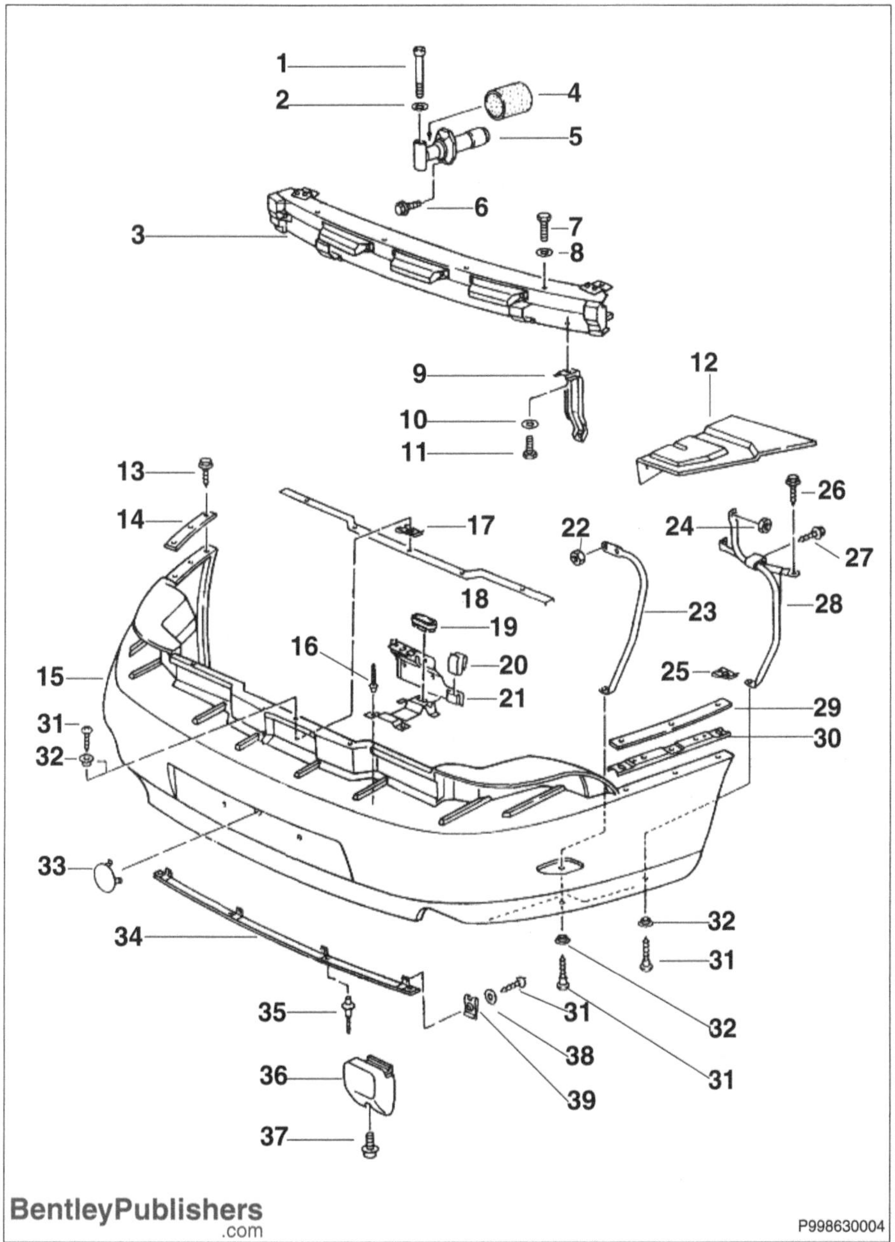

1. Fastener, M12
2. Washer
3. Bumper
4. Heat shield
5. Impact absorber
6. Fastener, M8
7. Fastener, M5
8. Washer
9. Rear support
10. Washer
11. Screw, M5
12. Support
13. Screw
14. Piping
15. Bumper cover
16. Rivet
17. Speed nut
18. Retaining strip
19. Guide
20. Retaining clip
21. Support
22. Nut, M5
23. Bracket
24. Nut, M6
25. Speed nut
26. Screw
27. Screw
28. Bracket
29. Piping
30. Retaining strip
31. Self-tapping screw
32. Spacer
33. Cover
34. Retaining strip
35. Rivet
36. Bumper guard
37. Fastener, M10
38. Washer
39. Speed nut

63

Rear bumper cover, removing and installing

– Raise vehicle and support safely. Remove rear wheels.

> ***WARNING—***
>
> • *Make sure the car is stable and well supported at all times. A floor jack is not adequate support.*

– Remove left and right rear turn signal lights. See **94 Exterior Lights** for more information.

◄ Remove trim cover fasteners (**arrows**). Remove trim cover.

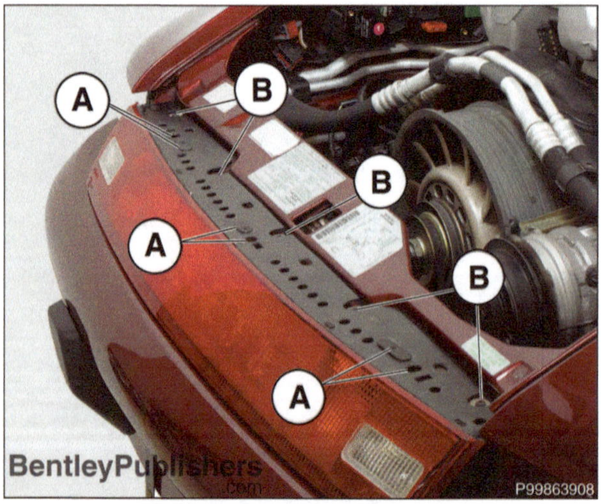

◄ Remove rear reflector fasteners (**A**). Remove reflector.

– Remove upper bumper fasteners (**B**).

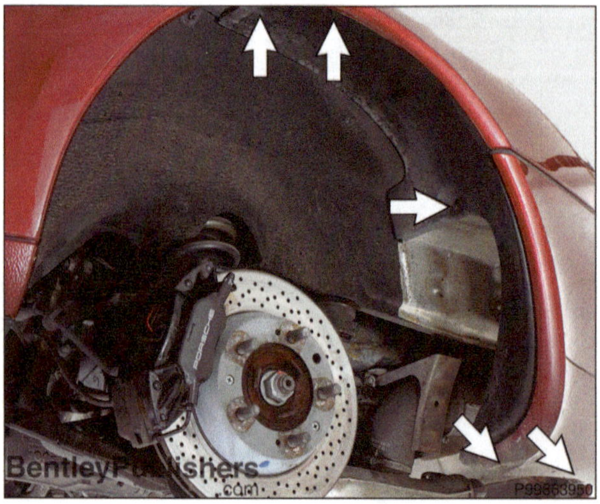

◄ Working in rear wheel wells, remove wheel housing liner and heat shield fasteners (**arrows**).

NOTE–

• *Left side shown. Right side is similar.*

– Peel out and remove both wheel housing liners.

P998631004A

◄ Working inside wheel wells, remove bumper fasteners (**arrows**) from both sides.

NOTE–

- *Left side shown. Right side is similar.*

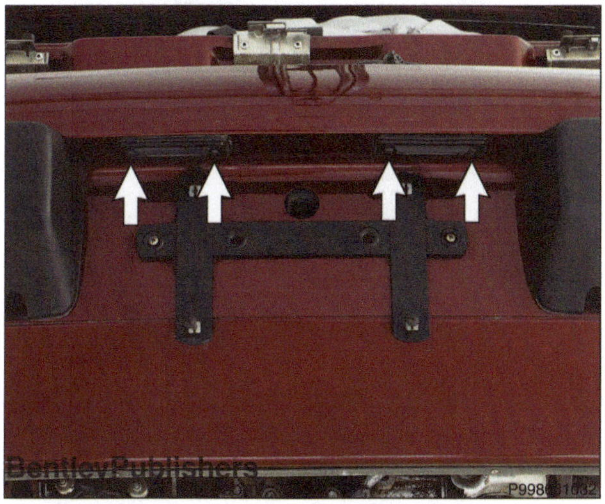

◄ Remove license plate light screws (**arrows**).

– Pull out lights and disconnect connectors.

NOTE–

- *For more information see* **94 Exterior Lights**.

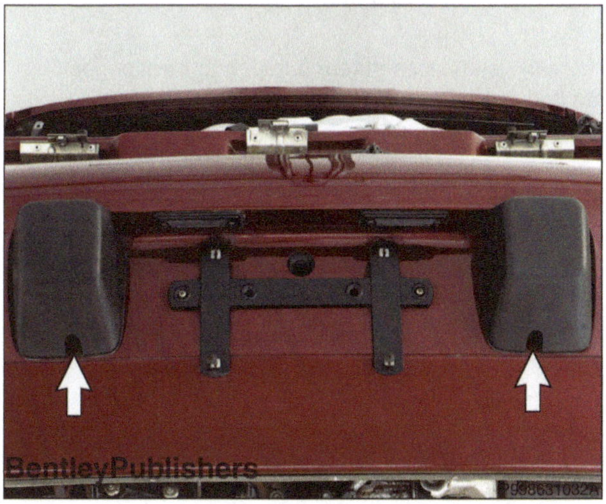

◄ Remove bumper guard mounting bolts (**arrows**).

– Remove bumper guards.

63

Rear bumper cover, removing and installing

< Remove heat shield fasteners (**arrows**) along bumper cover.

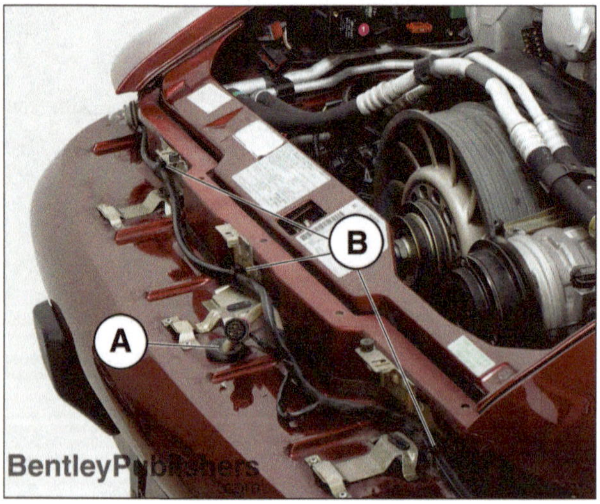

< Pull license plate light wiring with sealing grommet through bumper cover opening (**A**). Unclip wiring harness from retaining clips (**B**).

< Feed rear light wiring harness through cut-out (**arrow**) in right turn signal recess.

– With help from assistant, loosen rear bumper cover at edges then pull outward and carefully remove.

– Installation is reverse of removal.

Rear bumper, removing and installing

– Raise vehicle and support safely. Remove rear wheels.

> ***WARNING—***
> - *Make sure the car is stable and well supported at all times. A floor jack is not adequate support.*

– Remove rear bumper cover. See **Rear bumper cover, removing and installing** in this repair group.

◄ Remove upper heat shield fasteners (**arrows**).

– Remove heat shield and set aside.

◄ Remove lower heat shield bolts (**arrows**).

◄ Remove main bumper support bolts (**arrows**).

NOTE–
- *If necessary, push heat shield toward body to access bumper fasteners.*

– Lift bumper from impact absorbers.

– If necessary, remove impact absorber mounting bolts and lift out absorbers.

– Installation is reverse of removal.

63

64 Door Windows

GENERAL

This repair group covers door glass and power window regulator repairs.

For removal and installation of the doors, door panels and window frame, see **57 Doors and Locks**.

DOOR WINDOWS

Door vapor barrier, removing and installing

– Remove door trim panel. See **57 Doors and Locks**.

< Using plastic prying tool, lever off cover (**arrow**) and remove screw.

< Remove screw (**A**) and remove trim plate. Remove screws (**B**).

– Lift up upper door trim. Disconnect mirror switch and alarm light electrical connectors.

– Remove upper door trim.

64

Remove vapor barrier fasteners (**arrows**).

Carefully peel vapor barrier (**arrow**) from door.

Door window regulator and motor, removing and installing

- Remove door trim panel. See **57 Doors and Locks**.
- Remove door vapor barrier. See **Door vapor barrier, removing and installing** in this repair group.
- Lower window halfway into door. If necessary, reconnect window switch.

With help from an assistant, hold glass in position. Remove window regulator and motor bolts (**arrows**).

- Slide regulator guide wheels off of guide rail on glass.

Door window regulator and motor, removing and installing

< Unbolt window motor (**arrows**) from regulator prior to removal.

< Remove regulator, then motor through hole in door.

– Lower glass carefully into door.

– Installation is reverse of removal, noting the following:

  • Lubricate guide channel and regulator with white lithium grease.

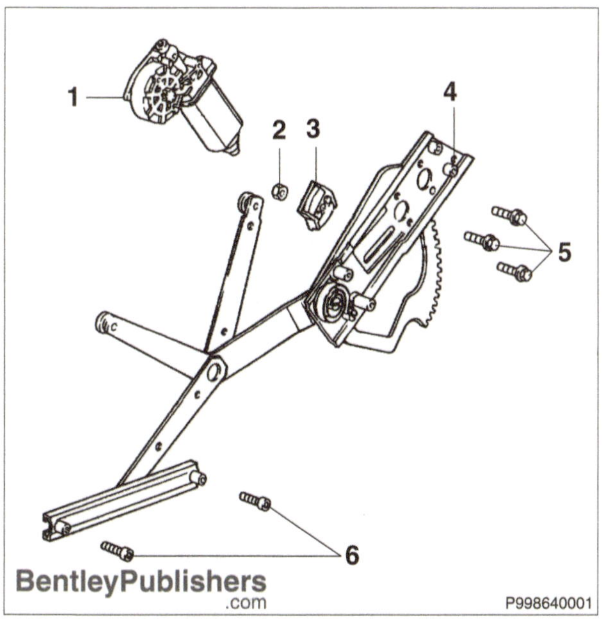

< Window regulator assembly

1. Window motor
2. Lock nut, M8 (replace with new)
3. Stop bracket
4. Window regulator assembly
5. Fastener, M8
6. Fastener, M6

Door window glass replacing

– **Coupe**: Remove door window frame. See **57 Doors and Locks**.

– Separate window regulator from glass. See **Door window regulator and motor, replacing and installing** in this repair group.

– Lower glass carefully into door.

◄ Using plastic prying tool, pry up outer weatherstrip clips (**arrows**) working from one end. Remove strip from door.

– Peel rubber inner weather strip from door.

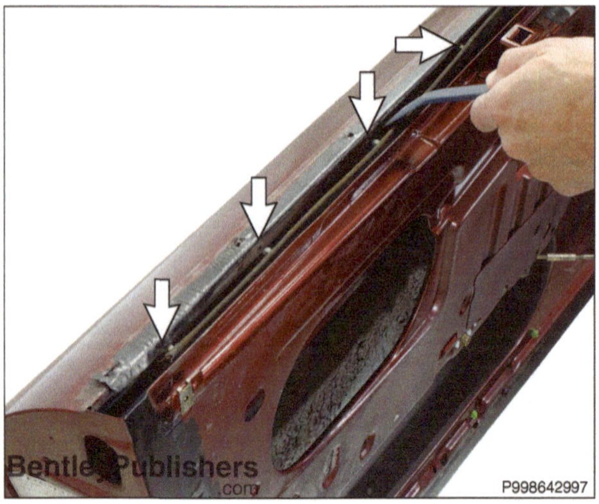

◄ Using plastic prying tool, lever felt-covered strip clips (**arrows**) away from door. Leave clip closest to front of door in place.

– Swing moulding up and out of way.

◄ **Cabriolet**: Pry off access cover (**arrow**) on upper door edge.

– Working through access opening, remove bolts holding window channel to door.

P998643004

◄ **Cabriolet**: Pull window channel (**arrow**) out of door.

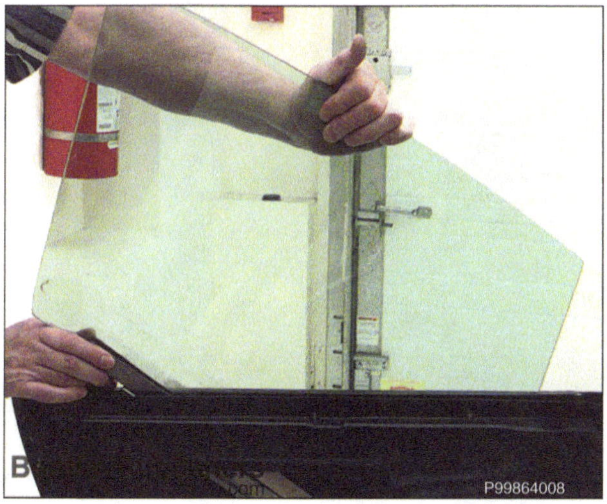
P99864008

◄ Carefully angle glass down in front and lift from door.

– Installation is reverse of removal.

> ***CAUTION—***
> • *If replacing a broken window, clean glass from window channels.*

Door window, adjusting

Window regulator adjustment should be done whenever the regulator has been removed from the door. The glass should contact the top of the window frame (or convertible top) squarely and should seat against the window seal uniformly. The window regulator has adjustments for in-out tilt and front-rear height parallelism.

– Remove door trim panel. See **57 Doors and Locks**.

◄ Loosen rear regulator guide rail bolts (**arrows**).

P998642700C

Door window, adjusting

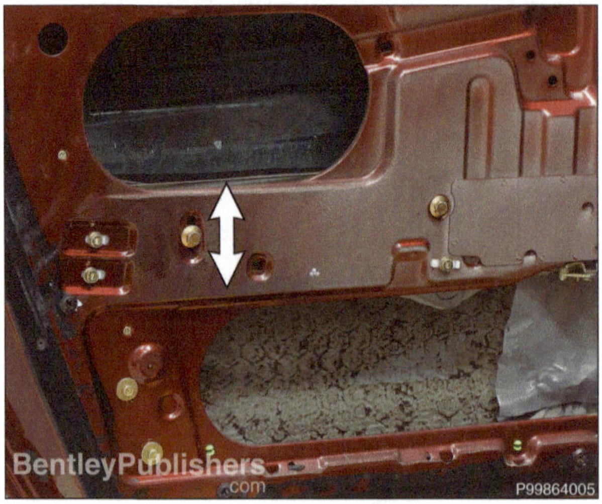

⋞ Move rear arm up or down in elongated slot (**arrow**) to adjust glass parallelism. Tighten bolt and check adjustment.

⋞ Loosen height (stop) adjusting bolt (**arrow**).

⋞ Reposition window stop left or right in elongated slot. Tighten bolt and check window height adjustment.

- Move stop toward rear to decrease amount of window travel.
- Move stop toward front to increase amount of window travel.

– Install door trim panel. See **57 Doors and Locks**.

66 Exterior Equipment

General

This repair group includes repair information for the outside rear view mirrors, emblems and body rocker covers.

Outside Rear View Mirrors

Mirror glass, motor, gaskets and outside plastic cover are all serviceable separately.

> ***CAUTION—***
> - *Mirror components should be at room temperature or above before removal to reduce the risk of damage.*

Mirror glass, replacing

> ***CAUTION—***
> - *Mirror glass replacement procedure can damage glass.*

– Adjust mirror so that bottom edge of glass is as far from housing as possible.

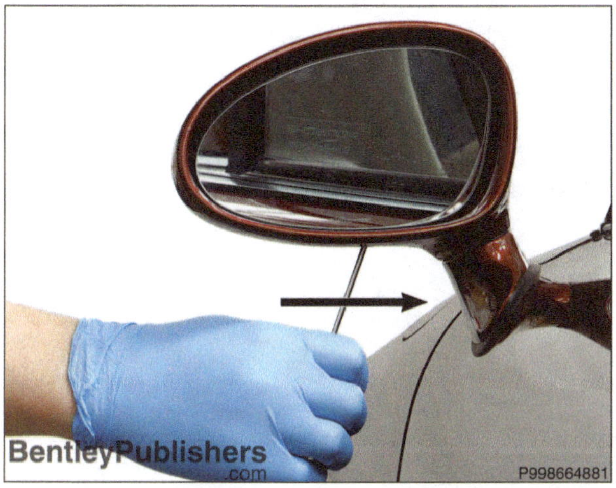

< Insert small flat blade screwdriver into hole in bottom of mirror housing to engage locking ring.

– Pivot locking ring in direction of **arrow** to release glass.

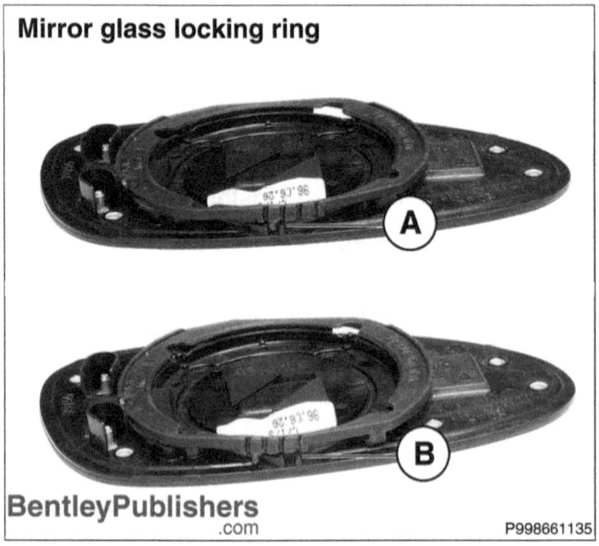

- Mirror glass locking ring.
  - Locked position (**A**)
  - Unlocked position (**B**)
- Using plastic prying tool, gently lever off mirror glass.
- Disconnect wires for mirror heater.
- Installation is reverse of removal, noting the following:
  - Match location of four mating pins to mounts in motor.
  - Turn lock-ring clockwise to lock into position.

Outside mirror components

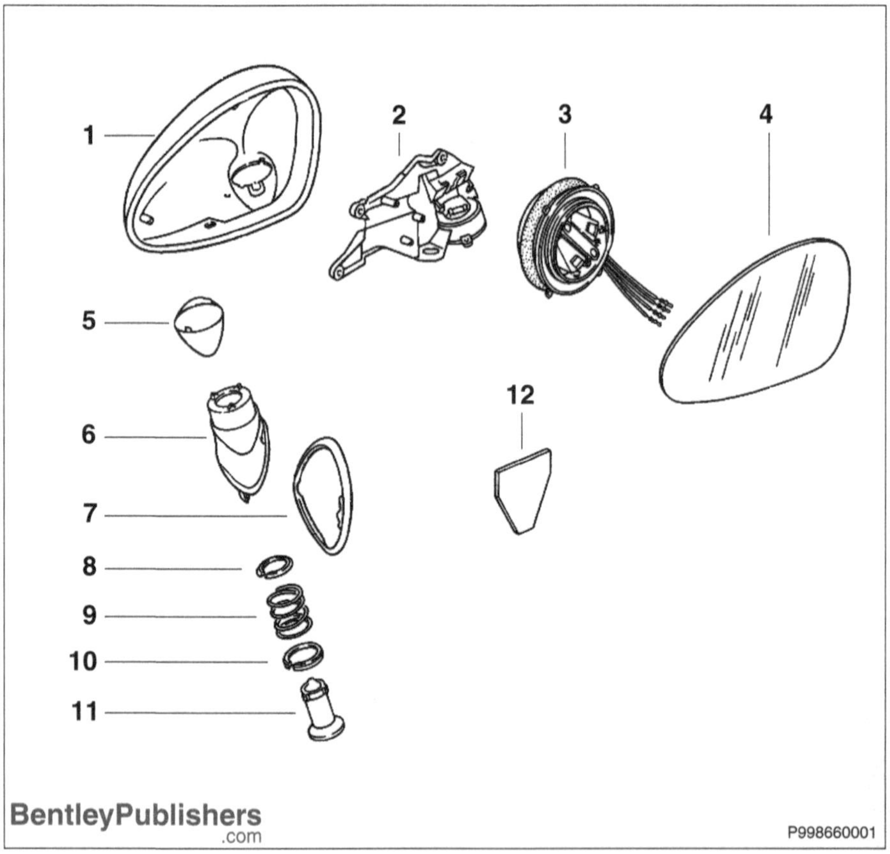

1. Mirror housing
2. Reinforcement plate
3. Mirror motor
4. Mirror glass
5. Gasket
6. Mirror base
7. Mirror pad
8. Upper centering washer
9. Compression spring
10. Lower centering washer
11. Bolt
12. Gasket

Outside mirror housing, removing and installing

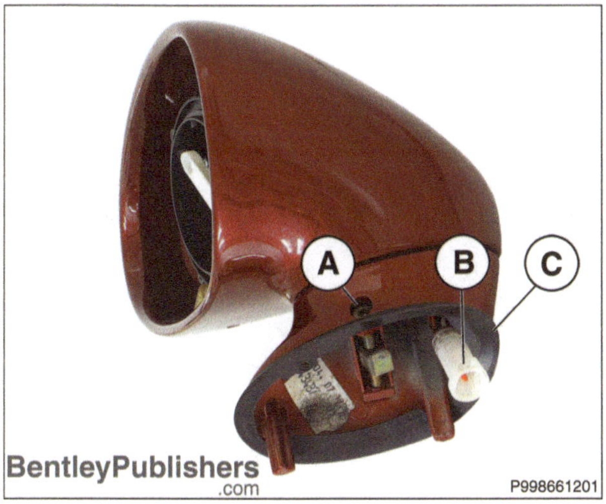

- Loosen 4 mm Allen bolt (**A**) on mirror base while supporting mirror assembly.
- Pull mirror away and disconnect electrical connector (**B**).
- Installation is reverse of removal, noting the following:
  - Ensure proper fit of rubber base gasket (**C**) to door.

Mirror heater, checking

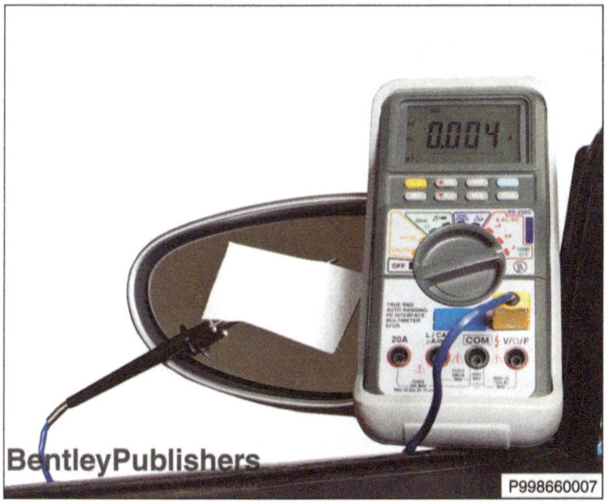

- Using a thermal sensor, measure temperature at center of mirror before turning ignition ON.
- Turn ignition ON to activate mirror heater circuit. Wait 1 minute.
- Temperature of glass should increase (warm to touch) if heater is functioning properly.

Emblems

Front emblem, removing and installing

- With trunk open, remove emblem fasteners (**arrows**) from behind.

Rear emblem, removing and installing

– Heat emblem and engine hood with heat gun until adhesive becomes soft and gummy.

> ***CAUTION—***
> *• Excessive heat can damage paint and clear coat.*

< Using dental floss or monofiliment fishing line, separate emblem from engine hood by sliding line under emblem using a sawing motion.

– Clean residual adhesive from emblem and engine hood using appropriate solvent.

– Emblem fastens to engine hood using double-sided trim tape.

Rocker Covers

Rockers covers are retaining to the body using multiple screws and plastic nuts. Some nuts are concealed. Additional trim may have to be removed to access fasteners.

Rocker cover components

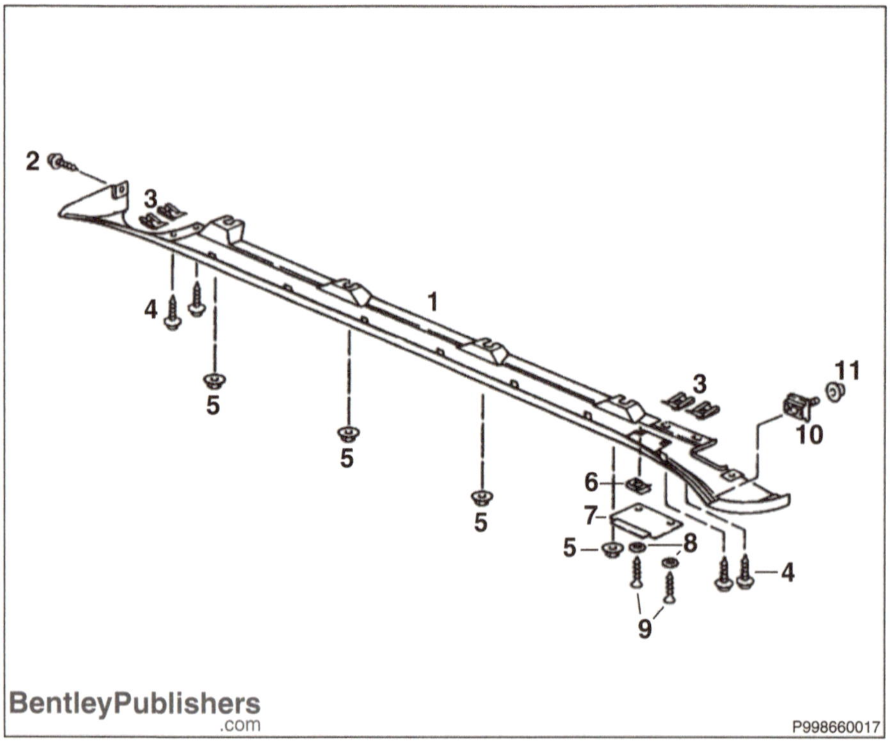

1. Rocker panel
2. Screw
3. Speed nut
4. Screw
5. Plastic nut
6. Speed nut
7. Cover
8. Washer
9. Self-tapping screw
10. Support
11. Plastic nut

69 Seat Belts, Airbags

General

This repair group covers seat belt and airbag component replacement.

Be sure to have airbag system diagnosis and component testing carried out by a Porsche workshop.

NOTE–

- *Special test equipment is required to retrieve airbag fault codes, diagnose system faults and reset airbag warning light. The warning light remains ON until problems are corrected and the fault memory cleared.*
- *A general bolt fastening torque table is given in* **03 Service and Maintenance**.

Warnings and Cautions

WARNING—

- *If the airbag warning light is ON, airbags will not deploy in the event of an accident. Be sure to have the airbag system inspected and repaired immediately.*
- *Airbags are deployed by explosive devices. Handled improperly or without adequate safeguards, these components are very dangerous. Serious injury may result if airbag service is attempted by persons unfamiliar with the Porsche airbag system and its approved service procedures.*
- *Disconnect battery and cover negative (–) battery terminal with insulating tape before starting diagnostic, troubleshooting or service work on airbag components.*
- *When handling a control module or other electronic component, use approved methods of dissipating electrostatic charge.*
- *After disconnecting battery, wait 1 minute before beginning work on airbag components.*
- *If an airbag is deployed due to an accident, Porsche specifies that certain airbag system components be replaced.*
- *Do not deploy an airbag prior to disposal. It should be deployed by a specialized disposal company. Contact a Porsche Center for the latest information on proper airbag disposal.*
- *When removing a deployed airbag unit, avoid contact with the skin; wear gloves. In case of skin contact, wash with water.*
- *Do not allow airbag system components to come in contact with cleaning solutions or grease.*
- *Do not subject airbag components to temperatures above 194°F (90°C).*
- *When reconnecting battery, make sure no one is inside the vehicle.*
- *Store an airbag unit with the padded side facing up. Do not leave an airbag unit unattended.*
- *If the airbag unit or airbag control module is dropped from a height of 0.5 meters (1.5 ft) or more, do not use component.*
- *Remove the airbag control module from the vehicle before performing any welding.*
- *Do not attempt to repair airbag components.*
- *Do not modify or repair seat belts. Do not modify seat belt mounting points.*
- *Do not bleach or dye seat belt webbing. Webbing that is severely faded or re-dyed does not meet strength requirements.*
- *Clean belts with luke-warm soap solution only.*
- *Periodically inspect seat belts for webbing defects such as cuts or pulled threads.*
- *Immediately after replacing a damaged or worn seat belt, destroy the old belt to prevent it from being reused.*

> ***CAUTION—***
> - *Prior to disconnecting battery, read battery disconnection cautions given in* **00 Warnings and Cautions**.
> - *To prevent marring trim when working on interior components, work with a plastic prying tool.*

Seat Belts

Front seat belt components

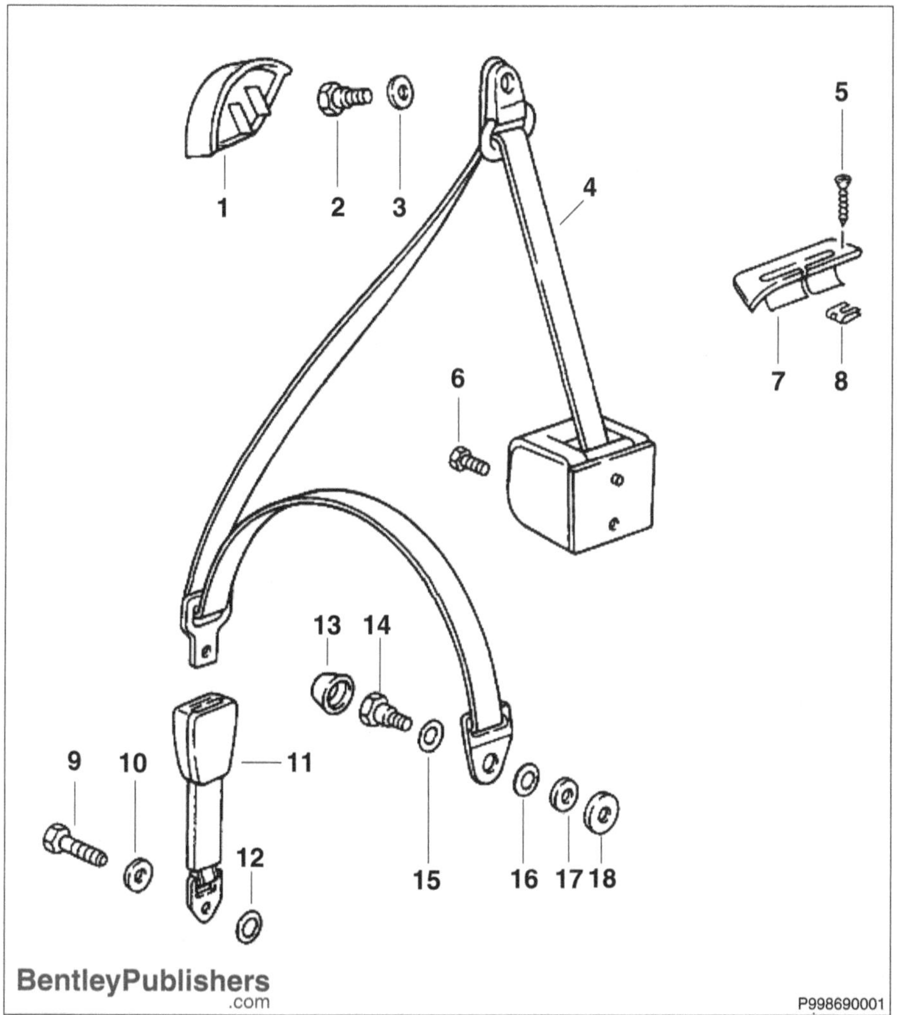

1. Cover
2. Screw
3. Washer
4. Front seat belt assembly
5. Self-tapping screw
6. Screw (Coupe / Targa)
7. Trim (Coupe / Targa)
8. Speed nut (Coupe / Targa)
9. Bolt
10. Washer
11. Seat belt buckle
12. Washer
13. Cap
14. Screw
15. Washer
16. Washer
17. Spacer sleeve
18. Spacer

Front seat belt, removing and installing (Cabriolet)

- Remove rear side panel fasteners (**arrows**).
- Pull back side panel and disconnect speaker electrical connector.
- Remove side panel.

- Remove lower seat belt fastener (**arrow**).

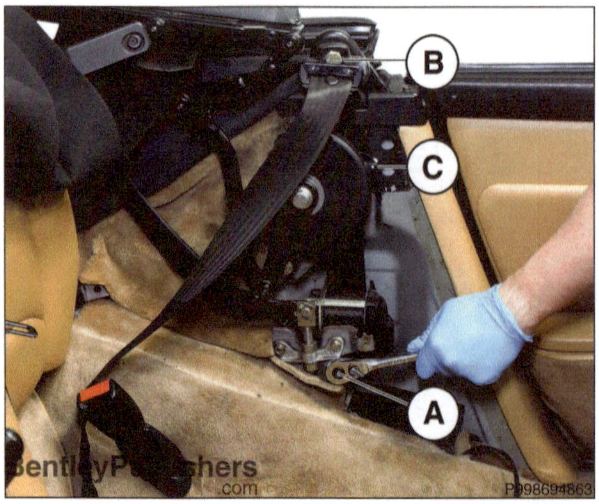

- Remove seat belt reel bolt (**A**).
- Remove cover over upper seat belt bolt (**B**), then remove bolt.
- Feed seat belt around bracket (**C**).
- Installation is reverse of removal.

Front seat belt, removing and installing (Coupe and Targa)

◄ Unclip seat belt trim cover (**A**).

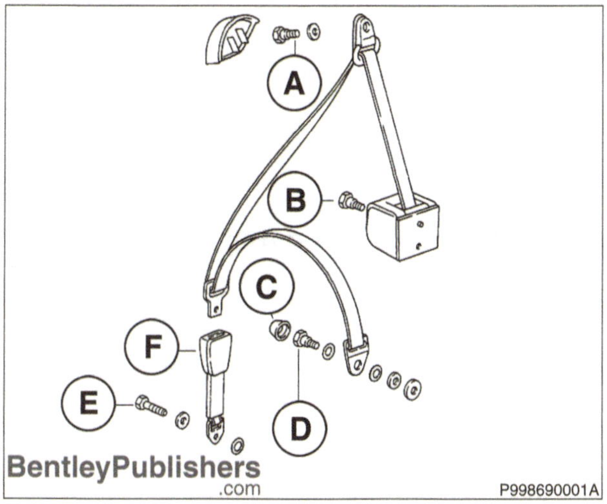

◄ Remove seat belt upper bolt (**A**).

- Remove cover (**C**) and seat belt bolt (**D**).
- Feed seat belt and upper mount through rear side panel.
- Remove rear side panel. See **70 Interior Trim**.
- Remove seat belt reel bolt (**B**).
- Remove seat belt assembly.
- Remove bolt (**E**) and seat belt receiver (**F**) if necessary.
- Installation is reverse of removal.

69

Rear seat belt components

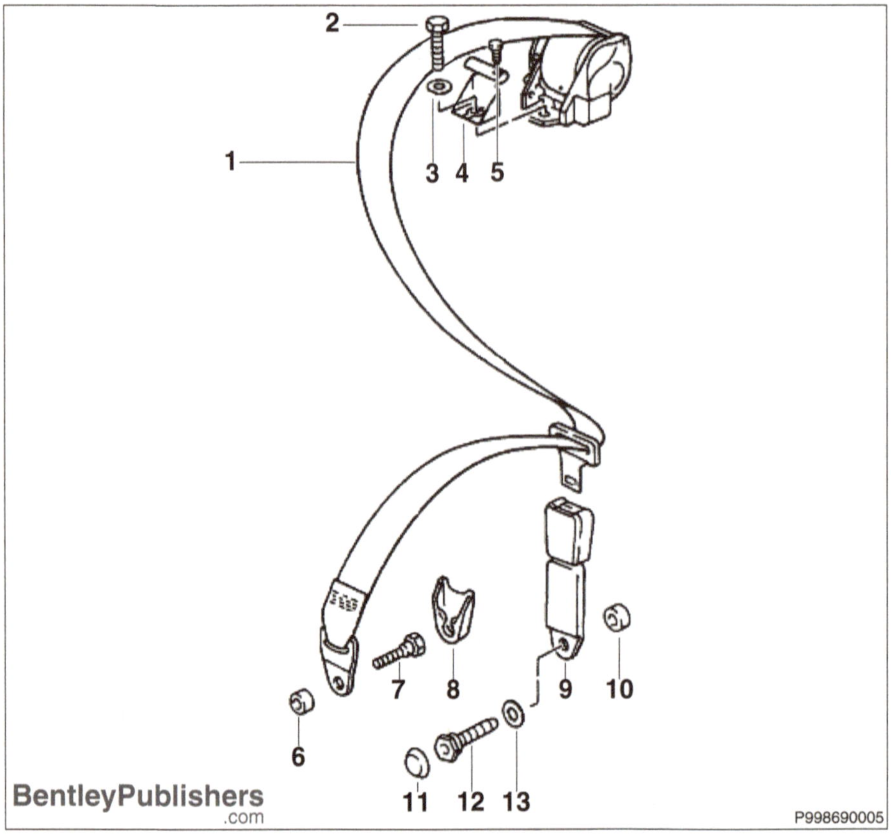

1. Seat belt
2. Screw
3. Washer
4. Bracket
5. Plug
6. Spacer sleeve
7. Screw
8. Cap
9. Seat belt buckle
10. Spacer sleeve
11. Cap
12. Screw
13. Washer

Rear seat belt, removing and installing (Cabriolet)

◀ Unclip lower seat belt trim cover (**arrow**).

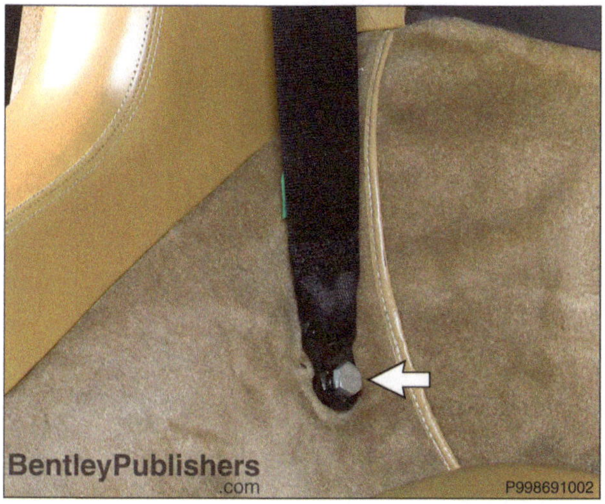

- ⮜ Remove bolt (**arrow**).
- – Remove seat backrest and seat belt frame cover. See **72 Seats**.

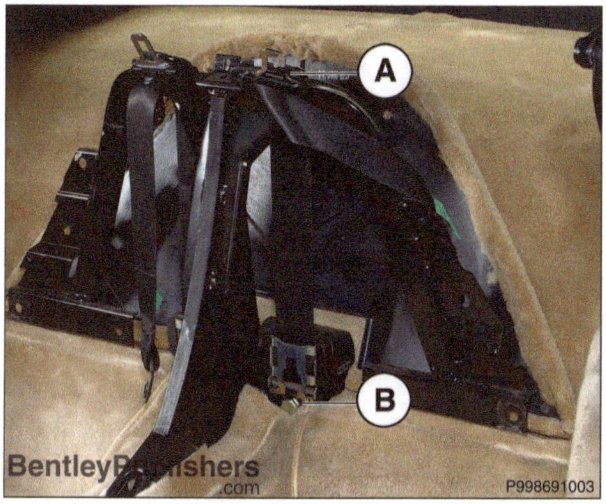

- ⮜ Feed seat belt through notch in upper guide (**A**).
- – Remove bolt (**B**), then remove seat belt.
- – Installation is reverse of removal.

Rear seat belt, removing and installing (Coupe and Targa)

- – Remove or fold down rear seat back. See **72 Seats**.

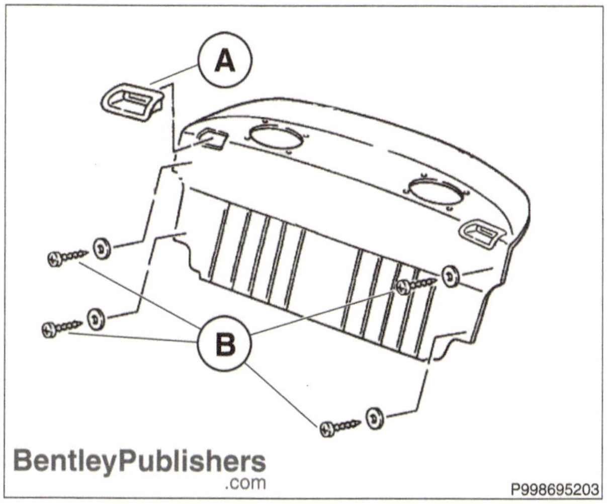

- ⮜ Remove rear deck panel fasteners (**B**).
- – Lever off seat belt trim (**A**) then slide down to seat belt buckle.

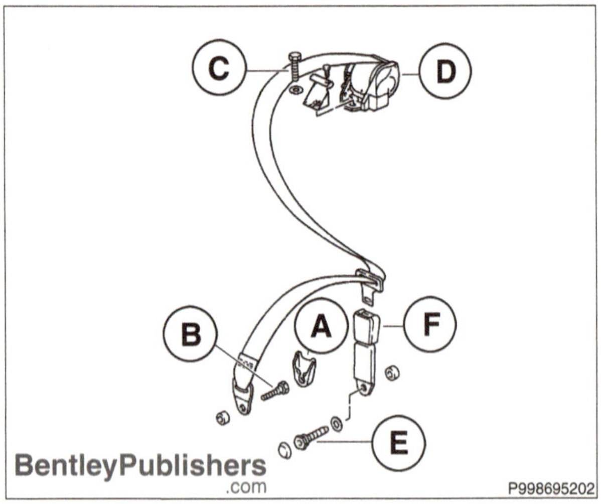

- Remove cover (**A**) then remove bolt (**B**).
- Feed seat belt through rear cover panel.
- Disconnect rear speakers and pull out rear cover panel.
- Remove bolt (**C**) then remove rear seat belt reel (**D**).
- Remove bolt (**E**) then remove seat belt receiver (**F**).
- Installation is reverse of removal.

AIRBAGS

Airbag system, general

993 models are equipped with driver and passenger airbags.

Faults codes stored in the airbag system cannot be erased until the relevant fault is remedied. Control unit replacement should be carried out if the system has been triggered three times.

Airbag component location

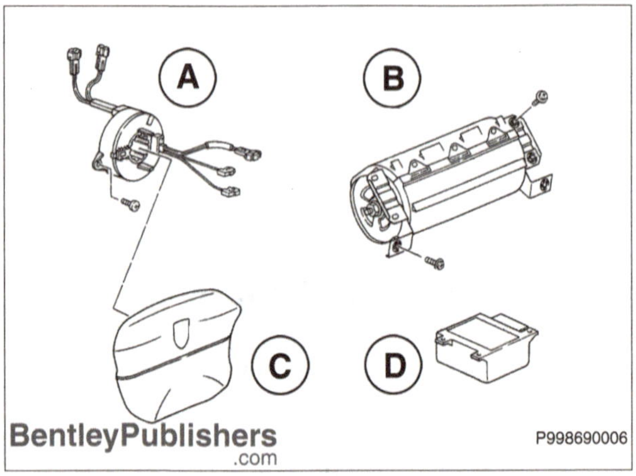

Airbag system components:
- Driver airbag contact spring (**A**) located behind steering wheel.
- Passenger airbag (**B**) located behind passenger side dashboard.
- Driver airbag (**C**) mounted to steering wheel.
- Airbag control unit (**D**) located under dashboard, towards center.

Airbag control unit

Airbag control unit controls/functions:
- Crash detection
- Ignition (deployment) of airbags
- System monitoring (fault detection and fault storage)
- Diagnosis
- Programming
- System readiness check (each time car is started)

Contact spring

The contact spring allows for a constant electrical connection to the horn and airbag contacts as the steering wheel turns. The contact spring is mounted behind the steering wheel.

When the steering wheel is removed, the contact spring automatically locks to prevent inadvertent repositioning. When removing steering wheel and contact spring, ensure that the wheels are pointing straight ahead.

Driver airbag

The driver airbag mounts directly to the face of the steering wheel. The driver airbag unit consists of a gas generator with retaining plate and a thermoplastic cushion. The airbag is made of a non-coated polyamide fabric.

During airbag deployment, the cover is torn apart and the airbag is fully expanded within 30 milliseconds

Passenger airbag

Passenger airbag, located behind passenger side dashboard, consists of a gas generator and a fabric airbag.

Airbag control unit, removing and installing

The airbag control unit is mounted in place with shear nuts. Special tools are required to remove the shear nuts.

– Disconnect negative (–) battery cable and cover battery terminal with insulating material.

> ***CAUTION—***
>
> - *Prior to disconnecting battery, read battery disconnection cautions in* **00 Warnings and Cautions**.

> ***WARNING—***
>
> - *After disconnecting battery, wait 1 minute before beginning work on airbag components.*
> - *Airbag control unit must not be moved while connector is plugged in.*

From under passenger dash, lift locking tab (**arrow**) and remove airbag electrical connector.

– Using a shear nut removal socket (Porsche special tool 9259), remove shear nuts on left and right side.

– Remove airbag module.

– Installation is reverse of removal, noting the following:
  - Ensure mounting location on body is clean.
  - Using new shear bolts, tighten until bolt head shears off.

Driver airbag, removing and installing

– Disconnect negative (–) battery cable and cover battery terminal with insulating material.

> ***CAUTION—***
> - *Prior to disconnecting battery, read battery disconnection cautions in* **00 Warnings and Cautions.**

> ***WARNING—***
> - *After disconnecting battery, wait 1 minute before beginning work on airbag components.*
> - *Airbag control unit must not be moved while connector is plugged in.*

< Using T30 Torx driver, remove airbag bolts (**arrows**).

< Fold airbag forward, then carefully lever off airbag connector (**arrow**). Remove airbag and store safely.

> ***WARNING—***
> - *Improper handling of airbag can cause serious injury. Store airbag with horn pad facing up. If stored facing down, accidental deployment can propel it violently into the air, causing injury.*

Installation is reverse of removal, noting the following:
- Replace self-locking airbag mounting fasteners.
- Take care to not pinch contact spring harnesses.
- Using a Porsche scan tool, check and reset fault memory.
- Observe torque tightening specifications.
- Check function of horn when completed.

Tightening torque	
Airbag to steering wheel, T30 Torx	10 Nm (7 ft-lb)

Contact spring, removing and installing

– Disconnect negative (–) battery cable and cover battery terminal with insulating material.

> ***CAUTION—***
>
> - *Prior to disconnecting battery, read battery disconnection cautions in* **00 Warnings and Cautions**.

> ***WARNING—***
>
> - *After disconnecting battery, wait 1 minute before beginning work on airbag components.*
> - *Airbag control unit must not be moved while connector is plugged in.*

– Ensure front wheels are facing straight ahead.

– Remove driver airbag. See **Driver airbag, removing and installing** in this repair group.

– Remove steering wheel. See **48 Steering**.

NOTE–

- Once steering wheel is removed, contact spring will lock to prevent unintentional turning. When steering wheel is refit, it will unlock automatically.

◄ Remove trim cover fasteners (**arrows**), then remove trim covers.

P998693978

◄ Remove contact spring fasteners (**arrows**).

– Slide contact spring from steering shaft.

NOTE–

- *To gain access to contact spring connector, climate control panel, radio and knee bar removal may be necessary.*
- *Removing climate control panel. See* **80 Heating**.
- *Removing radio. See* **91 Radio**.
- *Removing knee bar. See* **70 Interior Trim**.

69

- Disconnect contact spring electrical connector (**arrow**) and remove.
- Installation is reverse of removal. Note the following:
  - Contact spring will unlock once steering wheel is installed.
  - Remount airbag using new mounting screws. See **Driver airbag, removing and installing** in this repair group.
  - Align bosses on contact spring to detents in steering wheel back.

CAUTION—

- *Do not pinch airbag contact spring harnesses.*

- Using a Porsche scan tool, check and reset fault memory.

Passenger airbag, removing and installing

- Disconnect negative (–) battery cable and cover battery terminal with insulating material.

CAUTION—

- *Prior to disconnecting battery, read battery disconnection cautions in* **00 Warnings and Cautions**.

WARNING—

- *After disconnecting battery, wait 1 minute before beginning work on airbag components.*
- *Airbag control unit must not be moved while connector is plugged in.*

- Remove radio and mounting bracket. See **91 Radio**.
- Remove climate control panel. See **80 Heating**.
- Remove ashtray and knee bar with glove box. See **70 Interior Trim**.

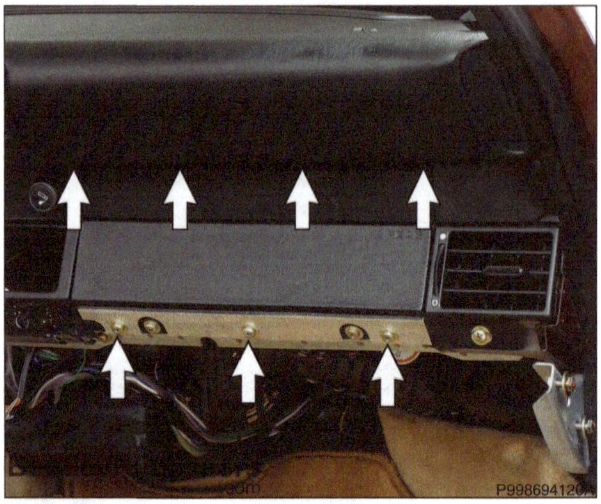

- Remove airbag flap fasteners (**arrows**).
- Pull down on flaps to remove.

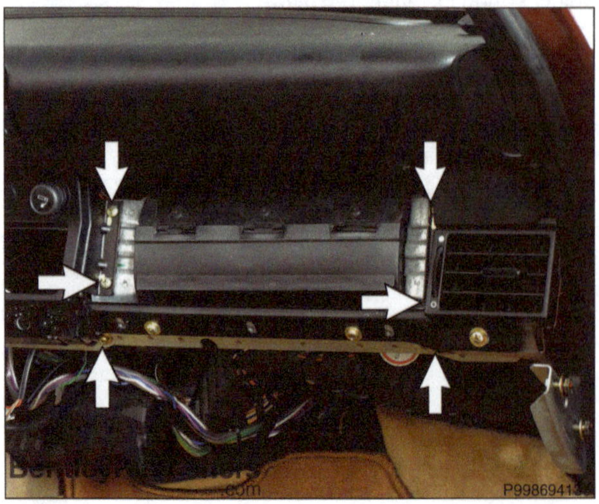

- Remove airbag mounting fasteners (**arrows**) then remove airbag.
- Disconnect orange airbag connector.
- Installation is reverse of removal, noting the following:
  - Replace airbag mounting fasteners.
  - Using a Porsche scan tool, check and reset fault memory.

70 Interior Trim

General

This repair group covers interior trim removal and installation. See the following repair groups for additional information:

- **57 Doors and Locks** for interior door panel removal.
- **72 Seats** for seat removal.
- **69 Seat Belts, Airbags** for airbag removal.
- **91 Radio** for radio removal.
- **48 Steering** for steering column cover service.
- **90 Instruments, Horns** for instrument removal.

Center Console

Center console, removing and installing

◄ Remove rubber insert (**arrow**).

Center console, removing and installing

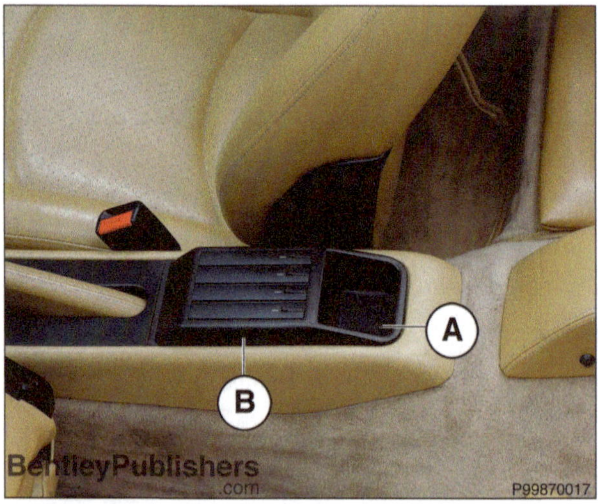

◄ Remove screw (**A**), then remove rear tray (**B**) from center console.

◄ Remove screw (**arrow**) underneath parking brake handle.

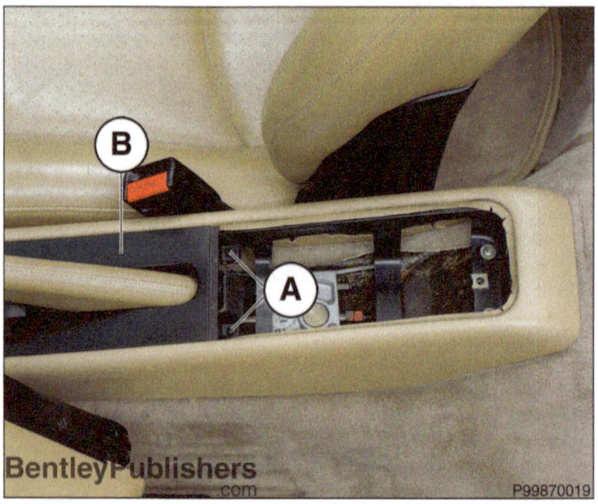

◄ Remove screws (**A**), then remove parking brake handle trim (**B**).

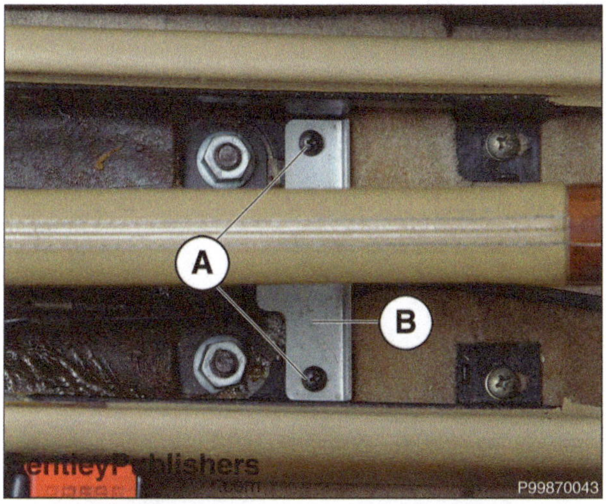

◀ Working underneath parking brake handle, remove screws (**A**), then remove reinforcement (**B**).

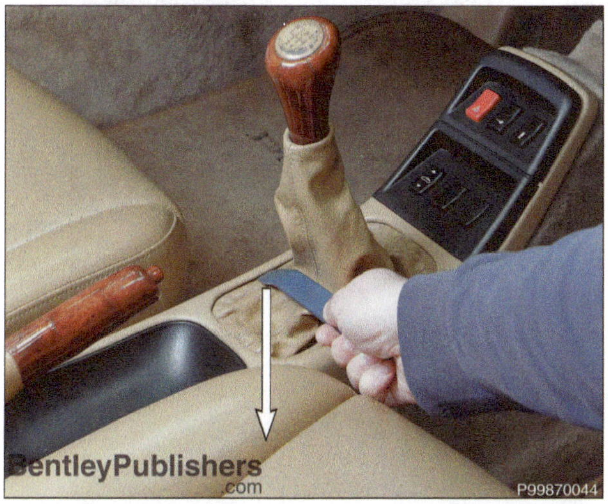

◀ Using a plastic prying tool, press down on left side of shift boot to disengage from center console.

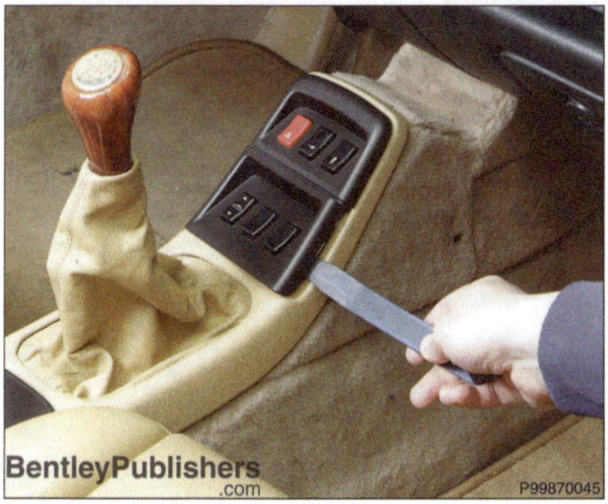

◀ Using a plastic prying tool, lever out lower the upper switch cover plates from center console.

Center console, removing and installing

◀ Disconnect electrical connectors (**arrows**) and remove switch covers.

NOTE–

- *Lower switch cover electrical connectors shown. Upper switch cover is similar.*

◀ Working at front of center console, remove screws (**arrows**).

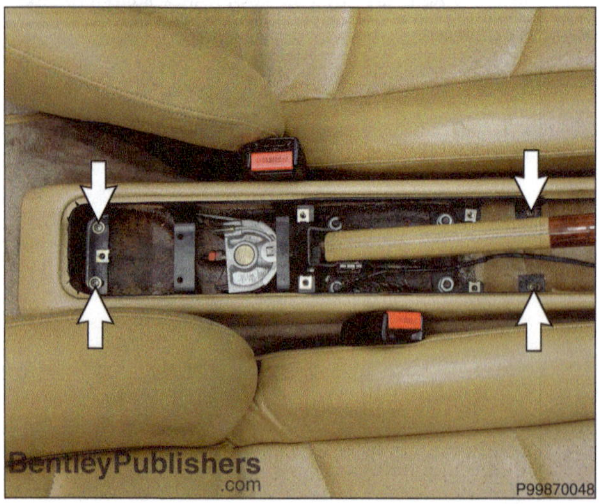

◀ Working at rear of center console, remove screws (**arrows**) .

- Lift center console straight up and remove from vehicle.
- Installation is reverse of removal.

Dashboard

Gauges and instruments are removed by prying them out of the dashboard cover. See **90 Instruments** for instrument removal.

A cracked or damaged dashboard cover can be replaced using the procedure given below.

Dashboard cover, removing and installing

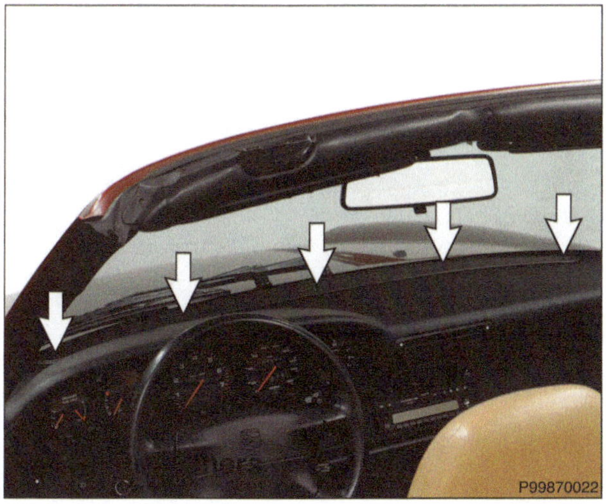

- Working at top of dashboard, remove defroster duct screws (**arrows**), then remove duct.

- Remove dashboard cover screws (**arrows**).

Dashboard cover, removing and installing

< Using a plastic prying tool, lever out and remove cover (**arrow**) on right side of dashboard.

– Repeat for left side.

< Working at right side of dashboard, remove screw (**arrow**).

– Repeat for left side.

< Using a small screwdriver, release locking tabs (**arrows**) inside vent and gently pull center vent from dashboard.

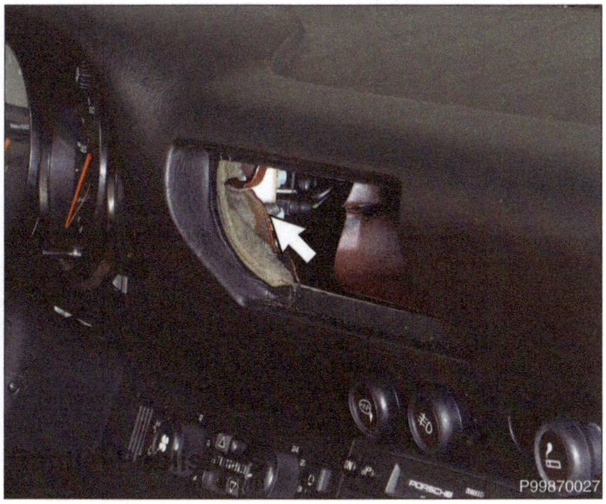
P99870027

◀ Working inside vent opening, remove nut and washer (**arrow**).

P99870028

◀ Carefully lift dashboard cover out in direction of **arrow**.

– Installation is reverse of removal.

Ash tray, removing and installing

P99870012

◀ Open ash tray and fold down spring-loaded bracket (**arrow**).

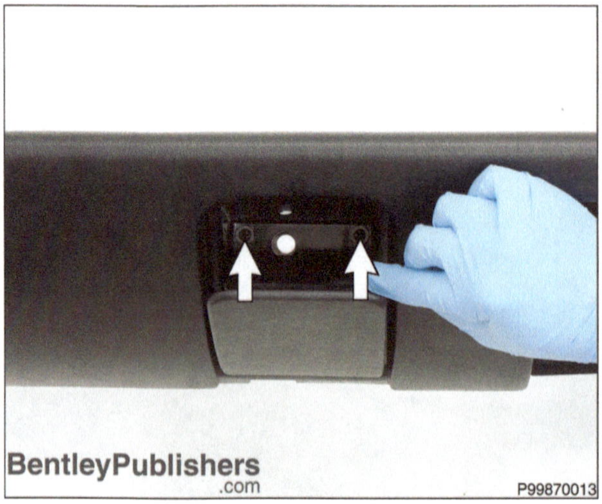

◄ Working behind spring-loaded bracket, remove screws (**arrows**), then remove ash tray from knee bar.

– Installation is reverse of removal.

Knee bar, removing and installing

– Disconnect negative (–) battery cable and cover with insulating material.

> ***CAUTION—***
>
> • *Prior to disconnecting the battery, read the battery disconnection notes given in* **03 Service and Maintenance**.

– Remove radio. See **91 Radio**.

– Remove climate control panel. See **80 Heating**.

Vehicles model year 1995

◄ Remove outer A/C vents from dashboard.

- Remove screw (**A**) from trim panel on side of dash.
- While pulling gently on A/C vent, release locking tabs (**arrows**) at top and bottom of vent with a small screwdriver.

NOTE–

• *Left side shown. Right side is similar.*

– Working inside vent opening, remove nuts and washers securing knee bar to dashboard.

Continued for all vehicles

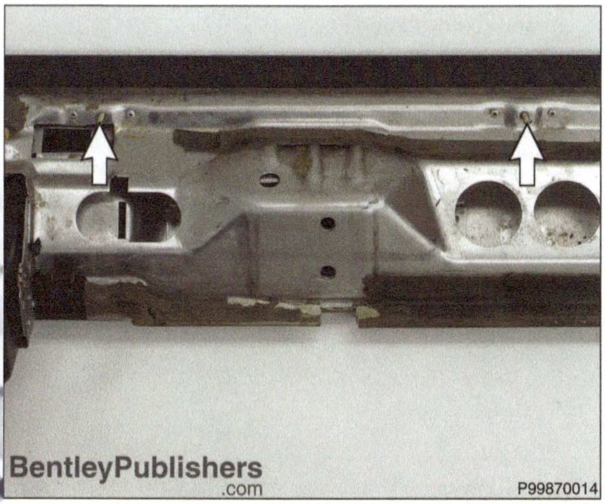

⋖ Continued for all vehicles: Working through radio and A/C control unit openings, remove nuts and washers (**arrows**) securing knee bar to dashboard .

NOTE–

- *Knee bar shown removed for clarity.*

– Remove ash tray. See **Ash tray, removing and installing** in this repair group.

⋖ Working in ash tray recess, remove bolts (**arrows**).

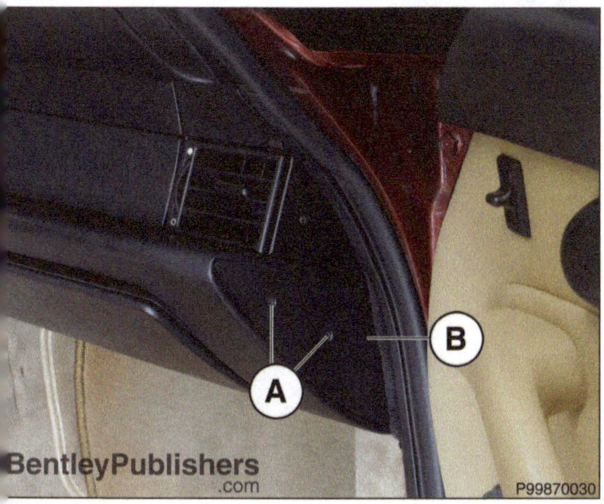

⋖ Working at right side of dashboard, remove screws (**A**), then remove cover (**B**).

– Repeat for left side.

- ◄ Working at right side of dashboard, remove bolts (**arrows**).
- Repeat for left side.
- Carefully lift out and remove knee bar.
- Installation is reverse of removal.

GLOVE COMPARTMENT

Glove compartment, removing and installing

- ◄ Working under passenger side of dashboard, remove screws (**arrows**) and washers.
- Open glove compartment.

- ◄ Carefully work glove compartment door hinges (**arrows**) out of dashboard and remove glove compartment door.

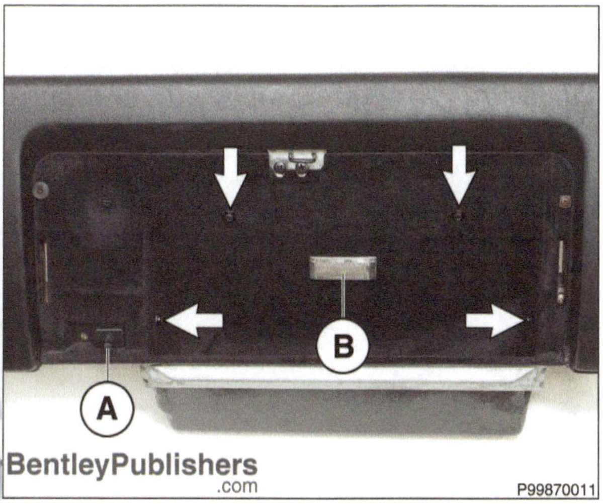

- Unclip glove compartment contact switch (**A**). Disconnect electrical connectors and remove switch.
- Unclip glove compartment light (**B**). Disconnect electrical connectors and remove light.
- Remove fasteners (**arrows**) and remove glove compartment from knee bar.
- Installation is reverse of removal.

Interior Panels

Rear side panel, removing and installing

Coupe

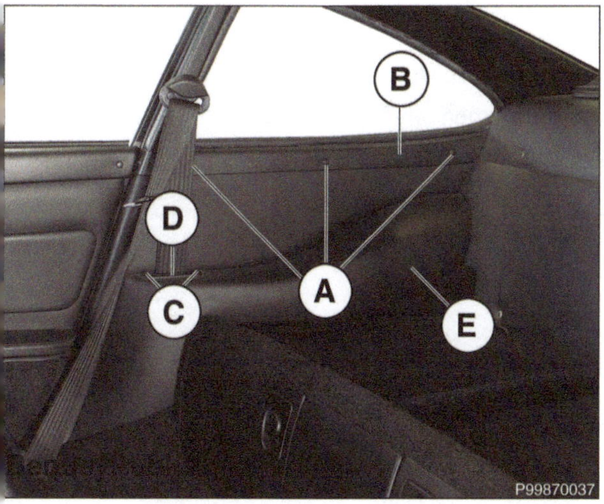

- Remove screws and washers (**A**) and remove upper trim piece (**B**).
- Remove screws (**C**) and pull seat belt guide (**D**) out of side panel.
- Disconnect seat belt from B-pillar. See **69 Seat Belts, Airbags**.
- Feed seat belt through side panel.
- Pull side panel (**E**) up and out of vehicle.
- Installation is reverse of removal.

Cabriolet

- Remove screws (**arrows**).
- Lift side panel up and out.
- Disconnect electrical connectors from speaker and remove rear side panel.
- Installation is reverse of removal.

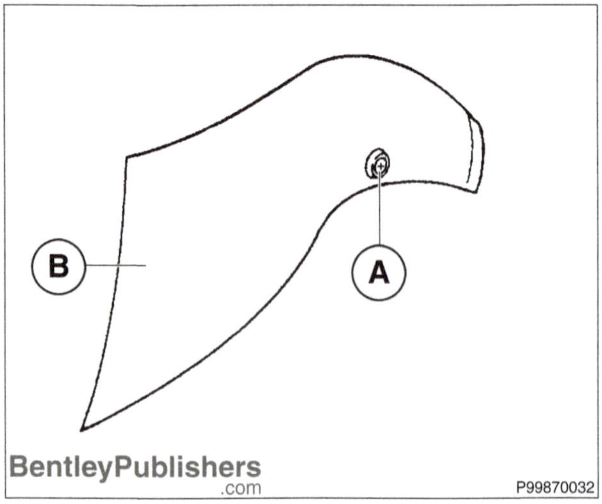

Targa

- Remove screw and washer (**A**). Lift out and remove side panel (**B**).
- Installation is reverse of removal.

Sun Visors

Sun visor, removing and installing

Coupe

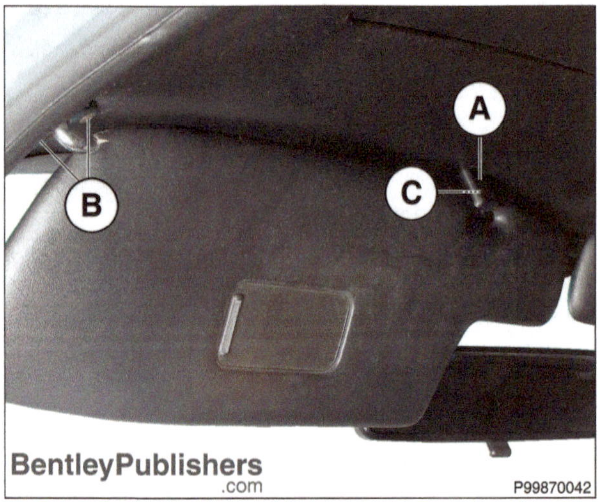

- Snap sun visor out of clip (**A**) and turn 90° to side.
- Remove screws (**B**), then remove sun visor.
- Remove screw (**C**) and clip (**A**).
- Installation is reverse of removal.

Cabriolet and Targa

- Remove screws (**arrows**), then remove sun visor.
- Installation is reverse of removal.

72 Seats

GENERAL

This repair group covers vehicle seats.

For information on seat belts, see **69 Seat Belts, Airbags**.

Front seats are available in comfort, sport and basic versions (electrically operated) and feature:

- Integral head rests
- Back rest release mechanism on both sides of seat
- Belt lock on seat frame
- Twin seat hinges
- Optional adjustable seat heater
- Optional lumbar support

Coupe and Targa models feature folding rear seats which double as a parcel shelf. The backs of the rear seats are carpet-covered, and a plastic molding is fitted to the front edge to prevent luggage from moving forward.

72

The Cabriolet rear seat is fixed in position and features center-mounted seat belt reels.

Seats

Front seat, removing and installing

NOTE–

- *If the power seat cannot be moved forward or backward with the power seat motors, see* **Power seat, moving manually** *later in this repair group to manually position the seat to access seat anchor bolts.*

Move seat to forward position. Remove mounting fasteners (**arrows**) at rear seat rails.

Move seat to rear position. Remove mounting fasteners (**arrow**) at front seat rails.

Vehicle with power or heated seat

– Disconnect negative (–) cable from battery and cover with insulating material.

> ***CAUTION—***
>
> - *Prior to disconnecting the battery, read the battery disconnection notes given in* **03 Service and Maintenance**.

– Tilt seat back.

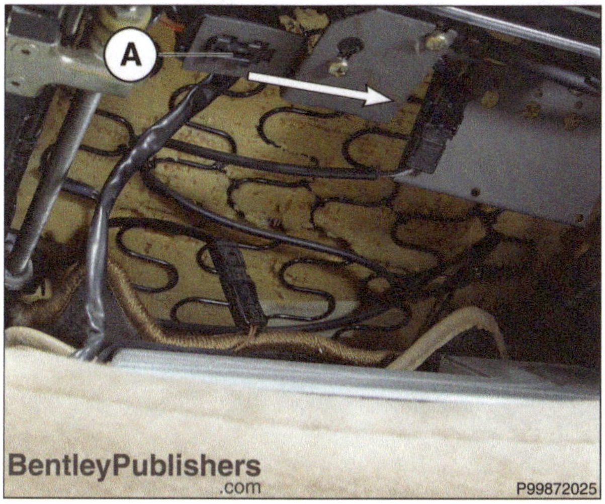

◄ Slide connector (**A**) in direction of **arrow** to disconnect from bracket.

◄ Disconnect electrical connectors (**arrows**).

NOTE–

- *Cut wire ties as necessary.*

Continued for all vehicles

– Remove seat from car.

NOTE–

- *Use a blanket to protect door sill from scuffing during seat removal.*

– Installation is reverse of removal, noting the following:

- Use wire ties or equivalent means to keep seat harness wiring from damage.
- Install all seat mounting bolts finger-tight before torquing.

Tightening torque	
Seat rail to body	23 Nm (17 ft-lb)

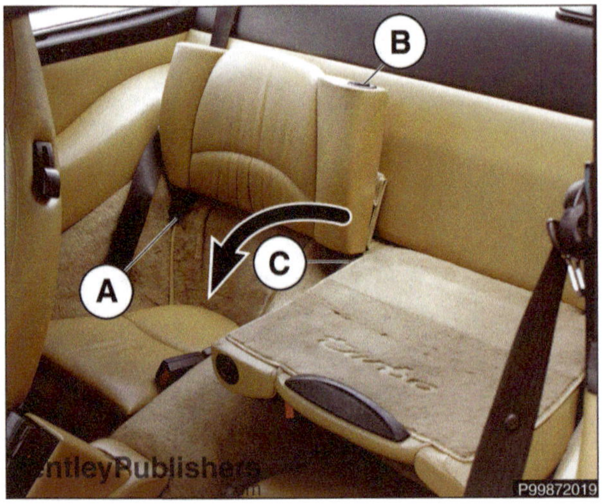

Rear seat backrests, removing and installing (Coupe and Targa)

NOTE–

- *Rear seat bottoms are glued in place.*

◄ Remove seat back fastener (**A**).

– Push button (**B**) and fold down seat back.

– Pull seat back out of mounting bracket (**C**) in direction of **arrow** to remove.

– Installation is reverse of removal.

Rear seat backrests, removing and installing (Cabriolet)

NOTE–

- *Rear seat bottoms are glued in place.*

– Close convertible top.

◄ Working behind rear seat, remove screws (arrows). Pull out carpeted trim panel to gain access to seat backrest nuts.

◄ Working behind rear seat backrests, remove nuts (**arrows**), then pull rear seat backrests out from front.

– Installation is reverse of removal.

Seat belt frame cover, removing and installing (Cabriolet)

– Remove rear seat backrests. See **Rear seat backrests, removing and installing** in this repair group.

◄ Pry off and remove cover (**arrow**).

NOTE–

- *Left side shown. Right side is similar.*

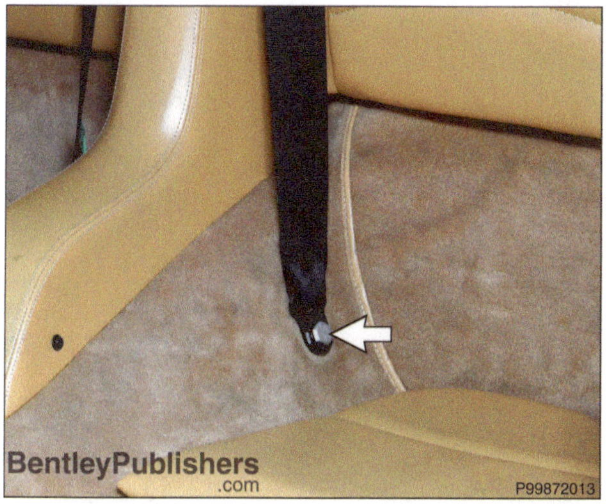

◄ Remove seat belt anchor bolt (**arrow**).

NOTE–

- *Left side shown. Right side is similar.*

◄ Using a plastic prying tool, lever off and remove screw cover (**arrow**).

NOTE–

- *Left side shown. Right side is similar.*

Seat belt frame cover, removing and installing (Cabriolet)

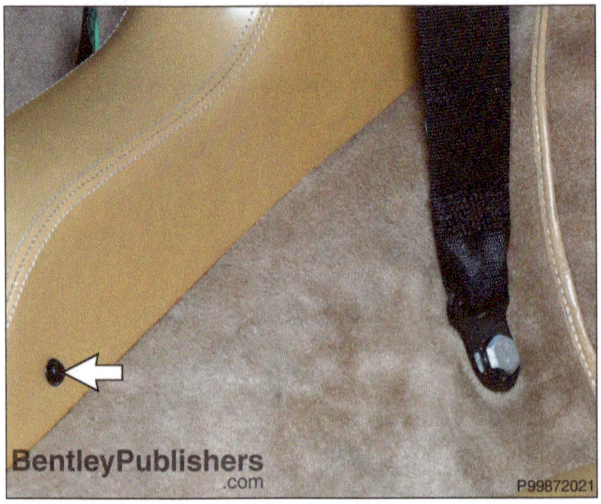

◀ Remove screw (**arrow**).

NOTE–

- *Left side shown. Right side is similar.*

◀ Remove screws (**arrows**).

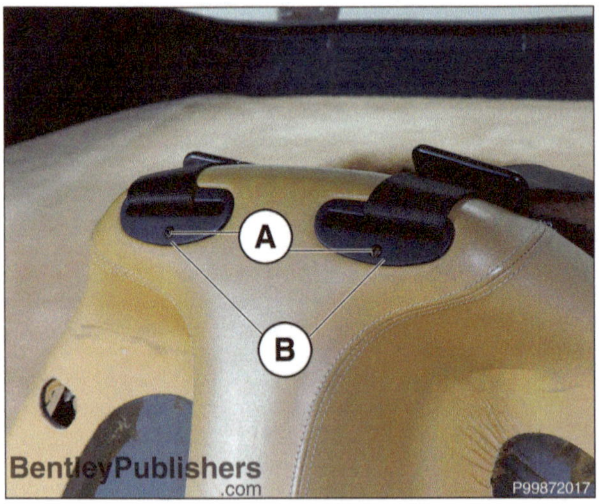

◀ Remove screws (**A**).

– Using a plastic prying tool, gently lever out seat belt guides (**B**) from seat belt frame cover.

– Feed seat belt guides and seat belts through seat belt frame cover.

< Working behind rear seat, remove nut (**arrow**).

NOTE–

- *Seat belt frame cover shown removed for clarity.*

– Lift out and remove belt frame cover from vehicle.

– Installation is reverse of removal.

Power seat, moving manually

The following procedure can be used to move the seat forward and aft to access the seat anchor bolts if the power seat is not working. A power drill and a seat drive cable (can use a damaged cable if one end of the cable is undamaged) are needed to carry out this procedure.

NOTE–

- *If power seat does not move and seat motors cannot be heard running when operating seat switch, check electrical connector under seat and fuse in central electric panel first. See* **97 Fuses, Relays, Component Locations**.
- *If seat does not move and motors can be heard running when operating seat switch, check drive cables and drive transmissions.*

CAUTION—

- *If power seat jams in place, do not operate seat in an attempt to free it up, as this may result in further damage to the power seat drives and cables*

< Working at front of seat, remove Torx fasteners (**arrows**) and remove trim piece.

72

< Carefully pry up and remove wire clip (arrow) at seat fore/aft drive motor.

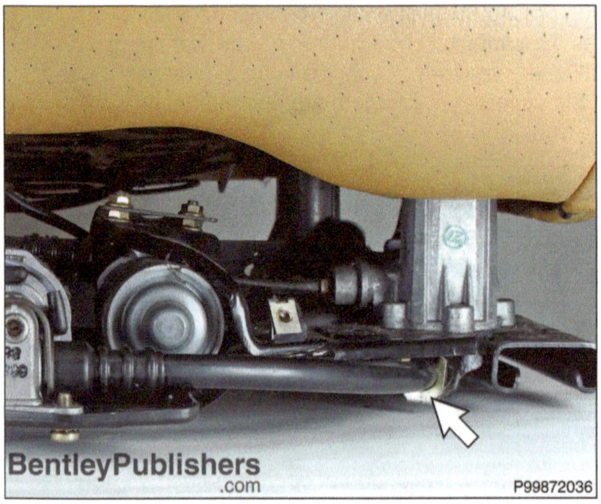

< Carefully lever off and remove clip (**arrow**) securing cable housing to seat frame.

– Pull drive cable with housing out of motor and seat rail adjuster.

– Repeat procedure for remaining fore/aft drive cable and cable housing.

– Inspect cable ends for damage. Replace frayed or unraveled cables.

< Insert end of drive cable into power drill and tighten chuck firmly onto drive cable.

NOTE–

- *If using a damaged cable, insert damaged (frayed) end of cable into driil, with undamaged end of cable facing out.*

P99872037

Insert undamaged square end of drive cable into fore/aft seat rail adjuster.

– Use power drill to move seat a small amount.

NOTE–

- *Spin seat rail adjuster clockwise to move seat forward and counterclockwise to move seat back.*
- *Move seat in small increments. Moving seat too much on one side can result in the seat jamming in the seat rail.*

– Move cable with drill to opposite fore/aft seat rail adjuster. Using drill, move seat a small amount in same direction.

P99872005

Repeat process on left and right sides until seat anchor bolts can be accessed.

– If necessary, repeat procedure to move seat to opposite end of tracks to access remaining bolts.

Power seat drive cable, replacing

This procedure describes replacing one fore/aft drive cable. Replacement of other seat drive cables is similar.

Although the fore/aft drive cables can be replaced with the seat installed, it is recommended that the seat be removed or unbolted and tilted rearward to gain access to the under side of the seat.

– Remove seat. See **Front Seat, removing and installing** in this repair group.

– Place seat upside down on a flat work surface.

P99872033

Working at front of seat, remove Torx fasteners (**arrows**) and remove trim piece.

Power seat drive cable, replacing

< Lever up and remove wire clip (**arrow**) at seat fore/aft drive motor.

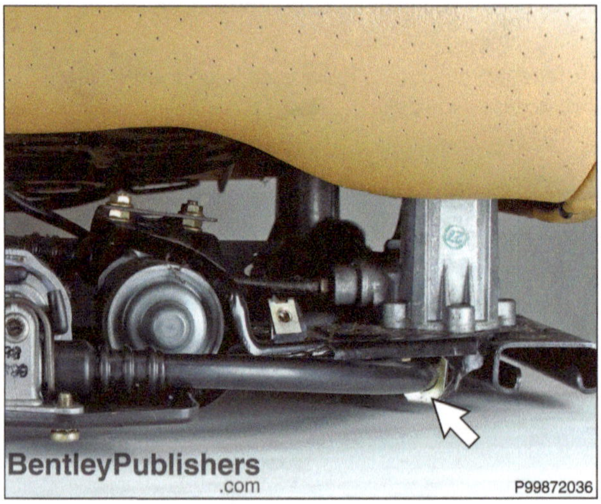

< Carefully lever off and remove clip (**arrow**) securing cable housing to seat frame.

– Pull drive cable with housing out of motor and fore/aft seat rail adjuster.

NOTE–

- *Inspect cable ends for damage. Replace cable if end is frayed.*

< Check that seat rails are evenly aligned on left and right sides. Measure distance **A** on left and right sides. If necessary, adjust rail position by turning threaded adjuster (**arrow**) by hand.

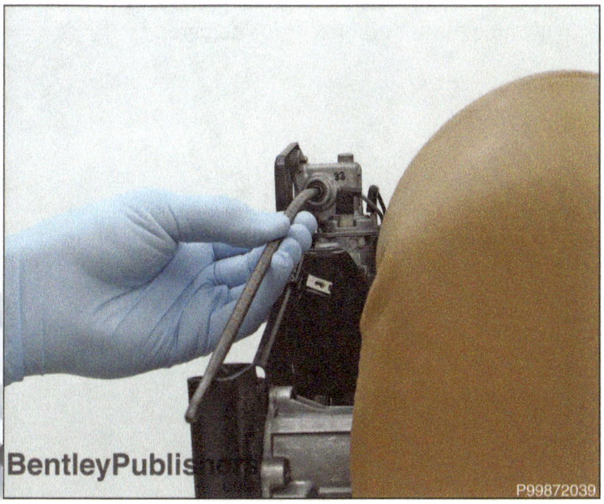

◄ Fit new seat fore/aft drive cable fully into drive motor.

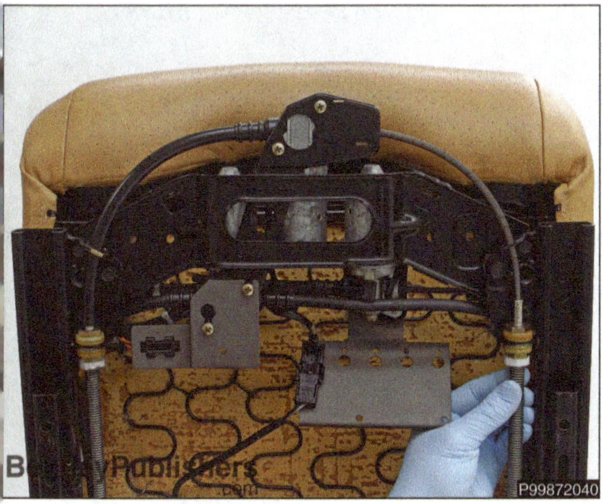

◄ Route cable through seat frame opening and into seat rail adjuster. Turn seat rail adjuster slightly by hand until square end of cable slides smoothly into square drive end of seat rail adjuster.

– Remove cable from rail adjuster and motor.

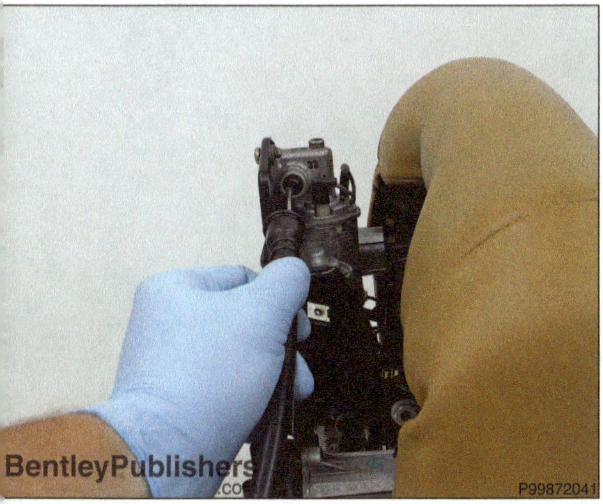

◄ Insert cable into cable housing and position housing with cable through seat frame and into position. Slide drive cable fully into drive motor.

Power seat drive cable, replacing

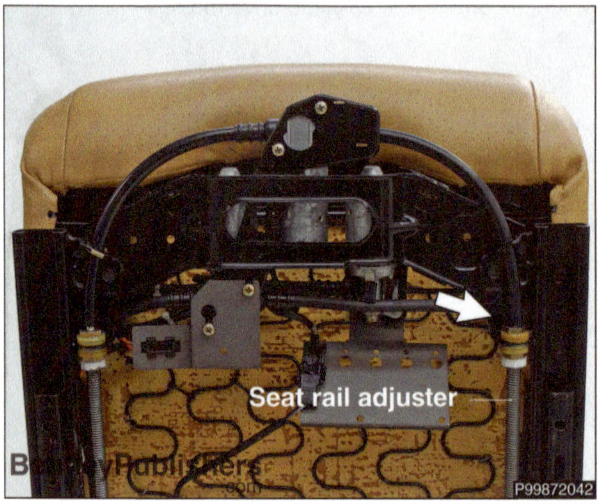

◄ Position housing onto seat rail adjuster (**arrow**), making sure drive cable slides fully into square end of seat rail adjuster.

◄ Install clip securing cable housing to seat frame (**arrow**).

◄ Install wire clip to seat motor, making sure clip secures cable housings to motor.

– Remainder of installation is reverse of removal.

80 Heating

General

This group covers heating system component replacement. Refer to the following groups for further information:

- **26 Exhaust System** for heat exchanger information.
- **87 Air-conditioning**
- **EWD Electrical wiring diagrams**

Air for the heating and ventilation system is brought in through the rear deck lid by the engine cooling fan (**5**). This air is then diverted via a rear electric blower fan (**4**) and sent through the left and right exhaust heat exchangers (**9**) where it is heated. Once heated, the air is pushed through ducts (**3**) by the rear blower motor and pulled by the left (**1**) and right (**2**) front blower fans behind the dashboard. Electrically-controlled flaps (**6**) open and close to allow heated air to pass through as needed.

There is a temperature sensor located in the climate control panel (**7**) that monitors the temperature in the cabin. The system uses fresh air mixed with heated and A/C air to maintain the proper temperature. If heated air is not required, the heater flaps close and heater air is vented to the atmosphere through air ducts (**8**) located under the engine.

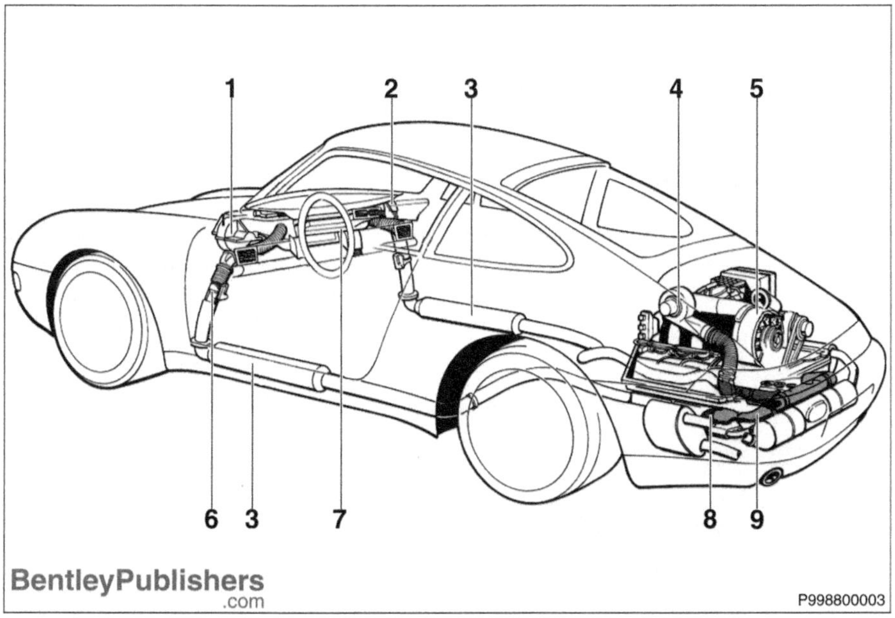

1. Blower motor, left
2. Blower motor, right
3. Heating duct
4. Rear blower motor
5. Engine cooling fan
6. Temperature mixing flap
7. Climate control panel
8. Atmospheric vent
9. Heat exchangers

Heating System Components

Heat exchangers

The heat exchangers are integral with the exhaust manifolds. For heat exchanger removal and installation, see **26 Exhaust System**.

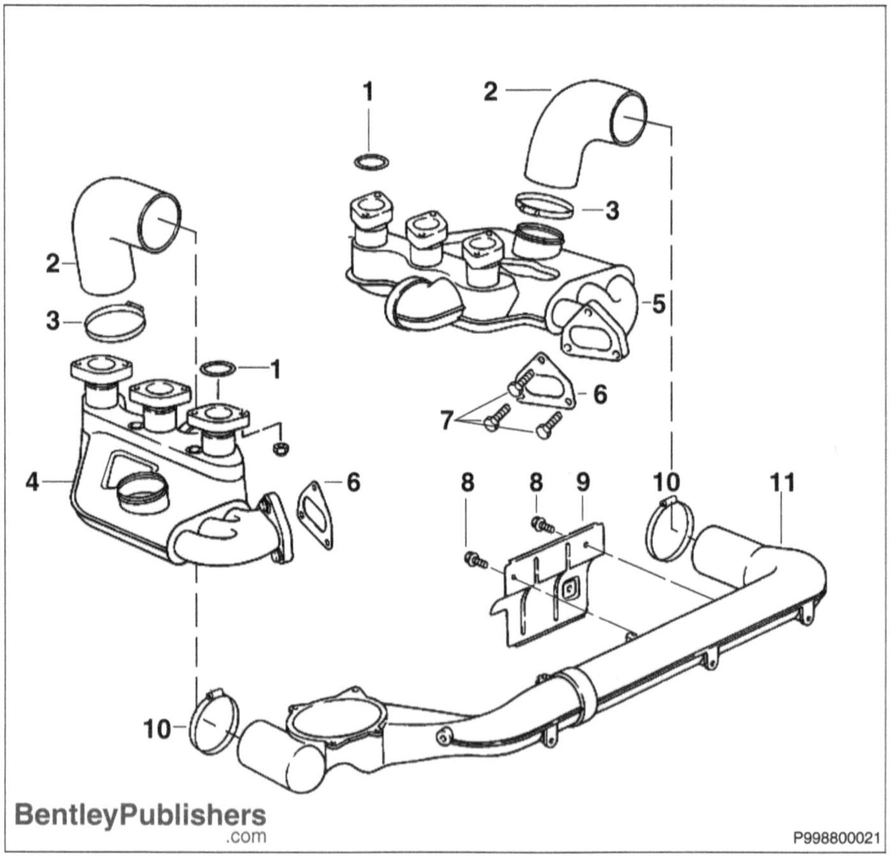

1. Exhaust manifold gasket
2. Heater hose
3. Hose clamp
4. Heat exchanger, left
5. Heat exchanger, right
6. Gasket
7. Bolt, M8
8. Screw, M6
9. Heat shield
10. Hose clamp
11. Heating tube

Climate control panel, removing and installing

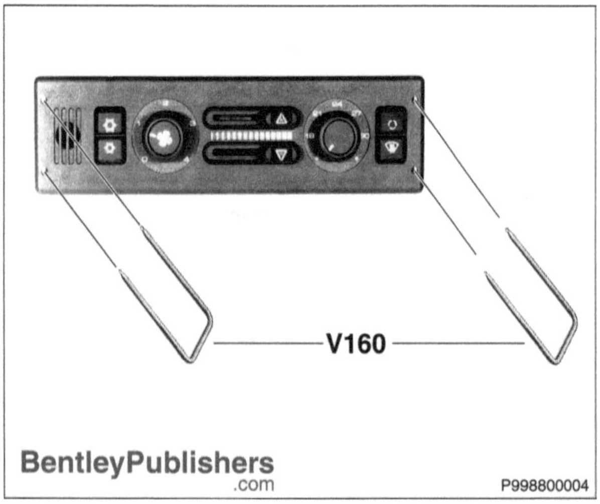

- ◄ Insert Porsche special tool **V160** into holes of control panel face to unlock. Lever out climate control panel.
- – Unplug electrical connectors from back of panel.
- – Installation is reverse of removal.

NOTE–

- *For information on climate control panel function, see* **87 Air-conditioning**.
- *To remove cage that holds control panel in place, pry locking tabs back on all sides and slide cage out of dashboard.*

Engine compartment blower motor ballast resistor

The rear blower motor ballast resistor (**arrow**) is clipped into the rear blower motor pipe in the left side of the engine bay. This resistor controls low speed (stage 1) of the rear blower motor.

NOTE–

- *If the blower motor runs on high speed (stage 2) only, the blower motor resistor is not functioning correctly.*

– Remove ballast resistor by unclipping from plastic pipe and disconnecting wiring terminals.

Inspect ballast resistor. Check that reed contact is firmly contacting terminal beneath it. Inspect contacts for burning or corrosion. Replace ballast resistor if any faults are found.

NOTE–

- *If the bimetallic thermoswitch feature of the resistor has been triggered, the contacts will not be touching each other. A concaved bimetallic plate is designed to electrically open the contacts if high current is present in the circuit.*

Reset bimetallic thermoswitch by inserting a thin object (i.e. needle) through edge of bimetallic thermoswitch and onto bimetallic plate. Contacts will be touching when bimetallic plate is reset

Engine compartment blower motor, removing and installing

◀ Remove fasteners (**arrows**).

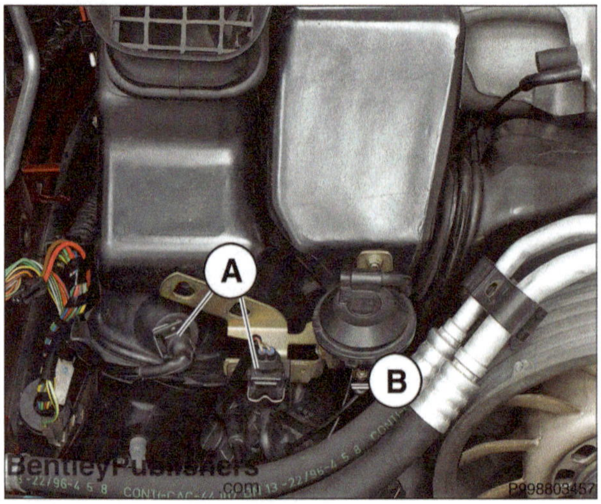

◀ Disconnect electrical connectors (**A**) and vacuum hose (**B**) at housing bracket.

◀ Lift off upper air duct (**arrow**).

Engine compartment blower motor, removing and installing

< Remove heating duct elbow hose clamp (**arrow**).

– Remove rubber duct.

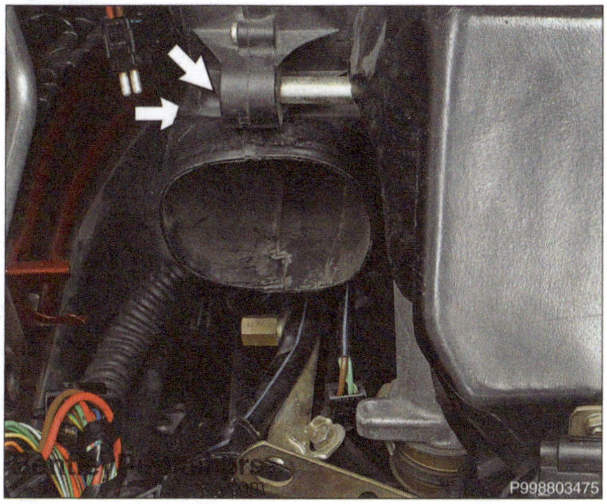

< Remove blower unit bolts (**arrows**).

– Disconnect blower fan motor electrical connector.

< Blower fan assembly shown removed for clarity.

- (**A**) Blower retaining bolts.

– Pull blower motor out of support in a rearward direction.

– Installation is reverse of removal. Note the following:

- Ensure proper fitting of hose connections and clamps on heating duct elbow.

Front blower motors, removing and installing

Special Porsche tools are required to pull the front blower motor from the blower housing. Read the procedure through before beginning repair.

– Disconnect negative (–) battery cable.

> ***CAUTION–***
> *• Prior to disconnecting the battery, read the battery disconnection cautions given in* **01 Warnings and Cautions.**

< Remove central electric panel cover (**A**). Remove rubber weather strip (**B**) and cowl panel cover (**C**).

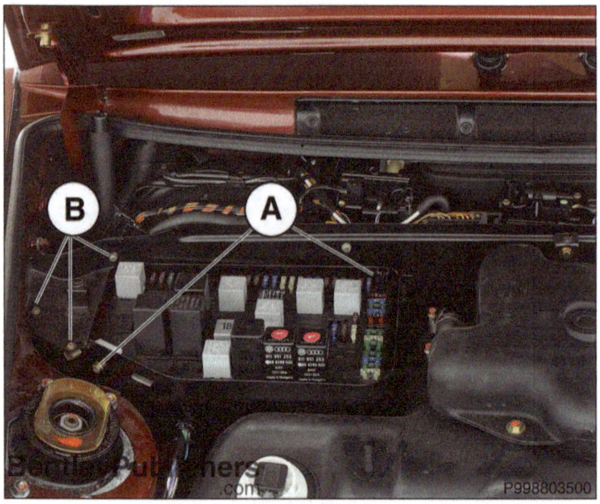

< Remove central electric panel fasteners (**A**) and wiring harness cover fasteners (**B**).

< Disconnect connector (**arrow**) from right side blower fan housing and slide out of bracket.

< Remove right side cross panel fasteners (**arrows**).

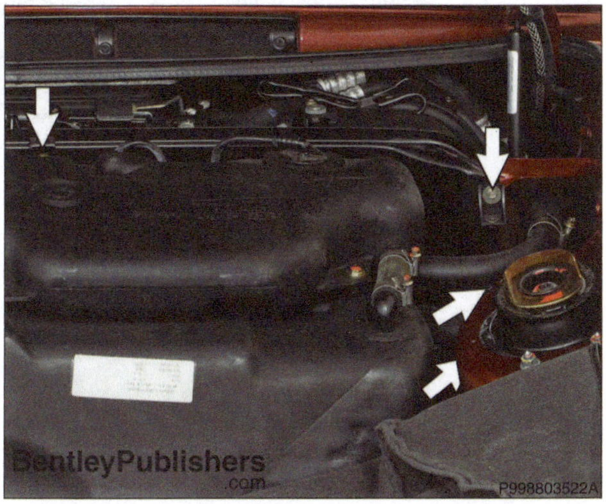

< Remove left side cross panel fasteners (**arrows**).

- Pull out cross panel.

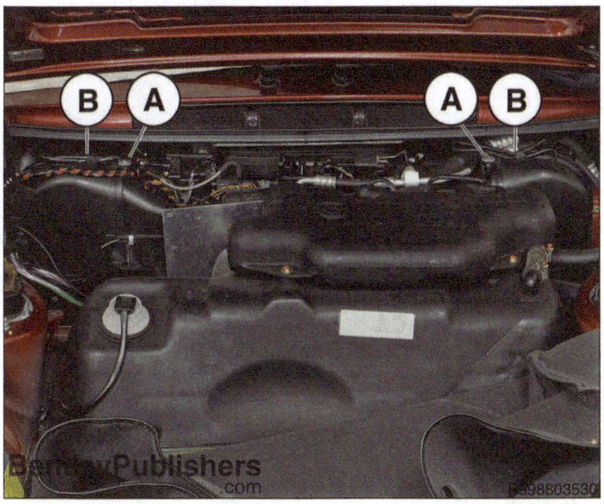

< Remove cabin filter housing retaining nuts (**A**).

- Remove clip (**B**) and wiring from cabin filter housings.
- Lift off cabin filter housing covers.

Remove cabin micro filters (**arrows**) from housing by lifting up out of slots.

Front blower fan motor (**arrow**) is now visible. There is a blower fan behind each cabin air filter.

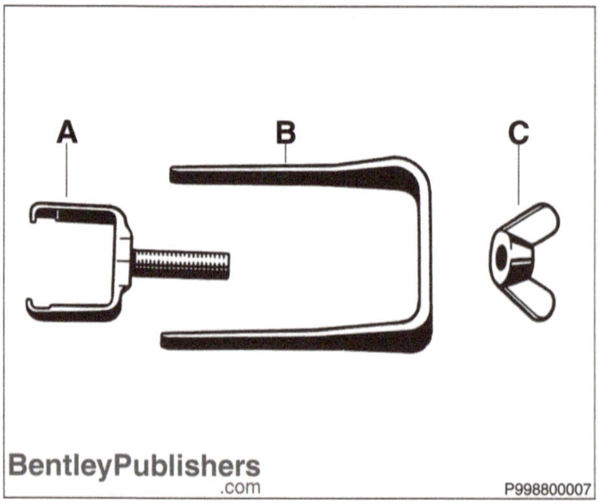

Engage Porsche special tool 9512 (**A**) onto fan motor.

◀ Push tool (**B**) through fan cage and onto cutouts (**arrows**) and tighten wing nut (**C**) until fan motor releases.

NOTE–

- *Note that special tools are installed off center to fan motor.*

– Pull fan assembly out of housing.

– Disconnect fan motor connector.

– Installation is reverse of removal. Note the following:
  - Ensure blower motor engages properly.
  - Ensure wires are not squeezed behind fan motor.
  - Check to see that blower fan turns freely.
  - Ensure cabin filter is seated in grooves in fan housing and fan housing cover.

Front blower motor resistor, removing and installing

– Disconnect negative (–) battery cable.

CAUTION–

- *Prior to disconnecting the battery, read the battery disconnection cautions given in* **01 Warnings and Cautions.**

◀ Pull back trunk carpeting and remove central electric panel cover (**A**). Remove rubber weather strip (**B**) and cowl panel cover (**C**).

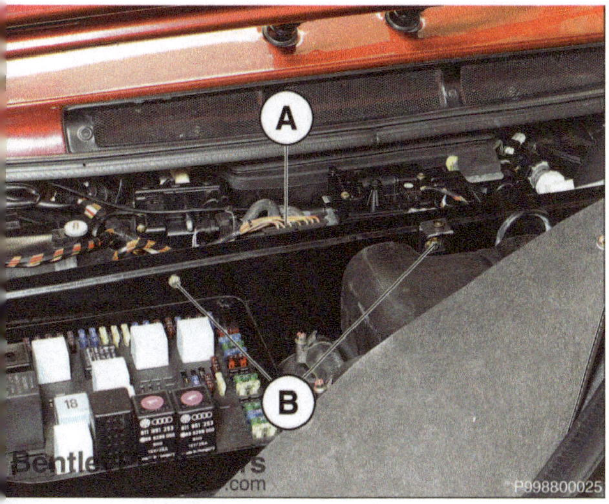

◀ Disconnect electrical connector (**A**). Counterhold nuts at rear of cross panel and remove bolts (**B**).

80

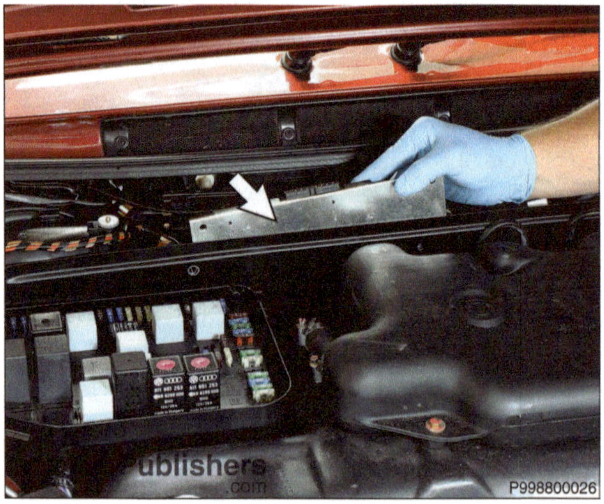

< Lift heater blower motor resistor (**arrow**) up and out.

– Installation is reverse of removal.

Heater flap and ducts

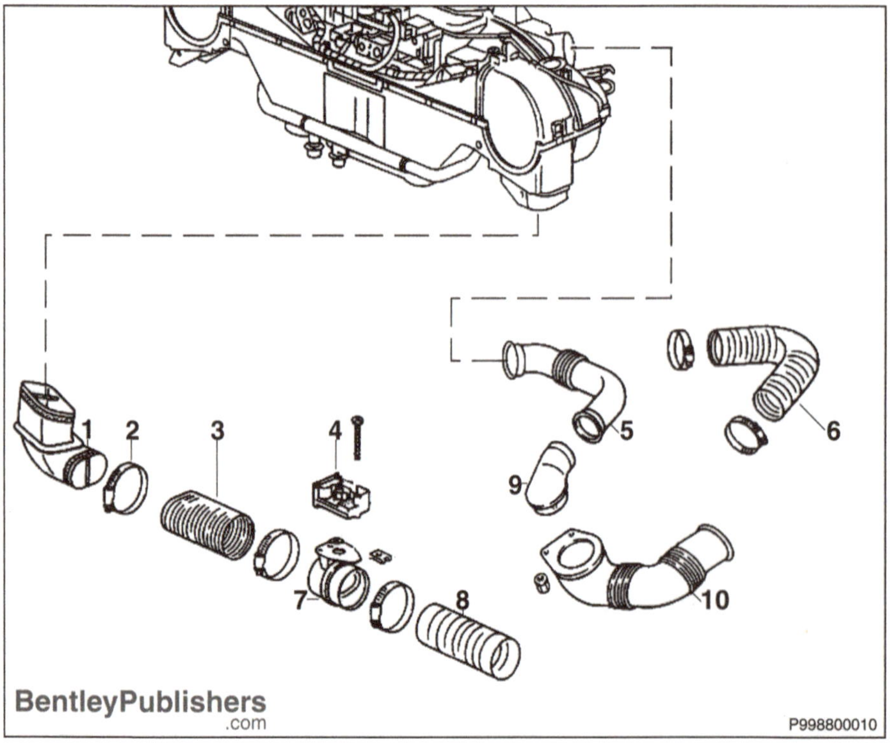

1. Footwell outlet vent
2. Hose clamp
3. Duct connecting hose
4. Heater flap motor
   - Opens and closes heater flap (**7**)
5. Air duct
6. Duct connecting hose
7. Temperature mixing flap assembly
8. Duct connecting hose
9. Elbow
10. Air duct

A/C and heater unit

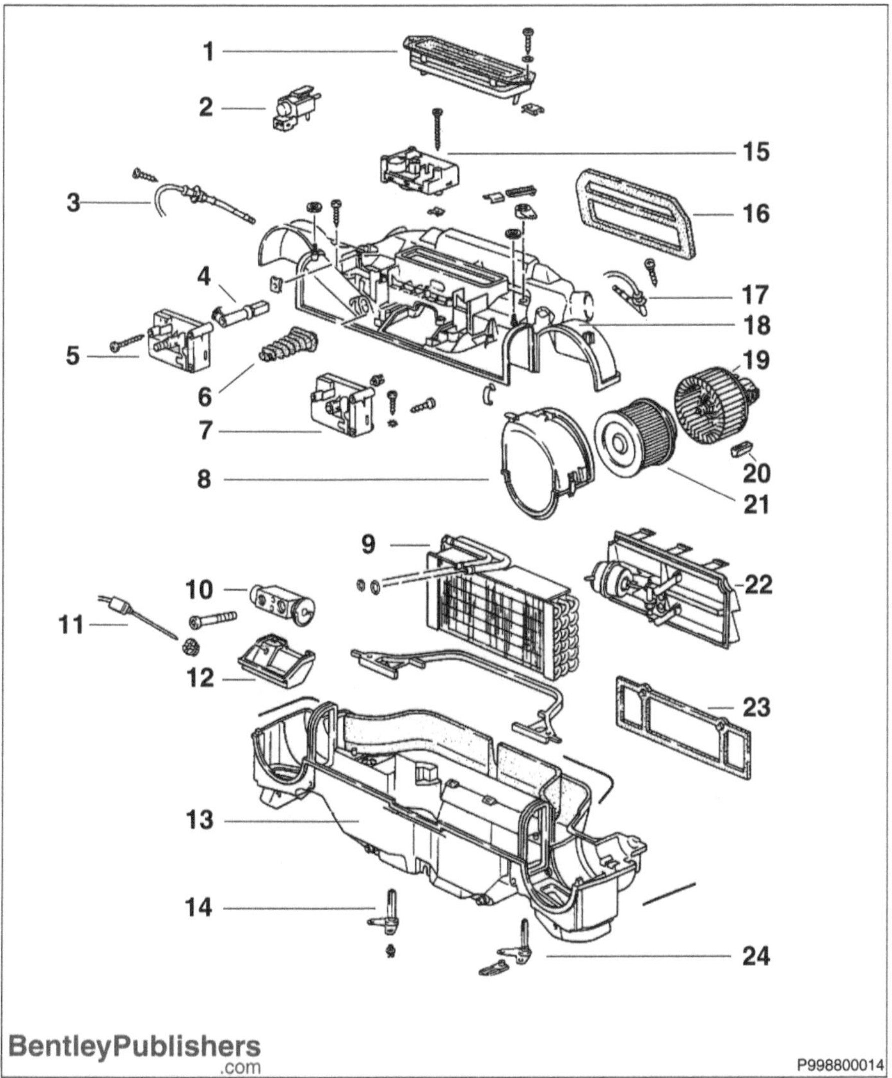

1. Upper bellows
2. Change over valve
3. Temperature mixing chamber repair kit
4. Joint
5. Defroster flap electric motor
6. Bellows
7. Temperature mixing sensor
8. Fan housing cover
9. A/C evaporator
10. A/C expansion valve
11. Temperature sensor
12. Expansion valve cover
13. A/C and heater unit housing, lower half
14. Lower connecting lever
15. Fresh air flap electric motor
16. Center duct gasket
17. Temperature mixing sensor
18. A/C and heater unit housing, upper half
19. Blower fan assembly, left side (right side not shown)
20. Rubber mounting
21. Cabin micro filter, left side (right side not shown)
22. Flap assembly
23. Lower gasket
24. Lower connecting lever

NOTE–

- *To remove A/C and heater unit, A/C system must be evacuated. For more information see* **87 Air-conditioning**.
- *To remove A/C and heater unit, fuel tank must be removed. See* **20 Fuel Storage and Supply**.
- *Procedure for removing and installing A/C and heater unit is not covered in this manual.*

87 Air-conditioning

General

This repair group covers interior ventilation and air-conditioning. For additional information, see:

- **03 Service and Maintenance** for A/C compressor drive belt and cabin micro filter replacement.
- **80 Heating** for blower fan removal and installation.
- **EWD Electrical Wiring Diagrams**.

Some A/C or heater system repair procedures require that the A/C refrigerant charge be evacuated. A/C system recharging procedures should be performed at a workshop with proper A/C machine and trained certified technicians.

A/C and heater control panel

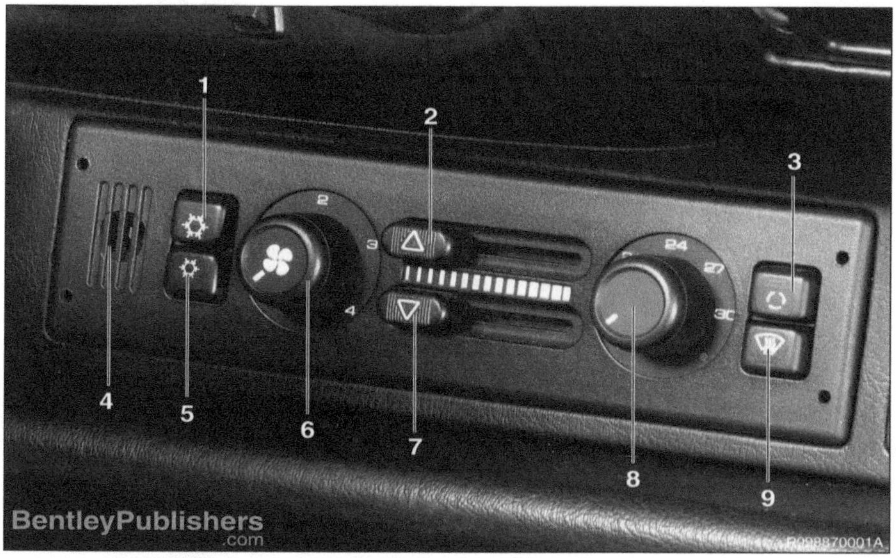

1. A/C switch (max A/C)
2. Air distribution slide control (upper)
   - Defroster vents and center vents
3. Air recirculation switch
4. Interior temperature sensor
5. A/C switch
6. Fan speed switch
7. Air distribution slide control (lower)
   - Footwell vents
8. Temperature selector
9. Defroster switch

87

A/C and heater unit

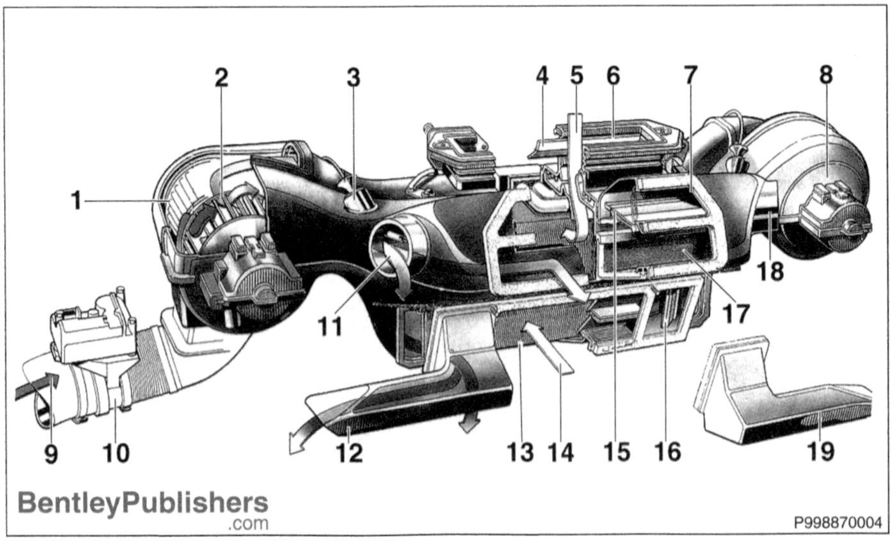

1. Cabin micro filter
2. Blower motor, left
3. Mixing chamber sensor, left
4. Fresh air flap
5. Fresh air intake
6. Fresh air intake
7. Defroster outlet
8. Blower motor, right
9. Heated air
10. Temperature mixing flap, left
11. Side duct assembly, left
12. Foot well outlet duct, left
13. Evaporator
14. Recirculated air
15. Defroster flap
16. Louvers for foot well outlet
17. Center duct outlet
18. Side duct assembly, right
19. Foot well outlet duct, right

A/C system warnings and cautions

WARNING—

- *Wear hand and eye protection (gloves and goggles) when working around the A/C system. If refrigerant oil comes in contact with your skin or eyes, immediately flush skin or eyes with cool water for 15 minutes and seek medical attention.*
- *Work in a well ventilated area. If you inhale refrigerant oil, breathe fresh air immediately and seek medical attention.*
- *Do not expose any component of the A/C system to high temperatures. Excessive heat could burst the system.*
- *Keep refrigerant away from open flames. Poisonous gas is produced if it burns. Do not smoke near refrigerant gases for the same reason.*
- *The A/C system is filled with refrigerant gas, which is under pressure. Pressurized refrigerant in the presence of oxygen may form a combustible mixture. Do not introduce compressed air into any refrigerant container.*
- *Electric welding near refrigerant hoses causes R134a to decompose. Discharge system before welding.*
- *Dispose of refrigerant oil as hazardous waste.*

CAUTION—
- *US law requires that any person who services a motor vehicle air-conditioner be properly trained and certified. Technicians must complete an approved recycling course to be certified.*
- *It is recommended that all A/C service be left to a Porsche work shop or other qualified A/C service facility.*
- *State and local governments may have additional requirements regarding A/C servicing. Always comply with state and local laws.*
- *Repair any A/C system which is known to be losing its charge.*
- *Do not top off a partially charged refrigerant system. Discharge system, evacuate and then recharge system.*
- *Refrigerant oil (PAG oil) is hygroscopic (water-absorbent). Do not store in open container.*
- *The mixture of refrigerant oil and refrigerant (R134a) attacks some metals and alloys (for example, copper) and breaks down certain hose materials. Use only hoses and pipes that are identified with a green mark (stripe) or the lettering "R134a".*
- *Immediately plug open connections on A/C components and pipes to prevent dirt and moisture contamination.*
- *Do not steam clean A/C condensers or evaporators. Use only cold water or compressed air.*
- *Do not store refrigerant oil in vicinity of flames, heat sources or strongly oxidizing materials.*
- *In case of fire, use carbon dioxide (CO2) extinguisher or sprayed water jet on A/C components.*
- *After refilling or testing refrigerant lines, replace protective caps. They serve as additional seals.*

A/C System Component Replacement

A/C system components

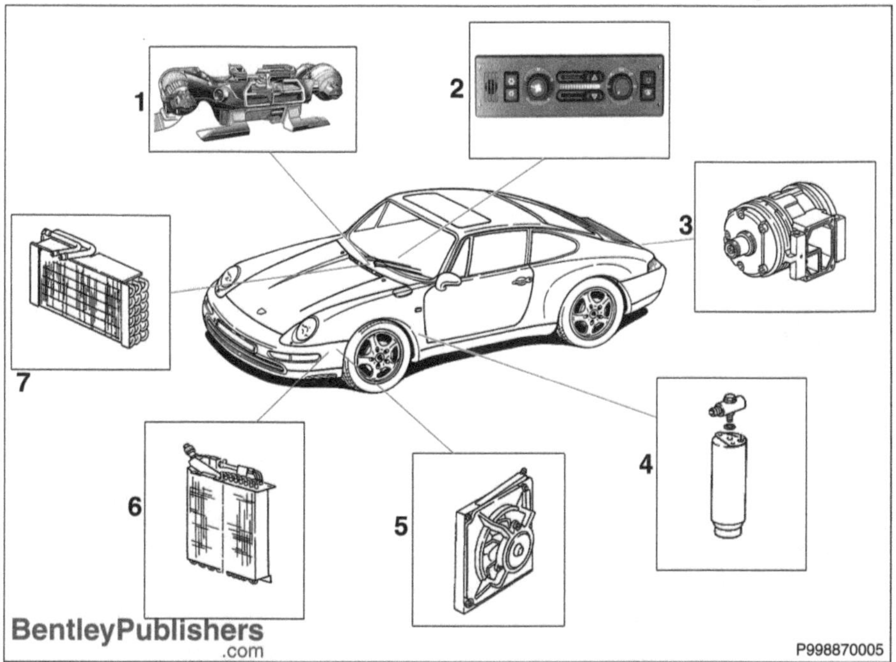

1. A/C and heater housing
   - Contains A/C evaporator
   - Contains left / right blower fans
2. A/C and heater control panel
   - Contains interior temperature sensor.
3. A/C compressor
   - Right side of engine compartment, belt driven
4. Receiver / dryer
5. Condenser fan
6. Condenser
   - Condenser fan (**5**) attached
7. Evaporator
   - In A/C and heater housing (**1**)
   - Expansion valve attaches to evaporator pipes

Refrigerant and refrigerant oil

The 993 air-conditioning system uses R134a refrigerant. Refrigerant oil is used to lubricate the A/C compressor.

A/C fluid capacities	
Refrigerant (R134a)	840 g (29.50 oz)
A/C lubricating oil (compressor)	140 ml (4.70 oz)

Compressor, removing and installing

This procedure covers removal and installation of the air-conditioning compressor only. Be sure to read **A/C system warnings and cautions** in this repair group before beginning work.

The belt-driven A/C compressor is easily accessible from the engine compartment.

Removing

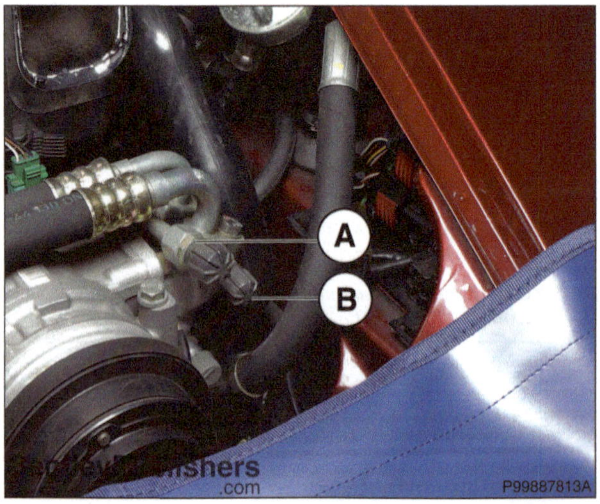

◄ A/C service ports are located to right of compressor.

- High pressure port (**A**)
- Low pressure port (**B**)

– Following manufacturer's instructions, connect an approved refrigerant recovery, recycle, recharge machine to the A/C system and remove refrigerant.

> ***WARNING—***
>
> - *Do not discharge/charge the A/C system without proper equipment and training. Damage to the vehicle and personal injury may result. Read warnings and cautions given earlier.*

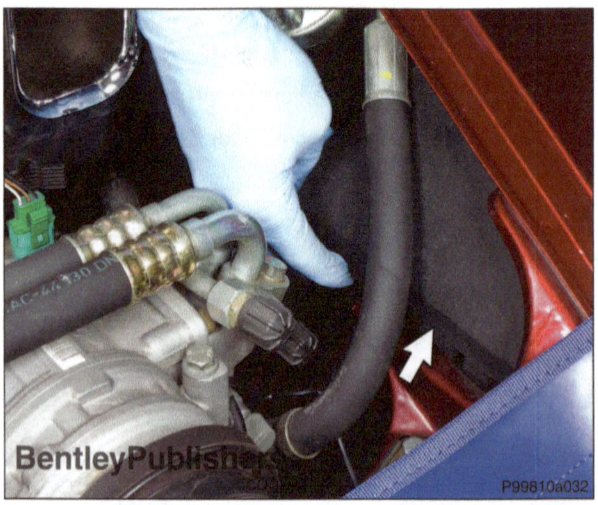

◄ Working in engine compartment on right side, remove plastic cover by unclipping bottom and pulling up (**arrow**).

Remove compressor electrical connector (**arrow**).

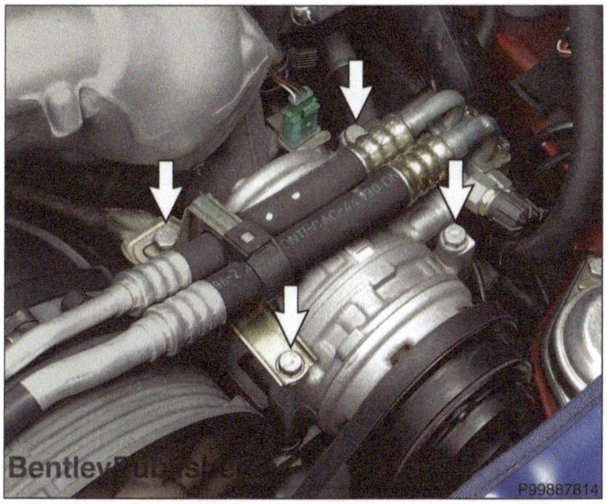

Loosen compressor mounting fasteners (**arrows**) but do not remove.

Release tensioning bolt lock nut and unscrew tensioning bolt (**inset**), which is located just below compressor on passenger side.

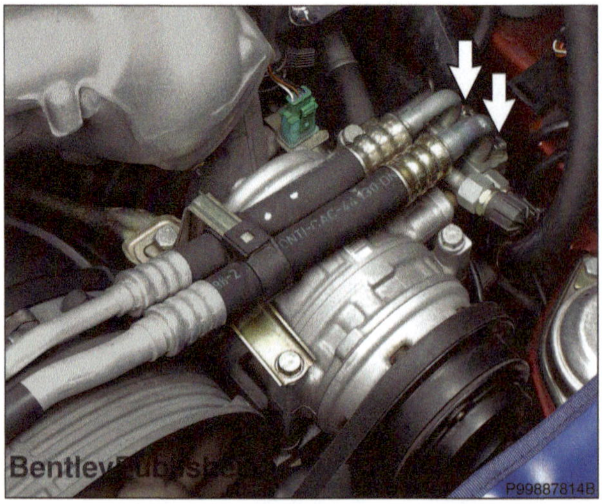

◄ Remove refrigerant line bolts (**arrows**) from A/C compressor. Plug ports and lines immediately.

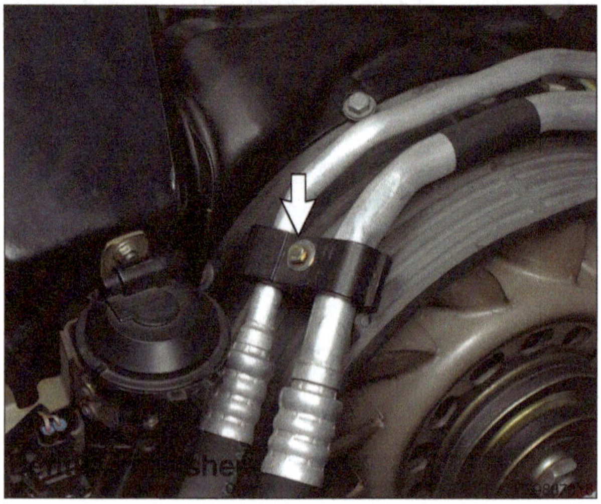

◄ Remove A/C line bracket bolt (**arrow**).

– Remove compressor drive belt. See **03 Service and Maintenance**.

– Remove compressor mounting fasteners. Move refrigerant lines aside.

– Remove compressor from engine.

Installing

Check that replacement compressor is filled with the correct volume of oil for the entire A/C system and is pressurized. Prior to installation:

- Open pressure cap to relieve pressure. Use shop towel to prevent oil spray.
- Pour approximately 50 ml of oil from new compressor into measuring glass. 90 ml should remain in compressor when installed.

– Install compressor to engine and install drive belt. See **03 Service and Maintenance**. Do not fully secure mounting bolts until drive belt is in place.

Tightening torque	
Compressor mounting bolts (M8)	28 Nm (20.5 ft-lb)

– Refit A/C lines to compressor. Replace all O-rings and coat with refrigerant oil.

Tightening torque	
Refrigerant pipe to compressor (M8)	23 Nm (17 ft-lb)

– Refit and adjust tensioning bolt and lock nut. See **03 Service and Maintenance** for belt tensioning recommendations.

– Install electrical connector to compressor.

– Recharge the A/C system following equipment manufacturer's instructions. Be sure to use correct type and volume of refrigerant oil. See **Refrigerant and refrigerant oil** in this repair group.

Condenser, removing and installing

The A/C condenser and condenser fan are located behind the left front wheel, behind the wheel housing liner. A/C condenser replacement requires that the A/C refrigerant be properly discharged (evacuated).

> ***WARNING—***
>
> - *Do not discharge/charge A/C system without proper equipment and training. Damage to the vehicle and personal injury may result. Read warnings and cautions given earlier.*

Removing

– Following manufacturer's instructions, connect an approved refrigerant recovery/recycling/recharging unit to A/C system and remove refrigerant.

– Raise vehicle and support safely. Remove left front wheel.

> ***WARNING—***
>
> - *Make sure vehicle is stable and well supported at all times. Use a professional automotive lift or jack stands. Place jack stands underneath structural chassis points, not underneath suspension parts. A floor jack is not adequate support.*

– Remove left front wheel housing liner. See **50 Fenders**.

– Remove lower section of front spoiler assembly.

◄ Separate ballast resistor electrical connector (**arrow**).

NOTE–

- *Bumper cover shown removed for clarity. Procedure can be performed with bumper cover in place.*

◄ Disconnect refrigerant pipes (**arrows**) from condenser. Immediately plug ports and pipes with suitable air-tight seal.

87

◀ Disconnect headlight washer hoses (**A**) from regulator valve.

– Disconnect condenser fan electrical connector (**B**) then pull away clips and wiring from fan.

– Remove vent hose (**C**) from clips.

◀ Remove condenser support bracket fasteners (**arrows**).

– Remove condenser and bracket from vehicle with fan by pulling whole unit downwards.

NOTE–

• *Rubber mounting pin secures top of condenser bracket.*

– With bracket removed, remove screws from support bracket and lift condenser out from above.

Installing

– Installation is reverse of removal Note the following:
- Replace O-rings and use clean refrigerant oil to lubricate.
- Uncap refrigerant lines just prior to refitting lines.
- Replace A/C oil that was drawn off (approximately 20 ml).
- Use proper tightening torques when installing A/C pipes.

Tightening torques	
5/8" - 18 UNF (unified fine thread)	17 Nm (12.5 ft-lb)
3/4" - 16 UNF (unified fine thread)	24 Nm (18.0 ft-lb)

– Recharge A/C system following manufacturer's instructions for A/C machine.

Condenser fan, removing and installing

The A/C condenser fan is located behind the left front wheel, behind the wheel house liner.

– Remove left front wheel housing liner. See **50 Fenders**.

◄ Disconnect condenser fan motor electrical connector (**arrow**).

NOTE–

- *Bumper cover shown removed for clarity. Procedure can be performed with bumper cover in place.*

◄ Disconnect headlight washer hoses (**A**) from regulator valve.

– Remove vent hose (**B**) from clips.

◄ Remove fan fasteners (**arrows**) then remove fan assembly from below.

– Remove fan from retaining bracket.

– Installation in reverse of removal. Note the following:

- After refitting, ensure that fan blades rotate without interference.

A/C condenser fan ballast resistor

The A/C condenser fan controls low speed operation of A/C condenser fan. This resistor is located behind the left front fender area on backside of A/C condenser bracket.

◄ The ceramic A/C condenser fan ballast resistor can be prone to failure (broken resistor pictured).

– Raise vehicle and support safely. Remove left front wheel.

> ***WARNING—***
>
> - *Make sure vehicle is stable and well supported at all times. Use a professional automotive lift or jack stands. Place jack stands underneath structural chassis points, not underneath suspension parts. A floor jack is not adequate support.*

– Remove left front wheel housing liner. See **50 Fenders**.

– Remove lower section of front spoiler assembly.

◄ Disconnect ballast resistor electrical connector (**arrow**).

NOTE–

- *Bumper cover shown removed for clarity. Procedure can be performed with bumper cover in place.*

◄ Disconnect headlight washer hoses (**A**) from regulator valve.

– Remove vent hose (**B**) from clips.

- Remove condenser support bracket fasteners (**arrows**).
- Pull condenser down to release from upper rubber retaining pin.
- Ballast resistor (**A**) is mounted to top of condenser bracket towards chassis.

NOTE–

- Using 4 mm Allen wrench, remove screw securing ballast resistor to bracket and remove resistor.
- Installation is reverse of removal.

A/C receiver/drier, replacing

The A/C receiver/drier is mounted in the rear of the left front wheel housing.

> ***CAUTION—***
>
> *• The receiver/drier must be replaced when:*
> *There is dirt in the A/C system.*
> *The compressor has seized or has been replaced.*
> *The condenser or evaporator are replaced.*
> *The A/C system is leaking and there is no more refrigerant*
> *The A/C system was open for 24 hours or more.*

- Remove left front wheel housing liner. See **50 Fenders**.
- Following manufacturer's instructions, connect an approved refrigerant recovery/recycling/recharging unit to A/C system and remove refrigerant.

> ***WARNING—***
>
> *• Do not discharge/charge the A/C system without proper equipment and training. Damage to the vehicle and personal injury may result. Read warnings and cautions given earlier.*

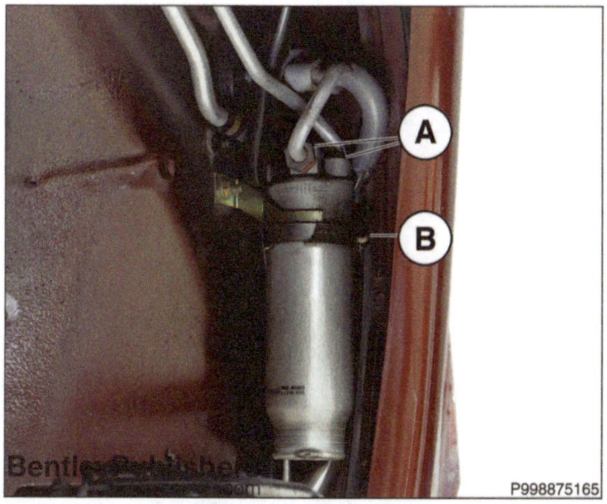

- Disconnect and plug refrigerant lines (**A**). Remove clamp (**B**) and remove receiver drier.
- Installation is reverse of removal. Note the following:
  - Be sure there is sufficient oil in receiver dryer.
  - Before recharging, evacuate system for at least 30 minutes.
  - Recharge system following equipment manufacturer's instructions.
  - Always replace O-rings when reconnecting refrigerant lines.

A/C evaporator and expansion valve

The integrated climate control system combines heating and air-conditioning functions in one housing behind the dashboard. The A/C evaporator is located between the two halves of the A/C and heater unit. This unit has to be removed to replace evaporator.

NOTE–

- *Replacement of the A/C and heater unit is not covered in this manual. For more information see* **80 Heating**.

A/C and heater unit housing

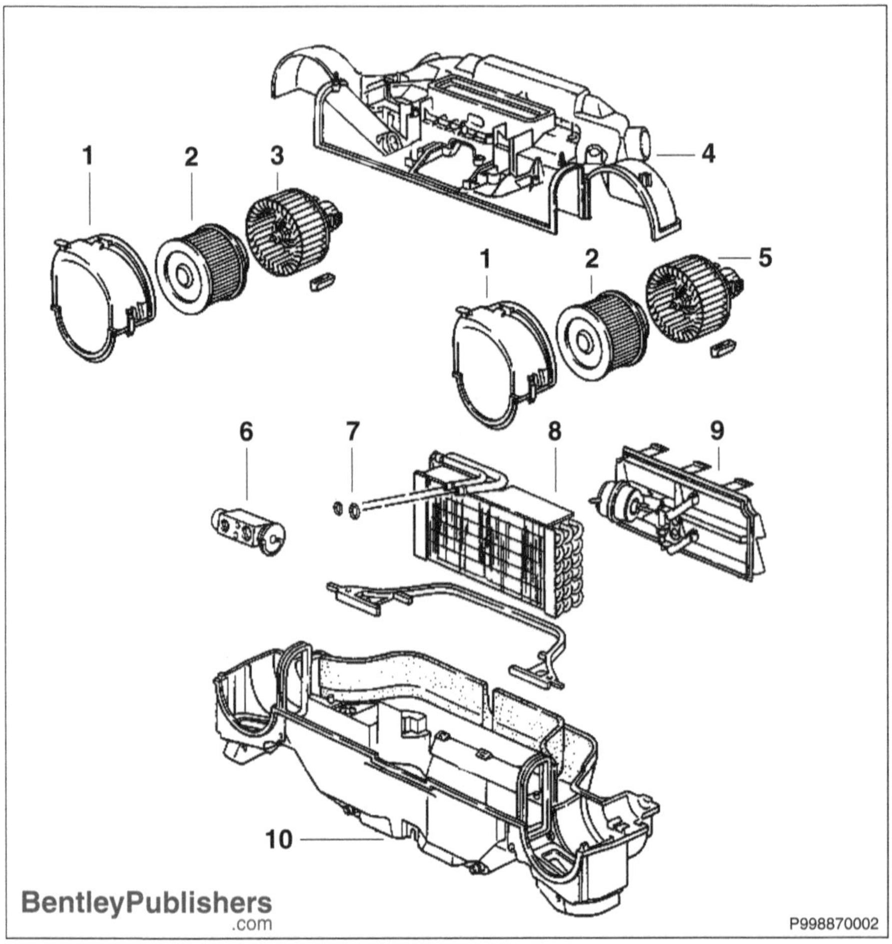

1. Fan housing cover
2. Cabin micro filter
3. Blower motor (right side)
4. Upper housing
5. Blower motor (left side)
6. A/C expansion valve
7. Expansion valve O-rings
8. A/C evaporator
9. Distribution flap
10. Lower housing

90 Instruments, Horns

General

This repair group covers removal and installation of gauges and warning lights, as well as instrument bulb replacement.

Electrical system safety precautions

See repair group **97 Fuses, Relays, Component Locations** for additional **Warnings and Cautions** before performing any work on vehicle electrical systems.

CAUTION—

- *Before disconnecting the battery, read battery disconnection cautions in* **00 Warnings and Cautions**.
- *Read and record control module fault memory before disconnecting the battery.*
- *Record radio station presets and obtain radio activation code before disconnecting the battery.*
- *Switch the ignition OFF and remove the negative (–) battery cable before removing any electrical components.*
- *Connect and disconnect electrical connectors and ignition test equipment leads only while the ignition is OFF.*
- *Only use a digital multimeter for electrical tests.*

Instrument layout

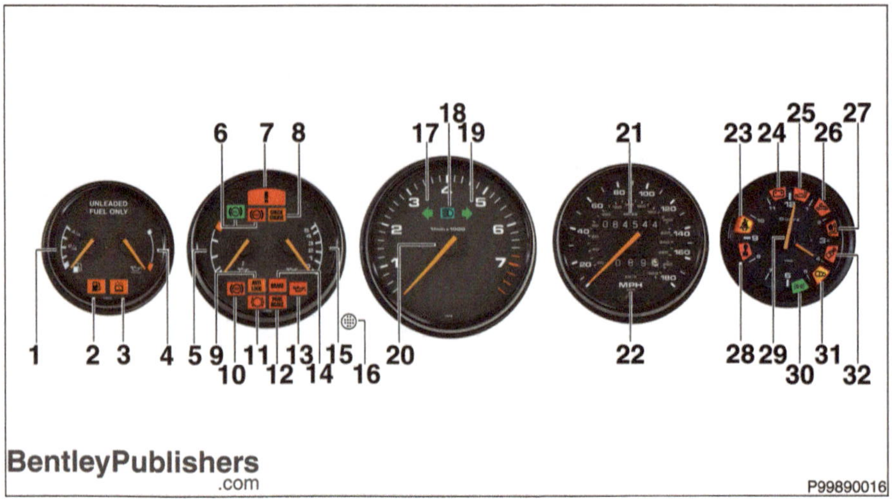

1. Fuel gauge
2. Low fuel level warning light
3. Low washer fluid warning light
4. Oil level gauge
5. Oil temperature gauge
6. ABD (automatic brake differential) lights
7. Central warning light
8. Check engine light
9. ABS warning light
10. Brake pressure warning light
    - Carrera 4, Carrera 4S, Turbo
11. Brake pad wear indicator
12. Parking brake light
13. Oil pressure warning light
14. Brake fluid warning light
15. Oil pressure gauge
16. Warning buzzer
17. Left turn signal
18. High beam indicator light
19. Right turn signal
20. Tachometer
21. Odometer
22. Speedometer
23. Seat belt warning light
24. Alternator warning light
25. Rear spoiler warning light
26. Convertible top warning light
27. Airbag warning light
28. Tiptronic emergency indicator light
29. Clock
30. Parking light indicator light
31. Engine fan belt warning light
32. Immobilizer warning light

Instruments

There are five electronic instruments (gauges) with various warning lights. Each instrument is mounted in a rubber retaining ring. Disconnect battery negative (–) cable and cover with insulating material before beginning work on instruments.

Fuel level / oil level gauge

P99890015

◄ The combination fuel level and oil level gauge monitors vehicle fuel and oil level. It also houses the low fuel and low washer fluid warning lights.

If the fuel level is low, the warning light will illuminate. Avoid high speed cornering if the low fuel level light is on. A blinking fuel level light indicates a fault in the system.

Oil level is sufficient when needle is horizontal with the engine running. Requirements for the oil level gauge to indicate correctly:

- vehicle level.
- engine warm (oil temperature in range of third marking on oil temperature gauge).
- engine idling.

If the washer fluid level falls to 2.1 quarts, the warning light will illuminate.

Oil temperature / oil pressure gauge

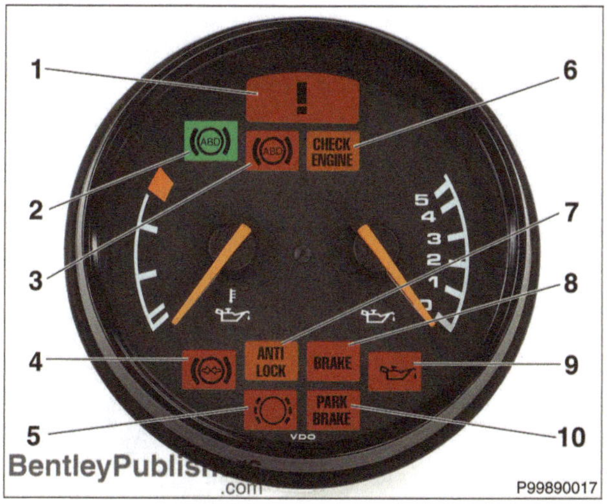

P99890017

◄ The combination oil temperature and pressure gauge monitors both oil temperature and pressure. It also houses the following lights:

1. Central warning light
2. ABD active light
3. ABD warning light
4. Brake pressure warning light (hydraulic brake boost)
   - Carrera 4, Carrera 4S, Turbo only
5. Brake pad wear indicator light
6. Check engine light
7. Antilock brake warning light
8. Brake fluid warning light
9. Engine oil pressure warning light
10. Parking brake indicator light

The central warning light and individual warning lights come on momentarily when ignition is switched ON. If a fault occurs, the central warning light comes on with the corresponding warning light(s). The cancel button in the center console can be used to cancel the central warning light.

Tachometer

The tachometer also houses the turn signal indicator lights, the high beam indicator light, and the on-board computer display (if equipped).

Speedometer/odometer

The speedometer indicates the vehicles speed in both miles per hour and kilometers per hour.

The upper odometer reads the total distance traveled.

The lower odometer records trip mileage, and can be reset to zero at any time by pressing the reset knob below.

- On vehicles equipped with a digital on-board computer the analog trip odometer is replaced by a digital display.

Clock

The 993 has an analog clock that also houses the following warning lights:

1. Tiptronic emergency program indicator light
2. Seat belt warning light
3. Alternator warning light
4. Rear spoiler warning light
   - Warning remains lit until vehicle speed exceeds 4 mph.
5. Convertible top indicator light
6. Airbag warning light
7. Vehicle immobilizer status indicator light
8. Engine fan belt warning light
9. Parking light indicator light

To set the clock, press the rocker switch (**arrow**).

- Press rocker switch briefly to set clock forward one minute.
- Press and hold rocker switch for rapid setting.

Instruments, removing and installing

Using plastic prying tool (**arrow**), lever instrument and rubber retaining ring out of dashboard.

NOTE–

- *Fuel level / oil level gauge removal shown. Removal of other instruments is similar.*

> ***CAUTION—***
>
> - *Use care not to damage dashboard or gauge.*

Label and disconnect electrical connectors (**arrows**).

– Installation is reverse of removal.

Instrument bulbs, replacing

– Remove instrument. See **Instruments, removing and installing** in this repair group.

– Turn instrument over.

◄ Using a small screwdriver, rotate bulb socket (**arrow**) 45° counterclockwise to unlock.

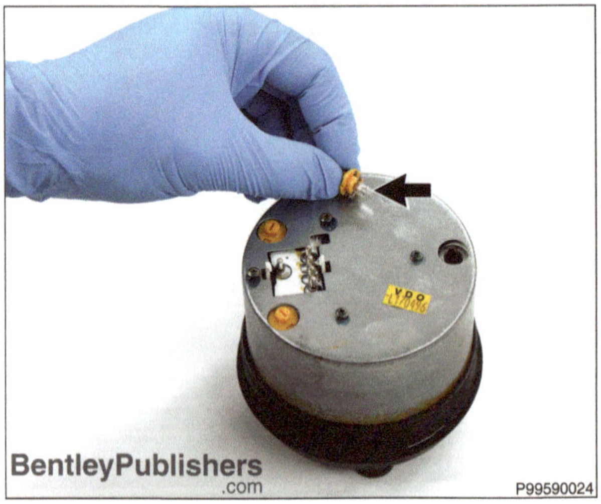

◄ Lift out bulb (**arrow**) to replace.

> ***CAUTION—***
>
> *• If bulb is dirty or has been handled by the glass end, clean bulb with soft, lint-free cloth and rubbing alcohol. Oils or dirt can reduce bulb lifespan.*

– Installation is reverse of removal.

Horns

The horn system consists of dual horns, horn relay, and horn buttons on the steering wheel. The electrical contact between the rotating steering wheel and the steering column is through the air bag contact spring. The horns are located below the right headlight and are accessible by removing the front bumper.

NOTE–

• *See* **69 Seat Belts, Airbags** *for contact spring replacement information.*

The horns are hard wired to ground. Pressing the horn button activates the horn relay to supply the horns with positive voltage.

Horns, removing and installing

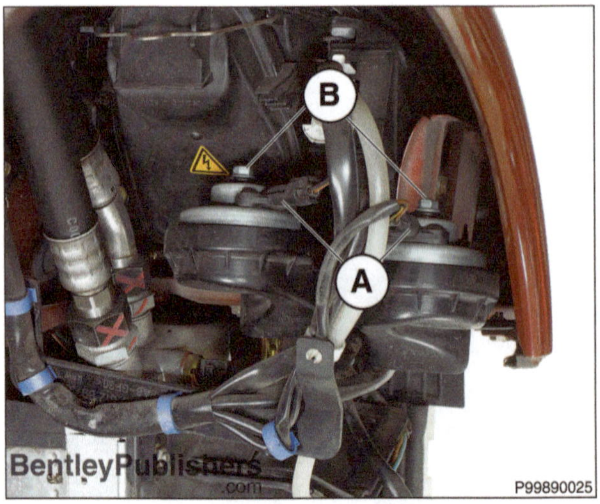

– Remove right front wheel housing liner. See **50 Fenders**.

◄ Unplug wiring harnesses from horns (**A**).

– Remove horn mounting bolts (**B**) and remove horns from bracket.

– Installation is reverse of removal.

91 Radio

GENERAL

Porsche 993 models came equipped with a variety of cassette and CD sound systems. This repair group covers radio removal and installation. For related information, see:

- **90 Instruments, Horns** for information on the instrument cluster.
- **EWD Electrical Wiring Diagrams** for radio wiring.

RADIO

Radio, removing and installing

Radio security code and radio station presets will be lost when radio is removed or power to the radio is interrupted. Radio security code can be obtained from an authorized Porsche dealer.

CAUTION—

- *Record the radio station presets and obtain the radio security code before disconnecting the radio.*

P99891001

Press release button (**arrow**) and remove radio face plate.

◄ Insert radio removal tools (**A**) in slots at base of radio.

– Lever tools in direction of **arrows** and slide radio out of dashboard far enough to access electrical connectors.

– Disconnect antenna and electrical connectors and remove radio from vehicle.

– Installation is reverse of removal. Note the following:

- Connect antenna and electrical connectors.
- Press radio into dashboard until it snaps into place.
- Set radio security code and presets recorded earlier. See **Radio security code, entering** in this repair group.

Radio security code, entering

Porsche 993 radios are equipped with an anti-theft security code. If the battery is disconnected or radio is removed, the radio code will need to be entered upon reconnecting the battery.

When battery is reconnected, the LCD panel will display the word "CODE".

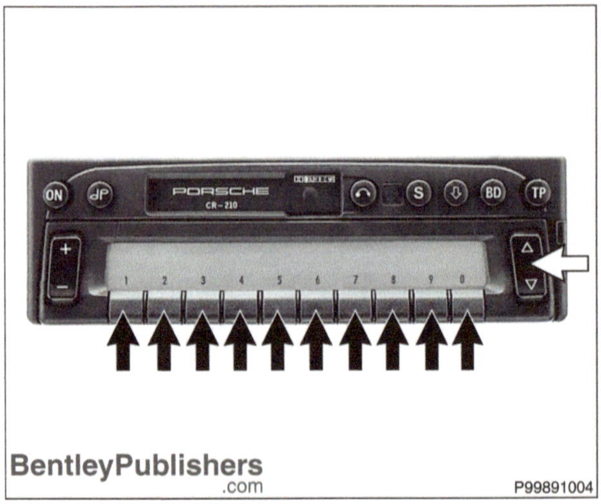

◄ Enter 4-digit radio code using buttons (**black arrows**) along bottom of LCD panel.

- After first digit is entered, the word "CODE" disappears and the first digit is displayed followed by three dashes.
- As digits are entered, the next line in sequence flashes.
- When all four numbers have been entered, the first digit flashes. At this time you can correct any error by entering the complete code.
- Once the correct code is displayed, enter by pushing the signal seek switch (**white arrow**).
- If wrong radio code is entered, radio will be locked into "WAIT" mode from 15 minutes to 24 hours, depending on how many times the wrong code was entered.
- If LCD panel displays the word "WAIT", turn ignition key to accessory position for a period of time.

92 Wipers and Washers

General

This repair group covers wiper blade service, wiper assembly repairs, and windshield washer repairs.

For additional information, see the following:

- **48 Steering** for steering column stalk switch assembly.
- **97 Fuses, Relays, Component Locations**.
- **EWD Electrical Wiring Diagrams**.

Cautions

CAUTION—

- *Switch ignition OFF and disconnect the negative (–) battery cable before removing any electrical components.*
- *Prior to disconnecting the battery, read battery disconnection cautions in* **00 Warnings and Cautions**.
- *Prior to disconnecting the battery, obtain radio security and record radio presets.*
- *Only use a digital multimeter for electrical tests.*

Wiper Arms and Blades

Wiper blade cleaning problems

Common problems with the windshield wipers include streaking or sheeting, water drops after wiping, and blade chatter.

Streaking is usually caused when wiper blades are coated with road film or car wash wax.

Drops that remain behind after wiping are usually caused by oil, road film, or diesel exhaust residue on the glass.

Chatter may be caused by dirty or worn blades, or by wiper arms that are out of alignment.

- Clean blades using soapy water. If cleaning does not cure problem, replace blades.
- Use an alcohol or ammonia solution or other non-abrasive cleaner to clean windshield.
- Clean blades and windshield as described above.
- Bend wiper arm so that there is even pressure along blade and the blade is perpendicular to windshield at rest.
- If problems persist, replace blades and wiper arms.

Wiper blade, replacing

- Turn ignition OFF and remove key.
- Pivot wiper arm off glass.

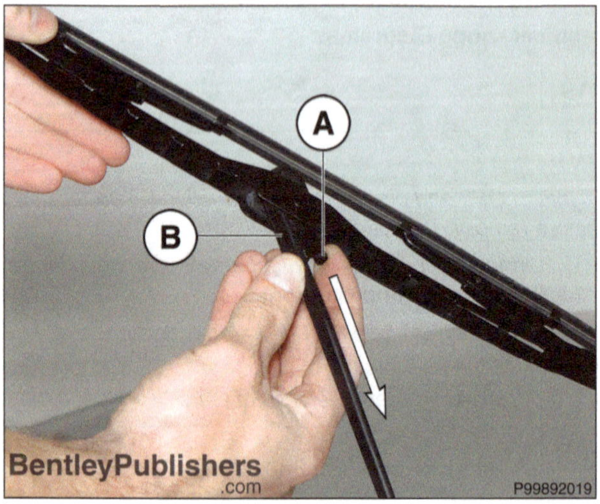

- Position wiper blade approximately perpendicular to wiper arm.
- Press retaining tab (**A**).
- Slide blade plastic retaining clip out of arm (**B**) in direction of **arrow**.
- Installation is reverse of removal. Install wiper blade clip to wiper arm until it clicks into position.

Wiper arm, removing and installing

– With wipers in parked position, turn ignition OFF and remove key.

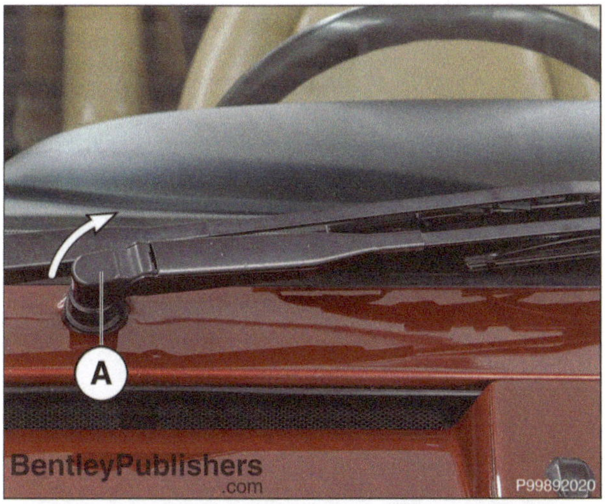

◀ Gently pivot cover (**A**) in direction of **arrow** to access wiper arm mounting nut.

◀ Remove wiper arm mounting nut (**arrow**) and washer.

– Mark position of wiper arm on shaft to aid installation.

◀ Remove wiper arm from shaft.

NOTE–

- *If the arm does not come off freely, gently lever the end of the arm back and forth to help loosen the arm at the splined fitting.*

– Installation is reverse of removal. Tighten mounting nut after adjusting position of arms.

Wiper and Washer Controls

Wiper and washer system fuse

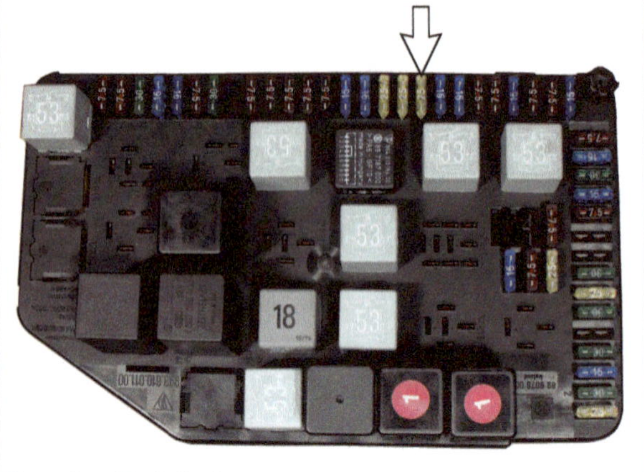

BentleyPublishers.com P99892023

Fuse is located in central electric panel in trunk.

Wiper and washer fuse	
Location	Central electric panel
Number	24
Rating	25 A

Wiper and washer system relays

BentleyPublishers.com P99892024

Relays are located in trunk, in central electric panel.

Wiper and washer relays	
Headlight washer (if equipped)	Relay R43
Intermittent wiper	Relay R42

Wiper and washer switch

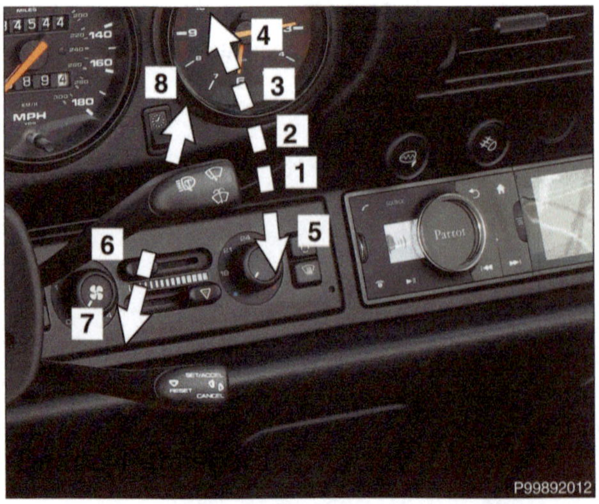

P99892012

Wiper stalk switch positions are as follows:

1. Wipers parked
2. Slow wipe
3. Regular wipe
4. Fast wipe
5. Intermittent wipe
6. Windshield washer pump only
7. Windshield washer pump and wipers
8. Headlight washer (if equipped)

The wiper and washer stalk switch is one component of an integrated steering column switch assembly which includes headlight dimmer, turn signal switches, and, if equipped, on-board computer controller and cruise control switch. Steering column stalk switch replacement is covered in **48 Steering**.

Turn knurled knob (**A**) to set wipe interval.

Wiper Repairs

Windshield wiper assembly components

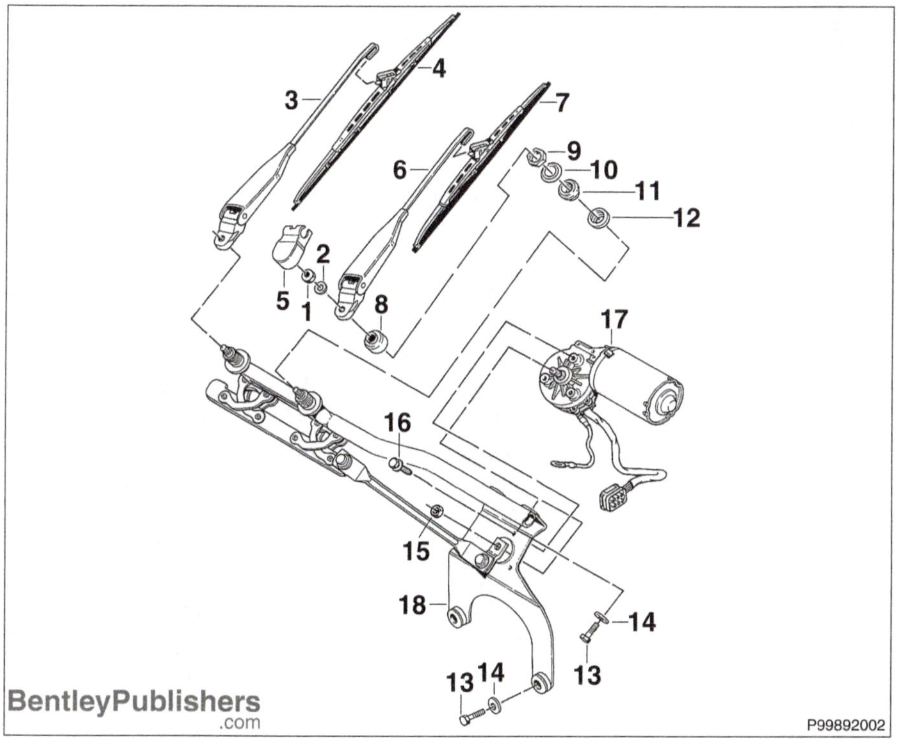

1. M8 nut
   - Tighten to 15 Nm (11 ft-lb)
2. Washer
3. Right wiper arm
4. Right wiper blade
5. Cap
6. Left wiper arm
7. Left wiper blade
8. Cap
9. Nut
   - Tighten to 10 + 1 Nm (7 + 0.7 ft-lb)
10. Flange washer
11. Rubber bushing
12. Rubber bushing
13. M6 x 22 mm fastener
    - Tighten to 10 Nm (7 ft-lb)
14. Washer
15. Nut
16. Fastener
    - Tighten to 11 Nm (8 ft-lb)
17. Wiper motor
18. Crank gear

Wiper motor, removing and installing

Read the procedure through before beginning. The A/C refrigerant charge system needs to be evacuated so that A/C lines can be disconnected and moved out of the way to access the wiper motor.

– Disconnect negative (–) battery cable and cover negative terminal with insulating material.

> ***CAUTION—***
>
> • *Prior to disconnecting the battery, read the battery disconnection cautions in* **00 Warnings and Cautions**.
>
> • *Be sure to have the radio anti-theft code before disconnecting the battery.*

– Remove steering wheel and steering column cover. See **48 Steering**.

– Remove instruments. See **90 Instruments, Horns**.

– Remove air guide tube inside dashboard.

◄ Disconnect electrical connector (**arrow**) from wiper motor.

– Support trunk lid and disconnect driver's side trunk lid strut. See **55 Trunk Lid, Engine Hood**.

– Vehicles with air conditioning: Evacuate refrigerant charge.

> ***WARNING—***
>
> • *Wear hand and eye protection (gloves and goggles) when working around the A/C system. If refrigerant oil comes in contact with your skin or eyes, immediately flush skin or eyes with cool water for 15 minutes and seek medical attention.*
>
> • *Work in a well ventilated area. If you inhale refrigerant oil, breathe fresh air immediately and seek medical attention.*
>
> • *Do not expose any component of the A/C system to high temperatures. Excessive heat could burst the system.*
>
> • *Keep refrigerant away from open flames. Poisonous gas is produced if it burns. Do not smoke near refrigerant gases for the same reason.*
>
> • *The A/C system is filled with refrigerant gas which is under pressure. Pressurized refrigerant in the presence of oxygen may form a combustible mixture. Do not introduce compressed air into any refrigerant container.*
>
> • *Electric welding near refrigerant hoses causes R134a to decompose. Discharge system before welding.*
>
> • *Dispose of refrigerant oil as hazardous waste.*

◄ Release clips (**A**) and remove central electric panel cover.

– Pull up rubber weather strip (**B**).

– Pull cowl panel cover (**C**) up and out of vehicle.

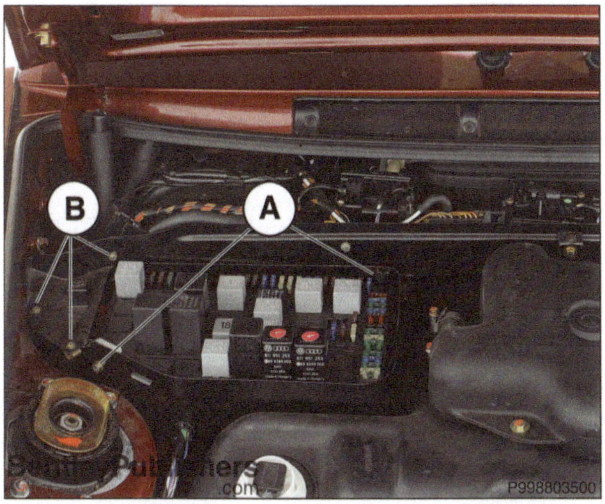

⊲ Remove central electric panel fasteners (**A**) and wiring harness cover fasteners (**B**).

⊲ Remove connector (**arrow**) from atop right side blower fan housing and slide out of bracket.

– Place protective cover on fender and lay central electric panel on fender.

⊲ Remove right side cross panel fasteners (**arrows**).

Wiper motor, removing and installing

◄ Remove left side cross panel fasteners (**arrows**).

– Pull out cross panel.

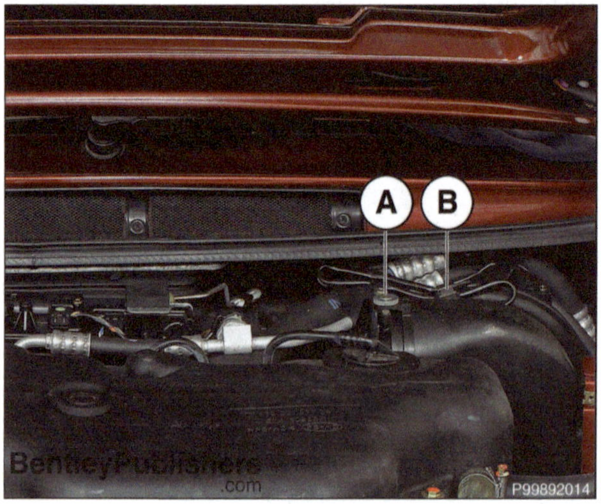

◄ Remove cabin filter housing retaining nut (**A**). Remove clip (**B**) and wiring from atop cabin filter housing.

– Disconnect electrical connector from fan housing lid.

– Lift off cabin filter housing cover.

– Disconnect refrigerant lines from expansion valve. Immediately plug all open connections. Loosen retaining clip.

– Remove front left wheel housing liner. See **50 Fenders**.

◄ Remove fastener (**arrow**) securing wheel housing refrigerant lines to trunk refrigerant lines. Immediately plug all open connections.

– Move refrigerant lines to side.

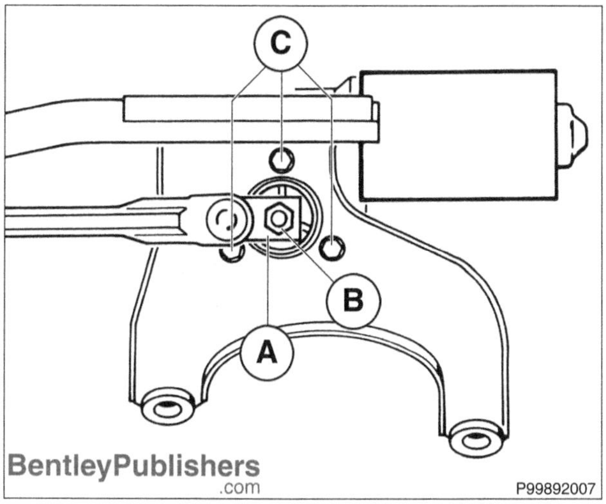

- Using a 21 mm open-ended wrench, counterhold wiper linkage (**A**) and remove wiper linkage to wiper motor fastener (**B**).
- Remove bolts (**C**) securing windshield wiper motor to bracket and withdraw windshield wiper motor through tachometer opening in dashboard.
- Installation is reverse of removal. Note the following:
  - Set wiper motor temporarily in place with a stud bolt and nut. Install two bolts, remove nut and stud, and install third bolt.
  - Connect vehicle battery and switch wiper motor on then off to return motor to park position.
  - Install wiper linkage. Cable and linkage must form a straight line.
  - Switch wiper motor on and check wiper position.
  - Disconnect negative (–) battery cable and cover with insulating material.
  - Vehicles with air conditioning: Install removed components. For A/C refrigerant lines, use new O-rings. Lubricate O-rings with refrigerant oil before installing. Recharge A/C system.
  - When installing airbag, replace airbag self-locking fasteners.

Tightening torques	
Wiper motor to linkage	22 Nm ± 2 Nm (16 ± 1.5 ft-lb)
Wiper motor to bracket	11 Nm (8 ft-lb)

Rear wiper assembly components

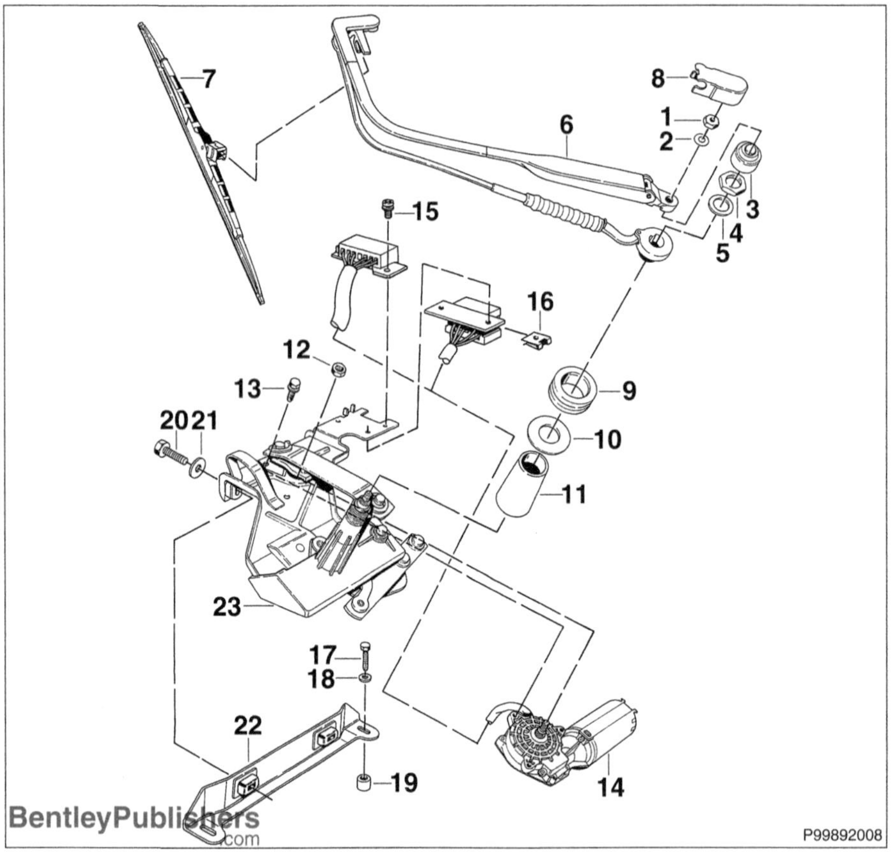

1. M8 nut
   - Tighten to 20 - 25 Nm (15 - 18 ft-lb)
2. Washer
3. Cap
4. M16 Nut
   - Tighten to 14 - 16 Nm (10 - 12 ft-lb)
5. Washer
6. Wiper arm
7. Wiper blade
8. Cap
9. Rubber bushing
10. Washer
11. Cover
12. Nut
    - Tighten to 10 Nm (7 ft-lb)
13. Screw
    - Tighten to 11-12 Nm (8 - 9 ft-lb)
14. Wiper motor
15. Self-tapping screw
16. Sheet metal nut
17. Screw
    - Tighten to 5 Nm (4 ft-lb)
18. Washer
19. Bushing
20. Screw
    - Tighten to 8 - 10 Nm (6 - 7 ft-lb)
21. Washer
22. Retaining bracket
23. Assembly frame

Washers

Windshield washer components

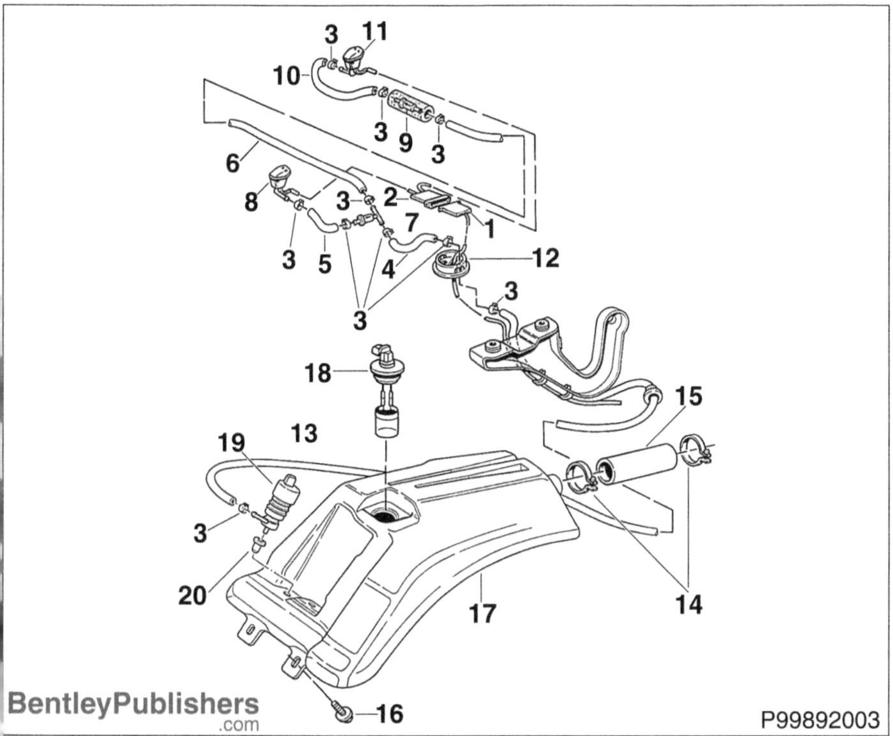

1. Wire harness electrical connector
2. Heated spray nozzle electrical connector
3. Hose clamp
4. Hose
5. Hose
6. Hose
7. Valve
8. Left hand spray nozzle
9. Valve
10. Hose
11. Right hand spray nozzle
12. Hose clamp
13. Hose
14. Hose clamp
15. Hose
16. Screw
17. Fluid reservoir
    - In left front wheel well. Remove wheel housing liner to access.
18. Fluid level sender
19. Windshield washer pump
20. Grommet

Headlight washer components

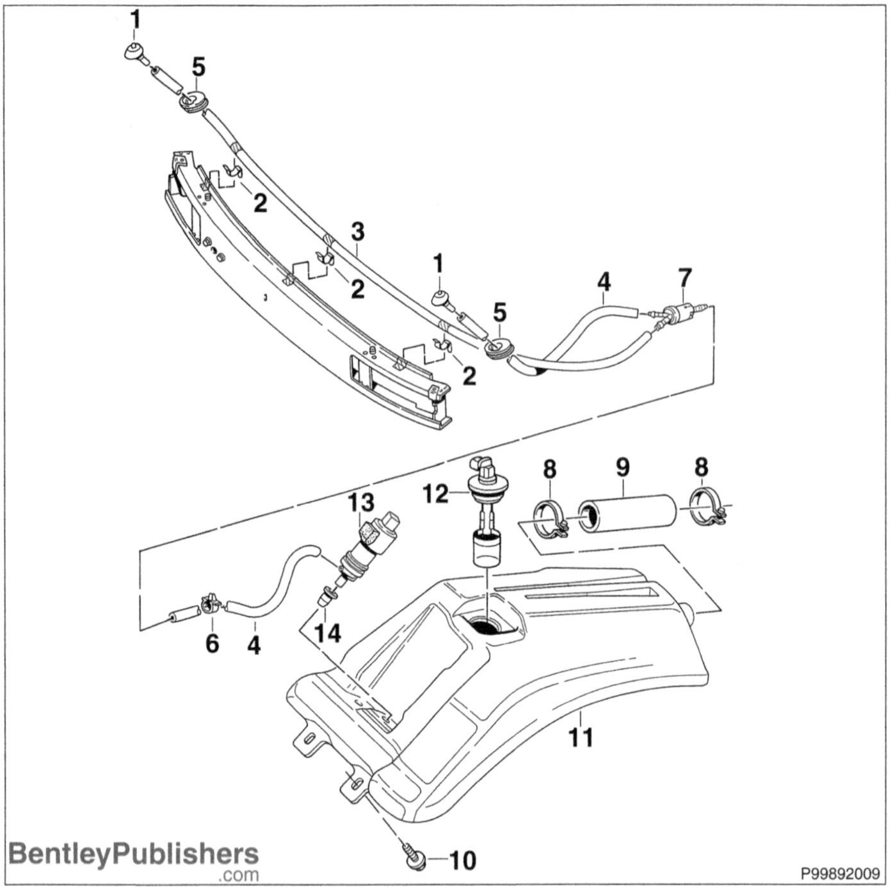

1. Spray nozzle
2. Clamp
3. Hose
4. Hose
5. Grommet
6. Bracket
7. Valve
8. Hose clamp
9. Hose
10. Screw
11. Fluid reservoir
12. Fluid level sender
13. Headlight washer pump
14. Grommet

94 Exterior Lights

GENERAL

This group covers exterior lighting repairs and bulb replacement.

For additional information, see:

- **34 Manual Transmission** for backup light switch.
- **37 Automatic Transmission** for gear position switch.
- **97 Fuses, Relays, Component Locations**
- **EWD Electrical Wiring Diagrams**

Headlight design

Porsche 993 model headlights integrate high and low beams in one headlight assembly.

The standard low beam uses a PES (poly ellipsoidal) headlight and an H1 bulb. The high beam is created with a VF (variable focus) reflector and a separate H1 bulb.

Litronic system

993 models were offered with an optional xenon headlight system known as Litronic.

To 1996: Xenon ignitor and bulb are serviced separately.

From 1997: Xenon ignitor was integrated into the headlight assembly.

Lighting troubleshooting

Many lighting problems can be traced to dirty or corroded sockets, or poorly fitting bulb contacts or connectors. Check that voltage is reaching the bulb and that the socket has a good connection to ground. See **EWD Electrical Wiring Diagrams** for more information.

In case of problems with one xenon headlight, swap over individual components (gas discharge bulb, ignition unit, control module) from the second headlight to isolate the defective component.

Exterior bulb applications

For convenience, exterior bulb applications for 993 models are listed in **Table a**.

Table a. Exterior bulb applications

Backup light	12V-21W
Brake lights/rear tail light	12V-21W/5W
Foglight	H3 12V-55 W
Headlight • Halogen (high and low) • Xenon (low beam)	 H1 12V-55 W DS2 12V-35W
License plate light	12V-5W (festoon)
Side marker light • Front	 12V-5W
Turn signal • Front • Rear (turn/parking)	 12V-21W 12V-21W

Warnings and Cautions

WARNING—

- *Xenon bulbs operate at high voltages. When working on xenon headlight components (control module, ignition unit, gas discharge lamp, drive motor), be sure to switch lights OFF.*
- *Operate xenon bulb only when installed in the headlight reflector.*
- *An operating light, especially a headlight bulb, becomes very hot. Allow bulbs to cool before working on the lighting system.*

> ***CAUTION—***
> - *Do not touch the glass of bulbs with bare skin. Dirt and skin oils may cause a bulb to fail prematurely. If necessary wipe bulb using a clean cloth dampened with rubbing alcohol.*
> - *Use only original equipment replacement bulbs. Non-original bulbs may cause false failure readings on the dashboard display.*
> - *Use only specified bulbs. A bulb with higher wattage may cause damage to bulb housing.*
> - *To avoid marring car paint or trim, work with a plastic prying tool or wrap screwdriver tip with tape.*
> - *Clean headlights with water and standard window cleaning agent.*

Lighting Controls

Light switch, removing and installing

- Disconnect negative (–) battery cable.

◄ Pull headlight knob (**arrow**) to remove.

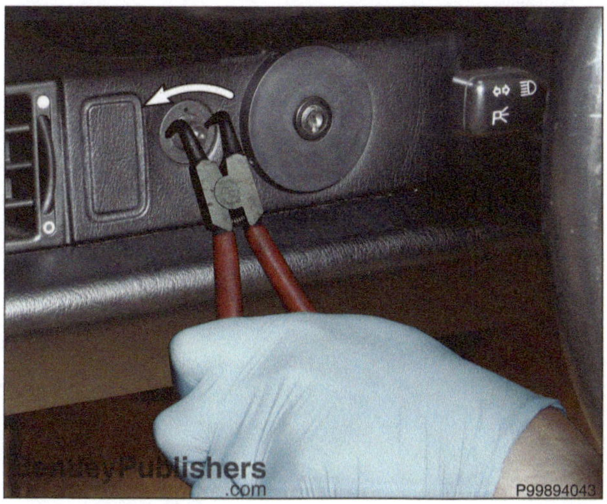

◄ Using snap ring pliers in retaining ring holes, turn threaded retaining ring counterclockwise to remove.

- Push light switch into dashboard.
- Label and disconnect wires.
- Remove switch from below dashboard.
- Installation is reverse of removal.

94

Hazard warning switch, removing and installing

– Disconnect negative (–) battery cable.

⊲ Using a plastic prying tool, lever out lower then upper switch cover plates from center console.

⊲ Disconnect electrical connectors (**arrows**) and remove switch covers.

NOTE–

- *Lower switch cover electrical connectors shown. Upper switch cover is similar.*

⊲ Unclip tabs (**arrows**) and disconnect hazard warning switch from rear of switch cover.

– Installation is reverse of removal.

Brake light switch, removing and installing

- Disconnect negative (–) battery cable.
- Working in pedal area, remove floor mat and peel carpet from pedal floor board.

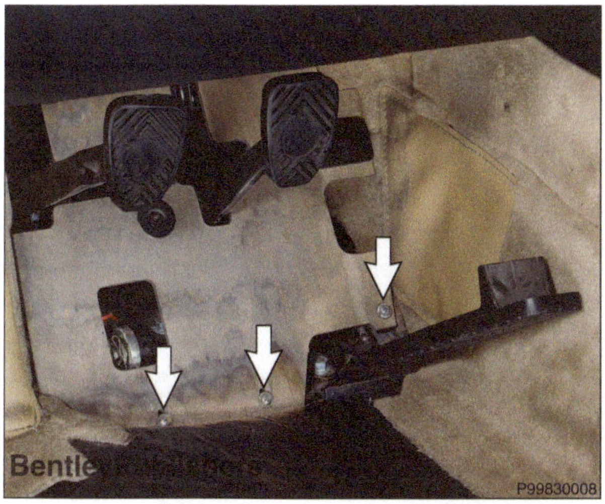

◄ Remove retaining screws at base of floor board (**arrows**).

- Pull accelerator pedal from top and off of linkage ball mount.
- Remove floor board.

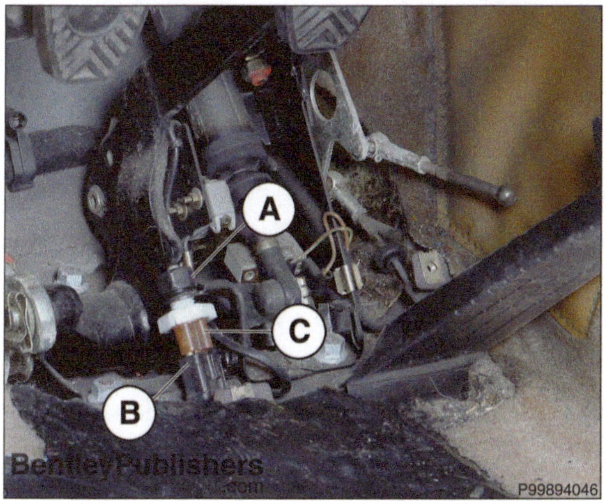

◄ Push brake pedal forward by hand and remove nut (**A**).

- Disconnect electrical connector (**B**) and remove brake light switch (**C**).

NOTE–

- *Vehicle with manual transmission shown. Vehicle with automatic transmission similar.*

- Installation is reverse of removal.

Lighting fuses

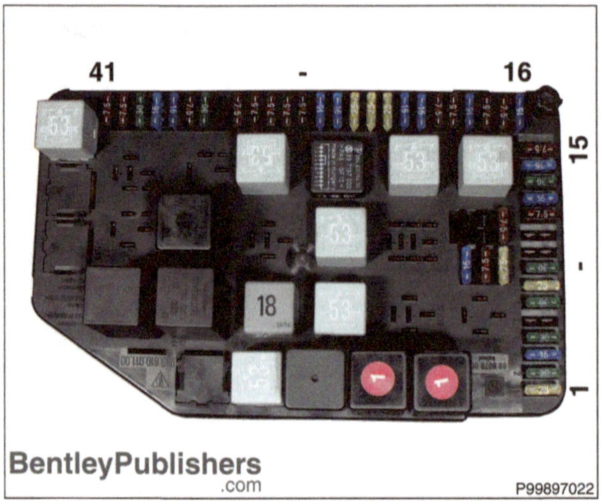

Lighting fuses are located along the perimeter of the central electric panel. Be sure to replace a bad fuse with one of equivalent rating.

Table b. Lighting system fuses

Protected circuit	Fuse	Rating in amps
Backup light (automatic transmission)	**15**	7.5
Backup light (manual transmission)	**14**	15
Brake light	**19**	15
Fog light	**14**	15
Front left turn signal	**31**	7.5
Front right turn signal	**32**	7.5
Left high beam	**20**	7.5
Left low beam	**22**	7.5
Left side marker light	**40**	7.5
Left side turn signal	**30**	7.5
License plate light	**36**	7.5
Rear fog light	**14**	15
Rear left turn signal	**30**	7.5
Right high beam	**21**	7.5
Right low beam	**23**	7.5
Right rear turn signal	**33**	7.5
Right side marker light	**41**	7.5
Right side turn signal	**33**	7.5

Front Lighting Repairs

Headlight aim, adjusting

Open front trunk and peel back trim on inner fender next to headlight assembly. Pull back adjuster access hole plug.

– Use 5 mm Allen wrench to adjust headlight aim:

1. Lateral adjustment: Turn upper adjuster
   - clockwise: Beam moves right.
   - counterclockwise: Beam moves left.
2. Height adjustment: Turn lower adjuster
   - clockwise: Beam moves up.
   - counterclockwise: Beam moves down.

Headlight assembly, removing and installing

– Disconnect negative (–) battery cable.

◀ Open trunk and pull back side trim on inner fender next to headlight assembly.

– Move headlight locking lever from position **A** to position **C** to unlock, and then to position **B**.

– Slide headlight forward out of fender. If necessary, turn headlight lock slightly to disengage.

◀ If removing headlight mount, carefully pry electrical connector (**arrow**) from bracket.

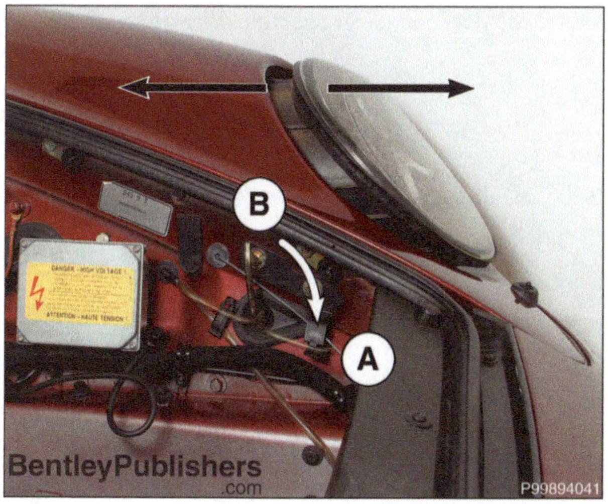

◀ When installing headlight, line up aligning pins on sides of headlight with headlight carrier plate rails. Press headlight firmly into fender, then turn headlight locking lever from position **B** to position **A**.

– Check level of headlight housing in relation to fender.

94

Headlight bulb, replacing

– Disconnect negative (–) battery cable.

– Remove headlight assembly. See **Headlight assembly, removing and installing** in this repair group.

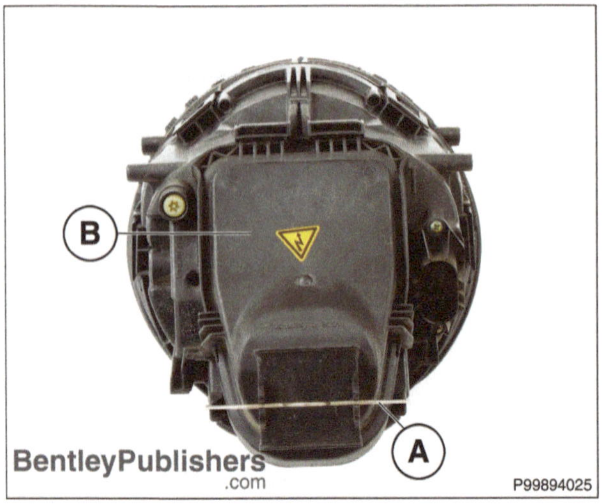

◄ Pivot retainer bar (**A**) down and remove cover (**B**).

◄ Halogen: Release retainer catch (**arrows**) and replace low beam bulb (**A**) or high beam bulb (**B**) as applicable.

◄ Litronic: Rotate electrical connector counterclockwise to remove.

< Litronic: Press retaining ring (**arrow**) and turn counterclockwise to remove.

< Lift out Litronic bulb (**arrow**) to remove.

– Installation is reverse of removal.

Front turn signal / parking light bulb, replacing

– Disconnect negative (–) battery cable and cover with insulating material.

< Remove screw (**A**). Pull turn signal lens (**B**) in direction of **arrow.**

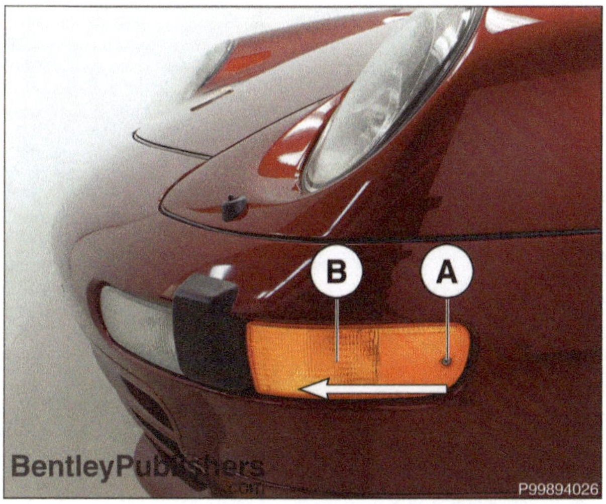

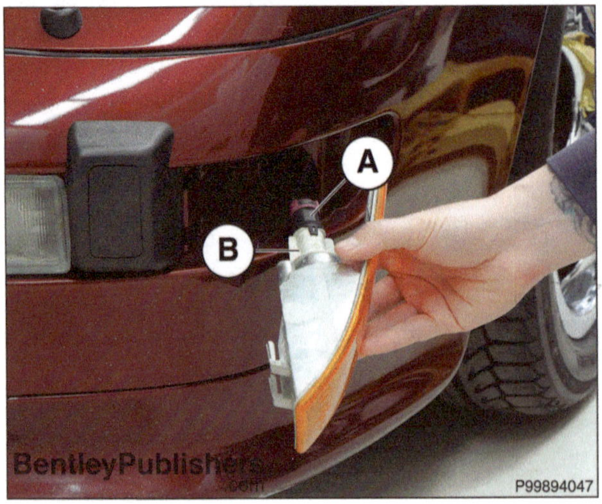

- Disconnect electrical connector (**A**).
- Twist bulb socket (**B**) counterclockwise to remove from lens.
- Twist bulb counterclockwise to remove from socket.
- Installation is reverse of removal.

Front foglight bulb, replacing

- Disconnect negative (–) battery cable and cover with insulating material.
- Using a plastic prying tool, gently lever cover off rubber buffer.

- Remove nuts and pull rubber buffer (**arrow**) forward to remove.

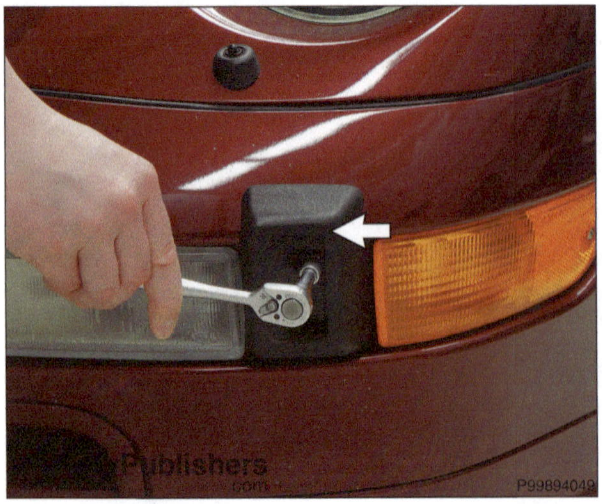

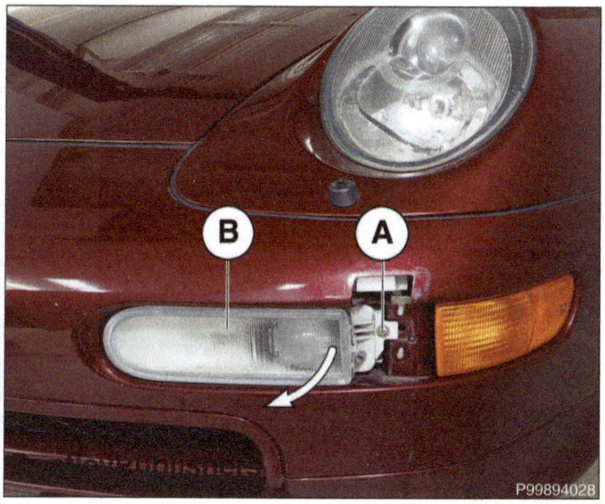

- Remove screw (**A**).
- Pivot out lens (**B**) and disconnect electrical connectors.

- Twist side marker light bulb (**arrow**) counterclockwise to remove.

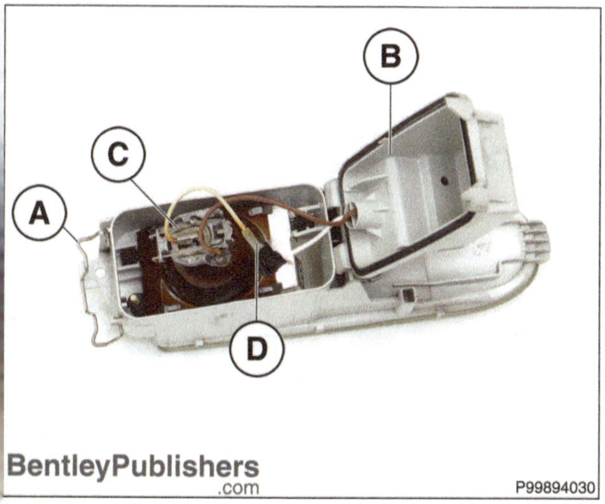

- Release retainer bar (**A**) and pivot cover (**B**) out.
- Release retaining bracket (**C**).
- Disconnect plug (**D**) on bulb cable and remove bulb.
- Installation is reverse of removal.

Rear Lighting Repairs

Taillights, servicing

– Disconnect negative (–) battery cable.

◀ Working on left side, remove screw (**arrow**) and slide rear side lens out of body.

– Repeat for right side.

◀ Disconnect electrical connectors (**arrows**) and remove side lens.

– Repeat for right side.

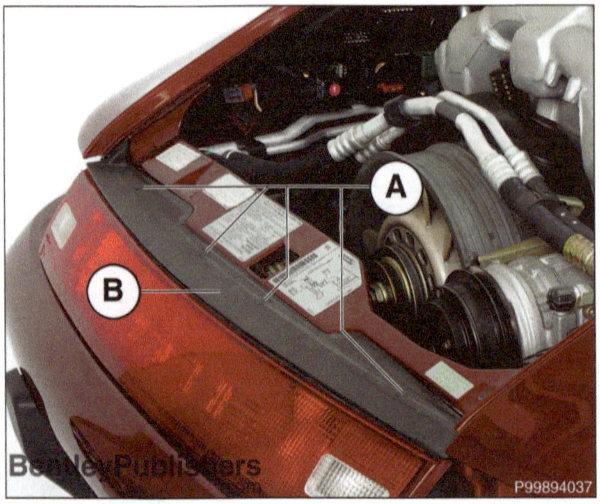

◀ Remove fasteners (**A**) and cover piece (**B**).

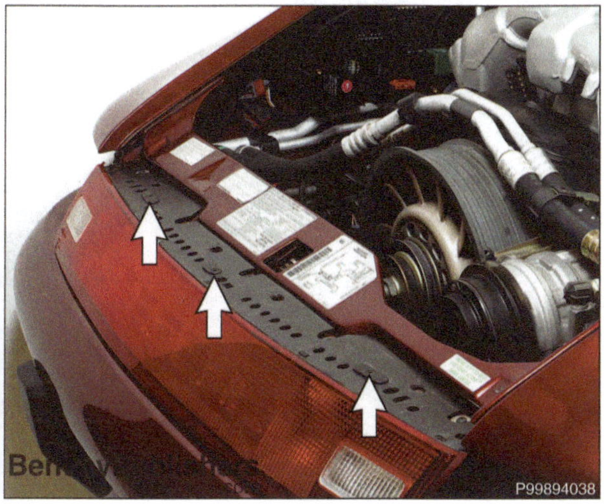

< Remove fasteners and support pieces (**arrows**).

- Slide center taillight lens out of body.
- Disconnect electrical connectors.

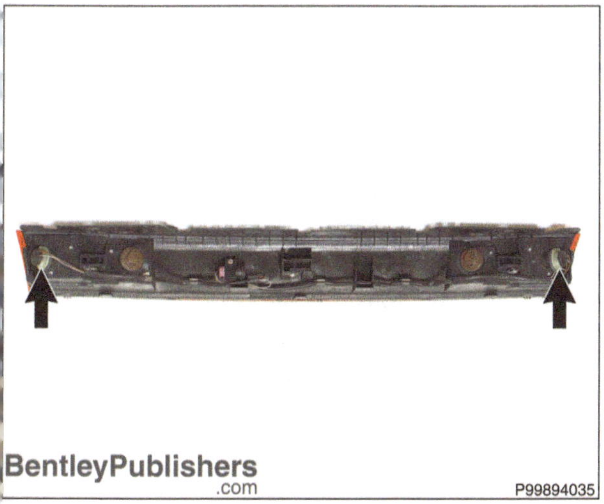

< Twist backup light bulbs (**arrows**) to remove from socket.

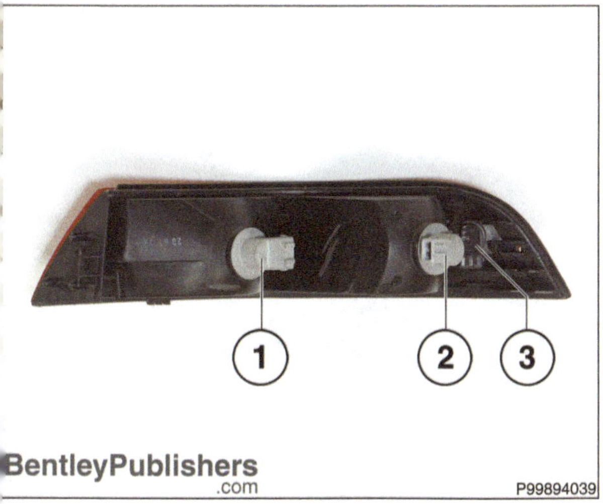

< Twist bulbs to remove from socket.

1. Side light/brake light
2. Turn signal
3. Side marker light

Center brake light, servicing

The center brake light bulbs are soldered in place. For service, it is recommended to contact a Porsche workshop.

License plate light bulb, removing and installing

– Disconnect negative (–) battery cable and cover with insulating material.

< Remove screws (**arrows**).

– Using a plastic prying tool, lever out and lower license plate light lenses.

< Disconnect electrical connectors (**arrows**).

– Carefully lever out bulb from contacts.

– Installation is reverse of removal, noting the following:

- Be sure rubber gasket is properly seated against bumper.

96 Interior Lights, Anti-theft

GENERAL

This repair group covers interior light bulb replacement as well as descriptions and repair information for the electronic immobilizer, central locking and alarm systems.

See the following repair groups for additional information:

- **57 Doors** for door locks
- **90 Instruments, Horns** for instrument cluster bulbs
- **94 Exterior Lights** for exterior bulbs

Warnings and Cautions

WARNING—

- *An operating light becomes very hot and will cause injury. Allow bulbs to cool before working on the lighting system.*

CAUTION—

- *Do not touch the glass of bulbs with bare skin. Dirt and skin oils may cause a bulb to fail prematurely. If necessary wipe bulb using a clean cloth dampened with rubbing alcohol.*
- *Use only original equipment replacement bulbs. Non-original bulbs may cause false failure readings on the dashboard display.*
- *Use only specified bulbs. A bulb with higher wattage may cause damage to bulb housing.*
- *To avoid marring car paint or trim, work with a plastic prying tool or wrap screwdriver tip with tape.*

Interior Lighting

The interior and trunk are illuminated by incandescent lights. The vehicle dome light is integrated into the infrared passenger compartment monitoring sensor.

For convenience, interior bulb applications are listed in **Table a**.

Table a. Interior bulb applications	
Engine compartment light	10 W (festoon)
Glove compartment light	3 W (festoon)
Roof lights	10 W (festoon)
Trunk light	10 W (festoon)

Trunk light bulb, replacing

– Disconnect negative (–) battery cable and cover with insulating material.

> ***CAUTION—***
> - *Prior to disconnecting the battery, read the battery disconnection cautions in* **00 Warnings and Cautions**.
> - *Be sure to have the radio anti-theft code before disconnecting the battery.*

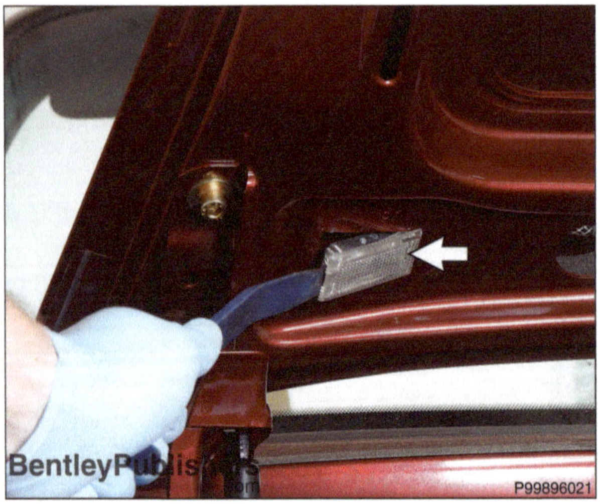

◄ Using a plastic prying tool or small flat blade screwdriver, carefully lever out trunk light lens (**arrow**).

◄ Carefully lever bulb (**arrow**) out of contacts.

– Installation is reverse of removal, noting the following:

- Carefully insert lens from one side first and press lens into place.

Engine compartment light bulb, replacing

– Disconnect negative (–) battery cable and cover with insulating material.

> ***CAUTION—***
>
> - *Prior to disconnecting the battery, read the battery disconnection cautions in* **00 Warnings and Cautions**.
> - *Be sure to have the radio anti-theft code before disconnecting the battery.*

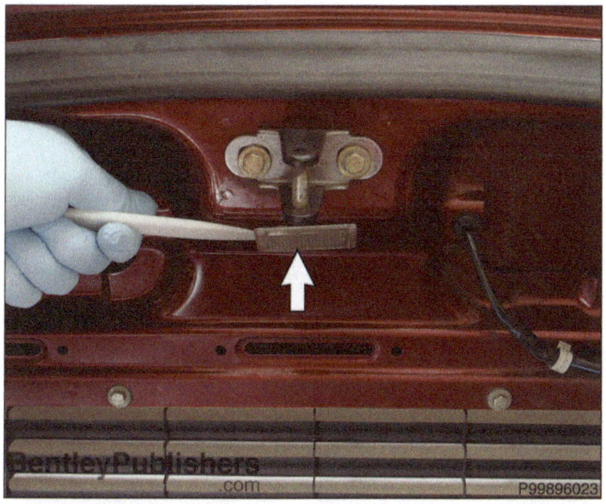

◄ Using a plastic prying tool or small flat blade screwdriver, carefully lever out engine compartment light lens (**arrow**).

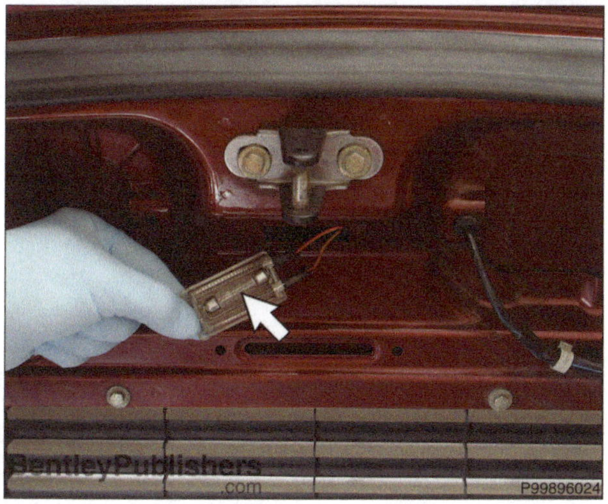

◄ Carefully lever bulb (**arrow**) out of contacts.

– Installation is reverse of removal, noting the following:

- Carefully insert lens from one side first and press lens into place.

Interior light bulb, replacing

– Disconnect negative (–) battery cable and cover with insulating material.

> ***CAUTION—***
>
> - *Prior to disconnecting the battery, read the battery disconnection cautions in* **00 Warnings and Cautions**.
> - *Be sure to have the radio anti-theft code before disconnecting the battery.*

◄ Using a plastic prying tool or small flat blade screwdriver, carefully lever out interior light lens (**arrow**).

– Carefully lever bulb out of contacts.

– Installation is reverse of removal, noting the following:

- Carefully insert lens from one side first and press lens into place.

NOTE–

- *Interior light is located in windshield frame between sun visors on Cabriolet and Targa models.*

Glove compartment light bulb, replacing

– Disconnect negative (–) battery cable and cover with insulating material.

> ***CAUTION—***
>
> - *Prior to disconnecting the battery, read the battery disconnection cautions in* **00 Warnings and Cautions**.
> - *Be sure to have the radio anti-theft code before disconnecting the battery.*

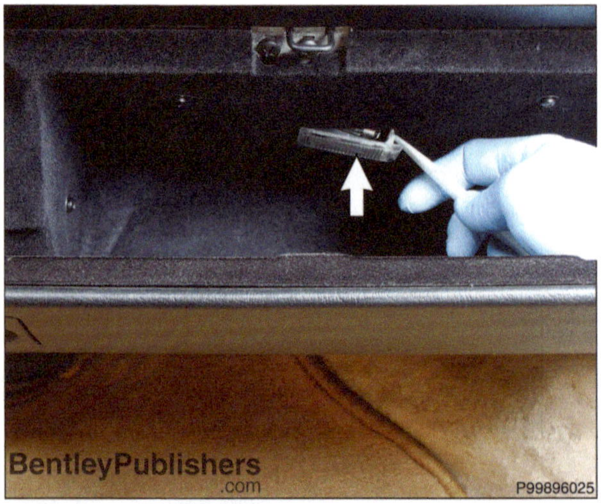

◄ Using a plastic prying tool or small flat blade screwdriver, carefully lever out glove compartment light lens (**arrow**).

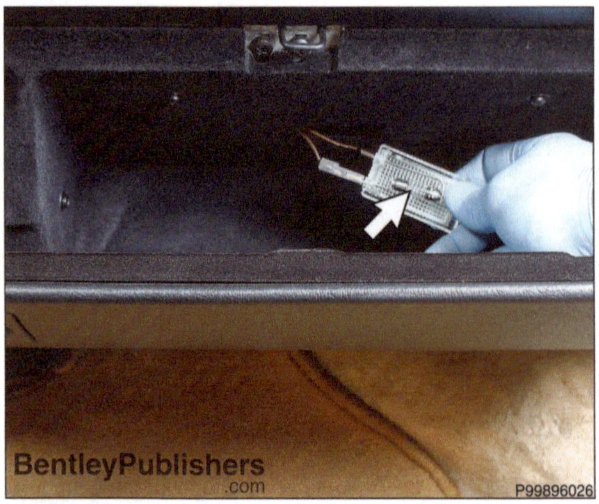

◄ Carefully lever bulb (**arrow**) out of contacts.

– Installation is reverse of removal, noting the following:

- Carefully insert lens from one side first and press lens into place.

ANTI-THEFT SYSTEM

Anti-theft system components

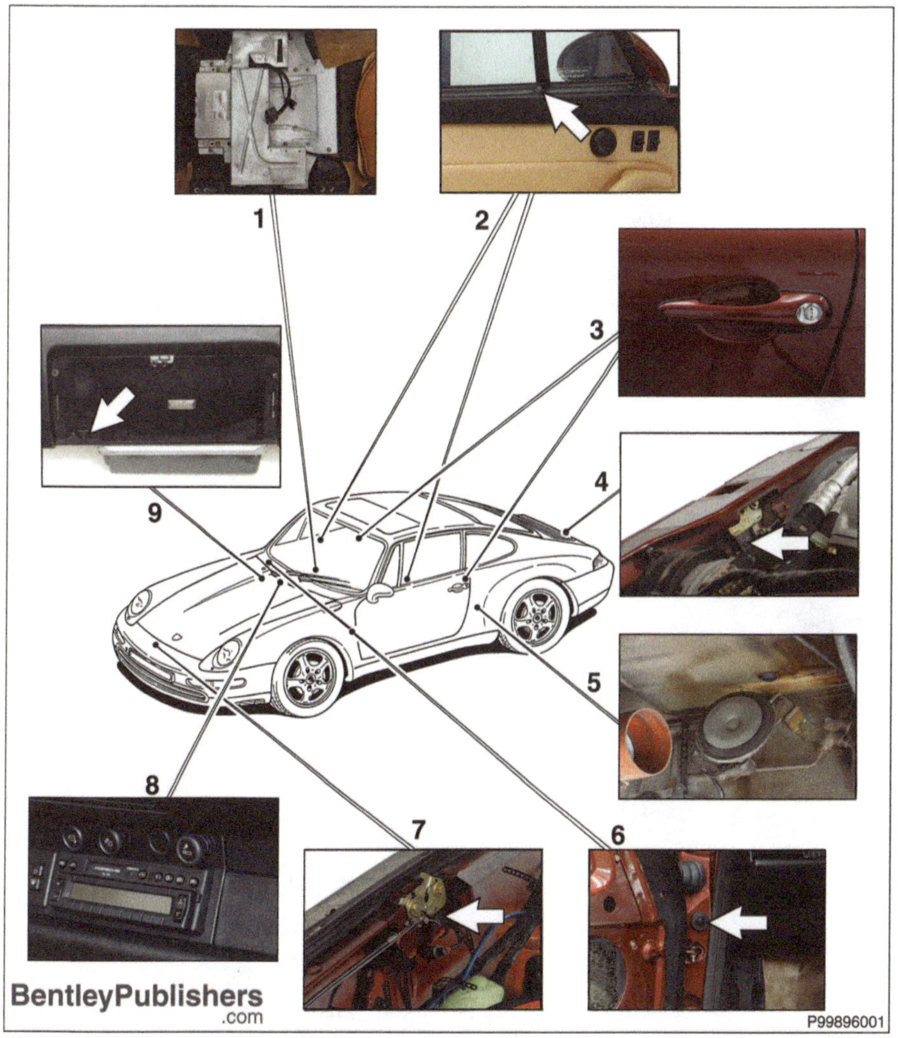

1. Anti-theft system control unit
   - Under driver's seat
2. Alarm readiness lamps
3. Door lock microswitch
4. Engine hood lock microswitch
5. Alarm horn
6. Door contact switches
7. Trunk lid lock microswitch
8. Radio alarm contact
9. Glove compartment switch

Vehicle security

The anti-theft system consists of a forced entry alarm, integrated central locking system, and electronic immobilizer (also referred to as anti-drive-off). The system performs the following functions:

- Electronic immobilization of DME
- Central locking
- Alarm system
- Infrared passenger compartment monitoring
- Radio anti-theft

NOTE–

- *In Porsche technical literature, the electronic immobilizer is referred to as anti-drive-off. The immobilizer, central locking and alarm share one central control module.*

The immobilizer is automatically activated when the alarm system is armed. If vehicle alarm is triggered, terminal 15 for the DME control unit is interrupted, disabling the ignition and fuel injection systems.

Central locking

Always activate central locking using remote key or central locking button on center console. In case of central locking system malfunction, locks can be operated manually using key in door lock cylinder. See **Door lock emergency operation** in this repair group.

If the door is not latched properly, the vehicle cannot be locked.

If the vehicle is unlocked using the keyless entry transmitter but none of the doors are opened within one minute, doors are re-locked automatically, and the alarm and immobilizer are reactivated.

Central locking switch (**arrow**) is located in center console switch panel. An indicator lamp inside switch illuminates when doors are locked.

Ignition must be switched ON to use central locking switch.

Doors are opened and electrically unlocked by pulling on inner door handle.

Alarm system

The alarm system is activated when the doors are locked. An LED in each door will flash to indicate that the alarm system is activated.

When the doors are unlocked, the alarm system is deactivated and the LEDs stop flashing.

By activating the key or remote transmitter twice, only the alarm is activated. Select this mode if people or animals are in the vehicle. The doors are locked but can be opened from inside using the central locking button.

The alarm monitors the following components:

- Both doors
- Trunk lid
- Engine compartment lid
- Glove compartment
- Radio
- Ignition

If the alarm is triggered, the alarm horn will sound and the interior and hazard warning lights will flash for approximately four minutes. The alarm LEDs will switch to a double flashing signal. Unlocking doors will deactivate alarm.

The alarm LEDs (**arrow**) are located in the top of the door panel.

Status of the alarm system is indicated by the alarm LEDs. See **Table b**.

Table b. Alarm LED status	
Alarm status	**LED signal**
Doors unlocked, alarm off	No flashing
Doors locked, alarm activated	Slow flashing
No fault condition detected during alarm system self-test	Fast flashing
Alarm system was switched on by locking door three times in rapid succession	Double flashing
Fault condition detected during alarm system self-test	Double flashing
Alarm system was triggered at least once	Double flashing

Passenger compartment monitor

The optional infrared passenger compartment monitor is integrated into the interior light, located between the sun visors. It consists of an infrared emitter and integrated interior light.

1. Infrared transmitter
2. Combined interior light lens and passenger compartment monitor button
3. Infrared receiver

If a person remains in the vehicle with the doors locked, deactivate the passenger compartment monitor.

– To deactivate passenger compartment monitor, press and hold button (**2**) for approximately three seconds.

A flashing signal indicates that passenger compartment monitor has been deactivated.

The monitor will flash again when doors are locked.

If the convertible top, Targa roof, or wind deflector is open, the monitor is automatically deactivated.

Remote key

◀ To 1995, vehicles equipped with immobilizer had a key with LED light and a separate remote control.

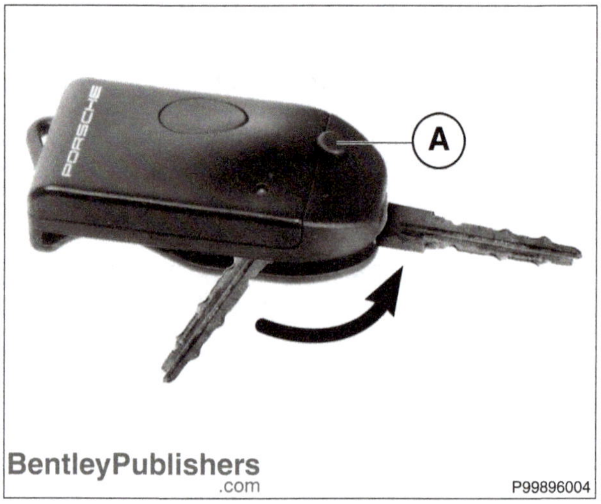

◀ From 1996, vehicles equipped with immobilizer had a remote control integrated with folding key.

– To unlock key from housing, press button **A** and pivot key out in direction of **arrow**. Press button **A** again to fold key back into housing.

◀ With key removed from ignition and doors closed, press remote button (**arrow**) to activate:

- central locking system.
- alarm system.
- immobilizer.

– Press button (**arrow**) again to deactivate immobilizer and alarm systems and unlock doors.

Remote key, replacing battery

NOTE–

- *Early remote key shown. Late remote key is similar.*

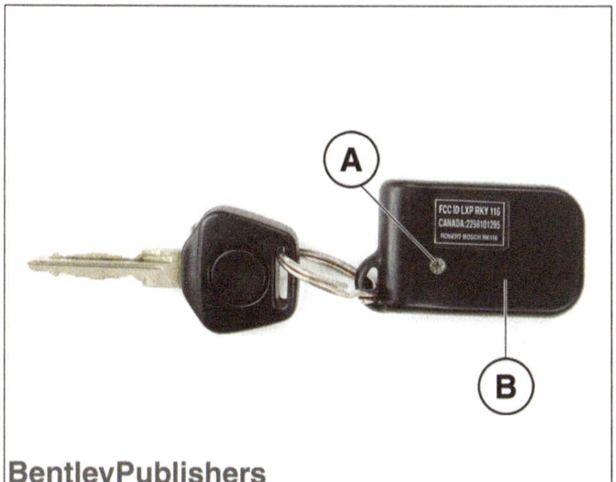

P99896018

- Remove fastener (**A**).
- Remove remote key cover (**B**).

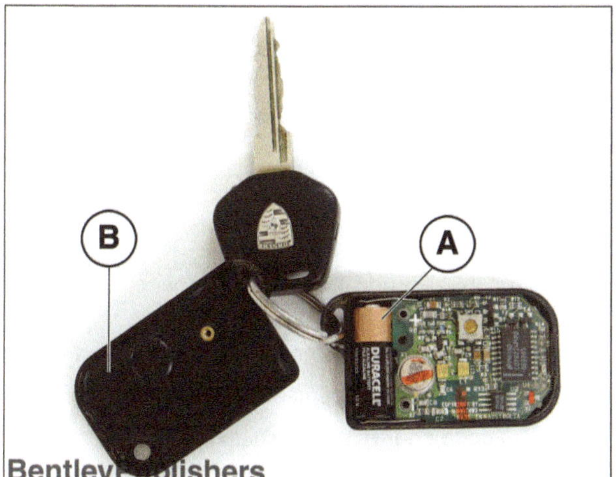

P99896019

- Replace battery (**A**) and install cover (**B**).

Table c. Remote key battery specifications	
Remote key	A23, 12V

Remote key, programming

Programming of the remote key requires a four digit code that is unique to the vehicle. Use this procedure to program a new remote. The code is supplied with the original documentation when the vehicle was new. An authorized Porsche Center can provide the code. Note that proof of vehicle ownership may be required.

NOTE–

- *The remote will not program if there are faults in the anti-theft system. Lock the car with the remote to check for proper LED operation. The LEDs in door should flash quickly for 10 seconds and then slow down to once per second. If the door LEDs double flash instead of the single slow flash, there are faults present in the anti-theft system. Faults must be corrected before the remote can be reprogrammed.*

P99896029

- Using fingernail or plastic prying tool, carefully pry out cap from door lock cylinder.
- Using key in lock cylinder, unlock door, lock door, and unlock door again within five seconds.
- Open door, insert ignition key and switch ignition ON within ten seconds.

P99896028

- Immobilizer warning lamp (**arrow**) in clock will illuminate and go out after fifteen seconds.
- Switch ignition OFF and ON within five seconds. Immobilizer warning light will illuminate and start flashing after fifteen seconds.
- Enter first digit of immobilizer code by switching ignition ON and OFF in rapid succession.
  - Start entering code within five seconds.
  - Each digit of immobilizer code represents number of ON / OFF switch repetitions.
  - Digit zero is entered by switching ignition on and off ten times.
- Wait for warning light acknowledgement.
- Enter next digit of code and wait for warning light acknowledgement.
- Repeat until four-digit code has been entered. When code has been entered properly, warning light will start flashing.

NOTE–

- *If warning light does not flash after code number has been entered, exit vehicle and lock door with key in lock cylinder. Repeat procedure.*
- *Additional remote keys must be programmed within one minute.*

- Press and hold remote key button until alarm readiness lamps in doors flash.
- Operate remote keys in succession until all remote keys have been programmed.

NOTE–

- *A maximum of four remote keys may be assigned to a vehicle at the same time.*
- *If no more remote control units are operated within 1 minute or the ignition is switched off, programming is stopped.*

Immobilizer, deactivating

Deactivating the immobilizer anti-theft system requires a four digit code that is unique to the vehicle. The code is supplied with the original documentation when the vehicle was new. An authorized Porsche Center can provide the code. Note that proof of vehicle ownership may be required.

In case of remote key, immobilizer, or central locking malfunction, the immobilizer system can be deactivated to allow starting of the vehicle.

NOTE–

- *Deactivating procedure must be completed within 100 seconds.*

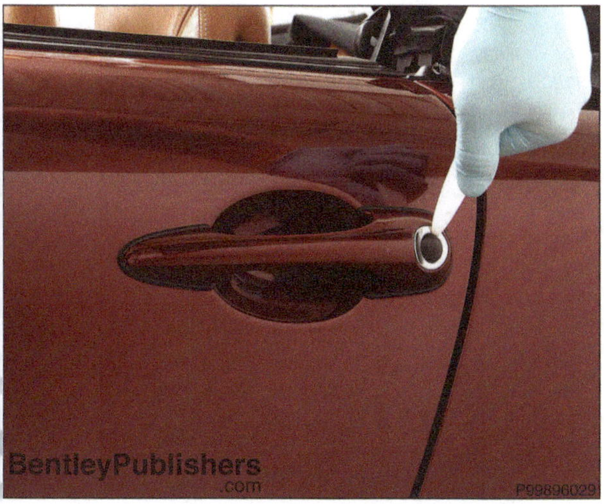

- Using fingernail or plastic prying tool, carefully pry out cap from door lock cylinder.
- Using key in lock cylinder, unlock door, lock door, and unlock door again within five seconds.
- Open door and switch ignition on within another ten seconds.

- Immobilizer warning lamp (**arrow**) in clock will illuminate and go out after fifteen seconds.
- Switch ignition OFF and ON within five seconds. Immobilizer warning light will illuminate and start flashing after fifteen seconds.
- Enter first digit of immobilizer code by switching ignition ON and OFF in rapid succession.
  - Start entering code within five seconds.
  - Each digit of immobilizer code represents number of ON / OFF switch repetitions.
  - Digit zero is entered by switching ignition on and off ten times.
- Wait for warning light acknowledgement.
- Enter next digit of code and wait for warning light acknowledgement.
- Repeat until four-digit code has been entered. When code has been entered properly, warning light will start flashing.
- Switch ignition on and off one more time.

NOTE–

- *If warning light does not flash after code has been entered, exit vehicle and lock door with key in lock cylinder. Repeat procedure.*

- Immobilizer is now deactivated. Start engine.
  - Alarm system and central locking are now unlocked.

Electronic immobilizer control module,removing and installing

- Disconnect negative (–) battery cable and cover with insulating material.
- Remove driver seat. See **72 Seats**.

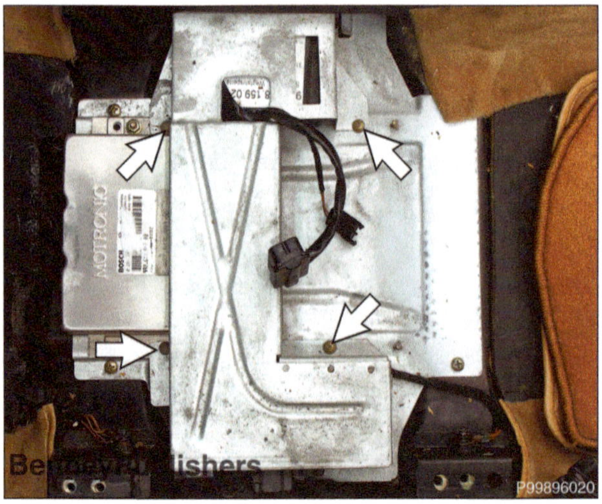

- Using a cross cut chisel, remove shear bolts (**arrows**) and remove control unit cover.
- Disconnect electrical connector and remove control unit.
- Installation is reverse of removal. Use new M6 x 13 mm shear bolts when installing.

97 Fuses, Relays, Component Locations

TABLES

General

This repair group provides a brief description of the electrical system and covers fuses, relays, ground locations and electrical component locations, primarily using photos or illustrations. Basic electrical system troubleshooting tips are also covered here.

Electrical system

Porsche 993 models are equipped with a 12-volt, direct current (dc), negative-ground electrical system. The voltage regulator maintains the voltage in the system at approximately the 12 vdc rating of the battery, and all circuits are grounded by direct or indirect connection to the negative (–) terminal of the battery.

The complex nature of 993 electrical systems requires a large number of fuses and electrical components. Locating the correct component in this array of equipment is an important first step in electrical diagnosis. Electrical equipment and accessories installed vary depending on model and model year. When working on electrical problems, use this component locator as well as wiring color codes from **EWD Electrical Wiring Diagrams** to identify the correct component(s).

Most electrical circuits are protected by fuses. Fuses for individual components are primarily housed in the central electric panel in the rear of the trunk, passenger side (**arrow**).

For a full listing of fuses and fuse locations, see **Fuse Locations and Ratings** in this repair group.

Battery notes

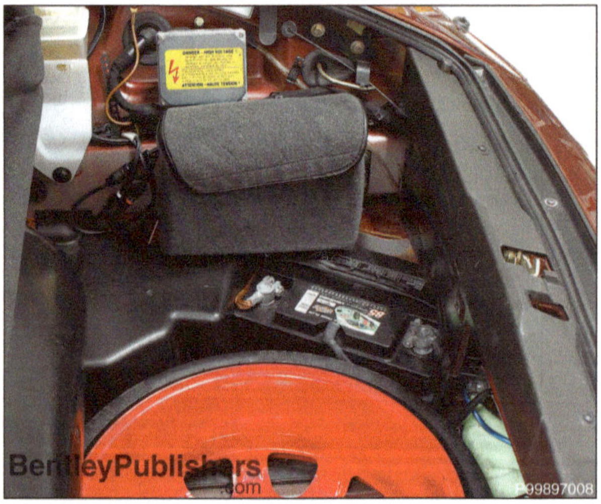

The battery is located in the front of the trunk on the left side.

If the battery is disconnected and reconnected, the following components may need to be reset:

- Clock
- On-board computer
- Radio anti-theft code. Be sure the radio code is available.

NOTE —

- *See* **91 Radio** *for the procedure on entering the radio security (anti-theft) code.*

Warning and Cautions

WARNING—

- *An airbag is an explosive device. Handle with extreme care. See warnings and cautions in* **69 Seat Belts, Airbags**.
- *The ignition system of the car operates at lethal voltages. If you have a weak heart or wear a pacemaker, do not expose yourself to ignition system electrical currents. Take extra precautions when working on the ignition system or when servicing the engine while it is running or the ignition key is ON. See* **28 Ignition System** *for additional ignition system warnings and cautions.*

CAUTION—

- *Before disconnecting the battery, read the battery disconnection cautions in* **00 Warnings and Cautions**.
- *Disconnecting the battery may erase control module stored DTC's. Read out and record all control module fault codes before disconnecting the battery.*
- *Record radio station presets and obtain the radio activation code before disconnecting the battery.*
- *Do not disconnect the battery with the engine running.*
- *Switch the ignition OFF and remove the negative (–) battery cable before removing any electrical components. Connect and disconnect electrical connectors and ignition test equipment leads only while the ignition is switched OFF.*
- *Relay and fuse positions are subject to change and may vary from car to car. If questions arise, an authorized Porsche dealer is the best source for the most accurate and up-to-date information.*
- *Use a digital multimeter for electrical tests. Switch the multimeter to the appropriate function and range before making test connections.*
- *Many control modules are static sensitive. Static discharge damages them permanently. Handle the modules using proper static prevention equipment and techniques.*
- *To avoid damaging harness connectors or relay panel sockets, use jumper wires with flat-blade connectors that are the same size as the connector or relay terminals.*
- *Do not try to start the engine of a car which has been heated above 176° F (80°C) (for example, in a paint drying booth). Allow it to cool to normal temperature.*
- *Disconnect the battery before doing any electric welding on the car.*
- *Do not wash the engine while it is running, or any time the ignition is ON.*
- *Choose test equipment carefully. Use a digital multimeter with at least 10 MW input impedance, or an LED test light. An analog meter (swing-needle) or a test light with a normal incandescent bulb may draw enough current to damage sensitive electronic components.*
- *Do not use an ohmmeter to measure resistance on solid state components such as control modules.*
- *Disconnect the battery before making resistance (ohm) measurements on a circuit.*

ELECTRICAL COMPONENT LOCATIONS

Component locations table

Table a is a cross-referenced listing of Porsche 993 electrical components. Figure and page numbers in columns 3 and 4 refer to photographs and illustrations in this repair group.

Keep in mind the following:

- Every component is not installed in every car.
- Due to changes in production, component locations may vary from what is illustrated. Consult your Porsche dealer for the latest information.
- Component location photos are from a 1997 Carrera Cabriolet model with standard transmission, unless noted otherwise.

Table a. Electrical component locations

Component	Location	Fig #	Page #
ABS control unit	In trunk, passenger side	**Fig. 9**	97-15
ABS hydraulic unit	In trunk, driver side	**Fig. 9**	97-15
ABS/ABD hydraulic unit	In trunk, driver side	**Fig. 9**	97-15
ABS relay	In trunk, passenger side	**Fig. 6**	97-13
ABS wheel speed sensors	At wheel carriers	**Fig. 17**	97-18
A/C compressor relay	Relay carrier plate in engine compartment	**Fig. 5**	97-12
Airbag control unit	Below dashboard in center of car		
Alarm (immobilizer) horn	Rear crossmember, left-hand side	**Fig. 13**	97-16
Alarm system / central locking system control unit	Under passenger seat		
Amplifier	Under passenger seat		
Antenna amplifier	In trunk		
Automatic transmission control unit	*see* Tiptronic transmission control unit		
Automatic transmission gear position switch	Right side of transmission		
Backup light relay (Tiptronic)	Central electric panel	**Fig. 4**	97-11
Backup light switch (manual transmission)	Right side of transmission	**Fig. 12**	97-16
Blower final stage (front blower motors)	In rear of trunk, on cabin microfilter housing, right side		
Blower motors, front (left, right)	Under cowl panel, left, right	**Fig. 7**	97-14
Blower motor, rear	In engine compartment, left side		
Blower resistor (rear blower motor, engine compartment)	In engine compartment, left side in blower air pipe		
Blower motor, rear, temperature sensor (NTC)	In duct to rear blower motor		
Brake booster pump (Carrera 4, Turbo)	In trunk		
Brake booster pump relay (Carrera 4, Turbo)	In trunk, passenger side		
Brake light switch	At brake pedal below floor panel		
Buzzer contact	On ignition lock		
Buzzer relay (1995)	In front of instruments, driver side		
Buzzer relay (from 1996)	In front of instrument panel		
Carbon canister shutoff valve	In left rear wheel housing		
Central electric panel (fuse/relay panel)	Rear of trunk, passenger side		97-10
Central informer (for instruments)	Behind knee bar, driver side, left.		
Clutch switch	Under floor panel by clutch pedal		

Table a. Electrical component locations

Component	Location	Fig #	Page #
Condenser blower fan (A/C)	In left front wheel housing	**Fig. 16**	97-17
Condenser blower fan relay (A/C)	Central electric panel	**Fig. 4**	97-11
Condenser blower fan resistor (A/C)	In front left wheel housing		
Convertible top control motor(s)	Behind rear seat		
Convertible top control unit	In dashboard, driver side		
Convertible top lock motor, left	At convertible top frame, left		
Convertible top lock motor, right	At convertible top frame, right		
Crankshaft position sensor	On left side of engine at flywheel end		
Cruise control control unit	Under passenger seat		
Cruise control actuator	Top left of engine under intake manifold	**Fig. 19**	97-19
Daytime running light relay - Canada	Central electric panel	**Fig. 4**	97-11
Diagnostic connector (OBD, OBD II)	In center console	**Fig. 20**	97-19
DME main relay	Central electric panel	**Fig. 4**	97-11
DME control module	Under driver seat		
Dual-tone horn	In right front wheel housing	**Fig. 15**	97-17
Dual-tone horn relay	Central electric panel	**Fig. 4**	97-11
ECM (Engine Control Module)	*see* DME control module		
Engine compartment blower	Top left side of engine		
Engine compartment blower relay	Engine compartment relay panel	**Fig. 5**	97-12
Engine compartment fuses	Engine compartment relay panel	**Fig. 2**	97-10
Engine speed sensor	*see* Crankshaft position sensor		
Engine temperature sensor	On engine, below secondary air pump	**Fig. 19**	97-19
Flasher (1995)	In center console		
Flasher (from 1996)	In front of instrument panel		
Fuel level sensor	In fuel tank, trunk	**Fig. 8** **Fig. 9**	97-14 97-15
Front fog light relay	Central electric panel	**Fig. 4**	97-11
Fuel pump	Beneath vehicle, right front (above access cover)	**Fig. 14**	97-17
Fuel pump relay	*see* DME main relay		
Fuse 46 (60 amp)	In trunk, behind ABS hydraulic unit	**Fig. 3**	97-10
Fuse 47 (60 amp)	In trunk, behind ABS hydraulic unit	**Fig. 3**	97-10
Fuse locations	Central electric panel Engine compartment relay panel	**Fig. 4** **Fig. 2**	97-11 97-10
Glass roof and sun blinds motors (Targa)	Rear window shelf		
Glass roof control unit (Targa)	*see* Targa top control unit		
Hall sensor, ignition	In distributor	**Fig. 18**	97-18
Headlight washer pump	By washer fluid reservoir, in front left wheel housing		
Headlight washer relay	Central electric panel	**Fig. 4**	97-11
Heater blower motor, front (left and right)	*see* Blower motor, front		
Heater blower motor, rear	*see* Blower motor, rear		
Heating and air conditioning control unit	On instrument panel		
Horns	*see* Dual tone horn		
Hydraulic boost pressure warning switch (Carrera 4)	In front of trunk	**Fig. 9**	97-15

Table a. Electrical component locations			
Component	**Location**	**Fig #**	**Page #**
Idle air control valve	On engine	**Fig. 18** **Fig. 19**	97-18 97-19
Ignition coil I	On engine	**Fig. 18**	97-18
Ignition coil II	On engine	**Fig. 18**	97-18
Ignition switching unit (1995 only)	Under driver seat		
Interior light with interior sensor	Front roof frame, center		
Instrument light control unit (from 1996)	Central electric panel	**Fig. 4**	97-11
Instrument light control unit (1995)	In trunk, driver side		
Intake air temperature sensor	In rubber shroud between mass air flow sensor and throttle housing		
Intermittent wiper control relay	Central electric panel	**Fig. 4**	97-11
Key buzzer	*see* Buzzer contact		
Kickdown switch	Under floor panel by accelerator pedal		
Knock sensor, cylinders 1-3	On engine, attached to knock sensor bridge on cylinders		
Knock sensor, cylinders 4-6	On engine, attached to knock sensor bridge on cylinders		
Lights on buzzer	In front of instruments		
Litronic headlight control unit (left)	In trunk, left	**Fig. 11**	97-16
Litronic headlight control unit (right)	In trunk, right	**Fig. 10**	97-15
Mass air flow sensor (to 1995)	Intake air duct		
Mass air flow sensor (from 1996)	Intake air housing		
DME control unit	*see* DME control module		
DME (MFI + DI) relay	*see* DME main relay		
Microswitch conv. top open/close limit	Behind side panel covering, right rear		
Microswitch, conv. top lock, left	Convertible top frame, in head bow, left		
Microswitch, conv. top lock, right	Convertible top frame, in head bow, right		
Microswitch, rear spoiler extended	On engine hood		
Microswitch, rear spoiler retracted	On engine hood		
Heater fan temperature sensor	*see* Blower motor, rear, temperature sensor (NTC)		
Oil cooler blower fan	In front right wheel housing	**Fig. 15**	97-17
Oil cooler blower relay	Central electric panel	**Fig. 4**	97-11
Oil cooler blower resistor	In front right wheel housing, top of oil cooler shroud.		
Oil level sensor	In rear right wheel housing, in engine oil tank		
Oil pressure sensor/switch	On engine, under secondary air pump		
Oil temperature sensor (for oil temp. gauge)	On engine, under intake manifold, right rear		
Oil temperature sensor (for oil cooler fan)	In front right wheel housing, in oil cooler		
Outside temperature sensor	In front bumper, left-hand side		
Oxygen sensor (1995)	In exhaust system, center		
Oxygen sensor, post-catalyst, cylinders 1-3 (1996 - 1998)	In exhaust system, left side, between header and catalytic converter		
Oxygen sensor, post-catalyst, cylinders 4-6 (1996 - 1998)	In exhaust system, right side, between header and catalytic converter		
Oxygen sensor, pre-catalyst, cylinders 1-3 (1996 - 1998)	In exhaust system, left side, between catalytic converter and muffler		

Table a. Electrical component locations			
Component	**Location**	**Fig #**	**Page #**
Oxygen sensor, pre-catalyst, cylinders 4-6 (1996 - 1998)	In exhaust system, right side, between catalytic converter and muffler		
Power window relay	Central electric panel	**Fig. 4**	97-11
P / N lock control unit	Central electric panel (1995)	**Fig. 4**	97-11
Pressure switch, air conditioning system	In rear of trunk	**Fig. 8** **Fig. 9**	97-14 97-15
Rear foglight cutout relay	Central electric panel	**Fig. 4**	97-11
Rear spoiler control unit	On right-hand side of transverse wall		
Rear window defogger control relay	Engine compartment relay panel	**Fig. 5**	97-12
Rear window wiper relay	On rear wiper motor		
Secondary air pump	Right rear of engine under intake manifold	**Fig. 19**	97-19
Secondary air pump relay	Engine compartment relay panel	**Fig. 5**	97-12
Starter	Transmission, right side		
Starter relay (Tiptronic transmission)	Central electric panel	**Fig. 4**	97-11
Sunroof motor	In roof above rear window, zippered compartment		
Sunroof relay	Central electric panel	**Fig. 4**	97-11
Fuel tank venting valve	On engine, top left	**Fig. 19**	97-19
Targa top control unit	Behind rear seat		
Terminal 15 E relay	Central electric panel	**Fig. 4**	97-11
Terminal X relay	Central electric panel	**Fig. 4**	97-11
Throttle position sensor	On throttle housing	**Fig. 19**	97-19
Tiptronic transmission control unit	Under driver seat beside DME		
Transverse acceleration sensor - Tiptronic (1995 only)	In center console		
V-belt sensor control	On rear of engine, middle	**Fig. 18**	97-18
Windshield washer fluid pump	In front left wheel housing		

Fuse Locations and Ratings

Vehicle power supply

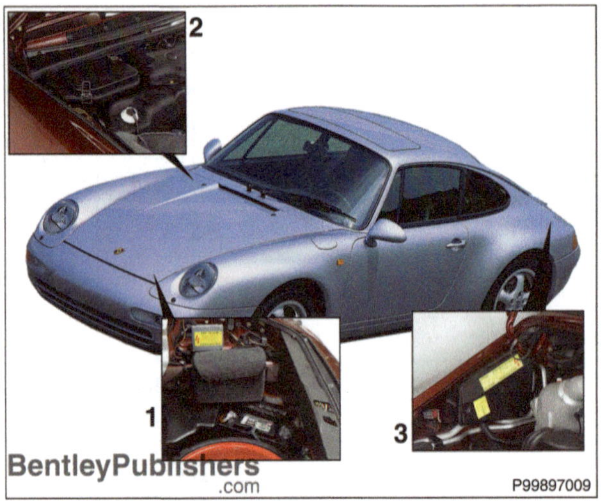

Main electrical power supply components for the Porsche 993:

1. Battery in front of trunk.
   - See page 97-2 in this repair group.
2. Central electric panel in rear of trunk.
   - Contains fuses and relays.
   - See **Fig. 1** on page 97-8 in this repair group.
3. Relay panel in left side of engine compartment.
   - Contains relays and fuses
   - See **Fig. 2** on page 97-10 and Fig. 5 on page 12 in this repair group.

Central electric panel

Main fuses and relays are behind cover in front of trunk, passenger side.

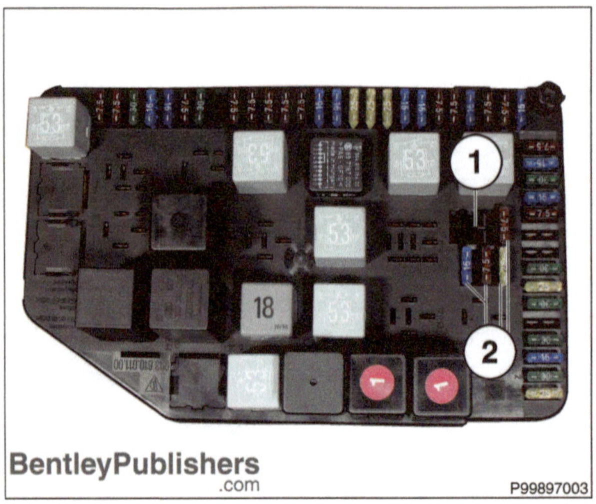

Fig. 1. Fuses in central electric panel

Fuses are organized along the perimeter of the central electric panel. Be sure to replace a bad fuse with one of same rating.

For fuse ratings, see **Table b** and **Table c**.

1. Fuse puller
2. Spare fuses

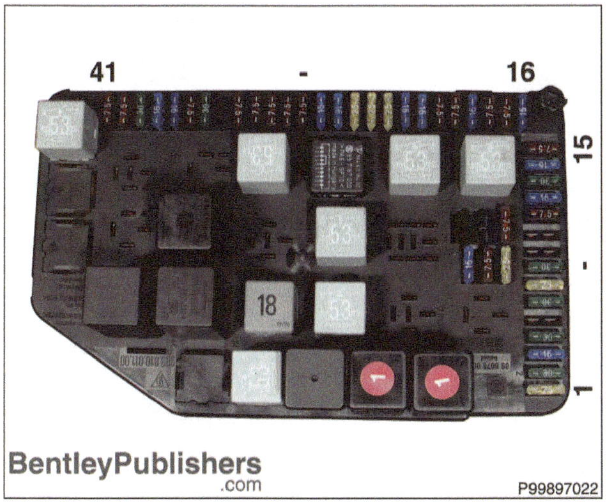

Table b. Fuses in central electric panel

Fuse	Rating in amps	Protected circuit
1	25	Passenger compartment blower
2	30	Oil cooler blower fan
3	15	Dual-tone horns
4	30	Power window regulator
5	–	Not used
6	30	Power seat, right
7	25	Cigarette lighter
8	30	Power seat, left
9	–	Not used
10	–	1995: Not used
	7.5	1996-1998: Mobile phone
11	7.5	1996: Interior lights
		1997-1998: Engine and luggage compartment light, glove box lamp
12	15	On-board computer, rear window wiper, warning flasher, clock
13	30	1995: Sunroof, power top
		1996-1998: Sunroof, motor for glass roof 1, power top, motor for roof blinds
14	15	Rear foglight, foglight, backup light (manual transmission)
15	7.5	Backup light (automatic transmission)
16	15	ABS
17	7.5	Heater, rear spoiler
18	7.5	Instruments, diagnostic connector, terminal 15E, central informer
19	15	Stop light, cruise control control unit, P / N lock (1995)
20	7.5	Left high beam
21	7.5	Right high beam
22	7.5	Left low beam
23	7.5	Right low beam
24	25	Windshield wipers, HCS
25	25	Seat heating
26	15	Fuel pump, heated oxygen sensor
27	15	Automatic transmission
28	15	Rear window defogger switch, rear window wiper, mirror adjustment
29	–	Not used
30	7.5	Rear left turn signal, left side turn signal
31	7.5	Front left turn signal
32	7.5	Front right turn signal
33	7.5	Right rear turn signal, Right side turn signal
34	–	Not used
35	–	1995: Not used
	30	1996-1998: Motor for glass roof 2
36	7.5	1995-1996: License plate lighting
		1997-1998: License plate lighting, instrument light
37	15	Rear spoiler
38	15	Radio amplifier
39	30	A/C condenser blower fan
40	7.5	Left side marker light
41	7.5	Right side marker light

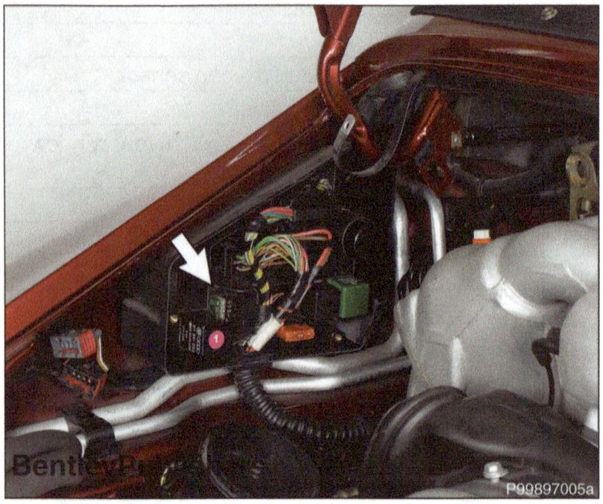

Fig. 2. Fuses in engine compartment relay panel

Fuses are located in engine compartment relay panel (**arrow**). Be sure to replace a bad fuse with one of same rating.

Table c. Fuses in engine compartment relay panel		
Fuse	Rating in amps	Protected circuit
42	30	Blower relay
43	25	Rear window defogger relay
44	7.5	A/C compressor relay
48	40	Secondary air pump relay

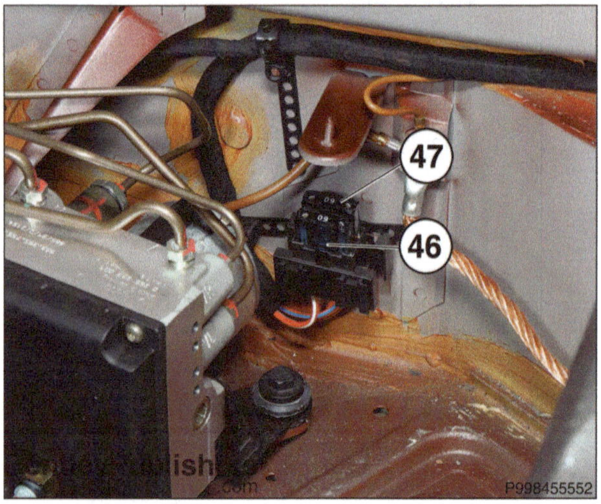

Fig. 3. Fuses in trunk

Fuses are located in trunk behind ABS hydraulic unit. Replace a blown fuse with one of same rating.

Table d. Fuses in trunk		
Fuse	Rating in amps	Protected circuit
46	60	ABS
47	60	Hydraulic booster pump (models with ABD) DME

Component Location Illustrations

In the photos that follow, the subject vehicle is a 1997 Carrera Cabriolet model with standard transmission.

Central electric panel

CAUTION—

- *Relay positions are subject to change and may vary from car to car. If questions arise, please check up-to-date repair information.*

The central electric panel contains an array of fuses and relays. The panel is in the upper right side of the trunk.

Working in trunk, unclip and remove central electric panel lid.

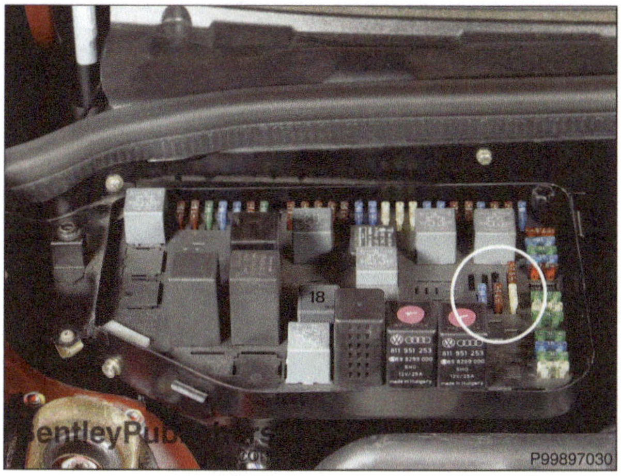

The central electric panel also contains spare fuses.

Fig. 4. Relays in central electric panel

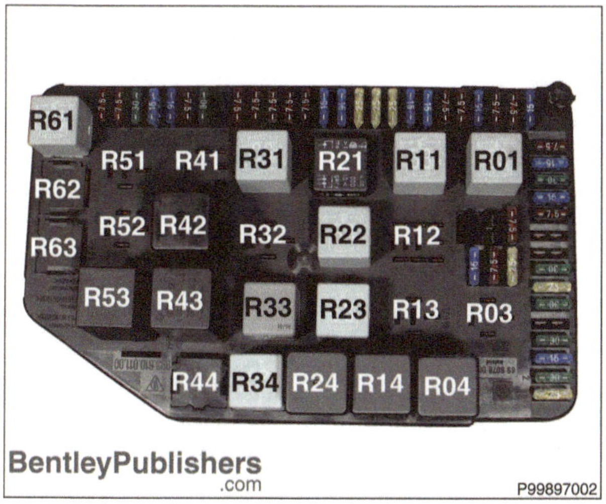

Relays in central electric panel. Relay application information is in **Table e**.

Table e. Central electric panel relays	
Position	**Function**
R01	Terminal X (II)
R03	Not used
R04	Oil cooler fan
R11	Terminal 15E
R12	1995-1996: Bridge adapter
	1997-1998: Not used
R13	Not used
R14	A/C condenser fan
R21	Sunroof
R22	Power window regulator
R23	Dual-tone horn
R24	Control unit P / N lock (1995) Instrument light control unit (1996–)
R31	Foglight
R32	1995: Not used
	1996-1998: Daytime running light control
R33	Terminal X
R34	Front fog lights
R41	Not used
R42	Wiper intermittent control
R43	Headlight washing
R44	Not used
R51	Backup light (automatic transmission)
R52	Not used
R53	DME main (MFI + DI) relay
R61	Starter relay
R62	Terminal 30 and terminal 15 fuse for telephone
R63	Daytime running light (Canada)

Engine compartment relay panel

> ***CAUTION—***
>
> *• Relay positions are subject to change and may vary from car to car. If questions arise, an authorized Porsche dealer is the best source for the most accurate and up-to-date information.*

The engine compartment relay panel contains a limited number of fuses and relays. The panel is in the left side of the engine compartment.

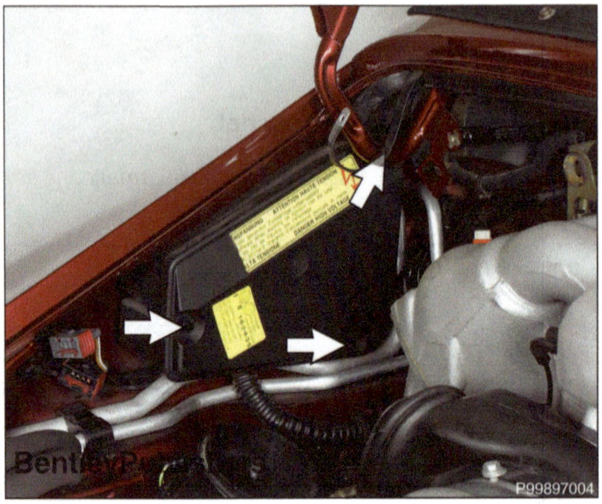

◄ Working in engine compartment, remove fasteners (**arrows**) and remove lid.

Fig. 5. Relays in engine compartment

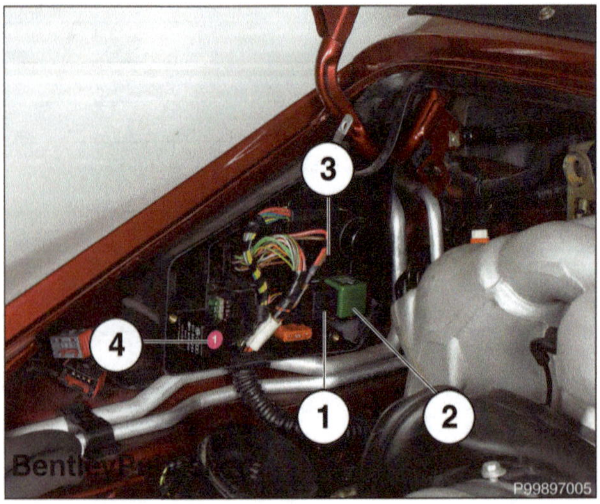

◄ Relays in engine compartment. Relay application information is in **Table f**.

Table f. Engine compartment relays	
Position	**Function**
1	Secondary air pump relay
2	A/C compressor relay
3	Rear window defogger control relay
4	Engine compartment blower relay

Trunk relay panel

> ***CAUTION—***
>
> - *Relay positions are subject to change and may vary from car to car. If questions arise, an authorized Porsche dealer is the best source for the most accurate and up-to-date information.*

The trunk relay panel contains several relays. The panel is in the right side of the trunk.

◄ Working in trunk, unclip and remove lid.

Fig. 6. Relays in trunk

◄ Relays in trunk. Relay application information is in **Table g**.

Table g. Relays in trunk	
Position	**Function**
R65	ABS hydraulic unit pump
R66	Hydraulic brake booster pump relay (Carrera 4)
R68	ABS hydraulic unit solenoids

Trunk components

Fig. 7. A/C and heater housing

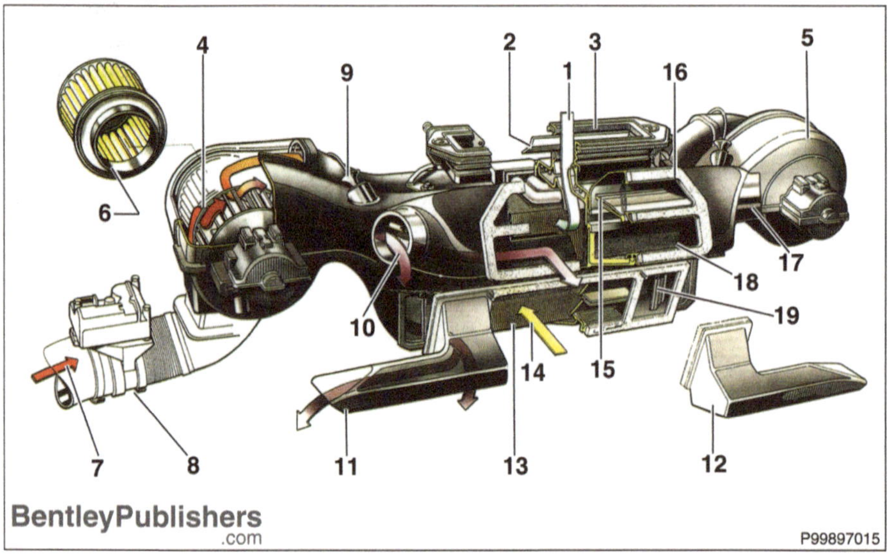

1. Fresh air intake
2. Fresh air flap
3. Fresh air intake
4. Blower, left
5. Blower, right
6. Cabin air particle filter
7. Heated air output
8. Temperature mixing flap, left
9. Mixing chamber sensor, left
10. Side nozzles assembly, left
11. Footwell outlet nozzle, left
12. Footwell outlet nozzle, right
13. A/C evaporator
14. Circulation air
15. Defroster flap
16. Defroster nozzle outlet
17. Side nozzle assembly, right
18. Center nozzle outlet
19. Footwell outlet

Fig. 8. Trunk overview, C2 (trim panels removed)

NOTE–

• 1997 C2 model shown.

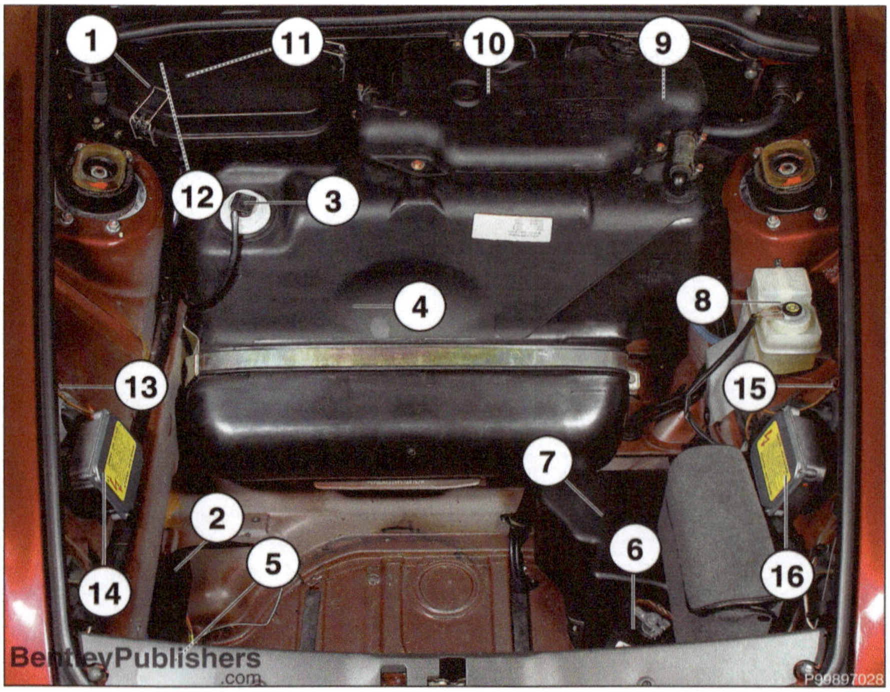

1. Central electric panel
2. Trunk relay panel
3. Fuel gauge sending unit
4. Fuel tank
5. ABS control unit
6. Battery
7. ABS/ABD hydraulic unit
8. Brake fluid reservoir
9. Left blower motor
10. A/C system pressure switch
11. Blower final stage
12. Right blower motor
13. Ground point 2
14. Right Litronic headlight control unit
15. Ground point 1
16. Left Litronic headlight control unit

Fig. 9. Trunk overview, C4 (trim panels removed)

NOTE–

- *1995 C4 model shown.*

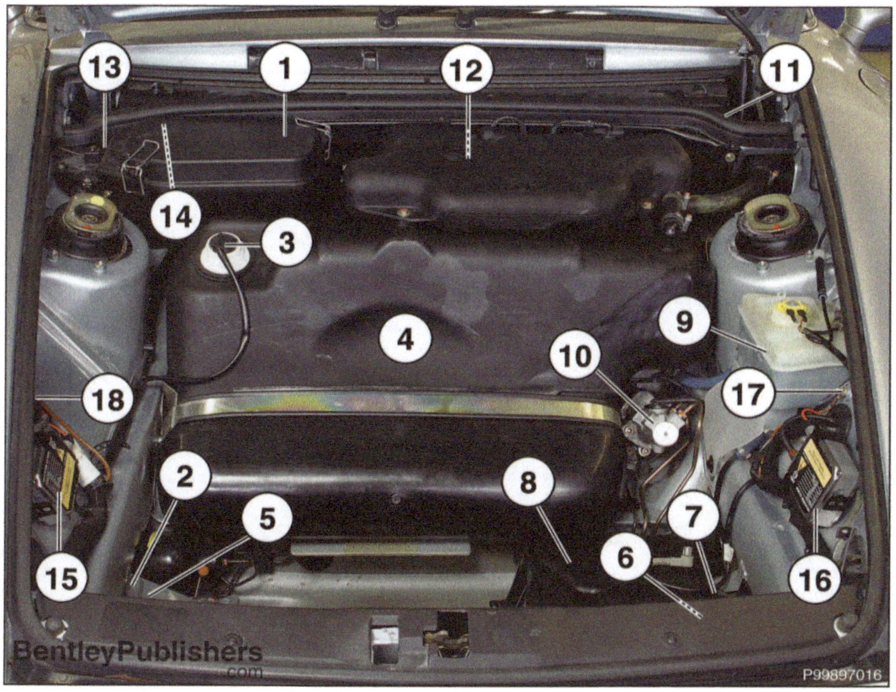

1. Central electric panel
2. Trunk relay panel
3. Fuel gauge sending unit
4. Fuel tank
5. ABS control unit
6. Hydraulic boost pressure switch
7. Battery
8. ABS/ABD hydraulic unit
9. Brake fluid reservoir
10. Brake master cylinder
11. Left blower motor
12. A/C system pressure switch
13. Blower final stage
14. Right blower motor
15. Right Litronic headlight control unit
16. Left Litronic headlight control unit
17. Ground point 1
18. Ground point 2

Fig. 10. Trunk, right front

Trim panels removed:

1. Litronic headlight control unit
2. Ground point 2
3. Fuel level sending unit
4. ABS control module
5. Relay panel

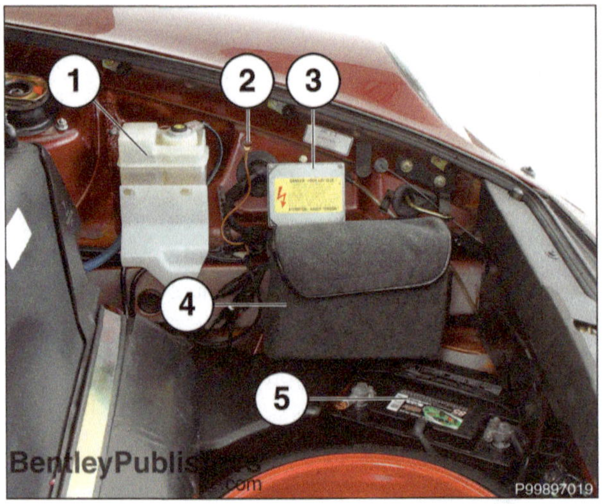

Fig. 11. Trunk, left front

Trim panels removed:

1. Brake fluid reservoir
2. Ground point 1
3. Litronic headlight control unit
4. CD changer
5. Battery

Components underneath vehicle

Fig. 12. Underneath rear of vehicle

Right side of manual transmission:

1. Back-up light switch

Fig. 13. Underneath rear of vehicle

Behind underbody trim panel, left side of rear crossmember.

1. Alarm horn

Fig. 14. Underneath front of vehicle

Behind fuel pump cover:

1. Fuel pump

Fig. 15. In right front wheel housing

Behind wheel housing liner:

1. Horns
2. Auxiliary oil cooler fan

Fig. 16. In left front wheel housing

Behind wheel housing liner:

1. A/C condenser fan

97

Fig. 17. At wheel carrier

Rear wheel carrier shown. Front is similar.

1. ABS wheel speed sensor

NOTE–

- *Rear wheel carrier shown. Front is similar.*

Engine components

Fig. 18. Engine compartment overview

Engine and transmission removed for clarity.

1. Ignition coils
2. Twin ignition distributor
3. Engine cooling fan
4. Alternator
5. Idle speed control
6. Vario-ram actuator
7. Power steering pump
8. Belt tension monitor

Fig. 19. Engine, rear

Engine and transmission removed for clarity

1. Oil filter
2. Secondary air pump
3. Throttle body with throttle position sensor
4. Idle speed control
5. Vario-ram actuator
6. Fuel tank vent valve
7. Engine compartment relay panel connector
8. Engine temperature sensor
9. Cruise control actuator

Diagnostic connection

Fig. 20. Diagnostic connection

◀ Left side of center console, driver footwell.

Ground points

Fig. 21. Ground point 1

Left side of trunk, near brake fluid reservoir.

Fig. 22. Ground point 2

Right side of trunk.

Fig. 23. Ground points 4 and 5

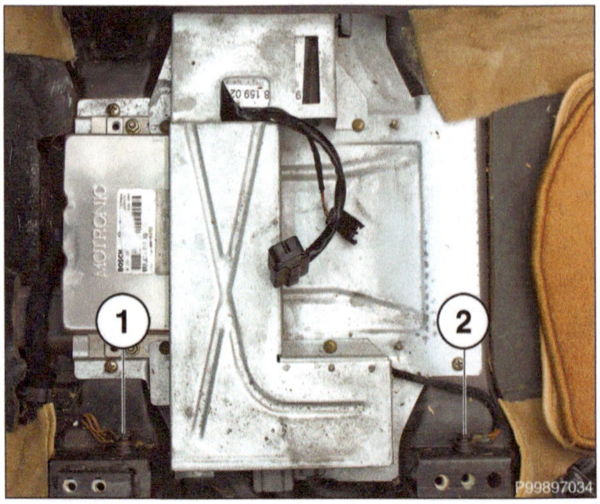

Under driver's seat.

1. Ground point 4
2. Ground point 5

Fig. 24. Ground point 10

Left side of trunk, near battery.

Grounds

Grounds are distributed throughout the vehicle body. Many are found under the interior carpets or behind trim panels. Several component grounds are often ganged. Ground positions vary among models. Lugs and connectors attached to ground are susceptible to damage and corrosion. Clean or renew as necessary.

Ground points, 1995

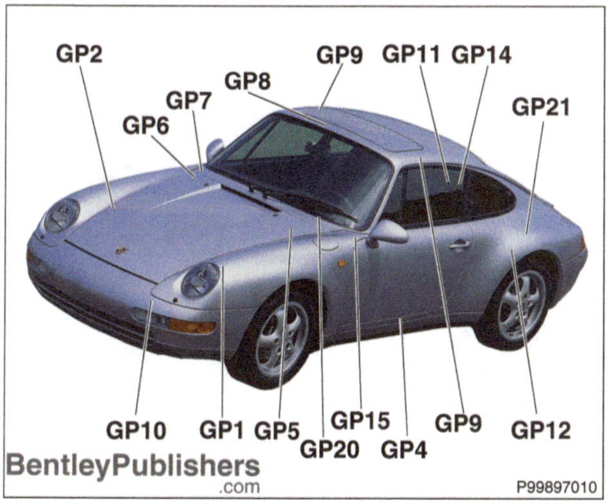

Fig. 25. 1995 grounds

Ground locations for 1995 vehicles are described in **Table h**. Figure and page numbers in table columns 3 and 4 refer to photographs and illustrations in this repair group.

Ground applications for individual electrical circuits for 1995 models are in **Table j** and **Table k**.

Table h. 1995 ground points

No.	Location	Fig #	Page #
GP 1	In trunk, near left wheel housing	**Fig. 21**	97-20
GP 2	In trunk, near right wheel housing	**Fig. 22**	97-20
GP 4	Ground for power consumers, under driver seat	**Fig. 23**	97-20
GP 5	On floor, under dashboard		
GP 6	In trunk, on right side member		
GP 7	In trunk, on left side member		
GP 8	Top of windshield frame, center		
GP 9	Left side of roof		
GP 9	Right side of roof		
GP 10	At battery, in trunk, left	**Fig. 24**	97-21
GP 11	Ground strap, in right rear of engine compartment		
GP 12	Left rear of engine compartment		
GP 14	Right rear of engine compartment		
GP 15	Bottom of left A-pillar		
GP 20	Seat cross member under passenger seat		
GP 21	Left side of crank case		

97

Ground points, 1996 - 1998

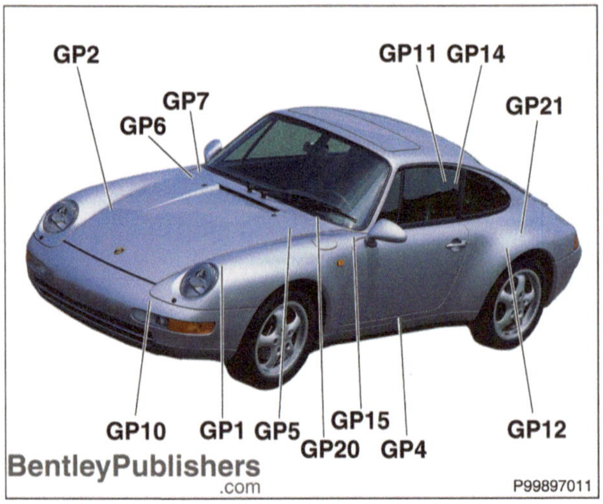

Fig. 26. 1996 - 1998 grounds

Ground locations for 1996 - 1998 vehicles are listed in **Table i**. Figure and page numbers in table columns 3 and 4 refer to photographs and illustrations in this repair group.

Ground applications for individual electrical circuits for 1996 through 1998 models are in **Table l** and **Table m**.

Table i. 1996-1998 ground points

No.	Location	Fig #	Page #
GP 1	In trunk, near left wheel housing	**Fig. 21**	97-20
GP 2	In trunk, near right wheel housing	**Fig. 22**	97-20
GP 4	Ground for power consumers, under driver seat	**Fig. 23**	97-20
GP 5	On floor, behind driver seat	**Fig. 23**	97-20
GP 6	In trunk, on right side member		
GP 7	In trunk, on left side member		
GP 10	At battery, in trunk, left	**Fig. 24**	97-21
GP 11	Ground strap, in right rear of engine compartment		
GP 12	Left rear of engine compartment		
GP 14	Right rear of engine compartment		
GP 15	Bottom of left A-pillar		
GP 20	Seat cross member under passenger seat		
GP 21	Left side of crank case		

Table j. 1995 (early) ground applications

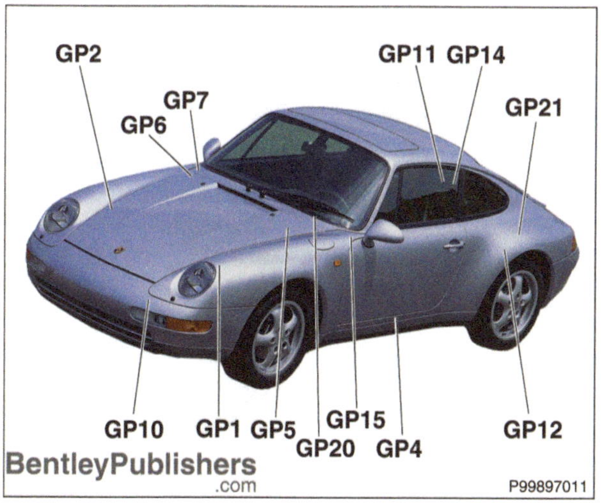

Circuit	GP 1	GP 2	GP 4	GP 5	GP 6	GP 7	GP 8	GP 9	GP 10	GP 11	GP 12	GP 14	GP 15	GP 20	GP 21
ABS control unit		*													
A/C compressor relay											*				
Airbag contact spring												*			
Airbag control unit													*		
Alarm horn													*		
Alarm system / central locking system control unit														*	
Antenna amplifier						*									
Battery									*						
Booster pump motor (Carrera 4)		*													
Booster pump relay (Carrera 4)		*													
Brake fluid level control switch	*														
Center console light switch, center panel													*		
Center console light switch, top panel													*		
Central informer												*	*		
Central locking button													*		
Cigarette lighter					*								*		
Clock												*			
Clock adjusting button												*			
Clutch switch			*												
Code lock				*											
Combined steering column switch													*		
Condenser blower	*														
Convertible top control unit													*		
Cruise control drive														*	
Cruise control control unit														*	
Diagnosis plug socket					*										
Driver seat			*												
Driver's switch, power window - pass. side					*										
Engine compartment blower															*

Ground points and applications

Table j. 1995 (early) ground applications

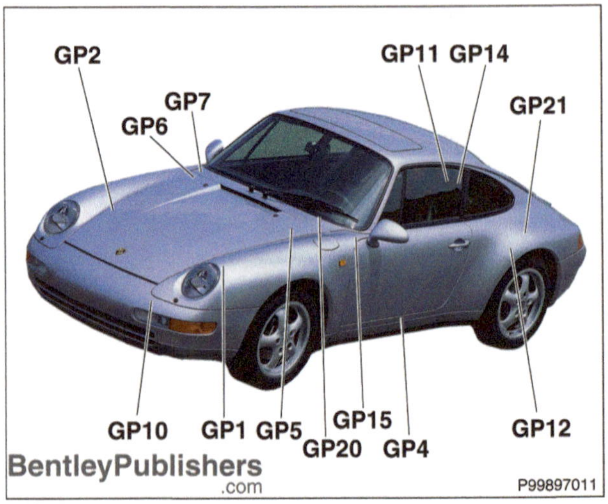

Circuit	GP 1	GP 2	GP 4	GP 5	GP 6	GP 7	GP 8	GP 9	GP 10	GP 11	GP 12	GP 14	GP 15	GP 20	GP 21
Flasher													*		
Foglight switch					*										
Fuel filler	*														
Fuel pump		*													
Generator															*
Glove compartment contact					*										
Buzzer relay													*		
Ground strap, starter-to-body										*					
Hazard light switch													*		
Headlight, left	*														
Headlight washer nozzle	*														
Heated washer nozzle, right													*		
High mounted stop light, convertible												*			
Ignition control unit				*											
Interior light - coupe, left								*							
Interior light - coupe, right								*							
Interior light - convertible, Targa							*								
Instrument cluster												*			
Instrument light control unit												*			
Kickdown switch				*											
Light switch					*							*			
Luggage compartment lock microswitch		*													
P / N lock													*		
Mass air flow sensor				*											
DME control unit				*											
Oil cooler blower					*										
Oil level sensor												*			
Passenger seat														*	
P / N lock control unit					*										
Radio									*						

Table j. 1995 (early) ground applications

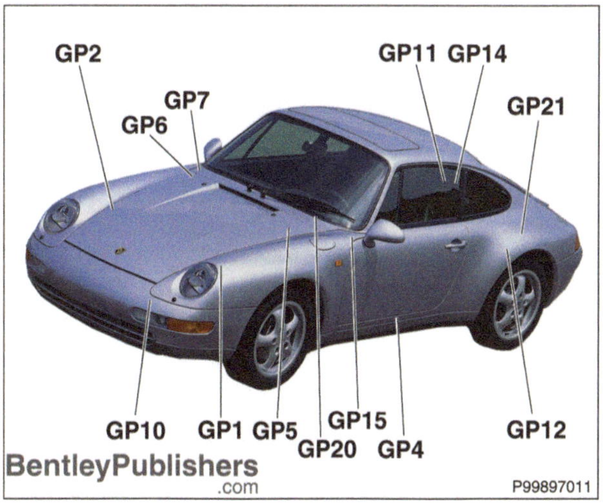

Circuit	GP 1	GP 2	GP 4	GP 5	GP 6	GP 7	GP 8	GP 9	GP 10	GP 11	GP 12	GP 14	GP 15	GP 20	GP 21
Rear foglight switch					*										
Rear package tray drive												*			
Rear package tray switch												*	*		
Rear spoiler control unit														*	
Rear spoiler extended microswitch														*	
Rear window defogger (coupe)												*			
Rear window defogger relay											*				
Rear window defogger switch					*										
Rear window wiper												*			
Rear window wiper relay												*			
Right headlight					*										
Safety belt lock			*												
Secondary air pump				*											*
Selector lever switch													*		
Side turn signal - left													*		
Side turn signal - right					*										
Speedometer												*	*		
Starter relay					*										
Sunroof switch													*		
Tachometer with on-board unit													*		
Tachometer without on-board unit													*		
Tailgate lock microswitch												*			
Tiptronic transmission control unit			*												
Transmission selector lever switch				*											
Dual-tone horn					*										
Warning buzzer light					*										
Wheel carrier, front left	*														
Wheel carrier, front right		*													
Windshield washer fluid pump	*														
Windshield wiper motor												*			

Table k. 1995 (late) ground applications

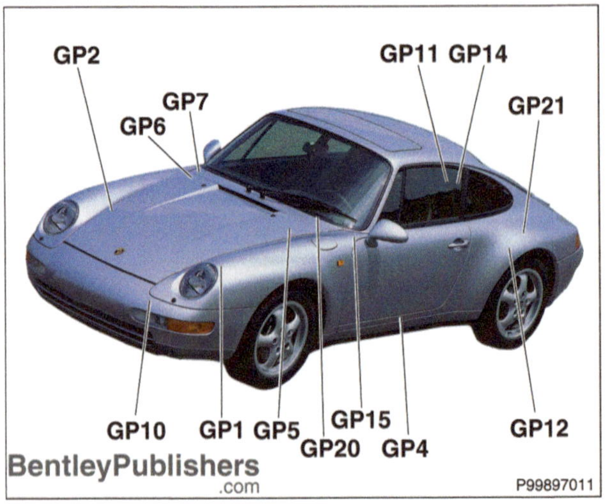

Circuit	GP 1	GP 2	GP 4	GP 5	GP 6	GP 7	GP 8	GP 9	GP 10	GP 11	GP 12	GP 14	GP 15	GP 20	GP 21
ABS control unit		*													
A/C compressor relay											*				
A/C and heater control unit					*										
Airbag contact spring													*		
Airbag control unit													*		
Alarm horn													*		
Alarm system / central locking system control unit														*	
Antenna amplifier						*									
Anti-drive-off control unit (anti-theft)			*												
Ashtray light					*										
Battery															
Blower final stage					*										
Booster pump motor		*													
Booster pump relay		*													
Brake fluid level control switch	*														
Center console light switch, center panel															
Center console light switch, top panel													*		
Central informer													*		
Central locking button													*		
Cigarette lighter					*										
Clock													*		
Clock adjusting button													*		
Clutch switch			*												
Combined steering column switch													*		
Condenser blower	*														
Convertible top control unit													*		
Cruise control drive														*	
Cruise control control unit														*	

Table k. 1995 (late) ground applications

Circuit	GP 1	GP 2	GP 4	GP 5	GP 6	GP 7	GP 8	GP 9	GP 10	GP 11	GP 12	GP 14	GP 15	GP 20	GP 21
Diagnosis plug socket									*				*		
Driver seat			*												
Engine compartment blower															*
Flasher													*		
Foglight switch					*										
Fuel filler	*														
Fuel pump		*													
Generator (Alternator)															*
Glove compartment contact					*										
Buzzer relay													*		
Ground strap, starter-to-body									*	*					
Hazard light switch													*		
Headlight, left	*														
Headlight washer nozzle	*														
Heated washer nozzle, right													*		
High mounted stop light, convertible												*			
Ignition control unit				*											
Interior light - coupe, left								*							
Interior light - coupe, right								*							
Interior light - convertible							*								
Instrument cluster													*		
Instrument light control unit												*	*		
Kickdown switch (Tiptronic)			*												
License plate light, left												*			
Light switch													*		
Luggage compartment lock microswitch		*													
Magnet for P / N lock													*		
Mass air flow sensor				*											
Microswitch, roof, right												*			

97

Ground points and applications

Table k. 1995 (late) ground applications

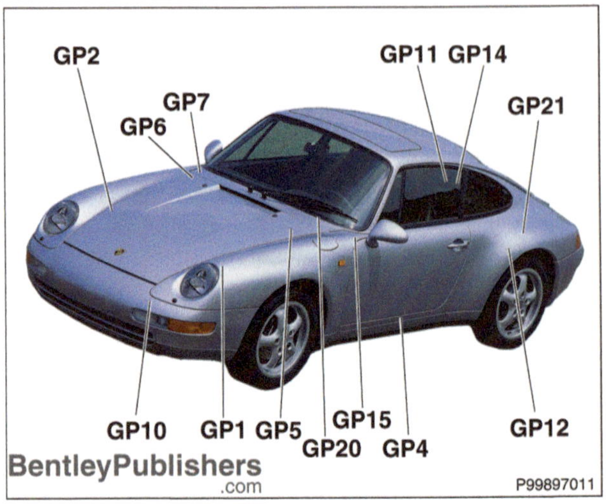

Circuit	GP 1	GP 2	GP 4	GP 5	GP 6	GP 7	GP 8	GP 9	GP 10	GP 11	GP 12	GP 14	GP 15	GP 20	GP 21
DME control unit			*	*											
Mirror heating switch					*										
Oil cooler blower					*										
Oil level sensor												*			
Passenger seat														*	
P / N lock control unit					*										
Power window motor switch, pass. side					*										
Pressure warning switch		*													
Radio									*						
Rear foglight switch					*										
Rear package tray drive												*			
Rear package tray switch												*			
Rear spoiler control unit														*	
Rear spoiler extended microswitch														*	
Rear window defogger (coupe)												*			
Rear window defogger relay											*				
Rear window heating button					*										
Rear window wiper												*			
Rear window wiper relay												*			
Right headlight					*										
Safety belt lock			*												
Secondary air pump															*
Selector lever switch			*										*		
Side turn signal - right					*										
Speedometer													*		
Starter relay					*										
Sunroof switch													*		
Suppressor capacitor															*
Tachometer with on-board unit													*		

Table k. 1995 (late) ground applications

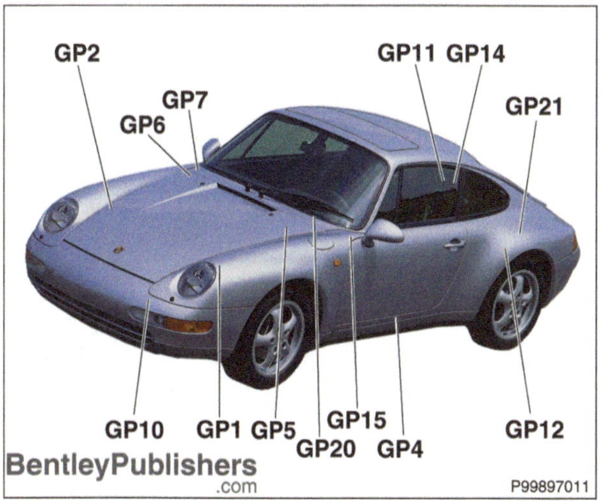

Circuit	GP 1	GP 2	GP 4	GP 5	GP 6	GP 7	GP 8	GP 9	GP 10	GP 11	GP 12	GP 14	GP 15	GP 20	GP 21
Tachometer without on-board unit													*		
Tailgate lock microswitch												*			
Tiptronic transmission control unit				*											
Dual-tone horn					*										
Wheel carrier, front left	*														
Wheel carrier, front right		*	*												
Windshield washer fluid pump	*														
Windshield wiper motor													*		

Ground points and applications

Table I. 1996 ground applications

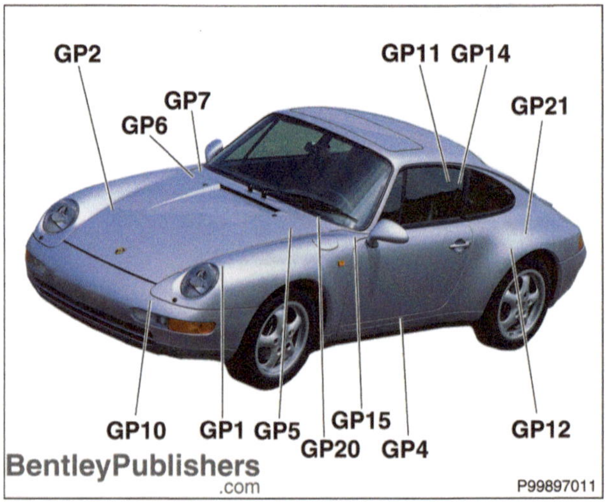

Circuit	GP 1	GP 2	GP 4	GP 5	GP 6	GP 7	GP 10	GP 11	GP 12	GP 14	GP 15	GP 20	GP 21
ABS control unit		*											
A/C compressor relay									*				
A/C and heater control unit					*								
Airbag contact spring											*		
Airbag control unit											*		
Alarm horn											*		
Alarm system / central locking system control unit												*	
Antenna amplifier						*							
Anti-drive-off control unit (anti-theft)			*										
Ashtray light					*								
Blind switch											*		
Blower final stage					*								
Booster pump motor		*											
Brake fluid level control switch	*												
Center console light switch, center panel											*		
Center console light switch, top panel											*		
Central informer											*		
Central locking button											*		
Cigarette lighter					*								
Clock											*		
Clock adjusting button											*		
Clutch switch			*										
Combined steering column switch											*		
Condenser blower	*												
Convertible top control unit											*		
Cruise control drive												*	
Cruise control control unit												*	
Diagnosis plug socket							*				*		
Driver seat			*										

Table I. 1996 ground applications

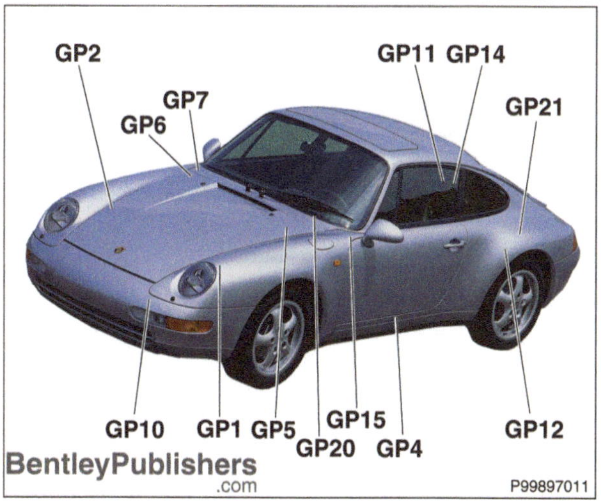

Circuit	GP 1	GP 2	GP 4	GP 5	GP 6	GP 7	GP 10	GP 11	GP 12	GP 14	GP 15	GP 20	GP 21
E-box, mobile phone											*		
Engine compartment blower													*
Flasher											*		
Foglight switch					*								
Fuel filler	*												
Fuel pump		*											
Generator (Alternator)													*
Glass roof control unit										*			
Glass roof switch											*		
Glove compartment contact					*								
Buzzer relay											*		
Ground strap, battery-to-body							*						
Ground strap, starter-to-body								*					
Hazard light switch											*		
Headlight, left	*												
Headlight washer nozzle	*												
Heated washer nozzle, right											*		
High mounted stop light, convertible										*			
Ignition control unit				*									
Interior light - coupe, right										*			
Interior light interior sensor - convertible											*		
Interior light - convertible											*		
Instrument cluster										*	*		
Instrument light control unit											*		
Kickdown switch			*										
Light switch											*		
Luggage compartment lock microswitch		*											
Mass air flow sensor				*									
Microswitch, roof, right										*			

Ground points and applications

Table I. 1996 ground applications

Circuit	GP 1	GP 2	GP 4	GP 5	GP 6	GP 7	GP 10	GP 11	GP 12	GP 14	GP 15	GP 20	GP 21
DME control unit			*	*									
Mirror heating switch, convertible					*								
Oil cooler blower					*								
Passenger seat												*	
Power window motor switch, passenger side					*								
Pressure warning switch		*											
Radio							*						
Rear foglight switch					*								
Rear package tray drive										*			
Rear package tray switch										*			
Rear spoiler control unit												*	
Rear spoiler extended microswitch												*	
Rear window defogger (coupe)										*			
Rear window defogger relay									*				
Rear window defogger switch													
Rear window heating button					*								
Rear window wiper										*			
Rear window wiper relay										*			
RPM counter with on-board unit											*		
RPM counter without on-board unit											*		
Safety belt lock			*										
Selector lever switch			*								*		
Side turn signal - left											*		
Side turn signal - right					*								
Speedometer											*		
Starter relay					*								
Sunroof switch											*		
Suppressor capacitor													*
Tailgate lock microswitch										*			

Table I. 1996 ground applications

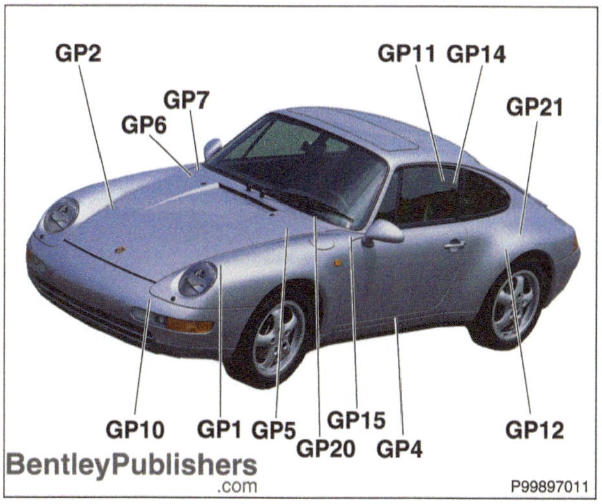

Circuit	GP 1	GP 2	GP 4	GP 5	GP 6	GP 7	GP 10	GP 11	GP 12	GP 14	GP 15	GP 20	GP 21
Tiptronic transmission control unit				*							*		
Dual-tone horn					*								
Washer fluid level switch	*												
Wheel carrier, front left	*												
Wheel carrier, front right		*	*										
Windshield washer fluid pump	*												
Windshield wiper motor											*		

Ground points and applications

Table m. 1997-1998 ground applications

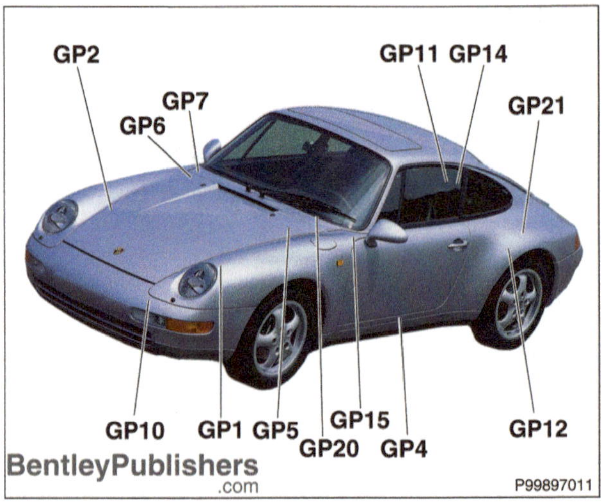

Circuit	GP 1	GP 2	GP 4	GP 5	GP 6	GP 7	GP 10	GP 11	GP 14	GP 15	GP 20	GP 21
ABS control unit		*										
A/C compressor relay									*			
A/C and heater control unit					*							
Airbag contact spring										*		
Airbag control unit										*		
Alarm horn										*		
Alarm system / central locking system control unit											*	
Antenna amplifier						*						
Anti-drive-off control unit (anti-theft)			*									
Ashtray light					*							
Battery							*					
Blind switch									*			
Blower final stage					*							
Booster pump motor		*										
Brake fluid level control switch	*											
Center console light switch, top panel										*		
Central informer										*		
Central locking button										*		
Child seat recognition (passenger seat)										*		
Cigarette lighter					*							
Clock										*		
Clock adjusting button										*		
Clutch switch			*									
Clutch switch interlock			*									
Combined steering column switch										*		
Condenser blower	*											
Convertible top control unit										*		
Cruise control drive											*	
Cruise control control unit											*	

Table m. 1997-1998 ground applications

Circuit	GP 1	GP 2	GP 4	GP 5	GP 6	GP 7	GP 10	GP 11	GP 14	GP 15	GP 20	GP 21
Diagnosis plug socket										*		
Driver seat			*									
E-box, mobile phone										*		
Engine compartment blower												*
Flasher										*		
Foglight switch					*							
Fuel filler	*											
Fuel pump		*										
Generator												*
Glass roof control unit									*			
Glass roof switch									*			
Glove compartment contact					*							
Buzzer relay										*		
Ground strap, battery-to-body							*					
Ground strap, starter-to-body								*				
Hazard light switch										*		
Headlight, left	*											
Headlight washer nozzle	*											
Heated washer nozzle, right										*		
High mounted stop light, convertible									*			
Ignition control unit				*								
Interior light - coupe, right									*			
Interior light interior sensor - convertible										*		
Interior light - convertible										*		
Instrument cluster									*	*		
Instrument light control unit										*		
Kickdown switch			*							*		
Light switch										*		
Luggage compartment lock microswitch		*										

Ground points and applications

Table m. 1997-1998 ground applications

Circuit	GP 1	GP 2	GP 4	GP 5	GP 6	GP 7	GP 10	GP 11	GP 14	GP 15	GP 20	GP 21
Mass air flow sensor				*								
Microswitch, roof, right									*			
DME control unit			*	*								
Mirror heating switch, convertible					*							
Oil cooler blower					*							
Passenger seat											*	
Power window motor switch, passenger side					*							
Pressure warning switch		*										
Radio							*					
Rear foglight switch					*							
Rear molding									*			
Rear package tray drive									*			
Rear package tray switch									*			
Rear spoiler control unit											*	
Rear spoiler extended microswitch											*	
Rear window defogger (coupe)									*			
Rear window defogger relay									*			
Rear window heating button					*							
Rear window wiper relay									*			
Tachometer with on-board unit										*		
Tachometer without on-board unit										*		
Safety belt lock			*									
Selector lever switch			*							*		
Side marker light - right					*							
Side turn signal - left										*		
Side turn signal - right					*							
Speedometer										*		
Sunroof switch										*		
Tailgate lock microswitch									*			

Table m. 1997-1998 ground applications

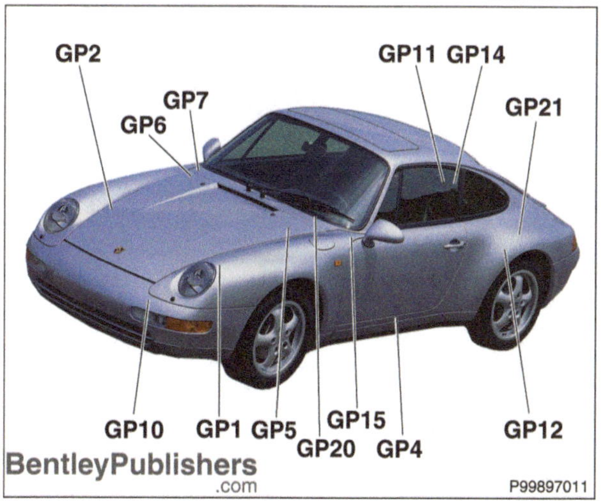

Circuit	GP 1	GP 2	GP 4	GP 5	GP 6	GP 7	GP 10	GP 11	GP 14	GP 15	GP 20	GP 21
Tiptronic transmission control unit				*						*		
Dual-tone horn					*							
Washer fluid level switch	*											
Wheel carrier, front left	*											
Wheel carrier, front right		*	*									
Windshield washer fluid pump	*											
Windshield wiper motor										*		

ELECTRICAL TROUBLESHOOTING

Four things are required for current to flow in any electrical circuit:

- Voltage source.
- Wires or connections to transport voltage.
- Load or device that uses electricity.
- Connection to ground.

Most problems can be found using a digital multimeter (volt / ohm / ammeter) to check the following:

- Voltage supply.
- Breaks in the wiring (infinite resistance / no continuity).
- A path to ground that completes the circuit.

Electric current is logical in its flow, always moving from the voltage source toward ground. Electrical faults can usually be located through a process of elimination. When troubleshooting a complex circuit, separate the circuit into smaller parts. General tests outlined below may be helpful in finding electrical problems. The information is most helpful when used with wiring diagrams.

Be sure to analyze the problem. Use wiring diagrams to determine the most likely cause. Get an understanding of how the circuit works by following the circuit from ground back to the power source.

When making test connections at connectors and components, use care to avoid spreading or damaging the connectors or terminals. Some tests may require jumper wires to bypass components or connections in the wiring harness. When connecting jumper wires, use blade connectors at the wire ends that match the size of the terminal being tested. The small internal contacts are easily spread apart, and this can cause intermittent or faulty connections that can lead to more problems.

Voltage and voltage drop

Wires, connectors, and switches that carry current are designed with very low resistance so that current flows with a minimum loss of voltage. A voltage drop is caused by higher than normal resistance in a circuit. This additional resistance actually decreases or stops the flow of current. Excessive voltage drop can cause problems ranging from dim headlights to sluggish wipers. Some common sources of voltage drops are corroded or dirty switches, dirty or corroded connections or contacts and loose or corroded ground wires and ground connections.

A voltage drop test is a good test to perform if current is flowing through the circuit but the circuit is not operating correctly. A voltage drop test helps pinpoint a corroded ground strap or a faulty switch. Normally, there should be less than 1 V drop across most wires or closed switches. Voltage drop across a connector or short cable varies but should not exceed 0.3 V.

A voltage drop test is generally more accurate than a simple resistance check because the resistances involved are often too small to measure with most ohmmeters. For example, a resistance as small as 0.02 W results in a 3 V drop in a typical 150 amp starter circuit. (150 amps x 0.02 W = 3 V)

Keep in mind that voltage with ignition key ON and voltage with engine running are not the same. With ignition ON and engine OFF, fully charged battery voltage is approximately 12.6 V. With engine running (charging voltage), normal voltage is approximately 14.0 V. Measure voltage at battery with ignition ON and then with engine running to get exact measurements.

Voltage, measuring

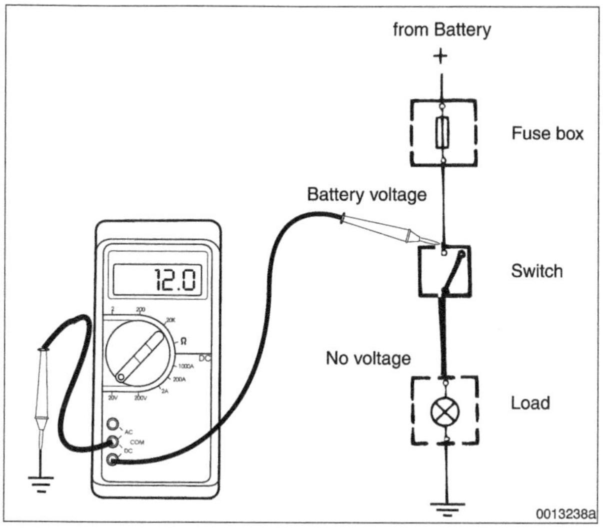

- Connect digital multimeter negative lead to a reliable ground point on vehicle. The negative (–) battery terminal is a good ground point.
- Connect digital multimeter positive lead to point in circuit you wish to measure.
- Check that voltage reading does not deviate more than 1 V from voltage at battery. If voltage drop is more than this, check for a corroded connector or loose ground wire.

Voltage drop, testing

Check voltage drop only when there is a load on the circuit, such as when operating the starter motor or turning on the headlights. Use a digital multimeter to ensure accurate readings.

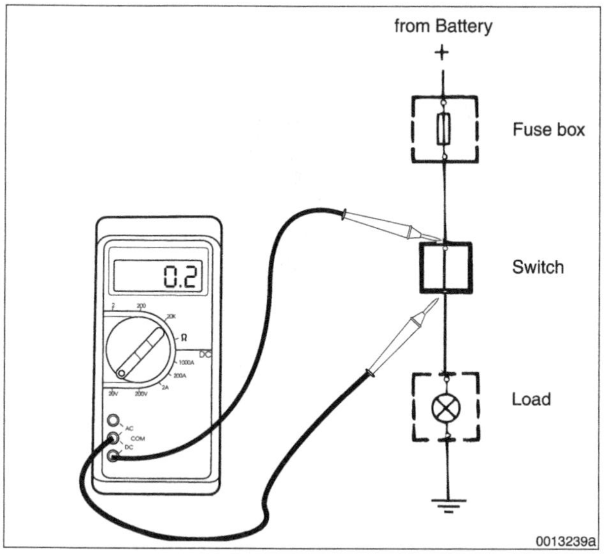

- Connect digital multimeter positive lead to positive (+) battery terminal or a positive power supply close to battery source.
- Connect digital multimeter negative lead to other end of cable or switch being tested.
- With power switched ON and circuit working, meter shows voltage drop (difference between two points). This value should not exceed 1 V.
- Maximum voltage drop in an automotive circuit, as recommended by the Society of Automotive Engineers (SAE), is as follows:
  - 0 V for small wire connections
  - 0.1 V for high current connections
  - 0.2 V for high current cables
  - 0.3 V for switch or solenoid contacts
- On longer wires or cables, the drop may be slightly higher. In any case, a voltage drop of more than 1.0 V usually indicates a problem.

Continuity, checking

Use a continuity test to check a circuit or switch. Because most automotive circuits are designed to have little or no resistance, a circuit or part of a circuit can be easily checked for faults using an ohmmeter. An open circuit or a circuit with high resistance does not allow current to flow. A circuit with little or no resistance allows current to flow easily.

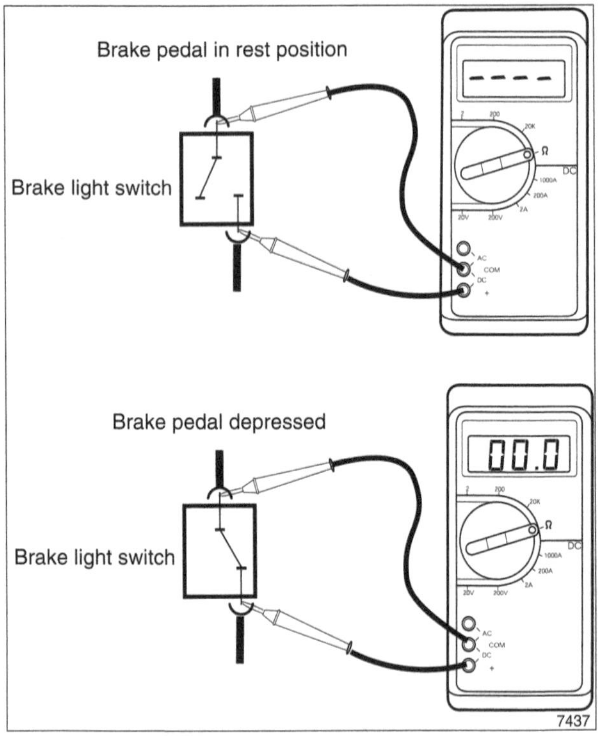

When checking continuity, switch ignition OFF. On circuits that are powered at all times, disconnect battery. Using the appropriate wiring diagram, test circuit for faulty connections, wires, switches, relays and engine sensors by checking for continuity.

< Example: Test brake light switch for continuity:

- With brake pedal in rest position (switch open) there is no continuity (infinite Ω).
- With pedal depressed (switch closed) there is continuity (0 Ω).

Short circuits, testing

Short circuits are exactly what the name implies. Current in the circuit takes a shorter path than it was designed to take. The most common short that causes problems is a short to ground where the insulation on a positive (+) wire wears away and the metal wire is exposed. When the wire rubs against a metal part of the vehicle or other ground source, the circuit is shorted to ground. If the exposed wire is live (positive battery voltage), a fuse blows or the circuit may be damaged.

Short circuits vary in nature and are often difficult to locate. They can be found using a logical approach based on knowledge of the current path.

Use a digital multimeter to locate short circuits.

CAUTION—

- *In circuits protected with high rating fuses (25 amp and greater), wires or circuit components may be damaged before the fuse blows. Check for wiring damage before replacing fuses of this rating. Also, check for correct fuse rating.*

Testing with ohmmeter

- Remove blown fuse from circuit and disconnect cables from battery. Disconnect harness connector from circuit load or consumer.

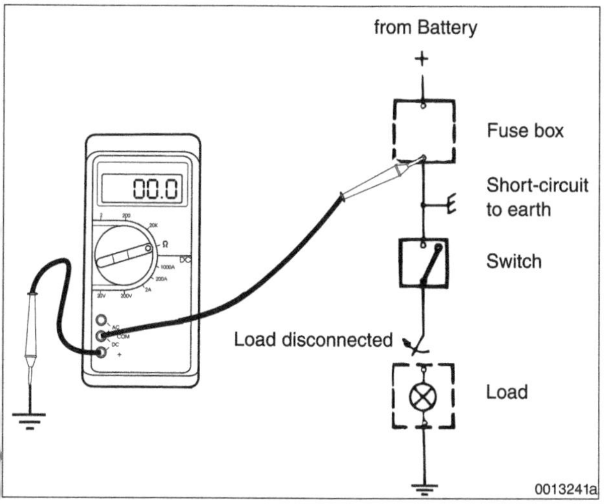

< Using an ohmmeter, connect one test lead to load side of fuse terminal (terminal leading to circuit) and other test lead to ground.

- If there is continuity to ground, there is a short to ground.
- If there is no continuity, work from wire harness nearest to fuse and relay panel and move or wiggle wires while observing meter. Continue to move down harness until meter displays a reading. This is the location of the short to ground.
- Visually inspect wire harness at this point for any faults. If no faults are visible, carefully slice open harness cover or wire insulation for further inspection. Repair any faults found.

Testing with voltmeter

- Remove blown fuse from circuit. Disconnect harness connector from circuit load or consumer.

NOTE

• Most fuses power more than one consumer. Be sure all consumers are disconnected when checking for a short circuit.

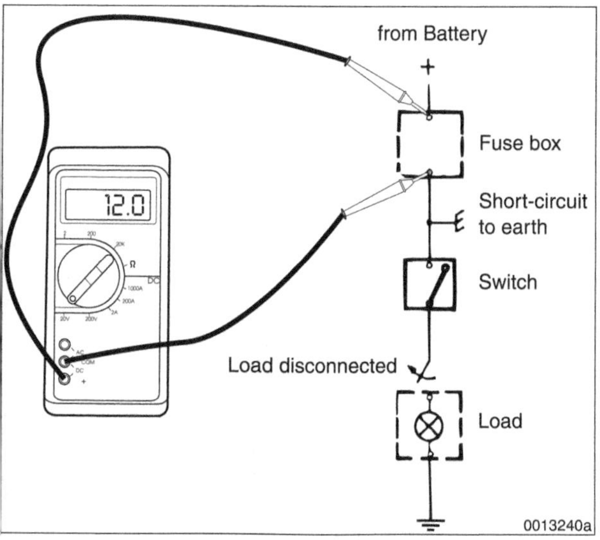

< Using a digital multimeter, connect test leads across fuse terminals. Make sure power is present in circuit. If necessary switch ignition ON.

- If voltage is present at voltmeter, there is a short to ground.
- If voltage is not present, work from wire harness nearest to fuse and relay panel and move or wiggle wires while observing meter. Continue to move down harness until meter displays a reading. This is the location of the short to ground.
- Visually inspect wire harness at this point for any faults. If no faults are visible, carefully slice open harness cover or wire insulation for further inspection. Repair any faults found.

EWD Electrical Wiring Diagrams

(continued on next page)

TABLES

General

The wiring schematics given in this section detail electrical circuit information for select model years. Note that electrical and electronic circuits may have more than one source of power and/or ground. In many cases the ground may also be a switched ground. Take time to study the schematics of the entire system prior to circuit troubleshooting.

Fuses, grounds and component locations are covered in **97 Fuses, Relays, Component Locations**.

Cautions

> ***CAUTION—***
> - *Airbags and pyrotechnic seat belt reels (where applicable) are deployed by explosive devices. Handled improperly or without adequate safeguards, these components are very dangerous. Before starting any work, refer to the warnings and cautions in* **69 Seat Belts, Airbags**.
> - *Prior to disconnecting the battery, read the battery disconnection cautions in* **00 Warnings and Cautions**.
> - *Connect and disconnect ignition system wires, multiple connectors and ignition test equipment leads only while the ignition is switched OFF. Keep clothing, hands, and feet dry if possible.*
> - *Switch a test meter to the appropriate function and range before making test connections.*
> - *Switch the ignition OFF and disconnect the negative (–) battery cable before removing electrical components.*

Wiring Overview

Each diagram shows the wiring, connectors, terminals, and electrical or electronic components of the circuit. It also identifies the wires by color or terminal coding. The information contained under this heading is organized to help navigate and better understand the wiring diagrams contained in this repair group.

Using wiring diagrams

Using the wiring diagrams for the first time may require some interpretation. Abbreviations, acronyms and not-so-obvious conventions are used throughout the diagrams. Many of the conventions used in the diagrams are detailed in the sample wiring diagram below.

Wiring color codes

Wire insulation colors in this section are abbreviated, depending on model year (early/late), acronyms may vary.

BLACK	BK	BK
BROWN	BR	BN
RED	RE	RD
ORANGE	OR	OG
YELLOW	YE	YE
GREEN	GN	GN
BLUE	BL	BU
VIOLET	VI	VT
GREY	GR	GY
WHITE	WT	WH
PINK	RS	PK
GOLD	-	GD
TURQUOISE	TK	TQ
SILVER	-	SR

PB04EWD005

Wiring diagram conventions

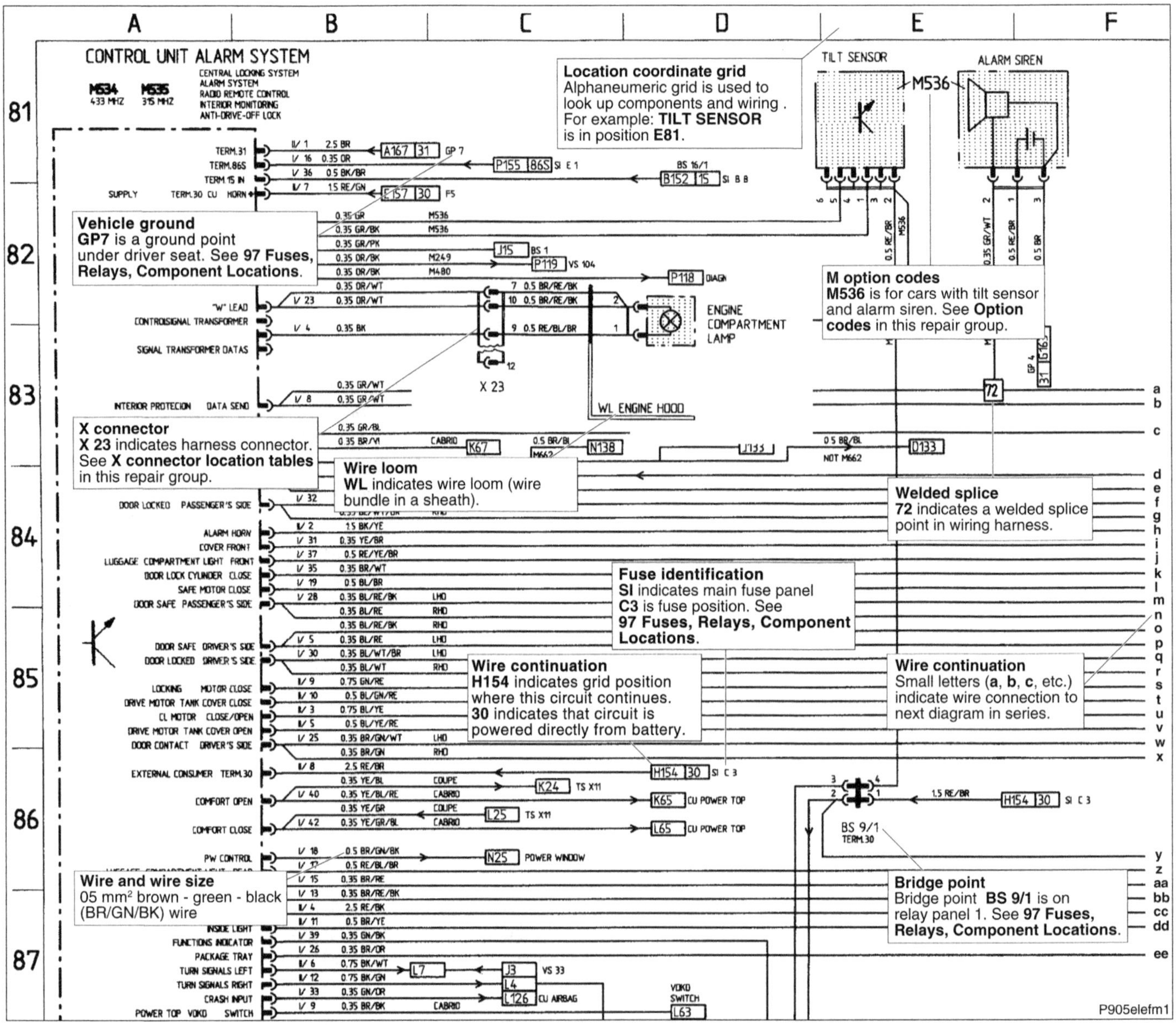

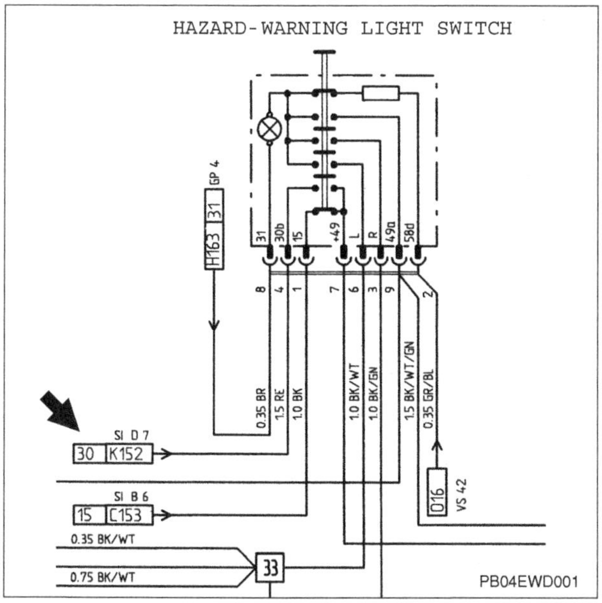

Circuit and terminal descriptions

Porsche designates electrical circuits, including junctions and grounds, with unique designations, most of which follow the German DIN standard. See Table a.

For example, if a terminal is labeled 30 (**arrow**), it tells you that positive (+) voltage is supplied to that terminal at all times directly from the battery. The sample schematic also shows a few other commonly used terminal designations, such as terminal 15 and terminal 31.

Table a. Electrical terminal designations (according to DIN 72 552 standard)

Number	Circuit description
1	Low voltage switched terminal of coil
4	High voltage center terminal of coil
X	Originates at ignition switch. Supplies power when ignition switch is in PARK, RUN or START position.
XE	Originates at ignition switch. Disrupts power when ignition switch is in START position.
15	Originates at ignition switch. Supplies power when ignition switch is in RUN or START position.
30	Battery positive (+) voltage. Supplies power whenever battery is connected. (Not dependent on ignition switch position, unfused)
31	Ground, battery negative (–) terminal
50	Starter control, supplies power from battery to starter solenoid when ignition switch is in START position only.
54	Stop light terminals
56	Headlight terminals
56a / 56b	High beam / low beam terminals
56d	Headlight flasher contact terminal
57	Parking lights terminal
58	Side marker lights, taillights, license plate lights, instrument panel lights
58d	Instrument panel lights
61	Alternator charge indicator lamp (generator D+)
85	Ground side (–) or relay coil
86	Power-in side (+) of relay coil
87	Power when relay coil is energized (usually terminal 87 connects to terminal 30 power when relay is energized)
D+	Alternator warning light and field energizing circuit

Porsche abbreviations and acronyms

Acronyms and abbreviations used in the wiring diagrams and in this manual are summarized in **Table b**.

Table b. Porsche abbreviations and acronyms

Number	Circuit description	Number	Circuit description
A/C	air-conditioning	ins.c	instrument cluster
ABD	automatic brake differential	LED	light emitting diode
ABS	antilock brakes	LF	low frequency
ADL	additional driving lights	lhd	left-hand drive
BB	brake booster	M5.2	Motronic 5.2
BPAU	brake booster	MY	model year
BL	backup light	NO	number
CE	central electric	NTC	negative temperature coefficient
CLOCK	clock signal	PL	plug
CLS	central locking system	PIN	pin
CO	carbon monoxide	POT	potentiometer
CP	connecting point	psd	Porsche limited slip differential
CU	control unit	PWm	pulse width modulation
CWL	central warning light	RA	headlight cleaning system
DATA	data signal	RA	rear axle
DEF	defrost	RDK	tire pressure control
DG	engine speed sensor	RFL	rear fog light
DI	electric ignition system with knock control	RHD	right hand drive
DP	disconnecting point	RL	rear left
DSP	digital sound package	RR	rear right
ESO	engine speed sensor output	RoW	rest of world
ess	engine speed sensor	SA	Saudi Arabia
ETC	electrical transmission control	SAFETY	safety lamp
FA	front axle	service	service
FL	front left	SG	control unit
FOG	fog light	SCS	combined steering column switch
FR	front right	service	service
GP	ground point	SI	fuse
GND	ground	SILA	safety lamp
gr	glass roof	sl	safety lamp
HCS	headlight cleaning system	SS	speed sensor
HF	high frequency	TDPS	tank differential pressure sensor
HFM	hot-film mass air flow sensor	TE	terminal
hvac	headlight vertical aim control	TEL.MUTE	radio muting control
inj. v	injector valve	TI	injection time
inj. valve	fuel injector	TN	engine speed signal
trp	twin-coil rotary positioner	ust	911 Turbo
turbo	911 turbo	WL	wiring loom (wiring harness)
TV	throttle valve	wp	weld point
TVS	throttle valve sensor	WW	worldwide

Option codes

Table c lists option codes.These codes are referred to in the wiring diagrams. For a more exhaustive listing of option codes, see **02 Product Familiarization**.

Table c. Option codes

Code	Definition	Number	Definition
M 002	RS road version	M 487	Norway
M 003	RS racing safety equipment	M 490	Audio option package
M 004	RS turbo version	M 513	Lumbar support seat right
M 113	Daytime running lights (Canada)	M 520	Interior monitoring
M 139	Adjustable seat heating, left side	M 530	Anti-theft (Immobilizer)
M 193	Japan version	m 545	Large tank
M 215	Saudi Arabia version	M 553	USA / Canada version
M 222	Anti-skid regulator (ASR)	M 562	Airbags, driver and passenger
M 224	Automatic brake differential	M 564	Without airbag
M 249	Automatic transmission (Tiptronic)	M 573	Air-conditioning system
M 339	C4 4WD drive	M 576	Without rear fog light
M 340	Adjustable seat heating, right side	M 586	Lumbar support seat left
M 383	Sport seat left electric height adjustment	M 601	Litronic (Xenon) headlights
M 387	Sport seat right electric height adjustment	M 602	High mounted stop light
M 419	Rear package tray	M 605	Headlight vertical aim control
M 425	Rear window wiper	M 614	Telephone preparation
M 437	Comfort seat, left	M 615	Cordless phone
M 438	Comfort seat, right	M 632	Recognition seat assignment passenger
M 439	Power top	M 650	Sunroof
M 441	Radio preparation	M 651	Power windows
M 451	Limited radio preparation	M 659	On-board computer
M 454	Cruise control	M 680	Digital sound package
M 480	6-speed manual transmission	M 692	CD-changer

X Connectors

Porsche identifies harness connectors using the letter X followed by a number. To see an example of an X-connector in a wiring diagram, look at grid coordinate N6 (1995 diagrams). There you will see X25, which is a 10-pin connector.

Table d. X-Connectors, 1995

Connector	Location, additional information	Connector	Location, additional information
X 1/1	In trunk, right, 10 pins	X 28	On steering column, 6 pins
X 1/2	In trunk, right, 10 pins	X 29	Engine compartment, right, 2 pins
X 3	Engine compartment, left, 14 pins	X 31	Below parcel tray, passenger's side, 4 pins
X 4/1	Under driver's seat, 14 pins	X 32	Under dashboard, driver's side, 2 pins
X 4/2	Under driver's seat, 14 pins	X 33	In trunk, left, 2 pins
X 11/1	Under dashboard, driver's side, 14 pins	X 34	In trunk, right, 2 pins
X 11/2	Under dashboard, driver's side, 10 pins	X 36	Behind passenger's side rear side panel, 4 pins
X 12/1	Under dashboard, passenger's side, 12 pins	X 37	Under driver's seat, 8 pins
X 12/2	Under dashboard, passenger's side, 10 pins	X 38	Under passenger's seat, 8 pins
X 20	Under dashboard, passenger's side, 14 pins	X 39	On trunk lid hinge, right, 6 pins
X 24	Engine compartment, right, 10 pins	X 60	Engine compartment, left, 21 pins
X 25	Engine compartment, right, 10 pins	X 65	Driver's door, 6 pins
X 26	Behind passenger's side rear side panel, 12 pins	X 66	Passenger's door, 6 pins
X 27	Below parcel tray, passenger side, 4 pins	X 69	In trunk, right, 14 pins

Table e. X-Connectors, 1995/2

Connector	Location, additional information	Connector	Location, additional information
X 1/1	In trunk, right, 10 pins	X 29	Engine compartment, right, 2 pins
X 1/2	In trunk, right, 10 pins	X 31	Below parcel tray, passenger's side, 4 pins
X 3	Engine compartment, left, 14 pins	X 32	Under dashboard, driver's side, 6 pins
X 4/1	Under driver's seat, 14 pins	X 33	In trunk, left, 2 pins
X 4/2	Under driver's seat, 14 pins	X 34	In trunk, right, 2 pins
X 11/1	Under dashboard, driver's side, 14 pins	X 36	Behind passenger's side rear side panel, 4 pins
X 11/2	Under dashboard, driver's side, 10 pins	X 37	Under driver's seat, 8 pins
X 12/1	Under dashboard, passenger's side, 12 pins	X 38	Under passenger's seat, 8 pins
X 12/2	Under dashboard, passenger's side, 10 pins	X 39	On trunk lid hinge, right, 6 pins
X 20	Under dashboard, passenger's side, 14 pins	X 60	Engine compartment, left, 21 pins
X 24	Engine compartment, right, 10 pins	X 65	Driver's door, 6 pins
X 25	Engine compartment, right, 10 pins	X 66	Passenger's door, 6 pins
X 26	Behind passenger's side rear side panel, 12 pins	X 69	In trunk, right, 14 pins
X 28	On steering column, 6 pins		

Table f. X-Connectors, 1996

Connector	Location, additional information	Connector	Location, additional information
X 1/1	In trunk, right, 10 pins	X 29	Engine compartment, right, 2 pins
X 1/2	In trunk, right, 10 pins	X 31	Below parcel tray, passenger's side, 8 pins
X 3	Engine compartment, left, 14 pins	X 32	Under dashboard, driver's side, 6 pins
X 4/1	Under driver's seat, 14 pins	X 33	In trunk, left, 2 pins
X 4/2	Under driver's seat, 14 pins	X 34	In trunk, right, 2 pins
X 11/1	Under dashboard, driver's side, 14 pins	X 36	Behind passenger's side rear side panel, 4 pins
X 11/2	Under dashboard, driver's side, 10 pins	X 37	Under driver's seat, 8 pins
X 12/1	Under dashboard, passenger's side, 12 pins	X 38	Under passenger's seat, 8 pins
X 12/2	Under dashboard, passenger's side, 10 pins	X 39	On trunk lid hinge, right, 6 pins
X 20	Under dashboard, passenger's side, 14 pins	X 60	Engine compartment, left, 21 pins
X 24	Engine compartment, right, 10 pins	X 65	Driver's door, 6 pins
X 25	Engine compartment, right, 10 pins	X 66	Passenger's door, 6 pins
X 26	Behind passenger's side rear side panel, 12 pins	X 69	In trunk, right, 14 pins
X 28	On steering column, 6 pins		

Table g. X-Connectors, 1997-1998

Connector	Location, additional information	Connector	Location, additional information
X 1/1	In trunk, right, 10 pins	X 26	Behind passenger's side rear side panel, 12 pins
X 1/2	In trunk, right, 10 pins	X 28	On steering column, 6 pins
X 3	Engine compartment, left, 14 pins	X 29	Engine compartment, right, 2 pins
X 4/1	Under driver's seat, 14 pins	X 31	Below parcel tray, passenger's side, 8 pins
X 4/2	Under driver's seat, 14 pins	X 32	Under dashboard, driver's side, 6 pins
X 4/3	Under driver's seat, 10 pins	X 36	Behind passenger's side rear side panel, 4 pins
X 11/1	Under dashboard, driver's side, 14 pins	X 37	Under driver's seat, 8 pins
X 11/2	Under dashboard, driver's side, 10 pins	X 38	Under passenger's seat, 8 pins
X 12/1	Under dashboard, passenger's side, 12 pins	X 39	On trunk lid hinge, right, 6 pins
X 12/2	Under dashboard, passenger's side, 10 pins	X 60	Engine compartment, left, 21 pins
X 20	Under dashboard, passenger's side, 14 pins	X 65	Driver's door, 6 pins
X 24	Engine compartment, right, 10 pins	X 66	Passenger's door, 6 pins
X 25	Engine compartment, right, 10 pins	X 69	In trunk, right, 14 pins

Sheet 1 - 1995

Lights (page 1 of 3)

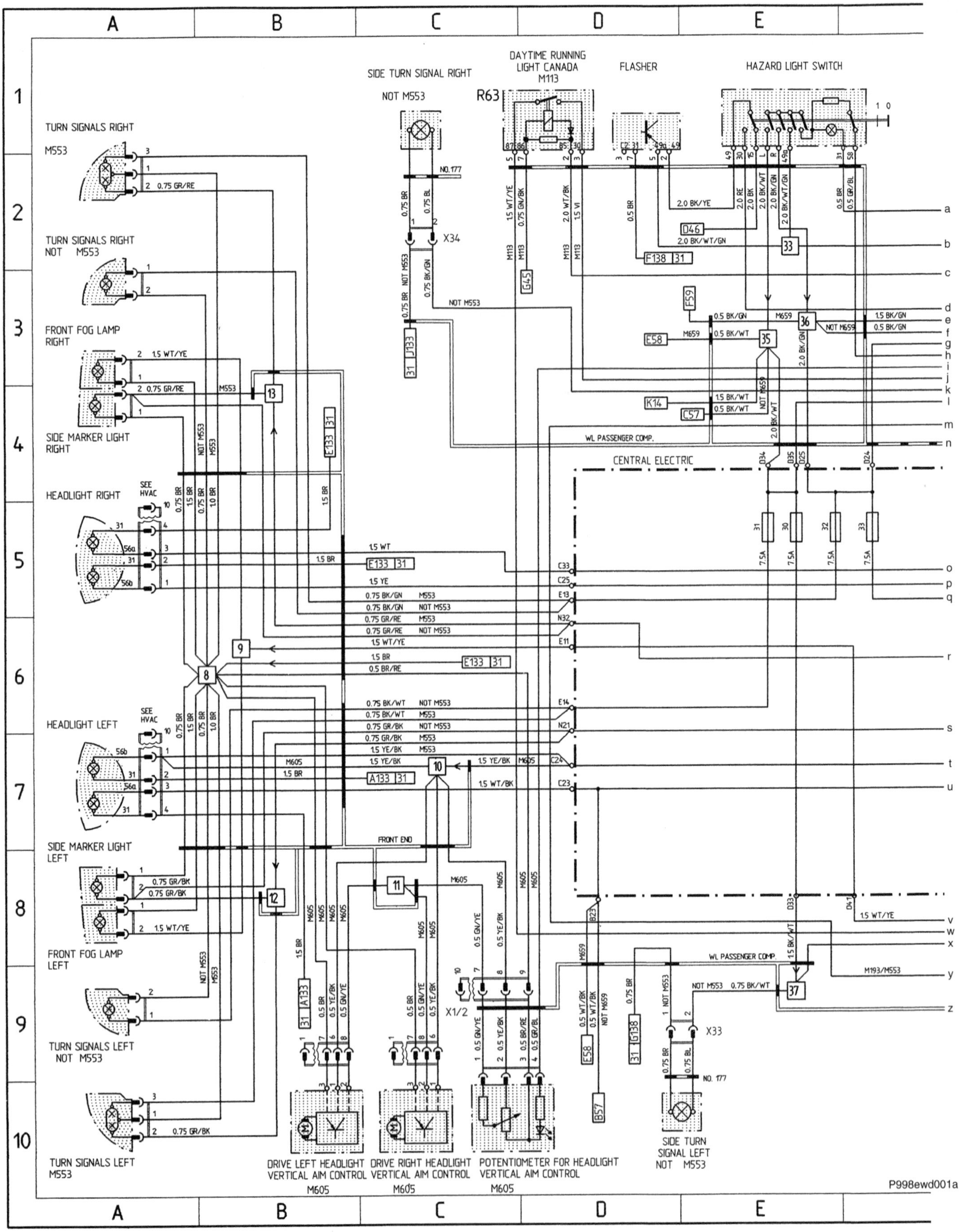

Lights
(page 2 of 3)

Sheet 1 - 1995

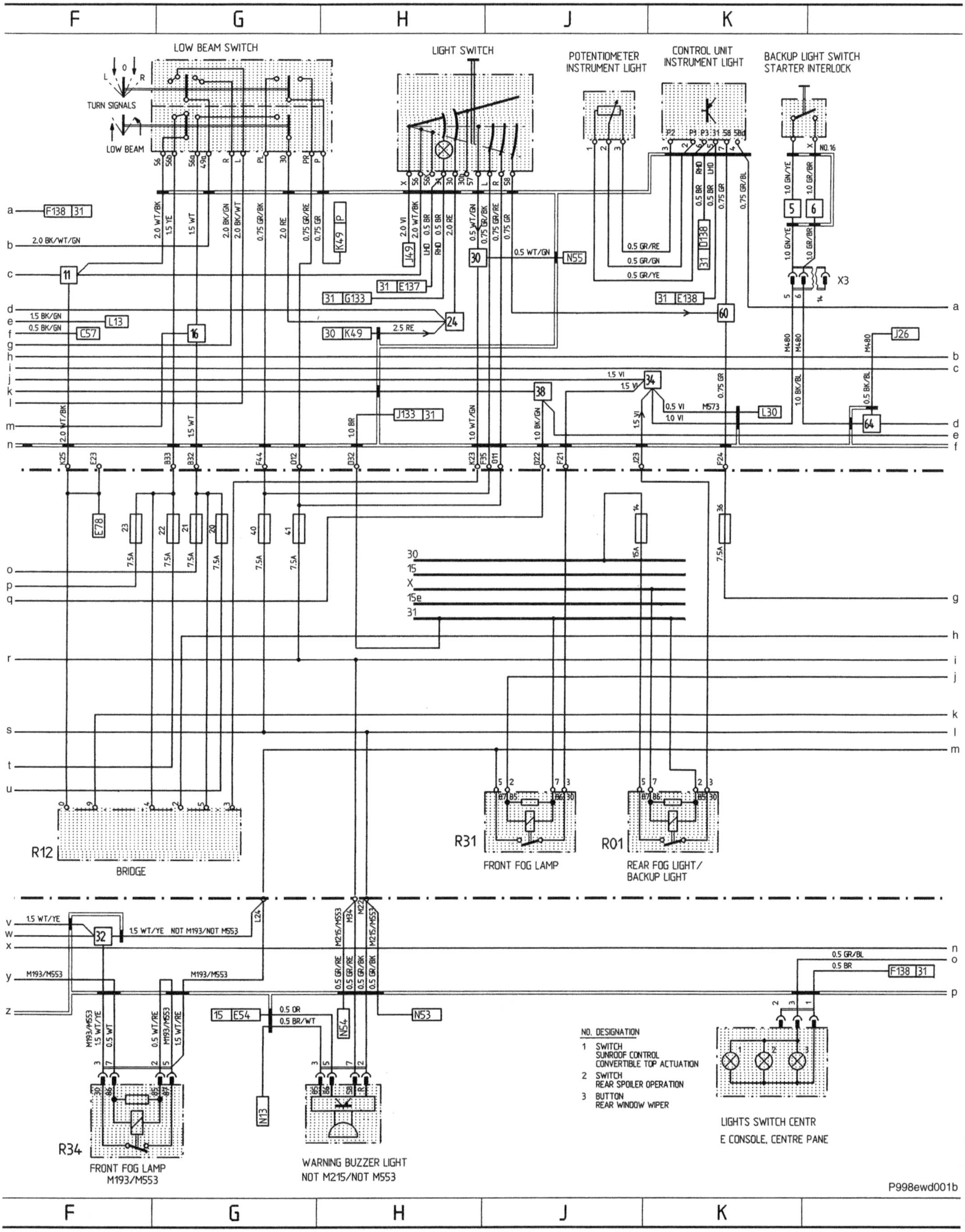

P998ewd001b

EWD

Sheet 1 - 1995

Lighting (page 3 of 3)

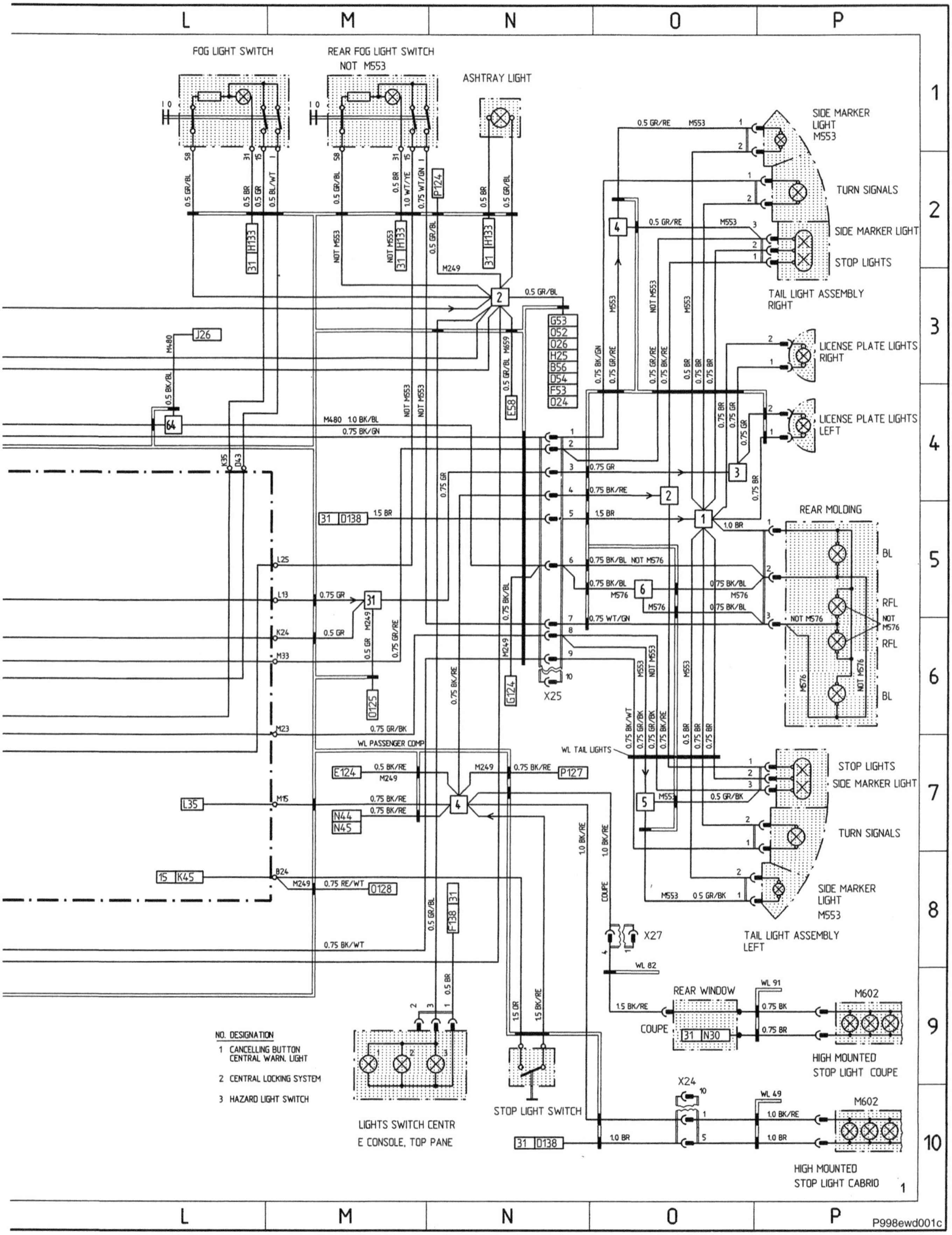

P998ewd001c

Seats, alarm system, central locking, mirror, power windows interior lighting (page 1 of 3)

Sheet 2 - 1995

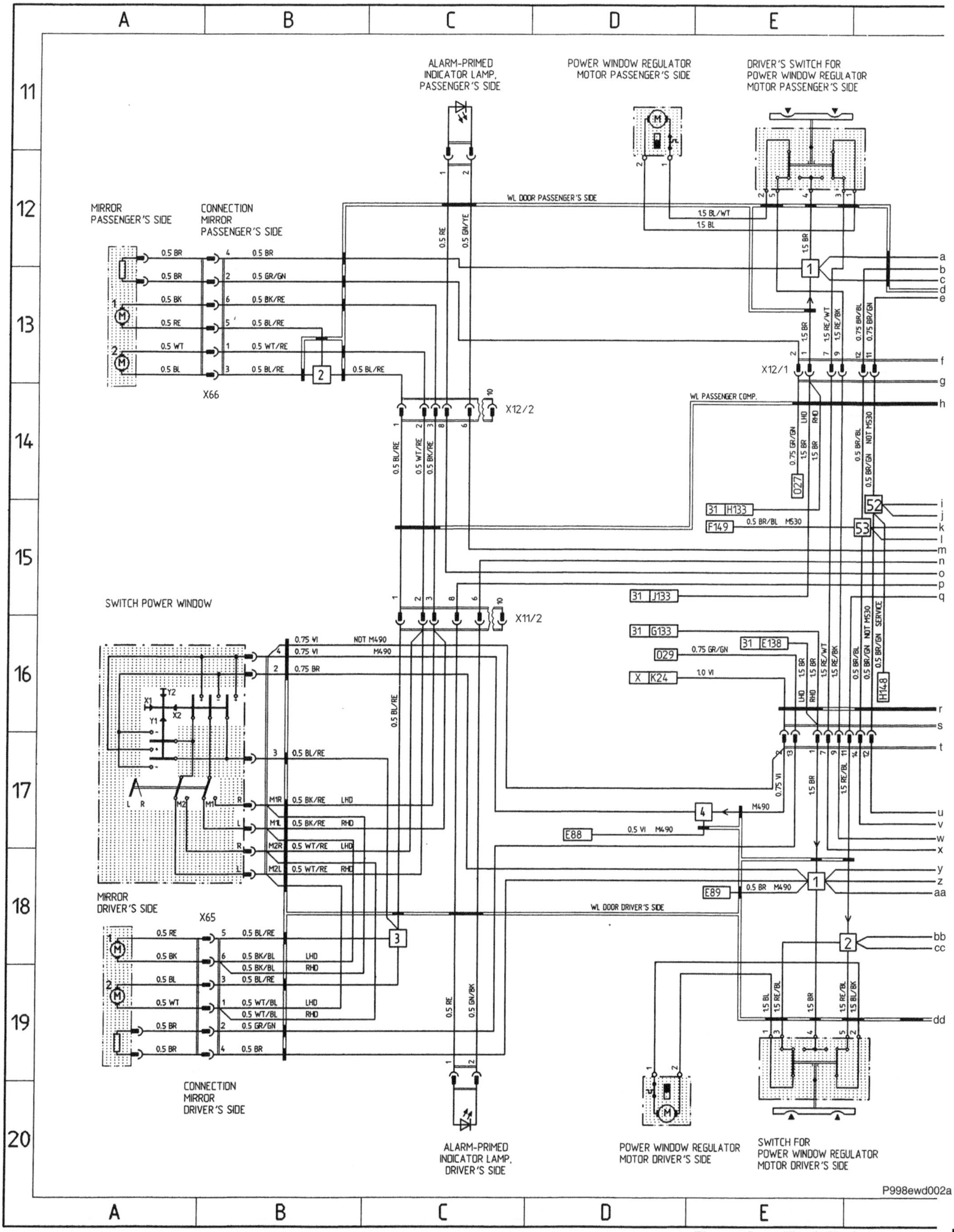

EWD

Sheet 2 - 1995 Seats, alarm system, central locking, mirror, power windows (page 2 of 3)

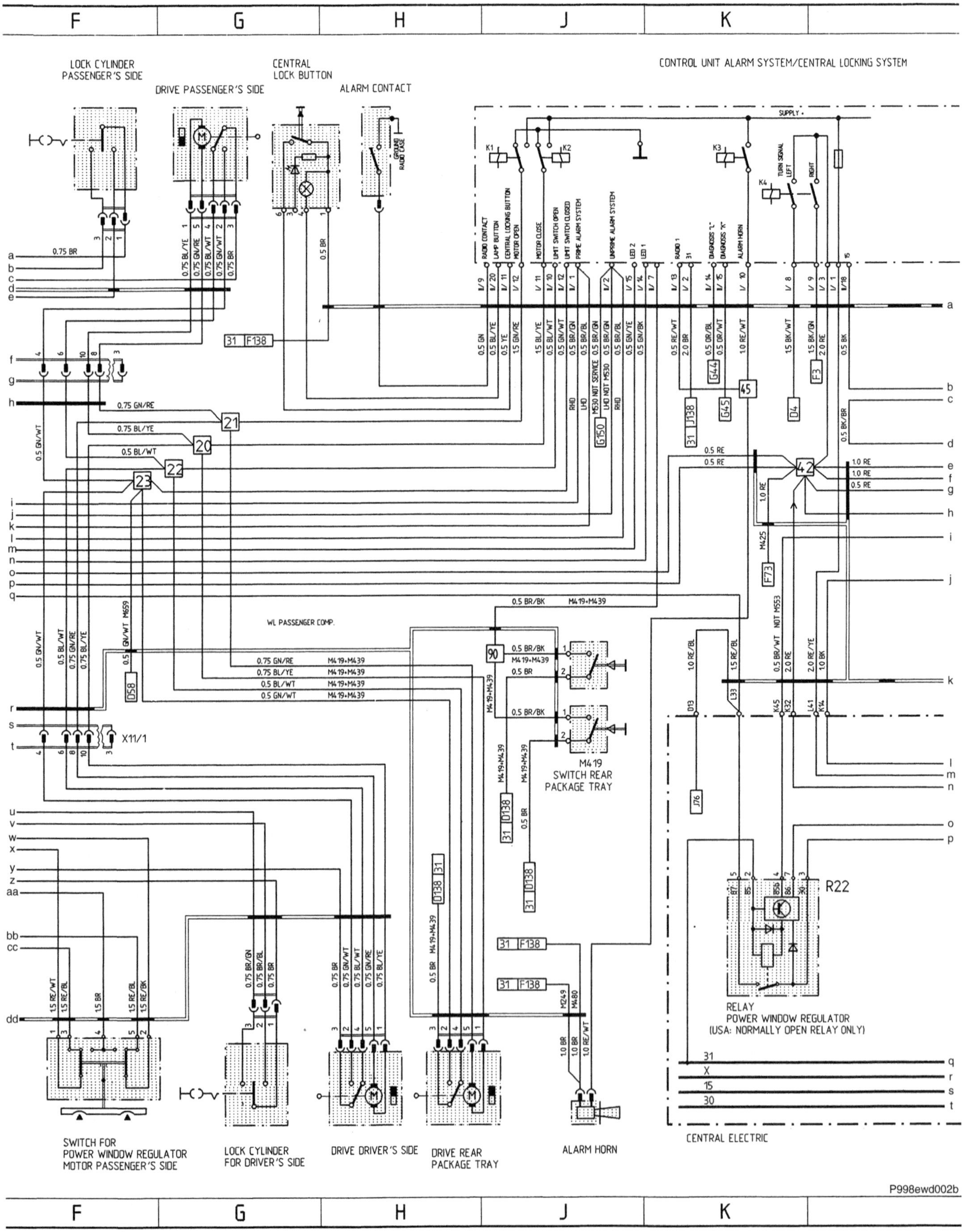

Seats, alarm system, central locking, mirror, power windows interior lighting (page 3 of 3)

Sheet 2 - 1995

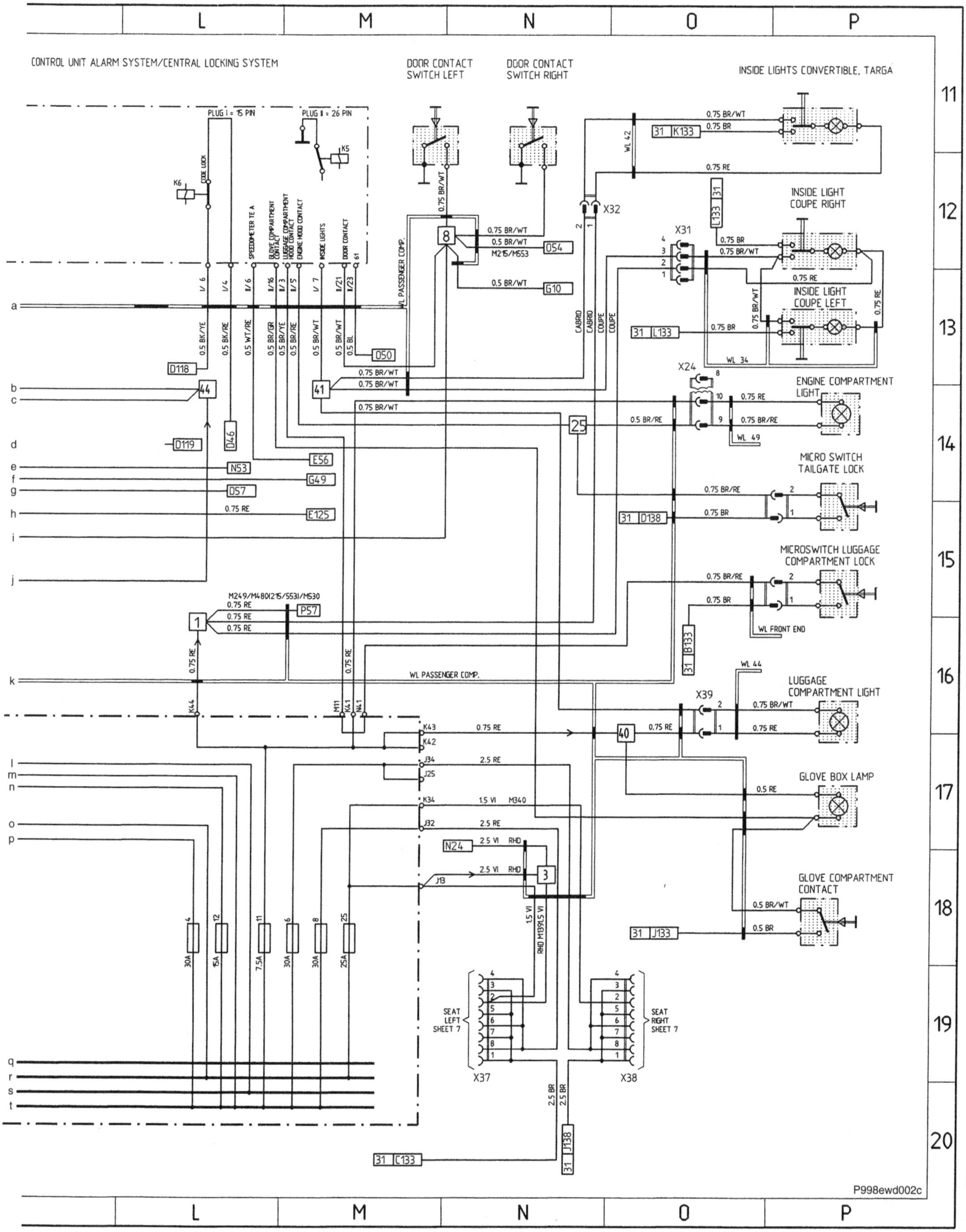

Sheet 3 - 1995

Heater, A/C
(page 1 of 3)

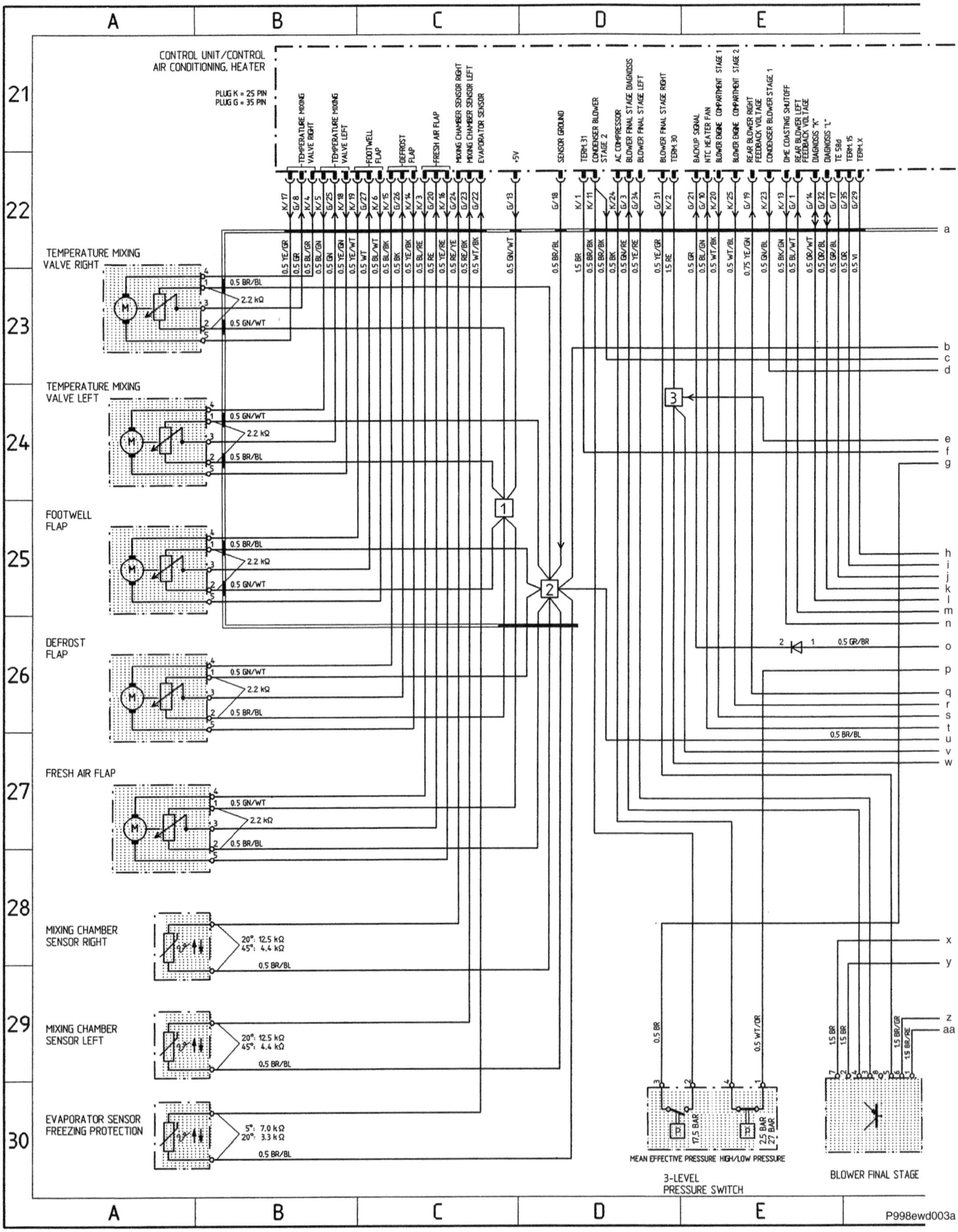

Heater, A/C (page 2 of 3)

Sheet 3 - 1995

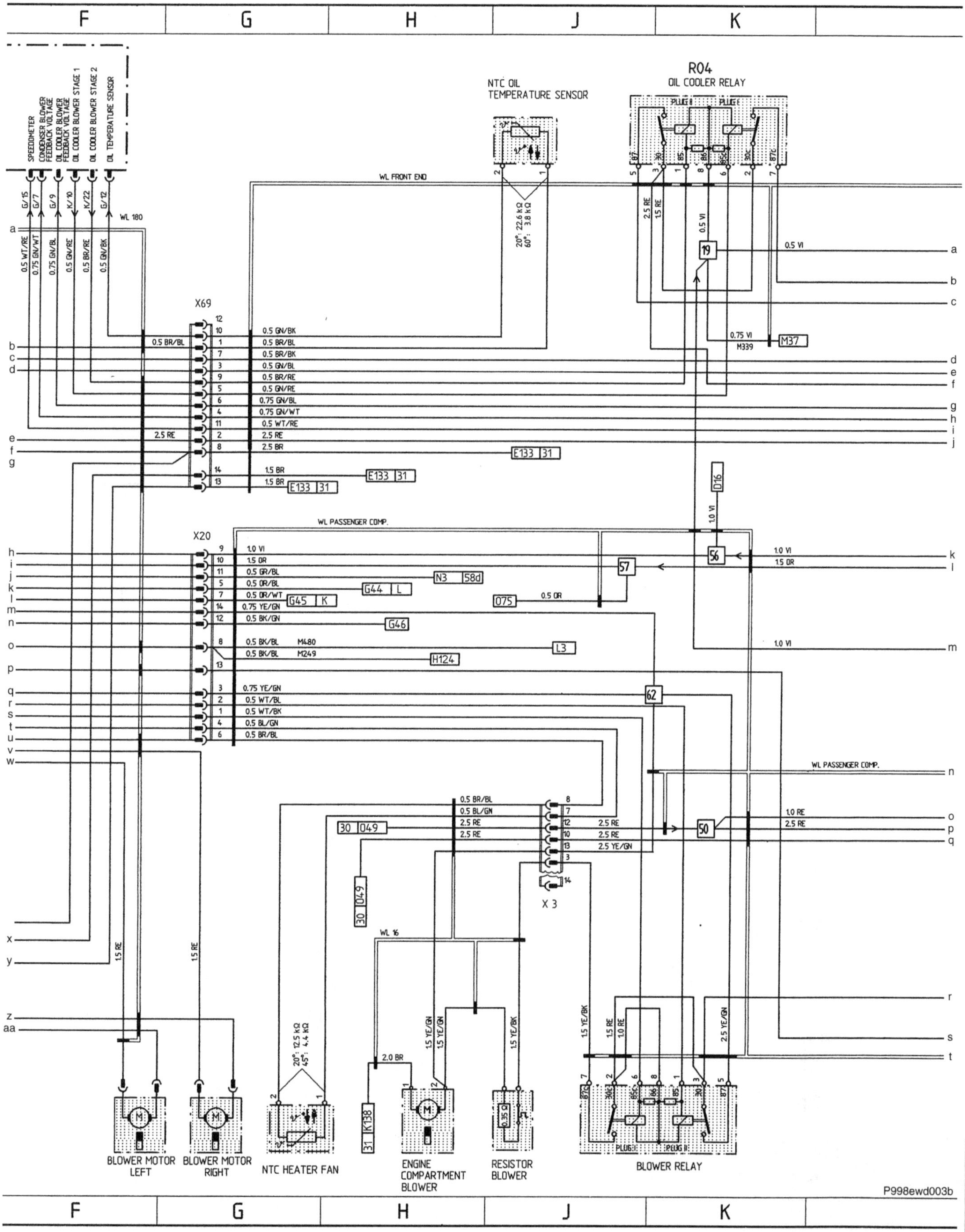

Sheet 3 - 1995

Heater, A/C
(page 3 of 3)

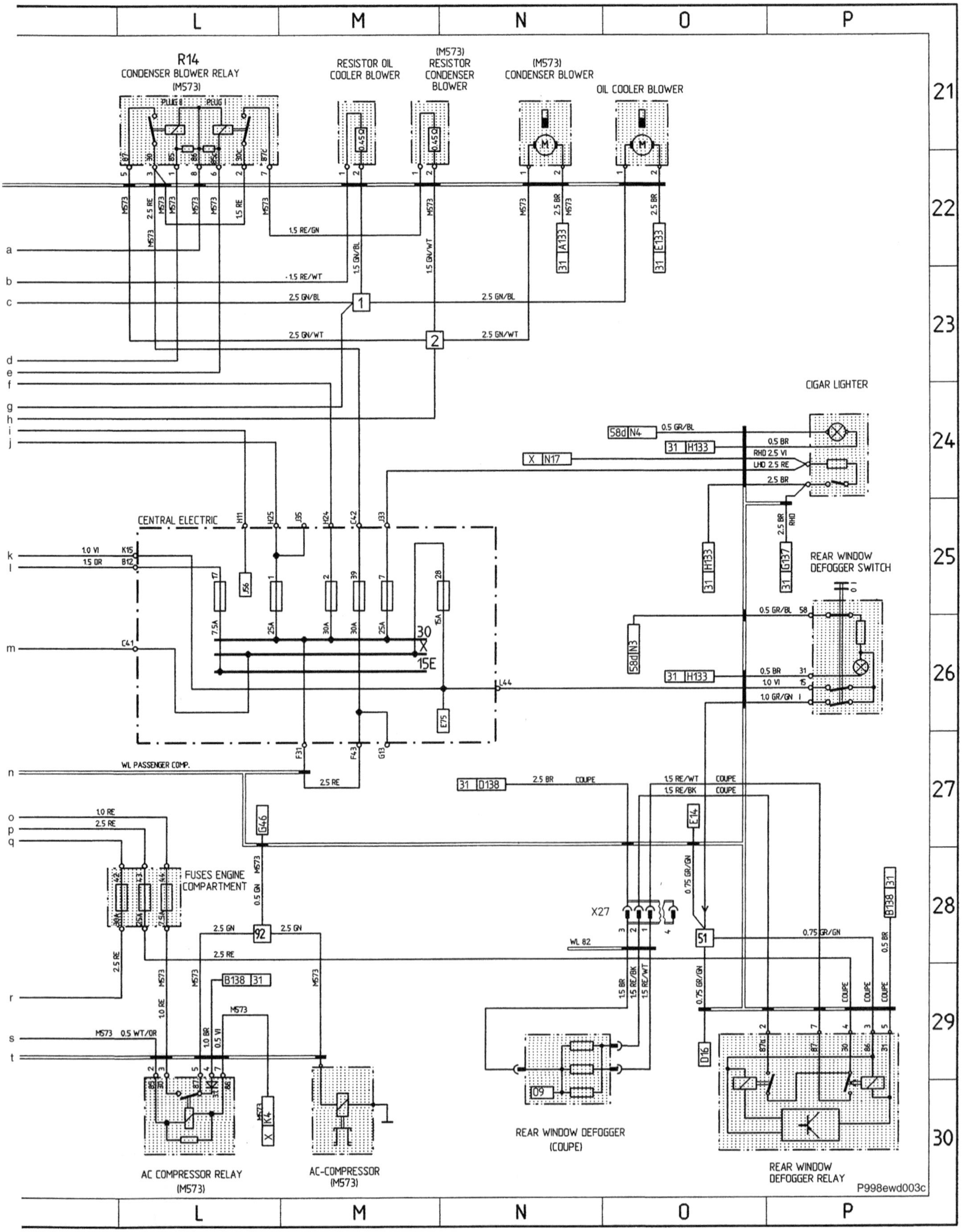

ABS
(page 1 of 3)

Sheet 4 - 1995

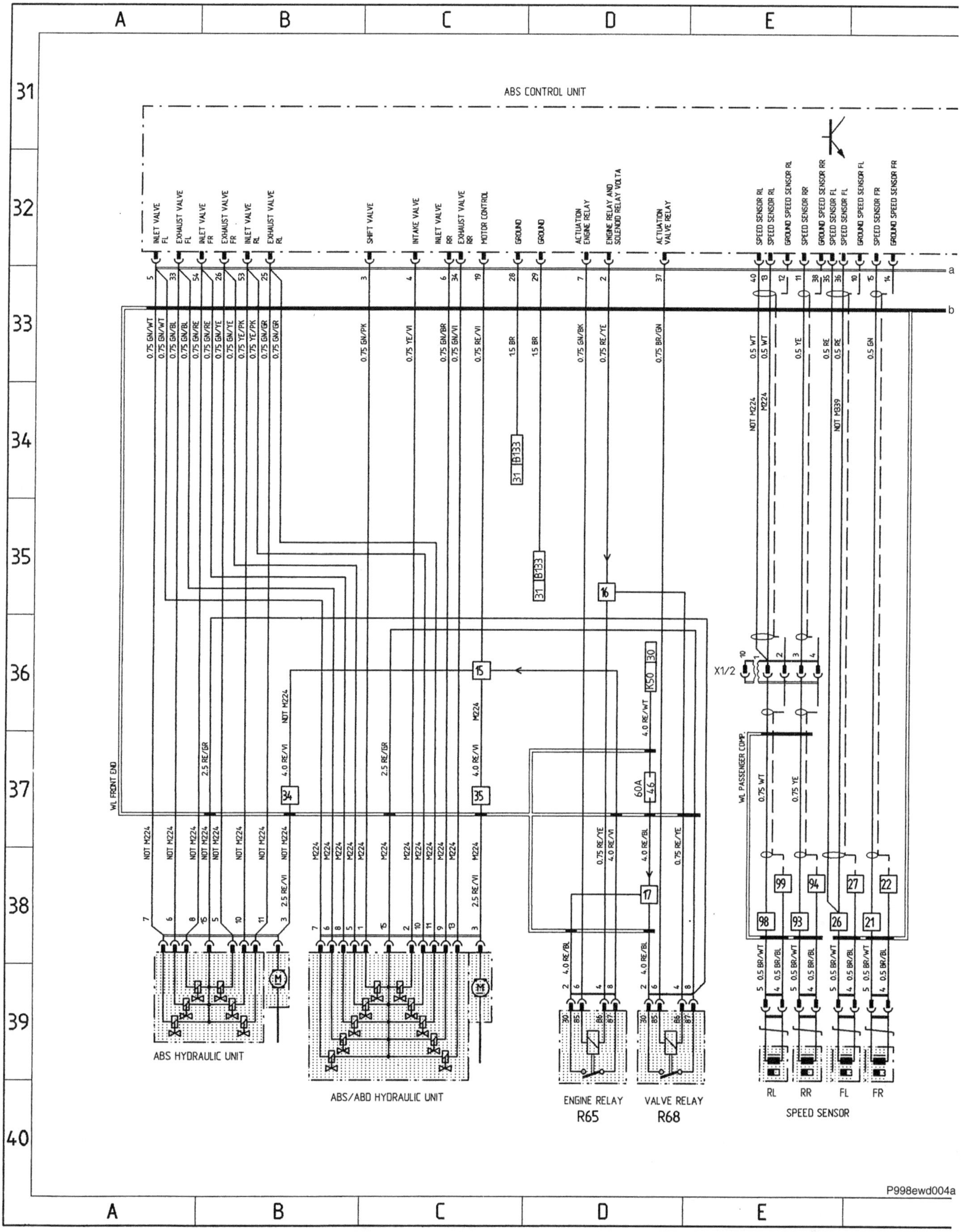

P998ewd004a

Sheet 4 - 1995

ABS
(page 2 of 3)

F	G	H	J	K	

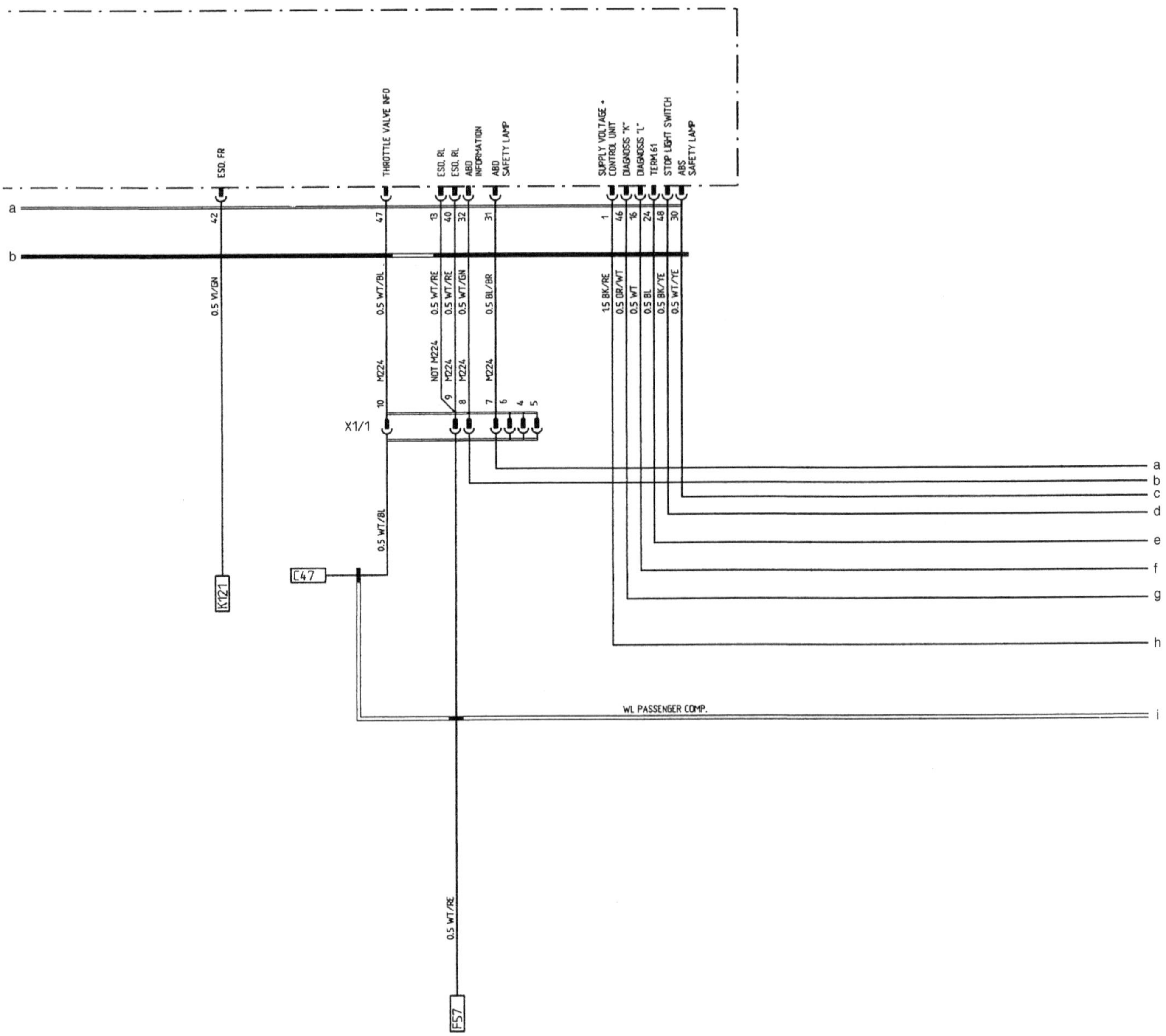

P998ewd004b

F	G	H	J	K	

ABS
(page 3 of 3)

Sheet 4 - 1995

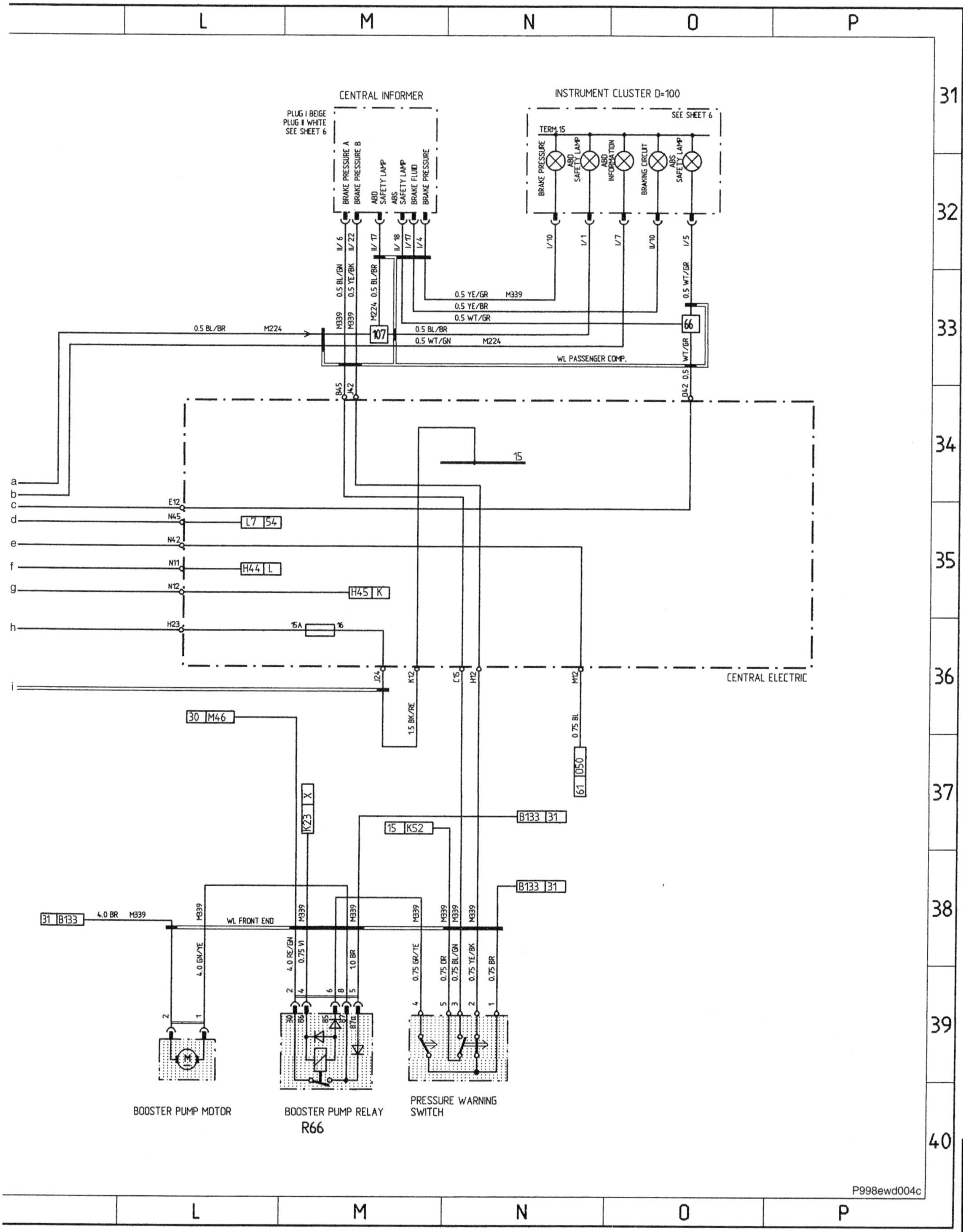

EWD

Sheet 5 - 1995 **Engine electrics, fuel, ignition system (DME M 2.10) (page 1 of 3)**

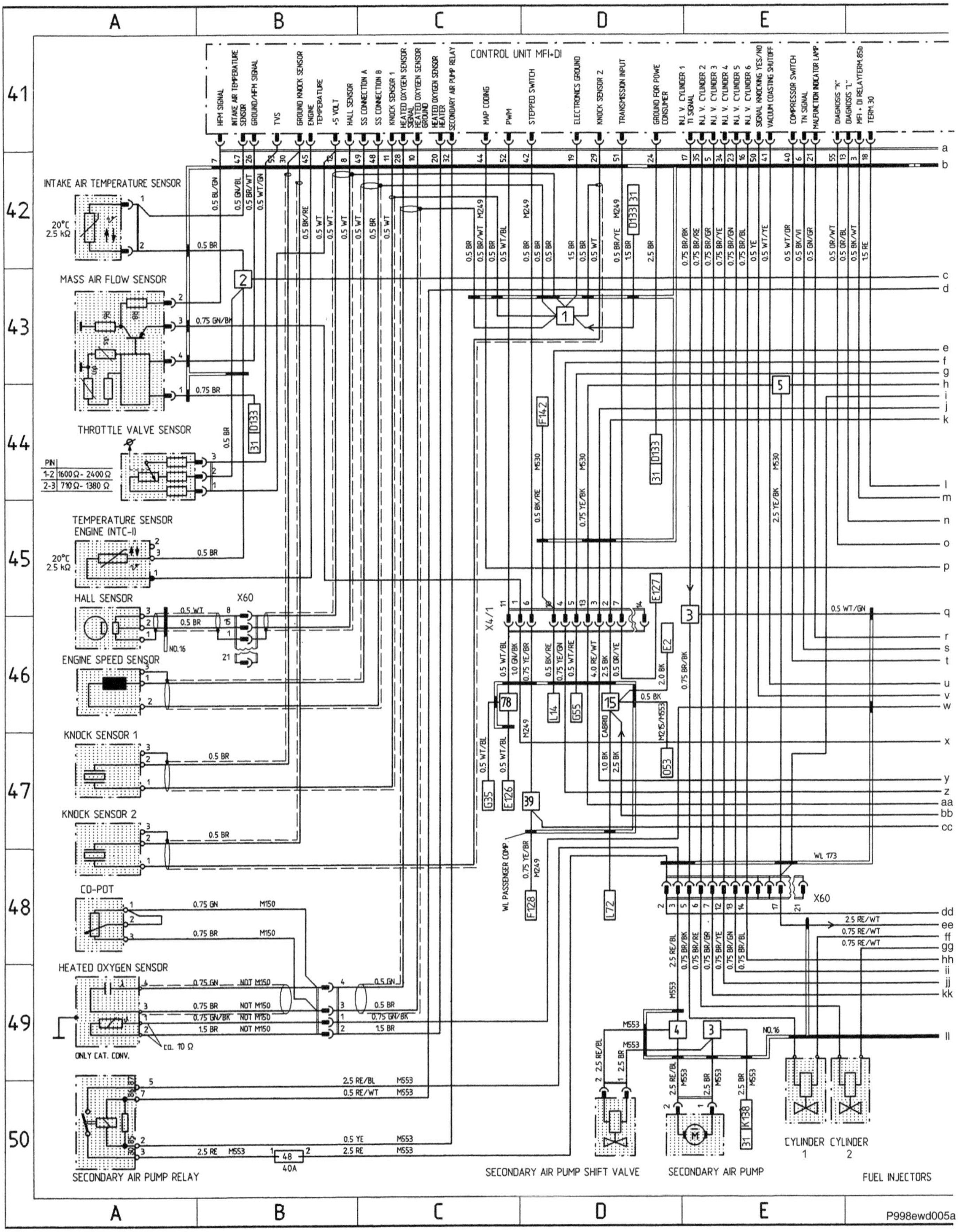

Engine electrics, fuel, ignition system (DME M 2.10) (page 2 of 3)

Sheet 5 - 1995

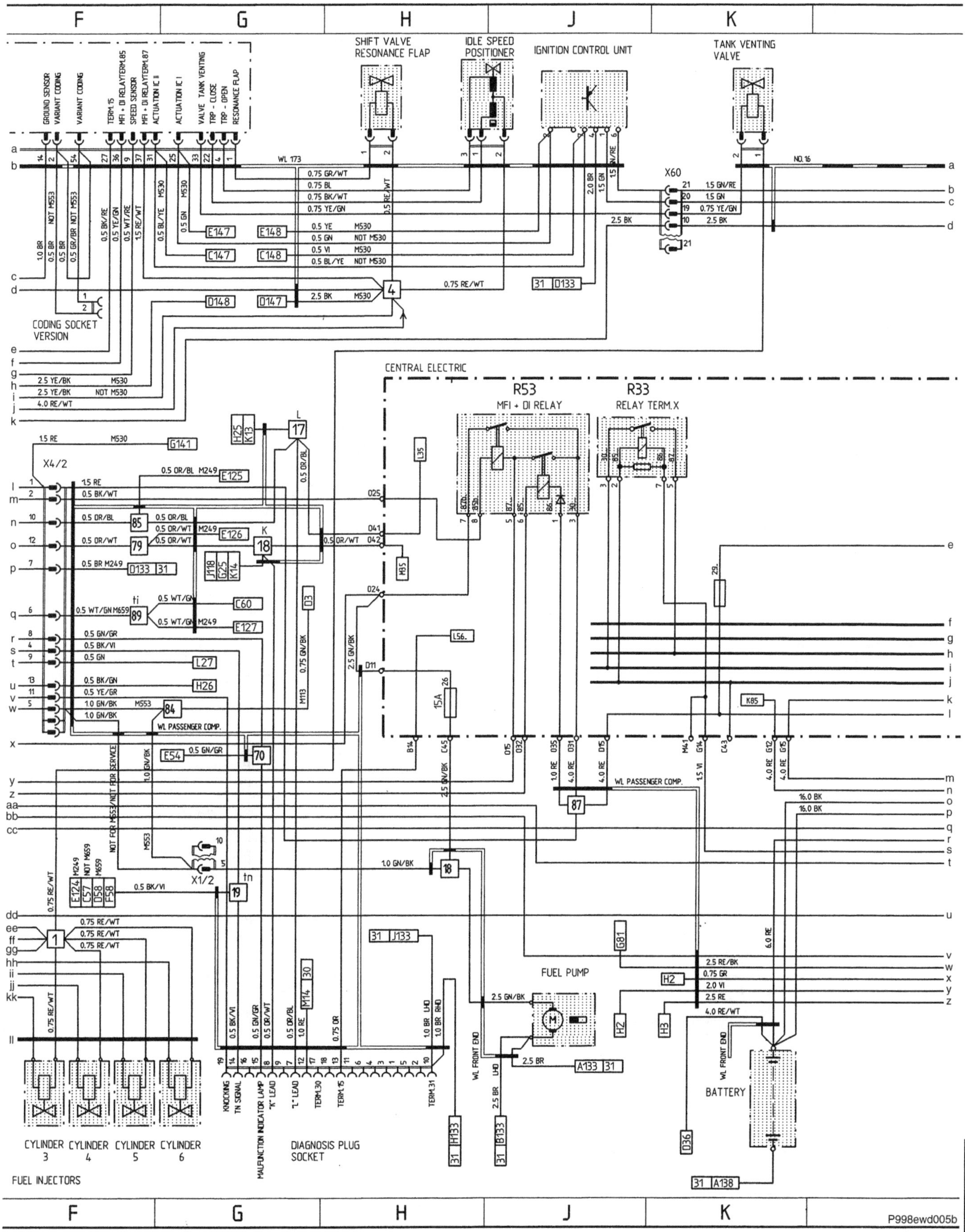

Sheet 5 - 1995

Engine electrics, fuel, ignition system (DME M 2.10) (page 3 of 3)

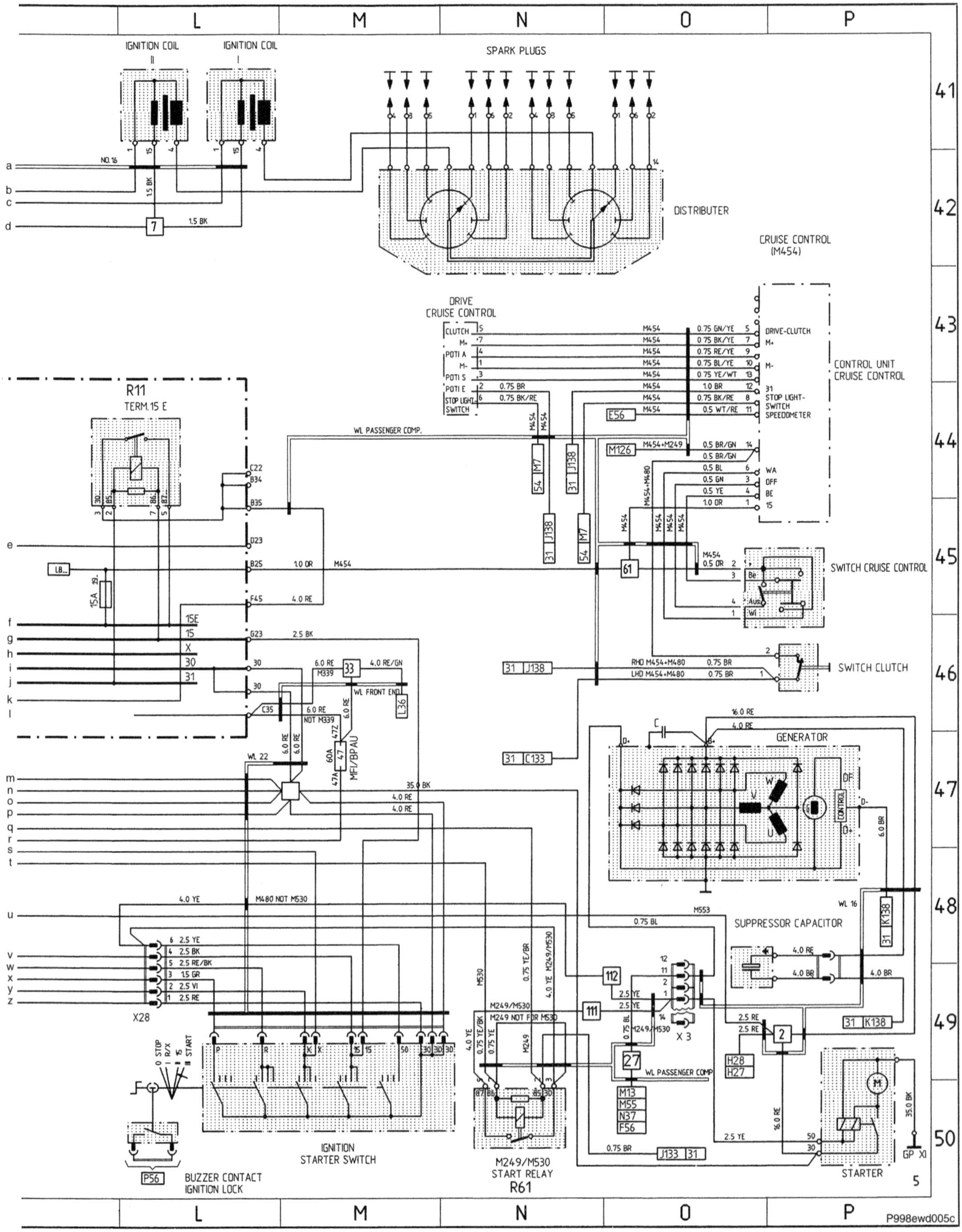

Instruments and sensor (page 1 of 3)

Sheet 6 - 1995

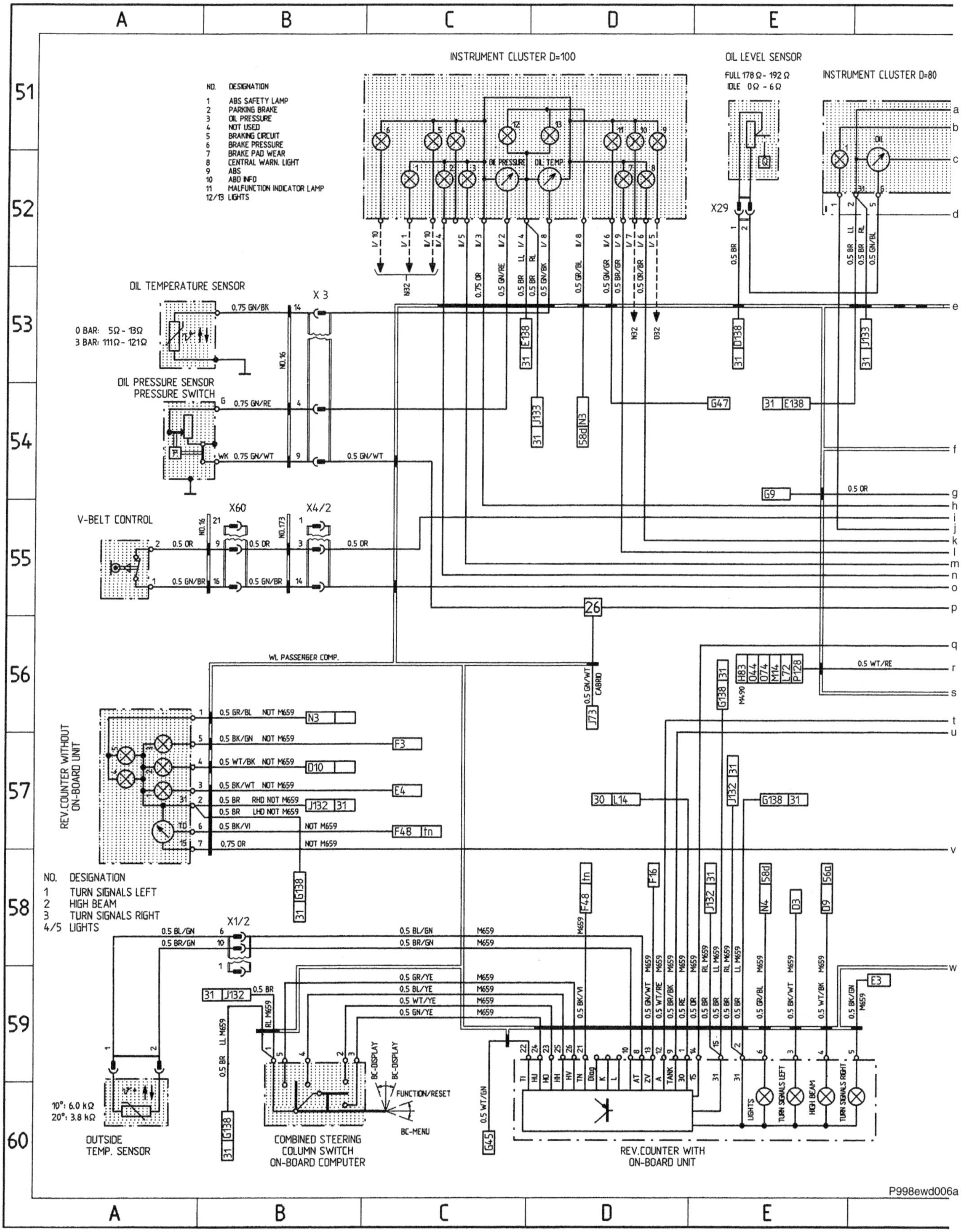

Sheet 6 - 1995

Instruments and sensor (page 2 of 3)

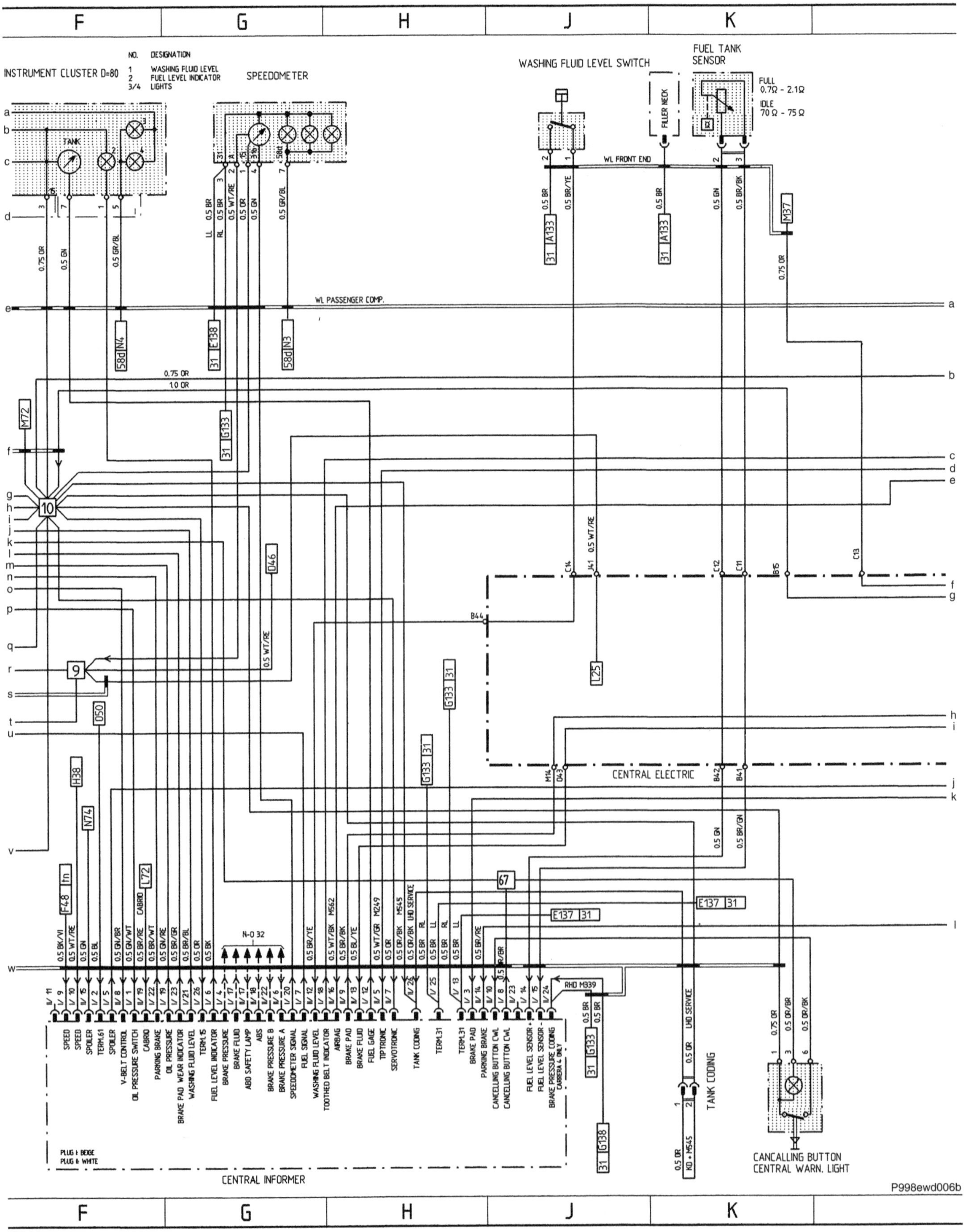

Instruments and sensor (page 3 of 3)

Sheet 6 - 1995

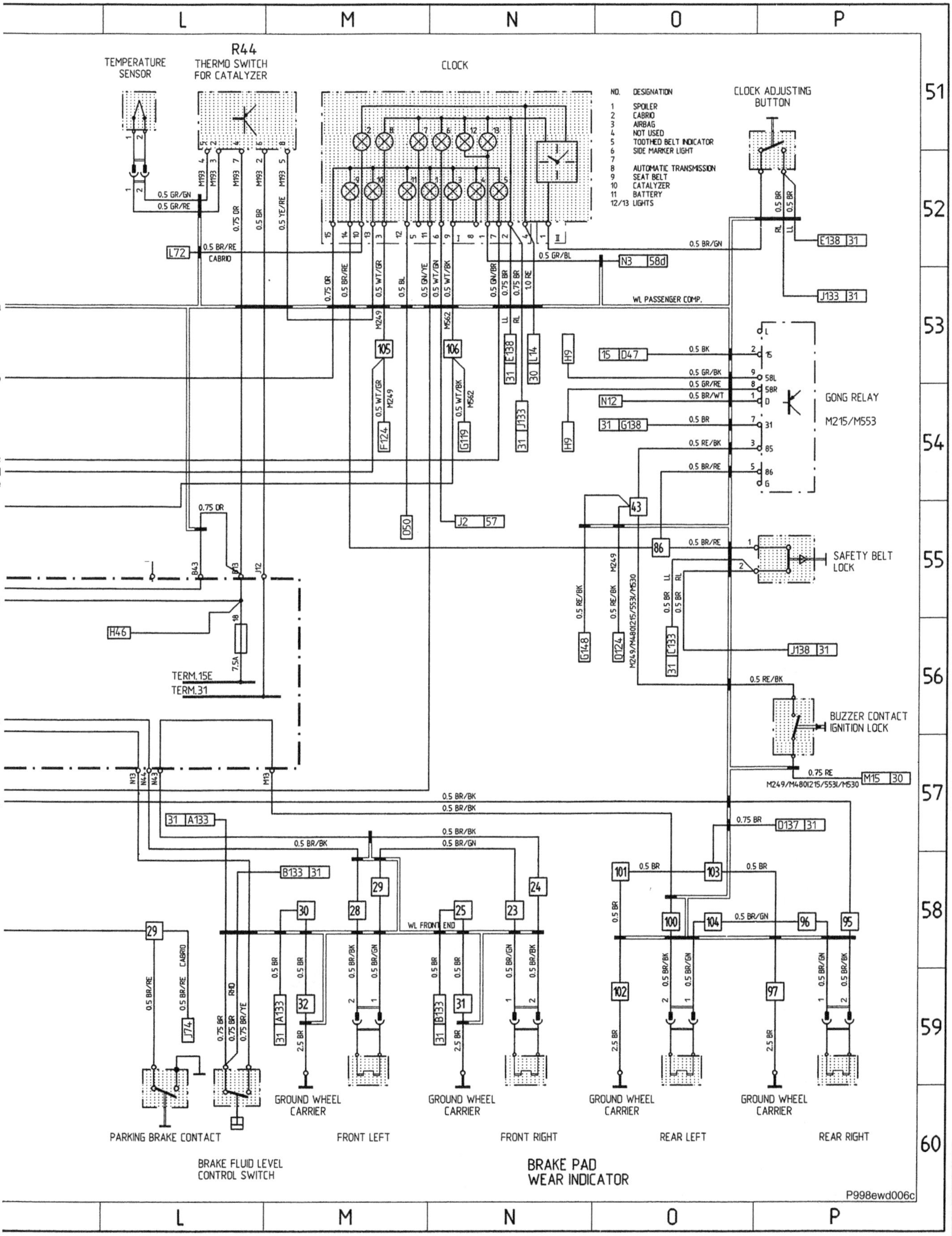

Sheet 7 - 1995

Seats
(page 1 of 2)

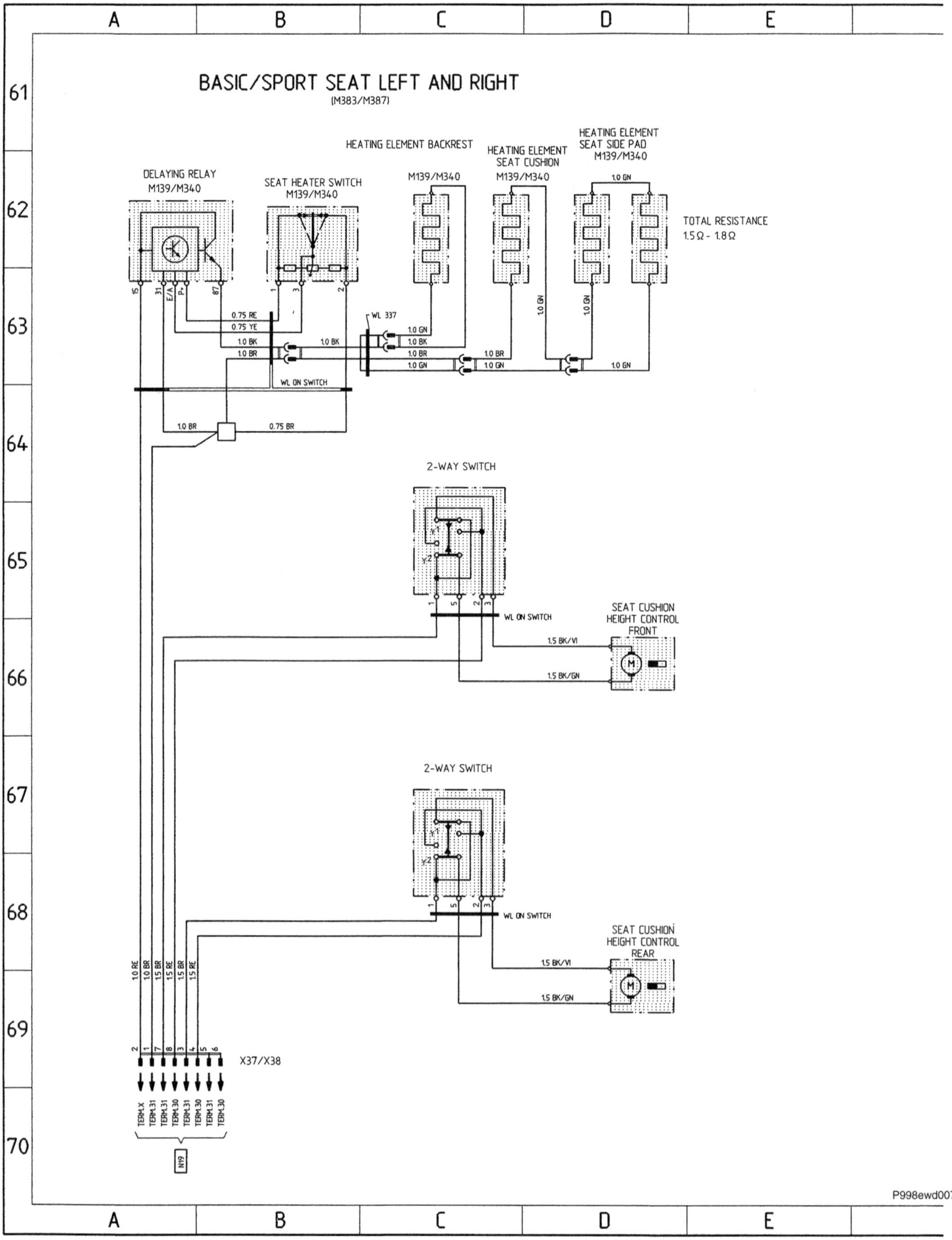

Seats
(page 2 of 2)

Sheet 7 - 1995

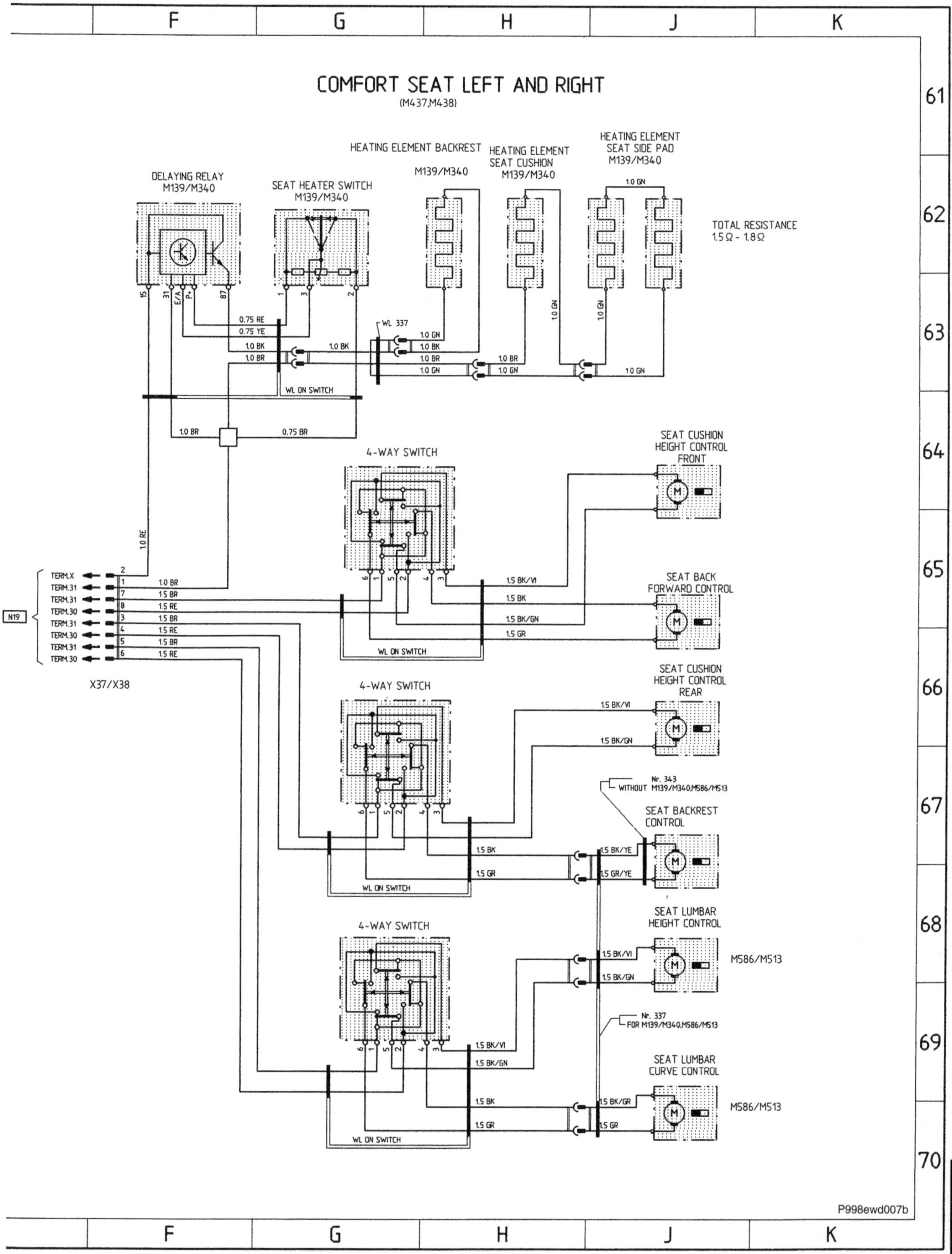

EWD

Sheet 8 - 1995 Rear spoiler, two-tone horns, wipe and wash cleaners, sunroof, convertible top (page 1 of 3)

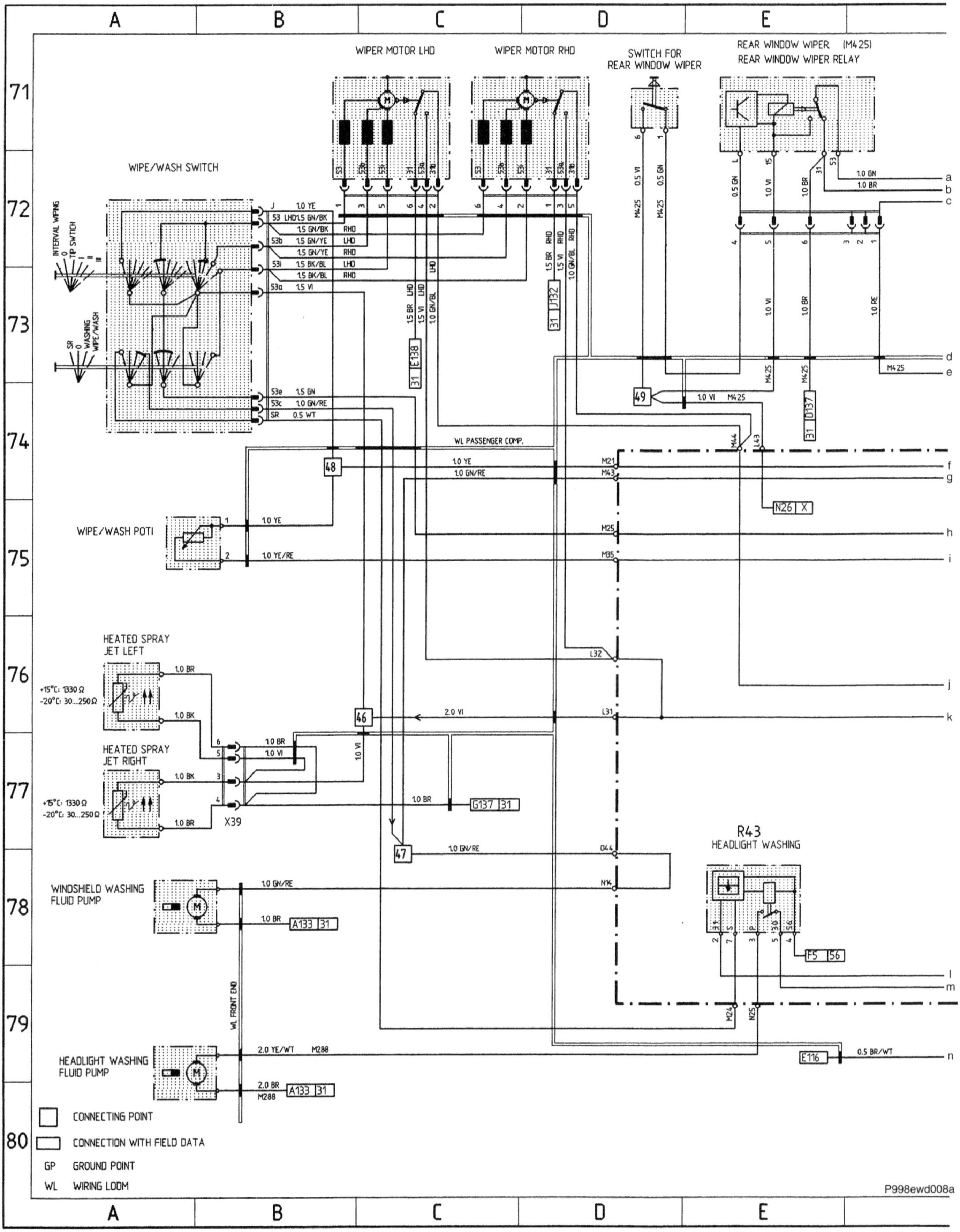

☐	CONNECTING POINT
▭	CONNECTION WITH FIELD DATA
GP	GROUND POINT
WL	WIRING LOOM

P998ewd008a

Rear spoiler, two-tone horns, wipe and wash cleaners, sunroof, convertible top (page 2 of 3)

Sheet 8 - 1995

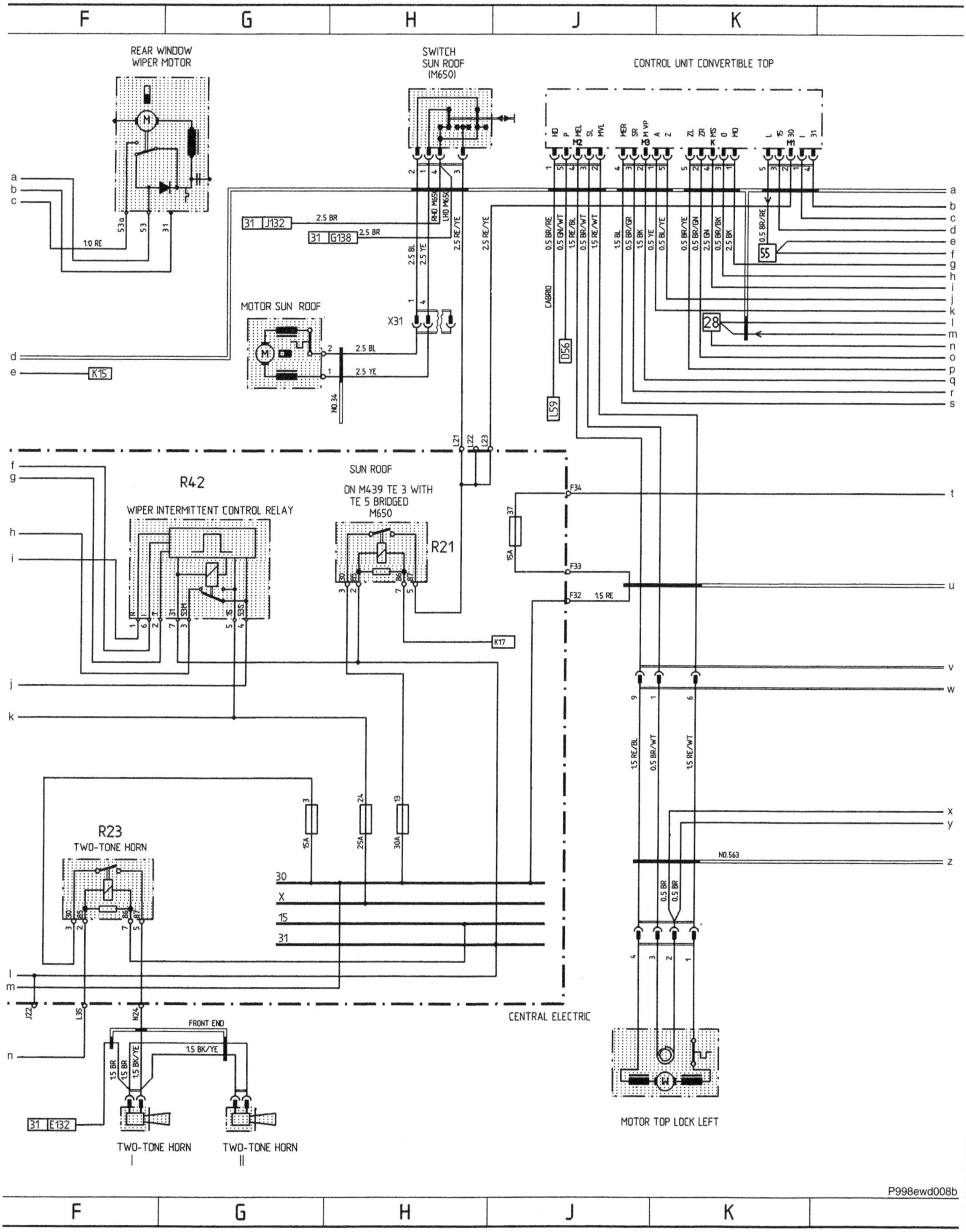

Sheet 8 - 1995 Rear spoiler, two-tone horns, wipe and wash cleaners, sunroof, convertible top (page 3 of 3)

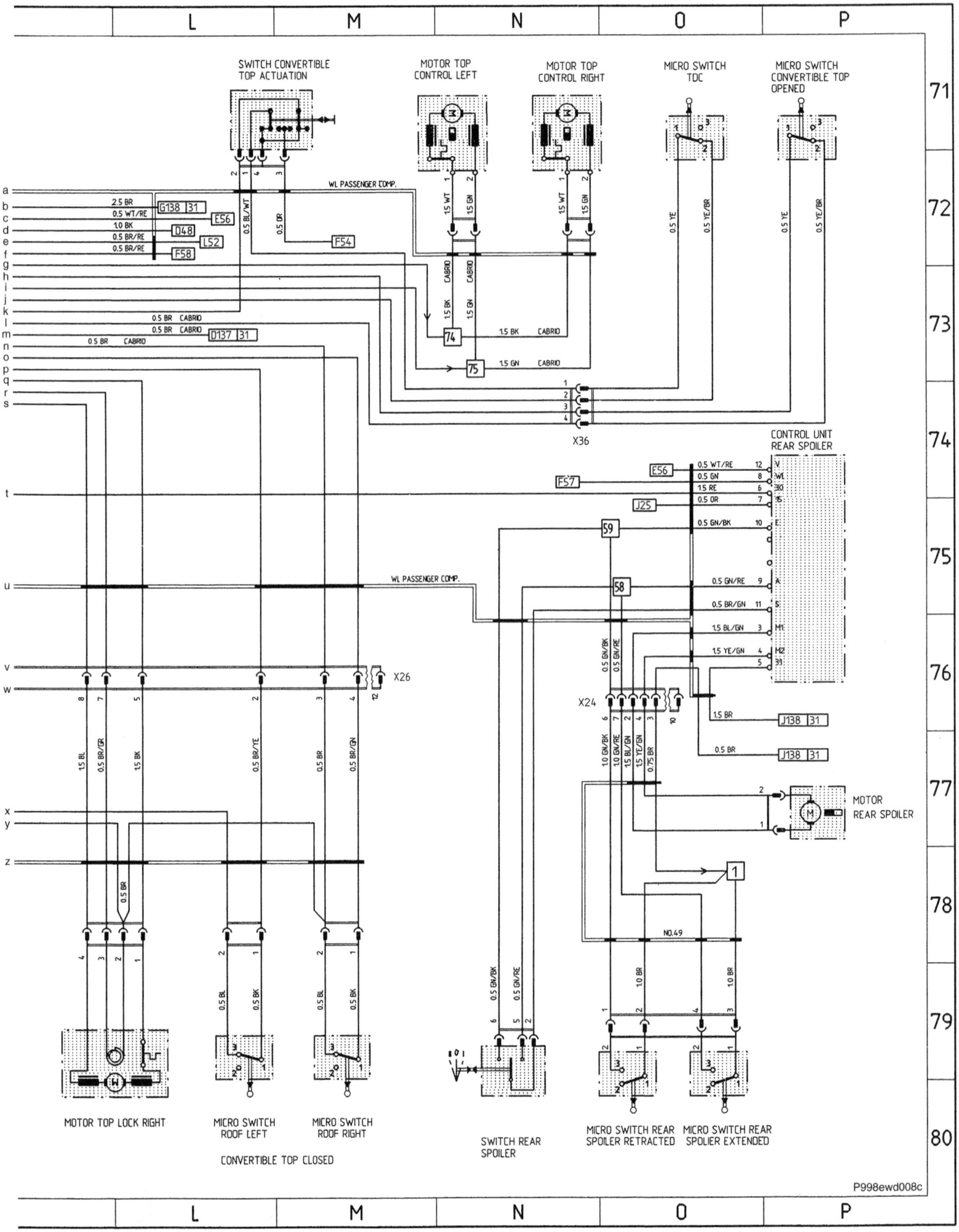

Radio, telephone (page 1 of 3)

Sheet 9 - 1995

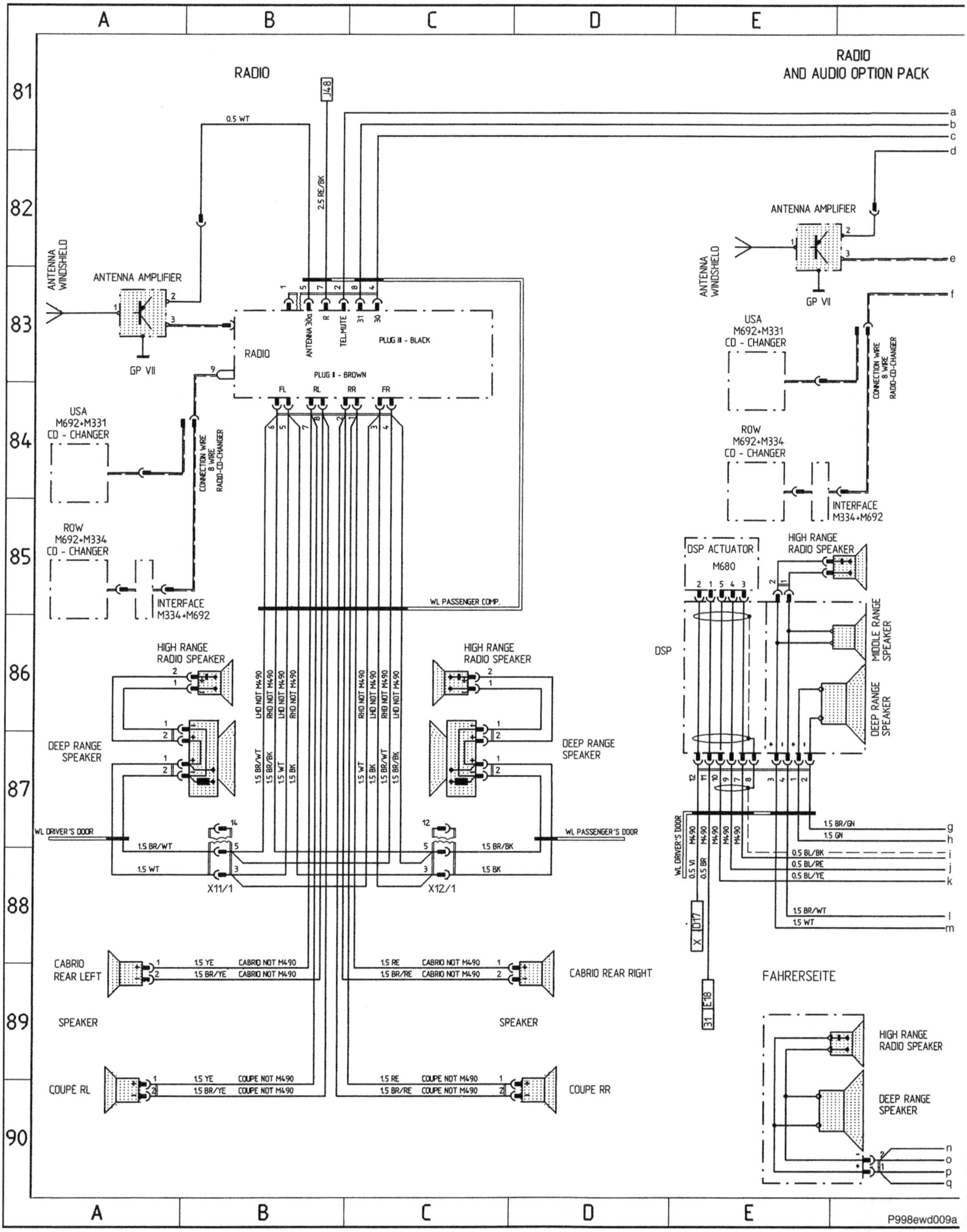

EWD

Sheet 9 - 1995

Radio, telephone (page 2 of 3)

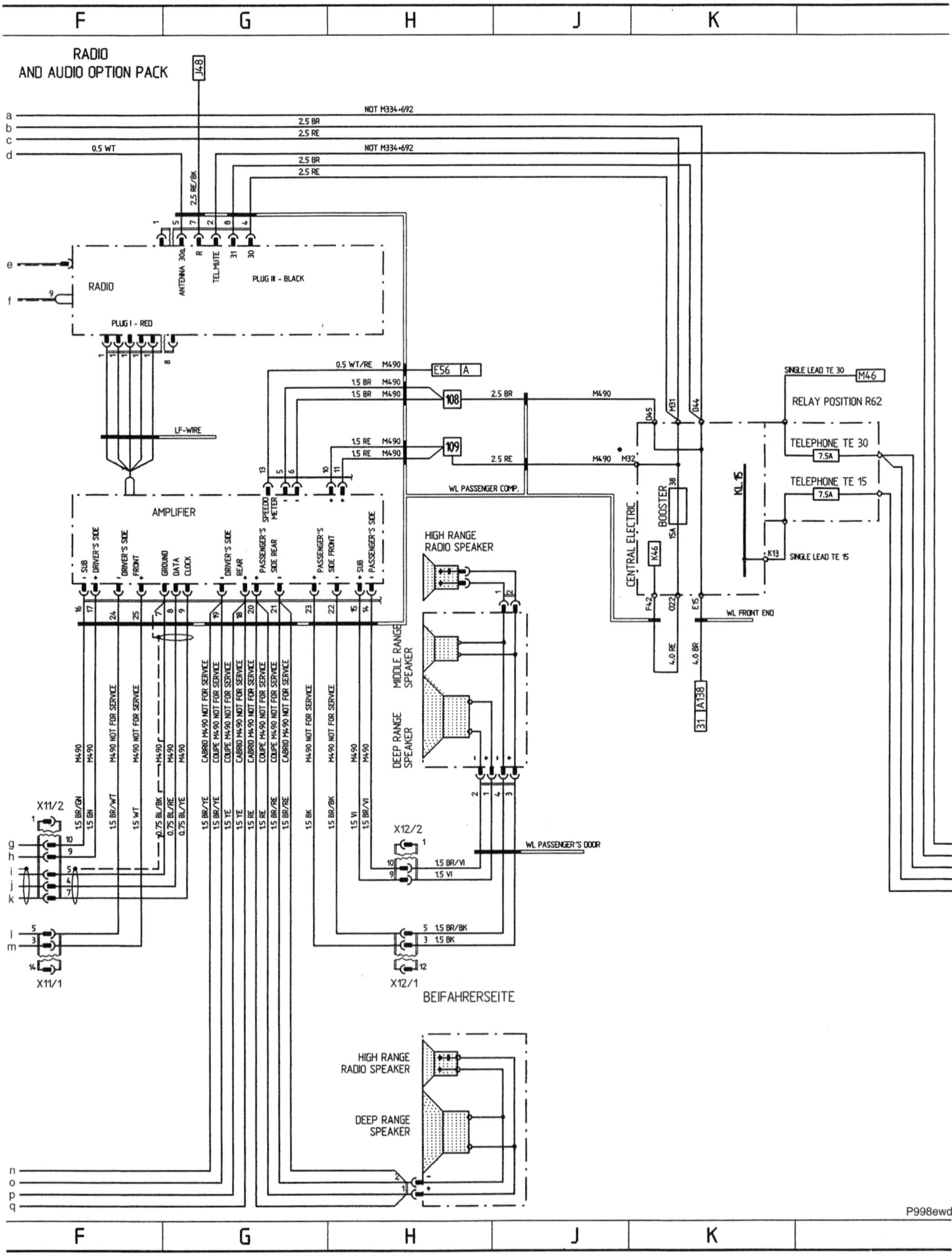

P998ewd009b

Radio, telephone
(page 3 of 3)

Sheet 9 - 1995

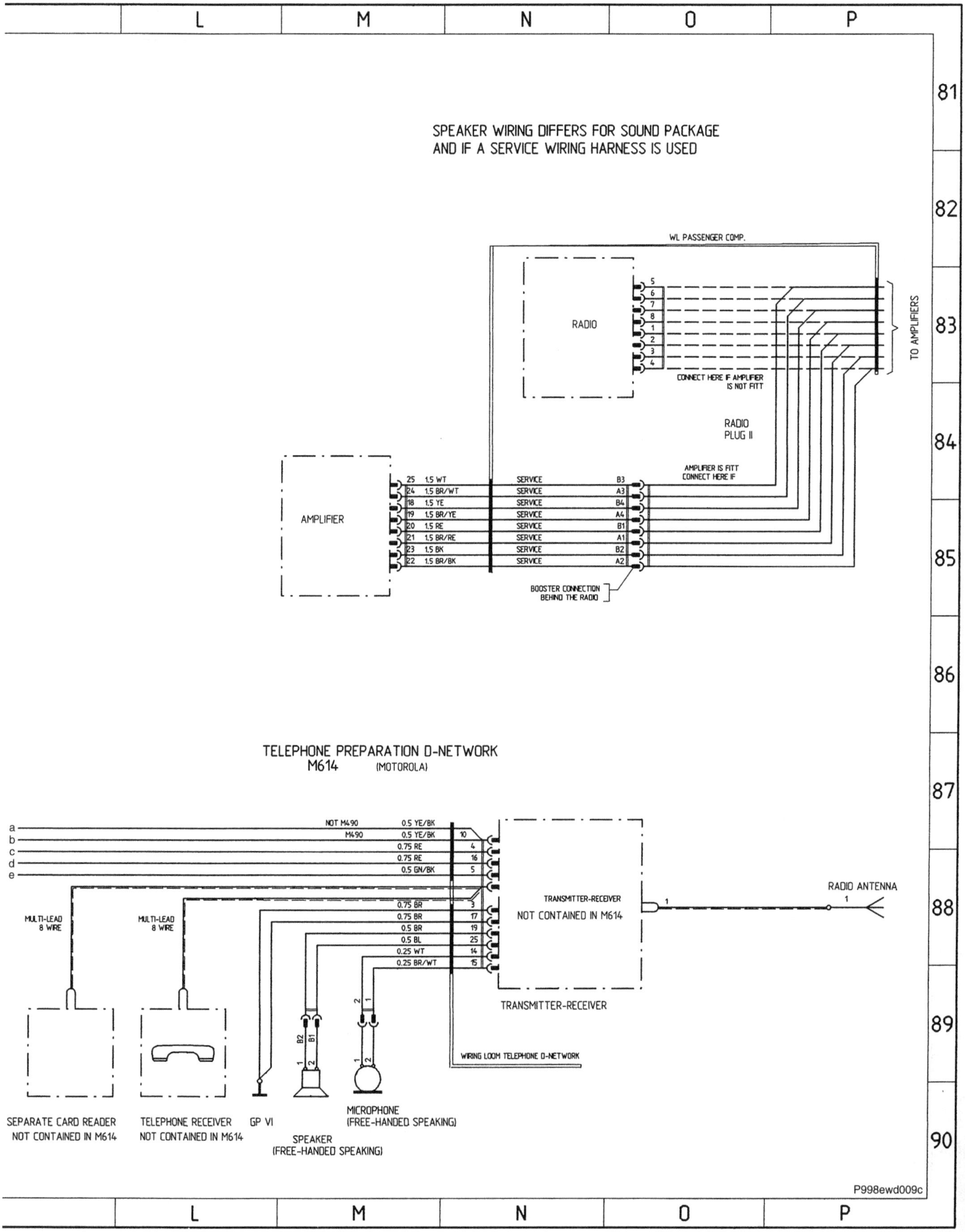

EWD

Sheet 10 - 1995

Foglights
(page 1 of 1)

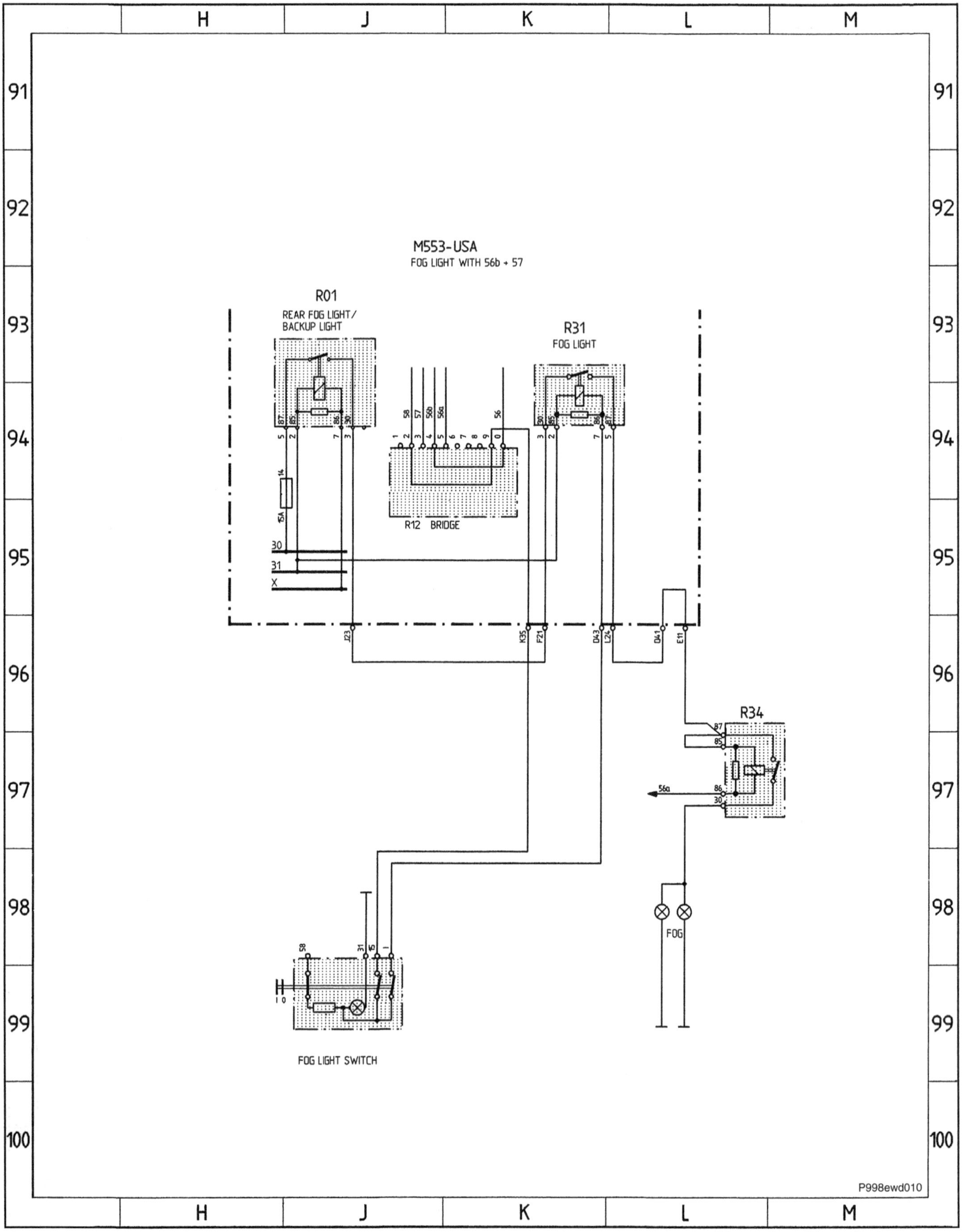

Central electric (page 1 of 3)

Sheet 11 - 1995

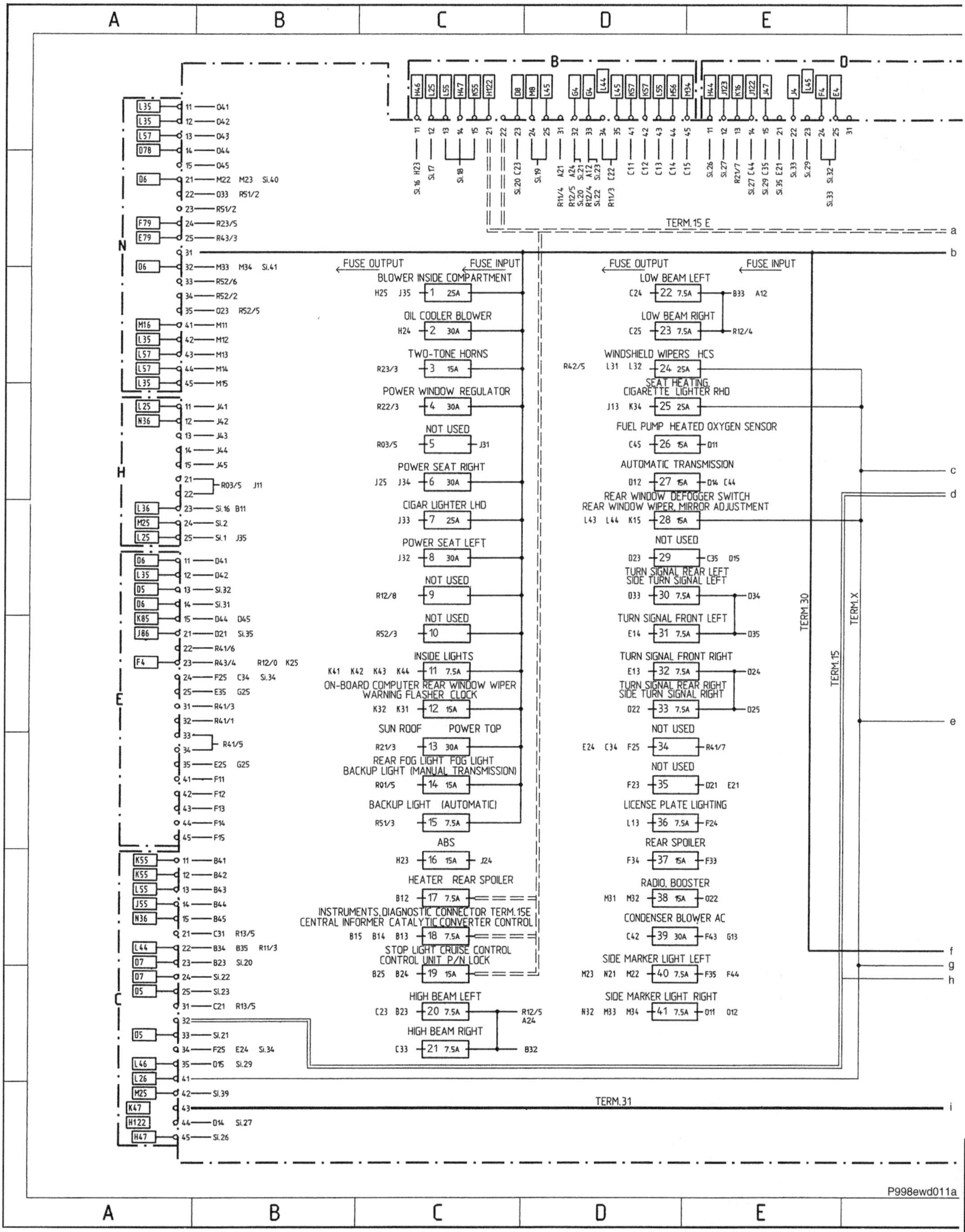

Sheet 11 - 1995

Central electric (page 2 of 3)

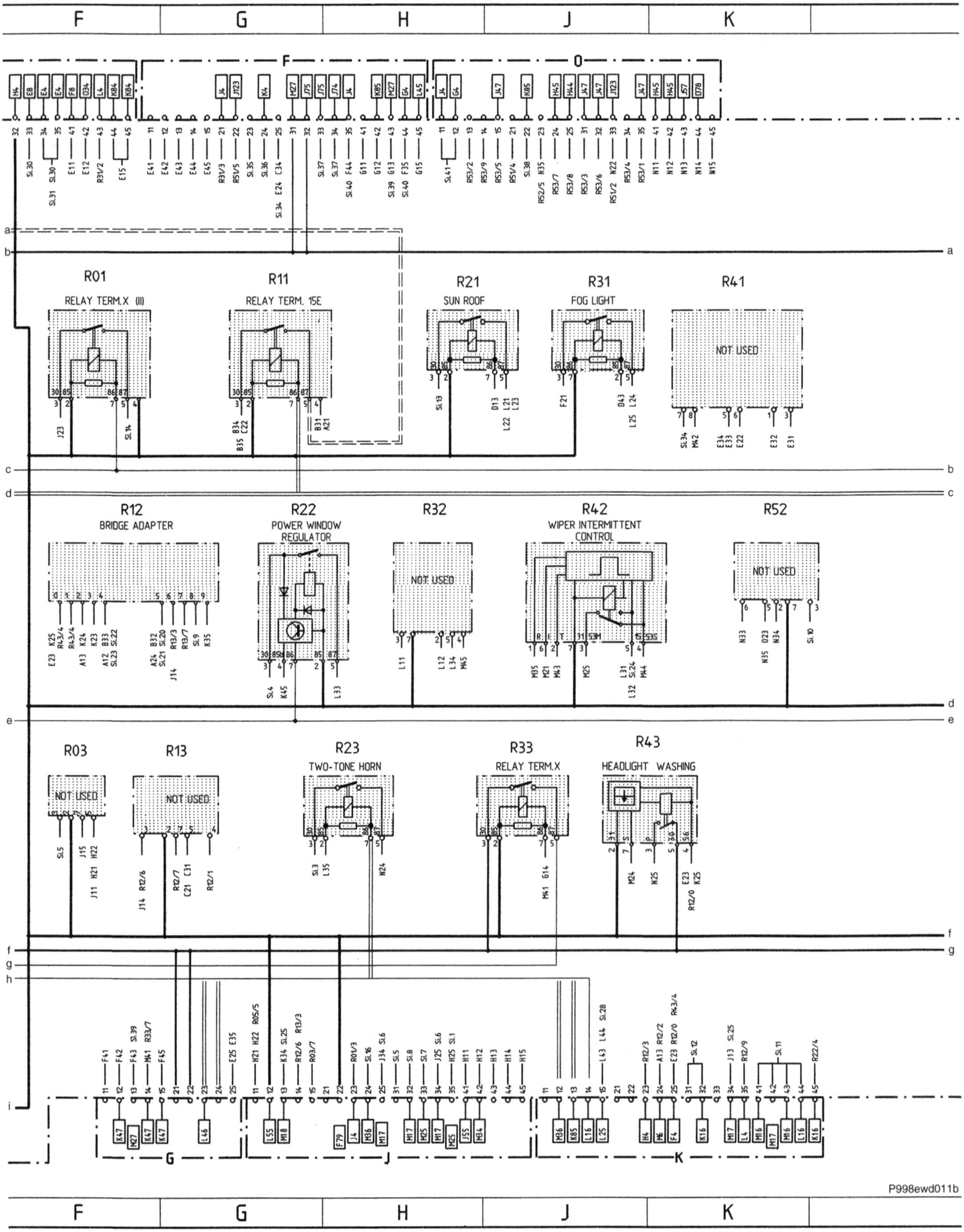

P998ewd011b

Central electric (page 3 of 3)

Sheet 11 - 1995

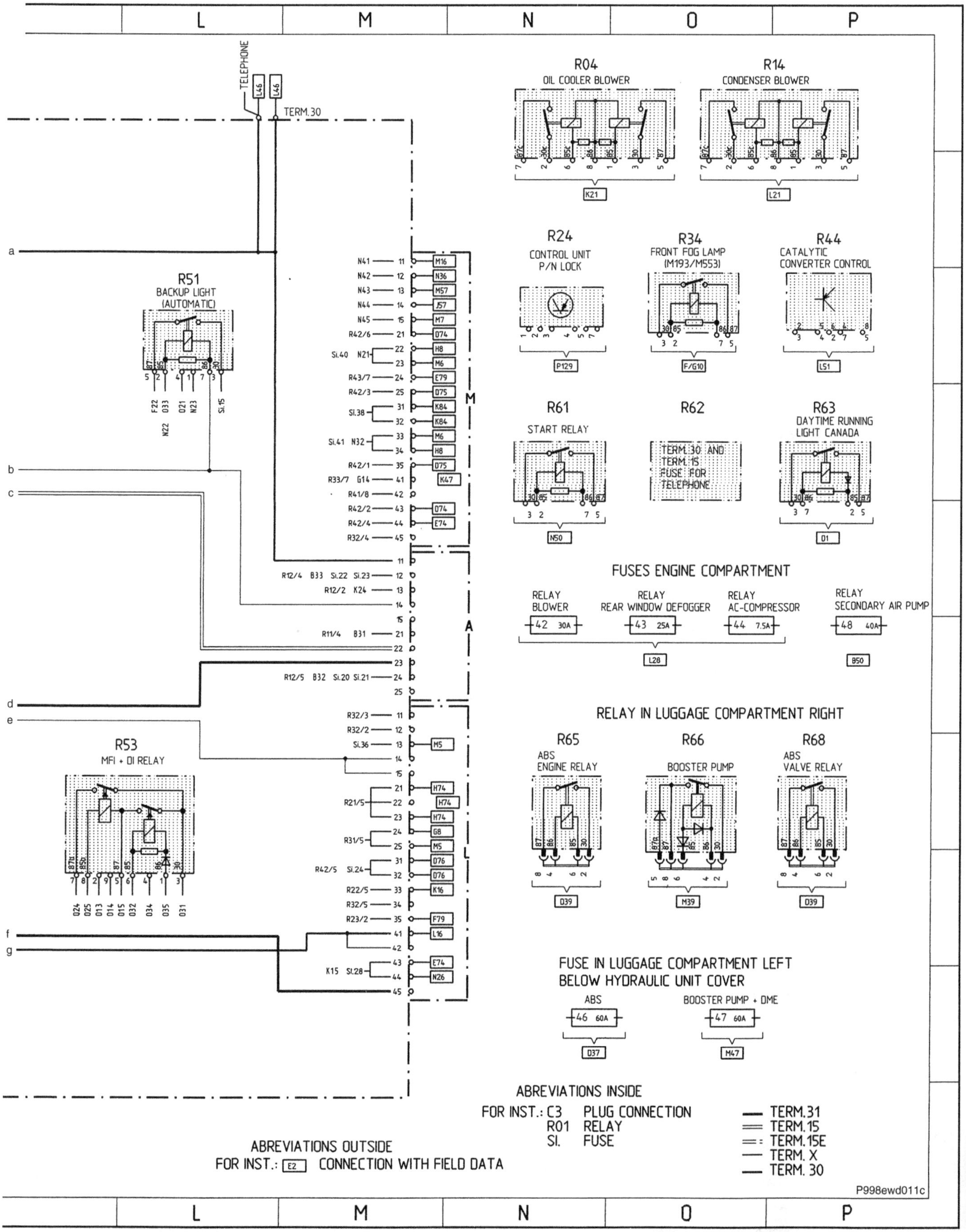

Sheet 12 - 1995

Airbag
(page 1 of 1)

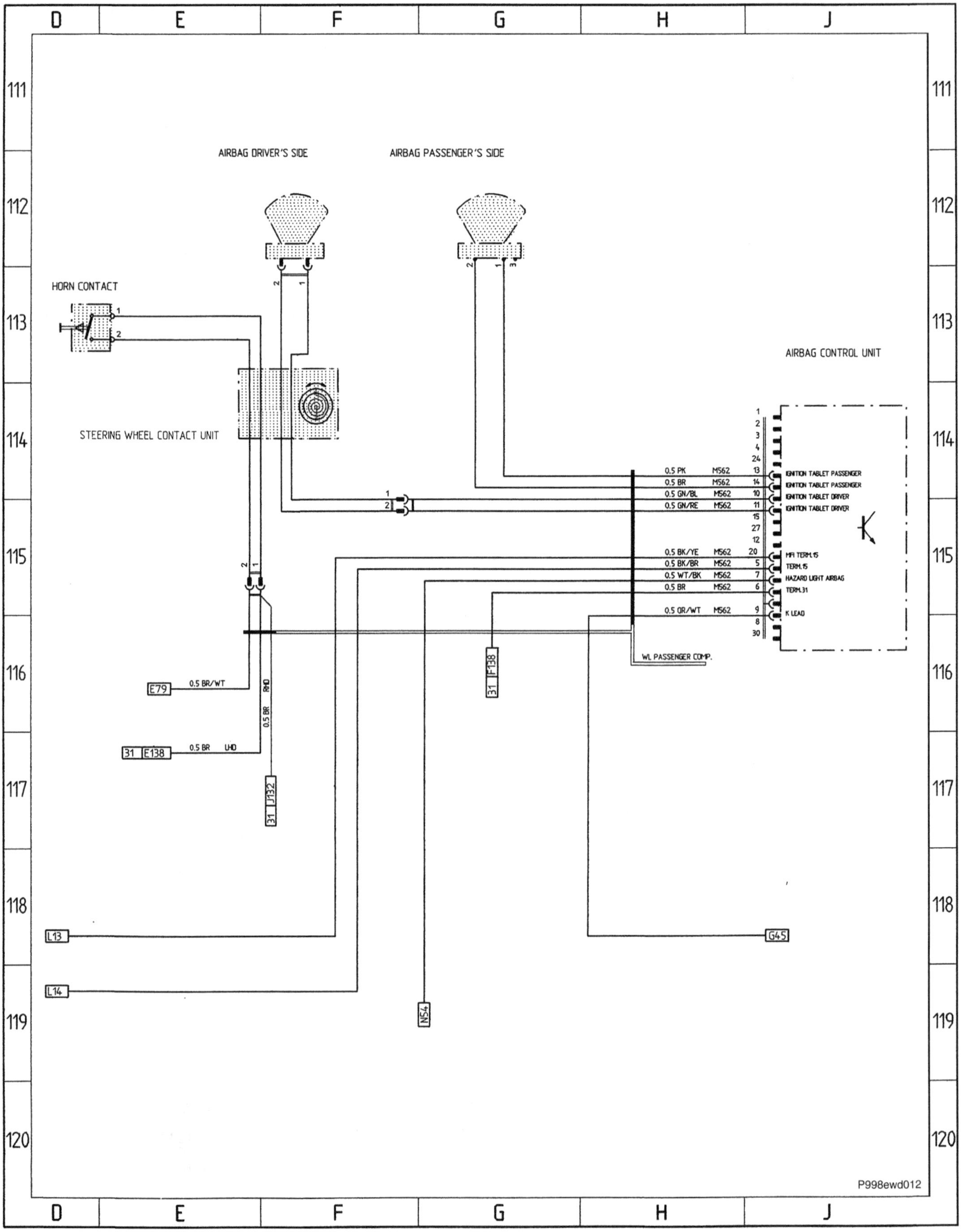

Tiptronic transmission (page 1 of 3)

Sheet 13 - 1995

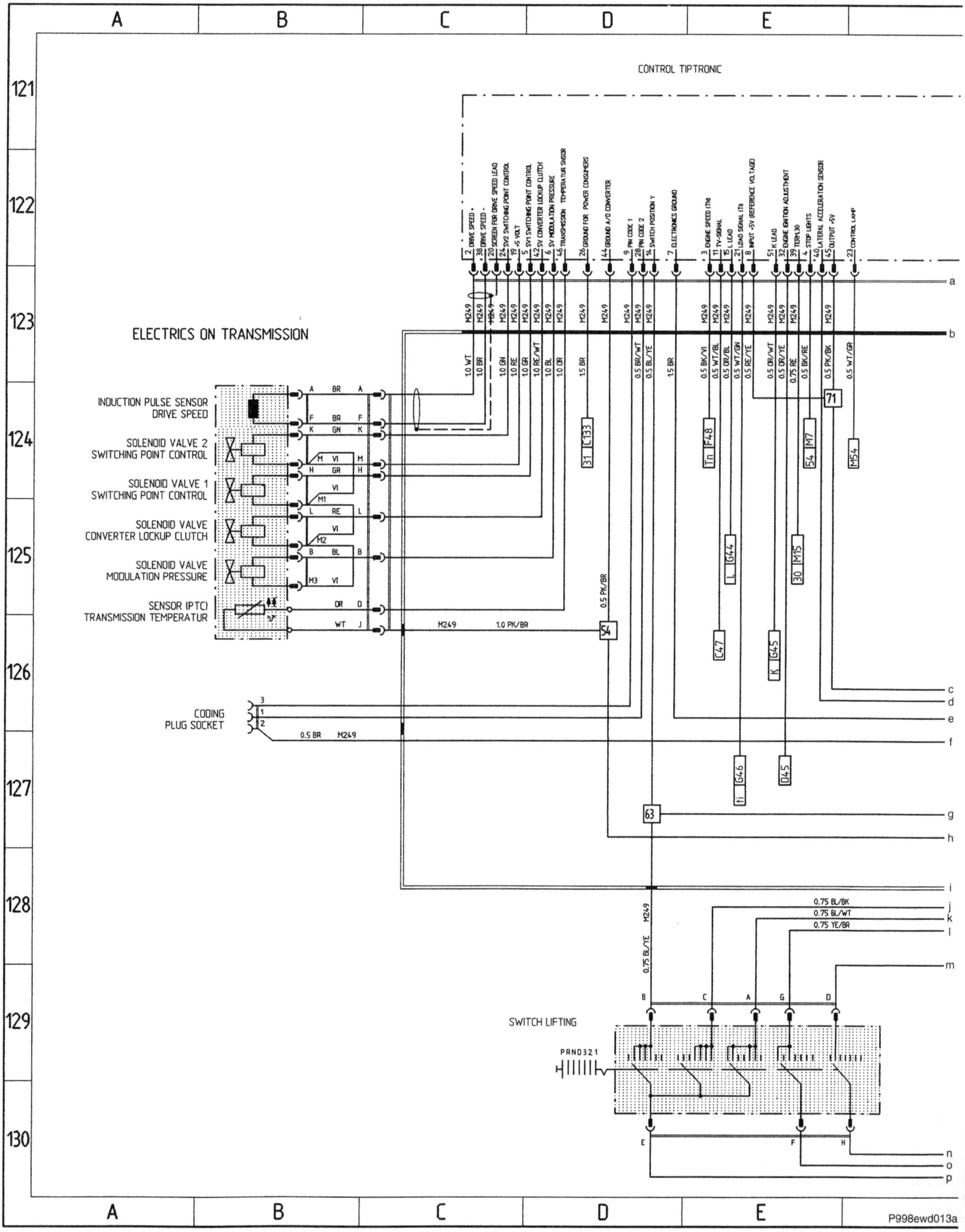

P998ewd013a

Sheet 13 - 1995

Tiptronic transmission (page 2 of 3)

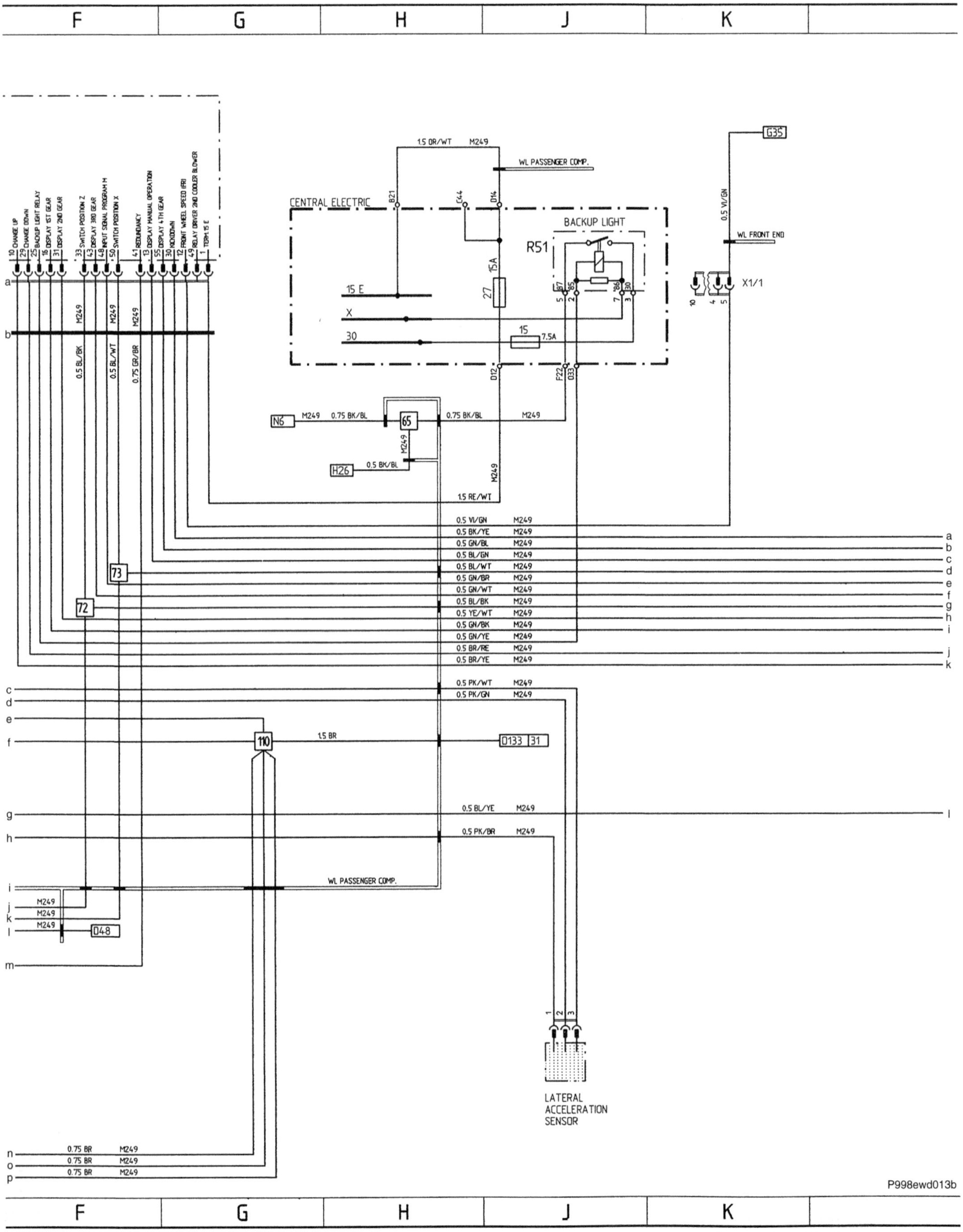

Tiptronic transmission (page 3 of 3)

Sheet 13 - 1995

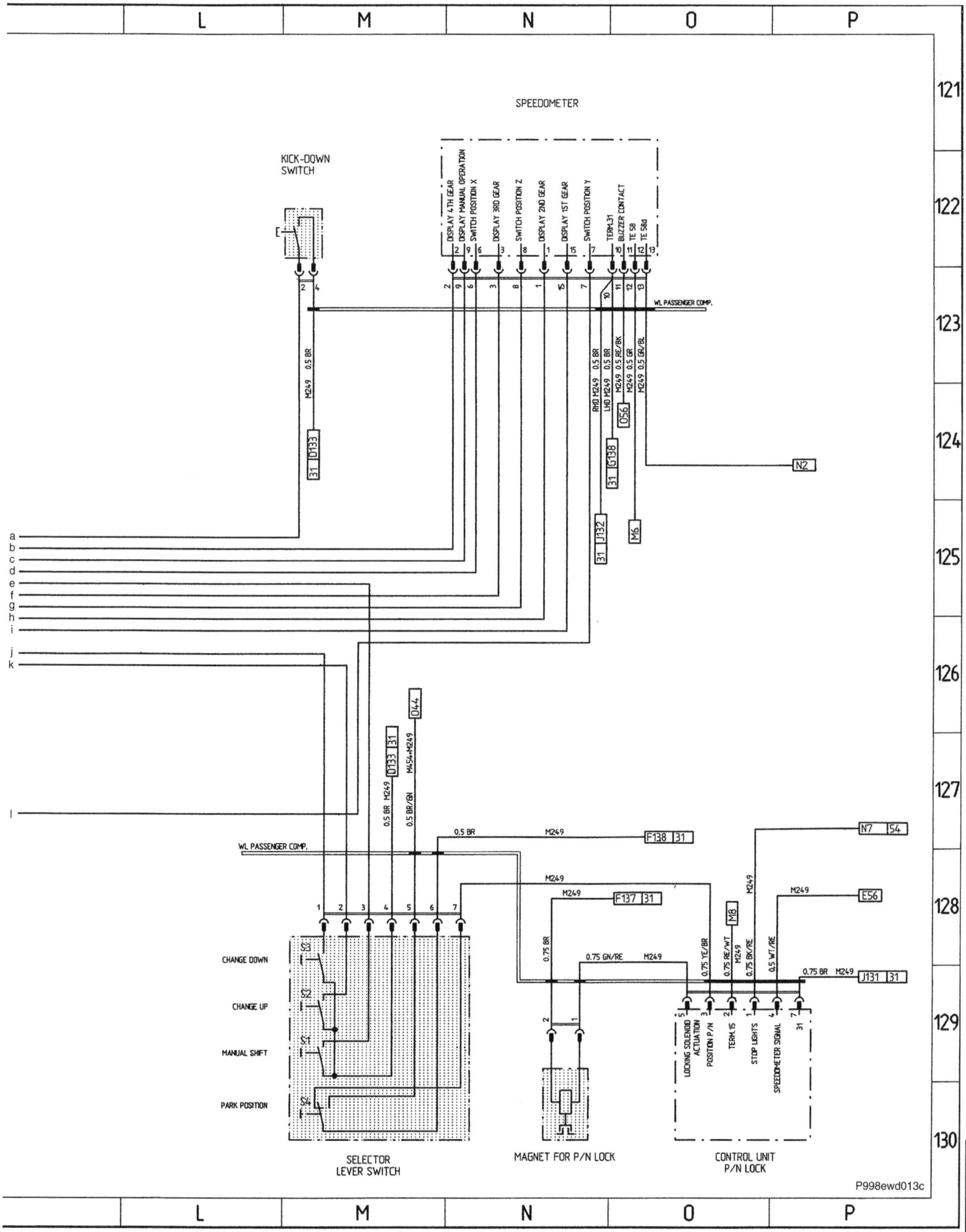

EWD

Sheet 14 - 1995

Ground points
(page 1 of 3)

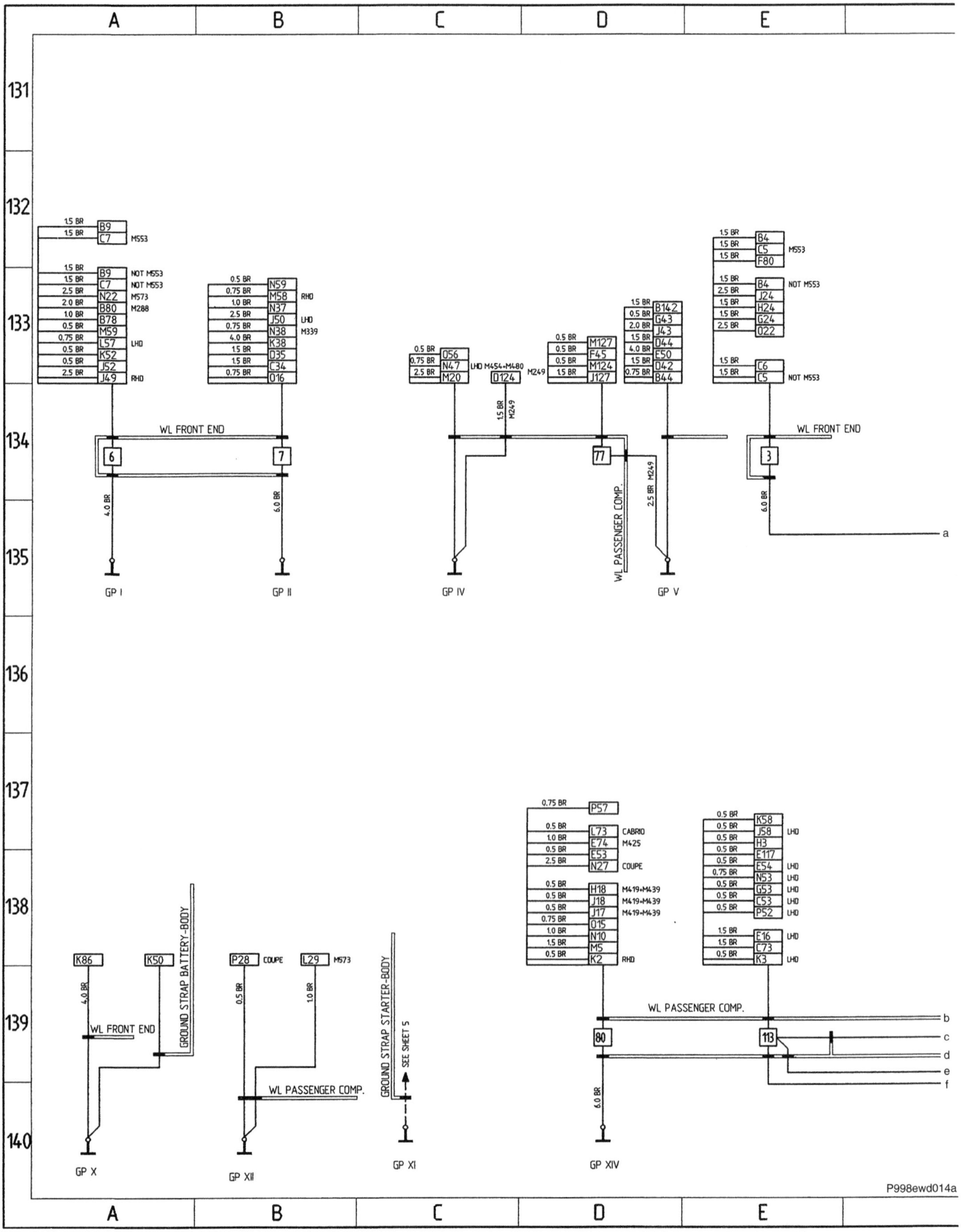

Ground points (page 2 of 3)

Sheet 14 - 1995

F	G	H	J	K	

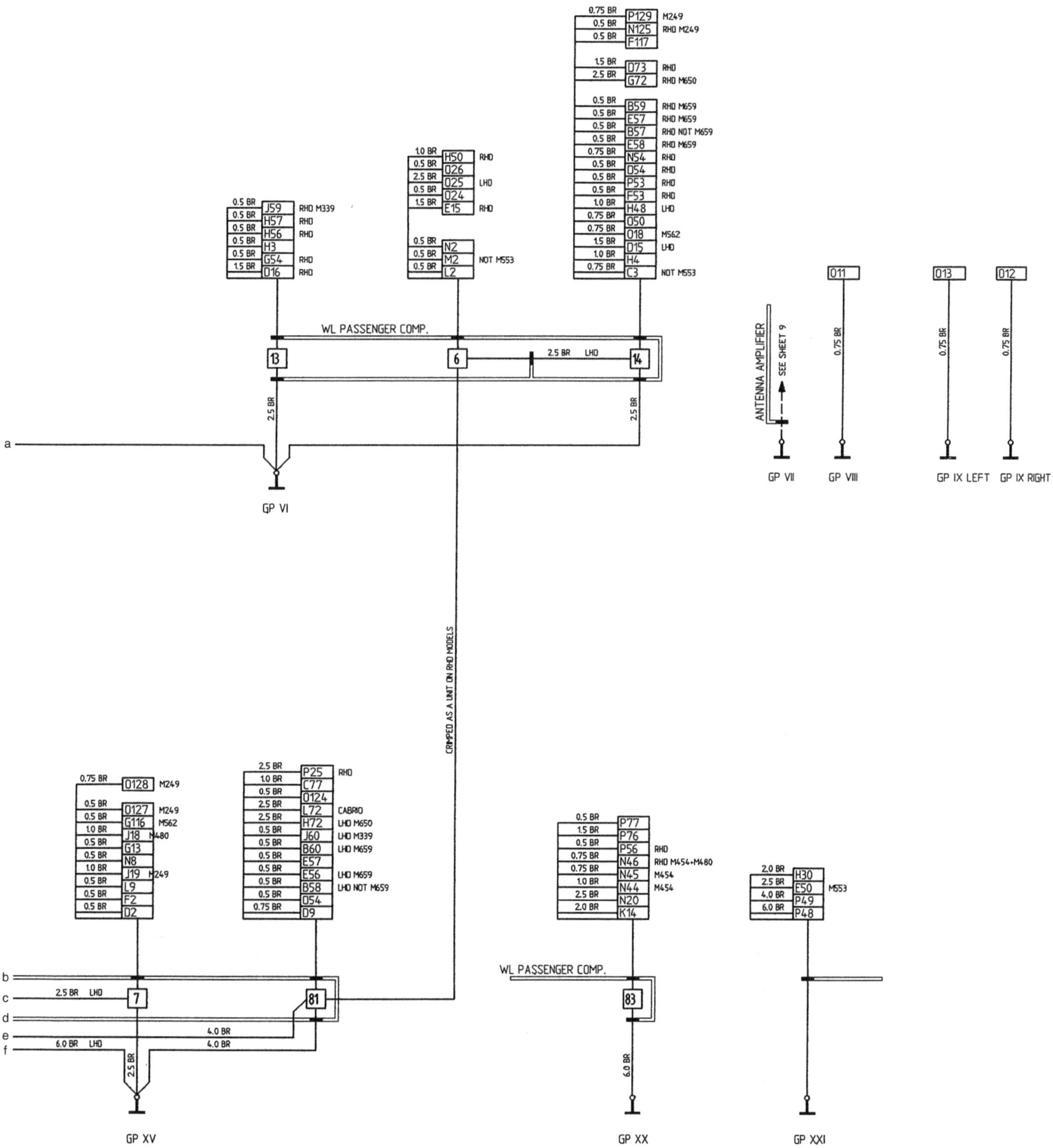

P998ewd014b

F	G	H	J	K	

EWD

Sheet 14 - 1995

Ground points
(page 3 of 3)

L	M	N	O	P

CODE	NOTE	POSITION IN VEHICLE LHD
GP I	LUGGAGE COMPARTMENT, WHEEL HOUSING LEFT	2cP
GP II	LUGGAGE COMPARTMENT, WHEEL HOUSING RIGHT	2cM
GP IV	GROUND FOR POWER CONSUMERS, UNDER SEAT LEFT	11eP
GP V	MOTOR ELEC./TIPTRONIK	8eP
GP VI	CENTRAL ELECTRIC, ON SIDE MEMBER RIGHT	6bL
GP VII	ANTENNA AMPLIFIER ON SIDE MEMBER LEFT	7bL
GP VIII	INSIDE LIGHT CONVERTIBLE, ROOF CENTRE FRONT	9aN
GP IX	INSIDE LIGHT, ROOF RIGHT	11aL
GP IX	INSIDE LIGHT, ROOF LEFT	11aP
GP X	BATTERY, LUGGAGE COMPARTMENT LEFT	3eP
GP XI	GROUND STRAP (STARTER, TRANSMISSION, ENGINE, BODY)	14cM
GP XII	ENGINE COMPARTMENT REAR LEFT	15cQ
GP XIV	ENGINE COMPARTMENT REAR RIGHT	15cL
GP XV	INSTRUMENTS, SIDE MEMBER LEFT	7bQ
GP XX	SEAT CROSS MEMBER, UNDER SEAT RIGHT	8eL
GP XXI	GROUND FOR POWER CONSUMERS, CONNECTION CRANK CASE	15cP

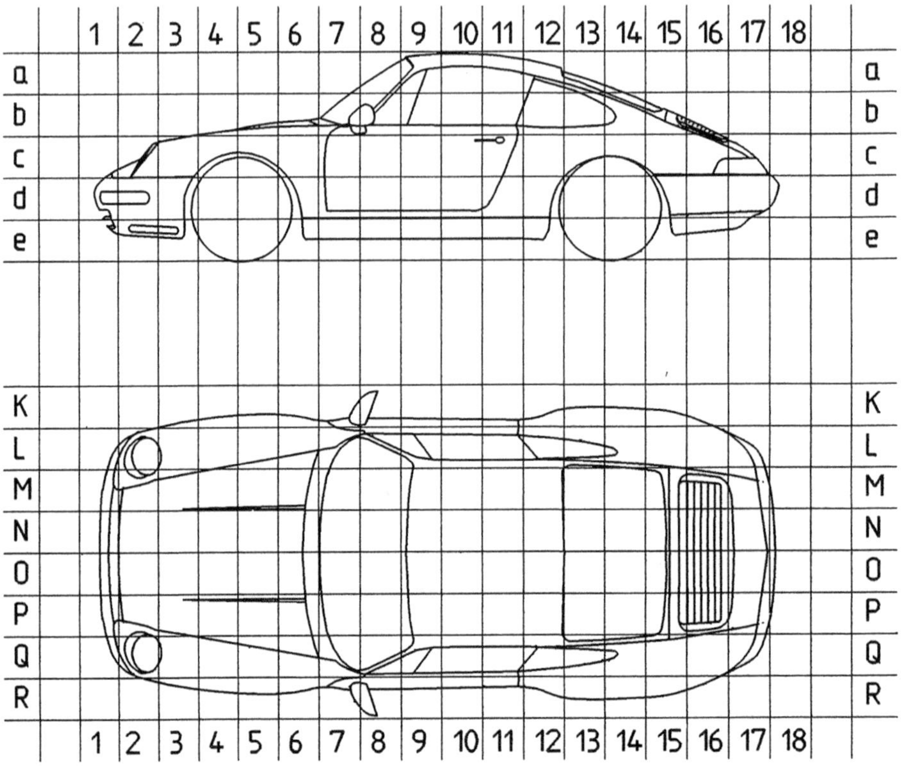

131 132 133 134 135 136 137 138 139 140

P998ewd014c

L	M	N	O	P

Construction components (page 1 of 3)

Sheet 15 - 1995

A | B | C | D | E

CONSTRUCTION COMPONENTS

DESIGNATION, FUNCTION	POSITION IN VEHICLE LHD	POSITION IN VEHICLE RHD	NOTE	FIELD IN WIRING DIAGRAM
ALARM HORN	13dP	13dP	REAR LH CROSSMEMBER	J20
ANTENNA AMPLIFIER	7bL	7bQ	IN LUGGAGE COMPARTMENT	A83,E82
ABS/ABD HYDRAULIC UNIT	3dP	3dP	IN LUGGAGE COMPARTMENT	C39
ABS RELAY	4cL	4cL	IN LUGGAGE COMPARTMENT RIGHT	D39
AMPLIFIER	9eM	9eM	UNDER THE SEAT RIGHT	F-H85, M85
BLOWER FINAL STAGE	6bN	6bN	IN LUGGAGE COMPARTMENT	F30
BLOWER ENGINE COMPARTMENT	16cO	16cO	ON ENGINE	H30
BLOWER MOTOR RIGHT	6cM	6cM	IN LUGGAGE COMPARTMENT	G30
BLOWER MOTOR LEFT	6cP	6cP	IN LUGGAGE COMPARTMENT	F30
BRAKE BOOSTER PUMP	1dO	1dO	IN LUGGAGE COMPARTMENT	L39
BACKUP LIGHT SWITCH	14dN	14dN	BY TRANSMISSION RIGHT	K/L1
BUZZER CONTACT	7cP	7cM	ON IGNITION LOCK	P56/57
CLUTCH SWITCH CRUISE CONTROL	6dP	6dN	BY CLUTCH PEDAL UNDER FLOOR PANEL	P46
CONTROL RELAY REAR WINDOW DEFOGGER	16cP	16cP	CARRIER PLATE IN ENGINE COMPARTMENT	O/P29/30
CONTROL UNIT ABS	2dM	2dM	IN LUGGAGE COMPARTMENT	A-J32
CONTROL UNIT AIRBAG	6cO	6cO	ON INTRUMENT PANEL	J114/116
CONTROL UNIT ALARM SYSTEM/CENTRAL LOCKING SYSTEM	9eM	9eM	UNDER THE SEAT RIGHT	J-M12
CONTROL UNIT TIPTRONIC TRANSMISSION	10dP	10dP	UNDER THE SEAT LEFT	C-G122
CONTROL UNIT CONVERTIBLE TOP	6cP	6cO	ON INTRUMENT PANEL	J/K71
CONTROL UNIT INSTRUMENT LIGHT	7bQ	7bR	IN LUGGAGE COMPARTMENT LEFT	K1
CONTROL UNIT REAR SPOILER	7cM	7cP	ON RH SIDE OF TRANSVERSE WALL	P74-76
CONTROL UNIT HEATING, AIR CONDITIONING	7cO	7cN	ON INTRUMENT PANEL	B-F21
CONTROL UNIT P/N LOCK	6cM	6cM	CENTRAL ELECTRIC	O/P129/130
CONTROL UNIT CRUISE CONTROL	9eN	9eN	UNDER THE SEAT RIGHT	P43/44
CENTRAL ELECTRIC	6cM	6cM	IN LUGGAGE COMPARTMENT	—
CENTRAL INFORMER	7cP	7cM	ON FIREWALL	M32
CONDENSER BLOWER	2dQ	2dQ	IN WHEEL HOUSING FRONT LEFT	N21
DIAGNOSIS CONNECTION	7cL	7cQ	IN PASSENGER'S FOOTWELL	G/H50
DRIVE CRUISE CONTROL	15cO	15cO	ON ENGINE	N43/44
ENGINE SPEED SENSOR	15cP	15cP	ON ENGINE	A46
FLASHER	7dN	7dN	IN CENTRE CONSOLE	D1
FUEL PUMP	6eM	6eP	BEHIND THE BASE	J49
FUSES ENGINE COMPARTMENT	16cP	16cP	CARRIER PLATE IN ENGINE COMPARTMENT	L28
FUSE IN LUGGAGE COMPARTMENT	4cP	4cP	BELOW HYDRAULIC UNIT COVER	D37/M47
FUEL LEVEL SENSOR	5cM	5cP	IN FUEL TANK	K51
GONG RELAY	7bQ	—	LH SIDE IN FRONT OF INSTRUMENTS	P53/54
HALL SENSOR	17cO	17cO	ON DISTRIBUTOR	A46
HYDRAULIC UNIT ABS	3dP	3dP	IN LUGGAGE COMPARTMENT	A/B39
HEADLIGHT CLEANER PUMP	3cQ	3cQ	BY WASHING FLUID RESERVOIR IN WHEEL HOUSING FL	A79/80
HEADLIGHT CLEANER RELAY	6cM	6cM	CENTRAL ELECTRIC	E78
IDLE SPEED POSITIONER	15bN	15bN	ON ENGINE	H/J41
IGNITION CONTROL UNIT	11dP	11dP	UNDER THE SEAT LEFT	J41
IGNITION COIL I	17cP	17cP	ON ENGINE	L41
IGNITION COIL II	17cP	17cP	ON ENGINE	L41
KICK-DOWN SWITCH	6dO	6dM	BY ACCELERATOR PEDAL UNDER FLOOR PANEL	M122
KNOCK SENSOR 1	15cP	15cP	ON ENGINE	A47
KNOCK SENSOR 2	16cN	16cN	ON ENGINE	A47
LATERAL ACCELERATION SENSOR TIPTRONIC TRANSMISSION	7dN	7dN	IN CENTRE CONSOLE	J129
MASS AIR FLOW SENSOR	16bO	16bO	ON ENGINE	A43
MICRO SWITCH ROOF RIGHT	9aN	9aN	AT CONVERTIBLE TOP FRAME	M79
MICRO SWITCH ROOF LEFT	9aO	9aO	AT CONVERTIBLE TOP FRAME	L79
MICRO SWITCH REAR SPOLIER EXTENDED	16bP	16bP	ON TAILGATE	O79
MICRO SWITCH REAR SPOILER RETRACTED	16bP	16bP	ON TAILGATE	O79
MICRO SWITCH TDC	12cL	12cL	BEHIND SIDE PANEL COVERING	O71
MICROSWITCH FOR TOP CONTROL	12cL	12cL	BEHIND SIDE PANEL COVERING	P71

P998ewd015a

A | B | C | D | E

EWD

Sheet 15 - 1995

Construction components (page 2 of 3)

DESIGNATION, FUNCTION	POSITION IN VEHICLE LHD	RHD	NOTE	FIELD IN WIRING DIAGRAM
MFI + DI RELAY	6cM	6cM	CENTRAL ELECTRIC	J44
MFI + DI CONTROL UNIT	9eP	9eP	UNDER THE SEAT LEFT	B-G41
MOTOR SUN ROOF	12aN	12aN	IN ROOF	G73
MOTOR TOP CONTROL RIGHT	15cN	15cN	BEHIND REAR SEAT	N71
MOTOR TOP CONTROL LEFT	15cO	15cO	BEHIND REAR SEAT	N71
MOTOR TOP LOCK RIGHT	9aN	9aN	AT CONVERTIBLE TOP FRAME	K/L79
MOTOR TOP LOCK LEFT	9aO	9aO	AT CONVERTIBLE TOP FRAME	J/K79
NTC HEATER FAN	16cP	16cP	ON ENGINE	G30
NTC TEMPERATURE FOR INTAKE AIR	15cN	15cN	ON ENGINE	A42
OIL PRESSURE SENSOR/SWITCH	14cN	14cN	ON ENGINE	A54
OIL COOLER BLOWER	2dL	2dL	IN WHEEL HOUSING FRONT RIGHT	O21
OIL LEVEL GAGE	12cK	12cK	IN WHEEL HOUSING REAR RIGHT	E51
OIL TEMPERATURE SENSOR FOR INSTRUMENT	14cN	14cN	ON ENGINE	A53
OIL TEMPERATURE SENSOR FOR OIL COOLER BLOWER	2dL	2dL	IN WHEEL HOUSING FRONT RIGHT	J21
OUTSIDE TEMP. SENSOR	2dP	2dP	FRONT LH BUMPER	A60
PRESSURE WARNING SWITCH	1dO	1dO	IN LUGGAGE COMPARTMENT	M/N39
PRESSURE SWITCH AIR CONDITIONING SYSTEM	6bO	6bO	IN LUGGAGE COMPARTMENT	E30
REAR WINDOW WIPER RELAY	15bO	15bO	ON REAR WINDOW WIPER MOTOR	E71
REAR FOG LIGHT CUTOUT RELAY	17dN	17dN	BEHIND REAR PANEL	H108
RELAY CONDENSER BLOWER	6cM	6cM	CENTRAL ELECTRIC	L21
RELAY AC COMPRESSOR	16cP	16cP	CARRIER PLATE IN ENGINE COMPARTMENT	L30
RELAY TWO-TONE HORN	6cM	6cM	CENTRAL ELECTRIC	F78
RELAY POWER WINDOW REGULATOR	6cM	6cM	CENTRAL ELECTRIC	K18/19
RELAY BLOWER ENGINE COMPARTMENT	16cP	16cP	CARRIER PLATE IN ENGINE COMPARTMENT	J/K30
RELAY STARTER TIPTRONIC-TRANSMISSION	6cM	6cM	CENTRAL ELECTRIC	N50
RELAY BACKUP LIGHT	7bQ	7bQ	IN LUGGAGE COMPARTMENT LEFT	J122
RELAY THERMO SWITCH CATALYTIC CONVERTER	6cM	6cM	CENTRAL ELECTRIC	L/M51
RELAY TERM.X	6cM	6cM	CENTRAL ELECTRIC	J/K44
RELAY TERM.15 E	6cM	6cM	CENTRAL ELECTRIC	K/L44
RELAY RFL - BACKUP LIGHT	6cM	6cM	CENTRAL ELECTRIC	J/K7, A-M94
RELAY FRONT FOG LAMP	6cM	6cM	CENTRAL ELECTRIC	J7, C-O94
RELAY OIL COOLER BLOWER	6cM	6cM	CENTRAL ELECTRIC	K21
RELAY BRAKE BOOSTER PUMP	4cL	4cL	IN LUGGAGE COMPARTMENT RIGHT	M39
RELAY SUN ROOF	6cM	6cM	CENTRAL ELECTRIC	H75
RELAY DAYTIME RUNNING LIGHT CANADA	6cM	6cM	CENTRAL ELECTRIC	D1
RELAY FOG LIGHT SWITCH OFF USA/JAPAN	6cM	6cM	CENTRAL ELECTRIC	H-P/94, F/G10
RESISTOR OIL COOLER BLOWER	2dL	2dL	IN WHEEL HOUSING FRONT RIGHT	M21
RESISTOR BLOWER	16dP	16dP	IN ENGINE COMPARTMENT	J30
RESISTOR FOR BLOWER CONDENSER	2dQ	2dQ	IN WHEEL HOUSING FRONT LEFT	M21
STOP LIGHT SWITCH	6dP	6dM	AT THE BRAKE PEDAL BELOW THE FLOOR PANEL	N9
SWITCH LIFTING	12dO	12dO	ON GEARBOX	D-F129
SECONDARY AIR PUMP RELAY	17cP	17cP	ON CARRIER ON LH SIDE IN ENGINE COMPARTMENT	A50
SHIFT VALVE RESONANCE FLAP	15bN	15bN	ON ENGINE	H41
STARTER	14dN	14dN	BY CLUTCH HOUSING RIGHT	P49/50
SECONDARY AIR PUMP	16cN	16cN	ON ENGINE	D/E50
THROTTLE VALVE SENSOR	15bN	15bN	ON THROTTLE VALVE ASSEMBLY	A44
TANK VENTING VALVE	16cP	16cP	ON ENGINE	K41
TEMPERATURE MOTOR	15cP	15cP	ON ENGINE	A45
TWO-TONE HORN I	3cL	3cL	IN WHEEL HOUSING FRONT RIGHT	F80
TWO-TONE HORN II	3cL	3cL	IN WHEEL HOUSING FRONT RIGHT	G80
V-BELT CONTROL	17dO	17dO	ON ENGINE REAR MIDDLE	A55
WINDSHIELD WASHING FLUID PUMP	3cQ	3cQ	BY WASHING FLUID RESERVOIR IN WHEEL HOUSING FL	A78
WIPER INTERMITTENT CONTROL RELAY	6cM	6cM	CENTRAL ELECTRIC	F/G75

P998ewd015b

Construction components
(page 3 of 3)

Sheet 15 - 1995

	L	M	N	O	P

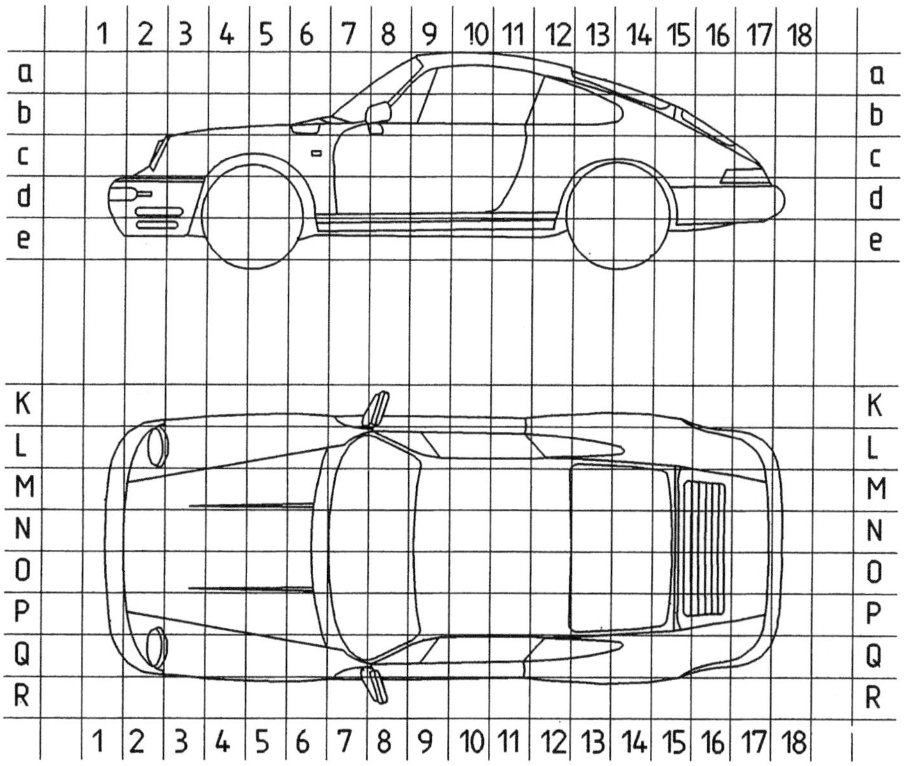

P998ewd015c

	L	M	N	O	P

Sheet 16 - 1995

Connectors and abbreviations (page 1 of 2)

A	B	C	D	E	F

PLUG CONNECTIONS

CODE	NUMBER OF PINS	DESIGNATION, FUNCTION	POSITION IN VEHICLE LHD	RHD	NOTE	FIELD
X 1/1	10	CONNECTION WL PASSENGER COMPARTMENT/ WL FRONT END	6cL	6cL	IN LUGGAGE COMPARTMENT RIGHT	G/H34, K123
X 1/2	10	CONNECTION WL PASSENGER COMPARTMENT/ WL FRONT END	6cL	6cL	IN LUGGAGE COMPARTMENT RIGHT	C9,E36,G48,B58-59
X 3	14	CONNECTION WL PASSENGER COMPARTMENT/ WL MOTOR	16cP	16cP	CARRIER PLATE IN ENGINE COMPARTMENT	K-L3,J27-28,D48-49,B53-54
X4/1	14	CONNECTION WL PASSENGER COMPARTMENT/ WL MFI + DI	9eP	9eP	UNDER THE SEAT LEFT	CD45-46,G147
X4/2	14	CONNECTION WL PASSENGER COMPARTMENT/ WL MFI + DI	9eP	9eP	UNDER THE SEAT LEFT	F44-46,B55
X11/1	14	CONNECTION WL PASSENGER COMPARTMENT/ WL DOOR DRIVER'S SIDE	6cQ	6cL	UNDER THE INSTRUMENT PANEL DIVER'S SIDE	E-F16-17,B87-88, F88
X11/2	10	CONNECTION WL PASSENGER COMPARTMENT/ WL DOOR DRIVER'S SIDE	6cQ	6cL	UNDER THE INSTRUMENT PANEL DIVER'S SIDE	C16 ,F87-88
X12/1	12	CONNECTION WL PASSENGER COMPARTMENT/ WL DOOR PASSENGER'S SIDE	6cL	6cQ	UNDER THE INSTRUMENT PANEL PASSENGER'S SIDE	E-F13,C87-88 ,H88
X12/2	10	CONNECTION WL PASSENGER COMPARTMENT/ WL DOOR PASSENGER'S SIDE	6cL	6cQ	UNDER THE INSTRUMENT PANEL PASSENGER'S SIDE	C14,H87/88
X20	14	CONNECTION WL PASSENGER COMPARTMENT/ WL HEATER, AIR CONDITIONER	6cM	6cM	UNDER INSTRUMENT PANEL RIGHT	G25-26
X24	10	CONNECTION WL PASSENGER COMPARTMENT/ WL TAILGATE	16cL	16cL	BOX RH SIDE OF ENGINE COMPARTMENT	O10,O13-14,O76
X25	10	CONNECTION WL PASSENGER COMPARTMENT/ WLTAIL LIGHTS	16cL	16cL	BOX RH SIDE OF ENGINE COMPARTMENT	N4-6
X26	12	CONNECTION WL PASSENGER COMPARTMENT/ WLCONVERTIBLE TOP	12cL	12cL	BEHIND SIDE PANEL COVERING	J-M76
X27	4	CONNECTION WL PASSENGER COMPARTMENT/ WLREAR WINDOW	15bM	15bM	BELOW THE PARCEL TRAY	O9,O28
X28	6	CONNECTION WL PASSENGER COMPARTMENT/ WLIGNITION LOCK	6cP	6cM	ON STEERING COLUMN	L48-49
X29	2	CONNECTION WL PASSENGER COMPARTMENT/ OIL TANK SENSOR	16cL	16cL	BOX RH SIDE OF ENGINE COMPARTMENT	E52
X31	4	CONNECTION WL PASSENGER COMPARTMENT/ WLINSIDE LIGHTS COUPE	15bM	15bM	BELOW THE PARCEL TRAY	O12-13,H73
X32	2	CONNECTION WL PASSENGER COMPARTMENT/ WLINSIDE LIGHTS CONVERTIBLE	6cQ	6cQ	UNDER INSTRUMENT PANEL LEFT	N-O12
X33	2	CONNECTION WL PASSENGER COMPARTMENT/ WLSIDE TURN SIGNAL LEFT	7bQ	7bQ	IN LUGGAGE COMPARTMENT ON SIDE MEMBER LEFT	D-E9
X34	2	CONNECTION WL PASSENGER COMPARTMENT/ WLSIDE TURN SIGNAL RIGHT	7bL	7bL	IN LUGGAGE COMPARTMENT ON SIDE MEMBER RIGHT	C2
X36	4	CONNECTION WL PASSENGER COMPARTMENT/ SWITCH LIFTING CONVERTIBLE TOP	12cL	12cL	BEHIND SIDE PANEL COVERING	N74
X37	8	CONNECTION WL PASSENGER COMPARTMENT/ WLSEAT LEFT	9eP	9eN	UNDER THE DRIVER SEAT	N19,A70,E65-66
X38	8	CONNECTION WL PASSENGER COMPARTMENT/ WLSEAT RIGHT	9eN	9eP	UNDER THE PASSEGER SEAT	N19,A70,E65-66
X39	6	CONNECTION WL PASSENGER COMPARTMENT/ WLLUGGAGE COMPARTMENT HOOD	7bL	7bL	ON RH SIDE OF LUGGAGE COMP. LID HINGE	O16,B77
X60	21	CONNECTION WL MOTOR/ WL MFI + DI	16cP	16cP	CARRIER PLATE IN ENGINE COMPARTMENT	B46,D-E48,K42,B55
X65	6	CONNECTION WL DOOR DRIVER'S SIDE/ WL OUTSIDE MIRROR LEFT	8cR	8cK	DRIVER'S DOOR	B18-19
X66	6	CONNECTION WL DOOR PASSENGER'S SIDE/ WL OUTSIDE MIRROR RIGHT	8cK	8cR	PASSENGER'S DOOR	B13
X69	14	CONNECTION WL FRONT END/ WL HEATER, AIR CONDITIONER	6cM	6cM	IN LUGGAGE COMPARTMENT RIGHT	G23-24

M - NUMBERS

M 113 CANADA VERSION (DAYTIME RUNNING LIGHT)
M 139 ADJUSTABLE SEAT HEATING, LEFT SEAT
M 193 JAPAN VERSION
M 215 SAUDI-ARABIA VERSION
M 224 AUTOMATIC BRAKE DIFFERENTIAL
M 249 AUTOMATIC TRANSMISSION (TIPTRONIC)
M 339 FWD
M 340 ADJUSTABLE SEAT HEATING, RIGHT SEAT
M 383 SPORT SEAT LEFT ELECTRIC HEIGHT ADJUSTMENT
M 387 SPORT SEAT RIGHT ELECTRIC HEIGHT ADJUSTMENT
M 419 REAR PACKAGE TRAY
M 425 REAR WINDOW WIPER
M 437 COMFORT SEAT LEFT
M 438 COMFORT SEAT RIGHT
M 439 ELECTRIC CONVERTIBLE TOP
M 454 CRUISE CONTROL
M 480 TRANSMISSION
M 487 NORWAY
M 490 AUDIO OPTION PACK
M 513 LUMBAR SUPPORT SEAT RIGHT
M 553 USA - CANADA VERSION
M 562 AIRBAG
M 573 AIR CONDITIONING SYSTEM
M 576 WITHOUT REAR FOG LIGHT
M 586 LUMBAR SUPPORT SEAT LEFT
M 602 HIGH MOUNTED STOP LIGHT
M 605 HEADLIGHT VERTICAL AIM CONTROL
M 614 TELEPHONE PREPARATION
M 650 SUN ROOF
M 659 ON-BOARD COMPUTER
M 680 DIGITAL SOUND PACKAGE

CONNECTION

Z999|99

FIELD INDICATION

TERMINAL ID

P IGNITION SWITCH POSITION 0
X IGNITION SWITCH POSITION I+II
15 IGNITION SWITCH POSITION II+III
R IGNITION SWITCH POSITION I+II+III
K DIAGNOSIS K
L DIAGNOSIS L
30 PERMANENT POSITIVE
31 GROUND
54 STOP LIGHT
56 LIGHT SWITCH POSITION II
56a/56b HIGH BEAM/LOW BEAM
57 LIGHT SWITCH POSITION I
58 LIGHT SWITCH POSITION I+II
58R/58L SIDE MARKER LIGHT RIGHT/LEFT
58d INSTRUMENT LIGHTS
Ti INJECTION SIGNAL
Tn ROTATIONAL SPEED SIGNAL
61 GENERATOR D+

P998ewd016a

A	B	C	D	E	F

Connectors and abbreviations (page 2 of 2)

Sheet 16 - 1995

G	H	J	K	L	M	N

ABBREVIATIONS

CODE	MEANING	CODE	MEANING	CODE	MEANING
ABD	AUTOMATIC BRAKE DIFFERENTIAL	FR	FRONT RIGHT	RHD	RIGHT-HAND DRIVE
ABS	ANTILOCK BRAKING SYSTEM	FOG	FOG LIGHT	RL	REAR LEFT
AC	AIR CONDITIONING SYSTEM	GP	GROUND POINT	RR	REAR RIGHT
ADL	ADDITIONAL DRIVING LIGHTS	HF	HIGH FREQUENCY	RDK	TIRE PRESSURE CONTROL
BB	BRAKE BOOSTER	HFM	HOT-FILM AIR MASS METER	ROW	REST OF WORLD
BL	BACKUP LIGHT	HVAC	HEADLIGHT VERTICAL AIM CONTROL	RFL	REAR FOG LIGHT
CLOCK	CLOCK SIGNAL	HCS	HEADLIGHT CLEANING SYSTEM	SA	SAUDI ARABIA
CO	CARBON MONOXIDE	INS.C	INSTRUMENT CLUSTER	SCS	COMBINED STEERING COLUMN SWITCH
CWL	CENTRAL WARN. LIGHT	INJ.V	INJECTOR VALVE	SERVICE	SERVICE
CLS	CENTRAL LOCKING SYSTEM	IC	IGNITION CIRCUIT	SL	SAFETY LAMP
CE	CENTRAL ELECTRIC	LF	LOW FREQUENCY	SS	SPEED SENSOR
CP	CONNECTING POINT	LED	LIGHT-EMITTING DIODE	ESS	ENGINE SPEED SENSOR
CL	CLOSE	LHD	LEFT-HAND DRIVE	TV	THROTTLE VALVE
CU	CONTROL UNIT	MFI + DI	DIGITAL ENGINE ELECTRONICS	TVS	THROTTLE VALVE SENSOR
DATA	DATA SIGNAL	NO	NUMBER	TE	TERMINAL
DEF	DEFROST	NTC	NEGATIVE TEMPERATURE COEFFICIENT	TEL.MUTE	RADIO MUTING CONTROL
DSP	DIGITAL SOUND PACKAGE	OP	OPEN	TI	INJECTION TIME
D I	ELECTRIC IGNITION SYSTEM WITH KNOCK CONTROL	PIN	PIN	TN	SPEED
DP	DISCONNECTING POINT	POT	POTENTIOMETER	TRP	TWIN-COIL ROTARY POSITIONER
ETC	ELECTRICAL TRANSMISSION CONTROL	PSD	PORSCHE LIMITED SLIP DIFFERENTIAL	USA	USA
ESO	ENGINE-SPEED SENSOR OUTPUT	PWM	PULSE WIDTH MODULATION	WL	WIRING LOOM
FA	FRONT AXLE	PL	PLUG	WP	WELD POINT
FL	FRONT LEFT	RA	REAR AXLE	WW	WORLDWIDE

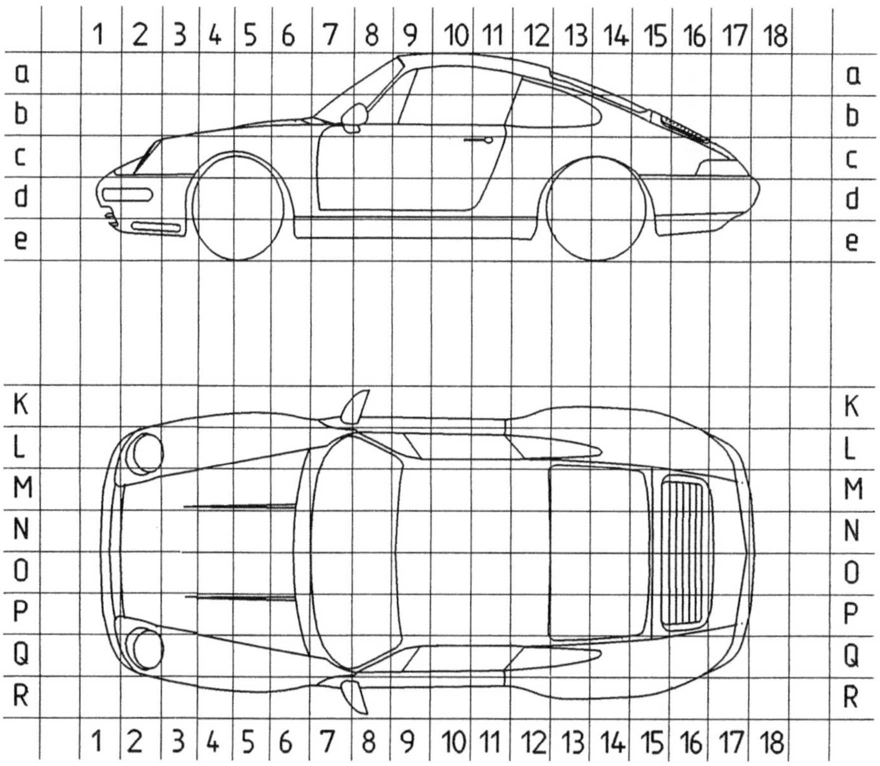

P998ewd016b

G	H	J	K	L	M	N

EWD

Sheet 17 - 1995

Anti-theft (option code M530)
(page 1 of 1)

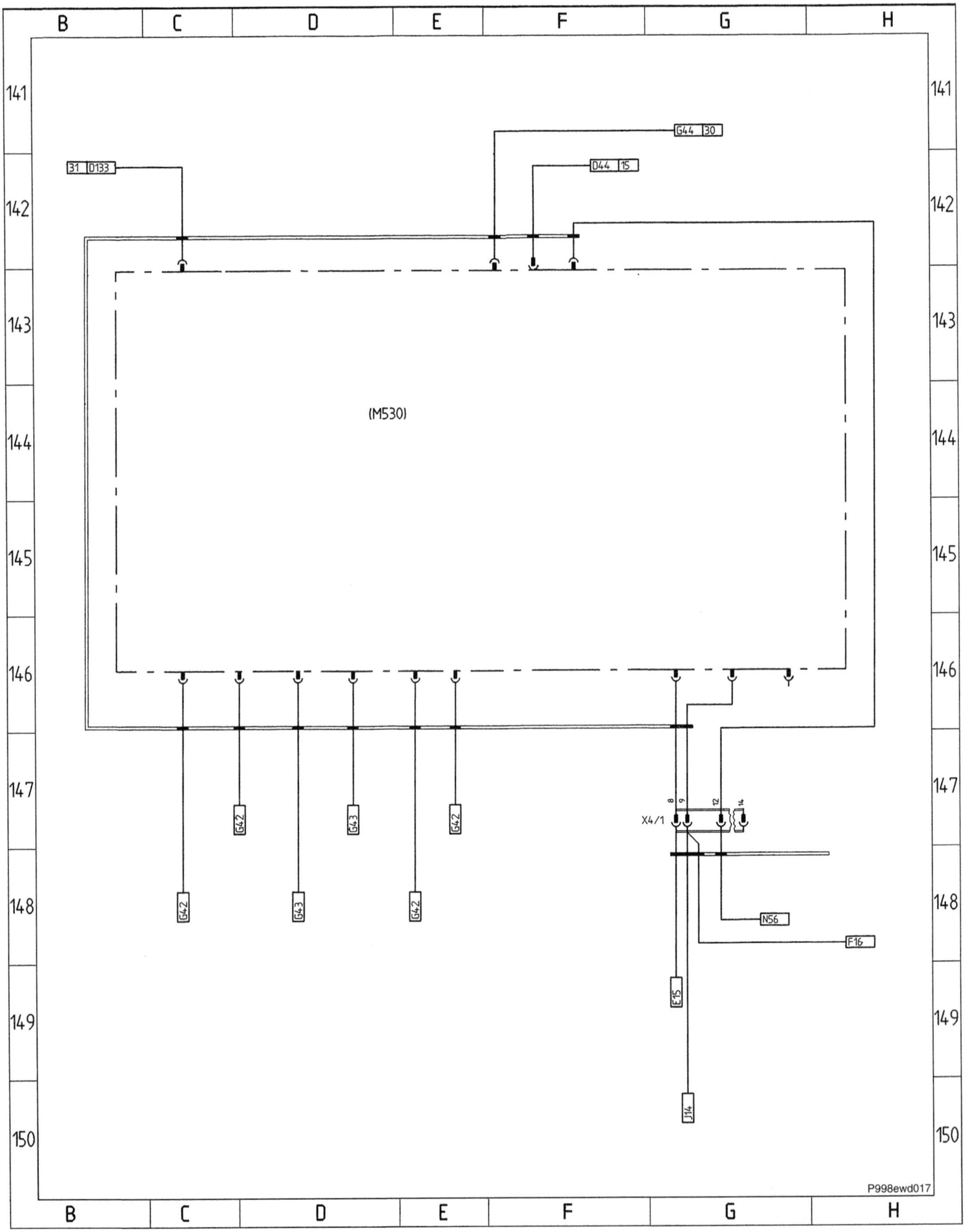

Lights
(page 1 of 3)

Sheet 1 -1995/2

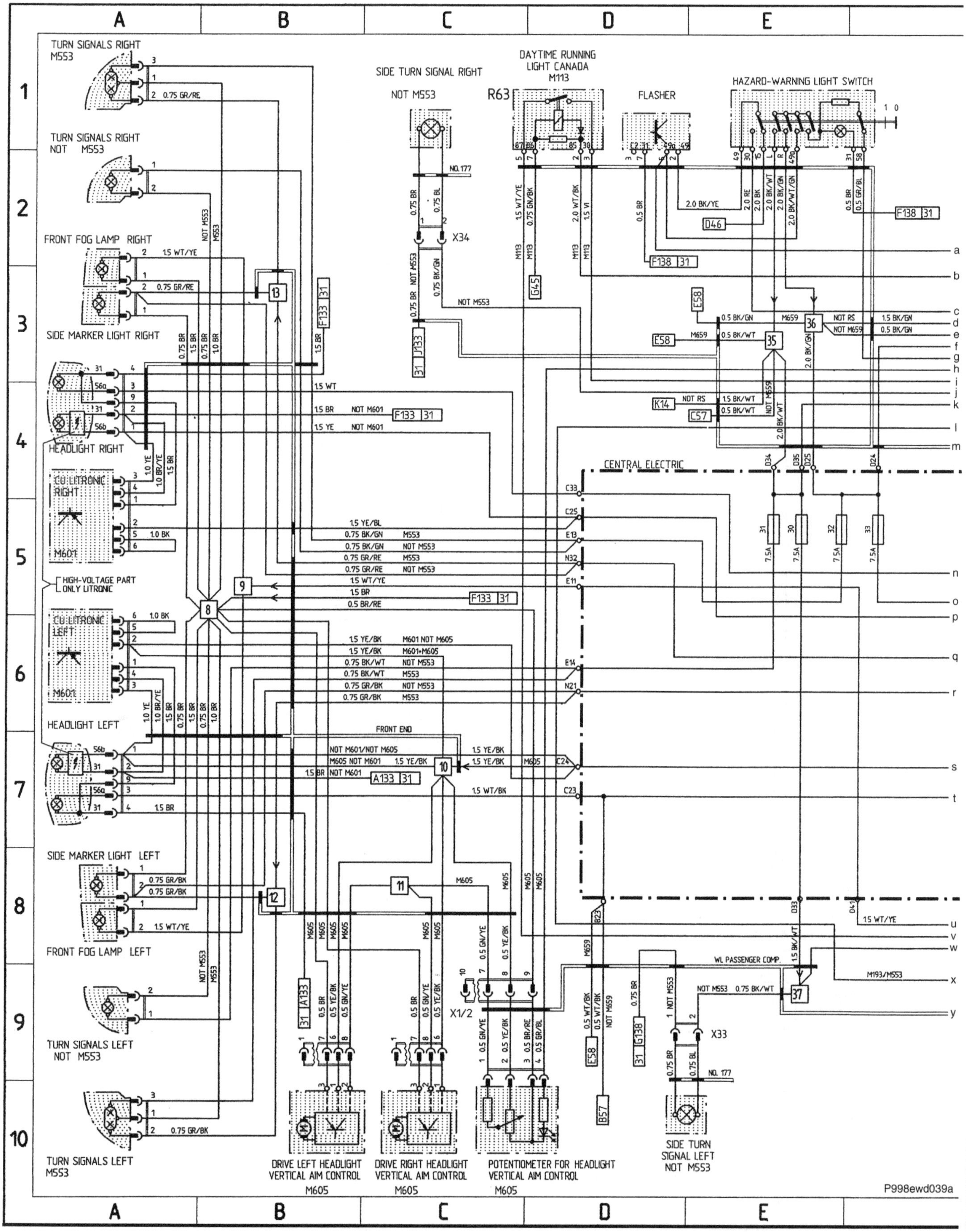

Sheet 1 - 1995/2

Lights
(page 2 of 3)

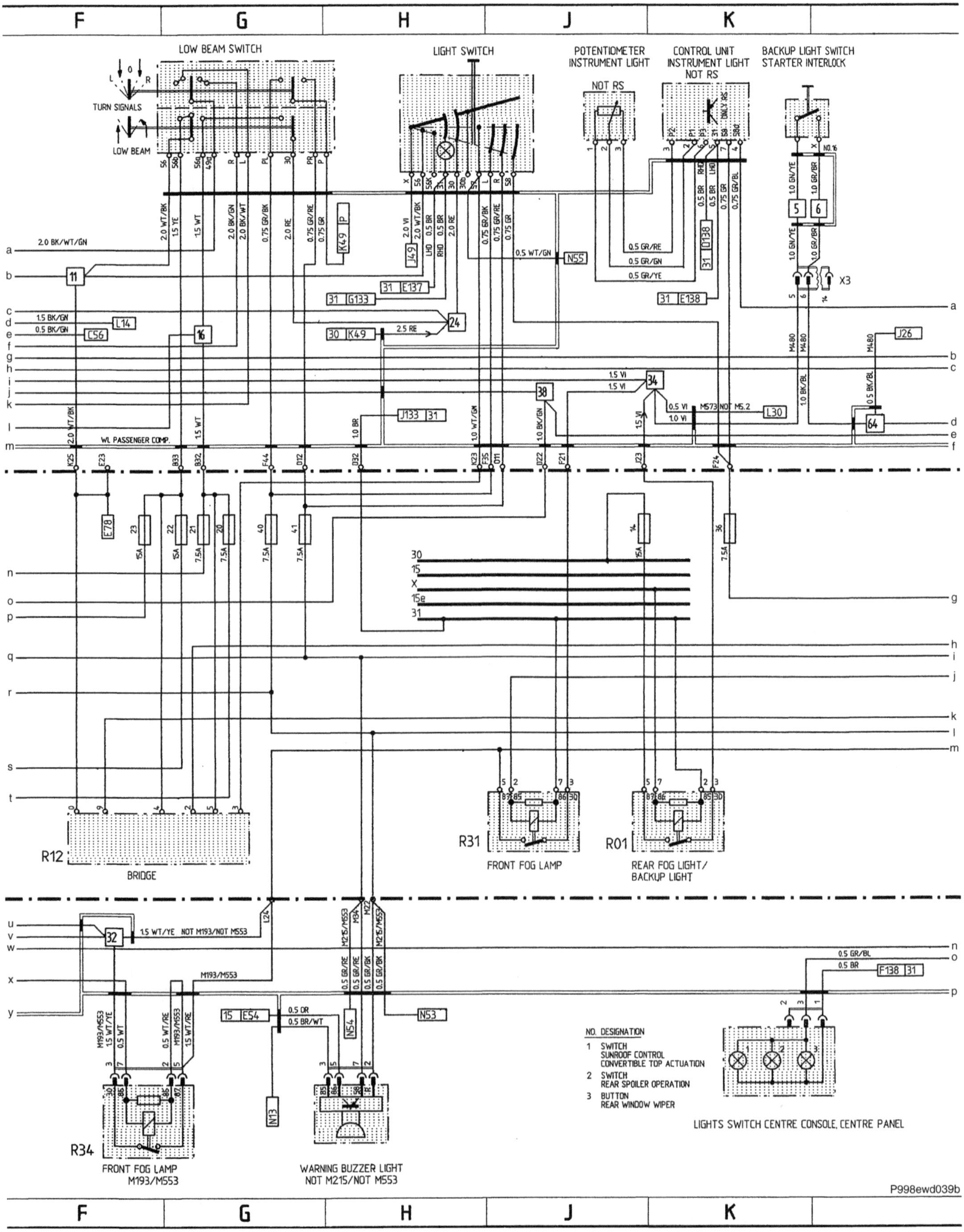

Lights
(page 3 of 3)

Sheet 1 - 1995/2

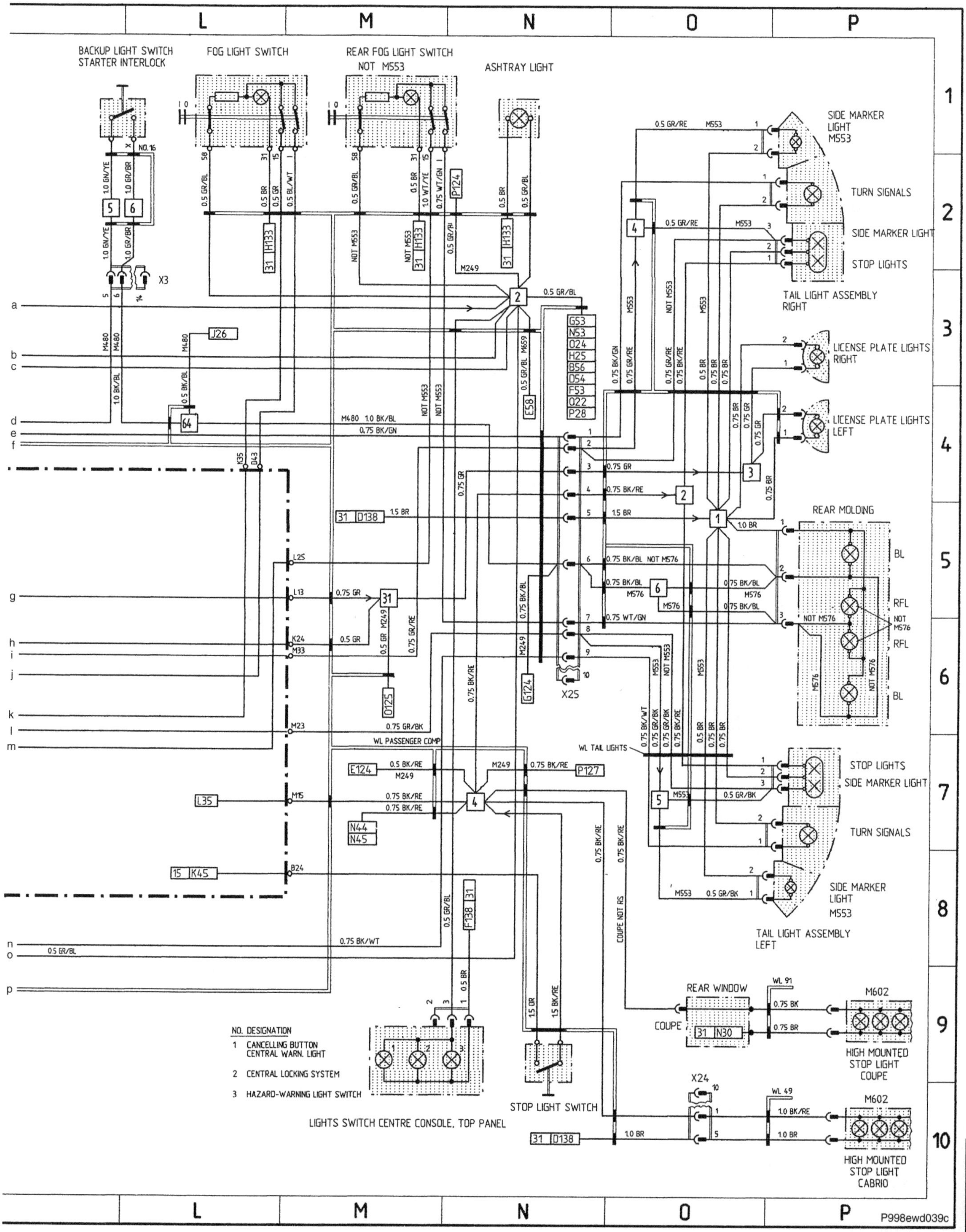

Sheet 2 - 1995/2

Seats, alarm system, central locks, mirror, power window regulator, insid e lights (page 1 of 3)

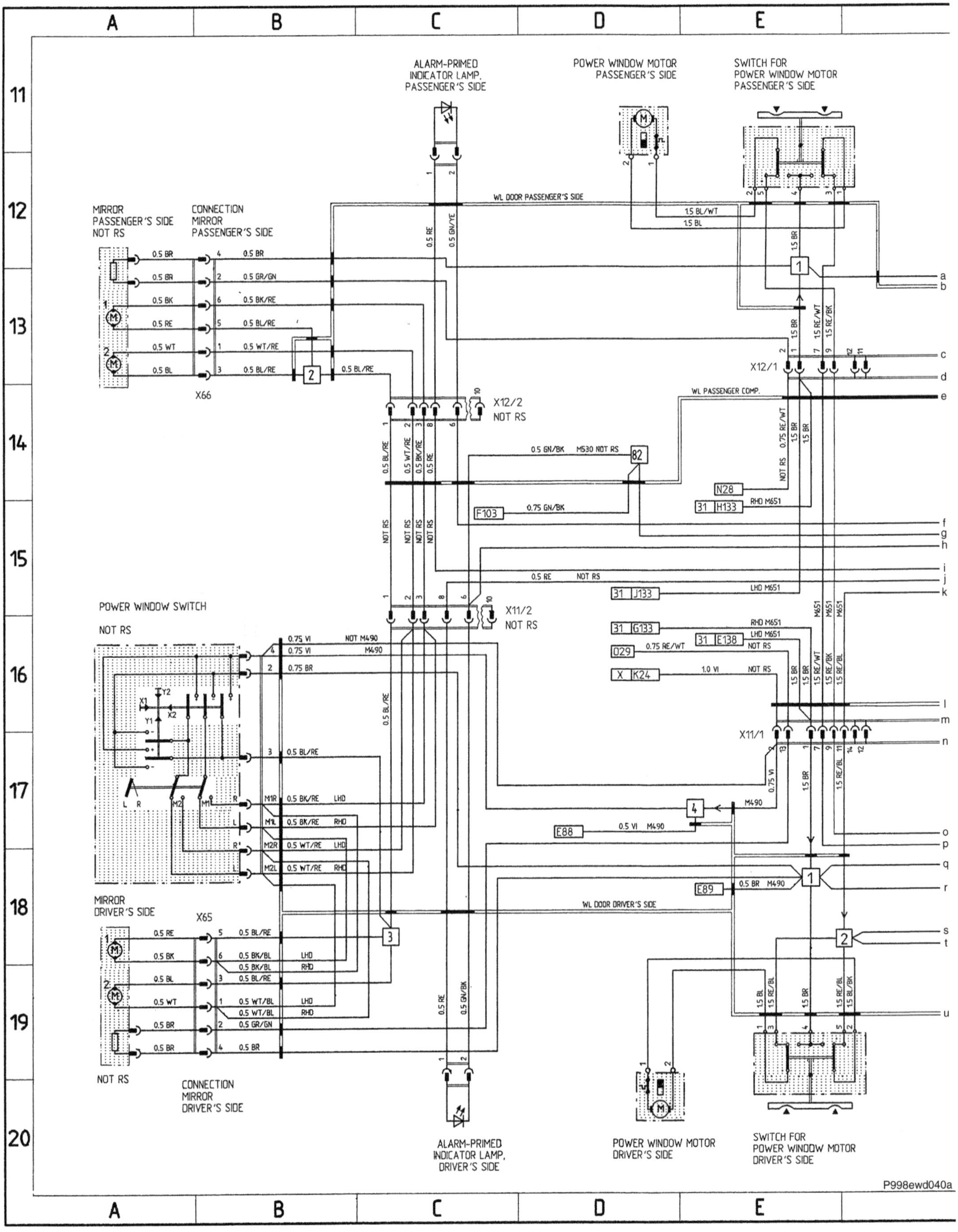

Seats, alarm system, central locks, mirror, power window regulator, inside lights (page 2 of 3)

Sheet 2 - 1995/2

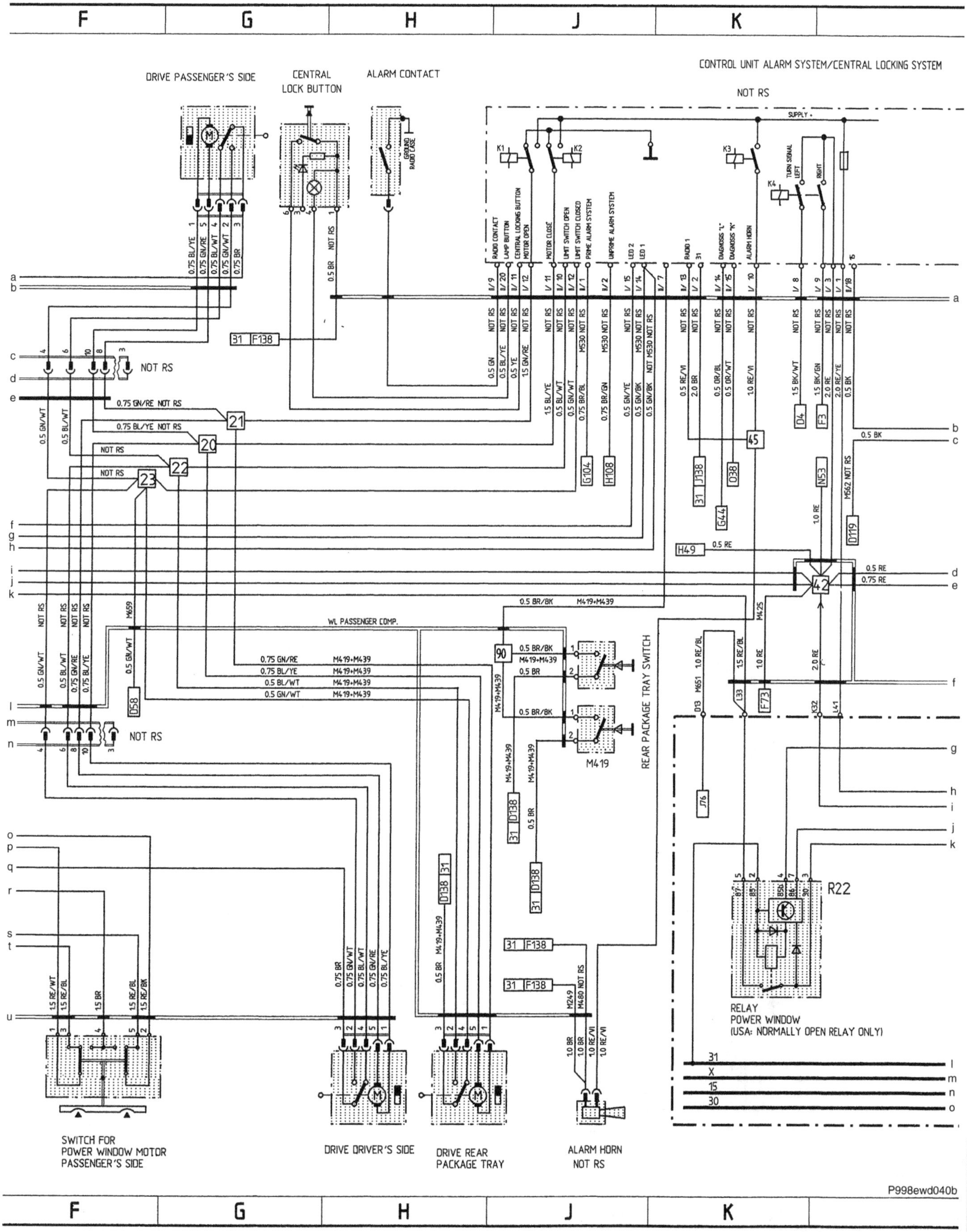

Sheet 2 - 1995/2

Seats, alarm system, central locks, mirror, power window regulator, inside lights (page 3 of 3)

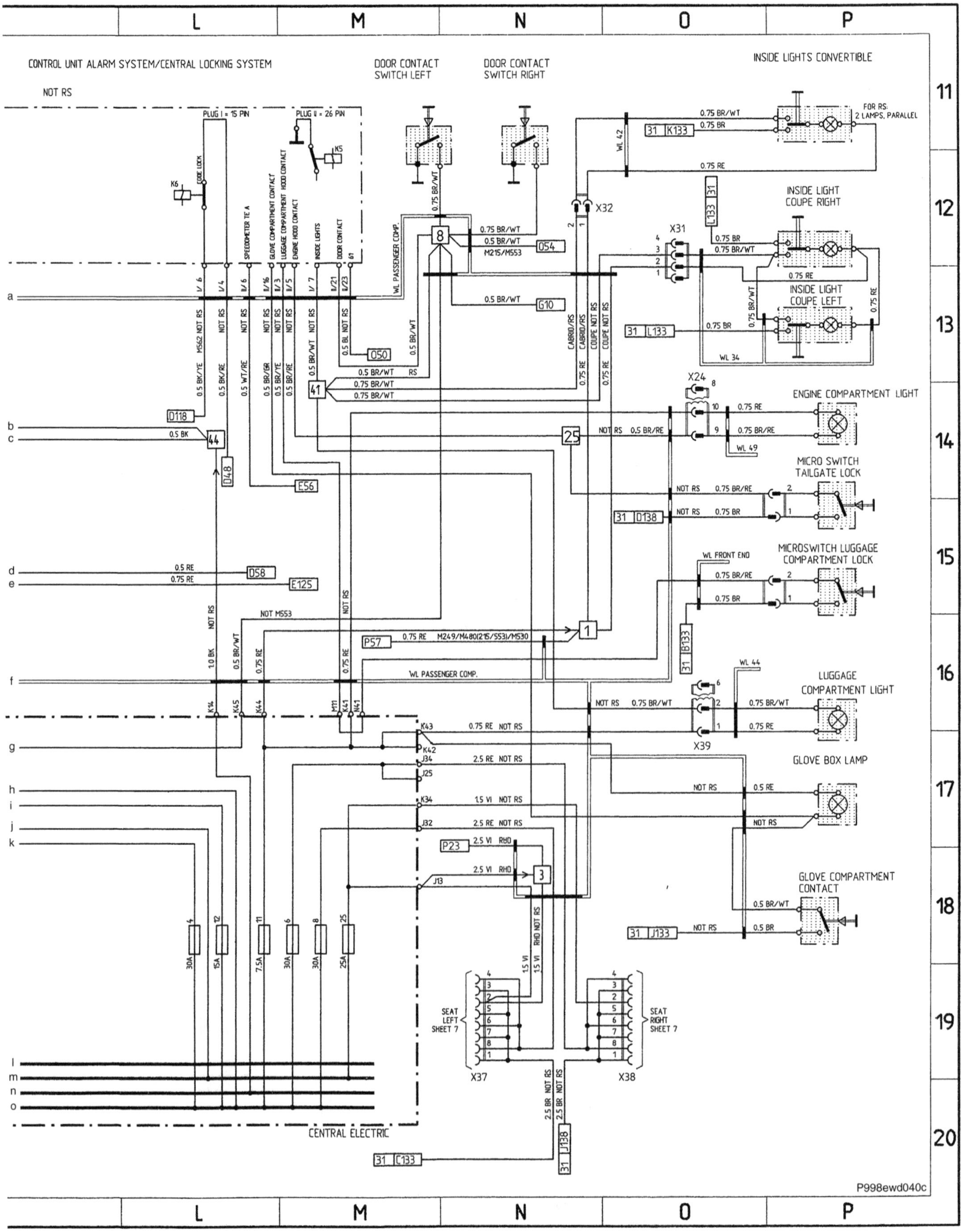

Heater, A/C (page 1 of 3)

Sheet 3 - 1995/2

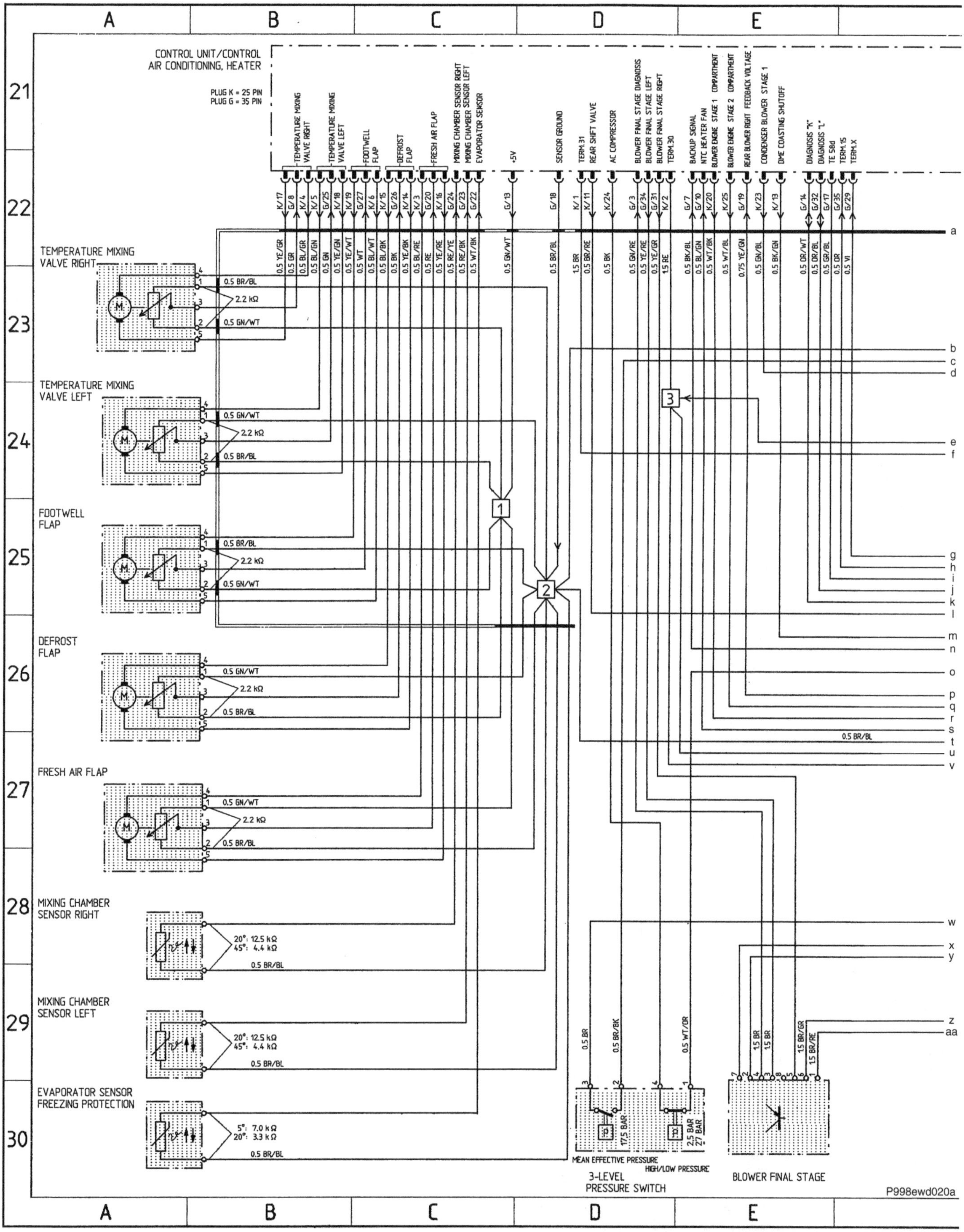

Sheet 3 - 1995/2

Heater, A/C (page 2 of 3)

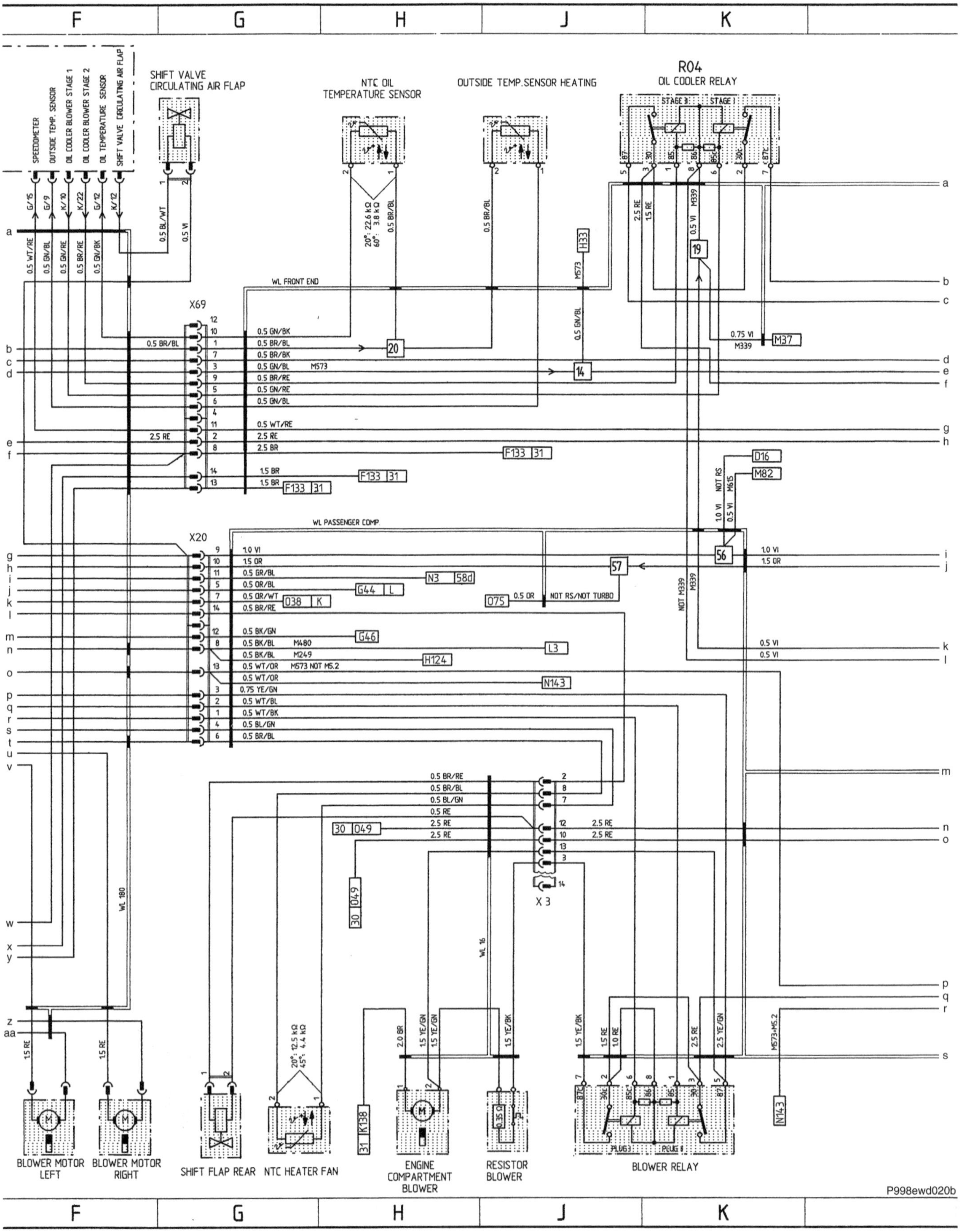

Heater, A/C (page 3 of 3)

Sheet 3 - 1995/2

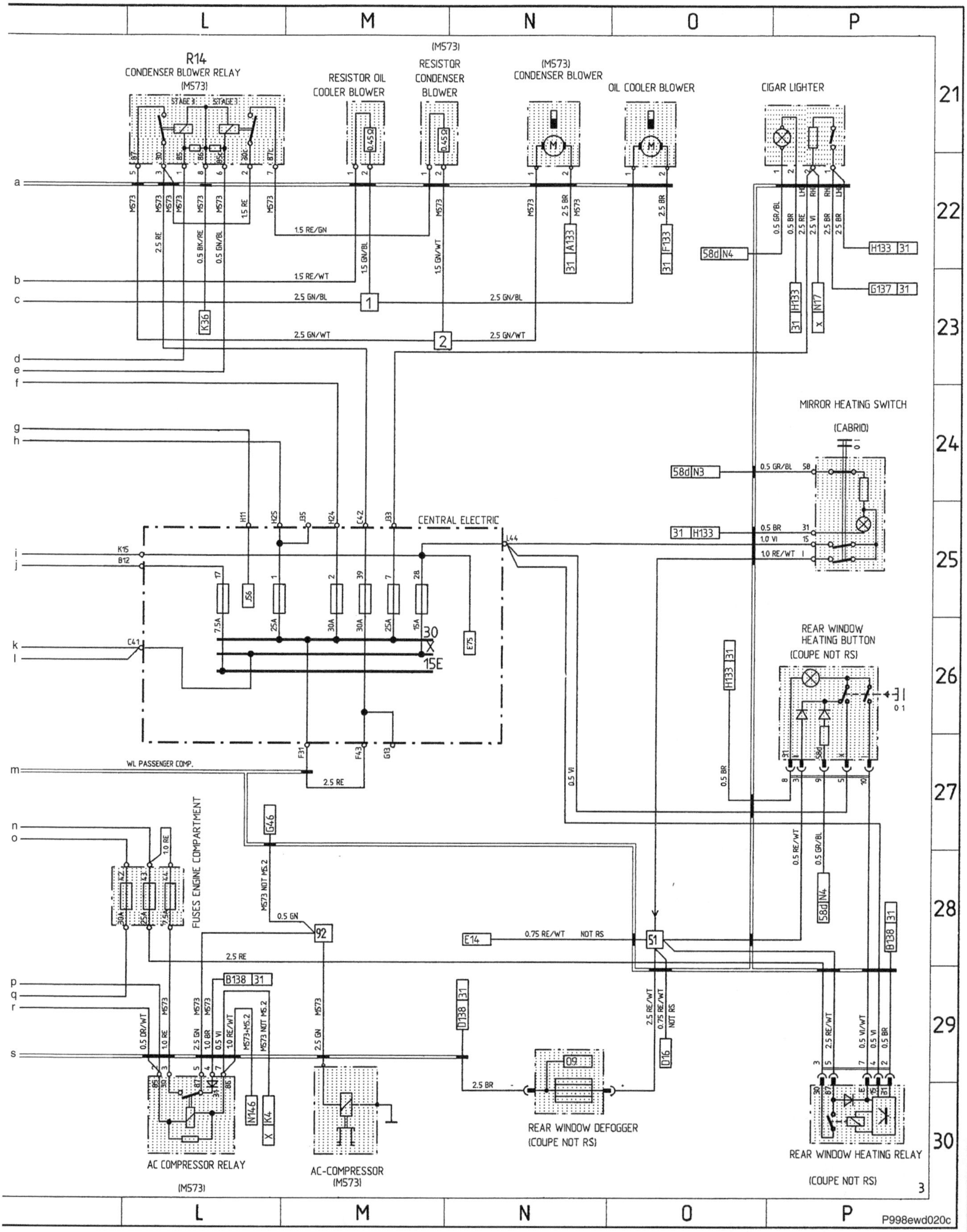

Sheet 4A - 1995/2

ABS
(page 1 of 3)

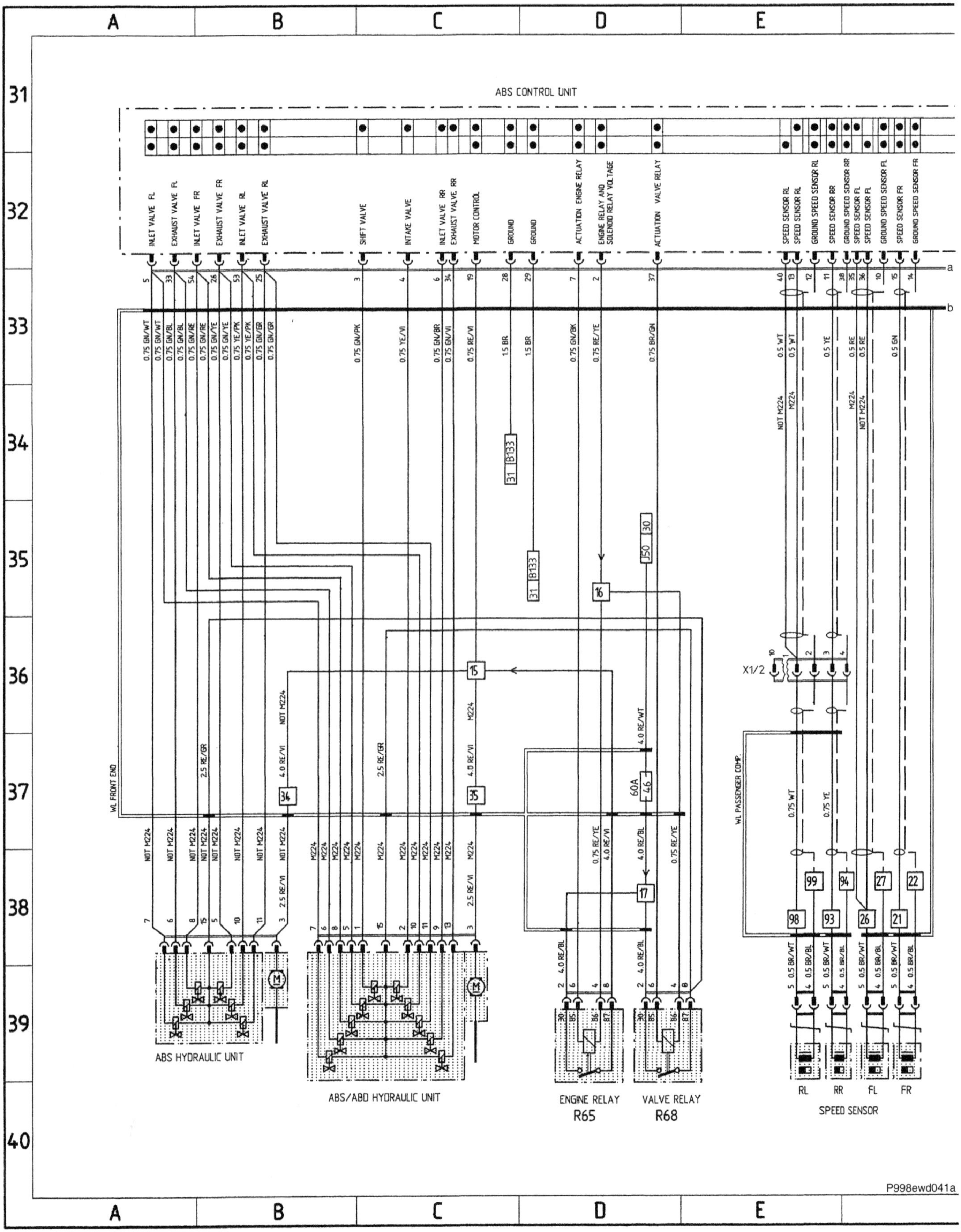

ABS
(page 2 of 3)

Sheet 4A - 1995/2

F	G	H	J	K	

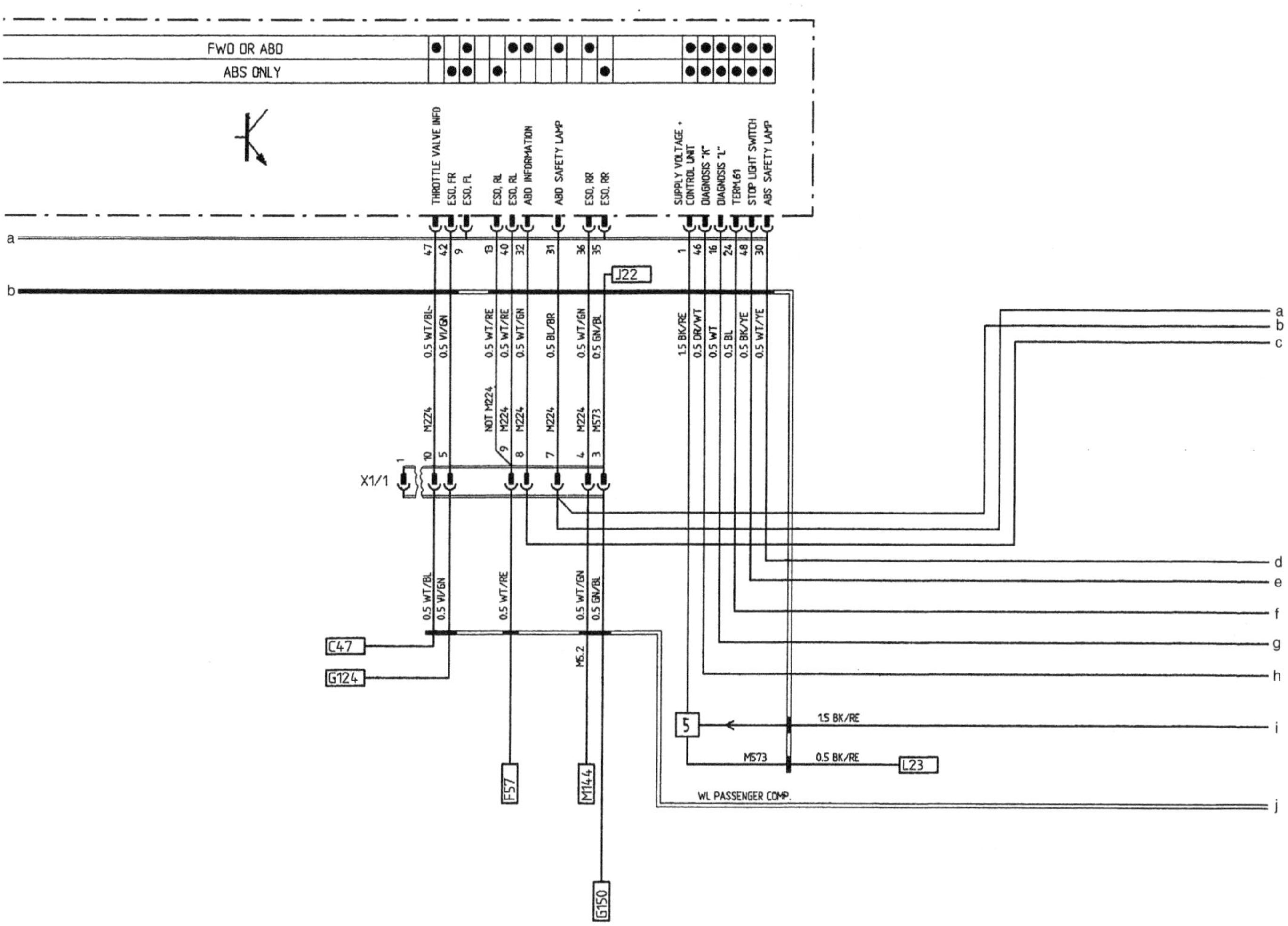

P998ewd041b

F	G	H	J	K	

EWD

Sheet 4A - 1995/2

ABS (page 3 of 3)

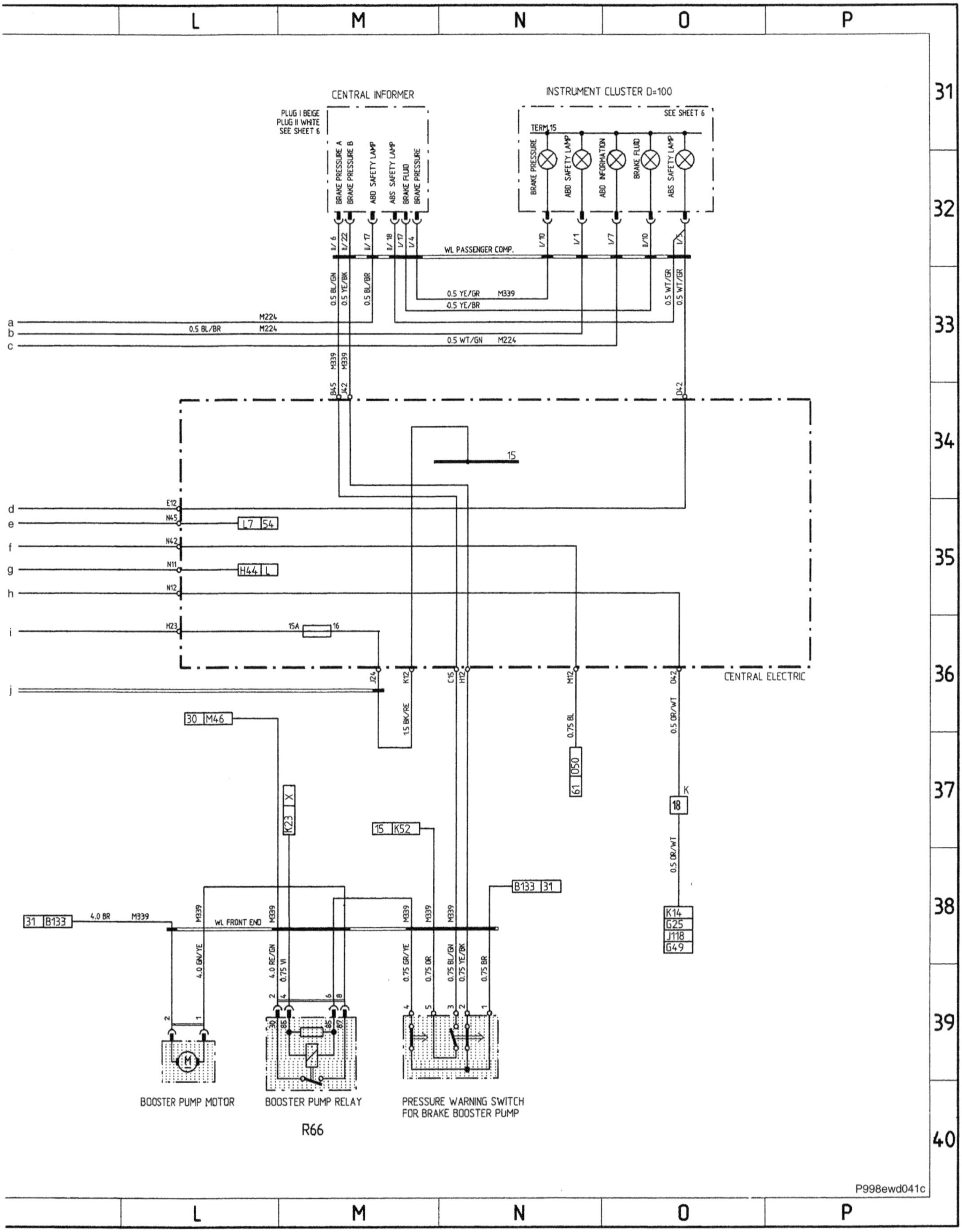

ABS (RS)
(page 1 of 3)

Sheet 4B - 1995/2

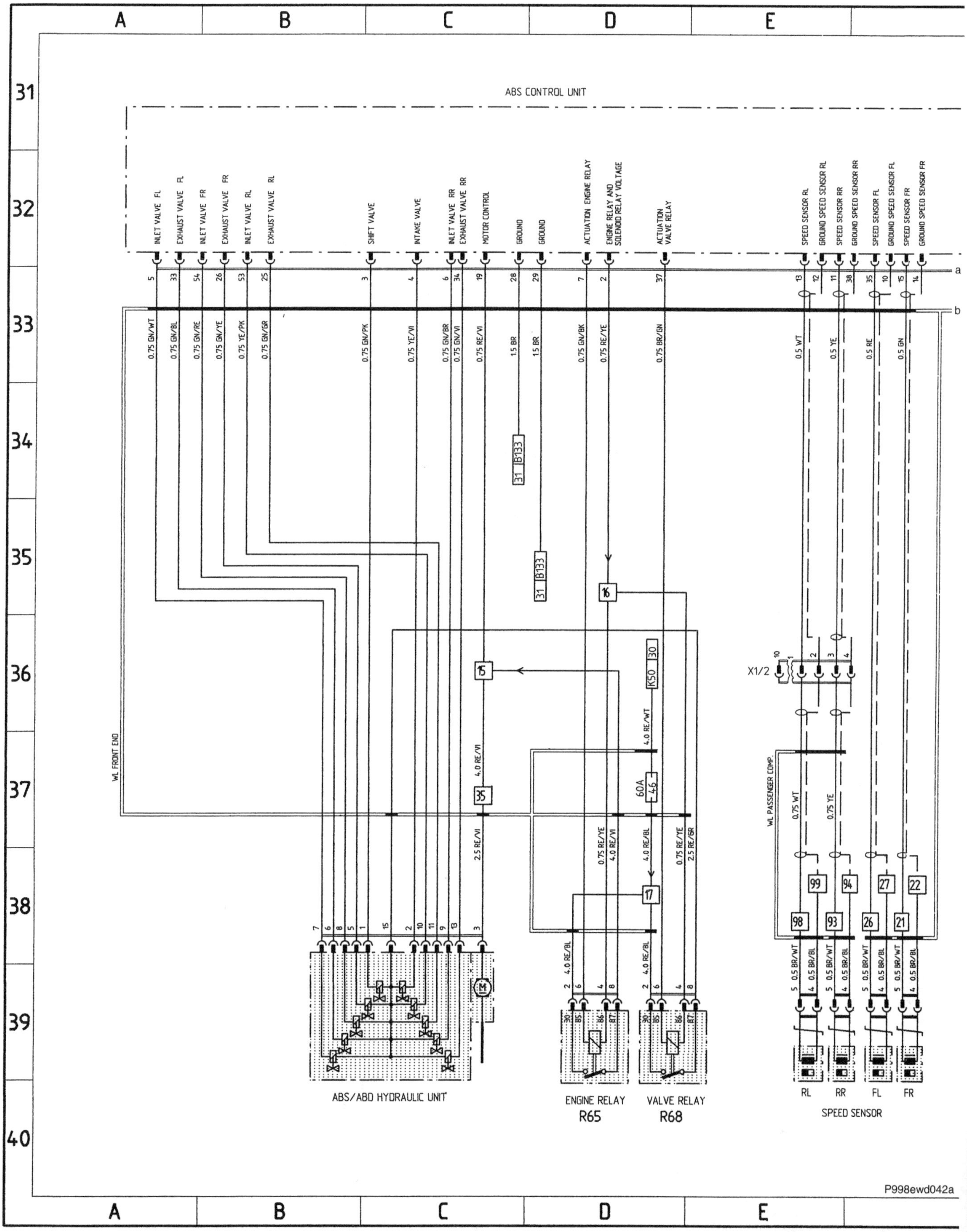

P998ewd042a

Sheet 4B - 1995/2

ABS (RS) (page 2 of 3)

F	G	H	J	K

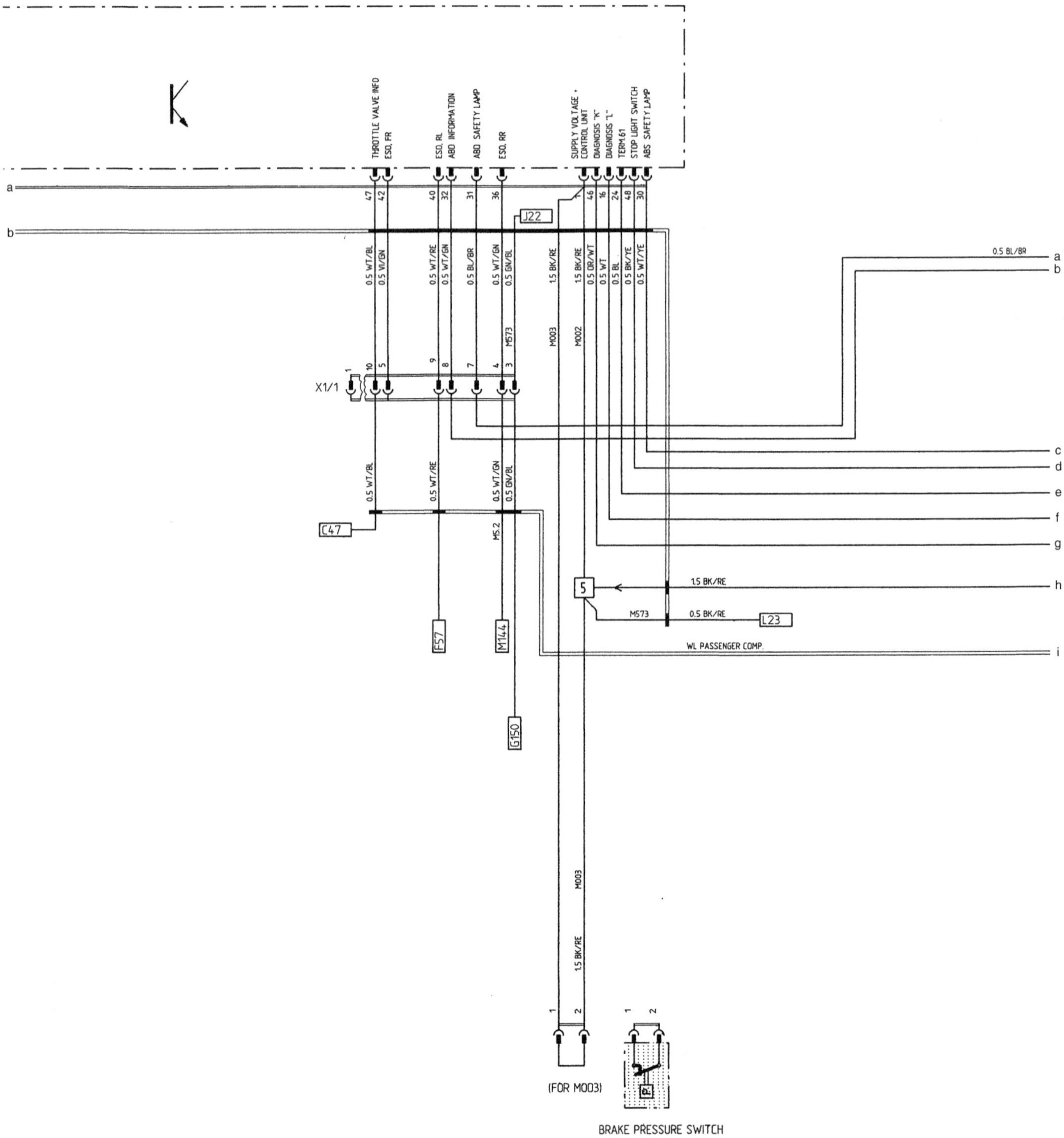

P998ewd042b

F	G	H	J	K

ABS (RS)
(page 3 of 3)

Sheet 4B - 1995/2

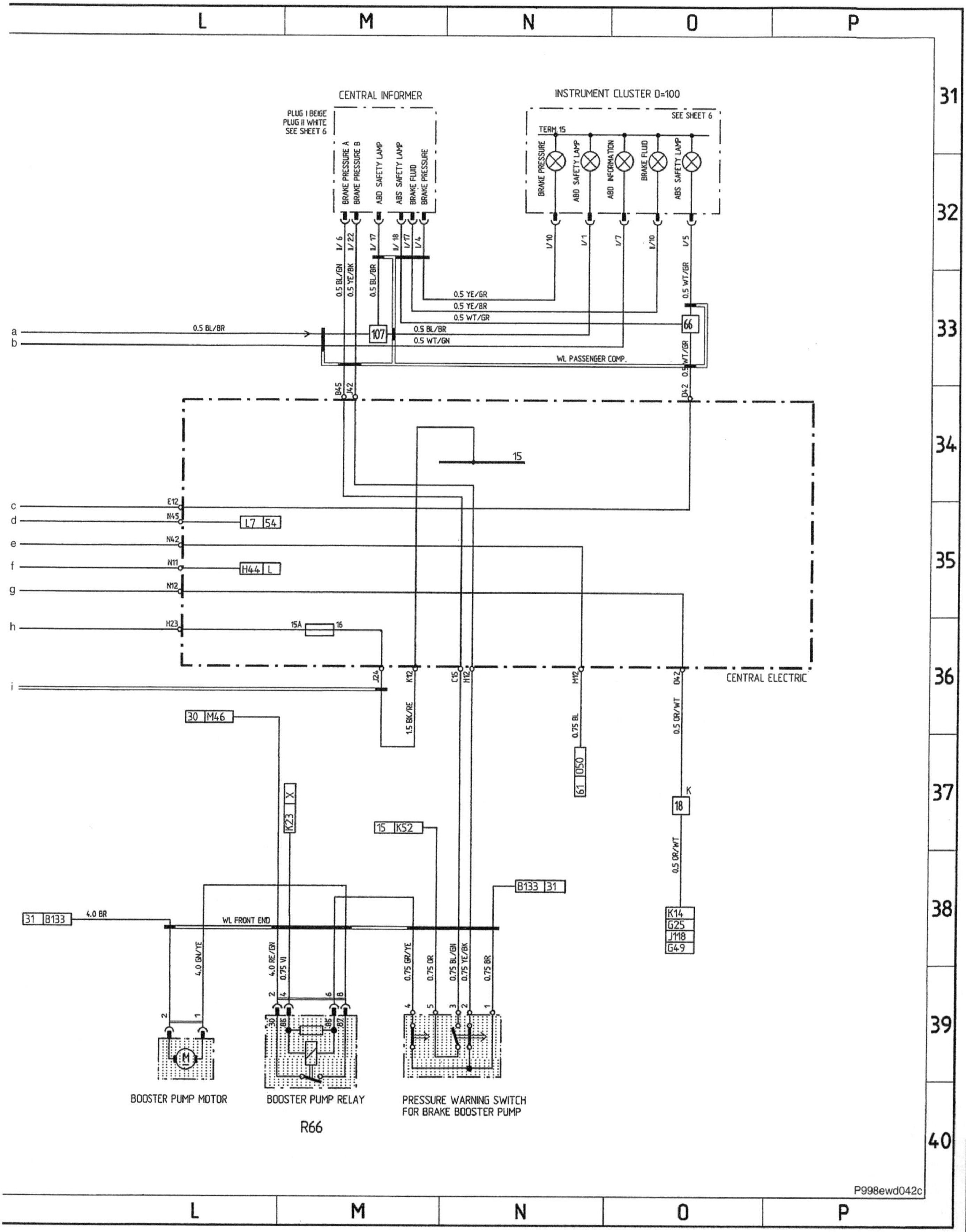

EWD

Sheet 5A - 1995/2

Engine electrics, fuel, ignition system (DME M 2.10) (page 1 of 3)

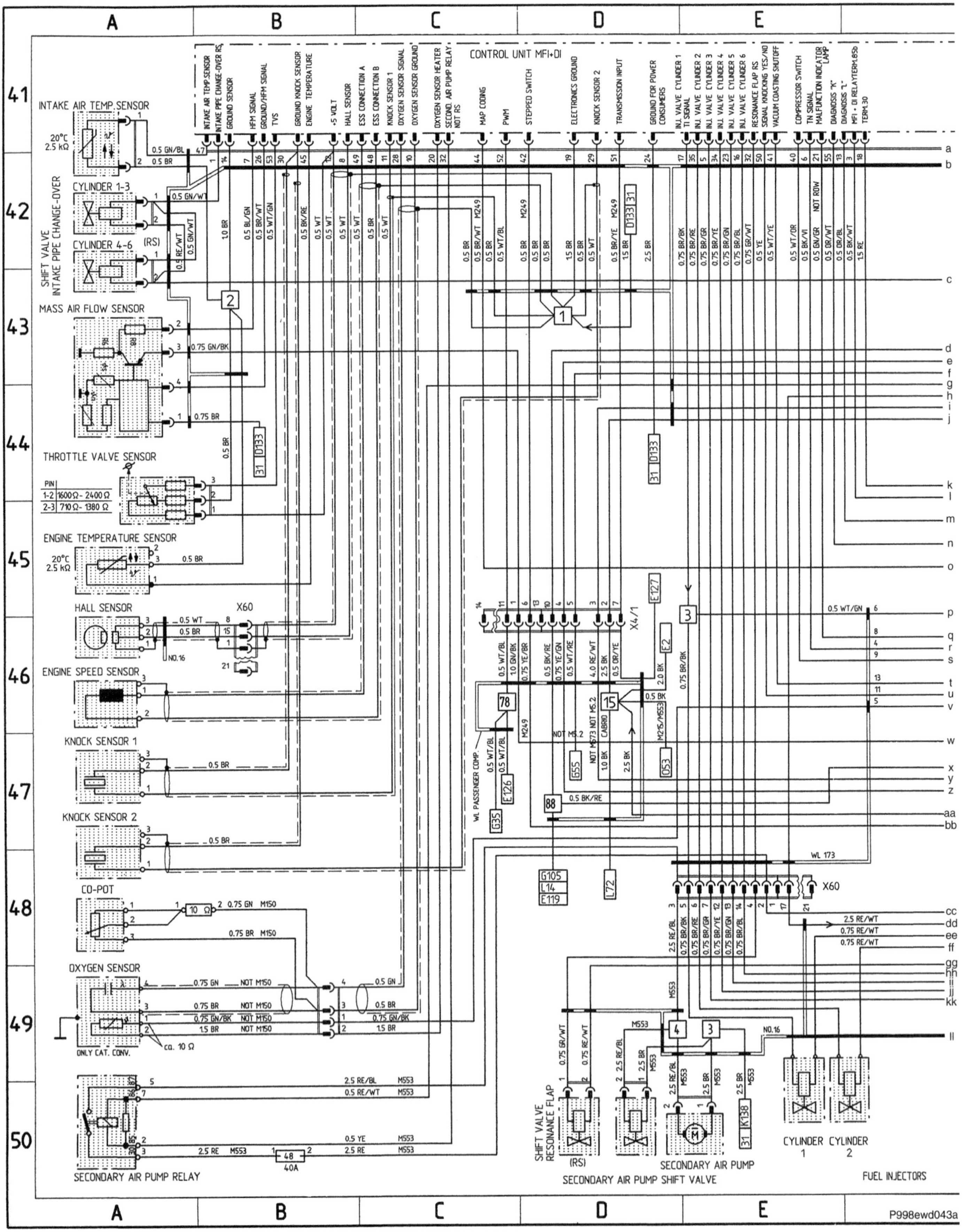

Engine electrics, fuel, ignition system (DME M 2.10) (page 2 of 3)

Sheet 5A - 1995/2

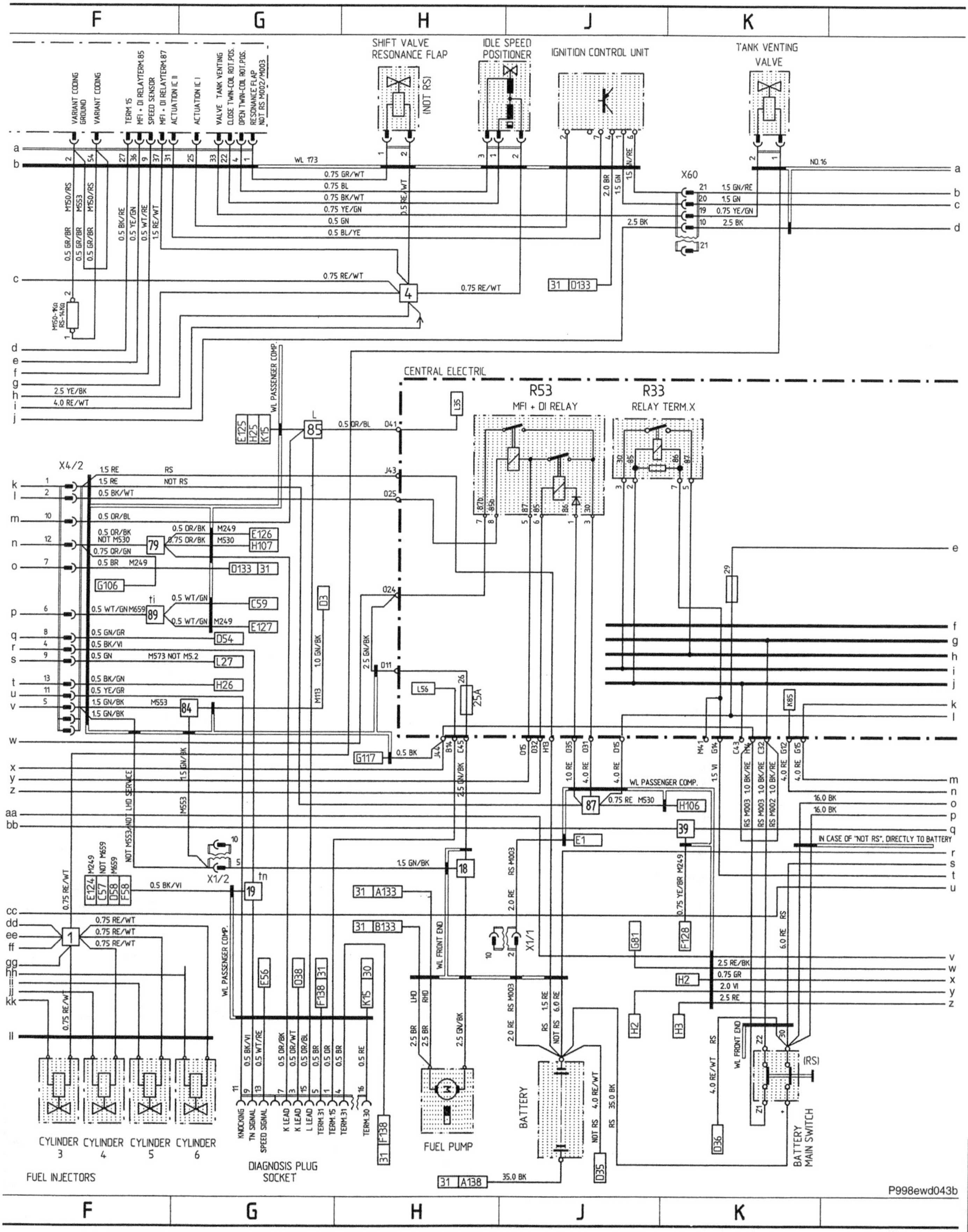

Sheet 5A - 1995/2

Engine electrics, fuel, ignition system (DME M 2.10) (page 3 of 3)

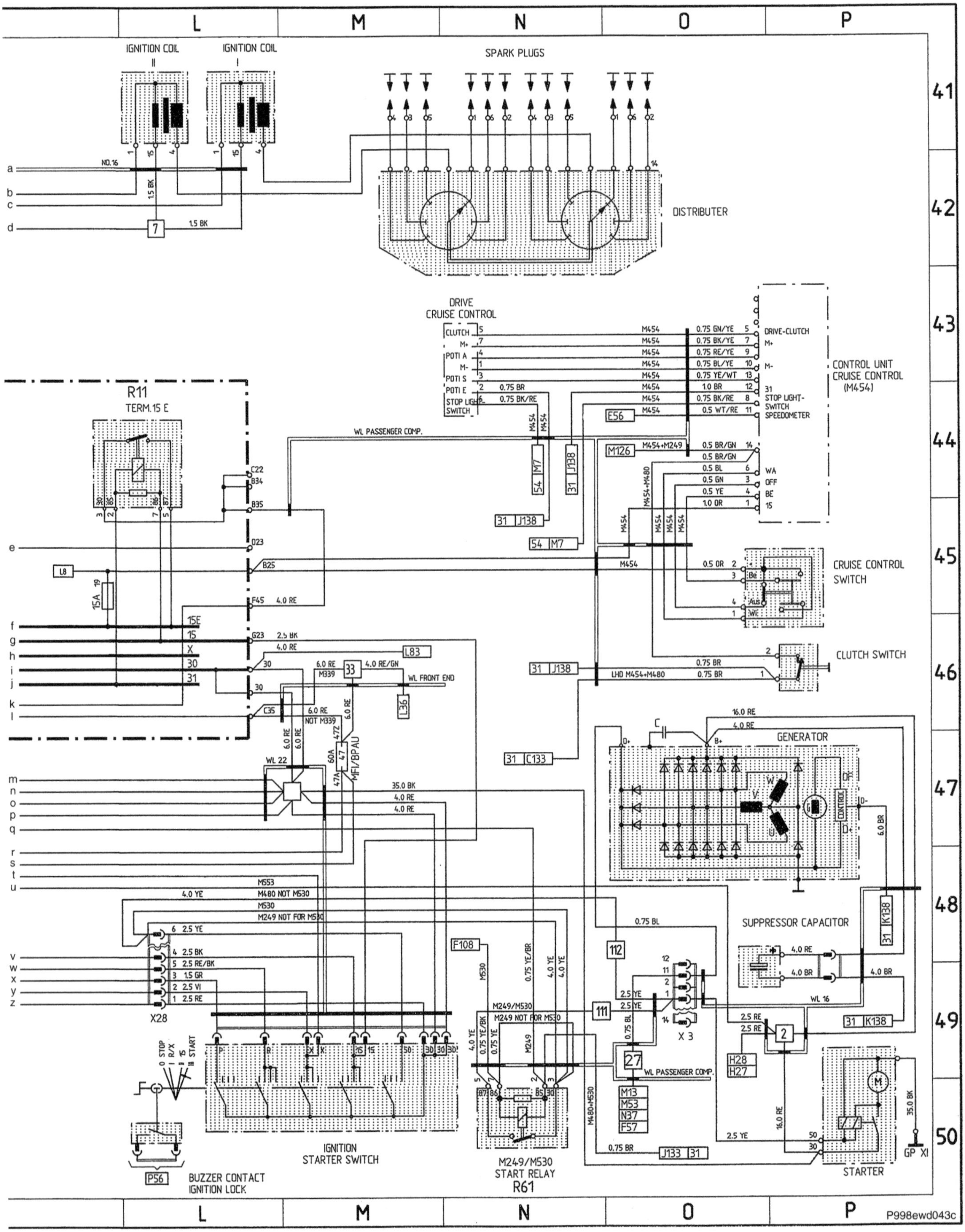

Instruments and sensor (page 1 of 3)

Sheet 6 - 1995/2

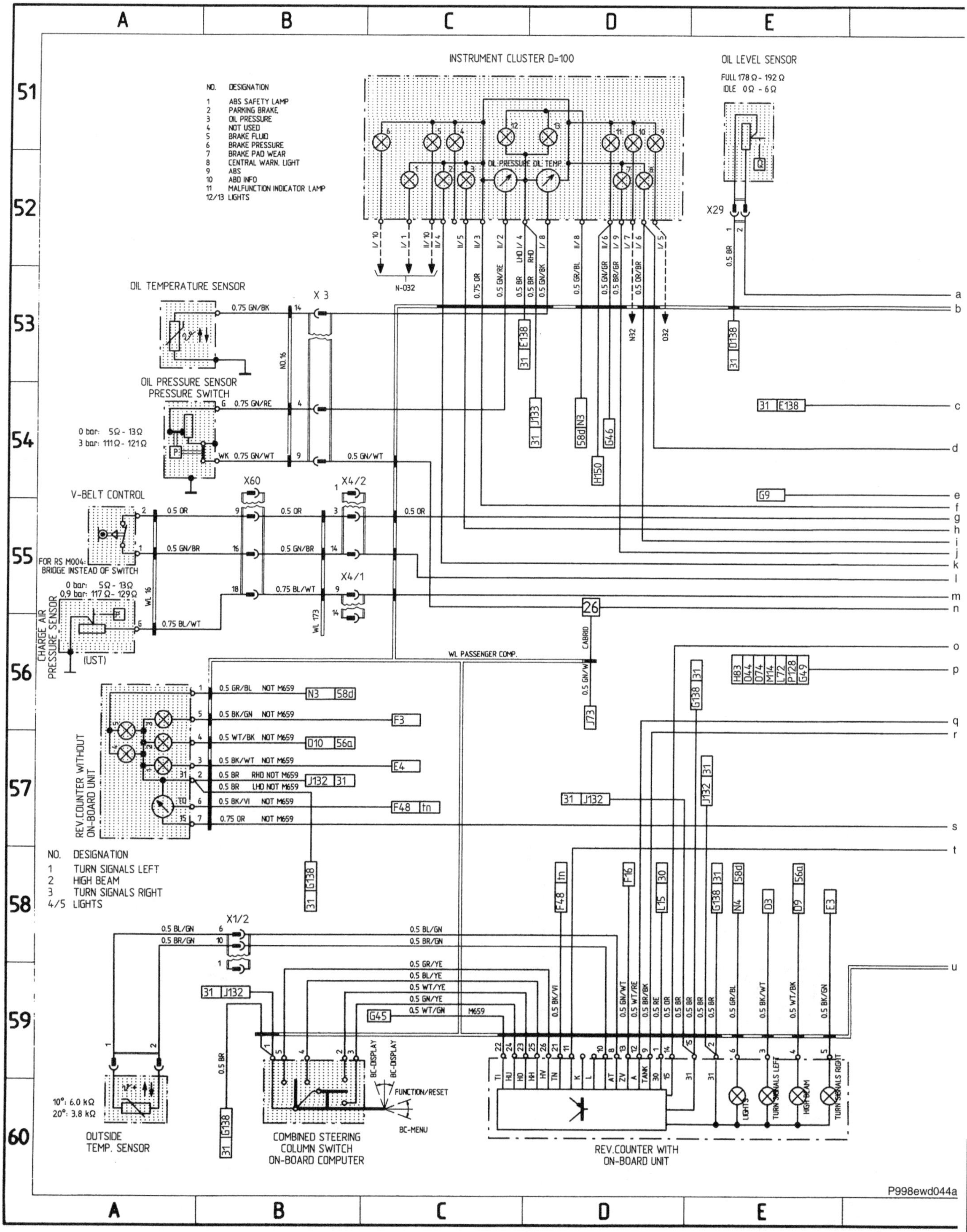

Sheet 6 - 1995/2

Instruments and sensor (page 2 of 3)

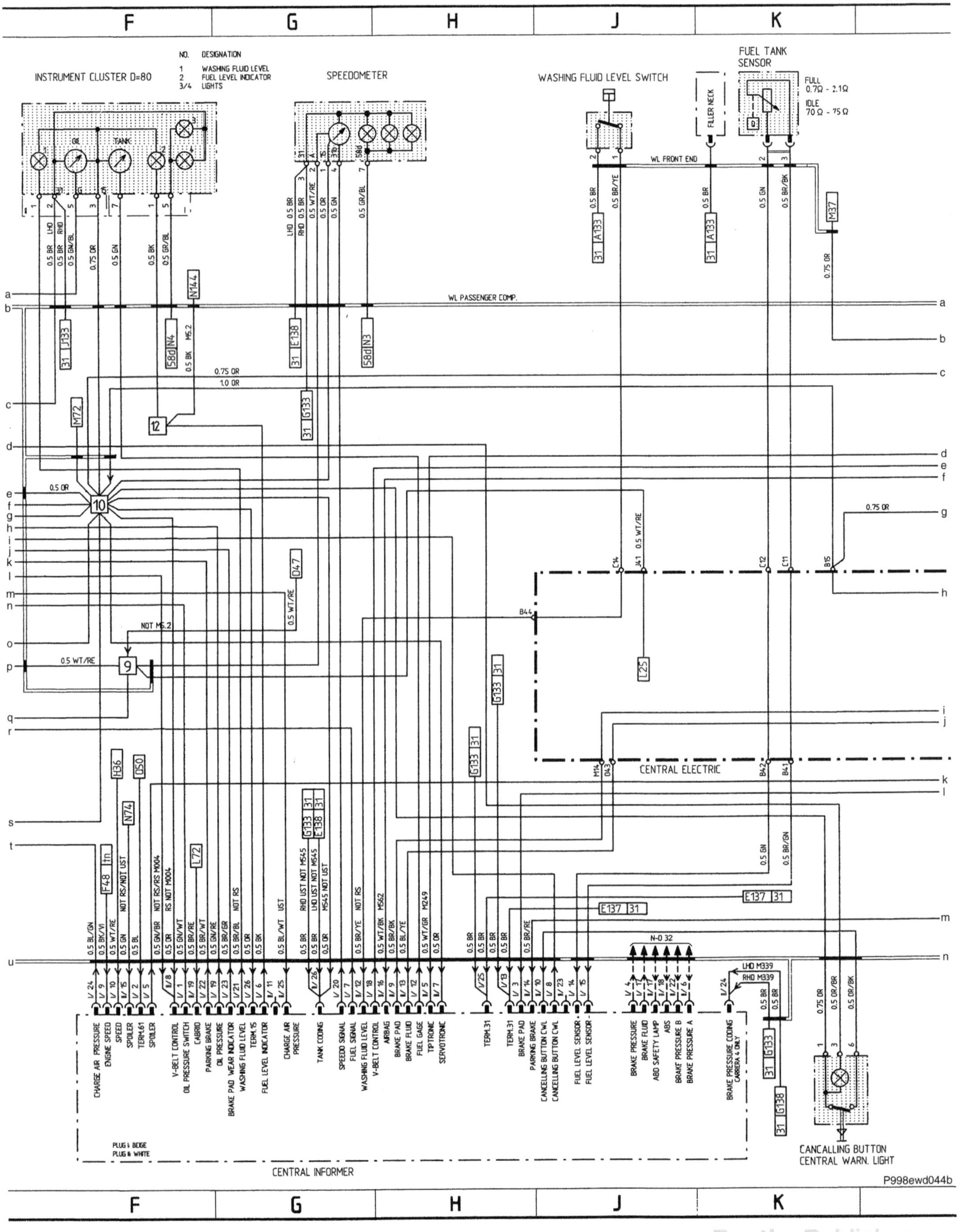

P998ewd044b

F G H J K

Instruments and sensor (page 3 of 3)

Sheet 6 - 1995/2

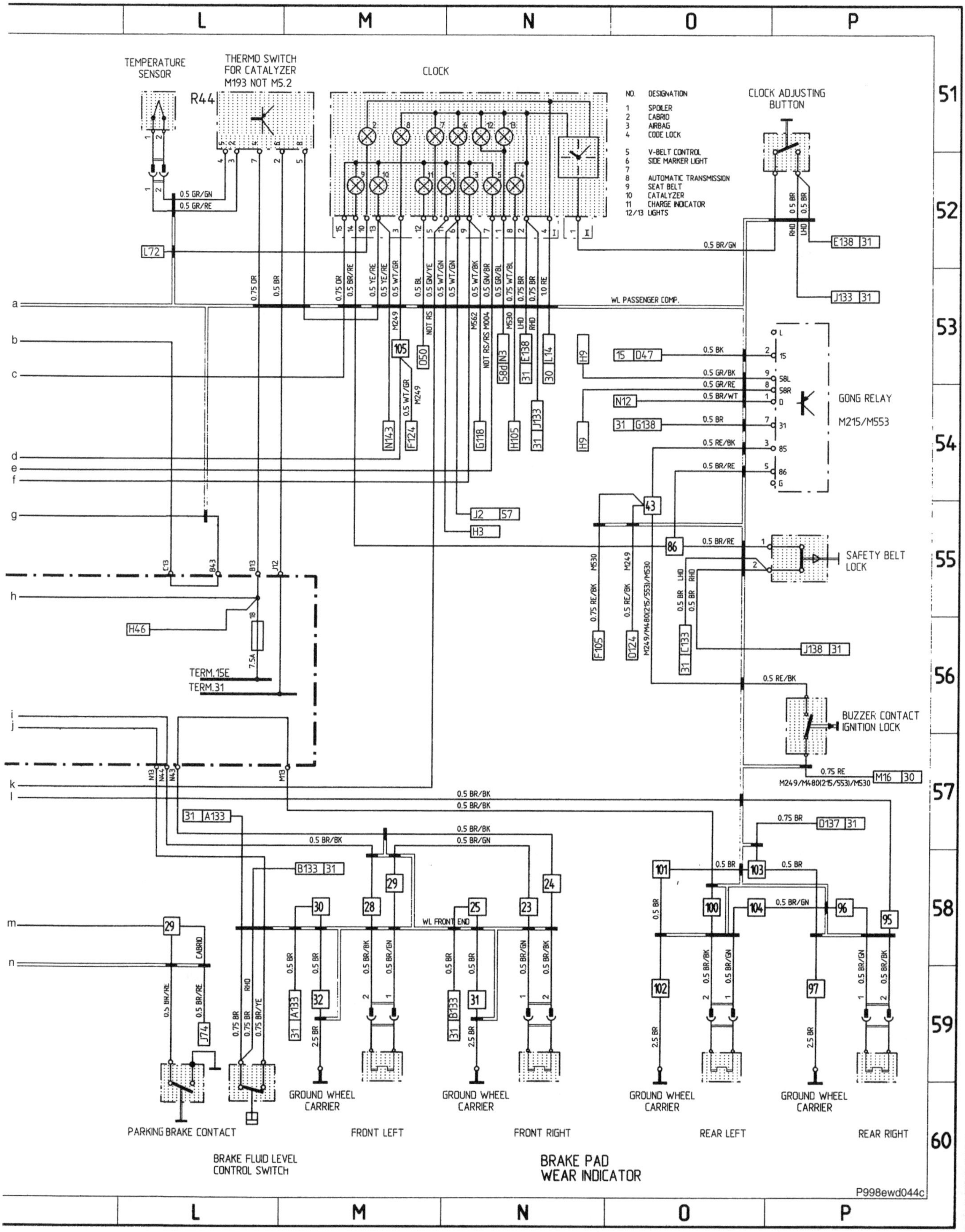

EWD

Sheet 7 - 1995/2

Seats
(page 1 of 2)

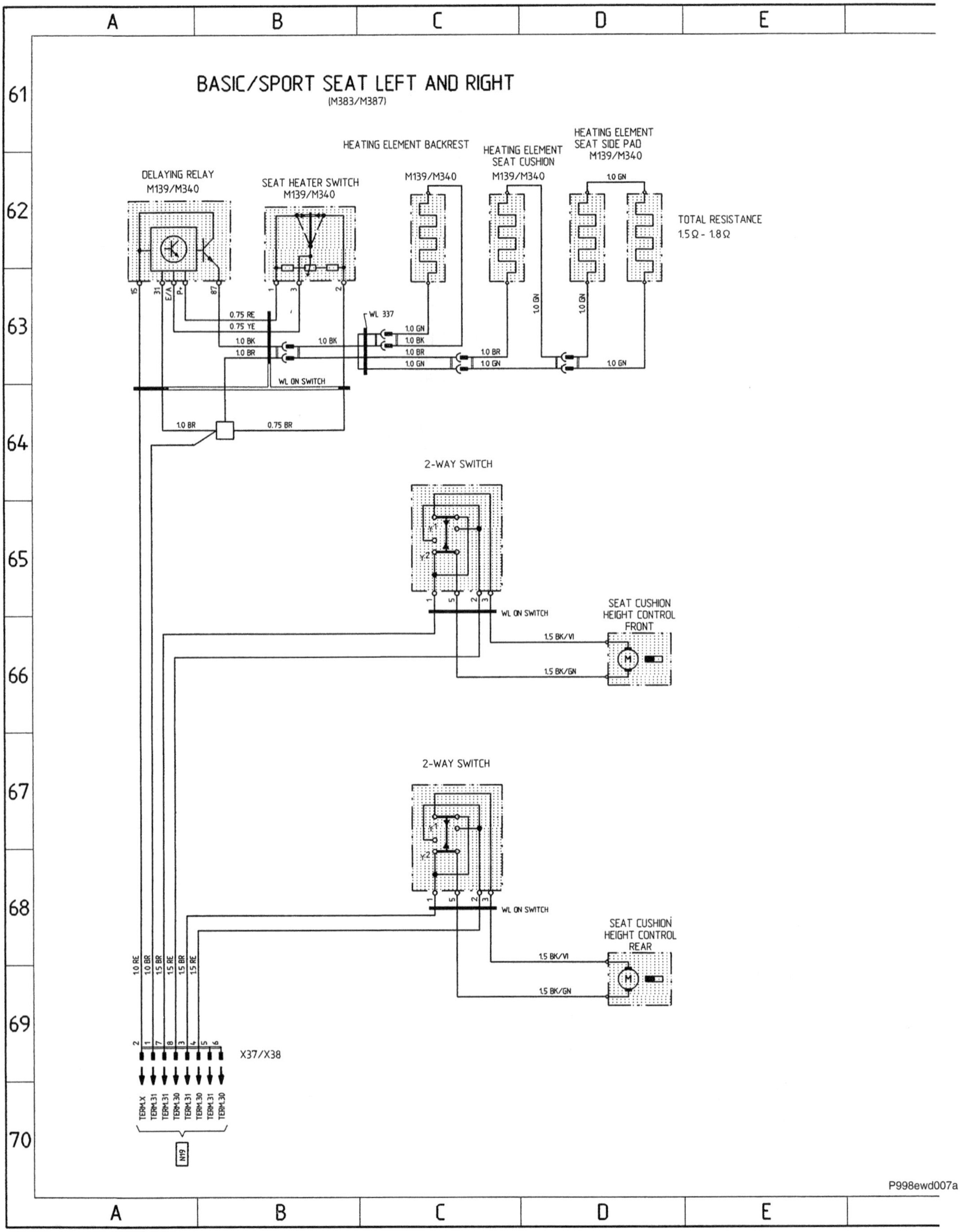

Seats (page 2 of 2)

Sheet 7 - 1995/2

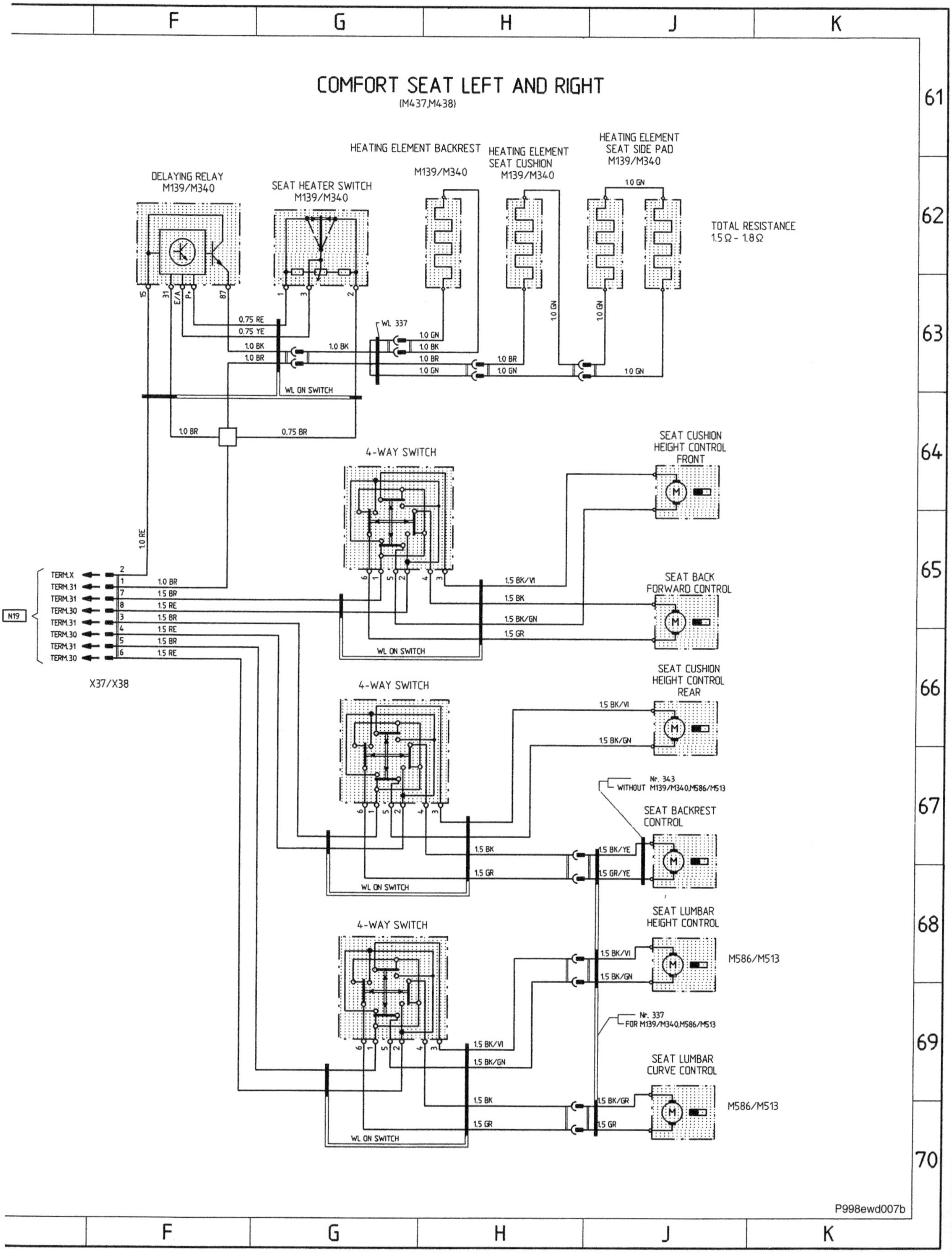

Sheet 8 - 1995/2 Rear spoiler, two-tone horns, wipe-and-wash cleaners, sunroof, convertible top (page 1 of 3)

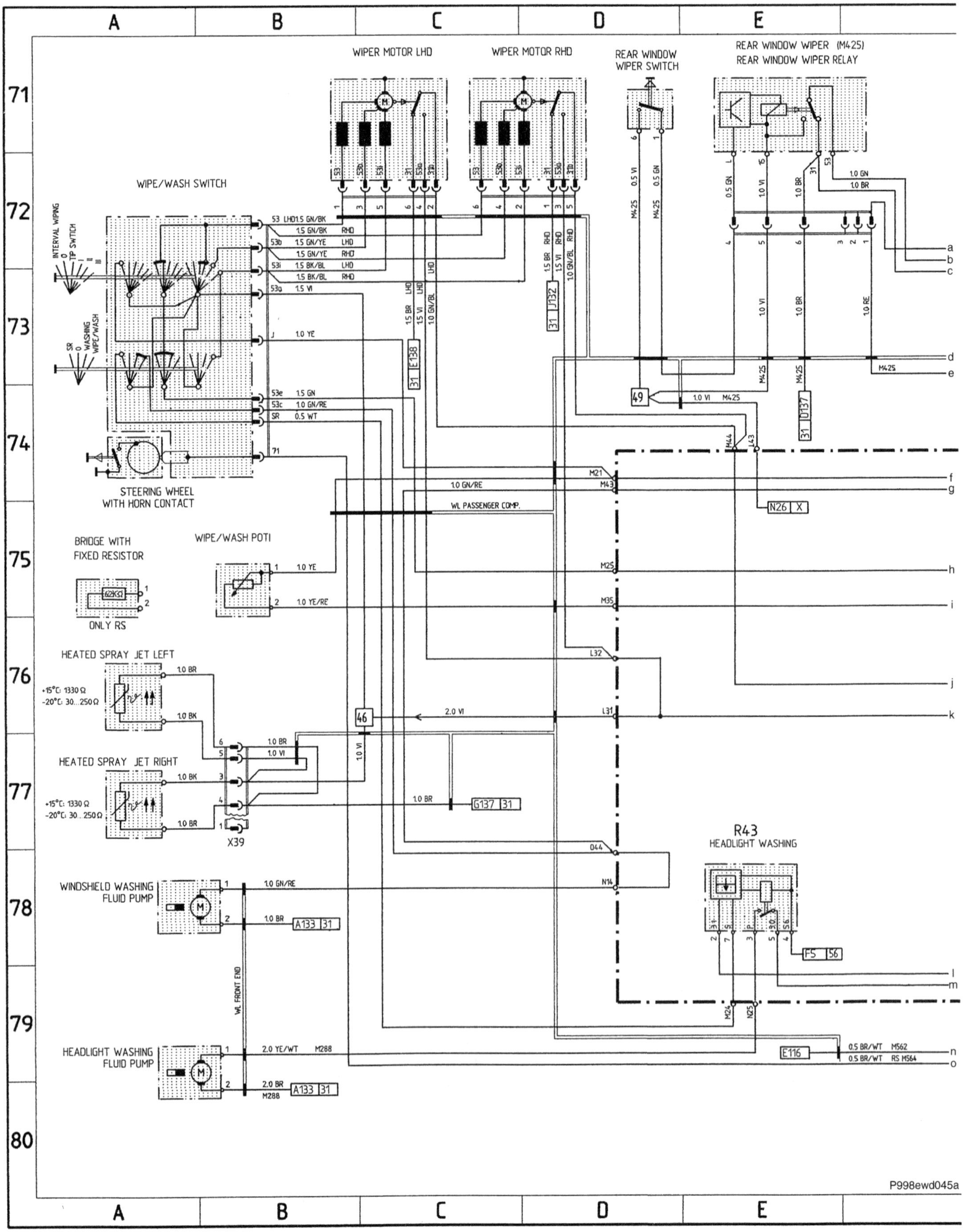

Rear spoiler, two-tone horns, wipe-and-wash cleaners, sunroof, convertible top (page 2 of 3) **Sheet 8 - 1995/2**

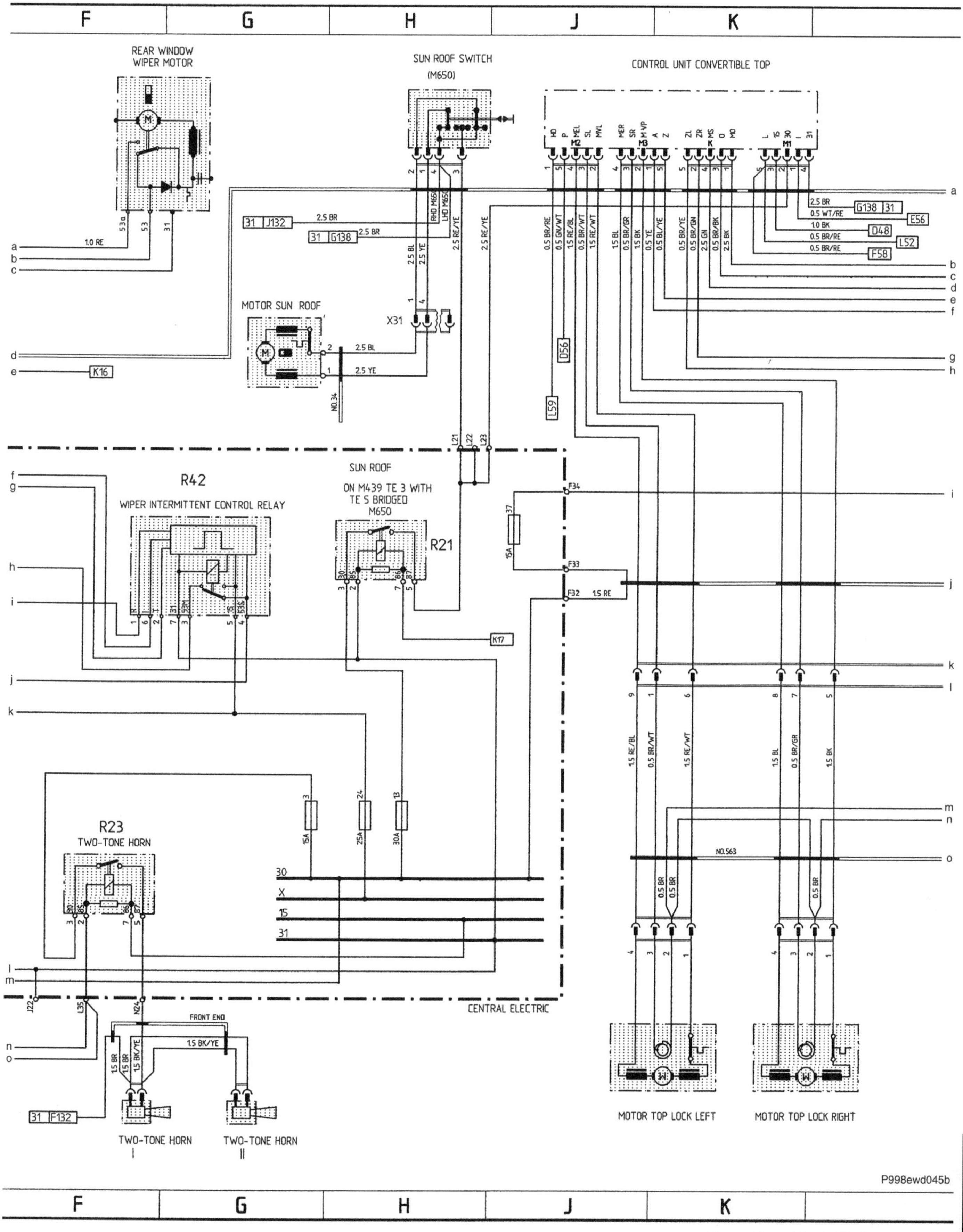

Sheet 8 - 1995/2 Rear spoiler, two-tone horns, wipe-and-wash cleaners, sunroof, convertible top (page 3 of 3)

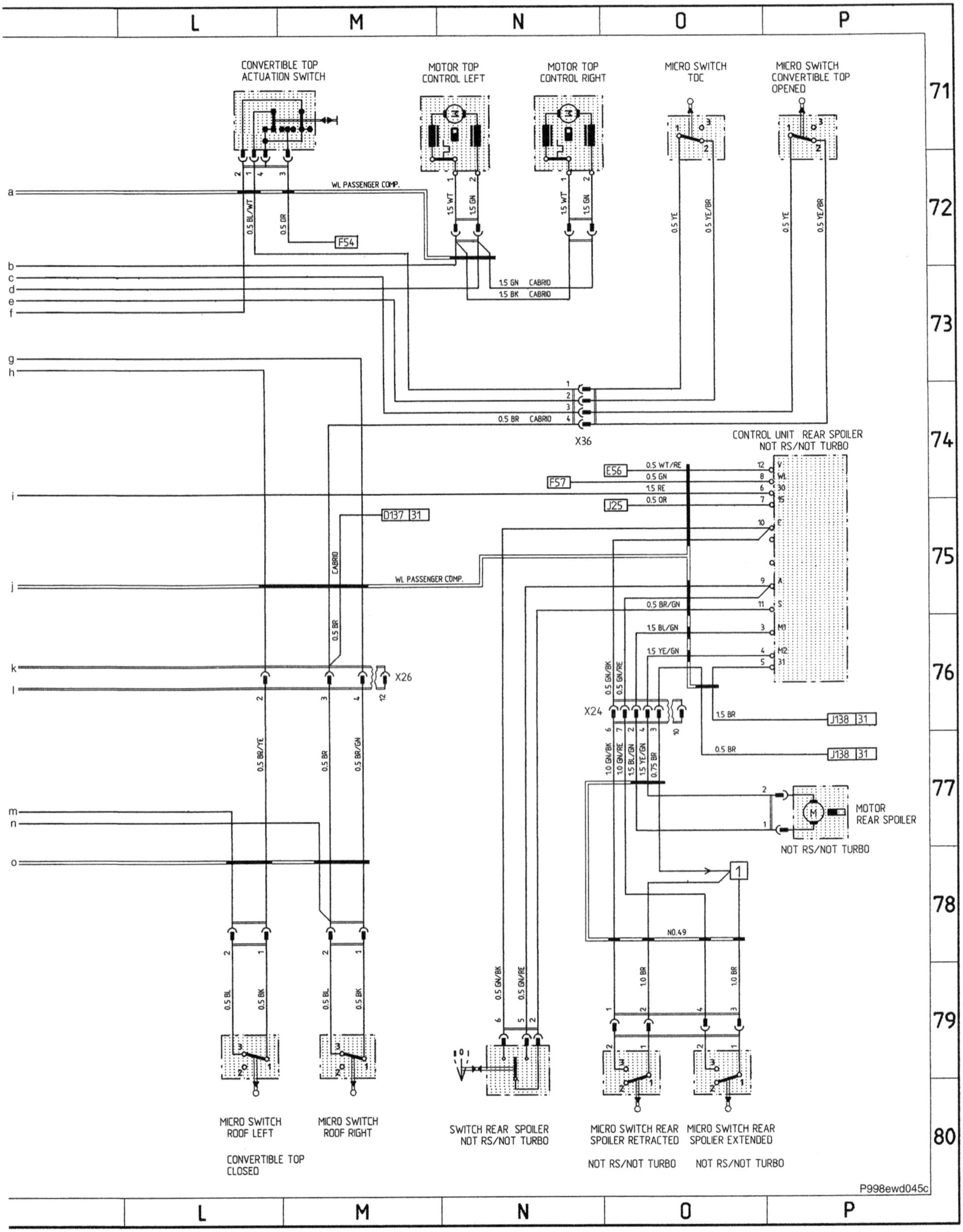

Radio, telephone (page 1 of 3)

Sheet 9 - 1995/2

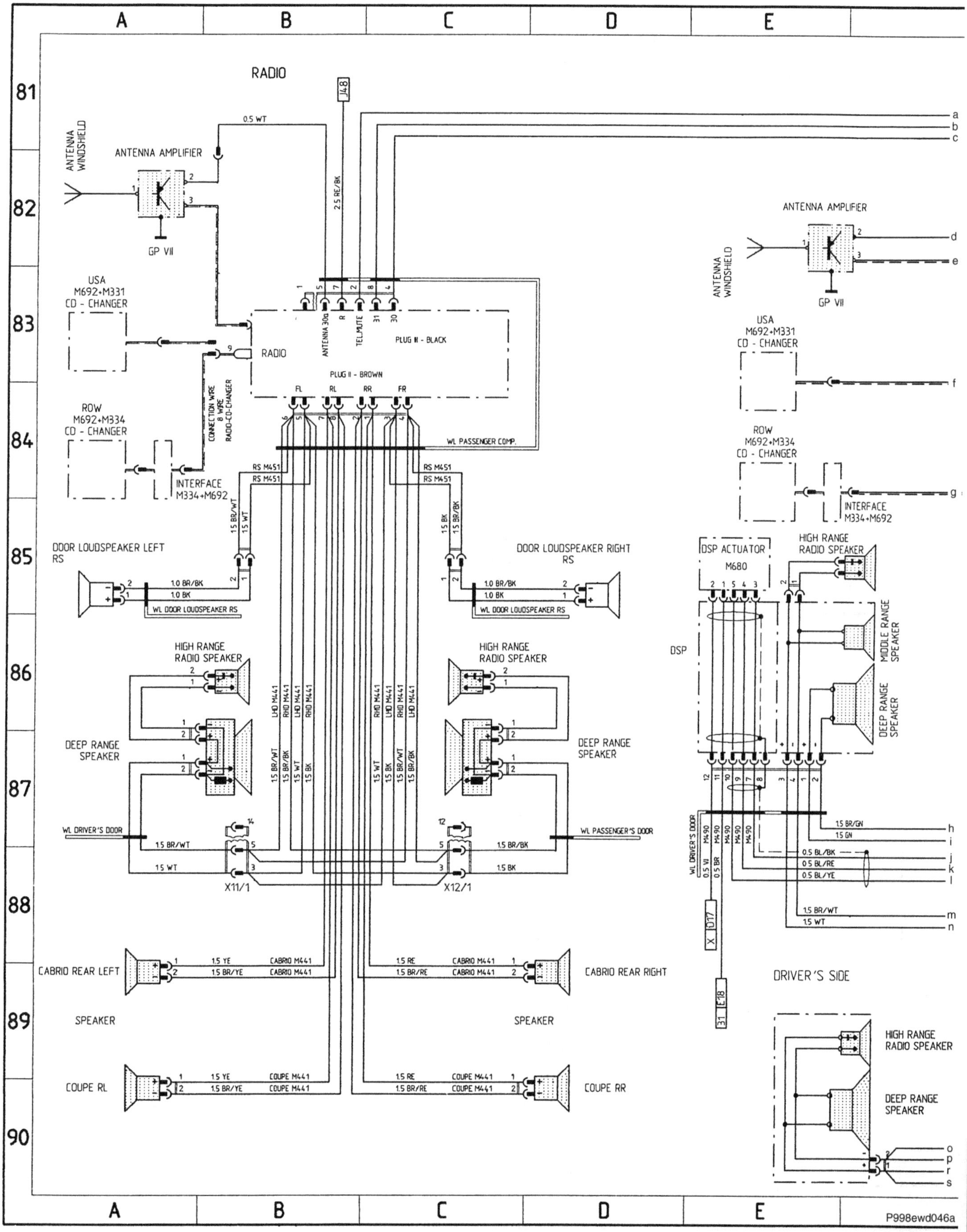

Sheet 9 - 1995/2

Radio, telephone (page 2 of 3)

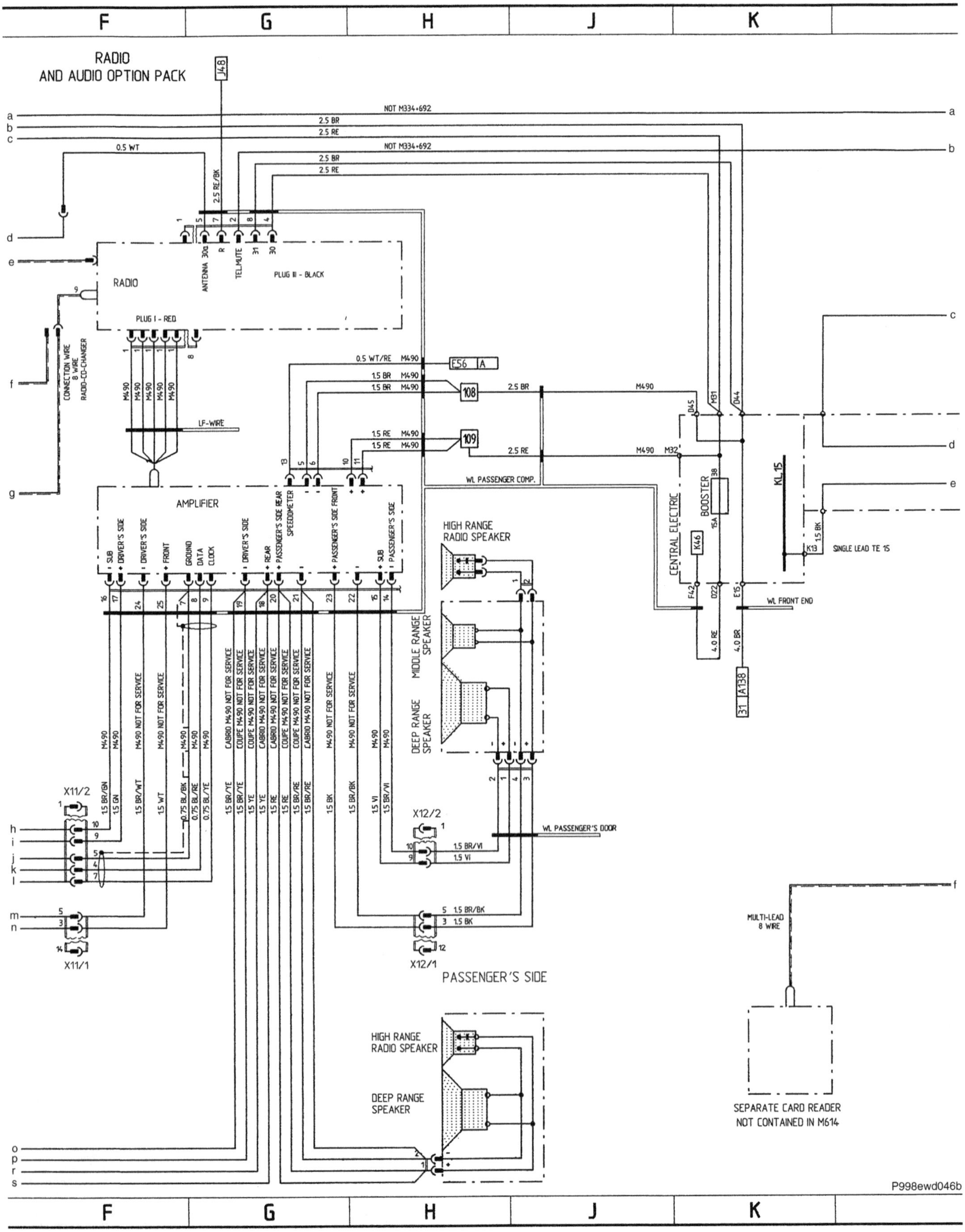

Radio, telephone (page 3 of 3)

Sheet 9 - 1995/2

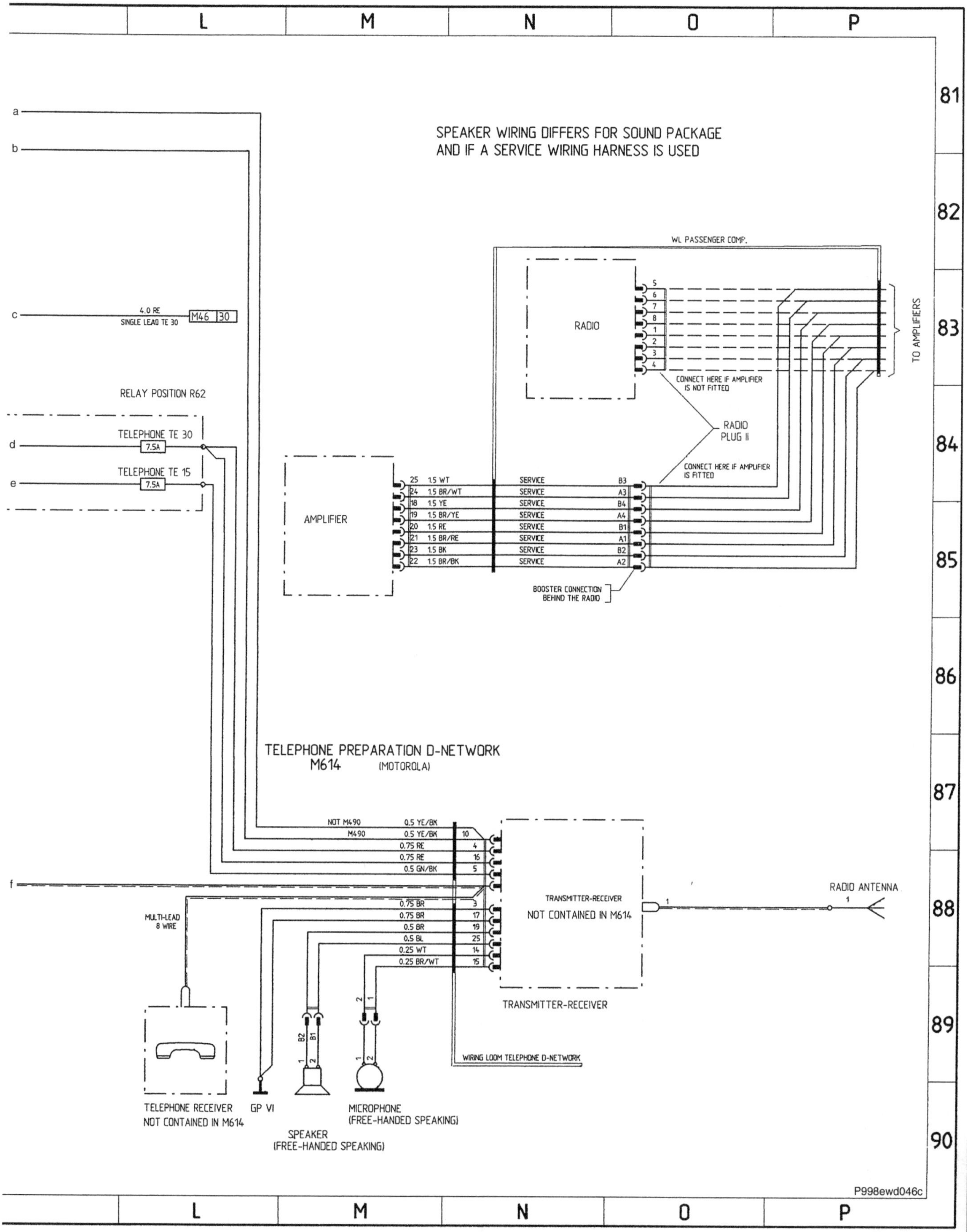

Sheet 10 - 1995/2

Foglight, rear foglight (page 1 of 1)

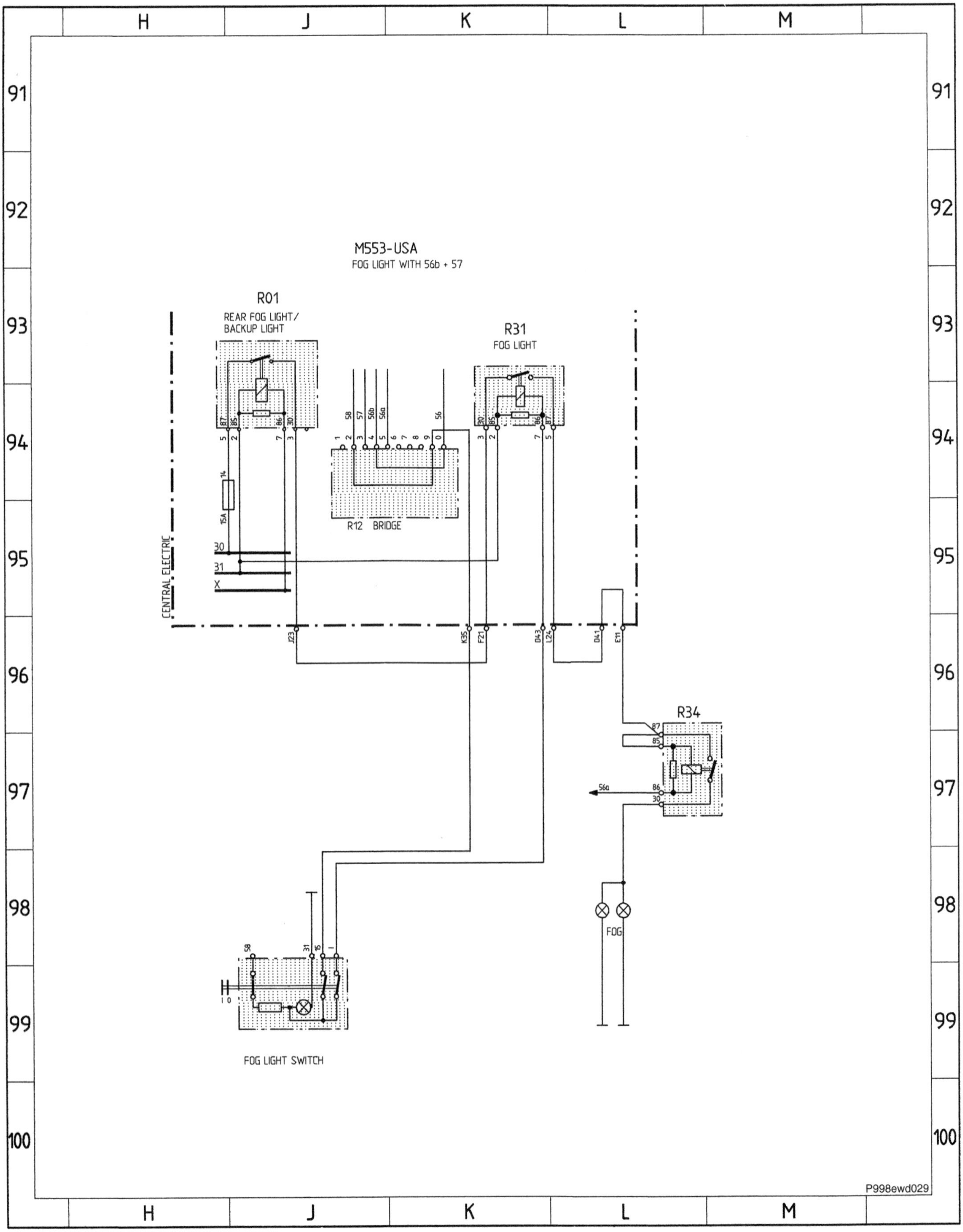

Anti-theft (option code M530) (page 1 of 1)

Sheet 11 - 1995/2

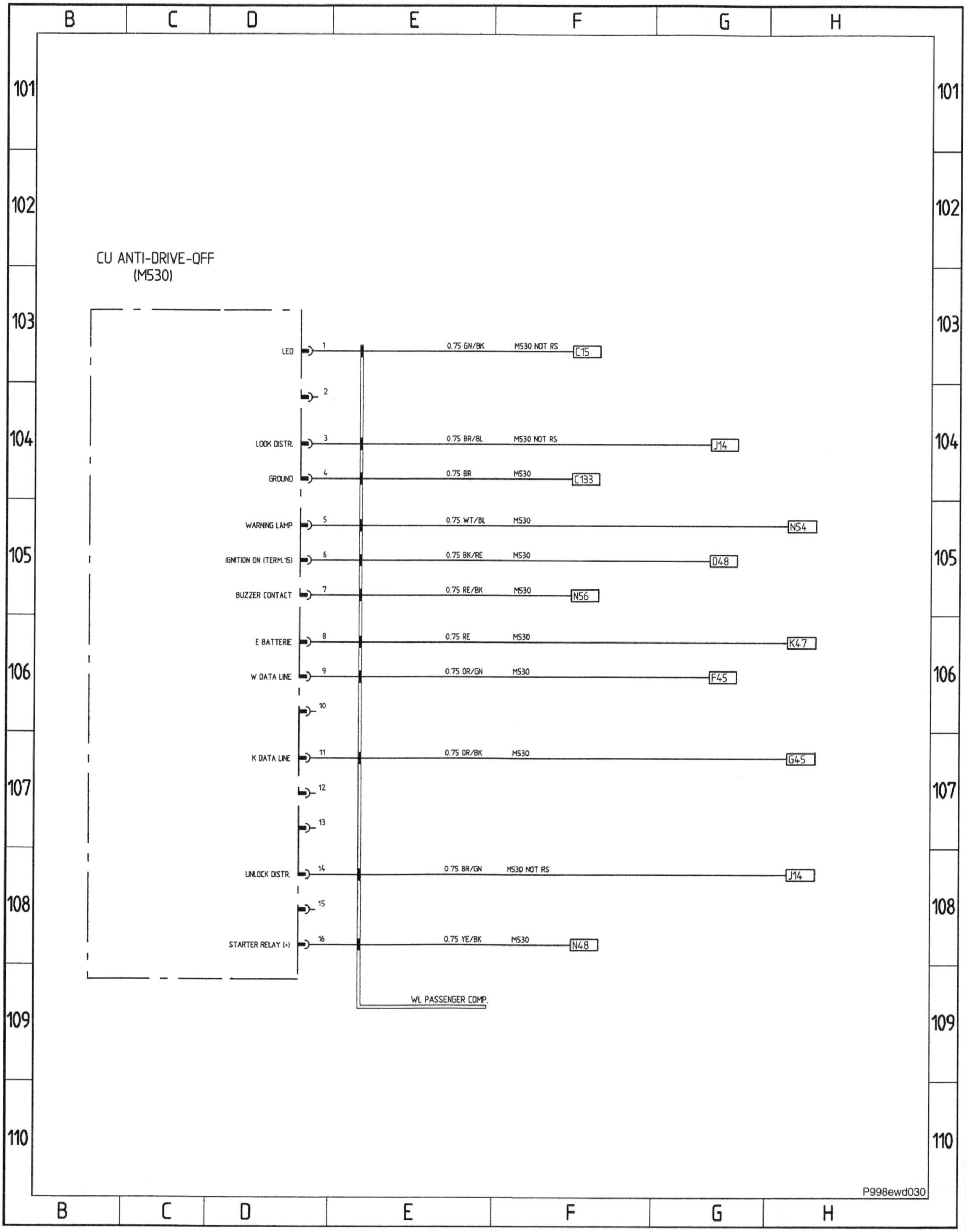

EWD

Sheet 12 - 1995/2

Airbag
(page 1 of 1)

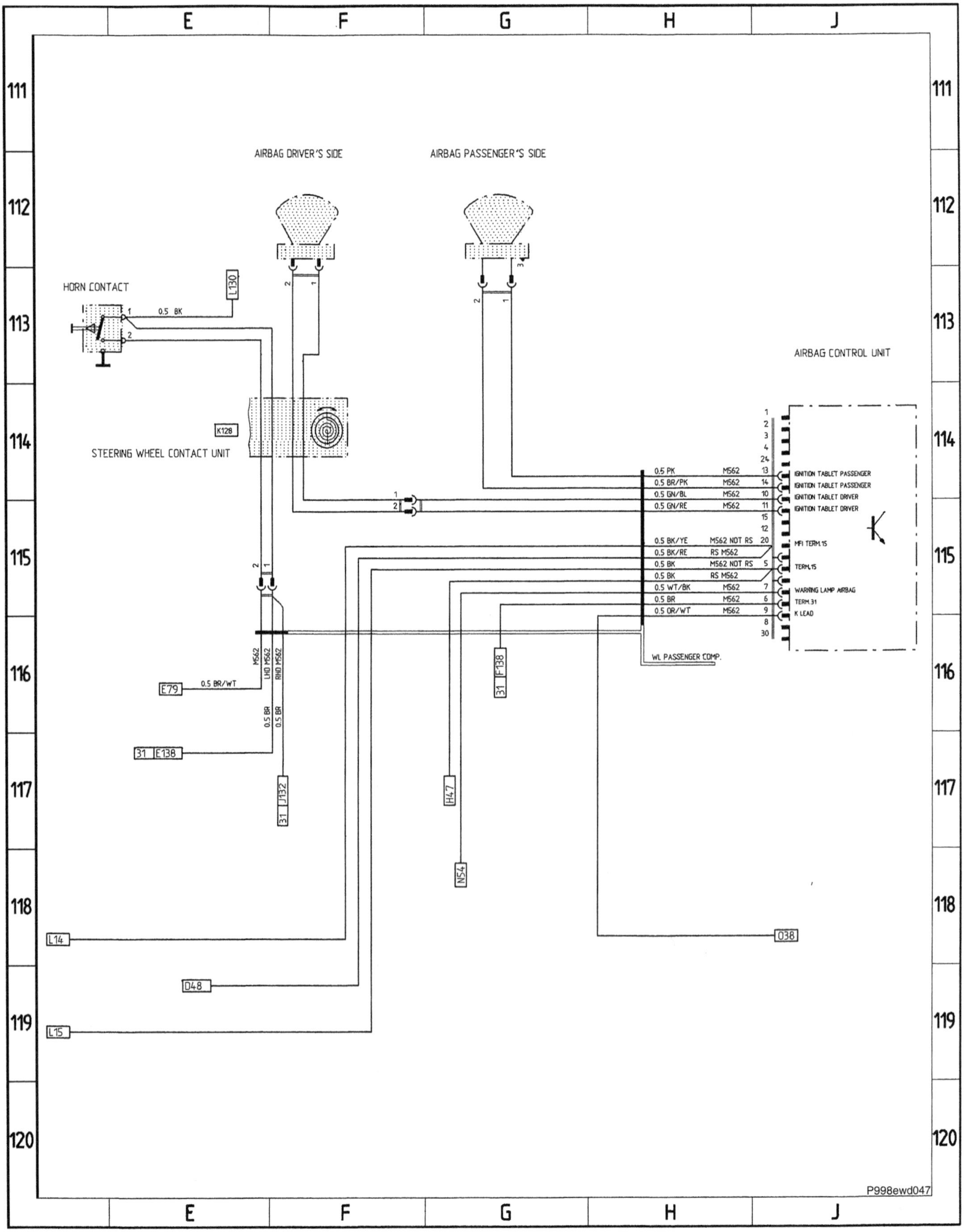

Tiptronic transmission
(page 1 of 3)

Sheet 13 - 1995/2

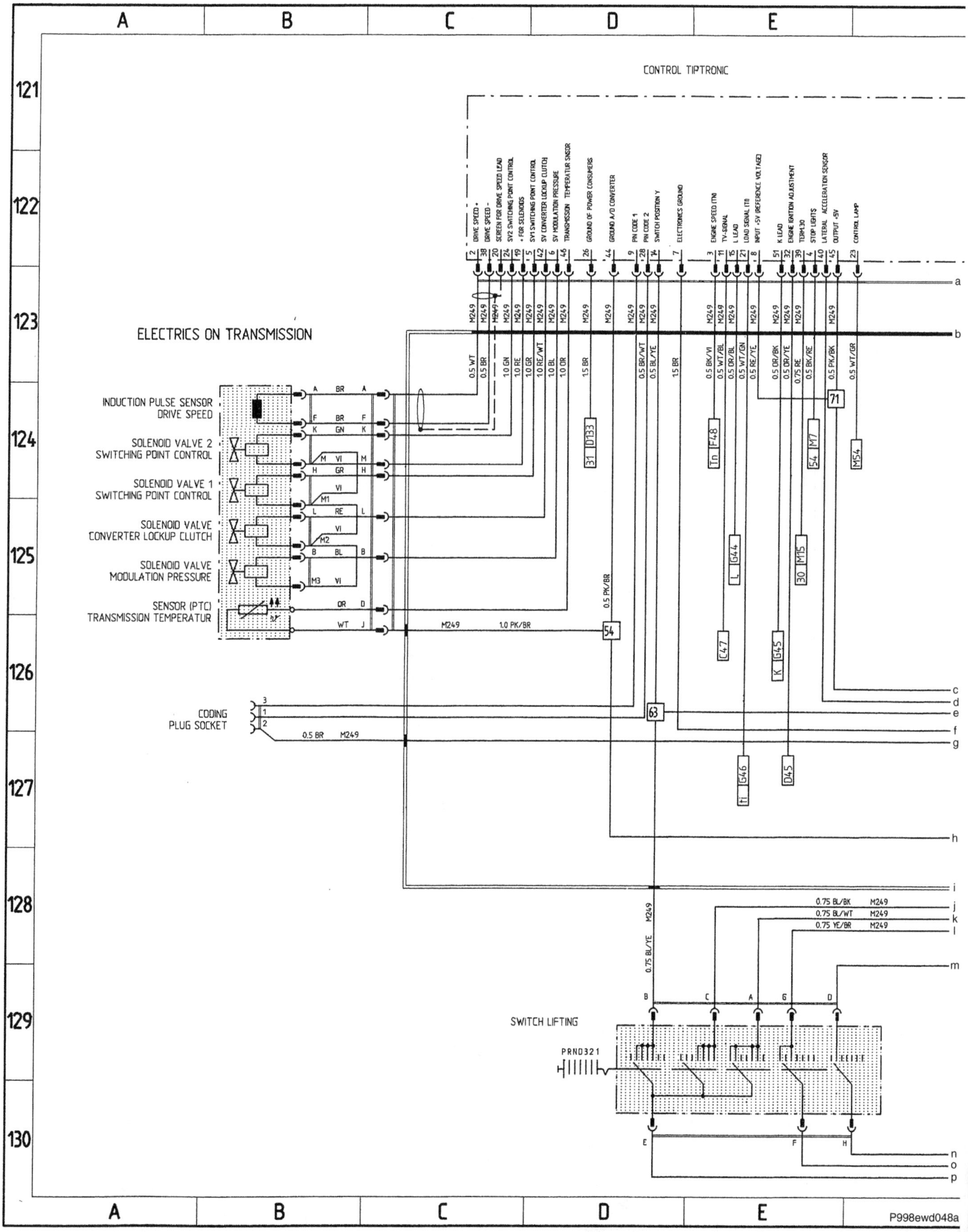

Sheet 13 - 1995/2

Tiptronic transmission (page 2 of 3)

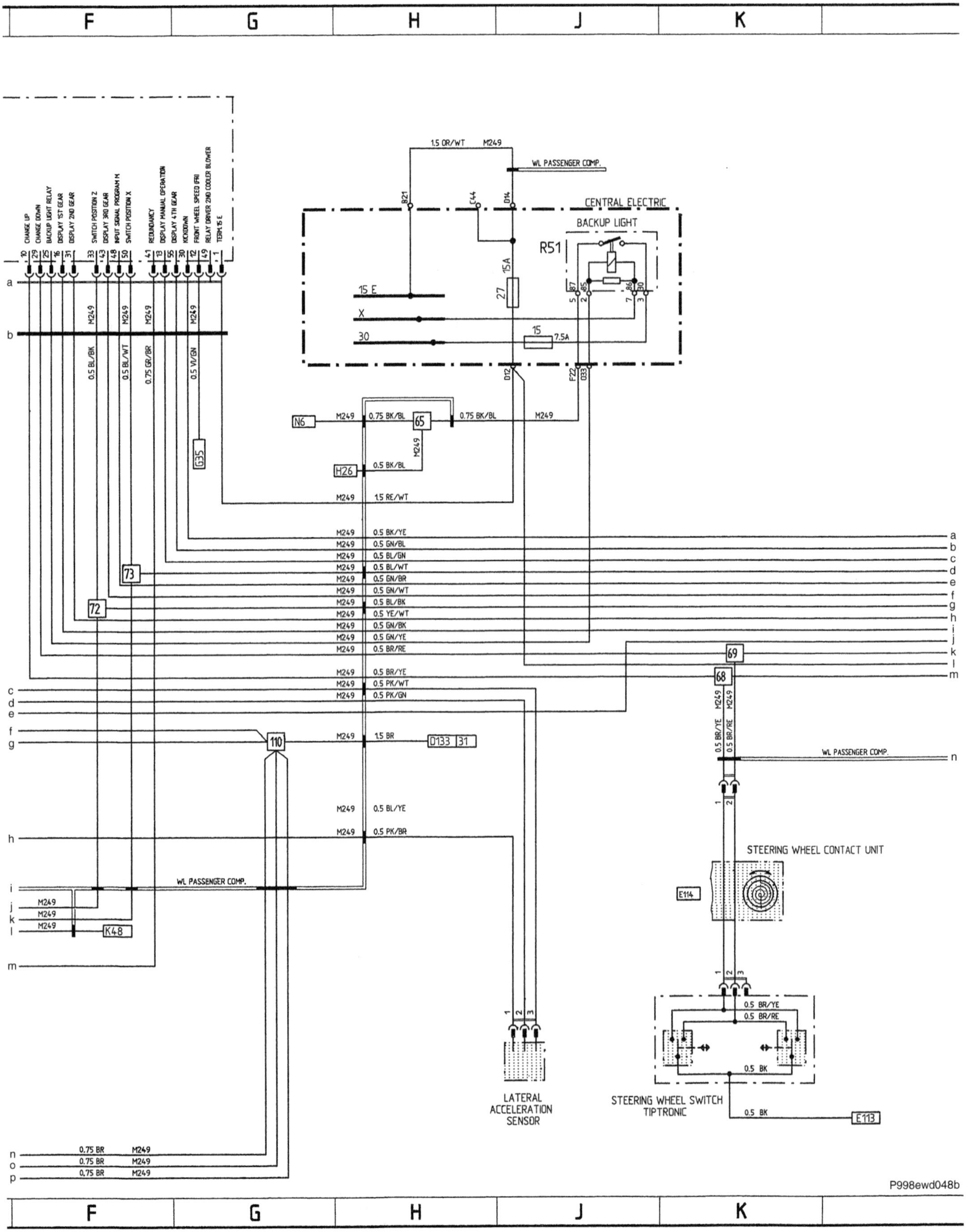

P998ewd048b

Tiptronic transmission (page 3 of 3)

Sheet 13 - 1995/2

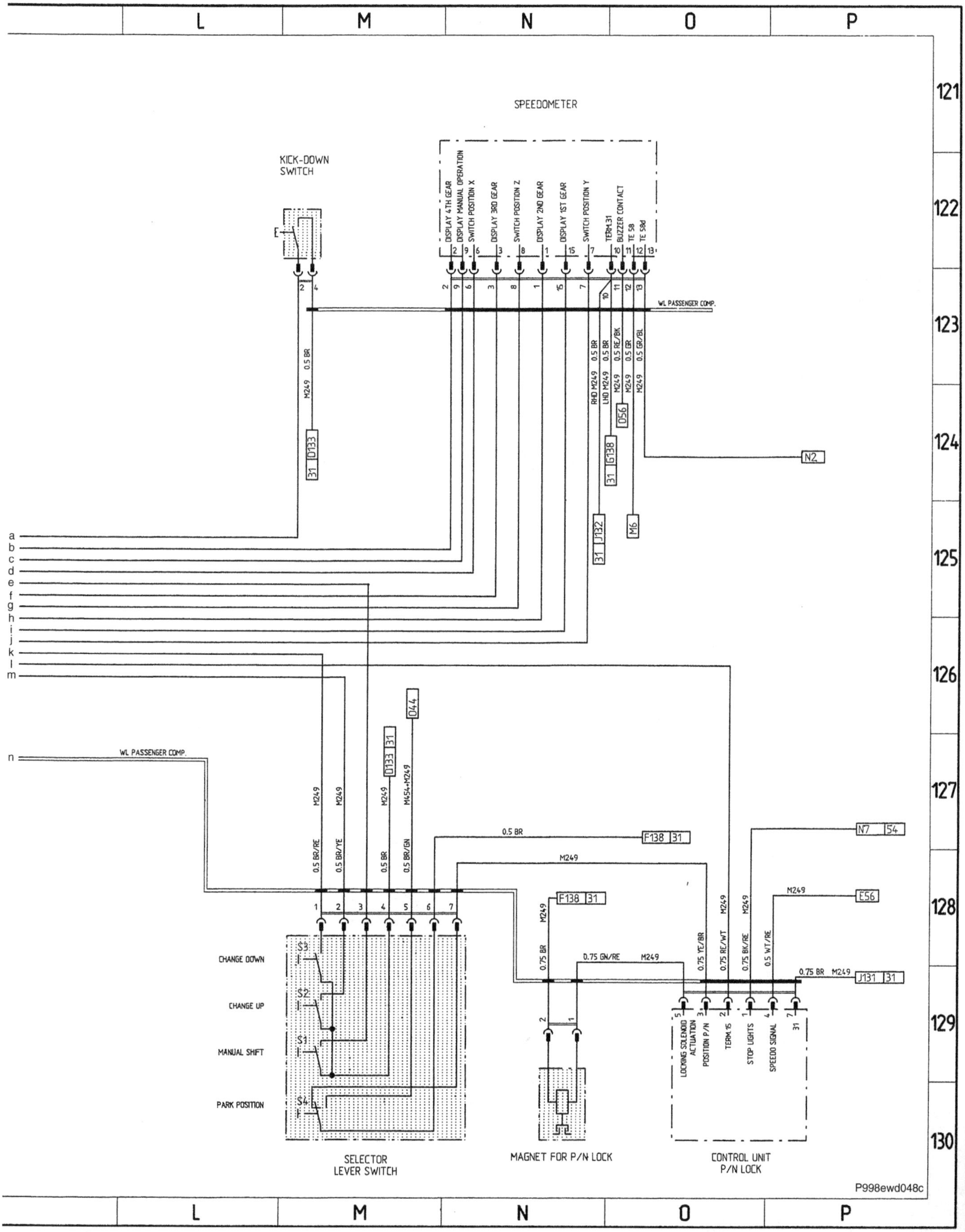

Sheet 14 - 1995/2

Ground points (page 1 of 3)

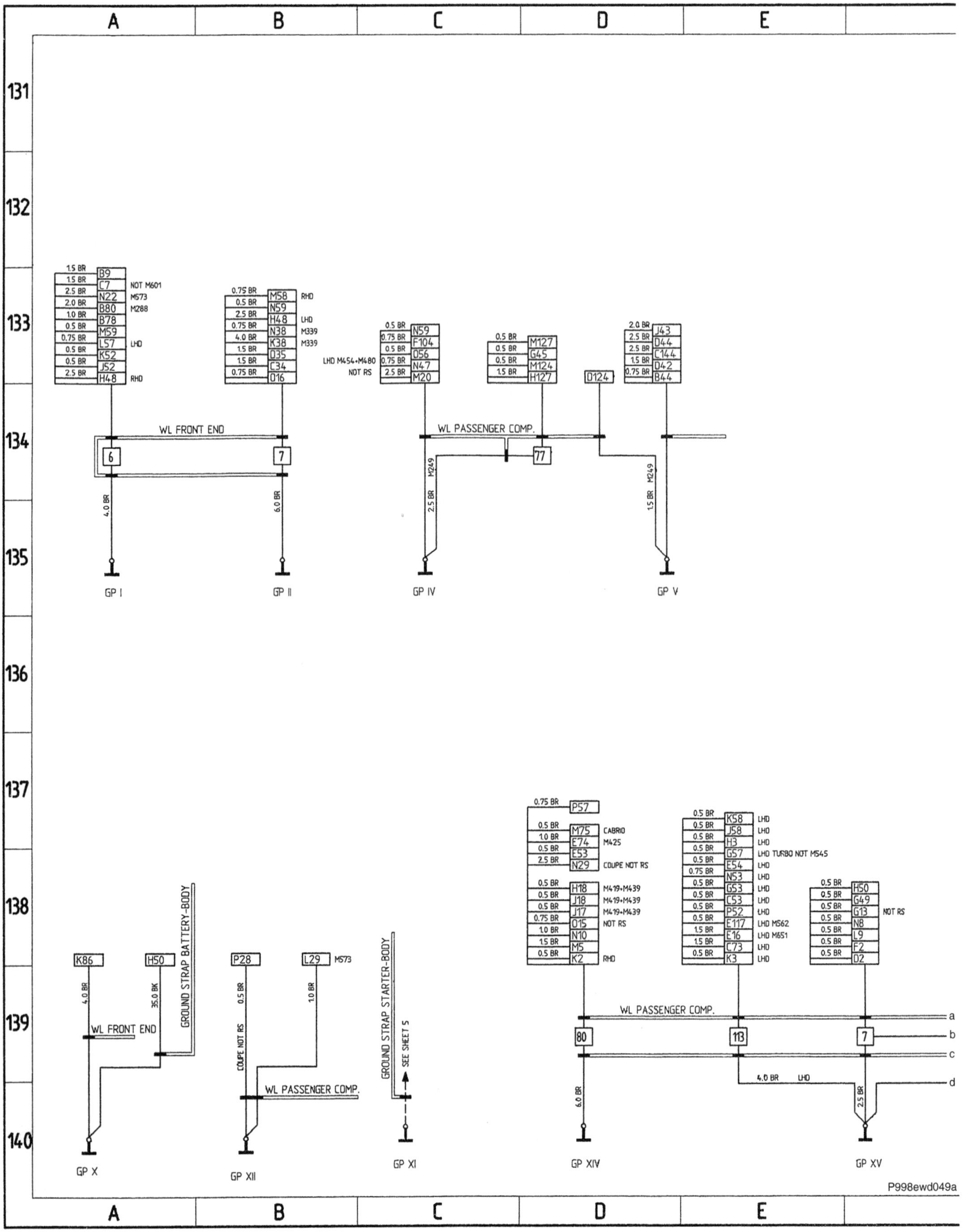

Ground points (page 2 of 3)

Sheet 14 - 1995/2

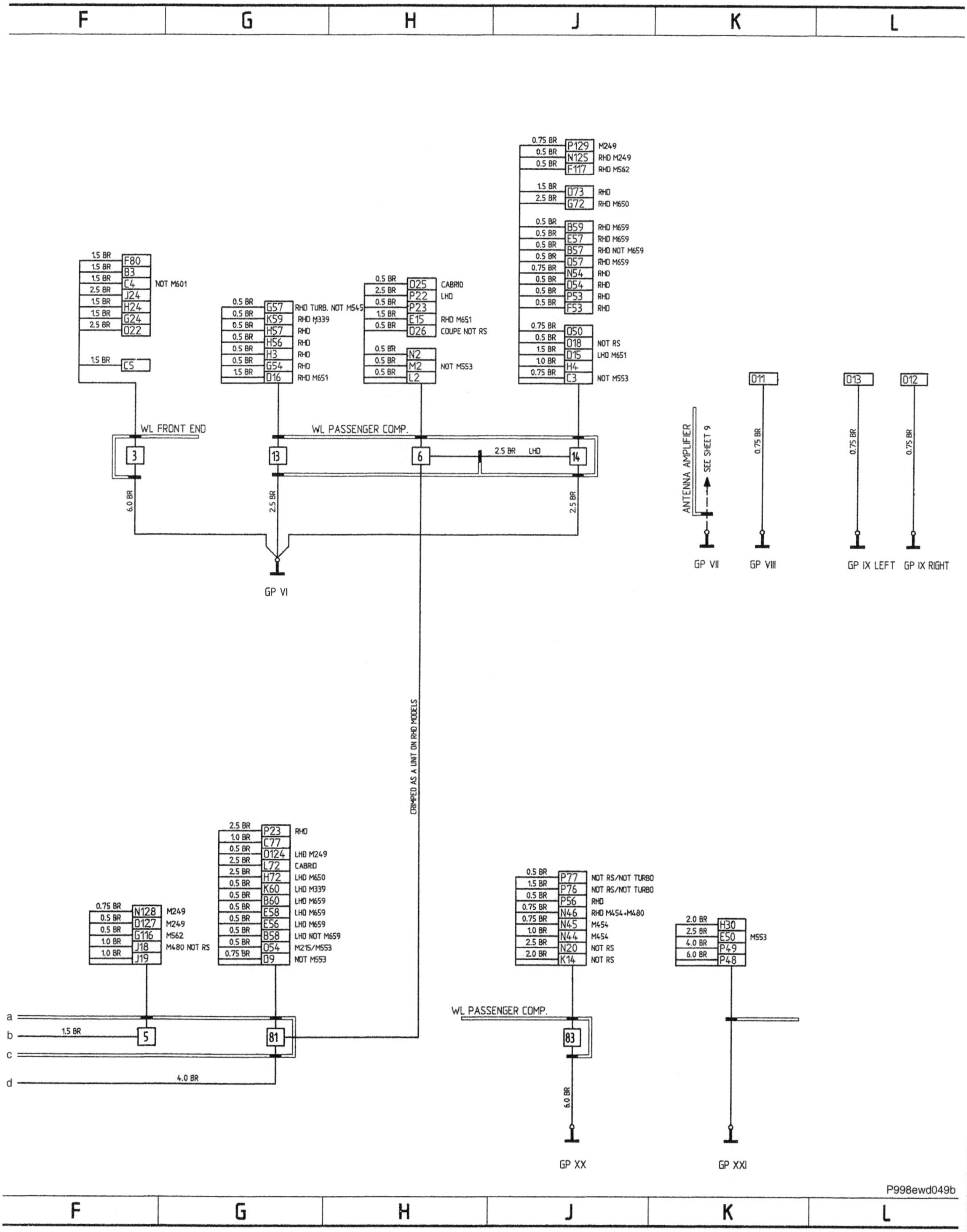

EWD

Sheet 14 - 1995/2

Ground points (page 3 of 3)

L	M	N	O	P

131

CODE	NOTE	POSITION IN VEHICLE LHD
GP I	LUGGAGE COMPARTMENT, WHEEL HOUSING LEFT	2cP
GP II	LUGGAGE COMPARTMENT, WHEEL HOUSING RIGHT	2cM
GP IV	GROUND OF POWER CONSUMERS, UNDER SEAT LEFT	11eP
GP V	ENGINE ELECTRONICS/TIPTRONIC	8eP
GP VI	CENTRAL ELECTRIC, ON SIDE MEMBER RIGHT	6bL
GP VII	ANTENNA AMPLIFIER ON SIDE MEMBER LEFT	7bL
GP VIII	INSIDE LIGHT CONVERTIBLE, ROOF CENTRE FRONT	9aN
GP IX	INSIDE LIGHT, ROOF RIGHT	11aL
GP IX	INSIDE LIGHT, ROOF LEFT	11aP
GP X	BATTERY, LUGGAGE COMPARTMENT LEFT	3eP
GP XI	GROUND STRAP (STARTER, TRANSMISSION, ENGINE, BODY)	14cM
GP XII	ENGINE COMPARTMENT REAR LEFT	15cO
GP XIV	ENGINE COMPARTMENT REAR RIGHT	15cL
GP XV	INSTRUMENTS, SIDE MEMBER LEFT	7bO
GP XX	SEAT CROSS MEMBER, UNDER SEAT RIGHT	8eL
GP XXI	GROUND OF POWER CONSUMERS, CAMSHAFT CASING CYL.3	15cP

132
133
134
135

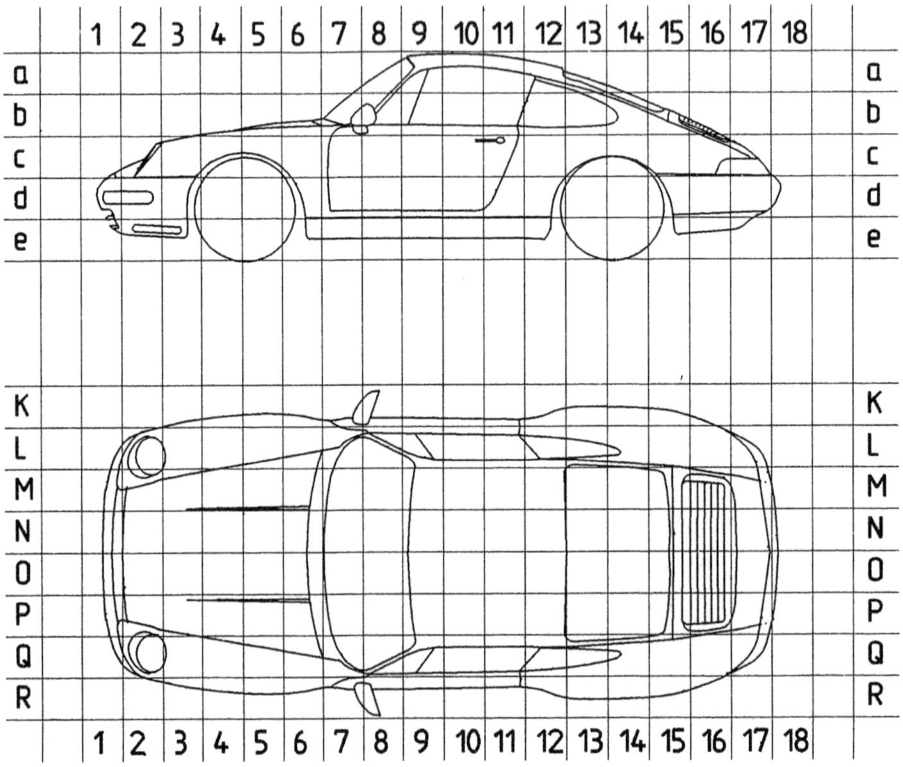

136
137
138
139
140

P998ewd049c

L	M	N	O	P

Construction components (page 1 of 3)

Sheet 15 - 1995/2

A | B | C | D | E | F

CONSTRUCTION COMPONENTS

DESIGNATION, FUNCTION	POSITION IN VEHICLE LHD	RHD	NOTE	FIELD IN WIRING DIAGRAM
ALARM HORN	13dP	13dP	REAR LH CROSSMEMBER	J20
ANTENNA AMPLIFIER	7bL	7bQ	IN LUGGAGE COMPARTMENT	A83,E82
ABS/ABD HYDRAULIC UNIT	3dP	3dP	IN LUGGAGE COMPARTMENT	C39
ABS RELAY	4cL	4cL	IN LUGGAGE COMPARTMENT RIGHT	D39
AMPLIFIER	9eM	9eM	UNDER THE SEAT RIGHT	F-H85, M85
BLOWER FINAL STAGE	6bN	6bN	IN LUGGAGE COMPARTMENT	F30
BLOWER ENGINE COMPARTMENT	16cO	16cO	ON ENGINE	H30
BLOWER MOTOR RIGHT	6cM	6cM	IN LUGGAGE COMPARTMENT	G30
BLOWER MOTOR LEFT	6cP	6cP	IN LUGGAGE COMPARTMENT	F30
BRAKE BOOSTER PUMP	1dO	1dO	IN LUGGAGE COMPARTMENT	L39
BACKUP LIGHT SWITCH	14dN	14dN	BY TRANSMISSION RIGHT	K/L1
CONDENSER BLOWER	2dQ	2dQ	IN WHEEL HOUSING FRONT LEFT	N21
CLUTCH SWITCH CRUISE CONTROL	6dP	6dN	BY CLUTCH PEDAL UNDER FLOOR PANEL	P46
CONTROL RELAY REAR WINDOW DEFOGGER	16cP	16cP	CARRIER PLATE IN ENGINE COMPARTMENT	O/P29/30
CONTROL UNIT ABS	2dM	2dM	IN LUGGAGE COMPARTMENT	A-J32
CONTROL UNIT AIRBAG	6cO	6cO	ON INTRUMENT PANEL	J114/116
CONTROL UNIT ALARM SYSTEM/CENTRAL LOCKING SYSTEM	9eM	9eM	UNDER THE SEAT RIGHT	J-M12
CONTROL UNIT TIPTRONIC TRANSMISSION	10dP	10dP	UNDER THE SEAT LEFT	C-G122
CONTROL UNIT CONVERTIBLE TOP	6cP	6cO	ON INTRUMENT PANEL	J/K71
CONTROL UNIT INSTRUMENT LIGHT	7bQ	7bR	IN LUGGAGE COMPARTMENT LEFT	K1
CONTROL UNIT REAR SPOILER	7cM	7cP	ON RH SIDE OF TRANSVERSE WALL	P74-76
CONTROL UNIT HEATING, AIR CONDITIONING	7cO	7cN	ON INTRUMENT PANEL	B-F21
CONTROL UNIT P/N LOCK	6cM	6cM	CENTRAL ELECTRIC	O/P129/130
CONTROL UNIT CRUISE CONTROL	9eN	9eN	UNDER THE SEAT RIGHT	P43/44
CHARGE AIR TEMPERATURE SENSOR	16cN	16cN	ON ENGINE	A142
CENTRAL ELECTRIC	6cM	6cM	IN LUGGAGE COMPARTMENT	—
CENTRAL INFORMER	7cP	7cM	ON FIREWALL	M32
DIAGNOSIS CONNECTION	7cL	7cQ	IN PASSENGER'S FOOTWELL	G50
DRIVE CRUISE CONTROL	15cO	15cO	ON ENGINE	N43/44
ENGINE SPEED SENSOR	15cP	15cP	ON ENGINE	A46
ENGINE TEMPERATURE SENSOR	15cP	15cP	ON ENGINE	A45
FLASHER	7dN	7dN	IN CENTRE CONSOLE	D1
FUEL PUMP	6eM	6eP	BEHIND THE BASE	H50
FUSES ENGINE COMPARTMENT	16cP	16cP	CARRIER PLATE IN ENGINE COMPARTMENT	L28
FUSE IN LUGGAGE COMPARTMENT	4cP	4cP	BELOW HYDRAULIC UNIT COVER	D37/M47
FUEL LEVEL SENSOR	5cM	5cP	IN FUEL TANK	K51
GONG RELAY	7bQ	—	LH SIDE IN FRONT OF INSTRUMENTS	P53/54
HALL SENSOR	17cO	17cO	ON DISTRIBUTOR	A46
HYDRAULIC UNIT ABS	3dP	3dP	IN LUGGAGE COMPARTMENT	A/B39
HEADLIGHT CLEANER PUMP	3cQ	3cQ	BY WASHING FLUID RESERVOIR IN WHEEL HOUSING FL	A79/80
HEADLIGHT CLEANER RELAY	6cM	6cM	CENTRAL ELECTRIC	E78
IDLE SPEED POSITIONER	15bN	15bN	ON ENGINE	H/J41
IGNITION CONTROL UNIT	11dP	11dP	UNDER THE SEAT LEFT	J41
IGNITION COIL I	17cP	17cP	ON ENGINE	L41
IGNITION COIL II	17cP	17cP	ON ENGINE	L41
KICK-DOWN SWITCH	6dO	6dM	BY ACCELERATOR PEDAL UNDER FLOOR PANEL	M122
KNOCK SENSOR CYL.1-3	15cP	15cP	ON ENGINE	A47
KNOCK SENSOR CYL.4-6	16cN	16cN	ON ENGINE	A47
LATERAL ACCELERATION SENSOR TIPTRONIC TRANSMISSION	7dN	7dN	IN CENTRE CONSOLE	J129
LIGHTS ON BUZZER	7cP	7cM	ON IGNITION LOCK	P56/57
MFI + DI RELAY	6cM	6cM	CENTRAL ELECTRIC	J44
MFI + DI CONTROL UNIT	9eP	9eP	UNDER THE SEAT LEFT	B-G41
MASS AIR FLOW SENSOR	16bO	16bO	ON ENGINE	A43
MICRO SWITCH ROOF RIGHT	9aN	9aN	AT CONVERTIBLE TOP FRAME	M79
MICRO SWITCH ROOF LEFT	9aO	9aO	AT CONVERTIBLE TOP FRAME	L79
MICRO SWITCH REAR SPOILER EXTENDED	16bP	16bP	ON TAILGATE	O79
MICRO SWITCH REAR SPOILER RETRACTED	16bP	16bP	ON TAILGATE	O79

P998ewd050a

A | B | C | D | E

EWD

Sheet 15 - 1995/2

Construction components (page 2 of 3)

DESIGNATION, FUNCTION	POSITION IN VEHICLE LHD	RHD	NOTE	FIELD IN WIRING DIAGRAM
MICRO SWITCH TDC	12cL	12cL	BEHIND SIDE PANEL COVERING	O71
MICROSWITCH FOR TOP CONTROL	12cL	12cL	BEHIND SIDE PANEL COVERING	P71
MOTOR SUN ROOF	12aN	12aN	IN ROOF	G73
MOTOR TOP CONTROL RIGHT	15cN	15cN	BEHIND REAR SEAT	N71
MOTOR TOP CONTROL LEFT	15cO	15cO	BEHIND REAR SEAT	N71
MOTOR TOP LOCK RIGHT	9aN	9aN	AT CONVERTIBLE TOP FRAME	K/L79
MOTOR TOP LOCK LEFT	9aO	9aO	AT CONVERTIBLE TOP FRAME	J/K79
NTC HEATER FAN	16cP	16cP	ON ENGINE	G30
NTC TEMPERATURE FOR INTAKE AIR	15cN	15cN	ON ENGINE	A42
OUTSIDE TEMP. SENSOR	2dP	2dP	FRONT LH BUMPER	A60
OXYGEN SENSORS CYL.1-3	16dP	16dP	ON ENGINE	F150
OXYGEN SENSORS CYL.4-6	16dM	16dM	ON ENGINE	D150
OIL PRESSURE SENSOR/SWITCH	14cN	14cN	ON ENGINE	A54
OIL COOLER BLOWER	2dL	2dL	IN WHEEL HOUSING FRONT RIGHT	O21
OIL LEVEL SENSOR	12cK	12cK	IN WHEEL HOUSING REAR RIGHT	E51
OIL TEMPERATURE SENSOR FOR INSTRUMENT	14cN	14cN	ON ENGINE	A53
OIL TEMPERATURE SENSOR FOR OIL COOLER BLOWER	2dL	2dL	IN WHEEL HOUSING FRONT RIGHT	J21
PRESSURE TRANSDUCER, CHARGE AIR PRESSURE	16bM	16bM	ON ENGINE	A56
PRESSURE SWITCH AIR CONDITIONING SYSTEM	6bO	6bO	IN LUGGAGE COMPARTMENT	E30
PRESSURE WARNING SWITCH	1dO	1dO	IN LUGGAGE COMPARTMENT	M/N39
REAR FOG LIGHT CUTOUT RELAY	17dN	17dN	BEHIND REAR PANEL	H108
REAR WINDOW WIPER RELAY	15bO	15bO	ON REAR WINDOW WIPER MOTOR	E71
RESISTOR BLOWER	16dP	16dP	IN ENGINE COMPARTMENT	J30
RESISTOR FOR BLOWER CONDENSER	2dQ	2dQ	IN WHEEL HOUSING FRONT LEFT	M21
RESISTOR OIL COOLER BLOWER	2dL	2dL	IN WHEEL HOUSING FRONT RIGHT	M21
RELAY CONDENSER BLOWER	6cM	6cM	CENTRAL ELECTRIC	L21
RELAY AC COMPRESSOR	16cP	16cP	CARRIER PLATE IN ENGINE COMPARTMENT	L30
RELAY TWO-TONE HORN	6cM	6cM	CENTRAL ELECTRIC	F78
RELAY POWER WINDOW	6cM	6cM	CENTRAL ELECTRIC	K18/19
RELAY BLOWER ENGINE COMPARTMENT	16cP	16cP	CARRIER PLATE IN ENGINE COMPARTMENT	J/K30
RELAY STARTER TIPTRONIC-TRANSMISSION	6cM	6cM	CENTRAL ELECTRIC	N50
RELAY BACKUP LIGHT	7bQ	7bQ	IN LUGGAGE COMPARTMENT LEFT	J122
RELAY THERMO SWITCH CATALYTIC CONVERTER	6cM	6cM	CENTRAL ELECTRIC	L/M51
RELAY TERM.X	6cM	6cM	CENTRAL ELECTRIC	J/K44
RELAY TERM.15 E	6cM	6cM	CENTRAL ELECTRIC	K/L44
RELAY RFL - BACKUP LIGHT	6cM	6cM	CENTRAL ELECTRIC	J/K7, A-M94
RELAY FRONT FOG LAMP	6cM	6cM	CENTRAL ELECTRIC	J7, C-094
RELAY OIL COOLER BLOWER	6cM	6cM	CENTRAL ELECTRIC	K21
RELAY BRAKE BOOSTER PUMP	4cL	4cL	IN LUGGAGE COMPARTMENT RIGHT	M39
RELAY SUN ROOF	6cM	6cM	CENTRAL ELECTRIC	H75
RELAY DAYTIME RUNNING LIGHT CANADA	6cM	6cM	CENTRAL ELECTRIC	D1
RELAY FOG LIGHT SWITCH OFF USA/JAPAN	6cM	6cM	CENTRAL ELECTRIC	H-P/94, F/G10
SECONDARY AIR PUMP RELAY	17cP	17cP	ON CARRIER ON LH SIDE IN ENGINE COMPARTMENT	A50
STOP LIGHT SWITCH	6dP	6dM	AT THE BRAKE PEDAL BELOW THE FLOOR PANEL	N9
SWITCH LIFTING	12dO	12dO	ON GEARBOX	D-F129
SHIFT VALVE RESONANCE FLAP	15bN	15bN	ON ENGINE	H41
STARTER	14dN	14dN	BY CLUTCH HOUSING RIGHT	P49/50
SECONDARY AIR PUMP	16cN	16cN	ON ENGINE	D/E50
THROTTLE VALVE SENSOR	15bN	15bN	ON THROTTLE VALVE ASSEMBLY	A44
TIMING VALVE	16cM	16cM	ON ENGINE	L150
TANK VENTING VALVE	16cP	16cP	ON ENGINE	K41
TWO-TONE HORN I	3cL	3cL	IN WHEEL HOUSING FRONT RIGHT	F80
TWO-TONE HORN II	3cL	3cL	IN WHEEL HOUSING FRONT RIGHT	G80
V-BELT CONTROL	17dO	17dO	ON ENGINE REAR MIDDLE	A55
WINDSHIELD WASHING FLUID PUMP	3cQ	3cQ	BY WASHING FLUID RESERVOIR IN WHEEL HOUSING FL	A78
WIPER INTERMITTENT CONTROL RELAY	6cM	6cM	CENTRAL ELECTRIC	F/G75

P998ewd050b

Construction components (page 3 of 3)

Sheet 15 - 1995/2

	L	M	N	O	P

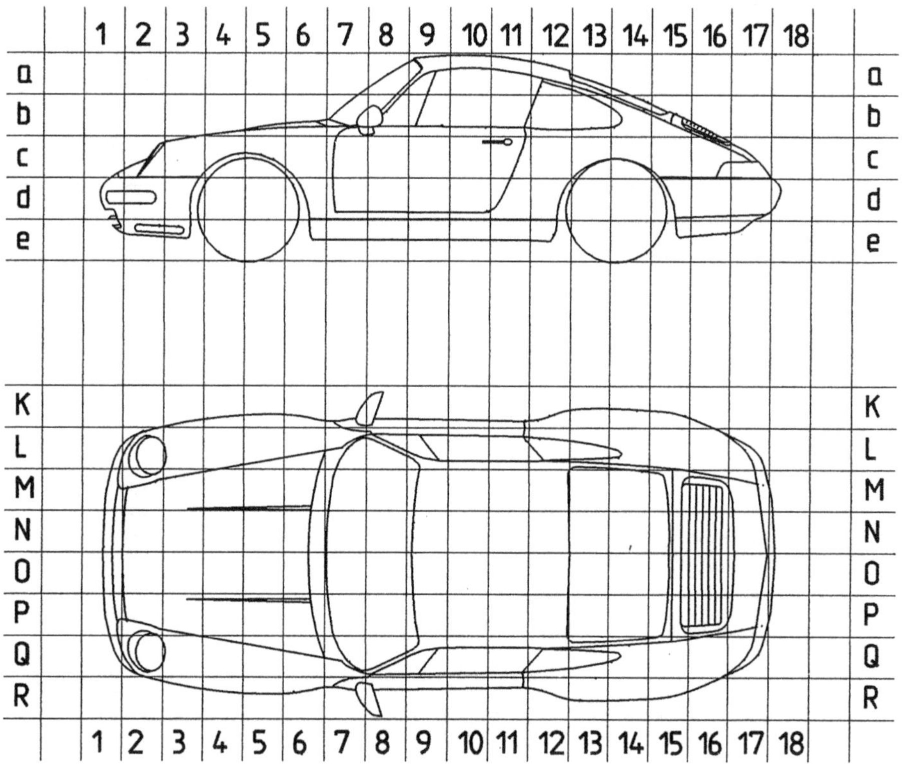

P998ewd050c

	L	M	N	O	P

Sheet 16 - 1995/2

Connectors and abbreviations (page 1 of 2)

PLUG CONNECTIONS

CODE	NUMBER OF PINS	DESIGNATION, FUNCTION	POSITION IN VEHICLE LHD	RHD	NOTE	FIELD
X 1/1	10	CONNECTION WL PASSENGER COMPARTMENT/ WL FRONT END	6cL	6cL	IN LUGGAGE COMPARTMENT RIGHT	G/H34,K123
X 1/2	10	CONNECTION WL PASSENGER COMPARTMENT/ WL FRONT END	6cL	6cL	IN LUGGAGE COMPARTMENT RIGHT	C9,E36,G48,B58/59
X 3	14	CONNECTION WL PASSENGER COMPARTMENT/ WL MOTOR	16cP	16cP	CARRIER PLATE IN ENGINE COMPARTMENT	K/L3,J27/28,048/49,B53/54
X4/1	14	CONNECTION WL PASSENGER COMPARTMENT/ WL MFI + DI	9eP	9eP	UNDER THE SEAT LEFT	C/D45/46,B55,L143
X4/2	14	CONNECTION WL PASSENGER COMPARTMENT/ WL MFI + DI	9eP	9eP	UNDER THE SEAT LEFT	F45/46,B55,G/H148
X11/1	14	CONNECTION WL PASSENGER COMPARTMENT/ WL DOOR DRIVER'S SIDE	6cQ	6cL	UNDER THE INSTRUMENT PANEL DIVER'S SIDE	E/F16/17,B87/88, F88
X11/2	10	CONNECTION WL PASSENGER COMPARTMENT/ WL DOOR DRIVER'S SIDE	6cQ	6cL	UNDER THE INSTRUMENT PANEL DIVER'S SIDE	C16 ,F87/88
X12/1	12	CONNECTION WL PASSENGER COMPARTMENT/ WL DOOR PASSENGER'S SIDE	6cL	6cQ	UNDER THE INSTRUMENT PANEL PASSENGER'S SIDE	E/F13,C87/88 ,H88
X12/2	10	CONNECTION WL PASSENGER COMPARTMENT/ WL DOOR PASSENGER'S SIDE	6cL	6cQ	UNDER THE INSTRUMENT PANEL PASSENGER'S SIDE	C14,H87/88
X20	14	CONNECTION WL PASSENGER COMPARTMENT/ WL HEATER, AIR CONDITIONER	6cM	6cM	UNDER INSTRUMENT PANEL RIGHT	G25/26
X24	10	CONNECTION WL PASSENGER COMPARTMENT/ WL TAILGATE	16cL	16cL	BOX RH SIDE OF ENGINE COMPARTMENT	010,013/14,076
X25	10	CONNECTION WL PASSENGER COMPARTMENT/ WL TAIL LIGHTS	16cL	16cL	BOX RH SIDE OF ENGINE COMPARTMENT	N4-6
X26	12	CONNECTION WL PASSENGER COMPARTMENT/ WL CONVERTIBLE TOP	12cL	12cL	BEHIND SIDE PANEL COVERING	J-M76
X28	6	CONNECTION WL PASSENGER COMPARTMENT/ WL IGNITION LOCK	6cP	6cM	ON STEERING COLUMN	L48/49
X29	2	CONNECTION WL PASSENGER COMPARTMENT/ OIL TANK SENSOR	16cL	16cL	BOX RH SIDE OF ENGINE COMPARTMENT	E52
X31	4	CONNECTION WL PASSENGER COMPARTMENT/ WL INSIDE LIGHTS COUPE	15bM	15bM	BELOW THE PARCEL TRAY	012/13,H73
X32	2	CONNECTION WL PASSENGER COMPARTMENT/ WL INSIDE LIGHTS CONVERTIBLE	6cQ	6cQ	UNDER INSTRUMENT PANEL LEFT	N/012
X33	2	CONNECTION WL PASSENGER COMPARTMENT/ WL SIDE TURN SIGNAL LEFT	7bQ	7bQ	IN LUGGAGE COMPARTMENT ON SIDE MEMBER LEFT	D/E9
X34	2	CONNECTION WL PASSENGER COMPARTMENT/ WL SIDE TURN SIGNAL RIGHT	7bL	7bL	IN LUGGAGE COMPARTMENT ON SIDE MEMBER RIGHT	C2
X36	4	CONNECTION WL PASSENGER COMPARTMENT/ SWITCH LIFTING CONVERTIBLE TOP	12cL	12cL	BEHIND SIDE PANEL COVERING	N74
X37	8	CONNECTION WL PASSENGER COMPARTMENT/ WL SEAT LEFT	9eP	9eN	UNDER THE DRIVER SEAT	N19,A70,E65/66
X38	8	CONNECTION WL PASSENGER COMPARTMENT/ WL SEAT RIGHT	9eN	9eP	UNDER THE PASSEGER SEAT	N19,A70,E65/66
X39	6	CONNECTION WL PASSENGER COMPARTMENT/ WL LUGGAGE COMPARTMENT HOOD	7bL	7bL	ON RH SIDE OF LUGGAGE COMP. LID HINGE	016,B77
X60	21	CONNECTION WL MOTOR/ WL MFI + DI	16cP	16cP	CARRIER PLATE IN ENGINE COMPARTMENT	B46,E48,K42,B55,M142,M148
X65	6	CONNECTION WL DOOR DRIVER'S SIDE/ WL OUTSIDE MIRROR LEFT	8cR	8cK	DRIVER'S DOOR	B18/19
X66	6	CONNECTION WL DOOR PASSENGER'S SIDE/ WL OUTSIDE MIRROR RIGHT	8cK	8cR	PASSENGER'S DOOR	B13
X69	14	CONNECTION WL FRONT END/ WL HEATER, AIR CONDITIONER	6cM	6cM	IN LUGGAGE COMPARTMENT RIGHT	G23/24

M - NUMBERS

M 113	CANADA VERSION (DAYTIME RUNNING LIGHT)	M 487	NORWAY
M 139	ADJUSTABLE SEAT HEATING, LEFT SEAT	M 490	AUDIO OPTION PACK
M 193	JAPAN VERSION	M 513	LUMBAR SUPPORT SEAT RIGHT
M 215	SAUDI-ARABIA VERSION	M 530	CODE LOCK
M 224	AUTOMATIC BRAKE DIFFERENTIAL	M 545	LARGE TANK
M 249	AUTOMATIC TRANSMISSION (TIPTRONIC)	M 553	USA / CANADA VERSION
M 339	FWD	M 562	AIRBAG
M 340	ADJUSTABLE SEAT HEATING, RIGHT SEAT	M 564	WITHOUT AIRBAG
M 383	SPORT SEAT LEFT ELECTRIC HEIGHT ADJUSTMENT	M 573	AIR CONDITIONING SYSTEM
M 387	SPORT SEAT RIGHT ELECTRIC HEIGHT ADJUSTMENT	M 576	WITHOUT REAR FOG LIGHT
M 419	REAR PACKAGE TRAY	M 586	LUMBAR SUPPORT SEAT LEFT
M 425	REAR WINDOW WIPER	M 601	LITRONIC
M 437	COMFORT SEAT LEFT	M 602	HIGH MOUNTED STOP LIGHT
M 438	COMFORT SEAT RIGHT	M 605	HEADLIGHT VERTICAL AIM CONTROL
M 439	ELECTRIC CONVERTIBLE TOP	M 614	TELEPHONE PREPARATION
M 441	RADIO PREPARATION	M 650	POWER SUNROOF
M 451	LIMITED RADIO PREPARATION	M 651	POWER WINDOWS
M 454	CRUISE CONTROL	M 659	ON-BOARD COMPUTER
M 480	TRANSMISSION	M 680	DIGITAL SOUND PACKAGE

CONNECTION ---- Z999 99

FIELD INDICATION

TERMINAL ID

P	IGNITION SWITCH POSITION 0	56	LIGHT SWITCH POSITION II
X	IGNITION SWITCH POSITION I+II	56a/56b	HIGH BEAM/LOW BEAM
15	IGNITION SWITCH POSITION II+III	57	LIGHT SWITCH POSITION I
R	IGNITION SWITCH POSITION I+II+III	58	LIGHT SWITCH POSITION I+II
K	DIAGNOSIS K	58R/58L	SIDE MARKER LIGHT RIGHT/LEFT
L	DIAGNOSIS L	58d	INSTRUMENT LIGHTS
30	PERMANENT POSITIVE	Ti	INJECTION SIGNAL
31	GROUND	Tn	ROTATIONAL SPEED SIGNAL
54	STOP LIGHT	61	GENERATOR D+

P998ewd051a

Connectors and abbreviations (page 2 of 2)

Sheet 16 - 1995/2

F	G	H	J	K	L

ABBREVIATIONS

CODE	MEANING	CODE	MEANING	CODE	MEANING
ABD	AUTOMATIC BRAKE DIFFERENTIAL	FOG	FOG LIGHT	RDK	TIRE PRESSURE CONTROL
ABS	ANTILOCK BRAKING SYSTEM	GP	GROUND POINT	ROW	REST OF WORLD
AC	AIR CONDITIONING SYSTEM	HCS	HEADLIGHT CLEANING SYSTEM	RHD	RIGHT-HAND DRIVE
ADL	ADDITIONAL DRIVING LIGHTS	HF	HIGH FREQUENCY	RFL	REAR FOG LIGHT
BPAU	BRAKE BOOSTER	HFM	HOT-FILM AIR MASS METER	SA	SAUDI ARABIA
BL	BACKUP LIGHT	HVAC	HEADLIGHT VERTICAL AIM CONTROL	SAFETY LAMP	SAFETY LAMP
CL	CLOSE	INJ. VALVE	FUEL INJECTOR	SERVICE	SERVICE
CLS	CENTRAL LOCKING SYSTEM	INS.C	INSTRUMENT CLUSTER	SCS	COMBINED STEERING COLUMN SWITCH
CU	CONTROL UNIT	IC	IGNITION CIRCUIT	SS	SPEED SENSOR
CWL	CENTRAL WARN. LIGHT	LED	LIGHT-EMITTING DIODE	THROTTLE	THROTTLE VALVE
CE	CENTRAL ELECTRIC	LHD	LEFT-HAND DRIVE	TVS	THROTTLE VALVE SENSOR
CP	CONNECTING POINT	LF	LOW FREQUENCY	TE	TERMINAL
CLOCK	CLOCK SIGNAL	MFI + DI	DIGITAL ENGINE ELECTRONICS	TEL.MUTE	RADIO MUTING CONTROL
CO	CARBON MONOXIDE	M5.2	MOTRONIC 5.2	TI	INJECTION TIME
DATA	DATA SIGNAL	NO	NUMBER	TN	SPEED
DEF	DEFROST	NTC	NEGATIVE TEMPERATURE COEFFICIENT	TURBO	911 TURBO
DSP	DIGITAL SOUND PACKAGE	OP	OPEN	TRP	TWIN-COIL ROTARY POSITIONER
DP	DISCONNECTING POINT	PIN	PIN	USA	USA
DI	ELECTRIC IGNITION SYSTEM WITH KNOCK CONTROL	POT	POTENTIOMETER	WL	WIRING LOOM
ESO	ENGINE-SPEED SENSOR OUTPUT	PSD	PORSCHE LIMITED SLIP DIFFERENTIAL	WP	WELD POINT
ESS	ENGINE SPEED SENSOR	PWM	PULSE WIDTH MODULATION	WW	WORLDWIDE
ETC	ELECTRICAL TRANSMISSION CONTROL	PL	PLUG		
FA	FRONT AXLE	RA	REAR AXLE		
FL	FRONT LEFT	RL	REAR LEFT		
FR	FRONT RIGHT	RR	REAR RIGHT		

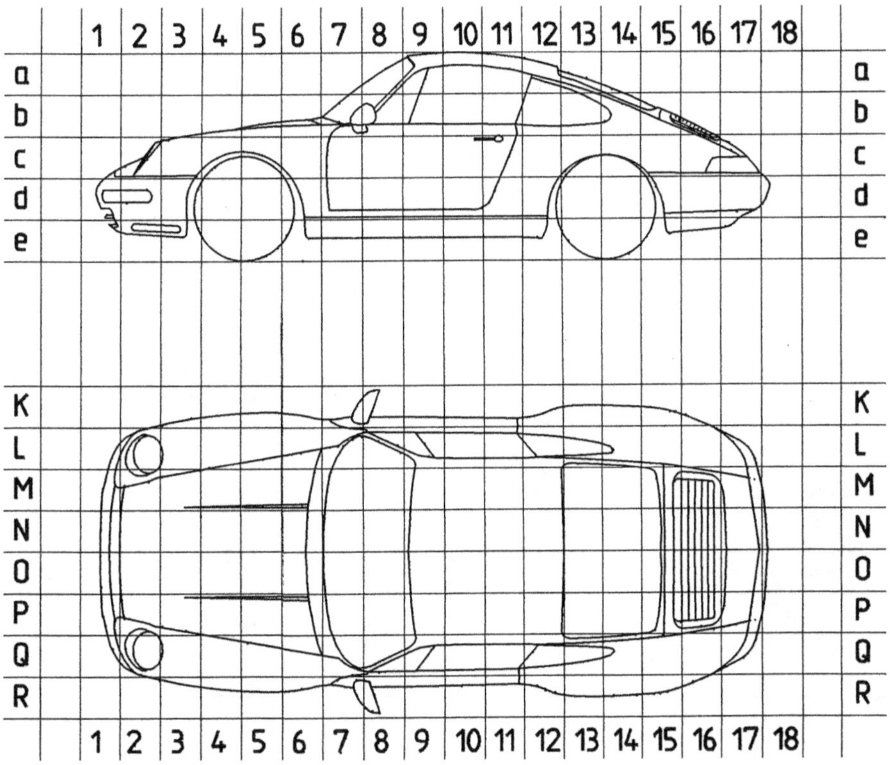

P998ewd051b

F	G	H	J	K	L

EWD

Sheet 17 - 1995/2

Central electric (page 1 of 3)

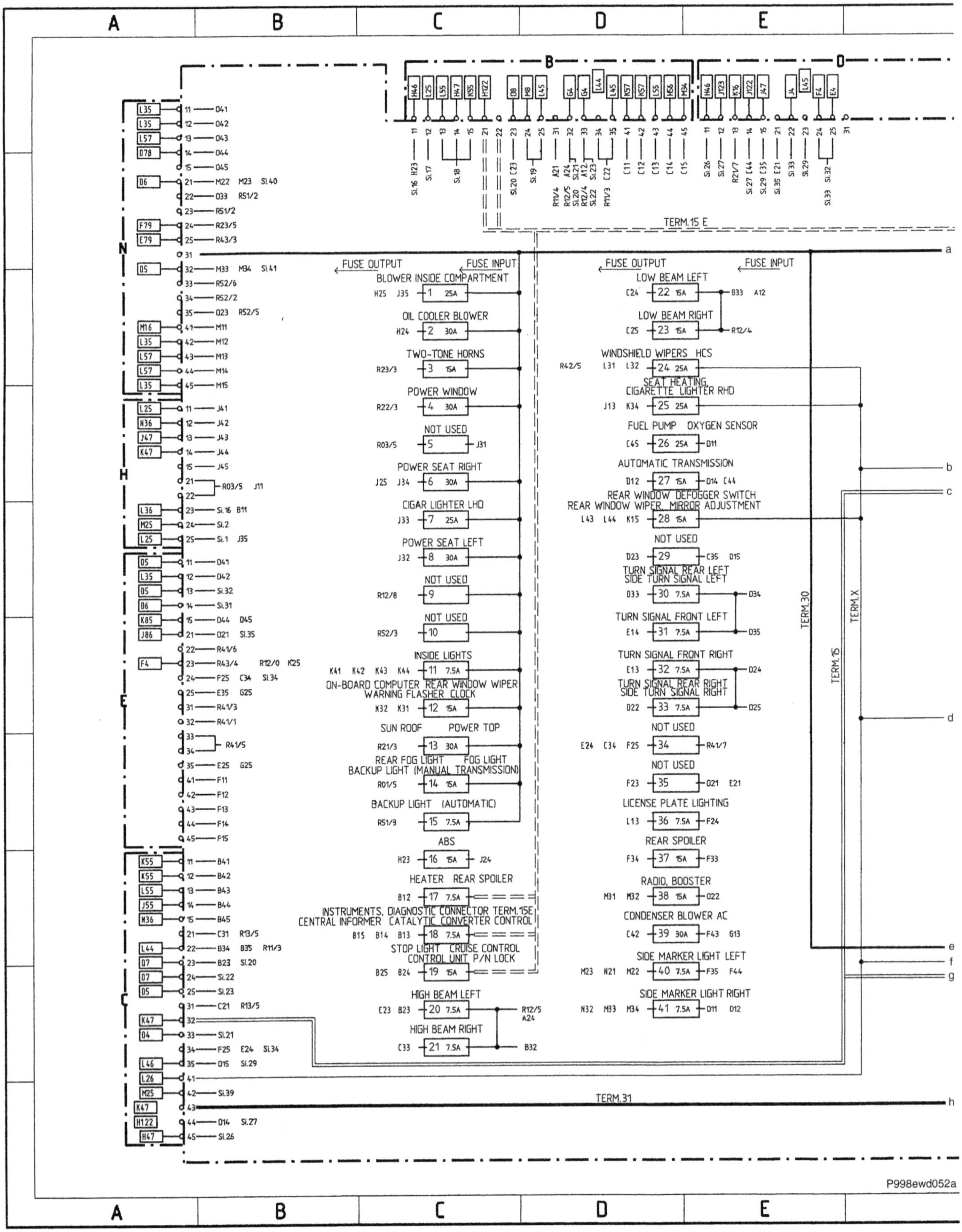

P998ewd052a

Central electric (page 2 of 3)

Sheet 17 - 1995/2

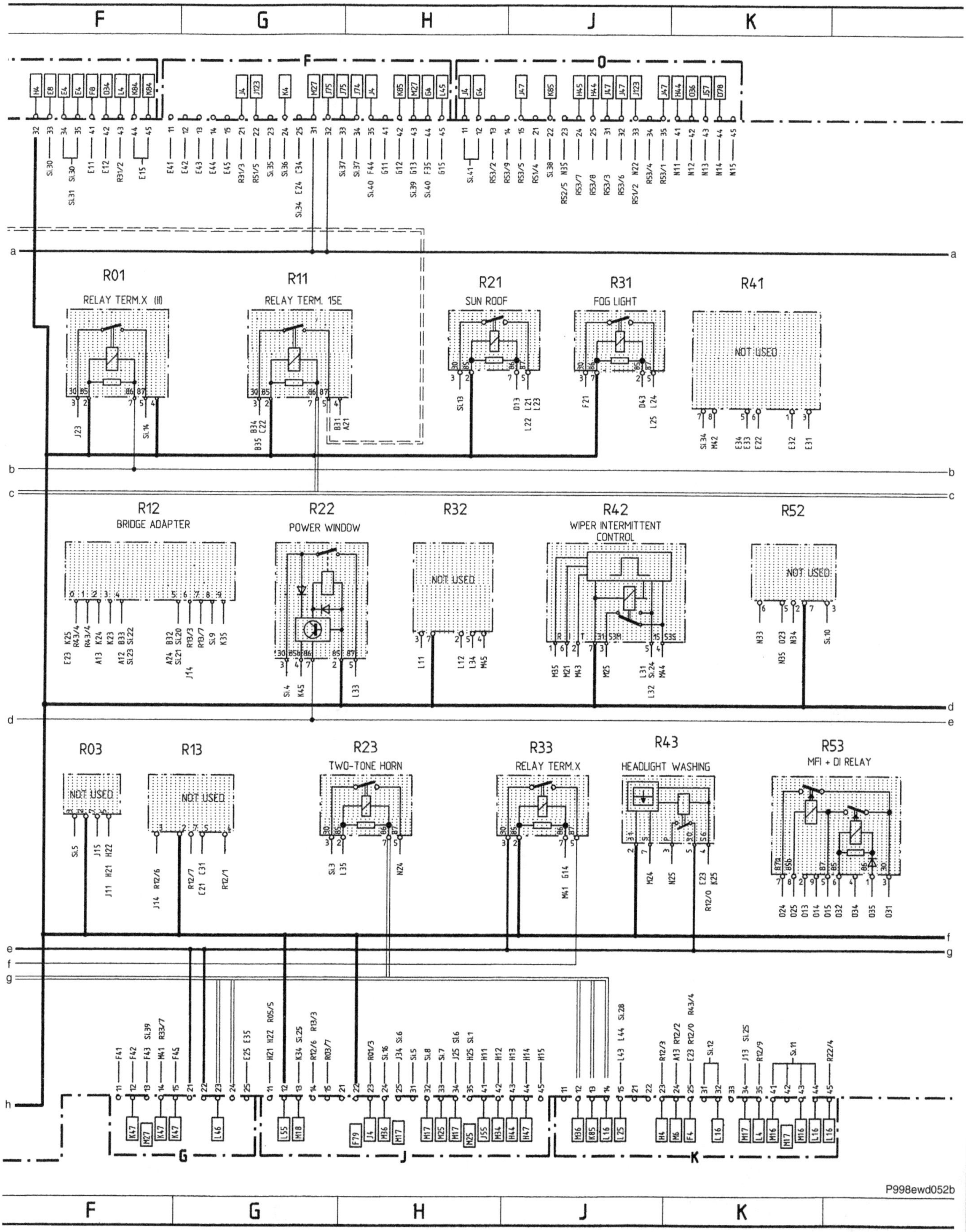

P998ewd052b

Sheet 17 - 1995/2

Central electric (page 3 of 3)

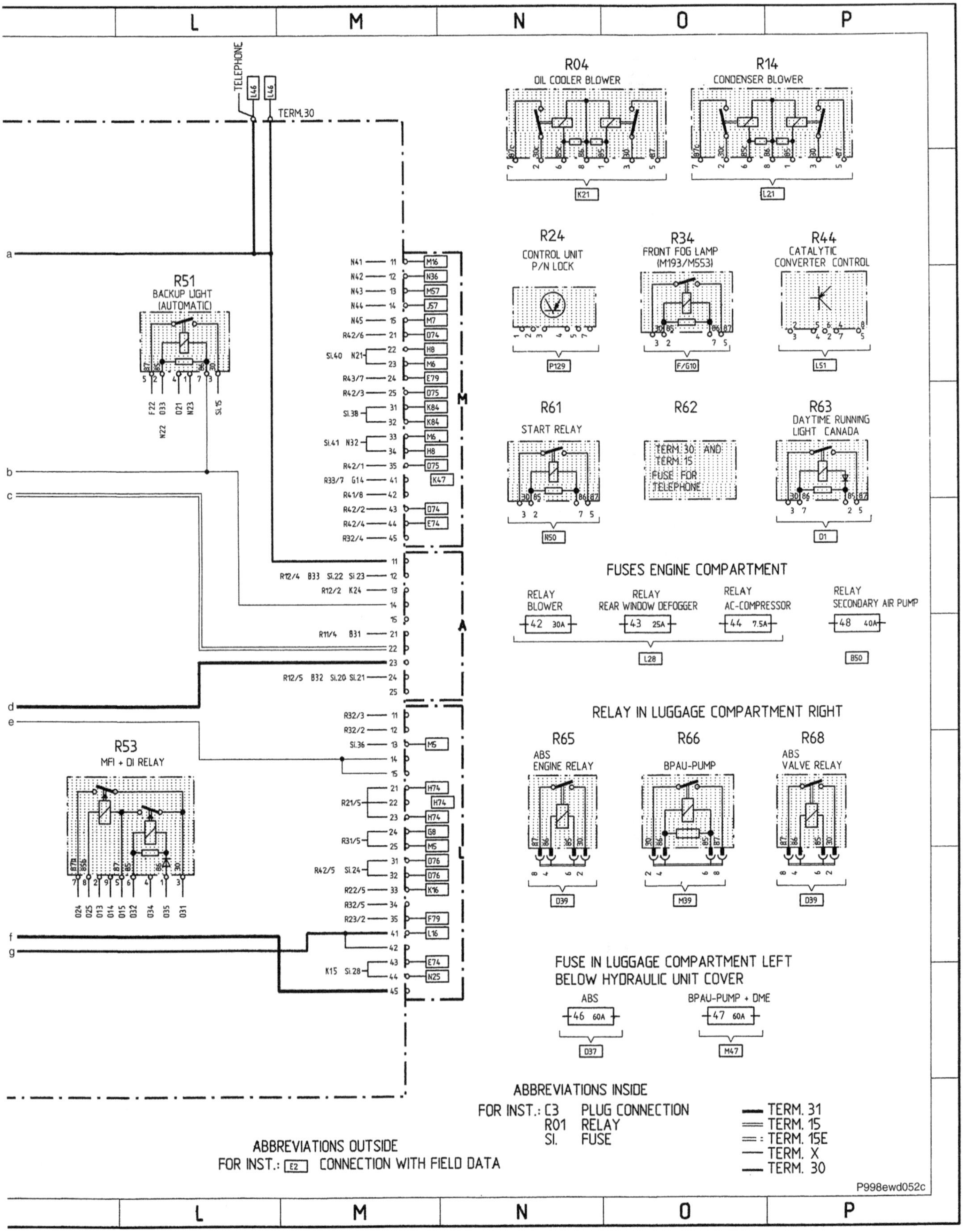

Lights
(page 1 of 3)

Sheet 1 - 1996

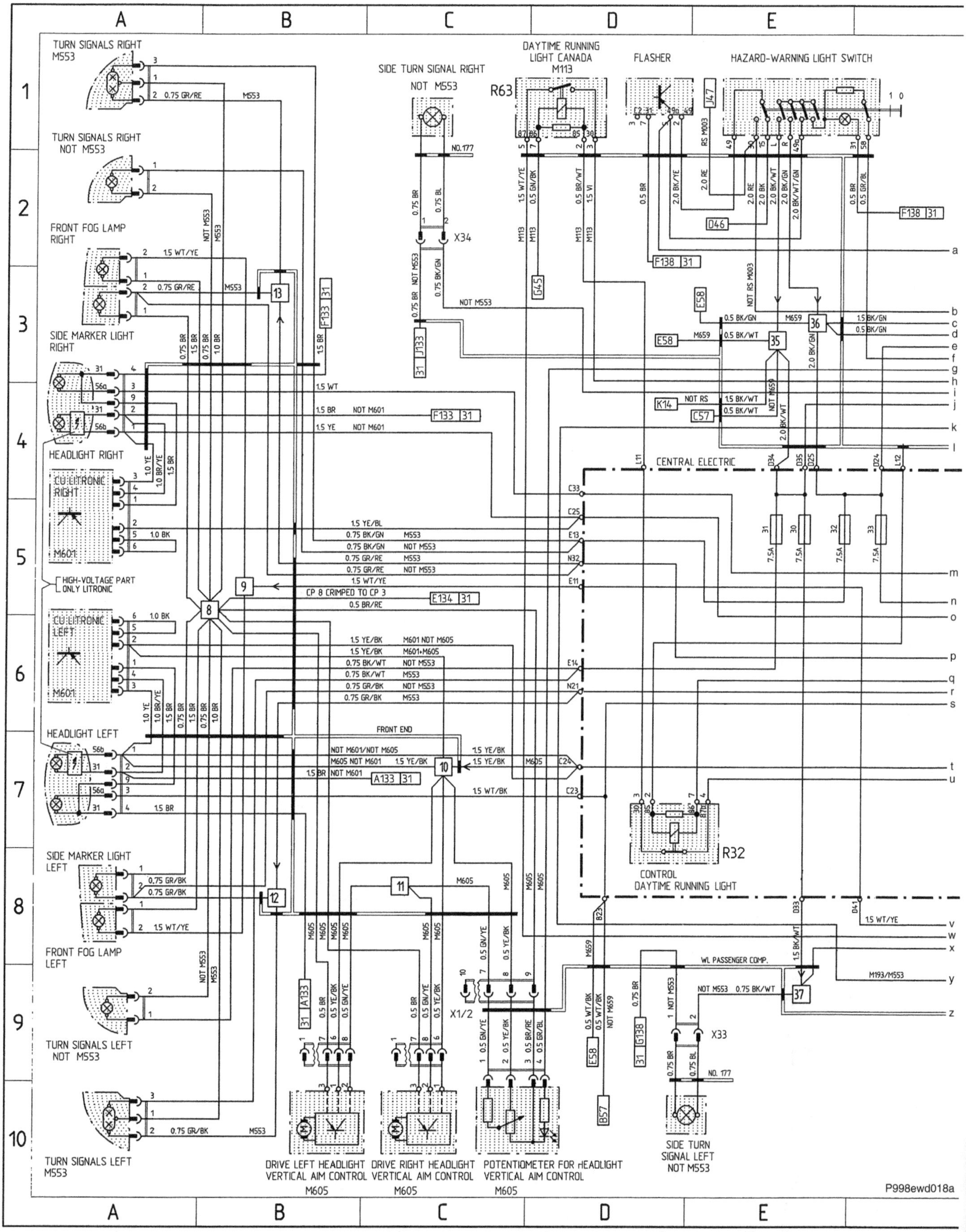

EWD

Sheet 1 - 1996

Lights
(page 2 of 3)

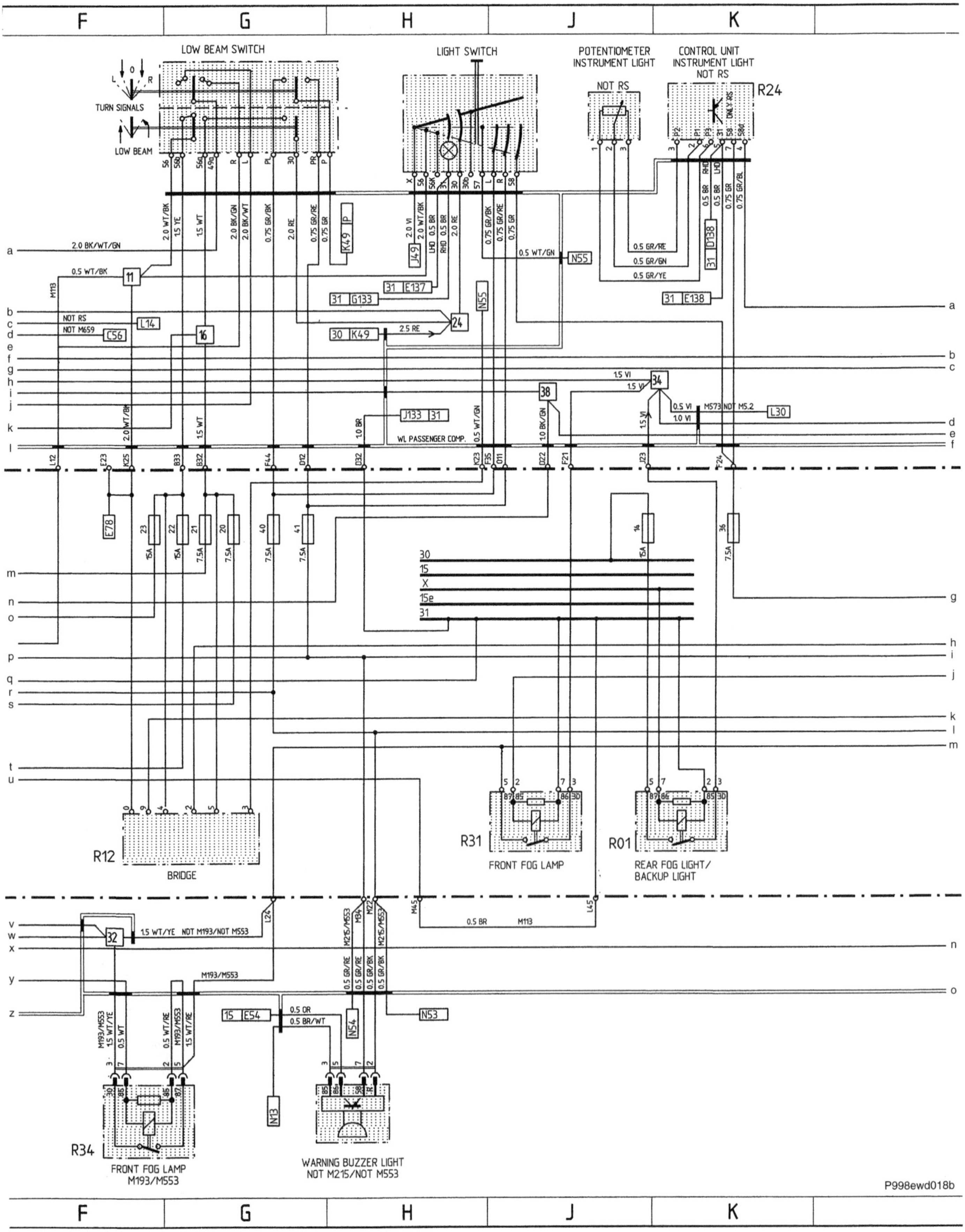

Lights
(page 3 of 3)

Sheet 1 - 1996

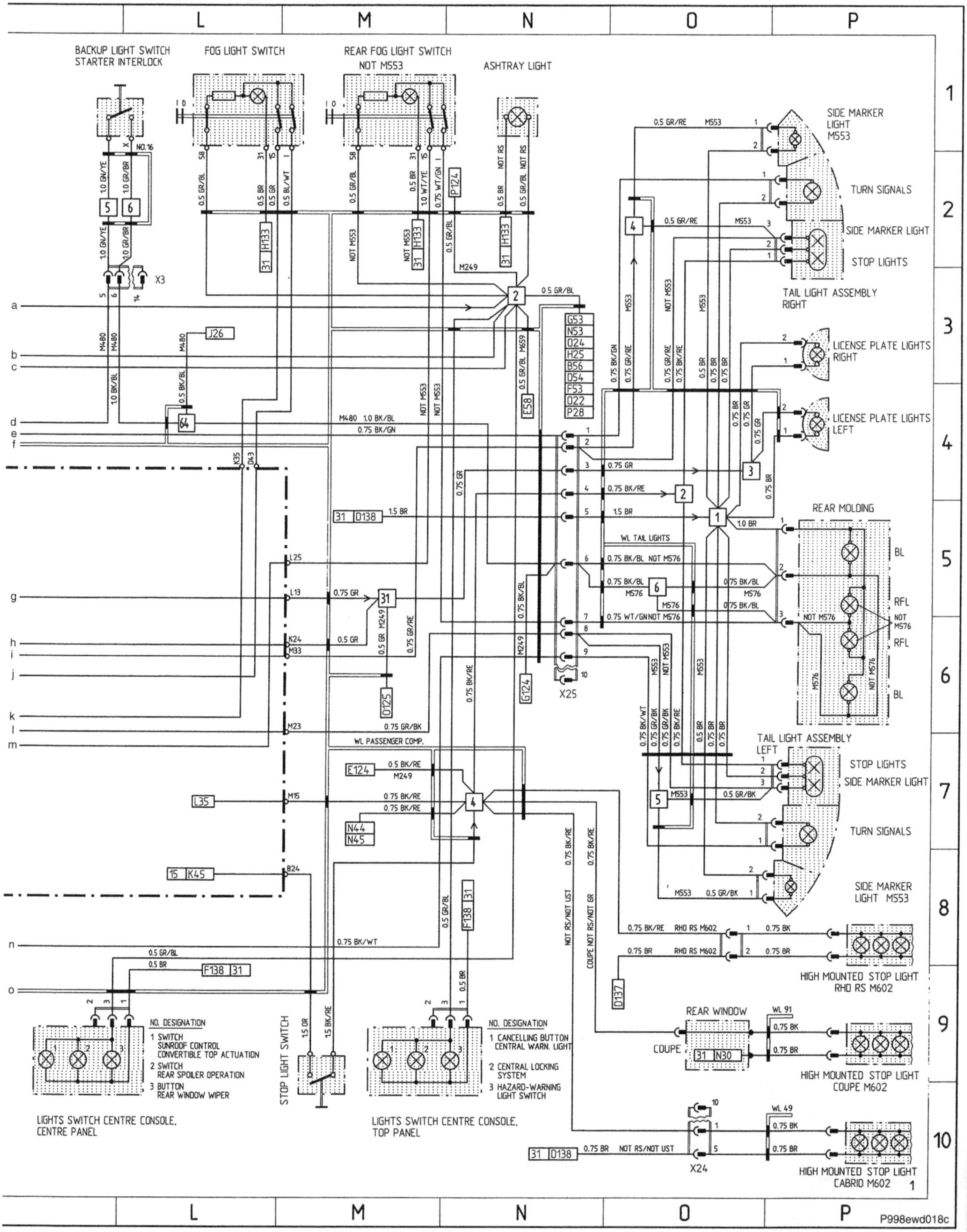

Sheet 2 - 1996

Seats, alarm system, central locks, mirror, power window regulator, inside lights (page 1 of 3)

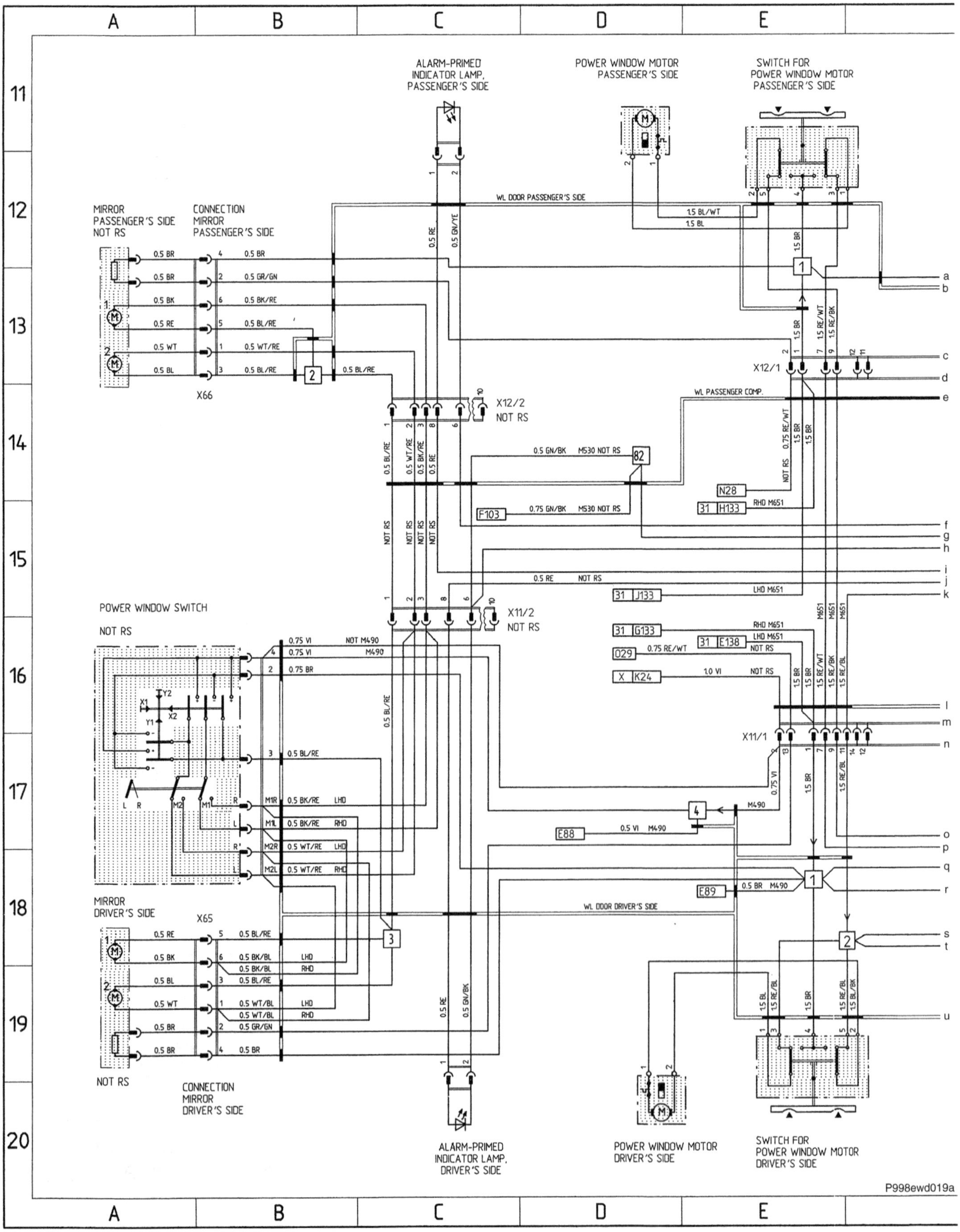

P998ewd019a

Seats, alarm system, central locks, mirror, power window regulator, inside lights (page 2 of 3)

Sheet 2 - 1996

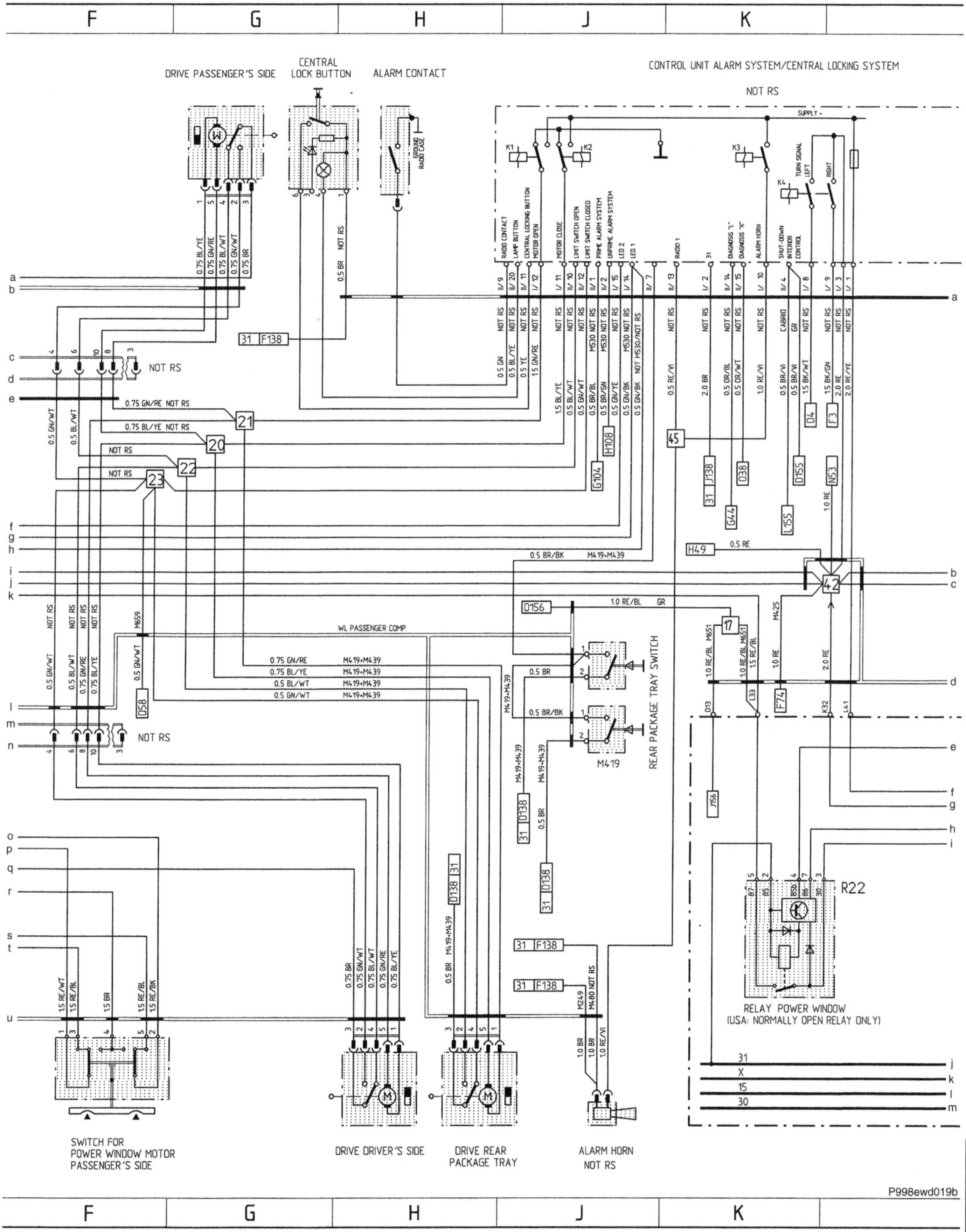

F G H J K

P998ewd019b

Sheet 2 - 1996

Seats, alarm system, central locks, mirror, power window regulator, inside lights (page 3 of 3)

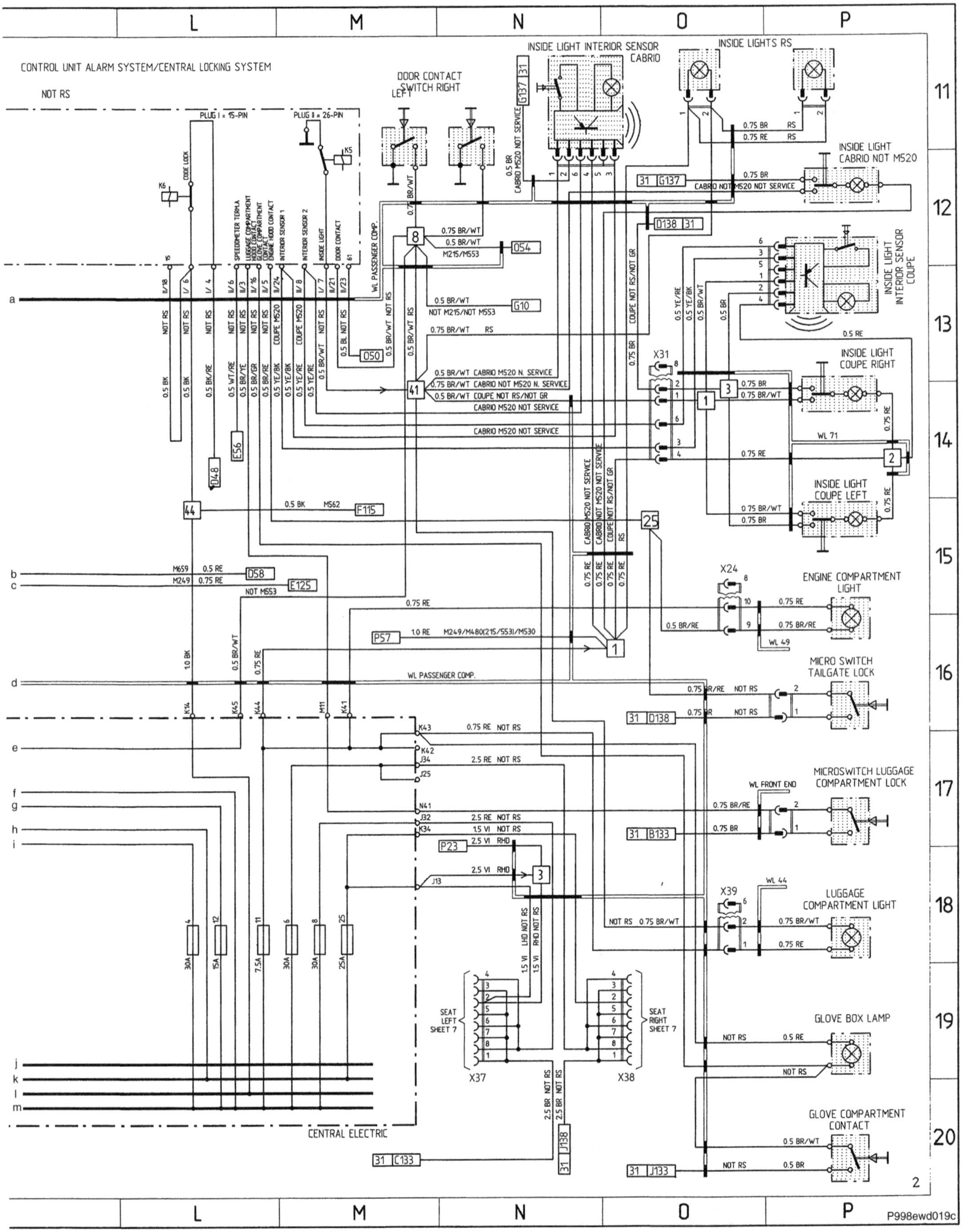

Heater, A/C (page 1 of 3)

Sheet 3 - 1996

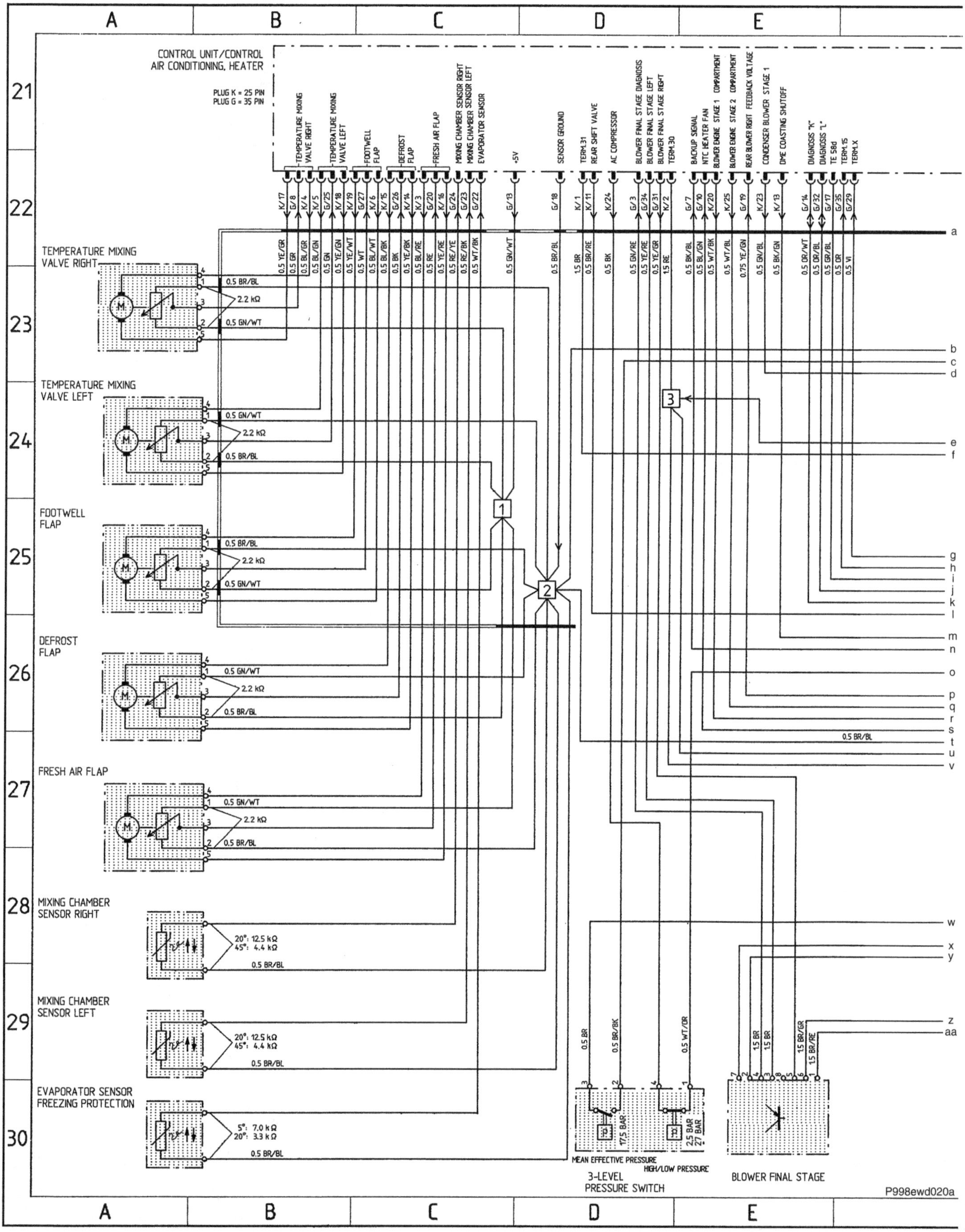

EWD

Sheet 3 - 1996

Heater, A/C (page 2 of 3)

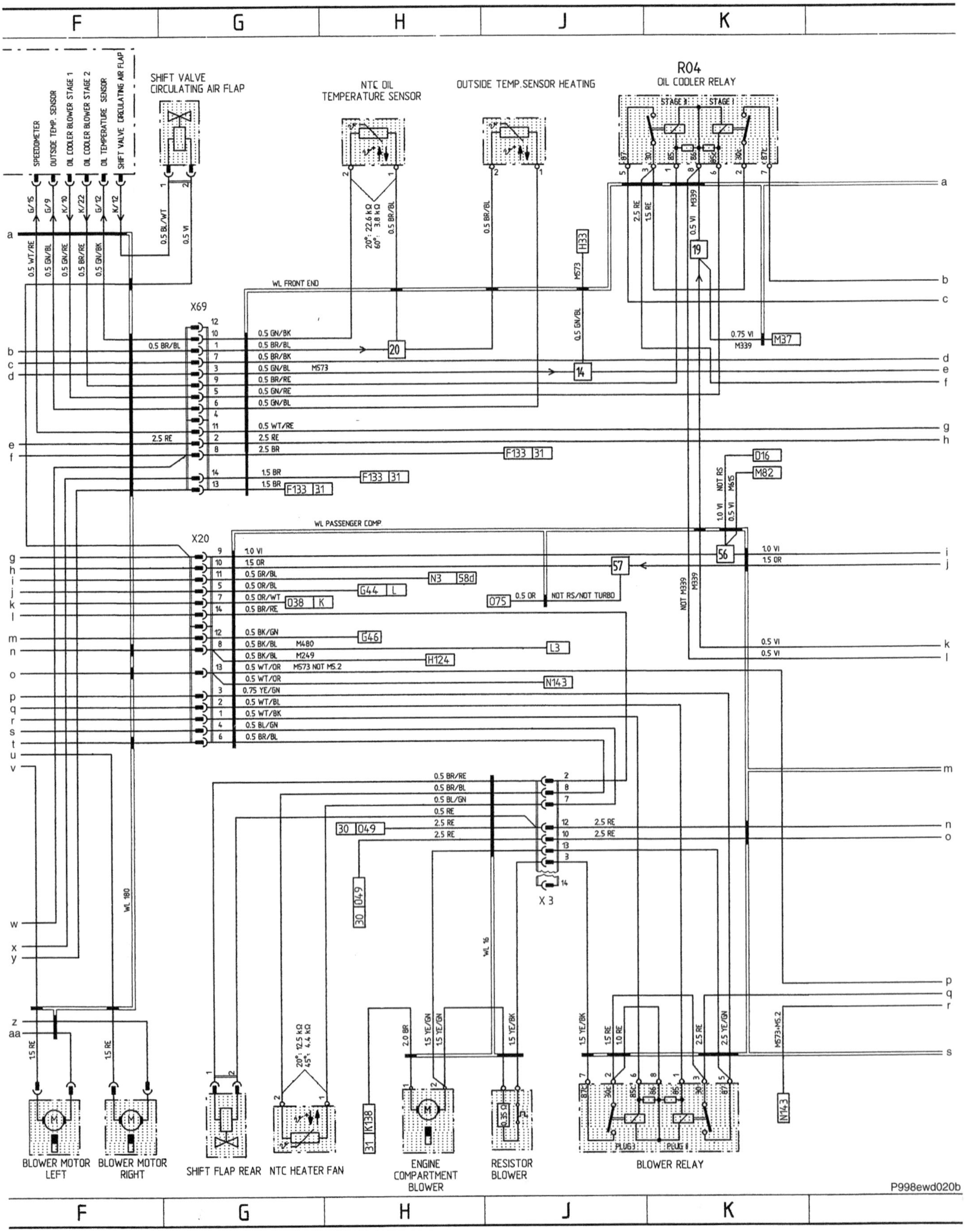

Heater, A/C (page 3 of 3)

Sheet 3 - 1996

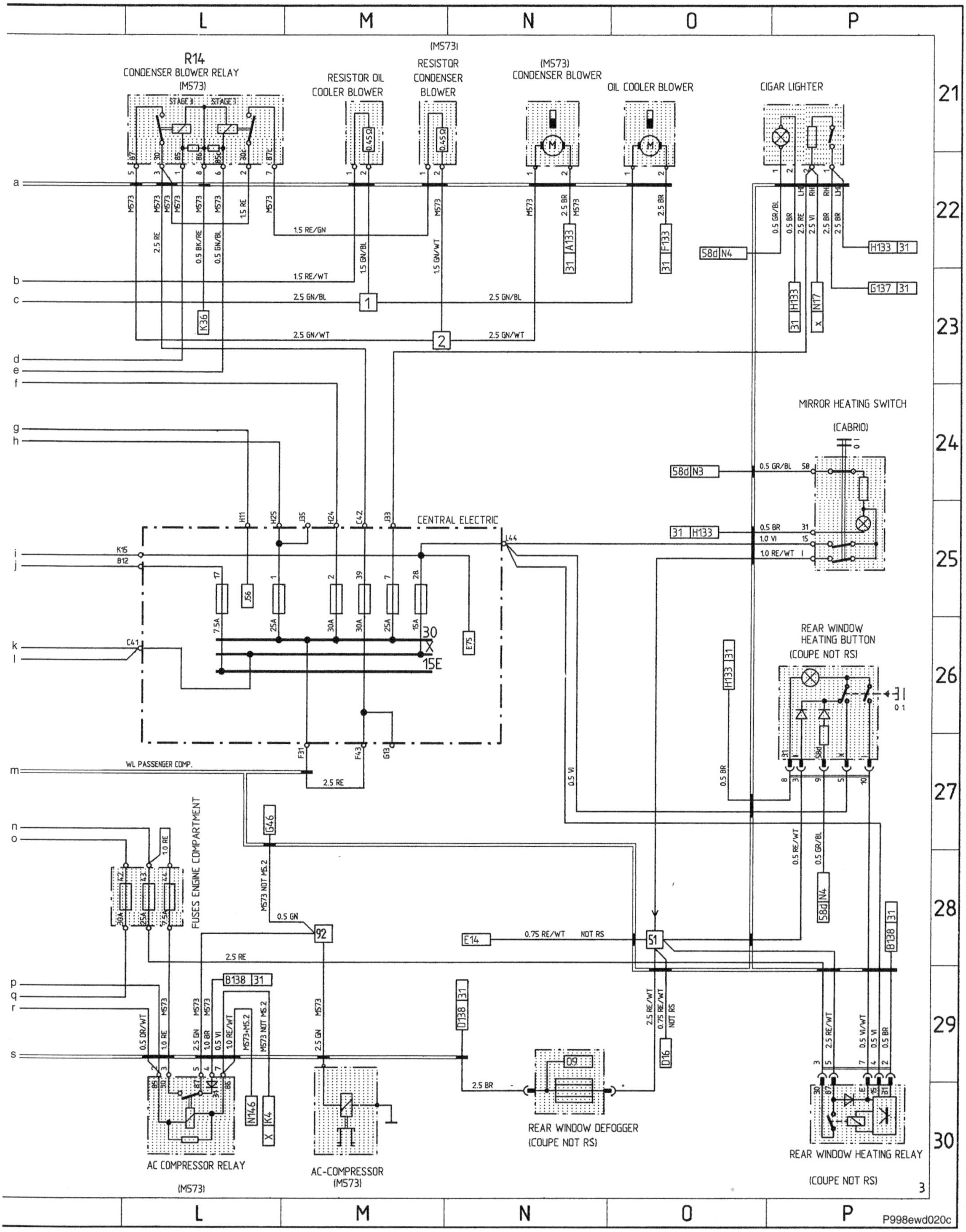

EWD

Sheet 4A - 1996

ABS
(page 1 of 3)

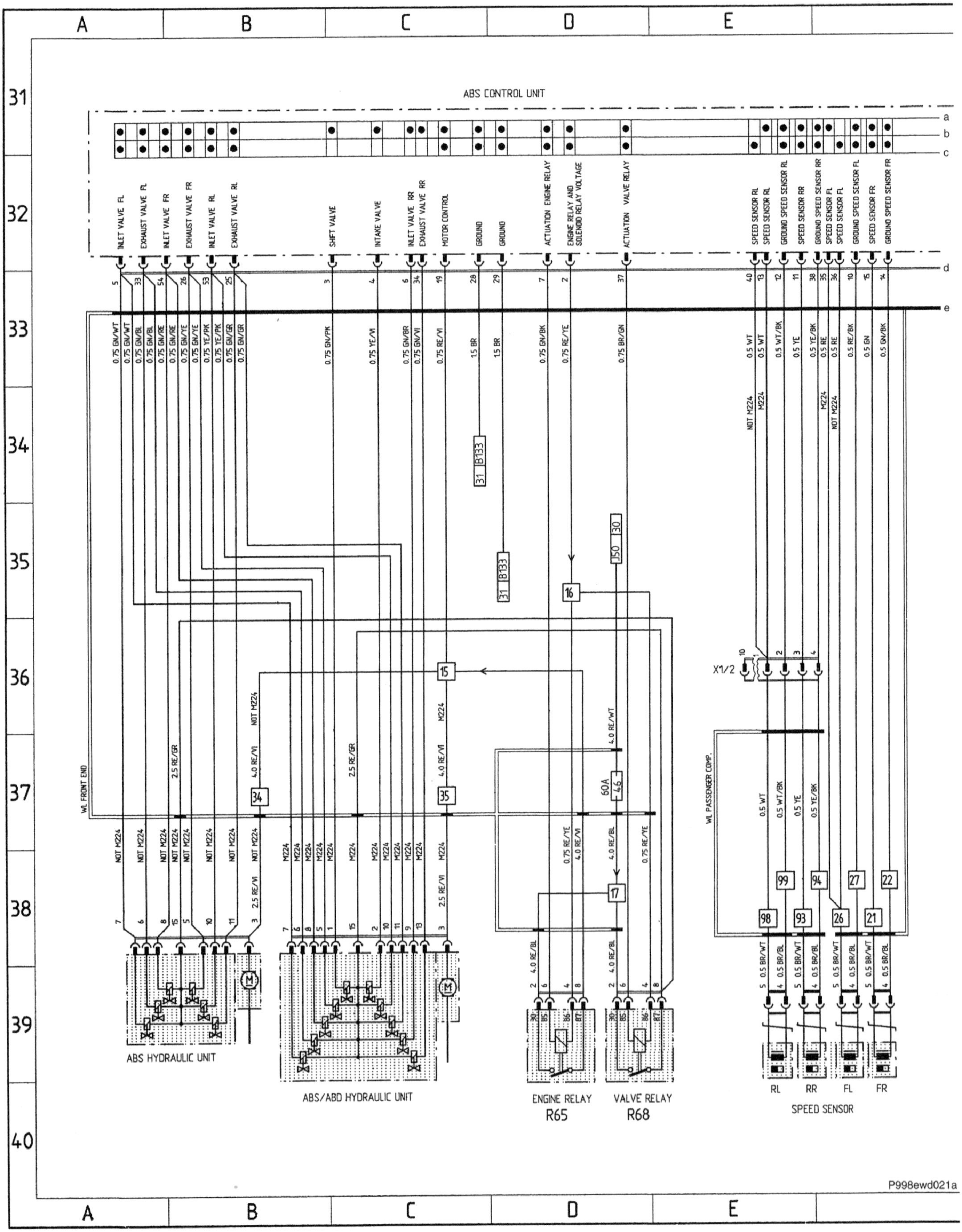

P998ewd021a

ABS
(page 2 of 3)

Sheet 4A - 1996

F	G	H	J	K	

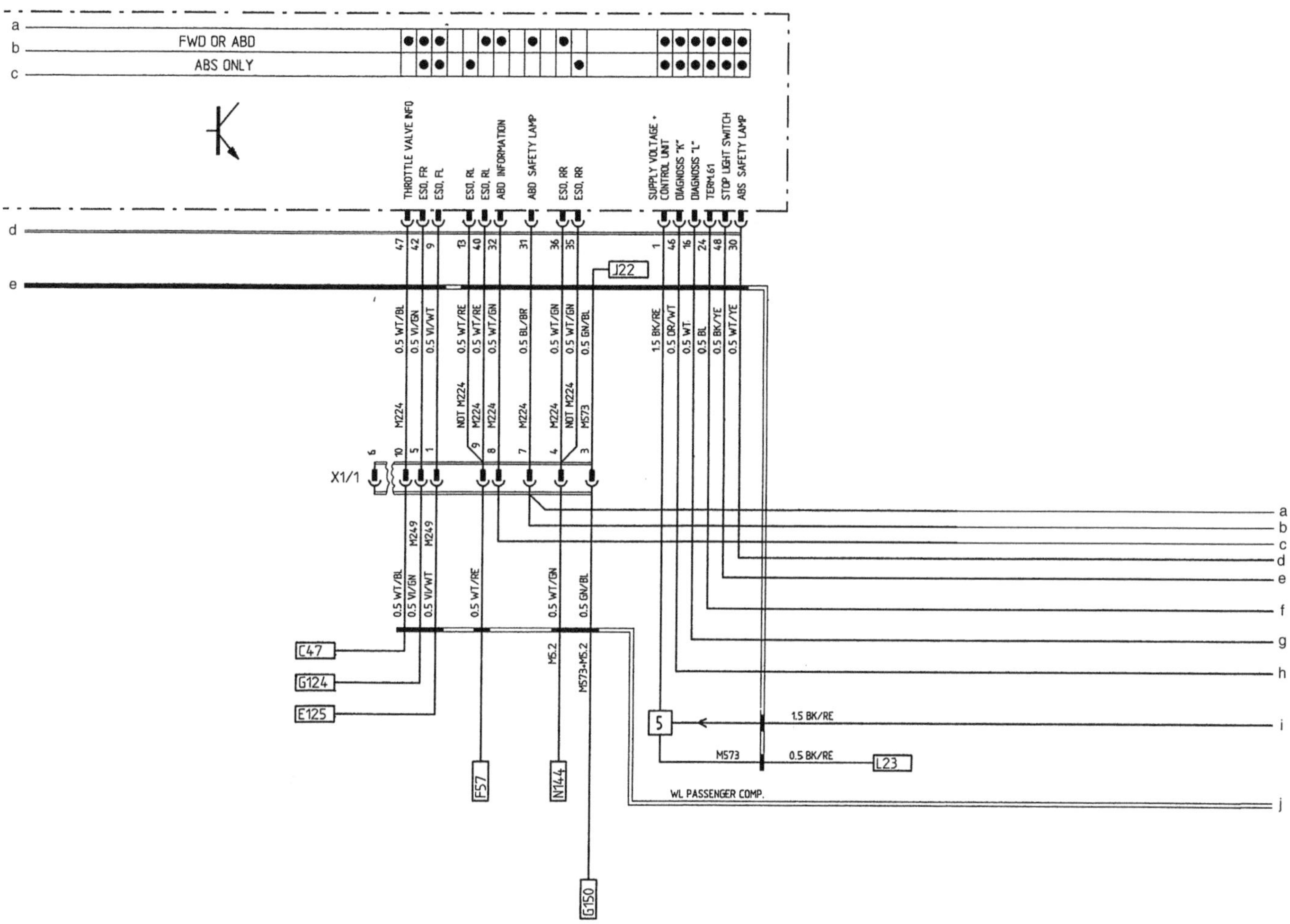

F	G	H	J	K	

P998ewd021b

EWD

Sheet 4A - 1996

ABS
(page 3 of 3)

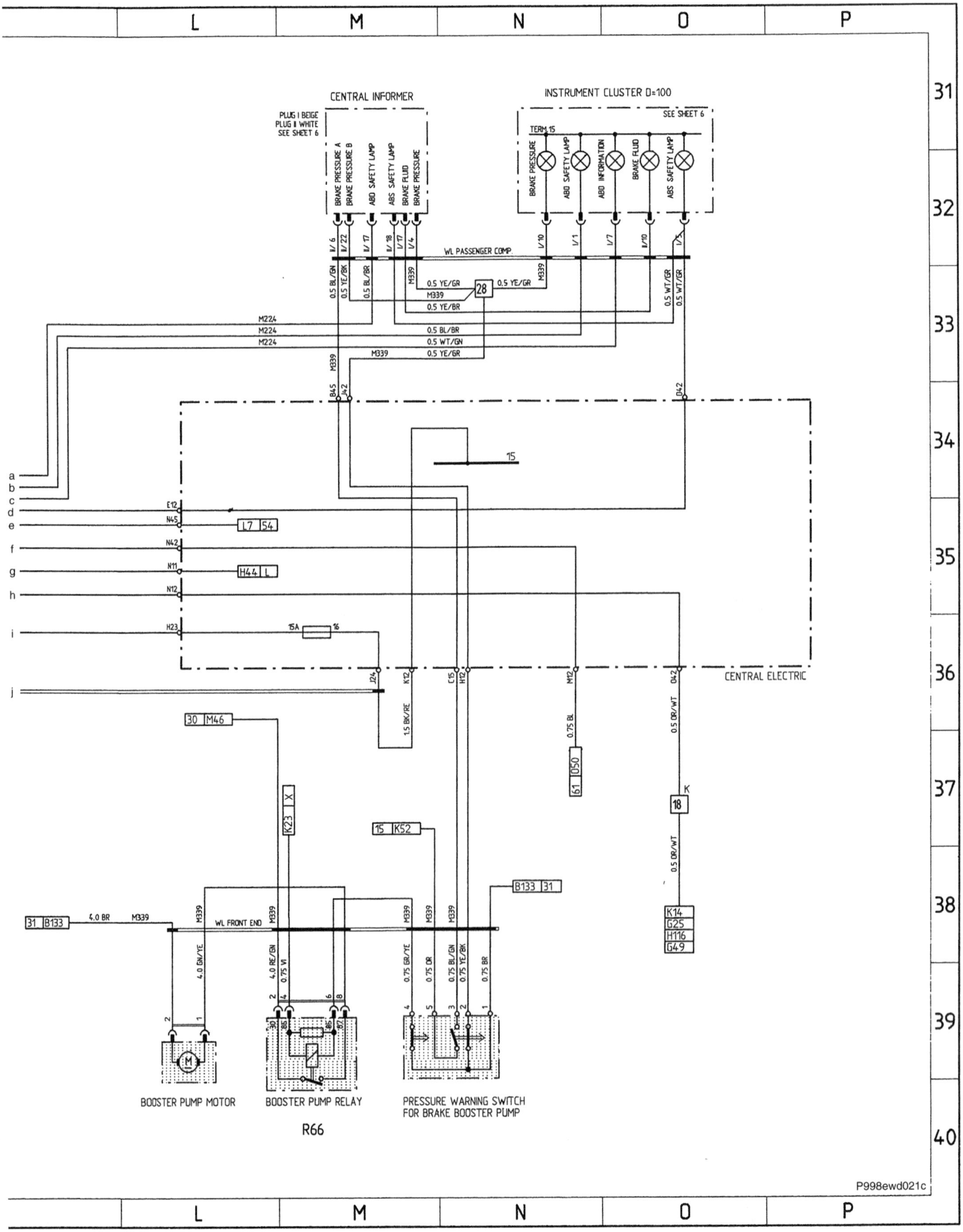

ABS (RS)
(page 1 of 3)

Sheet 4B - 1996

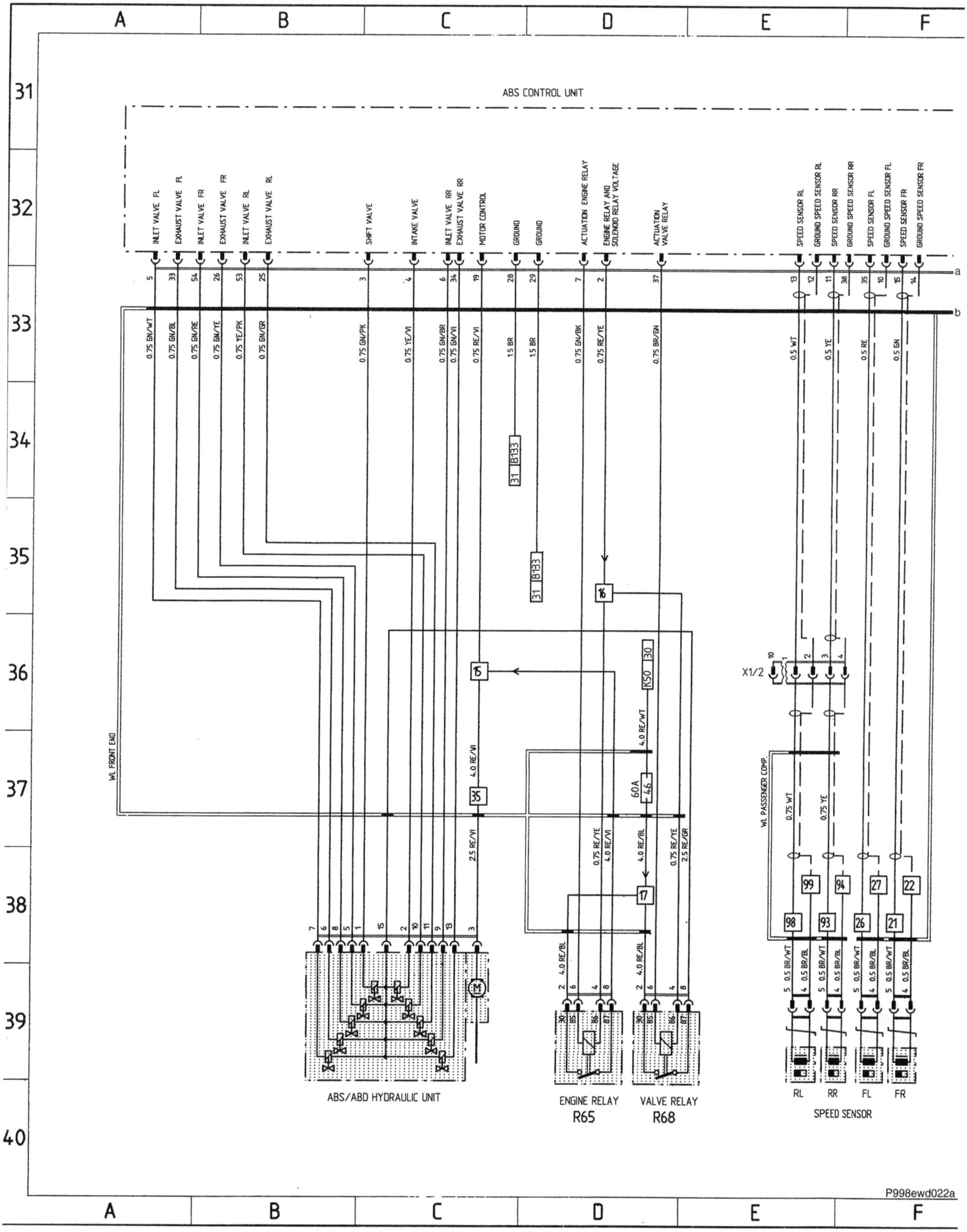

EWD

Sheet 4B - 1996

ABS (RS)
(page 2 of 3)

	G	H	J	K	

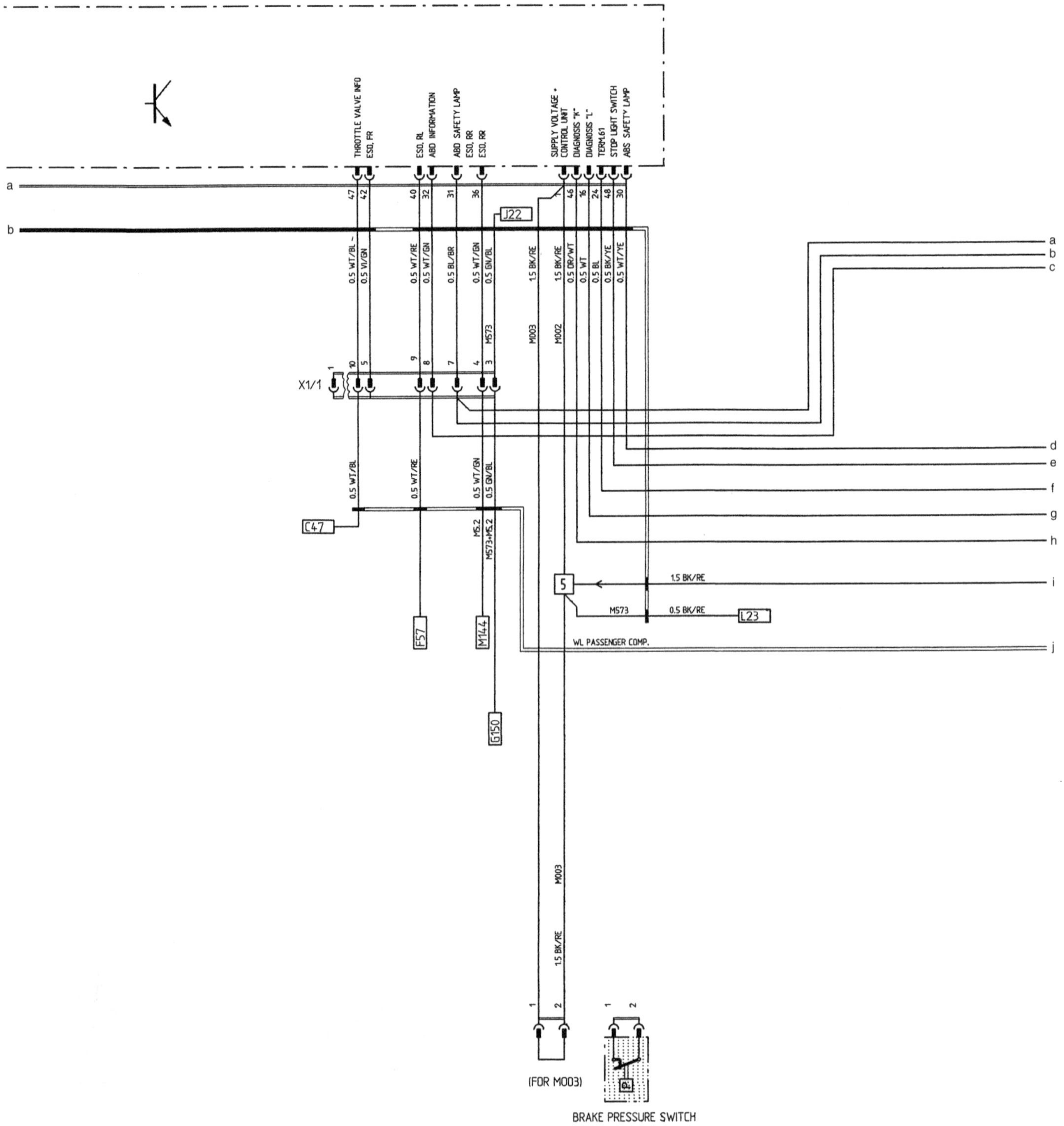

P998ewd022b

	G	H	J	K	

ABS (RS)
(page 3 of 3)

Sheet 4B - 1996

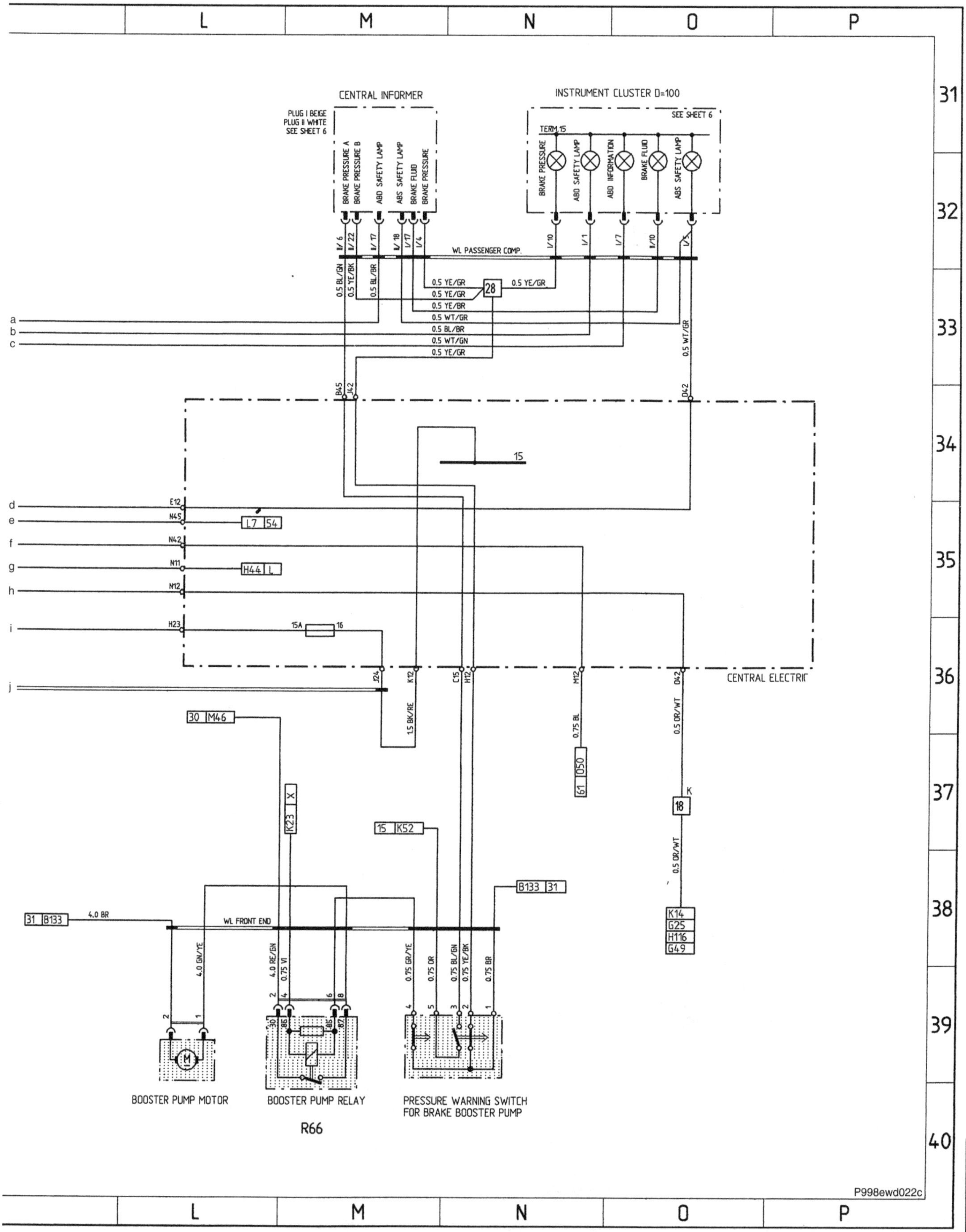

EWD

Sheet 5A - 1996 Engine electrics, ignition system (DME M 2.10, ROW) (page 1 of 3)

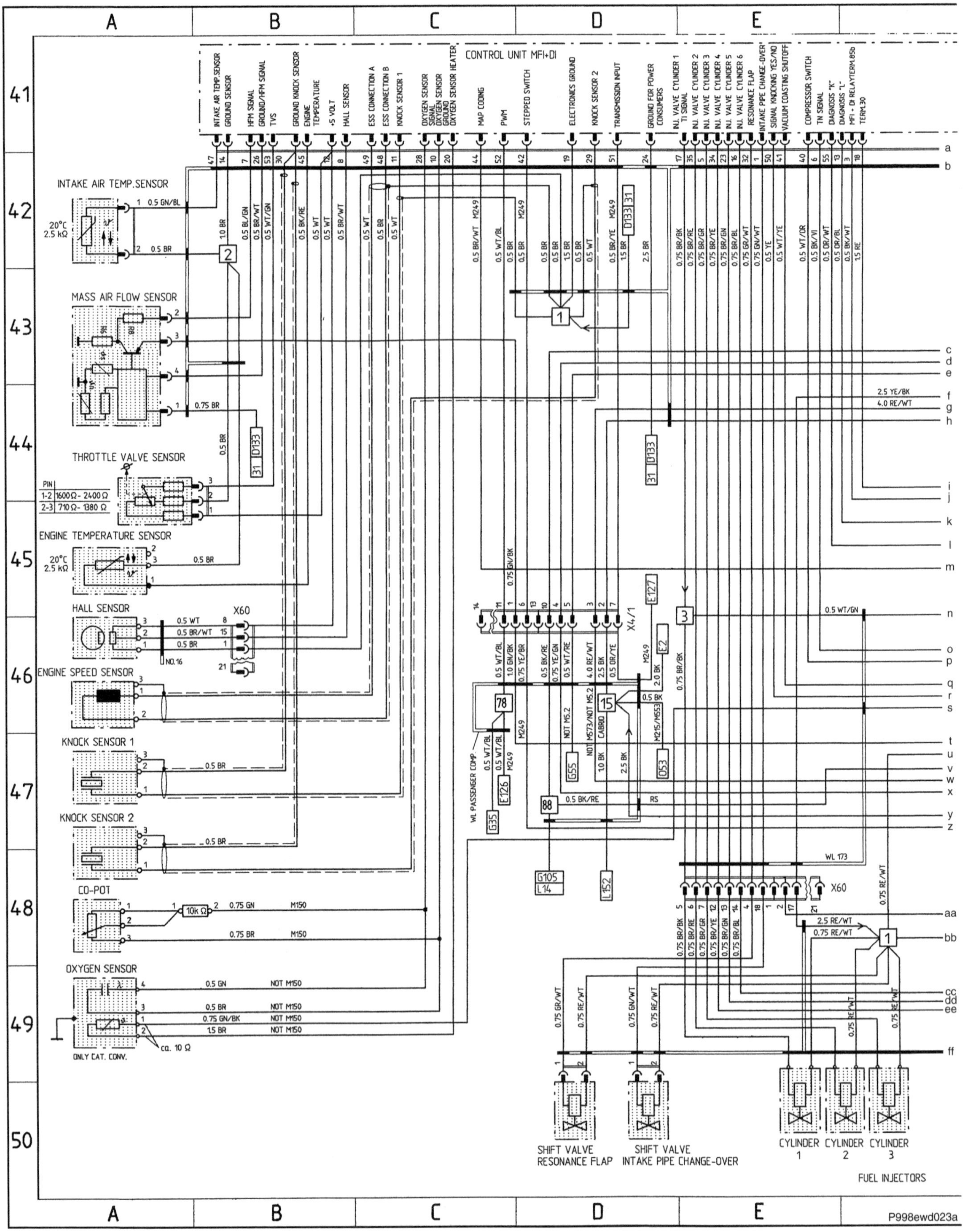

P998ewd023a

Engine electrics, ignition system (DME M 2.10, ROW) (page 2 of 3)

Sheet 5A - 1996

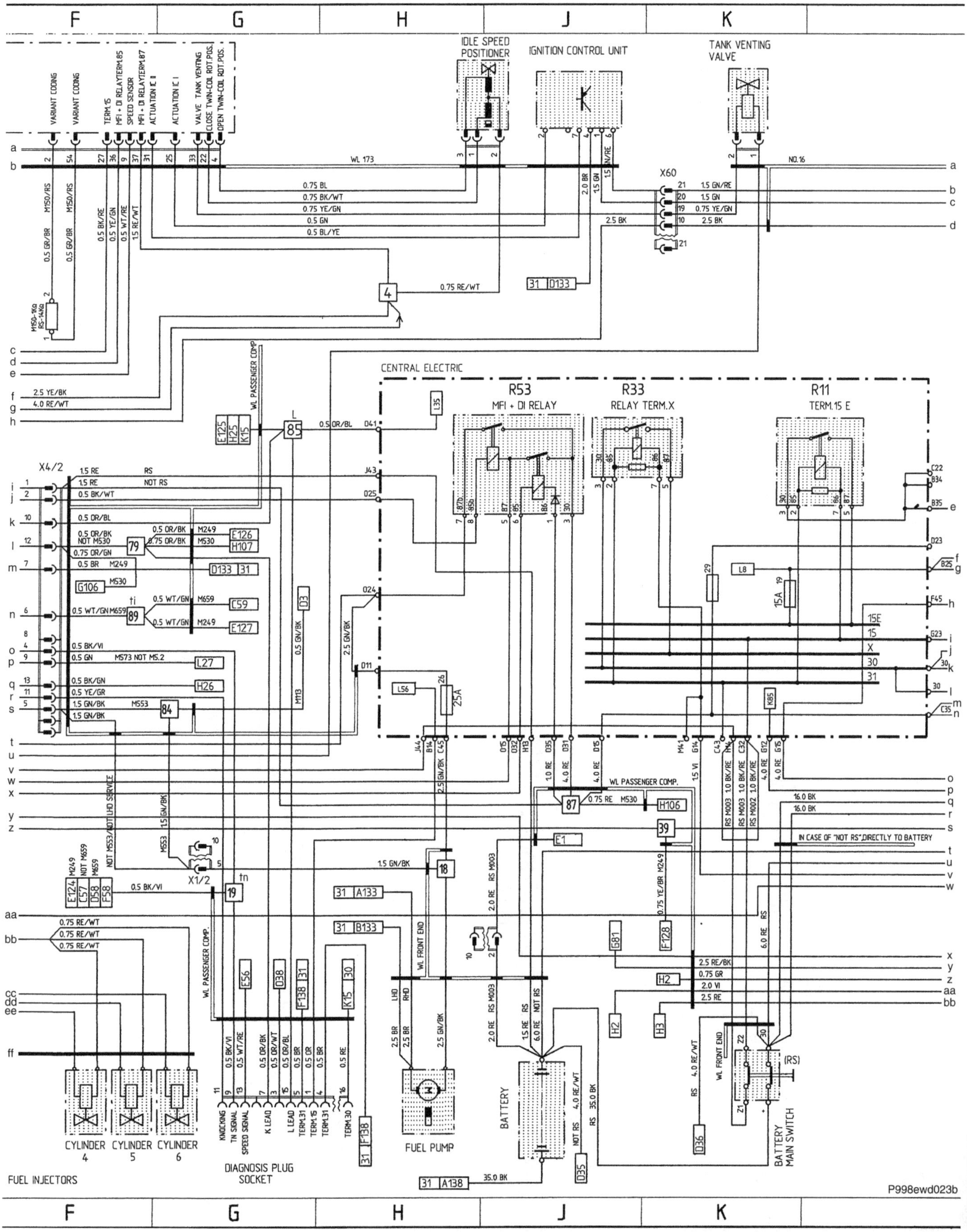

Sheet 5A - 1996 Engine electrics, ignition system (DME M 2.10, ROW)
(page 3 of 3)

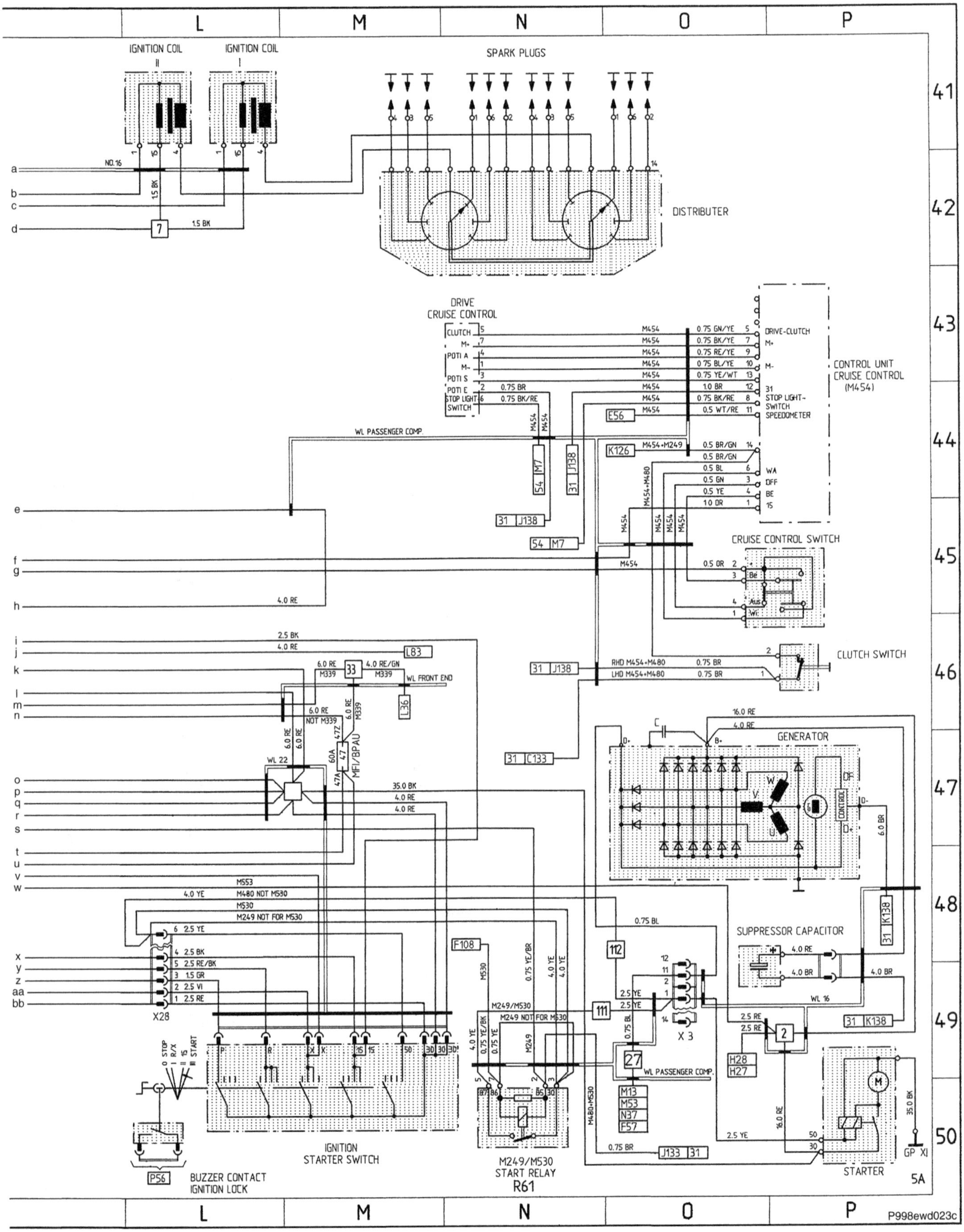

Engine electrics, ignition system (DME M 5.2, US) (page 1 of 3)

Sheet 5B - 1996

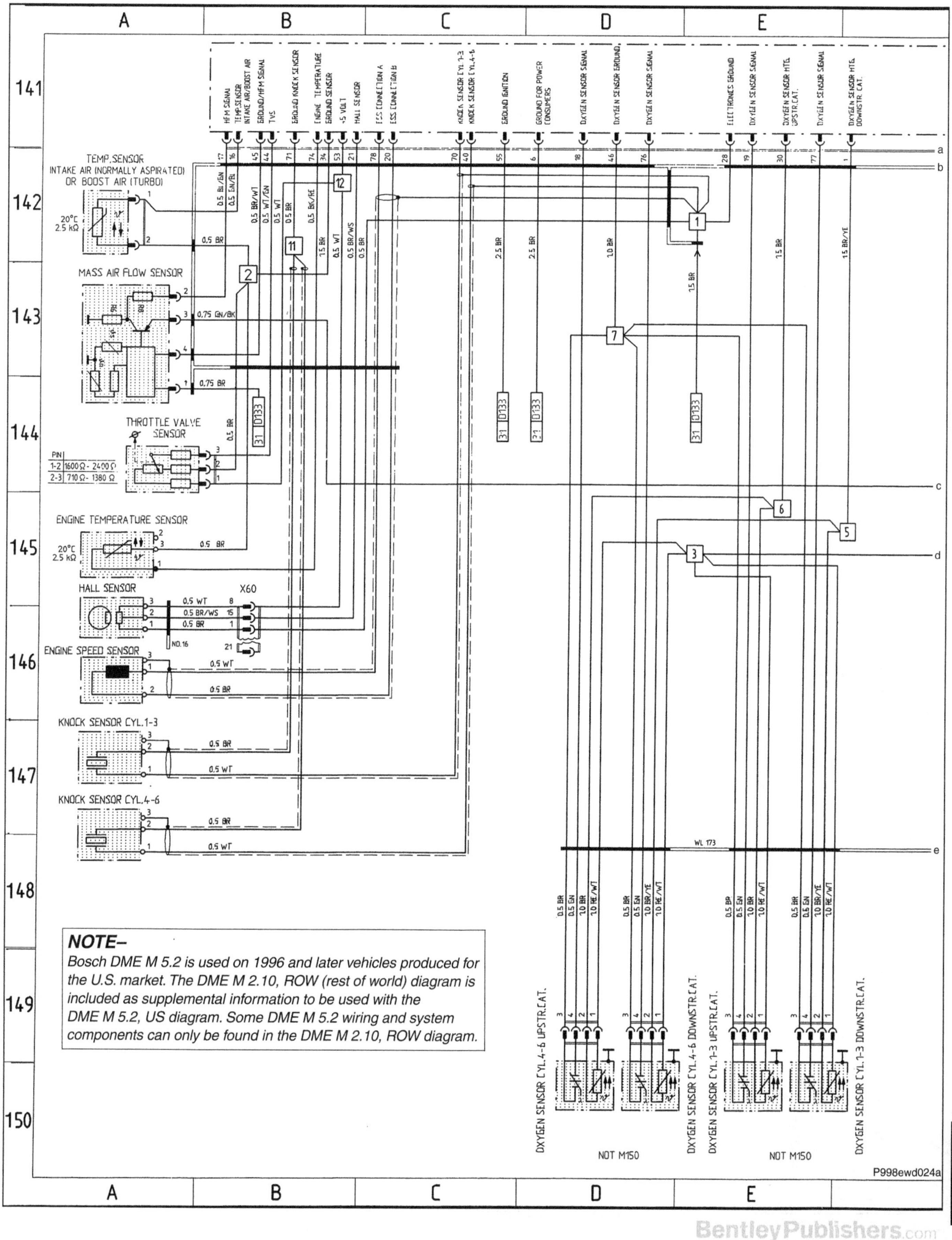

NOTE–

Bosch DME M 5.2 is used on 1996 and later vehicles produced for the U.S. market. The DME M 2.10, ROW (rest of world) diagram is included as supplemental information to be used with the DME M 5.2, US diagram. Some DME M 5.2 wiring and system components can only be found in the DME M 2.10, ROW diagram.

Sheet 5B - 1996

Engine electrics, ignition system (DME M 5.2, US) (page 2 of 3)

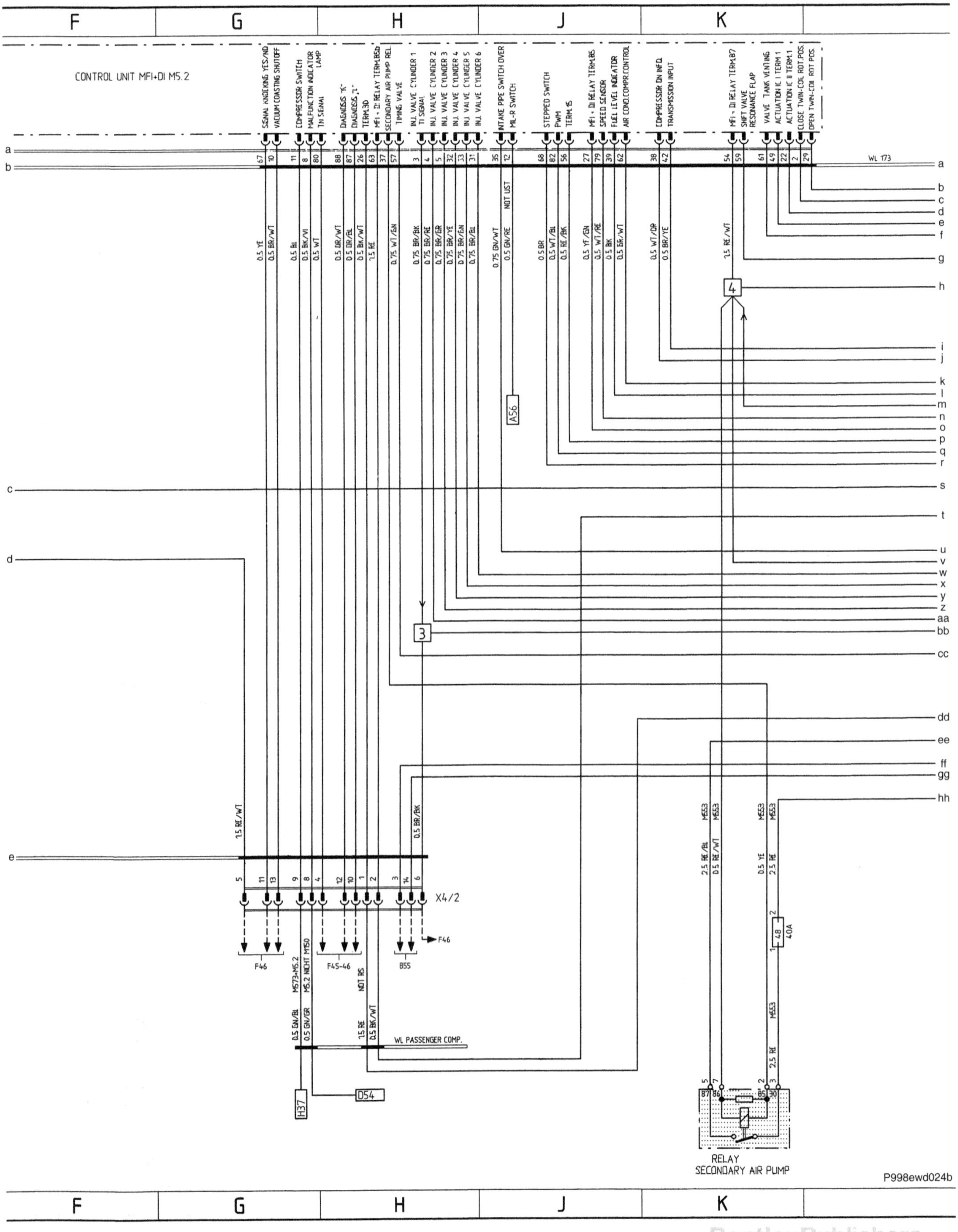

P998ewd024b

Engine electrics, ignition system (DME M 5.2, US) (page 3 of 3)

Sheet 5B - 1996

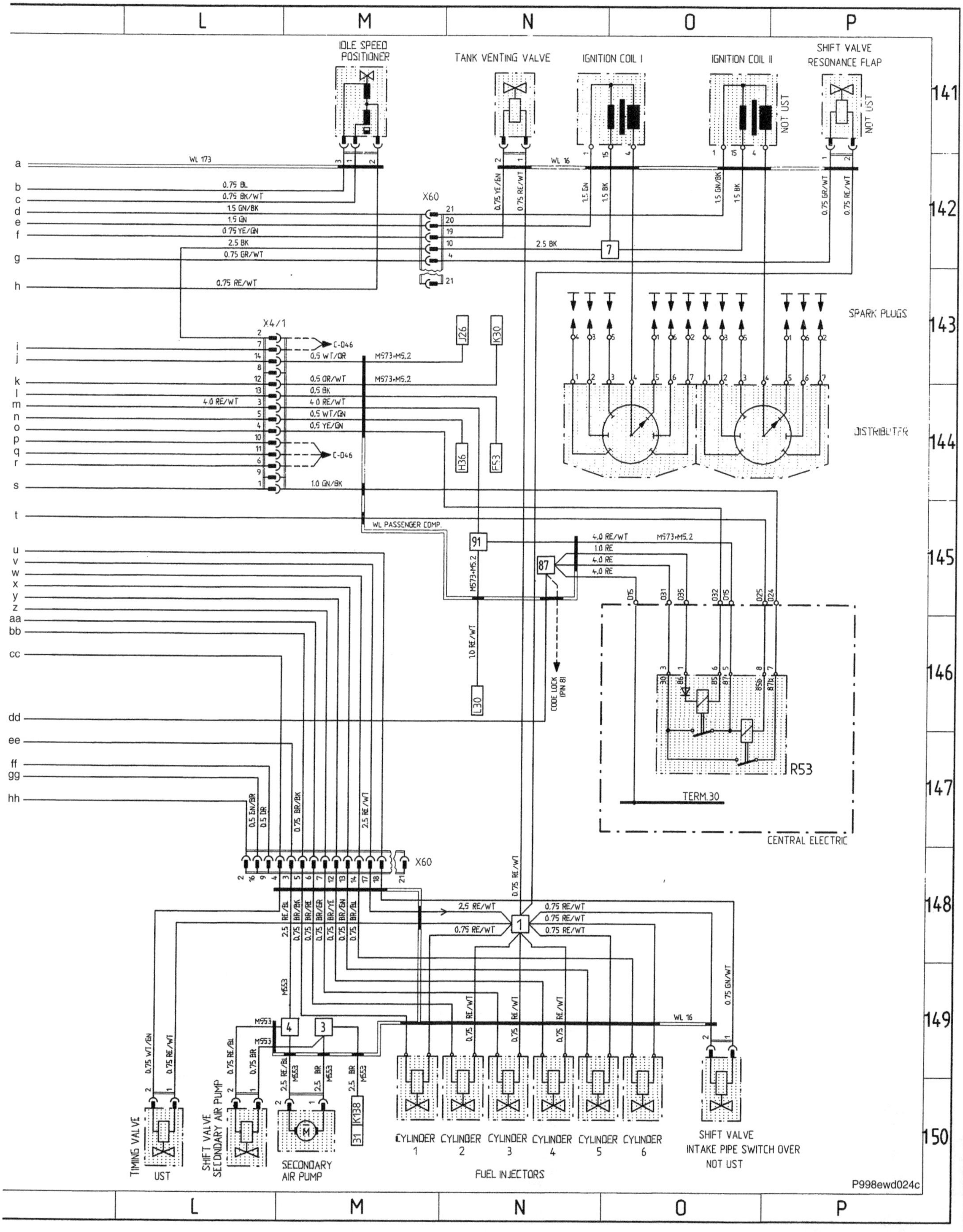

Sheet 6 - 1996

Instruments and sensor (page 1 of 3)

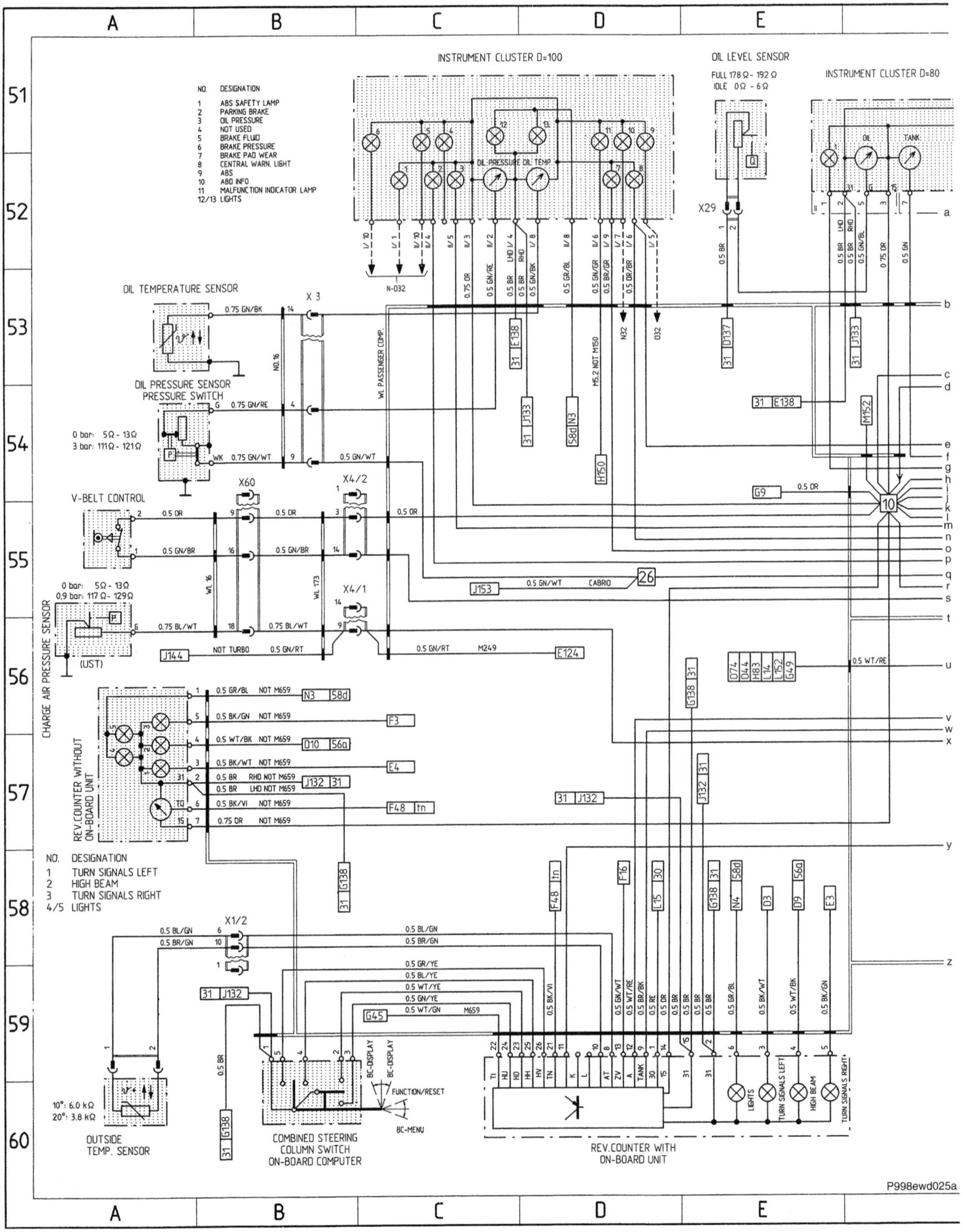

Instruments and sensor
(page 2 of 3)

Sheet 6 - 1996

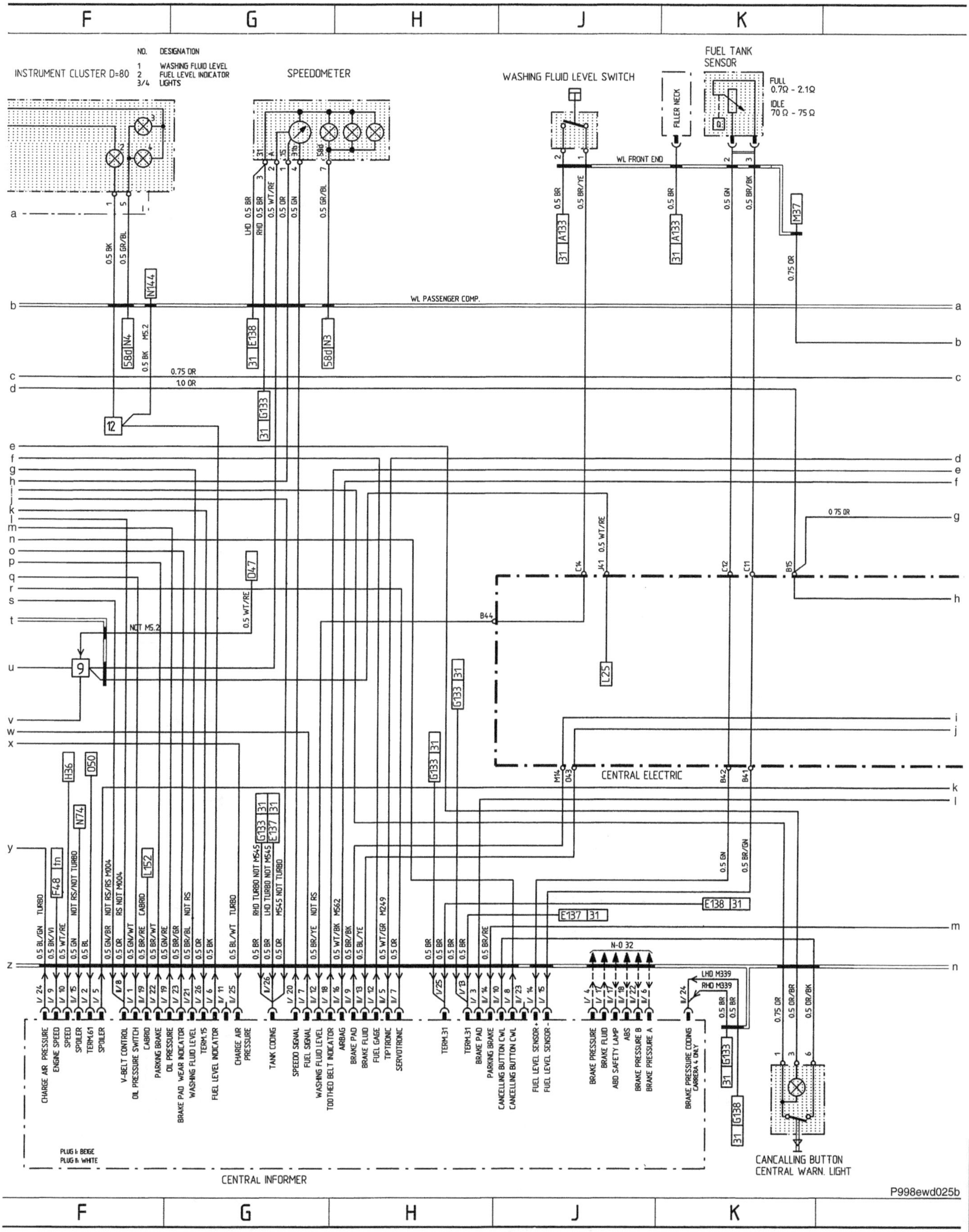

P998ewd025b

EWD

Sheet 6 - 1996

Instruments and sensor (page 3 of 3)

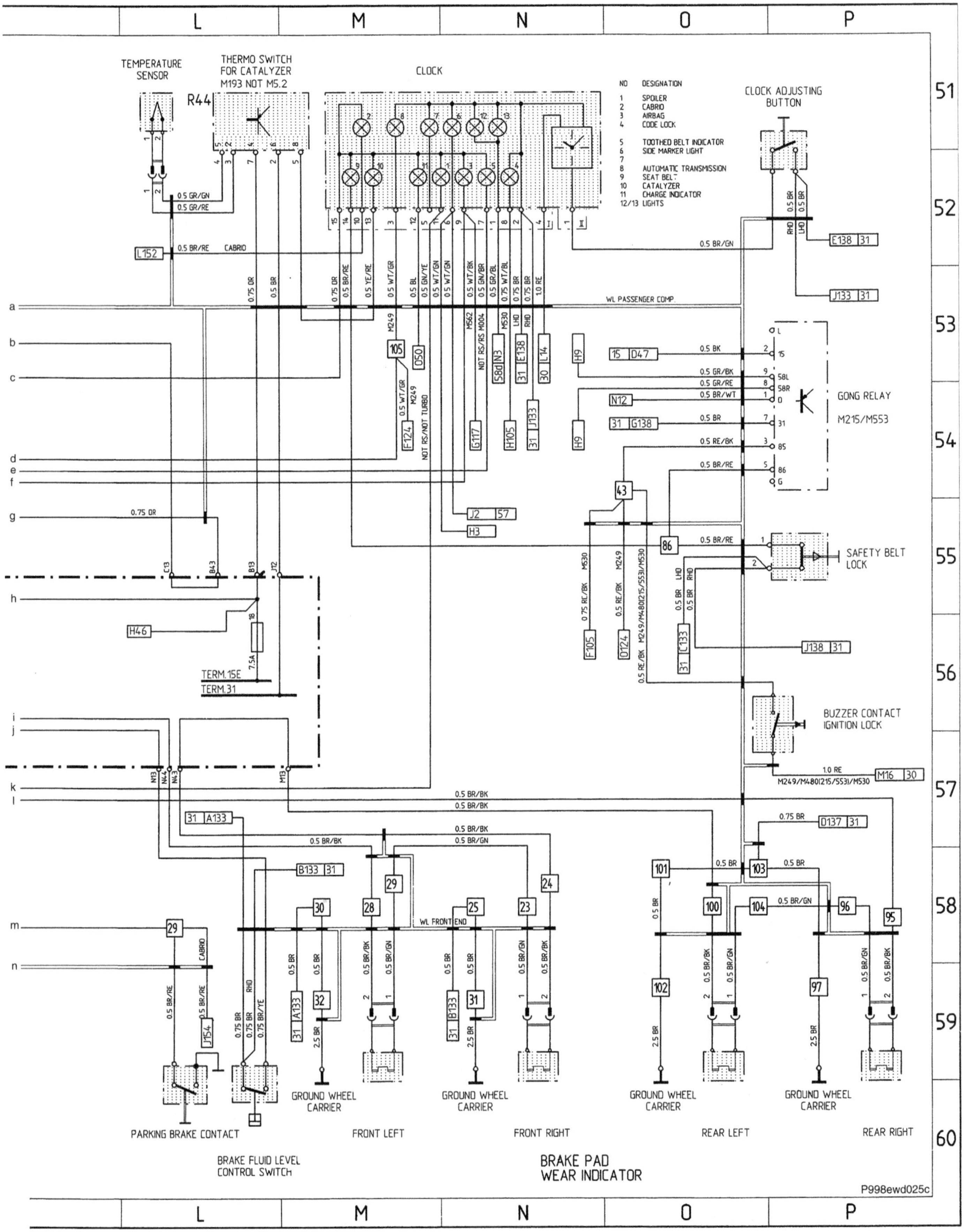

Seats, differing cable layout for service (page 1 of 3)

Sheet 7 - 1996

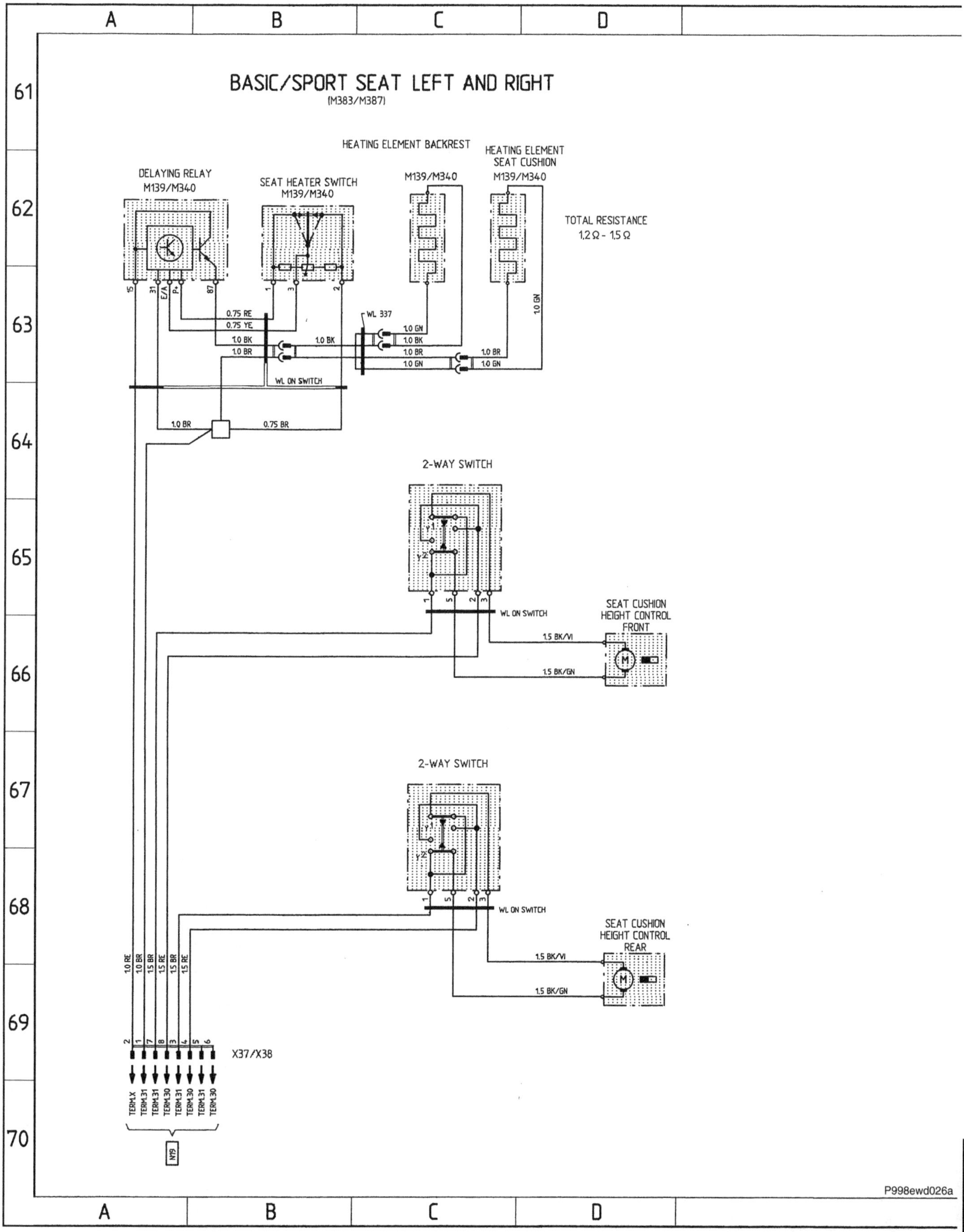

P998ewd026a

Sheet 7 - 1996

Seats, differing cable layout for service (page 2 of 3)

E	F	G	H	J	K

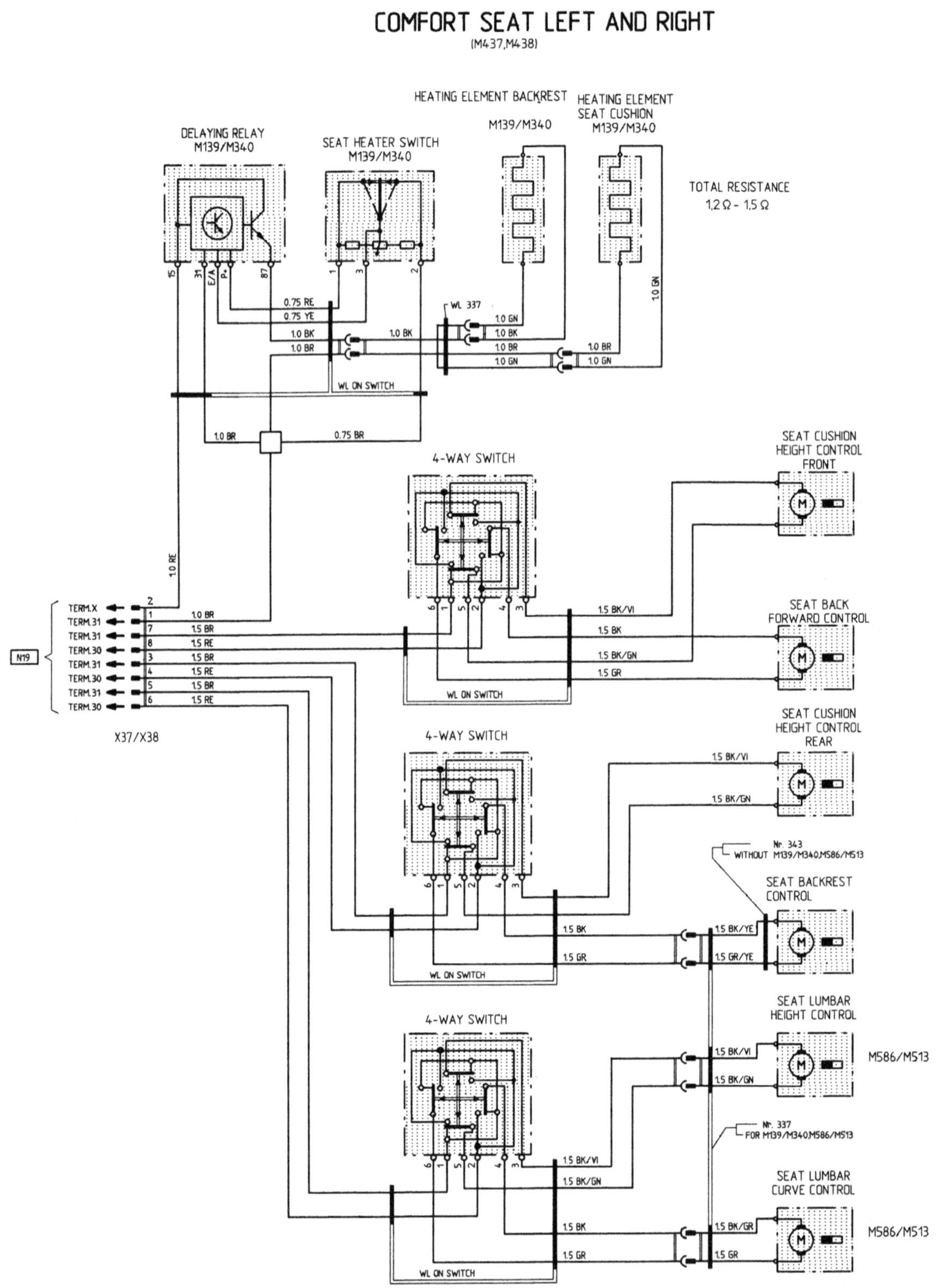

P998ewd026b

E	F	G	H	J	K

Seats, differing cable layout for service (page 3 of 3)

Sheet 7 - 1996

L	M	N	O	P

61

SPEAKER WIRING DIFFERS FOR SOUND PACKAGE
AND IF A SERVICE WIRING HARNESS IS USED

62

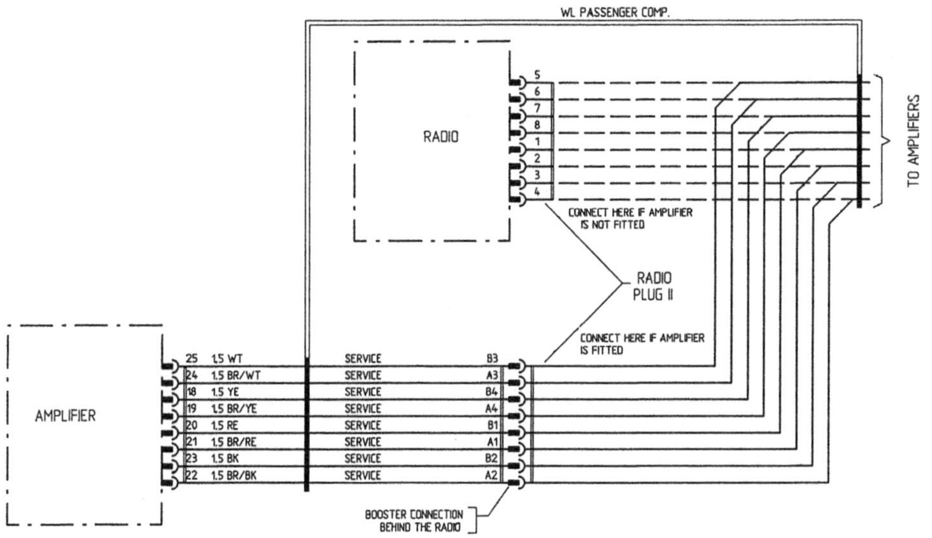

63

64

65

66

67

68

69

70

P998ewd026c

L	M	N	O	P

EWD

Electrical Wiring Diagrams EWD-126

Sheet 8 - 1996

Rear spoiler, two-tone horns, wipe-and-wash cleaners (page 1 of 3)

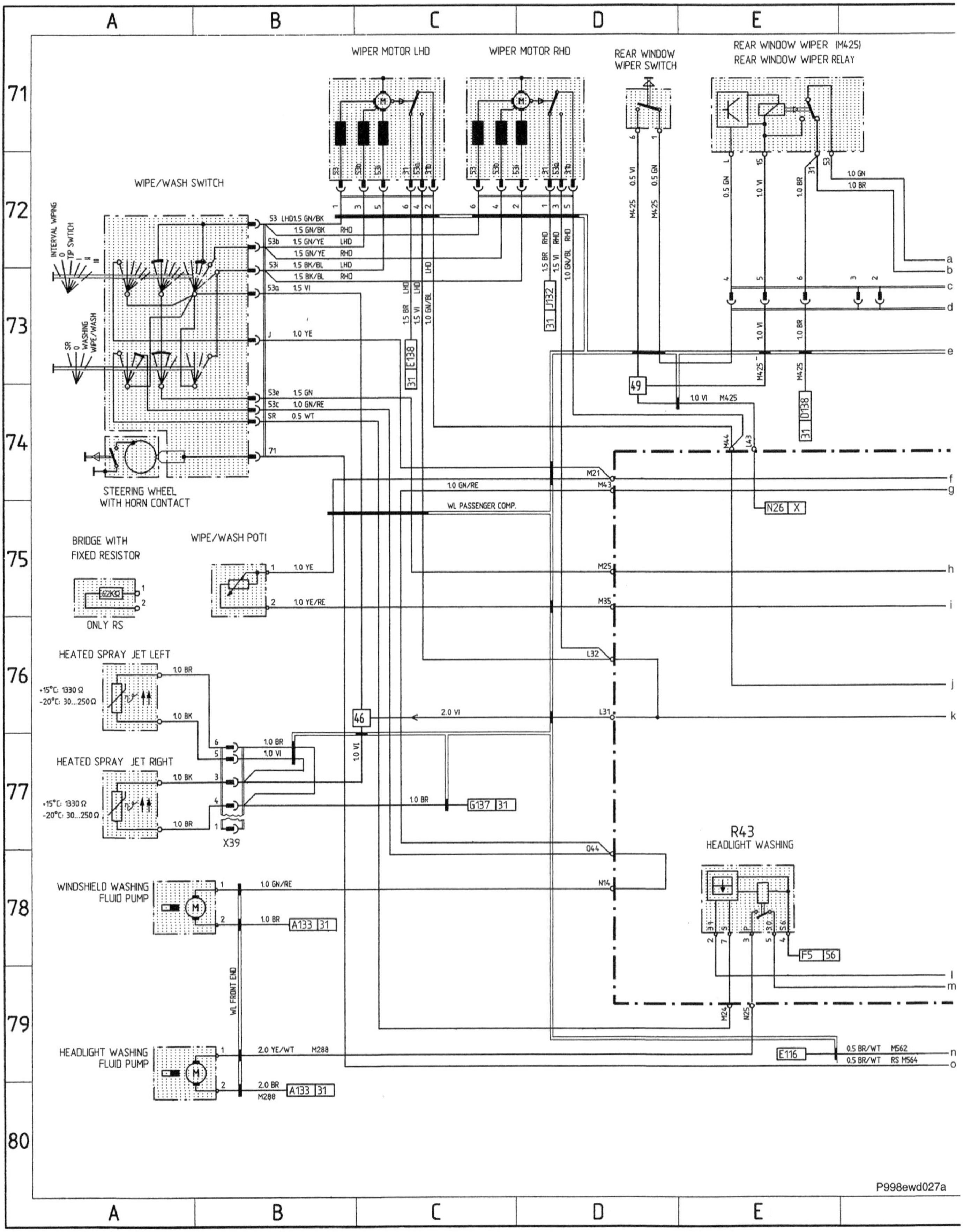

Rear spoiler, two-tone horns, wipe-and-wash cleaners (page 2 of 3)

Sheet 8 - 1996

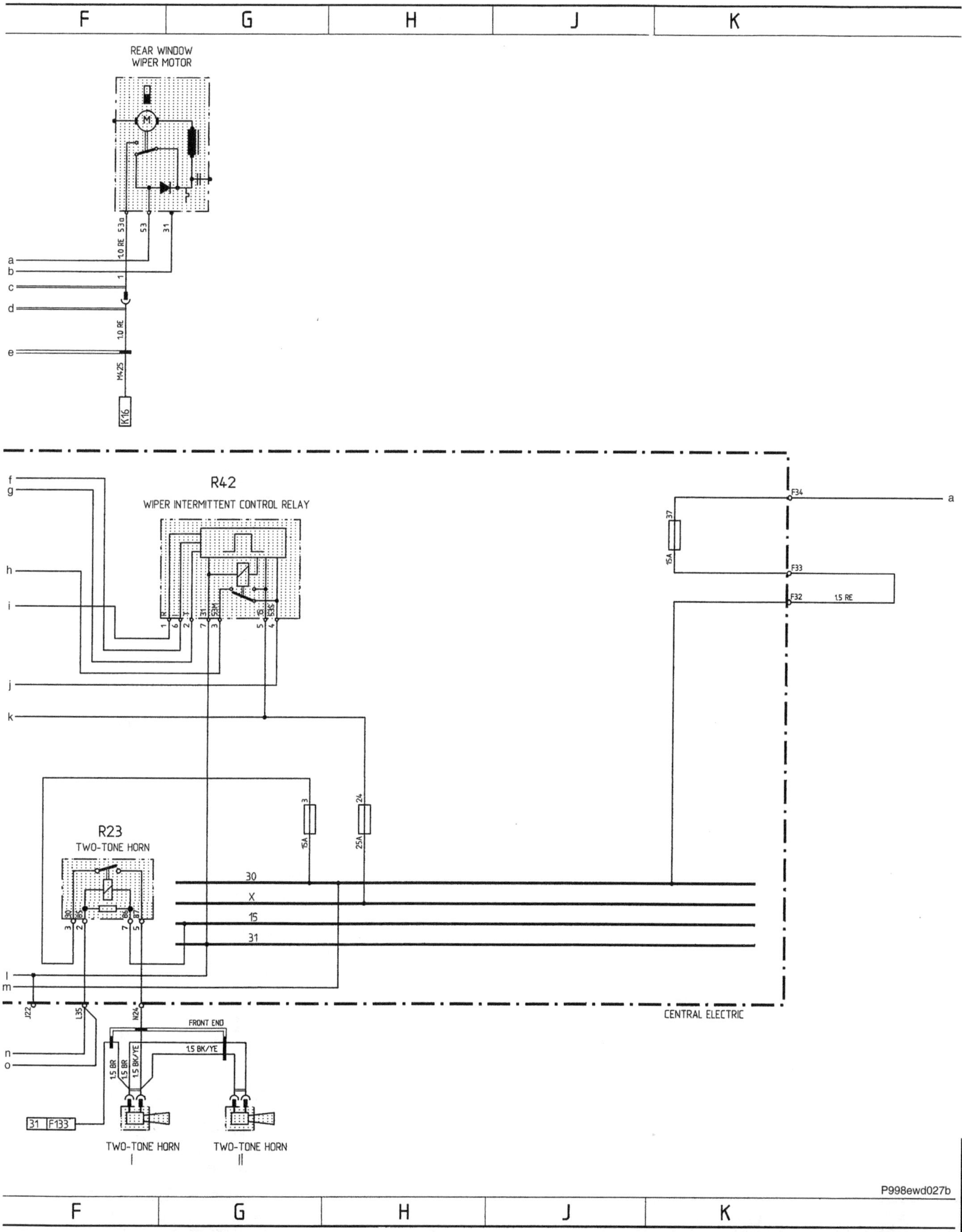

EWD

Sheet 8 - 1996 **Rear spoiler, two-tone horns, wipe-and-wash cleaners (page 3 of 3)**

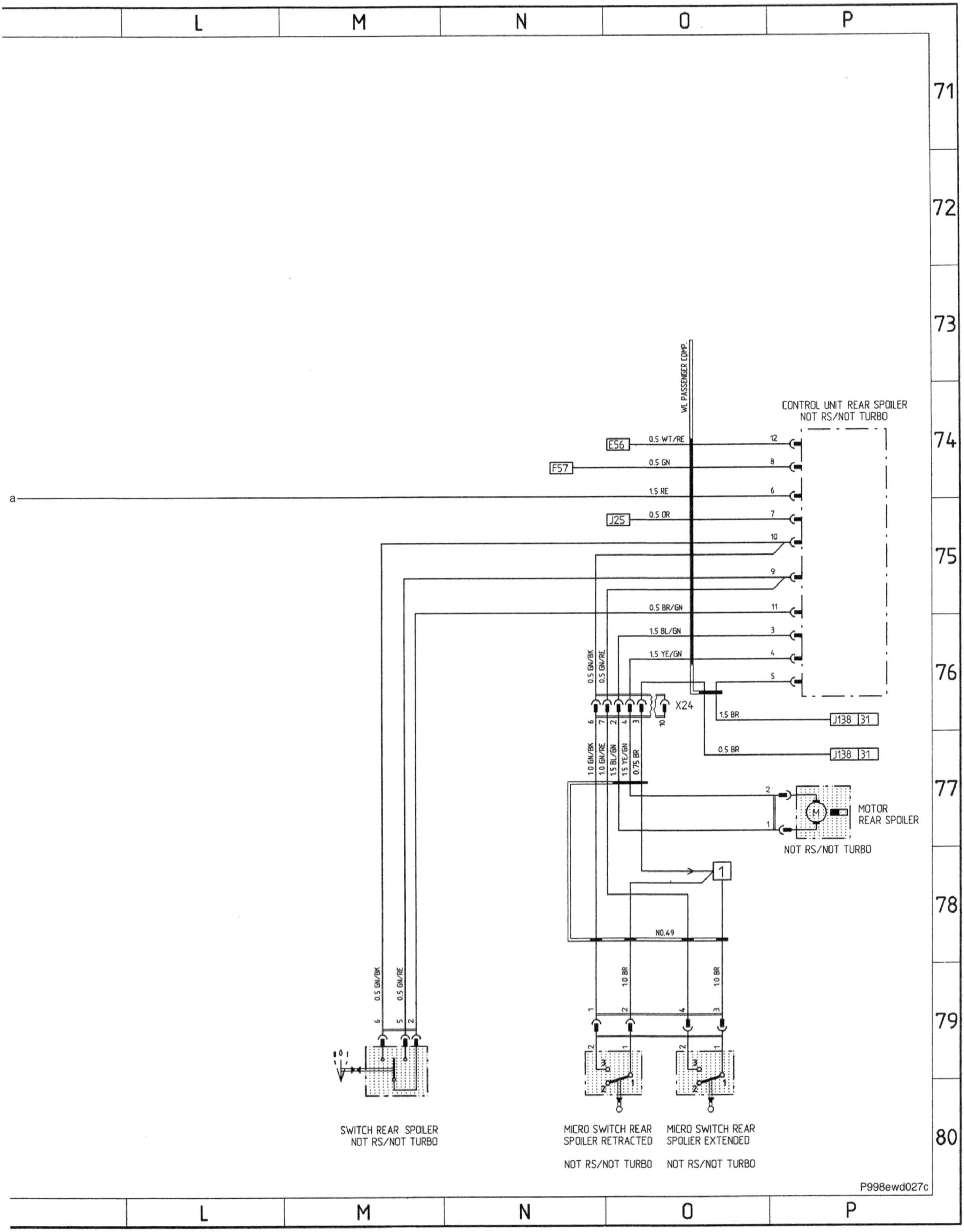

Radio, telephone (page 1 of 3)

Sheet 9 - 1996

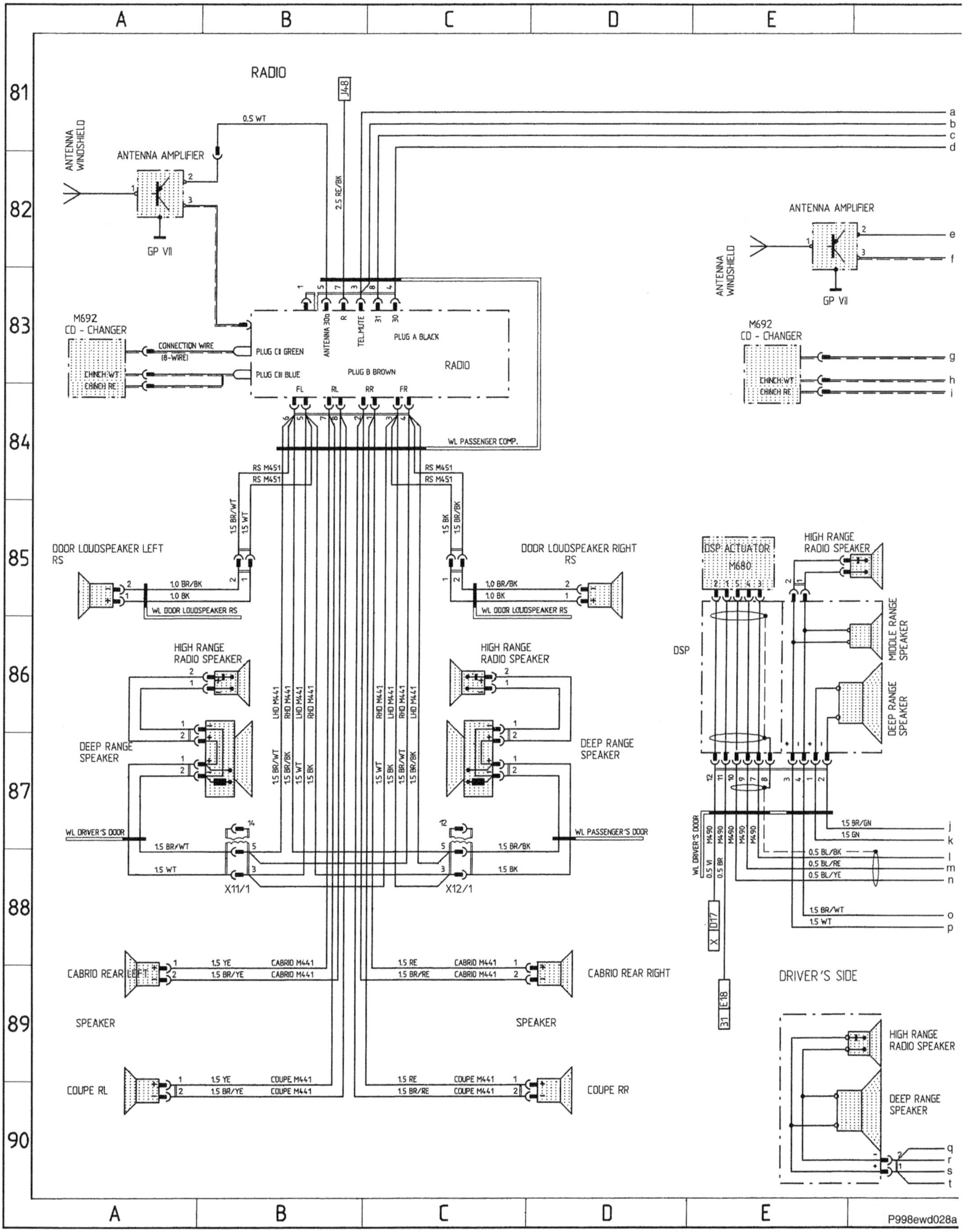

Sheet 9 - 1996

Radio, telephone (page 2 of 3)

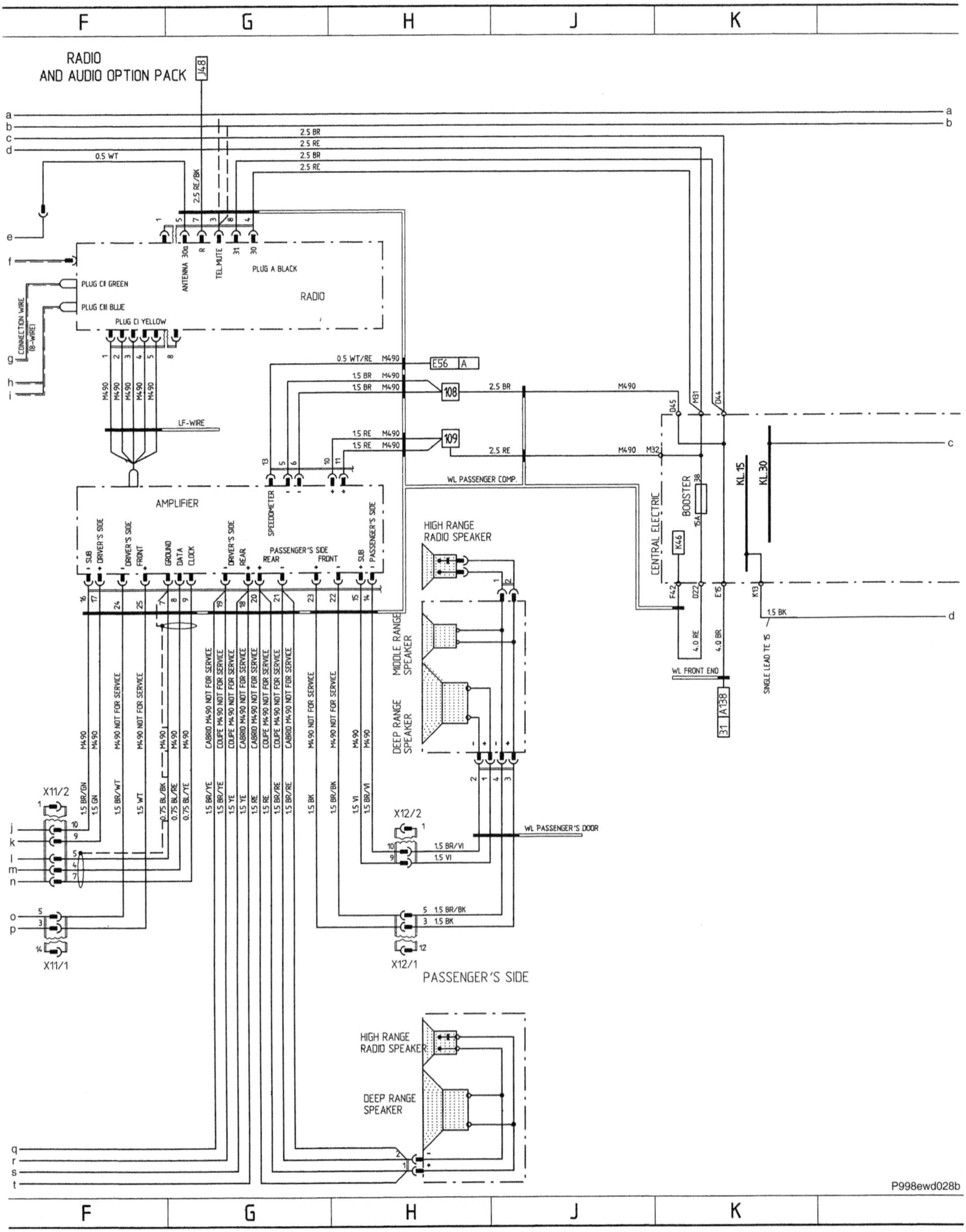

Radio, telephone
(page 3 of 3)

Sheet 9 - 1996

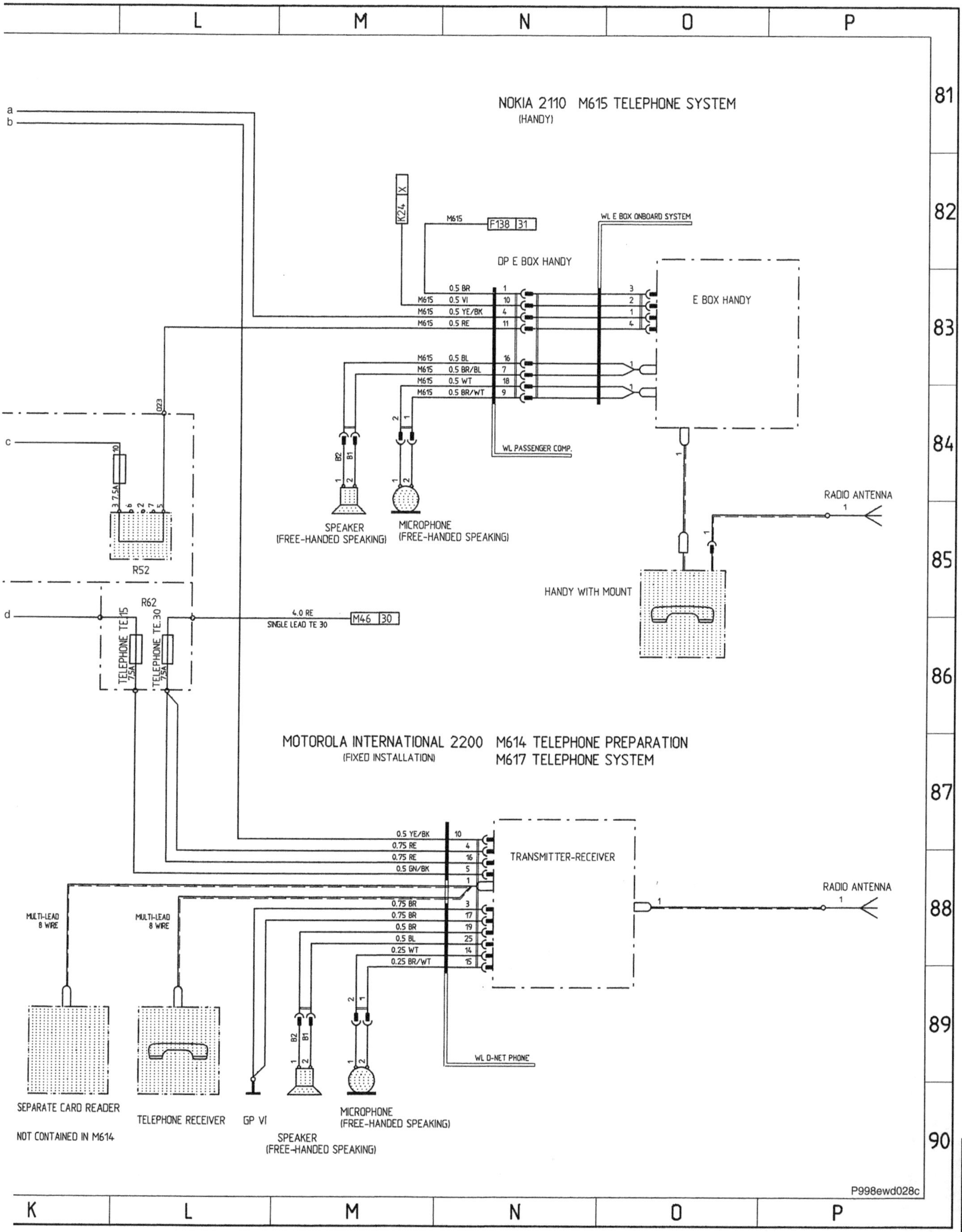

EWD

Sheet 10 -1996

Foglights
(page 1 of 1)

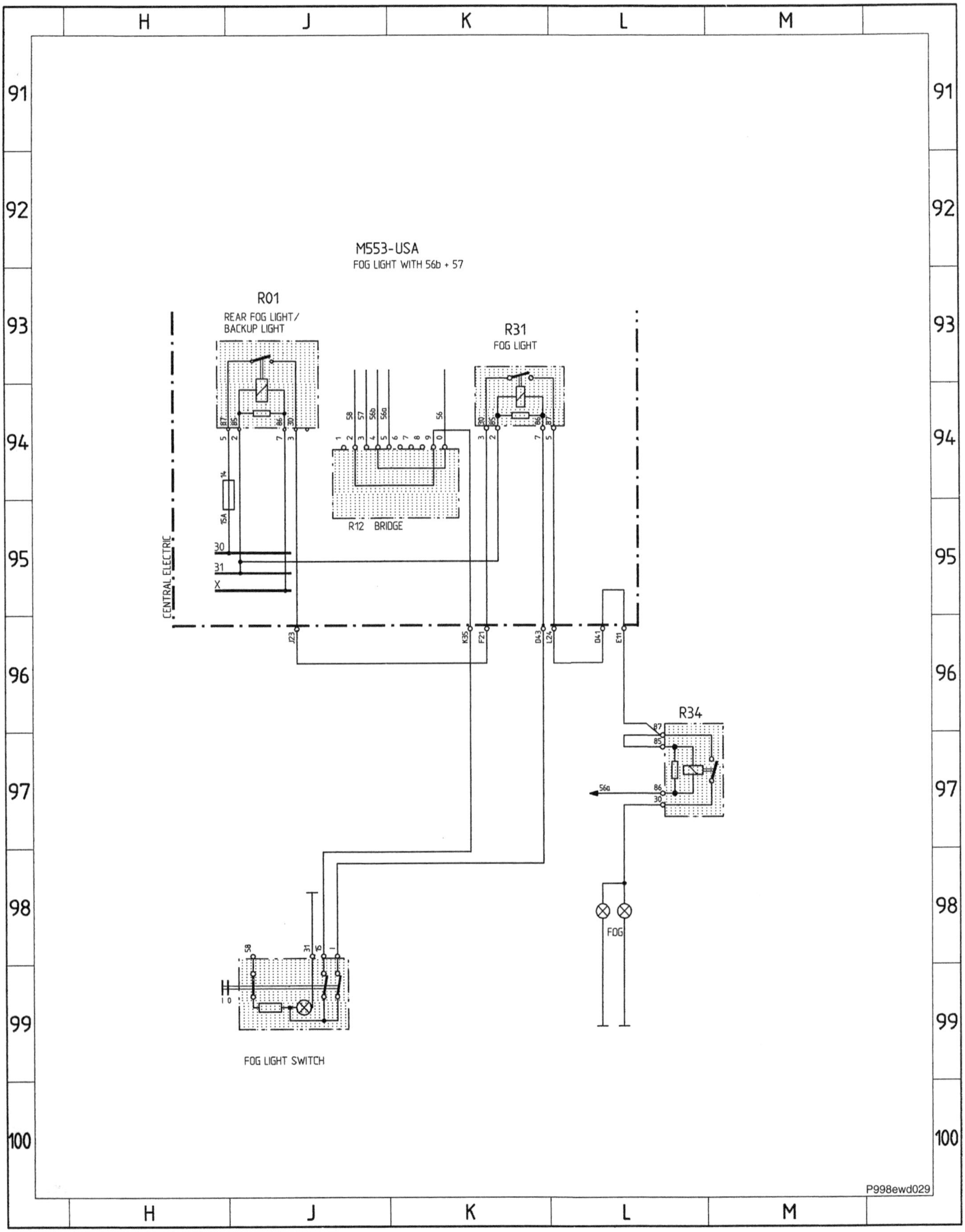

Anti-theft (option code M530) (page 1 of 1)

Sheet 11 - 1996

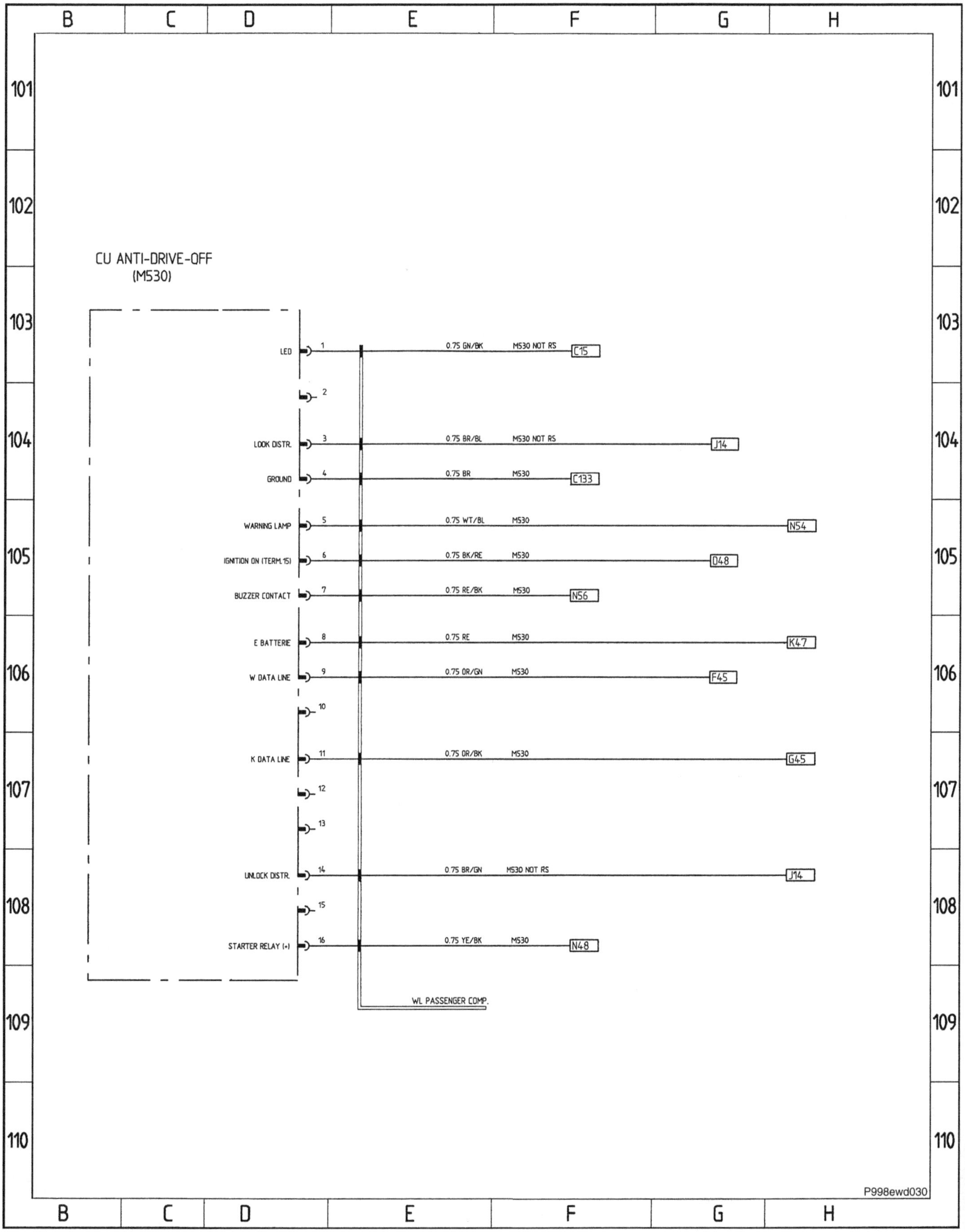

EWD

Sheet 12 - 1996

Airbag
(page 1 of 1)

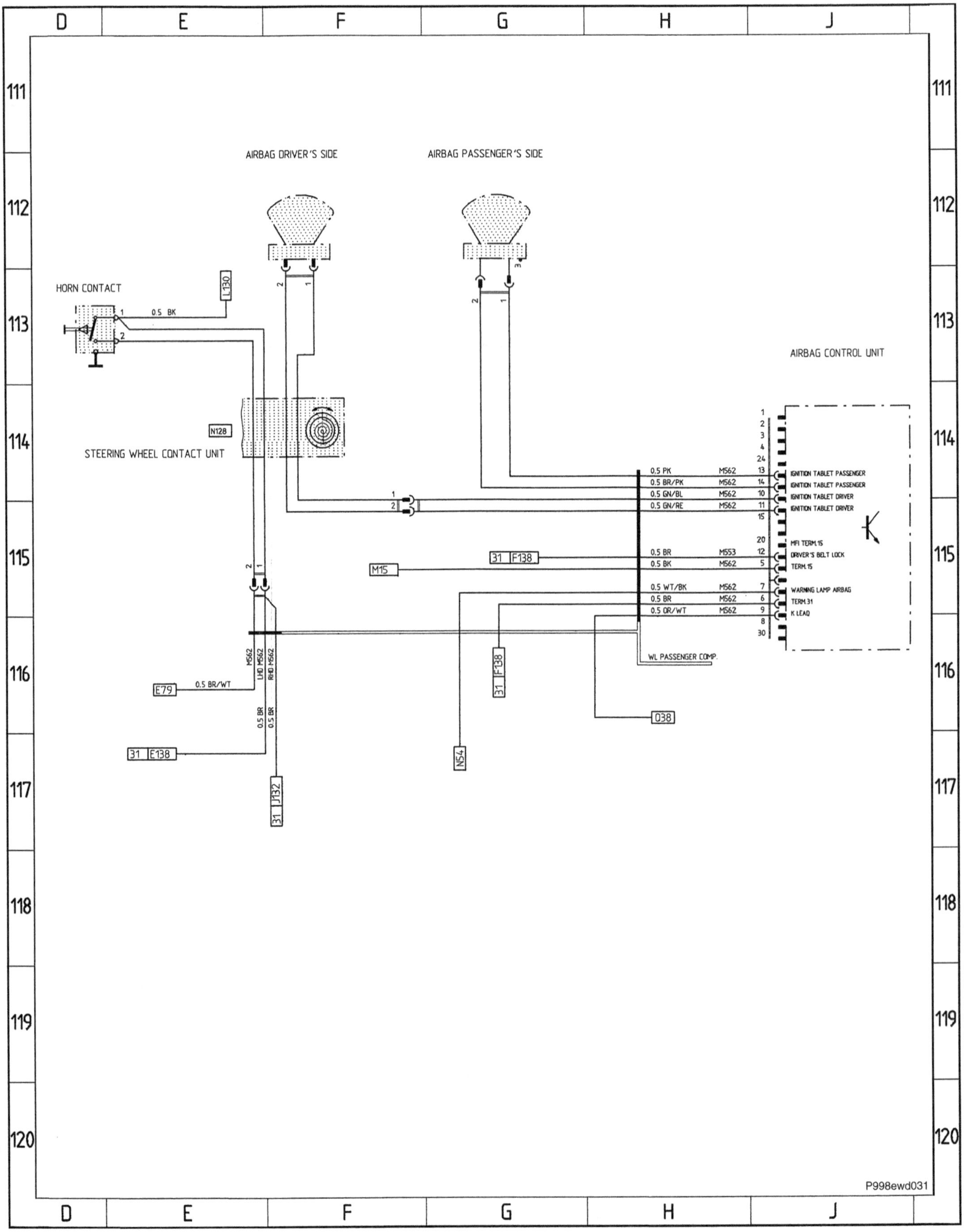

Tiptronic transmission (page 1 of 3)

Sheet 13 -1996

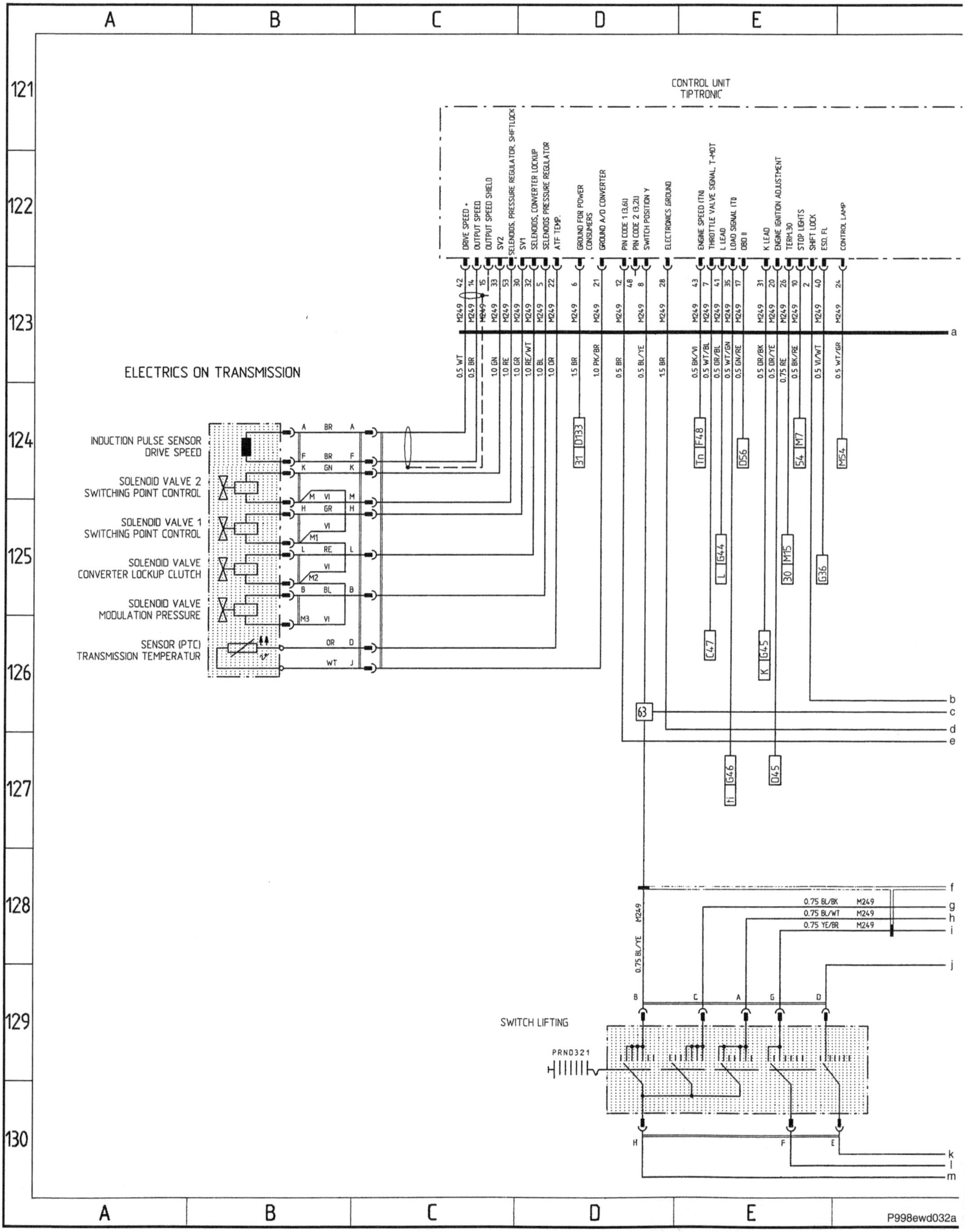

A B C D E

121 122 123 124 125 126 127 128 129 130

P998ewd032a

Sheet 13 -1996

Tiptronic transmission (page 2 of 3)

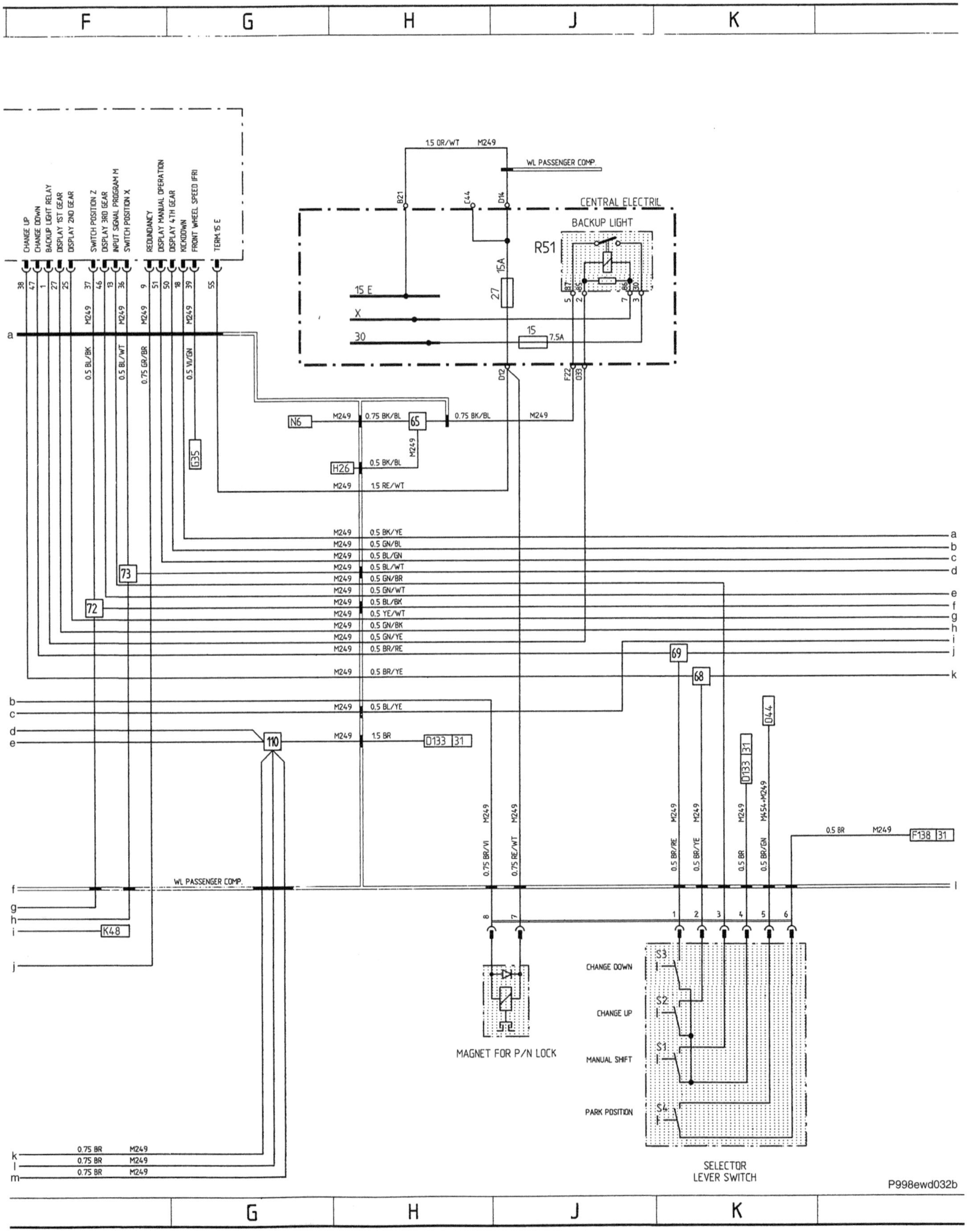

P998ewd032b

Tiptronic transmission (page 3 of 3)

Sheet 13 -1996

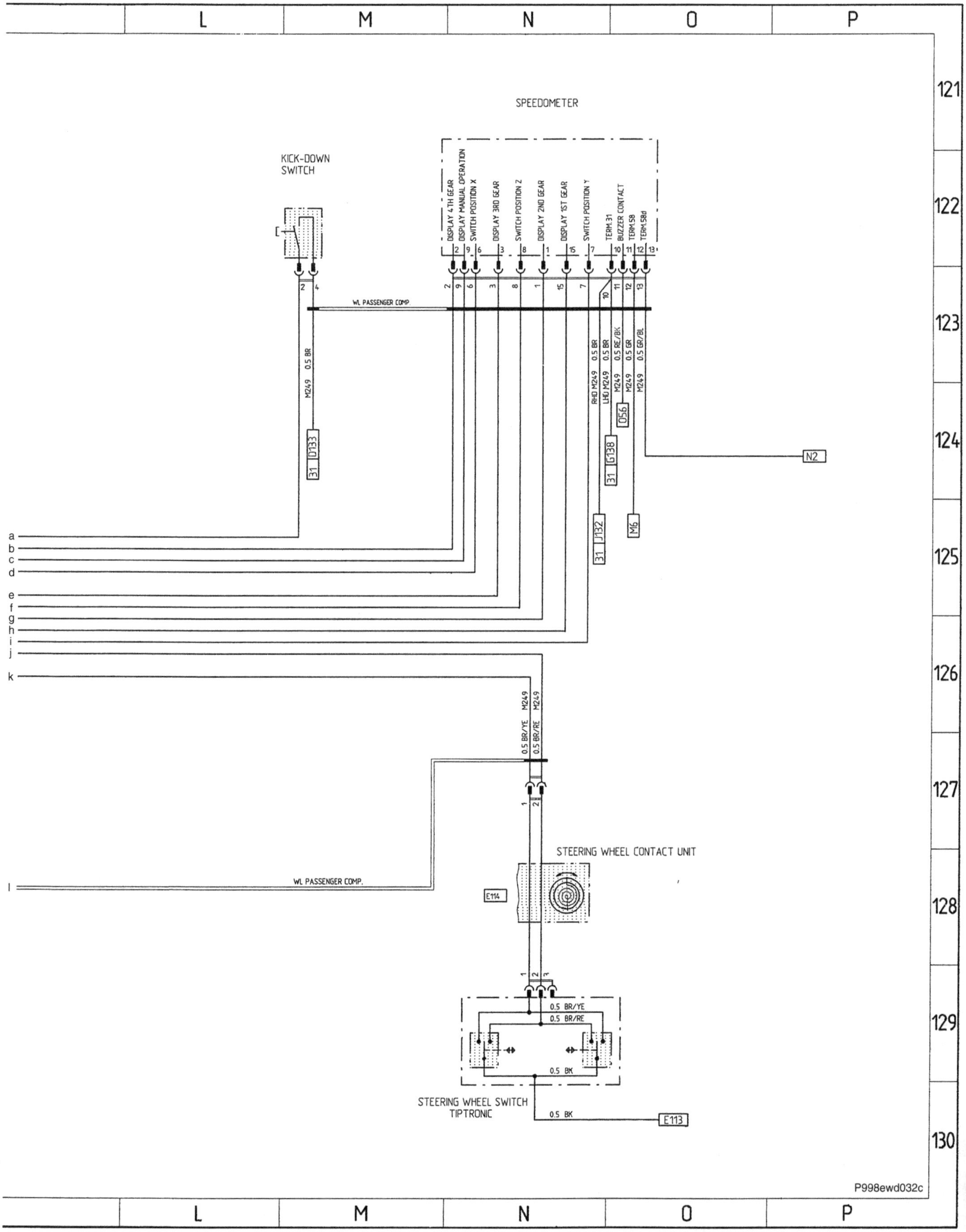

EWD

Sheet 14 - 1996

Ground points (page 1 of 3)

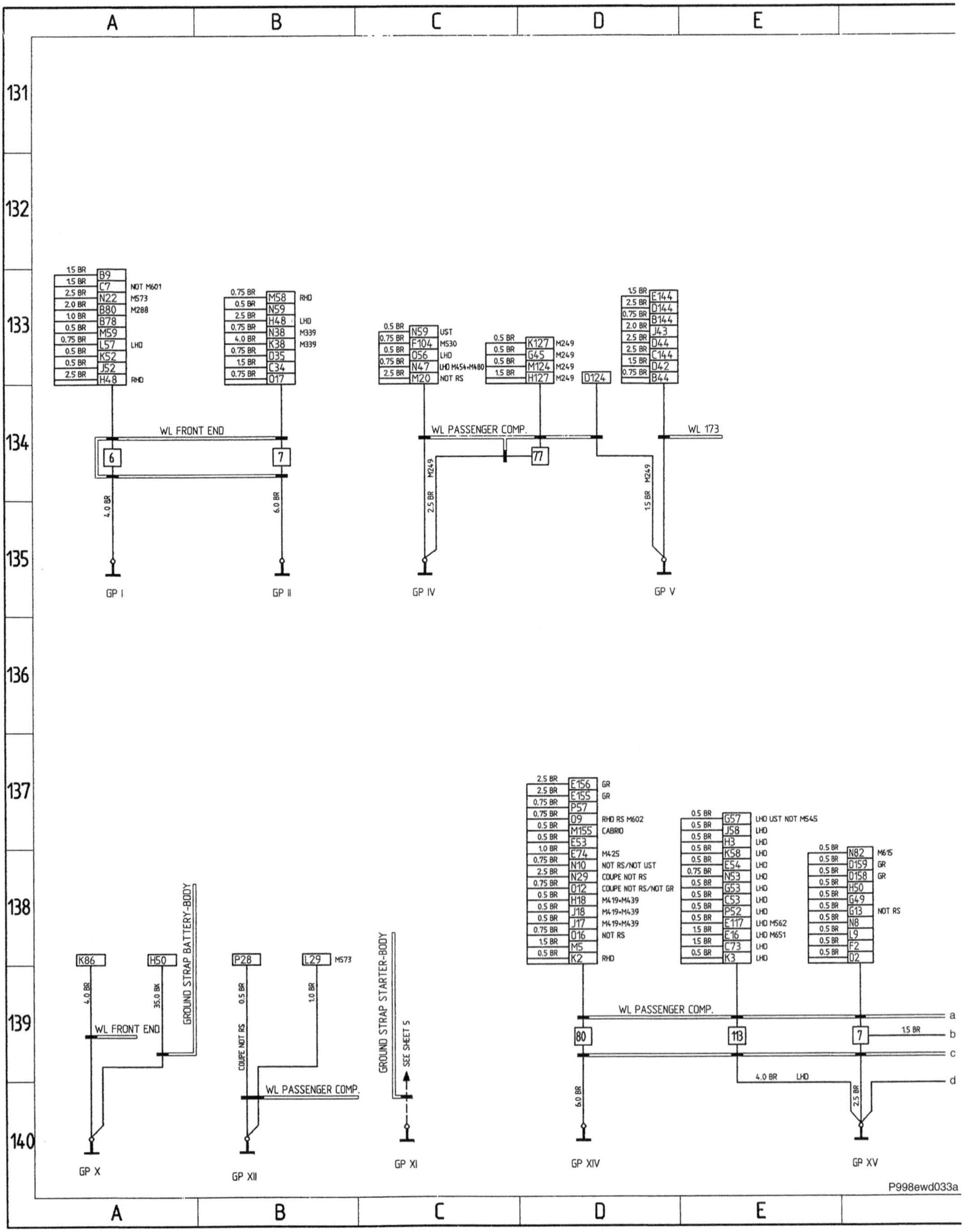

Ground points (page 2 of 3)

Sheet 14 - 1996

F	G	H	J	K

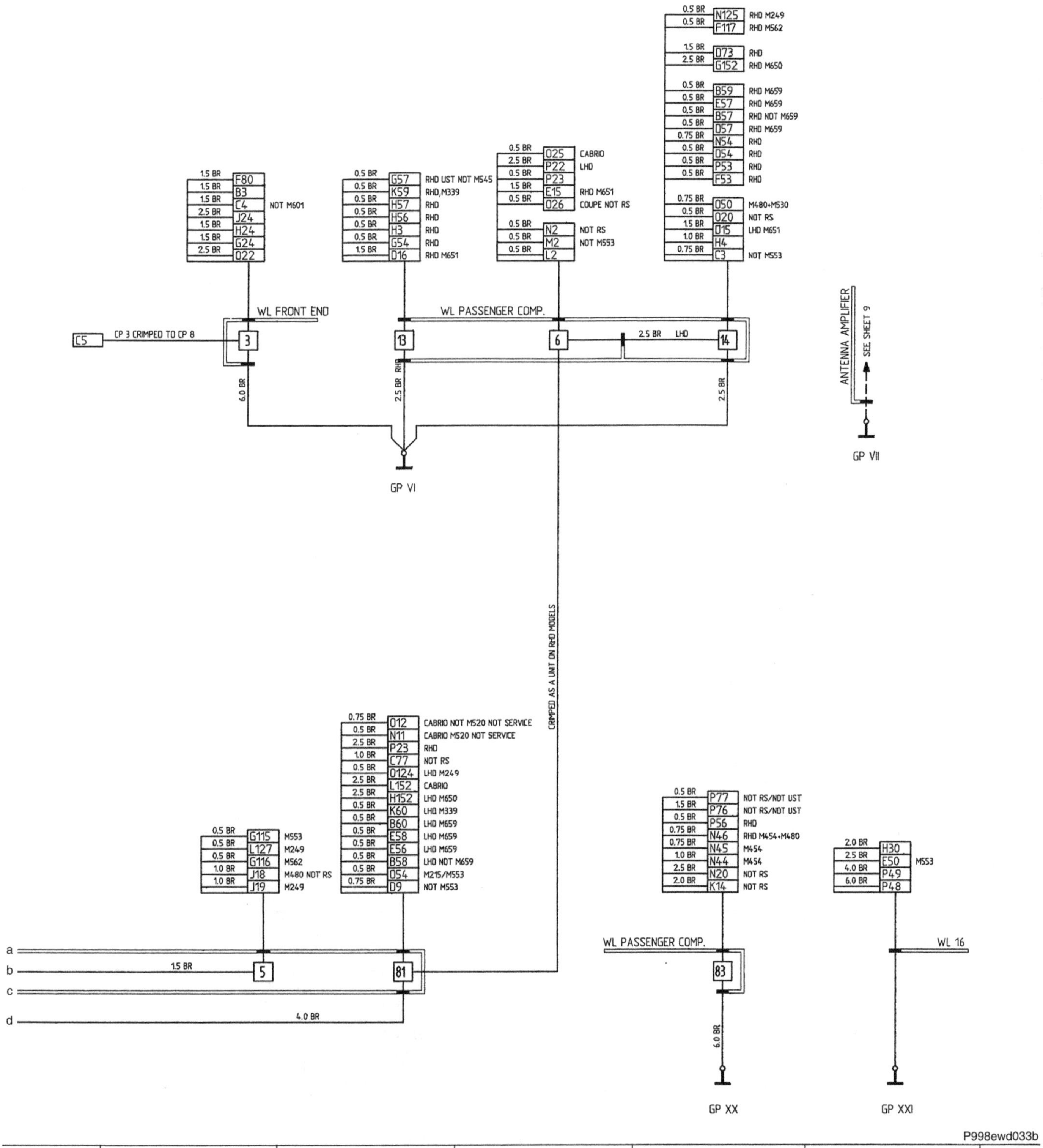

F	G	H	J	K

Sheet 14 - 1996

Ground points
(page 3 of 3)

CODE	NOTE	POSITION IN VEHICLE LHD
GP I	LUGGAGE COMPARTMENT, WHEEL HOUSING LEFT	2cP
GP II	LUGGAGE COMPARTMENT, WHEEL HOUSING RIGHT	2cM
GP IV	GROUND OF POWER CONSUMERS, UNDER SEAT LEFT	8eP
GP V	ENGINE ELECTRONICS/TIPTRONIC	11eP
GP VI	CENTRAL ELECTRIC, ON SIDE MEMBER RIGHT	6bL
GP VII	ANTENNA AMPLIFIER ON SIDE MEMBER LEFT	7bL
GP X	BATTERY, LUGGAGE COMPARTMENT LEFT	3eP
GP XI	GROUND STRAP (STARTER, TRANSMISSION, ENGINE, BODY)	14cM
GP XII	ENGINE COMPARTMENT REAR LEFT	15cQ
GP XIV	ENGINE COMPARTMENT REAR RIGHT	15cL
GP XV	INSTRUMENTS, SIDE MEMBER LEFT	7bQ
GP XX	SEAT CROSS MEMBER, UNDER SEAT RIGHT	8eL
GP XXI	GROUND OF POWER CONSUMERS, CAMSHAFT CASING CYL.3	15cP

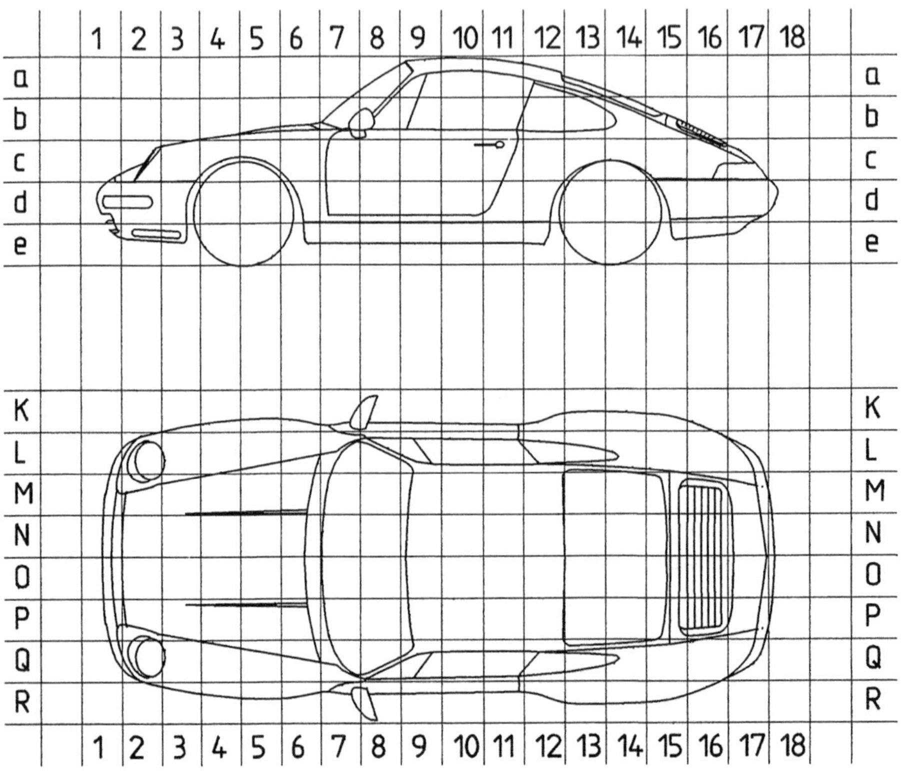

P998ewd033c

Convertible top, sunroof, glass roof (page 1 of 3)

Sheet 15 - 1996

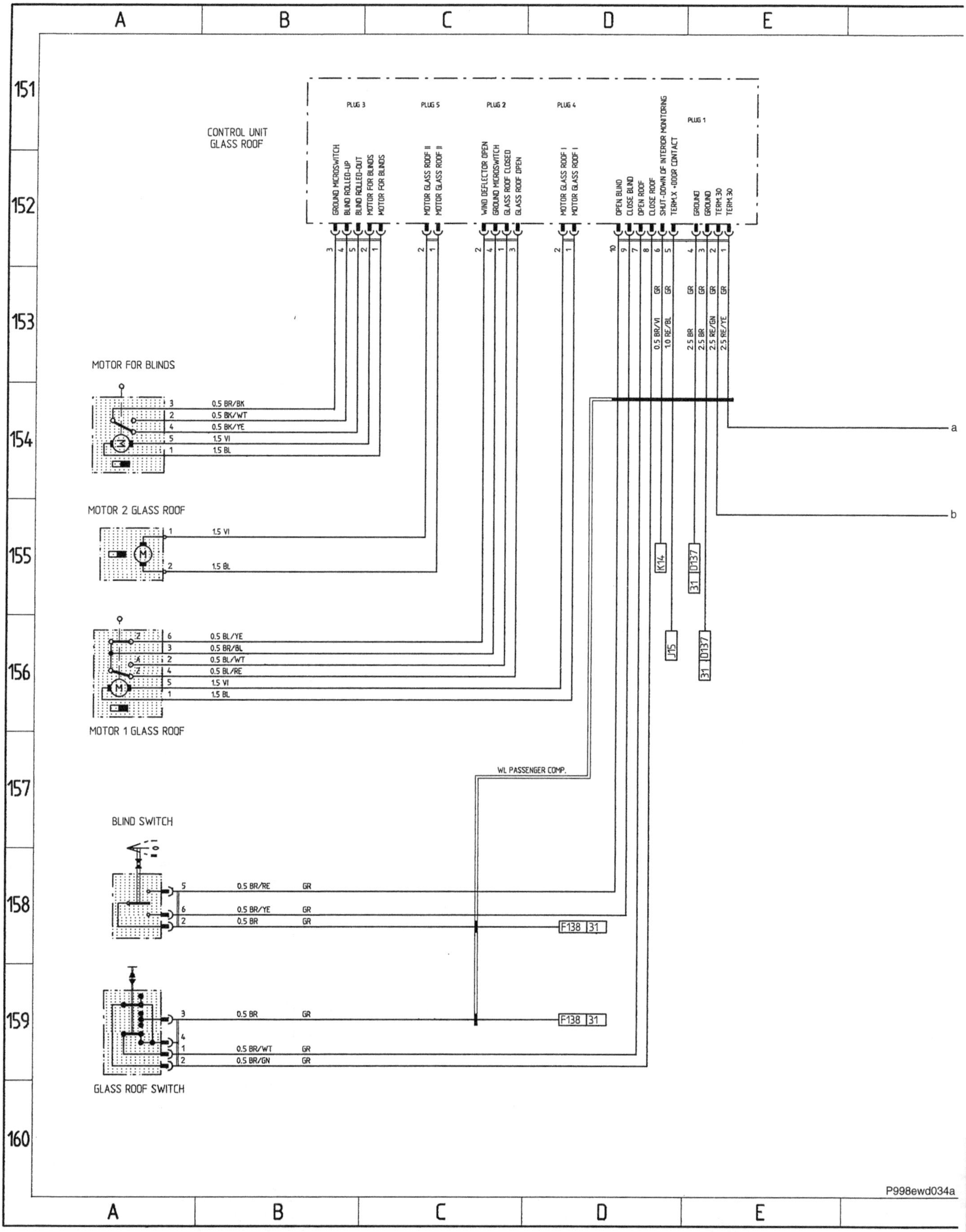

EWD

Sheet 15 - 1996

Convertible top, sunroof, glass roof (page 2 of 3)

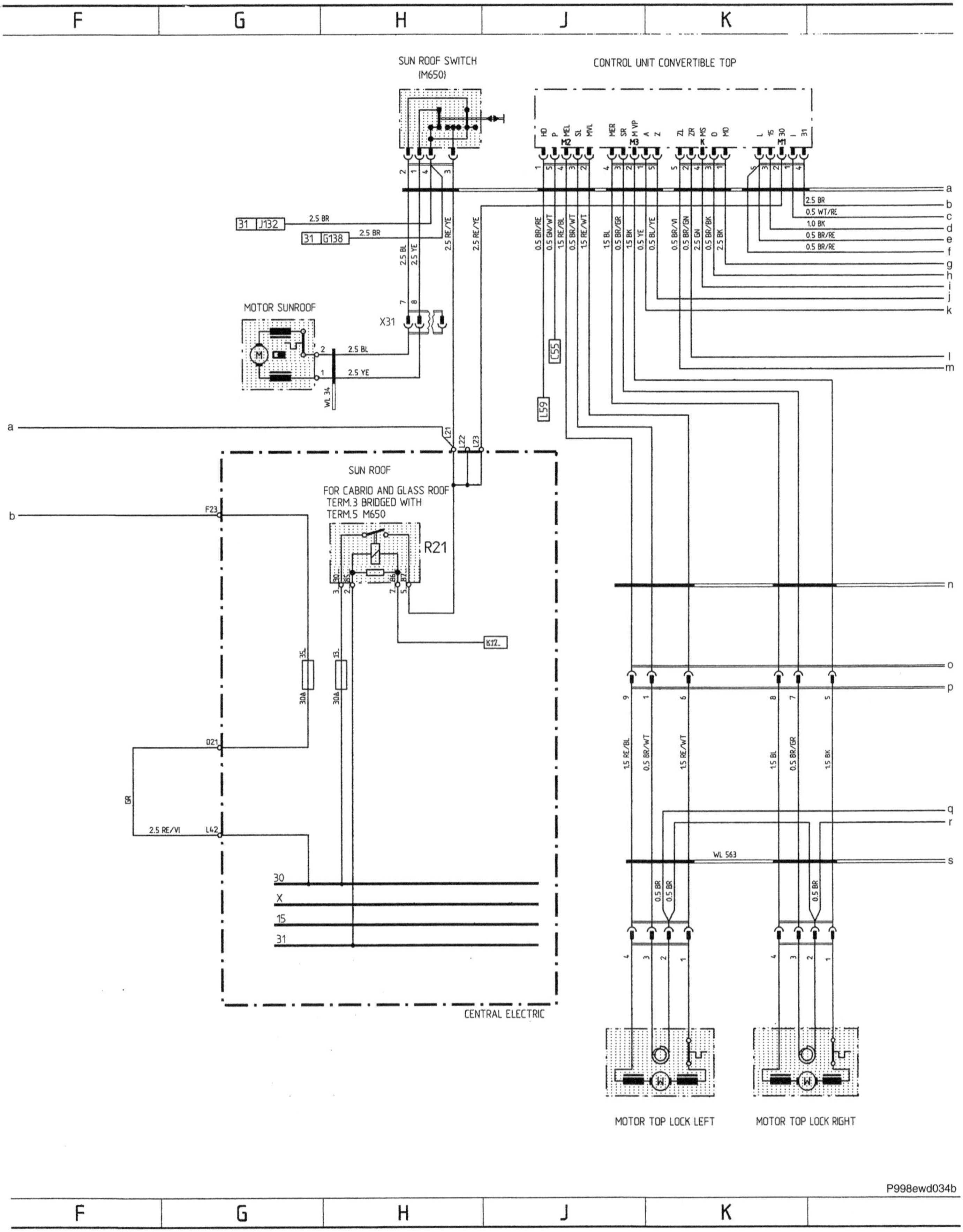

Convertible top, sunroof, glass roof (page 3 of 3)

Sheet 15 - 1996

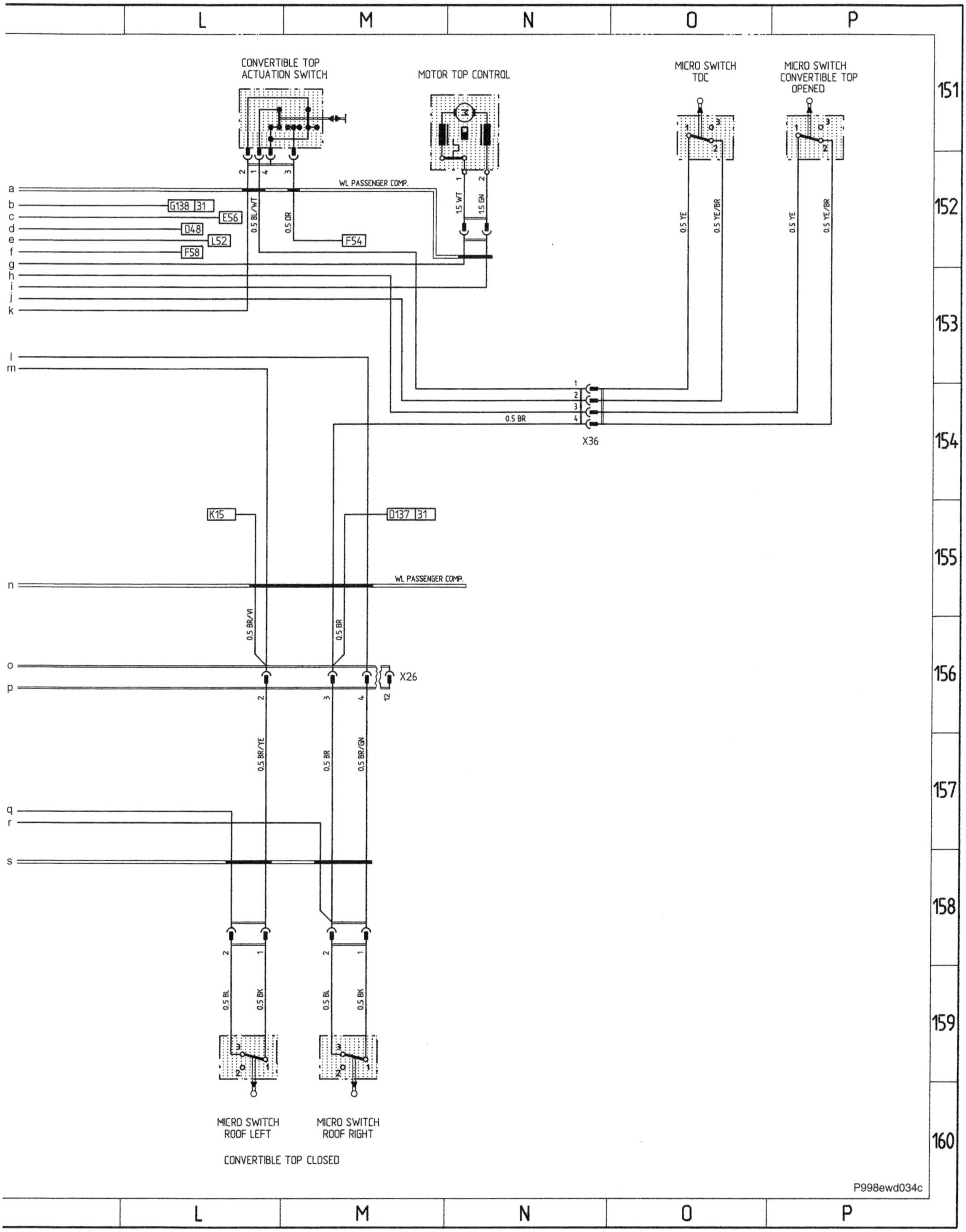

Sheet 16 - 1996

Construction components (page 1 of 3)

A | B | C | D | E

CONSTRUCTION COMPONENTS

DESIGNATION, FUNCTION	POSITION IN VEHICLE LHD	RHD	NOTE	FIELD IN WIRING DIAGRAM
ALARM HORN	13dP	13dP	REAR LH CROSSMEMBER	J20
ANTENNA AMPLIFIER	7bL	7bQ	IN LUGGAGE COMPARTMENT	A83,E82
OUTSIDE TEMP. SENSOR	2dP	2dP	FRONT LH BUMPER	A60
FLASHER	6cQ	6cL	IN FRONT OF INSTRUMENTS	D1
STOP LIGHT SWITCH	6dP	6dM	AT THE BRAKE PEDAL BELOW THE FLOOR PANEL	N9
DIAGNOSIS CONNECTION	8cO	8cO	IN CENTRE CONSOLE	G50
MFI + DI RELAY	6cM	6cM	CENTRAL ELECTRIC	J44
MFI + DI CONTROL UNIT	9eP	9eP	UNDER THE SEAT LEFT	B-G41
ENGINE SPEED SENSOR	15cP	15cP	ON ENGINE	A46
THROTTLE VALVE SENSOR	15bN	15bN	ON THROTTLE VALVE ASSEMBLY	A44
PRESSURE TRANSDUCER, CHARGE AIR PRESSURE	16bM	16bM	ON ENGINE	A56
PRESSURE SWITCH AIR CONDITIONING SYSTEM	6bO	6bO	IN LUGGAGE COMPARTMENT	E30
PRESSURE WARNING SWITCH	1dO	1dO	IN LUGGAGE COMPARTMENT	M/N39
E BOX HANDY	7cN	7cN	ON RH SIDE OF TRANSVERSE WALL	O83
TWO-TONE HORN I	3cL	3cL	IN WHEEL HOUSING FRONT RIGHT	F80
TWO-TONE HORN II	3cL	3cL	IN WHEEL HOUSING FRONT RIGHT	G80
BLOWER FINAL STAGE	6bN	6bN	IN LUGGAGE COMPARTMENT	F30
BLOWER ENGINE COMPARTMENT	16cO	16cO	ON ENGINE	H30
BLOWER MOTOR RIGHT	6cM	6cM	IN LUGGAGE COMPARTMENT	G30
BLOWER MOTOR LEFT	6cP	6cP	IN LUGGAGE COMPARTMENT	F30
GONG RELAY	6cQ	—	LH SIDE IN FRONT OF INSTRUMENTS	P53/54
HALL SENSOR	17cO	17cO	ON DISTRIBUTOR	A46
REAR WINDOW WIPER RELAY	15bO	15bO	ON REAR WINDOW WIPER MOTOR	E71
HYDRAULIC UNIT ABS	3dP	3dP	IN LUGGAGE COMPARTMENT	A/B39
ABS/ABD HYDRAULIC UNIT	3dP	3dP	IN LUGGAGE COMPARTMENT	C39
INSIDE LIGHT WITH INTERIOR SENSOR	9aN	9aN	FRONT ROOF FRAME	N11
V-BELT CONTROL	17dO	17dO	ON ENGINE REAR MIDDLE	A55
KICK-DOWN SWITCH	6dO	6dM	BY ACCELERATOR PEDAL UNDER FLOOR PANEL	M122
KNOCK SENSOR CYL.1-3	15cP	15cP	ON ENGINE	A47
KNOCK SENSOR CYL.4-6	16cN	16cN	ON ENGINE	A47
CONDENSER BLOWER	2dQ	2dQ	IN WHEEL HOUSING FRONT LEFT	N21
FUEL PUMP	6eM	6eP	BEHIND THE BASE	H50
CLUTCH SWITCH CRUISE CONTROL	6dP	6dN	BY CLUTCH PEDAL UNDER FLOOR PANEL	P46
OXYGEN SENSOR CYL.1-3	16dP	16dP	ON ENGINE	F150
OXYGEN SENSOR CYL.4-6	16dM	16dM	ON ENGINE	D150
IDLE SPEED POSITIONER	15bN	15bN	ON ENGINE	H/J41
MASS AIR FLOW SENSOR	16bO	16bO	ON ENGINE	A43
MICRO SWITCH ROOF RIGHT	9aN	9aN	AT CONVERTIBLE TOP FRAME	M159
MICRO SWITCH ROOF LEFT	9aO	9aO	AT CONVERTIBLE TOP FRAME	L159
MICRO SWITCH REAR SPOILER EXTENDED	16bP	16bP	ON TAILGATE	O79
MICRO SWITCH REAR SPOILER RETRACTED	16bP	16bP	ON TAILGATE	O79
MICRO SWITCH TDC	12cL	12cL	BEHIND SIDE PANEL COVERING	O151
MICROSWITCH FOR TOP CONTROL	12cL	12cL	BEHIND SIDE PANEL COVERING	P151
MOTOR SUNROOF	12aN	12aN	IN ROOF	G153
MOTOR TOP CONTROL RIGHT	15cN	15cN	BEHIND REAR SEAT	N151
MOTOR TOP CONTROL LEFT	15cO	15cO	BEHIND REAR SEAT	N151
MOTOR TOP LOCK RIGHT	9aN	9aN	AT CONVERTIBLE TOP FRAME	K/L159
MOTOR TOP LOCK LEFT	9aO	9aO	AT CONVERTIBLE TOP FRAME	J/K159
MOTORS GLASS ROOF AND SUN BLINDS	14bO	14bO	REAR WINDOW SHELF	A154-156
REAR FOG LIGHT CUTOUT RELAY	17dN	17dN	BEHIND REAR PANEL	H108
NTC HEATER FAN	16cP	16cP	ON ENGINE	G30
NTC TEMPERATURE FOR INTAKE AIR	15cN	15cN	ON ENGINE	A42
OIL PRESSURE SENSOR/SWITCH	14cN	14cN	ON ENGINE	A54
OIL COOLER BLOWER	2dL	2dL	IN WHEEL HOUSING FRONT RIGHT	O21
OIL LEVEL SENSOR	12cK	12cK	IN WHEEL HOUSING REAR RIGHT	E51
OIL TEMPERATURE SENSOR FOR INSTRUMENT	14cN	14cN	ON ENGINE	A53
OIL TEMPERATURE SENSOR FOR OIL COOLER BLOWER	2dL	2dL	IN WHEEL HOUSING FRONT RIGHT	J21
SWITCH LIFTING	12dO	12dO	ON GEARBOX	D-F129

P998ewd035a

A | B | C | D | E

Construction components (page 2 of 3)

Sheet 16 - 1996

DESIGNATION, FUNCTION	POSITION IN VEHICLE LHD	RHD	NOTE	FIELD IN WIRING DIAGRAM
BRAKE BOOSTER PUMP	1dO	1dO	IN LUGGAGE COMPARTMENT	L39
RELAY FOR REAR WINDOW DEFOGGER	16cP	16cP	CARRIER PLATE IN ENGINE COMPARTMENT	O/P29/30
ABS RELAY	4cL	4cL	IN LUGGAGE COMPARTMENT RIGHT	D39
RELAY CONDENSER BLOWER	6cM	6cM	CENTRAL ELECTRIC	L21
RELAY AC COMPRESSOR	16cP	16cP	CARRIER PLATE IN ENGINE COMPARTMENT	L30
RELAY TWO-TONE HORN	6cM	6cM	CENTRAL ELECTRIC	F78
RELAY POWER WINDOW	6cM	6cM	CENTRAL ELECTRIC	K18/19
RELAY BLOWER ENGINE COMPARTMENT	16cP	16cP	CARRIER PLATE IN ENGINE COMPARTMENT	J/K30
RELAY STARTER TIPTRONIC-TRANSMISSION	6cM	6cM	CENTRAL ELECTRIC	N50
RELAY BACKUP LIGHT	7bQ	7bQ	IN LUGGAGE COMPARTMENT LEFT	J122
RELAY THERMO SWITCH CATALYTIC CONVERTER	6cM	6cM	CENTRAL ELECTRIC	L/M51
RELAY TERM.X	6cM	6cM	CENTRAL ELECTRIC	J/K44
RELAY TERM.15 E	6cM	6cM	CENTRAL ELECTRIC	K/L44
RELAY RFL BACKUP LIGHT	6cM	6cM	CENTRAL ELECTRIC	J/K7, A-M94
RELAY FRONT FOG LAMP	6cM	6cM	CENTRAL ELECTRIC	J7, C-O94
RELAY OIL COOLER BLOWER	6cM	6cM	CENTRAL ELECTRIC	K21
RELAY BRAKE BOOSTER PUMP	4cL	4cL	IN LUGGAGE COMPARTMENT RIGHT	M39
RELAY SUN ROOF	6cM	6cM	CENTRAL ELECTRIC	H155
SECONDARY AIR PUMP RELAY	17cP	17cP	ON CARRIER ON LH SIDE IN ENGINE COMPARTMENT	A50
DAYTIME RUNNING LIGHT CONTROL RELAY	6cM	6cM	CENTRAL ELECTRIC	D8
RELAY DAYTIME RUNNING LIGHT CANADA	6cM	6cM	CENTRAL ELECTRIC	D1
RELAY FOG LIGHT SWITCH OFF USA/JAPAN	6cM	6cM	CENTRAL ELECTRIC	H-P/94, F/G10
BACKUP LIGHT SWITCH	14dN	14dN	BY TRANSMISSION RIGHT	K/L1
SHIFT VALVE RESONANCE FLAP	15bN	15bN	ON ENGINE	H41
WINDSHIELD WASHING FLUID PUMP	3cQ	3cQ	BY WASHING FLUID RESERVOIR IN WHEEL HOUSING FL	A78
FUSES ENGINE COMPARTMENT	16cP	16cP	CARRIER PLATE IN ENGINE COMPARTMENT	L28
FUSE IN LUGGAGE COMPARTMENT	4cP	4cP	BELOW HYDRAULIC UNIT COVER	D37/M47
HEADLIGHT CLEANER PUMP	3cQ	3cQ	BY WASHING FLUID RESERVOIR IN WHEEL HOUSING FL	A79/80
HEADLIGHT CLEANER RELAY	6cM	6cM	CENTRAL ELECTRIC	E78
STARTER	14dN	14dN	BY CLUTCH HOUSING RIGHT	P49/50
DRIVE CRUISE CONTROL	15cO	15cO	ON ENGINE	N43/44
CONTROL UNIT ABS	2dM	2dM	IN LUGGAGE COMPARTMENT	A-J32
CONTROL UNIT AIRBAG	6cO	6cO	ON INTRUMENT PANEL	J114/116
CONTROL UNIT ALARM SYSTEM/CENTRAL LOCKING SYSTEM	9eM	9eM	UNDER THE SEAT RIGHT	J-M12
CONTROL UNIT CONVERTIBLE TOP	6cP	6cO	ON INTRUMENT PANEL	J/K151
CONTROL UNIT GLASS ROOF	13cO	13cO	EMERGENCY SEAT WELL	D151-152
CONTROL UNIT REAR SPOILER	7cM	7cP	ON RH SIDE OF TRANSVERSE WALL	P74-76
CONTROL UNIT HEATING, AIR CONDITIONING	7cO	7cN	ON INTRUMENT PANEL	B-F21
CONTROL UNIT INSTRUMENT LIGHT	6cM	6cM	CENTRAL ELECTRIC	K1
CONTROL UNIT CRUISE CONTROL	9eN	9eN	UNDER THE SEAT RIGHT	P43/44
CONTROL UNIT TIPTRONIC TRANSMISSION	10dP	10dP	UNDER THE SEAT LEFT	C-G122
SECONDARY AIR PUMP	16cN	16cN	ON ENGINE	D/E50
LIGHTS ON BUZZER	6cQ	6cL	IN FRONT OF INSTRUMENTS	P56/57
TIMING VALVE	16cM	16cM	ON ENGINE	L150
FUEL LEVEL SENSOR	5cM	5cP	IN FUEL TANK	K51
TANK VENTING VALVE	16cP	16cP	ON ENGINE	K41
CHARGE AIR TEMPERATURE SENSOR	16cN	16cN	ON ENGINE	A142
ENGINE TEMPERATURE SENSOR	15cP	15cP	ON ENGINE	A45
RESISTOR BLOWER	16dP	16dP	IN ENGINE COMPARTMENT	J30
RESISTOR FOR BLOWER CONDENSER	2dQ	2dQ	IN WHEEL HOUSING FRONT LEFT	M21
RESISTOR OIL COOLER BLOWER	2dL	2dL	IN WHEEL HOUSING FRONT RIGHT	M21
AMPLIFIER	9eM	9eM	UNDER THE SEAT RIGHT	F-H85, M85
WIPER INTERMITTENT CONTROL RELAY	6cM	6cM	CENTRAL ELECTRIC	F/G75
CENTRAL ELECTRIC	6cM	6cM	IN LUGGAGE COMPARTMENT	—
CENTRAL INFORMER	7cP	7cM	ON FIREWALL	M32
IGNITION CONTROL UNIT	11dP	11dP	UNDER THE SEAT LEFT	J41
IGNITION COIL I	17cP	17cP	ON ENGINE	L41
IGNITION COIL II	17cP	17cP	ON ENGINE	L41

P998ewd035b

Sheet 16 - 1996

Construction components
(page 3 of 3)

	L	M	N	O	P

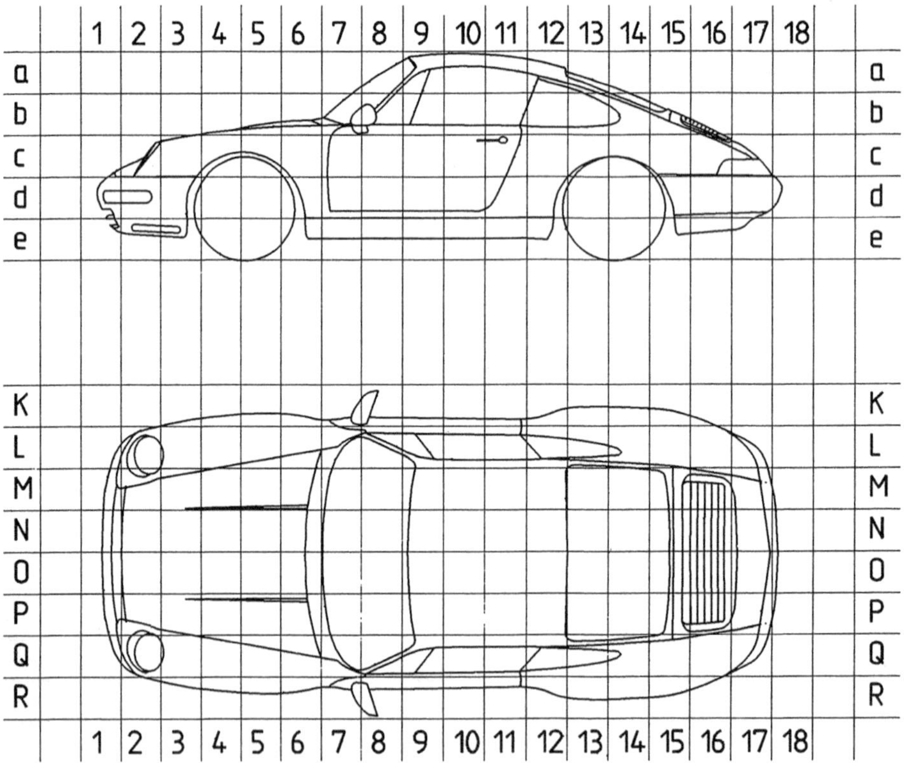

P998ewd035c

	L	M	N	O	P

Connectors and abbreviations (page 1 of 2)

Sheet 17 - 1996

A	B	C	D	E	F

PLUG CONNECTIONS

CODE	NUMBER OF PINS	DESIGNATION, FUNCTION	POSITION IN VEHICLE LHD	RHD	NOTE	FIELD IN WIRING DIAGRAM
X 1/1	10	CONNECTION WL PASSENGER COMPARTMENT/ WL FRONT END	6cL	6cL	IN LUGGAGE COMPARTMENT RIGHT	G/H34,H/J48
X 1/2	10	CONNECTION WL PASSENGER COMPARTMENT/ WL FRONT END	6cL	6cL	IN LUGGAGE COMPARTMENT RIGHT	C9,E36,G48,B58/59
X 3	14	CONNECTION WL PASSENGER COMPARTMENT/ WL MOTOR	16cP	16cP	CARRIER PLATE IN ENGINE COMPARTMENT	K/L3,J27/28,048/49,B53/54
X4/1	14	CONNECTION WL PASSENGER COMPARTMENT/ WL MFI + DI	9eP	9eP	UNDER THE SEAT LEFT	C/D45/46,B55,L143
X4/2	14	CONNECTION WL PASSENGER COMPARTMENT/ WL MFI + DI	9eP	9eP	UNDER THE SEAT LEFT	F45/46,B55,G/H148
X11/1	14	CONNECTION WL PASSENGER COMPARTMENT/ WL DOOR DRIVER'S SIDE	6cQ	6cL	UNDER THE INSTRUMENT PANEL DIVER'S SIDE	E/F16/17,B87/88, F88
X11/2	10	CONNECTION WL PASSENGER COMPARTMENT/ WL DOOR DRIVER'S SIDE	6cQ	6cL	UNDER THE INSTRUMENT PANEL DIVER'S SIDE	C16 ,F87/88
X12/1	12	CONNECTION WL PASSENGER COMPARTMENT/ WL DOOR PASSENGER'S SIDE	6cL	6cQ	UNDER THE INSTRUMENT PANEL PASSENGER'S SIDE	E/F13,C87/88 ,H88
X12/2	10	CONNECTION WL PASSENGER COMPARTMENT/ WL DOOR PASSENGER'S SIDE	6cL	6cQ	UNDER THE INSTRUMENT PANEL PASSENGER'S SIDE	C14,H87/88
X20	14	CONNECTION WL PASSENGER COMPARTMENT/ WL HEATER, AIR CONDITIONER	6cM	6cM	UNDER INSTRUMENT PANEL RIGHT	G25/26
X24	10	CONNECTION WL PASSENGER COMPARTMENT/ WL TAILGATE	16cL	16cL	BOX RH SIDE OF ENGINE COMPARTMENT	010,015/16,076
X25	10	CONNECTION WL PASSENGER COMPARTMENT/ WL TAIL LIGHTS	16cL	16cL	BOX RH SIDE OF ENGINE COMPARTMENT	N4-6
X26	12	CONNECTION WL PASSENGER COMPARTMENT/ WL CONVERTIBLE TOP	12cL	12cL	BEHIND SIDE PANEL COVERING	J-M156
X28	6	CONNECTION WL PASSENGER COMPARTMENT/ WL IGNITION LOCK	6cP	6cM	ON STEERING COLUMN	L48/49
X29	2	CONNECTION WL PASSENGER COMPARTMENT/ OIL TANK SENSOR	16cL	16cL	BOX RH SIDE OF ENGINE COMPARTMENT	E52
X31	8	CONNECTION WL PASSENGER COMPARTMENT/ WL INSIDE LIGHT COUPE	15bM	15bM	BELOW THE PARCEL TRAY	014,H153
X32	6	CONNECTION WL PASSENGER COMPARTMENT/ WL INSIDE LIGHT CONVERTIBLE	6cQ	6cQ	UNDER INSTRUMENT PANEL LEFT	N/012
X33	2	CONNECTION WL PASSENGER COMPARTMENT/ WL SIDE TURN SIGNAL LEFT	7bQ	7bQ	IN LUGGAGE COMPARTMENT ON SIDE MEMBER LEFT	D/E9
X34	2	CONNECTION WL PASSENGER COMPARTMENT/ WL SIDE TURN SIGNAL RIGHT	7bL	7bL	IN LUGGAGE COMPARTMENT ON SIDE MEMBER RIGHT	C2
X36	4	CONNECTION WL PASSENGER COMPARTMENT/ SWITCH LIFTING CONVERTIBLE TOP	12cL	12cL	BEHIND SIDE PANEL COVERING	N154
X37	8	CONNECTION WL PASSENGER COMPARTMENT/ WL SEAT LEFT	9eP	9eN	UNDER THE DRIVER SEAT	N19,A70,E65/66
X38	8	CONNECTION WL PASSENGER COMPARTMENT/ WL SEAT RIGHT	9eN	9eP	UNDER THE PASSEGER SEAT	N19,A70,E65/66
X39	6	CONNECTION WL PASSENGER COMPARTMENT/ WL LUGGAGE COMPARTMENT HOOD	7bL	7bL	ON RH SIDE OF LUGGAGE COMP. LID HINGE	018,B77
X60	21	CONNECTION WL MOTOR/ WL MFI + DI	16cP	16cP	CARRIER PLATE IN ENGINE COMPARTMENT	B46,E48,K42,B55,M142,M148
X65	6	CONNECTION WL DOOR DRIVER'S SIDE/ WL OUTSIDE MIRROR LEFT	8cR	8cK	DRIVER'S DOOR	B18/19
X66	6	CONNECTION WL DOOR PASSENGER'S SIDE/ WL OUTSIDE MIRROR RIGHT	8cK	8cR	PASSENGER'S DOOR	B13
X69	14	CONNECTION WL FRONT END/ WL HEATER, AIR CONDITIONER	6cM	6cM	IN LUGGAGE COMPARTMENT RIGHT	G23/24

M-NUMBERS

M002	RS ROAD VERSION	M487	NORWAY
M003	RS RACING SAFETY EQUIPMENT	M490	AUDIO OPTION PACK
M004	RS TURBO VERSION	M513	LUMBAR SUPPORT SEAT RIGHT
M113	CANADA VERSION (DAYTIME RUNNING LIGHT)	M520	INTERIOR MONITORING
M139	ADJUSTABLE SEAT HEATING, LEFT SEAT	M530	CODE LOCK
M193	JAPAN VERSION	M545	LARGE TANK
M215	SAUDI-ARABIA VERSION	M553	USA / CANADA VERSION
M224	AUTOMATIC BRAKE DIFFERENTIAL	M562	AIRBAG
M249	AUTOMATIC TRANSMISSION (TIPTRONIC)	M564	WITHOUT AIRBAG
M339	FWD	M573	AIR CONDITIONING SYSTEM
M340	ADJUSTABLE SEAT HEATING, RIGHT SEAT	M576	WITHOUT REAR FOG LIGHT
M383	SPORT SEAT LEFT ELECTRIC HEIGHT ADJUSTMENT	M586	LUMBAR SUPPORT SEAT LEFT
M387	SPORT SEAT RIGHT ELECTRIC HEIGHT ADJUSTMENT	M601	LITRONIC
M419	REAR PACKAGE TRAY	M602	HIGH MOUNTED STOP LIGHT
M425	REAR WINDOW WIPER	M605	HEADLIGHT VERTICAL AIM CONTROL
M437	COMFORT SEAT LEFT	M614	TELEPHONE PREPARATION
M438	COMFORT SEAT RIGHT	M615	CORDLESS PHONE
M439	ELECTRIC CONVERTIBLE TOP	M650	POWER SUNROOF
M441	RADIO PREPARATION	M651	POWER WINDOWS
M451	LIMITED RADIO PREPARATION	M659	ON-BOARD COMPUTER
M454	CRUISE CONTROL	M680	DIGITAL SOUND PACKAGE
M480	TRANSMISSION	M692	CD-CHANGER

CONNECTION

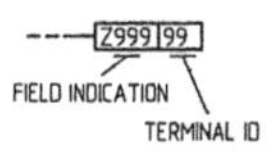

P	IGNITION SWITCH POSITION	0	56	LIGHT SWITCH POSITION	II
X	IGNITION SWITCH POSITION	I+II	56a/56b	HIGH BEAM/LOW BEAM	
15	IGNITION SWITCH POSITION	II+III	57	LIGHT SWITCH POSITION	I
R	IGNITION SWITCH POSITION	I+II+III	58	LIGHT SWITCH POSITION	I+II
K	DIAGNOSIS K		58R/58L	SIDE MARKER LIGHT RIGHT/LEFT	
L	DIAGNOSIS L		58d	INSTRUMENT LIGHTS	
30	PERMANENT POSITIVE		Ti	INJECTION SIGNAL	
31	GROUND		Tn	ROTATIONAL SPEED SIGNAL	
54	STOP LIGHT		61	GENERATOR D+	

P998ewd036a

A	B	C	D	E	F

EWD

Sheet 17 - 1996

Connectors and abbreviations (page 2 of 2)

G	H	J	K	L

ABBREVIATIONS

CODE	MEANING	CODE	MEANING	CODE	MEANING
ABD	AUTOMATIC BRAKE DIFFERENTIAL	SERVICE	SERVICE	CL	CLOSE
ABS	ANTILOCK BRAKING SYSTEM	INS.C	INSTRUMENT CLUSTER	SA	SAUDI ARABIA
AC	AIR CONDITIONING SYSTEM	TE	TERMINAL	CU	CONTROL UNIT
BPAU	BRAKE BOOSTER	SCS	COMBINED STEERING COLUMN SWITCH	SILA	SAFETY LAMP
CLOCK	CLOCK SIGNAL	WL	WIRING LOOM	WP	WELD POINT
CO	CARBON MONOXIDE	LED	LIGHT-EMITTING DIODE	HCS	HEADLIGHT CLEANING SYSTEM
DATA	DATA SIGNAL	LHD	LEFT-HAND DRIVE	PL	PLUG
DEF	DEFROST	HVAC	HEADLIGHT VERTICAL AIM CONTROL	TEL.MUTE	RADIO MUTING CONTROL
SS	SPEED SENSOR	M5.2	MOTRONIC 5.2	TI	INJECTION TIME
ESO	ENGINE-SPEED SENSOR OUTPUT	GP	GROUND POINT	TN	ENGINE SPEED
ESS	ENGINE SPEED SENSOR	LF	LOW FREQUENCY	DP	DISCONNECTING POINT
TV	THROTTLE VALVE	FOG	FOG LIGHT	USA	USA
TVS	THROTTLE VALVE SENSOR	NO.	NUMBER	UST	911 TURBO
MFI + DI	DIGITAL ENGINE ELECTRONICS	RFL	REAR FOG LIGHT	FA	FRONT AXLE
DSP	DIGITAL SOUND PACKAGE	NTC	NEGATIVE TEMPERATURE COEFFICIENT	FL	FRONT LEFT
ETC	ELECTRICAL TRANSMISSION CONTROL	OP	OPEN	FR	FRONT RIGHT
INJ. VALVE	FUEL INJECTOR	PIN	PIN	CP	CONNECTING POINT
D I	ELECTRIC IGNITION SYSTEM WITH KNOCK CONTROL	POT	POTENTIOMETER	WW	WORLDWIDE
GR	GLASS ROOF	PSD	PORSCHE LIMITED SLIP DIFFERENTIAL	CE	CENTRAL ELECTRIC
RA	REAR AXLE	PWM	PULSE WIDTH MODULATION	ADL	ADDITIONAL HIGH BEAM
HF	HIGH FREQUENCY	RDK	TIRE PRESSURE CONTROL	IC	IGNITION CIRCUIT
HFM	HOT-FILM AIR MASS METER	ROW	REST OF WORLD	CLS	CENTRAL LOCKING SYSTEM
RL	REAR LEFT	BL	BACKUP LIGHT	TRP	TWIN-COIL ROTARY POSITIONER
RR	REAR RIGHT	RHD	RIGHT-HAND DRIVE	CWL	CENTRAL WARN. LIGHT

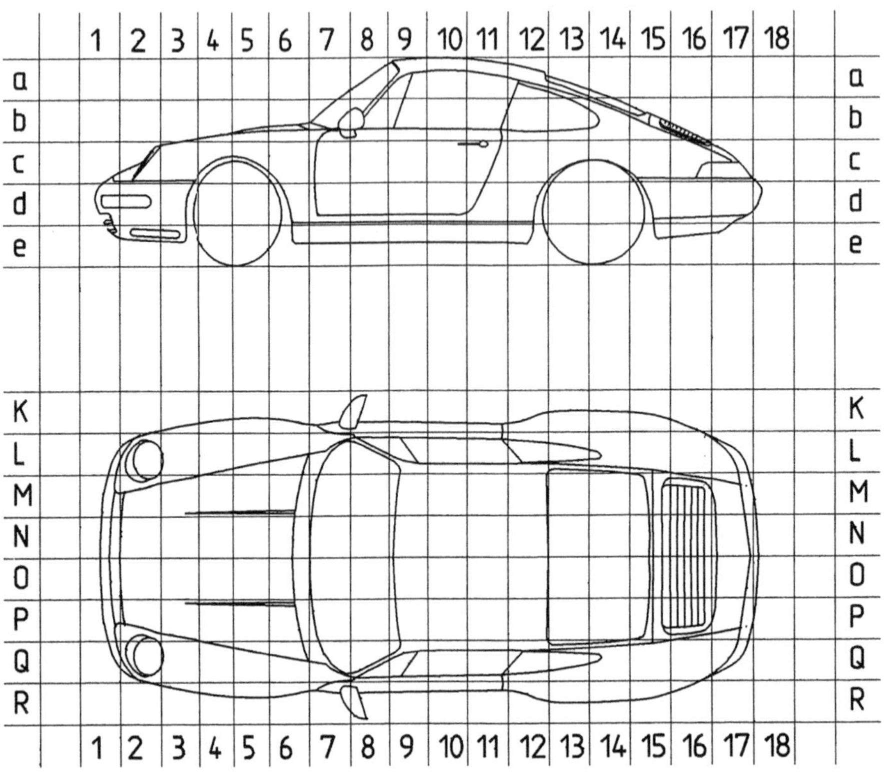

P998ewd036b

G	H	J	K	L

Central electric (page 1 of 3)

Sheet 18 - 1996

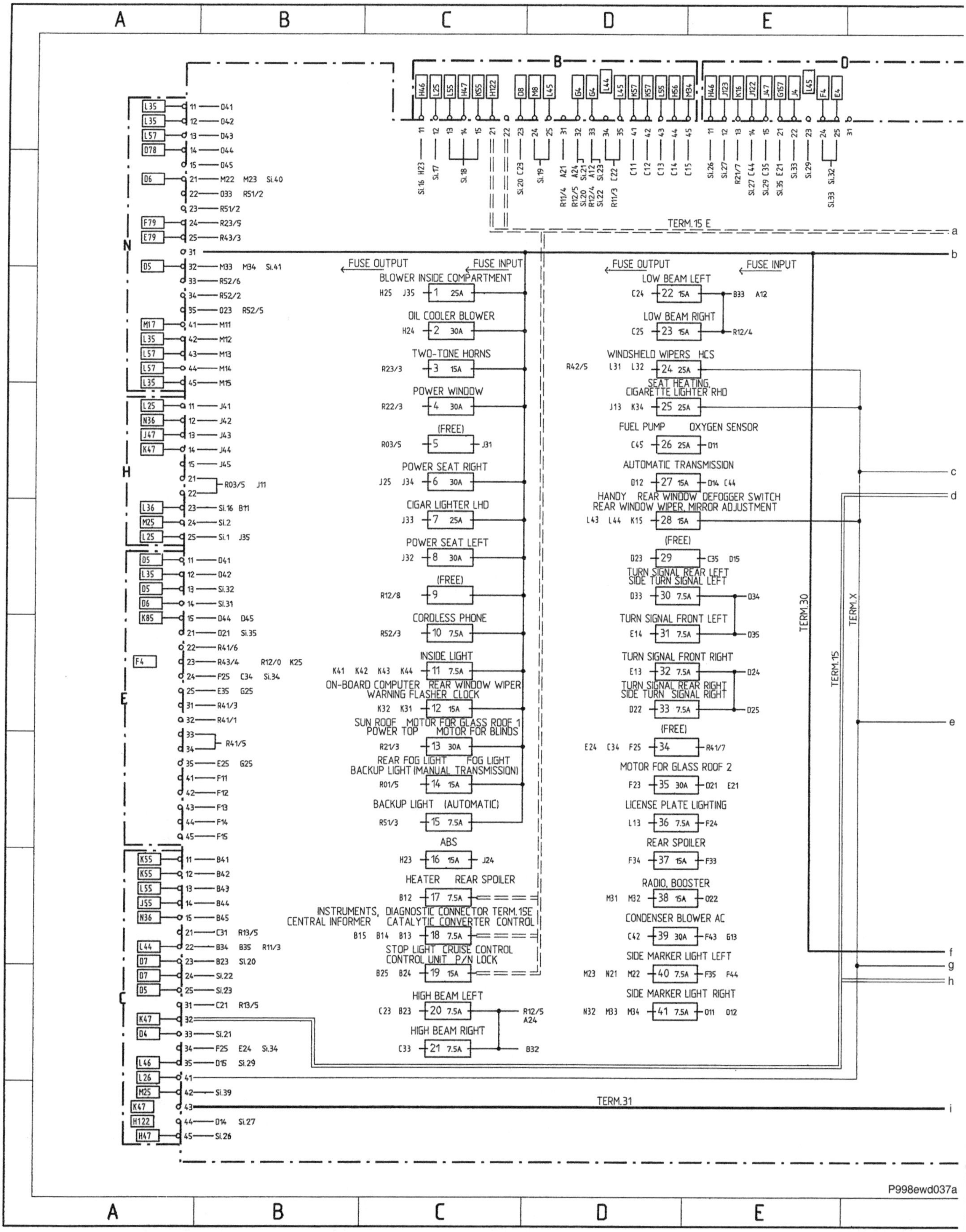

Sheet 18 - 1996

Central electric (page 2 of 3)

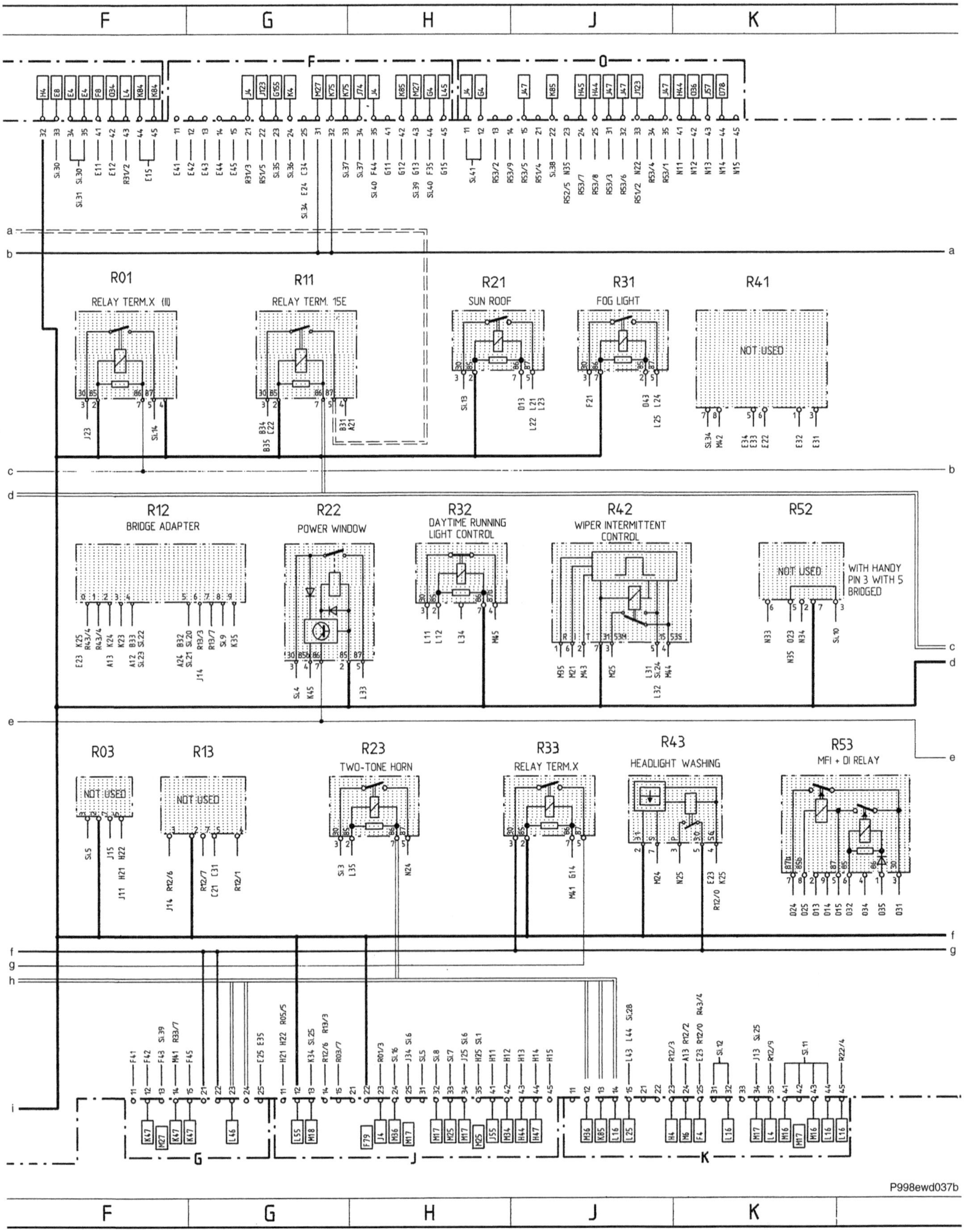

Central electric (page 3 of 3)

Sheet 18 - 1996

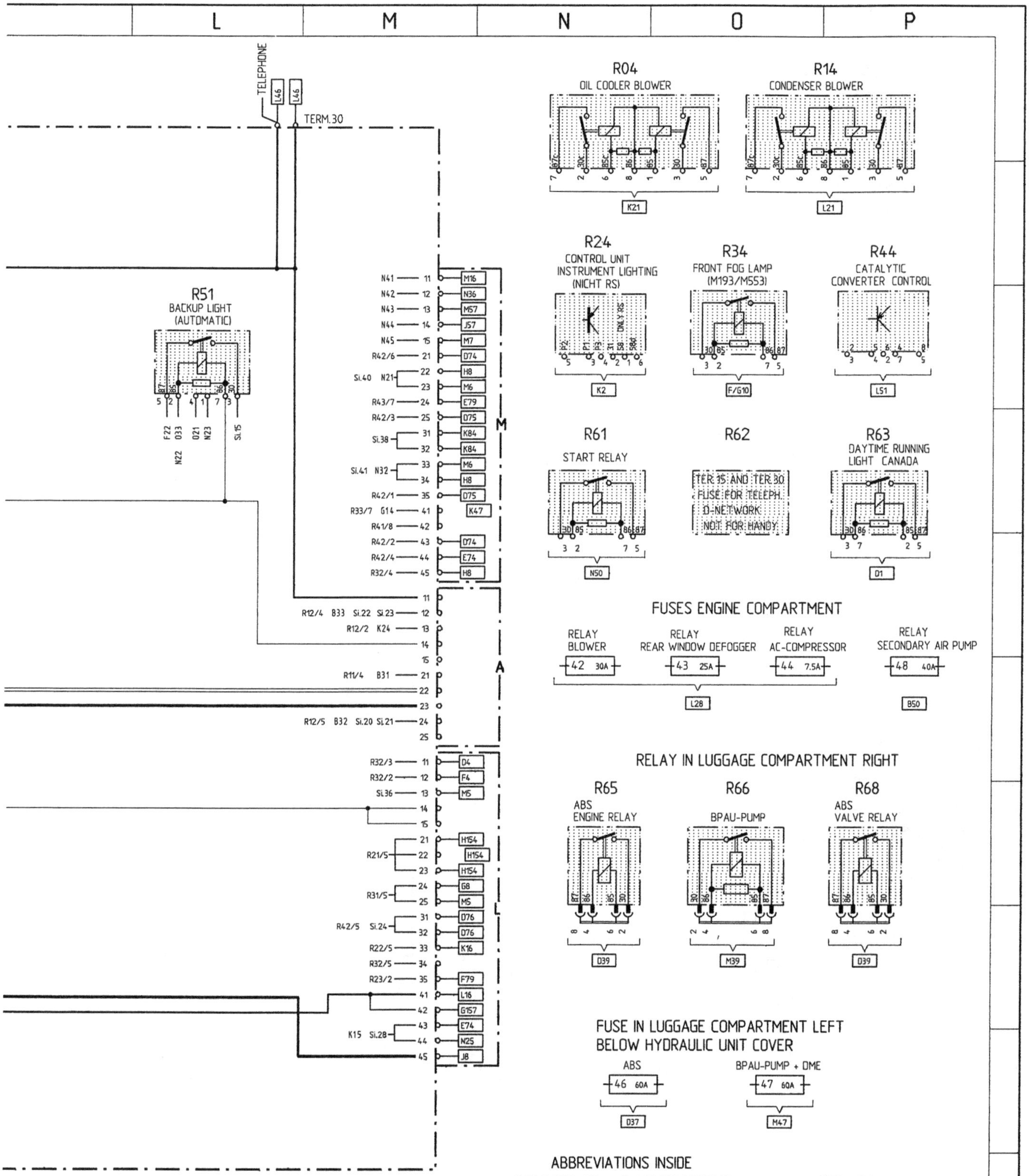

ABBREVIATIONS INSIDE
FOR INST.: C3 PLUG CONNECTION
R01 RELAY
SI. FUSE

ABBREVIATIONS OUTSIDE
FOR INST.: E2 CONNECTION WITH FIELD DATA

— TERM.31
═ TERM.15
═: TERM.15E
— TERM. X
— TERM. 30

P998ewd037c

K L M N O P

EWD

Sheet 1 - 1997/1998

Lights (page 1 of 3)

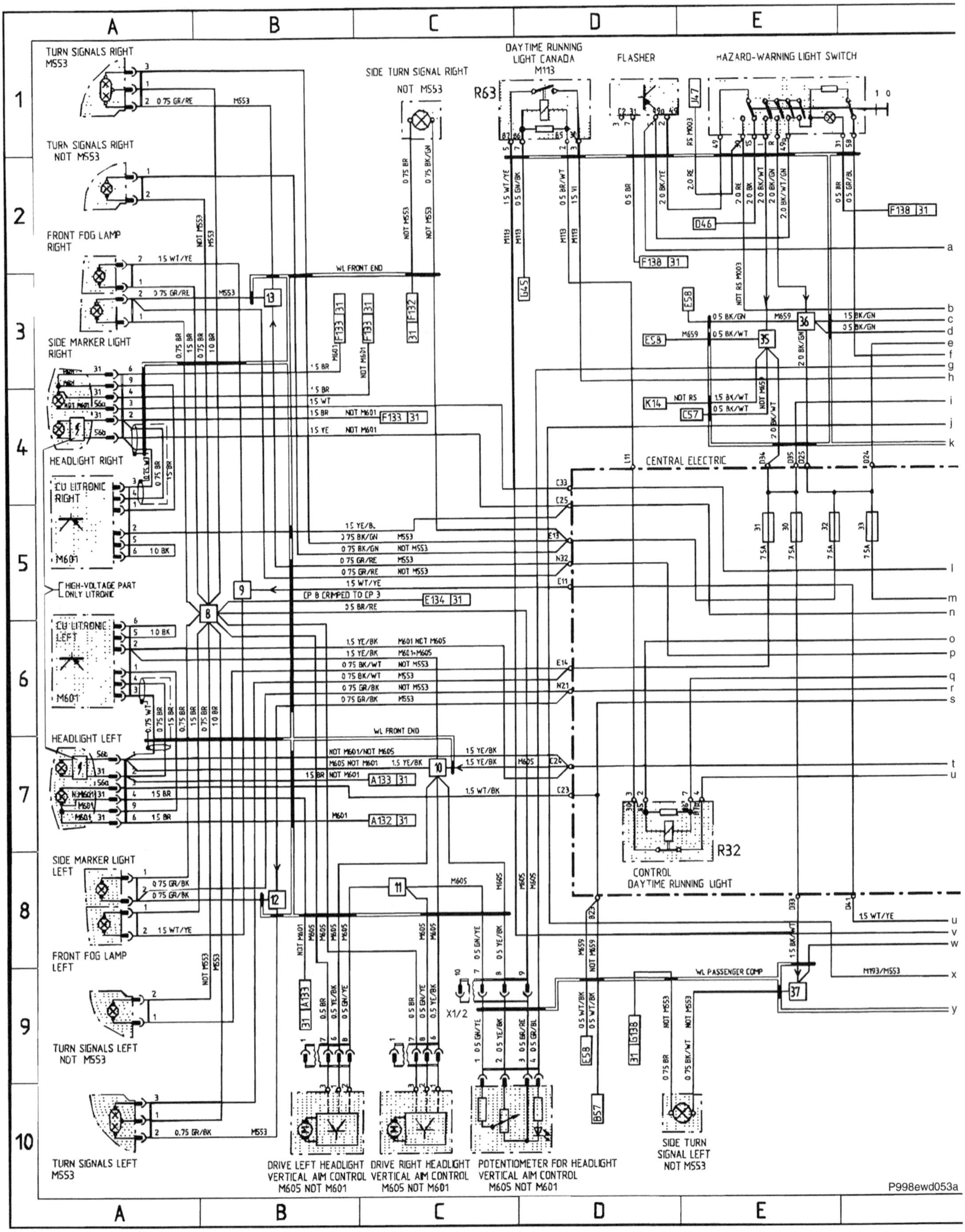

Lights (page 2 of 3)

Sheet 1 - 1997/1998

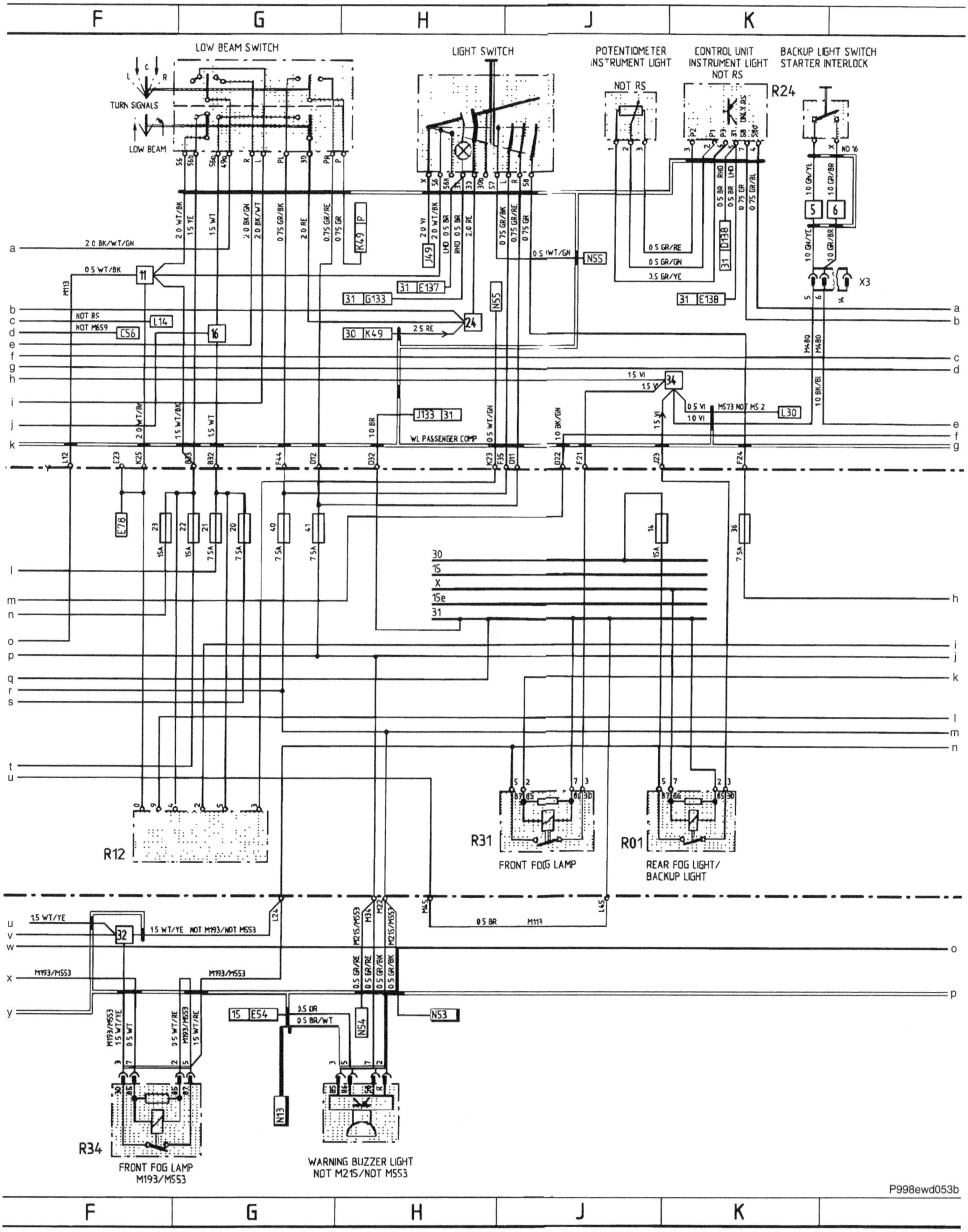

Sheet 1 - 1997/1998

Lights (page 3 of 3)

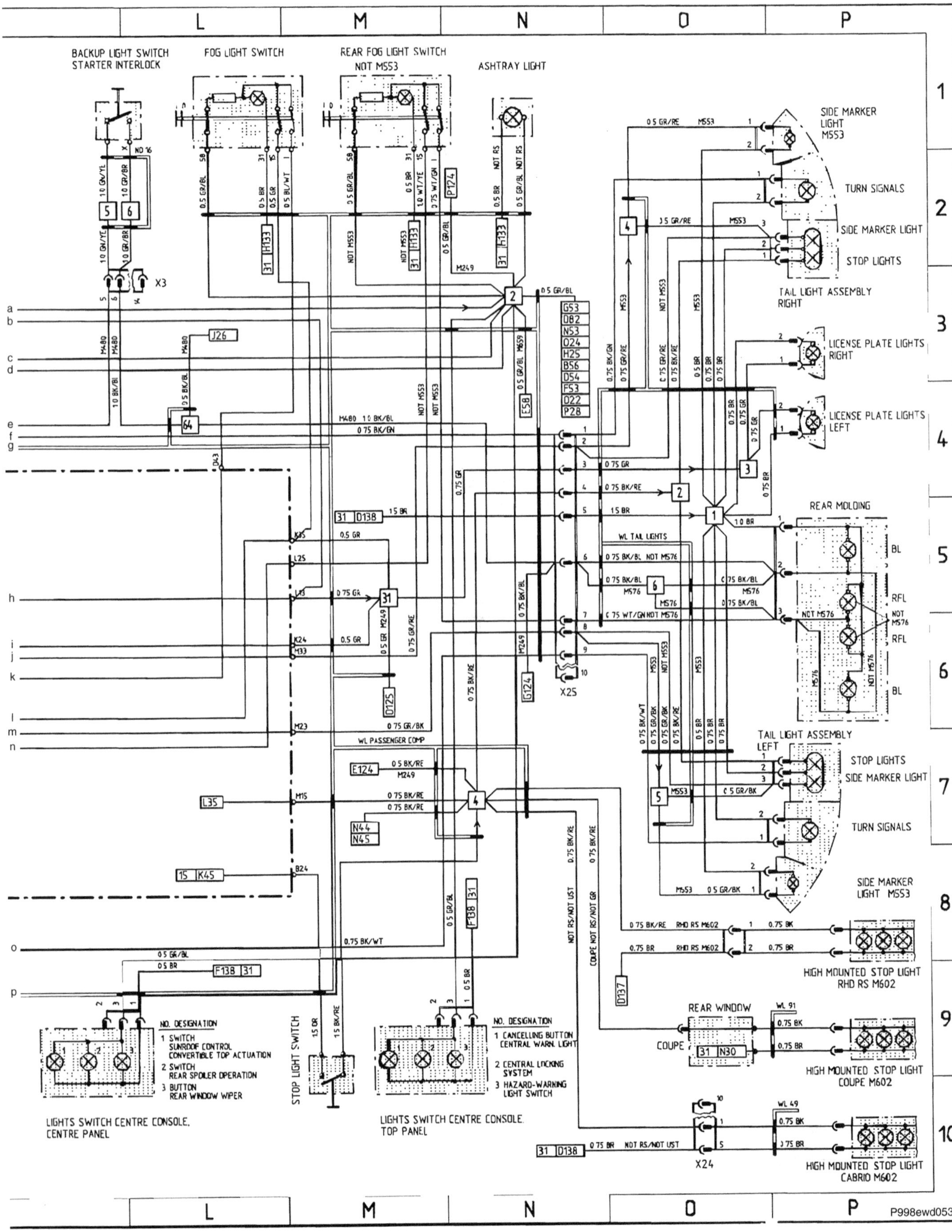

Seats, alarm system, central locks, mirror, power window regulator, inside lights (page 1 of 3)

Sheet 2 - 1997/1998

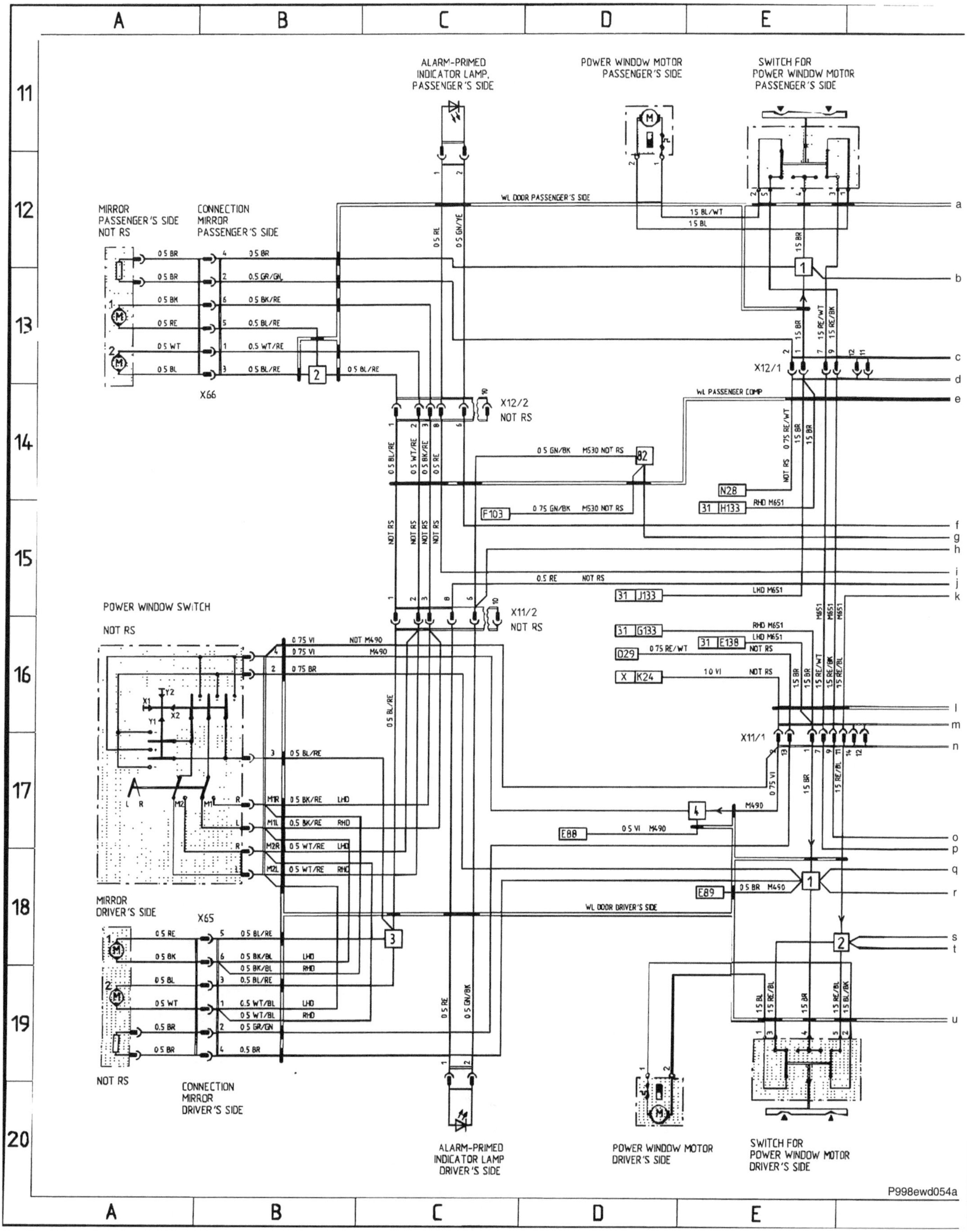

Sheet 2 - 1997/1998

Seats, alarm system, central locks, mirror, power window regulator, inside lights (page 2 of 3)

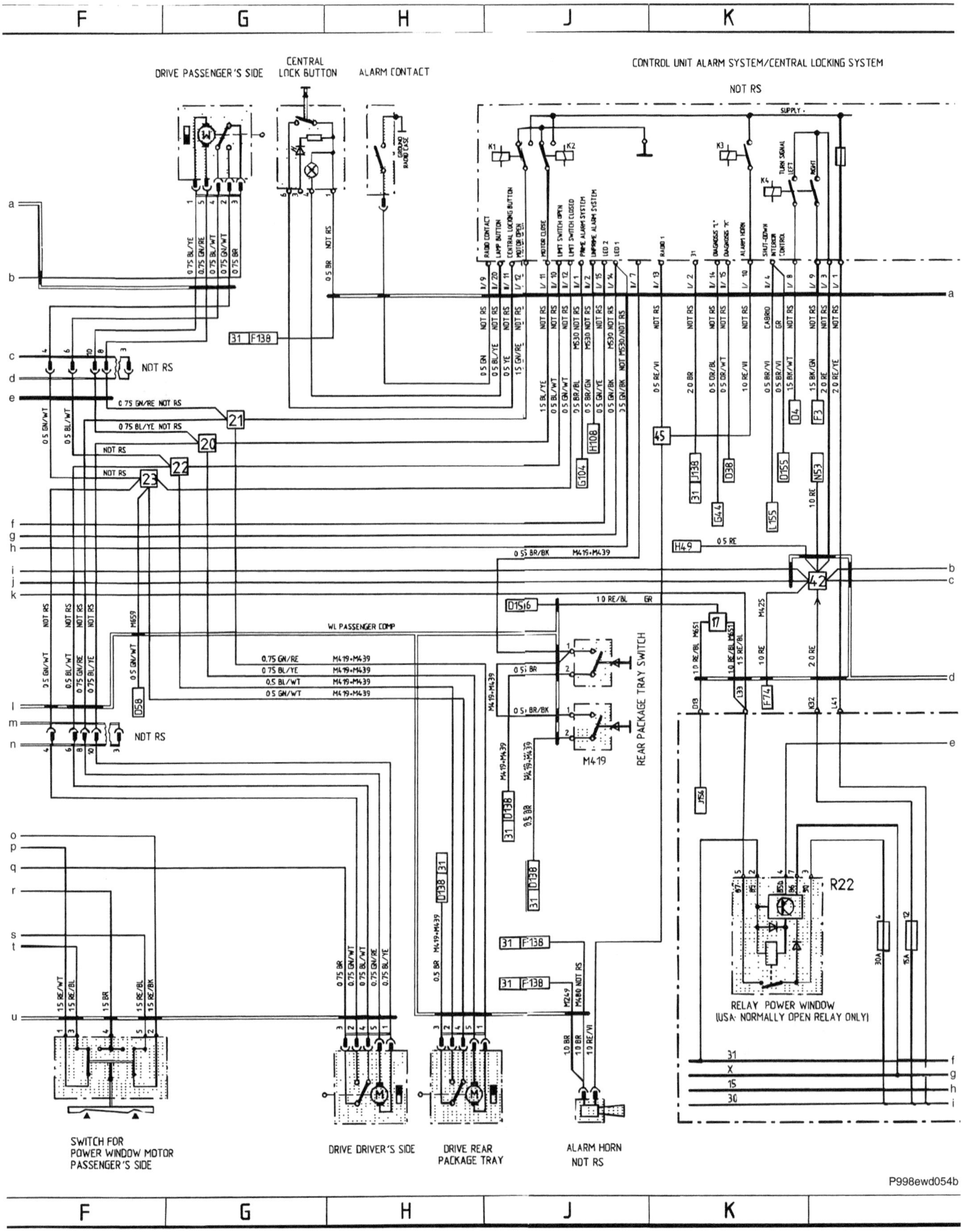

P998ewd054b

Seats, alarm system, central locks, mirror, power window regulator, inside lights (page 3 of 3)

Sheet 2 - 1997/1998

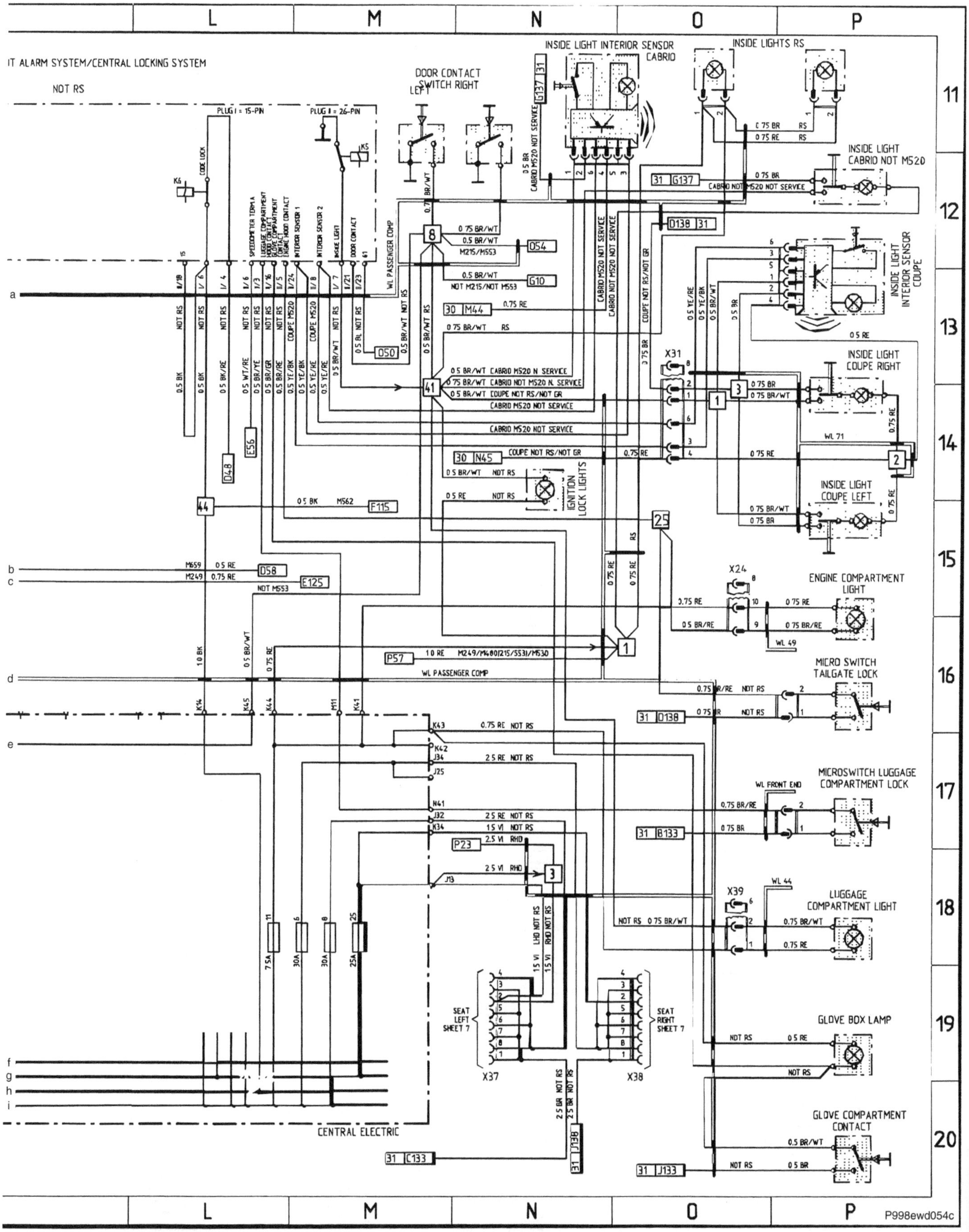

EWD

Sheet 3 - 1997/1998

Heater, A/C (page 1 of 3)

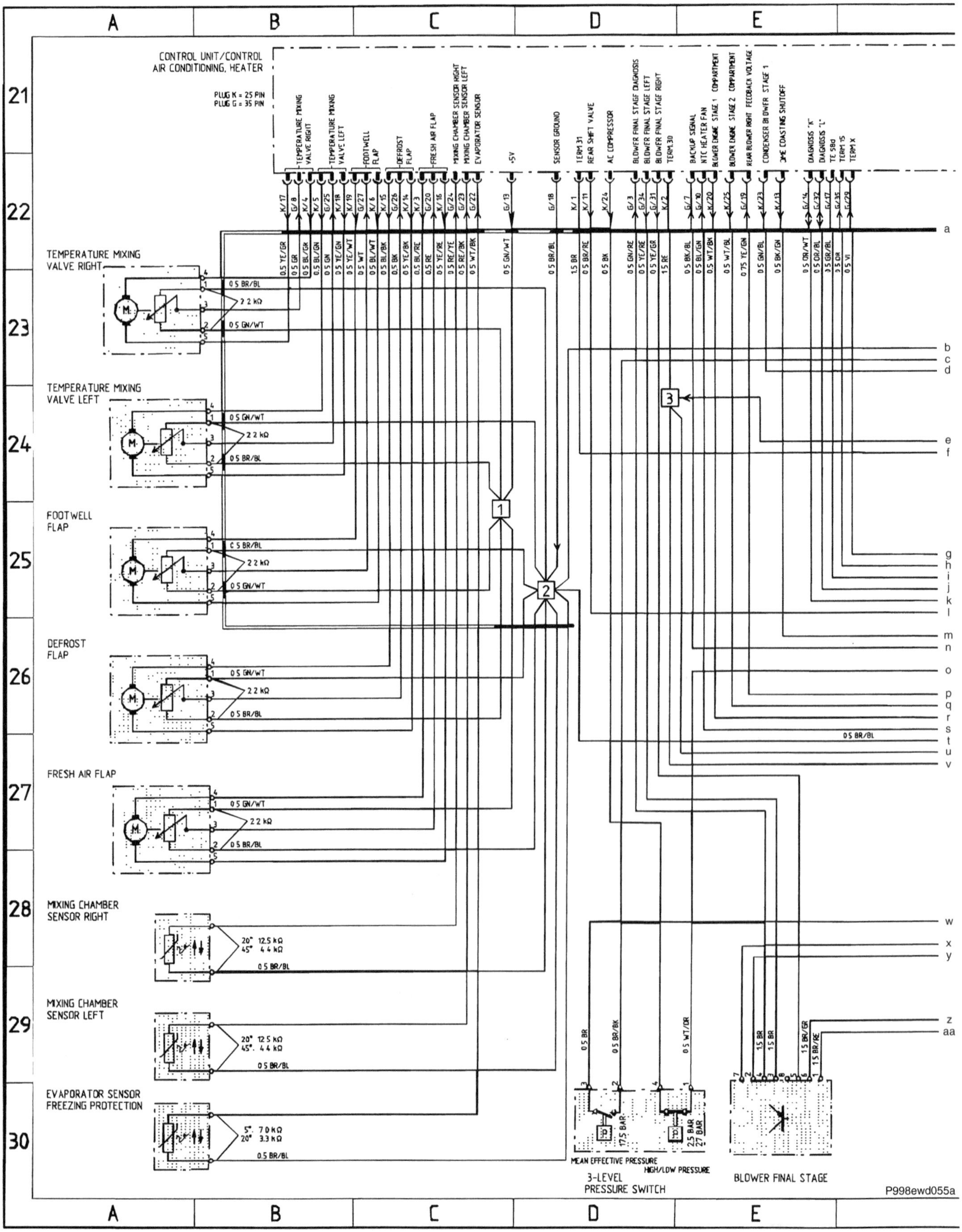

Heater, A/C (page 2 of 3)

Sheet 3 - 1997/1998

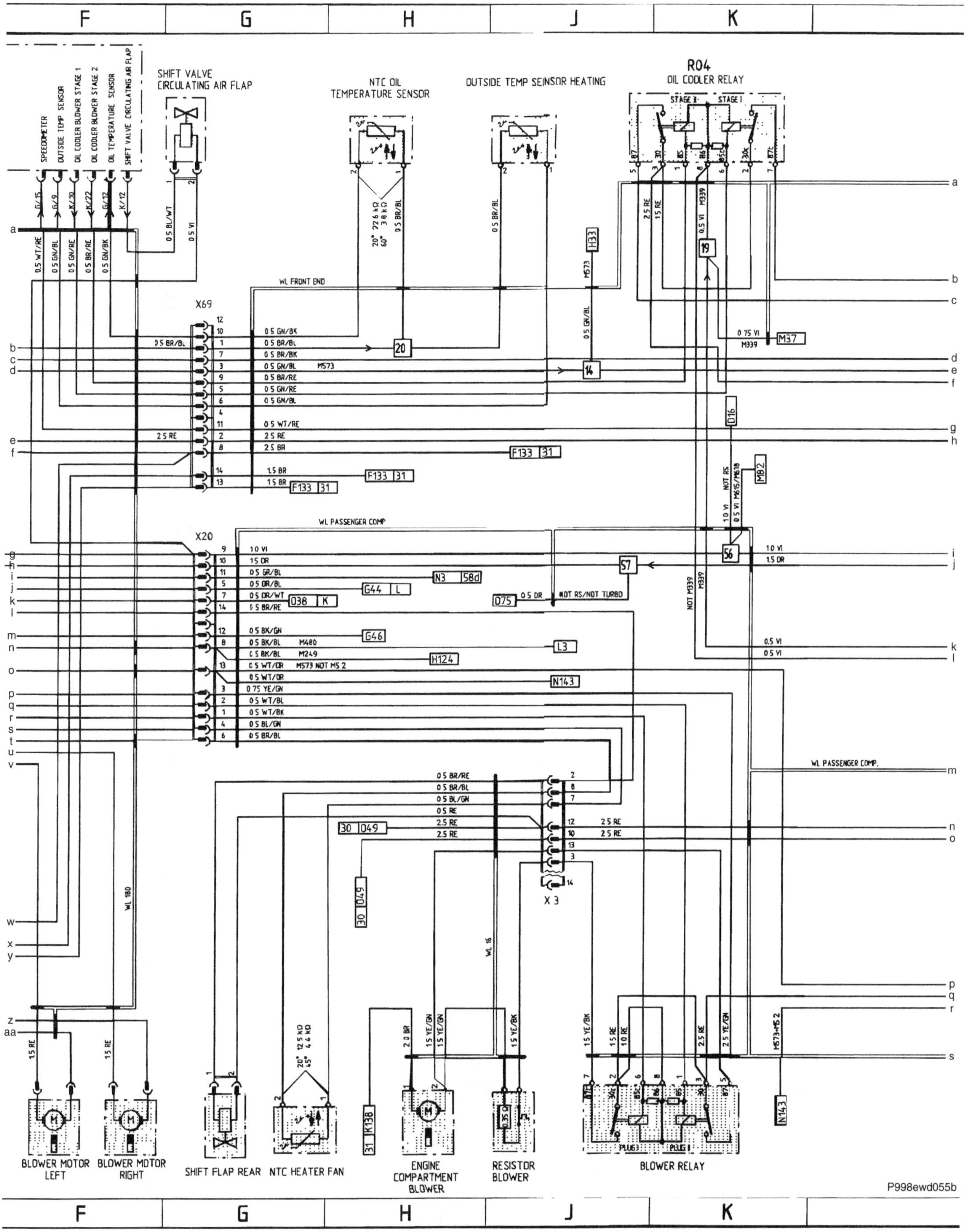

Sheet 3 - 1997/1998

Heater, A/C (page 3 of 3)

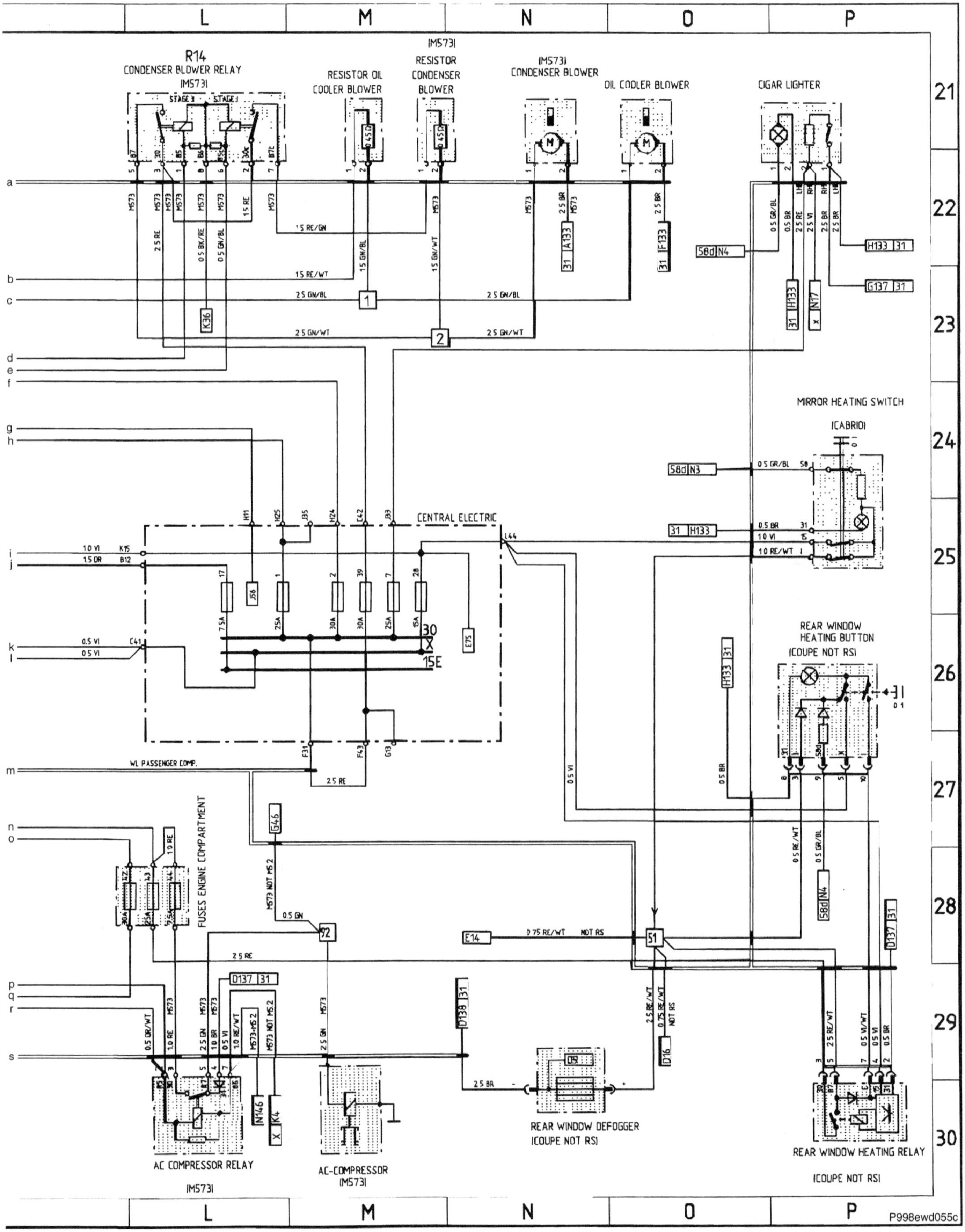

ABS
(page 1 of 3)

Sheet 4A - 1997/1998

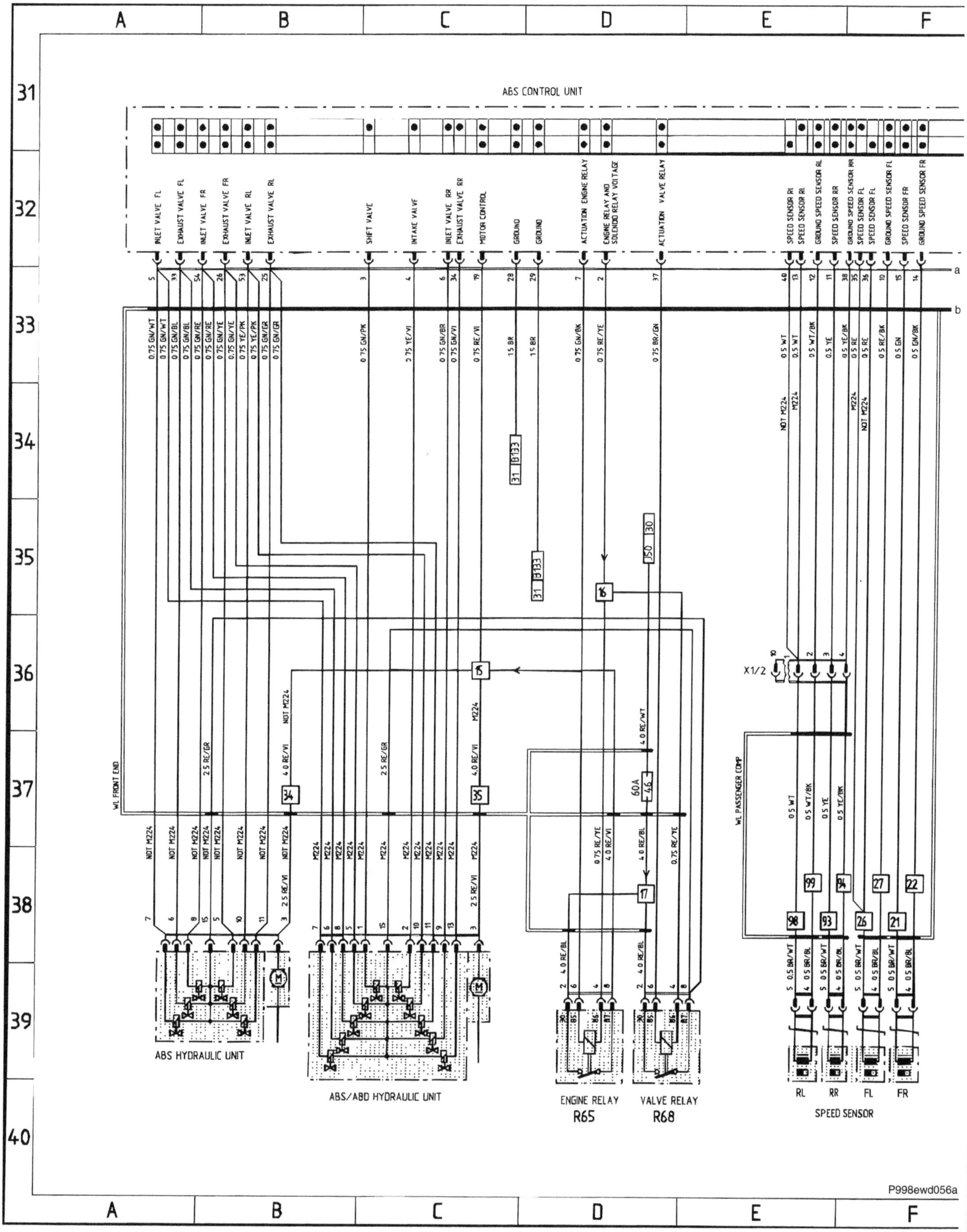

EWD

Sheet 4A - 1997/1998

ABS (page 2 of 3)

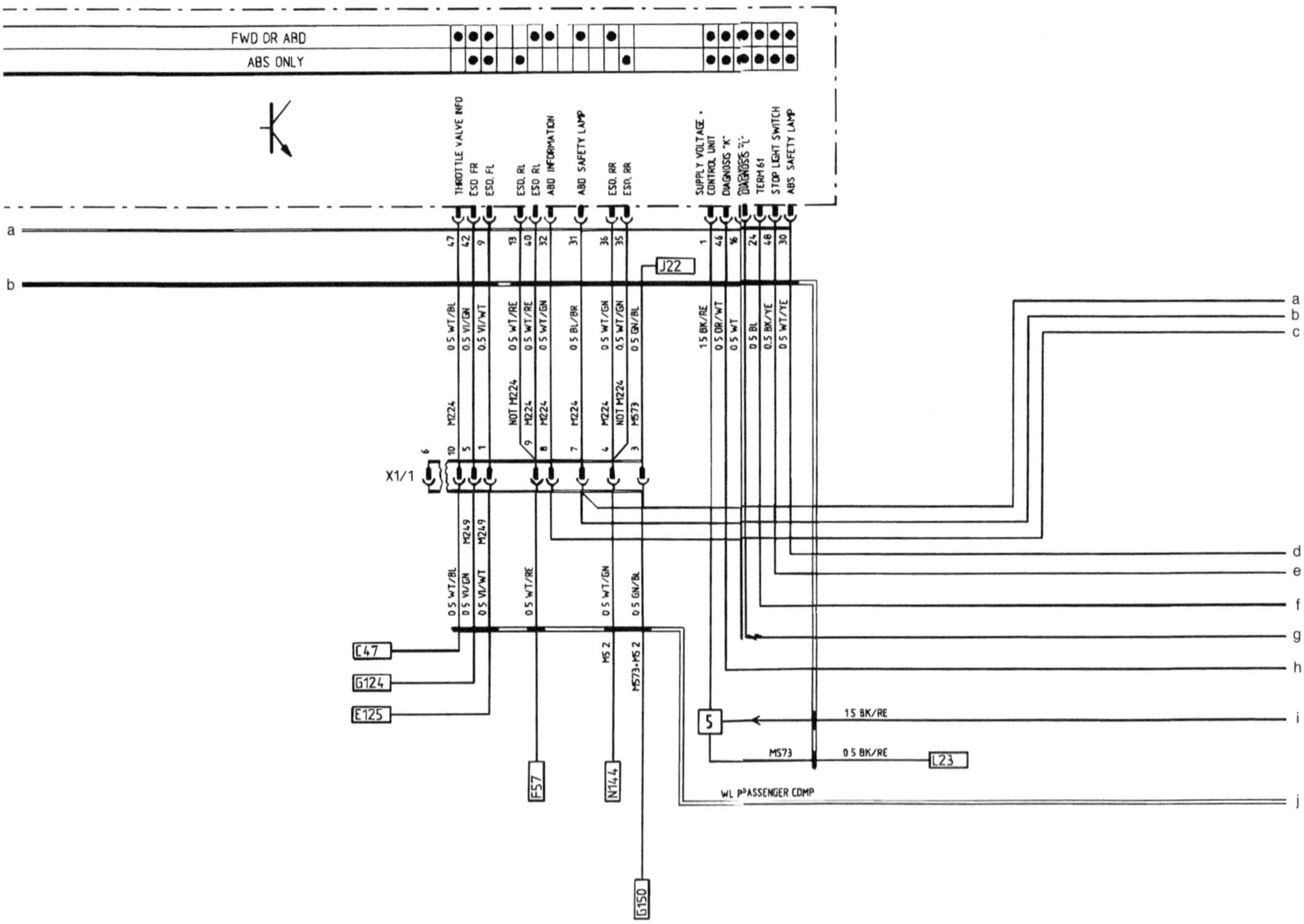

P998ewd056b

ABS
(page 3 of 3)

Sheet 4A - 1997/1998

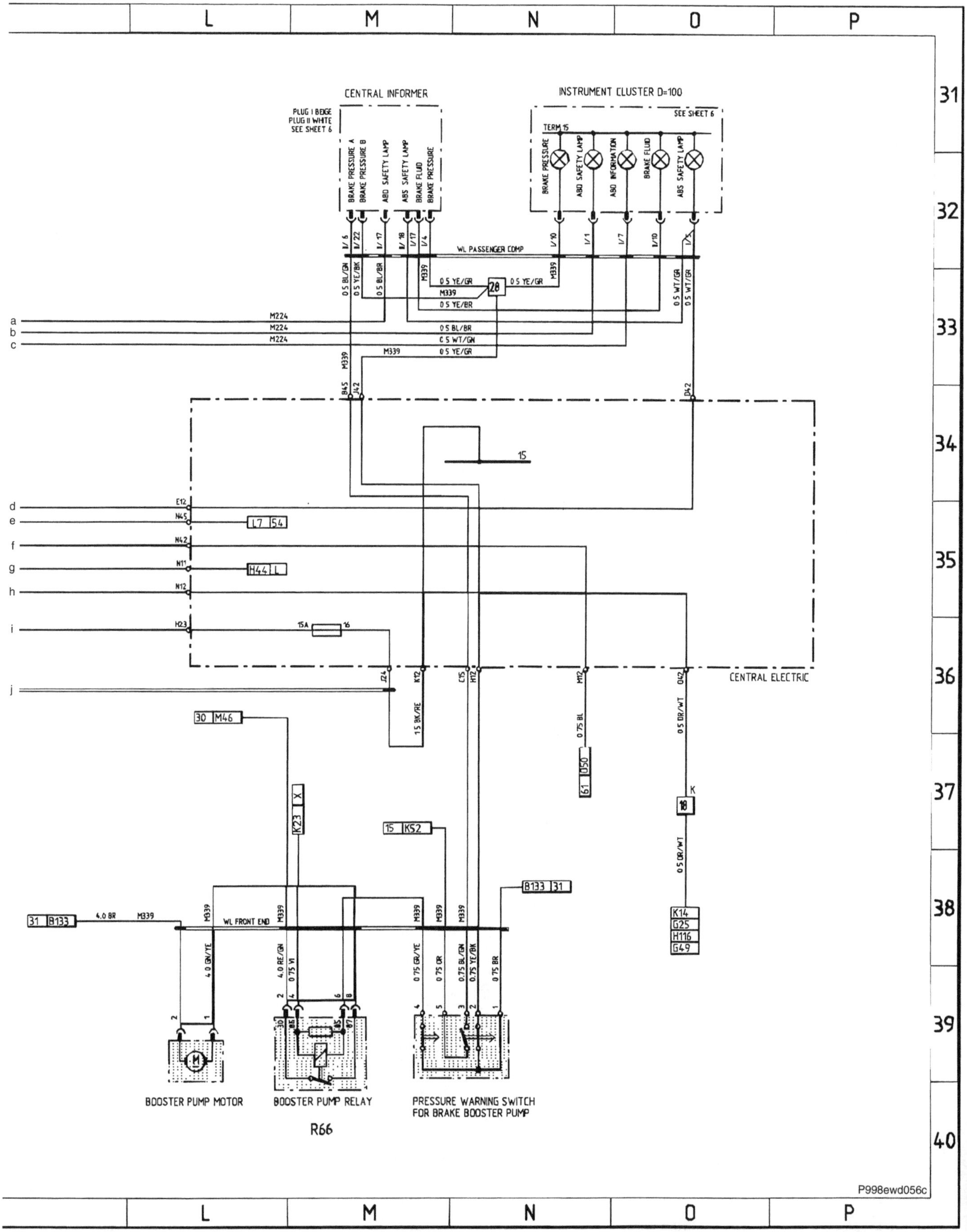

P998ewd056c

Sheet 4A - 1997/1998

ABS (RS) (page 1 of 3)

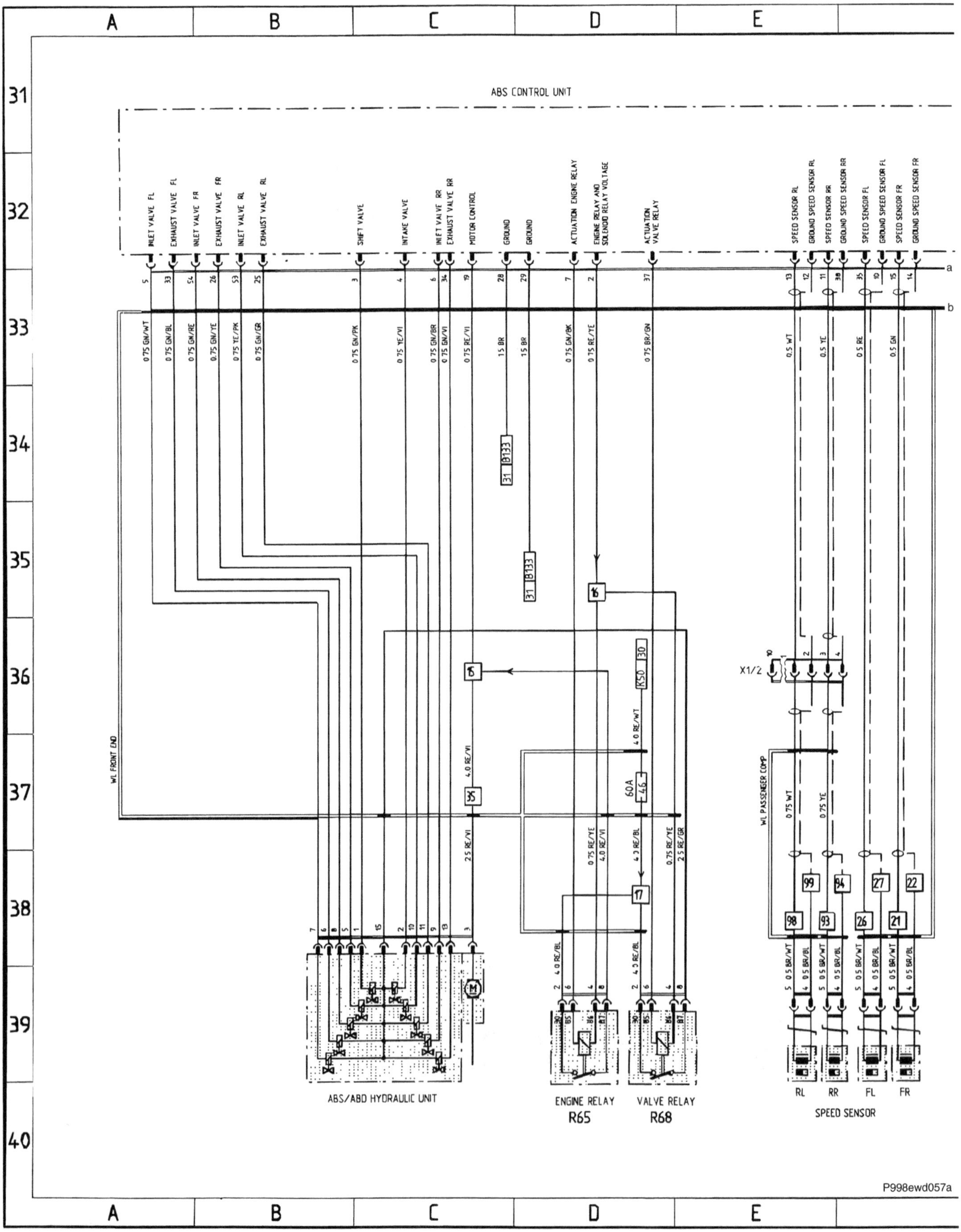

ABS (RS) (page 2 of 3)

Sheet 4A - 1997/1998

F	G	H	J	K	

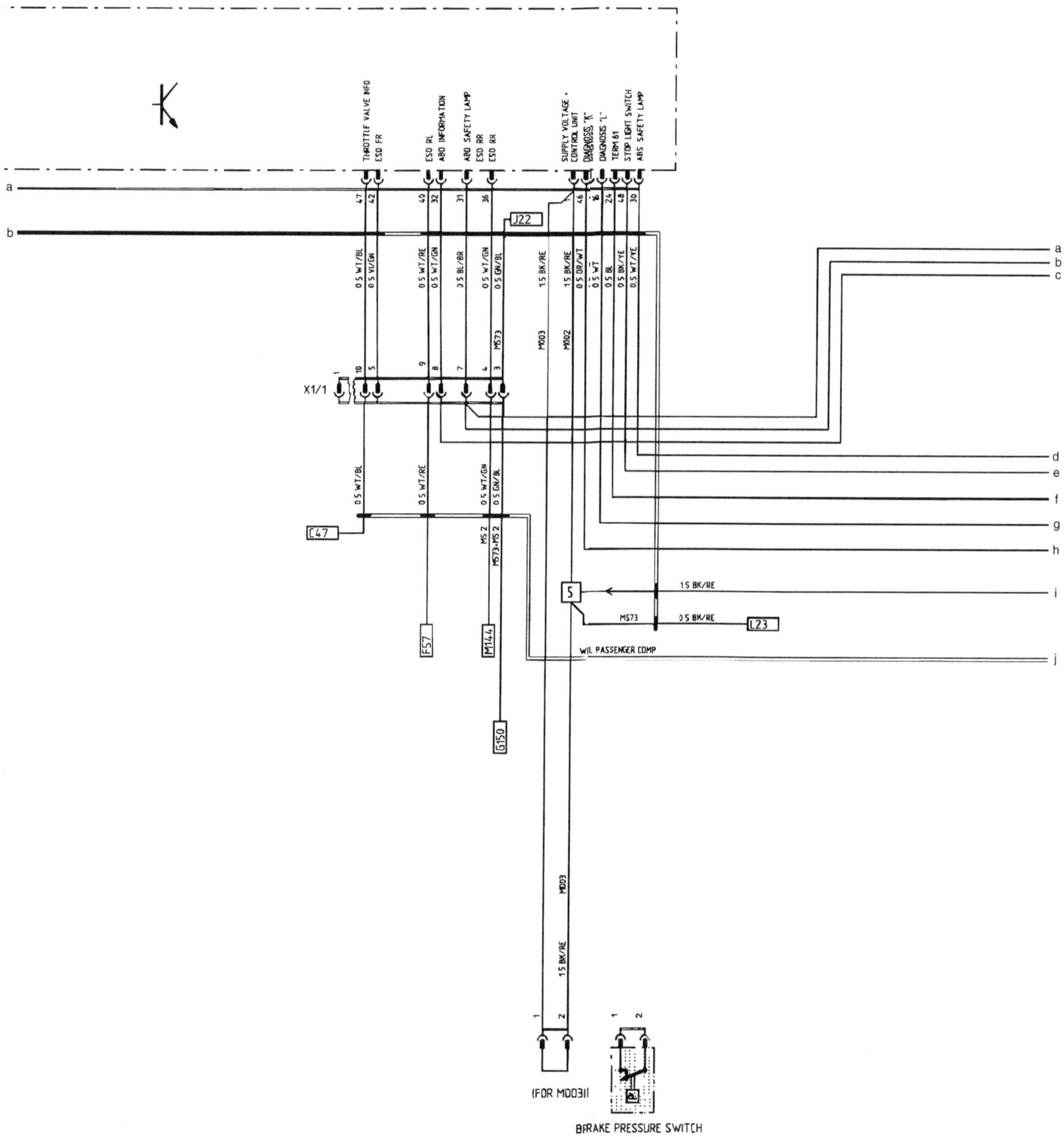

P998ewd057b

F	G	H	J	K	

Sheet 4A - 1997/1998

ABS (RS)
(page 3 of 3)

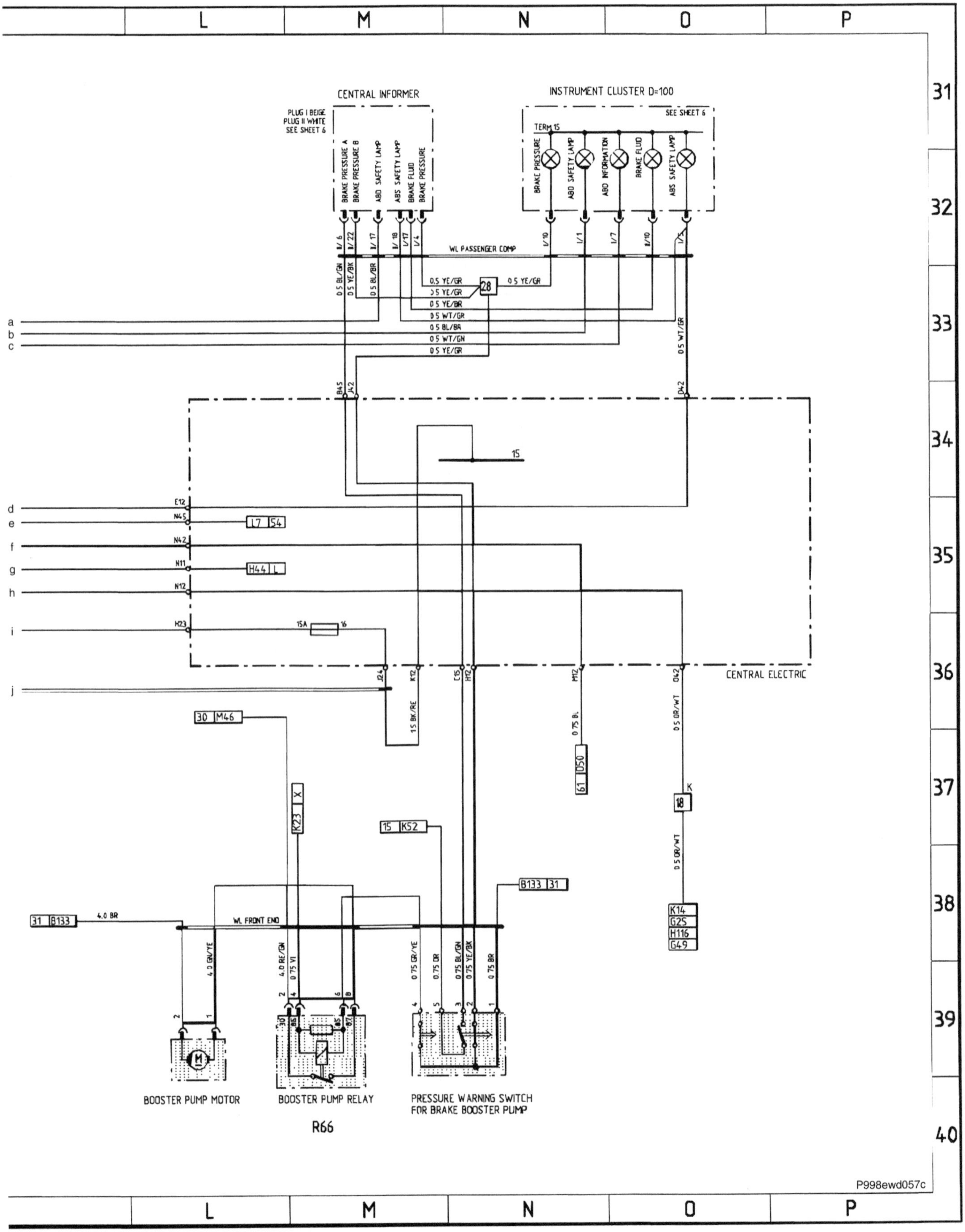

Engine electrics, ignition system (DME M 2.10, ROW) (page 1 of 3)

Sheet 5A - 1997/1998

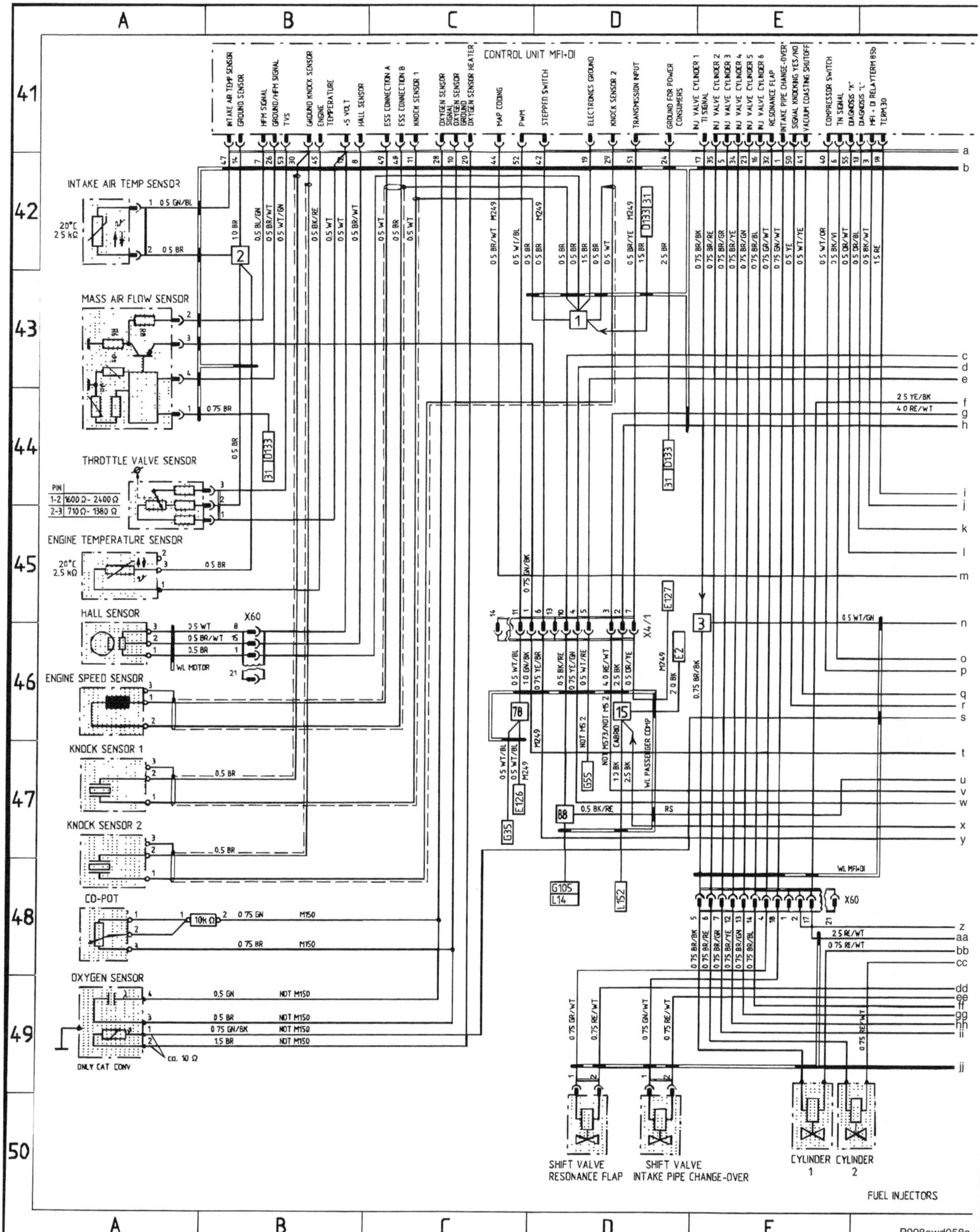

P998ewd058a

Sheet 5A - 1997/1998

Engine electrics, ignition system (DME M 2.10, ROW) (page 2 of 3)

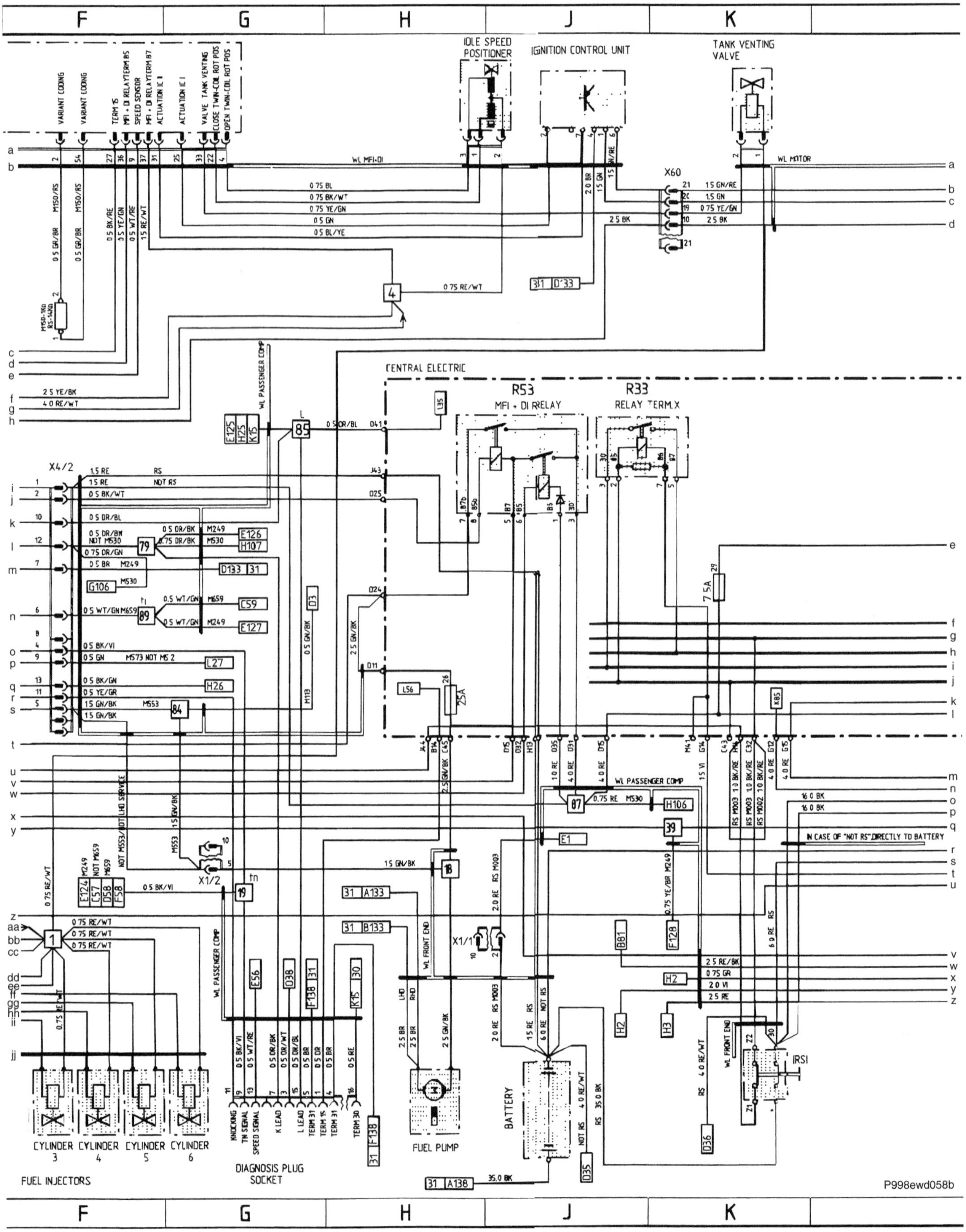

Engine electrics, ignition system (DME M 2.10, ROW) (page 3 of 3)

Sheet 5A - 1997/1998

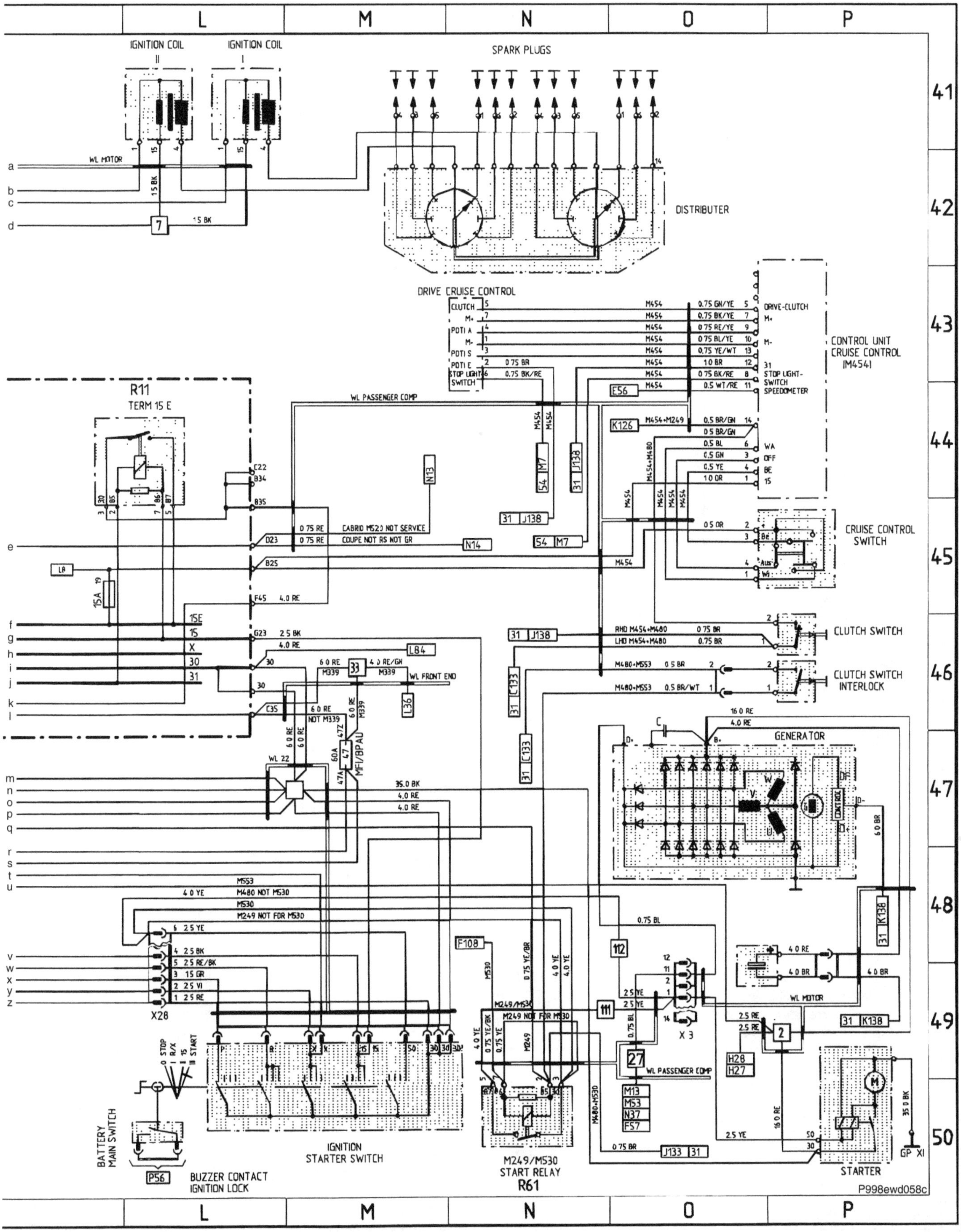

Sheet 5B - 1997/1998

Engine electrics, ignition system (DME M 5.2, US) (page 1 of 3)

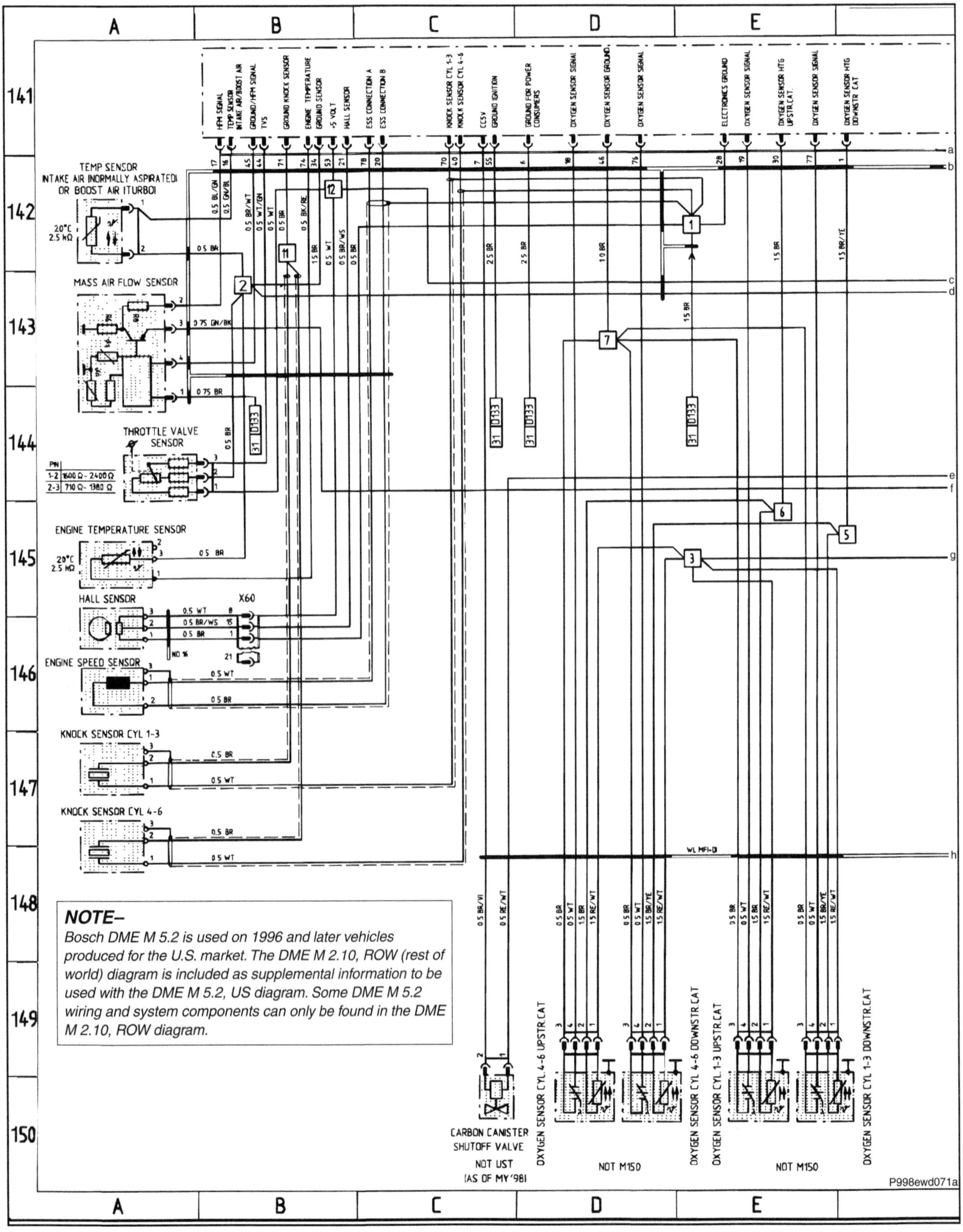

NOTE–

Bosch DME M 5.2 is used on 1996 and later vehicles produced for the U.S. market. The DME M 2.10, ROW (rest of world) diagram is included as supplemental information to be used with the DME M 5.2, US diagram. Some DME M 5.2 wiring and system components can only be found in the DME M 2.10, ROW diagram.

Engine electrics, ignition system (DME M 5.2, US) (page 2 of 3)

Sheet 5B - 1997/1998

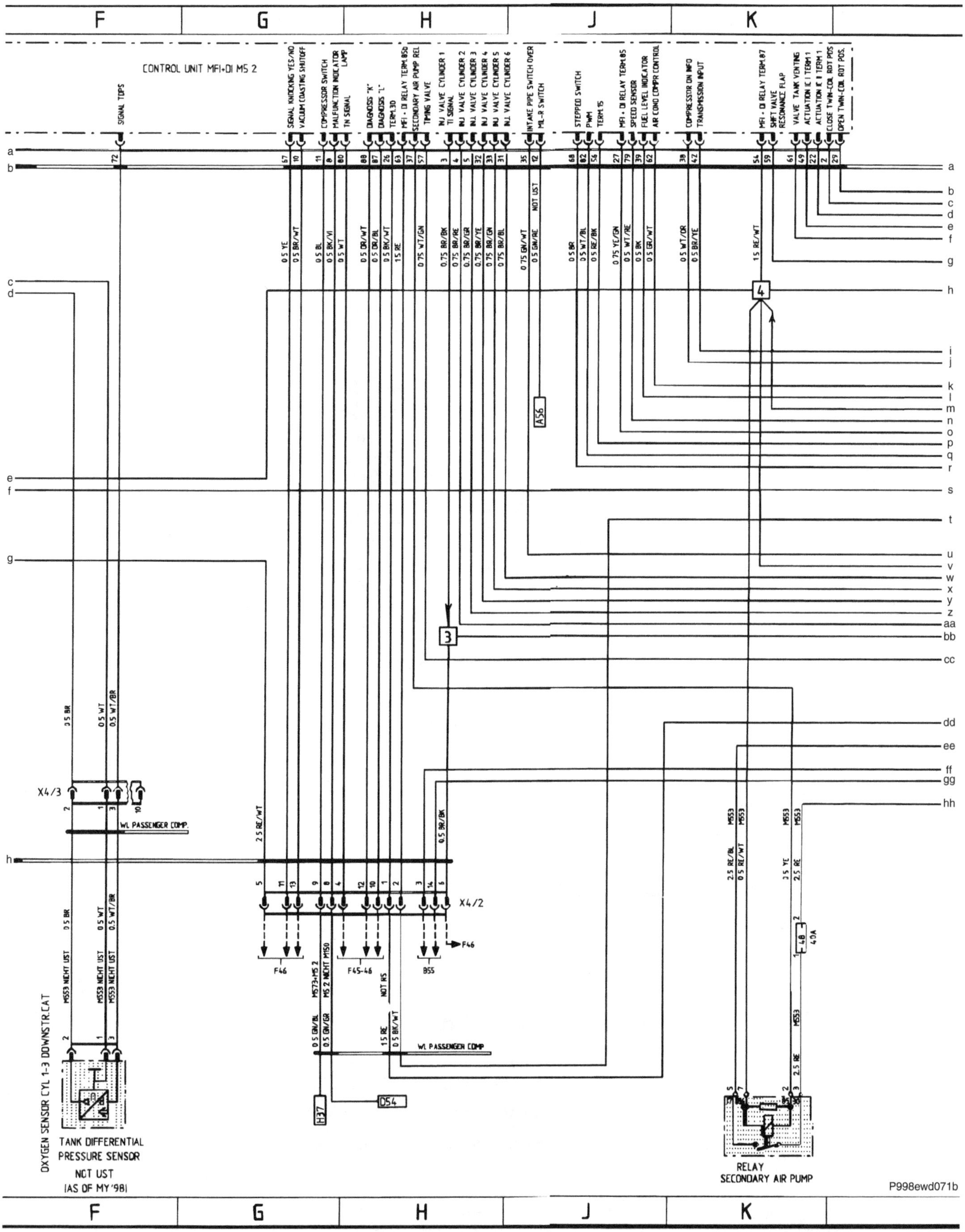

Sheet 5B - 1997/1998

Engine electrics, ignition system (DME M 5.2, US) (page 3 of 3)

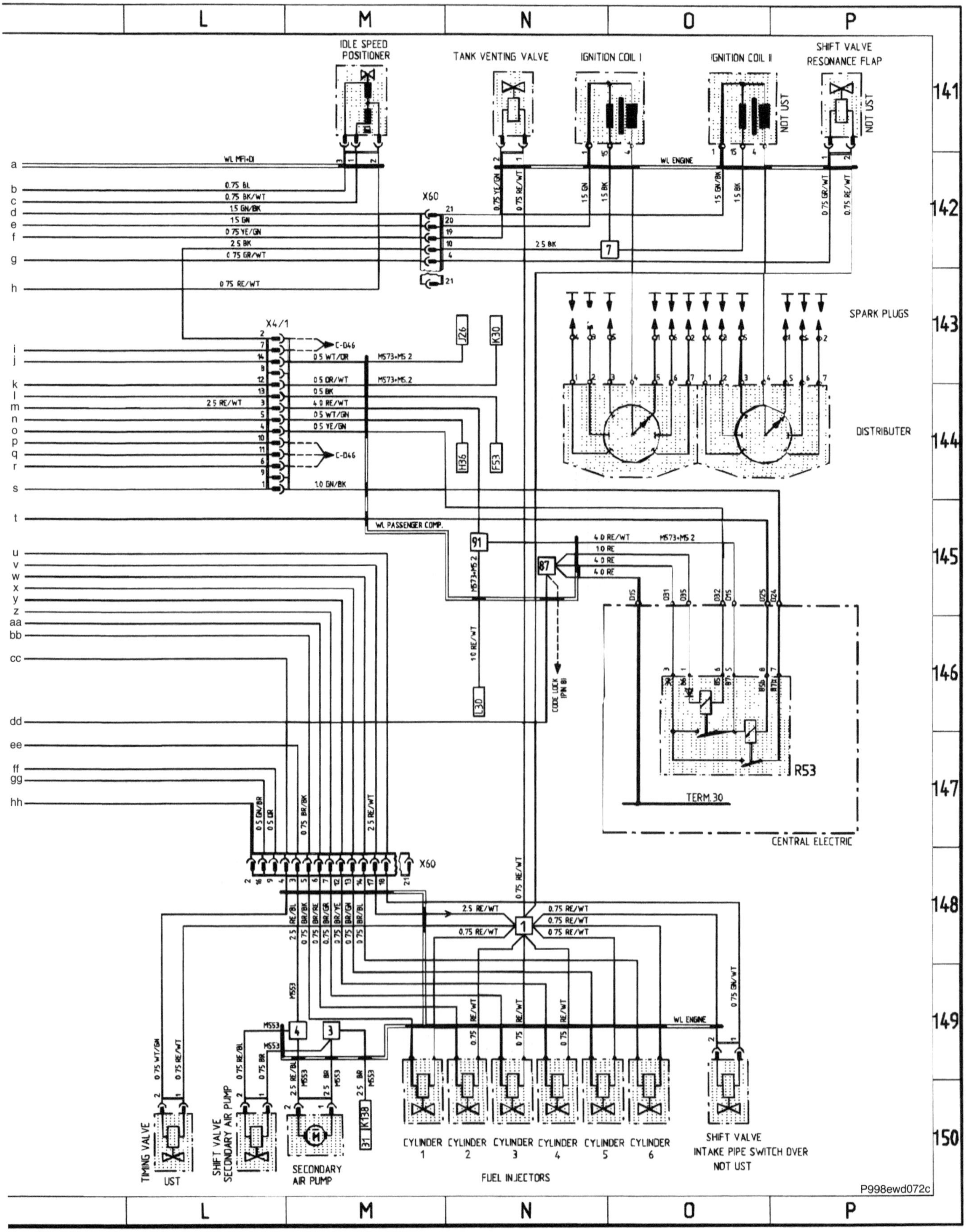

Instruments and sensor (page 1 of 3)

Sheet 6 - 1997/1998

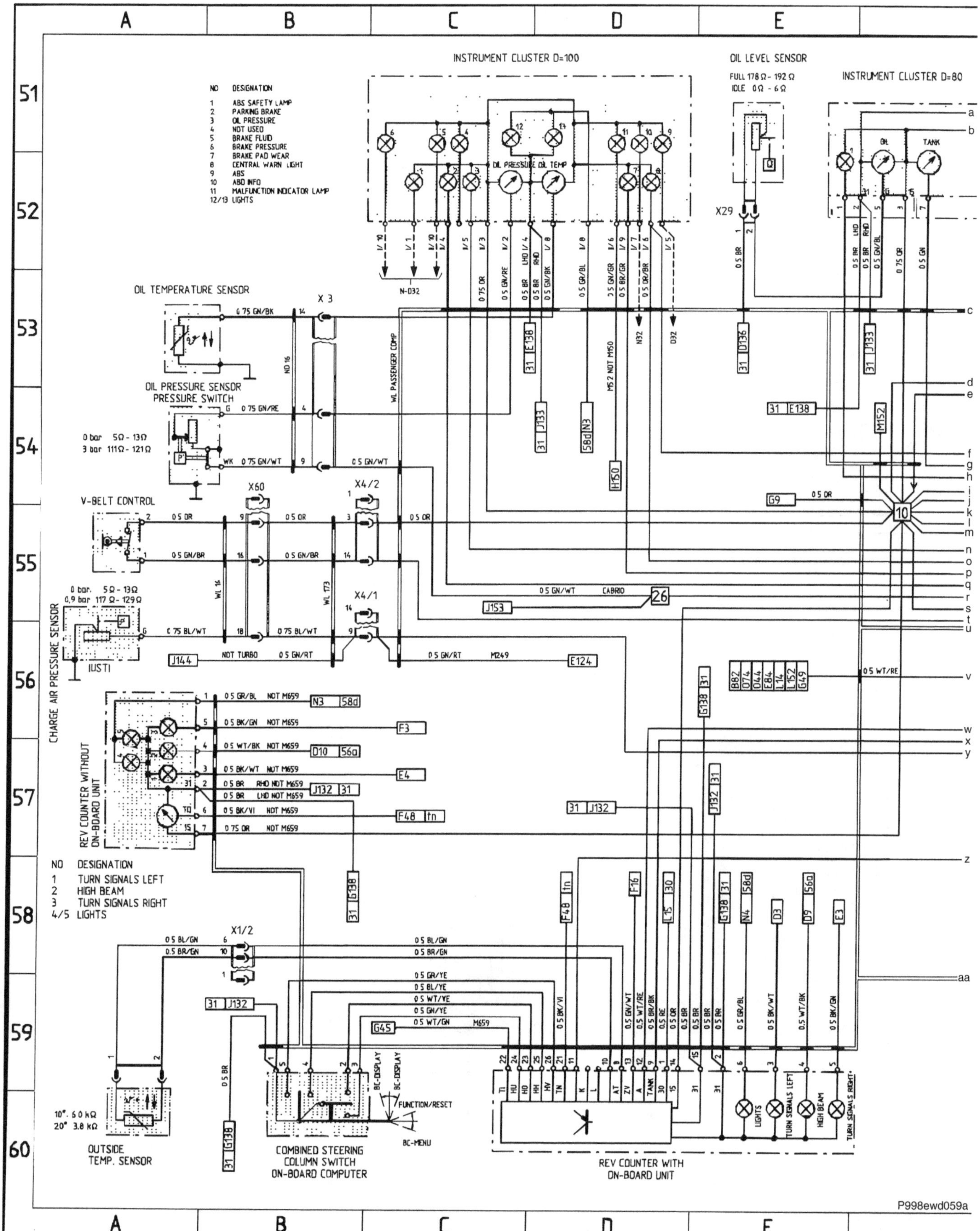

EWD

Sheet 6 - 1997/1998

Instruments and sensor (page 2 of 3)

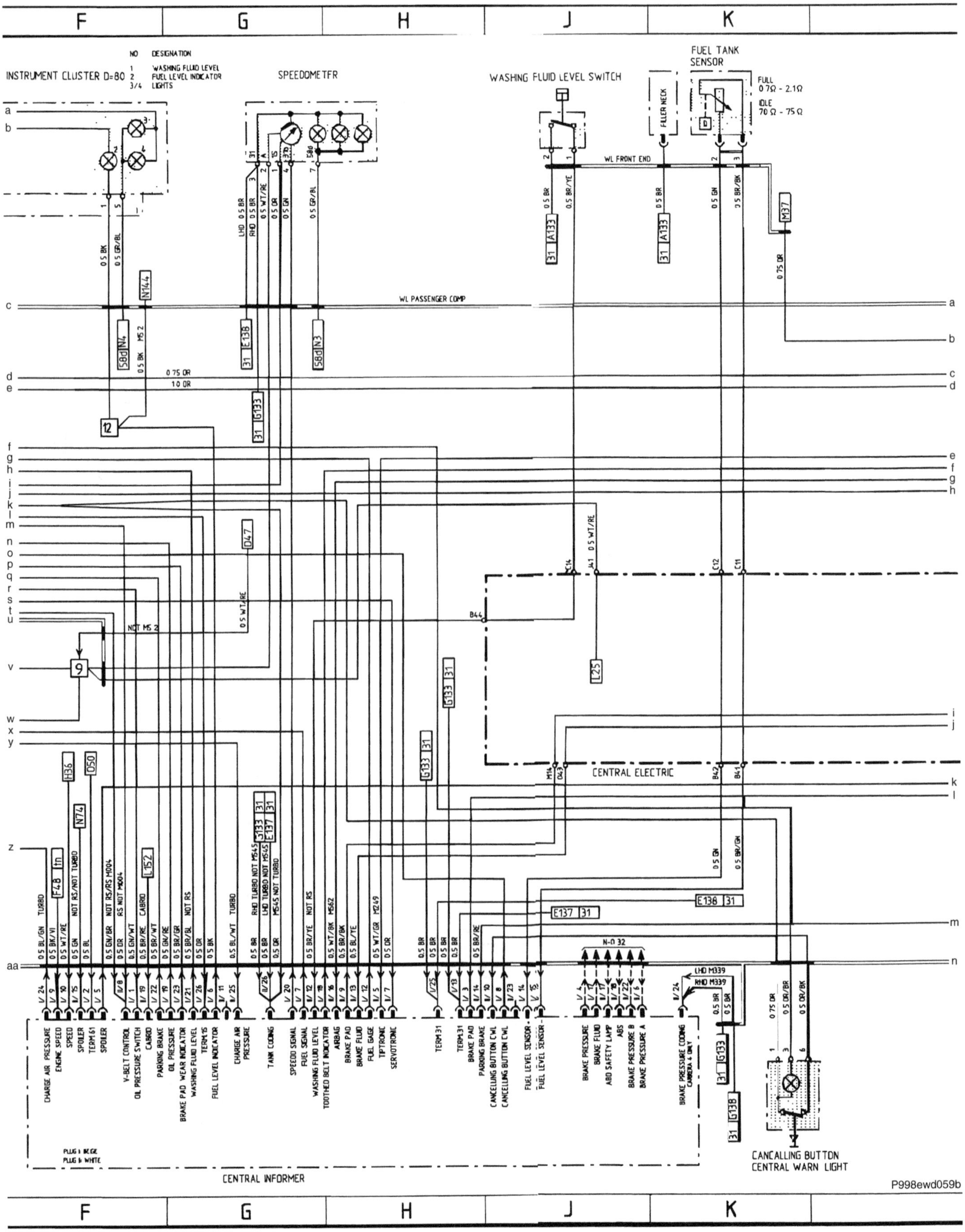

P998ewd059b

Instruments and sensor (page 3 of 3)

Sheet 6 - 1997/1998

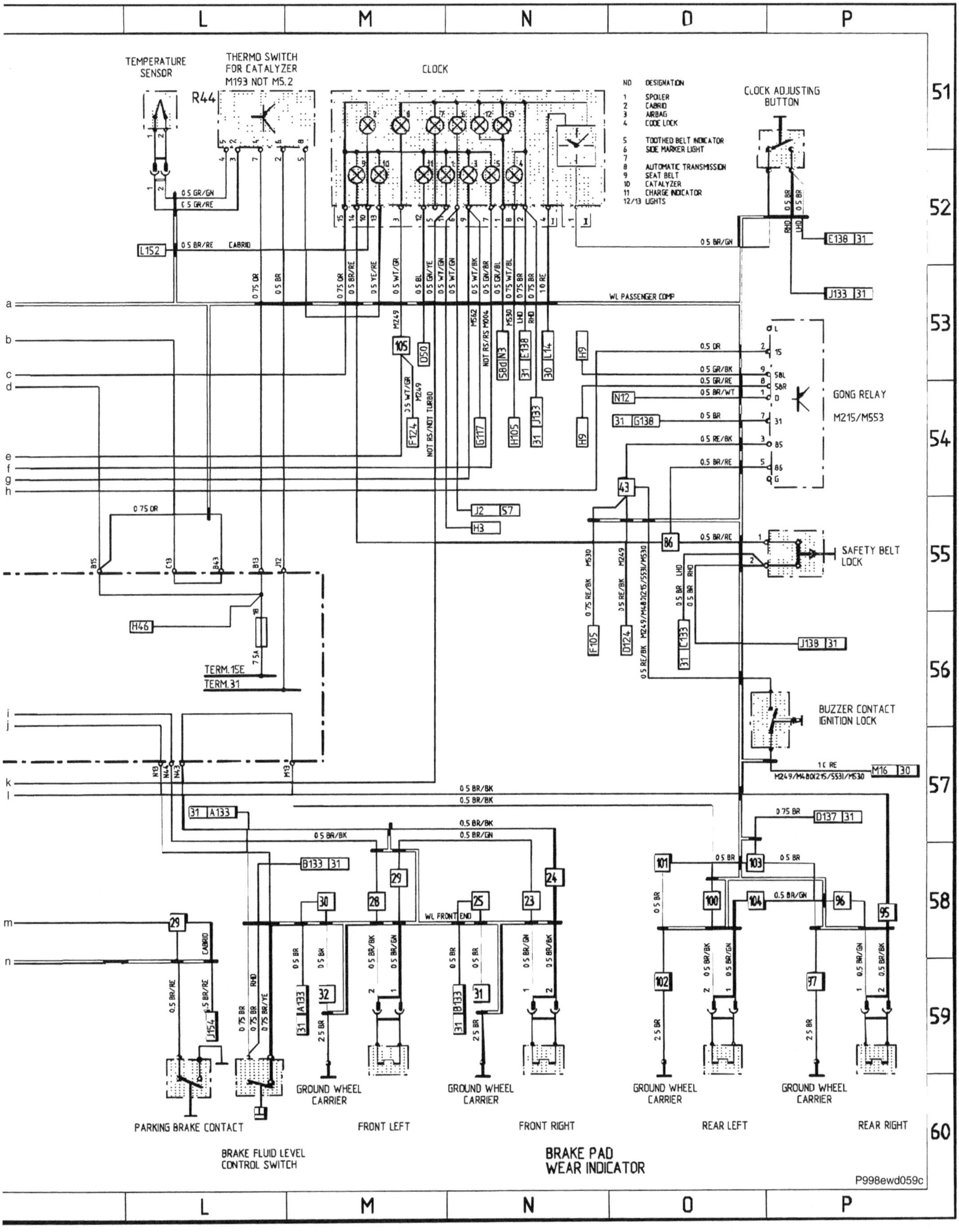

EWD

Sheet 7 - 1997/1998

Seats, differing cable layout for service (page 1 of 3)

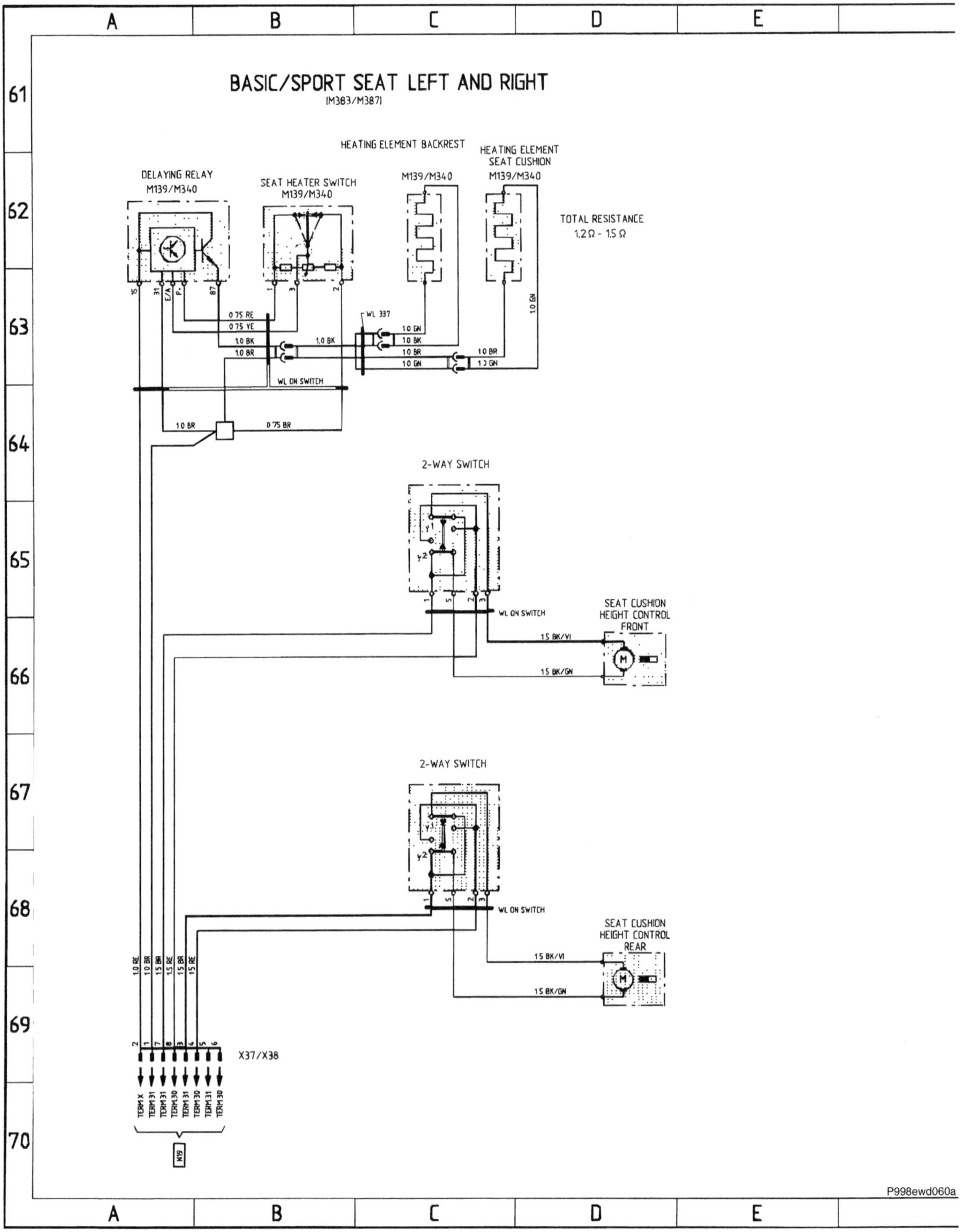

P998ewd060a

Seats, differing cable layout for service (page 2 of 3)

Sheet 7 - 1997/1998

F	G	H	J	K

COMFORT SEAT LEFT AND RIGHT
(M437,M438)

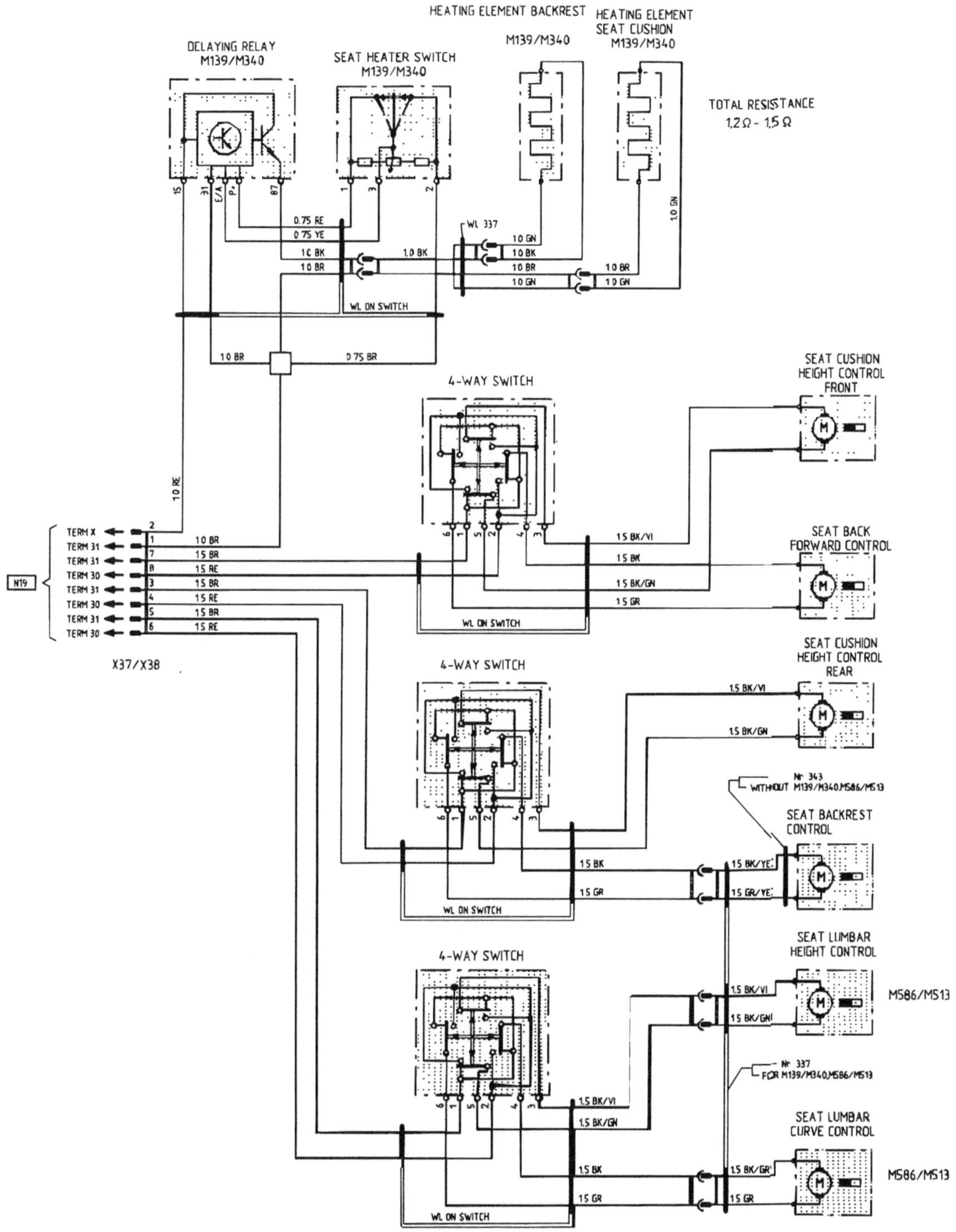

P998ewd060b

F	G	H	J	K

EWD

Sheet 7 - 1997/1998

Seats, differing cable layout for service (page 3 of 3)

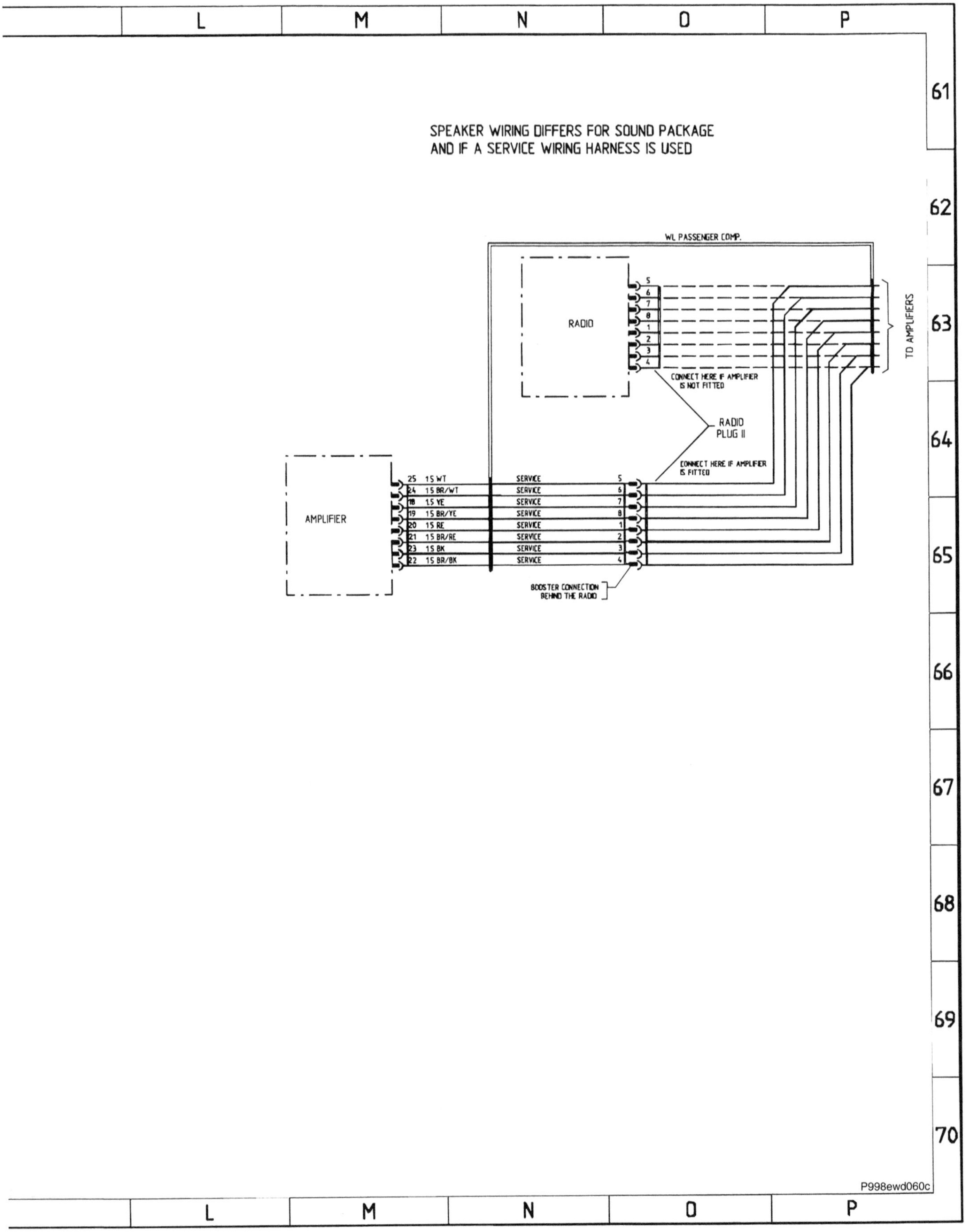

Rear spoiler, two-tone horns, wipe and wash cleaners (page 1 of 3)

Sheet 8 - 1997/1998

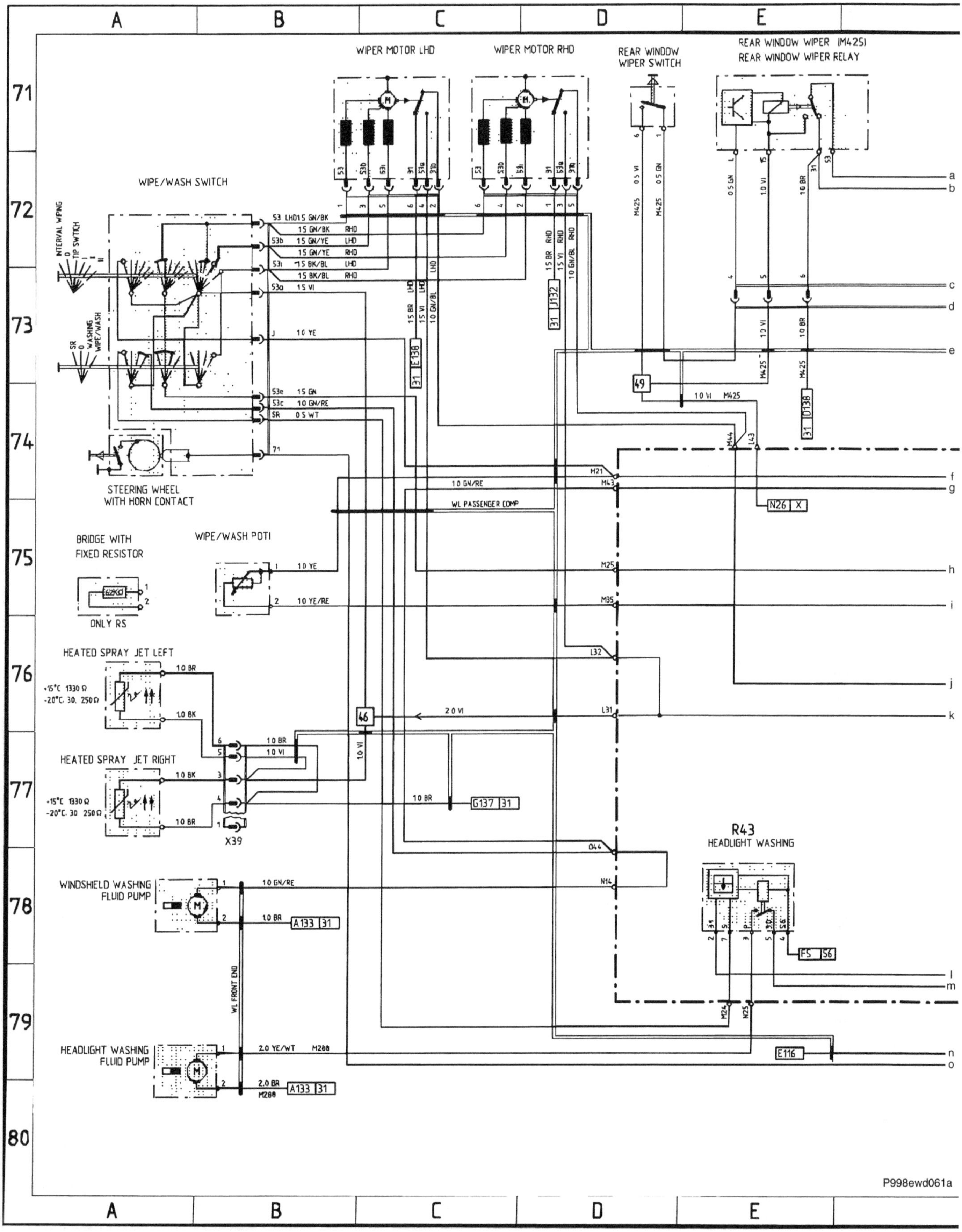

Sheet 8 - 1997/1998

Rear spoiler, two-tone horns, wipe and wash cleaners (page 2 of 3)

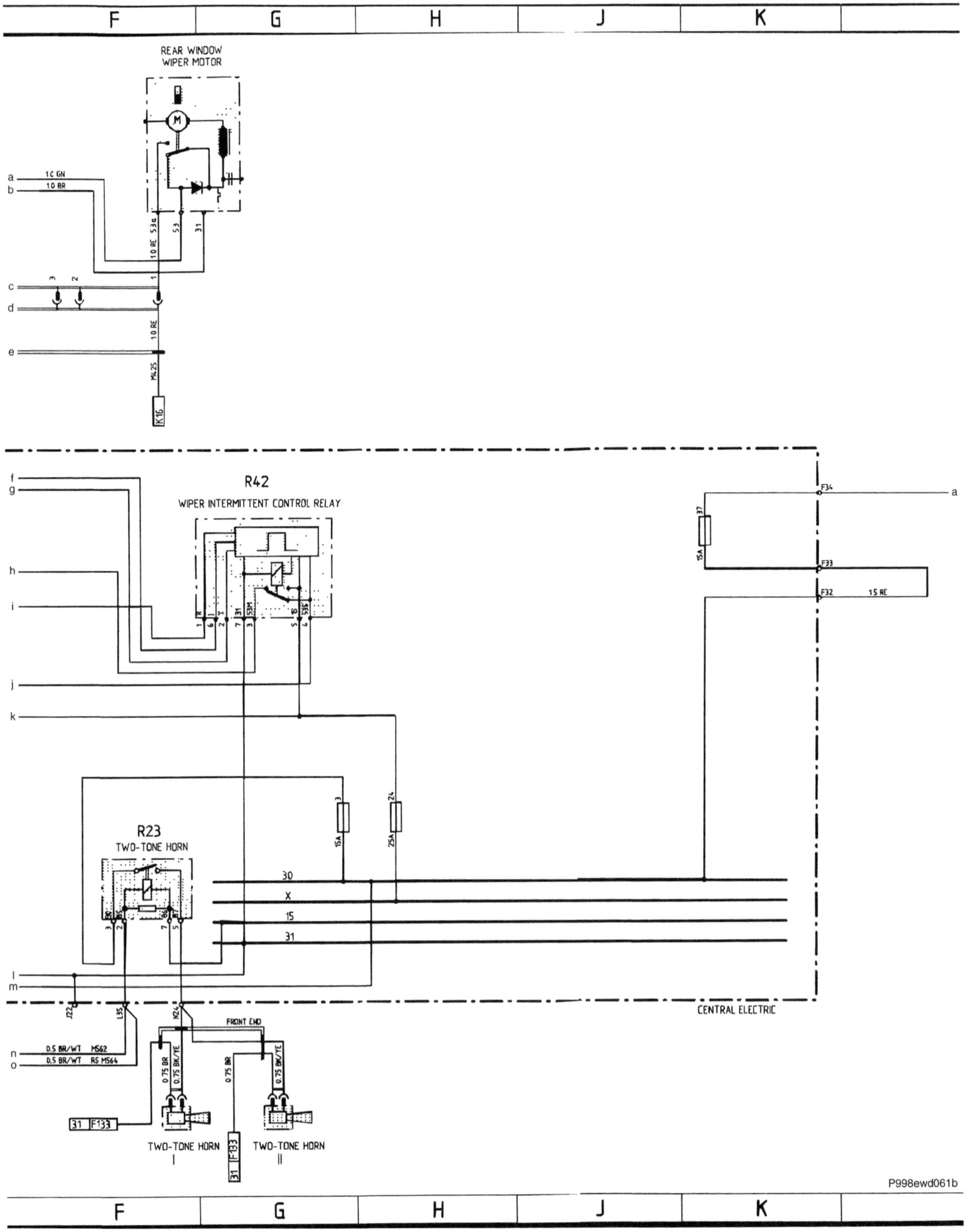

P998ewd061b

Rear spoiler, two-tone horns, wipe and wash cleaners (page 3 of 3)

Sheet 8 - 1997/1998

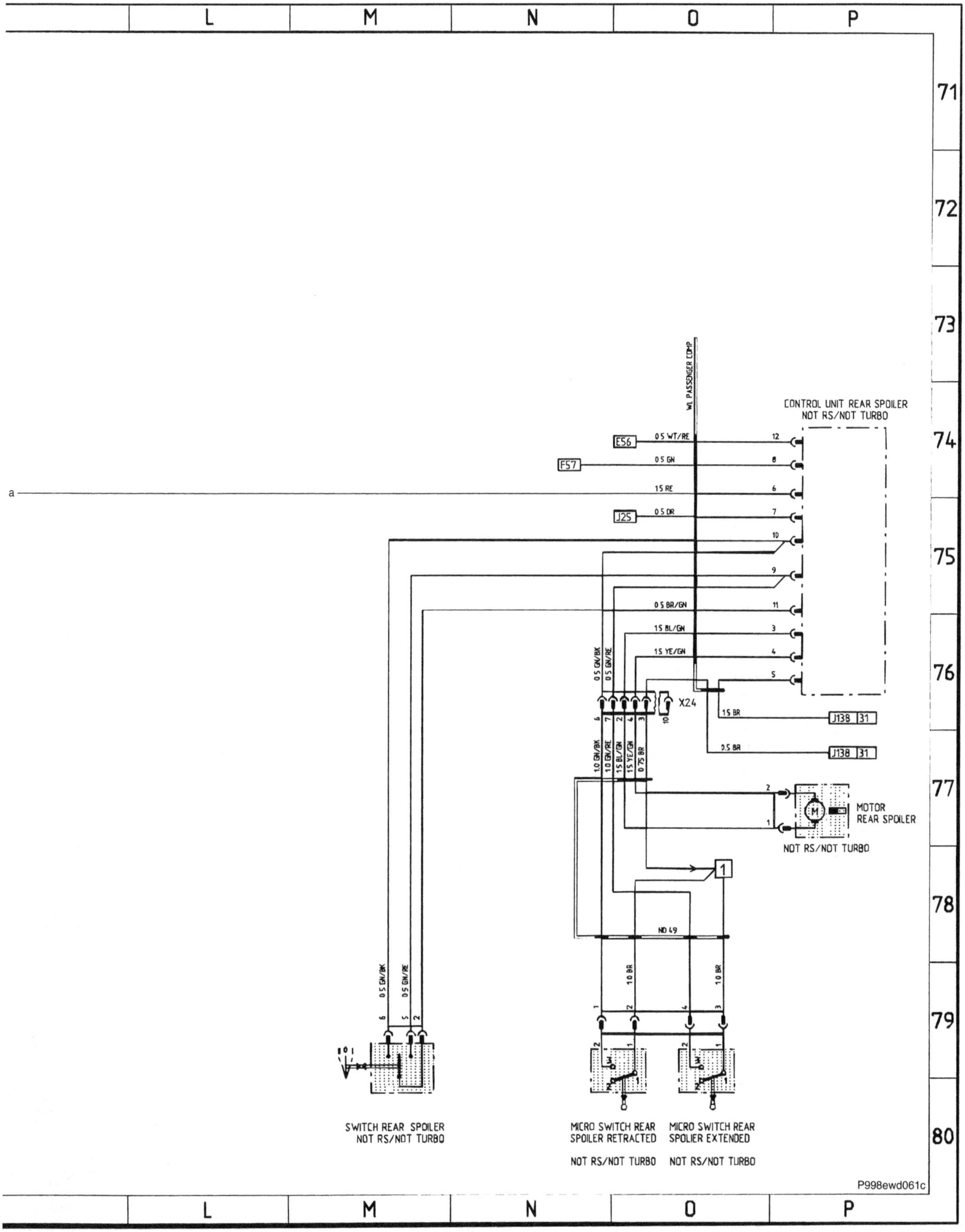

EWD

Sheet 9 - 1997/1998

Radio, telephone
(page 1 of 3)

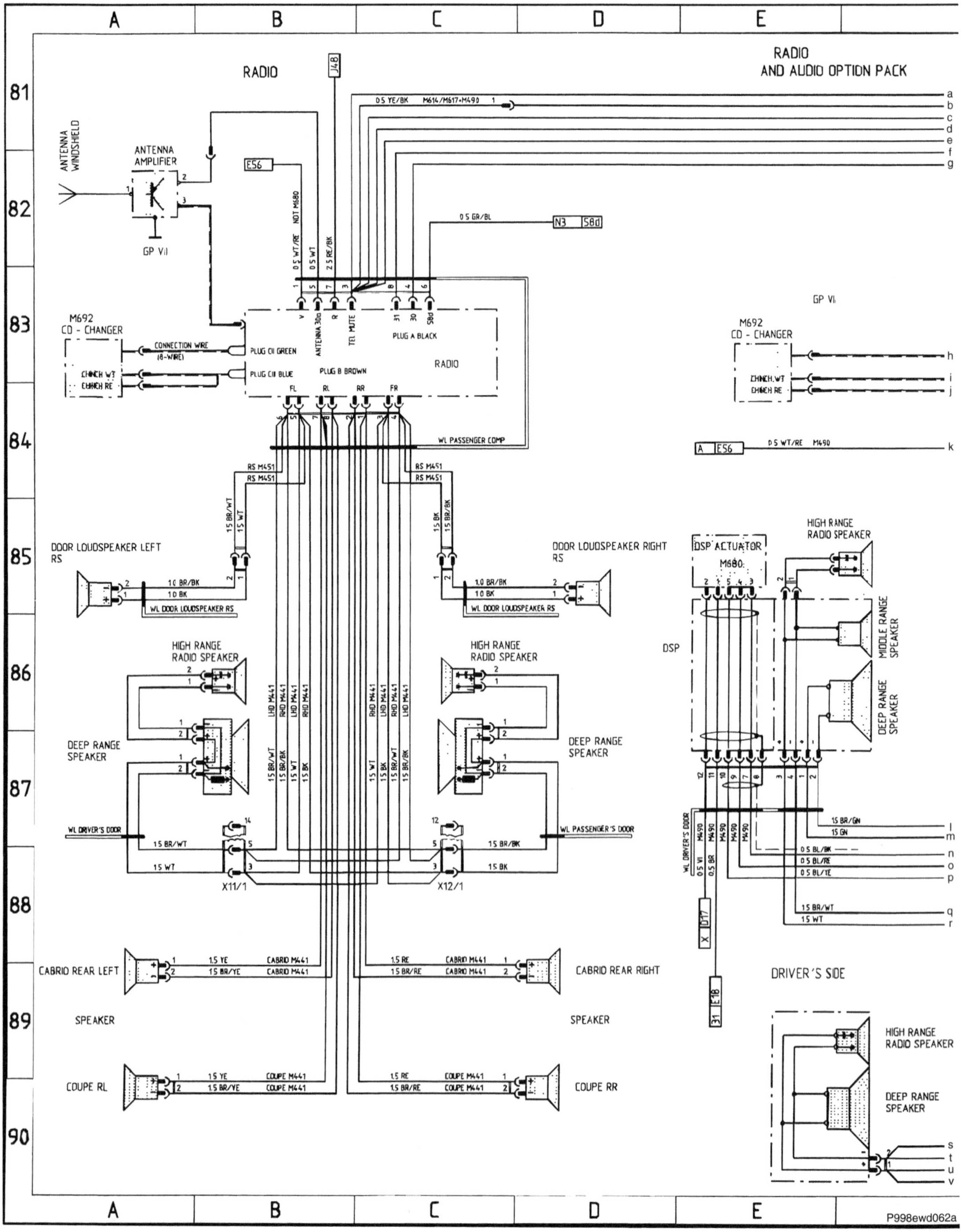

Radio, telephone (page 2 of 3)

Sheet 9 - 1997/1998

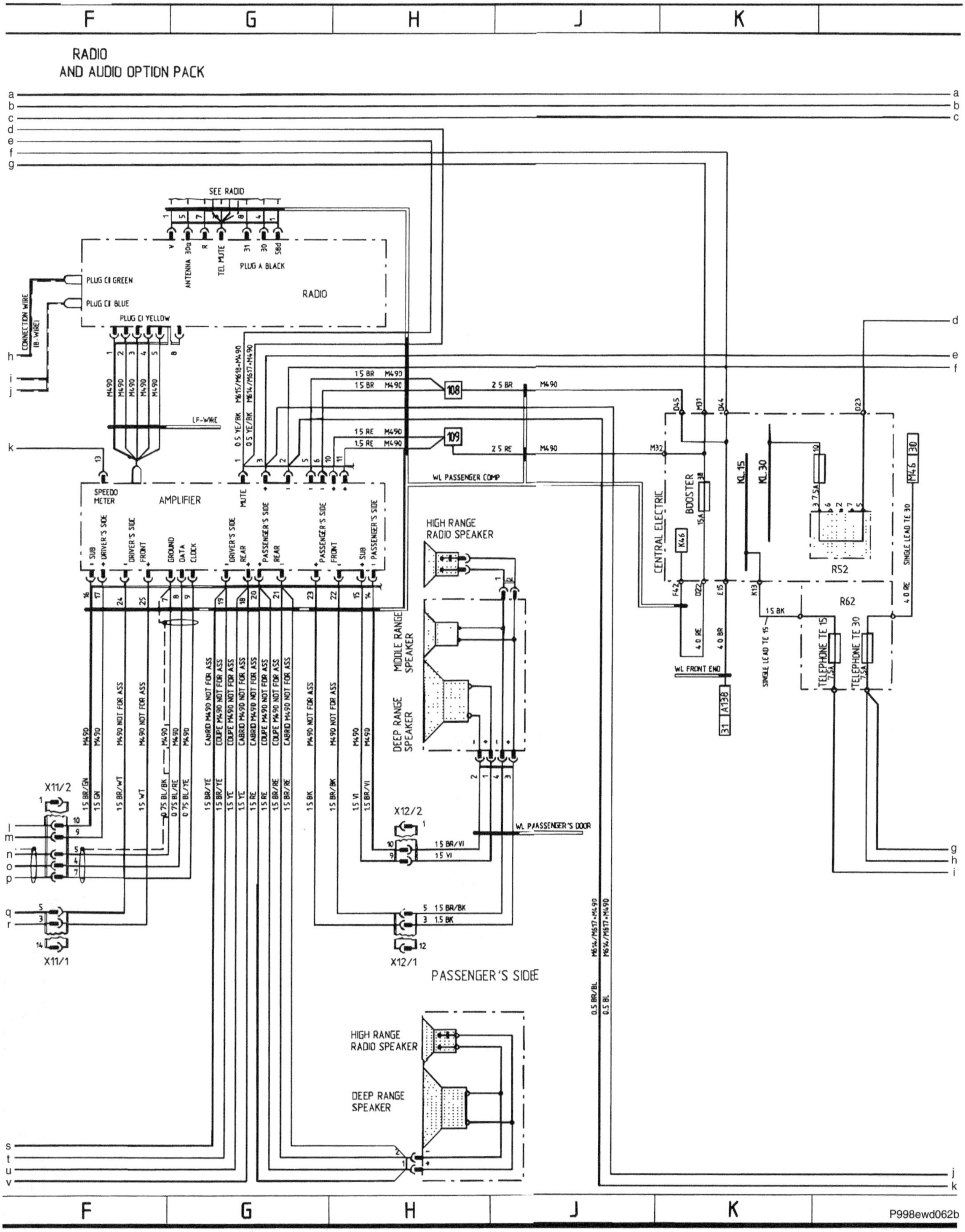

P998ewd062b

Sheet 9 - 1997/1998

Radio, telephone (page 3 of 3)

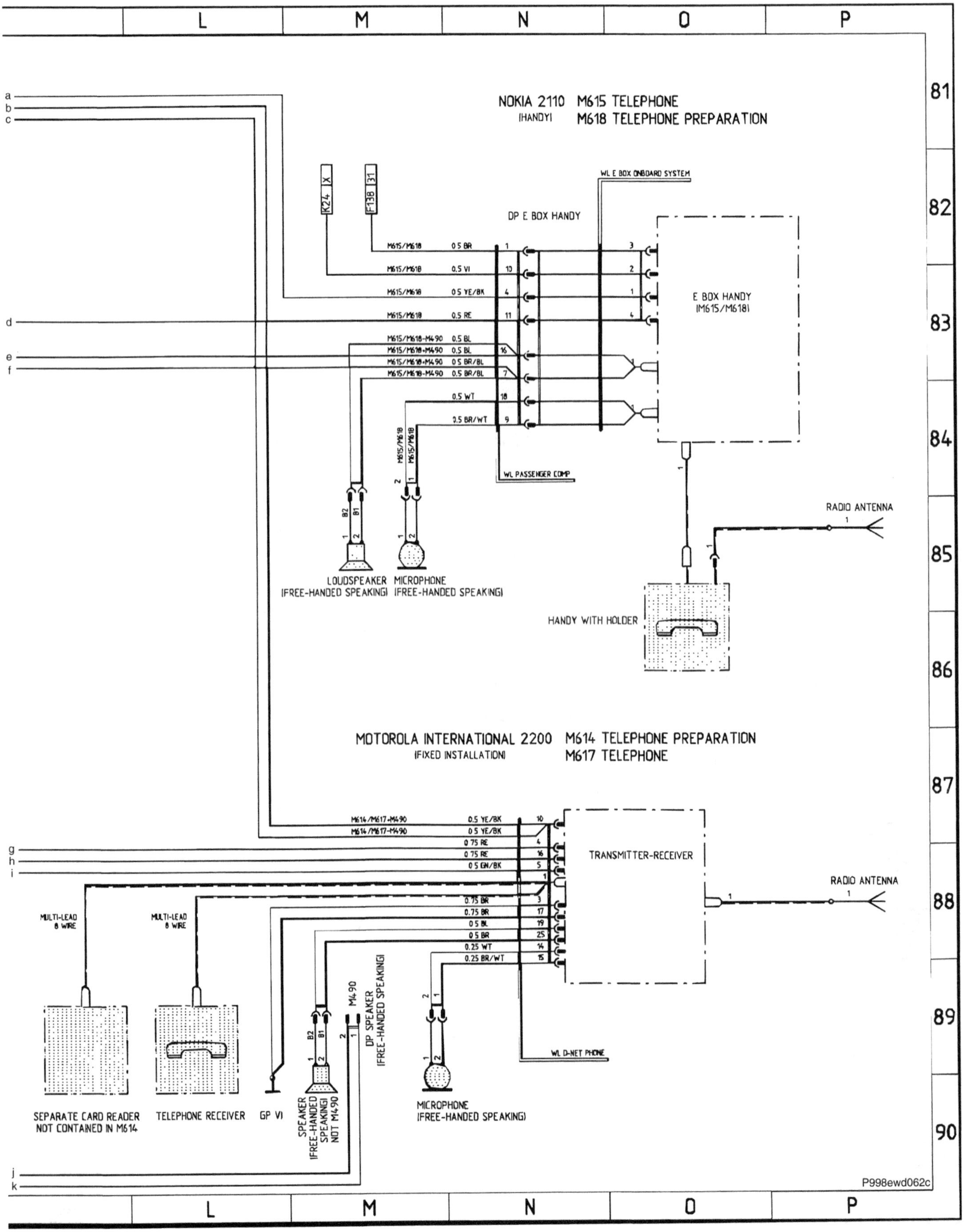

Anti-theft (option code M530) (page 1 of 1)

Sheet 11 - 1997/1998

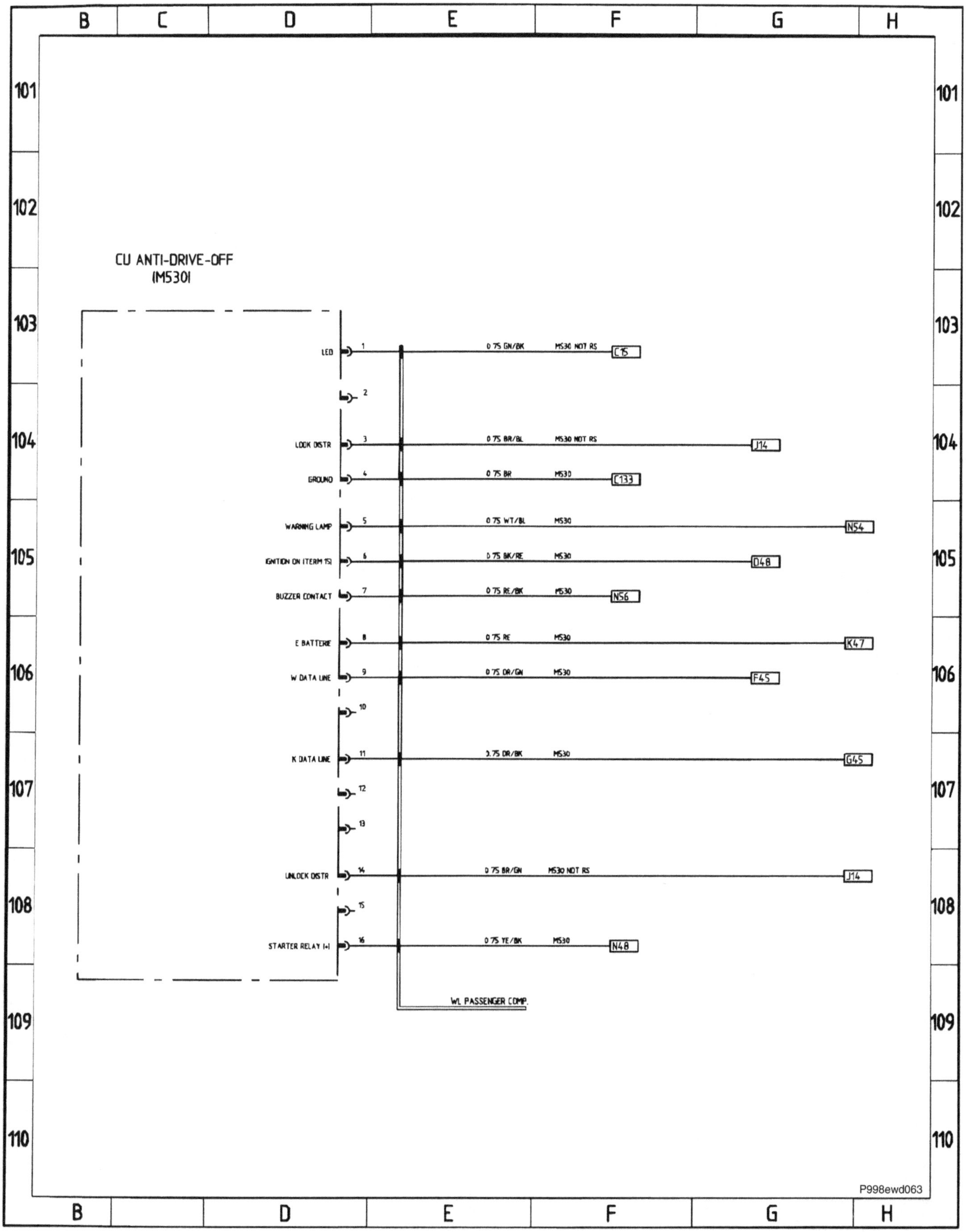

EWD

Sheet 12 - 1997/1998

Airbag
(page 1 of 1)

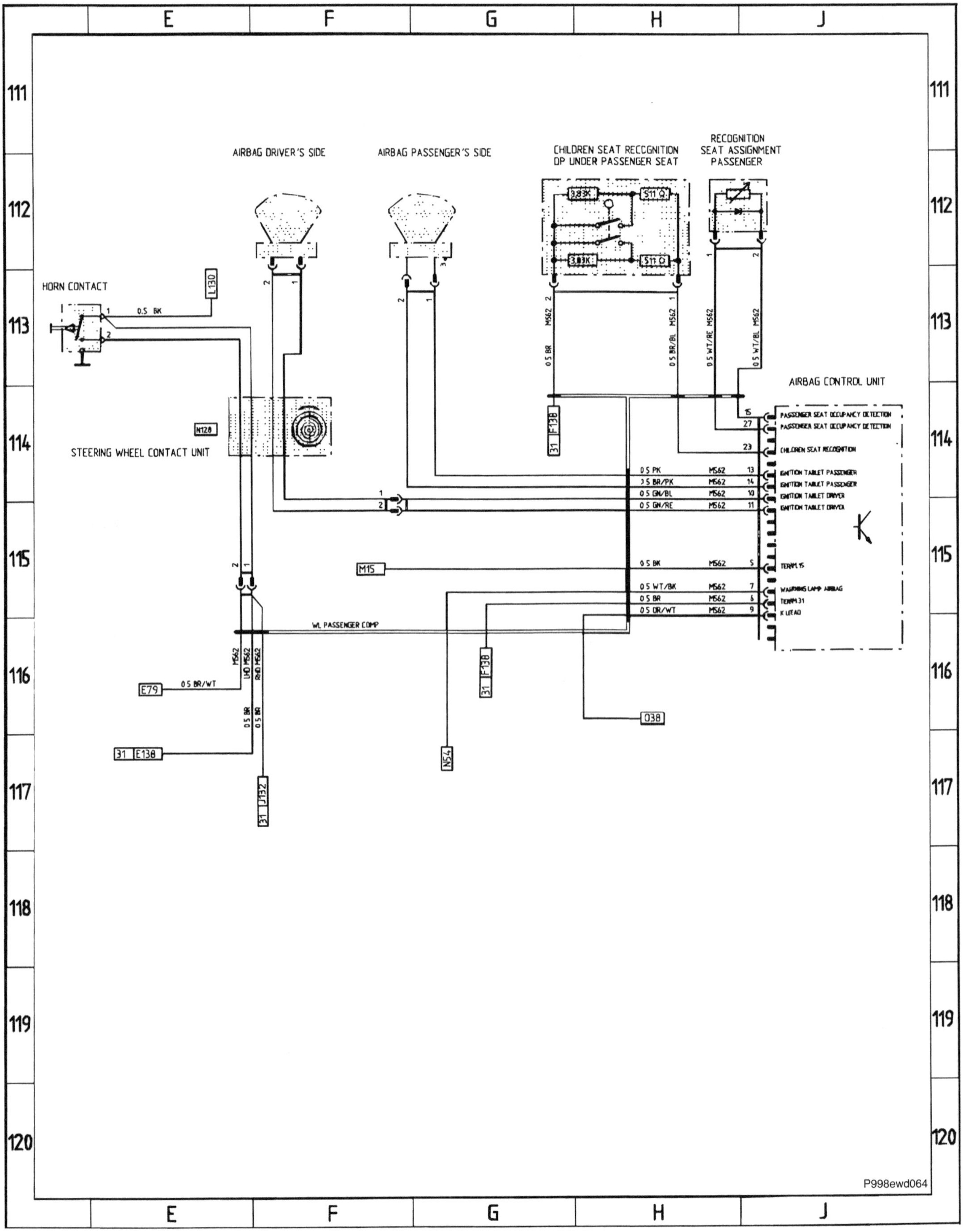

P998ewd064

Tiptronic-transmission (page 1 of 3)

Sheet 13 - 1997/1998

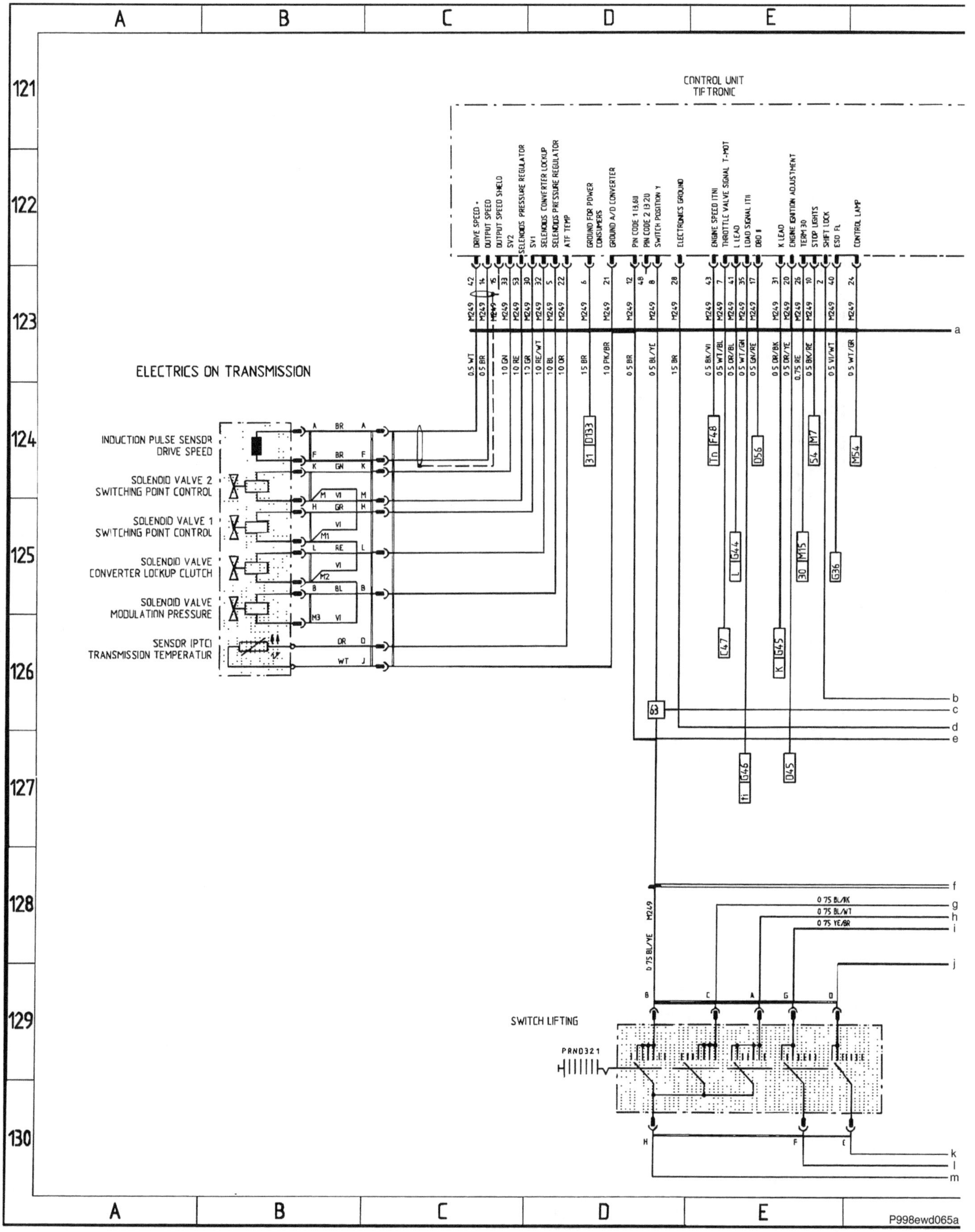

Sheet 13 - 1997/1998

Tiptronic-transmission (page 2 of 3)

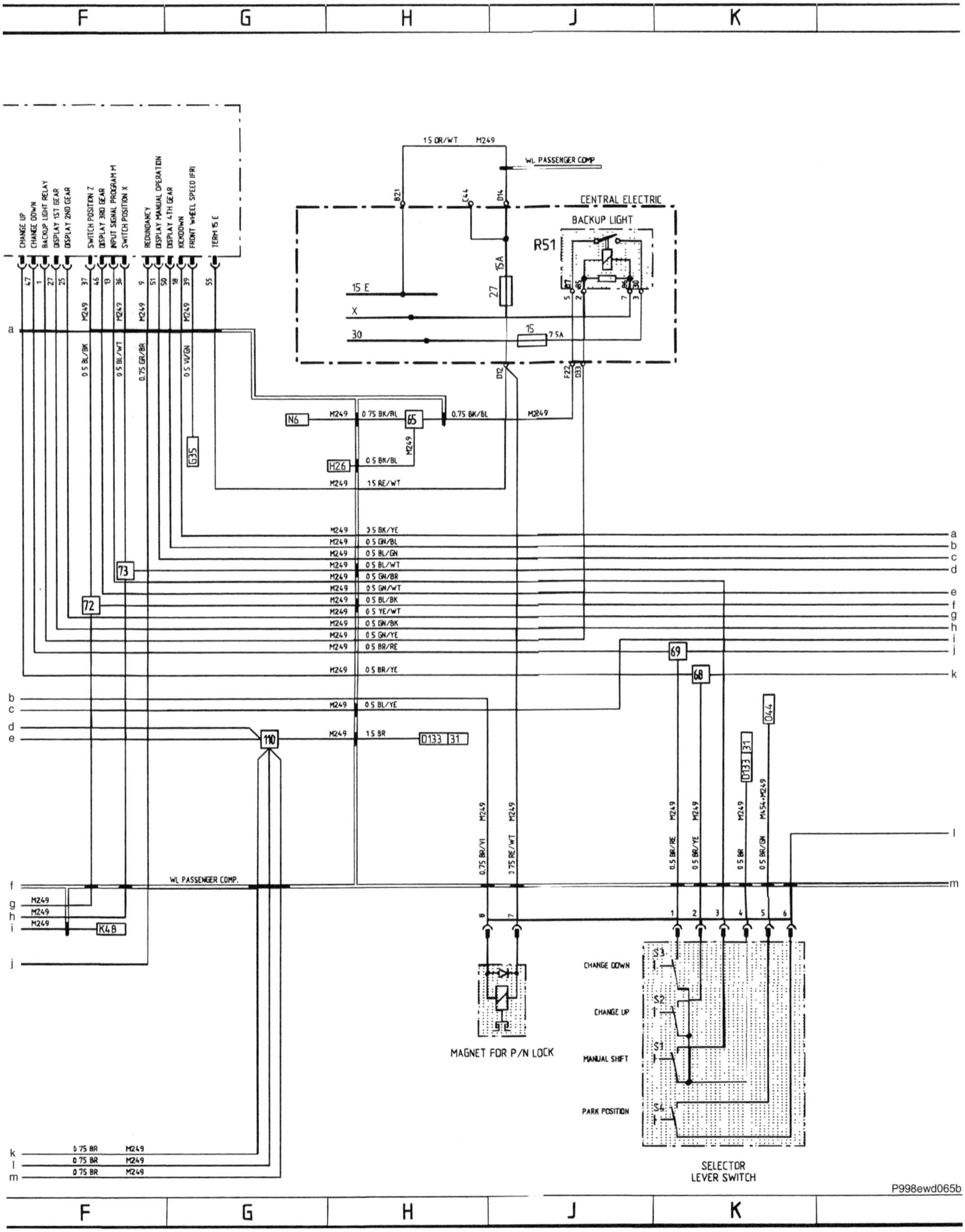

Tiptronic-transmission (page 3 of 3)

Sheet 13 - 1997/1998

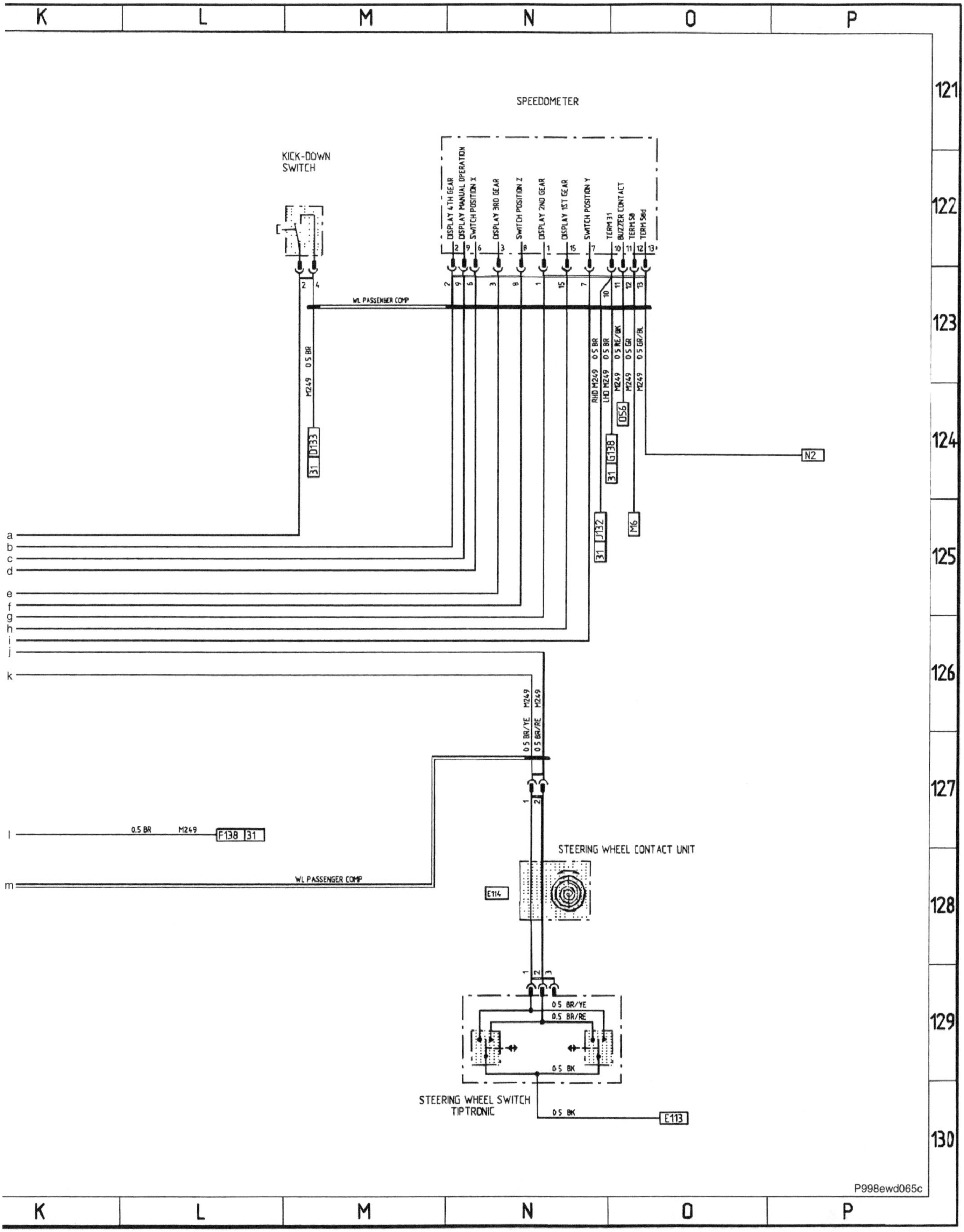

Sheet 14 - 1997/1998

Ground points (page 1 of 3)

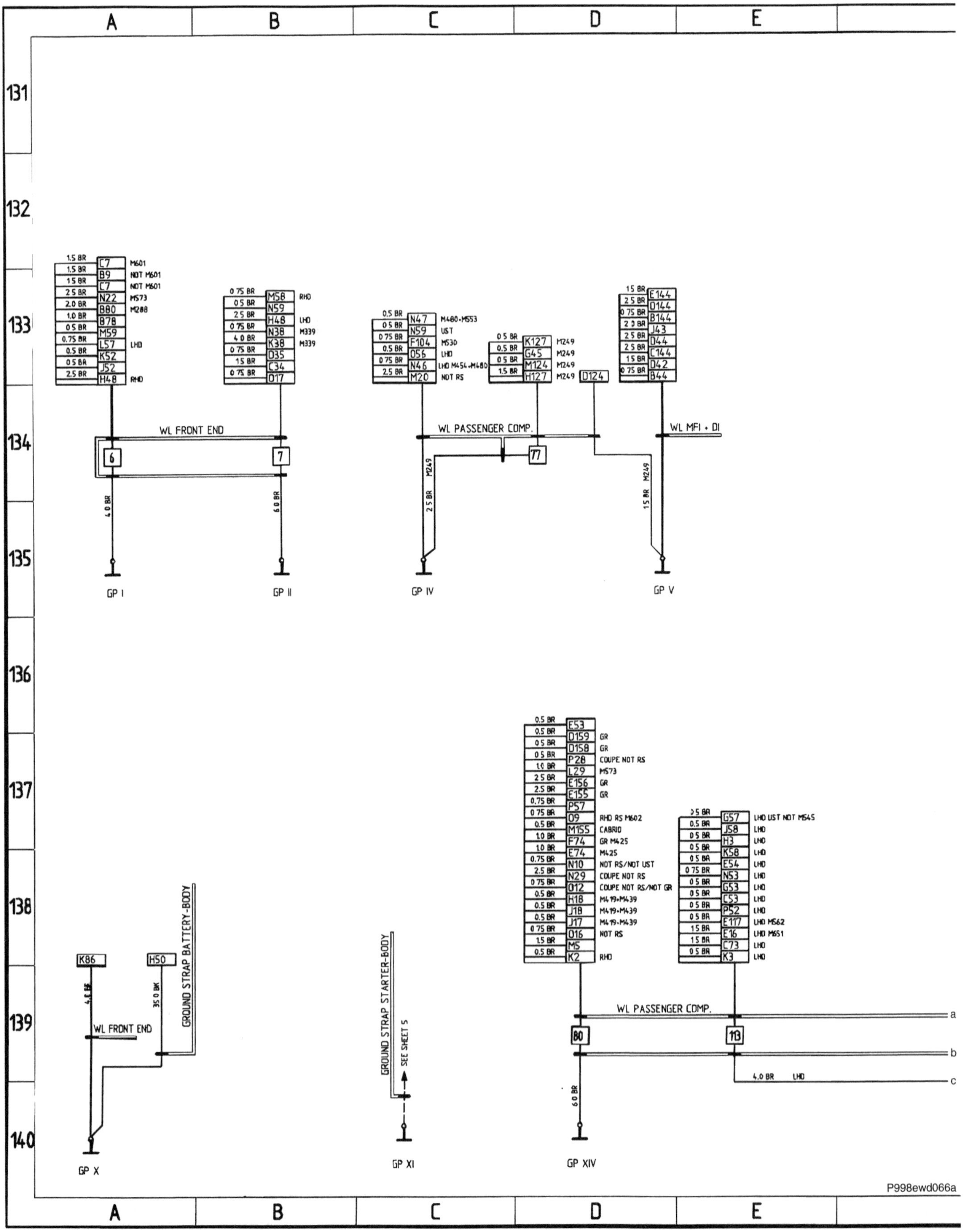

Ground points (page 2 of 3)

Sheet 14 - 1997/1998

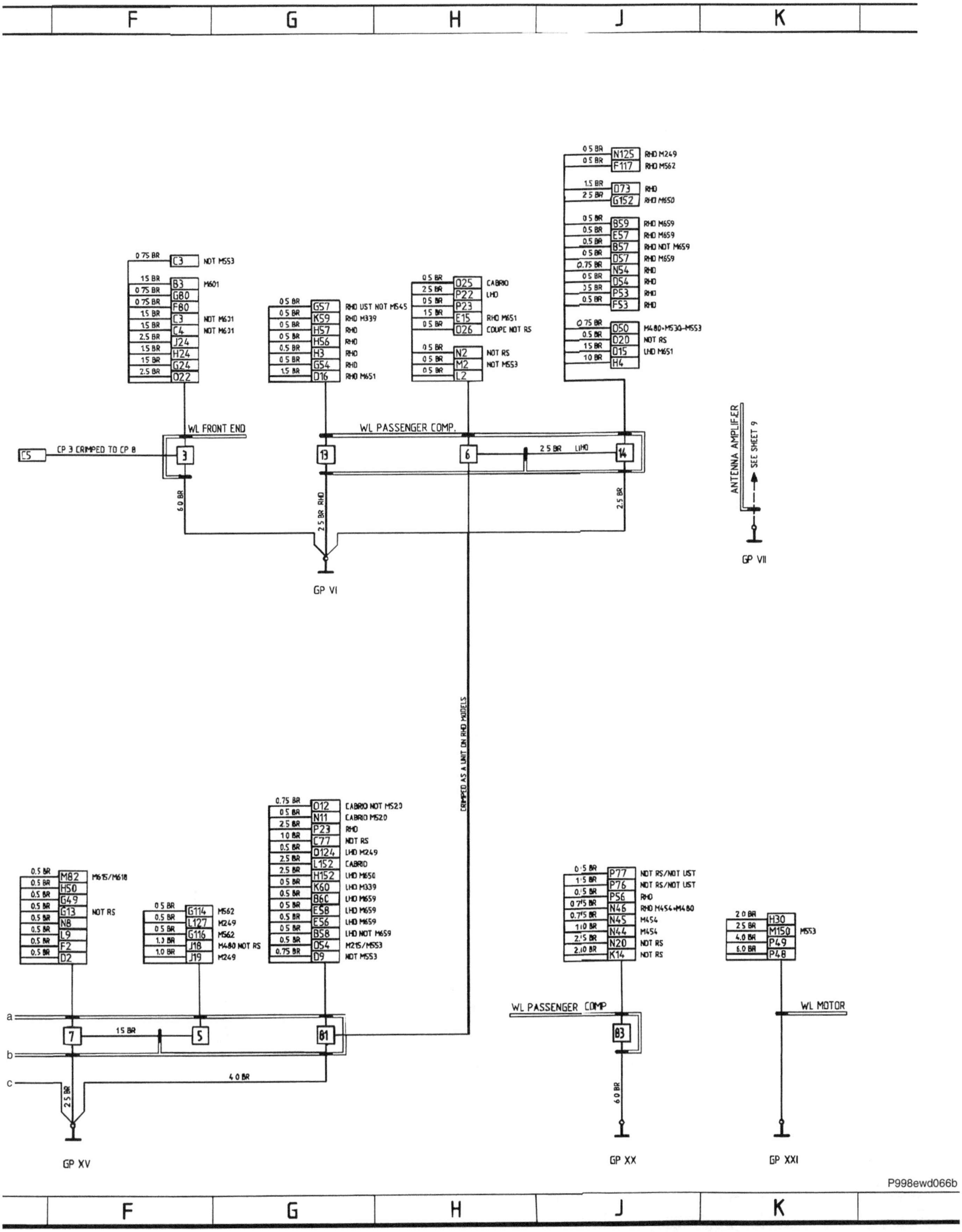

**Sheet 14 -
1997/1998**

Ground points
(page 3 of 3)

K	L	M	N	O	P

CODE	NOTE	POSITION IN VEHICLE LHD
GP I	LUGGAGE COMPARTMENT, WHEEL HOUSING LEFT	2cP
GP II	LUGGAGE COMPARTMENT, WHEEL HOUSING RIGHT	2cM
GP IV	GROUND OF POWER CONSUMERS, UNDER SEAT LEFT	8eP
GP V	ENGINE ELECTRONICS/TIPTRONIC	11eP
GP VI	CENTRAL ELECTRIC, ON SIDE MEMBER RIGHT	6bL
GP VII	ANTENNA AMPLIFIER ON SIDE MEMBER LEFT	7bL
GP X	BATTERY, LUGGAGE COMPARTMENT LEFT	3eP
GP XI	GROUND STRAP (STARTER, TRANSMISSION, ENGINE, BODY)	14cM
GP XII	ENGINE COMPARTMENT REAR LEFT	15cQ
GP XIV	ENGINE COMPARTMENT REAR RIGHT	15cL
GP XV	INSTRUMENTS, SIDE MEMBER LEFT	7bQ
GP XX	SEAT CROSS MEMBER, UNDER SEAT RIGHT	8eL
GP XXI	GROUND OF POWER CONSUMERS, CAMSHAFT CASING CYL 3	15cP

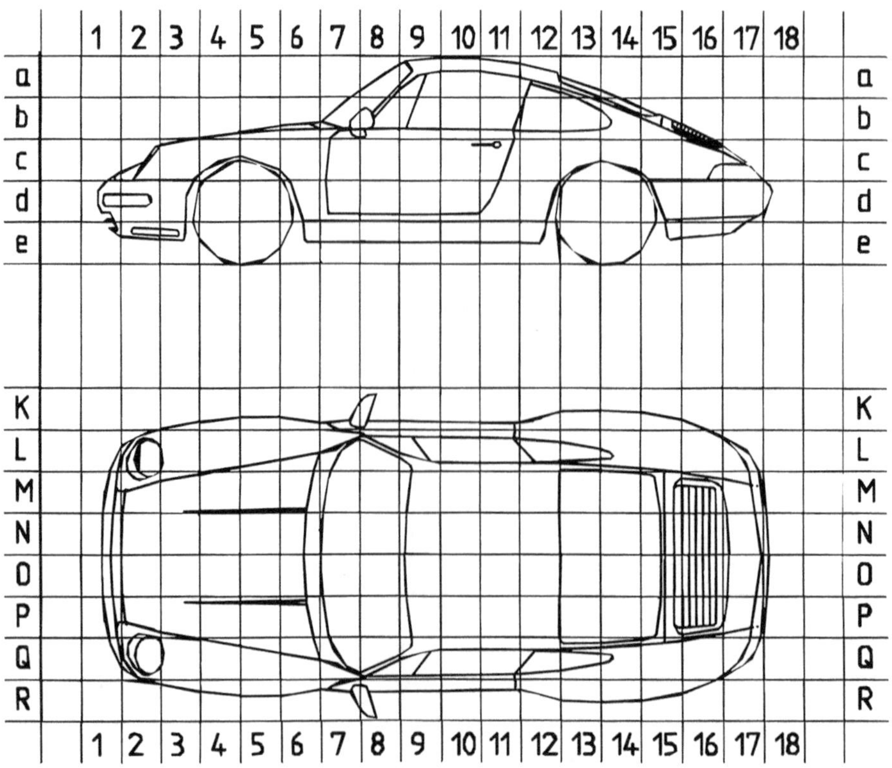

131 132 133 134 135 136 137 138 139 140

P998ewd066c

K	L	M	N	O	P

Convertible top, sunroof, glass roof (page 1 of 3)

Sheet 15 - 1997/1998

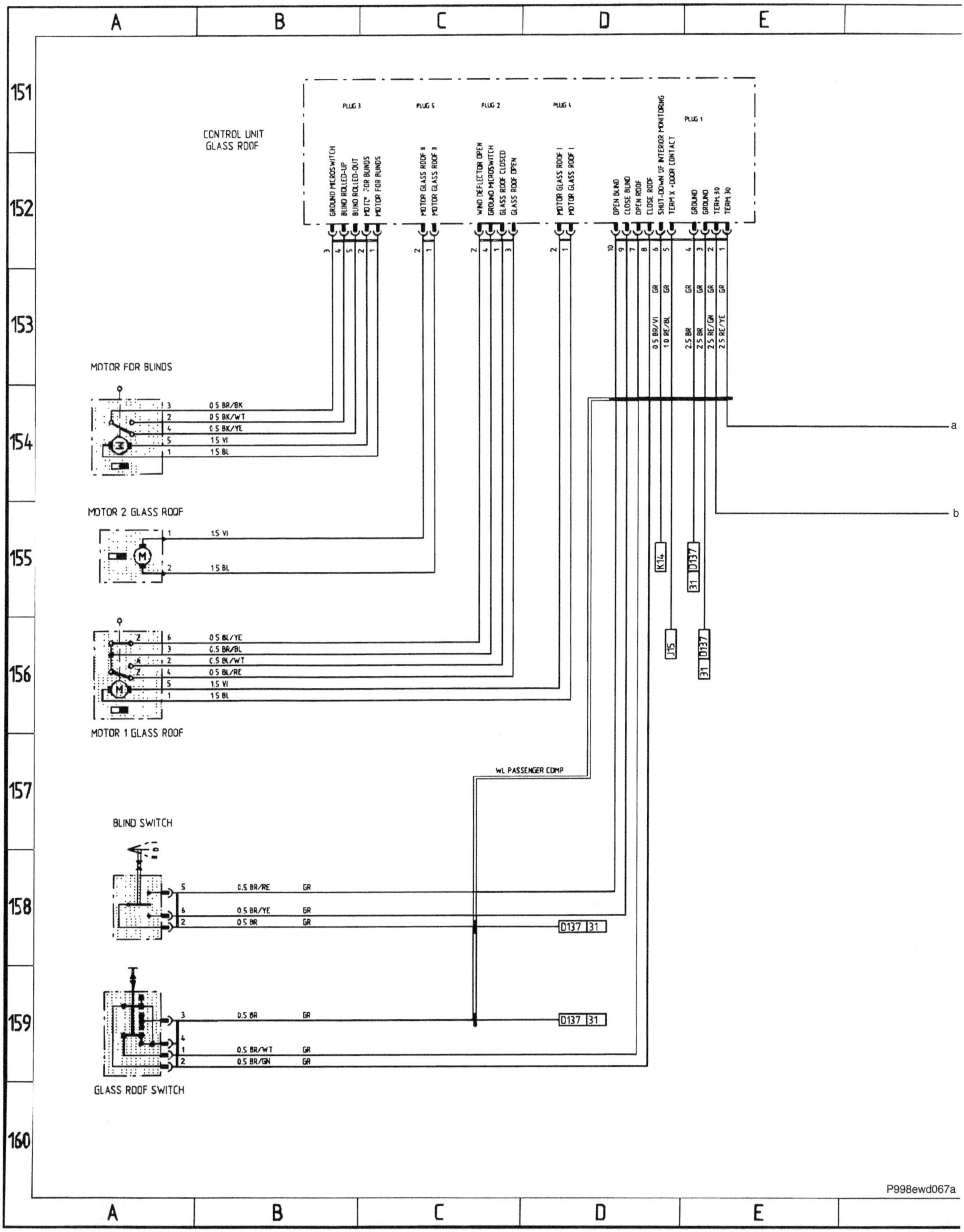

Sheet 15 - 1997/1998

Convertible top, sunroof, glass roof (page 2 of 3)

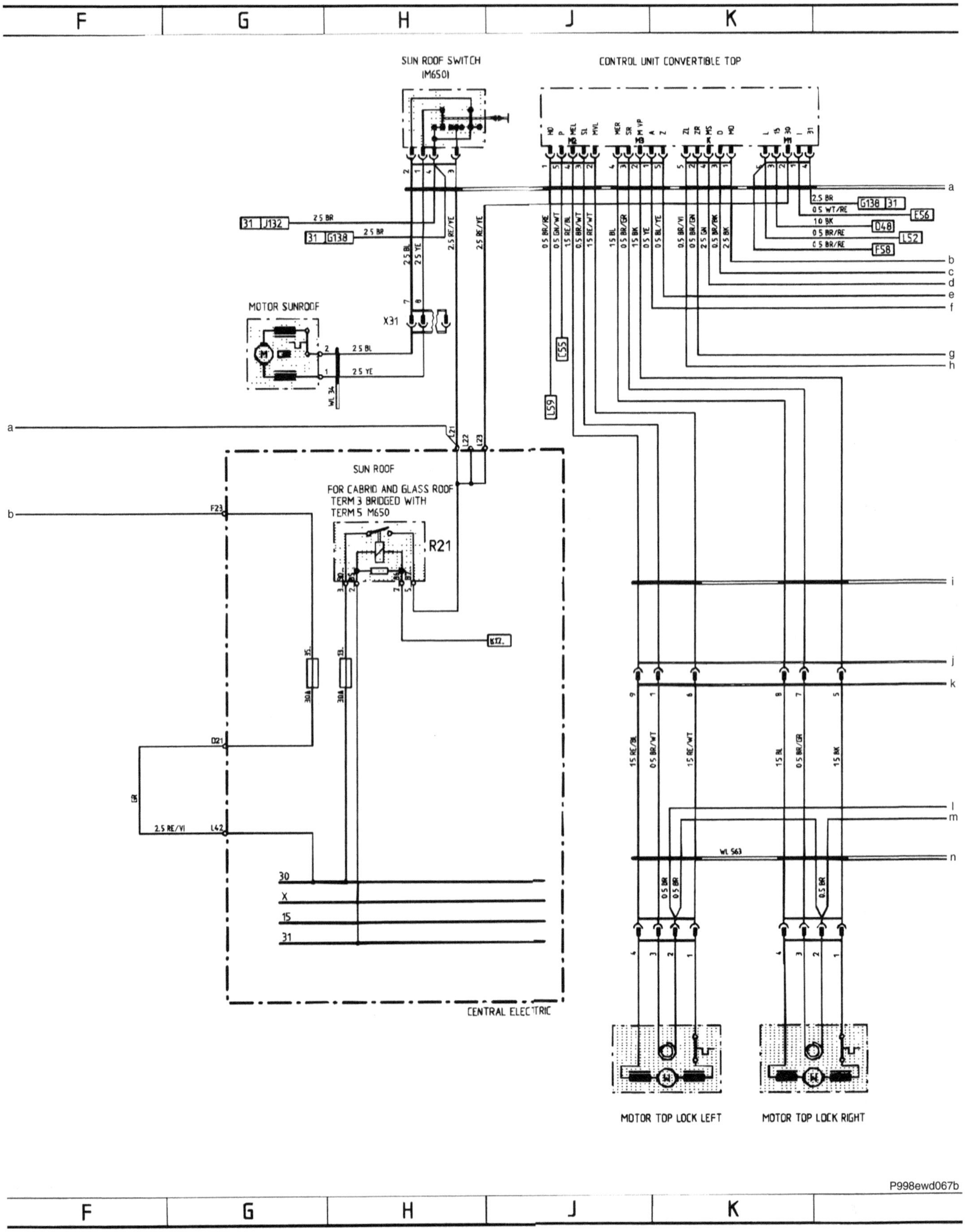

Convertible top, sunroof, glass roof (page 3 of 3)

Sheet 15 - 1997/1998

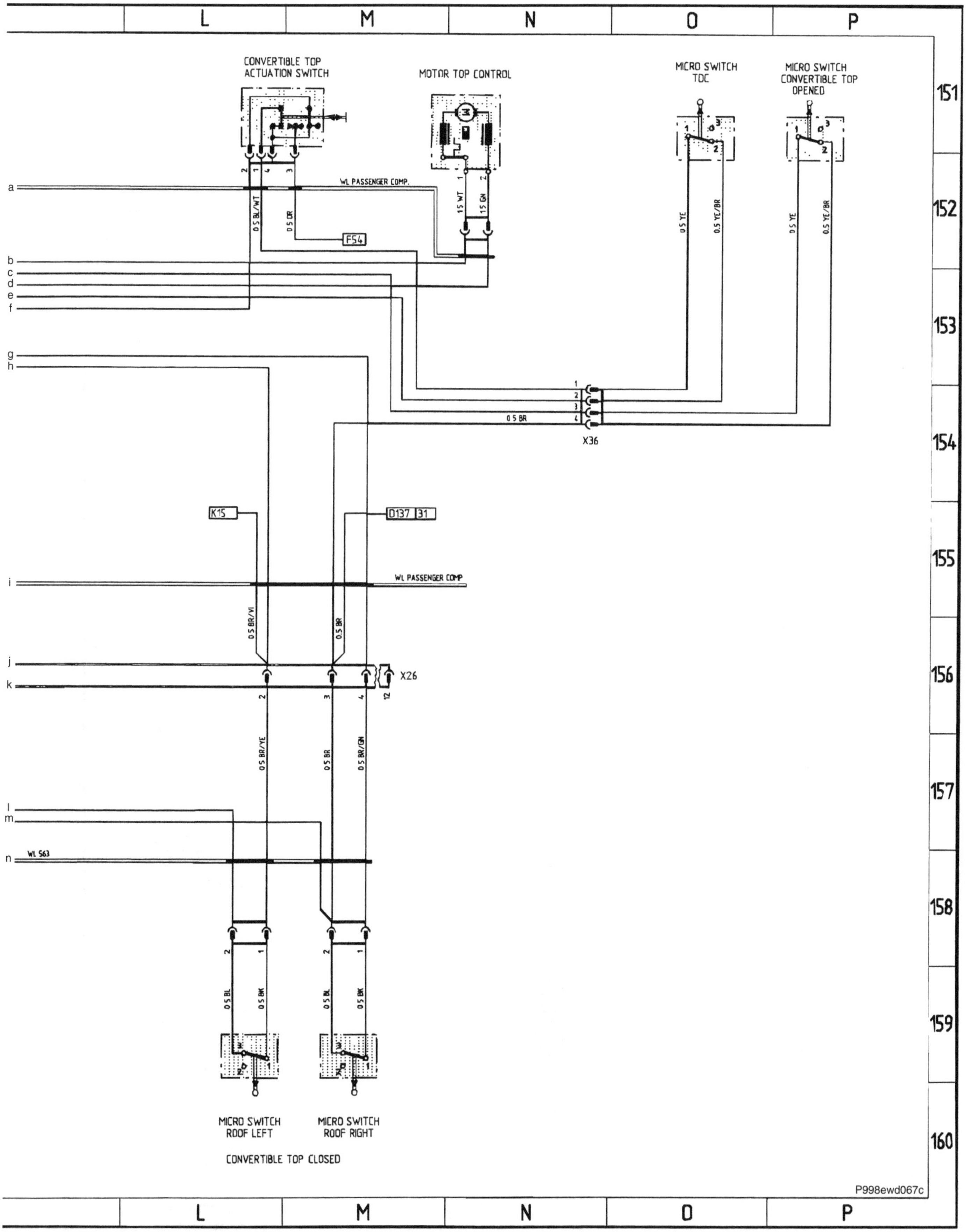

Sheet 16 - 1997/1998

Construction components (page 1 of 3)

A B C D E

CONSTRUCTION COMPONENTS

DESIGNATION, FUNCTION	POSITION IN VEHICLE LHD	RHD	NOTE	FIELD IN WIRING DIAGRAM
CARBON CANISTER SHUTOFF VALVE	13cR	—	IN WHEEL HOUSING REAR LEFT	C150
ALARM HORN	13dP	13dP	REAR LH CROSSMEMBER	J20
ANTENNA AMPLIFIER	7bL	7bQ	IN LUGGAGE COMPARTMENT	A83,E82
OUTSIDE TEMP. SENSOR	2dP	2dP	FRONT LH BUMPER	A60
FLASHER	6cQ	6cL	IN FRONT OF INSTRUMENTS	D1
STOP LIGHT SWITCH	6dP	6dM	AT THE BRAKE PEDAL BELOW THE FLOOR PANEL	N9
DIAGNOSIS CONNECTION	8cO	8cO	IN CENTRE CONSOLE	G50
MFI + DI RELAY	6cM	6cM	CENTRAL ELECTRIC	J44
MFI + DI CONTROL UNIT	9eP	9eP	UNDER THE SEAT LEFT	B-G41
ENGINE SPEED SENSOR	15cP	15cP	ON ENGINE	A46
THROTTLE VALVE SENSOR	15bN	15bN	ON THROTTLE VALVE ASSEMBLY	A44
PRESSURE TRANSDUCER, CHARGE AIR PRESSURE	16bM	16bM	ON ENGINE	A56
PRESSURE SWITCH AIR CONDITIONING SYSTEM	6bO	6bO	IN LUGGAGE COMPARTMENT	E30
PRESSURE WARNING SWITCH	1dO	1dO	IN LUGGAGE COMPARTMENT	M/N39
E BOX HANDY	7cN	7cN	ON RH SIDE OF TRANSVERSE WALL	O83
TWO-TONE HORN I	3cL	3cL	IN WHEEL HOUSING FRONT RIGHT	F80
TWO-TONE HORN II	3cL	3cL	IN WHEEL HOUSING FRONT RIGHT	G80
BLOWER FINAL STAGE	6bN	6bN	IN LUGGAGE COMPARTMENT	F30
BLOWER ENGINE COMPARTMENT	16cO	16cO	ON ENGINE	H30
BLOWER MOTOR RIGHT	6cM	6cM	IN LUGGAGE COMPARTMENT	G30
BLOWER MOTOR LEFT	6cP	6cP	IN LUGGAGE COMPARTMENT	F30
GONG RELAY	6cQ	—	LH SIDE IN FRONT OF INSTRUMENTS	P53/54
HALL SENSOR	17cO	17cO	ON DISTRIBUTOR	A46
REAR WINDOW WIPER RELAY	15bO	15bO	ON REAR WINDOW WIPER MOTOR	E71
HYDRAULIC UNIT ABS	3dP	3dP	IN LUGGAGE COMPARTMENT	A/B39
ABS/ABD HYDRAULIC UNIT	3dP	3dP	IN LUGGAGE COMPARTMENT	C39
INSIDE LIGHT WITH INTERIOR SENSOR	9aN	9aN	FRONT ROOF FRAME	N11
V-BELT CONTROL	17dO	17dO	ON ENGINE REAR MIDDLE	A55
KICK-DOWN SWITCH	6dO	6dM	BY ACCELERATOR PEDAL UNDER FLOOR PANEL	M122
KNOCK SENSOR CYL 1-3	15cP	15cP	ON ENGINE	A47
KNOCK SENSOR CYL.4-6	16cN	16cN	ON ENGINE	A47
CONDENSER BLOWER	2dQ	2dQ	IN WHEEL HOUSING FRONT LEFT	N21
FUEL PUMP	6eM	6eP	BEHIND THE BASE	H50
CLUTCH SWITCH CRUISE CONTROL	6dP	6dN	BY CLUTCH PEDAL UNDER FLOOR PANEL	P46
OXYGEN SENSOR CYL 1-3	16dP	16dP	ON ENGINE	F150
OXYGEN SENSOR CYL 4-6	16dM	16dM	ON ENGINE	D150
IDLE SPEED POSITIONER	15bN	15bN	ON ENGINE	H/J41
MASS AIR FLOW SENSOR	16bO	16bO	ON ENGINE	A43
MICRO SWITCH ROOF RIGHT	9aN	9aN	AT CONVERTIBLE TOP FRAME	M159
MICRO SWITCH ROOF LEFT	9aO	9aO	AT CONVERTIBLE TOP FRAME	L159
MICRO SWITCH REAR SPOILER EXTENDED	16bP	16bP	ON TAILGATE	O79
MICRO SWITCH REAR SPOILER RETRACTED	16bP	16bP	ON TAILGATE	O79
MICRO SWITCH TDC	12cL	12cL	BEHIND SIDE PANEL COVERING	O151
MICROSWITCH FOR TOP CONTROL	12cL	12cL	BEHIND SIDE PANEL COVERING	P151
MOTOR SUNROOF	12aN	12aN	IN ROOF	G153
MOTOR TOP CONTROL RIGHT	15cN	15cN	BEHIND REAR SEAT	N151
MOTOR TOP CONTROL LEFT	15cO	15cO	BEHIND REAR SEAT	N151
MOTOR TOP LOCK RIGHT	9aN	9aN	AT CONVERTIBLE TOP FRAME	K/L159
MOTOR TOP LOCK LEFT	9aO	9aO	AT CONVERTIBLE TOP FRAME	J/K159
MOTORS GLASS ROOF AND SUN BLINDS	14bO	14bO	REAR WINDOW SHELF	A154-156
REAR FOG LIGHT CUTOUT RELAY	17dN	17dN	BEHIND REAR PANEL	H108
NTC HEATER FAN	16cP	16cP	ON ENGINE	G30
NTC TEMPERATURE FOR INTAKE AIR	15cN	15cN	ON ENGINE	A42
OIL PRESSURE SENSOR/SWITCH	14cN	14cN	ON ENGINE	A54
OIL COOLER BLOWER	2dL	2dL	IN WHEEL HOUSING FRONT RIGHT	O21
OIL LEVEL SENSOR	12cK	12cK	IN WHEEL HOUSING REAR RIGHT	E51
OIL TEMPERATURE SENSOR FOR INSTRUMENT	14cN	14cN	ON ENGINE	A53
OIL TEMPERATURE SENSOR FOR OIL COOLER BLOWER	2dL	2dL	IN WHEEL HOUSING FRONT RIGHT	J21
SWITCH LIFTING	12dO	12dO	ON GEARBOX	O-F129

P998ewd068a

A B C D E

Construction components (page 2 of 3)

Sheet 16 - 1997/1998

DESIGNATION, FUNCTION	POSITION IN VEHICLE LHD	RHD	NOTE	FIELD IN WIRING DIAGRAM
BRAKE BOOSTER PUMP	1dO	1dO	IN LUGGAGE COMPARTMENT	L39
RELAY FOR REAR WINDOW DEFOGGER	16cP	16cP	CARRIER PLATE IN ENGINE COMPARTMENT	O/P29/30
ABS RELAY	4cL	4cL	IN LUGGAGE COMPARTMENT RIGHT	D39
RELAY CONDENSER BLOWER	6cM	6cM	CENTRAL ELECTRIC	L21
RELAY AC COMPRESSOR	16cP	16cP	CARRIER PLATE IN ENGINE COMPARTMENT	L30
RELAY TWO-TONE HORN	6cM	6cM	CENTRAL ELECTRIC	F78
RELAY POWER WINDOW	6cM	6cM	CENTRAL ELECTRIC	K18/19
RELAY BLOWER ENGINE COMPARTMENT	16cP	16cP	CARRIER PLATE IN ENGINE COMPARTMENT	J/K30
RELAY STARTER TIPTRONIC-TRANSMISSION	6cM	6cM	CENTRAL ELECTRIC	N50
RELAY BACKUP LIGHT	7bQ	7bQ	IN LUGGAGE COMPARTMENT LEFT	J122
RELAY THERMO SWITCH CATALYTIC CONVERTER	6cM	6cM	CENTRAL ELECTRIC	L/M51
RELAY TERM.X	6cM	6cM	CENTRAL ELECTRIC	J/K44
RELAY TERM 15 E	6cM	6cM	CENTRAL ELECTRIC	K/L44
RELAY RFL BACKUP LIGHT	6cM	6cM	CENTRAL ELECTRIC	J/K7, A-M94
RELAY FRONT FOG LAMP	6cM	6cM	CENTRAL ELECTRIC	J7, C-O94
RELAY OIL COOLER BLOWER	6cM	6cM	CENTRAL ELECTRIC	K21
RELAY BRAKE BOOSTER PUMP	4cL	4cL	IN LUGGAGE COMPARTMENT RIGHT	M39
RELAY SUN ROOF	6cM	6cM	CENTRAL ELECTRIC	H155
SECONDARY AIR PUMP RELAY	17cP	17cP	ON CARRIER ON LH SIDE IN ENGINE COMPARTMENT	A50
DAYTIME RUNNING LIGHT CONTROL RELAY	6cM	6cM	CENTRAL ELECTRIC	D8
RELAY DAYTIME RUNNING LIGHT CANADA	6cM	6cM	CENTRAL ELECTRIC	D1
RELAY FOG LIGHT SWITCH OFF USA/JAPAN	6cM	6cM	CENTRAL ELECTRIC	H-P/94, F/G10
BACKUP LIGHT SWITCH	14dN	14dN	BY TRANSMISSION RIGHT	K/L1
SHIFT VALVE RESONANCE FLAP	15bN	15bN	ON ENGINE	H41
WINDSHIELD WASHING FLUID PUMP	3cQ	3cQ	BY WASHING FLUID RESERVOIR IN WHEEL HOUSING FL	A78
FUSES ENGINE COMPARTMENT	16cP	16cP	CARRIER PLATE IN ENGINE COMPARTMENT	L28
FUSE IN LUGGAGE COMPARTMENT	4cP	4cP	BELOW HYDRAULIC UNIT COVER	D37/M47
HEADLIGHT CLEANER PUMP	3cQ	3cQ	BY WASHING FLUID RESERVOIR IN WHEEL HOUSING FL	A79/80
HEADLIGHT CLEANER RELAY	6cM	6cM	CENTRAL ELECTRIC	E78
STARTER	14dN	14dN	BY CLUTCH HOUSING RIGHT	P49/50
DRIVE CRUISE CONTROL	15cO	15cO	ON ENGINE	N43/44
CONTROL UNIT ABS	2dM	2dM	IN LUGGAGE COMPARTMENT	A-J32
CONTROL UNIT AIRBAG	6cO	6cO	ON INTRUMENT PANEL	J114/116
CONTROL UNIT ALARM SYSTEM/CENTRAL LOCKING SYSTEM	9eM	9eM	UNDER THE SEAT RIGHT	J-M12
CONTROL UNIT CONVERTIBLE TOP	6cP	6cO	ON INTRUMENT PANEL	J/K151
CONTROL UNIT GLASS ROOF	13cO	13cO	EMERGENCY SEAT WELL	O151-152
CONTROL UNIT REAR SPOILER	7cM	7cP	ON RH SIDE OF TRANSVERSE WALL	P74-76
CONTROL UNIT HEATING, AIR CONDITIONING	7cO	7cN	ON INTRUMENT PANEL	B-F21
CONTROL UNIT INSTRUMENT LIGHT	6cM	6cM	CENTRAL ELECTRIC	K1
CONTROL UNIT CRUISE CONTROL	9eN	9eN	UNDER THE SEAT RIGHT	P43/44
CONTROL UNIT TIPTRONIC TRANSMISSION	10dP	10dP	UNDER THE SEAT LEFT	C-G122
SECONDARY AIR PUMP	16cN	16cN	ON ENGINE	D/E50
LIGHTS ON BUZZER	6cQ	6cL	IN FRONT OF INSTRUMENTS	P56/57
TIMING VALVE	16cM	16cM	ON ENGINE	L150
TANK DIFFERENTIAL PRESSURE SENSOR	6cN	—	ON FUEL TANK	F150
FUEL LEVEL SENSOR	5cM	5cP	IN FUEL TANK	K51
TANK VENTING VALVE	16cP	16cP	ON ENGINE	K41
CHARGE AIR TEMPERATURE SENSOR	16cN	16cN	ON ENGINE	A142
ENGINE TEMPERATURE SENSOR	15cP	15cP	ON ENGINE	A45
RESISTOR BLOWER	16dP	16dP	IN ENGINE COMPARTMENT	J30
RESISTOR FOR BLOWER CONDENSER	2dQ	2dQ	IN WHEEL HOUSING FRONT LEFT	M21
RESISTOR OIL COOLER BLOWER	2dL	2dL	IN WHEEL HOUSING FRONT RIGHT	M21
AMPLIFIER	9eM	9eM	UNDER THE SEAT RIGHT	F-H85, M85
WIPER INTERMITTENT CONTROL RELAY	6cM	6cM	CENTRAL ELECTRIC	F/G75
CENTRAL ELECTRIC	6cM	6cM	IN LUGGAGE COMPARTMENT	—
CENTRAL INFORMER	7cP	7cM	ON FIREWALL	M32
IGNITION CONTROL UNIT	11dP	11dP	UNDER THE SEAT LEFT	J41
IGNITION COIL I	17cP	17cP	ON ENGINE	L41
IGNITION COIL II	17cP	17cP	ON ENGINE	L41

P998ewd068b

Sheet 16 - 1997/1998

Construction components (page 3 of 3)

	L	M	N	O	P

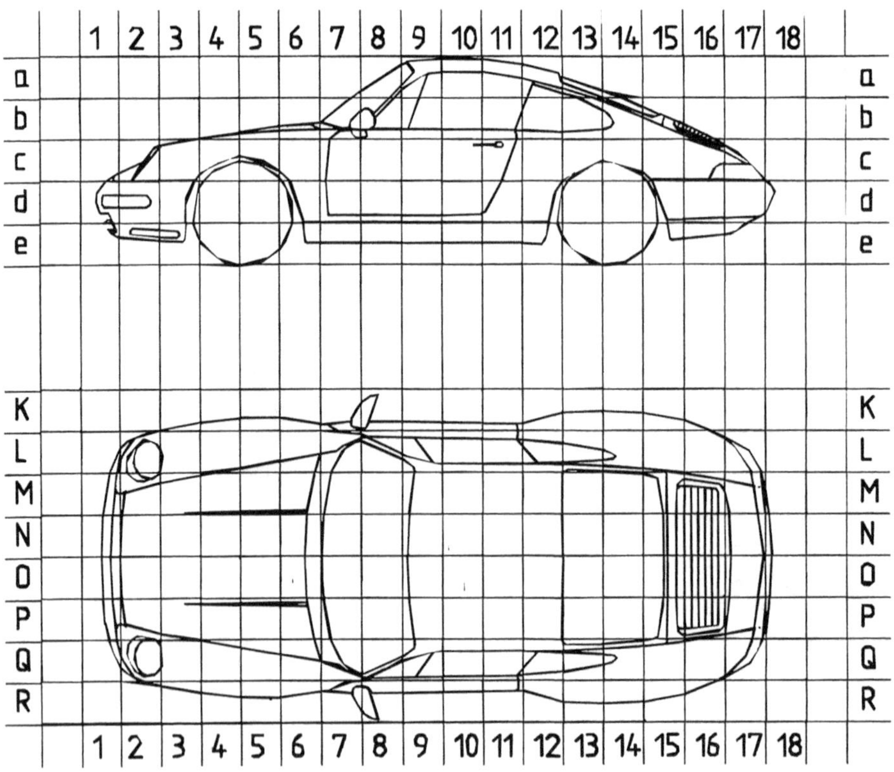

P998ewd068c

	L	M	N	O	P

Connectors and abbreviations (page 1 of 2)

Sheet 17 - 1997/1998

A	B	C	D	E

PLUG CONNECTIONS

CODE	NUMBER OF PINS	DESIGNATION, FUNCTION	POSITION IN VEHICLE LHD	RHD	NOTE	FIELD IN WIRING DIAGRAM
X 1/1	10	CONNECTION WL PASSENGER COMPARTMENT/ WL FRONT END	6cL	6cL	IN LUGGAGE COMPARTMENT RIGHT	G/H34,H/J48
X 1/2	10	CONNECTION WL PASSENGER COMPARTMENT/ WL FRONT END	6cL	6cL	IN LUGGAGE COMPARTMENT RIGHT	C9,E36,G48,B58/59
X 3	14	CONNECTION WL PASSENGER COMPARTMENT/ WL MOTOR	16cP	16cP	CARRIER PLATE IN ENGINE COMPARTMENT	K/L3,J27/28,O48/49,B53/54
X4/1	14	CONNECTION WL PASSENGER COMPARTMENT/ WL MFI + DI	9eP	9eP	UNDER THE SEAT LEFT	C/D45/46,B55,L143
X4/2	14	CONNECTION WL PASSENGER COMPARTMENT/ WL MFI + DI	9eP	9eP	UNDER THE SEAT LEFT	F45/46,B55,G/H148
X4/3	10	CONNECTION WL PASSENGER COMPARTMENT/ WL MFI + DI	9eP	9eP	UNDER THE SEAT LEFT	F147
X11/1	14	CONNECTION WL PASSENGER COMPARTMENT/ WL DOOR DRIVER'S SIDE	6cQ	6cL	UNDER THE INSTRUMENT PANEL DIVER'S SIDE	E/F16/17,B87/88, F88
X11/2	10	CONNECTION WL PASSENGER COMPARTMENT/ WL DOOR DRIVER'S SIDE	6cQ	6cL	UNDER THE INSTRUMENT PANEL DIVER'S SIDE	C16,F87/88
X12/1	12	CONNECTION WL PASSENGER COMPARTMENT/ WL DOOR PASSENGER'S SIDE	6cL	6cQ	UNDER THE INSTRUMENT PANEL PASSENGER'S SIDE	E/F13,C67/88,H88
X12/2	10	CONNECTION WL PASSENGER COMPARTMENT/ WL DOOR PASSENGER'S SIDE	6cL	6cQ	UNDER THE INSTRUMENT PANEL PASSENGER'S SIDE	C14,H87/88
X20	14	CONNECTION WL PASSENGER COMPARTMENT/ WL HEATER, AIR CONDITIONER	6cM	6cM	UNDER INSTRUMENT PANEL RIGHT	G25/26
X24	10	CONNECTION WL PASSENGER COMPARTMENT/ WL TAILGATE	16cL	16cL	BOX RH SIDE OF ENGINE COMPARTMENT	O10,O15/16,O76
X25	10	CONNECTION WL PASSENGER COMPARTMENT/ WL TAIL LIGHTS	16cL	16cL	BOX RH SIDE OF ENGINE COMPARTMENT	N4-6
X26	12	CONNECTION WL PASSENGER COMPARTMENT/ WL CONVERTIBLE TOP	12cL	12cL	BEHIND SIDE PANEL COVERING	J-M156
X28	6	CONNECTION WL PASSENGER COMPARTMENT/ WL IGNITION LOCK	6cP	6cM	ON STEERING COLUMN	L48/49
X29	2	CONNECTION WL PASSENGER COMPARTMENT/ OIL TANK SENSOR	16cL	16cL	BOX RH SIDE OF ENGINE COMPARTMENT	E52
X31	8	CONNECTION WL PASSENGER COMPARTMENT/ WL INSIDE LIGHT COUPE	15bM	15bM	BELOW THE PARCEL TRAY	O14,H153
X32	6	CONNECTION WL PASSENGER COMPARTMENT/ WL INSIDE LIGHT CONVERTIBLE	6cQ	6cQ	UNDER INSTRUMENT PANEL LEFT	N/O12
X36	4	CONNECTION WL PASSENGER COMPARTMENT/ SWITCH LIFTING CONVERTIBLE TOP	12cL	12cL	BEHIND SIDE PANEL COVERING	N154
X37	8	CONNECTION WL PASSENGER COMPARTMENT/ WL SEAT LEFT	9eP	9eN	UNDER THE DRIVER SEAT	N19,A70,E65/66
X38	8	CONNECTION WL PASSENGER COMPARTMENT/ WL SEAT RIGHT	9eN	9eP	UNDER THE PASSEGER SEAT	N19,A70,E65/66
X39	6	CONNECTION WL PASSENGER COMPARTMENT/ WL LUGGAGE COMPARTMENT HOOD	7bL	7bL	ON RH SIDE OF LUGGAGE COMP LID HINGE	O18,B77
X60	21	CONNECTION WL MOTOR/ WL MFI + DI	16cP	16cP	CARRIER PLATE IN ENGINE COMPARTMENT	B46,E48,K42,B55,M142,M148
X65	6	CONNECTION WL DOOR DRIVER'S SIDE/ WL OUTSIDE MIRROR LEFT	8cR	8cK	DRIVER'S DOOR	B18/19
X66	6	CONNECTION WL DOOR PASSENGER'S SIDE/ WL OUTSIDE MIRROR RIGHT	8cK	8cR	PASSENGER'S DOOR	B13
X69	14	CONNECTION WL FRONT END/ WL HEATER, AIR CONDITIONER	6cM	6cM	IN LUGGAGE COMPARTMENT RIGHT	G23/24

M-NUMBERS

M002	RS ROAD VERSION
M003	RS RACING SAFETY EQUIPMENT
M004	RS TURBO VERSION
M113	CANADA VERSION (DAYTIME RUNNING LIGHT)
M139	ADJUSTABLE SEAT HEATING, LEFT SEAT
M193	JAPAN VERSION
M215	SAUDI-ARABIA VERSION
M224	AUTOMATIC BRAKE DIFFERENTIAL
M249	AUTOMATIC TRANSMISSION (TIPTRONIC)
M339	FWD
M340	ADJUSTABLE SEAT HEATING, RIGHT SEAT
M383	SPORT SEAT LEFT ELECTRIC HEIGHT ADJUSTMENT
M387	SPORT SEAT RIGHT ELECTRIC HEIGHT ADJUSTMENT
M419	REAR PACKAGE TRAY
M425	REAR WINDOW WIPER
M437	COMFORT SEAT LEFT
M438	COMFORT SEAT RIGHT
M439	ELECTRIC CONVERTIBLE TOP
M441	RADIO PREPARATION
M451	LIMITED RADIO PREPARATION
M454	CRUISE CONTROL
M480	TRANSMISSION
M487	NORWAY
M490	AUDIO OPTION PACK
M513	LUMBAR SUPPORT SEAT RIGHT
M520	INTERIOR MONITORING
M530	CODE LOCK
M545	LARGE TANK
M553	USA / CANADA VERSION
M562	AIRBAG
M564	WITHOUT AIRBAG
M573	AIR CONDITIONING SYSTEM
M576	WITHOUT REAR FOG LIGHT
M586	LUMBAR SUPPORT SEAT LEFT
M601	LITRONIC
M602	HIGH MOUNTED STOP LIGHT
M605	HEADLIGHT VERTICAL AIM CONTROL
M614	TELEPHONE PREPARATION
M615	CORDLESS PHONE
M632	RECOGNITION SEAT ASSIGNMENT PASSENGER
M650	POWER SUNROOF
M651	POWER WINDOWS
M659	ON-BOARD COMPUTER
M680	DIGITAL SOUND PACKAGE
M692	CD-CHANGER

CONNECTION

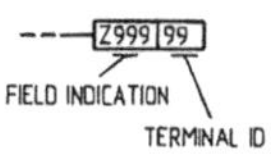

P	IGNITION SWITCH POSITION	0	56	LIGHT SWITCH POSITION	II
X	IGNITION SWITCH POSITION	I+II	56a/56b	HIGH BEAM/LOW BEAM	
15	IGNITION SWITCH POSITION	II+III	57	LIGHT SWITCH POSITION	I
R	IGNITION SWITCH POSITION	I+II+III	58	LIGHT SWITCH POSITION	I+II
K	DIAGNOSIS K		58R/58L	SIDE MARKER LIGHT RIGHT/LEFT	
L	DIAGNOSIS L		58d	INSTRUMENT LIGHTS	
30	PERMANENT POSITIVE		Ti	INJECTION SIGNAL	
31	GROUND		Tn	ROTATIONAL SPEED SIGNAL	
54	STOP LIGHT		61	GENERATOR D+	

P998ewd069a

A	B	C	D	E

Sheet 17 - 1997/1998

Connectors and abbreviations (page 2 of 2)

G H J K L

ABBREVIATIONS

CODE	MEANING	CODE	MEANING	CODE	MEANING
ABD	AUTOMATIC BRAKE DIFFERENTIAL	FL	FRONT LEFT	PWM	PULSE WIDTH MODULATION
ABS	ANTILOCK BRAKING SYSTEM	FOG	FOG LIGHT	RA	REAR AXLE
AC	AIR CONDITIONING SYSTEM	FR	FRONT RIGHT	RDK	TIRE PRESSURE CONTROL
ADL	ADDITIONAL HIGH BEAM	GP	GROUND POINT	RFL	REAR FOG LIGHT
ASS	AFTER SALES SERVICE	GR	GLASS ROOF	RHD	RIGHT-HAND DRIVE
BL	BACKUP LIGHT	HCS	HEADLIGHT CLEANING SYSTEM	RL	REAR LEFT
BPAU	BRAKE BOOSTER	HF	HIGH FREQUENCY	ROW	REST OF WORLD
CCSV	CARBON CANISTER SHUTOFF VALVE	HFM	HOT-FILM AIR MASS METER	RR	REAR RIGHT
CE	CENTRAL ELECTRIC	HVAC	HEADLIGHT VERTICAL AIM CONTROL	SA	SAUDI ARABIA
CL	CLOSE	IC	IGNITION CIRCUIT	SCS	COMBINED STEERING COLUMN SWITCH
CLOCK	CLOCK SIGNAL	INJ VALVE	FUEL INJECTOR	SILA	SAFETY LAMP
CLS	CENTRAL LOCKING SYSTEM	INS C	INSTRUMENT CLUSTER	SS	SPEED SENSOR
CO	CARBON MONOXIDE	LED	LIGHT-EMITTING DIODE	TDPS	TANK DIFFERENTIAL PRESSURE SENSOR
CP	CONNECTING POINT	LF	LOW FREQUENCY	TE	TERMINAL
CU	CONTROL UNIT	LHD	LEFT-HAND DRIVE	TEL MUTE	RADIO MUTING CONTROL
CWL	CENTRAL WARN LIGHT	MFI + DI	DIGITAL ENGINE ELECTRONICS	TI	INJECTION TIME
DATA	DATA SIGNAL	MY	MODELYEAR	TN	ENGINE SPEED
DEF	DEFROST	M5 2	MOTRONIC 5.2	TRP	TWIN-COIL ROTARY POSITIONER
D I	ELECTRIC IGNITION SYSTEM WITH KNOCK CONTROL	NO	NUMBER	TV	THROTTLE VALVE
DP	DISCONNECTING POINT	NTC	NEGATIVE TEMPERATURE COEFFICIENT	TVS	THROTTLE VALVE SENSOR
DSP	DIGITAL SOUND PACKAGE	OP	OPEN	USA	USA
ESO	ENGINE-SPEED SENSOR OUTPUT	PIN	PIN	UST	911 TURBO
ESS	ENGINE SPEED SENSOR	PL	PLUG	WL	WIRING LOOM
ETC	ELECTRICAL TRANSMISSION CONTROL	POT	POTENTIOMETER	WP	WELD POINT
FA	FRONT AXLE	PSD	PORSCHE LIMITED SLIP DIFFERENTIAL	WW	WORLDWIDE

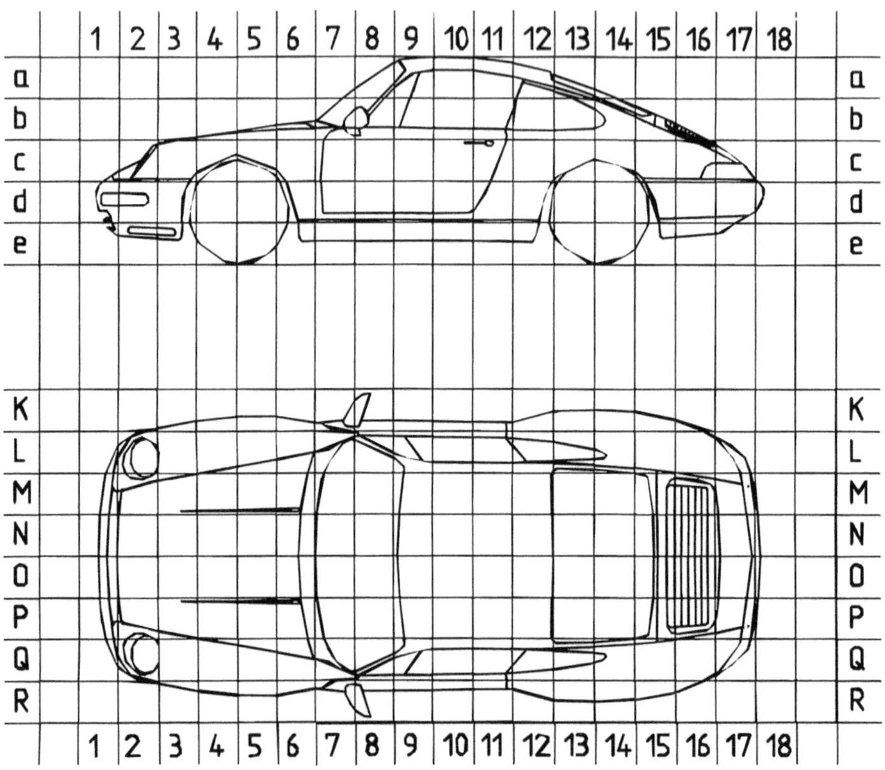

P998ewd069b

G H J K L

Central electric (page 1 of 3)

Sheet 18 - 1997/1998

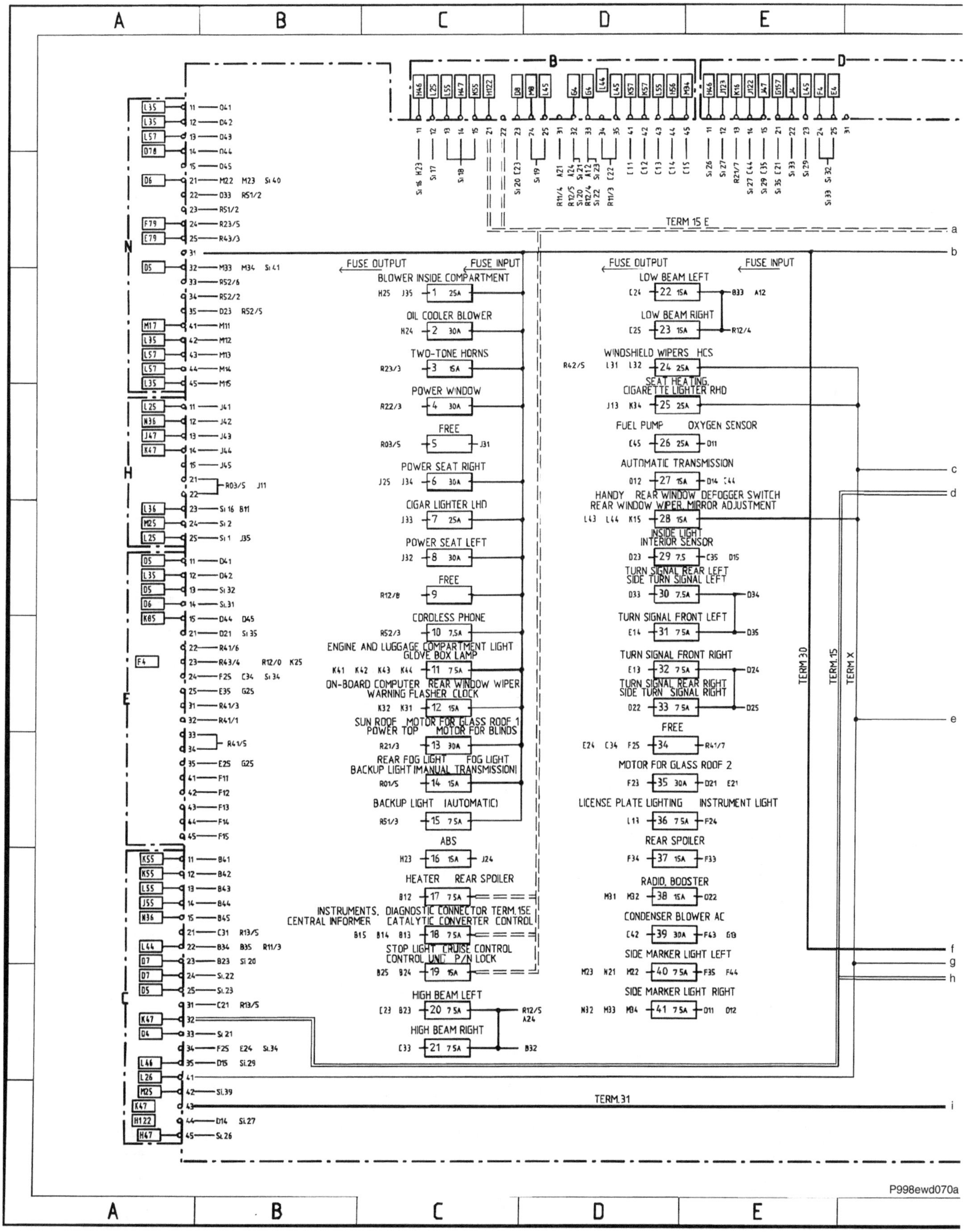

EWD

Sheet 18 - 1997/1998

Central electric (page 2 of 3)

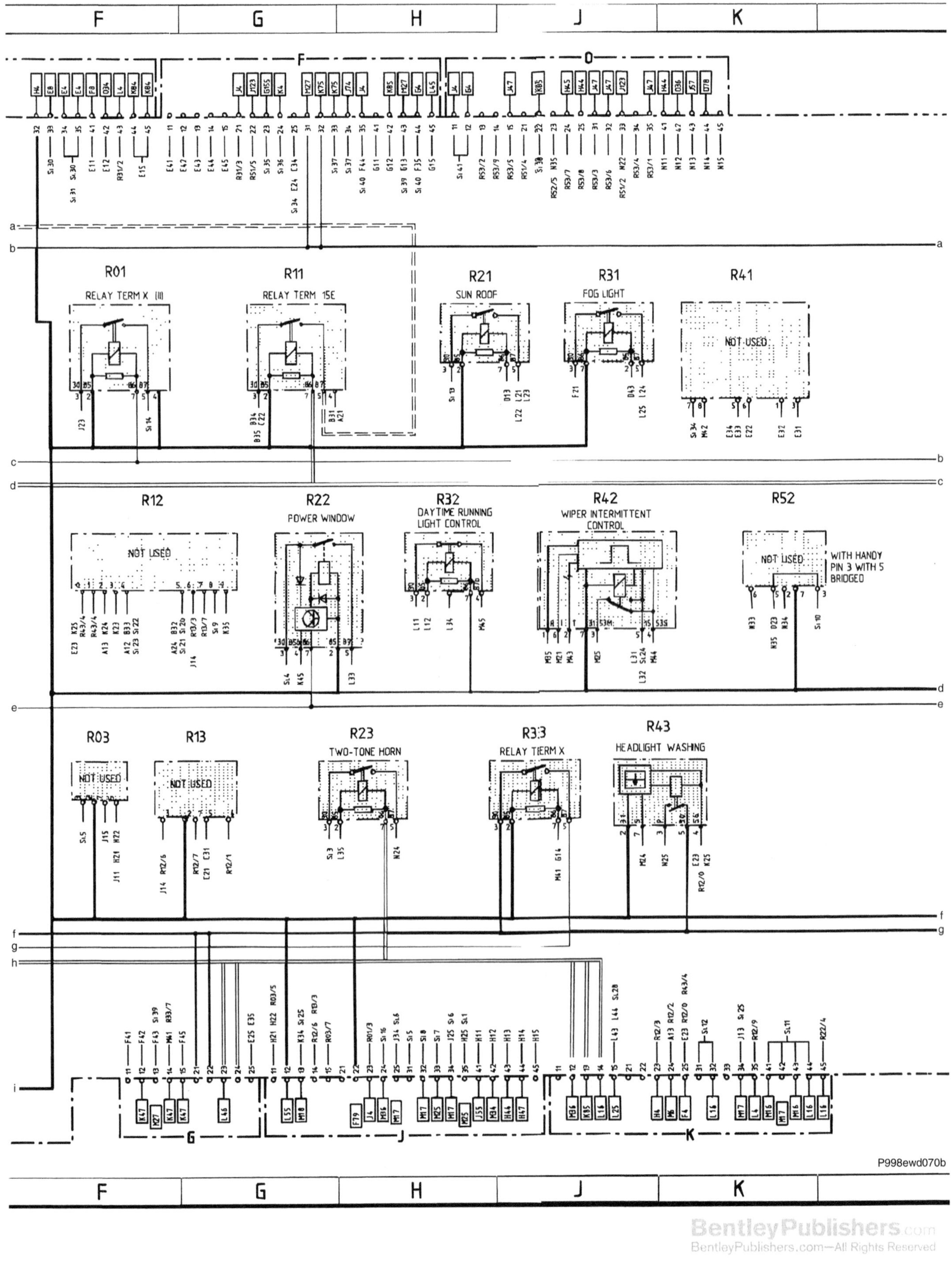

Central electric
(page 3 of 3)

Sheet 18 - 1997/1998

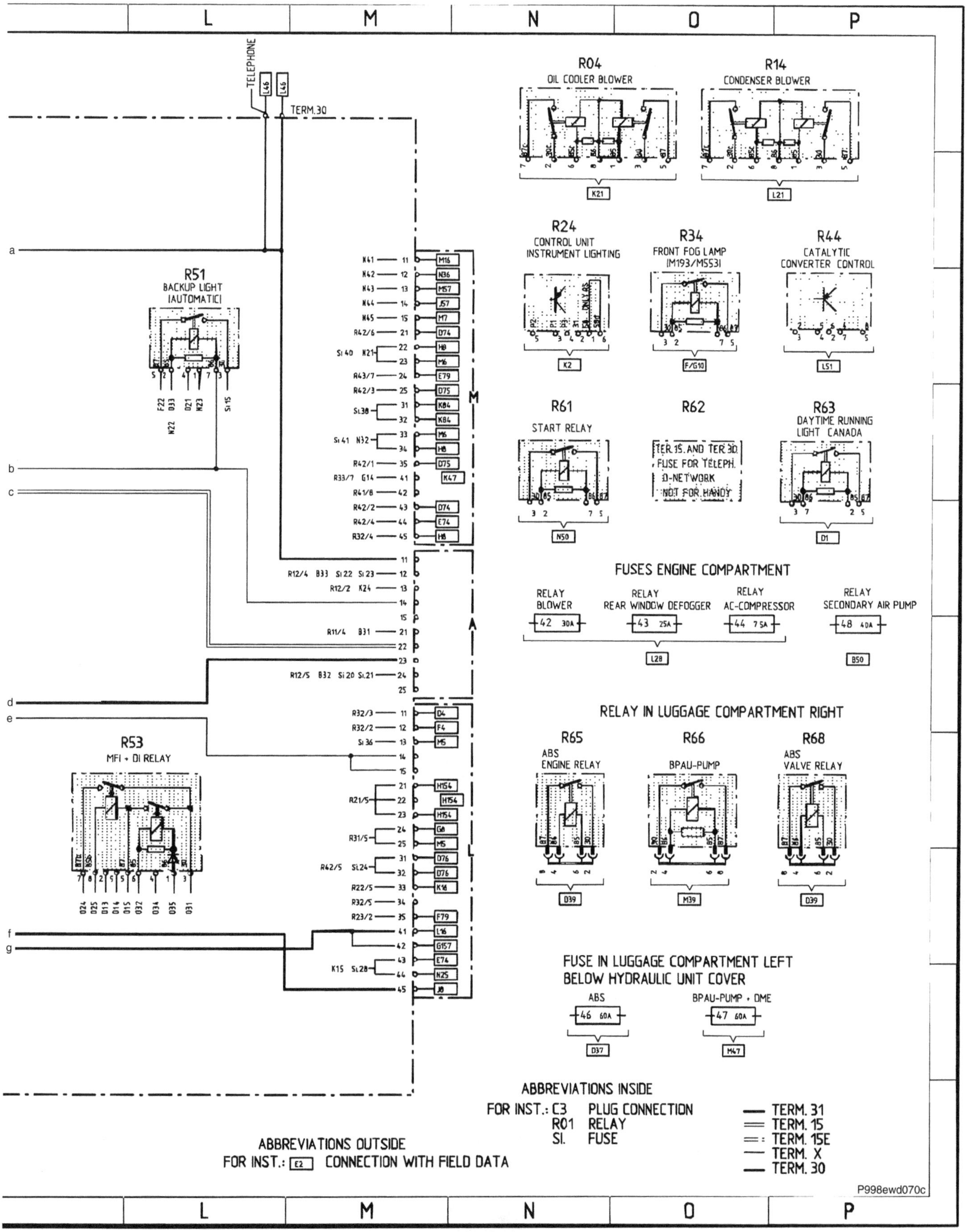

ABBREVIATIONS INSIDE
FOR INST.: C3 PLUG CONNECTION
R01 RELAY
SI. FUSE

TERM. 31
TERM. 15
TERM. 15E
TERM. X
TERM. 30

ABBREVIATIONS OUTSIDE
FOR INST.: E2 CONNECTION WITH FIELD DATA

P998ewd070c

L M N O P

EWD

OBD On-Board Diagnostics

GENERAL

This repair group covers the fundamentals and equipment requirements of On-Board Diagnostics (OBD II) as they apply to Porsche 993 vehicles. Also given here are diagnostic fault codes tables for DME (ODD I and OBD II) and Tiptronic transmissions.

Fault memory, accessing

A Porsche diagnostic scan tool can be used to access fault codes and OBD data.

Porsche 993 systems capable of on-board self-diagnosis:

- Engine management (DME M 2.10 and DME M 5.2)
- Airbag
- ABS/ABD
- Automatic (Tiptronic) transmission
- Alarm system
- Heater system

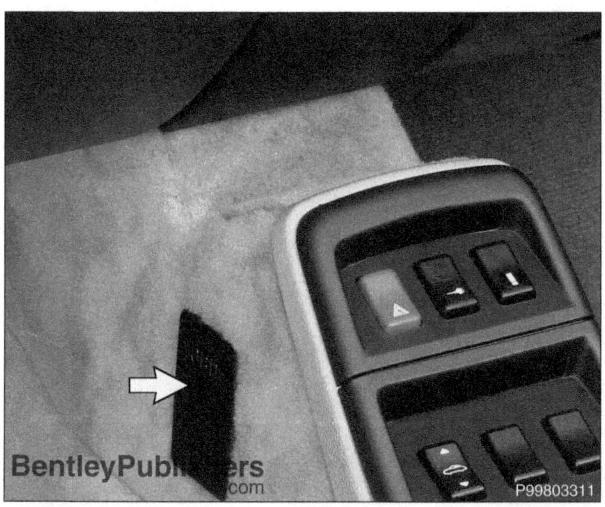

The diagnostic scan tool is connected via the diagnostic port. Remove cover (**arrow**) to access port. Port is located in driver's footwell on left side of center console.

NOTE–

- *1996 - 1998 models are equipped with an advanced OBD II (DME M 5.2) emissions-related diagnostic system. 1995 models are equipped with OBD I (DME M 2.10) .*
- *For OBD I vehicles, emissions-related fault information is displayed with blink codes through the Check Engine light. See* **DME 2.10 Fault Code Diagnosis**.
- *For OBD II vehicles, a generic OBD II compatible diagnostic scan tool can be used to retrieve basic emissions-related DTCs. A list of DTCs can be found in* **OBD On-Board Diagnostics**.

On-Board Diagnostics (OBD II)

OBD II information below applies to 1996 through 1998 models with DME M 5.2.

OBD II standards were developed by the SAE (Society of Automotive Engineers) and CARB (California Air Resources Board). OBD II is the second generation of on-board self-diagnostic equipment requirements. Since 1996, the standards have been applied to all passenger vehicles sold in the United States.

On-board diagnostic capabilities are incorporated into the hardware and software of the engine control module (ECM) to monitor components that can affect vehicle emissions.

Emission-influencing components are checked by a diagnostic routine (called a monitor) to verify that it is functioning properly. If a problem or malfunction is detected, the diagnostic executive built into the OBD II system stores a fault code and may illuminate the malfunction indicator light (MIL) on the instrument panel.

OBD II scan tool equipment is needed to access the fault memory and OBD II data.

OBD II system standards consist mainly of the following:

- Standardized 16-pin OBD II diagnostic connector
- Standardized diagnostic trouble codes (DTCs) for all vehicle manufacturers
- Trouble codes must be accessible via a generic diagnostic scan tool
- Saving of operating conditions data (freeze frame) at time of malfunction
- Standardized nomenclature and abbreviations for components and systems
- Testing of three-way catalytic converter efficiency
- Engine misfire recognition
- Monitoring of fuel tank ventilation system
- Monitoring of secondary air injection system
- Monitoring of oxygen sensor system adaptation
- Monitoring of all emissions-relevant components
- Malfunction indicator lamp (MIL) and fault memory
- Protection against manipulation of control modules
- Readiness codes
- Standardized readout of operating data such as rpm, temperature, etc.

P998OBD001

Malfunction indicator light (MIL)

Porsche 993 vehicles are equipped with a malfunction indicator light (MIL) (**arrow**).

OBD II software causes the MIL to illuminate when emission levels exceed 1.5 times Federal standards.

MIL illuminates under the following conditions:

- Ignition key ON before cranking (bulb check function).
- Engine management system fault detected for two consecutive OBD II drive cycles. See Drive cycle in this repair group.
- Catalyst-damaging fault detected.
- Component malfunction causes emissions to exceed 1.5 times OBD II standards.
- Manufacturer-defined specifications exceeded.
- Implausible input signal.
- Misfire faults.
- Leak in evaporative system.
- Oxygen sensors observe no purge flow from purge valve evaporative system.
- ECM fails to enter closed-loop operation within specified time.
- ECM or automatic transmission control module (TCM) in “limp home” mode.

Additional MIL information:

- A fault code is stored within the ECM upon the first occurrence of a fault in the system being checked.
- Two complete consecutive drive cycles with the fault present illuminate the MIL. The exception to the two-fault requirement is a catalyst-damaging fault, which illuminates the MIL immediately.
- If the second drive cycle was not complete and the fault was not checked, the ECM counts the third drive cycle as the next consecutive drive cycle. The MIL illuminates if the system is checked and the fault is still present.
- Once the MIL is illuminated, it remains illuminated until the vehicle completes three consecutive drive cycles without detecting a fault.
- An existing fault code is cleared from memory automatically when the vehicle completes 80 consecutive drive cycles without the fault being detected.

Freeze frame data

OBD II systems are required to save the following operating conditions data at the time of malfunction, for the first and most recent occurrence of a fault:

- Engine load
- Engine temperature
- Engine rpm
- Intake air temperature
- Throttle angle
- Fuel tank level
- Voltage supply
- Counter for hours of operation
- Oxygen sensor status and adaptation values
- Ignition angle
- Air mass value
- Engine rpm
- Vehicle speed

Scan tool

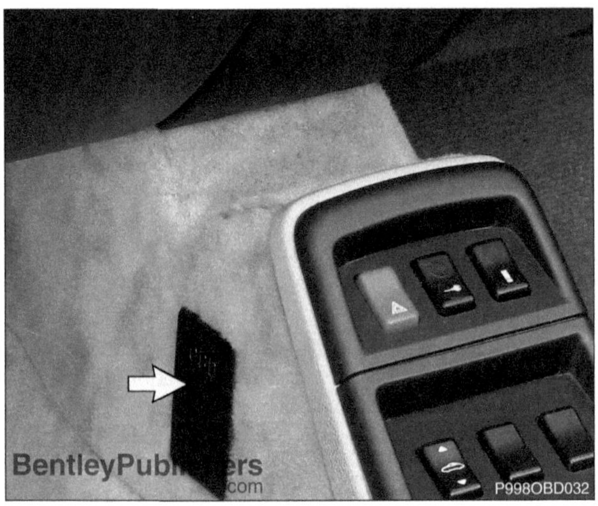

Diagnostics begin by connecting a scan tool to the 16-pin OBD II plug inside the vehicle (**arrow**).

Diagnostic scan tools available at the time of this printing include the Porsche factory tools and a small number of aftermarket Porsche-specific tools.

In addition to professional scan tools, inexpensive generic OBD II scan tool software programs and handheld units are readily available. Though limited, they are nonetheless powerful diagnostic tools. These tools read live data streams and freeze frame data as well as a host of other valuable diagnostic data.

Diagnostic monitors

Diagnostic monitors run tests and checks on specific emission control systems, components, and functions.

A complete drive cycle is required for the tests to be valid. See **Drive cycle** in this repair group. The diagnostic monitor signals the ECM of the loss or impairment of the signal or component and determines if a signal or sensor is faulty based on 3 conditions:

- Signal or component shorted to ground
- Signal or component shorted to B+
- Signal or component missing (open circuit)

The OBD II system monitors the emission control systems that are installed. Emission control systems vary by vehicle model and year. OBD II software monitors the following:

- Oxygen sensors
- Catalysts
- Engine misfire
- Fuel tank evaporative control system
- Secondary air injection
- Fuel system

Oxygen sensor monitoring: Four oxygen sensors are used on the DME M 5.2 system. When driving conditions allow, response rate and switching time of each oxygen sensor is monitored. The oxygen sensor heater function is also monitored. The OBD II system differentiates between pre-catalyst and post-catalyst oxygen sensors and reads each one individually. In order for the oxygen sensor to be effectively monitored, the system must be in closed loop operation.

Catalyst monitoring: This strategy monitors the two heated oxygen sensors per bank of cylinders. It compares the oxygen content going into the catalytic converter to the oxygen leaving the converter.

The diagnostic software knows that most of the oxygen should be used up during the oxidation phase. If it sees higher than programmed values, a fault is set and the MIL illuminates.

Misfire monitor: This strategy monitors crankshaft speed fluctuations and determines if an engine misfire occurs by monitoring variations in speed between each crankshaft sensor trigger point. This strategy is so finely tuned that it can determine the severity of the misfire.

The system determines if a misfire is occurring, as well as other pertinent misfire information such as:

- Specific cylinder(s)
- Severity of the misfire event
- Emissions relevant or catalyst damaging

Misfire detection is an on-going monitoring process that is only disabled under certain limited conditions.

Fuel system monitoring: This monitor looks at the fuel delivery needed (long / short term fuel trim) for proper engine operation based on programmed data. If too much or not enough fuel is delivered over a predetermined time, a DTC is set and the MIL is turned on.

Fuel trim refers to adjustments to base fuel schedule. Long-term fuel trim refers to gradual adjustments to the fuel calibration adjustment as compared to short term fuel trim. Long term fuel trim adjustments compensate for gradual changes that occur over time.

Fuel system monitoring monitors the calculated injection time (ti) in relation to engine speed, load, and pre-catalytic converter oxygen sensor(s) signals.

Using this data, the system optimizes fuel delivery for all engine operating conditions.

Evaporative system monitoring: This monitor checks the fuel storage system and related fuel lines for leaks. It can detect very small leaks anywhere in the system.

Drive cycle

The OBD II drive cycle is an important concept in understanding OBD II requirements. The purpose of the drive cycle is to run all of the emission-related on-board diagnostics over a broad range of driving conditions.

A drive cycle is considered complete when all of the diagnostic monitors have run their tests without interruption. For a drive cycle to be initiated, the vehicle must be started cold and brought up to 70° C (158° F) and at least 40° C (104° F) above its original starting temperature.

Readiness codes

Inspection / maintenance (I/M) readiness codes are mandated as part of OBD II. The readiness code is stored after complete diagnostic monitoring of specified components and systems is carried out. The readiness code function was designed to prevent manipulating an I/M emission test procedure by clearing faults codes or disconnecting the ECM or battery.

Readiness codes indicate whether the OBD II system is actually ready to monitor the various emission control systems on the vehicle. The vehicle must complete a drive cycle to set readiness codes. The code is binary:

- 0 for ready
- 1 for not ready

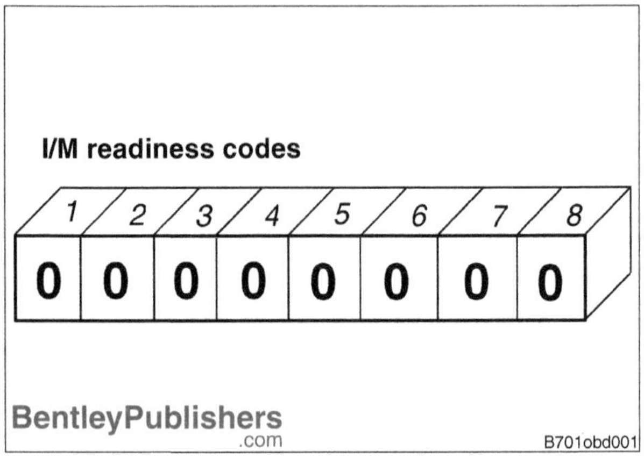

The parameters which are monitored for readiness are:

1. Catalyst efficiency
2. Catalyst heating
3. Fuel tank evaporative control
4. Secondary air injection
5. A/C refrigerant (not applicable)
6. Oxygen sensors
7. Oxygen sensor heaters
8. Exhaust gas recirculation (not applicable)

When all zeros are displayed, the system has established readiness. Readiness codes can be displayed using Porsche and aftermarket scan tools.

Readiness codes are set to 1 (not ready) in the following cases:

- The battery or ECM is disconnected.
- DTCs are erased after completion of repairs and a drive cycle is not completed.

DTC example: P 0 3 0 6

- P: A powertrain problem
- 0: SAE sanctioned ('generic')
- 3: Related to an ignition system / misfire
- 06 Misfire detected at cylinder #6

DTCs provide a freeze frame or snap-shot of a vehicle performance or emissions fault at the moment that the fault first occurs. This information is accessible through generic scan tools.

Freeze frame data contains, but isn't limited to, the following information:

- Engine load (calculated)
- Engine rpm
- Short and long-term fuel trim
- Vehicle speed
- Coolant temperature
- Intake manifold pressure
- Open / closed loop operation
- Fuel pressure (if available)

OBD II Diagnostic Trouble Codes

Diagnostic Trouble Codes (DTCs)

OBD II DTCs apply to 1996 - 1998 vehicles. Each digit represents a specific value. When the MIL illuminates it indicates that a DTC has been stored:

- DTCs are stored as soon as they occur, whether or not the MIL illuminates.
- DTCs store and display a time stamp.
- DTCs record the current fault status: Present, not currently present, or intermittent.

DTC digit interpretation	
1st digit	
P	powertrain
B	body
C	chassis
2nd digit	
0	SAE
1	Porsche
3rd digit	
0	total system
1	fuel
2	air ratio
3	ignition system or misfire
4	additional emission controls
5	vehicle speed & idle control
6	control module & initiating systems
7	transmission

Table a. OBDII diagnostic trouble codes (from 1996)

SAE Code	Porsche code(s)	Component or system	Fault conditions	Fault area
P0000	51, 56	Tiptronic manual program switch	• short circuit to ground • no manual program	• wiring • transmission control unit • selector lever switch • Porsche code 51 indicates fault in manual program switch • Porsche code 56 indicates fault in instrument cluster control
P0101	115	Mass air flow sensor	• fault signal implausible	• mass air flow sensor faulty
P0102	115	Mass air flow sensor	• below lower limit	• break in wiring short to ground
P0103	115	Mass air flow sensor	• above upper limit	• short to B+
P0112	124	Intake air temperature circuit - low input Intake air temperature circuit - below lower limit	• below lower limit	• Intake air temperature sensor faulty • short to ground • break in wiring
P0113	124	Intake air temperature circuit - high input Intake air temperature circuit - above upper limit	• upper limit exceeded	• short to B+
P0115	123	ECT Engine coolant temperature sensor signal implausible	• malfunctions in the ECT sensor or when recording engine temperature • short circuit to ground	• wiring • ECT sensor • ECM
P0117	123	ECT Engine coolant temperature malfunction	• malfunctions in the ECT sensor or when recording engine temperature • short circuit to ground	• wiring • ECT sensor • ECM
P0122	117	Throttle position (TP) sensor - entry signal too low	• below lower limit	• interruption or short circuit in wiring • TP sensor faulty • ECM faulty
P0123	117	Throttle position (TP) sensor - entry signal too high	• above upper limit	• interruption or short circuit in wiring • TP sensor faulty • ECM faulty
P0130	10	Oxygen sensor ahead of catalytic converter short circuit of wires or limited voltage increase (cylinders 1-3)	• crack in ceramic material • reference air polluted • short circuit signal wire to sensor ground • When P0130 is stored, P1115 will also be stored (this is not an actual fault)	• oxygen sensor • wiring harness • ECM

Table a. OBDII diagnostic trouble codes (from 1996)

SAE Code	Porsche code(s)	Component or system	Fault conditions	Fault area
P0131	10	Short circuit to ground at signal wire of oxygen sensor ahead of catalytic converter or incorrect voltage (cylinders 1-3)	• short circuit of signal wire to ECM ground or housing ground (P1115 and P1119 may also be stored - these are not actual faults) • short circuit of sensor ground to ECM ground or housing ground (when this fault occurs, faults for all four oxygen sensors are stored, even though only one is likely faulty) • short circuit to ground of ECM final stage	• wiring harness • oxygen sensor • ECM
P0132	10	Short circuit to plus at oxygen sensor ahead of catalytic converter (cylinders 1-3)	• oxygen sensor short circuit to plus • wiring harness short circuit to plus • ECM final stage short circuit to plus • when this fault occurs, faults for all four oxygen sensors are stored, even though only one is likely faulty • this fault may cause misfires	• wiring harness • oxygen sensor • ECM
P0133	15	Aging of oxygen sensor ahead of three way catalytic converter (TWC) (cylinders 1-3)	• estimated values are exceeded by 1.5 x	• oxygen sensor
P0134	10	Interruption of signal for oxygen sensor ahead of catalytic converter (cylinders 1-3)	• interruption of signal wire or ground wire of one oxygen sensor • oxygen sensor heating not functioning • ECM final stage faulty • interruption of the common sensor ground • When P0134 is stored, P1115 may also be stored • P0134 is also stored as P0154 if there is an interruption in the common sensor ground	• wiring harness • oxygen sensor • ECM
P0136	12	Oxygen sensor behind catalytic converter short circuit of wires (cylinders 1-3)	• short circuit signal wire to sensor ground (short circuit of wires) • ECM final stage faulty • When P0136 is stored, P1117 will also be stored	• wiring harness • oxygen sensor • ECM

OBD

Table a. OBDII diagnostic trouble codes (from 1996)				
SAE Code	**Porsche code(s)**	**Component or system**	**Fault conditions**	**Fault area**
P0137	12	Short circuit to ground at signal wire of oxygen sensor behind catalytic converter or incorrect voltage (cylinders 1-3)	• short circuit of signal wire to ECM ground or housing ground (P1115 and P1119 may also be stored - these are not actual faults) • short circuit of sensor ground to ECM ground or housing ground (when this fault occurs, faults for all four oxygen sensors are stored, even though only one is likely faulty) • short circuit to ground of ECM final stage	• wiring harness • oxygen sensor • ECM
P0138	12	Short circuit to plus at oxygen sensor behind catalytic converter (cylinders 1-3)	• oxygen sensor short circuit to plus • wiring harness short circuit to plus • ECM final stage short circuit to plus • when this fault occurs, faults for all four oxygen sensors are stored, even though only one is likely faulty • this fault may cause misfires	• wiring harness • oxygen sensor • ECM
P0139	17	Aging of oxygen sensor after three way catalytic converter (TWC) (cylinders 1-3)	• oxygen sensor value does not change toward rich or lean	• oxygen sensor
P0140	12	Interruption of signal for oxygen sensor behind catalytic converter (cylinders 1-3)	• interruption of signal wire or ground wire of one oxygen sensor • oxygen sensor heating not functioning • ECM final stage faulty • interruption of the common sensor ground • when P0140 is stored, P1117 may also be stored	• wiring harness • oxygen sensor • ECM
P0150	18	Oxygen sensor ahead of catalytic converter short circuit or wires or limited voltage increase (cylinders 4-6)	• crack in ceramic material • reference air polluted • short circuit signal wire to sensor ground • when P0150 is shedded, a non-shedded fault P1121 will also be stored (it is not an actual fault)	• oxygen sensor • wiring harness • ECM

Table a. OBDII diagnostic trouble codes (from 1996)

SAE Code	Porsche code(s)	Component or system	Fault conditions	Fault area
P0151	18	Short circuit to ground at signal wire of oxygen sensor ahead of catalytic converter or incorrect voltage (cylinders 4-6)	• short circuit of signal wire to ECM ground or housing ground (P1115 and P1119 may also be stored - these are not actual faults) • short circuit of sensor ground to ECM ground or housing ground (when this fault occurs, faults for all four oxygen sensors are stored, even though only one is likely faulty) • short circuit to ground of ECM final stage	• wiring harness • oxygen sensor • ECM
P0152	18	Short circuit to plus at oxygen sensor ahead of catalytic converter (cylinders 4-6)	• oxygen sensor short circuit to plus • wiring harness short circuit to plus • ECM final stage short circuit to plus • when this fault occurs, faults for all four oxygen sensors are stored, even though only one is likely faulty • this fault may cause misfires	• wiring harness • oxygen sensor • ECM
P0153	21, 22	Aging of oxygen sensor ahead of three way catalytic converter (TWC) (cylinders 4-6)	• estimated values are exceeded by 1.5 x	• oxygen sensor
P0154	18	Interruption of signal for oxygen sensor ahead of catalytic converter (cylinders 4-6)	• interruption of signal wire or ground wire of one oxygen sensor • oxygen sensor heating not functioning • ECM final stage faulty • When P0154 is stored, P1119 may also be stored • P0154 is also stored as P0134 if there is an interruption in the common sensor ground	• wiring harness • oxygen sensor • ECM
P0156	20	Oxygen sensor behind catalytic converter short circuit of wires (cylinders 4-6)	• short circuit signal wire to sensor ground (short circuit of wires) • ECM final stage faulty • When P0156 is stored, P1121 will also be stored	• wiring harness • oxygen sensor • ECM

Table a. OBDII diagnostic trouble codes (from 1996)

SAE Code	Porsche code(s)	Component or system	Fault conditions	Fault area
P0157	20	Short circuit to ground at signal wire of oxygen sensor behind catalytic converter or incorrect voltage (cylinders 4-6)	• short circuit of signal wire to ECM ground or housing ground (P1115 and P1119 may also be stored - these are not actual faults) • short circuit of sensor ground to ECM ground or housing ground (when this fault occurs, faults for all four oxygen sensors are stored, even though only one is likely faulty) • short circuit to ground of ECM final stage	• wiring harness • oxygen sensor • ECM
P0158	20	Short circuit to plus at oxygen sensor behind catalytic converter (cylinders 4-6)	• oxygen sensor short circuit to plus • wiring harness short circuit to plus • ECM final stage short circuit to plus • when this fault occurs, faults for all four oxygen sensors are stored, even though only one is likely faulty • this fault may cause misfires	• wiring harness • oxygen sensor • ECM
P0159	23	Aging of oxygen sensor after three way catalytic converter (TWC) (cylinders 4-6)	• oxygen sensor value does not change toward rich or lean	• oxygen sensor
P0160	20	Interruption of signal for oxygen sensor behind catalytic converter (cylinders 4-6)	• interruption of signal wire or ground wire of one oxygen sensor • oxygen sensor heating not functioning • ECM final stage faulty • when P0160 is stored, P1121 may also be stored	• wiring harness • oxygen sensor • ECM
P0300	62, 75	Misfire recognized	• Misfire of several cylinders	• contaminated, faulty, or worn spark plug • ignition wires faulty • faulty high tension insulation • acceleration restrictions • loose connections
			• combustion misfire	• mixture too rich • mixture too lean • Porsche code 62 indicates emission-related misfire • Porsche code 75 indicates catalyst-damaging misfire

Table a. OBDII diagnostic trouble codes (from 1996)				
SAE Code	**Porsche code(s)**	**Component or system**	**Fault conditions**	**Fault area**
P0301	50, 63	Cylinder 1 - misfire detected	• Misfire of cylinder 1	• contaminated, faulty, or worn spark plug • ignition wires faulty • faulty high tension insulation • acceleration restrictions • loose connections
			• combustion misfire	• mixture too rich • mixture too lean • Porsche code 50 indicates emission-related misfire • Porsche code 63 indicates catalyst-damaging misfire
P0302	51, 64	Cylinder 2 - misfire detected	• Misfire of cylinder 2	• contaminated, faulty, or worn spark plug • ignition wires faulty • faulty high tension insulation • acceleration restrictions • loose connections
			• combustion misfire	• mixture too rich • mixture too lean • Porsche code 51 indicates emission-related misfire • Porsche code 64 indicates catalyst-damaging misfire
P0303	52, 65	Cylinder 3 - misfire detected	• Misfire of cylinder 3	• contaminated, faulty, or worn spark plug • ignition wires faulty • faulty high tension insulation • acceleration restrictions • loose connections
			• combustion misfire	• mixture too rich • mixture too lean • Porsche code 52 indicates emission-related misfire • Porsche code 65 indicates catalyst-damaging misfire
P0304	53, 66	Cylinder 4 - misfire detected	• Misfire of cylinder 4	• contaminated, faulty, or worn spark plug • ignition wires faulty • faulty high tension insulation • acceleration restrictions • loose connections
			• combustion misfire	• mixture too rich • mixture too lean • Porsche code 53 indicates emission-related misfire • Porsche code 66 indicates catalyst-damaging misfire

Table a. OBDII diagnostic trouble codes (from 1996)

SAE Code	Porsche code(s)	Component or system	Fault conditions	Fault area
P0305	54, 67	Cylinder 5 - misfire detected	• Misfire of cylinder 5	• contaminated, faulty, or worn spark plug • ignition wires faulty • faulty high tension insulation • acceleration restrictions • loose connections
			• combustion misfire	• mixture too rich • mixture too lean • Porsche code 54 indicates emission-related misfire • Porsche code 67 indicates catalyst-damaging misfire
P0306	55, 68	Cylinder 6 - misfire detected	• Misfire of cylinder 6	• contaminated, faulty, or worn spark plug • ignition wires faulty • faulty high tension insulation • acceleration restrictions • loose connections
			• combustion misfire	• mixture too rich • mixture too lean • Porsche code 55 indicates emission-related misfire • Porsche code 68 indicates catalyst-damaging misfire
P0326	210	Knock sensor 1 - signal implausible	• signal not plausible	• knock sensor loose • knock sensor bridge loose or damaged • wiring break or short circuit • connector corrosion • knock sensor faulty
P0331	211	Knock sensor 2 - signal implausible	• signal not plausible	• knock sensor loose • knock sensor bridge loose or damaged • wiring break or short circuit • connector corrosion • knock sensor faulty
P0336	111	Crankshaft position sensor - not in specified range	• No CKP sensor signal when engine is running	• break or short circuit in wiring • CKP sensor faulty • ECM faulty
P0341	112	Camshaft position (CMP) sensor signal implausible	• signal implausible	• no power supply • break in wiring • short circuit to B+ • short circuit to ground • CMP sensor faulty
P0410	80	Secondary air injection signal implausible (cylinders 1-3)	• secondary air injection pump is not triggered • secondary air injection pump does not work • air supply lines restricted • electric change-over valve does not work • air change-over valve does not work	• wiring harness or fuse 48 • relay • ECM • air supply lines • secondary air injection (AIR) pump • electric change-over valve • air change-over valve

Table a. OBDII diagnostic trouble codes (from 1996)				
SAE Code	Porsche code(s)	Component or system	Fault conditions	Fault area
P0420	40	TWC conversion too low (cylinders 1-3)	• at least three speed/load ranges are detected as faulty	• catalytic converter
P0430	45	TWC conversion too low (cylinders 4-6)	• at least three speed/load ranges are detected as faulty	• catalytic converter
P0441	93	Fuel tank ventilation malfunction	• no change at oxygen sensing and no change of the duty cycle of the IAC valve when evaporative emission (EVAP) canister purge valve opens	• faulty EVAP canister purge valve • blocked fresh air line • leaking EVAP canister purge valve • leaking line between EVAP canister and air intake system
P0444	98	EVAP canister purge valve - open circuit	• circuit open	• wiring • EVAP canister purge valve • ECM
P0445	98	EVAP canister purge valve - short circuit	• short circuit to B+ • short circuit to ground	• wiring harness • ECM
P0501	120	Vehicle speed sensor (VSS) not in specified range	• no speed signal	• break in wiring or short circuit from ECM to ABS control module • break in wiring or short circuit from right rear wheel speed sensor ABS control module • right rear wheel speed sensor faulty • pulse wheel for speed sendor damaged or contaminated with brake dust • ABS control module faulty
P0506	32	Idle air control - engine speed too low	• idle speed too low	• IAC valve is sticking • bypass line kinked or blocked • air cleaner blocked
P0507	32	Idle air control - engine speed too high	• idle speed too low	• IAC valve is sticking • air intake system leaking
P0602	44	Control unit defective (watchdog)	• Tiptronic control unit defective • restricted driving program	• Tiptronic control unit defective
P0603	49	Control unit (EEPROM)	• Tiptronic control unit defective	• Tiptronic control unit defective
P0604	101, 102, 104	ECM faulty	• ECM faulty	• ECM • Porsche code 101 indicates ECM internal RAM fault • Porsche code 101 indicates ECM external RAM fault • Porsche code 101 indicates DTC memory fault
P0605	103	ECM faulty	• ECM faulty	• ECM
P0605	42	Control unit (checksum)	• Tiptronic control unit defective • restricted driving program	• Tiptronic control unit defective

Table a. OBDII diagnostic trouble codes (from 1996)

SAE Code	Porsche code(s)	Component or system	Fault conditions	Fault area
P0705	38	Multi-function switch (start) - signal implausible	• discontinuity, short circuit to ground and to plus • restricted driving program	• wiring • multi-function switch • control module
P0706	35	Multi-function switch - signal implausible	• discontinuity, short circuit to ground and to plus • restricted driving program	• wiring • multi-function switch • Tiptronic control module
P0710	37	Temperature sensor, transmission	• discontinuity/short circuit to plus, short circuit to ground • default value 60° C	• wiring • temperature sensor • Tiptronic control unit
P0721	45	Downshift protection - signal implausible	• No speed signal or wrong speed signals • restricted driving program	• speed sensor • Transmission (slipping clutch)
P0722	36	Speed sensor, transmission - signal implausible	• discontinuity, short circuit to ground and to plus • restricted driving program	• wiring • Tiptronic control unit • speed sensor
P0725	46	RPM limiter - signal implausible	• No speed signal or incorrect speed signals • restricted driving program	• speed sensor • Transmission (slipping clutch)
P0727	21	RPM limiter <- ECM control unit	• discontinuity/short circuit to ground, to plus • restricted driving program	• wiring • Tiptronic control unit
P0731	71	Gear monitoring, 1st gear - signal implausible	• Mechanical/hydraulic fault in transmission (e.g. solenoid valve does not operate despite correct electrical triggering)	• solenoid valve • valve chamber • transmission
P0732	72	Gear monitoring, 2nd gear - signal implausible	• Mechanical/hydraulic fault in transmission (e.g. solenoid valve does not operate despite correct electrical triggering)	• solenoid valve • valve chamber • transmission
P0733	73	Gear monitoring, 3rd gear - signal implausible	• Mechanical/hydraulic fault in transmission (e.g. solenoid valve does not operate despite correct electrical triggering)	• solenoid valve • valve chamber • transmission
P0734	74	Gear monitoring, 4th gear - signal implausible	• Mechanical/hydraulic fault in transmission (e.g. solenoid valve does not operate despite correct electrical triggering)	• solenoid valve • valve chamber • transmission
P0743	33	Solenoid valve, torque converter clutch	• discontinuity/short circuit to ground, to plus • restricted driving program	• wiring • transmission wiring harness • solenoid valve • control unit
P0748	34	Pressure regulator	• discontinuity/short circuit to ground, to plus • restricted driving program	• wiring • transmission wiring harness • solenoid valve • control unit
P0753	31	Solenoid valve 1	• discontinuity/short circuit to ground, to plus • restricted driving program	• wiring • solenoid valve • control unit

Table a. OBDII diagnostic trouble codes (from 1996)

SAE Code	Porsche code(s)	Component or system	Fault conditions	Fault area
P0758	32	Solenoid valve 2	• discontinuity/short circuit to ground, to plus • restricted driving program	• wiring • solenoid valve • control unit
P1102	13	Oxygen sensor heating 1 ahead of catalytic converter - above upper limit	• heating current too high	• heating resistance too low • short to plus
P1105	14	Oxygen sensor heating 1 behind catalytic converter - above upper limit	• heating current too high	• heating resistance too low • short to plus
P1107	5	Oxygen sensor heating 2 ahead of catalytic converter - above upper limit	• heating current too high	• heating resistance too low • short to plus
P1110	4	Oxygen sensor heating 2 behind catalytic converter - above upper limit	• heating current too high	• heating resistance too low • short to plus
P1115	13	Oxygen sensor heating 1 ahead of catalytic converter - below lower threshold	• heating current too low	• transition resistance • heating resistance too high • interruption in the heating circuits • interruption in the sensor signal wire • short circuit to plus (final stage disconnected)
P1117	14	Oxygen sensor heating 1 behind catalytic converter - below lower threshold	• heating current too low	• transition resistance • heating resistance too high • interruption in the heating circuits • interruption in the sensor signal wire • short to B+ (End-stage switched off)
P1119	5	Oxygen sensor heating 2 ahead of catalytic converter - below lower threshold	• heating current too low	• transition resistance • heating resistance too high • interruption in the heating circuits • interruption in the sensor signal wire • short circuit to plus (final stage disconnected)
P1121	4	Oxygen sensor heating 2 behind catalytic converter - below lower threshold	• heating current too low	• transition resistance • heating resistance too high • interruption in the heating circuits • interruption in the sensor signal wire • short to B+ (End-stage switched off)
P1123	27	Oxygen sensing adaptation range 1 (cylinders 1-3) lean limit	• fuel/air mixture is so rich that the control is up to the lean limit	• fuel pressure too high • fuel injector leaking
P1124	27	Oxygen sensing adaptation range 1 (cylinders 1-3) enrichment limit	• fuel/air mixture is so lean that the control is at the enrichment limit	• intake system leaking • fuel pressure too low • fuel injectors contaminated • volume supply of fuel pump too low

OBD

Table a. OBDII diagnostic trouble codes (from 1996)

SAE Code	Porsche code(s)	Component or system	Fault conditions	Fault area
P1125	35	Oxygen sensing adaptation range 1 (cylinders 4-6) lean limit	• fuel/air mixture is so rich that the control is up to the lean limit	• fuel pressure too high • fuel injector leaking
P1126	35	Oxygen sensing adaptation range 1 (cylinders 4-6) enrichment limit	• fuel/air mixture is so lean that the control is at the enrichment limit	• intake system leaking • fuel pressure too low • fuel injectors contaminated • volume supply of fuel pump too low
P1127	26	Oxygen sensing adaptation range 2 (cylinders 1-3) lean limit	• fuel/air mixture is so rich that the control is up to the lean limit	• fuel pressure too high • fuel injector leaking
P1128	26	Oxygen sensing adaptation range 2 (cylinders 1-3) enrichment limit	• fuel/air mixture is so lean that the control is at the enrichment limit	• intake system leaking • fuel pressure too low • fuel injectors contaminated • volume supply of fuel pump too low
P1129	34	Oxygen sensing adaptation range 2 (cylinders 4-6) lean limit	• fuel/air mixture is so rich that the control is up to the lean limit	• fuel pressure too high • fuel injector leaking
P1130	34	Oxygen sensing adaptation range 2 (cylinders 4-6) enrichment limit	• fuel/air mixture is so lean that the control is at the enrichment limit	• intake system leaking • fuel pressure too low • fuel injectors contaminated • volume supply of fuel pump too low
P1140	121	Load calculation - signal implausible	• signal implausible	• improper adjustment of operating cable for cruise control • throttle position sensor • false air
P1213	150	Injector circuit cylinder 1 - short to B+	• upper threshold exceeded	• fuel injector (coil) short circuit • short circuit to B+ in wiring
P1214	151	Injector circuit cylinder 2 - short to B+	• upper limit exceeded	• fuel injector (coil) short circuit • short circuit to B+ in wiring
P1215	152	Injector circuit cylinder 3 - short to B+	• upper limit exceeded	• fuel injector (coil) short circuit • short circuit to B+ in wiring
P1216	153	Injector circuit cylinder 4 - short to B+	• upper limit exceeded	• fuel injector (coil) short circuit • short circuit to B+ in wiring
P1217	154	Injector circuit cylinder 5 - short to B+	• upper limit exceeded	• fuel injector (coil) short circuit • short circuit to B+ in wiring
P1218	155	Injector circuit cylinder 6 - short to B+	• upper limit exceeded	• fuel injector (coil) short circuit • short circuit to B+ in wiring
P1225	150	Fuel injector, cylinder 1 - short to ground	• below lower threshold	• short circuit to ground in wiring
P1226	151	Fuel injector, cylinder 2 - short to ground	• below lower limit	• short circuit to ground in wiring

Table a. OBDII diagnostic trouble codes (from 1996)

SAE Code	Porsche code(s)	Component or system	Fault conditions	Fault area
P1227	152	Fuel injector, cylinder 3 - short to ground	• below lower limit	• short circuit to ground in wiring
P1228	153	Fuel injector, cylinder 4 - short to ground	• below lower limit	• short circuit to ground in wiring
P1229	154	Fuel injector, cylinder 5 - short to ground	• below lower limit	• short circuit to ground in wiring
P1230	155	Fuel injector, cylinder 6 - short to ground	• below lower limit	• short circuit to ground in wiring
P1237	150	Fuel injector, cylinder 1 - interruption	• interruption	• fuel injector (coil) interruption • interruption in wiring
P1238	151	Fuel injector, cylinder 2 - interruption	• interruption	• fuel injector (coil) interruption • interruption in wiring
P1239	152	Fuel injector, cylinder 3 - interruption	• interruption	• fuel injector (coil) interruption • interruption in wiring
P1240	153	Fuel injector, cylinder 4 - interruption	• interruption	• fuel injector (coil) interruption • interruption in wiring
P1241	154	Fuel injector, cylinder 5 - interruption	• interruption	• fuel injector (coil) interruption • interruption in wiring
P1242	155	Fuel injector, cylinder 6 - interruption	• interruption	• fuel injector (coil) interruption • interruption in wiring
P1386	222	Knock control test pulse - signal implausible	• Internal ECM checking faulty	• ECM
P1411	208	Secondary air injection signal implausible (cylinders 4-6)	• secondary air injection pump is not triggered • secondary air injection pump does not work • air supply lines restricted • electric change-over valve does not work • air change-over valve does not work	• wiring harness or fuse 48 • relay • ECM • air supply lines • secondary air injection (AIR) pump • electric change-over valve • air change-over valve
P1455	170	A/C compressor control - interruption	• interruption/no signal	• interruption in the wiring
P1456	170	A/C compressor control - upper limit exceeded	• upper limit exceeded	• short circuit in wiring to B+
P1457	170	A/C compressor control - below lower limit	• below lower limit	• short circuit to ground in wiring
P1458	138	A/C compressor signal - implausible operating range	• A/C compressor is triggered, although the climate control is switched off	• short circuit to ground in wiring • A/C compressor is triggered • interruption in wiring for climate control readiness
P1501	167	End-stage fuel pump relay - below lower limit	• below lower limit	• ECM relay faulty • short circuit to ground • ECM faulty

Table a. OBDII diagnostic trouble codes (from 1996)

SAE Code	Porsche code(s)	Component or system	Fault conditions	Fault area
P1502	167	End-stage fuel pump relay - upper limit exceeded	• upper limit exceeded	• ECM relay faulty • short circuit to B+ • ECM faulty
P1510	168	Idle control valve opening coil - open	• interruption	• wiring interruption • IAC valve faulty • ECM faulty
P1512	175	Intake manifold resonance flap - short circuit to B+	• upper limit exceeded	• change-over valve (solenoid) short circuit • short circuit to B+ in the cable
P1513	168	Idle control valve opening coil - open	• below lower limit	• short circuit to ground • IAC valve faulty • ECM faulty
P1514	168	Idle control valve opening coil - open	• upper limit exceeded	• short circuit to ground • IAC valve faulty • ECM faulty
P1515	175	Intake manifold resonance flap - short circuit to ground	• value below lower limit	• short circuit to ground in the cable
P1516	175	Intake manifold resonance flap - open circuit	• open circuit	• change-over valve (solenoid) open circuit • discontinuity in the cable
P1541	167	End-stage fuel pump relay - interruption	• interruption	• wiring interruption • ECM relay faulty • ECM faulty
P1551	169	Idle air control (IAC) valve - interruption	• interruption	• wiring interruption • IAC valve faulty • ECM faulty
P1552	169	Idle air control (IAC) valve closing coil - below lower limit	• below lower limit	• short circuit to ground • IAC valve faulty • ECM faulty
P1553	169	Idle air control (IAC) valve - upper limit exceeded	• upper limit exceeded	• short circuit to ground • IAC valve faulty • ECM faulty
P1570	39	Immobilizer - active Immobilizer implausible operating range	• implausible operating range	• wiring between ECM and immobilizer control module interrupted • short circuit to ground or short circuit to B+
P1571	39	Immobilizer - open Immobilizer - interruption / no signal	• implausible operating range	• immobilizer control module faulty
P1585		Misfire with empty fuel tank	• Misfire and fuel tank empty	• fuel tank empty
P1593	209	Intake manifold length tuning 2 (cylinders 4-6) - open circuit	• open circuit	• change-over valve (solenoid) open circuit • discontinuity in the cable
P1594	209	Intake manifold length tuning 2 (cylinders 4-6) - short circuit to ground	• value below lower limit value	• short circuit to ground in the cable
P1595	209	Intake manifold length tuning 2 (cylinders 4-6) - short circuit to B+	• upper limit value exceeded	• short circuit to B+ in the cable
P1600	107	Power supply - below lower limit	• battery voltage below 10 V	• battery faulty • transition resistance

Table a. OBDII diagnostic trouble codes (from 1996)

SAE Code	Porsche code(s)	Component or system	Fault conditions	Fault area
P1601	107	Power supply - upper limit exceeded	• battery voltage above 16 V	• voltage regulator in alternator faulty
P1602	108	Power supply - interruption	• terminal 30 missing • presence of P1602 does not necessarily indicate a fault • P1602 is stored when battery or ECM are disconnected	• wiring interrupted • battery was disconnected • ECM connector was removed
P1610	164	MIL activated externally - open circuit	• interruption in wiring • no signal	• wiring harness • no voltage at Tiptronic control unit • speed signal from ECM missing
P1611	164	MIL activated externally - short circuit to ground	• short circuit to ground	• wiring harness • Tiptronic control unit
P1614	164	MIL activated externally - short circuit to B+	• short circuit to B+	• wiring harness • Tiptronic control unit
P1640	105	ECM faulty	• ECM faulty	• ECM
P1691	165	Malfunction indication light (MIL), check engine, interruption	• interruption	• interruption in wiring
P1692	165	MIL check engine - below lower limit	• below lower limit	• short circuit to ground
P1693	165	MIL check engine - upper limit exceeded	• upper limit exceeded	• short circuit to B+
P1703	59	Switch, R position	• discontinuity/short circuit to plus, to ground • restricted driving program	• wiring • multi-function switch • Tiptronic control unit
P1704	53	Kickdown switch	• short circuit to ground • no kickdown	• wiring • kickdown switch • Tiptronic control unit
P1710	55	Speed signal (FR) - signal implausible	• discontinuity/short circuit to plus, to ground • no manual program	• ABS control unit • Tiptronic control unit • wiring harness
P1715	62	Speed signal (FL) - signal implausible	• discontinuity/short circuit to plus, to ground • no manual program	• ABS control unit • Tiptronic control unit • wiring harness
P1746	43	Control unit defective (relay)	• Tiptronic control unit defective • restricted driving program	• Tiptronic control unit defective
P1750	11	Voltage > control unit - signal implausible	• discontinuity • short circuit to plus • short circuit to ground • loose contact • voltage exceeds the valid range • restricted driving program	• fuse • wiring • relay • control unit • alternator • controller
P1761	63	Shiftlock P/N - discontinuity/short circuit to ground	• discontinuity/short circuit to ground	• wiring • Tiptronic control unit
P1762	63	Shiftlock P/N - short circuit to plus	• short circuit to plus	• wiring • solenoid for P/N lock

Table a. OBDII diagnostic trouble codes (from 1996)				
SAE Code	**Porsche code(s)**	**Component or system**	**Fault conditions**	**Fault area**
P1765	25	Throttle valve signal <-DTC memory of ECM control unit Reading out ECM - signal implausible	• no signal from ECM control unit • default throttle value 9% • no kickdown	• wiring • throttle-valve potentiometer • ECM control unit
P1766	25	Throttle valve signal <- ECM control unit	• discontinuity/short circuit to ground, short circuit to plus • default throttle value 9% • no kickdown	• wiring • Tiptronic control unit
P1770	22	Load signal <- ECM control unit	• discontinuity/short circuit to ground, short circuit to plus • hard shifting	• wiring • Tiptronic control unit
P1781	24	Ignition angle intervention - discontinuity/short circuit to ground	• discontinuity/short circuit to ground • hard shifting	• wiring • Tiptronic control unit
P1782	24	Ignition angle intervention - short circuit to plus	• short circuit to plus • hard shifting	• wiring • ECM control unit
P1782	135	Engine engagement / normal engine torque - short circuit to ground	• wiring from Tiptronic control unit pin 20 to ECM pin 42 has ground potential for longer than approximately 2.6 seconds	• wiring harness • Tiptronic control unit

DME M 2.10 Fault Diagnosis

DME connector pinout (55-pin)

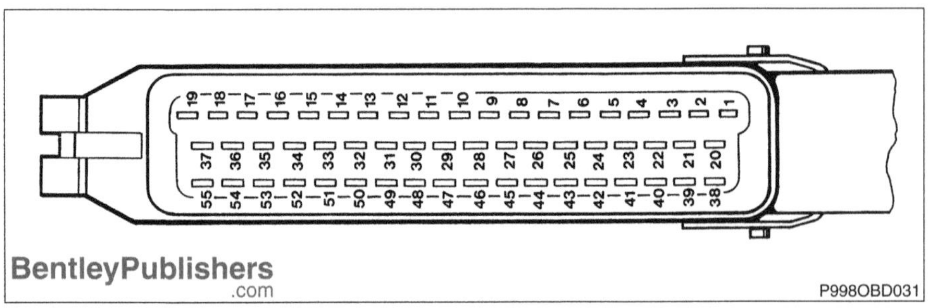

1. Resonance flap
2. Not used
3. Ground, DME relay 85 b
4. Intake manifold flap open
5. Cylinder 3 fuel injector
6. RPM
7. Mass air flow sensor signal
8. Hall sender signal
9. Vehicle speed
10. Ground oxygen sensor
11. Knock sensor 1
12. Positive supply: throttle potentiometer, Hall sender
13. Diagnosis L wire
14. Ground supply: throttle potentiometer, NTC sensors
15. Not used
16. Cylinder 6 fuel injector
17. Cylinder 1 fuel injector
18. Battery positive (+)

DME connector pinout (55-pin)

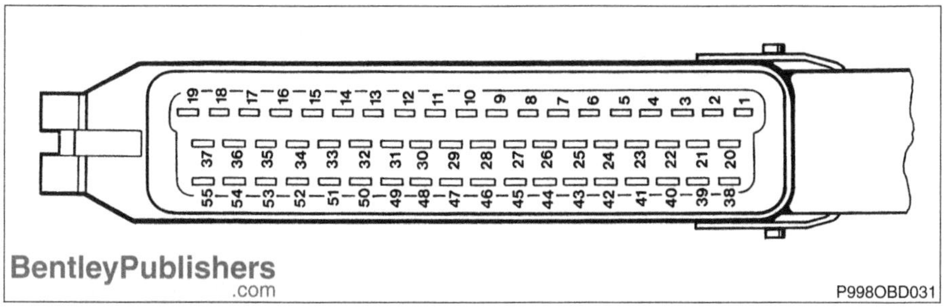

19. Ground, electronics shield

20. Oxygen sensor, heating

21. Check engine light

22. Intake manifold flap close

23. Cylinder 5 fuel injector

24. Ground: final stages

25. Final stage, ignition circuit 1

26. Ground, mass air flow sensor

27. Terminal 15

28. Oxygen sensor

29. Knock sensor 2

30. Ground, knock sensors

31. Final stage, ignition circuit 2

32. Secondary air pump

33. Fuel tank vent

34. Cylinder 4 fuel injector

35. Cylinder 2 fuel injector

36. Ground, DME relay 85

37. DME relay 87

38. Not used

39. Not used

40. AC compressor clutch

41. Heater control unit

42. Selector lever switch

43. Not used

44. Tiptronic coding

45. Engine NTC II

46. Not used

47. Intake air temperature sensor

48. Engine speed sensor –

49. Engine speed sensor +

50. Engine knock: yes or no

51. Transmission input

52. Throttle position to transmission

53. Throttle position sensor

54. Characteristic map switch

55. Diagnosis K wire

DME M 2.10 fault (blink) codes, accessing

For 1995 model vehicles, fault (blink) codes can be checked with or without a Porsche scan tool.

– To check fault (blink) codes without scan tool:
  - Turn ignition ON. Do not start engine.
  - Press and hold accelerator to floor for three to five seconds. Wait for MIL to flash once and release accelerator.
  - MIL will flash to indicate fault codes one digit at a time, starting with a single flash, then with a slight pause between digits. For example, fault code 23 will be displayed as a single flash, a slight pause, two flashes, a slight pause, then three flashes.
  - Press accelerator to floor then release to display next fault code.
  - Once all fault codes have been displayed, MIL will flash once.

Table b. DME M 2.10 fault code diagnosis (1995)

Test point	Fault code	Item	Possible faults, remedies, remarks
1a		Power supply for DME control unit (V)	Permanent positive (terminal 30). If there is no B+ at terminal 30, fault memory is cleared Test procedure: • Using test leads, connect voltmeter to terminal 24 (-) and terminal 18 (+) of the control unit plug. Display: Battery voltage • No display: Check current flow and ground paths in accordance with wiring diagram. 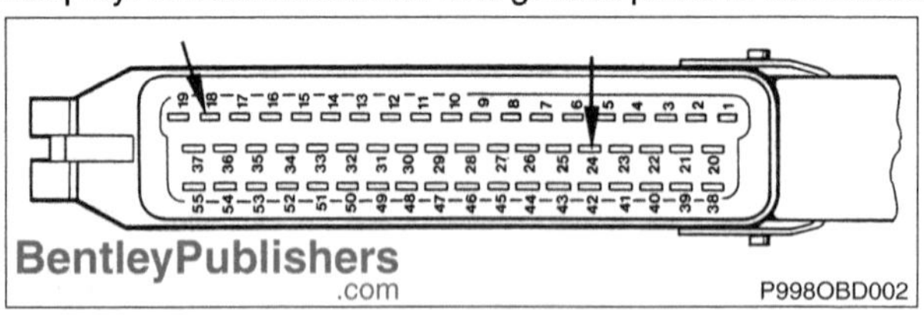
1b	1_11	Power supply for DME control unit (V) too high/ too low	Power supply via terminal 15 • Using test leads, connect a voltmeter to terminal 24 (-) and terminal 27 (+) of the control unit. • Turn ignition ON. Display: Battery voltage • No display: Check current flow according to wiring diagram 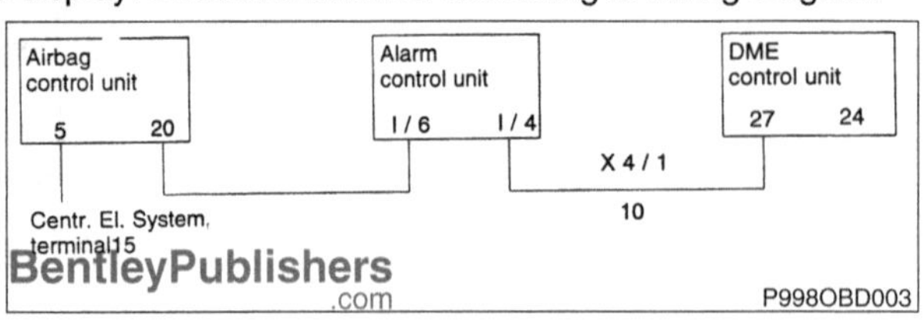

Table b. DME M 2.10 fault code diagnosis (1995)

Test point	Fault code	Item	Possible faults, remedies, remarks
2	1_14	Engine temperature sensor (ECTS II/Ω)	Using Porsche scan tool, read engine temperature directly under menu item "Actual values", If no plausible display: • Connect ohmmeter test leads to terminal 45 and terminal 14 of the disconnected DME control unit plug. • Display at: 0°C = 4.4 -6.8 kΩ 15-30°C = 1.4-3.6 kΩ 40°C = 1-1.3 kΩ 80°C = 250 -390 Ω 100°C = 160 -210 Ω • If the above values are not obtained, check directly at engine temperature sensor. ***NOTE–*** *• Temperature sensor 2 informs the control unit of the engine temperature. It provides additional fuel in the cold starting and warm-up stages of engine operation.* Open circuit (∞ Ω): • The DME control unit adjusts to a value pre-set in the control unit that approximately corresponds to that of the engine at operating temperature. When the engine is warm, enrichment by the faulty temperature sensor 2 does not occur (emergency running program). This results in starting problems when the engine is cold (no cold start enrichment). Short circuit to ground: • When engine is cold: No engine pickup, too lean, engine stops. No effect if engine is at operating temperature. • Replacement value is applicable to both types of fault. 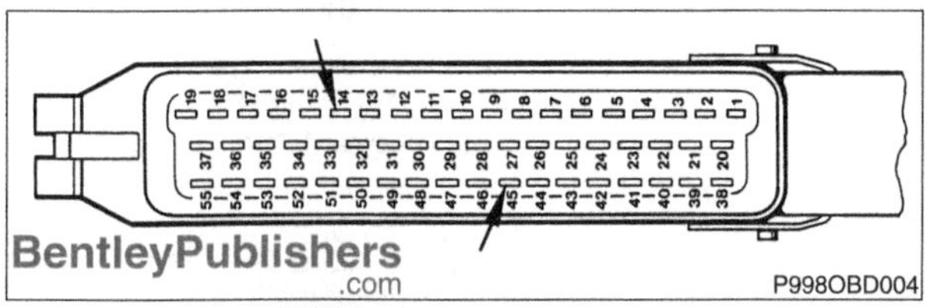
3	1_15 1_16	Throttle potentiometer	Using Porsche scan tool, read throttle angle directly in the "Actual values" menu item. If no plausible display is obtained, check power supply. • Connect 3-pin adapter lead VW 1501 between throttle potentiometer and disconnected plug. • Connect voltmeter between lead No.1 and No.2. • Ignition ON = display: approximately 5 V (Power supply of throttle potentiometer) • No display: Check according to wiring diagram • Display approximately 0.5 V: Connect voltmeter to lead No.2 and No.3. Operate throttle. Voltage should now increase to approximately 4.3 V

OBD

Table b. DME M 2.10 fault code diagnosis (1995)			
Test point	**Fault code**	**Item**	**Possible faults, remedies, remarks**
4	1_18	RPM signal	Run test using an oscilloscope. Connect and adjust shop oscilloscope according to manufacturer's instructions. • Connect oscilloscope test lead with terminal 49 and terminal 48 of the disconnected DME control unit plug. • Start engine. Sine wave fluctuations of 3 V min. must now be displayed. An intermittently higher amplitude indicates the reference mark signal. 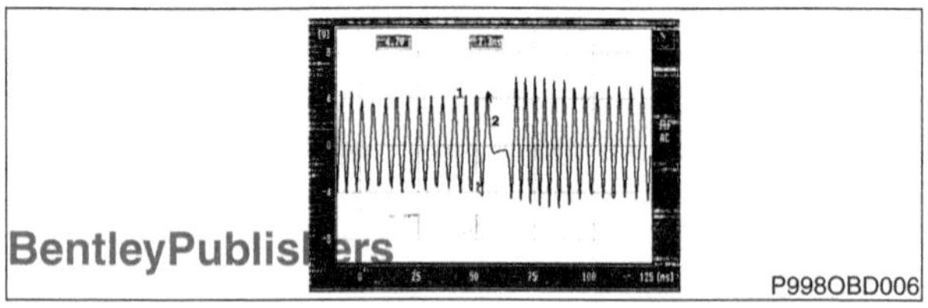 P998OBD006 **1.** RPM signal **2.** Reference mark signal • If the voltage signal is too low (< 3V), the gap between sensor and ring gear may be adjusted incorrectly. • Sensor gap: Between ring gear and sensor 1.0 ± 0.2 mm • Using Porsche scan tool, the reference mark signal may be read directly in the "Input signals" menu item. Test option • Check rpm signal with test lead VW 1501. • Connect test lead between rpm sender and disconnected control unit plug. Connect oscilloscope to test lead. • On vehicles with immobilizer, pull starter relay (R61) off the central electrical system and bridge plug-in contacts 30 and 87 (identifications 3 and 5 on central electrical system) with fuse-protected test lead. 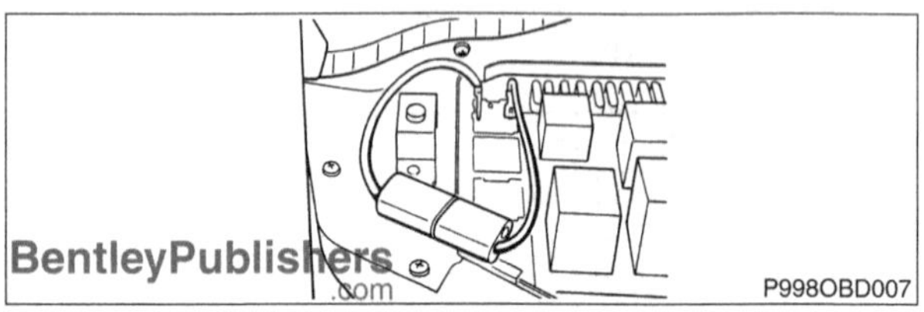 P998OBD007 • Start engine • Sine wave fluctuations of 3 V min. must now be displayed. An intermittently higher amplitude indicates the reference mark signal. 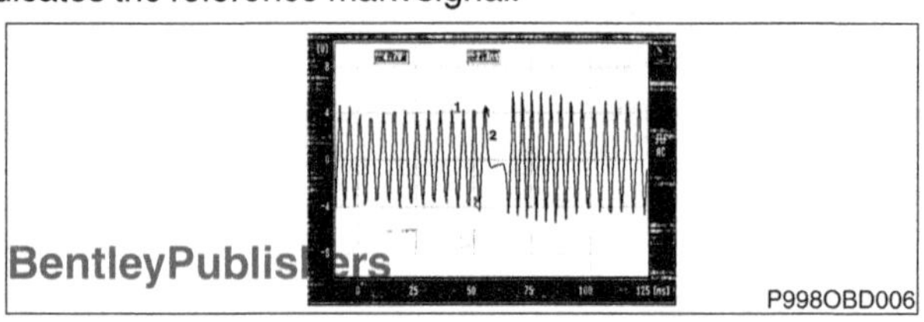 P998OBD006

Table b. DME M 2.10 fault code diagnosis (1995)

Test point	Fault code	Item	Possible faults, remedies, remarks
5	1_19	Speed signal/ speedometer	Using Porsche scan tool, the speed signal may be read directly in the Actual values menu item. As an alternative, check with an oscilloscope. • To do so, connect oscilloscope to terminal 9 and terminal 24 of the control unit plug. • Turn left front wheel manually. • The following signal must now be displayed: 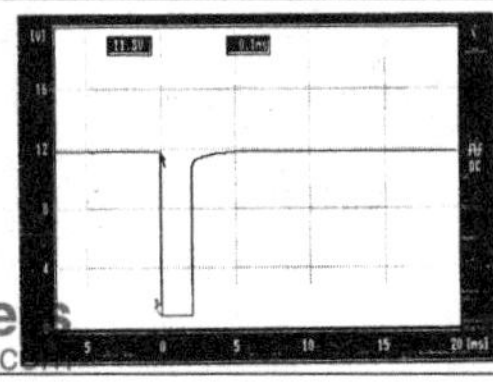 P998OBD008

<table>
<tr><th colspan="4">Table b. DME M 2.10 fault code diagnosis (1995)</th></tr>
<tr><th>Test point</th><th>Fault code</th><th>Item</th><th>Possible faults, remedies, remarks</th></tr>
<tr><td>6</td><td>1_21</td><td>Hot wire mass air flow sensor (V/Ω)</td><td>
Voltage supply (V)
• Turn plug at hot wire mass air flow sensor counterclockwise and disconnect.
• Connect voltmeter to terminals 1 and 3.
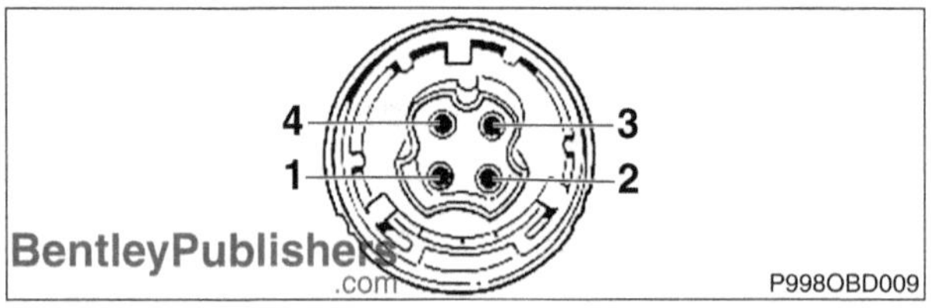

Bridge DME relay
• Pull DME relay (R53) off the central electrical system and bridge plug-in contacts 30 and 87 b (identifications 3 and 7 on central electrical system) with fuse-protected test lead. The fuel pump should be heard running.

• Ignition ON:
• Display: Voltage approximately 10.8 Volt
• No display: check in accordance with wiring diagram
• Reconnect plug.
The mass air flow sensor signal may be tested directly in the “Actual values” menu item using Porsche scan tool.
• To check the signal, remove upper air cleaner section and start engine.
• Display: approximately 1.0 Volt
• Blow against hot wire. This must cause the System Tester display to change.
Checking the hot wire signal (V)
• Connect plug to mass air flow sensor.
• Pull off DME plug.
• Bridge DME relay.
• Connect DME plug terminal 36 to ground (e.g. door stop).
• Connect Voltmeter to DME plug terminal 7 and 26 (ground)
• Display: 160 mV
• Blow against hot wire in mass air flow sensor and observe voltmeter. Voltage must change.
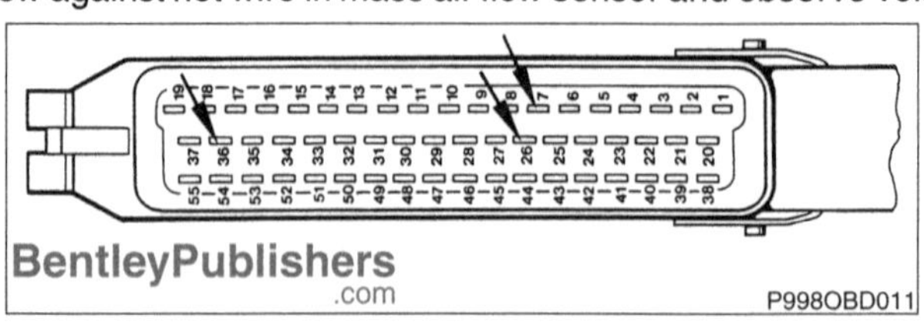

</td></tr>
</table>

Table b. DME M 2.10 fault code diagnosis (1995)

Test point	Fault code	Item	Possible faults, remedies, remarks
7	1_22	Oxygen sensor (V) (sensor signal)	Checking the sensor signal Using Porsche scan tool, read oxygen sensor signal directly under the "Actual values" menu item. • Diagnosis can only be carried out if an engine temperature of 70° C has been reached for more than 1 minute. • If not, disconnect oxygen sensor plug. 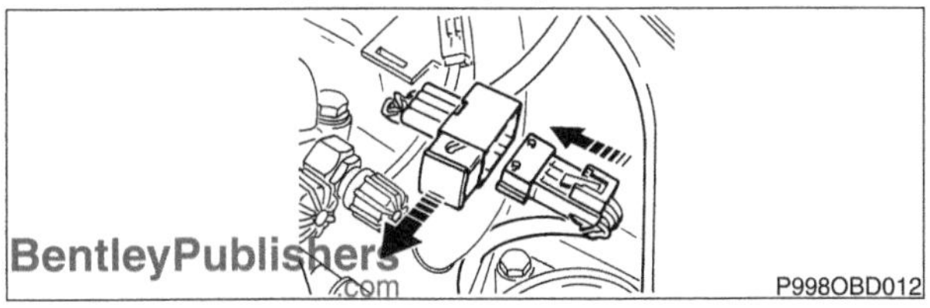 P998OBD012 • Connect digital voltmeter with terminals 3 and 4 at the sensor end. 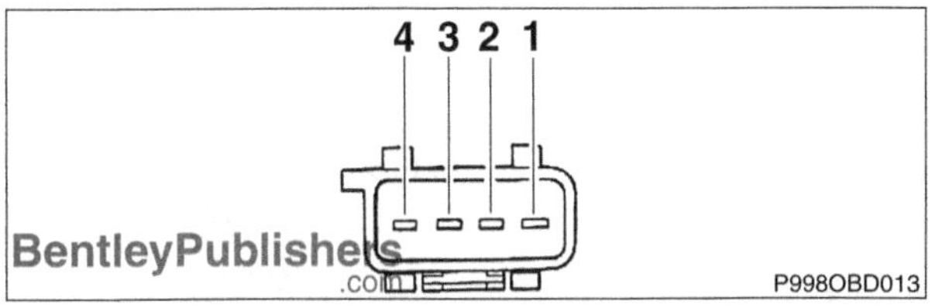 P998OBD013 • Start engine and allow to warm up so that the oxygen sensor reaches its operating temperature. When the mixture is enriched, e.g. during acceleration, a change in the voltage signal must be displayed. • Display: approximately 150 mV -900 mV (acc. to mixture composition) Oxygen sensor heating test • In case of delayed regulation response, check the oxygen sensor heating. Connect voltmeter to terminals 1 (+) and 2 of disconnected sensor plug on the control unit end with engine running. • Battery voltage must be present. 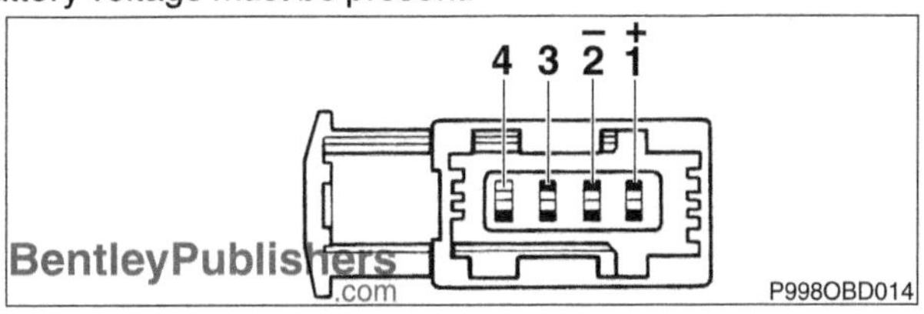 P998OBD014
8	1_23	Oxygen regulation stop	The oxygen regulator cannot operate within its control range if extreme problems of mixture preparation occur, e.g. due to an excessively lean setting because of unmetered air, or due to an excessively rich setting because of a faulty injector valve. The oxygen regulator then moves up to the stop position. Oxygen regulation too rich: • Check intake system for leaks Oxygen regulation too lean: • Check fuel pressure • Check injector valves for leaks
8	1_24	Oxygen sensor short to + / short to ground	If the control unit detects an oxygen sensor voltage signal of more than 1.4 V or less than 0.1 V, the control unit switches to operation without oxygen sensor. (Short to ground or open circuit). If regulation does not work and the sensor voltage is O.K., use Porsche scan tool to check control unit coding before replacing it.

Table b. DME M 2.10 fault code diagnosis (1995)

Test point	Fault code	Item	Possible faults, remedies, remarks
9	1_25	Intake air temperature sensor	Using Porsche scan tool, read intake air temperature directly under "Actual values" menu item. • If no display or unplausible value, connect terminals 47 and 14 of disconnected DME control unit plug to ohmmeter. 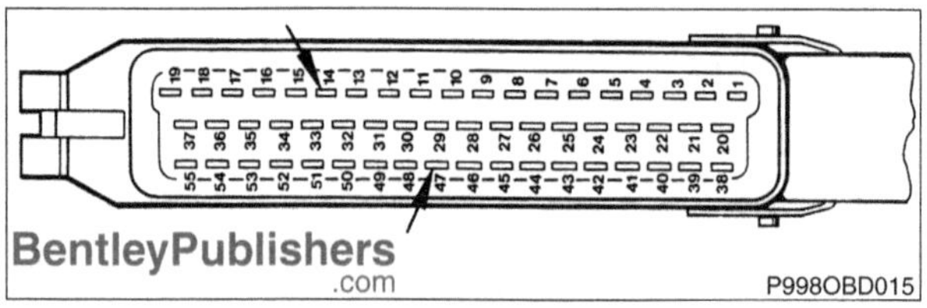 Display at: 0° C = 4.4 - 6.8 kΩ 15 - 30° C= 1.4 - 3.6 kΩ 40° C = 1 - 1.3 kΩ
10	1_26	Ignition timing change	Using Porsche scan tool, read ignition timing change signal directly in the "Input signals" menu item. On Tiptronic vehicles, ignition is retarded when a gear change is made. • When the ignition timing change fault message occurs, check wiring continuity between DME plug terminal 51 and Tiptronic connector terminal 32. 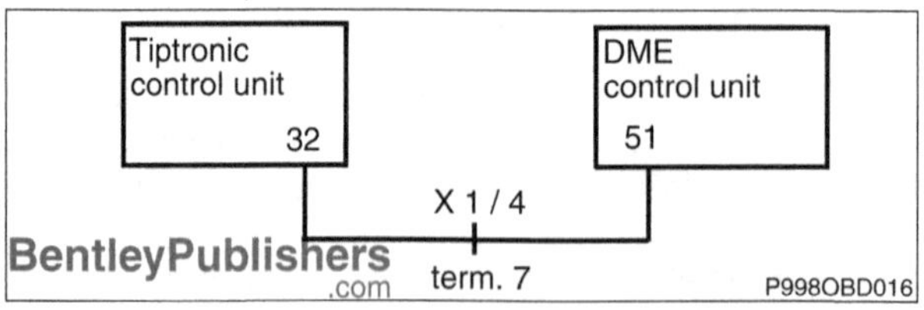 • This fault causes the Tiptronic transmission to operate in emergency mode. For test to be valid, a test drive is required since the signal is only displayed for a very short time.

Table b. DME M 2.10 fault code diagnosis (1995)

Test point	Fault code	Item	Possible faults, remedies, remarks
11 12	1_27 1_28	Idle stabilizer	The idle stabilizer is designed as a twin-winding actuator with one opening winding and one closing winding. • Using Porsche scan tool, read actuation of the idle stabilizer directly in the "Drive links" menu item. If no audible pulse is present, check the following: • Voltage supply Connect voltmeter with disconnected idle stabilizer plug terminal 2 and engine ground. Ignition ON. Display: Battery voltage Configuration of terminals: Terminal 1 Closing winding signal Terminal 2 Battery voltage Terminal 3 Opening winding signal 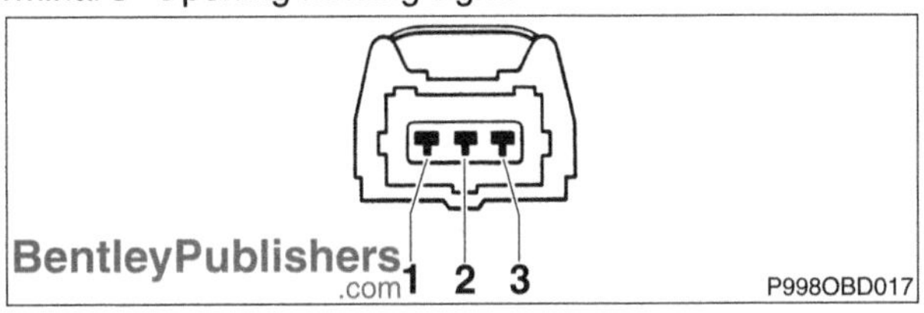 No display: • Check power supply in accordance with wiring diagram. • Check the control signal • Using Porsche scan tool, read actuation of the idle stabilizer directly in the "Drive links" menu item. Operational check: Switch on loads in idle mode. • % display must change, idle speed remains constant. • If not: Connect 3-pin adapter lead VW 1501 between idle stabilizer and plug port. Connect oscilloscope to the adapter lead terminal 1 (closing winding) as well as terminal 3 (opening winding). Make sure the wiring connectors are not shorted to vehicle ground to avoid short circuits. • With engine running, the following display must appear: 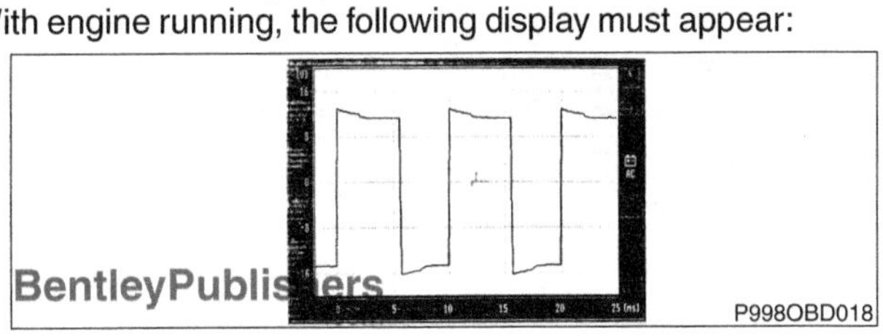 **1.** Signal (terminal 1, opening closing signal) **2.** Signal (terminal 3, opening winding signal) • If no audible pulse is detected even though voltage is present and a signal is applied, replace the idle stabilizer.
13	1_31	Knock sensor 1	Mounting of knock sensor (observe torque and type of screw). • Check wiring harness and electrical connector in accordance with wiring diagram. • Reconnecting the plugs helps to eliminate contact resistances. • Replace knock sensor. • If the knock sensor is faulty, ignition timing is retarded by 6° on crankshaft at a certain engine load.
14	1_32	Knock sensor 2	Mounting of knock sensor (observe torque and type of screw). • Check wiring harness and electrical connector in accordance with wiring diagram. • Reconnecting the plugs helps to eliminate contact resistances. • Replace knock sensor. • If the knock sensor is faulty, ignition timing is retarded by 6° on crankshaft.
15	1_33	Control unit faulty (knock computer)	Ignition timing is retarded by 6° on the crankshaft for all cylinders from a certain engine load if this fault occurs. Replace control unit.

OBD

Table b. DME M 2.10 fault code diagnosis (1995)

Test point	Fault code	Item	Possible faults, remedies, remarks
16	1_34	Hall signal	The hall sender is connected to the control unit for voltage supply (approximately 5 V) and ground connection. The signal is fed to the control unit via the 0 line. In case of error display: Check voltage supply, ground connection and signal line in accordance with the wiring diagram. Hall sender plug terminals — Control unit plug terminals 1 to 19 2 to 8 3 to 12
17	1_36	Idle CO potentiometer	On vehicles without catalytic converter that show fault code 36. start by testing the control unit coding with Porsche scan tool "Actual values" menu item. • If the coding is O.K., check power supply for CO potentiometer and potentiometer signal in accordance with the wiring diagram.
18	1_41	Control unit faulty	If this fault is detected by the control unit, maximum engine speed is limited to 6000 RPM 6 minutes after starting the engine. This is done to protect the engine.
19	1_42	Fuel pump relay (DME relay)	Start engine. • The fuel pump must operate during engine starting. If not: • Switch off ignition. Disconnect control unit plug. Use a test lead to connect terminal 36 and terminal 24 of the control unit plug. Using an additional test lead, connect terminal 3 of DME plug to ground (e.g. door stop). The pump must run. If not: • Check in accordance with wiring diagram 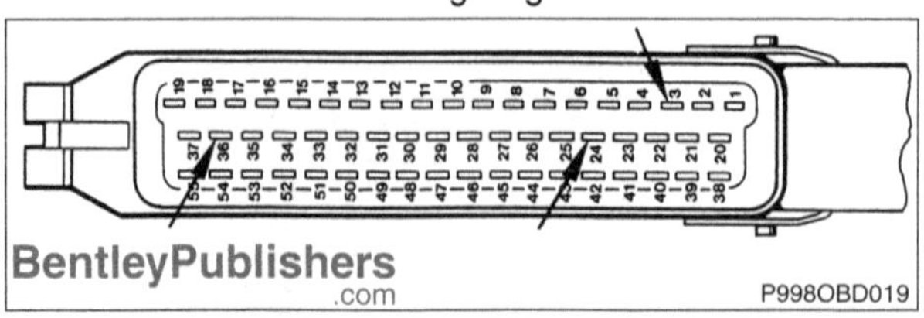

Table b. DME M 2.10 fault code diagnosis (1995)

Test point	Fault code	Item	Possible faults, remedies, remarks
20	1_43	Tank ventilation valve	When the engine runs at operating temperature, the tank ventilation valve is opened for a longer or shorter period as a function of the load. The opening period is determined by a ground pulse from the control unit. Activation test • To activate the tank ventilation valve directly, use Porsche scan tool, selecting "Drive links" menu item. • If there is no audible pulse from the tank venting valve, check voltage supply at the terminal of the tank ventilation valve plug and body ground according to the wiring diagram. • Ignition ON. • Display: Battery voltage  P998OBD020 Control signal test • Connect DME test lead (Bosch No.1 684 463 093) between tank ventilation valve and plug connection. Connect and adjust engine tester according to manufacturer's instructions. • The tank ventilation valve is not activated permanently. • Testing must be performed within 7 minutes after starting the engine at operating temperature. Then interrupt activation of ventilation valve for approximately 75 seconds, continue afterwards. • Start engine and accelerate. With the engine at operating temperature. The following display must be visible on the tester: 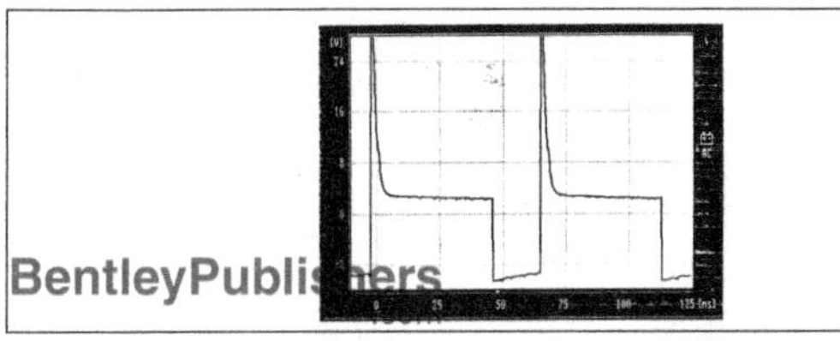 P998OBD021 • The signal becomes wider as the air throughput increases. • If there is no signal, check path in accordance with wiring diagram.
21	1_45	Check engine light	A ground signal is fed from the control unit terminal 21 to the "Check Engine lamp", causing this lamp to come on when an emission control component fails. • If the "Check Engine" warning lamp fails, this fault is stored in the fault memory. • Check by supplying ground to disconnected DME control unit plug terminal 21 and switching ignition ON. The Check Engine lamp must now come on. 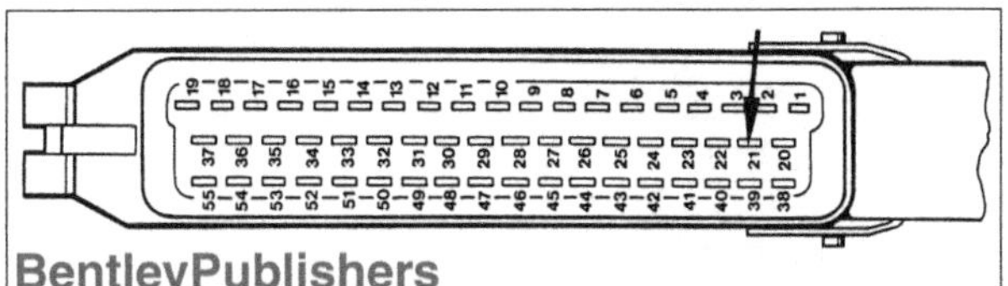 P998OBD022 • To replace a faulty display lamp, always use the specified bulb.

Table b. DME M 2.10 fault code diagnosis (1995)

Test point	Fault code	Item	Possible faults, remedies, remarks
22 23 24 25 26 27	1_51 1_52 1_53 1_54 1_55 1_56	Injector valves (V/Ω)	Using Porsche scan tool, check injector valves directly in “Drive links” or “Actuator active” menu item. • Selective injection allows each injector valve to be actuated individually. In case of the Drive links test point, a rather weak switching noise of the injector valves is audible. • In case of the Actuator active menu item, each individual injector valve may be isolated with the engine running. Power supply Disconnect valve plug, connect voltmeter to the injector valve plug terminal (refer to drawing) and ground (engine). • Ignition ON. • Display: Battery voltage 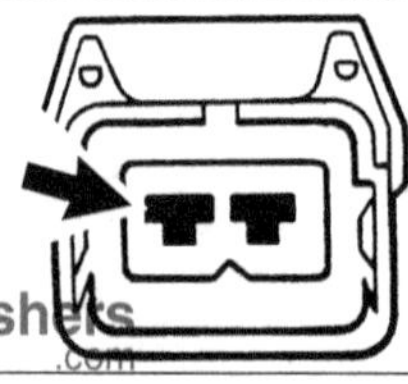 P998OBD020 • If no battery voltage is displayed, check according to wiring diagram. Checking coil resistance of injector valves • Disconnect valve plug. Check coil resistance at injector valve terminal contacts with an ohmmeter. • Test value: approximately 16 Ω Injection output stage • Adjust oscilloscope according to manufacturer’s instructions. Connect Bosch test lead (1 684 463 093) between injector valve and plug. Connect oscilloscope according to manufacturer’s instructions with the test lead. • Make sure the tester leads are not grounded in any way. • Start engine. If the injection output stage operates correctly or if the tester connections are correct, respectively, the following signal must be displayed: 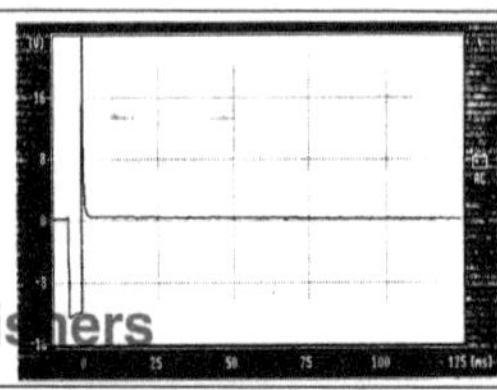 P998OBD023

Table b. DME M 2.10 fault code diagnosis (1995)

Test point	Fault code	Item	Possible faults, remedies, remarks
28	1_67 1_68 1_69	Ignition final stages	The ignition circuits may be checked using the "Actuator active" menu item on the Porsche scan tool (with engine running). Check control signal for the ignition final stages. Ignition circuit I • Connect test lead plus of the oscilloscope to terminal 2 and test lead minus to terminal 4 (ground) of the disconnected final stage plug (ignition control unit). 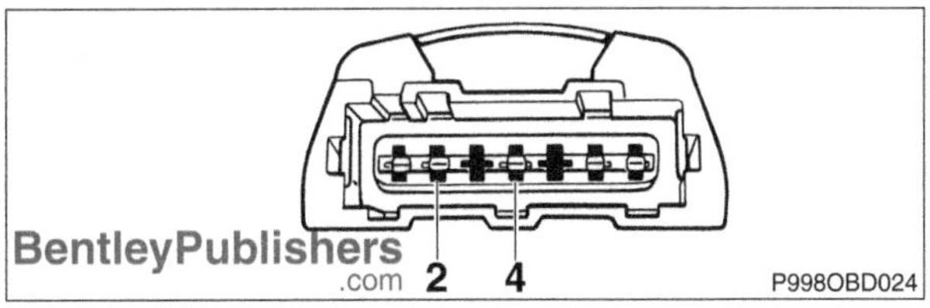 P998OBD024 • Operate starter. The oscilloscope must display the activating signal of the DME control unit. 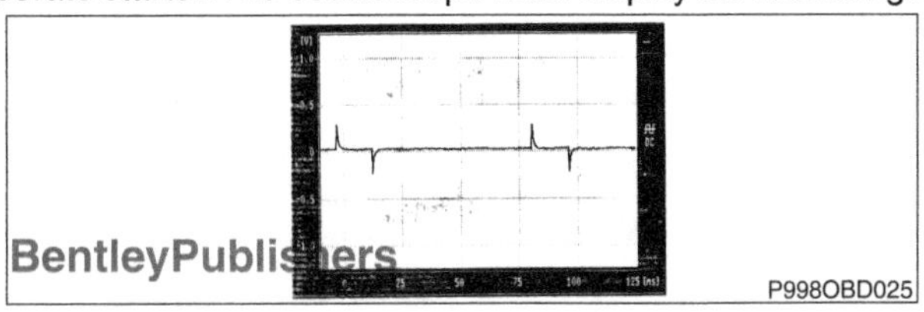 P998OBD025 Ignition circuit II • Connect test lead plus of the oscilloscope to terminal 7 and test lead minus to terminal 7 (ground) of the disconnected final stage plug (ignition control unit). 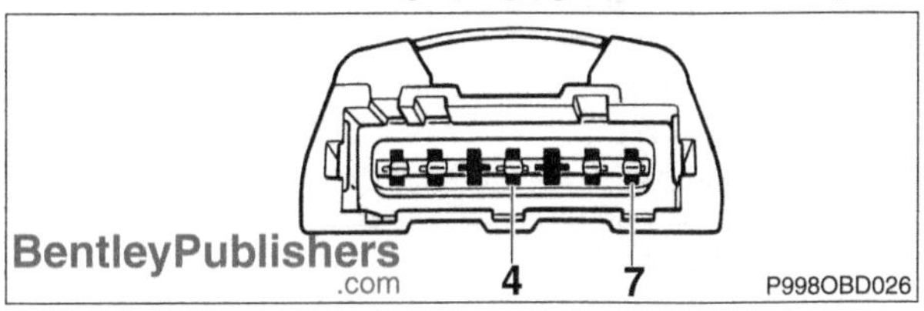 P998OBD026 • Operate starter. The activating signal must be displayed as for ignition circuit 1 Checking control signal for ignition coils Ignition circuit I • Plug connected to ignition control unit. Push back rubber sleeve and connect test lead plus to terminal 1. • Ground test lead minus 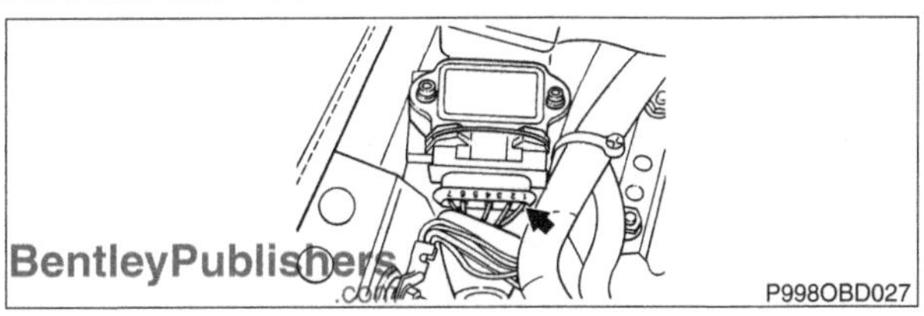 P998OBD027 **continued on next page**

Table b. DME M 2.10 fault code diagnosis (1995)

Test point	Fault code	Item	Possible faults, remedies, remarks
28	1_67 1_68 1_69	Ignition final stages	• Start engine. Oscilloscope shows primary pattern of ignition coil (ignition circuit I) 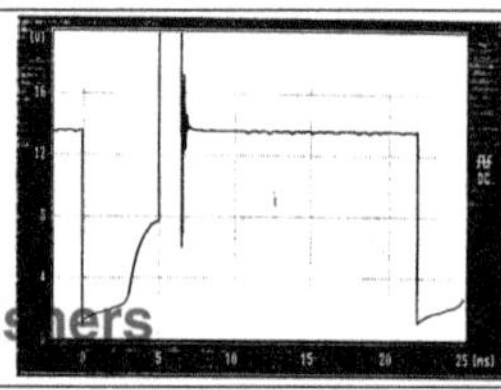 P998OBD028 Ignition circuit" • Plug connected to ignition control unit. Push back rubber sleeve and connect test lead plus to terminal 7. Ground test lead minus. • Start engine. Oscilloscope shows primary pattern of ignition coil (ignition circuit II). 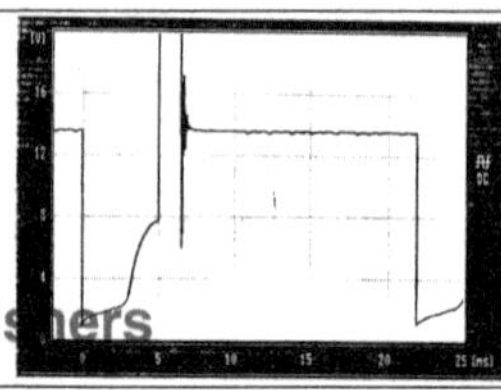 P998OBD028
29	1_44	Secondary air pump relay	The secondary air pump may be tested using the "Actuator active" menu item on the Porsche scan tool. Conditions for testing: • Cylinder head temperature > 80° C (> 176° F) After selecting "Actuator active", the auxiliary air pump will be activated after a waiting period of approximately 10 seconds. This will be indicated by the * symbol shown in the upper right hand corner of the display. Activation of the pump is limited to 20 seconds. AIR pump * Control U probe 1.10 0.01 V Return: N P998OBD029 When the pump is running, sensor voltage tends approaches 0 V and oxygen sensor regulation approaches 1.20. Activating the pump • If oxygen sensor regulation and sensor voltage do not change after the pump activating symbol has appeared on the display, check pump activation and the relay according to the wiring diagram. Activating on-off valve • If the pump activating symbol is displayed, the pump is audibly running, but the oxygen sensor regulation does not approach 1.2 and the sensor voltage does not approach 0.1, the fault lies in the valve area or wiring to valve.

Tiptronic Fault Diagnosis

Tiptronic connector pinout (1995)

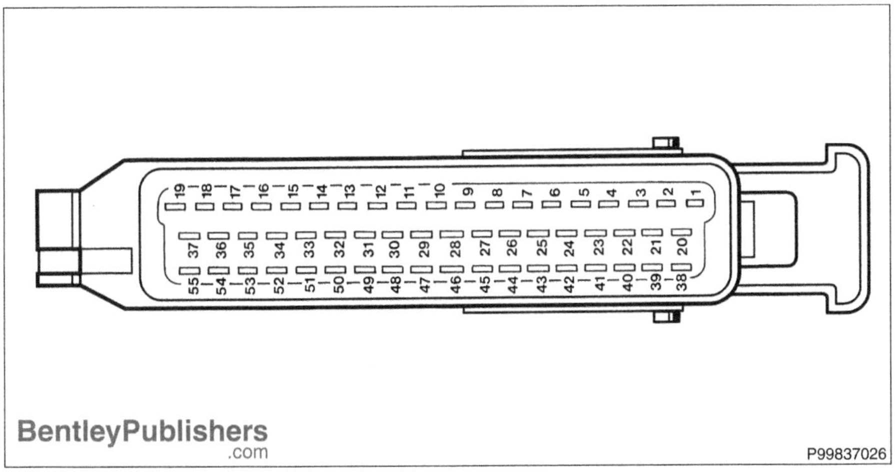

1. Terminal 15 E
2. Output rpm +
3. Engine rpm (TN)
4. Stop light
5. Solenoid valve 1
6. Solenoid valve modulating pressure
7. Electronics - Ground
8. Input +5 V (reference voltage)
9. Pin coding 1
10. Upshift
11. Throttle signal
12. Front wheel speed (FR)
13. Manual program display
14. Gear position switch
15. L wire
16. 1st gear display
17. Not used
18. Not used
19. + for solenoid valves
20. Shield / output rpm
21. Load signal (TI)
22. Not used
23. Warning lamp
24. Solenoid valve 2
25. Reverse light relay
26. Power ground
27. Not used
28. Pin coding 2
29. Downshift
30. Kickdown
31. 2nd gear display
32. Change of ignition timing
33. Position Z switch
34. Not used
35. Not used
36. Not used
37. Not used
38. Output rpm –
39. Terminal 30

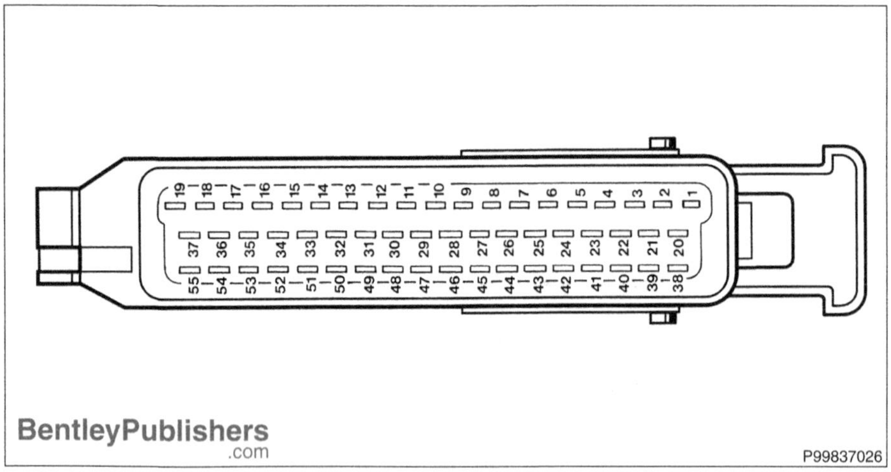

40. Transverse acceleration sensor

41. Redundancy

42. Lockup clutch solenoid valve

43. 3rd gear display

44. Ground (ATF temperature sensor/ transverse acceleration sensor)

45. Output + 5V

46. ATF - Temperature sensor

47. Not used

48. Input signal (manual program)

49. Not used

50. Position X switch

51. K wire

52. Not used

53. Not used

54. Not used

55. 4th gear display

Tiptronic fault diagnosis (1995)

Table c. Tiptronic fault code diagnosis (1995)

Test point	Fault code	Item	Result of fault	Possible faults
1		Supply voltage (Terminal 30)	No diagnosis possible	Open circuit, short to ground, loose contact • Open circuit/short to ground: Check fuse No. 12 • Fuse faulty: Pull connector off transmission control unit. Check pin 39 for short to ground. • Short to ground? Check wiring according to wiring diagram • No short to ground: Transmission control unit faulty • Fuse OK: Check voltage according to wiring diagram
2	11	Supply voltage (Terminal 15)	Tiptronic warning lamp does not come on, transmission is in emergency mode	Open circuit, short to positive, short to ground, loose contact, voltage outside of valid range **1.** Tiptronic warning lamp does not light when ignition is OFF. • Check fuse No. 27 • Measure voltage at fuse: No voltage: Check wire to relay R11 or operation of relay. • Measure voltage at transmission control unit connector pin 1: No voltage: Check wiring according to wiring diagram. **2.** Voltage outside of valid range Fault code 11: signal implausible • Low voltage? U < 11 V Check battery, alternator, regulator. • OK? Check for contact resistance at junctions, deformation and corrosion of contacts. • Excessive voltage? Check vehicle power supply (alternator, regulator) U > 14.5 V with engine running.

Table c. Tiptronic fault code diagnosis (1995)

Test point	Fault code	Item	Result of fault	Possible faults
3	13	Supply voltage, drive links	Tiptronic warning lamp comes on, transmission is in emergency mode	Possible faults: Transmission relay in control unit cannot pick up or drop out • Pull off transmission connector. Measure voltage at pin M (system voltage). 1. No voltage: Pull off transmission control unit connector. • Check wiring of transmission control unit pin 19 to transmission connector pin M. • Wiring OK: Transmission control unit faulty. 2. Voltage cannot be switched off: Fault lamp does not come on when fault occurs (e.g. connector pulled off at throttle potentiometer). • Pull off transmission control unit. • No voltage at pin M: Transmission control unit faulty. • Voltage at pin M: Short to positive in wiring (Transmission - transmission control unit).
4	14	Supply voltage, sensor 5 V	Tiptronic warning lamp comes on, transmission is in emergency mode	Possible faults: Open circuit, short to ground, short to positive • Voltage supply 5 V ± 0.5 V 1. Check voltage at transverse acceleration sensor pin 3 with ignition ON. 2. Pull connector off transmission control unit. Ignition OFF. • Check wiring from transmission control unit connector pin 45 to transverse acceleration sensor pin 3 for continuity. Short to ground and short to positive from transverse acceleration sensor pin 3 to pin 7 (ground) and pin 1 (15).
5	15	RPM signal (engine)	Tiptronic warning lamp comes on, transmission is in emergency mode	Possible faults: Open circuit/short to ground, short to positive, loose contact 1. Signal check • Check signal with Porsche scan tool. (Engine operation). • Check signal with oscilloscope at transmission control unit connector pin 3. • If signal OK: Transmission control unit faulty. • Signal shape 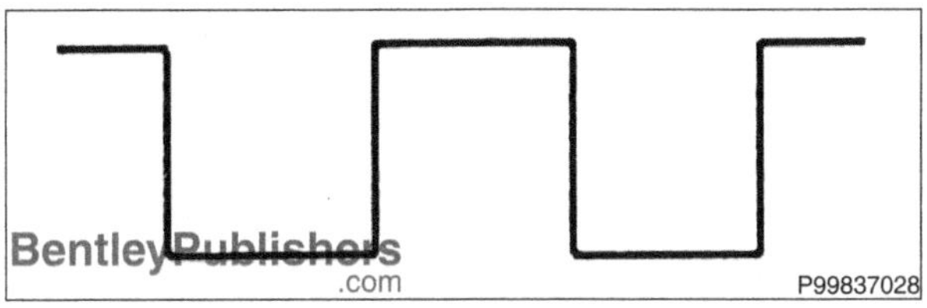 P99837028 2. Open circuit, short to ground, short to positive • Check wiring to wiring diagram. • Check transmission control unit connector pin 3 to DME connector pin 6 for continuity, short to ground and short to positive. (Ground at transmission control unit = pin 7, positive at transmission control unit = pin 1) • Connections OK: DME control unit faulty

OBD

Table c. Tiptronic fault code diagnosis (1995)

Test point	Fault code	Item	Result of fault	Possible faults
6	22	Load signal	Tiptronic warning lamp comes on, transmission is in emergency mode	Possible faults: Open circuit, short to ground, short to positive, loose contact **1.** Signal check • Check signal with Porsche scan tool (injection time) with engine running. • Check signal with oscilloscope at transmission control unit connector pin 21. 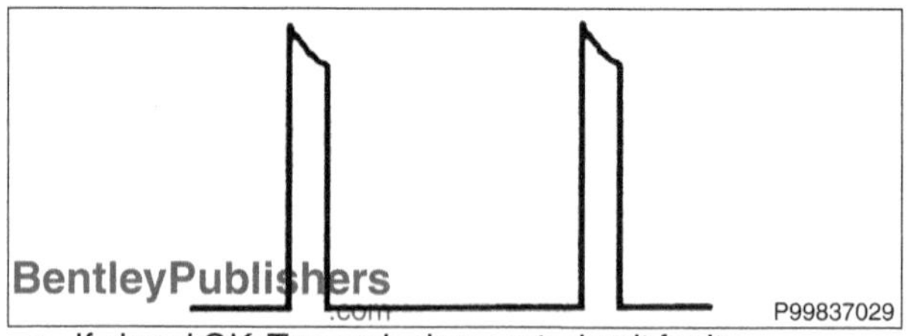 • If signal OK: Transmission control unit faulty. • Signal shape: TI signal **2.** Open circuit, short to ground, short to positive • Check wiring according to wiring diagram. Check transmission control unit connector pin 21 to DME connector pin 17 for continuity, short to ground and short to positive. (Ground at transmission control unit connector = pin 7, positive at transmission control unit connector = pin 1) • Connections OK: DME control unit faulty.
7	24	Change of ignition timing	Tiptronic warning lamp comes on, transmission is in emergency mode	Possible faults: Open circuit, short to ground, short to positive **1.** Check operation with Porsche scan tool (drive links). • Idle speed must drop. If it does not, check as follows: **2.** Check voltage drop • Remove cover of plugged-in connector X4/1. • Switch ignition ON. • Measure voltage at pin 7. Specification: approximately 5 V. • Trigger signal with Tester. Voltage must drop. • Operation OK: Check wiring to DME control unit (Connector X 4/1 pin 7 to DME control unit pin 51) • Wiring OK: DME control unit faulty • Operation not OK: Check wiring to transmission control unit (connector X4/1 pin 7 to transmission control unit connector pin 32). • Wiring OK: Transmission control unit faulty. • Note: Check engine to body ground strap • 1 = Connector X 4/1 (below left seat): 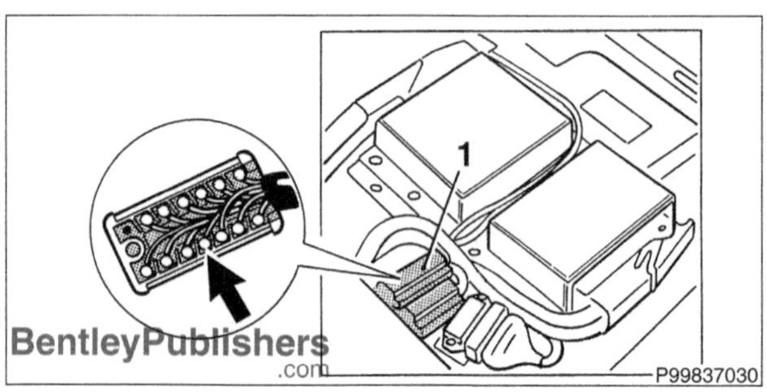

Table c. Tiptronic fault code diagnosis (1995)

Test point	Fault code	Item	Result of fault	Possible faults
8	25	Throttle Potentiometer	Tiptronic warning lamp comes on, transmission is in emergency mode	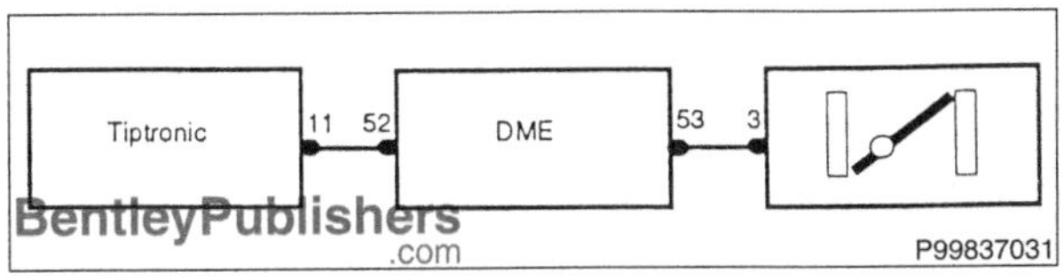 Detectable faults: Open circuit/short to ground, short to positive The "Throttle potentiometer" fault is also stored in the DME. The fault must be corrected according to the DME test plan (Test point 3). Tiptronic test: • Check display % with Porsche scan tool (Actual values menu). • If no display: **1.** Measure voltage signal at transmission control unit connector pin 11 with voltmeter. Vary from 1 V at idle to approximately 10 V at full throttle (Ignition ON) . • If no voltage: **2.** Check wiring according to wiring diagram. • Check transmission control unit connector pin 11 to DME pin 52 for continuity, short to ground and short to positive (ground at transmission control unit connector =pin 7, positive at transmission control unit connector = pin 1). • If continuity OK = DME control unit faulty. • If voltage OK = Transmission control unit faulty.
9	31	Solenoid valve 1	Tiptronic warning lamp comes on, transmission is in emergency mode	Possible faults: Open circuit, short to ground, short to positive **1.** Check operation with Porsche scan tool (Drive link test). • Acoustical test: Triggering cycle of solenoids is audible as a clicking noise near transmission. **2.** Pull connector off transmission control unit. • Measure resistance between pin 5 and pin 19. Specification: 34 Ω, tolerance 10 % • Check that pins 5 and 19 are free from shorts to ground. • Check OK: Transmission control unit faulty. **3.** Pull off connector at transmission. • Check wiring (transmission control unit connector pin 5 to transmission connector pin H and transmission control unit connector pin 19 to transmission connector pin M). • Wiring OK: Solenoid valve faulty. ***NOTE–*** *• If a "sporadic solenoid fault" occurs, check connections for pushed-back pins.*

Table c. Tiptronic fault code diagnosis (1995)

Test point	Fault code	Item	Result of fault	Possible faults
10	32	Solenoid valve 2	Tiptronic warning lamp comes on, transmission is in emergency mode	Possible faults: Open circuit, short to ground, short to positive **1.** Check operation with Porsche scan tool (Drive link test). • Acoustical test: Triggering cycle of solenoids is audible as a clicking noise near transmission. **2.** Pull off connector at transmission control unit. • Measure resistance between pin 24 and pin 19: Specification: 34 Ω, tolerance 10 % • Check that pins 24 and 19 are free from shorts to ground: Check OK: Transmission control unit faulty. **3.** Also pull off connector from transmission. • Check wiring (transmission control unit connector pin 24 to transmission connector pin K and transmission control unit connector pin 19 to transmission connector pin M). • Wiring OK: Solenoid valve faulty.
	33	Solenoid valve, torque converter, clutch	Tiptronic warning lamp comes on, transmission is in emergency mode	Possible faults: Open circuit, short to ground, short to positive **1.** Check operation with Porsche scan tool (Drive link test). • Acoustical test: Triggering cycle of solenoids is audible as a clicking noise near transmission. **2.** Pull off connector at transmission control unit. • Measure resistance between pin 42 and pin 19: Specification: 34 Ω, tolerance 10 % • Check that pins 42 and 19 are free from shorts to ground: Check OK: Transmission control unit faulty **3.** Also pull off connector from transmission. • Check wiring (transmission control unit connector pin 42 to transmission connector pin K and transmission control unit connector pin 19 to transmission connector pin M). • Wiring OK: Solenoid valve faulty.
12	34	Pressure regulator	Tiptronic warning lamp comes on, transmission is in emergency mode	Possible faults: Open circuit, short to ground, short to positive **1.** Pull off connector from transmission control unit. • Measure resistance between pin 6 and pin 19: Specification: 6 Ω ± 2 Ω • Check that pins 6 and 19 are free from shorts to ground: Check OK: Transmission control unit faulty. **2.** Also pull off connector from transmission. • Check wiring (Transmission control unit connector pin 6 to transmission connector pin B and transmission control unit connector pin 19 to transmission connector pin M). • Wiring OK: Pressure regulator faulty.

<table>
<tr><th colspan="5">Table c. Tiptronic fault code diagnosis (1995)</th></tr>
<tr><th>Test point</th><th>Fault code</th><th>Item</th><th>Result of fault</th><th>Possible faults</th></tr>
<tr><td>13</td><td>35</td><td>Selector lever switch</td><td>Tiptronic warning lamp comes on, transmission is in emergency mode</td><td>
Possible faults: Open circuit, short to ground, short to positive
1. Check operation with Porsche scan tool.
• Adjustment of Porsche scan tool:
Menu: “Input signals” = Selector lever/Manual program
• Shift through selector lever positions P, R, N, D, 3, 2, 1 one after another.
• Compare:
Selector lever position
Position indicator in speedometer
Position indicator on Porsche scan tool

NOTE–
• Due to its design, the display on Porsche scan tool appears with a certain delay. Shift through positions sufficiently slowly.

• No matching: Check wiring according to wiring diagram, check position switch with/without wiring.
2. Pull off transmission control unit connector, switching ignition OFF.
• Use ohmmeter to check transmission control unit connector pin 7 for ground.
• If no ground is present, refer to wiring diagram MP IV.
3. Measure position switch according to table 1.
• Ohmmeter display – = No continuity
0 = Continuity
Table 1
<table>
<tr><th></th><th>Pin 14</th><th>Pin 33</th><th>Pin 50</th><th>Pin 41 -> to ground</th></tr>
<tr><td>P</td><td>–</td><td>–</td><td>0</td><td>0</td></tr>
<tr><td>R</td><td>0</td><td>–</td><td>0</td><td>0</td></tr>
<tr><td>N</td><td>0</td><td>–</td><td>–</td><td>–</td></tr>
<tr><td>D</td><td>0</td><td>0</td><td>–</td><td>–</td></tr>
<tr><td>3</td><td>0</td><td>0</td><td>0</td><td>–</td></tr>
<tr><td>2</td><td>–</td><td>0</td><td>0</td><td></td></tr>
<tr><td>1</td><td>–</td><td>0</td><td>–</td><td></td></tr>
</table>
• Measurements OK: Control unit faulty.
• Measurements present, but incorrect sequence?
• Check cable from selector lever switch to position switch:
Is cable adjusted correctly?
Repair instructions: See Shift lever cable, adjusting, in this repair group.
Open circuit, short to ground, short to positive?
4. Check position switch without connections:
Pull connector off switch.
Check switch for continuity according to Table 2.
continued on next page
</td></tr>
</table>

<table>
<tr><th colspan="5">Table c. Tiptronic fault code diagnosis (1995)</th></tr>
<tr><th>Test point</th><th>Fault code</th><th>Item</th><th>Result of fault</th><th>Possible faults</th></tr>
<tr><td>13</td><td>35</td><td>Selector lever switch</td><td>Tiptronic warning lamp comes on, transmission is in emergency mode</td><td>Ohmmeter display
– = No continuity
0 = Continuity
Table 2
<table>
<tr><th></th><th>Pin E-B</th><th>Pin E-C</th><th>Pin E-A</th><th>Pin F-G</th><th>Pin H-D</th></tr>
<tr><td>P</td><td>–</td><td>–</td><td>0</td><td>0</td><td>0</td></tr>
<tr><td>R</td><td>0</td><td>–</td><td>0</td><td>–</td><td>0</td></tr>
<tr><td>N</td><td>0</td><td>–</td><td>–</td><td>0</td><td>–</td></tr>
<tr><td>D</td><td>0</td><td>0</td><td>–</td><td>–</td><td></td></tr>
<tr><td>3</td><td>0</td><td>0</td><td>0</td><td></td><td></td></tr>
<tr><td>2</td><td>–</td><td>0</td><td>0</td><td></td><td></td></tr>
<tr><td>1</td><td>–</td><td>0</td><td>–</td><td></td><td></td></tr>
</table></td></tr>
<tr><td>14</td><td>36</td><td>Speed sensor, transmission</td><td>Tiptronic warning lamp comes on, transmission is in emergency mode</td><td>Possible faults: Open circuit, short to ground, short to positive, loose contact

NOTE–
• Check with Porsche scan tool (Actual values). Store fault as of engine speed = 2,800 rpm.
1. Pull off connector from transmission control unit.
• Measure resistance between pin 2 and pin 38:
Specification: approximately 350 Ω
• Check that pin 2 and pin 38 are free from shorts to ground:
Check OK: Transmission control unit faulty
2. Also pull connector off transmission.
• Check continuity:
Transmission control unit connector pin 2 to pin A
Pin 38 to pin F
Short to ground from pin 2 and pin 38 to pin 7 (ground)
Pin 2 and pin 38 to pin 20 (screening)
Short to positive from pin 2 and pin 38 to pin 1 (positive)

NOTE–
• If it is to be assumed that intermittent faults linked to the output speed occur, check the screening connection at pin 20 after opening the connector housing
• Wiring OK: Check sensor and wiring in transmission.
• Possible transmission faults: ATF strainer clogged, air intake if oil starvation occurs etc., power transmission interrupted</td></tr>
</table>

Table c. Tiptronic fault code diagnosis (1995)				
Test point	**Fault code**	**Item**	**Result of fault**	**Possible faults**
15	37	Transmission temperature sensor	If a fault occurs, a replacement value 60° C is assumed	Possible faults: Open circuit, short to ground, short to positive, corrupted signal **NOTE–** *• Existing faults are stored only when the engine is running.* **1.** Check temperature with Porsche scan tool (Actual values). **2.** Pull off connector from transmission control unit: • Measure resistance between pin 46 and pin 44. Specification: approximately 1.00 kΩ/20° C 1.15 kΩ/40° C 1.30 kΩ/60° C • Check wiring: Repair short circuit. • Check OK: Transmission control unit faulty. Replace control unit. **NOTE–** *• If the display "Temperature sensor short to positive" and/or "Transverse acceleration sensor short to positive" appears, control unit may be faulty.*
16	38	Selector lever switch (for starting)	Tiptronic warning lamp comes on, transmission is in emergency mode	Specification: Engine can be started in P or N positions only Actual status: Engine can also be started in selector lever positions other than P and N Possible faults: Open circuit, short to ground, short to positive Start is possible except in P and N **1.** Check adjustment of bowden cable to position switch. • If required, run electrical test as described under Test point 13, item 4. • No start possible? **2.** Pull starter relay R61 from central electric panel. • Ground must be present at relay base terminal 85 when the selector lever is set to P and N. • Ground is present: Check start relay and ignition switch with related wiring. • No ground: Check position switch and wiring to start relay. **3.** Check position switch as described under Test point 13, item 4. • Position switch OK • Check continuity: a) Position switch pin G to starter relay R61. b) Junction X4/1 pin 6 to DME connector pin 42. • Wiring OK: Transmission-related starting requirements are met.
17	42	Control unit faulty	Tiptronic warning lamp comes on, transmission is in emergency mode	Replace control unit.
18	43	Control unit faulty	Tiptronic warning lamp comes on, transmission is in emergency mode	Replace control unit.

OBD

Table c. Tiptronic fault code diagnosis (1995)

Test point	Fault code	Item	Result of fault	Possible faults
19	44	Control unit faulty	Tiptronic warning lamp comes on, transmission is in emergency mode	Replace control unit.
20	45	Downshift protection	Tiptronic warning lamp comes on, transmission is in emergency mode	Possible faults: • Corrupted rpm signal from output or engine, or incorrect output/engine rpm ratio • Faulty transmission (Clutch slips) 1. Check inductive rpm sensor, Test point 14. 2. Check transmission.
21	46	Rev limiter	Tiptronic warning lamp comes on, transmission is in emergency mode	Possible faults: • Corrupted rpm signal from output or engine, or incorrect output/engine rpm ratio • Faulty transmission (Clutch slips) 1. Check inductive rpm sensor, Test point 14. 2. Check transmission.
22	51	Manual program switch	No manual program available	Detectable faults: Short to ground 1. Check operation with Porsche scan tool (Input signals). 2. Pull off connector from transmission control unit. • Check ground at pin 26. • Check connection from transmission control unit connector pin 48 to selector lever switch/pin 3. • Connection OK Replace selector lever switch plate. • Switch and connection OK: Transmission control unit faulty.
23	52	Up/down shift tip switch	No manual program available	Detectable faults: Short to ground 1. Check operation with Porsche scan tool (Input signals). 2. Pull off connector from transmission control unit. • Check ground at pin 26. • Check connection from transmission control unit connector / pin 10 and selector lever switch/pin 2 (upshift) or connection from transmission control unit connector/pin 29 to selector lever switch/pin 1 (downshift). • Connection OK: Replace selector lever switch unit. • Switch and connection OK: Transmission control unit faulty.
24	53	Kickdown switch	No kickdown shift	Detectable faults: Short to ground 1. A) Acoustical switch test. B) Check operation with Porsche scan tool (Input signals test). 2. Remove and check kickdown switch. ***NOTE–*** *• Adjust switch after replacement* 3. Pull off connector from transmission control unit. • Check connection from transmission control unit connector pin 30 to kickdown switch pin 2. • Check OK: Transmission control unit faulty.

Table c. Tiptronic fault code diagnosis (1995)

Test point	Fault code	Item	Result of fault	Possible faults
25	54	Transverse acceleration sensor	Upshifting in curves is not prevented	Detectable faults: Open circuit/short to ground. short to positive **1.** Check sensor with Porsche scan tool (Actual values) • (Nominal value approximately 0 g with vehicle on a level surface. The sensor can be checked by raising the vehide on one side.) **2.** Pull connector off the sensor. • Check sensor supply (5 V) at pin 3 and ground at pin 1. **3.** Check wiring according to wiring diagram. • Check continuity: Pin 45/transmission control unit connector to pin 3/sensor Pin 40/transmission control unit connector to pin 2/sensor Pin 44/transmission control unit connector to pin 1/sensor • Short to ground: Pin 2 and pin 3/sensor to pin 7/transmission control unit connector (ground) • Short to positive: Pin 1, 2 and 3/sensor to pin 1/transmission control unit connector (positive) • Wiring OK: Replace sensor. • Fault persists: Transmission control unit faulty.
26	55	Speed signal 1 (ABS)	Upshifting is not prevented, no manual program	Detectable faults: Open circuit/short to ground/short to positive **1.** Signal comes from ABS. ABS OK? **2.** Check signal with Porsche scan tool, raising the vehide and rotating right front wheel manually. **3.** Check wiring according to wiring diagram. • Check continuity: Pin 12/transmission control unit connector to ABS control unit connector pin 42 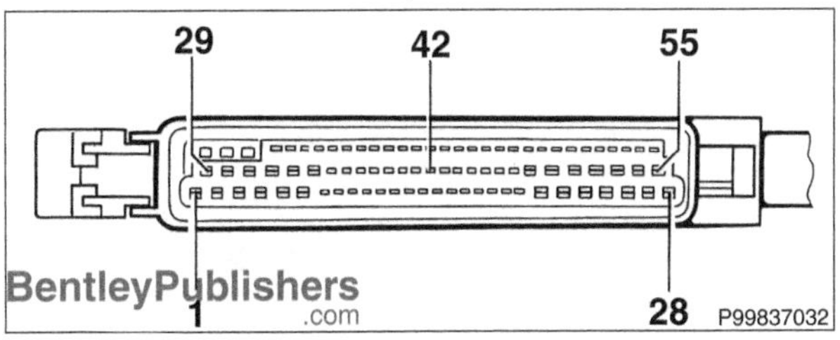 • Short to ground: Pin 12 to pin 7/transmission control unit connector (ground) • Short to positive: Pin 12 to pin 1/transmission control unit connector (positive) **4.** Check signal with oscilloscope (transmission control unit connector pin 12) (ignition ON). • Speed signal 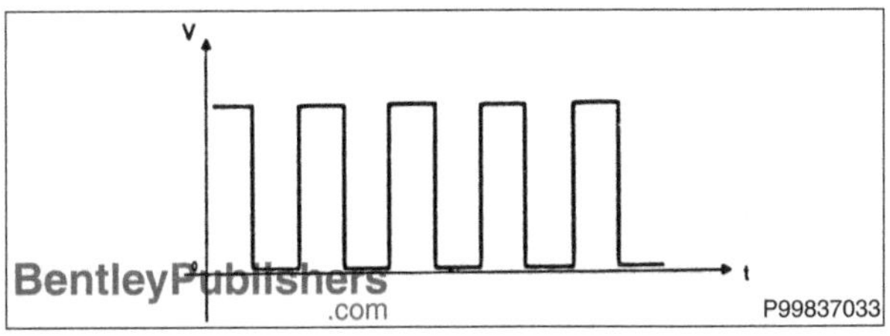 • Signal present: Transmission control unit faulty. • No signal: ABS faulty.

Table c. Tiptronic fault code diagnosis (1995)

Test point	Fault code	Item	Result of fault	Possible faults
			No manual program	
			No downshift during braking	
27	56	Combi-instrument input	No diagnosis possible	Detectable faults: Open circuit/short to ground, short to positive 1. Switch manual program on and off. • Check display in instrument. 2. Pull off transmission control unit connector. Connect pin 13 to ground. • Display OK: EGS faulty 3. Check wiring of transmission control unit pin 13 to speedometer pin 9. • Wiring OK: Speedometer faulty
28	59	R-position switch	Tiptronic warning lamp comes on, transmission is in emergency mode	Possible faults: Open circuit, short to ground, signal unplausible 1. Ground must be present at pin 41 when selector lever switch is in position P, R (N is possible due to overlaps). • No ground: Check wiring from transmission control unit connector pin 41 to position switch connector D. • Wiring OK: Check position switch as under Test point 13 item 4.
29	60	Reverse light relay	No diagnosis possible	Detectable faults: Open circuit/short to ground, short to positive 1. Trigger reverse light relay with Porsche scan tool (Drive links). The relay and the reverse lights must switch on and off in two second intervals. (Acoustical test. Relay is located in Central Electrical System R51). 2. Pull off connector from transmission control unit. • Short-circuit pin 25 to ground (pin 7). • Does reverse light relay pick up? • Transmission control unit faulty 3. Pull off reverse light relay. • Check voltage at terminal 30 of relay base. • Check connection from transmission control unit connector/pin 25 to relay base/terminal 85. • Wiring OK: Relay faulty
30	XX	Unknown fault code	Fault memory	Check ground points

Tiptronic connector pinout (from 1996)

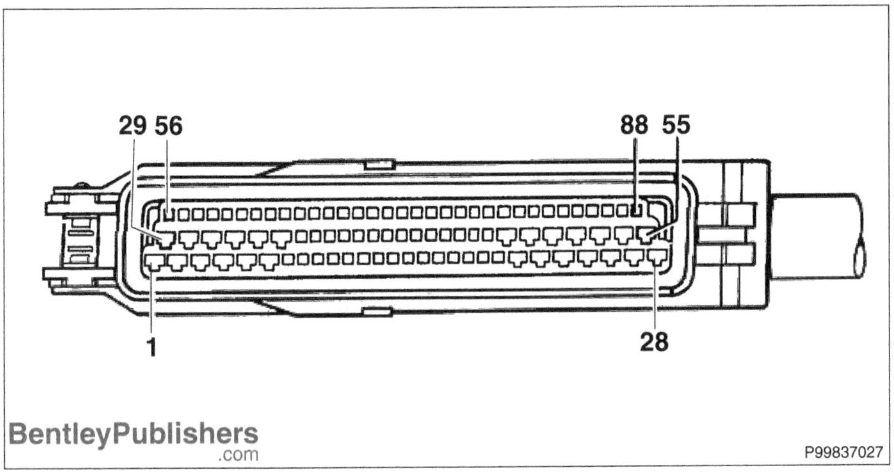

1. Relay, reverse light
2. Shiftlock
3. Not used
4. Not used
5. Solenoid valve/pressure regulator
6. Power ground
7. Throttle-valve signal, engine temperature
8. Switch position Y
9. Redundancy
10. Brake light
11. Not used
12. Pin coding 1 (3, 61)
13. Input signal, manual program
14. Output speed
15. Shield, output speed
16. Not used
17. OBD II fault lamp
18. Kickdown
19. Not used
20. Engine intervention
21. Sensor ground
22. ATF temperature
23. Not used
24. Fault lamp
25. Display, 2nd gear
26. Terminal 30
27. Display, 1st gear
28. Electronics ground
29. Not used
30. Solenoid valve 1
31. K wire
32. Solenoid valve, torque converter lockup clutch
33. Solenoid valve 2
34. Not used
35. Load signal (TR)
36. Switch position X
37. Switch position Z
38. Upshifting
39. Front wheel speed, right
40. Front wheel speed , left

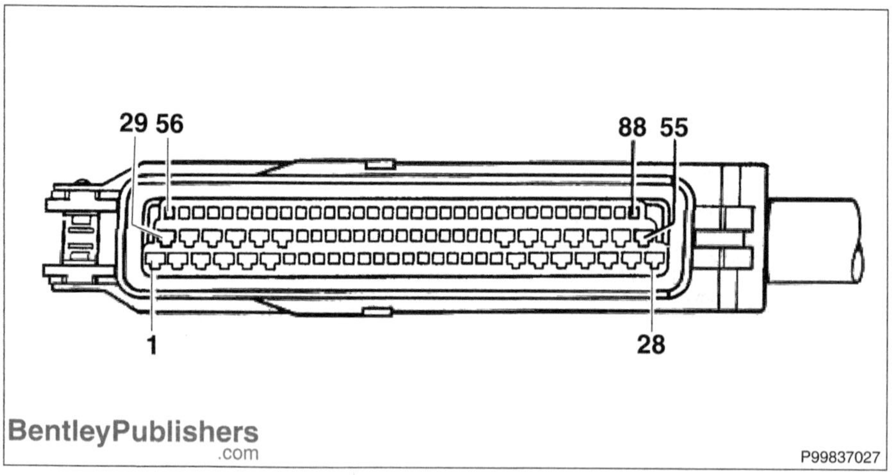

41. L wire

42. Output speed +

43. Engine speed (TN)

44. Not used

45. Not used

46. Display, 3rd gear

47. Downshifting

48. Not used

49. Not used

50. Display, 4th gear

51. Display, manual program

Tiptronic fault code diagnosis (from 1996)

Table d. Tiptronic fault code diagnosis (from 1996)

Test point	Fault code	Title	Causes	Possible faults
1	11	Voltage <ECM control unit	Tiptronic warning lamp does not come on, transmission is in emergency mode	Possible faults: Open circuit, short to positive, short to ground, loose contact, voltage outside of valid range 1. Tiptronic warning lamp does not light when ignition is OFF • Check fuse No. 27 • Measure voltage at fuse. • No voltage: Check wire to relay R11 (Terminal 15E) or operation of relay. • Measure voltage at transmission control unit connector pin 55. • No voltage: Check wiring according to wiring diagram 2. Voltage outside of valid range • Fault code 11 Signal unplausible • Low voltage? U < 11 V • Check battery, alternator, regulator. • OK? Check for contact resistance at junctions, deformation and corrosion of contacts. • Excessive voltage? Check power supply of vehicle (alternator, regulator) U > 14.5 V with engine running.

Table d. Tiptronic fault code diagnosis (from 1996)

Test point	Fault code	Title	Causes	Possible faults
2	21	RPM signal <ECM control unit	Tiptronic warning lamp comes on, transmission is in emergency mode	Possible faults: Open circuit/short to ground, short to positive, loose contact **1.** Signal check • Check signal with Porsche scan tool. (Engine operation). • Check signal with oscilloscope at transmission control unit connector pin 43. • If signal OK: Transmission control unit faulty • Signal shape 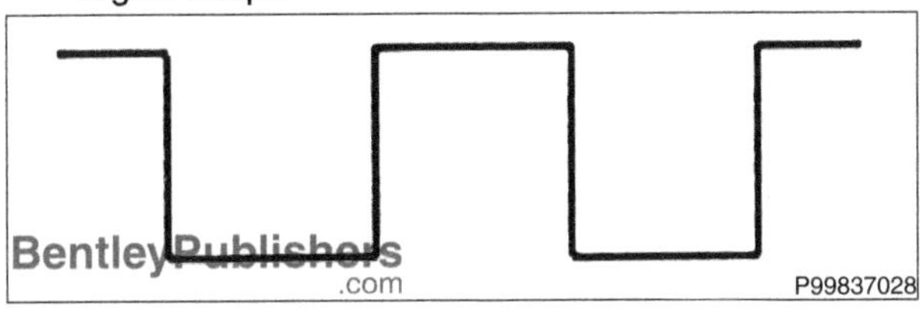 **2.** Open circuit, short to ground, short to positive • Check wiring to wiring diagram. • Check transmission control unit connector pin 43 to DME connector pin 6 for continuity, short to ground and short to positive. (Ground at transmission control unit = pin 28, positive at transmission control unit = pin 55) • Connections OK: DME control unit faulty
3	22	Load signal <ECM control unit	Tiptronic warning lamp comes on, transmission is in emergency mode Hard shifting	Possible faults: Open circuit, short to ground, short to positive, loose contact **1.** Signal check • Check signal with Porsche scan tool (injection time) with engine running • Check signal with oscilloscope at transmission control unit connector pin 35. • If signal OK: Transmission control unit faulty • Signal shape: TI signal 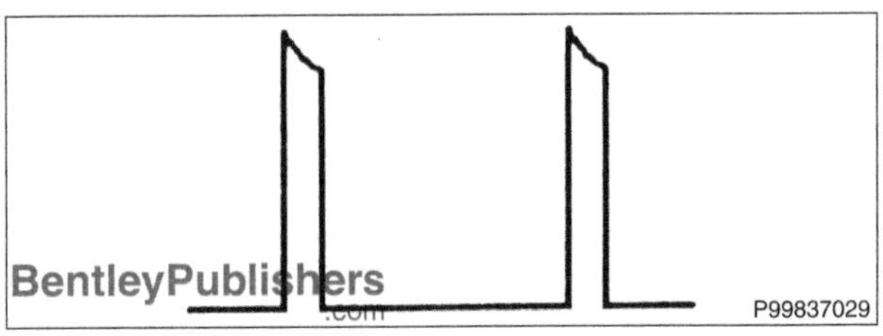

Table d. Tiptronic fault code diagnosis (from 1996)

Test point	Fault code	Title	Causes	Possible faults
4	24	Ignition-angle intervention (plus)	Tiptronic warning lamp comes on, transmission is in emergency mode Hard shifting	Possible faults: Open circuit, short to ground, short to positive 1. Check operation with Porsche scan tool (drive link test). Idle speed must drop. If it does not, check as follows: 2. Check voltage drop: • Remove cover of plugged-in connector X4/1. • Switch ignition OFF. • Measure voltage at pin 7. Specification: approximately 5 V. • Trigger signal with Tester. Voltage must drop. • Operation OK: Check wiring to DME control unit (Connector X 4/1 pin 7 to DME control unit pin 51). • Wiring OK: DME control unit faulty. • Operation not OK: Check wiring to transmission control unit (connector X4/1 pin 7 to transmission control unit connector pin 20). • Wiring OK: Transmission control unit faulty ***NOTE–*** *• Check engine to body ground strap* • Connector X 4/1 (below left seat) 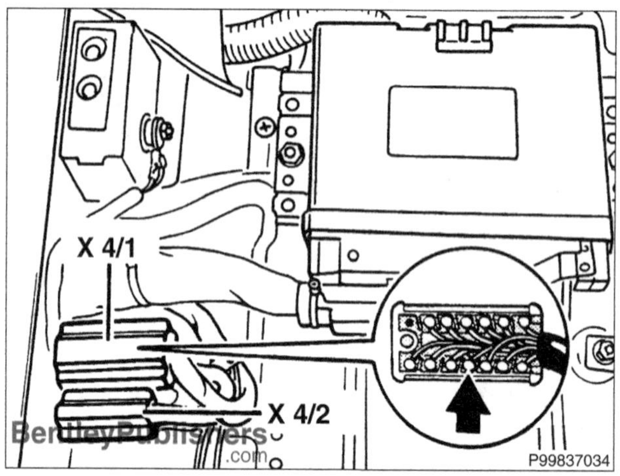

Table d. Tiptronic fault code diagnosis (from 1996)

Test point	Fault code	Title	Causes	Possible faults
5	25	Throttle-valve signal <ECM control unit Read out DTC memory	Tiptronic warning lamp comes on, transmission is in emergency mode Default throttle valve value 9% No kickdown	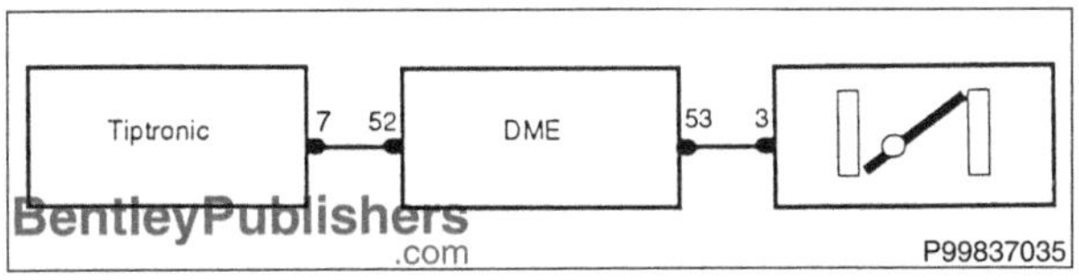 Detectable faults: Open circuit/short to ground, short to positive "Throttle potentiometer" fault is also stored in DME, fault must be corrected according to DME test plan. • Tiptronic test: Check display % with Porsche scan tool (Actual values menu) If no display: **1.** Measure voltage signal at transmission control unit connector pin 7 with voltmeter. Vary from 1 V at idle to approximately 10 V at full throttle (Ignition ON) If no voltage: **2.** Check wiring according to wiring diagram. • Check transmission control unit connector pin 7 to DME pin 52 for continuity, short to ground and short to positive (ground at transmission control unit connector = pin 28, positive at transmission control unit connector = pin 55). • If continuity OK = DME control unit faulty • If voltage OK = Transmission control unit faulty

Table d. Tiptronic fault code diagnosis (from 1996)

Test point	Fault code	Title	Causes	Possible faults
6	31	Solenoid valve 1	Tiptronic warning lamp comes on, transmission is in emergency mode	Possible faults: Open circuit, short to ground, short to positive **1.** Check operation with Porsche scan tool (Drive link test). • Acoustical test: Triggering cycle of solenoids is audible as a clicking noise near the transmission. **2.** Pull connector off transmission control unit. • Measure resistance between pin 30 and pin 53. Specification: 34 Ω, tolerance 10 %. • Check that pins 30 and 53 are free from shorts to ground. • Check OK: Transmission control unit faulty. **3.** Also pull off connector at transmission. • Check wiring (transmission control unit connector pin 30 to transmission connector pin H and transmission control unit connector pin 53 to transmission connector pin M). 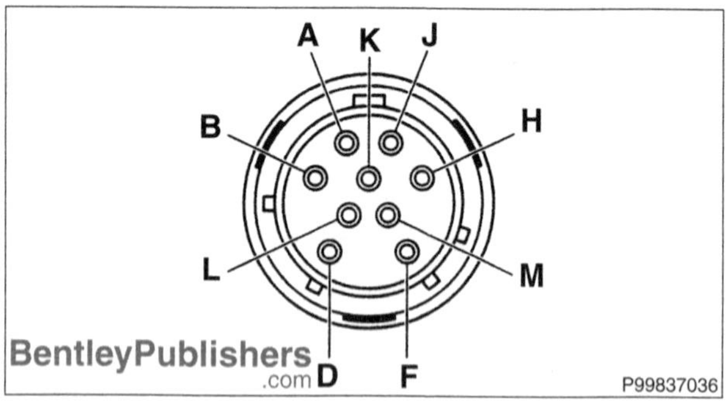 • Wiring OK: Solenoid valve faulty ***NOTE–*** *• If a "sporadic solenoid fault" occurs, check connections for pushed-back pins.*

Table d. Tiptronic fault code diagnosis (from 1996)

Test point	Fault code	Title	Causes	Possible faults
7	32	Solenoid valve 2	Tiptronic warning lamp comes on, transmission is in emergency mode	Possible faults: Open circuit, short to ground, short to positive **1.** Check operation with Porsche scan tool (Drive link test). • Acoustical test: Triggering cycle of solenoids is audible as a clicking noise near the transmission. **2.** Pull off connector at transmission control unit. • Measure resistance between pin 33 and pin 53. Specification: 34 Ω, tolerance 10 %. • Check that pins 33 and 53 are free from shorts to ground: Check OK: Transmission control unit faulty **3.** Also pull off connector from transmission. • Check wiring (transmission control unit connector pin 33 to transmission connector pin K and transmission control unit connector pin 53 to transmission connector pin M). 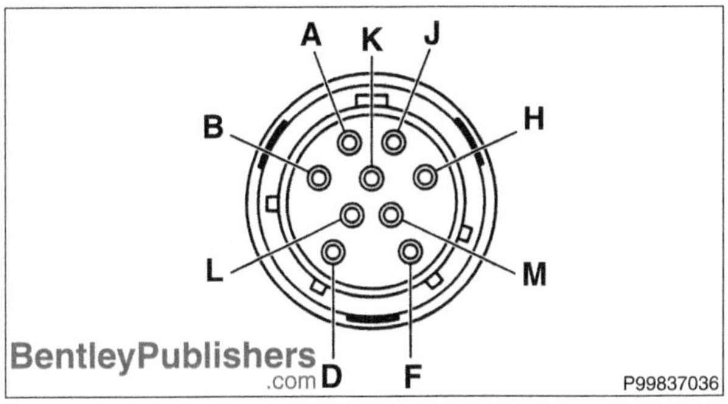 • Wiring OK: Solenoid valve faulty

<table>
<tr><th colspan="5">Table d. Tiptronic fault code diagnosis (from 1996)</th></tr>
<tr><th>Test point</th><th>Fault code</th><th>Title</th><th>Causes</th><th>Possible faults</th></tr>
<tr><td>8</td><td>33</td><td>Solenoid valve, torque converter clutch</td><td>Tiptronic warning lamp comes on, transmission is in emergency mode</td><td>Possible faults: Open circuit, short to ground, short to positive
1. Check operation with Porsche scan tool (Drive link test)
• Acoustical test:
Triggering cycle of solenoids is audible as a clicking noise near the transmission.
2. Pull off connector from transmission control unit. Measure resistance between pin 32 and pin 53. Specification: 34 Ω, tolerance 10 %.
• Check that pins 32 and 53 are free from shorts to ground.
Check OK: Transmission control unit faulty
3. Also pull off connector from transmission.
• Check wiring (transmission control unit connector pin 32 to transmission connector pin L and transmission control unit connector pin 53 to transmission connector pin M).
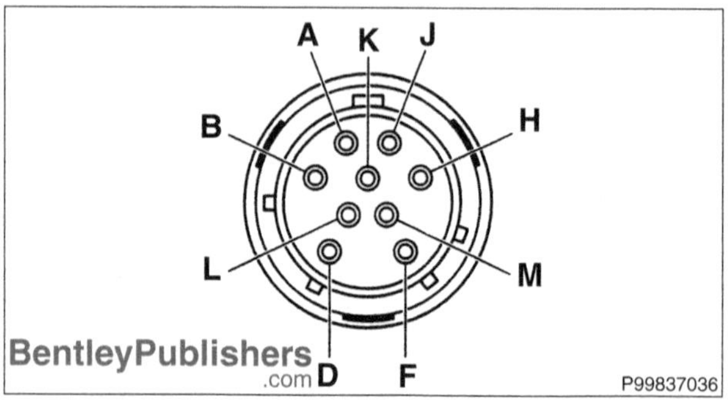

• Wiring OK: Solenoid valve faulty</td></tr>
<tr><td>9</td><td>34</td><td>Pressure regulator</td><td>Tiptronic warning lamp comes on, transmission is in emergency mode</td><td>Possible faults: Open circuit, short to ground, short to positive
1. Check operation with Porsche scan tool (Drive link test).
• Acoustical test:
Triggering cycle of solenoids is audible as a clicking noise near the transmission.
2. Pull off connector from transmission control unit. Measure resistance between pin 5 and pin 53. Specification: 6 Ω :± 2 Ω
• Check that pins 5 and 53 are free from shorts to ground. Check OK: Transmission control unit faulty
3. Also pull off connector from transmission.
• Check wiring (transmission control unit connector pin 5 to transmission connector pin B and transmission control unit connector pin 53 to transmission connector pin M).
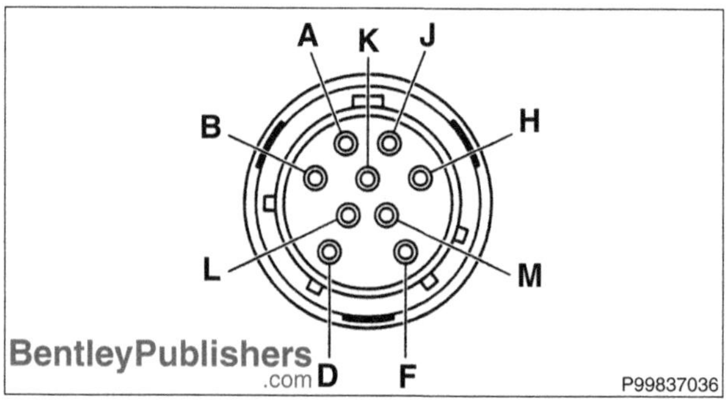

• Wiring OK: Solenoid valve faulty</td></tr>
</table>

<table>
<tr><th colspan="5">Table d. Tiptronic fault code diagnosis (from 1996)</th></tr>
<tr><th>Test point</th><th>Fault code</th><th>Title</th><th>Causes</th><th>Possible faults</th></tr>
<tr><td>10</td><td>35</td><td>Multi-function switch</td><td>Tiptronic warning lamp comes on, transmission is in emergency mode</td><td>
Possible faults: Open circuit, short to ground, short to positive
1. Check operation with Porsche scan tool.
• Adjustment of Porsche scan tool:
• Menu: “Input signals” = Selector lever/Manual program
• Shift through selector lever positions P, R, N, 0, 3, 2, 1 one after another.
• Compare:
Selector lever position
Position indicator in speedometer
Position indicator on Porsche scan tool
NOTE–
• Due to its design, the display on the Porsche scan tool appears with a certain delay - shift through positions sufficiently slowly.
• No matching: Check wiring according to wiring diagram, check position switch with/without wiring.
2. Pull off transmission control unit connector.
• Use ohmmeter to check transmission control unit connector pin 28 for ground.
• If no ground is present, refer to wiring diagrams.
3. Measure position switch according to table 1.
• Ohmmeter display – = No continuity
0 = Continuity
• Table 1:
<table>
<tr><th></th><th>Pin 8</th><th>Pin 37</th><th>Pin 36</th><th>Pin 9 -> to ground</th></tr>
<tr><td>P</td><td>–</td><td>–</td><td>0</td><td>0</td></tr>
<tr><td>R</td><td>0</td><td>–</td><td>0</td><td>0</td></tr>
<tr><td>N</td><td>0</td><td>–</td><td>–</td><td>–</td></tr>
<tr><td>D</td><td>0</td><td>0</td><td>–</td><td>–</td></tr>
<tr><td>3</td><td>0</td><td>0</td><td>0</td><td>–</td></tr>
<tr><td>2</td><td>–</td><td>0</td><td>0</td><td></td></tr>
<tr><td>1</td><td>–</td><td>0</td><td>–</td><td></td></tr>
</table>
• Measurements OK: Control unit faulty
• Measurements present, but incorrect sequence?
• Check cable from selector lever switch to position switch
• Is cable adjusted correctly?
• Repair instructions: See Shift lever cable, adjusting, in this repair group. Open circuit. short to ground, short to positive?
4. Check position switch without connections
• Pull connector off the switch.
• Check switch for continuity according to Table 2.
continued on next page
</td></tr>
</table>

<table>
<tr><th colspan="5">Table d. Tiptronic fault code diagnosis (from 1996)</th></tr>
<tr><th>Test point</th><th>Fault code</th><th>Title</th><th>Causes</th><th>Possible faults</th></tr>
<tr><td>10</td><td>35</td><td>Multi-function switch</td><td>Tiptronic warning lamp comes on, transmission is in emergency mode</td><td>
• Ohmmeter display — – = No continuity, 0 = Continuity
• Table 2
<table>
<tr><th></th><th>Pin H-B</th><th>Pin H-C</th><th>Pin H-A</th><th>Pin F-G</th><th>Pin E-D</th></tr>
<tr><td>P</td><td>–</td><td>–</td><td>0</td><td>0</td><td>0</td></tr>
<tr><td>R</td><td>0</td><td>–</td><td>0</td><td>–</td><td>0</td></tr>
<tr><td>N</td><td>0</td><td>–</td><td>–</td><td>0</td><td>–</td></tr>
<tr><td>D</td><td>0</td><td>0</td><td>–</td><td>–</td><td>–</td></tr>
<tr><td>3</td><td>0</td><td>0</td><td>0</td><td>–</td><td>–</td></tr>
<tr><td>2</td><td>–</td><td>0</td><td>0</td><td>–</td><td>–</td></tr>
<tr><td>1</td><td>–</td><td>0</td><td>–</td><td>–</td><td>–</td></tr>
</table>
• Fault? Replace gear position switch. See Automatic transmission gear position switch, removing and installing, in this repair group.
</td></tr>
<tr><td>11</td><td>36</td><td>Speed sensor Transmission</td><td>Tiptronic warning lamp comes on, transmission is in emergency mode</td><td>
Possible faults: Open circuit, short to ground, short to positive, loose contact
NOTE–
• Check with Porsche scan tool (Actual values). Store fault as of engine speed = 2,800 rpm.
1. Pull off connector from transmission control unit.
• Measure resistance between pin 14 and pin 42. Specification: approximately 350 Ω
• Check that pin 14 and pin 42 are free from shorts to ground. Check OK: Transmission control unit faulty
2. Pull connector off transmission.
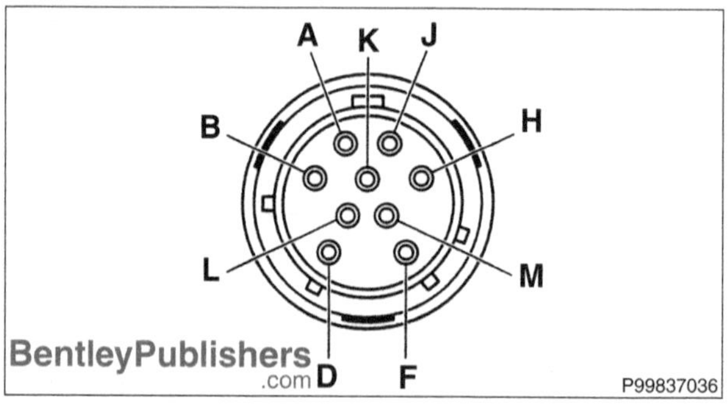

• Check continuity:
Transmission control unit connector pin 14 to pin F
Pin 42 to pin A
Short to ground from pin 14 and pin 42 to pin 28 (ground)
Pin 14 and pin 42 to pin 15 (screening)
Short to positive from pin 14 and pin 42 to pin 55 (positive)
</td></tr>
</table>

Table d. Tiptronic fault code diagnosis (from 1996)

Test point	Fault code	Title	Causes	Possible faults
12	37	Temperature sensor, transmission	If a fault occurs, a replacement value 60° C is assumed	Possible faults: Open circuit, short to ground, short to positive, corrupted signal *NOTE–* *• Existing faults are stored only when the engine is running.* **1.** Check temperature with Porsche scan tool (Actual values). **2.** Pull off connector from transmission control unit. • Measure resistance between pin 21 and pin 22. • Specification: approximately 1.00 kΩ/20° C 1.15 kΩ/40° C 1 .30 kΩ/60° C • Check wiring. • Check OK: Transmission control unit faulty
13	38	Multi-function switch (start)	Tiptronic warning lamp comes on, transmission is in emergency mode	Specification: Engine can be started in P or N positions only Actual status: Engine can also be started in selector lever positions other than P and N Possible faults: Open circuit, short to ground, short to positive Start is possible except in P and N **1.** Check adjustment of bowden cable to position switch. • If required, run electrical test as described under Test point 10, item 4. • No start possible? **2.** Pull start relay R61 from central electric panel. • Ground must be present at relay base terminal 85 when the selector lever is set to P and N. • Ground is present: Check start relay and ignition switch with related wiring. • No ground: Check position switch and wiring to start relay. **3.** Check position switch as described under Test point 10, item 4. • Position switch OK • Check continuity: Position switch pin G to start relay R61. Junction X4/1 pin 6 to DME connector pin 42 Wiring OK Transmission-related starting requirements are met.
14	42	Control unit defective (checksum)	Tiptronic warning lamp comes on, transmission is in emergency mode	Replace control unit.
15	43	Control unit defective (relay)	Tiptronic warning lamp comes on, transmission is in emergency mode	Replace control unit.
16	45	Control unit defective (watchdog)	Tiptronic warning lamp comes on, transmission is in emergency mode	Replace control unit.

Table d. Tiptronic fault code diagnosis (from 1996)

Test point	Fault code	Title	Causes	Possible faults
17	46	RPM limiter	Tiptronic warning lamp comes on, transmission is in emergency mode	Possible faults: Corrupted rpm signal from output or engine, or incorrect output/ engine rpm ratio Faulty transmission (Clutch slips) **1.** Check inductive rpm sensor, Test point 11. **2.** Check transmission.
18	46	Downshift protection	Tiptronic warning lamp comes on, transmission is in emergency mode	Faulty transmission (Clutch slips) **1.** Check inductive rpm sensor, Test point 11. **2.** Check transmission.
19	49	Control unit defective (EEPROM)	No reaction	Replace control unit.
20	51	Manual program switch	No manual program available	Detectable faults: Short to ground **1.** Check operation with Porsche scan tool (Input signals) **2.** Pull off connector from transmission control unit. • Check connection from transmission control unit connector pin 13 to selector lever switch/pin 3 . • Connection OK: Replace selector lever switch plate. • Switch and connection OK: Transmission control unit faulty
21	53	Switch, kickdown	No manual program available No kickdown	Detectable faults: Short to ground **1.** A) Acoustical switch test B) Check operation with Porsche scan tool (Input signals test). **2.** Remove and check kickdown switch ***NOTE–*** *• Adjust switch after replacement* **3.** Pull off connector from transmission control unit. • Check connection from transmission control unit connector pin 18 to kickdown switch pin 2. • Check OK: Transmission control unit faulty

Table d. Tiptronic fault code diagnosis (from 1996)

Test point	Fault code	Title	Causes	Possible faults
22	55	Speed signal (FR)	Upshifting is not prevented, no manual program	Detectable faults: Open circuit,short to ground,short to positive **1.** Signal comes from ABS. ABS OK? **2.** Check signal with Porsche scan tool, raising the vehicle and rotating right-hand front wheel manually. **3.** Check wiring according to wiring diagram 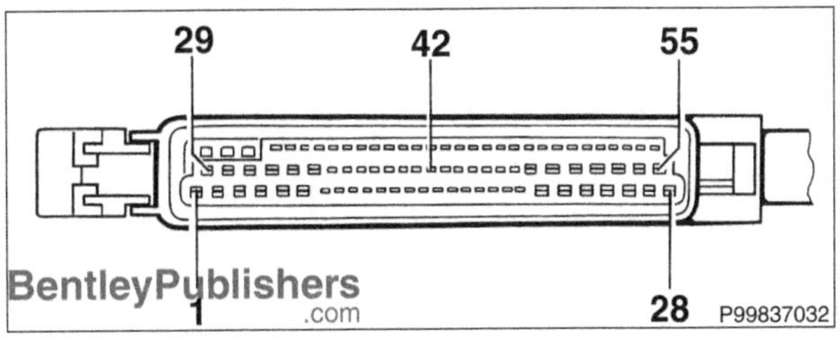 • Check continuity: Transmission control unit connector pin 39 to ABS control unit connector pin 42 • Short to ground: Transmission control unit connector pin 39 to pin 28 (ground) • Short to positive: Transmission control unit connector pin 39 to pin 55 (positive) **4.** Check signal with oscilloscope (transmission control unit connector pin 38) (ignition ON). • Speed signal 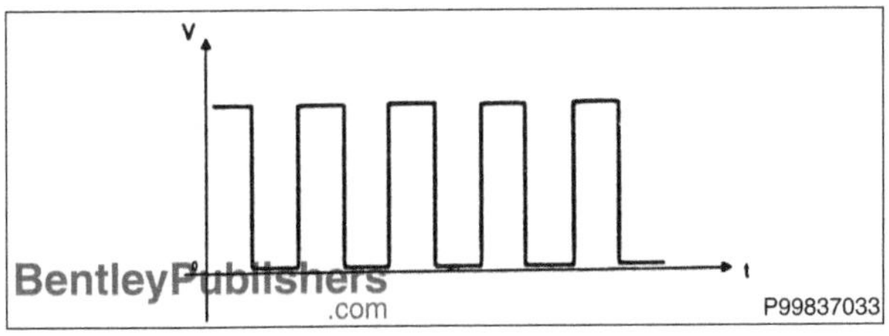 • Signal present: Transmission control unit faulty. • No signal: ABS faulty
23	56	Instrument cluster control	No diagnosis possible	Detectable faults: Open circuit/short to ground, short to positive **1.** Switch the manual program on and off with the engine running. • The position display "0" must go out when the manual range is engaged. **2.** Pull off the transmission control unit connector and connect pin 51 to ground. The position display/automatic mode must go out when the engine is not running. • If OK: Transmission control unit connector is defective. **3.** Check wiring of transmission control unit pin 51 to speedometer pin 9. • Wiring OK: Speedometer faulty
24	59	Switch, R position	Tiptronic warning lamp comes on, transmission is in emergency mode	Possible faults: Open circuit, short to ground, signal unplausible **1.** Ground must be present at pin 9 when selector lever switch is in position P, R (N is possible due to overlaps). • No ground: Check wiring from transmission control unit connector pin 41 to position switch connector D. • Wiring OK Check position switch as under Test point 10 item 4.

OBD

Table d. Tiptronic fault code diagnosis (from 1996)

Test point	Fault code	Title	Causes	Possible faults
25	62	Speed signal (FL)	Upshifting is not prevented, no manual program	Detectable faults: Open circuit/short to ground/short to positive 1. Signal comes from ABS. ABS OK? 2. Check signal with Porsche scan tool, raising the vehicle and rotating left-hand front wheel manually. 3. Check wiring according to wiring diagram. 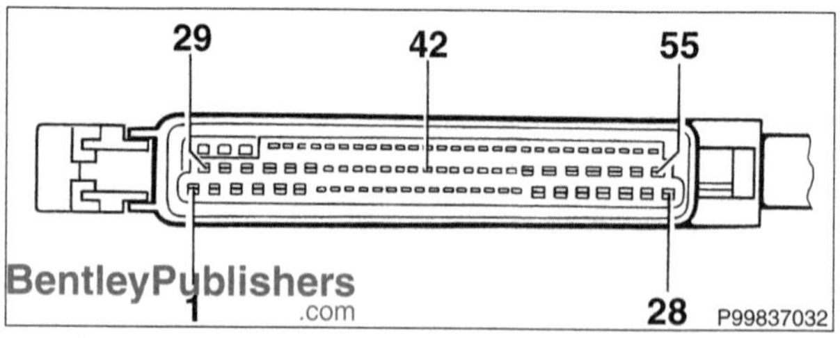 • Check continuity: Transmission control unit connector pin 40 to ABS control unit connector pin 9. • Short to ground: Transmission control unit connector pin 40 to pin 28 (ground) • Short to positive: Transmission control unit connector pin 40 to pin 55 (positive) 4. Check signal with oscilloscope (transmission control unit connector pin 40) (ignition ON) • Speed signal 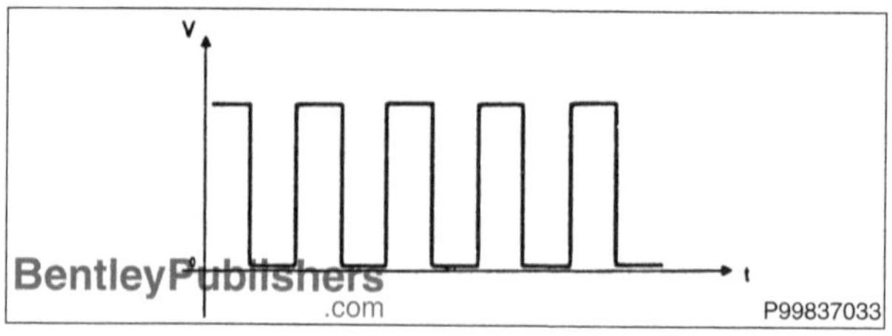 • Signal present: Transmission control unit faulty • No signal: ABS faulty
26	63	Shiftlock P/N ground	No reaction	Fault possibilities: Discontinuity, short circuit to ground, short circuit to plus ***NOTE–*** *• When the ignition is switched ON, a drive position can be selected from selector lever position P or N only when the brake pedal is pressed additionally.* 1. Positive potential must be present at pin 10 of the transmission control unit connector when the ignition is switched ON and the brake is applied. • If this is not the case = Check fuse 19, brake light switch and the wiring of the central electric system. 2. Connect pin 2 of the transmission control unit connector to ground. • The solenoid should operate (the activating pulse can be heard near the selector lever operating mechanism). • If this is not the case = Check fuse 27 and wiring in accordance with the wiring diagram. • If fuse 27 and wiring are OK = Solenoid defective. 3. If points 1 and 2 are OK = Control unit defective.

Table d. Tiptronic fault code diagnosis (from 1996)				
Test point	**Fault code**	**Title**	**Causes**	**Possible faults**
27	71	Gear monitoring 1st gear	No reaction	Fault possibility: Signal implausible ***NOTE–*** *• The gear monitoring function for 1st through 4th gears monitors the ratio of engine speed/output speed. In the event of setpoint deviations, the transmission control unit detects whether shifting was performed mechanically/hydraulically.* If a fault is stored, this indicates a mechanical/hydraulic fault in the transmission. **1.** Is Fault code 36 (Speed sensor) stored? • If yes = Remedy the fault in accordance with test point 11. **2.** If Fault code 36 is not stored = Check transmission.
28	72	Gear monitoring 2nd gear	No reaction	Fault possibility: Signal implausible ***NOTE–*** *• The gear monitoring function for 1st through 4th gears monitors the ratio of engine speed/output speed. In the event of setpoint deviations, the transmission control unit detects whether shifting was performed mechanically/hydraulically.* If a fault is stored, this indicates a mechanical/hydraulic fault in the transmission. **1.** Is Fault code 36 (Speed sensor) stored? • If yes = Remedy the fault in accordance with test point 11. **2.** If Fault code 36 is not stored = Check transmission.
29	73	Gear monitoring 3rd gear	No reaction	Fault possibility: Signal implausible ***NOTE–*** *• The gear monitoring function for 1st through 4th gears monitors the ratio of engine speed/output speed. In the event of setpoint deviations, the transmission control unit detects whether shifting was performed mechanically/hydraulically.* If a fault is stored, this indicates a mechanical/hydraulic fault in the transmission. **1.** Is Fault code 36 (Speed sensor) stored? • If yes = Remedy the fault in accordance with test point 11. **2.** If Fault code 36 is not stored = Check transmission.
30	74	Gear monitoring 4th gear	No reaction	Fault possibility: Signal implausible ***NOTE–*** *• The gear monitoring function for 1st through 4th gears monitors the ratio of engine speed/output speed. In the event of setpoint deviations, the transmission control unit detects whether shifting was performed mechanically/hydraulically.* If a fault is stored, this indicates a mechanical/hydraulic fault in the transmission. **1.** Is Fault code 36 (Speed sensor) stored? • If yes = Remedy the fault in accordance with test point 11. **2.** If Fault code 36 is not stored = Check transmission.

WARNING

Your common sense, good judgement and general alertness are crucial to safe and successful service work. Before attempting any work on your Porsche, be sure to read **00 Warnings and Cautions** *and the copyright page at the front of the manual. Review these warnings and cautions each time you prepare to work on your car. Please also read any warnings and cautions that accompany the procedures in the manual.*

A

B

C

WARNING

Your common sense, good judgement and general alertness are crucial to safe and successful service work. Before attempting any work on your Porsche, be sure to read **00 Warnings and Cautions** *and the copyright page at the front of the manual. Review these warnings and cautions each time you prepare to work on your car. Please also read any warnings and cautions that accompany the procedures in the manual.*

WARNING

Your common sense, good judgement and general alertness are crucial to safe and successful service work. Before attempting any work on your Porsche, be sure to read **00 Warnings and Cautions** *and the copyright page at the front of the manual. Review these warnings and cautions each time you prepare to work on your car. Please also read any warnings and cautions that accompany the procedures in the manual.*

F

WARNING

Your common sense, good judgement and general alertness are crucial to safe and successful service work. Before attempting any work on your Porsche, be sure to read **00 Warnings and Cautions** *and the copyright page at the front of the manual. Review these warnings and cautions each time you prepare to work on your car. Please also read any warnings and cautions that accompany the procedures in the manual.*

WARNING

Your common sense, good judgement and general alertness are crucial to safe and successful service work. Before attempting any work on your Porsche, be sure to read **00 Warnings and Cautions** *and the copyright page at the front of the manual. Review these warnings and cautions each time you prepare to work on your car. Please also read any warnings and cautions that accompany the procedures in the manual.*

O

P

R

S

WARNING

Your common sense, good judgement and general alertness are crucial to safe and successful service work. Before attempting any work on your Porsche, be sure to read **00 Warnings and Cautions** *and the copyright page at the front of the manual. Review these warnings and cautions each time you prepare to work on your car. Please also read any warnings and cautions that accompany the procedures in the manual.*

Notes